HOLT World Geography Today

TEXAS EDITION

HOLT, RINEHART AND WINSTON

A Harcourt Education Company

Austin • Orlando • Chicago • New York • Toronto • London • San Diego

The Authors

Prof. Robert J. Sager is Chair of Earth Sciences at Pierce College in Lakewood, Washington. Prof. Sager received his B.S. in geology and geography and M.S. in geography from the University of Wisconsin and holds a J.D. in international law from Western State University College of Law. He is the coauthor of several geography and earth science textbooks and has written many articles and educational media programs on the geography of the Pacific. Prof. Sager has received several National Science Foundation study grants and has twice been a recipient of the University of Texas NISOD National Teaching Excellence Award. He is a founding member of the Southern California Geographic Alliance and former president of the Association of Washington Geographers.

Prof. David M. Helgren is Director of the Center for Geographic Education at San Jose State University in California, where he is also Chair of the Department of Geography. Prof. Helgren received his Ph.D. in geography from the University of Chicago. He is the coauthor of several geography textbooks and has written many articles on the geography of Africa. Awards from the National Geographic Society, the National Science Foundation, and the L. S. B. Leakey Foundation have supported his many field research projects. Prof. Helgren is a former president of the California Geographical Society and a founder of the Northern California Geographic Alliance.

Editorial
Sue Miller, *Director*
Robert Wehnke, *Managing Editor*
Diana Holman Walker, *Senior Editor*
Daniel M. Quinn, *Pupil's Edition Project Editor*
Lissa B. Anderson, *Editor*
Andrew Miles, *Editor*
Jarred Prejean, *Associate Editor*

Technology Resources
Robert Scott Hrechko, *Internet Editor*
Annette Saunders, *Internet Editor*

Fact Checking
Bob Fullilove, *Editor*
Jenny Rose, *Associate Editor*

Copyediting
Julie Beckman, *Senior Copy Editor*
Katelijne A. Lefevere, *Copy Editor*

Sue Minkler, *Assistant Editorial Coordinator*
Gina Rogers, *Administrative Assistant*

Permissions
Kimberly Feden, *Permissions Editor*

Electronic Publishing
Robert Franklin, *EP Manager*
JoAnn Stringer, *Project Coordinator*
Christopher Lucas, *Project Coordinator*
Juan Baquera, *Production Artist*
Jim Gaile, *Production Artist*
Heather Jernt, *Production Artist*
Lana Kaupp, *Production Artist*
Kim Orne, *Production Artist*
Nanda Patel, *Production Artist*
Susan Savkov, *Production Artist*
Patty Zepeda, *Production Artist*
Barry Bishop, *Quality Control*
Sally Dewhirst, *Quality Control*
Sarah Willis, *Quality Control*
Angela Priddy, *Quality Control*

Pre-Press and Manufacturing
Leanna Ford, *Production Coordinator*
Antonella Posterino, *Production Coordinator*

New Media
Lydia Doty, *Senior Project Manager*
Armin Gutzmer, *Senior Manager, Training and Technical Support*

Art, Design and Photo

Book Design
Diane Motz, *Senior Design Director*
Robin Bouvette, *Senior Designer*
Mercedes Newman, *Designer*
Rina Ouellette, *Design Associate*
David Hernandez, *Designer*
Charlie Taliaferro, *Design Associate*
Lori Male, *Senior Designer*
Ed Diaz, *Design Associate*

Image Acquisitions
Joe London, *Director*
Tim Taylor, *Photo Research Supervisor*
Cindy Verheyden, *Senior Photo Researcher*
Stephanie Friedman, *Photo Researcher*
David Knowles, *Photo Researcher*
Elaine C. Tate, *Art Buyer Supervisor*
Angela Parisi, *Art Buyer*

Design New Media
Susan Michael, *Design Director*
Kimberly Cammerata, *Design Manager*
Grant Davidson, *Designer*

Media Design
Curtis Riker, *Design Director*
Richard Chavez, *Designer*

Graphic Services
Kristen Darby, *Manager*
Cathy Murphy, *Senior Image Designer*
Linda Wilbourn, *Image Designer*
Jane Dixon, *Image Designer*
Jeff Robinson, *Senior Ancillary Designer*

Cover Design
Pronk & Associates

While the chapter openers come from actual interviews, the young people's identities have been changed to protect their privacy.

Cover and Title Page photographs: Mount Shasta, California and background mountains, Texas Flag

Cover and Title Page Photo Credits: Mount Shasta, California, Artbase Inc.; Texas Flag, © Stockbyte

For acknowledgments, see page 797, which is an extension of the copyright page.

Printed in the United States of America

ISBN 0-03-065427-0

2 3 4 5 6 7 8 9 032 05 04 03

Academic Reviewers

Robin Elisabeth Datel
Instructor in Geography
California State University, Sacramento

Dennis Dingemans
Professor of Geography
University of California, Davis

Jeffrey Gritzner
Professor of Geography
The University of Montana

W. A. Douglas Jackson
Emeritus Professor, Geography and International Relations
Henry M. Jackson
School of International Studies
University of Washington

Robert B. Kent
Chair and Professor of Geography and Planning
University of Akron

Kwadwo Konadu-Agyemang
Professor of Geography and Planning
University of Akron

Nancy Lewis
Professor of Geography
University of Hawaii

Garth Andrew Myers
Associate Professor of Geography and African Studies
University of Kansas

Eric P. Perramond
Assistant Professor of Geography and Environmental Science
Stetson University

Bill Takizawa
Professor of Geography
San Jose State University

Brent Yarnal
Professor of Geography
Pennsylvania State University

Teacher Reviewers

Melissa Counihan
Rockdale High School
Rockdale, Texas

Tom Fischer
St. Clare School
St. Louis, Missouri

William Fisher
Bryan High School
Bryan, Texas

Steve Gargo
Appleton West High School
Appleton, Wisconsin

Lisa Haydel
Evergreen Junior High School
Houma, Louisiana

Nancy Lehmann-Carssow
Instructional Specialist
Lanier High School
Austin, Texas

C. Eugene Price
Heritage Junior-Senior High School
Monroeville, Indiana

Ron Scholten
Social Studies Department Chair
Providence High School
Burbank, California

Catharine Stoner
Wentzville High School
Wentzville, Missouri

Frank Thomas, Jr.
Hays County CISD
Austin, Texas

Susan W. Walker
Beaufort County Schools
Beaufort, South Carolina

Jane Young-Leatherman
Social Studies Department Chair
Durant High School
Durant, Oklahoma

Field Test Teachers

Amber Acuña
Luther Burbank High School
San Antonio, Texas

J. Mark Buffington
Mexico High School
Mexico, Missouri

Sandra Dawson-O'Bryan
Angleton High School
Angleton, Texas

Marie Ervin
James Madison High School
Houston, Texas

Patrick Haney
Carter High School
Fort Worth, Texas

Randy Kindschuh
Wisner-Pilger High School
Wisner, Nebraska

Jim Long
Holmes High School
San Antonio, Texas

Christine A. Moore
Fort Zumwalt South High School
St. Peters, Missouri

Carol Ragsdale
Central High School
West Helena, Arkansas

Juliann Warner
Martin High School
Arlington, Texas

HOLT

World Geography Today

TABLE OF CONTENTS

Inuit man and igloo in Canada

Parliament building in Budapest, Hungary

Vancouver, Canada

UNIT 2 The United States and Canada 138

Cornfield and farm, Minnesota

UNIT 3 Middle and South America 208

Lavender field, France

Tian Shan, Central Asia

Ancient stone heads, Turkey

UNIT 7

Africa 470

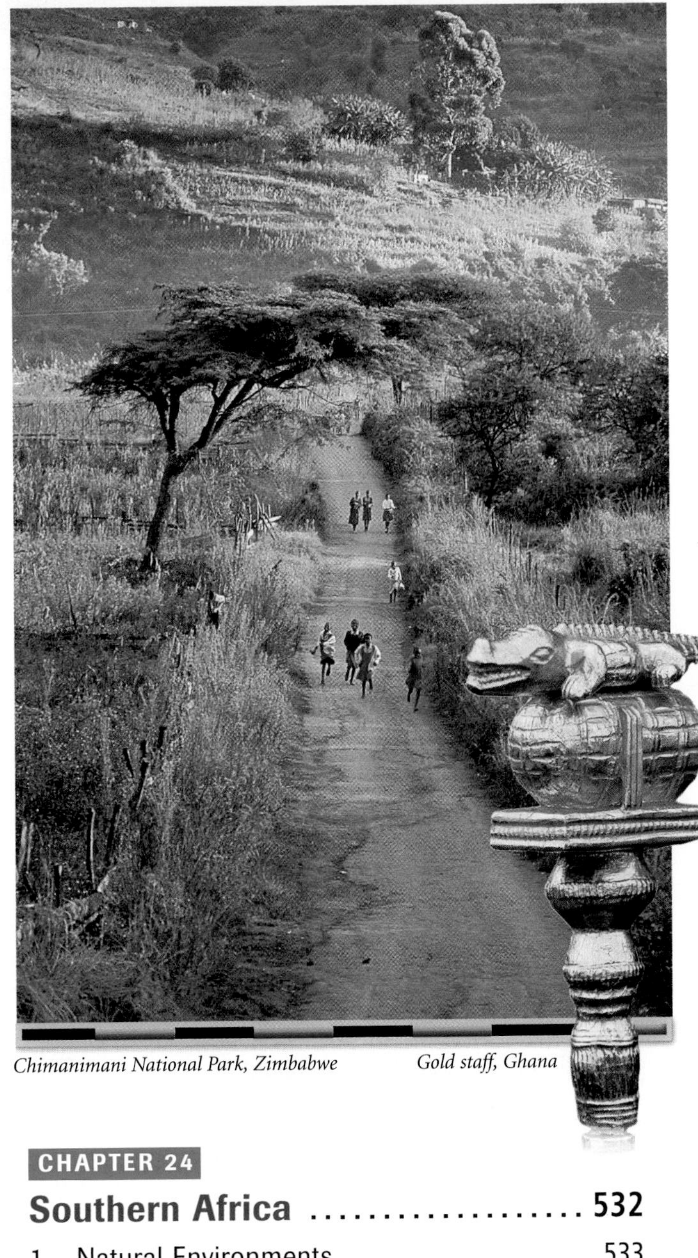

Chimanimani National Park, Zimbabwe

Gold staff, Ghana

Gorilla in Rwanda

Rice crops in Sri Lanka

UNIT 8 South Asia 552

UNIT 9 East and Southeast Asia .. 604

Korean fan dancers

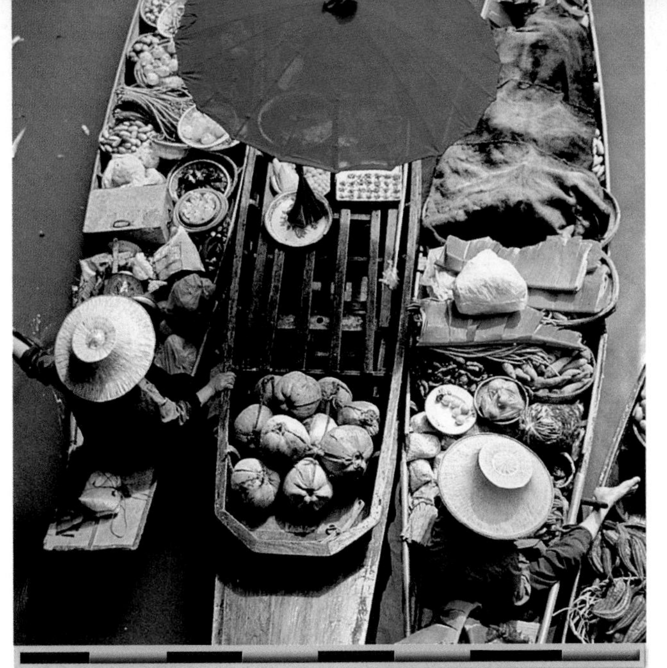

Food vendors in Bangkok, Thailand

REFERENCE SECTION

Aboriginal art, Australia

Geography and Your World

CASE STUDIES

Cities and Settlements

Geography for Life

Interdisciplinary Activities

connecting to *Literature*

FOCUS ON

Interdisciplinary Activities

FOCUS ON, cont.

Connecting to Other Disciplines

Technology Activities

Special Features

Technology Activities

internet connect

Skill-Building Activities

Skills Handbook

Skill Building: Using the Geographer's Tools

Skill-Building Activities

Skill Building for TAKS Workshops

Maps

Skill-Building Activities

Skill-Building Activities

Charts, Diagrams, and Graphs

How to Use Your Textbook

Use the chapter opener to preview the region you are about to study.

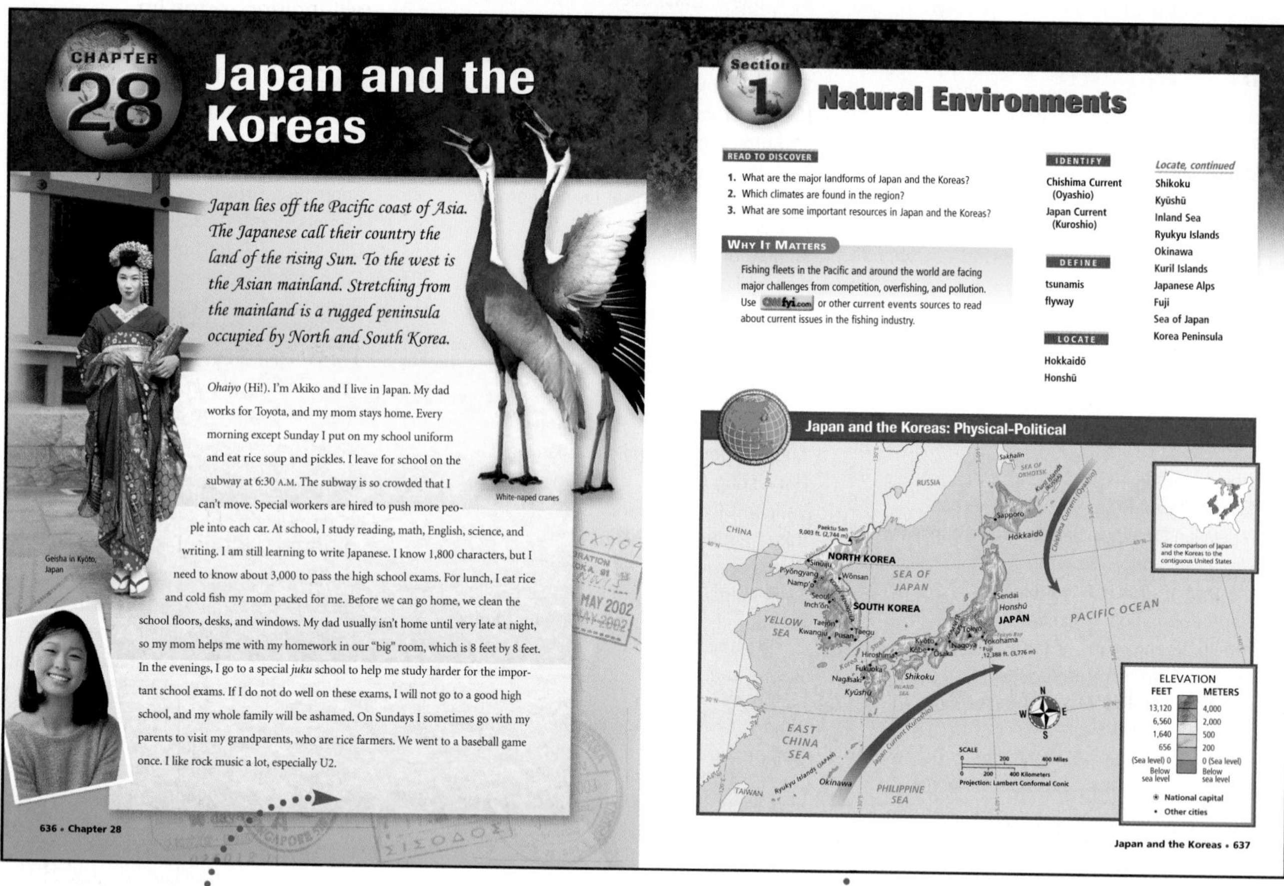

CHAPTER 28

Japan and the Koreas

Japan lies off the Pacific coast of Asia. The Japanese call their country the land of the rising Sun. To the west is the Asian mainland. Stretching from the mainland is a rugged peninsula occupied by North and South Korea.

Ohaiyo (Hi!). I'm Akiko and I live in Japan. My dad works for Toyota, and my mom stays home. Every morning except Sunday I put on my school uniform and eat rice soup and pickles. I leave for school on the subway at 6:30 A.M. The subway is so crowded that I can't move. Special workers are hired to push more people into each car. At school, I study reading, math, English, science, and writing. I am still learning to write Japanese. I know 1,800 characters, but I need to know about 3,000 to pass the high school exams. For lunch, I eat rice and cold fish my mom packed for me. Before we can go home, we clean the school floors, desks, and windows. My dad usually isn't home until very late at night, so my mom helps me with my homework in our "big" room, which is 8 feet by 8 feet. In the evenings, I go to a special *juku* school to help me study harder for the important school exams. If I do not do well on these exams, I will not go to a good high school, and my whole family will be ashamed. On Sundays I sometimes go with my parents to visit my grandparents, who are rice farmers. We went to a baseball game once. I like rock music a lot, especially U2.

Geisha in Kyōto, Japan

White-naped cranes

636 • Chapter 28

Section 1

Natural Environments

READ TO DISCOVER

1. What are the major landforms of Japan and the Koreas?
2. Which climates are found in the region?
3. What are some important resources in Japan and the Koreas?

WHY IT MATTERS

Fishing fleets in the Pacific and around the world are facing major challenges from competition, overfishing, and pollution. Use CNNfyi.com or other current events sources to read about current issues in the fishing industry.

IDENTIFY

Chishima Current (Oyashio)
Japan Current (Kuroshio)

DEFINE

tsunamis
flyway

LOCATE

Hokkaidō
Honshū

Locate, continued

Shikoku
Kyūshū
Inland Sea
Ryukyu Islands
Okinawa
Kuril Islands
Japanese Alps
Fuji
Sea of Japan
Korea Peninsula

Japan and the Koreas: Physical-Political

Japan and the Koreas • 637

An interview

with a student begins each regional chapter. These interviews give you a glimpse of what life is like for some people in the region you are about to study.

Chapter Map

The map at the beginning of Section 1 in regional chapters shows you the countries you will read about. You can use this map to identify country names and capitals and to locate physical features. These chapter maps will also help you create sketch maps in section reviews.

Use these built-in tools to read for understanding.

Read to Discover questions begin each section of *World Geography Today*. These questions serve as your guide as you read through the section. Keep them in mind as you explore the section content.

Why It Matters is an exciting way for you to make connections between what you are reading in your geography textbook and the world around you. Explore a topic that is relevant to our lives today by using **CNNfyi.com**.

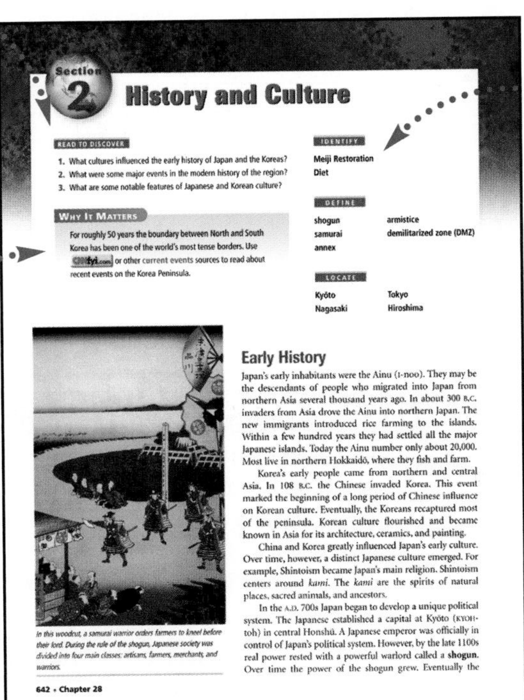

Section 2

History and Culture

READ TO DISCOVER

1. What cultures influenced the early history of Japan and the Koreas?
2. What were some major events in the modern history of the region?
3. What are some notable features of Japanese and Korean culture?

WHY IT MATTERS

For roughly 50 years the boundary between North and South Korea has been one of the world's most tense borders. Use CNNfyi.com or other current events sources to read about recent events on the Korea Peninsula.

IDENTIFY

Meiji Restoration
Diet

DEFINE

shogun
samurai
annex
armistice
demilitarized zone (DMZ)

LOCATE

Kyōto
Nagasaki
Tokyo
Hiroshima

In this woodcut, a samurai warrior orders farmers to kneel before their lord. During the rule of the shogun, Japanese society was divided into four main classes: artisans, farmers, merchants, and warriors.

Early History

Japan's early inhabitants were the Ainu (I-noo). They may be the descendants of people who migrated into Japan from northern Asia several thousand years ago. In about 300 B.C. invaders from Asia drove the Ainu into northern Japan. The new immigrants introduced rice farming to the islands. Within a few hundred years they had settled all the major Japanese islands. Today the Ainu number only about 20,000. Most live in northern Hokkaidō, where they fish and farm.

Korea's early people came from northern and central Asia. In 108 B.C. the Chinese invaded Korea. This event marked the beginning of a long period of Chinese influence on Korean culture. Eventually, the Koreans recaptured most of the peninsula. Korean culture flourished and became known in Asia for its architecture, ceramics, and painting.

China and Korea greatly influenced Japan's early culture. Over time, however, a distinct Japanese culture emerged. For example, Shintoism became Japan's main religion. Shintoism centers around *kami*. The *kami* are the spirits of natural places, sacred animals, and ancestors.

In the A.D. 700s Japan began to develop a unique political system. The Japanese established a capital at Kyōto (KYOH-toh) in central Honshū. A Japanese emperor was officially in control of Japan's political system. However, by the late 1100s real power rested with a powerful warlord called a **shogun**. Over time the power of the shogun grew. Eventually the

642 • Chapter 28

Identify and Define terms are introduced at the beginning of each section. The terms will be defined in context. They will include terms important to the study of geography and to the understanding of the region you are studying.

Our Amazing Planet features provide interesting facts about the region you are studying. Here you will learn about the origins of place-names and fascinating tidbits like the Earth's coldest place or largest living thing.

Interpreting the Visual Record features accompany many of the textbook's rich photographs. Pictures are one of the most important primary sources geographers can use to study our planet. These features invite you to analyze the images so that you can learn more about their content and their links to what you are studying in the section. Other captions ask you to interpret maps, graphs, and charts.

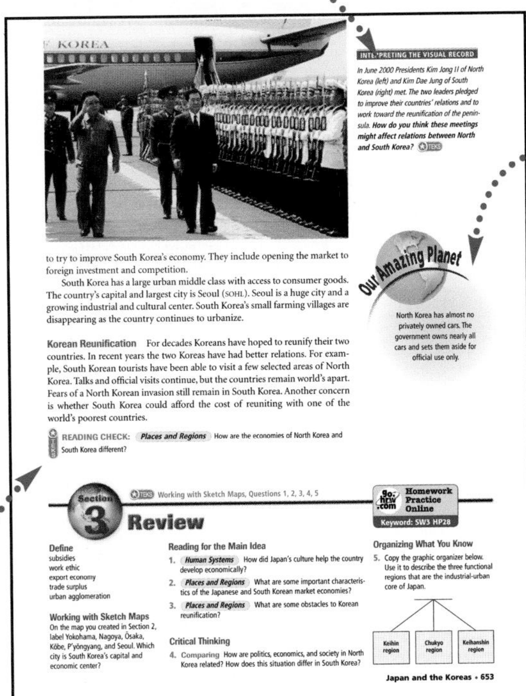

INTERPRETING THE VISUAL RECORD

In June 2000 Presidents Kim Jong Il of North Korea (left) and Kim Dae Jung of South Korea (right) met. The two leaders pledged to improve their countries' relations and to work toward the reunification of the peninsula. ***How do you think these meetings might affect relations between North and South Korea?*** TEKS

to try to improve South Korea's economy. They include opening the market to foreign investment and competition.

South Korea has a large urban middle class with access to consumer goods. The country's capital and largest city is Seoul (SOHL). Seoul is a huge city and a growing industrial and cultural center. South Korea's small farming villages are disappearing as the country continues to urbanize.

Korean Reunification For decades Koreans have hoped to reunify their two countries. In recent years the two Koreas have had better relations. For example, South Korean tourists have been able to visit a few selected areas of North Korea. Talks and official visits continue, but the countries remain world's apart. Fears of a North Korean invasion still remain in South Korea. Another concern is whether South Korea could afford the cost of reuniting with one of the world's poorest countries.

Our Amazing Planet

North Korea has almost no privately owned cars. The government owns nearly all cars and sets them aside for official use only.

READING CHECK: *Places and Regions* How are the economies of North Korea and South Korea different?

TEKS Working with Sketch Maps, Questions 1, 2, 3, 4, 5

Section 3

Review

go.hrw.com Homework Practice Online

Keyword: SW3 HP28

Define

subsidies
work ethic
export economy
trade surplus
urban agglomeration

Working with Sketch Maps

On the map you created in Section 2, label Yokohama, Nagoya, Ōsaka, Kōbe, P'yŏngyang, and Seoul. Which city is South Korea's capital and economic center?

Reading for the Main Idea

1. *Human Systems* How did Japan's culture help the country develop economically?
2. *Places and Regions* What are some important characteristics of the Japanese and South Korean market economies?
3. *Places and Regions* What are some obstacles to Korean reunification?

Critical Thinking

4. Comparing How are politics, economics, and society in North Korea related? How does this situation differ in South Korea?

Organizing What You Know

5. Copy the graphic organizer below. Use it to describe the three functional regions that are the industrial-urban core of Japan.

Keihin region
Chukyo region
Keihanshin region

Japan and the Koreas • 653

Reading Check questions appear often throughout the textbook to allow you to check your comprehension. As you read, pause for a moment to consider each Reading Check. If you have trouble answering the question, review the material that you just read.

Use these tools to pull together all of the information you have learned.

Reading for the Main Idea questions help review the main points you have studied in the section.

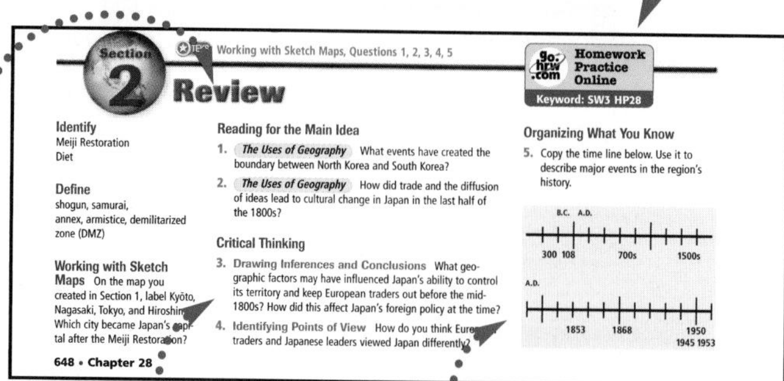

Section 2 Review

Working with Sketch Maps, Questions 1, 2, 3, 4, 5

go.hrw.com Homework Practice Online
Keyword: SW3 HP28

Identify
Meiji Restoration
Diet

Define
shogun, samurai, annex, armistice, demilitarized zone (DMZ)

Working with Sketch Maps On the map you created in Section 1, label Kyōto, Nagasaki, Tokyo, and Hiroshima. Which city became Japan's capital after the Meiji Restoration?

Reading for the Main Idea

1. *The Uses of Geography* What events have created the boundary between North Korea and South Korea?
2. *The Uses of Geography* How did trade and the diffusion of ideas lead to cultural change in Japan in the last half of the 1800s?

Critical Thinking

3. Drawing Inferences and Conclusions What geographic factors may have influenced Japan's ability to control its territory and keep European traders out before the mid-1800s? How did this affect Japan's foreign policy at the time?
4. Identifying Points of View How do you think European traders and Japanese leaders viewed Japan differently?

Organizing What You Know

5. Copy the time line below. Use it to describe major events in the region's history.

B.C. A.D.
300 108 700s 1500s
A.D.
1853 1868 1950 1945 1953

648 • Chapter 28

Homework Practice Online lets you log on to the go.hrw.com Web site to complete an interactive self-check of the material covered in the section.

Critical Thinking activities in section and chapter reviews allow you to explore a topic in greater depth and to build your skills.

Graphic Organizers will help you pull together important information from the section. You can complete the graphic organizer as a study tool to prepare for a test or writing assignment.

Locating Key Places activities ask you to identify and locate places you have read about in the chapter.

Using the Geographer's Tools, Geography for Life, and Building Social Studies Skills activities help you develop the skills you need to study geography and to answer standardized-test questions.

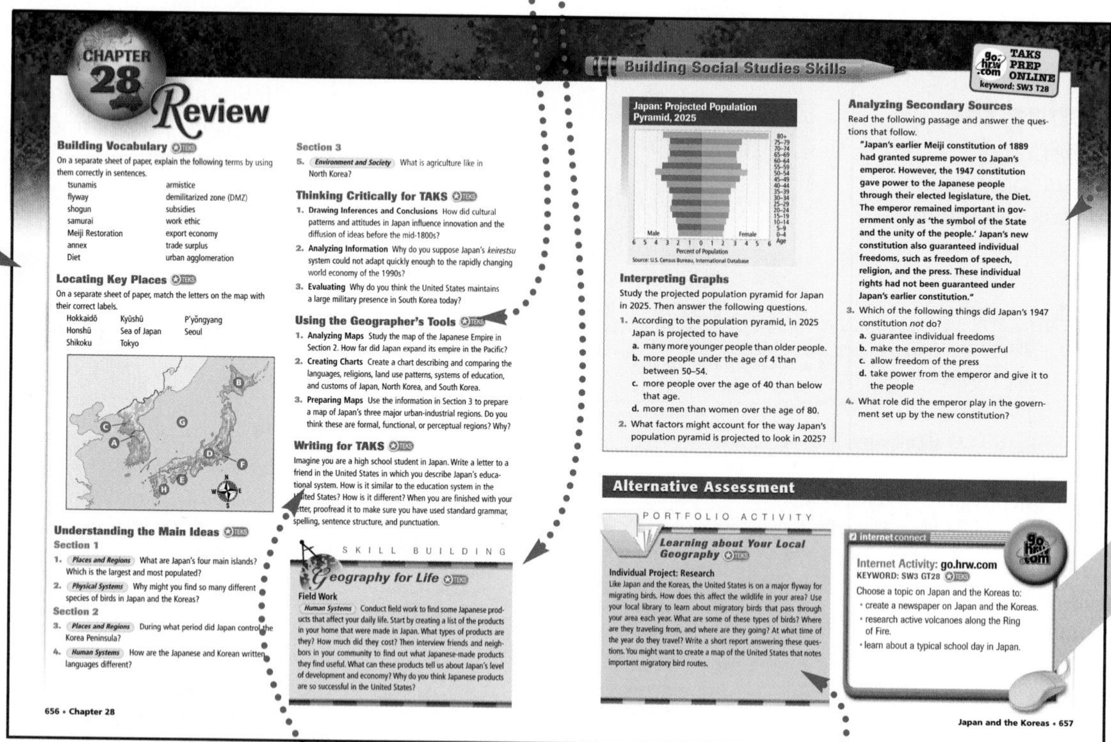

Chapter 28 Review

Building Vocabulary
On a separate sheet of paper, explain the following terms by using them correctly in sentences.

tsunamis, flyway, shogun, samurai, Meiji Restoration, annex, Diet, armistice, demilitarized zone (DMZ), subsidies, work ethic, export economy, trade surplus, urban agglomeration

Locating Key Places
On a separate sheet of paper, match the letters on the map with their correct labels.

Hokkaidō, Honshū, Shikoku, Kyūshū, Sea of Japan, Tokyo, P'yŏngyang, Seoul

Understanding the Main Ideas

Section 1

1. *Places and Regions* What are Japan's four main islands? Which is the largest and most populated?
2. *Physical Systems* Why might you find so many different species of birds in Japan and the Koreas?

Section 2

3. *Places and Regions* During what period did Japan control the Korea Peninsula?
4. *Human Systems* How are the Japanese and Korean written languages different?

Section 3

5. *Environment and Society* What is agriculture like in North Korea?

Thinking Critically for TAKS

1. Drawing Inferences and Conclusions How did cultural patterns and attitudes in Japan influence innovation and the diffusion of ideas before the mid-1800s?
2. Analyzing Information Why do you suppose Japan's *keiretsu* system could not adapt quickly enough to the rapidly changing world economy of the 1990s?
3. Evaluating Why do you think the United States maintains a large military presence in South Korea today?

Using the Geographer's Tools

1. Analyzing Maps Study the map of the Japanese Empire in Section 2. How far did Japan expand its empire in the Pacific?
2. Creating Charts Create a chart describing and comparing the languages, religions, land use patterns, systems of education, and customs of Japan, North Korea, and South Korea.
3. Preparing Maps Use the information in Section 3 to prepare a map of Japan's three major urban-industrial regions. Do you think these are formal, functional, or perceptual regions? Why?

Writing for TAKS
Imagine you are a high school student in Japan. Write a letter to a friend in the United States in which you describe Japan's educational system. How is it similar to the education system in the United States? How is it different? When you are finished with your letter, proofread it to make sure you have used standard grammar, spelling, sentence structure, and punctuation.

SKILL BUILDING
Geography for Life
Field Work
Human Systems Conduct field work to find some Japanese products that affect your daily life. Start by creating a list of the products in your home that were made in Japan. What types of products are they? How much did they cost? Then interview friends and neighbors in your community to find out what Japanese-made products they find useful. What can these products tell us about Japan's level of development and economy? Why do you think Japanese products are so successful in the United States?

656 • Chapter 28

Building Social Studies Skills

go.hrw.com TAKS PREP ONLINE keyword: SW3 T28

Japan: Projected Population Pyramid, 2025
Male Female
Percent of Population
Source: U.S. Census Bureau, International Database

Interpreting Graphs
Study the projected population pyramid for Japan in 2025. Then answer the following questions.

1. According to the population pyramid, in 2025 Japan is projected to have
   a. many more younger people than older people.
   b. more people under the age of 4 than between 50–54.
   c. more people over the age of 40 than below that age.
   d. more men than women over the age of 80.
2. What factors might account for the way Japan's population pyramid is projected to look in 2025?

Analyzing Secondary Sources
Read the following passage and answer the questions that follow.

"Japan's earlier Meiji constitution of 1889 had granted supreme power to Japan's emperor. However, the 1947 constitution gave power to the Japanese people through their elected legislature, the Diet. The emperor remained important in government only as 'the symbol of the State and the unity of the people.' Japan's new constitution also guaranteed individual freedoms, such as freedom of speech, religion, and the press. These individual rights had not been guaranteed under Japan's earlier constitution."

3. Which of the following things did Japan's 1947 constitution *not* do?
   a. guarantee individual freedoms
   b. make the emperor more powerful
   c. allow freedom of the press
   d. take power from the emperor and give it to the people
4. What role did the emperor play in the government set up by the new constitution?

Alternative Assessment

PORTFOLIO ACTIVITY
Learning about Your Local Geography
Individual Project: Research
Like Japan and the Koreas, the United States is on a major flyway for migrating birds. How does this affect the wildlife in your area? Use your local library to learn about migratory birds that pass through your area each year. What are some of these types of birds? Where are they traveling from, and where are they going? At what time of the year do they travel? Write a short report answering these questions. You might want to create a map of the United States that notes important migratory bird routes.

internet connect
Internet Activity: go.hrw.com
KEYWORD: SW3 GT28
Choose a topic on Japan and the Koreas to:
• create a newspaper on Japan and the Koreas.
• research active volcanoes along the Ring of Fire.
• learn about a typical school day in Japan.

Japan and the Koreas • 657

Writing about Geography activities let you practice your writing skills to explore in more detail topics you have studied in the chapter.

Portfolio Activities are exciting and creative ways to explore your local geography and to make connections to the region you are studying. Some activities ask you to work cooperatively, and others include projects to complete on your own.

Use these online tools to review and complete online activities.

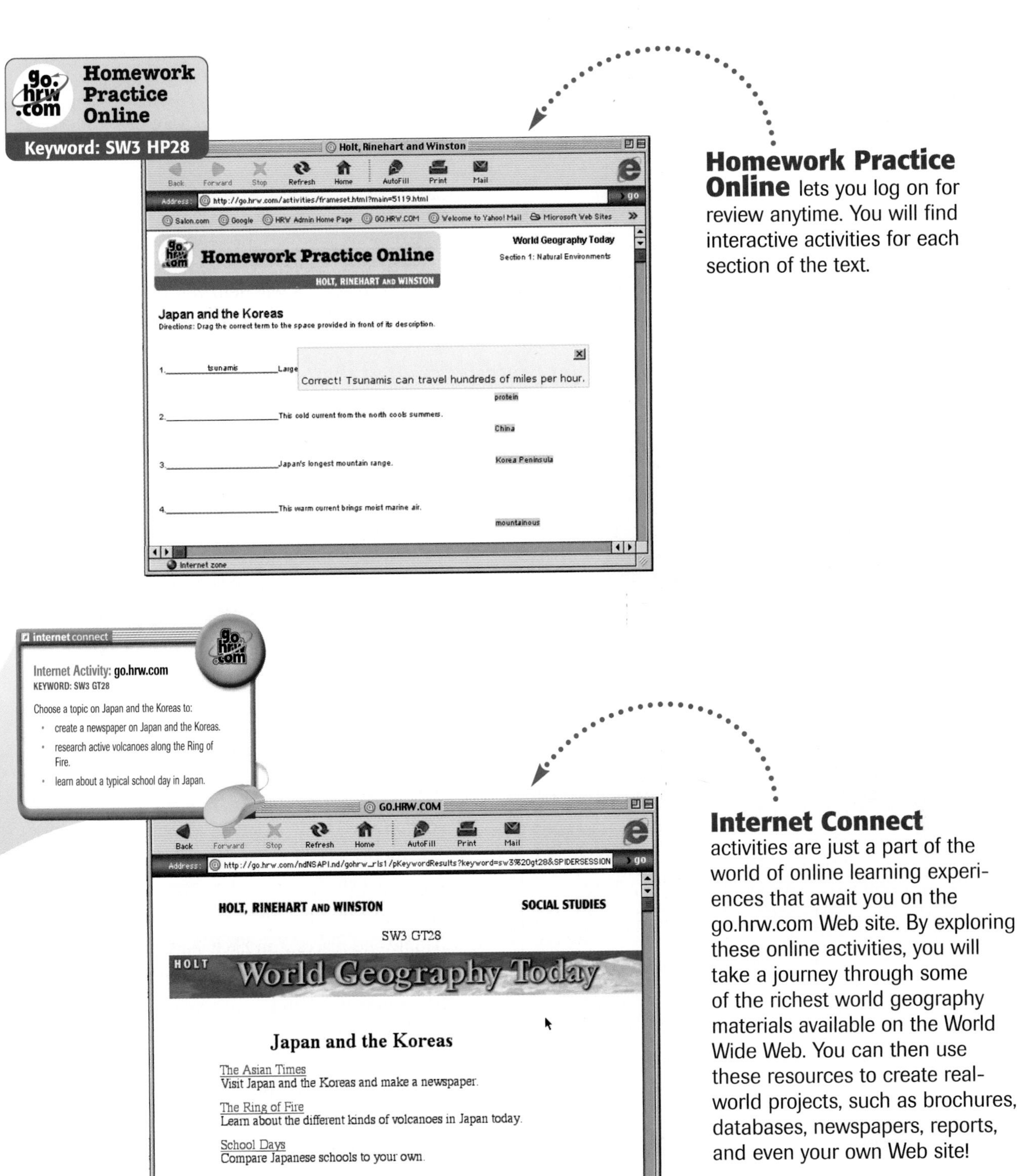

Homework Practice Online lets you log on for review anytime. You will find interactive activities for each section of the text.

Internet Connect activities are just a part of the world of online learning experiences that await you on the go.hrw.com Web site. By exploring these online activities, you will take a journey through some of the richest world geography materials available on the World Wide Web. You can then use these resources to create real-world projects, such as brochures, databases, newspapers, reports, and even your own Web site!

Why Geography Matters

Have you ever wondered. . .

why some places are deserts while other places get so much rain? What makes certain times of the year cooler than others? Why do some rivers run dry? Maybe you live near mountains and wonder what processes created them.

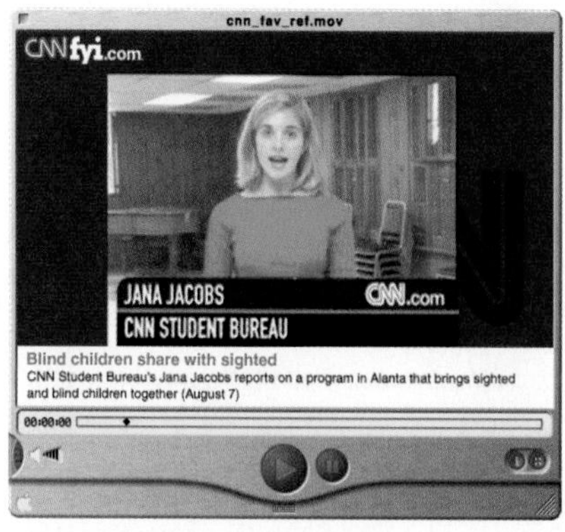

Do you know why the loss of huge forest areas in one part of the world can affect areas far away? Why does the United States have many different kinds of churches and other places of worship? Perhaps you are curious why Americans and people from other countries have such different points of view on many issues. The key to understanding questions and issues like these lies in the study of geography.

Student reporters contribute to **CNNfyi.com**.

Geography and Your World

All you need to do is watch or read the news to see the importance of geography. You have probably seen news stories about the effects of floods, volcanic eruptions, and other natural events on people and places. You likely have also seen how conflict and cooperation shape the relations between peoples and countries around the world. The Why It Matters feature beginning every section of World Geography Today uses the vast resources of **CNNfyi.com** or other current events sources to examine the importance of geography. Through this feature you will be able to draw connections between what you are studying in your geography textbook and events and conditions found around the world today.

My fall semester project testing lake water.

Geography and You

Anyone can influence the geography of our world. For example, the actions of individuals affect local environments. Some individual actions might pollute the environment. Other actions might contribute to efforts to keep the environment clean and healthy. Various other things also influence geography. For example, governments create political divisions, such as countries and states. The borders between these divisions influence the human geography of regions by separating peoples, legal systems, and human activities.

Governments and businesses also plan and build structures like dams, railroads, and airports, which change the physical characteristics of places. As you might expect, some actions influence Earth's geography in negative ways, others in positive ways. Understanding geography helps us evaluate the consequences of our actions.

Geography and Making Connections

When you think of the word geography, what comes to mind? Perhaps you simply picture people memorizing names of countries and capitals. Maybe you think of people studying maps to identify features like deserts, mountains, oceans, and rivers. These things are important, but the study of geography includes much more. Geography involves asking questions and solving problems. It focuses on looking at people and their ways of life as well as studying physical features like mountains, oceans, and rivers. Studying geography also means looking at why things are where they are and at the relationships between human and physical features of Earth.

The study of geography helps us make connections between what was, what is, and what may be. It helps us understand the processes that have shaped the features we observe around us, as well as the ways those features may be different tomorrow. In short, geography helps us understand the processes that have created a world that is home to more than 6 billion people and countless billions of other creatures.

Skills Handbook and Atlas

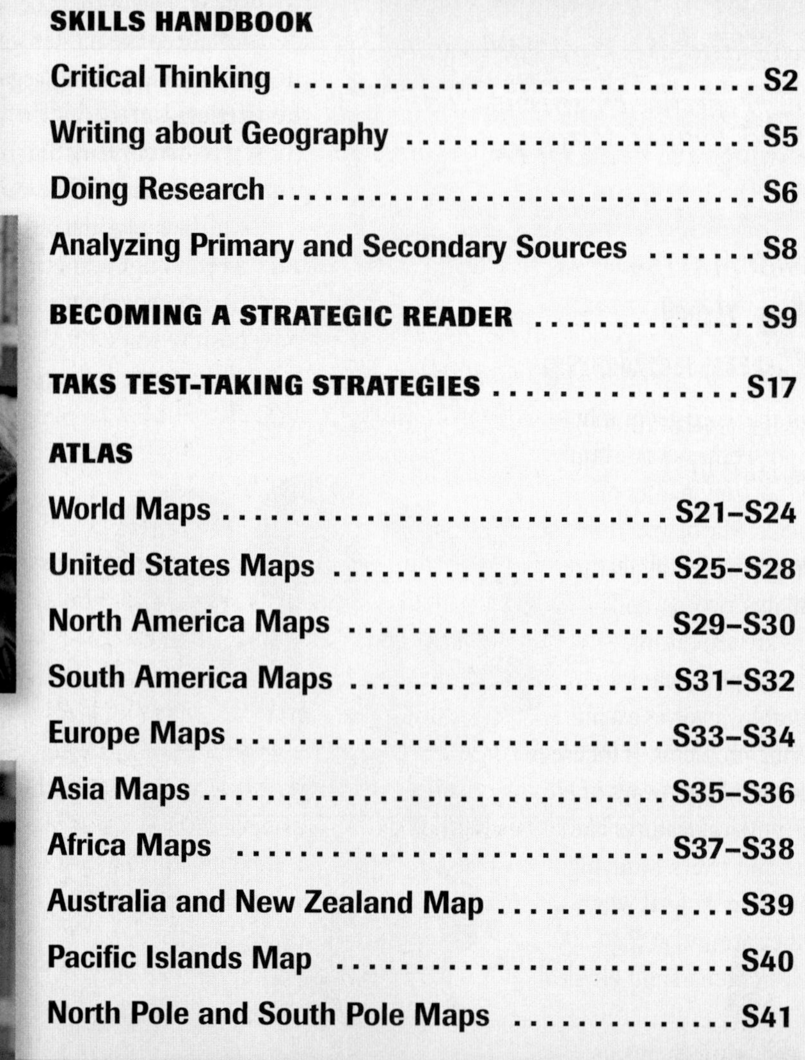

Critical Thinking

Throughout *World Geography Today, you are asked to think critically about some of the information you are studying. Critical thinking is the reasoned judgment of information and ideas. The development of critical thinking skills is essential to effective citizenship. Such skills empower you to exercise your civic rights and responsibilities as well as learn more about the world around you. Helping you develop critical thinking skills is an important goal of* World Geography Today. *The following critical thinking skills appear in the section reviews and chapter reviews of the textbook.*

Summarizing involves briefly restating information gathered from a larger body of information. Much of the writing in this textbook is summarizing. The geographical data in this textbook has been collected from many sources. Summarizing all the qualities of a region or country involves studying a large body of cultural, economic, geological, and historical information.

Finding the main idea is the ability to identify the main point in a set of information. This textbook is designed to help you focus on the main ideas in geography. The Read to Discover questions in each chapter help you identify the main ideas in each section. To find the main idea in any piece of writing, first read the title and introduction. These two elements may point to the main ideas covered in the text. Also, formulate questions about the subject that you think might be answered in the text. Having such questions in mind will focus your reading. Pay attention to any headings or subheadings, which may provide a basic outline of the major ideas. Finally, as you read, note sentences that provide additional details from the general statements that those details support. For example, a trail of facts may lead to a conclusion that expresses the main idea.

Mud mosque, Mali

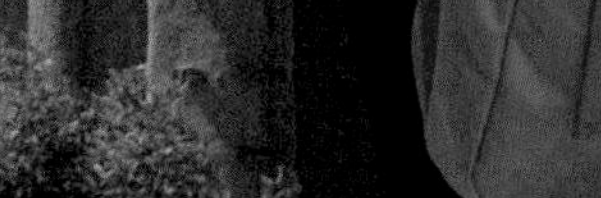

Comparing and contrasting involve examining events, points of view, situations, or styles to identify their similarities and differences. Comparing focuses on both the similarities and the differences. Contrasting focuses only on the differences. Studying similarities and differences between people and things can give you clues about the human and physical geography of a region.

Buddhist shrine, Myanmar

Stave church, Norway

Supporting a point of view involves identifying an issue, deciding what you think about it, and persuasively expressing your position. Your stand should be based on specific information. When taking a stand, state your position clearly and give reasons that support it.

Identifying points of view involves noting the factors that influence the outlook of an individual or group. A person's point of view includes beliefs and attitudes that are shaped by factors such as age, gender, race, and economic status. Identifying points of view helps us examine why people see things as they do. It also reinforces the realization that people's views may change over time or with a change in circumstances.

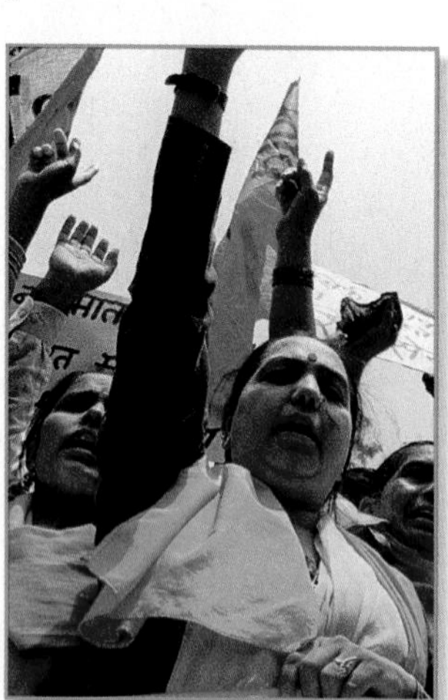
Political protest, India

Identifying bias is an important critical thinking skill in the study of any subject. When a point of view is highly personal or based on unreasoned judgment, it is considered biased. Sometimes, a person's actions reflect bias. At its most extreme, bias can be expressed in violent actions against members of a particular culture or group. A less obvious form of bias is a stereotype, or a generalization about a group of people. Stereotypes tend to ignore differences within groups.

Probably the hardest form of cultural bias to detect has to do with perspective, or point of view. When we use our own culture and experiences as a point of reference from which to make statements about other cultures, we are showing a form of bias called ethnocentrism.

Analyzing is the process of breaking something down into parts and examining the relationships between those parts. For example, to understand the processes behind forest loss, you might study issues involving economic development, the overuse of resources, and pollution.

Ecuador rainforest

Cleared forest, Kenya

Evaluating involves assessing the significance or overall importance of something. For example, you might evaluate the success of certain environmental protection laws or the effect of foreign trade on a society. You should base your evaluation on standards that others will understand and are likely to consider valid. For example, an evaluation of international relations after World War II might assess the political and economic tensions between the United States and the Soviet Union. Such an evaluation would also consider the ways those tensions affected other countries around the world.

Identifying cause and effect is part of interpreting the relationships between geographical events. A cause is any action that leads to an event; the outcome of that action is an effect. To explain geographical developments, geographers may point out multiple causes and effects. For example, geographers studying pollution in a region might note a number of causes.

Drawing inferences and drawing conclusions are two methods of critical thinking that require you to use evidence to explain events or information in a logical way. Inferences and conclusions are opinions, but these opinions are based on facts and reasonable deductions.

Drought in West Texas

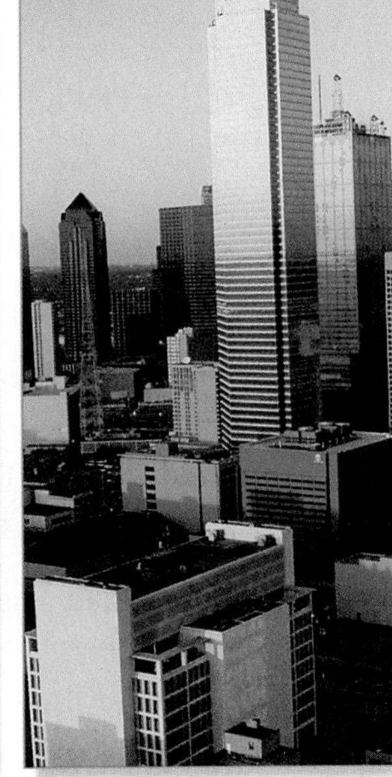

Dallas, Texas

For example, suppose you know that people are moving in greater and greater numbers to cities in a particular country. You also know that poor weather has hurt farming in outlying areas while industry has been expanding in cities. You might infer from this information some of the reasons for the increased migration to cities. You could conclude that poor harvests have pushed people to leave outlying areas. You might also conclude that the possibility of finding work in new industries may be pulling people to cities.

Making generalizations and predictions are two critical thinking skills that require you to form concise ideas from a large body of information. When you are asked to generalize, you must take into account many different pieces of information. You then form a unifying concept that can be applied to all of the pieces of information. Many times making generalizations can help you see trends. Looking at trends can help you form a prediction. Making a prediction involves looking at trends in the past and present and making an educated guess about how these trends will affect the future.

Communications technology, rural Brazil

SKILLS HANDBOOK

Writing about Geography

Writers have many different reasons for writing. In your study of geography, you might write to accomplish many different tasks. You might write a paragraph or short paper to express your own personal feelings or thoughts about a topic or event. You might also write a paper to inform your class about an event, person, place, or thing. Sometimes you may want to write in order to persuade or convince readers to agree with a certain statement or to act in a particular way.

You will find various kinds of questions at the end of each section, chapter, and unit throughout this textbook. Some questions will require in-depth answers. The following guidelines for writing will help you structure your answers so that they clearly express your thoughts.

The Writing Process

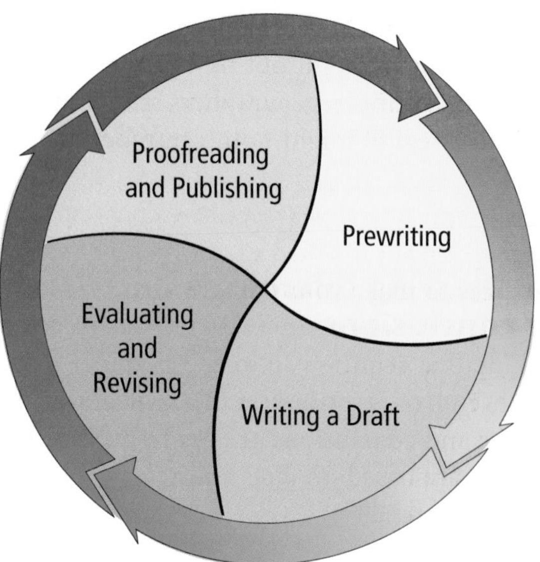

Prewriting Prewriting is the process of thinking about and planning what to write. It includes gathering and organizing information into a clear plan. Writers use the prewriting stage to identify their audience and purpose for what is to be written.

Often, writers do research to get the information they need. Research can include finding primary and secondary sources. You will read about primary and secondary sources later in this handbook.

Writing a Draft After you have gathered and arranged your information, you are ready to begin writing. Many paragraphs are structured in the following way:

- **Topic Sentence:** The topic sentence states the main idea of the paragraph. Putting the main idea into the form of a topic sentence helps keep the paragraph focused.
- **Body:** The body of a paragraph develops and supports the main idea. Writers use a variety of information, including facts, opinions, and examples, to support the main idea.
- **Conclusion:** The conclusion summarizes the writer's main points or restates the main idea.

Evaluating and Revising Read over your paragraphs and make sure you have clearly expressed what you wanted to say. Sometimes it helps to read your paragraphs aloud or to ask someone else to read it. Such methods help you identify rough or unclear sentences and passages. Revise the parts of your paragraph that are not clear or that stray from your main idea. You might want to add, cut, reorder, or replace sentences to make your paragraph as clear as possible.

Proofreading and Publishing Before you write your final draft, read over your paragraphs and correct any errors in grammar, spelling, sentence structure, or punctuation. Common mistakes include misspelled placenames, incomplete sentences, and improper use of punctuation, such as commas. You should use a dictionary and standard grammar guides to help you proofread your work.

After you have revised and corrected your draft, neatly rewrite or type your paper. Make sure your final version is clean and free of mistakes. The appearance of your final version can affect how your audience perceives and understands your writing.

Practicing the Skill

1. What are the steps in the writing process?
2. What elements are found in many paragraphs?
3. Write a paragraph or short paper about your community for a visitor. When you have finished your draft, review it and then mark and correct any errors in grammar, spelling, sentence structure, or punctuation. At the bottom of your draft, list key resources—such as a dictionary—that you used to check and correct your work. Then write your final draft. When you are finished with your work, use pencils or pens of different colors to underline and identify the topic sentence, body, and conclusion of your paragraph.

Doing Research

Research is at the heart of geographic inquiry. To complete a research project, you may need to use resources other than this textbook. For example, you may want to research specific places or issues not discussed in this textbook. You may also want to learn more about a certain topic that you have studied in a chapter. Following the guidelines below will help you plan and execute any research project you would like to undertake.

Planning The first step in approaching a research project is planning. Planning involves deciding on a topic and finding information about that topic.

- **Decide on a Topic.** Before starting any research project, you should decide on one topic. If you are working with a group, all group members should participate in choosing a topic. Sometimes a topic will be assigned to you, but at other times you may have to choose your own. Once you have settled on a topic, make sure you can find resources to help you research it.

- **Find Information.** In order to find a particular book, you need to know how libraries organize their materials. Libraries classify their books by assigning each book a call number that tells you its location. To find the call number, look in the library's card catalog. The card catalog lists books by author, by title, and by subject. Many libraries today have computerized card catalogs. Libraries often provide instructions on how to use their computerized card catalogs. If no instructions are available, ask a library staff member for help.

  Most libraries have encyclopedias, gazetteers, atlases, almanacs, and periodical indexes. Encyclopedias contain geographic, economic, and political data on individual countries, states, and cities. They also include discussions of historical events, religion, social and cultural issues, and much more. A gazetteer is a geographical dictionary that lists significant natural physical features and other places. An atlas contains maps and visual representations of geographic data. To find up-to-date facts, you can use almanacs, yearbooks, and periodical indexes. References like *The World Almanac and Book of Facts* include historical information and a variety of statistics. Periodical indexes, particularly *The Reader's Guide to Periodical Literature*, can help you locate informative articles published in magazines. *The New York Times Index* catalogs the newspaper articles published in the *New York Times*.

You may also want to find information on the World Wide Web. The World Wide Web is the part of the Internet where people put files called Web sites for other people to access. To search the World Wide Web, you must use a search engine. A search engine will provide you with a list of Web sites that contain keywords relating to your topic. Search engines also provide Web directories, which allow you to browse Web sites by subject.

Organizing Organization is key to completing research projects of any size. If you are working with a group, every group member should have an assigned task in researching, writing, and completing your project. You and all the group members should keep track of the materials that you used to conduct your research. Then compile those sources into a bibliography and turn it in with your research project.

In addition, information collected during research should be organized in an efficient way. A common method of organizing research information is to use index cards. If you have used an outline to organize your research, you can code each index card with the appropriate main idea number and supporting detail letter from the outline. Then write the relevant information on that card. You might also use computer files in the same way. These methods will help you keep track of what information you have collected and what information you still need to gather.

Some projects will require you to conduct original research. This original research might require you to interview people, conduct surveys, collect unpublished information about your community, or draw a map of a local place.

Before you do your original research, make sure you have all the necessary background information. Also, create a pre-research plan so that you can make sure all the necessary tools, such as research sources, are available.

Completing and Presenting Your Project Once you have completed your research project, you will need to present the information you have gathered in some fashion. Many times, you or your group will simply need to write a paper about your research. Research can also be presented in many other ways, however. For example, you could make an audio tape, a drawing, a poster board, a video, or a Web page to explain your research.

Practicing the Skill

1. What kinds of references would you need to research specific current events around the world?
2. Work with a group of four other students to plan, organize, and complete a research project on a topic of interest in your local community. For example, you might want to learn more about a particular individual or event that influenced your community's history. Other topics might include the economic features, physical features, and political features of your community.

Analyzing Primary and Secondary Sources

When conducting research, it is important to use a variety of primary and secondary sources of information. There are many sources of firsthand geographical information, including diaries, letters, editorials, and legal documents such as land titles. All of these are primary sources. Newspaper articles are also considered primary sources, although they are generally written after the fact. Other primary sources include personal memoirs and autobiographies, which people usually write late in life. Paintings and photographs of particular events, persons, places, or things make up a visual record and are also considered primary sources. Because they allow us to take a close-up look at a topic, primary sources are valuable geographic tools.

Secondary sources are descriptions or interpretations of events written after the events have occurred by persons who did not participate in the events they describe. Geography textbooks such as this one as well as biographies, encyclopedias, and other reference works are examples of secondary sources. Writers of secondary sources have the advantage of seeing what happened beyond the moment or place that is being studied. They can provide a perspective wider than that available to one person at a specific time.

How to Study Primary and Secondary Sources

1 **Study the Material Carefully.** Consider the nature of the material. Is it verbal or visual? Is it based on firsthand information or on the accounts of others? Note the major ideas and supporting details.

2 **Consider the Audience.** Ask yourself, "For whom was this message originally meant?" Whether a message was intended for the general public or for a specific private audience may have shaped its style or content.

3 **Check for Bias.** Watch for words or phrases that present a one-sided view of a person or situation.

4 **Compare Sources.** Study more than one source on a topic. Comparing sources gives you a more complete and balanced account of geographical events and their relationships to one another.

Practicing the Skill

1. What distinguishes secondary sources from primary sources?
2. What advantages do secondary sources have over primary sources?
3. Why should you consider the intended audience of a source?
4. Of the following, identify which are primary sources and which are secondary sources: a newspaper, a private journal, a biography, an editorial cartoon, a medieval tapestry, a deed to property, a snapshot of a family vacation, a magazine article about the history of Thailand, an autobiography. How might some of these sources prove to be both primary and secondary sources?

READING STRATEGIES

Becoming a Strategic Reader

by Dr. Judith Irvin

Everywhere you look, print is all around us. In fact, you would have a hard time stopping yourself from reading. In a normal day, you might read cereal boxes, movie posters, notes from friends, t-shirts, instructions for video games, song lyrics, catalogs, billboards, information on the Internet, magazines, the newspaper, and much, much more. Each form of print is read differently depending on your purpose for reading. You read a menu differently from poetry, and a motorcycle magazine is read differently than a letter from a friend. Good readers switch easily from one type of text to another. In fact, they probably do not even think about it, they just do it.

When you read, it is helpful to use a strategy to remember the most important ideas. You can use a strategy before you read to help connect information you already know to the new information you will encounter. Before you read, you can also predict what a text will be about by using a previewing strategy. During the reading you can use a strategy to help you focus on main ideas, and after reading you can use a strategy to help you organize what you learned so that you can remember it later. *World Geography Today* was designed to help you more easily understand the ideas you read. Important reading strategies employed in *World Geography Today* include:

A Tools to help you **preview and predict** what the text will be about

B Ways to help you **use and analyze visual information**

C Ideas to help you **organize the information** you have learned

A. Previewing and Predicting

How can I figure out what the text is about before I even start reading a section?

Previewing and **predicting** are good methods to help you understand the text. If you take the time to preview and predict before you read, the text will make more sense to you during your reading.

1 Usually, your teacher will set the purpose for reading. After reading some new information, you may be asked to write a summary, take a test, or complete some other type of activity.

Previewing and Predicting

step 1 Identify your purpose for reading. Ask yourself what you will do with this information once you have finished reading.

step 2 Ask yourself what is the main idea of the text and what are the key vocabulary words you need to know.

step 3 Use signal words to help identify the structure of the text.

step 4 Connect the information to what you already know.

2 As you preview the text, use **graphic signals** such as headings, subheadings, and boldface type to help you determine what is important in the text. Each section of *World Geography Today* opens by giving you important clues to help you preview the material.

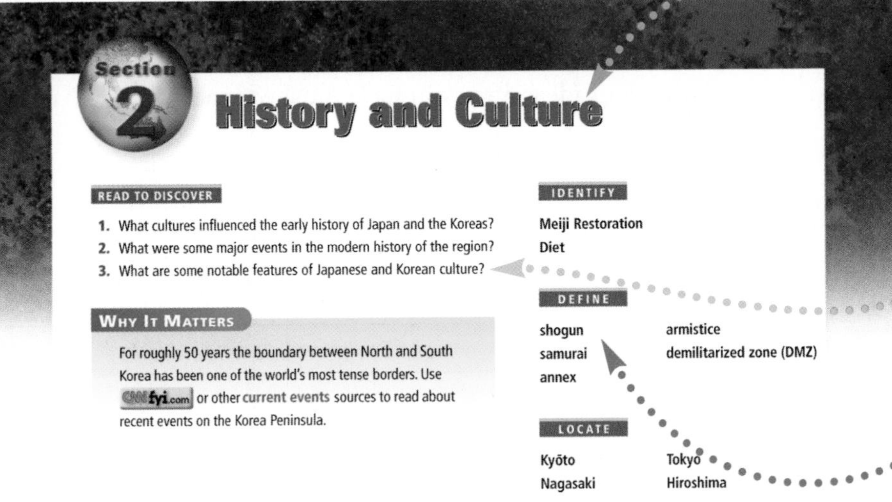

Looking at the section's **main heading** and subheadings can give you an idea of what is to come.

Read to Discover questions give you clues as to the section's main ideas.

Identify and Define terms let you know the key vocabulary you will encounter in the section.

3 Other tools that can help you in previewing are **signal words**. These words prepare you to think in a certain way. For example, when you see words such as *similar to, same as,* or *different from*, you know that the text will probably compare and contrast two or more ideas. Signal words indicate how the ideas in the text relate to each other. Look at the list below for some of the most common signal words grouped by the type of text structures they include.

SIGNAL WORDS

Cause and Effect	Compare and Contrast	Description	Problem and Solution	Sequence or Chronological Order
because since consequently this led to...so if...then nevertheless accordingly because of as a result of in order to may be due to for this reason not only...but	different from same as similar to as opposed to instead of although however compared with as well as either...or but on the other hand unless	for instance for example such as to illustrate in addition most importantly another furthermore first, second ...	the question is a solution one answer is	not long after next then initially before after finally preceding following on (date) over the years today when

4 Learning something new requires that you connect it in some way with something you already know. This means you have to think before you read and while you read. You may want to use a chart like this one to remind yourself of the information already familiar to you and to come up with questions you want answered in your reading. The chart will also help you organize your ideas after you have finished reading.

What I know	What I want to know	What I learned

B. Use and Analyze Visual Information

How can all the pictures, maps, graphs, and time lines with the text help me be a stronger reader?

Using visual information can help you understand and remember the information presented in *World Geography Today*. Good readers make a picture in their mind when they read. The pictures, charts, graphs, time lines, and diagrams that occur throughout *World Geography Today* are placed strategically to increase your understanding.

1 You might ask yourself questions like these:

Why did the writer include this image with the text?

What details about this image are mentioned in the text?

Analyzing Visual Information

step 1 As you preview the text, ask yourself how the visual information relates to the text.

step 2 Generate questions based on the visual information.

step 3 After reading the text, go back and review the visual information again.

step 4 Connect the information to what you already know.

2 After you have read the text, see if you can answer your own questions.

→ *Why is the train important?*

→ *What infrastructure is needed for this train to operate?*

→ *Why is the area around the rail line built up?*

B. Use and Analyze Visual Information *(continued)*

2 Maps, graphs, and charts help you organize information about a place. You might ask questions like these:

How does this map support what I have read in the text?

What does the information in these pie and bar graphs add to the text discussion?

→What is the purpose of this map?

→What special features does the map show?

→What do the colors, lines, and symbols on the map represent?

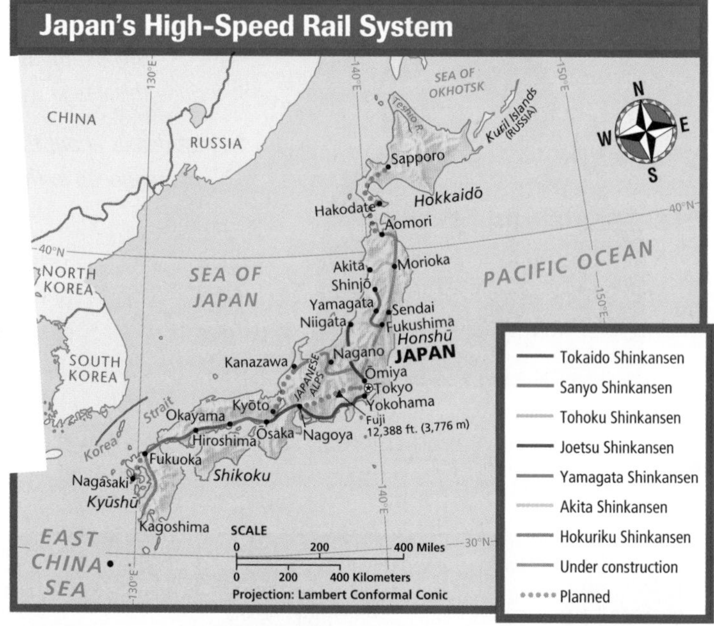

→What information is the writer trying to present with these graphs?

→Why did the writer use a pie graph and a bar graph to organize this information?

→Is information in the pie graph important to understanding the bar graph?

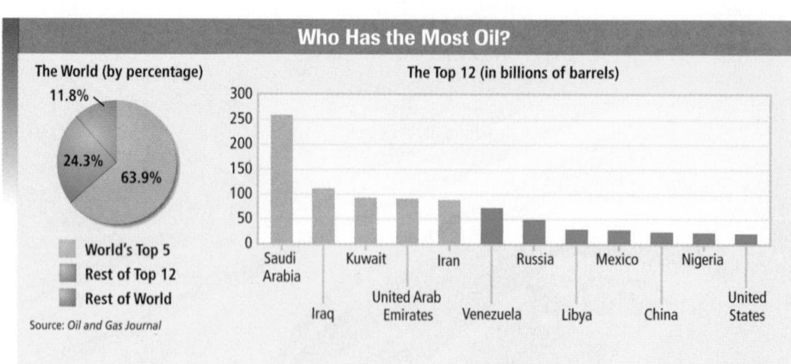

3 After reading the text, go back and review the visual information again.

4 Connect the information to what you already know.

C. Organize Information

Once I learn new information, how do I keep it all straight so that I will remember it?

To help you remember what you have read, you need to find a way of **organizing information.** Two good ways of doing this are by using graphic organizers and concept maps. **Graphic organizers** help you understand important relationships—such as cause-and-effect, compare/contrast, sequence of events, and problem/solution—within the text. **Concept maps** provide a useful tool to help you focus on the text's main ideas and organize supporting details.

Identifying Relationships

Using graphic organizers will help you recall important ideas from the section and give you a study tool you can use to prepare for a quiz or test or to help with a writing assignment. Some of the most common types of graphic organizers are shown below.

Cause and Effect

Events in history cause people to react in a certain way. Cause-and-effect patterns show the relationship between results and the ideas or events that made the results occur. You may want to represent cause-and-effect relationships as one cause leading to multiple effects,

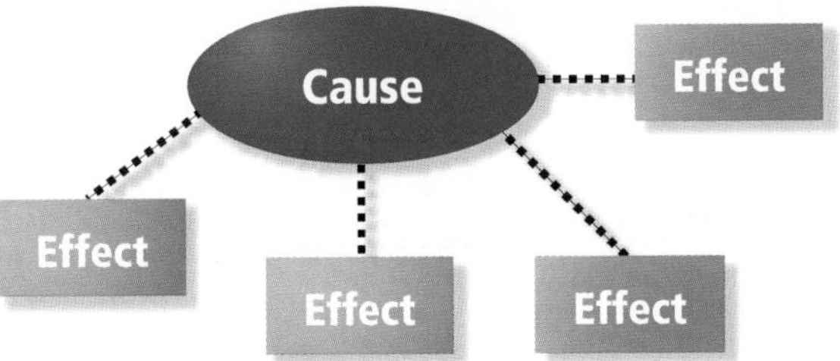

or as a chain of cause-and-effect relationships.

Cause 1 → Effect 1/ Cause 2 → Effect 2/ Cause 3 → Effect 3

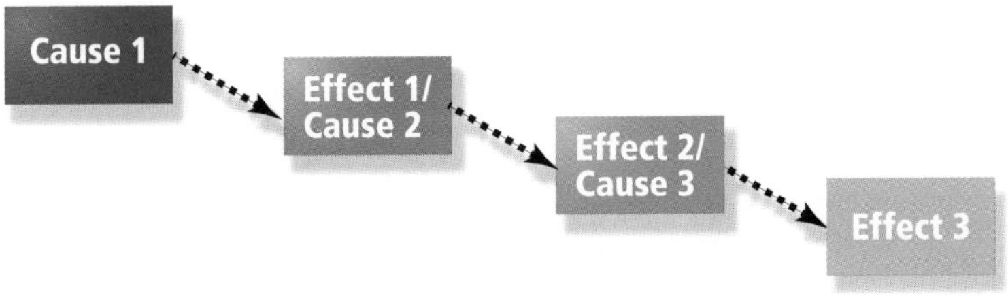

Constructing Graphic Organizers

step 1 Preview the text, looking for signal words and the main idea.

▼

step 2 Form a hypothesis as to which type of graphic organizer would work best to display the information presented.

▼

step 3 Work individually or with your classmates to create a visual representation of what you read.

C. Organize Information *(continued)*

Comparing and Contrasting

Graphic Organizers are often useful when you are comparing or contrasting information. Compare-and-contrast diagrams point out similarities and differences between two concepts or ideas.

Sequencing

Keeping track of dates and the order in which events took place is essential to understanding the history and geography of a place. Sequence or chronological-order diagrams show events or ideas in the order in which they happened.

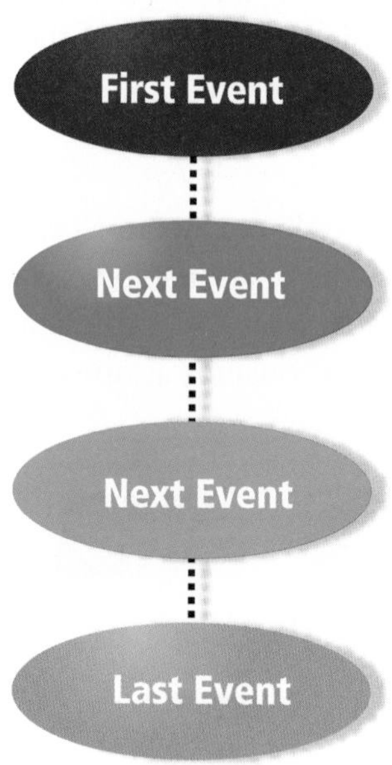

Problem and Solution

Problem-solution patterns identify at least one problem, offer one or more solutions to the problem, and explain or predict outcomes of the solutions.

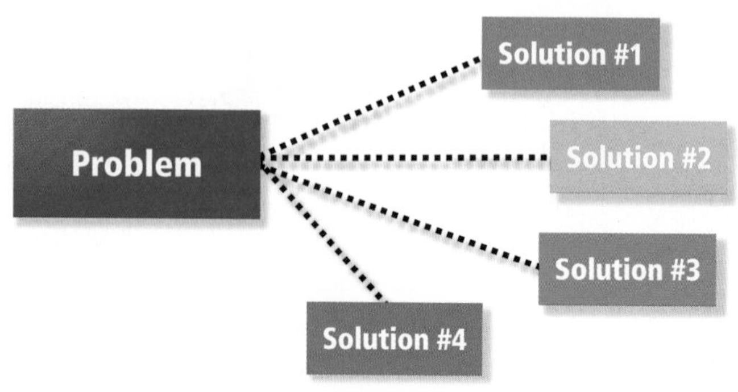

C. Organize Information *(continued)*

Identifying Main Ideas and Supporting Details

One special type of graphic organizer is the concept map. A concept map, sometimes called a semantic map, allows you to zero in on the most important points of the text. The map is made up of lines, boxes, circles, and/or arrows. It can be as simple or as complex as you need it to be to accurately represent the text.

Here are a few examples of concept maps you might use.

Main Idea

Supporting Idea 1 | Supporting Idea 2 | Supporting Idea 3

Detail 1 | Detail 2 | Detail 1 | Detail 2 | Detail 1 | Detail 2

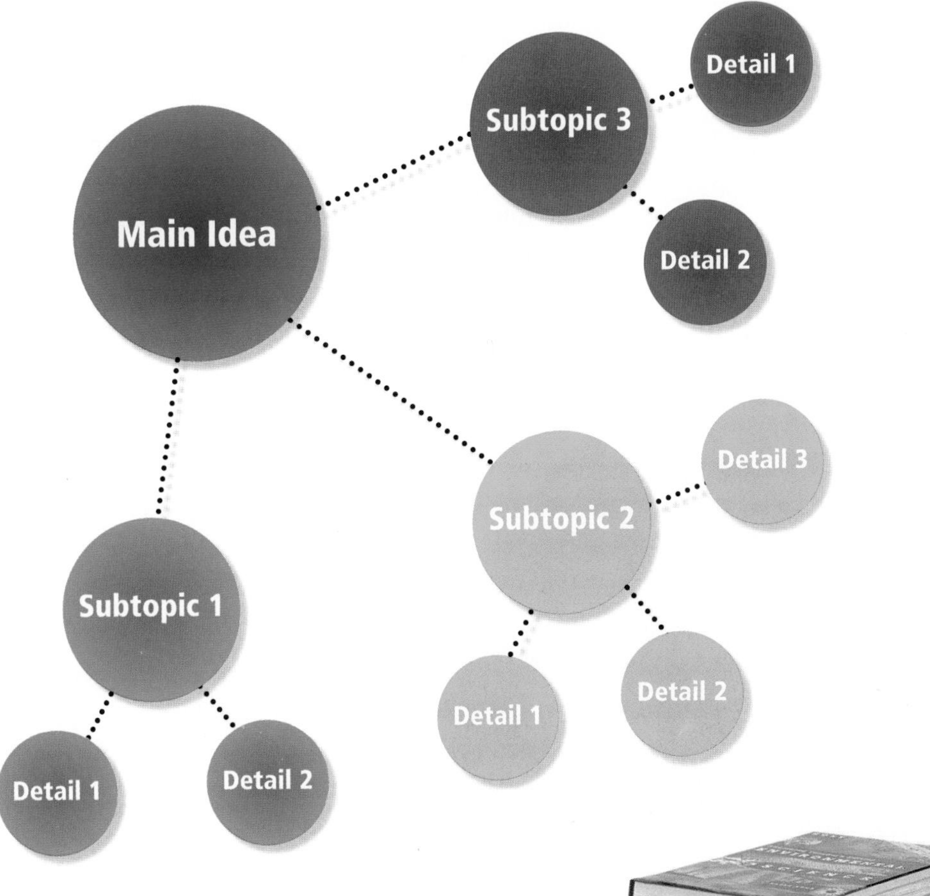

Constructing Concept Maps

step 1 Preview the text, looking at what type of structure might be appropriate to display on a concept map.

step 2 Taking note of the headings, bold-faced type, and text structure, sketch a concept map you think could best illustrate the text.

step 3 Using boxes, lines, arrows, circles, or any shapes you like, display the ideas of the text in the concept map.

TAKS Test-Taking Strategies

Every year in school, from grade 2 through grade 11, you will be asked to take the TAKS (Texas Assessment of Knowledge and Skills) Test. The test is designed to demonstrate the content and skills you have learned. It is important to keep in mind that in most cases the best way to prepare for the test is to pay close attention in class and take every opportunity to improve your general social studies, reading, writing, and mathematical skills.

Tips for Taking the Test

1. Be sure that you are well rested.
2. Be on time, and be sure that you have the necessary materials.
3. Listen to the teacher's instructions.
4. Read directions and questions carefully.
5. **DON'T STRESS!** Just remember what you have learned in class, and you should do well.

Practice the strategies at go.hrw.com

go.hrw.com
TAKS PREP ONLINE
keyword: SW3 T

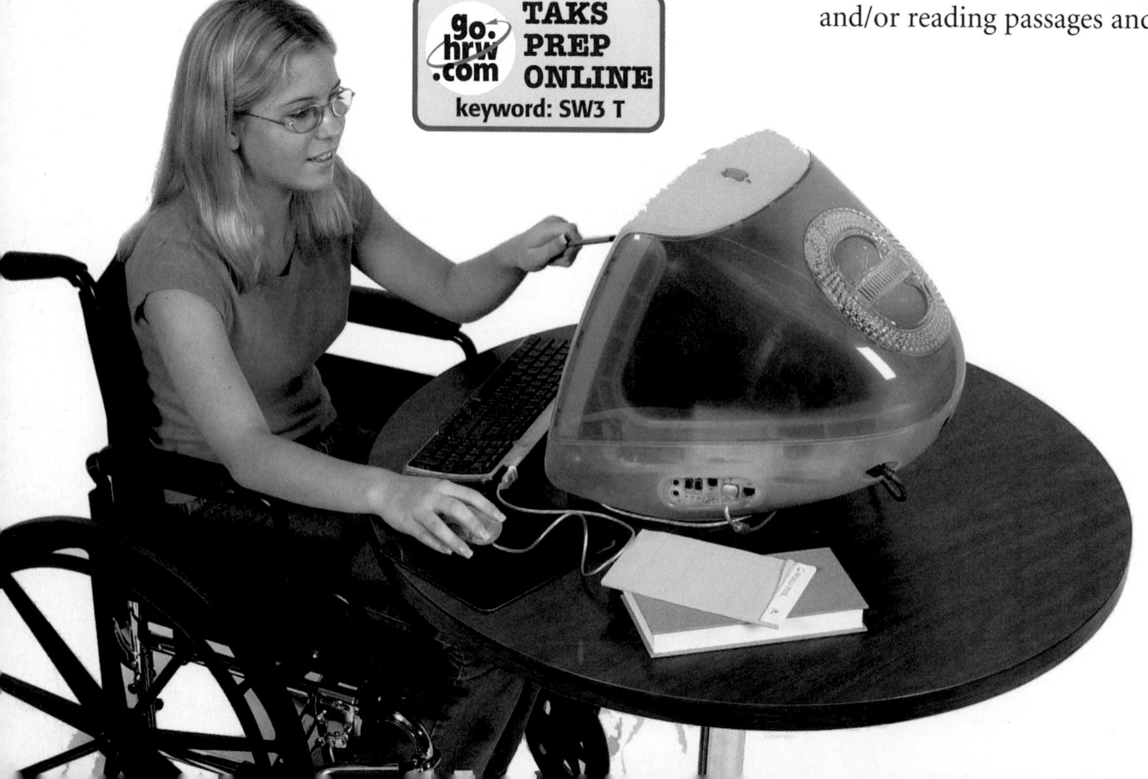

Tackling Social Studies

The social studies portions of the TAKS are designed to test your knowledge of the content and skills that you have been studying in one or more of your social studies classes. The objectives for TAKS are as follows:

1. Demonstrate an understanding of issues and events in history.
2. Demonstrate an understanding of geographic influences on historical issues and events.
3. Demonstrate an understanding of economic and social influences on historical issues and events.
4. Demonstrate an understanding of political influences on historical issues and events.
5. Use critical thinking skills to analyze social studies information.

The TAKS contains multiple-choice and may in the future contain open-ended questions also. The multiple-choice items will often be based on maps, tables, charts, graphs, pictures, cartoons, and/or reading passages and documents.

Tips for Answering Multiple-Choice Questions

1. If there is a written or visual piece accompanying the multiple-choice question, pay careful attention to the title, author, and date.
2. Then read through or glance over the content of the written or visual piece accompanying the question to familiarize yourself with it.
3. Next, read the multiple-choice question first for its general intent. Then reread it carefully, looking for words that give clues or can limit possible answers to the question. For example, words such as *most* or *best* tell you that there may be several correct answers to a question, but you should look for the most appropriate answer.
4. Read through the answer choices. Always read all of the possible answer choices even if the first one seems like the correct answer. There may be a better choice farther down in the list.
5. Reread the accompanying information (if any is included) carefully to determine the answer to the question. Again, note the title, author, and date of primary-source selections. The answer will rarely be stated exactly as it appears in the primary source, so you will need to use your critical thinking skills to read between the lines.
6. Think of what you already know about the time in history or person involved and use that to help limit the answer choices.
7. Finally, reread the question and selected answer to be sure that you made the best choice and that you marked it correctly on the answer sheet.

Strategies for Success

There are a variety of strategies you can prepare ahead of time to help you feel more confident about answering questions on the social studies TAKS. Here are a few suggestions:

1. Adopt an acronym—a word formed from the first letters of other words—that you will use for analyzing a document or visual piece that accompanies a question.

Helpful Acronyms

For a document, use SOAPS, which stands for

S Subject
O Overview
A Audience
P Purpose
S Speaker/author

For a picture, cartoon, map, or other visual piece of information, use OPTIC, which stands for

O Occasion (or time)
P Parts (labels or details of the visual)
T Title
I Interrelations (how the different parts of the visual work together)
C Conclusion (what the visual means)

2. Form visual images of maps and try to draw them from memory. The TAKS will most likely include maps showing many features, such as states, countries, continents, and oceans. Those maps may also show patterns in settlement and the size and distribution of cities. For example, in studying the United States, be able to see in your mind's eye such things as where the states and major cities are located. Know major physical features, such as the Mississippi River, the Appalachian and Rocky Mountains, the Great Plains, and the various regions of the United States, and be able to place them on a map. Such features may help you understand patterns in the distribution of population and the size of settlements.
3. When you have finished studying a geographic region or period in history, try to think of who or what might be important enough for the TAKS. You may want to keep your ideas in a notebook to refer to when it is almost time for the test.
4. The TAKS will likely test your understanding of the political, economic, and social processes that shape a region's history, culture, and geography. Questions may also ask you to

understand the impact of geographic factors on major events. For example, some may ask about the effects of migration and immigration on various societies and population change. In addition, questions may test your understanding of the ways humans interact with their environment.

5. For the skills area of the TAKS, practice putting major events and personalities in order in your mind. Sequencing people and events by dates can become a game you play with a friend who also has to take the TAKS. Always ask yourself "why" this event is important.
6. Follow the tips for TAKS reading below when you encounter a reading passage in social studies, but remember that what you have learned about history can help you in answering reading-comprehension questions.

Ready for Reading

The main goal of the reading sections of the TAKS Test is to determine your understanding of different aspects of a piece of writing. Basically, if you can grasp the main idea and the writer's purpose and then pay attention to the details and vocabulary so that you are able to draw inferences and conclusions, you will do well on the test.

Tips for Answering Multiple-Choice Questions

1. Read the passage as if you were not taking a test.
2. Look at the big picture. Ask yourself questions like, "What is the title?", "What do the illustrations or pictures tell me?", and "What is the writer's purpose?"
3. Read the questions. This will help you know what information to look for.
4. Reread the passage, underlining information related to the questions.
5. Go back to the questions and try to answer each one in your mind before looking at the answers.
6. Read all the answer choices and eliminate the ones that are obviously incorrect.

Types of Multiple-Choice Questions

1. **Main Idea** This is the most important point of the passage. After reading the passage, locate and underline the main idea.
2. **Significant Details** You will often be asked to recall details from the passage. Read the question and underline the details as you read, but remember that the correct answers do not always match the wording of the passage precisely.
3. **Vocabulary** You will often need to define a word within the context of the passage. Read the answer choices and plug them into the sentence to see what fits best.
4. **Conclusion and Inference** There are often important ideas in the passage that the writer does not state directly. Sometimes you must consider multiple parts of the passage to answer the question. If answers refer to only one or two sentences or details in the passage, they are probably incorrect.

Tips for Answering Short-Answer Questions

1. Read the passage in its entirety, paying close attention to the main events and characters. Jot down information you think is important.
2. If you cannot answer a question, skip it and come back later.
3. Words such as *compare, contrast, interpret, discuss,* and *summarize* appear often in short-answer questions. Be sure you have a complete understanding of each of these words.
4. To help support your answer, return to the passage and skim the parts you underlined.
5. Organize your thoughts on a separate sheet of paper. Write a general statement with which to begin. This will be your topic statement.
6. When writing your answer, be precise but brief. Be sure to refer to details in the passage in your answer.

Targeting Writing

On the TAKS Test you will occasionally be asked to write an essay. In order to write a concise essay, you must learn to organize your thoughts before you begin writing the actual composition. This keeps you from straying too far from the essay's topic.

Tips for Answering Composition Questions

1. Read the question carefully.
2. Decide what kind of essay you are being asked to write. Essays usually fall into one of the following types: persuasive, classificatory, compare/contrast, or "how to." To determine the type of essay, ask yourself questions like, "Am I trying to persuade my audience?", "Am I comparing or contrasting ideas?", or "Am I trying to show the reader how to do something?"
3. Pay attention to keywords, such as *compare, contrast, describe, advantages, disadvantages, classify,* or *speculate.* They will give you clues as to the structure that your essay should follow.
4. Organize your thoughts on a separate sheet of paper. You will want to come up with a general topic sentence that expresses your main idea. Make sure this sentence addresses the question. You should then create an outline or some type of graphic organizers to help you organize the points that support your topic sentence.
5. Write your composition using complete sentences. Also, be sure to use correct grammar, spelling, punctuation, and sentence structure.
6. Be sure to proofread your essay once you have finished writing.

Gearing Up for Math

On the TAKS Test you will be asked to solve a variety of mathematical problems that draw on the skills and information you have learned in class. If math problems sometimes give you difficulty, use the tips below to help you work through the problems.

Tips for Solving Math Problems

1. Decide what is the goal of the question. Read or study the problem carefully and determine what information must be found.
2. Locate the factual information. Decide what information represents key facts—the ones you must have to solve the problem. You may also find facts you do not need to reach your solution. In some cases, you may determine that more information is needed to solve the problem. If so, ask yourself, "What assumptions can I make about this problem?" or "Do I need a formula to help solve this problem?"
3. Decide what strategies you might use to solve the problem, how you might use them, and what form your solution will be in. For example, will you need to create a graph or chart? Will you need to solve an equation? Will your answer be in words or numbers? By knowing what type of solution you should reach, you may be able to eliminate some of the choices.
4. Apply your strategy to solve the problem and compare your answer to the choices.
5. If the answer is still not clear, read the problem again. If you had to make calculations to reach your answer, use estimation to see if your answer makes sense.

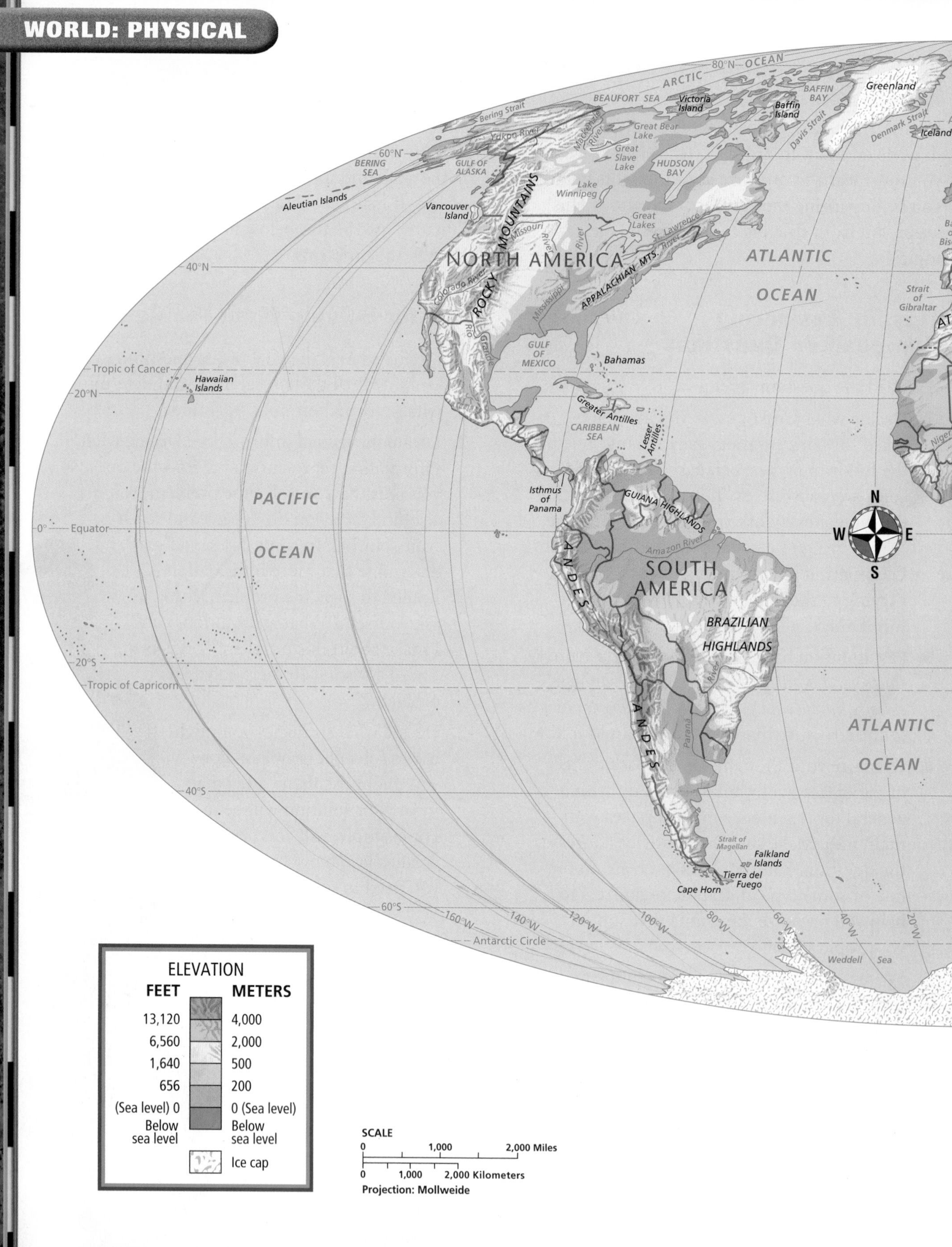
ARCTIC
80°N
OCEAN
BEAUFORT SEA
Victoria Island
BAFFIN BAY
Greenland
Baffin Island
Bering Strait
Yukon River
Mackenzie River
Great Bear Lake
Great Slave Lake
Davis Strait
Denmark Strait
Iceland
60°N
BERING SEA
GULF OF ALASKA
HUDSON BAY
Aleutian Islands
Lake Winnipeg
ROCKY MOUNTAINS
Vancouver Island
Missouri River
Great Lakes
St. Lawrence River
Bay of Biscay
NORTH AMERICA
APPALACHIAN MTS.
ATLANTIC
OCEAN
40°N
Colorado River
Mississippi River
Strait of Gibraltar
Rio Grande
GULF OF MEXICO
Bahamas
Tropic of Cancer
Hawaiian Islands
20°N
Greater Antilles
CARIBBEAN SEA
Lesser Antilles
Niger
PACIFIC
Isthmus of Panama
GUIANA HIGHLANDS
N
W
E
S
0°
Equator
OCEAN
Amazon River
ANDES
SOUTH AMERICA
BRAZILIAN HIGHLANDS
20°S
Tropic of Capricorn
Paraná
ATLANTIC
OCEAN
40°S
Strait of Magellan
Falkland Islands
Tierra del Fuego
Cape Horn
60°S
160°W
140°W
120°W
100°W
80°W
60°W
40°W
20°W
Antarctic Circle
Weddell Sea
ELEVATION
FEET
METERS
13,120
4,000
6,560
2,000
1,640
500
656
200
(Sea level) 0
0 (Sea level)
Below sea level
Below sea level
Ice cap
SCALE
0
1,000
2,000 Miles
0
1,000
2,000 Kilometers
Projection: Mollweide

ARCTIC OCEAN
North Cape
BARENTS SEA
KARA SEA
LAPTEV SEA
EAST SIBERIAN SEA
URAL MOUNTAINS
EUROPE
ALPS
BALTIC SEA
BLACK SEA
CASPIAN SEA
ARAL SEA
Lake Balkhash
ALTAY SHAN
GOBI
ASIA
Lake Baikal
SEA OF OKHOTSK
KAMCHATKA PENINSULA
Sakhalin
Hokkaidō
SEA OF JAPAN
Honshū
Shikoku
Kyūshū
EAST CHINA SEA
HIMALAYAS
THAR DESERT
MEDITERRANEAN SEA
ARABIAN PENINSULA
RED SEA
AFRICA
ARABIAN SEA
Bay of Bengal
Sri Lanka
Strait of Malacca
SOUTH CHINA SEA
Philippine Islands
Taiwan
Tropic of Cancer
PACIFIC OCEAN
MALAY PENINSULA
Borneo
Sumatra
Java
Sulawesi (Celebes)
New Guinea
Solomon Islands
Equator
Lake Victoria
Lake Tanganyika
INDIAN OCEAN
Madagascar
Mozambique Channel
KALAHARI DESERT
Cape of Good Hope
CORAL SEA
GREAT SANDY DESERT
AUSTRALIA
GREAT VICTORIA DESERT
GREAT DIVIDING RANGE
New Caledonia
Tropic of Capricorn
TASMAN SEA
North Island
South Island
Tasmania
ANTARCTICA
Iceland
Denmark Strait
KJØLEN MTS.
SCALE
0 250 500 750 Miles
0 250 500 750 Kilometers
Projection: Mollweide
NORTH SEA
BALTIC SEA
British Isles
ATLANTIC OCEAN
Bay of Biscay
ALPS
Danube River
Rhine River
Volga River
URAL MTS.
BLACK SEA
MEDITERRANEAN SEA
Strait of Gibraltar
Crete

WORLD: POLITICAL

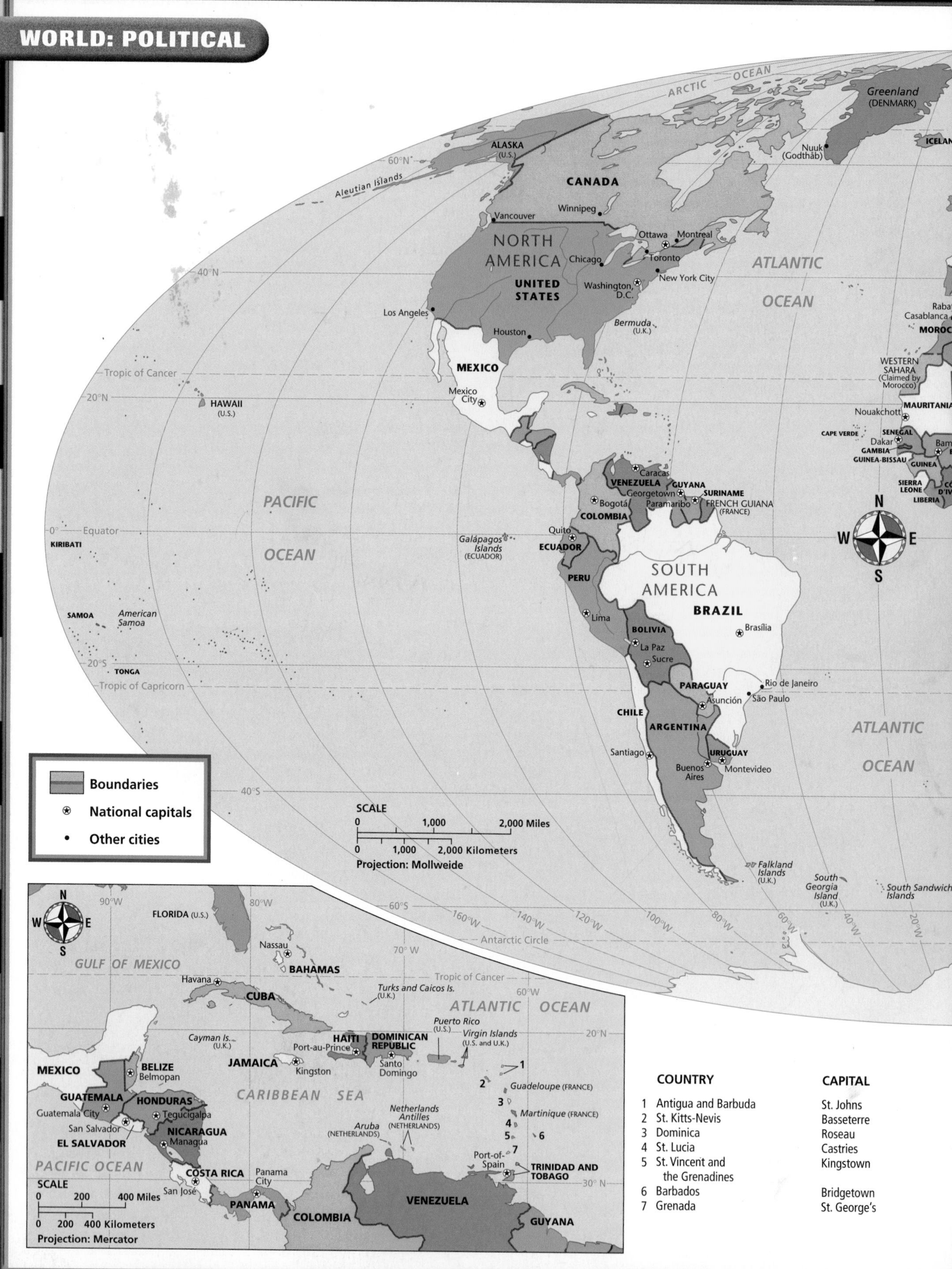

COUNTRY	CAPITAL
1 Antigua and Barbuda	St. Johns
2 St. Kitts-Nevis	Basseterre
3 Dominica	Roseau
4 St. Lucia	Castries
5 St. Vincent and the Grenadines	Kingstown
6 Barbados	Bridgetown
7 Grenada	St. George's

	COUNTRY	CAPITAL
1	Czech Republic	Prague
2	Slovakia	Bratislava
3	Slovenia	Ljubljana
4	Croatia	Zagreb
5	Bosnia and Herzegovina	Sarajevo
6	Macedonia	Skopje
7	Yugoslavia (Serbia and Montenegro)	Belgrade
8	Lithuania	Vilnius
9	Latvia	Riga
10	Estonia	Tallinn

THE UNITED STATES of AMERICA: PHYSICAL

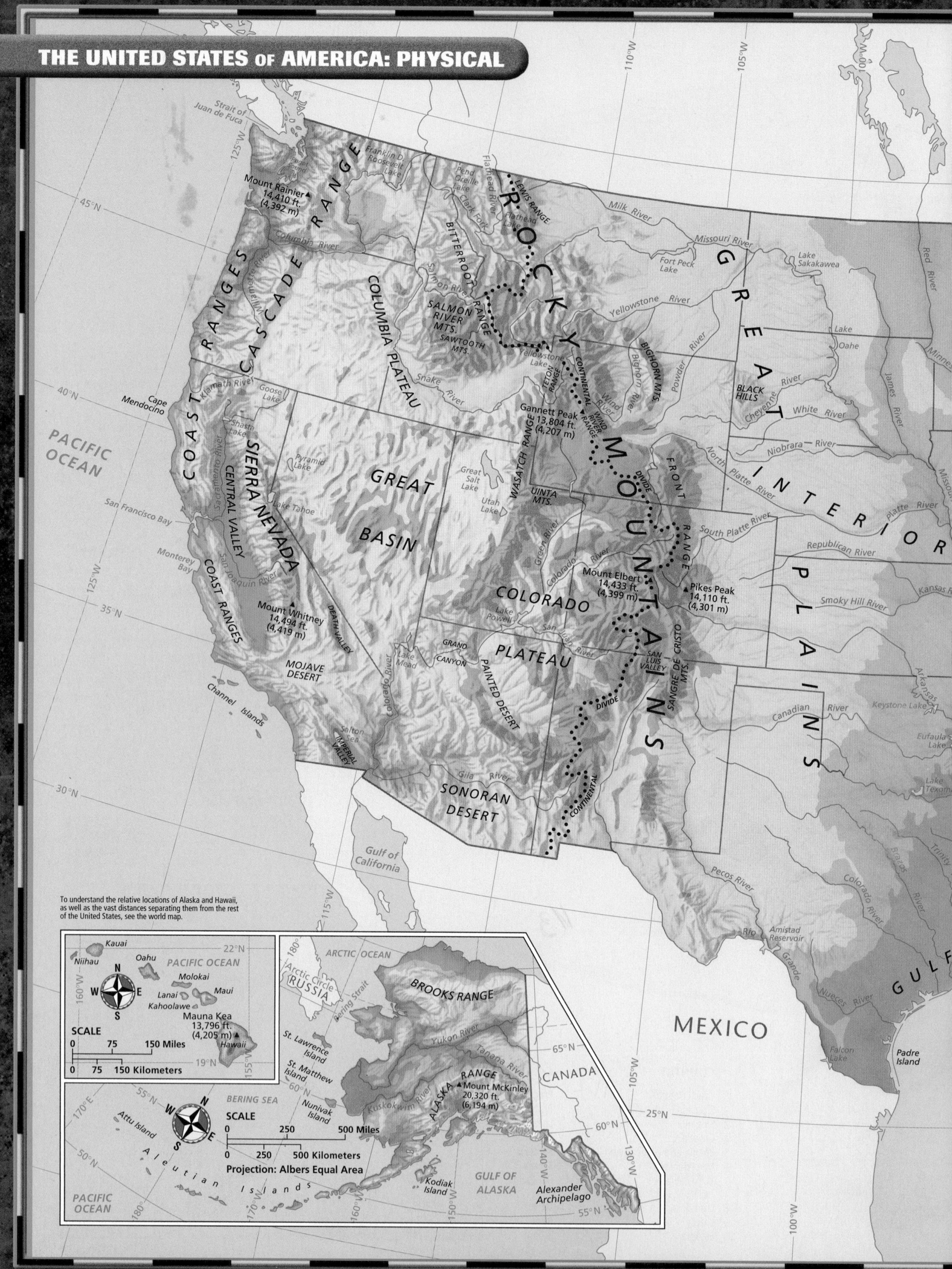

THE UNITED STATES OF AMERICA: PHYSICAL

THE UNITED STATES OF AMERICA: POLITICAL

THE UNITED STATES OF AMERICA: POLITICAL

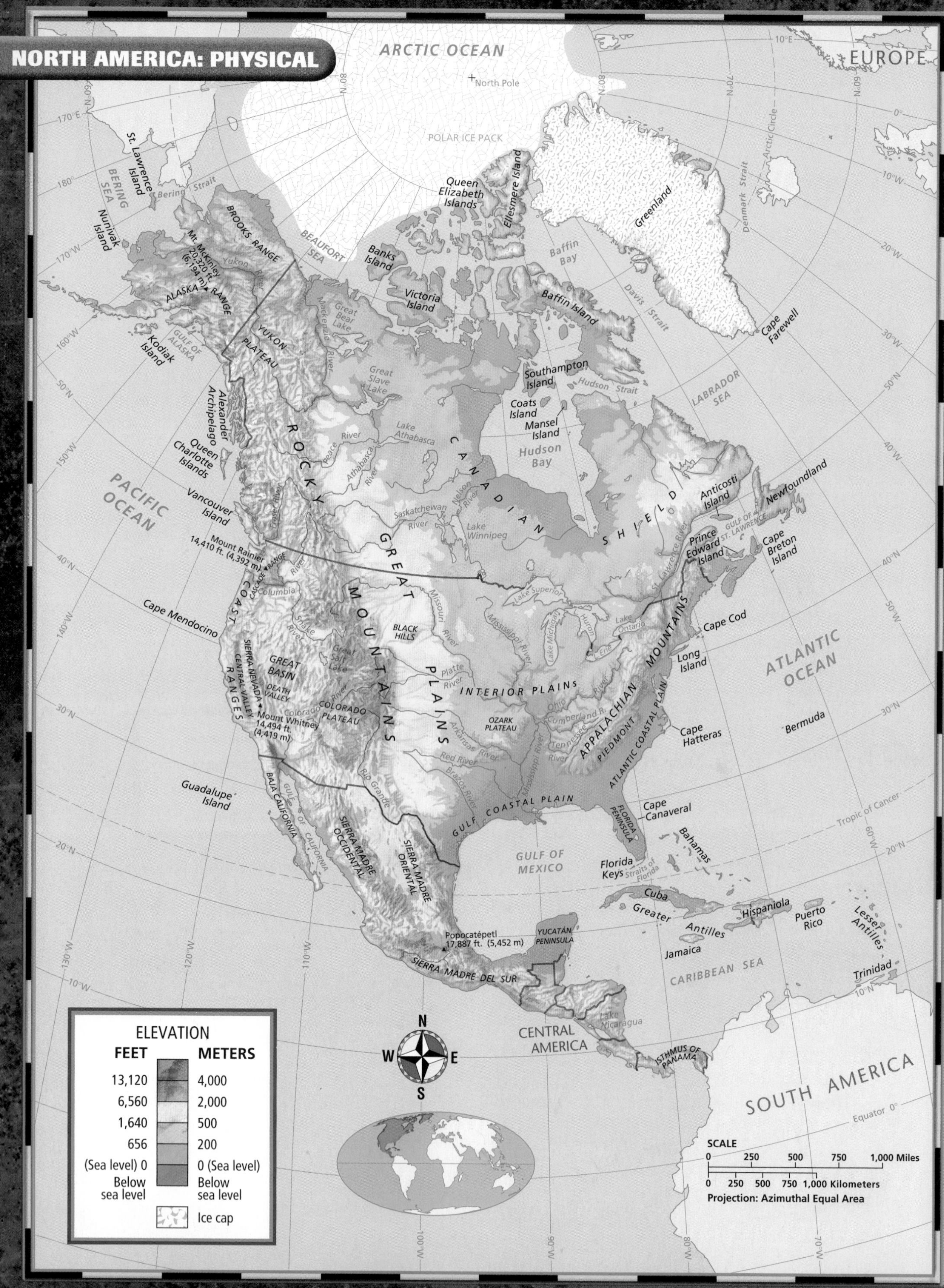
NORTH AMERICA: PHYSICAL
ARCTIC OCEAN
North Pole
POLAR ICE PACK
EUROPE
Greenland
Queen Elizabeth Islands
Ellesmere Island
Baffin Bay
Baffin Island
Banks Island
Victoria Island
BEAUFORT SEA
BERING SEA
Bering Strait
St. Lawrence Island
Nunivak Island
BROOKS RANGE
Mt. McKinley 20,320 ft. (6,194 m)
ALASKA RANGE
GULF OF ALASKA
Kodiak Island
YUKON PLATEAU
Great Bear Lake
Great Slave Lake
Southampton Island
Coats Island
Mansel Island
Hudson Bay
Hudson Strait
LABRADOR SEA
Davis Strait
Denmark Strait
Cape Farewell
Alexander Archipelago
Queen Charlotte Islands
PACIFIC OCEAN
Vancouver Island
ROCKY
MOUNTAINS
CANADIAN SHIELD
Lake Athabasca
Lake Winnipeg
Anticosti Island
Newfoundland
GULF OF ST. LAWRENCE
Prince Edward Island
Cape Breton Island
Mount Rainier 14,410 ft. (4,392 m)
CASCADE RANGE
COAST RANGES
Cape Mendocino
GREAT PLAINS
BLACK HILLS
Lake Superior
Lake Michigan
Lake Huron
Lake Erie
Lake Ontario
Cape Cod
Long Island
ATLANTIC OCEAN
GREAT BASIN
Great Salt Lake
SIERRA NEVADA
CENTRAL VALLEY
DEATH VALLEY
Mount Whitney 14,494 ft. (4,419 m)
COLORADO PLATEAU
INTERIOR PLAINS
OZARK PLATEAU
APPALACHIAN MOUNTAINS
PIEDMONT
ATLANTIC COASTAL PLAIN
Cape Hatteras
Bermuda
Guadalupe Island
BAJA CALIFORNIA
GULF OF CALIFORNIA
SIERRA MADRE OCCIDENTAL
SIERRA MADRE ORIENTAL
GULF COASTAL PLAIN
FLORIDA PENINSULA
Cape Canaveral
Tropic of Cancer
Bahamas
GULF OF MEXICO
Florida Keys
Straits of Florida
Cuba
Greater Antilles
Hispaniola
Puerto Rico
Lesser Antilles
Jamaica
YUCATÁN PENINSULA
Popocatépetl 17,887 ft. (5,452 m)
SIERRA MADRE DEL SUR
CARIBBEAN SEA
Trinidad
Lake Nicaragua
CENTRAL AMERICA
ISTHMUS OF PANAMA
SOUTH AMERICA
Equator 0°
ELEVATION
FEET
METERS
13,120
4,000
6,560
2,000
1,640
500
656
200
(Sea level) 0
0 (Sea level)
Below sea level
Below sea level
Ice cap
N
S
E
W
SCALE
0 250 500 750 1,000 Miles
0 250 500 750 1,000 Kilometers
Projection: Azimuthal Equal Area

NORTH AMERICA: POLITICAL

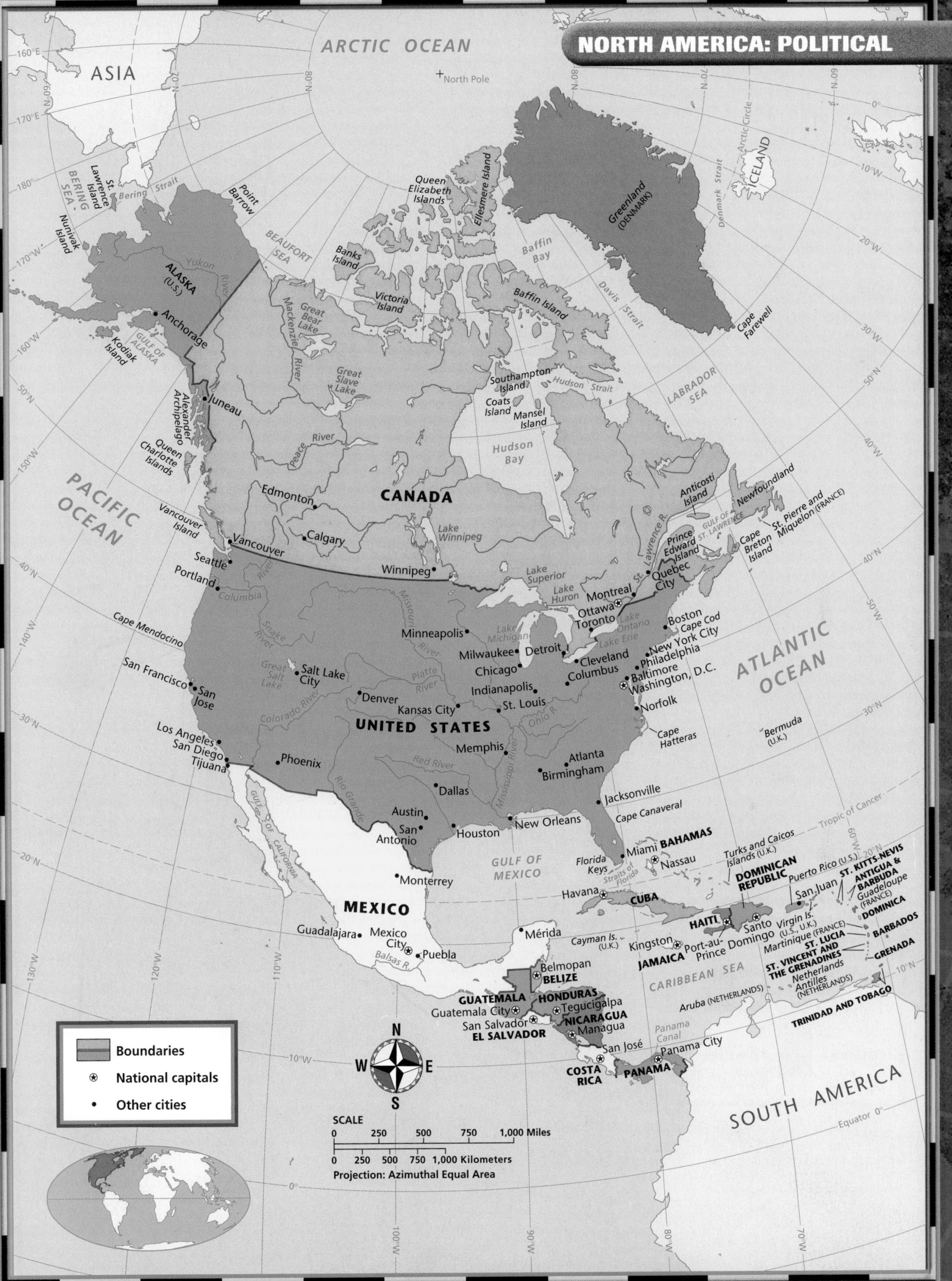

SOUTH AMERICA: PHYSICAL
CENTRAL AMERICA
CARIBBEAN SEA
Panama Canal
GULF OF PANAMA
Lake Maracaibo
Margarita Island
Tobago
Trinidad
Orinoco River Delta
Orinoco River
LLANOS
Meta River
Angel Falls
GUIANA HIGHLANDS
ATLANTIC OCEAN
Devil's Island
Cape Orange
Cauca River
Magdalena River
Mount Tolima 18,425 ft. (5,616 m)
Malpelo Island
ANDES
Amazon River Delta
Río Negro
Caquetá River
Japurá River
AMAZON BASIN
Amazon River
0° Equator
Equator 0°
Mount Chimborazo 20,561 ft. (6,267 m)
Galápagos Islands
GULF OF GUAYAQUIL
Marañón River
Juruá River
Purus
Madeira River
Tapajós River
Xingu River
Tocantins River
Parnaíba River
Ucayali River
Mount Huascarán 22,205 ft. (6,768 m)
Mamoré
Beni River
BRAZILIAN HIGHLANDS
MATO GROSSO PLATEAU
Araguaia River
São Francisco River
PACIFIC OCEAN
Lake Titicaca
Ancohuma Peak 20,958 ft. (6,388 m)
Lake Poopó
Pilcomayo
ATACAMA DESERT
GRAN CHACO
PLATEAU OF BRAZIL
Paraguay River
Tropic of Capricorn
San Ambrosio Island
San Félix Island
Salado River
Paraná River
Uruguay River
Mount Aconcagua 22,834 ft. (6,960 m)
Juan Fernández Islands
Río de la Plata
PAMPAS
Colorado River
SAN MATÍAS GULF
PATAGONIA
Chiloé Island
Chonos Archipelago
SAN JORGE GULF
Cape Tres Puntas
Bahía Grande
Strait of Magellan
Tierra Del Fuego
Falkland Islands
Cape Horn
South Georgia Island
ELEVATION
FEET
METERS
13,120 4,000
6,560 2,000
1,640 500
656 200
(Sea level) 0 0 (Sea level)
Below sea level Below sea level
SCALE
0 250 500 750 1,000 Miles
0 250 500 750 1,000 Kilometers
Projection: Azimuthal Equal Area
N
S
E
W
20°N
10°N
10°S
20°S
30°S
40°S
50°S
110°W
100°W
90°W
80°W
70°W
60°W
50°W
40°W
30°W
20°W

SOUTH AMERICA: POLITICAL
CENTRAL AMERICA
CARIBBEAN SEA
ATLANTIC OCEAN
PACIFIC OCEAN
Barranquilla
Cartagena
Caracas
Lake Maracaibo
VENEZUELA
Orinoco River
Georgetown
Paramaribo
Cayenne
GUYANA
SURINAME
FRENCH GUIANA (FRANCE)
Medellín
Bogotá
COLOMBIA
Cali
Malpelo Island (COLOMBIA)
Río Negro
Amazon River
Belém
Quito
ECUADOR
Guayaquil
Galápagos Islands (ECUADOR)
BRAZIL
PERU
Marañón River
Ucayali River
Trujillo
Recife
Callao
Lima
São Francisco River
Salvador
Lake Titicaca
BOLIVIA
La Paz
Arequipa
Brasília
Lake Poopó
Sucre
Belo Horizonte
PARAGUAY
Campinas
São Paulo
Rio de Janeiro
Asunción
Curitiba
Paraguay River
Tropic of Capricorn
San Ambrosio Island (CHILE)
San Félix Island (CHILE)
CHILE
Paraná River
Uruguay River
Pôrto Alegre
Córdoba
Rosario
URUGUAY
Valparaíso
Santiago
Juan Fernández Islands (CHILE)
Buenos Aires
Montevideo
Río de la Plata
ARGENTINA
Boundaries
National capitals
Other capitals
Other cities
SCALE
0 250 500 750 1,000 Miles
0 250 500 750 1,000 Kilometers
Projection: Azimuthal Equal Area
Strait of Magellan
Falkland Islands (U.K.)
Tierra del Fuego
South Georgia Island (U.K.)
Equator 0°
N S E W

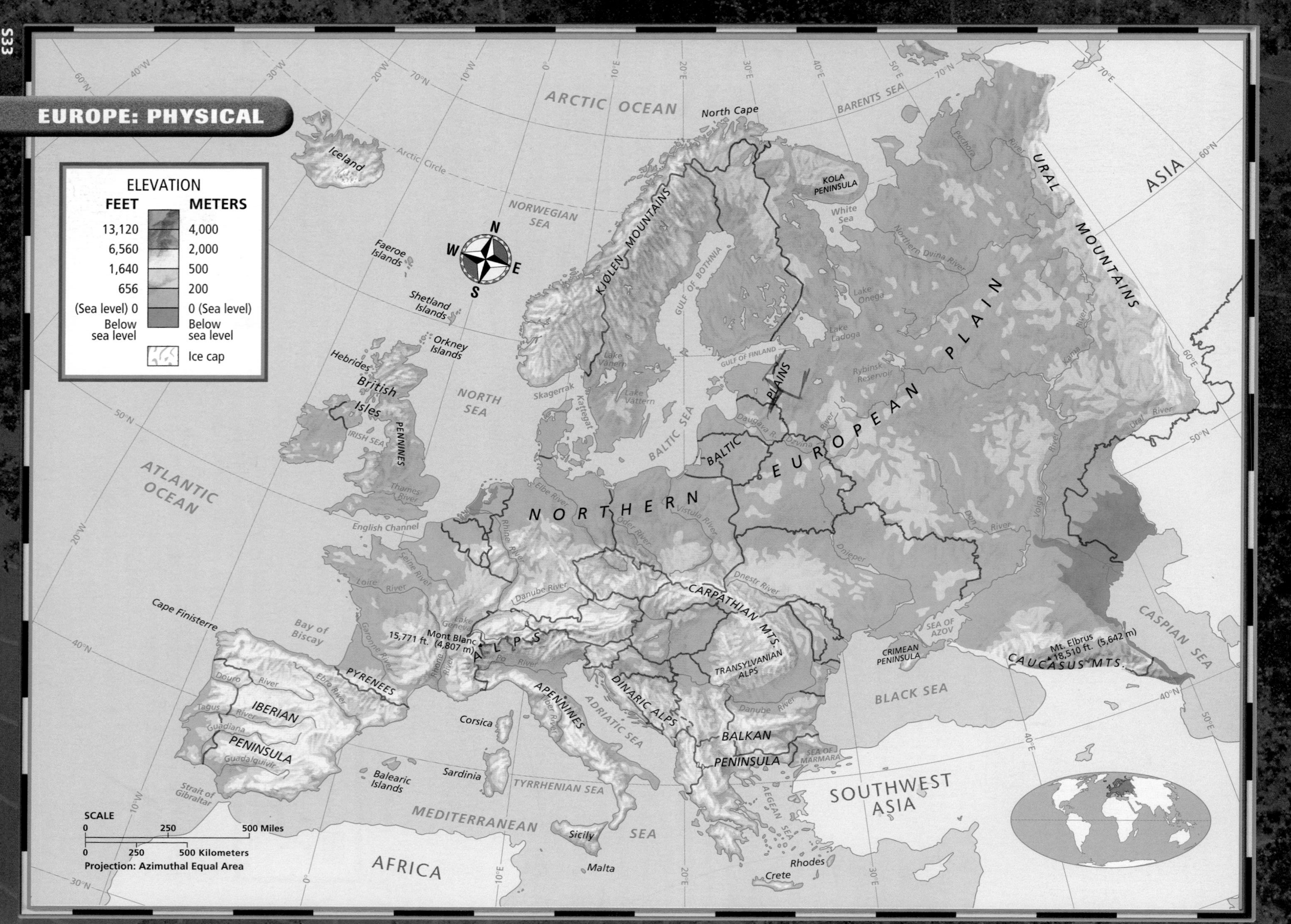
EUROPE: PHYSICAL
ELEVATION
FEET
METERS
13,120
4,000
6,560
2,000
1,640
500
656
200
(Sea level) 0
0 (Sea level)
Below sea level
Below sea level
Ice cap
ARCTIC OCEAN
North Cape
BARENTS SEA
Iceland
Arctic Circle
NORWEGIAN SEA
KOLA PENINSULA
White Sea
ASIA
URAL MOUNTAINS
Pechora River
Northern Dvina River
Lake Onega
Lake Ladoga
Faeroe Islands
Shetland Islands
Orkney Islands
Hebrides
British Isles
IRISH SEA
PENNINES
Thames River
NORTH SEA
Skagerrak
Kattegat
Lake Vänern
Lake Vättern
KJØLEN MOUNTAINS
GULF OF BOTHNIA
GULF OF FINLAND
BALTIC SEA
BALTIC PLAINS
Daugava R.
Dvina River
Rybinsk Reservoir
NORTHERN EUROPEAN PLAIN
Kama River
Ural River
Volga River
Don River
Dnieper
Dnestr River
ATLANTIC OCEAN
English Channel
Elbe River
Oder River
Vistula River
Rhine River
Seine River
Loire River
Danube River
Cape Finisterre
Bay of Biscay
Garonne River
Lake Geneva
Mont Blanc 15,771 ft. (4,807 m)
ALPS
Rhone River
Po River
CARPATHIAN MTS.
TRANSYLVANIAN ALPS
SEA OF AZOV
CRIMEAN PENINSULA
CASPIAN SEA
Mt. Elbrus 18,510 ft. (5,642 m)
CAUCASUS MTS.
BLACK SEA
PYRENEES
Douro River
Ebro River
Tagus River
Guadiana
Guadalquivir
IBERIAN PENINSULA
Corsica
APENNINES
Tiber River
ADRIATIC SEA
DINARIC ALPS
Danube River
BALKAN PENINSULA
SEA OF MARMARA
Sardinia
Balearic Islands
TYRRHENIAN SEA
Strait of Gibraltar
MEDITERRANEAN SEA
Sicily
Malta
AEGEAN SEA
Rhodes
Crete
SOUTHWEST ASIA
AFRICA
SCALE
0
250
500 Miles
0
250
500 Kilometers
Projection: Azimuthal Equal Area
N
S
E
W

EUROPE: POLITICAL
Boundaries
National capitals
Other cities
ARCTIC OCEAN
ATLANTIC OCEAN
NORTH SEA
BALTIC SEA
GULF OF BOTHNIA
GULF OF FINLAND
BARENTS SEA
WHITE SEA
BLACK SEA
CASPIAN SEA
MEDITERRANEAN SEA
ADRIATIC SEA
AEGEAN SEA
Bay of Biscay
English Channel
Strait of Gibraltar
ICELAND
Reykjavik
Faeroe Islands (DENMARK)
Shetland Islands
NORWAY
Bergen
Oslo
SWEDEN
Stockholm
Göteborg
FINLAND
Helsinki
North Cape
ESTONIA
Tallinn
LATVIA
Riga
LITHUANIA
Vilnius
RUSSIA
St. Petersburg
Moscow
Nizhniy Novgorod
URAL MOUNTAINS
ASIA
Ural River
Volga River
Don River
BELARUS
Minsk
POLAND
Warsaw
Kraków
UKRAINE
Kiev
Dnieper
MOLDOVA
Chişinău
ROMANIA
Bucharest
DENMARK
Copenhagen
UNITED KINGDOM
SCOTLAND
Edinburgh
NORTHERN IRELAND
Belfast
IRELAND
Dublin
British Isles
WALES
ENGLAND
Liverpool
London
Thames R.
Channel Islands (U.K.)
NETHERLANDS
The Hague
Amsterdam
BELGIUM
Brussels
LUXEMBOURG
Luxembourg
GERMANY
Hamburg
Elbe River
Berlin
Dresden
Cologne
Bonn
Munich
Danube River
Rhine River
FRANCE
Paris
Seine River
Loire River
Lyon
Rhone River
Marseille
PYRENEES
ANDORRA
Andorra la Vella
SWITZERLAND
Bern
Lake Geneva
LIECHTENSTEIN
Vaduz
ALPS
Milan
Po River
MONACO
Monaco
Corsica (FRANCE)
Sardinia (ITALY)
SAN MARINO
San Marino
ITALY
Rome
VATICAN CITY
Naples
Sicily
MALTA
Valletta
AUSTRIA
Vienna
CZECH REPUBLIC
Prague
SLOVAKIA
Bratislava
HUNGARY
Budapest
SLOVENIA
Ljubljana
CROATIA
Zagreb
BOSNIA & HERZEGOVINA
Sarajevo
YUGOSLAVIA
SERBIA
MONTENEGRO
Belgrade
Danube River
BULGARIA
Sofia
MACEDONIA
Skopje
ALBANIA
Tiranë
GREECE
Athens
Rhodes
Crete
PORTUGAL
Lisbon
Tagus River
SPAIN
Madrid
Valencia
Barcelona
Seville
Gibraltar (U.K.)
Balearic Islands (SPAIN)
AFRICA
SOUTHWEST ASIA
Arctic Circle
SCALE
0
250
500 Miles
0
250
500 Kilometers
Projection: Azimuthal Equal Area

ASIA: PHYSICAL

ASIA: POLITICAL

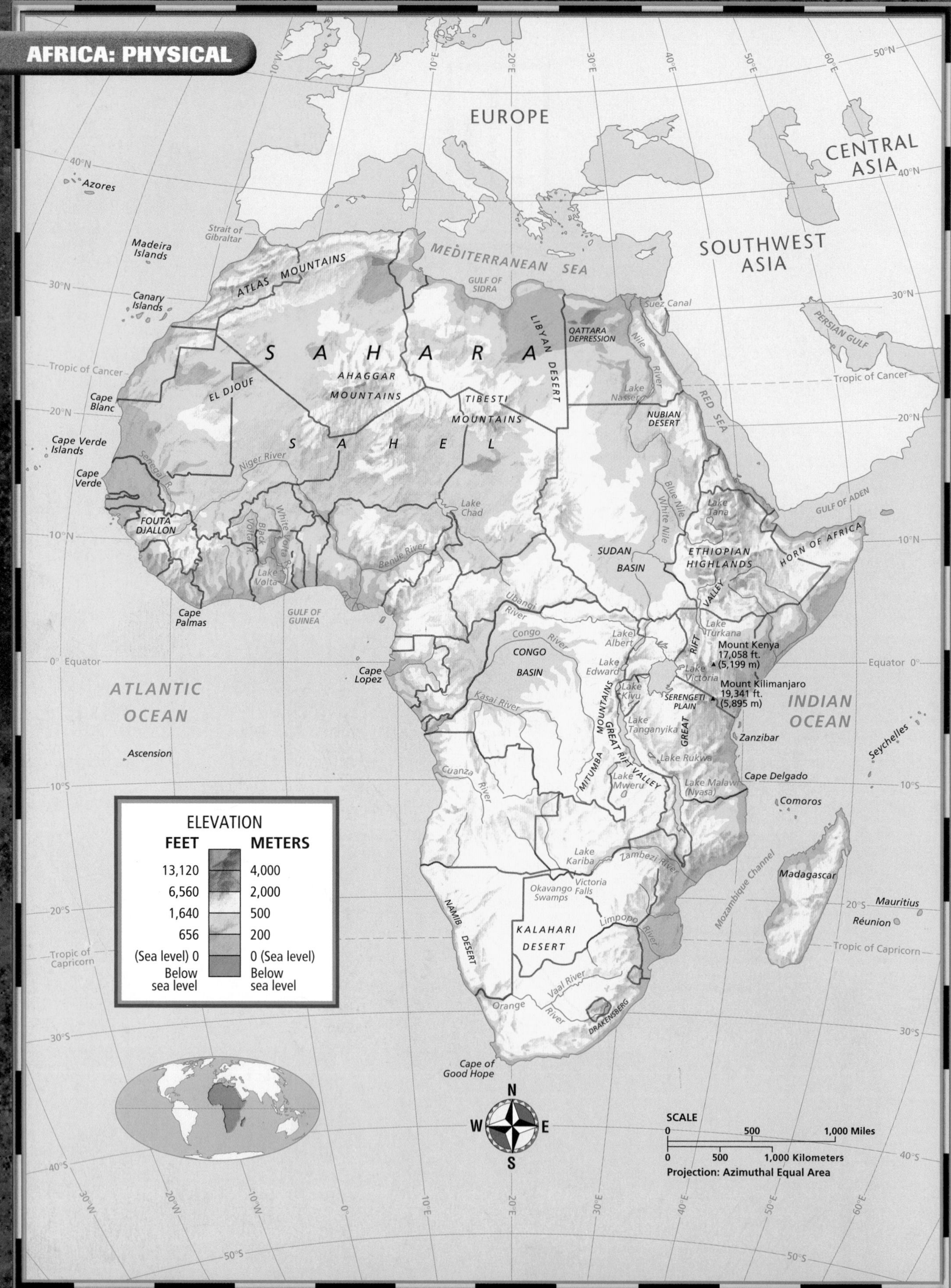
AFRICA: PHYSICAL
EUROPE
CENTRAL ASIA
SOUTHWEST ASIA
Azores
Madeira Islands
Strait of Gibraltar
Canary Islands
ATLAS MOUNTAINS
MEDITERRANEAN SEA
GULF OF SIDRA
Suez Canal
PERSIAN GULF
S A H A R A
AHAGGAR MOUNTAINS
EL DJOUF
LIBYAN DESERT
QATTARA DEPRESSION
Nile River
Lake Nasser
RED SEA
Tropic of Cancer
Cape Blanc
TIBESTI MOUNTAINS
NUBIAN DESERT
Cape Verde Islands
Cape Verde
S A H E L
Senegal R.
Niger River
Lake Chad
Blue Nile
White Nile
Lake Tana
GULF OF ADEN
FOUTA DJALLON
Black Volta R.
White Volta R.
Lake Volta
Benue River
SUDAN BASIN
ETHIOPIAN HIGHLANDS
HORN OF AFRICA
Cape Palmas
GULF OF GUINEA
Ubangi River
Congo River
CONGO BASIN
Lake Albert
Lake Turkana
GREAT RIFT VALLEY
Mount Kenya 17,058 ft. (5,199 m)
Lake Edward
Lake Victoria
Mount Kilimanjaro 19,341 ft. (5,895 m)
Cape Lopez
ATLANTIC OCEAN
INDIAN OCEAN
Equator
Lake Kivu
SERENGETI PLAIN
Kasai River
Lake Tanganyika
MITUMBA MOUNTAINS
Zanzibar
Seychelles
Ascension
Lake Rukwa
Cape Delgado
Cuanza River
Lake Mweru
Lake Malawi (Nyasa)
Comoros
ELEVATION
FEET
METERS
13,120
4,000
6,560
2,000
1,640
500
656
200
(Sea level) 0
0 (Sea level)
Below sea level
Below sea level
Lake Kariba
Zambezi River
Madagascar
Mozambique Channel
Victoria Falls
Okavango Swamps
NAMIB DESERT
Mauritius
Réunion
KALAHARI DESERT
Limpopo River
Tropic of Capricorn
Vaal River
Orange River
DRAKENSBERG
Cape of Good Hope
N
W
E
S
SCALE
0 500 1,000 Miles
0 500 1,000 Kilometers
Projection: Azimuthal Equal Area

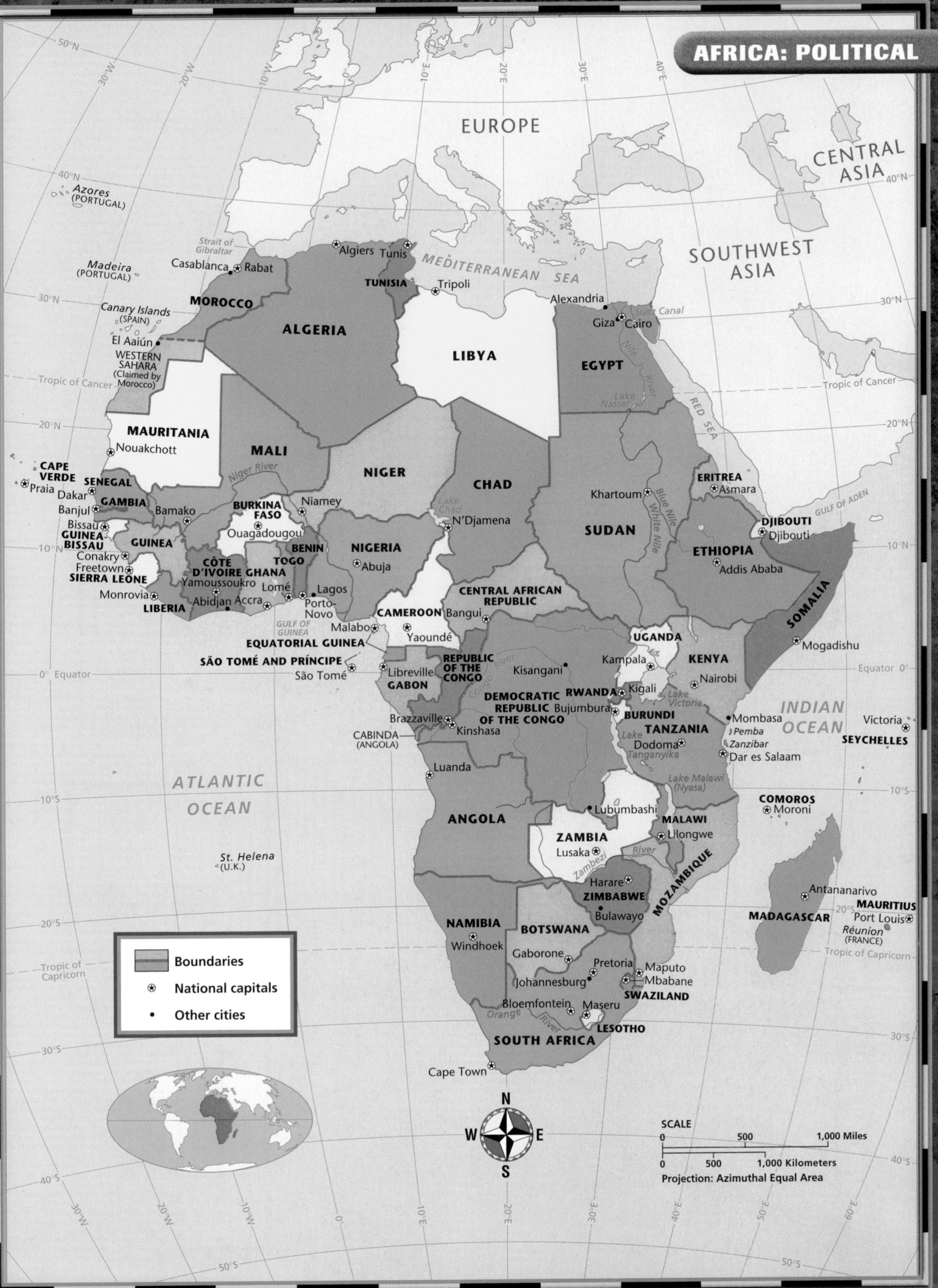
EUROPE
CENTRAL ASIA
SOUTHWEST ASIA
Azores (PORTUGAL)
Madeira (PORTUGAL)
Canary Islands (SPAIN)
Strait of Gibraltar
MEDITERRANEAN SEA
Casablanca
Rabat
Algiers
Tunis
TUNISIA
Tripoli
MOROCCO
ALGERIA
LIBYA
Alexandria
Suez Canal
Giza
Cairo
EGYPT
Nile River
Lake Nasser
RED SEA
El Aaiún
WESTERN SAHARA (Claimed by Morocco)
Tropic of Cancer
MAURITANIA
Nouakchott
MALI
Niger River
NIGER
CHAD
SUDAN
Khartoum
Blue Nile
White Nile
ERITREA
Asmara
GULF OF ADEN
DJIBOUTI
Djibouti
ETHIOPIA
Addis Ababa
SOMALIA
Mogadishu
CAPE VERDE
Praia
SENEGAL
Dakar
Banjul
GAMBIA
Bamako
BURKINA FASO
Ouagadougou
Niamey
Lake Chad
N'Djamena
Bissau
GUINEA-BISSAU
GUINEA
Conakry
Freetown
SIERRA LEONE
CÔTE D'IVOIRE
GHANA
TOGO
BENIN
NIGERIA
Abuja
Yamoussoukro
Monrovia
LIBERIA
Abidjan
Accra
Lomé
Lagos
Porto-Novo
GULF OF GUINEA
CENTRAL AFRICAN REPUBLIC
Bangui
CAMEROON
Yaoundé
Malabo
EQUATORIAL GUINEA
SÃO TOMÉ AND PRÍNCIPE
São Tomé
Libreville
GABON
REPUBLIC OF THE CONGO
Congo River
Kisangani
DEMOCRATIC REPUBLIC OF THE CONGO
UGANDA
Kampala
KENYA
Nairobi
RWANDA
Kigali
Lake Victoria
Bujumbura
BURUNDI
Mombasa
Pemba
Zanzibar
TANZANIA
Dodoma
Lake Tanganyika
Dar es Salaam
INDIAN OCEAN
Victoria
SEYCHELLES
Brazzaville
Kinshasa
CABINDA (ANGOLA)
Luanda
ATLANTIC OCEAN
Lake Malawi (Nyasa)
COMOROS
Moroni
Lubumbashi
ANGOLA
ZAMBIA
Lusaka
Zambezi River
MALAWI
Lilongwe
MOZAMBIQUE
St. Helena (U.K.)
Harare
ZIMBABWE
Bulawayo
Antananarivo
MADAGASCAR
MAURITIUS
Port Louis
Réunion (FRANCE)
NAMIBIA
Windhoek
BOTSWANA
Gaborone
Pretoria
Johannesburg
Maputo
Mbabane
SWAZILAND
Bloemfontein
Maseru
Orange River
LESOTHO
SOUTH AFRICA
Cape Town
Tropic of Capricorn
Equator
Boundaries
National capitals
Other cities
N
S
E
W
SCALE
0 500 1,000 Miles
0 500 1,000 Kilometers
Projection: Azimuthal Equal Area
50°N
40°N
30°N
20°N
10°N
0°
10°S
20°S
30°S
40°S
50°S
30°W
20°W
10°W
10°E
20°E
30°E
40°E
50°E
60°E

AUSTRALIA AND NEW ZEALAND

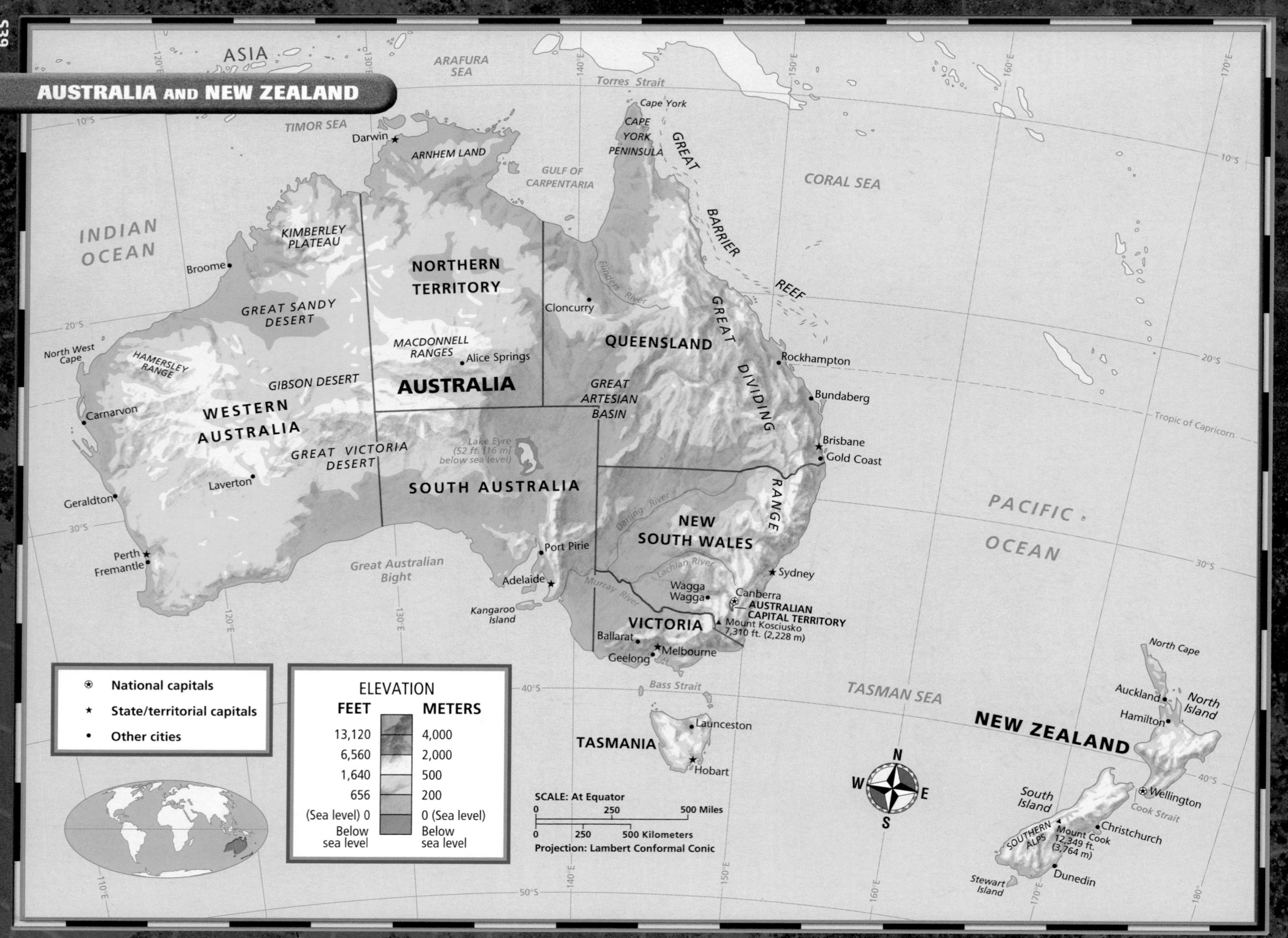

PACIFIC ISLANDS: POLITICAL

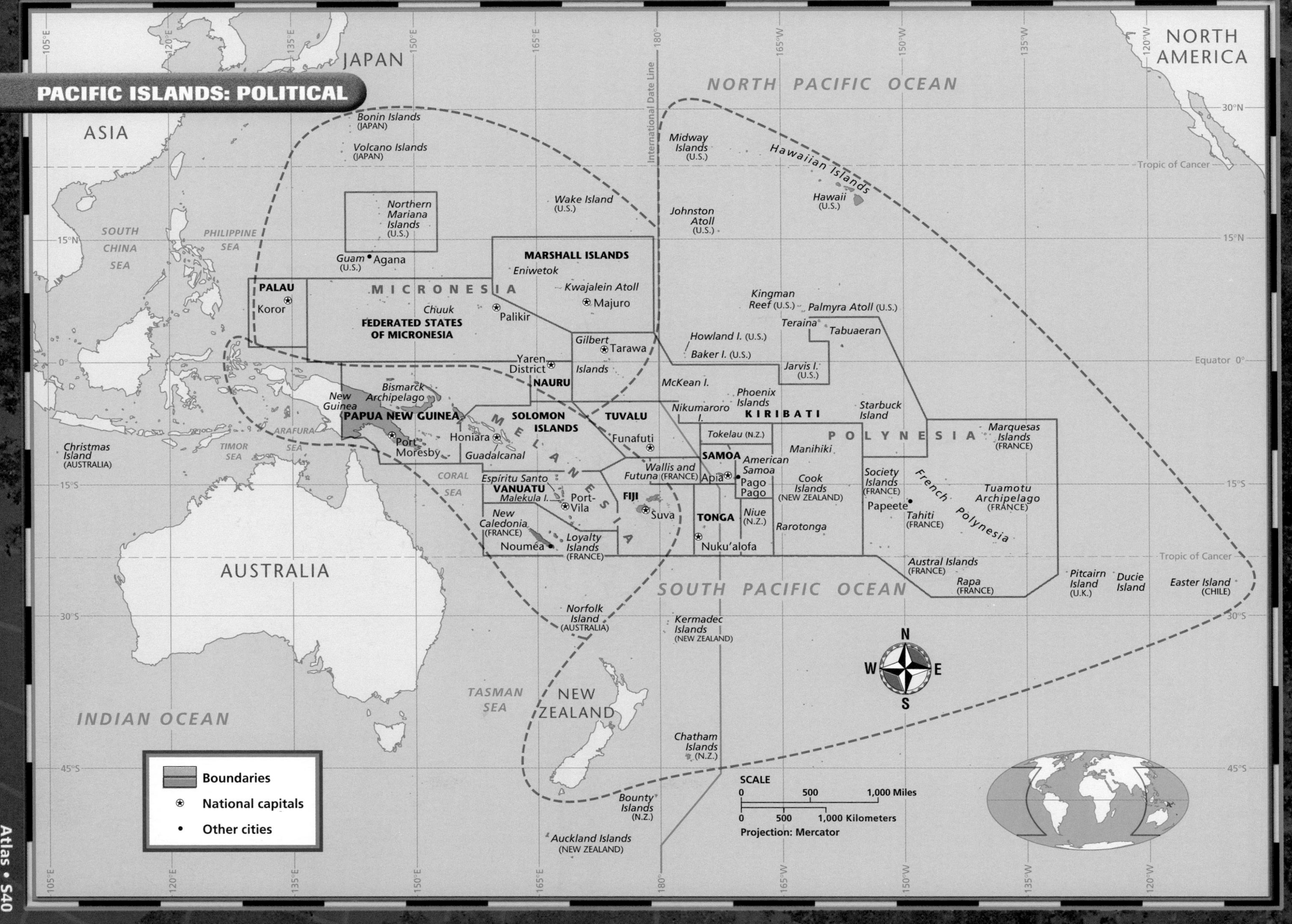

NORTH POLE

SOUTH POLE

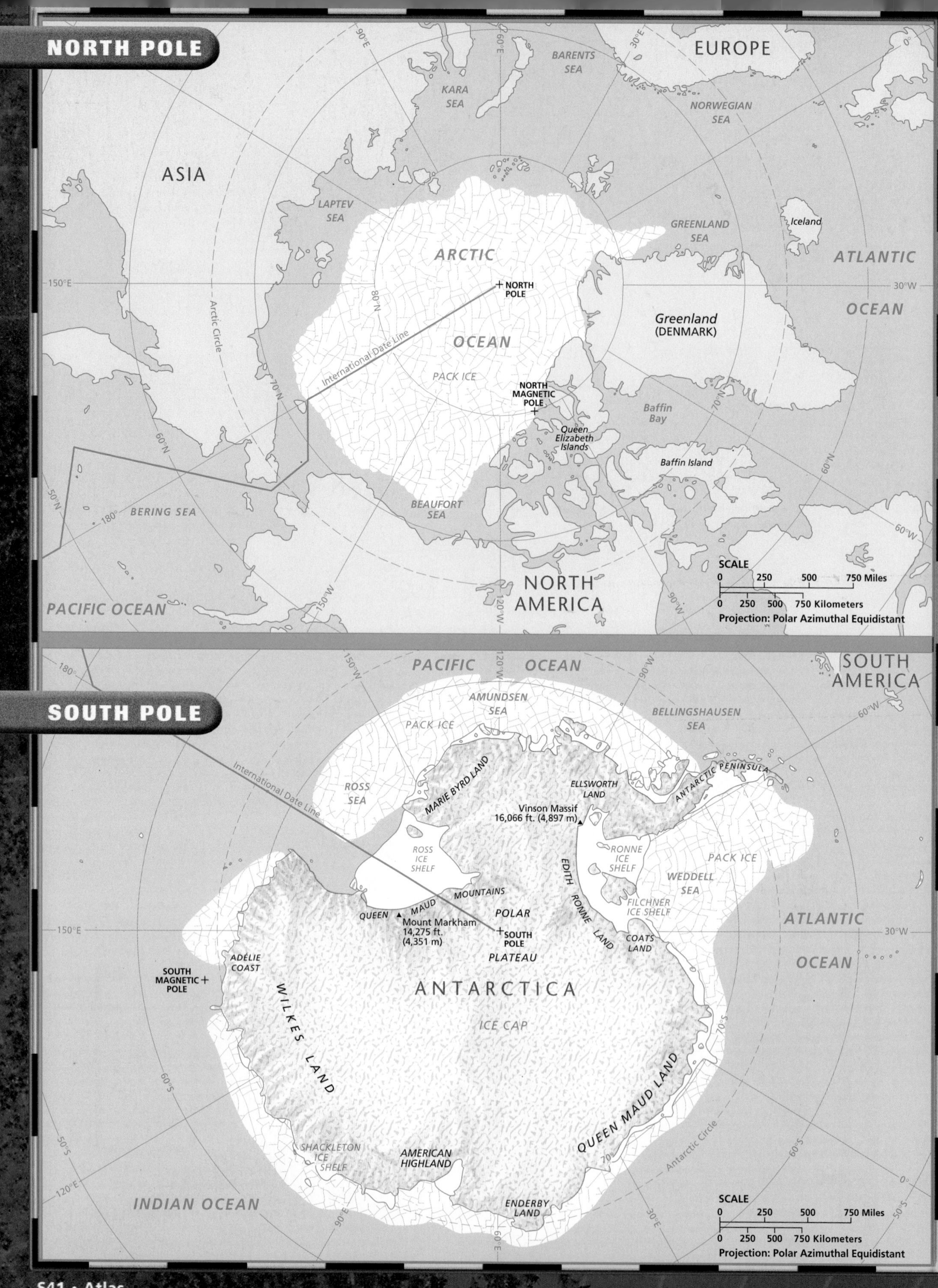

Texas Geography Today

TEXAS

Texas is located in the southern United States. The state shares a border with Mexico in the south and borders four U.S. states. New Mexico lies to the west, Oklahoma to the north, Arkansas to the northeast, and Louisiana to the east. Texas also has a long coastline on the Gulf of Mexico.

Bobcat

Section 1

Natural Environments

READ TO DISCOVER

1. What types of landforms are found in Texas?
2. What water resources does Texas have?
3. What are some characteristics of climates in Texas?
4. How does the state's physical environment affect plant and animal life?
5. How have Texans used the state's natural resources?

WHY IT MATTERS

Texans use water from many different sources. Overuse of some resources has caused water shortages. Use CNNfyi.com or other **current events** sources to learn more about water resources and water shortages in Texas.

LOCATE

Hill Country
Edwards Plateau
Pecos River
Guadalupe Peak
Red River
Brazos River
Colorado River
Neches River
Sabine River
Trinity River
Rio Grande

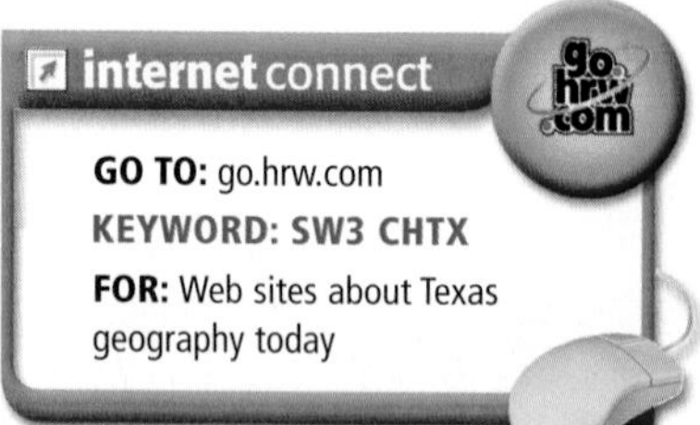

Landforms

Texas has as many different landscapes as some countries. Its landforms include canyons, islands, valleys, and hills that are extinct volcanoes. West Texas has mountains. However, plains, plateaus, and hills dominate the state. A plain is a flat or gently rolling area without a sharp rise or fall in elevation. Plateaus are elevated flatlands that drop sharply on one or more sides.

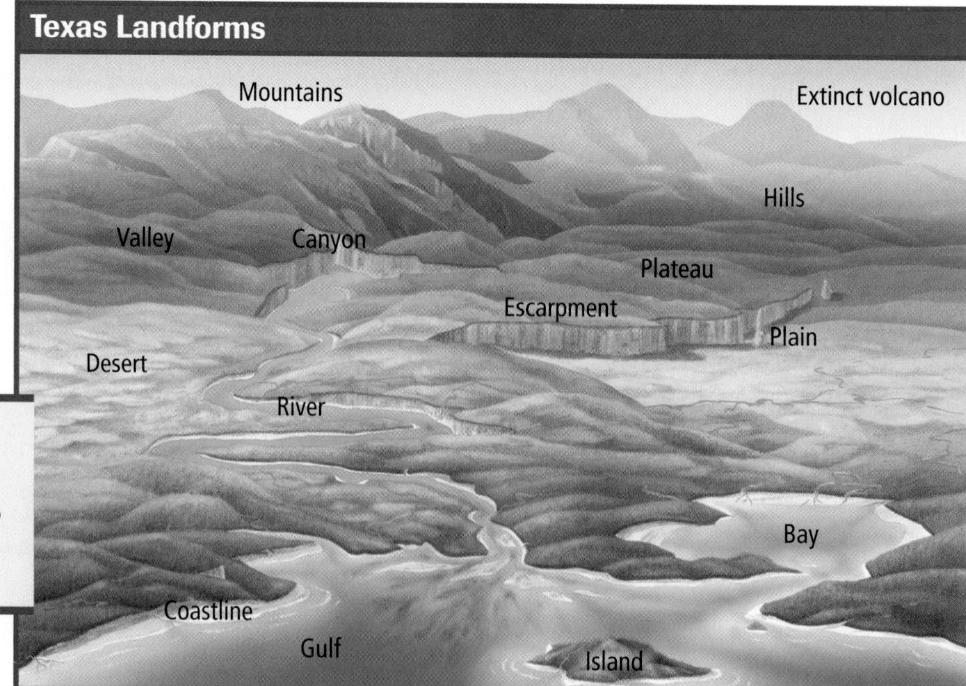

INTERPRETING THE DIAGRAM

Texas includes a wide variety of landforms and water features. ***Which of the illustrated physical features can be found in or near your community?***

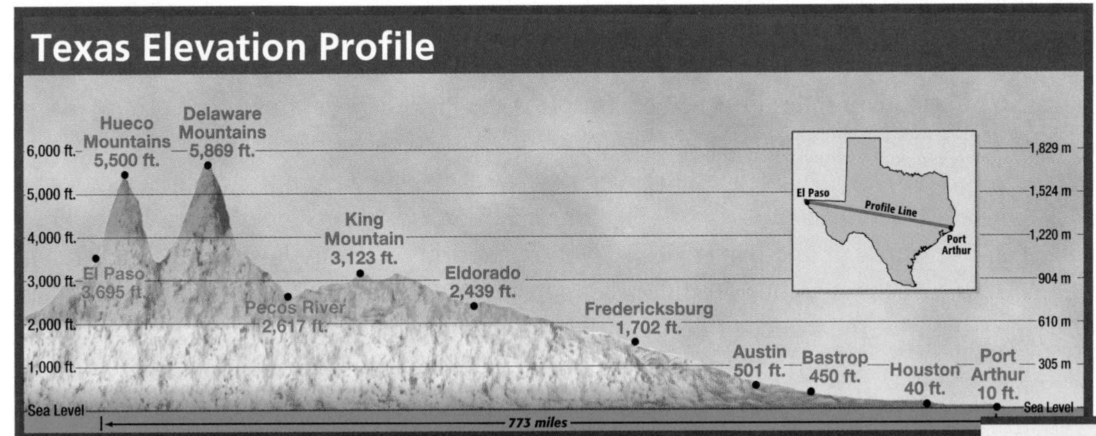

INTERPRETING THE DIAGRAM

This diagram shows changes in elevation along a line from West Texas to East Texas. ***What part of Texas has the greatest changes in elevation? What part of the state has the most constant elevations?*** TEKS

Anyone who has spent hour after hour driving across Texas knows that plains cover much of the state. In some areas, hills rise from the plains. Forests grow on the plains of East Texas. Just to the west of these forests lie gently rolling prairies, or treeless grasslands. Central Texas has more rugged hills, such as those found in the Hill Country. West of the Hill Country lies the Edwards Plateau. It is the state's largest plateau. The Edwards Plateau rises in elevation from east to west.

West of the Edwards Plateau the landscape becomes rougher and rockier. Several separate mountain ranges rise west of the Pecos River. They are all part of the Rocky Mountain system, which extends from Canada through the United States and into Mexico. The highest point in Texas, Guadalupe Peak, is part of the Guadalupe Range.

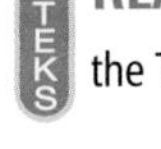

READING CHECK: ***Places and Regions*** How does the Texas landscape change from east to west?

Water

Texas has a wealth of lakes, rivers, and creeks. Bordering the state is the Gulf of Mexico. The Gulf and its bays are natural resources. Bays provide harbors for ships and nurseries for fish and other marine life. Both commercial and recreational fishers draw seafood from the Gulf. Shrimping is a multimillion dollar industry. Swimming and boating are popular Gulf Coast pastimes.

Texas has more than 40 rivers and 11,000 creeks and streams. (See the map.) The state's river systems can be divided into three groups. Some rivers and smaller streams in northern Texas flow into the Mississippi River. The Canadian River, which is known for the patches of quicksand along its course, and the Red River are among them.

A second group of rivers begins in Texas or one of its neighboring states and flows directly into the Gulf of Mexico. The Brazos, Colorado, Neches, Nueces, Sabine, and Trinity Rivers are in this group.

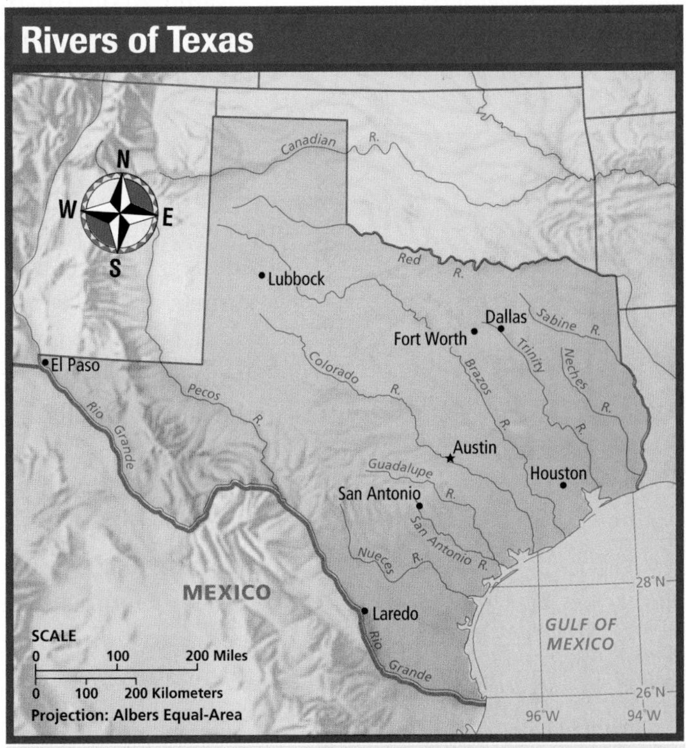

INTERPRETING THE MAP ***In which direction do the major rivers of Texas flow? How does the pattern of rivers appear to have affected the location of settlements?*** TEKS

Note that they flow generally northwest to southeast. The Colorado River travels some 600 miles (965 km) across Texas. Along the way it drains nearly 40,000 square miles (103,600 sq km). It is the longest river contained entirely within the state's boundaries. Far to the east, the Neches River flows across East Texas for 416 miles (669 km). The river takes its name from the Neches, a group of American Indians who lived nearby. For most of its length, the Neches flows through low-lying wooded land. The Trinity River lies just west of the Neches. It flows some 550 miles (885 km) across the prairies of North Texas, Dallas, and the woods of East Texas on its way to the Gulf.

A third group of Texas rivers consists of those in the Rio Grande system. The Rio Grande originates as a snow-fed stream high in the mountains of Colorado. From Colorado it flows through New Mexico before forming 1,254 miles (2,018 km) of the U.S.-Mexico border. However, drought and overuse have severely reduced the Rio Grande's volume. In fact, by 2001 the Rio Grande was not actually flowing all the way to the Gulf. For a while, a sandbar blocked the path of the weak stream.

READING CHECK: ***Places and Regions*** Into what groups can the river systems of Texas be divided?

Lakes Before people began shaping the landscape, Texas had few lakes. Caddo Lake in the state's northeast is the largest natural lake. Over the years, though, Texans have created hundreds of lakes by building dams across rivers. These dams control floods and create reservoirs. A reservoir is an artificial lake that stores water for human use, such as drinking water or for crop irrigation. A series of reservoirs called the Highland Lakes is located on the Colorado River. Lake Buchanan and Lake Travis are the largest of these reservoirs. These and the other Highland Lakes provide both water and recreational opportunities for Austin and other cities.

INTERPRETING THE VISUAL RECORD

A sunbather relaxes on the cliffs above Lake Travis, one of the Highland Lakes. ***How might the damming of the Colorado River have affected economic activities along the river's course?*** TEKS

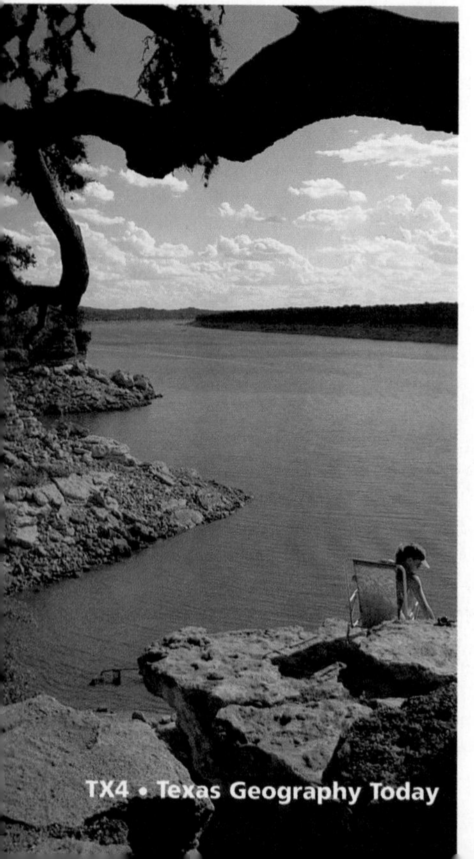

Aquifers Aquifers also provide water for Texas. An aquifer is a formation of rock, sand, or gravel that holds water underground. The process by which rainwater naturally refills an aquifer may be very slow. Water from the state's aquifers goes to farms, homes, and industries. Sometimes these water consumers clash over who should have access to an aquifer's water.

The Ogallala Aquifer is the largest underground water source in the state. It stretches from West Texas and New Mexico to South Dakota. The Ogallala is the largest aquifer in North America. It provides water for farming in the dry regions of the Texas Panhandle and High Plains. Almost 95 percent of the water pumped out of the aquifer is used for irrigation. Scientists are concerned that the amount of water pumped from the Ogallala is higher than the amount draining into it. To the south, the Edwards Aquifer provides water for Central Texas, including San Antonio. It is one of the biggest cities in the world to rely completely on a single groundwater source for its water supply. Cities and towns also draw heavily from the Gulf Coast Aquifer, which extends below 54 coastal counties. Because so much water has been pumped from this aquifer, in some places the ground has subsided, or sunk.

READING CHECK: ***Environment and Society*** Why have Texans created reservoirs and pumped water from aquifers?

Climate

Sometimes the weather in a certain Texas locale stays the same week after week or even month after month. Then, in a few minutes, it can change dramatically. However, the state's climates remain more constant. Climate is an area's pattern of weather over a long period of time. Temperature and precipitation are the main factors used to classify climates. You can identify the climate regions of Texas on the climate map of the United States and Canada in Unit 2.

Location is a key factor affecting Texas climates. The state lies in the lower middle latitudes, much closer to the equator than to the North Pole. The equator receives the Sun's most direct rays. As you might expect, Texas summers are hot. Afternoon temperatures often rise above 90°F (32°C). Late in the summer the temperature may climb above 110°F (43°C). Because buildings, cars, and pavement can raise air temperatures, cities are usually a few degrees hotter than the surrounding countryside.

Temperatures are often cooler near the Gulf of Mexico. Water cools and heats more slowly than land. Therefore, breezes that blow across the Gulf keep nearby land cooler in the summer than areas farther inland. In the winter the Gulf keeps the coast warmer. However, the Gulf Coast has higher humidity, which makes high temperatures there feel hotter. Humidity refers to the amount of moisture in the air.

Wind patterns also affect Texas climates. Some areas are hotter because the wind carries warm dry air from deserts farther west. The Rio Grande valley and north-central Texas are particularly affected by this pattern.

Winds can also blow in from the north, bringing cooler fall and winter temperatures to the Panhandle. Cold fronts called northers often sweep into the area. When a norther hits, the temperature can drop rapidly. Northers may bring howling winds, ice, and sometimes snow. Following a norther, the temperature is often low, but the sky is a clear blue. The blue sky may be the source of the term *blue norther*, a phrase often applied to these weather events. Northers can reach Central Texas and the Gulf Coast.

Elevation also affects temperature. Higher elevations are cooler because the air there is not as dense. Thin air does not hold heat as well as air at sea level. As a result, the mountains of West Texas are generally cooler, particularly at night, than lower elevations. Even within a small area, elevation makes a big difference. For example, temperatures in the mountains may be 20 degrees below those a few miles away on the desert floor.

In July 1979 Tropical Storm Claudette soaked an area near Alvin, just south of Houston, with 43 inches (109 cm) of rain in less than 24 hours. This remains the highest one-day rainfall total ever recorded in the United States.

A mild climate has helped make San Antonio a popular destination for tourists throughout the year. Here sightseers cruise along San Antonio's famous Riverwalk.

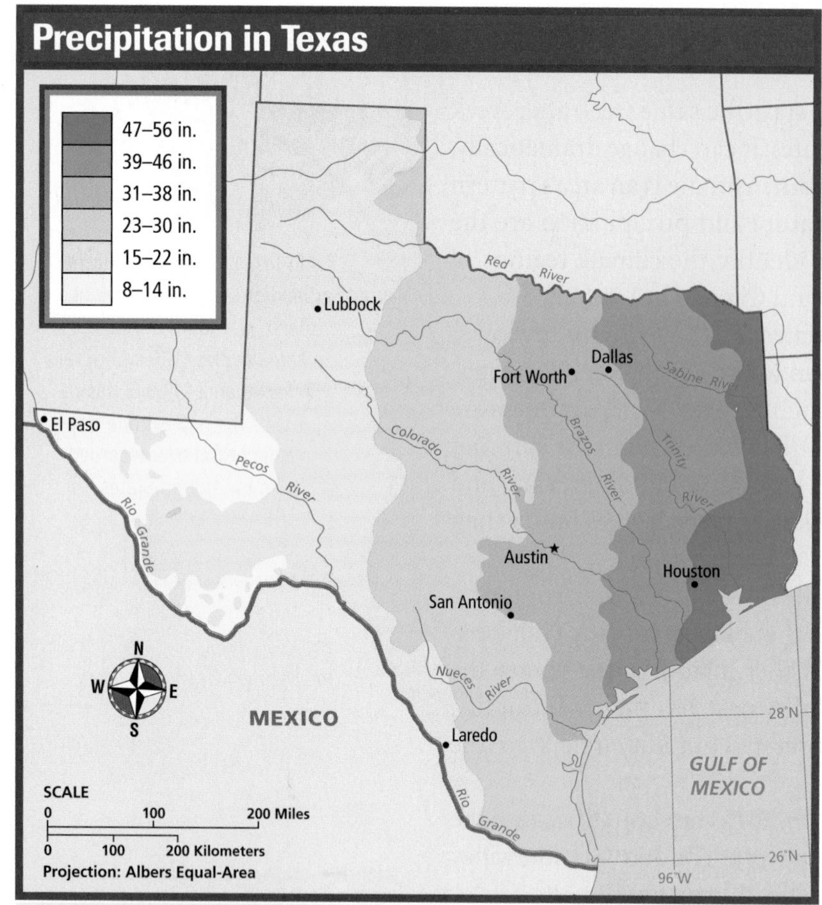

INTERPRETING THE MAP *Because they are scattered across the state, Texas cities vary greatly in the amount of precipitation they receive each year.* ***Which of the major cities receives the most rainfall? Which receives the least? How might this city's location help to make up for its lack of precipitation?***

Precipitation Precipitation is moisture falling as rain, snow, sleet, or hail. In Texas, the amount of precipitation increases from west to east. (See the map.) Average annual precipitation in West Texas is about 13 inches (33 cm) or less. In contrast, parts of East Texas may get more than 45 inches (114 cm) each year. East Texas gets more rain, mainly because it is close to the Gulf of Mexico. Warm moist air comes in from the Gulf. When this air passes over land it meets cooler air. As the air rises and cools, moisture condenses and may form raindrops.

North Texas receives some of its precipitation from winter snow. The Panhandle usually gets several heavy snows each year. The heaviest snowfall on record in Texas occurred in 1956. In that year the Amarillo area received more than 30 inches (76 cm) of snow.

Droughts and Floods When precipitation levels remain low for a long time, Texas goes through a period called a drought. Serious droughts have struck Texas in the 1890s, 1930s, 1950s, 1980s, and 1990s. Recent droughts have been severe. Heavy irrigation, increased water use by industry, and a growing population have placed great demands on water resources, including aquifers. As a result, water supplies in some places fall to extremely low levels during dry summer months. Droughts are particularly damaging to farms and ranches.

Sometimes a long drought is broken by heavy rainfall. Then floods are a threat. Within minutes, calm streams can become raging torrents. Rivers can overflow onto nearby land. Dams have helped control rivers. However, even swollen creeks can cause big problems in urban areas. For example, in May 1981, 10 inches (25 cm) of rain fell on Austin within two and a half hours. The ground was already soggy. Thirteen people were killed, mainly by creeks that surged over their banks onto city streets.

Severe Weather Texas gets its share of severe weather events, including tornadoes, hurricanes, and blizzards. Tornadoes are violent rotating storms that develop inside thunderstorms. Texas is on the southern edge of what is often called Tornado Alley. This area, the most tornado-prone in the world, extends from northwest Texas across Oklahoma and Kansas. Tornado winds can reach speeds of more than 300 miles (483 km) per hour. One account of an 1898 tornado described its effects. "[The tornado] blew cattle into the air, lodging them in trees, sucked all water from the Brazos River for a short distance and dumped a fifty-pound fish on dry land." A twister that struck Jarrell, Texas, in

1997 ripped the pavement from the highway. About 110 tornadoes struck Texas in 1998. Tornadoes have claimed hundreds of lives and caused millions of dollars in damage. The tornado season usually lasts from spring into summer. Tornadoes spawned by hurricanes can strike late into the fall.

Snow is almost unknown in South Texas, but the Panhandle usually gets more than 20 inches (50 cm) each year. Ice can break tree branches, which may then break power lines.

The hurricane season begins in June. These huge storms develop over the Atlantic Ocean and the Gulf of Mexico. Heavy rains and high tides usually accompany high winds. In 1900 up to 12,000 people died when a hurricane hit Galveston Island. This powerful storm still ranks as the most deadly natural disaster ever to strike the United States. A larger storm, Hurricane Carla, came ashore in 1961 near Port Lavaca. Winds rose to 175 miles (282 km) per hour, and storm tides reached 18.5 feet (5.6 m) above normal. Despite early warnings, 34 people died. Tropical storms that are not classified as hurricanes can also be deadly. In 2001, Tropical Storm Allison dumped about 3 feet (1 m) of rain over parts of Houston within days, and 22 people died in the floods.

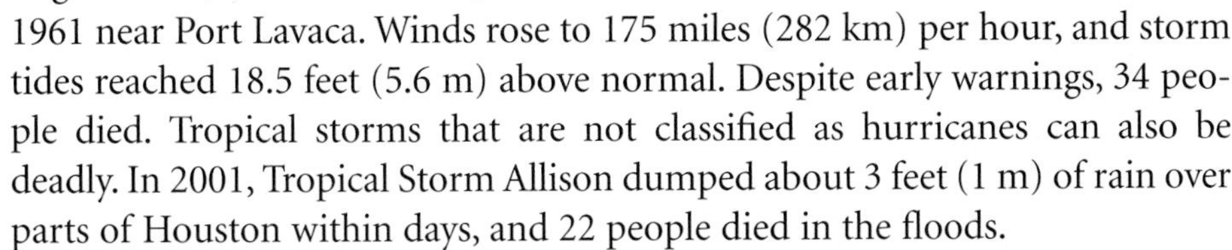

In winter, the Panhandle and north-central Texas are subject to blizzards, or large snowstorms. Blizzards bring strong winds, snow, and ice. They are particularly dangerous to cattle and other livestock. Damage done by the Great Blizzard of 1886–87 nearly destroyed the Texas cattle industry. With research and technology, scientists are better able to predict when and where severe weather will strike. These developments have saved countless lives.

TEKS **READING CHECK:** ***Physical Systems*** What are some hazardous environmental conditions that affect Texas?

Hurricane Allen swept over Corpus Christi in 1980. Strong winds and flooding caused millions of dollars' worth of property damage. A hurricane reaches the Texas coast about every three years.

Texas has more white-tailed deer—some 3 to 4 million—than any other state.

Plants and Animals

The diversity of Texas environments allows a range of plant life. A combination of climate, landforms, and soil conditions determine where certain plants can grow. For example, the state tree, the pecan, grows best in deep well-watered soil. As a result, it grows in most of the state's river valleys. Irrigation increases the range of the majestic pecan tree.

In contrast, only plants that survive long periods without water can grow in the dry lands of West Texas. Short grasses grow in clumps there. Cacti and other drought-tolerant plants such as yucca are common. Coniferous, or cone bearing, trees grow in the West Texas mountains. Coniferous trees commonly remain green all year. These hardy evergreen trees include junipers, piñon pines, and ponderosa pines.

The South Texas Plains get slightly more rain than West Texas. South Texas is often called brush country because shrubs and small trees grow there. Mesquite trees are scattered throughout the area. Ranchers often try to kill mesquites because the trees use precious water, reducing the amount of grass available to cattle. Oaks grow in places with more rain. Palm trees grow along the warm Gulf Coast of South Texas.

In the Panhandle, treeless plains stretch for miles. Town names such as Notrees and Levelland describe the landscape accurately. Soil conditions and annual rainfall do support many different kinds of grasses. However, farmland has replaced much of the Panhandle grasslands.

In most years the state's eastern third receives plenty of rain. The soil is also very fertile. As a result, a variety of trees, bushes, and grasses grow there easily. The East Texas Piney Woods are full of longleaf, shortleaf, and loblolly pine trees. Elm, hickory, and oak trees are also common.

The vegetation of Texas provides a home for many different animals. These include armadillos, bears, bobcats, coyotes, deer, javelinas, mountain lions, raccoons, skunks, and wild turkeys. Some creatures, notably opossums, raccoons, and white-tailed deer, have adapted to life in urban areas where food is available. Rivers and lakes shelter alligators, bass, catfish, and other animals. Among the Gulf's creatures are dolphins, oysters, redfish, and shrimp. Some 550 bird species have been seen in Texas.

People have destroyed the habitat of some animals. The ocelot and the whooping crane are just two species that might become extinct in Texas. An extinct species is one that has died out completely. In recent years the populations of some animals that were in danger of extinction have grown. For example, buffalo were hunted to the brink of extinction. Recently, the number of buffalo has increased dramatically.

A Texas horned lizard adorns a license plate that encourages conservation of the state's wildlife. The species is threatened with extinction in Texas.

READING CHECK: *Physical Systems* How does the climate of West Texas differ from that of East Texas? How does this affect plant life?

Natural Resources

Texas has a wealth of natural resources that have contributed to the state's growth. Its climate, soil, and water resources are the foundations of the state's agriculture, a significant part of the economy. In fact, the production of food and fibers such as cotton adds about $40 billion to the state's economy each year. More and more Texas farmlands, though, are owned and operated by large corporations instead of families.

Because East Texas has reliable rainfall and rich soils, farmers can grow a wide range of crops there. These include fruits, nuts, and vegetables. In the summer, East Texas produces large corn, tomato, and watermelon harvests. The moist climate also supports a timber industry. Trees are planted and harvested for many products. In about 20 counties along the southeastern Gulf Coast, farmers grow rice. Rivers provide irrigation for some of the rice fields.

Farmers in drier regions often use water from aquifers and rivers to irrigate their crops. For example, farmers in the Panhandle use the Ogallala Aquifer to irrigate vast wheat and cotton fields. In South Texas, farmers use irrigation to grow alfalfa, citrus fruits, cotton, melons, and vegetables. The mild winters allow farmers to harvest twice a year. Because the South Texas area north of Laredo produces crops all year, it is often called the Winter Garden.

Land Use in Texas

Grassland or prairie
Forest
Savanna
Cattle ranching
Cotton farming

SCALE 0 100 200 Miles; 0 100 200 Kilometers
Projection: Albers Equal-Area

INTERPRETING THE MAP *In the 1800s cattle ranching was concentrated in the Panhandle and West Texas. During the same period, cotton was grown only in East Texas.* ***How have these patterns changed? How might technology have allowed the spread of cotton production to other areas?*** TEKS

Livestock The natural resources of Texas have contributed to the growth of the livestock industry. Livestock includes cattle, chickens, goats, hogs, horses, sheep, and turkeys. Since early in the state's history, the grasslands have provided rich pastures for beef cattle. Where grass is sparse, the land supports fewer animals per acre. Cattle ranching is a big business. In 1999, Texas ranchers held some 14 million head of cattle. These herds were worth $7 billion. Dairy cattle numbered only about 340,000 the same year. Erath County, southwest of Dallas, accounted for almost 25 percent of the state's milk production.

Horses are still used on the state's huge ranches. However, urban and suburban areas where horses are ridden for recreation have the largest numbers. Yet Texas remains a leading state in the number of horses.

In the state's rockier regions, ranchers raise Angora goats. The Angora's silky hair is used to make mohair yarn, which is used in clothing. Texas produces nearly half of the world's mohair.

Cowhands herd longhorn cattle on the Y.O. Ranch, which has been a working ranch since 1880. Today's longhorns are descended from a mix of Spanish and English cattle.

After oil was discovered there in 1912, Burkburnett, near the Red River, became a boom town. By late 1918, wells were producing 7,500 barrels of oil a day. The town's population grew so much that 20 trains ran every day between Burkburnett and nearby Wichita Falls.

Energy Resources The energy resources of Texas are vital to the state's economy. These resources include oil, natural gas, and coal. Their use powers cars, industries, and homes.

Texas has a huge oil-drilling and refining industry. It began on January 10, 1901, when oil was discovered at Spindletop, near Beaumont. That day marked the beginning of the state's move from an agricultural to an industrial economy. Oil wells now operate in almost every Texas region. Refineries and processing plants convert the oil into fuel and other products. The production of oil and natural gas in Texas adds up to almost $20 billion a year. Thousands of people depend on jobs in the oil industry. In addition, revenue from oil wells on public land is a major source of funding for Texas public education. Universities, along with primary and secondary schools, benefit from these energy resources.

Oil, natural gas, and coal are nonrenewable resources. In other words, they are replaced very slowly by Earth's natural processes. To conserve nonrenewable resources, more people are becoming interested in renewable resources. These are resources that are easily replaced by Earth's natural processes. Some examples are the power of the Sun, or solar power, and wind power. Solar power can heat homes and generate power for other uses. Wind drives turbines that generate electricity. Some utility companies are starting to use renewable resources. For example, the city of Austin urges customers to support these alternatives. Some of the city's electricity is created by wind turbines in West Texas and solar power. Also, the city will burn the gas given off by the garbage in landfills. As oil and gas reserves are depleted, these power sources will probably be developed further.

Texas has a number of other natural resources that are important to industry. Sand and gravel go into concrete and other building materials. Texas is one of the world's leading sulfur producers. The state's salt deposits are also immense. Texas is a leading producer of helium, the gas that makes balloons rise. Other resources range from asphalt to gypsum.

Using Resources Wisely With a growing population, Texas uses more natural resources than ever. Great demands are placed on the air, land, and water. Forests may be lost if people do not plant trees to replace those cut down. Soil can wash away or lose its fertility if it is not conserved. Runoff from fields and streets can pollute rivers and streams. Car exhaust and factory fumes pollute the air.

No resource is more precious than water. Every year, huge amounts of water are pumped out of Texas aquifers. In fact, enough water is removed to cover some 11 million acres (4,455,000 ha) of land to a depth of one foot. However, only about half of this water is replaced naturally each year. Because aquifers refill slowly, water shortages may become more common. To protect

INTERPRETING THE VISUAL RECORD

Air pollution obscures the Houston skyline. In 1999 Houston was named the smoggiest city in the United States. In the foreground, ships move through the Houston Ship Channel, past petroleum processing and storage facilities. ***How does the photograph show both the source and results of air pollution?***

aquifers, Texans have established water conservation districts. These governing bodies try to establish better management of the state's aquifers. Sometimes, however, urban and rural interests conflict over control of the water conservation district. Governments, private citizens, and nonprofit organizations must work together to preserve the state's water and other natural resources.

✓ **READING CHECK:** *Environment and Society* What nonrenewable resources does Texas produce? What are the two main renewable resources being developed?

TEKS Questions 1, 2, 3, 4, 5

Section 1 Review

Homework Practice Online

Keyword: SW3 HPTX

Working with Sketch Maps

On a map of Texas that you draw or that your teacher provides, label the Hill Country, Edwards Plateau, Pecos River, Guadalupe Peak, Red River, Brazos River, Colorado River, Neches River, Sabine River, Trinity River, and Rio Grande. What and where is the largest underground water source in the state?

Reading for the Main Idea

1. *Environment and Society* What are some ways that Texans depend on, adapt to, and modify the physical environment?
2. *Physical Systems* What factors influence climate patterns in Texas?

Critical Thinking

3. **Analyzing Information** How might the development of renewable energy resources affect Texas?
4. **Drawing Inferences and Conclusions** How might policies such as the establishment of water conservation districts affect the state's geography and economy over time?

Organizing What You Know

5. Copy the chart below. Use it to explain how soil, water resources, forests, and oil contribute to the state's economy.

Resource	Importance to the state's economy
Soil	
Water resources	
Forests	
Oil	

Section 2

People and Culture

READ TO DISCOVER

1. Who were the first Texans?
2. Who lives in Texas today?
3. Where have Texans settled?
4. What factors have led to the growth of the Texas population?
5. What type of government does Texas have?

LOCATE

Houston
Dallas
San Antonio
Brownsville
El Paso
Waco

WHY IT MATTERS

Texas has a growing population. Many other places in the world are also experiencing a population boom. Use CNNfyi.com or other **current events** sources to learn more about population growth in Texas and other areas.

The First Texans

The first inhabitants of Texas arrived thousands of years ago. Some of them were hunter-gatherers, while others were farmers. Farming peoples were concentrated in eastern regions where rainfall was plentiful. The Caddo included about 25 tribal groups centered around the Red River. They raised corn and other crops, lived in villages, and made beautiful pottery. Native Americans who hunted game, such as buffalo, on the plains did not have permanent settlements. Among these tribes were the Comanche, Kiowa, Lipan Apache, Mescalero Apache, and Tonkawa.

When the first Spanish explorers came to eastern Texas, they encountered a people known as the Hasinai. The Spanish called the Hasinai by the group's word for friend or ally—*tejas* (TAY-hahs). The state takes its name from this word.

Native Americans armed with lances pursue buffalo across the Texas plains. The buffalo was a vital source of food. Its skin was used to make clothes and shelter.

After Europeans arrived, the American Indian population dropped rapidly. Disease and violence took a heavy toll. During the 1700s and 1800s some American Indians migrated to Texas from other parts of North America. Eventually, however, many Texas Indians were moved onto reservations. Today those who live on the state's three reservations form a small percentage of the 118,000 Texans who identify themselves as American Indians.

READING CHECK: *Human Systems* How did the arrival of Europeans affect Texas Indians?

Who Texans Are

Today there are almost 21 million Texas residents. Some Texans' ancestors lived in the area generations ago. Others are more recent arrivals. These groups' contributions to the state's diversity are reflected in dances, festivals, foods, languages, music, place-names, and in many other ways.

European Americans A slight majority of Texans trace their ancestry to people who came from Europe. Nearly 11 million Texans consider themselves non-Hispanic white. This number accounts for about 52 percent of the state's population. Many of this group's ancestors first came to Texas in the 1800s from the eastern United States. Immigrants who came directly from Europe also influenced Texas culture. Czech, French, German, Irish, and other European immigrants poured into Texas during the 1800s and early 1900s. Some of these groups maintain unique customs. For example, every year Texans of Scottish descent gather in Salado, north of Austin, for a "Gathering of the Clans." The event features piping, drumming, and Highland dancing.

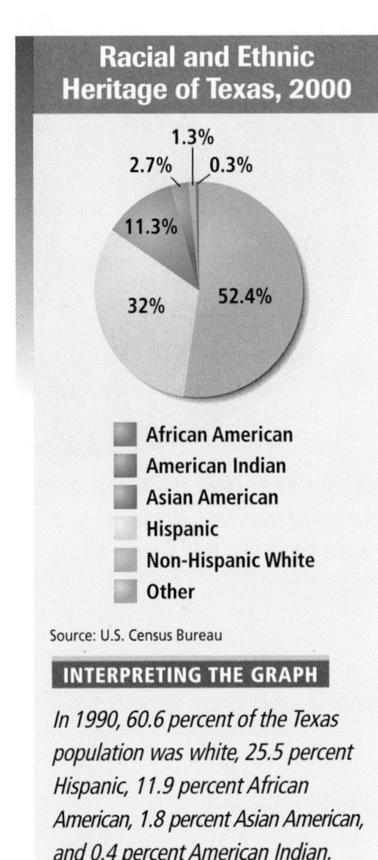

INTERPRETING THE GRAPH

In 1990, 60.6 percent of the Texas population was white, 25.5 percent Hispanic, 11.9 percent African American, 1.8 percent Asian American, and 0.4 percent American Indian. ***Which ethnic group's population has grown the fastest?***

Hispanics Some of the first European settlers came from Mexico or Spain. They introduced the Spanish language and Roman Catholic religion. Each year thousands of Mexican immigrants contribute to the growing Hispanic population. The term *Hispanic* describes someone of Latin American descent. A person of any race or ethnicity may be Hispanic. Use of the Spanish language is often the defining characteristic. More than 6.6 million Hispanics live in Texas. They make up about 32 percent of the state's population. Most Hispanic Texans or their ancestors came from Mexico, but others came from Cuba, Guatemala, or elsewhere in Central and South America. Aspects of Mexican culture, such as Mexican food, have become part of mainstream Texas culture.

African Americans African Americans also have a long history in Texas. Enslaved Africans were brought to Texas during the early and mid-1800s. Some enslaved African Americans came from elsewhere in the United States. During the 1900s the state's African American population grew. Nearly 2.4 million African Americans—or about 11.3 percent of the population—now live in Texas. African Americans have influenced music, religious practices, and other traditions. For example, musicians Scott Joplin and Huddie "Leadbelly" Ledbetter developed ragtime and blues, respectively. Zydeco, which came from African American communities in Louisiana, has further evolved in southeast Texas. African Americans have added new events to the calendar. June 19, or Juneteenth, celebrates the day on which the news reached Texas slaves that they had been freed.

Austinites celebrate Juneteenth with a parade. Juneteenth, which has been an official state holiday since 1979, is also celebrated in some places outside of Texas.

Indian children dance at a festival. In 1995 the largest Hindu temple complex in the United States was built west of Austin.

Asian Americans Within the past few decades more and more Asian Americans have made Texas their home. Nearly 560,000 Asian Americans live in the state. About 8 percent of immigrants to Texas in 1998 were Asians. Among the religions these immigrants follow are Buddhism, Hinduism, and Islam.

Two Asian groups in particular have influenced the state's cultural landscape. Texas has the second-largest Vietnamese community in the United States, after California. Vietnamese are concentrated in the Houston and Gulf Coast areas. Austin has a growing South Asian community. Because many are well-educated and speak English, these Indian immigrants have contributed to the city's high-tech industries.

TEKS **READING CHECK:** ***Human Systems*** What traditions have different ethnic groups brought to Texas?

Where Texans Have Settled

Historically, most Texans lived in rural areas. However, in the late 1800s people began migrating to cities in search of industrial jobs. By 1950, more Texans lived in cities than on farms and ranches. Texas cities have continued to grow. Now more than 84 percent of Texans live in cities. Houston, Dallas, and San Antonio are the three largest cities in Texas and are among the 10 largest in the United States.

People from Mexico have lived along the Rio Grande, from Brownsville to El Paso, for hundreds of years. Some families have been in the area since it was under Spanish rule or part of Mexico. Others came north to work on farms in the Rio Grande valley. A blend of Mexican and American influences has created a unique culture along the U.S.-Mexico border. This borderlands culture extends from California through Arizona and New Mexico to the Gulf of Mexico.

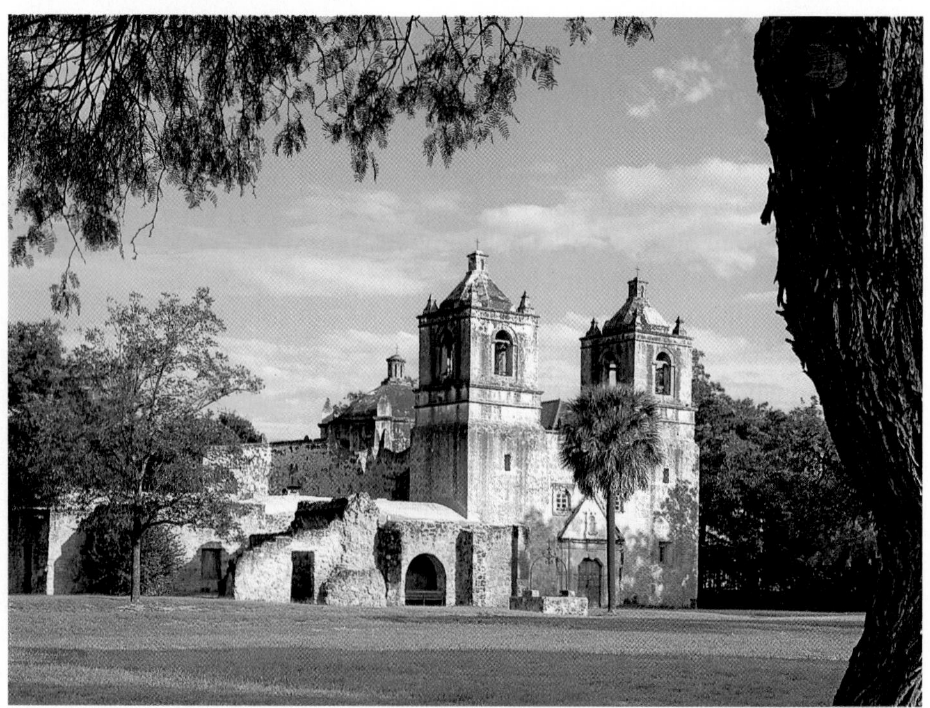

Catholic friars from Mexico established missions, like Mission Concepción in San Antonio, pictured here, to convert American Indians to Christianity. Settlements grew up around the missions.

Mexican influences remain strong elsewhere in South and West Texas also. Many businesses, cities, and cultural institutions have Spanish names, reflecting the role of Hispanics in their history and culture. San Antonio and El Paso have distinctly Mexican influences, as do numerous other towns. Spanish colonial architecture survives in many buildings. Tex-Mex food, which mixes northern Mexican and European cuisines, is popular. Today such Tex-Mex dishes as salsa and fajitas are standard fare across the United States.

Many settlers from the southern United States made their home in East Texas. The region's climates and soils were similar to what settlers had known in the American South. Some of these settlers fought in the revolution to make Texas independent from Mexico. When Texas became a republic, they set up a government similar to what already existed in the United States. They also allowed slavery. When Texas joined the United States, most Texans were already familiar with American ways. Today descendants of these early American settlers dominate the state's central and eastern regions.

INTERPRETING THE VISUAL RECORD

Fayette County, in Central Texas, boasts several churches with beautiful painted interiors. These Roman Catholic churches were built by Czech immigrants. Pictured is St. Mary's Catholic Church at High Hill. Also pictured is a descendant of the area's Czech settlers. ***How does this church's interior compare to other places of worship you have seen?***

European traditions flavor other areas. Czech immigrants settled in small Central Texas towns like West, which is near Waco. Customers drive for miles to buy Czech pastries called *kolaches* from local bakeries. German immigrants established Fredericksburg, New Braunfels, and Schulenburg, also in Central Texas. Some Germans of the Fredericksburg area built Sunday houses—small second homes near a church. A family would come to town for a weekend of shopping, socializing, and church attendance and stay at the modest home. These houses of one and one-half stories with an outside stair are distinctive features of the Gillespie County landscape. Other German practices also linger. German music festivals called *Saengerfests* attract enthusiastic audiences. Some of the original settlers' descendants speak German at home.

Norwegians settled in north-central Texas, particularly in Bosque County. Every year Clifton, which is west of Waco, holds a smorgasbord, a feast of Norwegian foods. Another group from Scandinavia, Swedes, first settled east of Austin at a town called New Sweden. Texans of Swedish descent now live throughout the state.

READING CHECK: ***Places and Regions*** What architectural styles and foods are part of the distinctive cultural landscape of cities like San Antonio and El Paso?

A Growing Population

Texas is becoming even more diverse as its population grows. To understand why the population is growing, it is important to learn about demography. Demography is a branch of geography that involves the study of human populations. Demographers study populations by analyzing statistics such as birthrate and death rate. The birthrate is the number of births per 1,000 people. The death rate is the number of deaths per 1,000 people. For example, in 1997 the birthrate in Texas was 17.39, while the death rate was 7.40. Therefore, far more people were being born than were dying. As a result, the population of Texas as a whole was growing. This is called natural increase. It accounts for more than half of the state's total population growth each year. Between 1991 and 1998 natural increase added more than 180,000 people to the state's population each year.

New Texans Immigration and migration are also significant factors in the growth of the Texas population. Immigration happens when people move to Texas from other countries. When people from other U.S. states move to Texas, the process is called migration.

Tens of thousands of immigrants arrive in Texas each year. For example, in 1998 alone more than 100,000 immigrants settled in the state. These newcomers arrive from many different countries, including Mexico, India, China, and Vietnam. In the 1990s immigration accounted for more than 20 percent of the state's total population growth.

Migration also adds to the state's growing population. From 1973 to 1983 a booming economy, particularly within the oil industry, helped draw more than 1.5 million people to Texas from other U.S. states. However, a drop in oil prices and a slowing economy led to declines in migration in the late 1980s. In the 1990s the economy recovered, encouraging further migration to Texas.

INTERPRETING THE VISUAL RECORD

The growing population of Texas is becoming more ethnically diverse, which in turn affects the state's economy and society. For example, these people in Houston work for a company that operates 11 Websites designed specifically for immigrants. ***How do you think the increasing ethnic diversity of Texas influences the development and spread of new ideas?*** TEKS

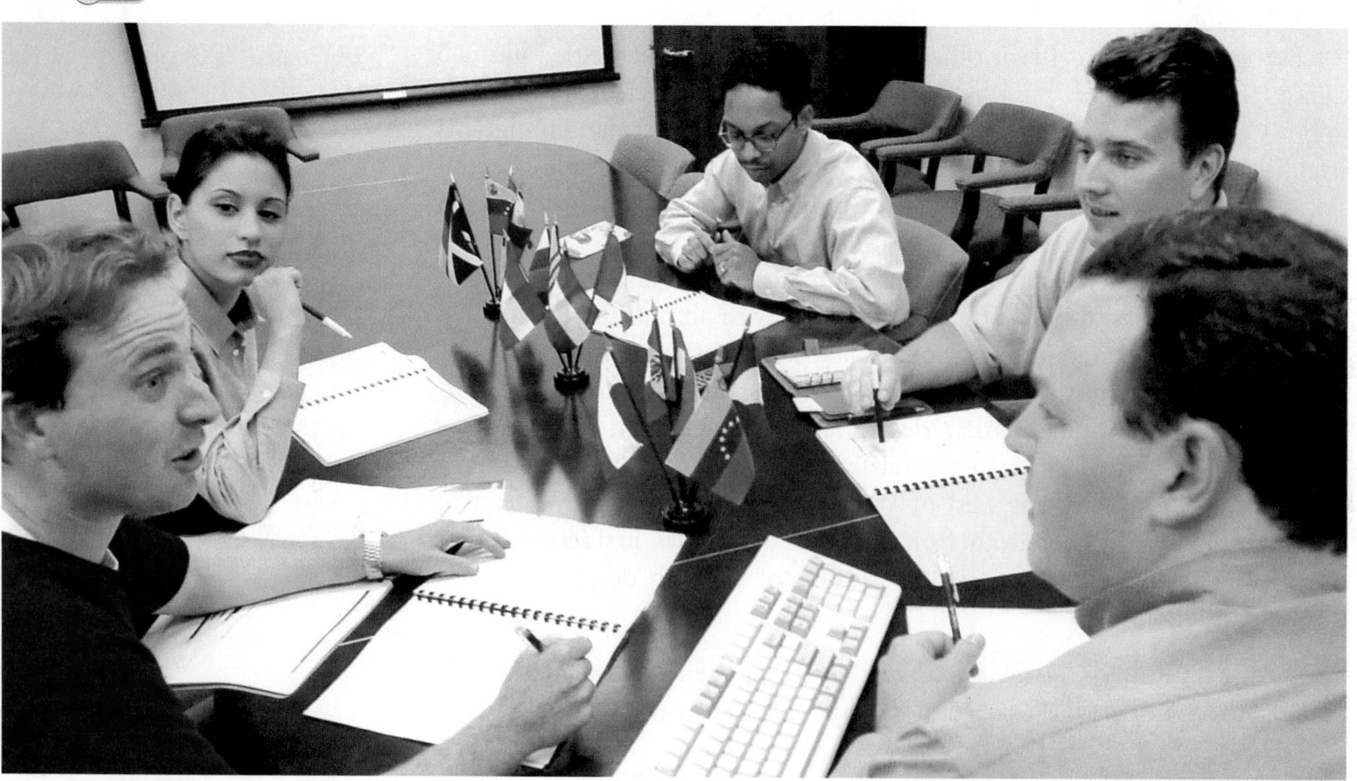

A dancer performs at Austin's First Annual Arirang Festival of Korean Culture in 2001. Koreans are one example of the many rapidly growing ethnic groups in Texas. In 1970 there were just 2,090 Koreans in the state. However, by 2000 that number had grown to 45,571. Many Koreans immigrated to Texas for economic opportunities, while others came later to join relatives.

A Changing State As the population of Texas continues to grow, the characteristics of the state's population are changing. Because so many people are born each year, the number of young people is increasing dramatically. In fact, about 6 million Texans are currently below the age of 18. As a result, Texas has a great need for schools and other services for its young population. At the same time, improvements in health care are allowing many people to live longer. As a result, the percentage of older people is increasing. In the coming years, this trend is expected to continue as the state's population as a whole gets older. This aging population will place greater demands on a wide range of government services and social programs such as retirement plans and medical care.

The population of Texas is also becoming more culturally diverse. This is largely the result of immigration and migration, which bring more and more cultures into the state. In addition to the state's largest ethnic groups—whites, Hispanics, and African Americans—many other groups are adding to the cultural diversity of Texas. For example, in the 2000 Census, thousands of Texans identified themselves as Vietnamese, Indian (South Asian), Chinese, Filipino, Korean, and Japanese.

Although the ethnic makeup of Texas is changing, Hispanics are still the state's second-largest ethnic group. Here, dancers perform a traditional Mexican dance in San Antonio.

TEKS **READING CHECK:** ***Human Systems*** What factors have contributed to population growth in Texas?

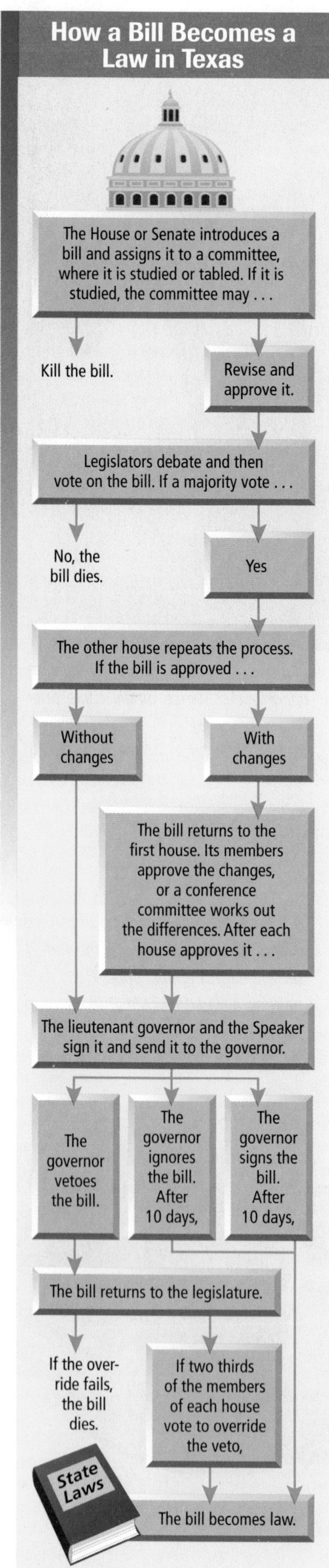

Texas Government

Like the other U.S. states, Texas has a democratic system of government. A democracy is a type of government in which citizens participate in free elections. In Texas every adult citizen can try to influence the development of public policies and decision-making processes. Governmental powers in Texas are divided among state, county, and city governments and special governing bodies such as school districts. Local issues, such as fire protection and neighborhood services, are handled by local governments. Statewide issues, such as managing state lands and parks, are handled by the state government.

The state government is organized in a way similar to the federal government of the United States. The state government has three branches—the legislative branch, executive branch, and judicial branch. The legislative branch writes bills, which are then sent to the governor. (See the diagram.) The executive branch administers the laws that are enacted. The governor heads the executive branch. He or she approves or vetoes the bills passed by the legislature. The judicial branch, including the state supreme court, is responsible for interpreting the laws enacted by the legislature and governor. Texas voters choose their legislators, governor, the heads of major executive departments, and judges in popular elections. In addition to the state government, Texans can turn to their local government to voice their opinions and concerns. Like the state government, local government officials are selected in popular elections.

People with a variety of political views and from all backgrounds participate in Texas government. For example, many people have different opinions about how much money their local or state government should raise through taxes. They may also have different views about how those governments should spend the tax money they raise. People with different viewpoints on taxes can try to convince government officials to promote the policies they think are best. Another example of a commonly debated

INTERPRETING THE VISUAL RECORD *City council meetings are public forums where local issues are debated.* ***What types of issues do you think are discussed at city council meetings?***

The Texas Capitol Building in Austin, built between 1882 and 1888, is the largest in the United States. Constructed of Texas limestone and granite, the building is topped by a statue representing liberty.

political issue involves road construction. As the populations of the state's cities grow, traffic congestion gets worse and worse. As a result, some people want to expand public transit systems, such as buses and light rail, to ease that congestion. However, other people think building new roads and highways would be a better approach to solving urban traffic problems. To resolve these issues, people can bring their opposing points of view to planning bodies and local governments, such as city councils. They can also organize to try to elect officials who share their points of view.

READING CHECK: *Human Systems* How do Texans with opposing viewpoints influence government policies and decision making? What are some examples of issues that people have different viewpoints on today?

TEKS Working with Sketch Maps, Questions 1, 2, 3, 4

Review

Homework Practice Online

Keyword: SW3 HPTX

Working with Sketch Maps On the map you created in Section 1, label Houston, Dallas, San Antonio, Brownsville, El Paso, and Waco. What are some towns that attracted many Czech or German immigrants?

Reading for the Main Idea

1. *Human Systems* Which immigrant group makes up about half a million Texans? How has the ability to speak English affected many members of this group?
2. *Human Systems* How are the characteristics of the state's population changing? What is causing these changes?

Critical Thinking

3. **Making Generalizations** How do you think distinctive cultural landscapes in various regions of Texas influence innovation and diffusion—the development and spread of cultural traits? Give some examples.
4. **Drawing Inferences and Conclusions** What do you think will happen to the population of Texas cities in the next two decades? Explain your answer.

Organizing What You Know

5. Copy the chart below. Use it to describe the main functions of the three branches of government in Texas.

Branch	Functions
Legislative	
Executive	
Judicial	

Section 3

The Regions of Texas

READ TO DISCOVER

1. Why is Texas considered a regional crossroads?
2. What types of features make the subregions of the Gulf-Atlantic Coastal Plain different?
3. How do the physical features of the interior plains affect industry in the region?
4. How have Texans adapted to the environment of the Great Plains?
5. How does the geography of the intermountain basins and plateaus affect human activities there?

WHY IT MATTERS

One of the most popular vacation areas in Texas is Big Bend National Park. Use CNNfyi.com or other **current events** sources to learn more about other popular parks in Texas and the United States.

LOCATE

Big Bend National Park
Tyler
Fort Worth
Texarkana
Beaumont
Corpus Christi
Galveston
Port Arthur
Bryan–College Station
Abilene
Wichita Falls
Lubbock
Amarillo
Odessa
Midland
Austin
San Angelo
Del Rio
Llano Basin
Marfa

The Hill Country is a region in Central Texas known for its colorful fields of bluebonnets, Indian paintbrush, and other wildflowers, which bloom each spring.

A Regional Crossroads

Texas is made up of many different regions. A region is an area that shares one or more characteristics that make it different from surrounding areas. The regions of Texas have distinct geographic and cultural patterns. Some geographers call Texas a regional crossroads because many different climatic, cultural, and physical regions meet there. Can you name some of the different regions that are found in Texas?

Regions in Texas Texas is part of a region in the southern United States called the Sun Belt. The Sun Belt's name refers to the warm climate that has attracted many new residents and industries to the region. Within Texas, some regions get their names from prominent geographic features. For example, the Redlands is an area of northeastern Texas that has reddish soils. Big Bend in West Texas is named after a large bend in the Rio Grande that wraps around the area. The Piney Woods in East Texas are named after the large pine forests that grow throughout the region.

Many other regions in Texas are named simply for their geographical location. For example, people living in Tyler say they are from East Texas. Other geographical regions include North Texas, South Texas, and West Texas. These regions have unclear boundaries because they are based partly on people's perceptions. Geographers describe these types of regions as perceptual regions. For example, people have different opinions about the exact boundaries of West Texas. Some argue that Fort Worth is in West Texas because the city

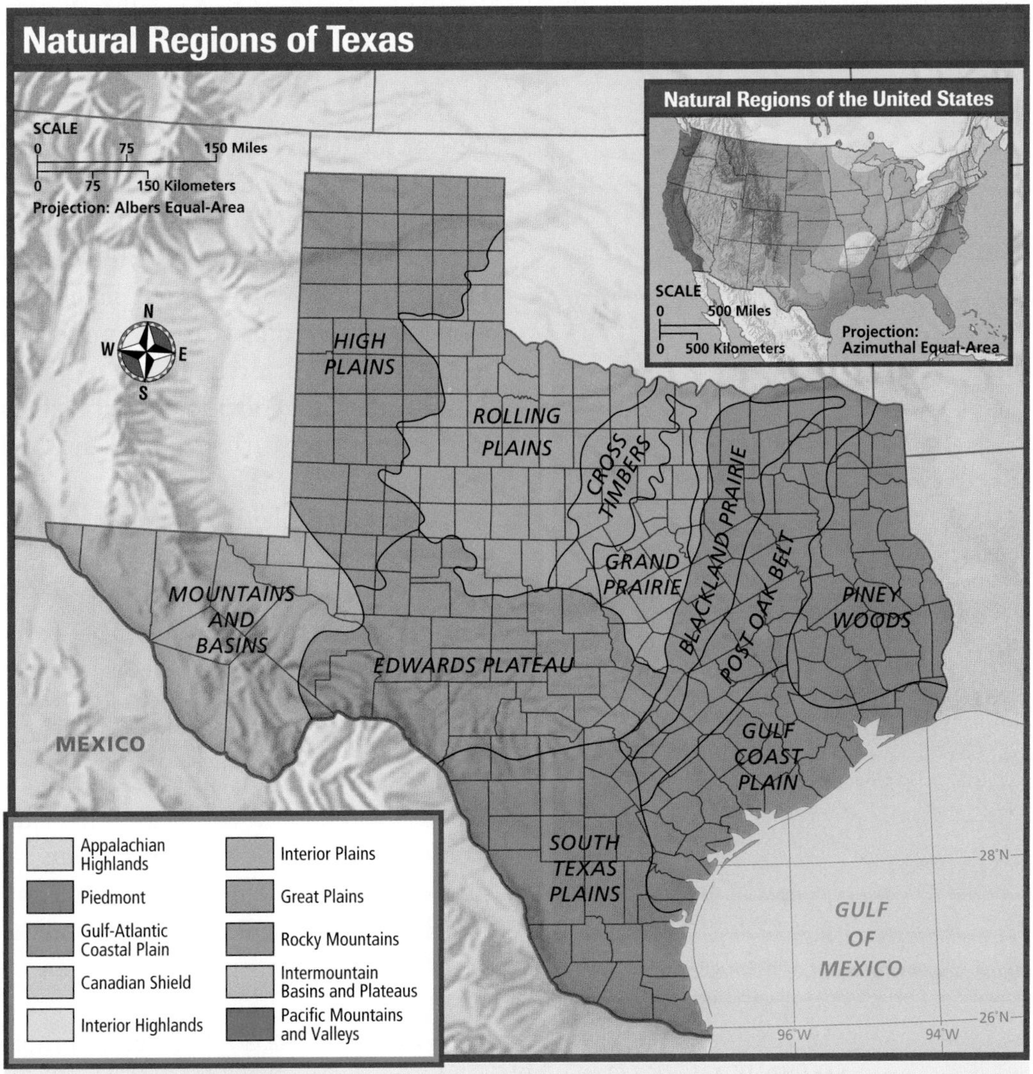

Natural Regions of Texas

SCALE 0 75 150 Miles
0 75 150 Kilometers
Projection: Albers Equal-Area

HIGH PLAINS
ROLLING PLAINS
CROSS TIMBERS
BLACKLAND PRAIRIE
GRAND PRAIRIE
POST OAK BELT
PINEY WOODS
MOUNTAINS AND BASINS
EDWARDS PLATEAU
GULF COAST PLAIN
SOUTH TEXAS PLAINS
MEXICO
GULF OF MEXICO
28°N
26°N
96°W
94°W

Natural Regions of the United States

SCALE 0 500 Miles
0 500 Kilometers
Projection: Azimuthal Equal-Area

Appalachian Highlands
Piedmont
Gulf-Atlantic Coastal Plain
Canadian Shield
Interior Highlands
Interior Plains
Great Plains
Rocky Mountains
Intermountain Basins and Plateaus
Pacific Mountains and Valleys

INTERPRETING THE MAP *The natural regions of Texas are part of larger regions that extend across North America.* ***Which four major natural regions found in the United States cover parts of Texas? Where is each located in the state?***

has close economic and historical ties to the region's cattle industry. However, others consider Fort Worth part of North Texas.

Texas can also be divided into natural regions. (See the map on this page.) Natural regions are defined by physical features and environmental conditions. These features may include climate types, landforms, plant life, rivers, and soils. Some geographers divide the continental United States into 10 major natural regions. Four of these natural regions—more than any other state—stretch into Texas. These regions include the Gulf-Atlantic Coastal Plain, the interior plains, the Great Plains, and an intermountain area of basins and plateaus. Each of these four natural regions has distinct landforms, climates, and vegetation patterns. Studying these regions can tell us a great deal about the diverse natural landscapes of Texas. These natural patterns, in turn, have had a major influence on the history, culture, and economy of the state.

READING CHECK: ***Places and Regions*** What physical features and environmental conditions define the natural regions of Texas?

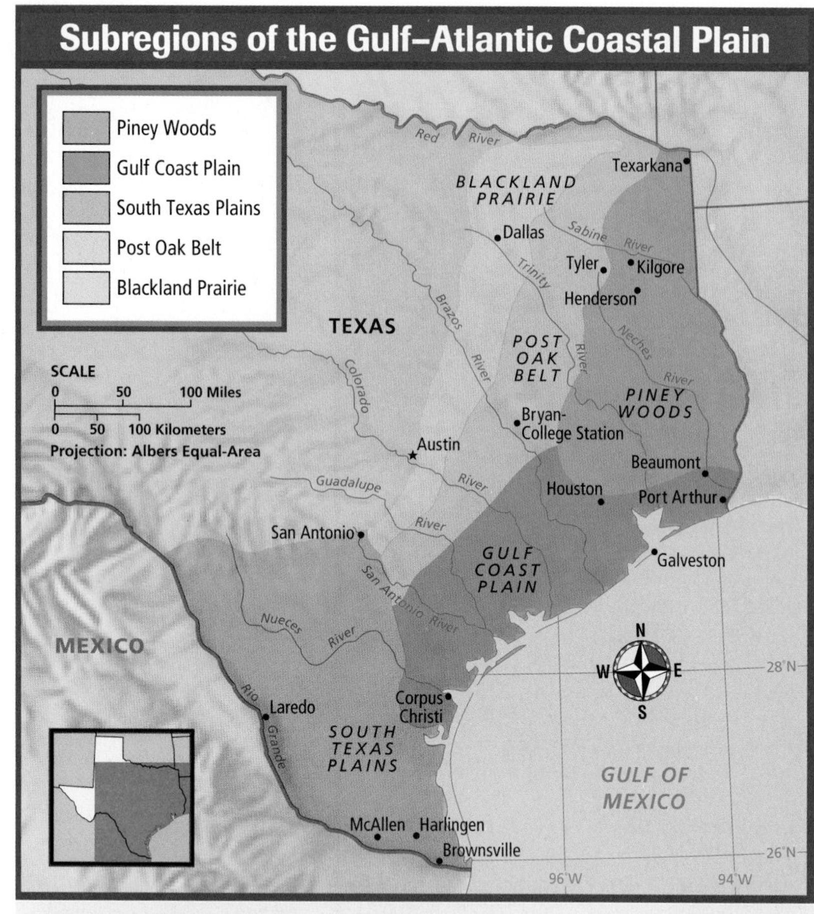

INTERPRETING THE MAP *The Gulf-Atlantic Coastal Plain covers much of eastern and southern Texas.* ***What do you think the environment of this region is like?***

The Gulf-Atlantic Coastal Plain

One of the largest regions in Texas is the Gulf-Atlantic Coastal Plain. (See the map.) This region stretches along the Gulf of Mexico and Atlantic coasts of the United States. In Texas the coastal plain runs along the state's entire coast. In some places it extends inland for hundreds of miles. Most of the Gulf-Atlantic Coastal Plain is low and marshy. Farther inland the land rises and becomes hilly.

The coastal plain contains the state's wettest region. Large amounts of rainfall and rich soils have led to the growth of dense forests in the eastern parts of the region. In the drier western areas, prairies and brush lands are more common. The coastal plain can be divided into five smaller subregions. In the east is the Piney Woods region. In the southeast is the Gulf Coast Plain. The South Texas Plains lie farther south. Stretching into Central and North Texas are the Post Oak Belt and the Blackland Prairie.

The Piney Woods The easternmost area of the Gulf-Atlantic Coastal Plain is known as the Piney Woods. The Piney Woods of Texas begin at the Texas-Louisiana border and extend west past the Trinity River. They stretch from the Red River in the north to the Gulf Coast Plain in the south.

Most of the land in the Piney Woods is made up of gently rolling hills. Elevations range from about 200 to 500 feet (60 to 150 m) above sea level. Humidity, temperatures, and rainfall are typically high. In fact, the Piney Woods have the wettest climate of any region in Texas. Most areas receive some 40 to 50 inches (102 to 127 cm) of rainfall each year. This rainfall helps support a large timber industry in heavily forested areas. Gum, hickory, oak, and pine trees are valuable sources of products such as wood pulp, which is used to make paper. Farming is also important to the region's economy. The wet climate and soils are ideal for producing many types of fruits and vegetables. In areas where farmers have cleared forests, tall grasses have grown. Farmers raise cattle in these open grassy areas. Oil is another of the region's natural resources. The oil industry contributes to the economy of cities such as Henderson and Kilgore.

Many people in the Piney Woods live in rural areas or small towns. Towns serve as local markets for farmers and for the lumber industry. The region's largest cities are Longview and Texarkana. Kilgore experienced a boom in the early 1930s, thanks to the oil-drilling business. These cities continue to prosper as their agriculture, oil, and timber industries grow.

The Gulf Coast Plain South and west of the Piney Woods lies the Gulf Coast Plain. This region extends south along the coast from the Sabine River to Corpus Christi Bay. Just offshore, a chain of barrier islands runs nearly the entire length of the Texas coast. These sandy islands are covered by brush and grasses. Shallow bays separate the islands from the mainland. On the mainland near the coast, the land is marshy. Inland, the lush grasslands of the plains are broken by scattered groups of trees.

INTERPRETING THE VISUAL RECORD

Padre Island off the southeastern coast of Texas is the state's largest barrier island and has the longest sand beach in the United States. Barrier islands help protect the mainland from ocean waves and tropical storms. ***What role do you think wave action plays in the formation of barrier islands?*** TEKS

Good soils and a favorable climate have made the Gulf Coast Plain a rich agricultural area. Farmers grow cotton, grains, and rice. The coastal grasslands also support one of the largest livestock industries in Texas. Even the Gulf waters are a valuable resource. Fishing and shrimping are major sources of income for many people in the area.

Another major offshore resource is oil. Offshore oil rigs provide jobs for many in the Gulf Coast region. The center of the Texas and U.S. oil-refining industry lies between Beaumont and Houston. The area also has a large petrochemical industry. Petrochemicals are chemicals made from oil and natural gas. They are used in many different industries.

Ports provide another valuable boost to the Gulf Coast's economy. Some of the region's largest cities are port cities, including Beaumont, Corpus Christi, Galveston, Houston, and Port Arthur. Farm and factory products of all kinds pass through these ports to other areas of the world. Goods from other countries enter Texas through these same cities. Houston lies about 50 miles (80 km) inland from the Gulf. However, the construction of the Houston Ship Channel in the early 1900s helped make Houston the third-largest seaport in the United States today.

READING CHECK: ***Places and Regions*** What natural resources are important in the Piney Woods and Gulf Coast Plain regions? What has helped make Houston a major seaport?

The South Texas Plains The South Texas Plains lie southwest of the Gulf Coast Plain. Near the Gulf of Mexico, the land is low and flat. To the north and west the land rises and becomes hillier. The climate is drier than the Gulf Coast Plain.

The South Texas Plains have many natural resources. Farming and ranching are the main contributors to the region's economy. One of the richest farming areas in the country—the Rio Grande valley, or simply the Valley—lies in this region. Temperatures there are warm for most of the year, and freezes are rare. The Valley has fertile soils from sediment deposited by the Rio Grande. Farmers grow large crops of citrus fruits, such as grapefruit and oranges. Aloe vera, sugarcane, and vegetables are other important crops. Ranches are found throughout much of the South Texas Plains. The largest is the huge King Ranch, which covers more than 1 million acres (405,000 ha). Oil and natural gas wells also dot the landscape.

The football teams of the University of Texas at Austin and Texas A&M University clash in College Station. These universities are two of the largest and most prestigious in Texas and are recognized nationally for their contributions to research and teaching.

Although the South Texas Plains region is largely rural, it does have several large cities. The largest is San Antonio. Founded in 1731 by the Spanish, the city is home to many historic buildings. It is also one of the state's most popular tourist destinations, featuring attractions such as the Alamo and the Riverwalk. In addition, military bases there employ many people. San Antonio is a retail trade center for South Texas and a major center for international trade with Mexico. Although San Antonio is the largest city in this region, Laredo is growing the fastest. Other border towns are also expanding rapidly.

The Post Oak Belt The Post Oak Belt lies west of the Piney Woods and Gulf Coast Plain. It stretches from the Red River in the north to just east of San Antonio in the south. The belt has a climate similar to that of the Piney Woods. However, it receives a little less rainfall, and the soil is sandier. The land of the Post Oak Belt is flat or gently rolling.

Some parts of the Post Oak Belt are covered in post oak, blackjack oak, elm, hickory, pecan, and walnut trees. However, early settlers cleared many areas for farming. Cotton is a major crop in the area. Other crops include corn and sorghum—a grain used to feed livestock. Farmers also raise cattle and hogs. Deposits of lignite, a type of soft coal, are also mined there.

As in the Piney Woods, most residents of the Post Oak Belt live in rural areas. However, a number of towns and small cities are scattered throughout the region. The largest cities are Tyler and Bryan–College Station. Manufacturing industries have developed in these cities. Food processing, furniture construction, and metalworking are some of the largest industries. Texas A&M University in College Station is a major educational institution and is the state's oldest public institution of higher education.

READING CHECK: ***Environment and Society*** How did early settlers modify the natural environment of the Post Oak Belt?

Tyler has long been famous for its roses. In the early 1900s a peach blight destroyed much of the region's fruit industry. As a result, many farmers began growing roses, which thrived in Tyler's warm climate. By the 1920s rose production had become a major business, and by the 1940s more than half of commercially grown rose bushes in the United States could be found within 10 miles (16 km) of Tyler.

Dallas was founded in 1841 near a natural ford, or crossing point, of the Trinity River. The city quickly developed into a service center for the surrounding rural area. Later it capitalized on its excellent location as a transportation center for moving products to manufacturing plants in the northern and eastern United States. Today the city's diverse economy includes manufacturing, transportation, and service industries.

The Blackland Prairie To the west of the Post Oak Belt is the Blackland Prairie. The climate there is mild like that of the Post Oak Belt. Prairies cover the landscape. Their rich black soils are ideal for farming. Farmers grow cotton, grains, and vegetables and raise cattle, chickens, and hogs.

The Blackland Prairie is one of the most heavily populated regions in Texas. Its cities include Garland, Grand Prairie, Mesquite, Plano, and Richardson, all of which are located around Dallas. The cities of Temple and Waco lie to the south.

Dallas lies in the northern part of the Blackland Prairie and is the second-largest city in Texas. Dallas has many industries. It is a center for banking and business in Texas and much of the southwestern United States. Many insurance and oil companies have their corporate headquarters in Dallas. The city is also a major center for the international cotton market. The Dallas economy depends on the manufacture of items ranging from computer electronics to missile parts and high-fashion clothing.

Like other large cities, the Dallas area is linked by an extensive transportation and transit system. This creates a functional region, or a region linked by the flow of goods or people. Nearby cities also make up part of a larger functional region. About 30 miles (48 km) to the west is Fort Worth. It is linked to Dallas by major highways and roads. Some people refer to Dallas, Fort Worth, and the smaller cities that lie between and around them as the Metroplex. The Metroplex is a major transportation hub. The Dallas–Fort Worth International Airport is located between these two cities. In addition, U.S. Interstate 35 runs through Dallas and Fort Worth. This highway starts at the Mexican border and runs north all the way to Duluth, Minnesota. Other roads continue from Duluth to Canada. Interstate 35 is a major transportation route, particularly for U.S.-Mexico trade.

READING CHECK: ***Human Systems*** What makes the Dallas transit system and the Dallas–Fort Worth Metroplex functional regions?

Subregions of the Interior Plains

INTERPRETING THE MAP *The interior plains occupy the heart of north-central Texas.* ***Based on the map, what types of landforms would you expect to find in this region?***

The Interior Plains

The interior plains region is one of the largest in the United States. It begins just west of the Appalachian Mountains and stretches from Canada as far south as the Colorado River in Texas. For the most part, this region is characterized by gently rolling prairies. These prairies are used for farming and ranching. Many of the country's farms are located in the region. In Texas there are three main subregions of the interior plains. They are the Grand Prairie, the Cross Timbers, and the Rolling Plains. (See the map.)

The Grand Prairie In Texas the Grand Prairie lies west of the Blackland Prairie and rests at a higher elevation. Some parts of the Grand Prairie are flat, while other areas have rolling hills. The climate of the Grand Prairie is similar to that of the Blackland Prairie. However, the layers of soil are thinner. Grasses, shrubs, and small trees are the major types of plants. These grasslands are an important natural resource. They are particularly well suited for raising cattle. Other livestock include goats, hogs, chickens, and sheep. The thin soil has limited farm production. Crops such as corn, oats, wheat, and other grains are grown mostly for animal feed.

Fort Worth is by far the largest city of the Grand Prairie. It is also the largest city in the Texas interior plains. Fort Worth plays a crucial role in the region's economy. It is a major processing and transportation center for livestock and farm products. The city has grain milling and storage facilities. With some of the busiest railyards in the country, Fort Worth is also a transportation center. Most important, however, are Fort Worth's manufacturing industries. The city specializes in the manufacture of airplanes, electronic equipment, and helicopters. Large aircraft manufacturing plants employ many people.

The Parker County Courthouse in Weatherford was completed in 1886 and is located at the exact geographic center of Parker County. The 254 county courthouses in Texas were built in a variety of architectural styles and include many excellent examples of late 1800s and early 1900s architecture.

The Cross Timbers The Cross Timbers surround the Grand Prairie on the west, north, and east. Two belts of forests make up the Cross Timbers—the Western Cross Timbers and the Eastern Cross Timbers. The Western Cross Timbers area lies west of the Grand Prairie and extends southward almost 200 miles (320 km) from the Red River to the Colorado River. The Eastern Cross Timbers are narrower, with an average width

Herders drive longhorn cattle to the Forth Worth Stockyards as part of the city's 150th anniversary celebrations. Fort Worth's history as a cattle marketing center draws thousands of tourists to the Stockyards area each year to enjoy its shops, restaurants, and Western atmosphere.

of less than 15 miles (24 km). They lie between the Grand Prairie and Blackland Prairie. The Eastern Cross Timbers extend southward from the Red River to the Brazos River.

The wooded landscape and sandy soil of the Cross Timbers absorb water well. As a result, the region is a good place for farming. Farmers have cleared the forest in many places to plant crops. These crops range from peanuts to corn, cotton, and hay. Ranchers also raise cattle, horses, and sheep. Since 1917 the region has provided Texas with energy resources. Oil and natural gas wells have been drilled in several of the region's counties.

The Eastern Cross Timbers run between Dallas and Fort Worth. Arlington and Denton are located in the Eastern Cross Timbers and are part of the Metroplex. These two cities have large manufacturing industries and excellent universities.

The Rolling Plains The Rolling Plains are in the westernmost part of the interior plains. The Red River borders the Rolling Plains on the north. The region extends west into the Panhandle and south beyond the Colorado River. The land of the Rolling Plains rises gradually from about 1,000 feet (305 m) in the east to 2,200 feet (670 m) in the west. Prairie grasses cover most of the hilly terrain. The shallow soil also supports mesquite trees, hardwoods, and brush. However, the region is mostly open prairies.

The Rolling Plains are particularly well suited for cattle ranching. Many of the state's largest ranches are located there. Steep valleys provide shelter for cattle, while grasslands and rivers provide food and water. In areas with poorer vegetation, ranchers raise sheep and goats. Where there is good farmland, farmers grow cotton, sorghum, and wheat. The region also has oil deposits.

The population of the Rolling Plains is small and largely rural. Small market towns are scattered throughout the area. Abilene and Wichita Falls are the largest cities. Both serve as distribution centers for the region. They also have meat and dairy processing facilities.

READING CHECK: ***Physical Systems*** How do the Rolling Plains differ from the Cross Timbers?

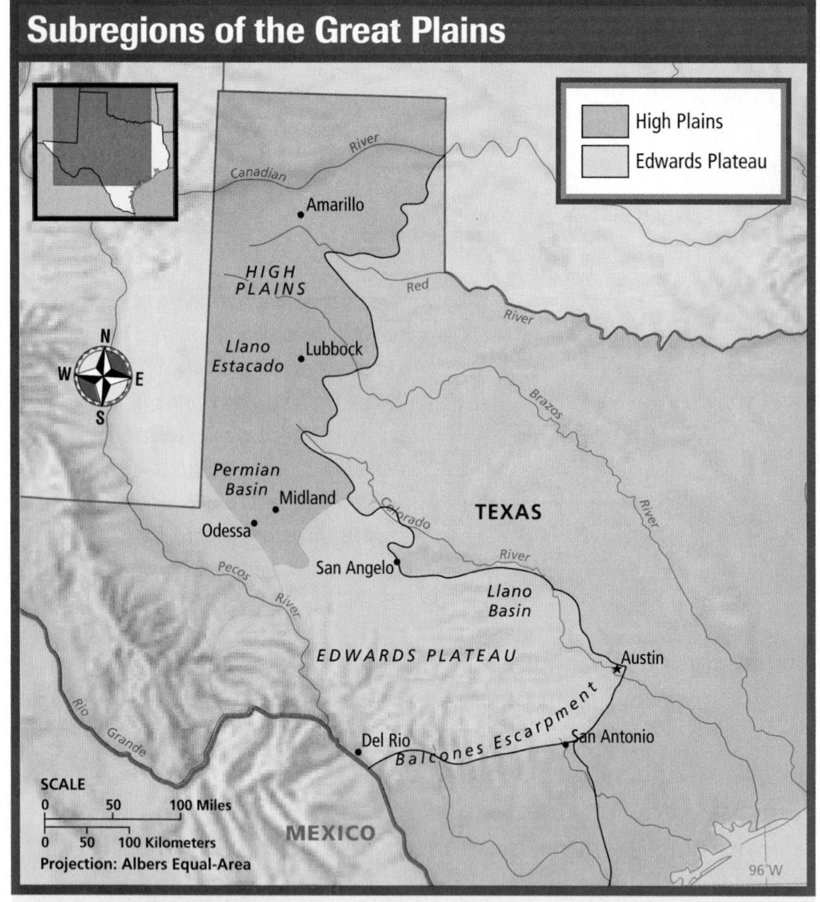

INTERPRETING THE MAP *The soils of the Great Plains are fertile. However, in Texas the region does not receive much precipitation, and irrigation supports agriculture.* ***Based on the map, how do you think the elevations of this region compare to other regions in Texas? Why?***

The Great Plains

The Great Plains extend from Canada to the Texas border. This region lies west of the interior plains and sits at the foot of the Rocky Mountains. In Texas the Great Plains stretch from the Panhandle to the Rio Grande. Vast grasslands cover the region, which is a productive farming area. To the south the landscape becomes more rugged. The Texas Great Plains are divided into the High Plains and the Edwards Plateau. (See the map.)

The High Plains The High Plains cover most of the Texas Panhandle, stretching almost as far south as the Pecos River. The eastern edge of the High Plains rises above the interior plains, separating the two regions. This also gives the High Plains its name. A hard layer of rock below the soil, known as the Caprock, is a notable physical feature. Erosion of the rock has created cliffs along the eastern and western edges of the High Plains.

The High Plains receive little regular rainfall. Early settlers there believed that the land was not suitable for farming. However, the region's rich grasslands attracted cattle ranchers. Today feedlots, where cattle are raised on feed rather than grass, are spread across the High Plains. In addition, large areas of grassland have been turned into farmland. Farmers pump water from the Ogallala Aquifer to irrigate their fields of cotton, sorghum, and wheat. However, overpumping of this groundwater is a growing concern in the region.

Large deposits of oil and natural gas lie in the High Plains. Many oil wells dot the area. The few cities there provide services for the ranching, farming, and oil industries. Lubbock and Amarillo serve as meat-processing and distribution centers. Odessa and Midland have several oil companies.

The Edwards Plateau Just south of the High Plains lies the Edwards Plateau. The Edwards Plateau extends east past the Colorado River and west past the Pecos River. The region's southern border is formed by the Rio Grande and Balcones Escarpment. The Balcones Escarpment is a limestone ridge that separates the Edwards Plateau from the Gulf-Atlantic Coastal Plain.

The Edwards Plateau is an elevated hilly area. A layer of limestone lies just below the surface. Erosion has washed away much soil, leaving only a thin layer above the limestone. Clumps of grass, shrubs, and cedar trees grow in this thin soil. Where the soil is deeper, short prairie grasses and mesquite trees grow. Ranchers raise cattle on these short grasses. Where the landscape is rockier, ranchers raise sheep and goats. They also lease land for recreation and hunting.

Much of the landscape of the Edwards Plateau is rugged. Nonetheless, Texans have settled and prospered in the region. The largest city, Austin, lies at the eastern edge of the plateau. Austin is the state capital. It is home to the University of Texas and high-tech industries and has a diverse economy. To the northwest is San Angelo, a distribution and manufacturing center. Along the Rio Grande, tourism and trade with Mexico have helped the city of Del Rio grow.

The Llano Basin lies in the northeastern area of the Edwards Plateau. A basin is a lower area surrounded by higher land. Erosion by the Colorado River and its tributaries created the Llano Basin. It lies some 1,000 feet (305 m) lower than the Edwards Plateau. There are few towns there. Most people are farmers or ranchers. Hunting and tourism also contribute to the region's economy.

The Governor's Mansion in Austin was built between 1854 and 1856 in the Greek Revival style.

READING CHECK: ***Physical Systems*** How has erosion shaped the landscape of the Great Plains?

The Intermountain Basins and Plateaus

The intermountain basins and plateaus region extends from Alaska to New Mexico and Texas. The region is called "intermountain" because it is located between the Rocky Mountains and Pacific mountain ranges. Basins and plateaus are the region's main landforms. Mountains and canyons are also found throughout the region, which has one of the most rugged and varied landscapes in Texas. It is also the state's driest region. The only major subregion there is called the Mountains and Basins. (See the map.)

The Mountains and Basins The Mountains and Basins subregion is bordered on the south and west by the Rio Grande. New Mexico marks its northern boundary. The region extends east to the Pecos River. The landscape of the Mountains and Basins includes mountains, plateaus, basins, and canyons. With its high mountains and low basins, the elevation of the area varies greatly. The climate is also extreme. Summers are very hot, and winters can be quite cold. Whatever the temperature, the region is almost always dry. The arid climate has limited plant life. Cacti, desert grasses, and shrubs grow in the dry rocky soil.

The dry climate has also limited human activities in the region. Farming and ranching are difficult. However, ranching is essential to the regional economy. Ranches must be large because the desert grasses and plants offer limited food for cattle, sheep, and goats. Some

The Intermountain Basins and Plateaus Region

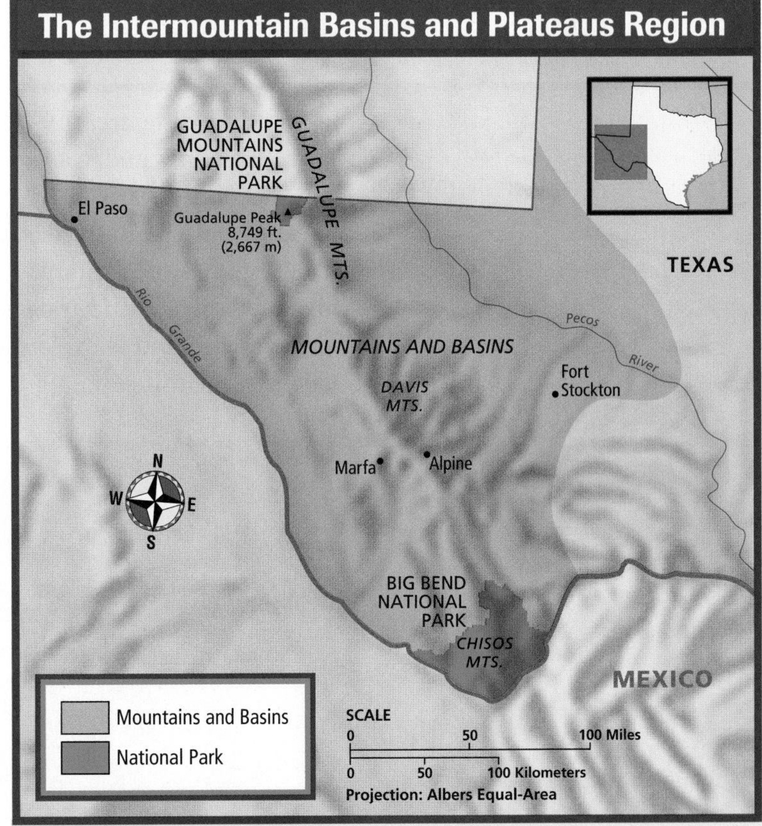

INTERPRETING THE MAP *The westernmost region of Texas includes the only mountains in the state. Several peaks in this area rise above 8,000 feet (2,440 m).* ***How do you think the area's rugged terrain has affected its population and economy?***

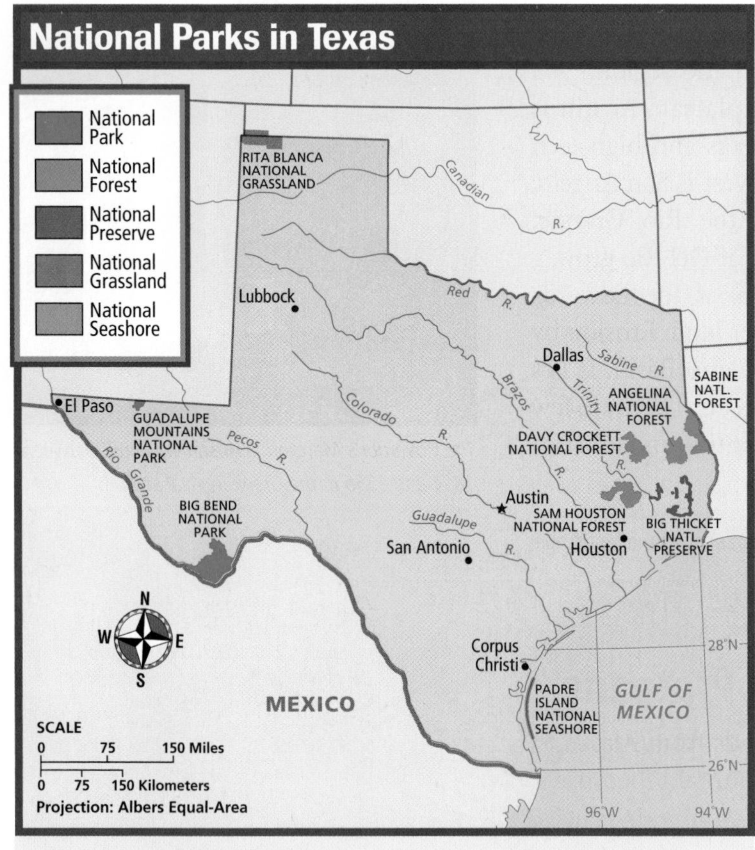

INTERPRETING THE MAP *In addition to its two national parks, Texas is home to a number of other protected areas, including four national forests.* ***In which natural region are these forests located?***

Texans have managed to farm by using irrigation. They grow alfalfa, cotton, pecans, and vegetables. Oil, silver, and sulfur are other natural resources that boost the local economy. In addition, the dramatic landscape has generated a tourist industry.

With limits on farming and ranching, the population in the region has remained small. It has only a few small towns. An exception is El Paso, one of the largest cities in Texas. El Paso is located along the Rio Grande in the westernmost corner of the state. Military bases and trade with Mexico have spurred the city's growth. Trade in the region is aided by U.S. Interstate 10, an important highway. Running through El Paso, Interstate 10 is a major east-west shipping route. Trucks often take it to avoid icy roads farther north.

National Parks Many businesses in El Paso have prospered by offering services to tourists. The beauty of the land is also a source of income for towns such as Alpine and Marfa. These towns serve as entry points to one of the most popular tourist areas in Texas.

Big Bend National Park is a popular camping and wildlife area in West Texas. It covers more than 1,250 square miles (3,240 sq km) and was the first national park in Texas. Big Bend includes an incredible variety of plants and animals. In fact, it is home to the greatest variety of bird species of all the U.S. national parks. More than 1,000 species of plants, 78 mammals, 56 reptiles, and 434 birds can be found inside the park's boundaries. These animals include the endangered peregrine falcon and Mexican long-nosed bat. Today Big Bend National Park is facing some difficult environmental challenges. Air pollution has dramatically

Big Bend's combination of desert plains and rocky mountain woodlands has allowed a unique group of plants and animals to flourish there. Several Big Bend species, such as the colima warbler and Mexican drooping juniper, are found nowhere else in the United States.

Guadalupe Mountains National Park contains Capitan Reef, one of the world's best examples of an ancient barrier reef. The reef was formed in shallow ocean waters more than 200 million years ago by decaying marine organisms. Over time, the area was pushed above Earth's surface and exposed by erosion.

reduced visibility in some areas, blocking the park's scenic views. In addition, the Rio Grande suffers from low water levels caused by water diversion. These low water levels have threatened rafting trips, a popular activity in the park. Pollution of the river is also a major problem.

Northwest of Big Bend National Park is Guadalupe Mountains National Park. The mountains of this range are actually the fossilized remains of an ancient reef. Guadalupe Peak, the highest point in Texas at 8,749 feet (2,667 m), is located within the park. Geologists study the park's deep canyons to learn about Earth's ancient history. The park is also a favorite destination of campers and hikers. Elk, mule deer, mountain lions, and black bears can be found at high elevations within the park.

One of the largest impact craters in the United States is located southwest of Odessa. It was formed thousands of years ago by iron meteorites that struck Earth. The impact shifted the limestone below the surface, forming a crater that covers about 10 acres (4 ha).

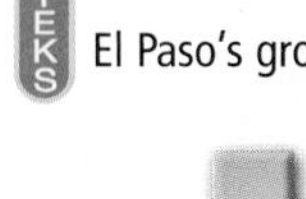

READING CHECK: *Environment and Society* What are two important sources of El Paso's growth?

TEKS Working with Sketch Maps, Questions 2, 3, 4, 5

Section 3 Review

Homework Practice Online
Keyword: SW3 HPTX

Working with Sketch Maps On the map you created in Section 2, label Big Bend National Park, Tyler, Fort Worth, Texarkana, Beaumont, Corpus Christi, Galveston, Port Arthur, Bryan–College Station, Abilene, Wichita Falls, Lubbock, Amarillo, Odessa, Midland, Austin, San Angelo, Del Rio, Llano Basin, and Marfa. Which city is the capital of Texas?

Reading for the Main Idea

1. *Physical Systems* Why is Texas considered a crossroads of natural regions?
2. *Places and Regions* What four natural regions of the United States are found in Texas? Which is the wettest? The driest?
3. *Environment and Society* How did the geographic features of the interior plains affect the economic life of the region?

Critical Thinking

4. **Drawing Inferences** How might the four natural regions of Texas be described as functional regions?

Organizing What You Know

5. Copy the chart below. Use it to explain where the subregions of the Gulf-Atlantic Coastal Plain are located and to describe their special characteristics.

Subregion	Location	Special features
Piney Woods		
Gulf Coast Plain		
South Texas Plains		
Post Oak Belt		
Blackland Prairie		

go.hrw.com TAKS PREP ONLINE keyword: SW3 TTX

TEXAS Review

Understanding the Main Ideas TEKS

Section 1

1. *Physical Systems* How do the climates of Texas affect the distribution of vegetation?

Section 2

2. *Human Systems* Is the population of Texas mostly urban or mostly rural? When and why did population distribution start to change?
3. *Human Systems* What ethnic groups make up much of the state's population?

Section 3

4. *Environment and Society* How have the physical features of the Edwards Plateau affected human activities in the region?
5. *Environment and Society* How did early settlers perceive farming possibilities in the High Plains of Texas? How have those perceptions changed? What has aided this change in perception?

Thinking Critically for TAKS TEKS

1. **Evaluating** What are some ways in which Texans depend on, adapt to, and modify the physical environment? What are some results of the modifications?
2. **Analyzing Information** What political, economic, social, and cultural features are most important to defining the character of Texas? Write a paragraph to describe the state's character.
3. **Drawing Inferences and Conclusions** How do you think the growing population of Texas might affect the use of natural resources?

Using the Geographer's Tools TEKS

1. **Analyzing Maps** Review the map of the natural regions of Texas. In which natural region is your community located? In which subregion is it located?
2. **Creating Maps** Use this textbook and other classroom and library resources to create a map of Texas. Locate and label the state's largest cities and capital.
3. **Creating Maps** Select a city in Texas. Use library resources to create a map of the city. Be sure to include the city's borders and voting precincts. How many precincts does the city have?

Writing for TAKS TEKS

Imagine that you have been asked to write an article about Texas for a tourism magazine. Choose one of the state's natural regions as your topic. Use the Internet and other resources to identify features you think visitors would find interesting. Then write your article. When you are finished with your article, proofread it to make sure you have used standard grammar, spelling, sentence structure, and punctuation.

Alternative Assessment

PORTFOLIO ACTIVITY

Learning about Your Local Geography TEKS

Group Project: Research

Plan, organize, and complete a research project about an issue being debated by your local city or county government. Assign members to report on differing points of view on the issue. Do cultural beliefs influence these views? What other influences are important? Track and describe the decision-making process used by your local government in developing policies related to the issue. Finally, create a panel discussion on the issue for your class. The panelists should include a moderator, members who have researched competing points of view, and members who have tracked the decision-making process of your local government.

internet connect

Internet Activity: go.hrw.com
KEYWORD: SW3 GTTX TEKS

Draw a map of Texas to show when and where different culture groups settled in Texas. Add pictures or phrases that describe each group's distinctive ways of life. Then write a report describing how these cultural patterns and landscapes seem to have influenced how new ideas, or innovations, evolved and spread, or diffused, throughout the state.

UNIT 1

The Geographer's World

Wulingyuan Scenic and Historic Interest Area, Hunan Province, China

CHAPTER 1 Studying Geography

Geography covers more topics than you might think. In this chapter you will learn about the wide scope of geography. You will also become familiar with some of the geographer's basic tools.

World map made by combining satellite images

Map of Africa from about 1650

Babylonian plan of the world, 600s B.C.

Themes and Essential Elements

READ TO DISCOVER

1. What are the two main branches of geography?
2. How do we use geography?
3. What are some ways we can organize our world and the study of geography?

DEFINE

geography
perspective
landscapes
cartography
meteorology
region
formal region
functional region
perception
perceptual regions

WHY IT MATTERS

Geography plays a major role in many international issues. Use CNNfyi.com or other **current events** sources to learn about a recent international news event that involves important geographical concepts.

What Is Geography?

Many people do not really know what **geography** is. Some people think geography is just memorizing lists of countries and state capitals. Other people think it is the study of rocks. Still others think geographers just look at maps and pictures of faraway places. However, there is much more to geography!

Geography is the study of everything on Earth, from rocks and rainfall to people and places. Geographers study how the natural environment influences people, how people's activities affect Earth, and how the world is changing. To do this, geographers look at many different things, including cities, cultures, plants, and resources. Geographers focus on where these things are or where related events happen.

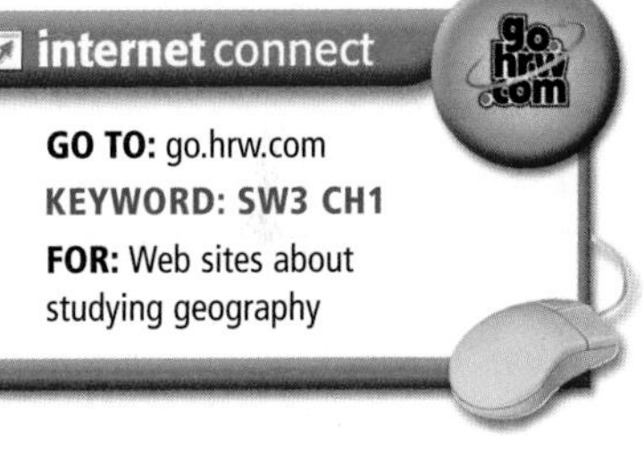

INTERPRETING THE VISUAL RECORD

Light shines through the shelter built by this Inuit man in Nunavut, Canada. ***How do you think the natural environment affects this person, and how do his activities affect Earth? How might his world be changing?*** TEKS

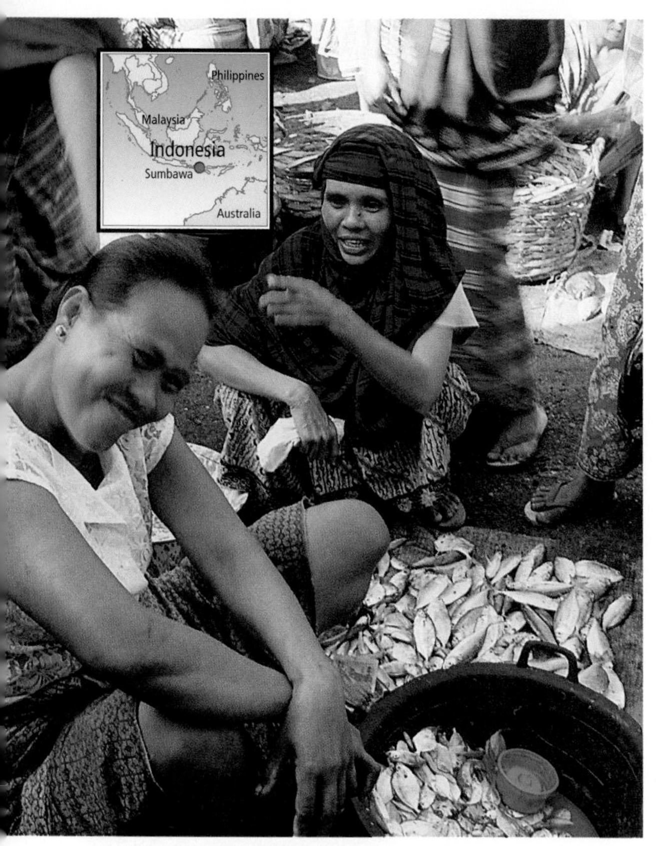

INTERPRETING THE VISUAL RECORD

Indonesian women sell fish at their local market. ***What might the photo indicate about the physical and human landscapes in which these women live?*** TEKS

Perspective—or the way a person looks at something—is an important part of learning about geography or any other subject. Geographers use a spatial perspective to study the world. That is, they look for patterns in where things are located on Earth and how they are arranged. Geographers then try to explain these patterns. They also look at a world that is shaped by **landscapes.** A landscape is the scenery of a place, including its physical, human, and cultural features. Geographers look at landscapes and try to explain what they see. For geographers the word *landscape* is almost magical. Each of us lives in a landscape. When studying world geography, we discover the amazing diversity of our world's landscapes.

Geography has two main branches—human geography and physical geography. Those who study human geography look at the distribution and characteristics of the world's people. They study where people live and work as well as their ways of life. They also look at how people make and trade things that they need to survive. The study of physical geography focuses on Earth's natural environments. These include Earth's landforms, water features, atmosphere, animals, plants, soils, and the processes that affect them. The interaction of people with their environment links human and physical geography together.

In this textbook you will use both human and physical geography to study the world. First, you will study an area's natural environment. Then you will learn about the area's human aspect and how it relates to the physical setting.

✓ **READING CHECK:** ***The Uses of Geography*** What are the two main branches of geography?

INTERPRETING THE VISUAL RECORD

Mount McKinley, the highest peak in North America, rises over Denali National Park and Preserve. Fireweed blooms in the foreground. ***What questions might physical and human geographers ask about this place?***

Who Uses Geography?

Now that you know what geography is, you might be wondering, who uses geography? You do. In fact, people all over the world use geography every day. We use it when we find our way to school or work and go on trips. We also use it when we watch the news on television and read about other countries. We think geographically every time we decide where to go and how to get there.

Most jobs require an understanding of geography. For example, a restaurant owner must find a good location for his or her business. Politicians need to know the geography of their districts. They must also understand the issues that are important to people there. In addition, a number of professions rely heavily on specially trained geographers.

Subfields of Geography Geography has many different subfields. One of the most well known is **cartography**—the study of maps and mapmaking. Maps are important because they help geographers study locations. Although some maps are still drawn by hand, computers have completely changed mapmaking. Computers store information from satellite images, photographs, and other sources. A cartographer then creates a map on a computer. Cartographers work for companies that publish maps, atlases, newspapers, magazines, and books. They also work for city planning agencies and other areas of government.

Another subfield of geography is **meteorology**—the study of weather. Meteorologists forecast how the weather will develop so that people know what to expect. You have probably watched these meteorologists, or weather forecasters, on local television.

Geographers at Work Many geographers work for governmental agencies. In fact, one of the largest employers of cartographers in the United States has been the United States Geological Survey (USGS). The USGS produces detailed maps of the whole country. Other agencies that hire geographers include the offices of most city, county, and state governments.

Many businesses hire geographers. Those geographers decide where to place new stores and plan shipping and trucking routes. They also help identify new markets. Geographers work in many different areas of business, such as tourism and travel and international sales.

Schools also hire geography teachers, who help people learn about the world. This knowledge is becoming more important as the different areas of the world become more closely linked. Geographic knowledge is also needed for good citizenship. Citizens who feel strongly about important geographic issues can try to influence public policies and decisions. Should we allow suburbs to be built over good farmland? Where should we put our garbage and dangerous materials? Helping citizens and governments find answers to these questions is the job of geographers.

A technician scans a map to create a digital version.

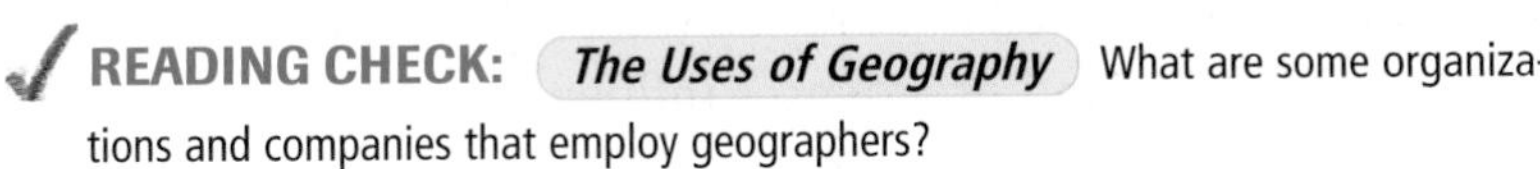

READING CHECK: ***The Uses of Geography*** What are some organizations and companies that employ geographers?

The word *geography* was first used by the ancient Greek geographer Eratosthenes (er-uh-TAHS-thuh-neez). It comes from two Greek words—*geo*, meaning "Earth," and *graphia*, which means "to describe."

γεωγραφία

How Do We Study Geography?

An important concept in geography is the idea of a **region**. A region is an area with one or more common features that make it different from surrounding areas. Cities, states, countries, and continents are examples. Organizing Earth's surface into smaller regions makes it easier to study our complex world.

Regions are defined by their physical and human features. Physical features include the kinds of climate, river systems, soils, and vegetation you find there. Human features include the languages, religions, and trade networks of an area. Sometimes the boundaries of a region are clear. For example, the United States is a political region with clear boundaries. In other places, the boundaries are harder to set. For example, the Corn Belt is a farming region in the midwestern United States. However, the Corn Belt does not have clearly set boundaries. It stretches across a number of states. Exactly where the Corn Belt begins and ends is not clear.

Regions can be any size. Countries, deserts, and mountain ranges are examples of large regions. Smaller regions include suburbs and neighborhoods. Regions can also be divided into smaller areas called subregions. For example, the Great Plains is a subregion within North America.

READING CHECK: ***Places and Regions*** What are some physical and human features that can define a region?

Types of Regions Geographers define regions in three basic ways. The first is a **formal region**. A formal region has one or more common features that make it different from surrounding areas. An example is the Sahel in Africa. This dry region lies between the Sahara, a vast desert to the north, and wetter forested areas to the south.

Formal regions can be based on almost any feature or combination of features. Those features might include population, income levels, crops, temperature, or rainfall. Physical features might define a formal region, such as the Rocky Mountains in the western United States. Economic features also might define such a region. For example, an industrial area in the northeastern and midwestern United States is also a formal region. This region was once called the Rust Belt because so many old factories there had shut down. Today new industries have revived the region's economy.

The second type of region is a **functional region**. These are made up of different places that are linked together and function as a unit. For example, a city transit system is a functional region. It includes many different places. However, the flow of people, trains, and buses link those places together.

Many functional regions are organized around a central point. Surrounding areas are linked to this point. For example, shopping malls are centers of functional regions linked to surrounding neighborhoods. Cities are also examples of these centers. They connect to suburbs, areas in the country, and industry, which all function together.

The third type of region involves human **perception**—our awareness and understanding of the environment around us. People view regions very differently. Our views are influenced not only by what is in a region but also by what is in us. Our ways of life and experiences influence how we perceive the world. Therefore, **perceptual regions** are regions that reflect human feelings

INTERPRETING THE VISUAL RECORD

The Piney Woods of East Texas qualifies as a formal region, mainly because of the pine trees that grow throughout the area. ***What other features might help define the Piney Woods?*** TEKS

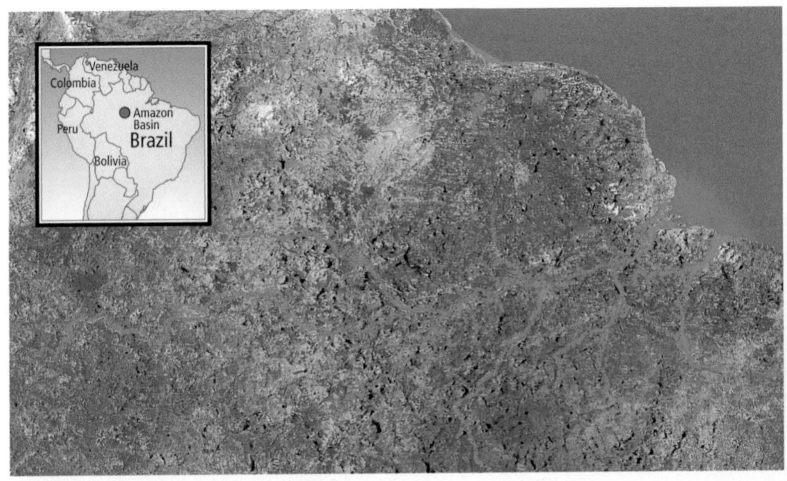

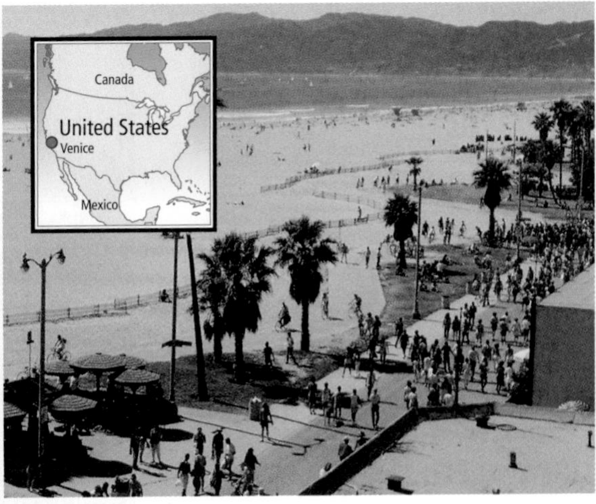

INTERPRETING THE VISUAL RECORD *The Amazon River system in South America, at the left, is a functional region. Places within it are linked by the river's flow. For many Americans, southern California, at the right, is a perceptual region.* ***What elements of southern California as a perceptual region are illustrated in this photo of the beach at Venice?*** TEKS

and attitudes. For example, "back home" is a perceptual region for most people. However, it may be hard to define exactly. The U.S. Midwest may be easier to define. The South—a region sometimes called Dixie—is another example. Many people perceive these areas to be distinct regions. These areas have their own special features that make them different from anywhere else. Yet people may view—or perceive—those features in differing ways.

READING CHECK: ***Places and Regions*** What are the three types of regions?

The Six Essential Elements The study of geography has long been organized according to five important themes. One theme, *location,* deals with the exact or relative spot of something on Earth. *Place* includes the physical and human features of a location. *Human-environment interaction* covers the ways people and environments interrelate with and affect each other. *Movement* involves how people and things change locations and the effects of these changes. *Region* organizes Earth into geographic areas with one or more shared characteristics. Another way to look at geography is to identify essential elements in its study. These six essential elements will be used throughout this textbook. They share many properties with the five themes of geography.

1. ***The World in Spatial Terms*** This element focuses on geography's spatial perspective. Crucial to this perspective is the use of maps to study people, places, and environments.

2. ***Places and Regions*** Our world has a vast number of unique places and regions. This element deals with the physical and human features of those places and how we define and perceive various regions.

3. ***Physical Systems*** Physical systems shape Earth's features. Geographers study earthquakes, mountains, rivers, volcanoes, weather patterns, and similar topics. They also study how plants and animals relate to these nonliving physical systems.

INTERPRETING THE VISUAL RECORD

Workers from an environmental organization called the Green Belt Movement teach schoolchildren in Kenya, East Africa, how to care for tree seedlings. ***If you were to label this photo with one of the six essential elements, which would you use? How does the essential element you chose relate to the photo?***

4. *Human Systems* People are central to geography. Our activities, movements, and settlements shape Earth's surface. The ways of life we follow and the things we produce and trade are part of the study of human systems. Geographers look at the causes and results of conflicts between peoples. The study of governments we set up and the features of cities and other settlements we live in are also part of this study.

5. *Environment and Society* Human actions, such as using oil or water, affect the environment. At the same time, Earth's physical systems—from weather to volcanoes—affect human activities. We depend on what the Earth provides to survive. The relationship between people and the environment is an important part of geography.

6. *The Uses of Geography* Geography helps us understand the relationships among people, places, and environments over time. Geography can help us to interpret the past and the present or to plan for the future.

✓ **READING CHECK:** *The Uses of Geography* What are six essential elements in the study of geography?

TEKS Question 1

Section 1 Review

Define geography, perspective, landscapes, cartography, meteorology, region, formal region, functional region, perception, perceptual regions

Reading for the Main Idea

1. *Places and Regions* What are examples of functional, formal, and perceptual regions?
2. *The Uses of Geography* What six essential elements are used to organize the study of geography?

Critical Thinking

3. **Contrasting** How is the study of human geography different from the study of physical geography?
4. **Making Generalizations and Predictions** What do you think a "geographical approach" to studying an issue might be?

Organizing What You Know

5. Create a graphic organizer like the one shown below. Use it to identify some of the jobs that geographers have.

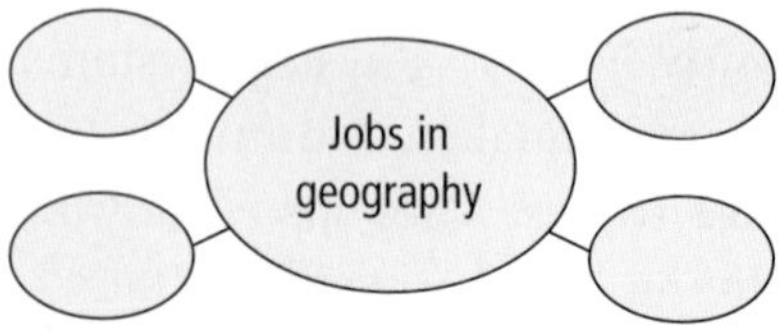

Skill Building: Using the Geographer's Tools

READ TO DISCOVER

1. How do geographers and mapmakers organize our world?
2. What kinds of special maps do geographers use?
3. How do geographers use climate graphs and population pyramids?

DEFINE

grid
latitude
longitude
equator
parallels
meridians
prime meridian
degrees
hemispheres
continents
atlas
map projections
great-circle route
compass rose
legend
contiguous
precipitation
topography
climate graphs
population pyramids

WHY IT MATTERS

News reporters often use maps to identify the locations of important events. Use CNNfyi.com or other **current events** sources to find out how maps are used in specific news stories.

Organizing the Globe

We begin our study of geography by looking at a globe. A globe is a scale model of Earth. It is useful for looking at the whole planet or at large areas of its land and water surface. One of the first things you will notice on the globe on this page is a pattern of lines. These lines circle the globe in east-west and north-south directions. This pattern is called a **grid**. The grid is made up of lines of **latitude** and **longitude**. Lines of latitude are drawn in an east-west direction. Lines of longitude are drawn in a north-south direction. The intersection of these imaginary lines helps us find the location of places.

Latitude and Longitude

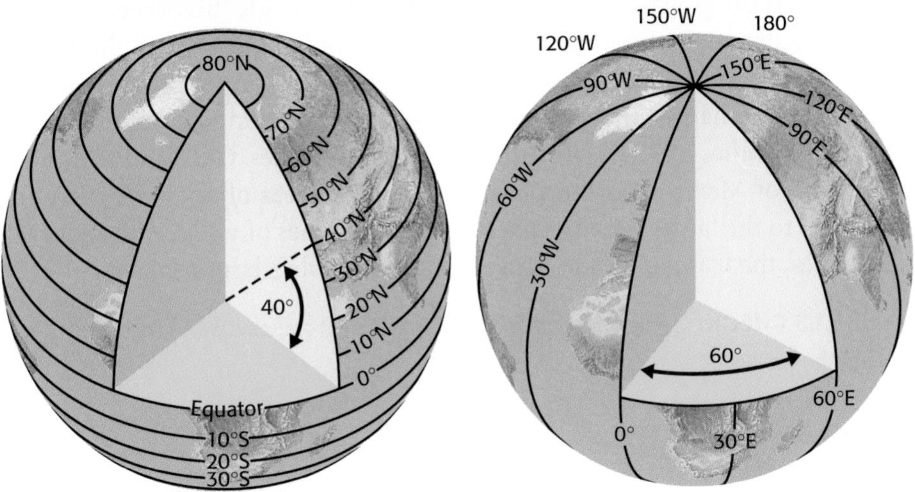

The illustration on the left shows lines of latitude. The north-south lines shown on the right are lines of longitude. Notice that lines of latitude are always the same distance apart.

NORTHERN HEMISPHERE

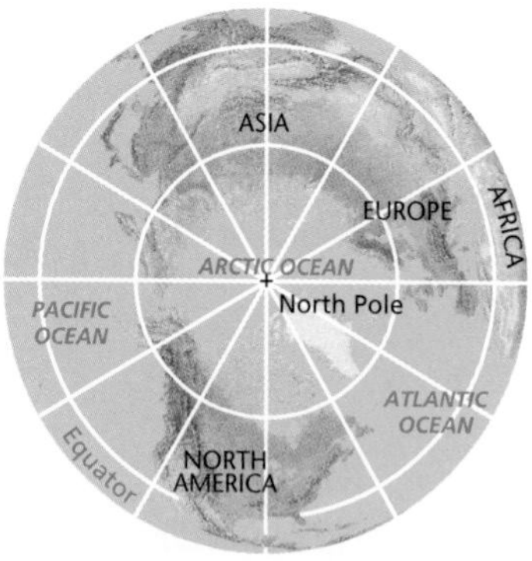

SOUTHERN HEMISPHERE

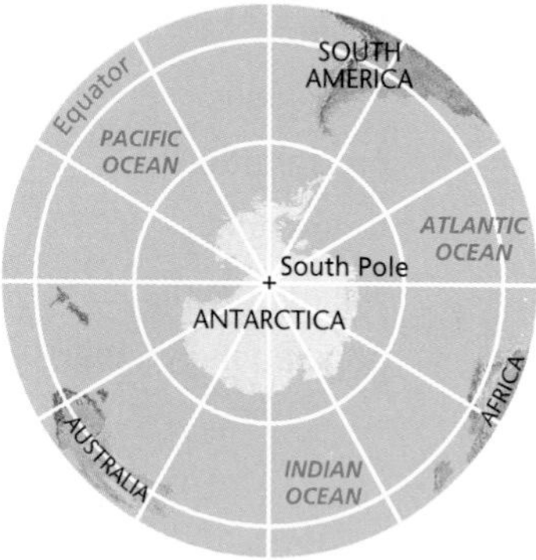

EASTERN HEMISPHERE

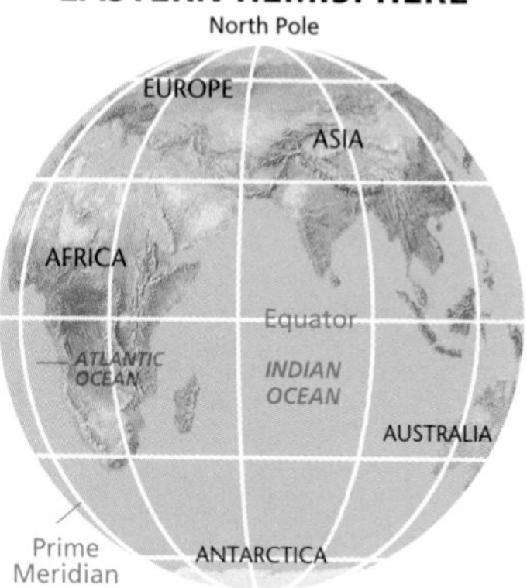

WESTERN HEMISPHERE

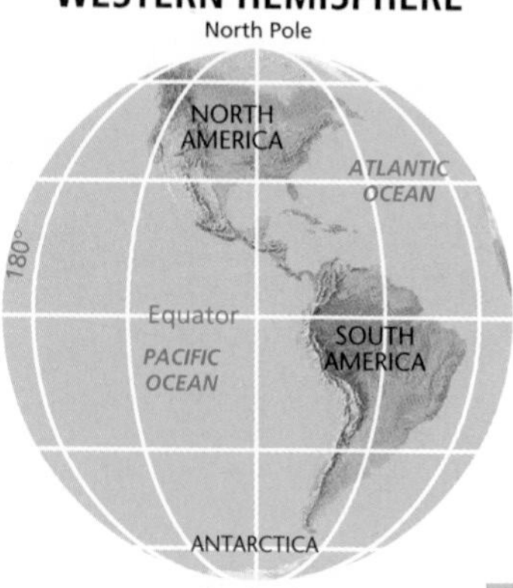

Lines of latitude measure distance north and south of the **equator**. The equator is an imaginary line that circles the globe halfway between Earth's North Pole and South Pole. Lines of latitude are also called **parallels**. This is because they are always parallel to the equator and each other. Lines of longitude are called **meridians**. They measure distance east and west of the **prime meridian**. This is an imaginary line drawn from the North Pole through Greenwich, England, to the South Pole. Parallels and meridians measure distances in **degrees**. The symbol for degrees is °. Degrees are further divided into minutes, for which the symbol is ′. There are 60 minutes in a degree.

As you can see on the globe on the previous page, parallels north of the equator are marked with an *N*. Those south of the equator are marked with an *S*. Lines of latitude range from 0°, for locations on the equator, to 90°N and 90°S, for locations at the North Pole and South Pole. Lines of longitude range from 0° on the prime meridian to 180° on a meridian in the mid–Pacific Ocean. Meridians west of the prime meridian to 180° are labeled with a *W*. Those east of the prime meridian are labeled with an *E*.

Hemispheres, Continents, and Oceans The globe's grid does more than help us locate places. Geographers also use those grid lines to organize the way we look at our world. For example, the equator divides the globe into two halves, or **hemispheres**. The half lying north of the equator is the Northern Hemisphere. The southern half is the Southern Hemisphere. The prime meridian and the 180° meridian divide the world into the Eastern Hemisphere and the Western Hemisphere. The prime meridian separates parts of Europe and Africa into two different hemispheres. To avoid this separation, some geographers divide the Eastern and Western Hemispheres in the Atlantic Ocean at 20°W. Doing so places all of Europe and Africa in the Eastern Hemisphere.

We can also organize our planet's land surface into seven large landmasses, called **continents**. There are seven continents: Africa, Antarctica, Asia, Australia, Europe, North America, and South America. Asia, the largest, is more than five times the size of Australia, the smallest. Landmasses smaller than continents and completely surrounded by water are called islands. Greenland is the world's largest island.

Geographers also organize Earth's water surface into separate areas. The largest area is the global ocean. Geographers further divide this ocean into four areas: the Atlantic Ocean, the Arctic Ocean, the Indian Ocean, and the Pacific Ocean. The Pacific is the largest ocean and the world's largest geographic feature. It is more than 12 times the size of the smallest ocean, the Arctic.

Smaller bodies of waters include seas, gulfs, and lakes. Gulfs and seas, such as the Gulf of Mexico and the Caribbean Sea, are areas of salt water that are connected to the larger oceans. Lakes are inland bodies of water. Although it is called a sea, the Caspian Sea in Asia is really the world's largest lake.

TEKS **SKILLS CHECK:** ***The World in Spatial Terms*** What are some ways geographers organize our world?

INTERPRETING THE MAPS ***In which hemispheres is the United States located? Which continents are located entirely within the Southern Hemisphere?***

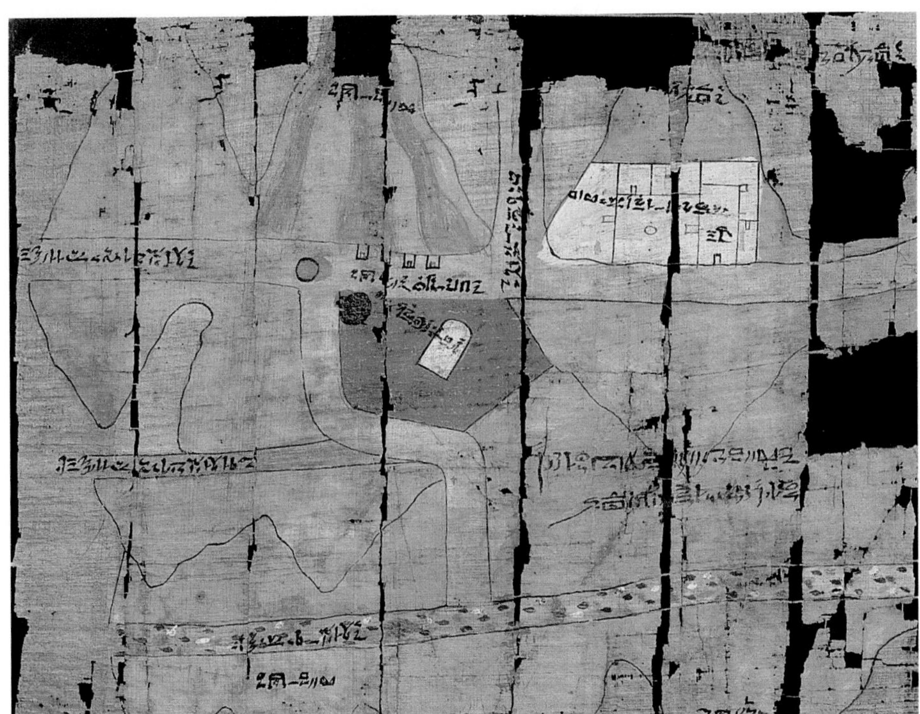

This map, which is more than 3,000 years old, shows an ancient mining and quarrying area in southeastern Egypt.

Making Maps

Globes are not the only useful visual tools for studying Earth. In organizing and identifying places in our world, geographers also use maps. Maps are flat representations of all or part of Earth's surface. A collection of maps in one book is called an **atlas**. You will find an atlas of world and regional maps at the front of this textbook.

Mapmakers have different ways of presenting our round Earth on flat maps. These different ways are called **map projections**. Because our planet is round, all flat maps—no matter their projection—have some distortion. For example, some maps do not show the true sizes of landmasses. This is particularly true at higher latitudes. For example, on some maps Greenland—which lies mostly within the Arctic Circle—might appear larger than Australia, a continent. However, those maps might be useful because they show true direction and true shapes. Other maps show size in true proportions but distort shapes. Mapmakers must choose the type of projection that is best for their purposes. The most common projections are cylindrical, conic, and flat-plane.

Map Projections Maps with cylindrical projections are designed as if a cylinder has been wrapped around the globe. The cylinder touches the globe only at the equator. The meridians are pulled apart and are parallel to each other instead of meeting at the poles. This causes landmasses near the poles to appear larger than they really are.

A Mercator map is a cylindrical projection. The Mercator map is useful for navigators because it shows true direction and shape. However, landmasses at high latitudes—such as Europe and North America—are exaggerated in size. They appear larger than they really are. Landmasses in lower latitudes may appear relatively smaller than they really are. (See the diagram.)

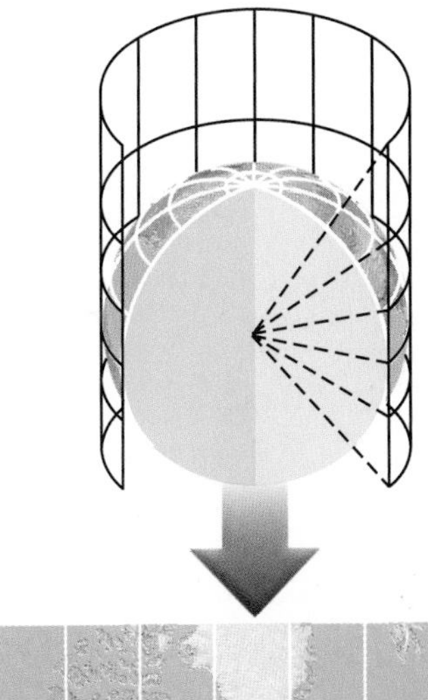

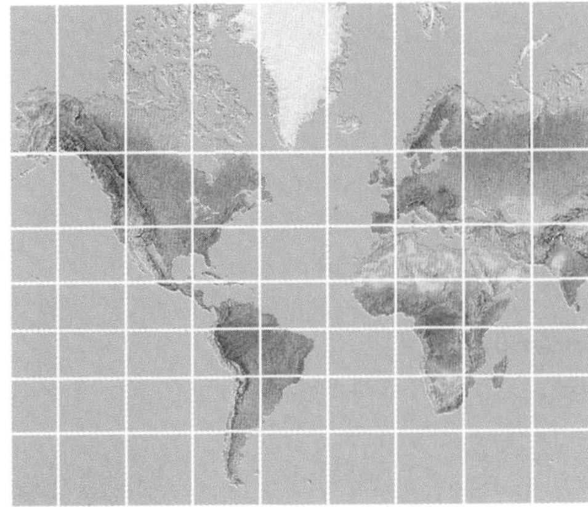

Cylindrical projection

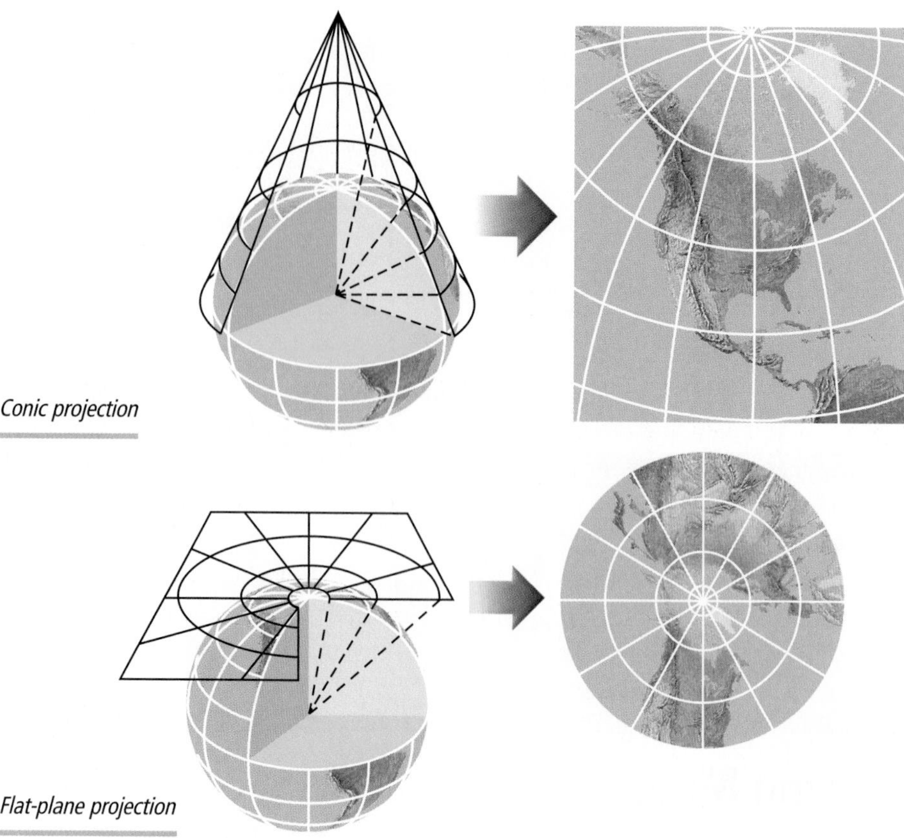
Conic projection

Flat-plane projection

Conic projections are designed as if a cone has been placed over the globe. A conic projection is most accurate along the lines of latitude where it touches the globe. It retains almost true shapes and sizes of landmasses along those locations. Conic projections are most useful for areas that have long east-west dimensions, such as the United States and Russia.

Flat-plane maps are those that appear to touch the globe at one point, such as the North Pole or the South Pole. A flat-plane projection is useful for showing true direction for airplane pilots and ship navigators. It also shows true area sizes, but it distorts shapes.

Great-Circle Route Drawing a straight line on a flat map will not show the shortest route between two places. Remember that maps represent a round world on a flat plane. The shortest route between any two places on the planet is called a **great-circle route**. (See the illustrations on this page.) Airline pilots and ship captains use great-circle routes to help them navigate. You can see

Using great-circle routes saves time and fuel for travelers. Use your atlas to find more examples of how great circle routes shorten other trips, such as those between Canada and Japan.

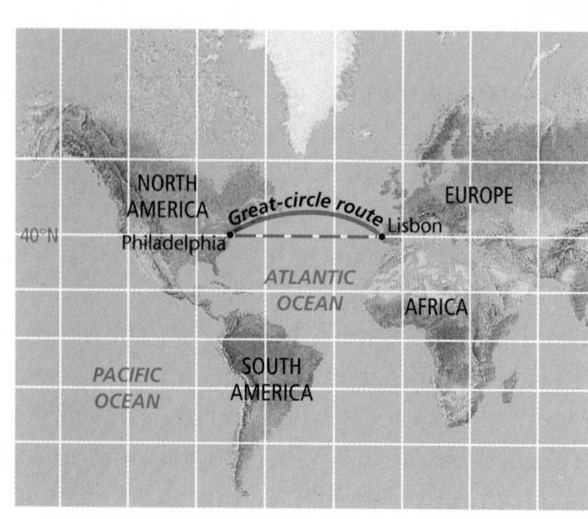

how a great-circle route shows the shortest distance between two points by using a round globe in your classroom or library.

SKILLS CHECK: ***The World in Spatial Terms*** What are three kinds of map projections that mapmakers use?

Understanding Map Elements

The study of geography involves more than simply looking at big places you see on a globe or map. It also involves looking at places at different scales—the size of an area and the level of detail that is shown. In fact, a map can show small areas, such as the floor plan of a building. It might show a neighborhood or a voting precinct. Maps can also show larger areas. For example, they might show whole cities, states, countries, continents, and oceans. You can see examples of maps at different scales throughout this textbook. For example, on this page are maps of Washington, D.C., and surrounding areas. Notice how details on each map change as the scale changes.

Washington, D.C.

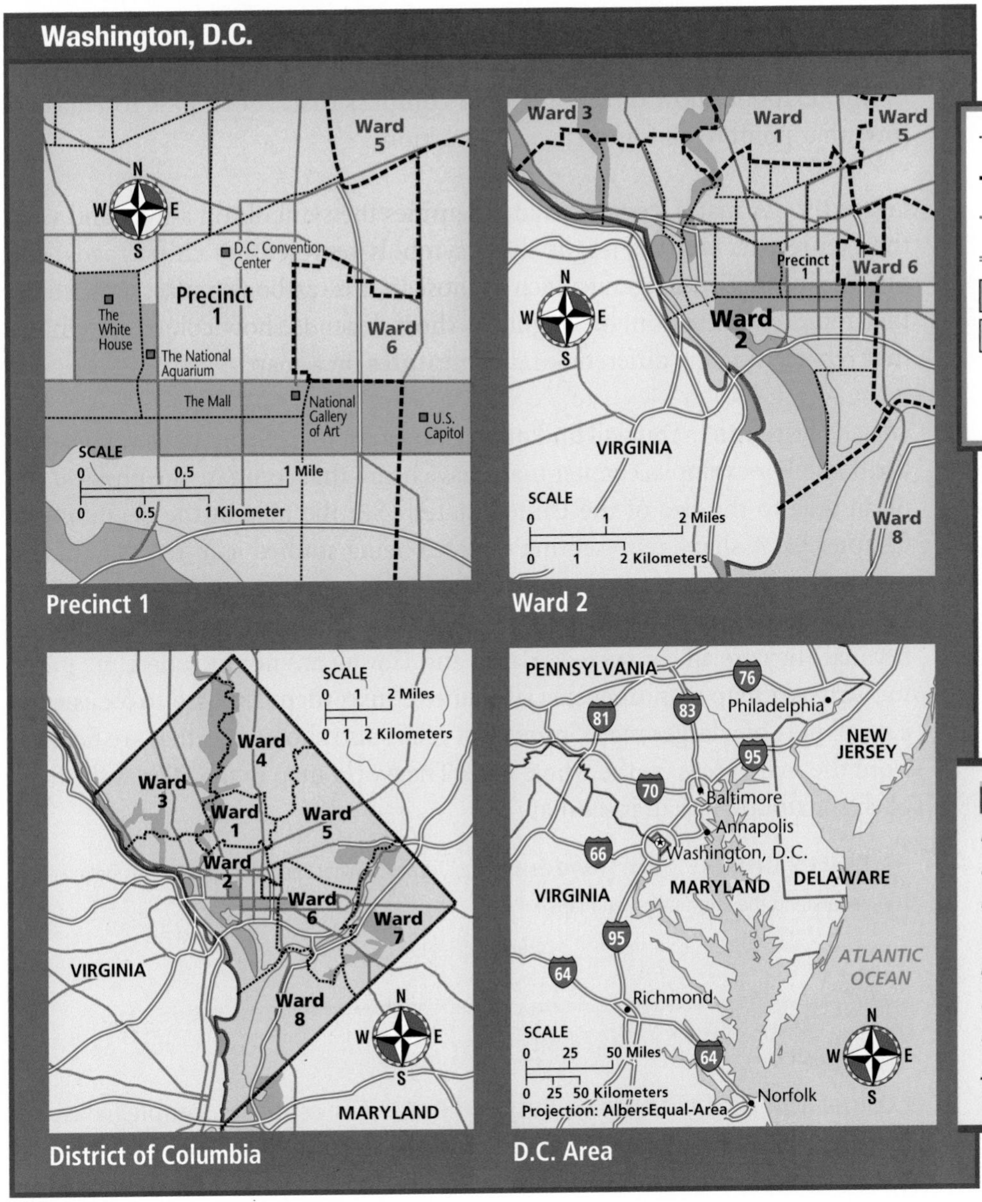

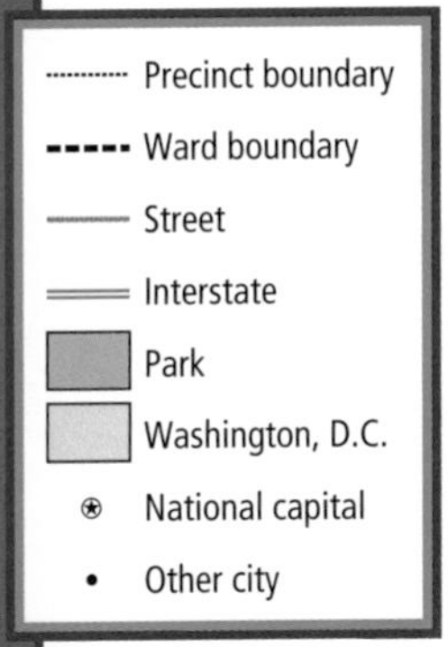

Precinct 1

Ward 2

District of Columbia

D.C. Area

- Precinct boundary
- Ward boundary
- Street
- Interstate
- Park
- Washington, D.C.
- National capital
- Other city

INTERPRETING THE MAPS *These maps show parts of the Washington, D.C., area at different scales. The Precinct 1 map has the largest scale.* ***What details might you find on a map with a scale larger than the Precinct 1 map? What might you learn from the D.C. Area map that you cannot learn from the other three maps?*** TEKS

People probably made maps even before written languages existed. However, the earliest known map is one discovered on a clay tablet at a place called Nuzi in what is now Iraq. The map dates to about 2500 B.C.

Maps may show different details depending on scale. However, they usually have the same basic elements. This is because maps, in some ways, are like messages sent out in code. Cartographers provide basic map elements to help us translate these codes. Thus, we can understand the information, or message, in a map. Almost all maps have several common elements. They include a distance scale, a directional indicator, and a key. The key is a guide that identifies symbols. You can see these elements on the Washington, D.C., maps.

Distance Scales A map's distance scale helps us determine real distances between points on a map. Remember that maps of small areas can show more detail than maps of large areas. Because of this, scales on those maps can indicate short distances. Some might show just one or two miles or kilometers. Others might show smaller distances—even just hundreds of feet or meters. Maps showing large areas, such as state and country maps, must have scales that indicate longer distances. Those scales might show distances in tens or even hundreds of miles or kilometers.

Directional Indicators A directional indicator shows which directions on a map are north, south, east, and west. Some mapmakers use a "north arrow," which points toward the North Pole. Most maps have north at the top. Maps in this textbook show direction with a **compass rose**. A compass rose has arrows that point to all four principal directions.

Legends A map's **legend**, or key, identifies the symbols on a map and what they represent. Legends might show symbols representing cities, roads, and other features. Some legends, such as those in this textbook's atlas, show colors that represent elevation, or height. In short, legends show colors or symbols that represent many different kinds of features on a map.

Other Elements You will find other important map elements as you use this textbook. For example, chapter maps have boxes that compare the physical size of an area to the size of the United States. (See the map at the beginning of Chapter 9.) A shape representing the area being studied is in red. It is placed over an outline of the **contiguous** United States. *Contiguous* means "connecting" or "bordering." The 48 states between Canada and Mexico are contiguous because they are all connected. Alaska and Hawaii are not included.

An inset map is another special element. Inset maps are used to focus in on a small part of a larger map. Some inset maps also show areas that are far away from the main areas on the whole map. The world map in the atlas at the front of this textbook has an inset map.

SKILLS CHECK: ***The World in Spatial Terms*** What special map elements help us understand the information on maps?

FOCUS ON HISTORY

Exploration and Changing Perceptions For centuries, people have used maps to reflect their perceptions of the world around them. For example, early Greeks created maps of the Mediterranean world. They had limited knowledge

INTERPRETING THE VISUAL RECORD

This map of India and the Pacific Ocean dates from 1570. Note the enlarged detail near the bottom of the page. ***How does this map show that sailors were uncertain about what dangers lay ahead for them? How do the distances and shapes shown on the map compare to those on today's maps?***

of what lay beyond that world. Some early European mapmakers even placed sea monsters at the edges of their maps. This reflected their uncertainty about what lay beyond the seas that surrounded the land areas. Early Chinese mapmakers placed China at the center of the world. They believed Chinese culture to be superior to all other cultures.

All of these limited perceptions began to change as sailors explored the outside world. The Chinese explorer Cheng Ho sailed to many places in Southeast Asia and eastern Africa in the 1400s. He took back to China his impressions of the people and places he encountered. About the same time, European explorers began sailing far from home. Soon European explorers sailed around Africa to reach Asia. However, some Europeans believed that a shorter route to Asia lay to the west—across the Atlantic Ocean.

In 1492 Christopher Columbus sailed westward from Spain. He and later explorers found two huge continents—the Americas—that lay between Europe and Asia. Knowledge of the Americas changed European perceptions of the size and features of their world. These changing perceptions drew more Europeans to the Americas and elsewhere. They also sparked a race for European colonies around the world. European countries became more powerful. However, local cultures in the Americas, Africa, and Asia were weakened or destroyed. Even parts of China—once seen by Chinese as the center of the world—fell under the influence of Europe.

READING CHECK: ***The Uses of Geography*** How did European perceptions of the world change after the voyages of Christopher Columbus? What changes did this bring to societies around the world?

Using Special-Purpose Maps

Geographers use many different kinds of maps. Many maps, for example, focus on certain kinds of information about a place or region. You will find

Climate Map, United States and Canada

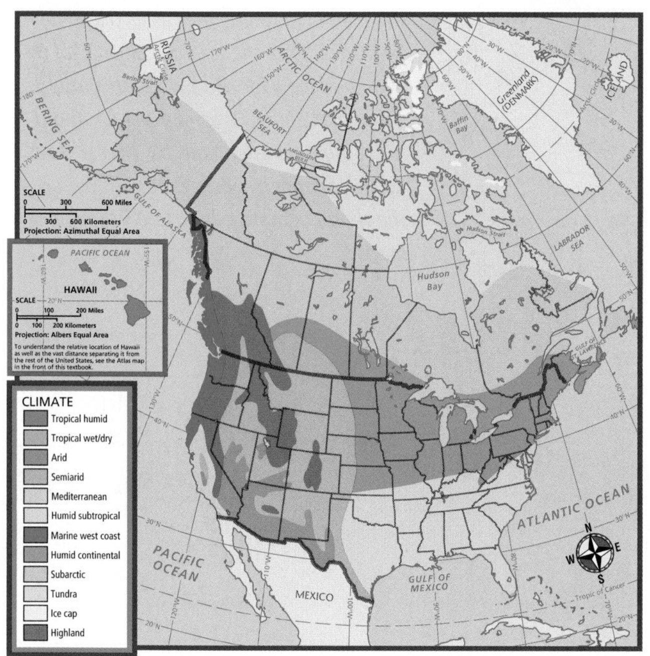

Precipitation Map, United States and Canada

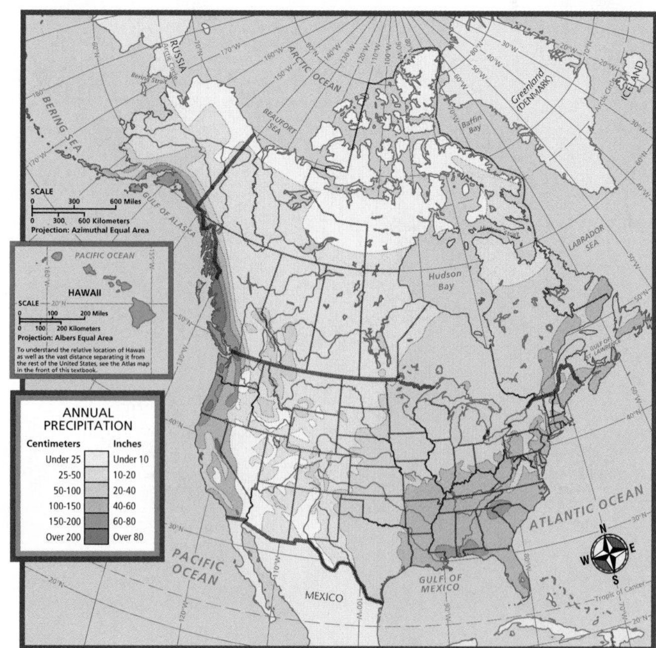

these special-purpose maps throughout this textbook. Some are political and physical atlas maps. You will find such maps at the front of this textbook and at the beginning of units and chapters. Political maps show the world's borders, cities, countries, states, and other political features. Physical maps show natural features like mountains, rivers, and other bodies of water. You will read about these kinds of features in Chapter 4. You will find symbols that identify the features on physical maps in legends. Of course, studying our world means more than looking at country borders, cities, and physical features. Each unit in this textbook opens with other special-purpose maps. These maps show climate, precipitation, population, and economic features.

Climate and Precipitation Maps Mapmakers use some maps to show weather patterns and atmospheric conditions. Climate maps use color to show the various climate regions of the world. You will read about climates in Chapter 3. Colors that identify climate types are found in a legend. However, boundaries between climate regions do not indicate a sudden change in average weather conditions. Instead, those boundaries mark areas of gradual change between climates. **Precipitation** maps are paired with climate maps at the beginning of units. The word *precipitation* refers to condensed droplets of water that fall as rain, snow, sleet, or hail. These kinds of maps show the average amount of precipitation that a region gets each year. Each map's legend uses colors to identify those amounts. By using the map's legend, you can see what areas get the most or least precipitation.

Population and Economic Maps Population maps give you a snapshot of the distribution of people in a region. You will read about population features in Chapter 5. Each color on a population map represents an average number of people living within a square mile or square kilometer. Sometimes symbols identify cities with populations of a certain size. The map's legend identifies these colors and symbols.

Population Map, United States and Canada

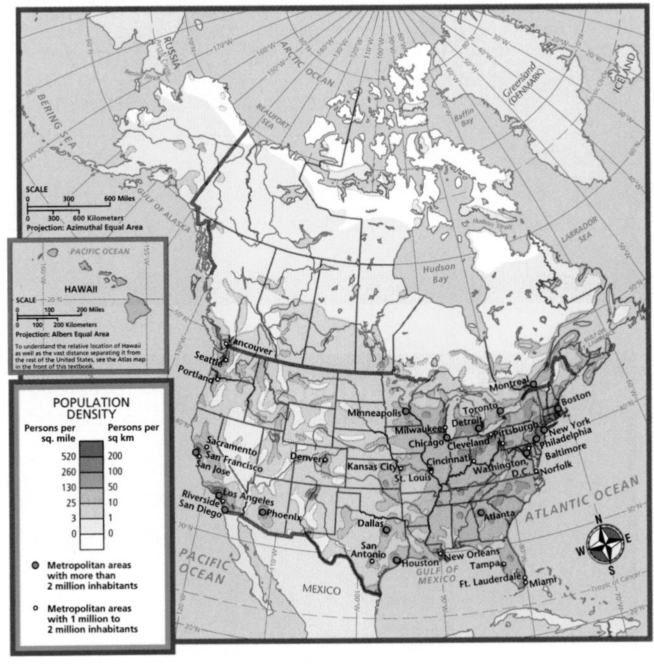

Land Use and Resources Map, United States and Canada

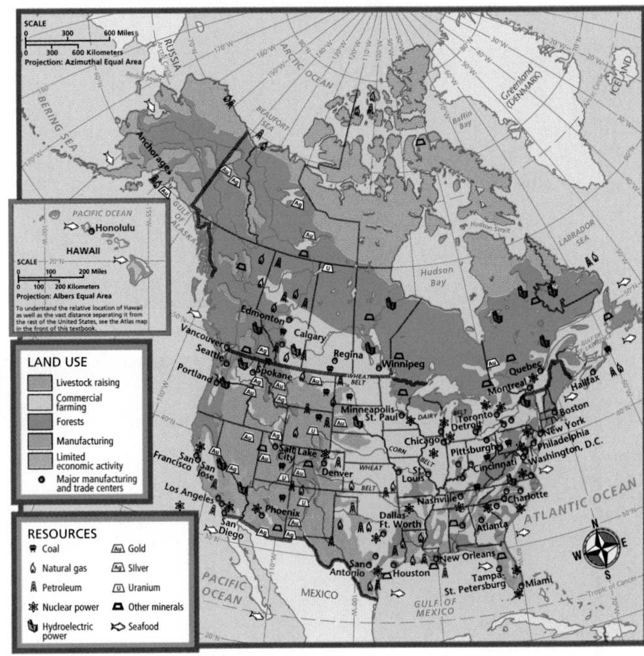

By looking at the population map of the United States in Unit 2 you can identify the most populated areas of the country. Those areas are mainly in the eastern half of the country, in the Midwest, and along the West Coast. In general, fewer people live in the interior and western states.

Economic maps show a region's important natural resources and the ways in which land is used. You will read about economic features in Chapter 6. Symbols show the location of resources, such as oil and gold. Colors show where land is used for farming or other economic activities.

SKILLS CHECK: ***The World in Spatial Terms*** How are colors used in climate, precipitation, population, and economic maps?

Elevation Profiles and Topographic Maps Some maps focus on an area's land features. You can see that each physical map in this textbook uses color to show land elevations. Elevation is the height of the land above sea level. Each color represents a different elevation.

In each unit atlas, you will also find an elevation profile like the one shown on the next page. An elevation profile shows a side view of a place or area. The profile shows the physical features of Guadalcanal, an island in the South Pacific. These features lie along a line from Point A to Point B in the elevation map below it.

Vertical (bottom to top) and horizontal (left to right) distances are calculated differently on elevation profiles. The vertical distance (such as the height of Mount Makarakomburu) is exaggerated when compared to the horizontal distance between Point A and Point B. This technique is called vertical exaggeration. If the vertical scale were not exaggerated, even tall mountains would appear as small bumps on an elevation profile.

The purpose of some maps is to show just the **topography**—or elevation, layout, and shapes—of the land. A special kind of topographical map is called a contour map. Contour maps provide a way of looking at the shapes of land in an

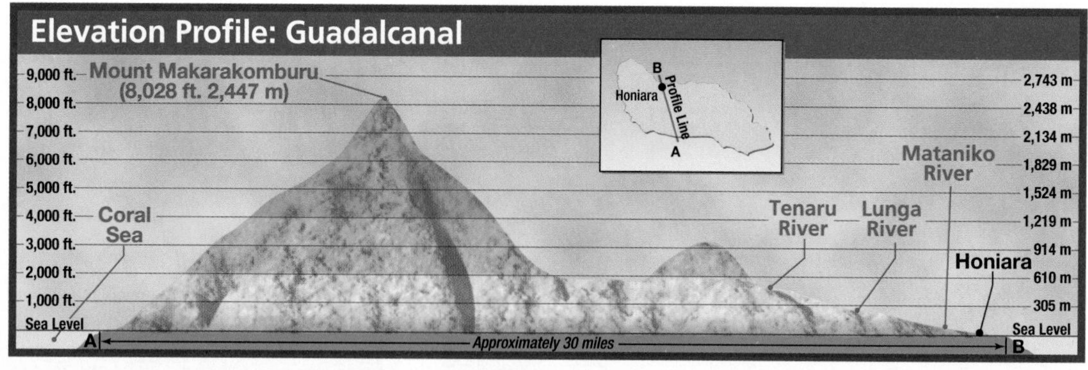

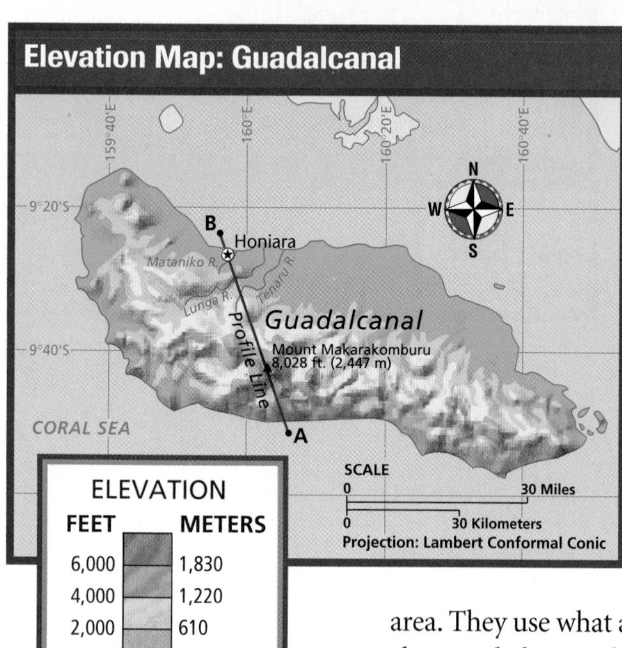

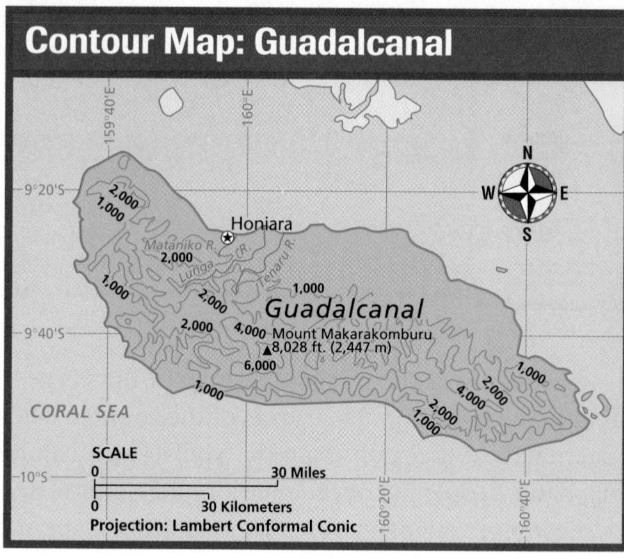

area. They use what are called contour lines to connect points of equal elevation above or below sea level. Elevation levels are written on the lines. The closer together the lines are, the steeper the land. For example, the land south of Mount Makarakomburu is steeper than the land on the northwestern end of the island.

SKILLS CHECK: ***The World in Spatial Terms*** Where is the largest, lowest, and flattest area of Guadalcanal?

Climate Graphs and Population Pyramids

Maps are not the only special tools geographers use to study the world around us. They also use many tools that you will find useful in economics, government, history, and other social studies. You can read more about bar graphs, charts, pie graphs, and tables in the Social Studies Skill-Building Handbook before Unit 1. In fact, you will find many charts, graphs, and tables in this textbook. You will also find two other common diagrams that show important geographic characteristics: **climate graphs** and **population pyramids**.

A climate graph shows the average temperatures and precipitation in a place. As you can see on the opposite page, along the left side of the climate graph is a range of average monthly temperatures. Along the right side is a range of average monthly precipitation amounts. The months of the year are labeled across the bottom. In this textbook's climate graphs, a red line shows the average monthly temperatures at the location. Green bars show average monthly precipitation amounts.

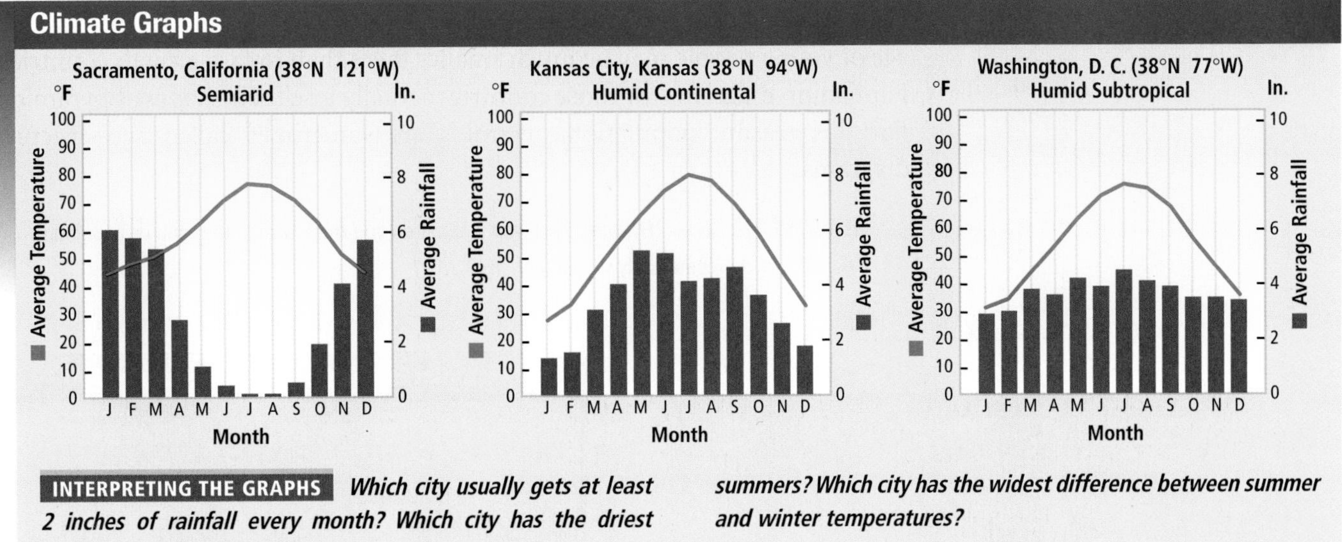

INTERPRETING THE GRAPHS *Which city usually gets at least 2 inches of rainfall every month? Which city has the driest summers? Which city has the widest difference between summer and winter temperatures?*

A population pyramid shows the percentages of males and females by age group in a country's population. As you can see in the pyramid on this page, these diagrams are split into two sides. Each bar on the left shows the percentage of a country's population that is male and of a certain age. The bars on the right show the same information for females. The percentages are labeled across the bottom of the diagram.

How do you think these population diagrams got their names? The base of a population pyramid shows the percentage of the youngest people in a country. The top shows the percentage of the oldest people. In many countries there are many more younger people than older people. Thus, the base of a country's diagram is often much wider than the top. The result is a diagram shaped like a pyramid.

Population pyramids help us understand population trends in countries. Countries that have large percentages of young people have populations that are growing rapidly. They have pyramids with very wide bases. Nigeria is an example of such a fast-growth country. On the other hand, in countries like

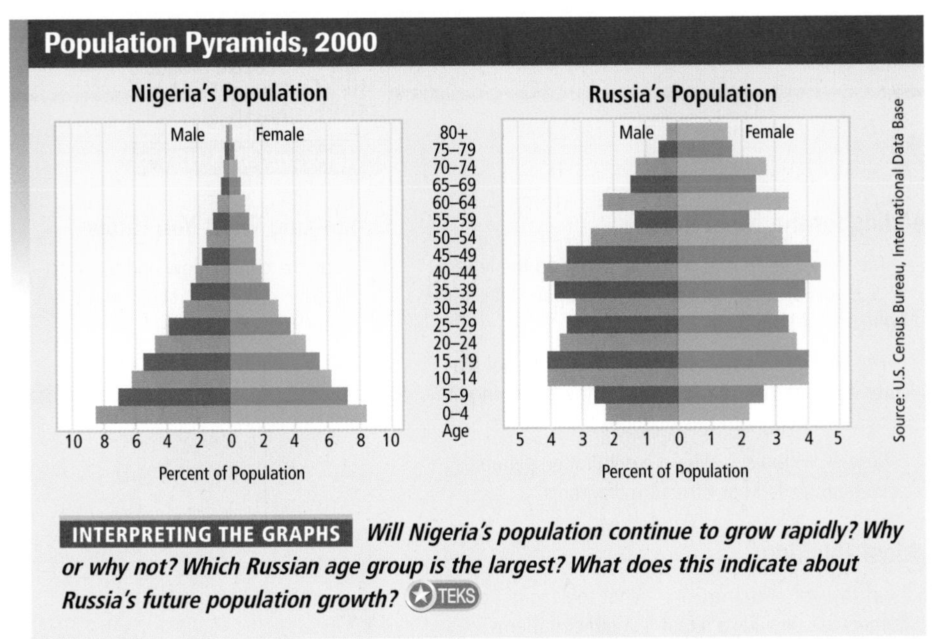

INTERPRETING THE GRAPHS *Will Nigeria's population continue to grow rapidly? Why or why not? Which Russian age group is the largest? What does this indicate about Russia's future population growth?* TEKS

Russia, populations are growing much more slowly or not at all. The percentage of young people may be much smaller there than in fast-growth countries. Population diagrams for those countries actually lose their shapes as pyramids. For this reason, population pyramids are sometimes called age-structure diagrams.

SKILLS CHECK: ***Human Systems*** What do you think a climate graph or population pyramid for your community would look like? Why?

Using Historical Maps

Studying geography often requires us to understand how human and physical processes have shaped a place or region over time. Geographers can use historical maps to study these processes and their effects. For example, the map here uses colors to show areas of North and South America that Europeans had settled and taken over before 1700. Students of geography can use this map to predict where various European languages are probably spoken in the Americas today.

Making Generalizations What else might a historical map like this one suggest about the human geography of North and South America today? TEKS

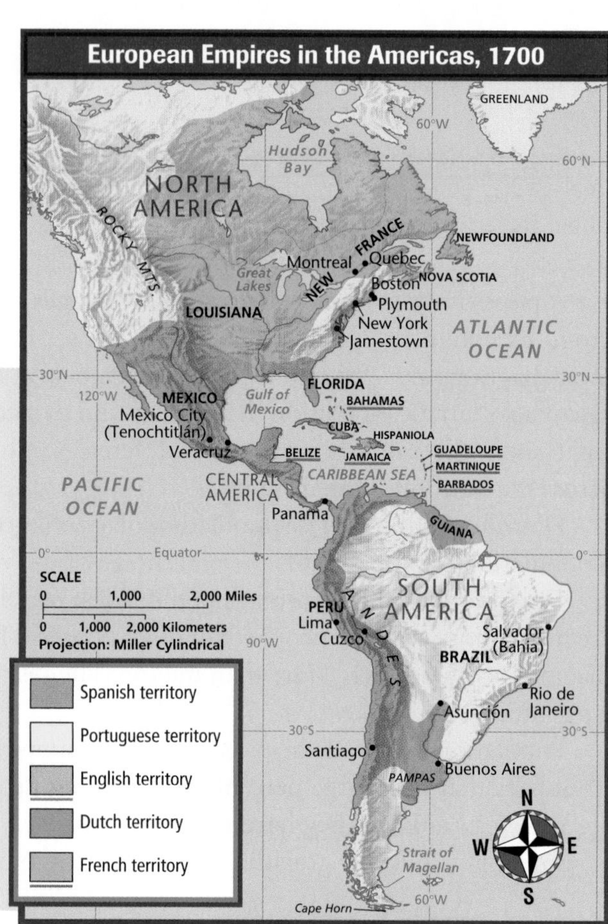

TEKS Questions 1, 2, 3, 4, 5,

Review

Define grid, latitude, longitude, equator, parallels, meridians, prime meridian, degrees, hemispheres, continents, atlas, map projections, great-circle route, compass rose, legend, contiguous, precipitation, topography, climate graphs, population pyramids

Working with Sketch Maps Sketch a map of the world. Label Earth's seven continents and the Arctic, Atlantic, Indian, and Pacific Oceans. In the margin, identify at least three bodies of water that are smaller than oceans.

Reading for the Main Idea

1. ***The World in Spatial Terms*** What do the letters *N, S, E,* and *W* mean when they accompany labels for latitude and longitude?
2. ***The World in Spatial Terms*** What regional features are found on special maps at the beginning of each unit throughout this textbook?
3. ***Human Systems*** Why are population pyramids sometimes called age-structure diagrams?

Critical Thinking

4. **Evaluating Information** What tools might geographers use to study a region's weather patterns?

Organizing What You Know

5. Copy the chart below and use it to describe cylindrical, conic, and flat-plane map projections.

Cylindrical	Conic	Flat-plane

Geography for Life

Geographic Information Systems

A geographic information system, or GIS, is a special kind of computer system. A GIS stores, displays, and maps locations and their features. Geographic information systems have become important tools in geography and many other fields. Among those fields are city planning, real estate, and environmental studies. Scientists also use them for a variety of tasks. For example, they can use a GIS to monitor natural hazards like volcanoes. They can also map crop growth and track the movements and locations of endangered species. Emergency workers can use a GIS to learn the shortest route to someone calling 911. Geographic information systems are also helpful in understanding the spread of disease and charting power outages. In short, they are valuable tools for many tasks. Today students can even earn a college degree in GIS technology.

How does a GIS work? The information collected for a map is called spatial data. With a GIS, different sets of this data are saved in a computer as "layers." (See the diagram.) These data layers can then be manipulated, compared, or combined. Cartographers can instantly display this data on a computer screen and get detailed information about each feature. They can also use a GIS to study features and their locations and to uncover relationships between them.

For example, imagine that you plan to open a restaurant. How would you find the best location for your new business? Using a GIS could help. With a GIS, you could find the locations of vacant buildings. You could then plot those located near major streets. You could also show the buildings' rental costs and information about the people who live nearby. That information might include how much money the people make. With this kind of information, a GIS could help you identify the best locations for your business.

Applying What You Know (TEKS)

1. **Summarizing** What is a GIS? What are some fields that use GIS technology?
2. **Problem Solving** What are some geographic problems and issues that could be studied using a geographic information system? What other kinds of spatial data might be useful?

Information Layering

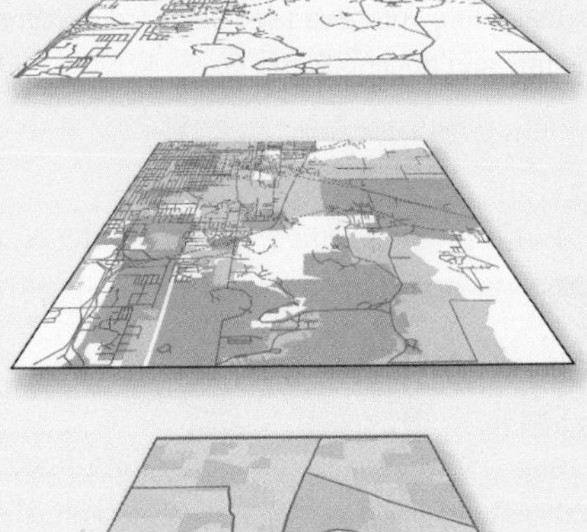

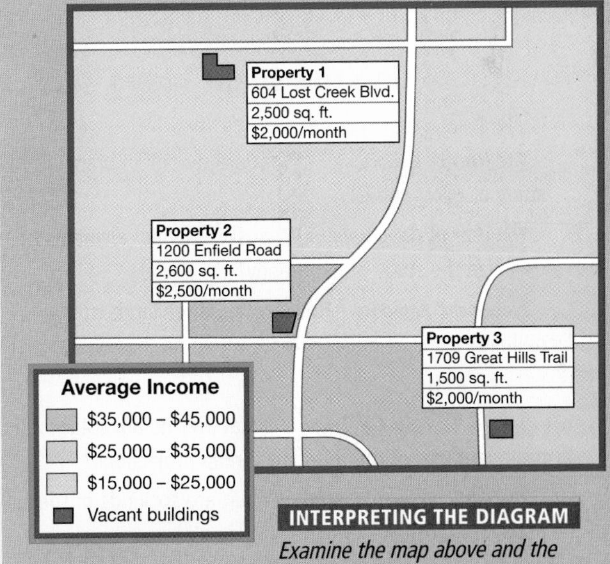

INTERPRETING THE DIAGRAM
Examine the map above and the layers at the left that would be combined in a GIS. ***Which of the three locations would you choose for a new restaurant? What influenced your choice?*** (TEKS)

Review

Building Vocabulary

On a separate sheet of paper, explain the following terms by using them correctly in sentences.

geography	perceptual regions	continents
cartography	equator	atlas
region	parallels	contiguous
formal region	meridians	topography
functional region	prime meridian	population pyramids

Locating Key Places

On a separate sheet of paper, match the letters on the map with their correct labels.

Africa	Europe	Atlantic Ocean
Antarctica	North America	Indian Ocean
Asia	South America	Pacific Ocean
Australia	Arctic Ocean	

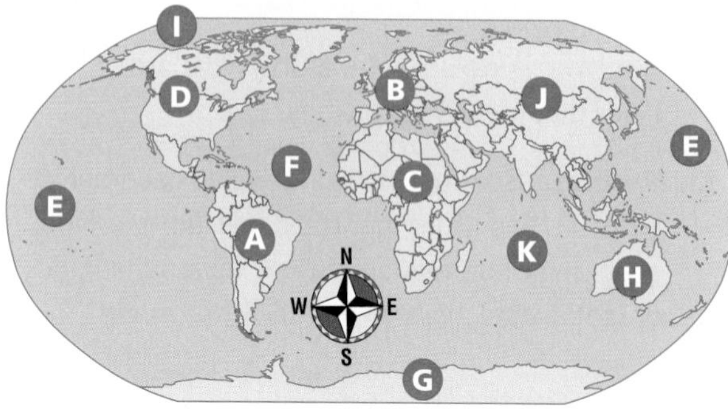

Understanding the Main Ideas

Section 1

1. *The Uses of Geography* What are two main branches in the study of geography?
2. *The Uses of Geography* What six essential elements help us organize the study of geography?
3. *Places and Regions* How are the three kinds of regions defined?

Section 2

4. *The World in Spatial Terms* What are the advantages and disadvantages of the three main map projections?
5. *The World in Spatial Terms* What are six kinds of special-purpose maps?

Thinking Critically for TAKS

1. **Analyzing Information** How are formal, functional, and perceptual regions different from each other?
2. **Evaluating** How would the six essential elements help you organize the study of your community's geography?
3. **Identifying Points of View** How might different perspectives affect the way different people perceive the region around your community?

Using the Geographer's Tools

Constructing a Population Pyramid Use the following percentages to construct a population pyramid for the United States. You can use the pyramids in Section 2 as a model. When you are finished, describe the population features that the diagram shows about the United States. Do you think the U.S. population is growing rapidly or slowly? Which age groups make up the largest parts of the population?

Age Groups	Male	Female
0–4	3.5%	3.3
5–9	3.7	3.5
10–14	3.7	3.5
15–19	3.7	3.5
20–24	3.4	3.3
25–29	3.2	3.3
30–34	3.5	3.6
35–39	4	4.1
40–44	4.1	4.1
45–49	3.5	3.7
50–54	3	3.2
55–59	2.3	2.5
60–64	1.8	2
65–69	1.6	1.9
70–74	1.4	1.8
75–79	1.1	1.6
80+	1.1	2.2

Writing for TAKS

Imagine that you are a geography teacher preparing for your first day of class. Prepare a short lecture describing geography and the kinds of things students will learn about during the course of their study. Include ways that students use geography in daily life and how they may use it in a future job. When you have finished your lecture, proofread it to make sure you have used standard grammar, spelling, sentence structure, and punctuation.

SKILL BUILDING

Geography for Life

Creating a Precinct Map

Environment and Society Use your library, the Internet, and other resources to locate information about voting precinct boundaries in your city or county. Then prepare a map showing those precincts and shade the precinct in which you live. Your map should also include notable human and physical features, such as major roads and rivers. In the margin of your map, note how those features may have helped shape precinct boundaries.

Building Social Studies Skills

Interpreting Maps

Study the map below. Then use the information from the map to help you answer the questions that follow.

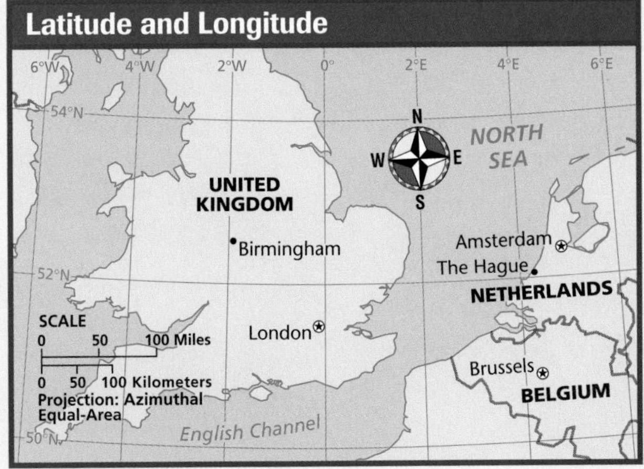

1. What is the approximate latitude and longitude of The Hague, capital of the Netherlands?
   a. 52°S, 8°E
   b. 52°N, 4°E
   c. 51°N, 1°W
   d. 54°N, 4°W
2. Imagine that your best friend went to Europe with her parents. Your friend decided to play a location game with you. She sent a postcard with the following message: "We flew in to a city near 52°N, 4°E. Two days later we went to a place close to 51°N, 4°E for a festival. Finally, we were near 52°N, 0° for three days to visit my mom's cousins." Write a few sentences in which you describe your best friend's trip, naming the cities she and her family visited.

Building Vocabulary

To build your vocabulary skills, answer the following questions.

3. *Perspective* is an important part of learning about geography.
   In which sentence does *perspective* have the same meaning as it does in the sentence above?
   a. This particular perspective is the east side of the building.
   b. The perspective from the lookout point was beautiful.
   c. From the child's perspective, the stranger looked 10 feet tall.
   d. Renaissance painters used new techniques to create perspective.
4. *Contiguous* means the same as
   a. bordering.
   b. distant.
   c. vertical.
   d. parallel.

Alternative Assessment

PORTFOLIO ACTIVITY

Learning about Your Local Geography TEKS

Individual Project: Researching Climate Statistics
Plan, organize, and complete a research project about climate conditions in your community or a nearby big city. Use the Internet, almanacs, and other resources to find information about average monthly temperatures and precipitation in the location you choose. Then use the data you collect to create a climate graph.

internet connect

Internet Activity: go.hrw.com
KEYWORD: SW3 GT1 TEKS

Choose a topic about studying geography:

Use the Internet to research geographic information systems (GIS) and how they can be applied to solve geographic and locational problems. Access the American Fact Finder mapmaker and create a map of your state that includes data on population and population density, education, economic and industrial development, and trade. When you create each map, locate your county and answer the questions posed about the data you have collected.

CHAPTER 2

Earth in Space

Earth is our home in the universe. In this chapter, you will learn about Earth's position in the solar system and relation to other objects in space. You will also learn about the Earth system—the interactions of Earth's land, water, air, and life.

Yerkes Observatory, Wisconsin

The Solar System

READ TO DISCOVER

1. What is Earth's position in the solar system?
2. How do rotation and revolution affect Earth?

DEFINE

solar system
planets
moons
satellite
solar energy
rotation
revolution

WHY IT MATTERS

Today scientists are making many exciting discoveries about the universe. Use CNNfyi.com or other **current events** sources to learn about how a recent scientific discovery has increased our knowledge of the universe.

Space and the Universe

If you look at the sky on a clear night, you can see thousands of stars. With a telescope you can see millions more. Beyond the telescope's view are trillions more. All these stars are part of the universe. The universe is made up of all existing things, including space and Earth. Most astronomers believe that the universe is 10 to 20 billion years old. Since its birth, the universe has been expanding continuously. It is unimaginable in size.

Space is filled with large objects called stars. Most stars are grouped together in huge clusters called galaxies. Many objects that look like individual stars to the naked eye are really billions of stars in a faraway galaxy. The Milky Way is the galaxy in which we live. The part of this galaxy that we can see from Earth looks like a bright milky streak across the night sky. The Sun is a medium-size star near the edge of the Milky Way.

When you look up at the night sky, do you ever wonder if there are places with life other than Earth? Scientists wonder about this and continue to search for clues. The vast universe holds unlimited possibilities for future discovery.

✓ **READING CHECK:** *Physical Systems* What are large groups of stars such as the Milky Way called?

This image of the Eagle Nebula shows an area where new stars are formed. The pillars of gas and dust contain material that condenses and ignites, forming stars.

The Planets The Sun and the group of bodies that revolve around it are called the **solar system**. The Sun's great size attracts the other objects through gravity. Besides the Sun, the solar system includes **planets**, asteroids, and comets. Planets are major bodies that orbit a star.

There are four inner and five outer planets in the solar system. The inner planets are closest to the Sun. They are Mercury, Venus, Earth, and Mars. The inner planets are terrestrial, meaning they

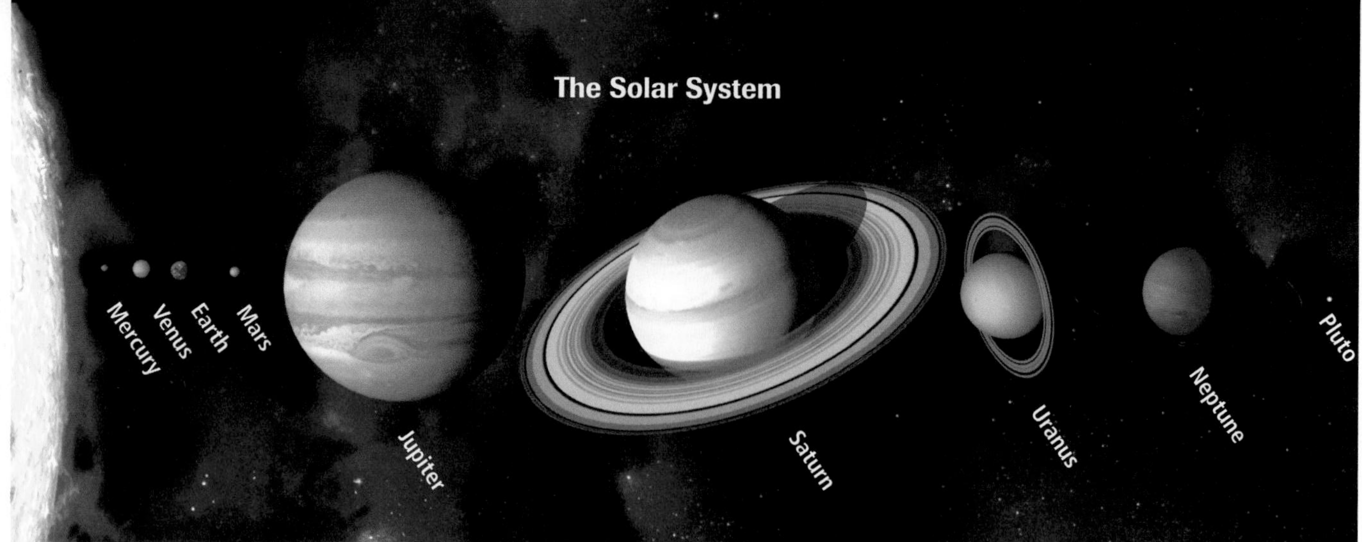

This diagram shows the nine planets and their relative sizes. However, the relative distance of each planet from the Sun is not accurate.

have a solid rocky surface. The outer planets are generally much larger and farther from the Sun. They are Jupiter, Saturn, Uranus, Neptune, and Pluto. The outer planets are mainly gaseous. An exception is Pluto, a small planet with a rocky core. Unlike stars, planets do not generate their own light. They are visible to us because they reflect sunlight.

Earth and the other planets look like they are perfectly round. However, this is not exactly correct. Earth's polar areas are slightly flattened. Around the equator, Earth is slightly bulged. This slight variation of a perfect sphere is called an oblate spheroid. *Oblate* means "flattened."

Moons are smaller objects that orbit a planet. A body that orbits a larger body can also be called a **satellite**. Moons are natural satellites. Earth has one moon. However, hundreds of human-made satellites also orbit Earth. Mercury and Venus are the only planets in the solar system with no moons. Saturn has at least 18.

GO TO: go.hrw.com
KEYWORD: SW3 CH2
FOR: Web sites about Earth in space

READING CHECK: ***Physical Systems*** What are the main bodies that make up the solar system?

Effects of the Moon and Sun on Tides

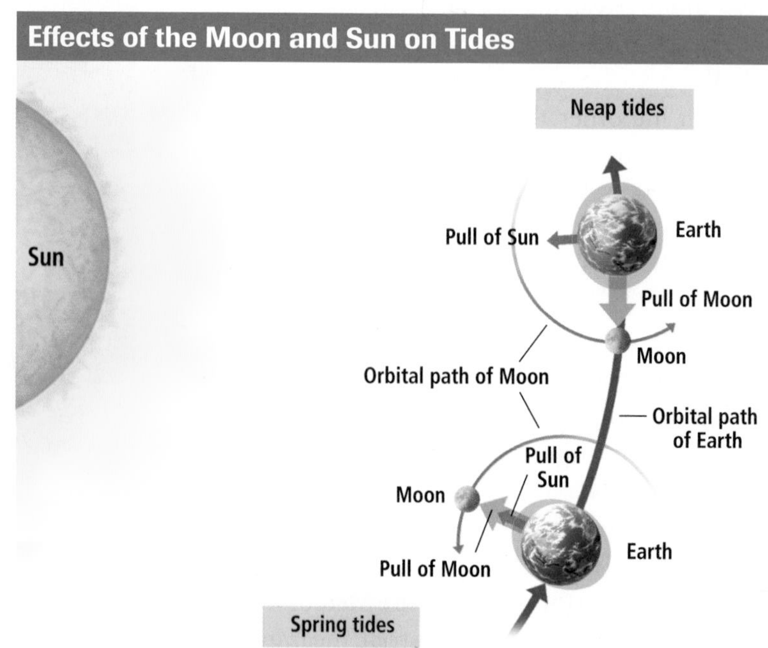

INTERPRETING THE DIAGRAM *Tides are higher than normal when the gravitational pull of the Moon and Sun combine. These tides, called spring tides, occur twice a month during the full moon and new moon. High tides are lower during neap tides, when the pull of the Sun is at a right angle to the pull of the Moon. Neap tides occur during quarter moons.* ***Why might knowing the schedule of high and low tides be important to people living in coastal areas?***

The Sun, Earth, and Moon The Sun is an average star in age, brightness, and size. It is small compared to huge stars called supergiants. However, compared to Earth, the Sun is very large. The diameter of Earth is about 8,000 miles (12,900 km). The diameter of the Sun is about 865,000 miles (1,390,000 km)—more than 100 times greater. The Sun operates like a giant thermonuclear reactor, releasing enormous amounts of energy.

Earth is the third planet from the Sun and the fifth-largest of the nine. Earth's orbit around the Sun is not a perfect circle. It is elliptical, or oval-shaped. Earth's orbit averages about 93 million miles (150 million km) from the Sun.

The Moon is about 240,000 miles (385,000 km) from Earth and is about one fourth the size of Earth. The Moon orbits Earth every 29.5 days, or about once every month. On clear nights you can see its surface. The Moon has a simple geography, with a barren volcanic surface and many craters. These craters were made by the impact of meteors and comets. Unlike Earth, the Moon has no air, water, or life.

The Sun, Earth, and Moon exert gravitational forces on one another that influence physical processes on Earth. The most obvious of these are the rise and fall of ocean tides. Tides are the response of the fluid ocean surface to the gravitational pull of the Moon and Sun. (See the diagram of tides.)

Between 1968 and 1972 NASA's Apollo missions landed astronauts on the Moon six times. In this photo, astronaut Harrison Schmitt studies a boulder in the Taurus-Littrow Valley. Schmitt made detailed descriptions of craters, boulders, and debris and helped bring back some 249 pounds (113 kg) of lunar material.

TEKS **READING CHECK:** ***Physical Systems*** What causes ocean tides?

Earth's Rotation, Revolution, and Tilt

Most of Earth's energy comes from the Sun. This type of energy is called **solar energy** and reaches Earth as light and heat. All life on Earth depends on solar energy. Solar energy affects weather, plants, animals, and human activities. It influences the clothes we wear, the homes we live in, the foods we grow and eat, and even which sports we play. Three different relationships between Earth and the Sun control how much solar energy is received at different locations. Do you know what these are?

Rotation Imagine that Earth has a rod running through it from the North Pole to the South Pole. This rod represents Earth's axis, and the planet spins around on it. One complete spin of Earth on its axis is one **rotation**, which takes 24 hours. Earth rotates in a west-to-east direction. We see the effects of Earth's rotation as the Sun "rising" in the east and "setting" in the west. To us, it appears that the Sun is moving across the sky. Actually, it is only Earth rotating on its axis.

Solar energy strikes only the half of Earth facing the Sun. If Earth did not rotate on its axis—creating day and night—only the half facing the Sun would receive solar energy. That side of Earth would be very hot. The half of the planet facing away from the Sun would always be dark and cold. Earth's rotation allows the entire planet's surface to receive the warming effects of daylight and the cooling effects of darkness.

Unique in the solar system, Earth is the only planet with liquid water at the surface, active mountain-building processes, and life.

Revolution In addition to rotating on its axis, Earth revolves around the Sun. It makes one elliptical orbit, or **revolution**, every 365 1/4 days—one Earth

Connecting to TECHNOLOGY

Astronomy

Astronomers study objects and matter beyond Earth. For most of history, people could see space only with the naked eye. However, with the help of modern technology, astronomers can now see far into space.

In the early 1600s the famous scientist Galileo used telescopes to study space. Telescopes magnify objects in space, making faint stars and galaxies visible. Telescopes have become central to astronomy. With the development of space-age technology, astronomy made even greater advances. For example, scientists sent out probes to collect new information about planets, moons, and other objects. The Hubble Space Telescope (shown below), built between 1978 and 1990, led to many exciting discoveries. The Hubble is the first powerful telescope ever placed in orbit around Earth, which gives it a much clearer view of space. Astronomers have used the Hubble to study black holes, galaxies, stars, and other distant objects.

Summarizing How has the development of modern technology advanced astronomy?

Hubble Space Telescope image of interacting spiral galaxies NGC 2207 and IC 2163

year. Each time you celebrate your birthday we have just completed another orbit around the Sun. For convenience, our calendars have 365 days in a year. To account for the one-fourth day gained each year, an extra day—February 29—is added to the calendar every four years. This year, one day longer than the previous three, is called a leap year.

Tilt If Earth's axis always pointed straight up and down in relation to the Sun, daylight hours would be the same at every location on Earth. Each day would consist of 12 hours of daylight and 12 hours of darkness. This would be true throughout the year. However, this is not the case because Earth's axis is tilted in relation to the Sun.

As Earth revolves around the Sun, its axis points toward the same spot in the sky. The North Pole points to a star known as the North Star. The position of the axis is fixed in respect to the North Star. Yet it is not fixed in relation to our Sun. Earth's orbit lies on a plane that runs from the center of the Sun to the center of the planet. Earth's axis is tilted 23 1/2 degrees from the perpendicular, or 90 degrees, to the plane of its orbit. Thus, as Earth revolves around the Sun, the North Pole points at times toward the Sun and at times away from the Sun. (See the diagram in Section 2.) The tilt of Earth on its axis affects the amount of solar energy that different places receive during the year.

TEKS **READING CHECK:** *Physical Systems* How do rotation, revolution, and tilt affect the amount of solar energy received at different locations on Earth?

Section 1 Review

Define
solar system
planets
moons
satellite
solar energy
rotation
revolution

Reading for the Main Idea

1. *Physical Systems* How are inner planets and outer planets different?
2. *Physical Systems* What is the geography of the Moon like?
3. *Physical Systems* What is the difference between rotation and revolution?

Critical Thinking

4. **Drawing Inferences and Conclusions** How do you think technologies developed to study space can affect people's lives on Earth?

Organizing What You Know

5. Create a chart like the one shown below. Use it to describe the Sun, Earth, and Moon.

Sun	Earth	Moon

Earth-Sun Relationships

READ TO DISCOVER

1. How does the angle of the Sun's rays affect the amount of solar energy received at different locations on Earth?
2. What are solstices and equinoxes?

WHY IT MATTERS

Many festivals and celebrations around the world take place during a solstice or equinox. Use CNNfyi.com or other **current events** sources to learn about a celebration at one of these times.

IDENTIFY

Tropic of Capricorn
Antarctic Circle
Arctic Circle
Tropic of Cancer

DEFINE

tropics
polar regions
solstice
equinox

Solar Energy and Latitude

As you know, different places on Earth receive different amounts of solar energy. Areas near the equator receive a lot of solar energy all year. These places are generally warm. We call these warm low-latitude areas near the equator the **tropics**. Other places get very little solar energy. These areas are at high latitudes and are cold most of the time. Because these areas surround the North and South Poles, we call them the **polar regions**. The areas between the tropics and the polar regions are called the middle latitudes. The amount of solar energy reaching these areas changes greatly during the year. They may be warm or cool, depending on the time of year.

The amount of solar energy that a place receives relates to the angle at which the Sun's rays strike Earth. Direct vertical solar rays heat Earth's surface more than angled rays. This is because the amount of solar energy in a direct ray is concentrated on a smaller area. The same amount of energy in an angled ray is spread over a larger area.

When the North Pole points toward the Sun, direct rays strike the Northern Hemisphere. (See the diagram.) Thus, the Northern Hemisphere receives more concentrated solar energy, making temperatures warmer. The length of time between sunrise and sunset also grows longer. At this time, the Southern Hemisphere receives more angled rays and is cooler.

When the North Pole tilts away from the Sun, the most direct rays strike the Southern Hemisphere. Now, the Southern Hemisphere receives more solar energy, experiences longer days, and thus has warmer temperatures. At this time, the Northern Hemisphere receives less solar energy and has cooler temperatures.

READING CHECK: *Physical Systems* Which latitudes receive the most solar energy throughout the year?

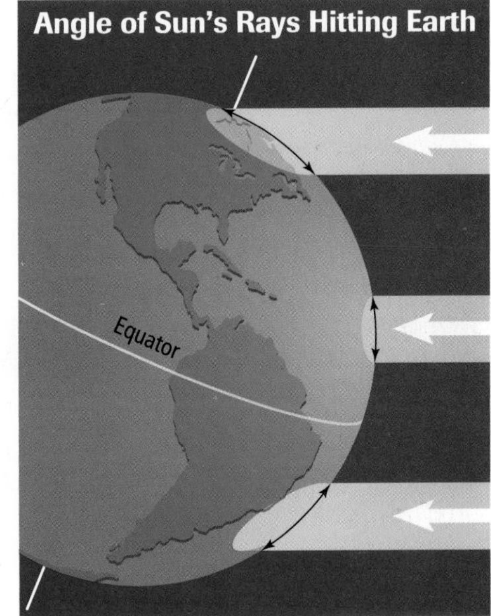

This diagram shows summer in the Northern Hemisphere. At this time, Earth tilts toward the Sun, and direct rays are more concentrated in the Northern Hemisphere. Meanwhile, in the polar regions and Southern Hemisphere, solar rays are spread over a larger area.

Our Amazing Planet

In the 200s B.C. the Greek astronomer Aristarchus proposed that Earth rotated on its axis—creating day and night—and revolved around the Sun each year.

The Seasons

We refer to the times of greater and lesser heat as the seasons. There are four general seasons: winter, spring, summer, and fall. Some regions, particularly the tropics, are warm year-round but have alternating wet and then dry seasons.

In each hemisphere, the Sun's energy is stronger during the summer. Daytime lasts longer. In the winter, daytime is shorter, and the Sun's energy is weaker. During spring and fall, the Sun's energy is more evenly distributed. At these times, daylight and darkness are closer to equal length. The tilt of Earth's axis causes the Northern and Southern Hemispheres to have opposite seasons at the same time of the year.

Solstices Twice during the year, Earth's poles tilt toward or away from the Sun more than at any other time. The time that Earth's poles point at their greatest angle toward or away from the Sun is called a **solstice**. Solstices occur each year about December 21 and June 21.

In the Northern Hemisphere, the December solstice has the fewest daylight hours of the year and is the first day of winter. The Southern Hemisphere on the same day has its greatest number of daylight hours, and it is the first day of summer. During the December solstice, the Sun's most direct rays strike Earth in the Southern Hemisphere along a parallel 23 1/2 degrees south of the equator. This parallel is called the **Tropic of Capricorn**. The South Pole is tilted toward the Sun and receives constant sunlight. All areas located south of the **Antarctic Circle** have 24 hours of daylight. The Antarctic Circle is the parallel 66 1/2 degrees south of the equator. Meanwhile, the area around the North Pole experiences constant darkness and is very cold. The parallel beyond which no

INTERPRETING THE DIAGRAM

As Earth revolves around the Sun, the tilt of the poles toward and away from the Sun causes the seasons to change. ***On which day is the North Pole pointed directly away from the Sun?***

The Seasons

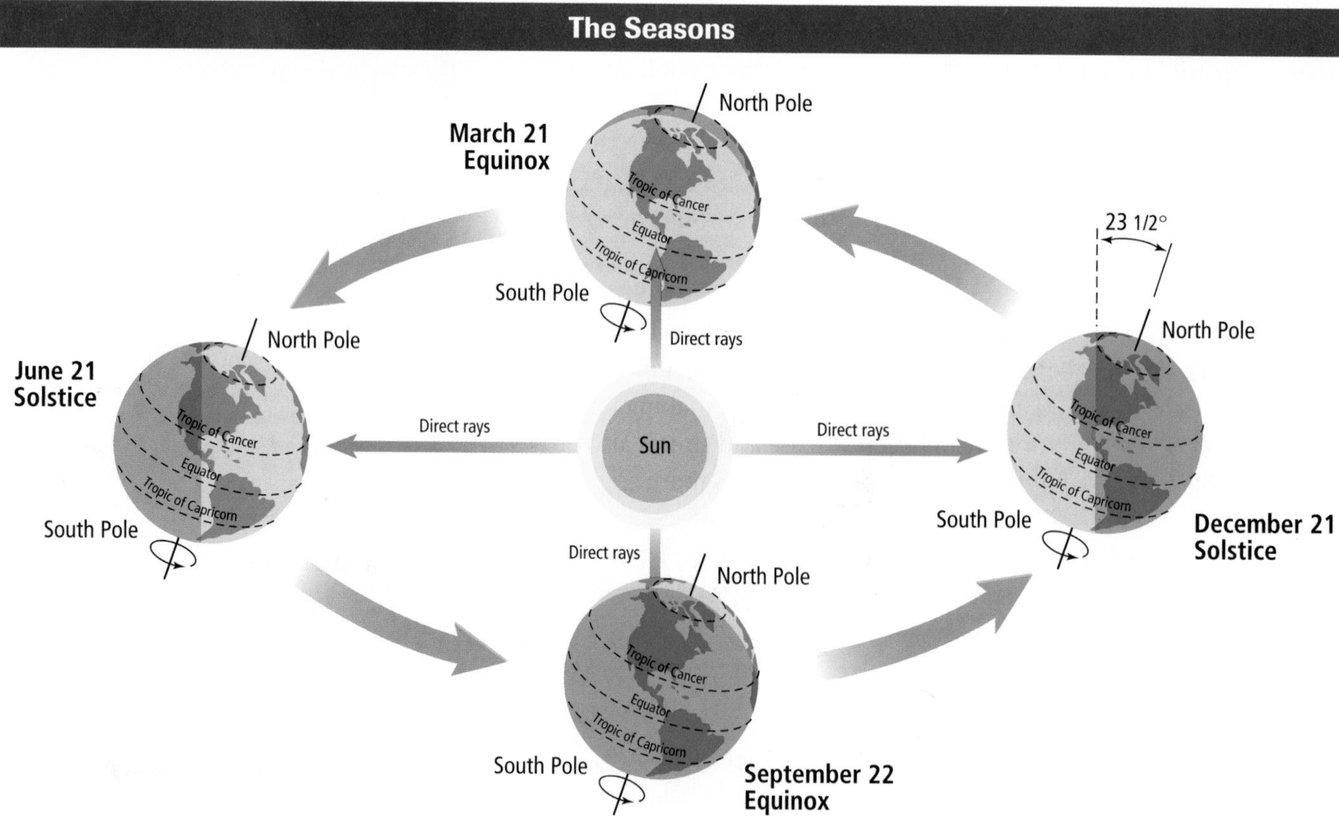

sunlight shines on this day is known as the **Arctic Circle**. It is located 66 1/2 degrees north of the equator.

During the June solstice, the Northern Hemisphere experiences the greatest number of daylight hours of the year and the first day of summer. On this day, the North Pole tilts toward the Sun. The Sun's direct rays are at their most northerly position, striking Earth at a line 23 1/2 degrees north of the equator. This line is called the **Tropic of Cancer**. If you traveled to Australia on the June solstice, it would be the first day of winter, which has the fewest daylight hours of the year. During the June solstice, the Sun never sets north of the Arctic Circle. Daylight lasts 24 hours. During this time, the opposite occurs south of the Antarctic Circle, where darkness lasts 24 hours.

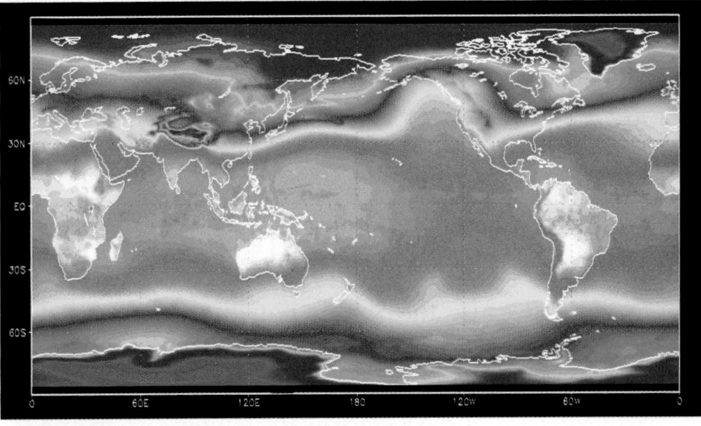

INTERPRETING THE MAP *This map shows Earth's average temperatures during December 19–25, 1999. Warmer temperatures are shown in red and yellow and colder temperatures are shown in purple and blue.* ***Where was the North Pole tilted in relation to the Sun when this data was collected? How did this affect the distribution of temperatures on Earth?*** TEKS

READING CHECK: *Physical Systems* When is the first day of winter in the Northern Hemisphere? In the Southern Hemisphere?

Equinoxes An **equinox** occurs twice each year when Earth's poles are not pointed toward or away from the Sun. *Equinox* means "equal night" in Latin. At this time, the direct rays of the Sun strike the equator, and both poles are at a 90 degree angle from the Sun. Both hemispheres receive an equal amount of sunlight—12 hours each.

Equinoxes occur on about March 21 and September 22. In the Northern Hemisphere, the March equinox marks the beginning of spring. To people living in the Southern Hemisphere, however, the March equinox marks the beginning of fall. The opposite situation occurs about September 22. Days between the solstices and equinoxes gradually become warmer or cooler, and daytime becomes longer or shorter, depending on where you live. This cycle is repeated each year, creating the four seasons.

READING CHECK: *Physical Systems* When is the first day of spring in the Northern Hemisphere? In the Southern Hemisphere?

TEKS Questions 2, 3, 5

Review

Identify

Tropic of Capricorn
Antarctic Circle
Arctic Circle
Tropic of Cancer

Define

tropics
polar regions
solstice
equinox

Reading for the Main Idea

1. *Places and Regions* Where are the tropics and polar regions found?
2. *Places and Regions* Where does the amount of solar energy Earth receives vary most during the year?

Critical Thinking

3. **Analyzing** When does winter occur on Earth? Explain your answer.
4. **Drawing Inferences and Conclusions** How do you think the tilt of Earth on its axis might affect life in the polar regions?

Organizing What You Know

5. Create a chart like the one shown below. Use it to describe the location, amount of solar energy received, and the time of year when solar energy is received for each region.

Tropics	Middle latitudes	Polar regions

Skill-Building Activity

Using a Time Zone Map

Because the Sun is not directly overhead every place on Earth at the same time, clocks are adjusted to reflect the difference in the Sun's position. Earth rotates on its axis once every 24 hours, so in one hour it makes 1/24 of a complete revolution. Since there are 360 degrees in a circle, we know that Earth turns 15 degrees of longitude each hour (360° ÷ 24 = 15°).

Earth turns in a west-to-east direction. As a result, the Sun rises first in New York, for example, and later in Los Angeles, which is farther west. If one place has the Sun directly overhead at noon, another place 15 degrees to the west will have the Sun directly overhead one hour later. The planet will have rotated 15 degrees during that hour. To account for this, we divide Earth into 24 time zones. Each time zone covers about 15 degrees of longitude. The time is an hour earlier for each 15 degrees you move westward on Earth. It is an hour later for each 15 degrees you move eastward. For example, if the Sun rises at 6:00 A.M. in New York, it is still only 3:00 A.M. in Los Angeles, three time zones to the west.

By international agreement, longitude is measured from the prime meridian, which passes through the Royal Greenwich Observatory in Greenwich (GREN-ich), England. Time is also measured from Greenwich and is called Greenwich mean time (GMT) or universal time (UT). For each time zone east of the prime meridian, clocks are set one hour ahead of GMT. For each time zone west of Greenwich, clocks are set one hour behind GMT. For example, when it is noon in London, England, it is 1:00 P.M. in Rome, Italy, one time zone to the east. At the same time, it is 7:00 A.M. in New York City, five time zones to the west.

Halfway around the planet from Greenwich is the international date line. It is a north-south line that runs through the Pacific Ocean. It generally follows the

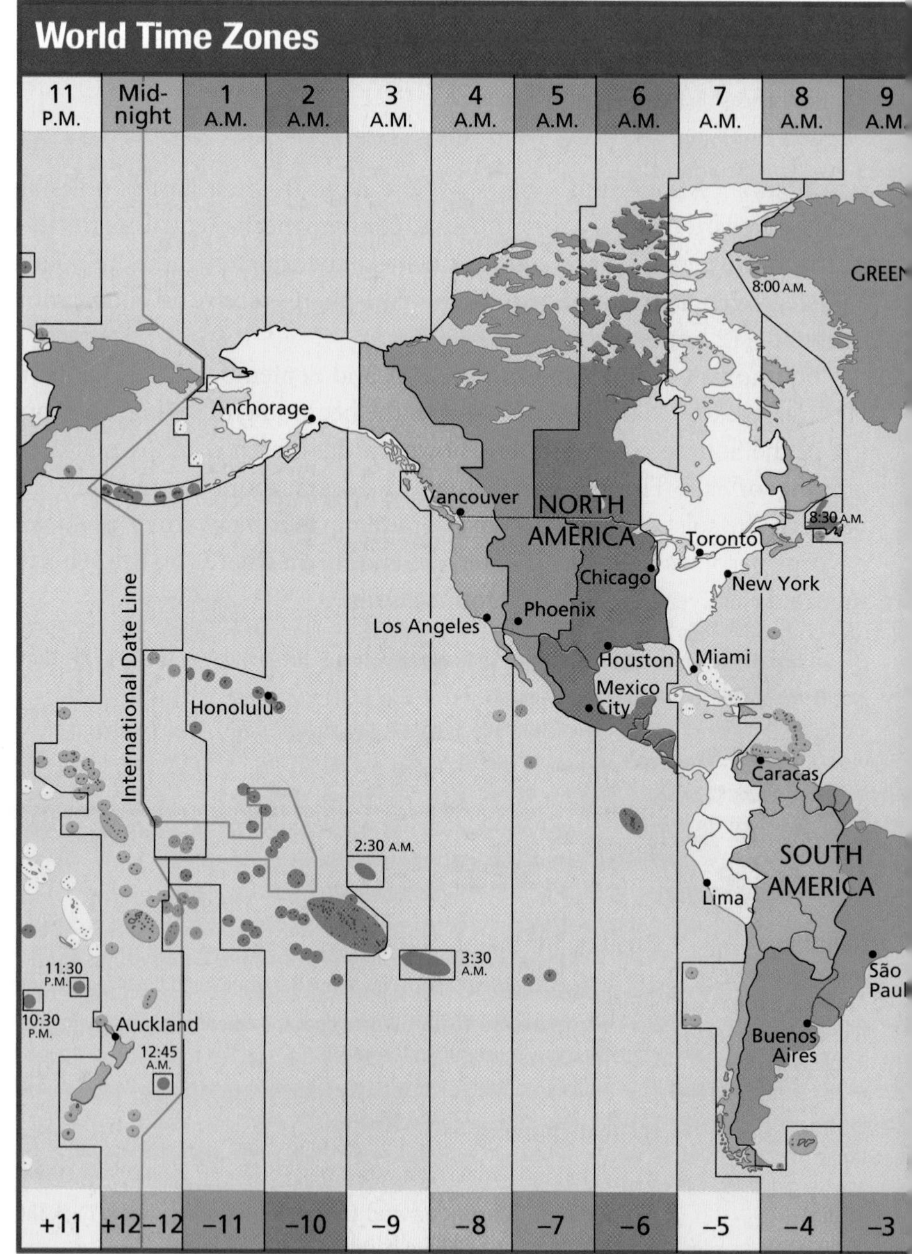

180° line of longitude. However, sometimes it leaves this line to avoid dividing island countries. When you cross the international date line, the date and day change. If you cross the date line from west to east, you gain a day. If you travel from east to west, you lose a day.

As you can see from the World Time Zones map, time zones do not follow meridians exactly. Instead, time zone lines often follow political boundaries. For example, in Europe and Africa many time zones follow country borders. The contiguous United States has four major time zones: eastern, central, mountain, and Pacific. Alaska and Hawaii are in separate time zones to the west. Some countries make local adjustments to the time in their time zones. For example, most of the United States has daylight savings time in the summer. People in places with daylight savings time adjust their clocks to have more daylight during the evening hours.

Practicing the Skill

1. In which time zone do you live? Check your time now. What time is it in New York?
2. How many time zones does China have?
3. If it is noon in New York, what time is it in London?

Section 3 The Earth System

READ TO DISCOVER

1. What are Earth's four spheres?
2. How is Earth's environment unique in the solar system?

DEFINE

atmosphere
lithosphere
hydrosphere
biosphere
environment

WHY IT MATTERS

Many global environmental issues are hotly debated today. Use CNNfyi.com or other **current events** sources to learn about the debate over one of those issues.

Earth's Four Spheres

Earth is a complex planet. Its different parts interact in a vast number of ways. The scale of some Earth interactions is so small that they are hard to notice. For example, ants and termites help mix decayed plant matter and soil in some places. Other interactions, such as rainfall and flooding, affect large regions.

Geographers call all these interactions the Earth system. A system is a group of different parts that interact to form a whole. Major parts of Earth can be viewed as separate from each other. However, they interact constantly. For example, when a volcano erupts, it not only affects the mountain where it is located but also the air, water, and life around it. Some large eruptions can affect global weather patterns. People are also part of the Earth system. Our actions affect Earth in many ways. For example, people convert large areas of Earth into farmland. This affects plant and animal life, soil fertility, and water use.

Kilauea volcano, on the island of Hawaii, is one of the largest, most active volcanoes in the world. Kilauea has been erupting steadily since 1983 and has generated numerous lava flows, which have added more than 500 acres (200 hectares) of land to the island. In 1986 Kilauea began releasing large amounts of sulfur dioxide gas into the air. This gas reacts chemically with particles in the air to produce volcanic smog, or "vog," which contaminates rainwater and causes a health hazard.

Geographers divide the Earth system into four major parts. Each part is called a sphere because it occupies a shell around the planet. The **atmosphere** is the envelope of gases that surrounds Earth. It is the least dense and outermost sphere, extending from Earth's surface into space. Earth's gravity holds the atmosphere around the planet. About 78 percent of Earth's atmosphere is a gas called nitrogen, and about 21 percent is oxygen. The rest is made up of carbon dioxide, ozone, and other gases. These gases and water vapor sustain life on Earth. The atmosphere also protects the planet from the Sun's harmful radiation.

The **lithosphere** is the solid crust of the planet. This outer crust includes rocks and soil. It forms Earth's continents, islands, and ocean floors.

The **hydrosphere** is all of Earth's water. Water covers about 70 percent of Earth's surface. The hydrosphere includes water in liquid, solid, and gaseous forms. Liquid water is found in oceans, lakes, rivers, and underground. Clouds and fog are made up of liquid droplets. Solid water, or ice, is found on both land and sea. Large amounts of ice are locked in glaciers in the polar regions. Earth is the only planet in the solar system known to have large amounts of surface water. Water is essential to all living organisms.

The **biosphere** is the part of Earth that includes all life forms. It includes all plants and animals. The biosphere overlaps the other three spheres. It extends from deep ocean floors to high in the atmosphere.

Earth's four major spheres are all interconnected. Each one affects the other. For example, the hydrosphere supplies people with water, which we need to live. It is also a home for plants and animals. The hydrosphere affects the lithosphere when rain breaks up rocks and washes them away. It also constantly interacts with the atmosphere, causing clouds and rain.

✓ **READING CHECK:** *Physical Systems* How are Earth's four spheres different from each other?

Earth's atmosphere from space

Earth's lithosphere—a volcano in Hawaii

Earth's hydrosphere—waves off the Oregon coast

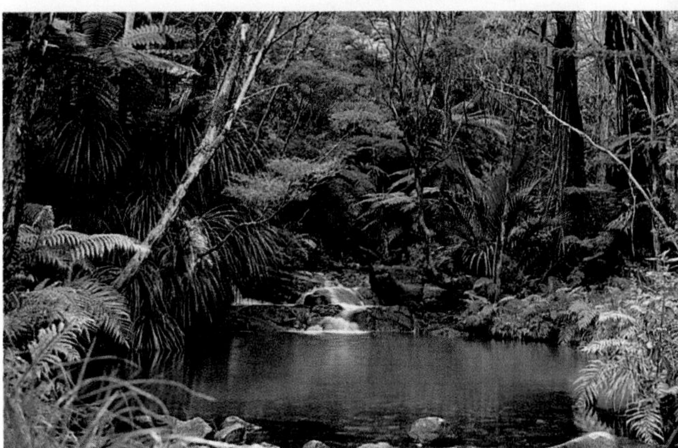

Earth's biosphere—a rain forest in New Zealand

Earth's Environment

Earth's four spheres make up the **environment**, or surroundings. The environment includes all the biological, chemical, and physical conditions that interact and affect life. Within our solar system, no other planet has an environment as complex as Earth's. Our closest neighbors, Venus and Mars, each have an atmosphere and a lithosphere. However, neither has a vast supply of liquid water. In recent years, scientists have looked far into space. They have discovered many other stars that have planets. In fact, scientists have found more than 50 planets outside our solar system. Do you think there is another planet in the universe that has the right environmental conditions for life?

Glen Canyon Dam, located in northern Arizona on the Colorado River, was completed in 1964 and created Lake Powell, a large reservoir that provides water and electricity to western states. Some environmental groups argue that the dam has disrupted the river's natural ecosystems and that Lake Powell should be drained. For some, the controversy surrounding the construction of the dam marked the beginning of the modern environmental movement.

Earth's environment is the key to our survival and quality of life. Therefore, we must be aware of how our activities affect the planet. One way we can learn more about Earth's environment is by studying environmental issues.

FOCUS ON CITIZENSHIP

Environmental Issues Many people today are concerned about the effects humans have on the environment. In fact, many geographers study a variety of environmental issues. These geographers study how to solve environmental problems at local, state, national, and international levels. Environmental studies can help us learn how to solve difficult environmental problems. They can also help us learn more about the world around us.

People often have different points of view about environmental issues. These different points of view can influence public policies and affect how decisions are made. For example, suppose government officials want to build a new dam. Some citizens might support the idea. A new dam could create jobs, prevent flooding, and increase water supplies. However, other people might oppose the construction of a new dam. They might be concerned that it would harm the environment. Some fish species might be threatened if they could not swim upriver to spawn. This, in turn, could affect the fishing industry. In addition, some land would be permanently flooded to protect other areas from occasional floods.

How might this problem or similar ones be resolved? Individuals and groups such as conservationists, farmers, and others may try to influence governmental decisions. This is called lobbying. Sometimes, legal challenges are made to a planned development. Other times, a group of citizens will force a vote on the issue. This allows the public to decide. Can you think of other ways to resolve difficult environmental issues?

READING CHECK: *Environment and Society* How can people with different points of view on environmental issues influence decision making?

TEKS Questions 2, 5

Section 3 Review

Define

atmosphere
lithosphere
hydrosphere
biosphere
environment

Reading for the Main Idea

1. *Physical Systems* How are Earth's four spheres connected to form a system?
2. *Environment and Society* How do different points of view on environmental issues influence public policies and decision-making processes? Give an example.

Critical Thinking

3. **Drawing Inferences and Conclusions** How does the atmosphere affect the other spheres?
4. **Making Generalizations and Predictions** How can human activities affect the Earth system?

Organizing What You Know

5. Copy the graphic organizer below. Use it to describe each of Earth's four spheres. In the center, describe ways the four spheres interrelate to influence Earth's environment.

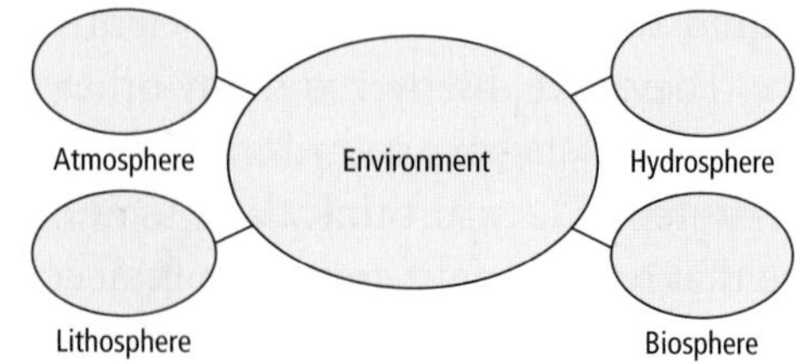

Geography for Life

Ice Ages and the Earth System

Throughout this textbook you will read about the forces and processes that have shaped the Earth system. Among the most important of these processes were Earth's ice ages. Ice ages are long periods of time during which thick ice sheets cover vast areas of land. During the ice ages, Earth's temperatures cooled for thousands or even millions of years, and ice sheets spread across the planet. During the most recent ice age, which ended about 10,000 years ago, ice sheets covered almost one third of Earth's present land area. (See the map of Pleistocene Glaciation.) Earth is currently experiencing a warm period between ice ages.

Ice ages greatly affect Earth's hydrosphere. For example, water from the oceans is frozen and locked in the expanding ice sheets. During the most recent ice age, enough water was frozen to lower global sea levels some 300 to 400 feet (90 to 120 m). As a result, more land was exposed at Earth's surface. The shapes and locations of coastlines differed greatly from how they look today. For example, the British Isles and mainland Europe were connected. North America and Asia were joined across what is now the Bering Strait. Rising ocean waters eventually covered those land bridges, however. Ocean levels rose because the warm period between ice ages causes ice to melt and ocean levels to rise. In fact, this process is partly responsible for the gradual rise in ocean levels today.

Some of the effects that ice ages have had on Earth's landscapes are easy to see. Slowly moving ice sheets carved and scraped the lithosphere, removing soil and forming holes. When the ice melted some of these holes filled with water and became lakes and swamps. The Great Lakes were formed in this way. Moving ice also wore down mountains and created great valleys.

In addition to their effects on the hydrosphere and lithosphere, ice ages also affected Earth's atmosphere and biosphere. Colder temperatures caused shifts in wind patterns. Ice-age temperature changes also caused a major redistribution of plants and animals. For example, expanding ice shifted the habitats of plant and animal species in the Northern Hemisphere from north to south. As a result, many species died off. During the last ice age, the distribution of human populations also changed. Most researchers believe that a land bridge across the Bering Strait allowed humans to move from eastern Asia to the Americas. Humans also moved across land links between areas in Europe, Asia, and Australia.

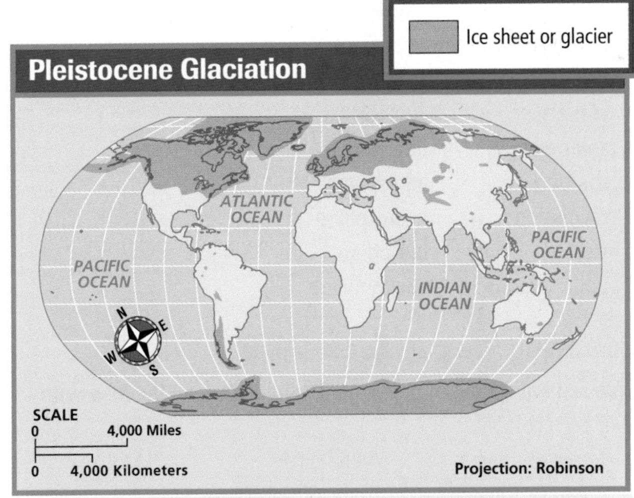

INTERPRETING THE MAP *The most recent ice age occurred mainly during the Pleistocene Epoch. This epoch began about 2 million years ago and ended about 10,000 years ago. Therefore, this ice age is known as the Pleistocene Ice Age, or Pleistocene Glaciation. During the Pleistocene Glaciation there were perhaps 12–16 periods of major ice advances. Between these cold periods were warmer times called interglacials.* ***Which parts of the world were covered with ice during the Pleistocene Glaciation?***

Applying What You Know

TEKS

1. **Summarizing** In what ways did ice ages affect the Earth system? How can some of those effects be seen today?

2. **Analyzing Information** In what ways might a new ice age affect Earth's biosphere?

CHAPTER 2 Review

Building Vocabulary TEKS

On a separate sheet of paper, explain the following terms by using them correctly in sentences.

solar system	solstice
planets	equinox
solar energy	atmosphere
rotation	lithosphere
revolution	hydrosphere
tropics	biosphere
polar regions	environment

Locating Key Places TEKS

On a separate sheet of paper, match the letters on the map with their correct labels.

Tropic of Cancer	Arctic Circle
Antarctic Circle	Tropic of Capricorn

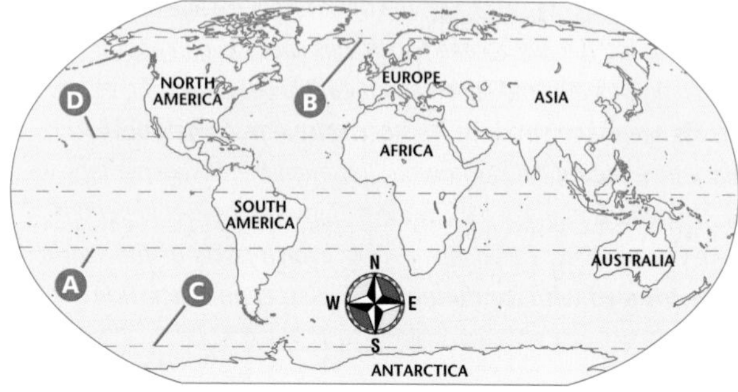

Understanding the Main Ideas TEKS

Section 1

1. *Physical Systems* What term describes Earth's shape?
2. *Physical Systems* How long does it take Earth to make one complete rotation on its axis? One complete revolution around the Sun?

Section 2

3. *Physical Systems* When do the Sun's most direct rays strike in the Northern Hemisphere? What effect does this have on temperatures there?
4. *Physical Systems* During the solstices, where do Earth's poles point in relation to the Sun?

Section 3

5. *Physical Systems* What is the Earth system?

Thinking Critically for TAKS TEKS

1. **Drawing Inferences and Conclusions** Suppose all you knew about a place was that it was located at 60° north latitude. What would this tell you about how much solar energy the place receives each year?
2. **Identifying Cause and Effect** How do Earth's revolution and tilt cause seasons?
3. **Making Generalizations and Predictions** If you lived in an area near the equator, would you experience all of the seasons? Explain.

Using the Geographer's Tools TEKS

1. **Analyzing Diagrams** Study the diagram of the seasons in Section 2. What is today's date? Using the diagram, estimate where Earth is on its path around the Sun.
2. **Interpreting Maps** Study the World Time Zones map that follows Section 2. Suppose a ship left Honolulu, Hawaii, at 10:00 P.M. on December 5 and traveled to Auckland, New Zealand. If the trip lasted six hours, what date and time would the ship arrive in Auckland?
3. **Preparing Diagrams** Using the information in this chapter, draw a diagram of Earth showing it tilted on its axis. Draw and label major lines of latitude, such as the equator, Tropics of Cancer and Capricorn, and Arctic and Antarctic Circles. How many degrees of latitude separate the Tropics of Cancer and Capricorn?

Writing for TAKS TEKS

Imagine you have been asked to help search for life beyond Earth. Write a short proposal outlining where you will search for life, what technologies you will use, and what you hope to find. When you are finished with your proposal, proofread it to make sure you have used standard grammar, spelling, sentence structure, and punctuation.

SKILL BUILDING

Geography for Life TEKS

Preparing Sketch Maps

Places and Regions Find a detailed surface map of the Moon from an atlas or other source. Use it to prepare a sketch map of the Moon's surface features and topography. When you are done, compare your map to a world map of Earth. What features are common on the Moon but not on Earth? How is the surface of the Moon different from the surface of Earth?

Building Social Studies Skills

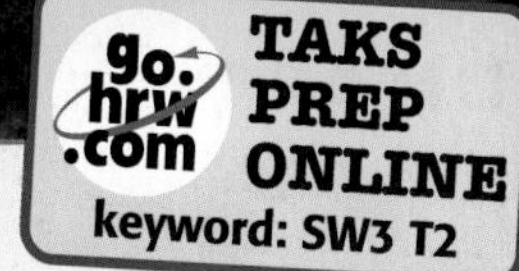

Sunrise and Sunset Times for Selected 2001 Dates in Sydney, Australia

Date	Sunrise	Sunset
March 21	6:59 A.M.	7:06 P.M.
June 21	6:58 A.M.	4:56 P.M.
September 22	5:45 A.M.	5:51 P.M.
December 21	5:43 A.M.	8:03 P.M.

Source: http://www.southernskies.com

Interpreting Charts

Study the chart above. Then use the information in the chart to help you answer the questions that follow.

1. On which date does summer begin in Sydney, Australia?
   a. March 21
   b. June 21
   c. September 22
   d. December 21
2. Look carefully at the times listed in the chart. How does this information relate to the tilt of Earth on its axis?

Analyzing Secondary Sources

Read the following passage and answer the questions.

"Telescopes have become central to astronomy. With the development of space-age technology, astronomy made even greater advances. For example, scientists sent out probes to collect new information about planets, moons, and other objects. The Hubble Space Telescope, built between 1978 and 1990, led to many exciting discoveries. The Hubble is the first powerful telescope ever placed in orbit around Earth, which gives it a much clearer view of space. Astronomers have used the Hubble to study black holes, galaxies, stars, and other distant objects."

3. The Hubble Space Telescope
   a. was built in the late 1800s.
   b. is located in Hawaii.
   c. is a probe sent out to study Mars.
   d. is the first powerful telescope ever placed in orbit around Earth.
4. Why might the Hubble's location give it such a clear view of space?

Alternative Assessment

PORTFOLIO ACTIVITY

Learning about Your Local Geography TEKS

Individual Project: Research

Plan, organize, and complete a research project on an environmental issue in your community. Check your local newspaper to find a current environmental issue. Then write a summary that explains what the issue is and describes different points of view about it. How might this issue be resolved? What role does public policy play in this issue? Have citizens tried to influence public policy regarding this issue? If so, how? How might a geographical perspective be used to help resolve this issue?

internet connect

Internet Activity: go.hrw.com
KEYWORD: SW3 GT2 TEKS

Choose a topic on Earth in space to:

- investigate the cause of seasons.
- explore recycling methods and create an action plan.
- report on the south polar regions and conditions in Antarctica.

CHAPTER 3 Weather and Climate

You have read about Earth in space and our planet's relationship to the Sun. In this chapter you will learn more about the Sun's energy and its effects on Earth. You will see how that energy affects atmospheric conditions and life on our planet.

Fog gathers over San Francisco Bay, California.

A tornado strikes Pampa, Texas.

Heavy snow blankets New York City.

Section 1 Factors Affecting Climate

READ TO DISCOVER

1. How does the Sun affect Earth's atmosphere?
2. How does atmospheric pressure distribute energy around the globe?
3. How do global wind belts affect weather and climate?
4. How do the oceans affect weather and climate?

WHY IT MATTERS

Has your state experienced a very hot summer or an unusually cold winter this year? Use CNNfyi.com or other **current events** sources to learn about the effects of weather extremes in the United States.

IDENTIFY

Gulf Stream

DEFINE

weather
climate
temperature
greenhouse effect
global warming
cyclones
prevailing winds
doldrums
front

The Sun and Latitude

The Sun plays the major role in Earth's **weather** and **climate** patterns. Weather is the condition of the atmosphere at a given time and place. Weather conditions in a geographic region over a long time are called climate. As you read earlier, solar energy heats Earth unevenly. The tilt of Earth as the planet revolves around the Sun is important. It determines which hemisphere receives the Sun's most direct rays at a given time of year. This process causes the changing seasons. Areas in the middle and high latitudes have distinct seasons. On the other hand, tropical locations in lower latitudes receive the most direct rays year-round. Thus, they are warm all year. Polar areas receive the least amount of energy from the Sun and are very cold all year.

What happens to the Sun's energy when it reaches Earth? About half is reflected back into space or absorbed by the atmosphere. Earth's surface absorbs the other half. Once absorbed, solar energy is converted into heat. The measurement of heat is called **temperature**.

Earth's atmosphere traps heat energy in a process called the **greenhouse effect**. Like the clear glass of a greenhouse, the atmosphere allows much sunlight to pass through. Earth's air then slows the rate at which the heat escapes into space. The greenhouse effect helps keep the planet warm.

Evidence seems to show that Earth has gotten warmer in recent decades. Most scientists believe that this process, called **global warming**, is caused by human activities. They point out that burning coal, natural gas, and oil adds carbon dioxide to the lower atmosphere. Because carbon dioxide absorbs heat, increased amounts

INTERPRETING THE VISUAL RECORD *Much of Texas suffered terrible heat and drought during 2000.* ***How might harsh weather affect how people view a region? How might their views differ from those held by people who live elsewhere?*** TEKS

During the Little Ice Age of the 1500s to 1800s, the average global temperature dropped about 1°C. Ice sheets advanced over Greenland farms, and the Baltic Sea and Thames River froze more often than they do now.

could enhance the greenhouse effect. These scientists urge us to reduce the production of carbon dioxide and other so-called greenhouse gases.

However, some people believe that global warming is part of a natural cycle. They believe the process may not be related entirely to human activities. This difference of opinion can affect public policies on the local, state, national, and international level. For example, citizens may convince a city council to ban gasoline-powered leaf blowers within city limits.

READING CHECK: *Physical Systems* How does Earth's atmosphere help keep the planet warm?

Atmospheric Pressure

Earth distributes the Sun's heat through a number of processes. These processes affect climate. Some of these effects occur in the atmosphere. Although you do not notice it, air in our atmosphere has weight. The air around you, extending to the top of the atmosphere, is always pushing on you. This force is called atmospheric pressure or air pressure. If you were climbing a mountain, as you moved higher there would be less air above you pushing down. The force pushing against you—the air pressure—would drop. At very high altitudes the air is too thin for humans to breathe. This is why high-flying aircraft are sealed and pressurized.

Air pressure also changes at different places on Earth's surface. Earth's unequal heating causes most of these changes. For example, when air is heated it expands, becomes less dense, and rises. This creates a low-pressure area. The rising air usually cools as it moves higher in the atmosphere. As the air rises and cools, the water vapor it carries may form clouds. These clouds may bring rain or even storms. For this reason, low pressure usually accompanies unstable weather conditions. All centers of low pressure are called **cyclones**. They can vary in intensity. Some might take the form of slight breezes and cloud cover. Others might become powerful storms with heavy rain and high winds.

These climbers use oxygen tanks as they scale Mount Everest in Nepal. Atmospheric pressure at the top of Mount Everest—at 29,035 feet, or 8,850 meters—is about one third what it is at sea level. Since the atmospheric pressure is so low at this altitude, about two-thirds less oxygen is available in each breath. As a result, climbers must force supplemental oxygen into their lungs.

On the other hand, cold air is dense and sinks toward Earth's surface. This process creates high-pressure areas. As air sinks, it heats and dries. Centers of high pressure usually bring stable, clear, and dry weather. However, they can also bring extreme heat in summer or bitter cold in winter.

On a global scale there are four major air pressure zones. They are the equatorial low, the subtropical highs, the subpolar lows, and the polar highs. Together they carry air back and forth between the equator and the Poles and between Earth's atmosphere and its surface. How does this work? Along the equator, the direct rays of the Sun cause warm air to rise, forming the equatorial low. This rising air cools in the upper atmosphere and flows toward the Poles. At about 30° latitude the cooled air begins to sink to the surface. This sinking causes the subtropical highs in each hemisphere. At the Poles, dense cold air sinks to the surface, causing the polar highs. The cold air then flows along the surface away from the Poles. At about 60° latitude the cold air forces warmer air flowing toward the Poles higher. This rising air forms the subpolar lows. (See the diagram.)

READING CHECK: ***Physical Systems*** What kind of weather is usually associated with an area of low pressure?

internet connect

GO TO: go.hrw.com
KEYWORD: SW3 CH3
FOR: Web sites about weather and climate

Global Wind Belts

Air pressure affects global wind patterns. Wind is the horizontal flow of air. Wind always flows from high to low pressure areas. For example, when air is released from a tire—an area of high pressure—it flows outward. Air will not flow into the tire unless a high-pressure hose pumps it in.

Pressure and Wind Systems

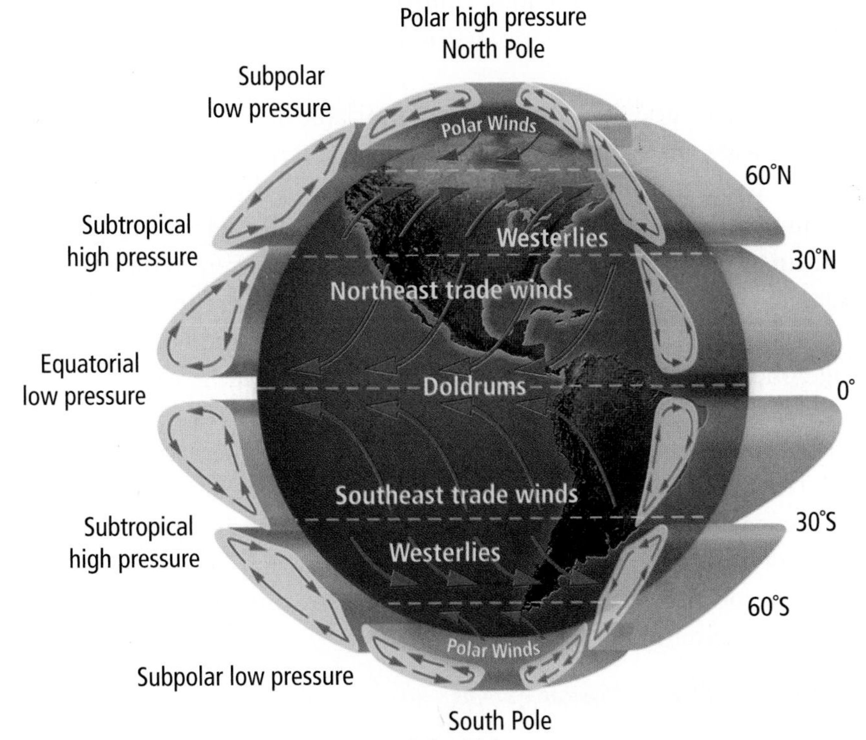

INTERPRETING THE DIAGRAM

Earth's rotation deflects the trade winds. Otherwise, they would flow more directly north or south. ***How do you think the trade winds affected exploration by European sailors?*** TEKS

A warm front heading northwest from the Gulf of Mexico brings rain to the plains of southern New Mexico. Cold fronts can lower the temperature rapidly—as much as 15°F within an hour.

Winds move heat and cold across the Earth's surface. This movement helps maintain a global energy balance. Some areas of the world have winds that blow from the same direction most of the time. These winds are known as **prevailing winds**. For example, the trade winds blow from the northeast and southeast toward the equator. They flow from the subtropical highs toward the equatorial low. These winds are so named because trading ships once used them to sail across the ocean.

Not all areas of the world lie in prevailing wind belts. The zone along the equator is calm, with no prevailing winds. This area is sometimes called the **doldrums**. Because the area has little wind, sailing ships could be caught there for long periods of time.

In the middle latitudes the prevailing winds are called westerlies. These winds flow generally from the west, from the subtropical highs toward the subpolar lows. Most of the contiguous United States is located in the westerlies. These winds carry most weather patterns and storms across the United States from west to east.

In the high latitudes the winds are more variable but come mainly from the east. These areas are subject to the cold polar easterlies. These strong winds blow from Arctic and Antarctic areas into the middle latitudes. In the United States, only Alaska is far enough north to be within this polar wind belt.

A **front** occurs when two air masses of widely different temperatures or moisture levels meet. Precipitation often occurs along these fronts. The warm westerlies meet the cold polar winds between about 40° and 60° latitude. This shifting zone where the cold and warm air masses meet is called the polar front.

There are also prevailing winds in the upper atmosphere, miles above the ground. The fastest of these high-speed westerly winds are called the jet streams. Wind speeds within jet streams can reach more than 300 miles (480 km) per hour. Usually there are two or three jet streams flowing in each hemisphere. Although we do not feel them directly, the jet streams move heat and steer major weather patterns.

READING CHECK: ***Physical Systems*** How is wind direction related to differences of air pressure?

Oceans and Currents

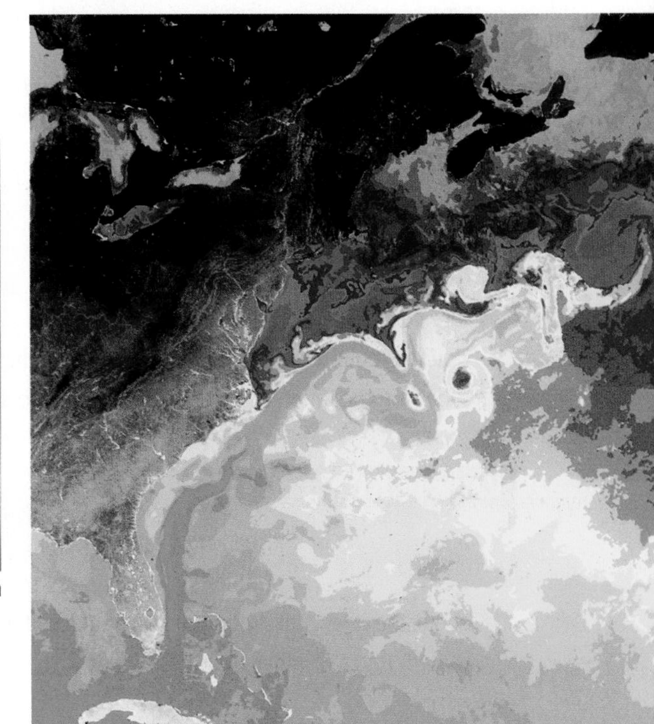

This infrared satellite image shows the Gulf Stream moving warm water from lower latitudes to higher latitudes. The dark red shape alongside Florida's east coast is the Gulf Stream.

Oceans also affect climate. Water heats and cools more slowly than land. Thus, land areas near oceans do not have such great temperature ranges as areas in the interior of continents. For example, Kansas City, Missouri, and San Francisco, California, are both located near the same latitude. However, Kansas City's winters are much colder and summers much warmer than those in San Francisco. This is because San Francisco lies on the Pacific coast. Kansas City is much farther from the ocean's moderating effects on temperatures.

Great rivers of seawater, called currents, are also important to climate. (See the map of world climate regions in Section 3.) Earth's winds and rotation as well as varying ocean temperatures create these ocean currents. The currents generally flow in circular paths. They move clockwise in the Northern Hemisphere and counterclockwise in the Southern Hemisphere.

Ocean currents move heat back and forth between the tropics and the polar regions. This movement helps maintain Earth's energy balance. Warm currents carry heated water from the tropics toward the cooler middle latitudes. The northward-flowing **Gulf Stream**, along the U.S. East Coast, is a good example of a warm current. Cool currents return cooled water from the middle and high latitudes toward the equator where it becomes warmer again. The southward-flowing California Current off the West Coast is a cool current. Cold ocean currents cool nearby land areas. Warm ocean currents make nearby land areas warmer. For example, consider the British Isles, which lie in high latitudes. You might expect to find cold climates there. However, a warm ocean current moderates the islands' climate.

TEKS **READING CHECK:** *Physical Systems* What are the main forces that create the ocean currents?

TEKS Questions 1, 2, 3, 4, 5

Review

Identify Gulf Stream

Define weather, climate, temperature, greenhouse effect, global warming, cyclones, prevailing winds, doldrums, front

Reading for the Main Idea

1. *Physical Systems* How do air pressure, global wind belts, and ocean currents affect Earth's energy balance?
2. *Physical Systems* How is air pressure affected when an area of Earth is heated?
3. *Places and Regions* Why are temperature ranges greater for places in the interior of continents than they are for places near oceans?

Critical Thinking

4. **Comparing** How do views on the causes of global warming differ among scientists and other people? Why is the issue important?

Organizing What You Know

5. Copy the word web shown below. Use it to identify major factors affecting climate discussed in this section.

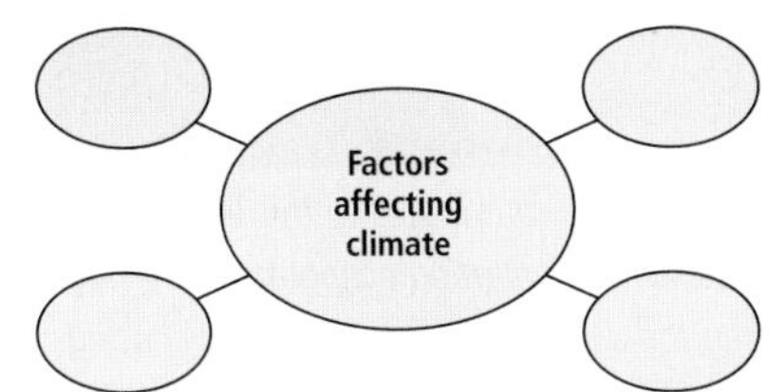

Section 2 Weather Factors

READ TO DISCOVER

1. What are the common forms of precipitation, and how are they formed?
2. How do mountains and elevation affect weather and climate?
3. What are the different types of storms, and how do they form?

DEFINE

evaporation
humidity
condensation
orographic effect
rain shadow
tornadoes
hurricanes
typhoons

WHY IT MATTERS

A long period of unusually low rainfall is called a drought. Use CNNfyi.com or other **current events** sources to learn about the effects of drought on human activities in the United States or other countries.

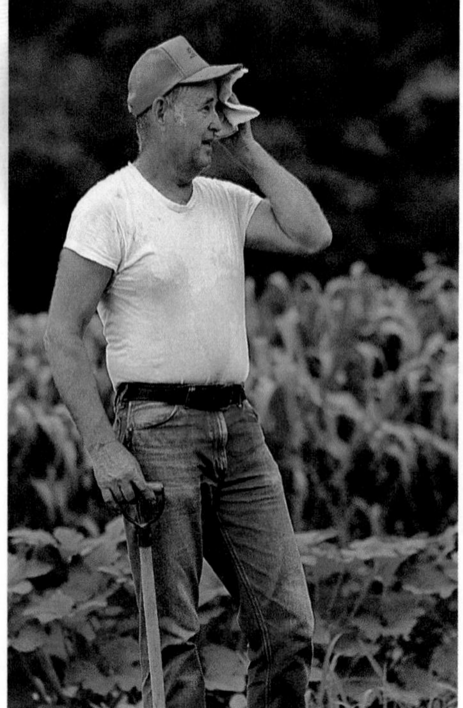

High humidity contributes to this Virginia gardener's discomfort. Perspiration evaporates very slowly when the air has a high water vapor content.

Precipitation

Water vapor plays an important role in many atmospheric processes. Without it, there would be no clouds, rain, or storms. The process by which water changes from a liquid to a gas is **evaporation**. Most water vapor that becomes rain is evaporated from the oceans. Some also evaporates from lakes, rivers, soils, and vegetation.

The amount of water vapor in the air is called **humidity**. The higher the temperature, the more water vapor the air can hold. When air cools, it will reach a temperature at which it cannot hold any more water vapor. At this point, **condensation** occurs. Condensation is the process by which water vapor changes from a gas into liquid droplets. Often you see condensation as clouds, dew, fog, or frost. If the condensed water droplets become large enough, they will fall as precipitation. There are four common forms of precipitation: rain, snow, sleet, and hail. Rain is, of course, a liquid. Snow is made up of generally six-sided ice crystals formed in the clouds. Sleet is rain that freezes as it falls. Hailstones are chunks of ice that form in storm clouds.

Precipitation is not evenly distributed around the world. It is generally highest in the persistent low-pressure zones. These zones are the equatorial low and the subpolar lows in the middle latitudes. Precipitation is generally lowest in the high-pressure zones. Those zones are the subtropical highs and the polar highs. Of course, precipitation also varies from season to season and from year to year in any given place.

READING CHECK: *Physical Systems* What are the four common forms of precipitation?

INTERPRETING THE VISUAL RECORD

Pictured on the left are mountains in the western part of the Sierra Nevada, in Sequoia National Park. At the right is part of the eastern slope of the same range, near Bishop, California. ***How does the vegetation of these places differ? What accounts for the difference?*** TEKS

Elevation and Mountain Effects

High elevation affects weather and climate. An increase in elevation—height on Earth's surface above sea level—causes a drop in temperature. The temperature drops 3.5°F per 1,000 feet of elevation (1°C/100 m). Thus, it may be warm at the base of a mountain while the summit is covered with snow or glaciers.

Mountains influence climates through the **orographic effect**. This effect occurs when moist air pushes against a mountain. The barrier forces the air to rise. The rising air cools and condenses, forming clouds and causing precipitation. As a result, the side of the mountain facing the wind receives a great deal of moisture. This side is known as the windward side. The side of the mountain facing away from the wind is the leeward side. As the air moves down the leeward side, it warms and dries. This drier area is called a **rain shadow**. Deserts are often located in rain shadows.

TEKS **READING CHECK:** ***Physical Systems*** How do high elevations and mountains influence weather and climate?

Skill-Building Activity

Reading a Weather Map

You have probably seen weather maps like this one on television or in the newspaper. Weather maps show atmospheric conditions either as they currently exist or as they are predicted to be in the future. Most weather maps have legends that explain what the colors and symbols on the map mean. The map on this page shows two cold fronts sweeping through the United States. Low-pressure systems are at the center of storms bringing rain and snow to the Northwest and upper Midwest. Notice that temperatures behind the cold front are significantly cooler than temperatures ahead of the front.

Analyzing Maps What forms of precipitation does this map show? TEKS

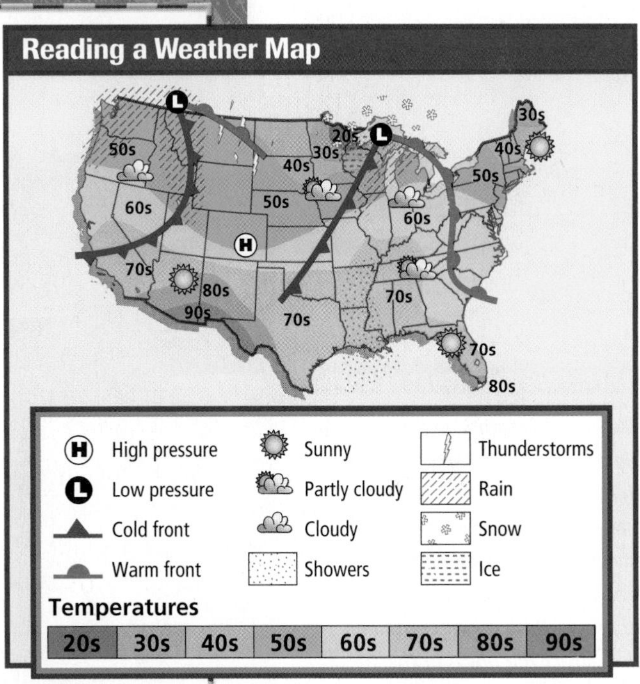

Our Amazing Planet

In the United States, lightning strikes the ground an estimated 30 million times a year. In an average year lightning kills more Americans than do tornadoes or hurricanes.

Storms

Storms are sudden and violent weather events. They can cause high winds, flooding, blowing snow, lightning, and turbulent seas. As a result, they can be extremely dangerous to human life and property. They also cause major problems for transportation.

Some very low-pressure cyclones that have rising unstable air become very large storms. They usually move from west to east in the middle latitudes, pushed by the jet stream flow. They are known as middle-latitude storms or extratropical cyclones. They can be huge, up to 1,000 miles (1,600 km) in diameter. These storms can travel across an entire continent or ocean. Most middle-latitude storms occur along a polar front. They form when cold dry polar air mixes with moist warm air from the tropics.

Middle-latitude storms may produce thunderstorms and **tornadoes**. These twisting spirals of air affect fairly small areas, but they can destroy almost anything in their path. The United States experiences more tornadoes than any other country.

Storms in the tropics differ from middle-latitude storms. Tropical cyclones are usually much smaller. Because there is no cold air present, they lack fronts. Also, they mainly travel westward, pushed by the trade winds. **Hurricanes** are the most powerful and destructive tropical cyclones. These rotating storms can bring heavy rain and winds higher than 155 miles per hour (249 kph). They begin over warm tropical seas. Those that strike the United States usually form in the tropical Atlantic Ocean. They become stronger as they move westward.

Hurricanes, which are called **typhoons** in the western Pacific Ocean, can create dangerously high waves. They may also produce thunderstorms and tornadoes. High seas brought by the storms can erode and destroy beaches and coastal areas. Sometimes large stretches of beach are swept away. Flooding can threaten people, property, and plant and animal life. Fortunately, hurricanes weaken as they move inland.

READING CHECK: ***Physical Systems*** What are examples of very violent middle-latitude and tropical storms?

INTERPRETING THE VISUAL RECORD

Hondurans look at a bridge destroyed during Hurricane Mitch in 1998. Hurricane Mitch has been labeled the deadliest Atlantic storm in two centuries. It caused massive damage throughout much of Central America and killed more than 9,000 people. ***How might continued rainfall affect the place in the photo?*** TEKS

Visible light image

Radar image

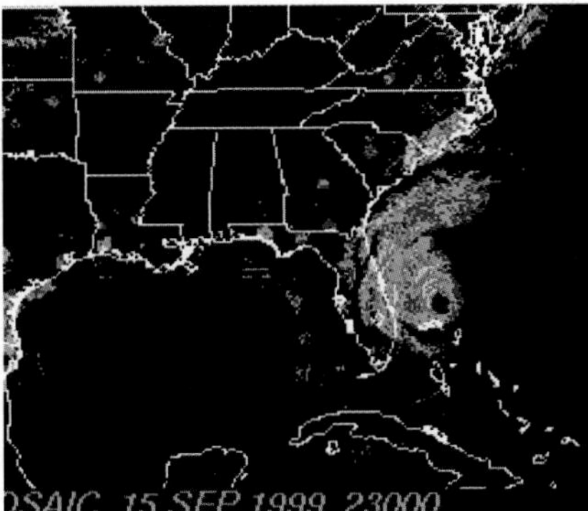

Infrared image

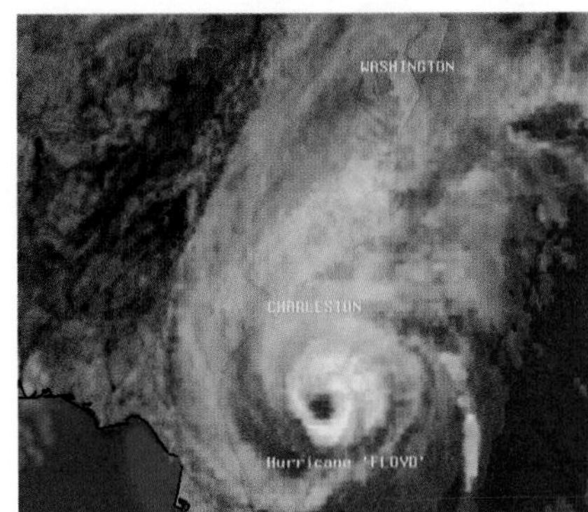

Hurricane Floyd, pictured above in three types of images, battered the eastern United States in mid-September 1999.

FOCUS ON SCIENCE, TECHNOLOGY, AND SOCIETY

Weather Satellites At one time, hurricanes could strike coastal areas with little warning. For example, in 1900 a hurricane caught residents of Galveston, Texas, unprepared. The storm killed thousands of people. Today satellites orbiting Earth help weather forecasters track the development of storms and save lives.

Weather satellites carry instruments that collect information about the Earth's surface and atmosphere. For example, they carry special cameras that make use of both visible light and invisible infrared light. Infrared sensors detect heat given off by clouds, land, and water. Radar can also be used to gather atmospheric data. With the information from these instruments, scientists can predict the path of a hurricane. They can also identify areas that may flood. People who live in affected areas can then leave before conditions worsen.

As technology improves, scientists can gather data on storms earlier in their development. Two satellites now use microwaves to peer through heavy clouds. They gather information on young hurricanes while the storms are still far out at sea. One of the satellites measures the sea's roughness. The other measures the rainfall in a given area.

For centuries, people looked to the skies for information about the weather. Now, scientists can use satellites to look down on Earth from above.

READING CHECK: *Physical Systems* How do weather satellites help people live safely in places affected by hurricanes?

TEKS Questions 1, 2, 3, 4, 5

Section 2 Review

Homework Practice Online
Keyword: SW3 HP3

Define

evaporation
humidity
condensation
orographic effect
rain shadow
tornadoes
hurricanes
typhoons

Reading for the Main Idea

1. *Physical Systems* How is temperature change related to condensation and precipitation?
2. *Physical Systems* What determines the direction of storm movement?

Critical Thinking

3. **Contrasting** How do storms in the tropics differ from storms in the middle latitudes?
4. **Analyzing Information** In what places near the equator might you expect to see snow and ice all year? Why?

Organizing What You Know

5. Using the weather map in the chapter as a guide, create a sketch map of the United States showing imaginary weather events. Use the standard weather map symbols. Then write a short description of the weather patterns you have displayed.

Section 3

Climate and Vegetation Patterns

READ TO DISCOVER

1. How do the two tropical climates differ?
2. What conditions are common in dry climates?
3. What climates are found in the middle latitudes?
4. What characterizes high-latitude and highland climates?

DEFINE

ecosystems
monsoon
savannas
arid
deciduous forests
coniferous forests
permafrost

WHY IT MATTERS

Scientists worry that climate changes could cause terrible problems in some parts of the world. Use CNNfyi.com or other **current events** sources to learn more about expected consequences of changes in global climate patterns.

Tropical Climates

All of the world's climates support a variety of **ecosystems**. An ecosystem is the community of plants and animals in an area. It also includes the nonliving parts of their environment, such as soil and climate. (See Geography for Life: World Ecosystems and Biomes in the next chapter.)

Among the world's most diverse ecosystems are those found in tropical climate regions. We will begin our discussion of world climate and vegetation patterns there. You can follow our discussion by studying the map and chart of world climate regions.

Tropical Humid Climate The tropical humid climate region is found in areas close to the equator. Those regions generally have warm temperatures and plentiful rainfall all year. They never have truly cold weather. Because the equator receives the Sun's direct rays all year, the area is always warm. As a result, warm air is always rising in the tropics. This continuous rising of warm unstable air brings almost daily thunderstorms and heavy rainfall.

Western Java, in Indonesia, has a tropical humid climate. This scene is from the Bogor Botanical Gardens. Some tropical plants, like the one growing on the tree trunk, need no more moisture than what is in the warm humid air.

Climate and Vegetation

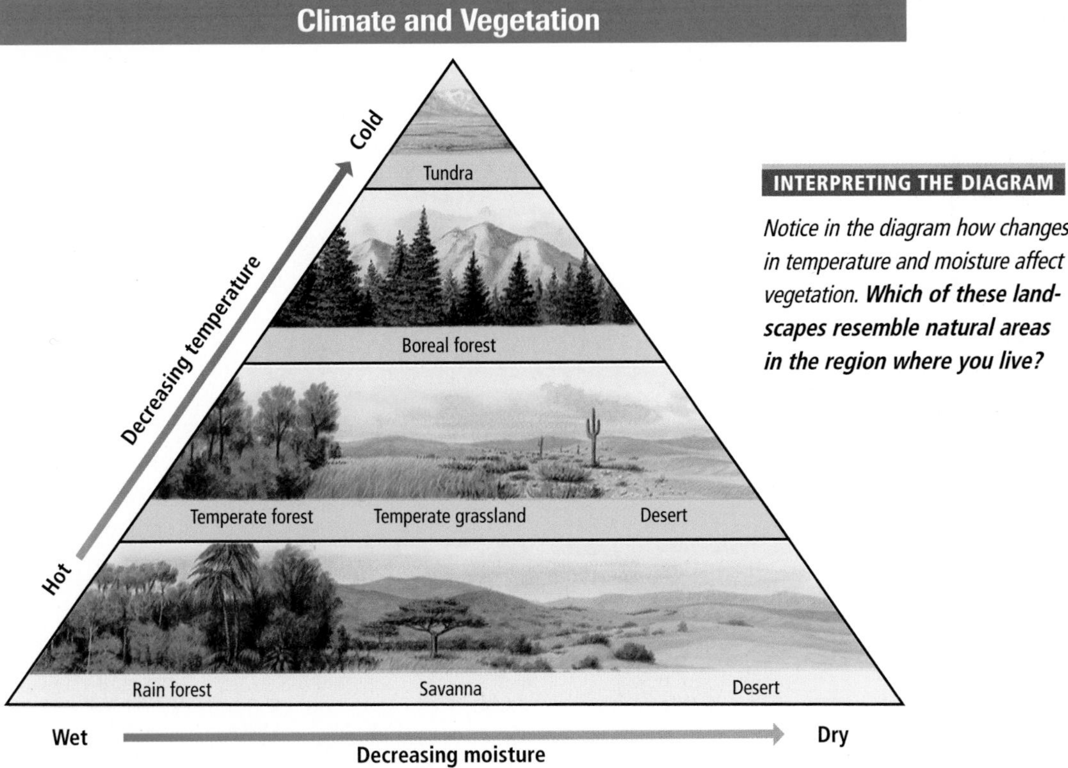

INTERPRETING THE DIAGRAM

Notice in the diagram how changes in temperature and moisture affect vegetation. ***Which of these landscapes resemble natural areas in the region where you live?***

In some tropical areas, rainfall is concentrated in one wet season. India and Southeast Asia have seasons of this type. During the summer months moist air flows into these areas from the warm ocean, a high-pressure area. The air flows to the hotter land, a low-pressure area. This air flow brings heavy rains to areas along the coast and inland. During the winter, dry air flows off the cooling continent (high-pressure areas). It flows toward the warm oceans (low-pressure areas). The air flow causes dry conditions on the continent. This wind system, in which winds completely reverse direction and cause seasons of wet and dry weather, is called the **monsoon**.

Warm temperatures and heavy rainfall create ideal conditions for plant growth. Thus, dense tropical rain forests thrive in tropical humid areas. These forests are the most complex land ecosystems in the world. Thousands of kinds of plants and animals live there.

Tropical Wet and Dry Climate Just to the north and south of the tropical humid climate is the tropical wet and dry climate. It is sometimes called the tropical savanna climate. This climate results from the seasonal change in the way the Sun's rays strike areas just north and south of the equator. During summer in these areas, the Sun's rays strike most directly. As a result, temperatures rise, creating low pressure and unstable, rising air. This, in turn, leads to heavy rainfall. During the winter the Sun's direct rays move to the opposite hemisphere. High pressure replaces low pressure. The high pressure system brings stable, cool, sinking air and a dry season. This alternating pattern of wet and dry seasons supports **savannas**. Savannas are areas of tropical grasslands, scattered trees, and shrubs.

READING CHECK: ***Physical Systems*** How does the tropical wet and dry climate differ from the tropical humid climate?

The World's Climate Regions

MONSOON AIR FLOW
- Wet monsoon
- Dry monsoon

MAJOR WORLD OCEAN CURRENTS
- Cool currents
- Warm currents

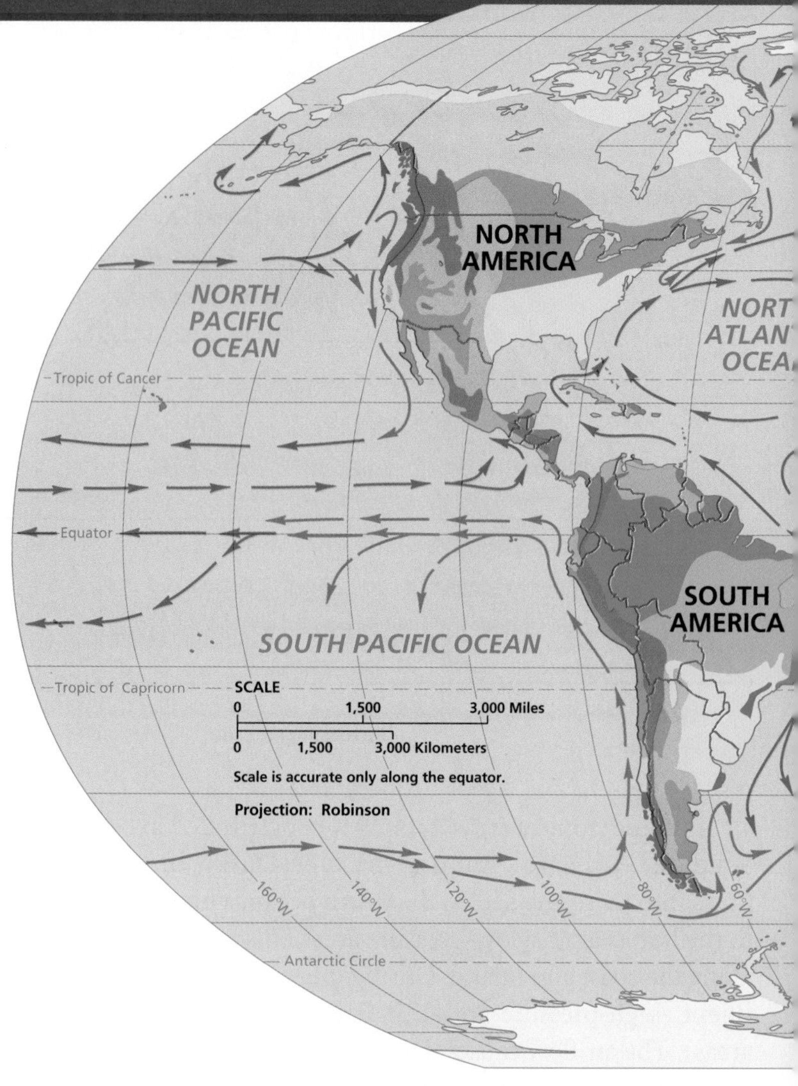

	Climate	Geographic Distribution	Major Weather Patterns	Vegetation
Tropical	TROPICAL HUMID	along equator; particularly equatorial South America, Congo Basin in Africa, Southeast Asia	warm and rainy year-round, with rain totaling anywhere from 65 to more than 450 in. (165–1,143 cm) annually; typical temperatures are 90°–95°F (32°–35°C) during the day and 65°–70°F (18°–21°C) at night	tropical rain forest
Tropical	TROPICAL WET AND DRY	between humid tropics and deserts; tropical regions of Africa, South and Central America, South and Southeast Asia, Australia	warm all year; distinct rainy and dry seasons; precipitation during the summer of at least 20 in. (51 cm); monsoon influences in some areas, such as South and Southeast Asia; summer temperatures average 90°F (32°C) during the day and 70°F (21°C) at night; typical winter temperatures are 75°–80°F (24°–27°C) during the day and 55°–60°F (13°–16°C) at night	tropical grassland with scattered trees
Dry	ARID	centered along 30° latitude; some middle-latitude deserts in interior of large continents and along western coasts; particularly Saharan Africa, Southwest Asia, central and western Australia, southwestern North America	arid; precipitation of less than 10 in. (25 cm) annually; sunny and hot in the tropics and sunny with great temperature ranges in middle latitudes; typical summer temperatures for lower-latitude deserts are 110°–115°F (43°–46°C) during the day and 60°–65°F (16°–18°C) at night, while winter temperatures average 80°F (27°C) during the day and 45°F (7°C) at night; in middle latitudes the hottest month averages 70°F (21°C)	sparse drought-resistant plants; many barren, rocky, or sandy areas
Dry	SEMIARID	generally bordering deserts and interiors of large continents; particularly northern and southern Africa, interior western North America, central and interior Asia and Australia, southern South America	semiarid; about 10–20 in. (25–51 cm) of precipitation annually; hot summers and cooler winters with wide temperature ranges similar to desert temperatures	grassland; few trees
Middle Latitudes	MEDITERRANEAN	west coasts in middle latitudes near cool ocean currents; particularly southern Europe, part of Southwest Asia, northwestern Africa, California, southwestern Australia, central Chile, south-western South Africa	dry sunny warm summers and mild wetter winters; precipitation averages 14–35 in. (35–90 cm) annually; typical temperatures are 75°–80°F (24–27°C) on summer days; the average winter temperature is 50°F (10°C)	scrub woodland and grassland
Middle Latitudes	HUMID SUBTROPICAL	east coasts in middle latitudes; particularly southeastern United States, eastern Asia, central southern Europe, southeastern parts of South America, South Africa, and Australia	hot humid summers and mild humid winters; precipitation year-round; coastal areas are in the paths of hurricanes and typhoons; precipitation averages 40 in. (102 cm) annually; typical temperatures are 75°–90°F (24°–32°C) in summer and 45°–50°F (7°–10°C) in winter	mixed forest

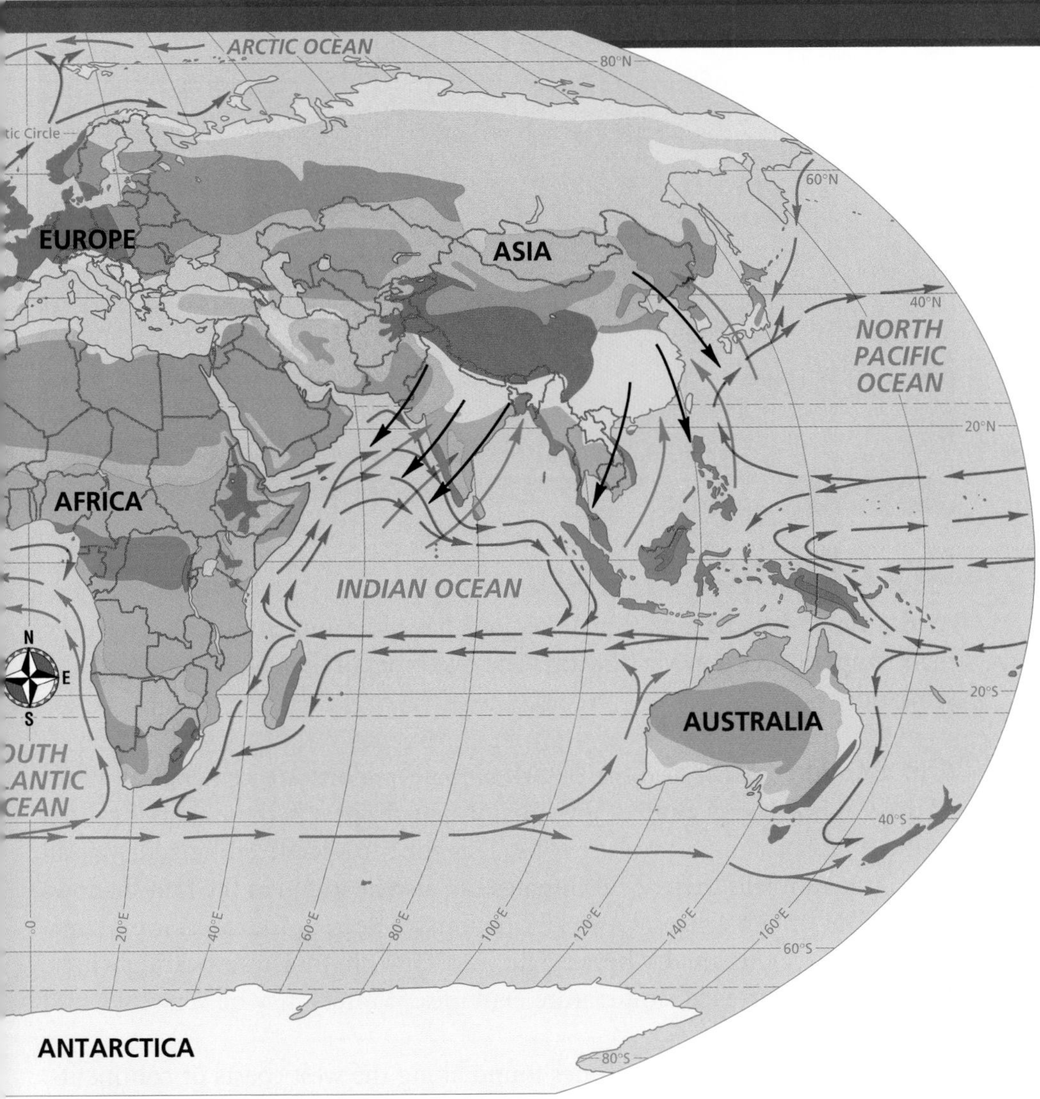

	Climate	Geographic Distribution	Major Weather Patterns	Vegetation
Middle Latitudes	MARINE WEST COAST	west coasts in upper-middle latitudes; particularly northwestern Europe and North America, southwestern South America, central southern South Africa, southeastern Australia, New Zealand	cloudy mild summers and cool rainy winters; strong ocean influence; precipitation averages 20–98 in. (51–250 cm) annually; westerlies bring storms, rain; average temperature in hottest month is usually between 60°F and 70°F (16°–21°C); average temperature in coolest month usually is above 32°F (0°C)	temperate evergreen forest
	HUMID CONTINENTAL	east coasts and interiors of upper-middle-latitude continents; particularly northeastern North America, northern and eastern Europe, northeastern Asia	four distinct seasons; long cold winters and short warm summers; precipitation amounts vary, usually 20–50 in. (51–127 cm) or more annually; average summer temperature is 75°F (24°C); average winter temperature is below freezing	mixed forest
High Latitudes	SUBARCTIC	higher latitudes of interior and east coasts of continents; particularly northern parts of North America, Europe, and Asia	extremes of temperature; long cold winters and short mild summers; low precipitation amounts all year; precipitation averages 5–15 in. (13–38 cm) in summer; temperatures in warmest month average 60°F (16°C) but can warm to 77°F (25°C); winter temperatures average below 0°F (–18°C)	northern evergreen forest
	TUNDRA	high-latitude coasts; particularly far northern parts of North America, Europe, and Asia, Antarctic Peninsula, subantarctic islands	cold all year; very long cold winters and very short cool summers; low precipitation amounts; precipitation average is 5–15 in. (13–38 cm) annually; warmest month averages less than 50°F (10°C); coolest month averages a little below 0°F (–18°C)	moss, lichens, low shrubs; permafrost bogs in summer
	ICE CAP	polar regions; particularly Antarctica, Greenland, Arctic Basin islands	freezing cold; snow and ice year-round; precipitation averages less than 10 in. (25 cm) annually; average temperatures in warmest month do not reach higher than freezing	no vegetation
	HIGHLAND	high mountain regions, particularly western parts of North and South America, eastern parts of Asia and Africa, southern and central Europe and Asia	greatly varied temperatures and precipitation amounts over short distances as elevation changes; prevailing wind patterns can affect rainfall on windward and leeward sides of highland areas	forest to tundra vegetation, depending on elevation

A boojum tree, at left, and saguaro cactus, at right, grow in the desert of Baja California, Mexico. Both of these unusual plants are well adapted to the desert's heat and lack of moisture. The boojum drops its leaves to save water, and the saguaro stores water in its fleshy trunk. Hurricanes sometimes sweep from the Pacific Ocean across Baja California. Then desert plants burst into bloom.

Dry Climates

Arid means "dry." All dry climate regions have low annual rainfall. However, their temperatures may vary greatly. The two types of dry climates are arid and semiarid.

Arid Climate Most arid, or desert, climate regions are centered at about 30 degrees north and south of the equator. The dryness in these areas is caused by the subtropical high-pressure zone. This zone has stable, sinking, dry air all year. Little rain falls there. Arid climates can also be found in the rain shadows of some mountain ranges. Other dry regions lie deep in the interior of continents. Such regions are dry because they are far from moisture-bearing winds. These areas experience temperature extremes. Winters may be very cold, and summers are very hot.

Small dry areas are sometimes found along the west coasts of continents. They lie where cool ocean currents create highly stable atmospheric conditions. Dry coastal deserts of this type are found along the west coasts of Australia, Mexico, South America, and southwestern Africa.

Plants and animals in these dry regions must be hardy. Most desert plants are small, and they may only sprout and flower after a rare rainfall. Soils tend to be thin and rocky.

Semiarid Climate The semiarid climate is a transition zone between the arid climate and the more humid climates. Semiarid climates receive more moisture than the deserts but less than the more humid areas. Where rainfall is heavy enough, grasses make up much of the plant life. Humans use many of these areas for growing grain. In the drier areas, plant life is more limited.

READING CHECK: *Physical Systems* What are four factors that can create dry climates?

Middle-Latitude Climates

Several types of climates are found in the middle latitudes. These are generally temperate climates. For the most part, they do not experience the extreme conditions found in tropical and high-latitude climates.

Mediterranean Climate The Mediterranean climate exists mainly in two kinds of areas. One is along coastal areas of southern Europe. The other is along west coasts of continents with cool ocean currents. Mediterranean climates do not usually extend far beyond inland mountain ranges. The stable sinking air of the subtropical high-pressure zone causes long, sunny, dry summers. During the mild winter, however, cool middle-latitude storms bring rain. The natural plant life of this climate is the Mediterranean scrub woodland. It includes short trees and shrubs with a few scattered large trees.

Like many other areas in southern Europe, Greece has a Mediterranean climate. This photo was taken on the island of Corfu.

Humid Subtropical Climate The humid subtropical climate is much more widespread than the Mediterranean climate. It is found on the eastern side of continents where there are warm ocean currents. The moist air flowing off the warm ocean waters greatly affects this climate area. Summers are hot and humid. Winters are mild, with occasional frost and some snow. Hurricanes or typhoons can be a danger here. The natural plant life of humid subtropical and other middle-latitude regions includes temperate forests. These forests can be divided into two main types. One type is **deciduous forests**. Trees there lose their leaves during part of the year. The other type is **coniferous forests**. Trees in those forests remain green all year.

Marine West Coast Climate The marine west coast climate is heavily influenced by oceans. This climate is generally found on the west coasts of continents in the upper middle latitudes. Temperatures in these areas are mild all year. Storms traveling across the oceans in the westerlies bring most of the rainfall. Winters are foggy, cloudy, and rainy. However, summers can be warm and sunny. This climate is widespread in northwestern Europe. Lowlands along the coast there allow cool moist ocean air to spread far inland. In some places, the rainy weather of this climate supports dense coniferous forests. These areas are called temperate rain forests.

Left: A coniferous forest grows in Banff National Park in the Canadian Rocky Mountains. Right: Deciduous trees in Portland, Oregon, change color in the fall before losing their leaves.

Humid Continental Climate The humid continental climate is found in the interiors and east coasts of upper-middle latitude continents. Invasions of both warm and cold air regularly affect these areas. The humid continental climate has the most changeable weather conditions. In fact, it experiences four distinct seasons. Because this climate type is situated along the polar front, storms bring rain throughout much of the year. Snow falls in winter. Precipitation is heavy enough to support forests.

TEKS **READING CHECK:** *Physical Systems* In areas with a Mediterranean climate, which season brings most of the rainfall?

High-Latitude and Highland Climates

There are three main types of high-latitude climates. They are subarctic, tundra, and ice cap climates.

Subarctic Climate The subarctic climate is located generally above 50° north latitude. However, a warm ocean current moderates the climate in areas above this latitude in northern Europe. The subarctic climate has long cold winters. Temperatures stay well below freezing for half of the year. The short summers can be warm, however. This climate also has the greatest annual temperature ranges in the world. Although severe, the climate supports vast evergreen forests. These northern forests are also called boreal forests.

The subarctic climate region is very large. It stretches across far northern North America, Europe, and Asia. In the Southern Hemisphere there is practically no land at these latitudes.

Tundra Climate Coastal areas in high latitudes have a tundra climate. These areas also have long winters. Temperatures are above freezing only during the short summers. In some places, water and soil below the ground's surface remain frozen throughout the year. The permanently frozen soil is called **permafrost**. During summer the permafrost makes it difficult for water from melting snow to seep into the ground. As a result, swamps and bogs form on the surface.

The tundra climate takes its name from the only kinds of vegetation that can survive there. Tundra vegetation is made up of lichens, mosses, herbs, and low shrubs. Trees cannot grow there. However, during the short summer the area bursts into flower.

INTERPRETING THE VISUAL RECORD

Tundra vegetation grows near Hudson Bay, Canada. ***Why might growing low to the ground be an advantage for these plants?*** TEKS

Guanaco feed on the plains of Torres del Paine National Park, Chile. Behind them rise the Andes. This long South American mountain range contains a variety of climates.

Ice Cap Climate Ice cap climates are found in Earth's polar regions. Those areas are always covered by huge flat masses of ice and snow. Most parts of Antarctica and Greenland are covered by ice caps. Few land plants and animals can survive in these climates. Hardy plant life may grow on exposed rocks. In contrast, the cold seas of the ice cap regions have rich marine ecosystems. Many birds, fish, and marine mammals live there. For example, seals and some whales feed on the fish of the Arctic Ocean. Polar bears feed on the seals.

Highland Climate Highland areas can have varying climates. This is partly because temperatures change with elevation. In addition, prevailing wind patterns can affect rainfall on windward and leeward sides of highland areas.

The lowest elevations of a mountain generally have a climate and vegetation similar to that of the surrounding area. Higher up the mountain, temperatures and air pressure are lower. The cooler temperatures limit the kinds of plants that can grow. No trees can grow above a certain level called the tree line. Climate conditions at the highest elevations are similar to those of the ice cap climate. Ice and snow are always present, and plant and animal life is scarce. Some mountains in the tropics have snow all year. For example, Africa's snowcapped Kilimanjaro lies only about 200 miles (322 km) south of the equator.

READING CHECK: *Physical Systems* Which of the polar climates supports large forests?

TEKS Questions 1, 2, 3, 4, 5

Section 3 Review

Homework Practice Online

Keyword: SW3 HP3

Define

ecosystems, monsoon, savannas, arid, deciduous forests, coniferous forests, permafrost

Reading for the Main Idea

1. *Physical Systems* Why are there wet and dry seasons in some tropical areas?
2. *Places and Regions* What kinds of precipitation levels and temperatures would you expect in arid climates?

Critical Thinking

3. **Drawing Inferences and Conclusions** Why do you think Mediterranean climates do not extend far inland past mountain ranges?
4. **Comparing** How do conditions on high mountaintops within highland climate regions resemble the ice cap climate?

Organizing What You Know

5. Create a graphic organizer like the one below. Add as many rows as you need to describe the vegetation that is typical in each of Earth's climate types.

Climate type	Vegetation

CASE STUDY

The Poles

Environment and Society Earth's axis intersects its surface at the North and South Poles. The Poles are the same distance from the equator, and both have ice cap climates. However, the Poles also have varying geographic features. They and their surrounding regions offer unique opportunities for expanding human knowledge about our world.

A Frozen Land

The South Pole is in the heart of Antarctica—a continent buried under snow and ice. It is the coldest, driest, windiest, and most isolated continent on Earth. Antarctica also has the highest average elevation of any continent. Russian scientists recorded the world's lowest temperature there, –128.6°F (–89.2°C). Even in the summer, the average temperature in Antarctica's interior stays far below freezing. The air is so cold that it cannot hold moisture. As a result, central Antarctica gets only 2 inches (5 cm) of precipitation per year. However, the ice has been building up for millions of years!

What does the bottom of the world look like? More than 95 percent of Antarctica's surface is ice. It is Earth's deep freeze, storing more than 90 percent of the planet's ice. Ice does not cover the Transantarctic Mountains, however. At 6,500 to 13,000 feet (2,000 to 4,000 m), these mountains break through the ice that blankets most of the continent. Mountains, plateaus, and valleys lie buried under the ice throughout the rest of the continent. Much of Antarctica's rock foundation lies under the frozen surface, some of it below sea level. If all of Antarctica's ice melted, parts of the continent would become island chains. In addition, without the weight of the ice pushing it downward, the underlying rock would rise.

A Frozen Sea

The North Pole lies in the middle of the Arctic Ocean instead of deep within a continent. Asia, Europe, and North America surround the Arctic Ocean. A permanent layer of sea ice covers the ocean's center. During the warmest months of the year, when the

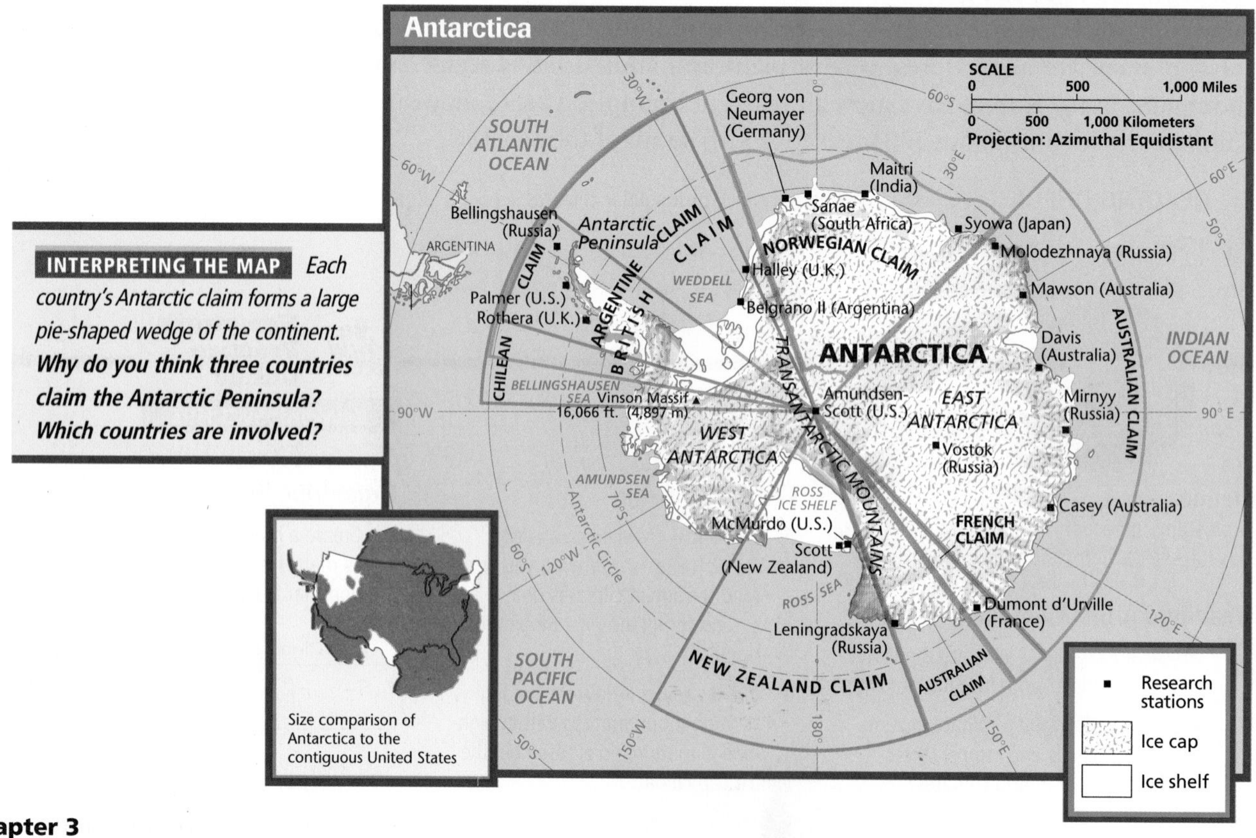

INTERPRETING THE MAP *Each country's Antarctic claim forms a large pie-shaped wedge of the continent.* ***Why do you think three countries claim the Antarctic Peninsula? Which countries are involved?***

Size comparison of Antarctica to the contiguous United States

temperature hovers near freezing, the edges of this frozen crust melt. When winter cold returns, so does the ice.

Compared to the stable and thick Antarctic ice cover, Arctic ice is very thin and loose. Antarctic ice is typically thousands of feet thick. In contrast, Arctic ice averages only 10 to 16 feet (3 to 5 m) in thickness. Arctic ice floats in large chunks on the ocean's surface. Currents, tides, and wind push and pull the ice. Cracks and ridges spread. Large sections of ice collide, combine, break up, and collide again.

INTERPRETING THE VISUAL RECORD *Bottom: The polar bear is at home in the Arctic Ocean and on land or ice. Top: At the North Pole, passengers from a Russian ship walk through all the time zones.* ***Through which hemispheres could these people walk?***

Research at the Poles

Scientists have conducted research at both Poles for many decades. During the 1950s and 1960s American and Soviet scientists mapped the Arctic Ocean's floor from research stations on the ice. However, because Arctic ice is always changing, these stations were not permanent.

In contrast, Antarctic research stations have been more durable. Some 29 countries have sponsored research projects in Antarctica. These projects address a wide range of topics. Some scientists search the ice for meteorites that provide information about our solar system. Others have compared air trapped in ancient ice bubbles with today's air. They learned that the use of fossil fuels has raised the amount of carbon dioxide in the air to the highest levels in human history. Some researchers concentrate on how animals survive in the frigid climate of Antarctica and the waters surrounding it.

A Harsh but Fragile Place

Antarctica is not immune to damage or change. For some time, people at research stations piled up trash and sewage and pushed it into the ocean. In addition, some energy companies have hoped to extract the continent's store of oil and minerals. Oil spills have already caused environmental damage. In 1991, 32 countries forged an agreement to protect Antarctica. The agreement forbids most activities in Antarctica that do not have a scientific purpose. It bans mining and drilling and limits tourism.

Still, Antarctica is changing. Satellite images show that the continent's ice sheet is shrinking. In 2001, scientists found that since 1992 about 8 trillion gallons of water had melted from a glacier in western Antarctica. That is enough to cover about 23 million acres of land with about 1 foot of water. Scientists believe this finding is further evidence of global warming.

Applying What You Know 

1. **Summarizing** What are some of the important differences in the physical geography of the North and South Poles?
2. **Identifying Cause and Effect** How have humans modified the physical environment of Antarctica? How may global warming be affecting Antarctica?

CHAPTER 3
Review

Building Vocabulary TEKS

On a separate sheet of paper, explain the following terms by using them correctly in sentences.

weather
condensation
doldrums
temperature
hurricanes
evaporation
cyclones
arid
prevailing winds
climate
rain shadow
front
greenhouse effect
monsoon
humidity

Locating Key Places TEKS

On a separate sheet of paper, match the letters on the map of Australia with their correct labels.

tropical humid climate
marine west coast climate
semiarid climate
arid climate
tropical wet and dry climate
humid subtropical climate
Mediterranean climate

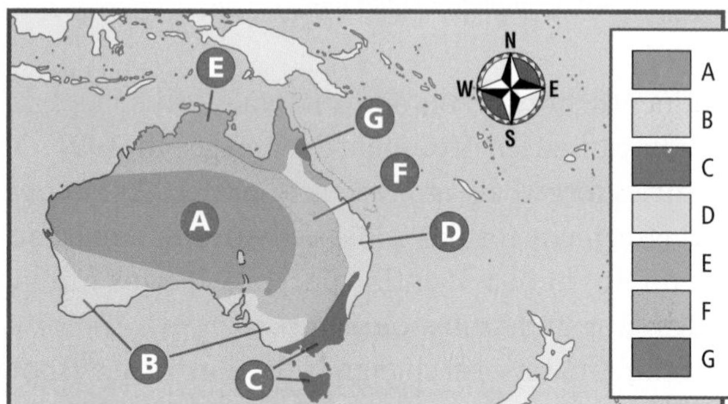

Understanding the Main Ideas TEKS

Section 1

1. *Physical Systems* How does latitude relate to climate?
2. *Physical Systems* How do atmospheric pressure zones and ocean currents affect Earth's energy balance?

Section 2

3. *Places and Regions* On a global scale, where is precipitation most common?
4. *Physical Systems* What effects can hurricanes have on local environments?

Section 3

5. *Places and Regions* Where will you find climates that are generally warm and wet all year? What creates dry weather conditions in arid regions?

Thinking Critically for TAKS TEKS

1. **Identifying Cause and Effect** Why is precipitation heavier in the persistent low-pressure zones of the world?
2. **Drawing Inferences and Conclusions** What climate factors do you think keep trees from growing in tundra regions?
3. **Analyzing Information** Around what lines of latitude are most arid regions centered? Why?

Using the Geographer's Tools TEKS

1. **Analyzing Diagrams** Study the diagram of Earth's pressure and wind systems. Then identify Earth's main prevailing wind patterns.
2. **Summarizing** Review the map and chart of Earth's climate regions. Create a word web for each climate type. Use the webs to describe the distribution and precipitation, temperature, and wind patterns of each climate. Then describe the factors that influence climate.
3. **Preparing Diagrams** Create two diagrams to show how the seasonal change in the way the Sun's rays strike the areas north and south of the equator helps create the tropical wet and dry climate. Draw one diagram to show how the climate is affected in the summer. The other diagram should show how the climate is affected in the winter.

Writing for TAKS TEKS

Conduct research on the latest scientific studies of global warming. You may want to concentrate on evidence from the polar regions. Pay particular attention to differences of opinion. Write a short newspaper article about your findings. When you are finished with your article, proofread it to make sure you have used standard grammar, spelling, sentence structure, and punctuation.

SKILL BUILDING

Geography for Life TEKS

Using Research Skills

Physical Systems Conduct research on three desert regions on different continents or at different latitudes. Use the climate and precipitation maps in this textbook's unit atlases to help you locate the deserts. Then write a report comparing and contrasting the three areas. Discuss the factors that create each arid climate and the range of temperatures and rainfall in each location. In addition, describe the plants and animals that live in each region.

Building Social Studies Skills

go.hrw.com TAKS PREP ONLINE keyword: SW3 T3

Interpreting Maps

Study the precipitation map below. Then use the information from the map to help you answer the questions that follow. Mark your answers on a separate sheet of paper.

Precipitation Map of New Zealand

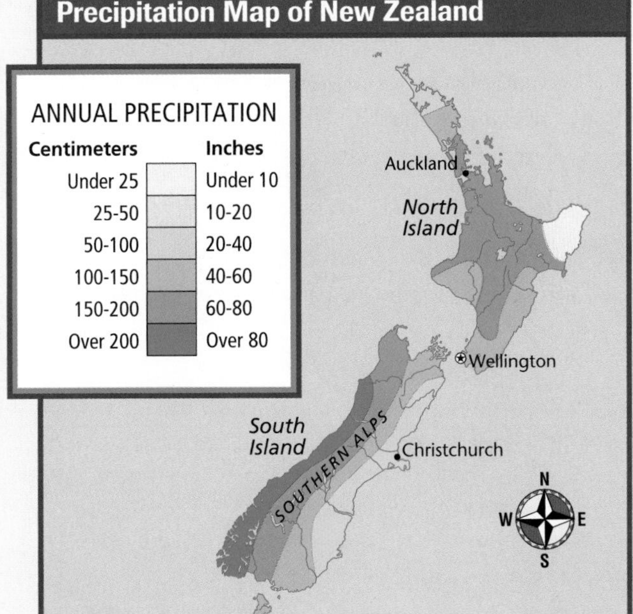

1. Of the following amounts of precipitation, which is the amount that falls in Wellington, according to the precipitation map?
   a. 19 cm
   b. 27 cm
   c. 88 cm
   d. 120 cm
2. Describe the difference in precipitation levels of the western and eastern coasts of New Zealand's South Island.

Using Language

The following passage contains mistakes in grammar, punctuation, or usage. Read the passage and then answer the following questions on a separate sheet of paper.

"[1] A humid tropical climate is warm all year. [2] A humid tropical climate is rainy all year. [3] People living in this climate do not see a change from summer to winter. [4] The heat in the tropics cause almost daily rainstorms."

3. Which word group contains an error in subject-verb agreement?
   a. 1
   b. 2
   c. 3
   d. 4
4. Write a sentence that effectively combines word groups 1 and 2.

Alternative Assessment

PORTFOLIO ACTIVITY

Learning about Your Local Geography TEKS

Group Project: Field Work

Within a group, use a barometer, rain gauge, and thermometer to keep a daily record of air pressure, rainfall, and temperature at your school. Track these values over several weeks. If possible, note how these factors change with the seasons. Then conduct research on weather records for your area and compare your observations to the official records. Write several generalizations about changes or lack of change in weather patterns.

internet connect

go.hrw.com

Internet Activity: go.hrw.com
KEYWORD: SW3 GT3 TEKS

Choose a topic on weather and climate to:

- use satellite maps to learn about weather conditions, fronts, and surface conditions around the world.
- create a poster to display research on hurricanes and the technology of hurricane tracking.
- investigate global warming.

CHAPTER 4 Landforms, Water, and Natural Resources

Earth has many different types of landforms. They are shaped by forces deep within the planet and by conditions on Earth's surface. In this chapter, you will learn about Earth's landforms, water features, and natural resources.

Eroded sandstone hills, Arizona

Turtle with creole fish, Pacific Ocean

Veins of gold and gold flakes

Landforms

READ TO DISCOVER

1. What physical processes inside Earth build up the land?
2. What physical processes on Earth's surface wear down the land?
3. How do these physical processes interact to create landforms?

DEFINE

core
mantle
magma
plate tectonics
continental drift
rift valleys
abyssal plains
continental shelves
trench
folds
faults
weathering
sediment
erosion
glaciers
plateau
alluvial fan
delta

WHY IT MATTERS

Earthquakes are among the natural forces that shape the land. They have changed landscapes in the United States in the distant past and in recent years. Use CNNfyi.com or other **current events** sources to learn about earthquakes and how they have affected different parts of the world.

internet connect

GO TO: go.hrw.com
KEYWORD: SW3 CH4
FOR: Web sites about landforms, water, and natural resources

Forces below Earth's Surface

Geology—the study of Earth's physical structures and the processes that have created them—is important to geographers. We live and build on hills, mountains, valleys, and other landforms. They can make travel easy or difficult. Landforms can also give us clues to what Earth was like in the past. Forces below Earth's surface are a key to the shaping of landforms.

Scientists have identified four important zones in Earth's interior, as you can see in the diagram. The planet's center is like a nuclear furnace, where decaying radioactive elements generate heat. At Earth's center, or **core**, both temperatures and pressures are very high. The core is divided into inner and outer layers. The inner core is solid. The outer core is mostly dense liquid metal, mainly iron and nickel. Beyond the core is the **mantle**—the zone that has most of Earth's mass. The uppermost layer is the crust. Although it is up to about 25 miles (40 km) thick, the crust is comparatively thin. Huge currents carry heat from the core through the mantle to the crust. Liquid rock within Earth is called **magma**. When this liquid rock spills out onto the surface it is called lava. Magma erupts from vents called volcanoes.

Internal Forces The theory of **plate tectonics** explains how forces within the planet create landforms. This theory views Earth's crust as divided into more than a dozen rigid, slow-moving plates. The plates can be compared to the cracked shell of a hard-boiled egg. Some plates are as large as a quarter of the planet, but others are only a few hundred miles across. The plates slowly move across the upper mantle, usually less than an inch per year. This process is called **continental drift**. Along the plate boundaries, the crust is subject to stresses that lead to melting, bending, and breaking. In the middle of

The Interior of Earth

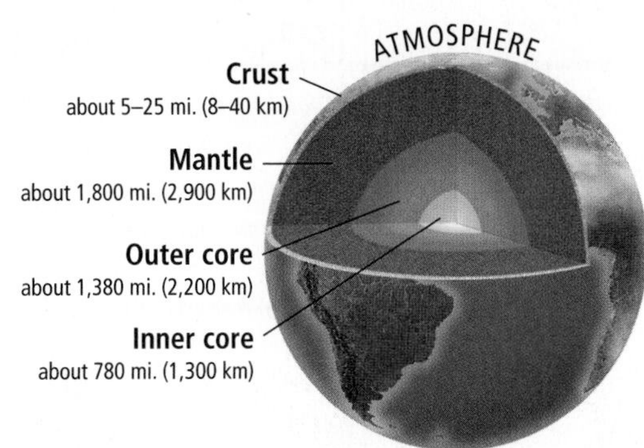

INTERPRETING THE DIAGRAM *Temperatures within Earth increase with depth. In the inner core, temperatures may reach as high as 12,000°F (7,000°C).* ***How far below the surface is Earth's inner core?***

Plate Tectonics

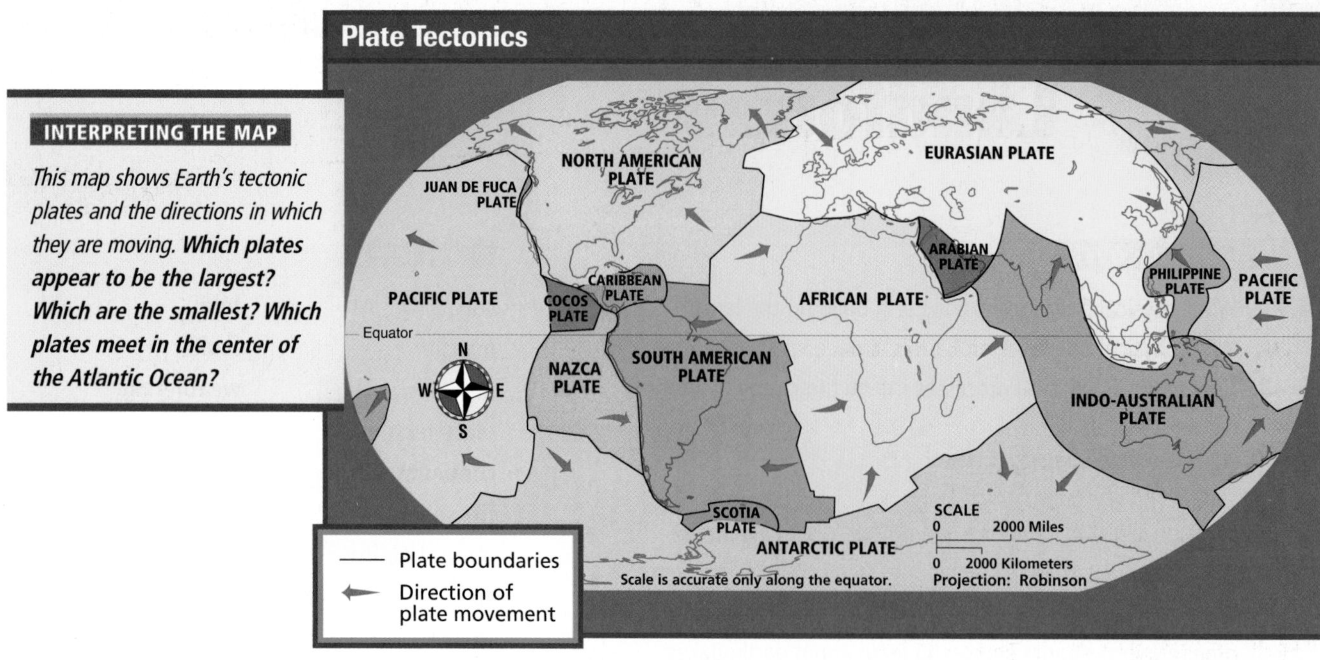

INTERPRETING THE MAP

This map shows Earth's tectonic plates and the directions in which they are moving. ***Which plates appear to be the largest? Which are the smallest? Which plates meet in the center of the Atlantic Ocean?***

plates, however, little tectonic activity takes place. Continental areas in the middle of plates are steadily eroded. Ocean floor areas in the middle of plates are steadily buried with sediment.

Volcanoes often form long rows and signal that a plate boundary is nearby. Earthquakes—sudden shakings of Earth's crust—take place when tectonic forces cause masses of rock inside the crust to break. Earthquakes are also common near plate boundaries.

Scientists use the theory of plate tectonics to explain the long history of Earth's surface. They believe that about 200 million years ago all the modern continents were part of one supercontinent called Pangaea (pan-JEE-uh). Pangaea then broke into two smaller supercontinents called Gondwana and Laurasia. These two, in turn, broke into the modern continents during the last 100 million years. The theory of plate tectonics helps explain the "fit" between the coastlines of Africa and South America. Rock formations that match up across the boundaries provide more evidence. The theory also helps geographers understand the origins of mountains and the landforms of the ocean floors.

Continental Drift

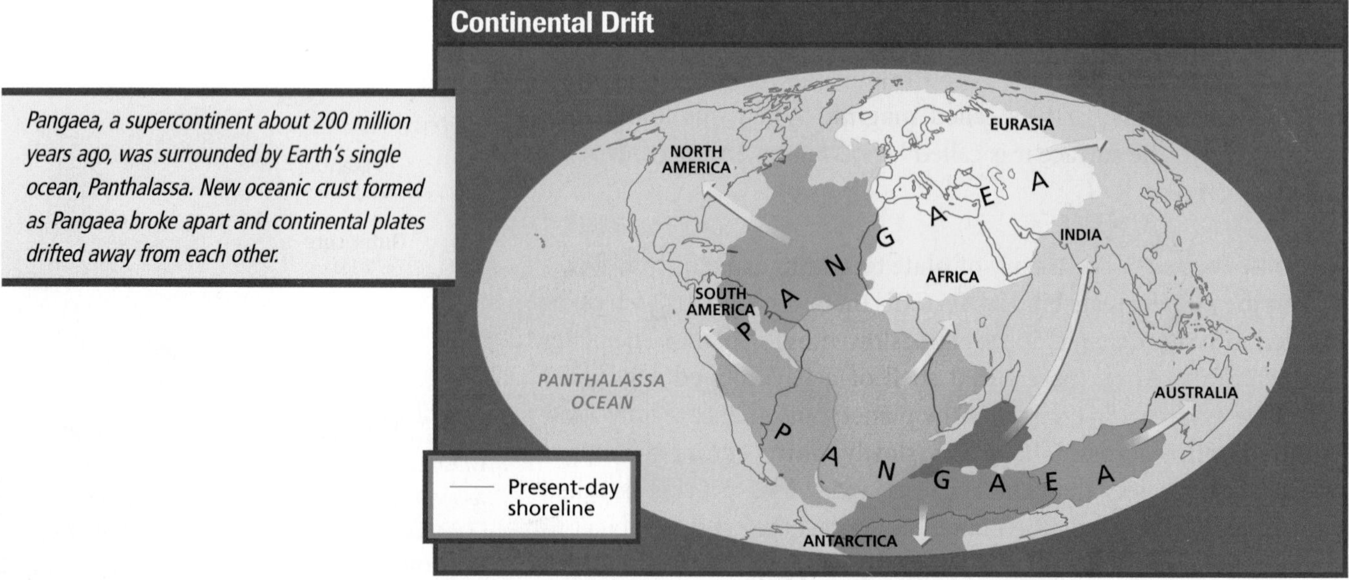

Pangaea, a supercontinent about 200 million years ago, was surrounded by Earth's single ocean, Panthalassa. New oceanic crust formed as Pangaea broke apart and continental plates drifted away from each other.

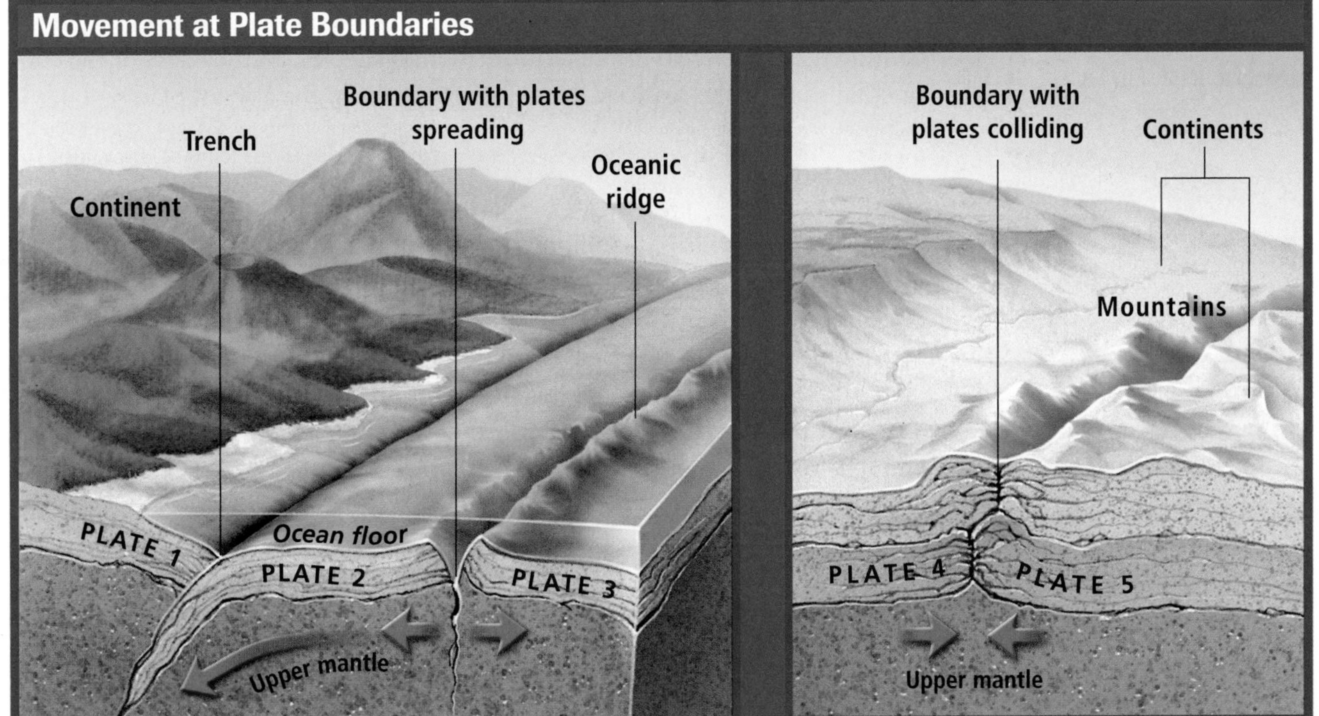

Left: Where Plate 2 pushes under Plate 1, a deep trench forms. Where Plate 2 and Plate 3 move apart, lava creates an oceanic ridge. Right: Where Plate 4 and Plate 5 collide, the crust is pushed up, forming a mountain range.

Plate Movement Three types of movements at plate boundaries are possible. First, the plates can move apart, or spread. Second, the plates can collide. Third, the plates can move laterally, slipping past each other. The movement of plates creates distinctive landforms.

Long ago in Earth's history the crust sorted itself into two layers of different kinds of rocks. The lower layer, made of heavier rock, is found on the ocean floors. Lying on top is a patchy layer of lighter rock. This layer makes up the continents. Nearly all spreading plate boundaries are found on the ocean floors. As fresh lava wells upward, this new crust pushes the plates apart. The rising heat also lifts the crust upward, building a chain of young volcanic mountains, called an oceanic ridge. Earth's oceanic ridges connect in a nearly continuous submarine mountain chain. This chain is nearly 40,000 miles (60,000 km) long and contains some of the world's highest peaks. The part of this chain that runs through the Atlantic Ocean is called the Mid-Atlantic Ridge. However, except for on a few islands, such as Iceland, these mountains are hidden beneath the waves.

A few spreading plate boundaries lie under continents. In these places, the crust stretches until it breaks, forming **rift valleys**. The biggest rift valleys are in eastern Africa.

Away from the oceanic ridges, rocks of the ocean floors gradually sink because they have no supporting heat below them. Here we find **abyssal plains**, the world's flattest and smoothest regions. In these areas a thick layer of mud, clay, and other materials has buried all features.

This map shows some of the landforms on the oceans' floors. Note how the Mid-Atlantic Ridge runs the length of the ocean and continues into the Indian Ocean.

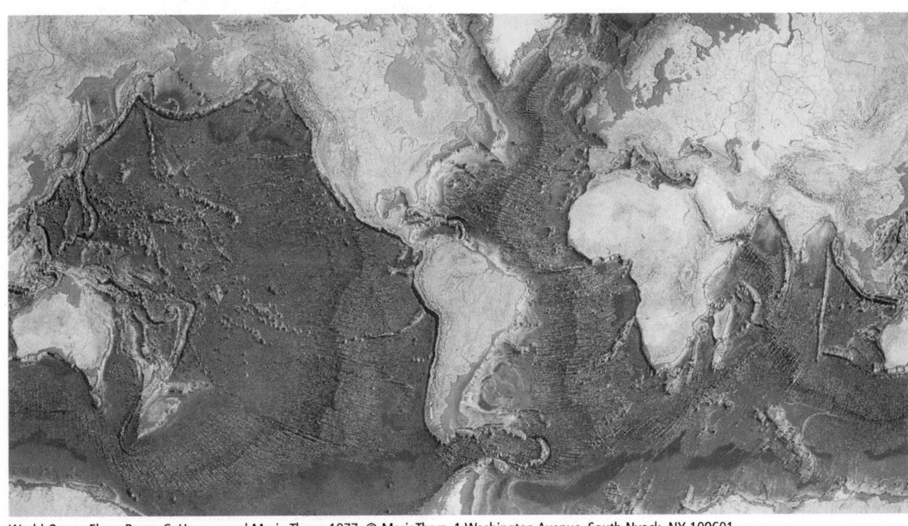

World Ocean Floor, Bruce C. Heezen and Marie Tharp, 1977, © MarieTharp, 1 Washington Avenue, South Nyack, NY 109601

Folds and Faults

Mountains formed by faults

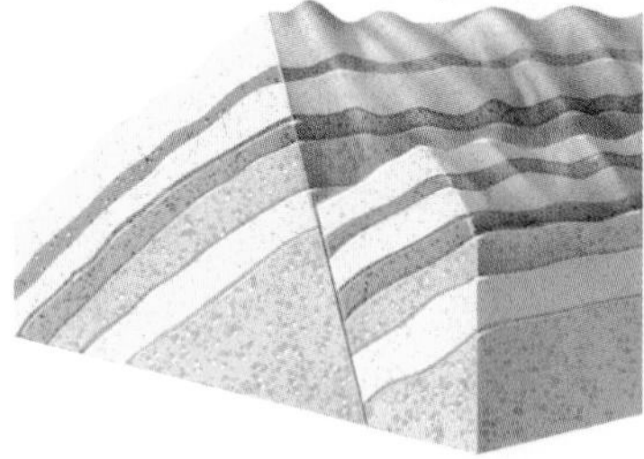

Mountains formed by folds

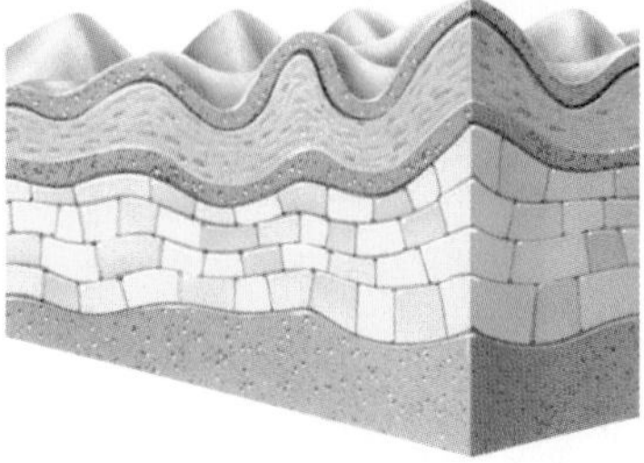

Top: Fault-block mountains form where parts of Earth's crust are pushed up along a fracture line, or fault. Bottom: Many large mountain ranges are created by folding, which occurs when two plates collide and their edges wrinkle.

The continental surface extends under the shallow ocean water around the continents. These areas are called **continental shelves**. At the edges of the continental shelves, the seafloor drops steeply down to the abyssal plains.

 **READING CHECK:** ***Physical Systems*** What physical process forms oceanic ridges?

When Plates Collide Colliding plate boundaries are found both on ocean floors and along continental edges. When two plates on the ocean floor collide, one slides underneath the other. This plate boundary is called a subduction zone, and the deep valley marking the plate collision is called a **trench**. The plate sliding downward generates heat as it grinds against the plate above it. This heat may produce a row of volcanoes. Some of these volcanoes rise high enough to become island chains. This process created places like the Tonga Trench and the nearby Tonga Islands in the western Pacific.

Sometimes one of the colliding plates is carrying a continent. In this case the heavier oceanic plate dives under the lighter continental plate. The squeezing of the continental plate causes volcanoes, **folds**, and **faults**. Folds are places where rocks have been compressed into bends. Faults are places where rock masses have broken apart and moved away from each other. (See the diagrams.)

It is also possible for two continental plates to collide. The result is a great long-lasting episode of mountain building with awesome folding, faulting, and volcanism. The Himalayas of Asia formed along such a plate boundary. There, scientists believe, a small continent marked by the triangle of modern India began crashing into Asia millions of years ago. The Himalayas, the mountains formed by this collision, are still growing. In fact, plate collisions have built most of the world's mountains.

Plates Moving Laterally When plates move laterally past each other, long fractures develop along the edges of both plates. The pressures along these boundaries are seldom uniform. While a little squeezing produces low mountains, a little spreading generates broad valleys. Earthquakes can be frequent in

Left: The San Andreas Fault cuts through the Carrizo Plain in California. Right: This freeway in Oakland, California, collapsed during a 1989 earthquake.

these areas. Again, most of these plate boundaries lie under the oceans. However, the San Andreas Fault in California is a famous example on land. Here a sliver of continental crust on the Pacific plate's edge is moving northwestward along the edge of North America. Meanwhile, the North American plate is being pushed westward by the oceanic ridge in the Atlantic Ocean. As this plate boundary occasionally skids along, earthquakes occur from Los Angeles to north of San Francisco.

READING CHECK: *Physical Systems* What physical processes created the Himalayas and the San Andreas Fault in California?

Forces on Earth's Surface

While tectonic forces are building up Earth's surface, other forces are wearing it down and making it more level. It is the interaction of these two kinds of forces that shapes the landforms we see.

Weathering and Erosion Rocks break and decay over time in a process called **weathering**. Weathering is usually slow and difficult to detect. However, even the hardest rocks will eventually wear down. Chemical processes cause some weathering. Substances in air and water react with the rock, creating new chemical compounds and slowly dissolving the rock. Weathering is also caused by physical processes that break rocks into smaller pieces. In desert areas, daytime heating and nighttime cooling can cause rocks to crack. In high mountains, repeated freezing and thawing of water inside a cracked rock can cause it to break even more. The roots of trees can pry rocks apart. Weathering breaks rock into smaller particles of gravel, sand, and mud called **sediment**.

Along with weathering, the other process changing landforms on Earth's surface is **erosion**. Erosion is the movement of surface material from one location to another. Water, wind, and ice cause erosion.

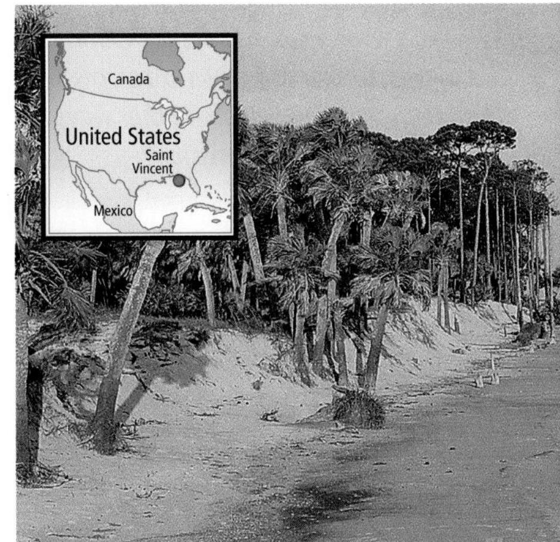

Water, Waves, and Wind Water is the most important force of erosion. Rainfall can cause rapid erosion where few plants protect the ground. Water erosion can begin as tiny channels on hillsides. If erosion is severe, a channel may grow into a gully. Running water can even carve deep canyons, such as the Grand Canyon. Rivers usually carry water and sediment from mountains in the center of continents all the way to the ocean. Wave action is another powerful force of erosion. During a storm, waves can tear away tons of beach sand in a few hours. Waves can also change shorelines slowly over many years.

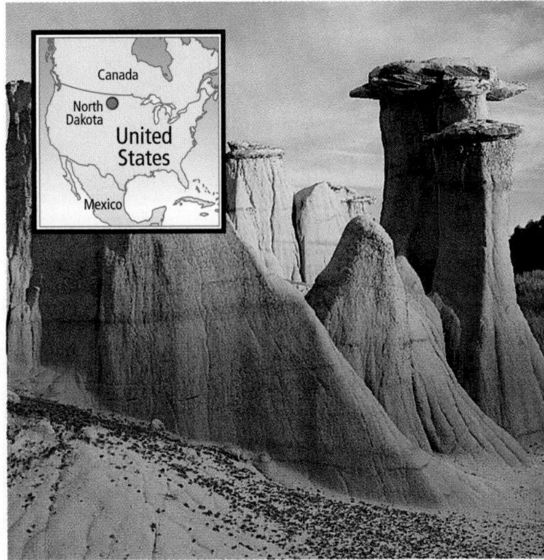

INTERPRETING THE VISUAL RECORD

Erosion wears away Earth's surface at an island beach off the Florida coast (top) and at Theodore Roosevelt National Park, North Dakota (bottom). ***What physical processes are causing erosion in these places?*** TEKS

Wind is another force that causes erosion. Plants protect most land surfaces from wind action. However, in dry lands, on beaches, and in places where people or animals have destroyed the vegetation, wind can cause significant erosion. Wind works in two ways. One is abrasion—that is, it blasts particles of sand against rock. The other way is by blowing sand and dust from one place to the next. Hills of wind-deposited sand are called dunes. Sand dunes are common on beaches and in deserts. Wind also lifts dust high into the atmosphere and transports it great distances. For example, dust from the Sahara in Africa is carried across the Atlantic Ocean to the Caribbean islands!

READING CHECK: *Physical Systems* How do wind and water shape the land?

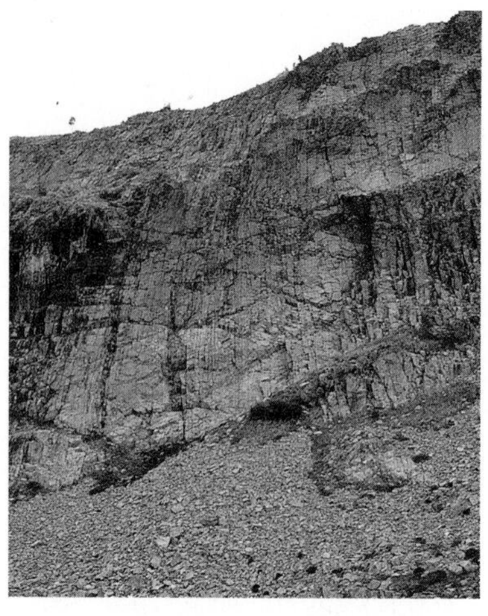

INTERPRETING THE VISUAL RECORD

Weathering is slowly wearing down this mountainside. Rock that has been shattered by frost and falls from a mountain forms what is called a talus slope. Gravity slowly pulls the rock down in a process called talus creep. ***In what other ways does gravity affect erosion?*** TEKS

The Power of Ice Thick masses of ice—called **glaciers**—also erode rock and move sediment. Glacier ice builds up when winter snows do not melt during the following summer. Glaciers may be great ice sheets covering whole regions. Today ice sheets more than two miles (three kilometers) thick cover most of Antarctica and Greenland. As these giant domes of ice sag with the pull of gravity, the edges of the ice sheets are pushed outward.

Glaciers are also found in the valleys of high mountains. These mountain glaciers flow slowly downhill, often sliding on a thin layer of liquid water at their bases. Glaciers can level anything in their paths. They can move rocks as big as houses over long distances. They can grind rocks into sediment as fine as flour. During the ice ages, glaciers reached to what are now St. Louis in North America and Moscow in Russia.

Glaciers can be found in high mountains all around the world, even on high mountains near the equator. Flowing downhill like slow rivers of ice, mountain glaciers carve great U-shaped valleys and sharp mountain peaks. The carving action of glaciers created many of the spectacular valleys we see around the world today. Yosemite Valley in California is a good example.

TEKS **READING CHECK:** ***Physical Systems*** What role do gravity and ice sheets play in shaping the land?

Shapes on the Land

One way to better understand landforms is to divide them into three groups. One group is built by tectonic processes. These landforms, such as mountains and some valleys, are created by volcanoes, folding, and faulting. Erosion and the depositing of sediment may be changing these landforms, rounding and smoothing their edges. Yet the basic landform is the result of tectonic processes.

A second type of landform is created by erosion. It is made of rock and has a thin layer of sediment or soil on the surface. The forces of erosion are slowly

Cyclists in Alaska descend to Black Rapids Glacier. In 1936 this glacier surged forward at the rate of 10 feet (3 m) per hour. It has slowed since then. The inset photo shows the face of a glacier. Here chunks may break off, or calve, and form icebergs.

lowering its surface. These landforms often reflect the hardness of the underlying rock. Harder rocks resist erosion and over time will stand above the surrounding land. One example of this landform is a **plateau**. A plateau is an elevated flatland that rises sharply above nearby land on at least one side. A plain—a nearly flat area—is often the final stage of a landscape wearing smooth.

A third kind of landform is formed by sediment deposited by ice, water, or wind. A sand dune in a desert is an example of this kind of landform. Another example is a floodplain. A floodplain is a landform of level ground built by sediment deposited by a river or stream.

The terrain in most regions is a jigsaw puzzle of many landforms. For example, a mountain range is formed by tectonic activity. Erosion may then form deep valleys between the mountains. The sediment eroded from the mountains may then be deposited at the mountains' bases. The result of this process can be an **alluvial fan**. This is a fan-shaped deposit of mud and gravel often found along the bases of mountains. Still later a stream may erode the sediments in the alluvial fan, carrying them all the way to a river mouth. There the sediment may move out into the ocean and sink, or the sediment may accumulate, building a **delta**. Eventually this sediment from the distant mountains could travel still farther. It might finally be deposited in an oceanic trench.

The location, shape, and size of landforms have influenced human settlement and transportation throughout history. For example, people tend to settle in flat areas where they can farm. People use rivers for water supplies and transportation. Many railroads and highways have been built along river valleys as well. People have also changed Earth's surface to suit their needs. Large machines can smooth ground for home construction, and explosives clear the way for roads. Governments build dams across rivers, turning valleys into lakes.

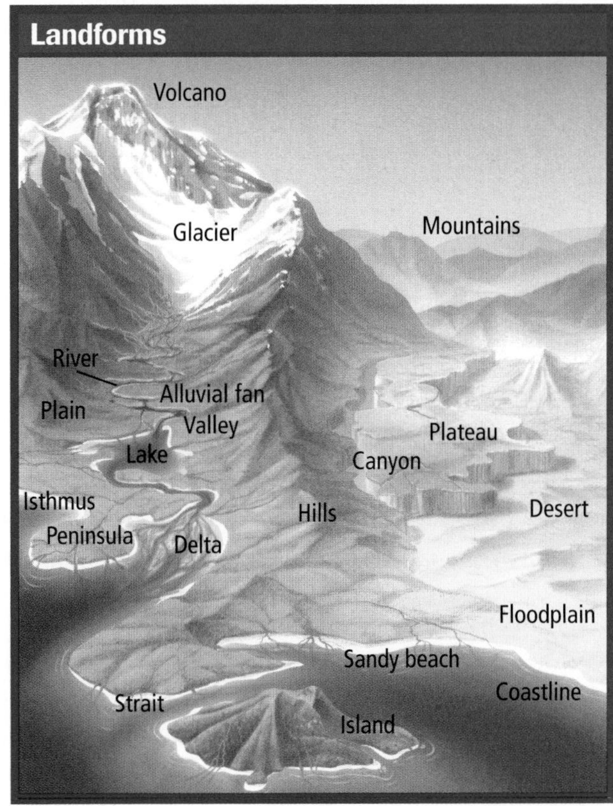

INTERPRETING THE VISUAL RECORD

Many of the landforms shown here can be found together in various regions of Earth. ***Which landforms can you identify where you live?***

READING CHECK: *Physical Systems* What are two kinds of landforms created by deposits of sediment?

TEKS Questions 1, 2, 3, 4, 5

Section 1 Review

Homework Practice Online
Keyword: SW3 HP4

Define
core, mantle, magma, plate tectonics, continental drift, rift valleys, abyssal plains, continental shelves, trench, folds, faults, weathering, sediment, erosion, glaciers, plateau, alluvial fan, delta

Reading for the Main Idea

1. *Physical Systems* What are some processes that shape landforms?
2. *Physical Systems* What are the three kinds of landforms?

Critical Thinking

3. **Analyzing Information** How might a plateau show the effects of both tectonic process and erosion?
4. **Drawing Inferences and Conclusions** Why do you think farming and raising livestock may lead to rapid erosion?

Organizing What You Know

5. Copy the table below. Use it to list the three types of tectonic plate boundaries, the landforms that result from each type, and an example of each type.

Type of plate boundary	Resulting landforms	Example

2 The Hydrosphere

READ TO DISCOVER

1. In what forms and where do we find water on Earth?
2. What are the causes and effects of floods?

DEFINE

desalinization
hydrologic cycle
headwaters
tributary
watershed
drainage basin
estuaries
wetlands
groundwater
water table

WHY IT MATTERS

Rainfall in any given area varies from year to year. Use CNNfyi.com or other current events sources to learn about record high and low rainfall levels in your area.

Water on Earth

Necessary for life, water is abundant on Earth. Yet 97 percent of the world's water is in oceans and is too salty for most uses. The salt in ocean water can be removed through **desalinization**. For example, some countries in the very dry region of Southwest Asia use this process to get freshwater. However, it is expensive because of the large amount of energy needed.

Less than 3 percent of the world's water is fresh. Most of this water is frozen in the ice caps of Antarctica and Greenland. The remainder is less than 1 percent of the world's water. This water is found in clouds, lakes, and rivers and in the ground. It is only this tiny proportion of the world's water that is available for human use.

The Hydrologic Cycle The amount of water on Earth stays much the same over time. However, its physical state is always changing, from gas (water vapor)

The Hydrologic Cycle

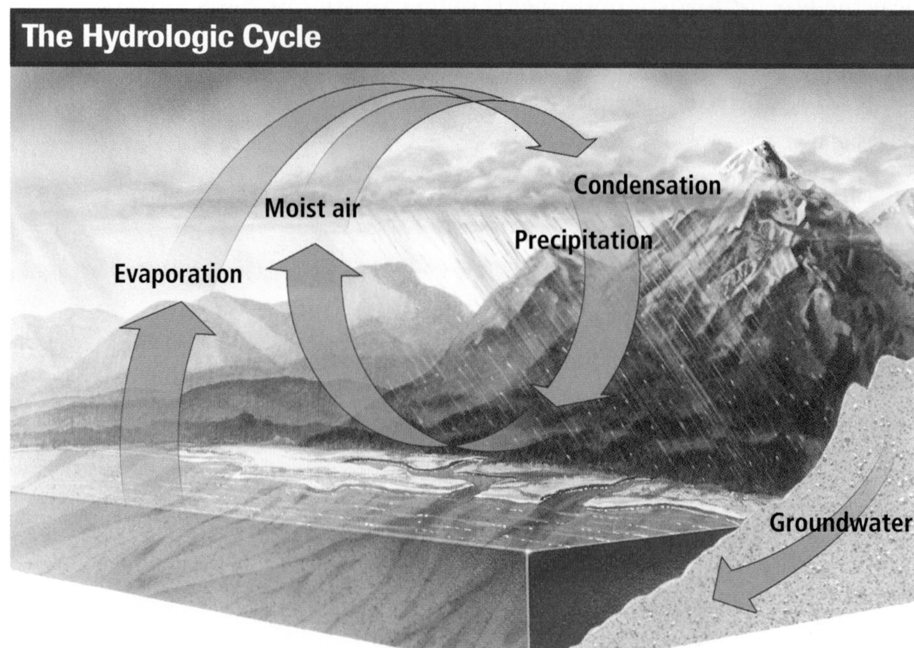

The circulation of water from one part of the hydrosphere to another is driven by solar energy. Following evaporation, water eventually returns to the surface in various forms of precipitation.

to liquid to solid (ice). The movement of water through the hydrosphere is called the **hydrologic cycle**. Solar energy, winds, and gravity drive this cycle. When heated by solar energy, water may change to vapor. This process is called evaporation. Most water in the atmosphere has evaporated from the ocean. As water vapor rises with the air, it cools and forms tiny liquid droplets in a process called condensation. Joining to form clouds, these droplets may grow to become raindrops heavy enough to fall to Earth. Water falling to the surface, either as rain, snow, or hail, is called precipitation. If it falls on land, water may be stored. It can build up in plants, in a river or stream, in a lake, or below the ground. Surface water either flows to the sea in a river or evaporates again and returns to the atmosphere.

READING CHECK: ***Physical Systems***
What forces drive the hydrologic cycle?

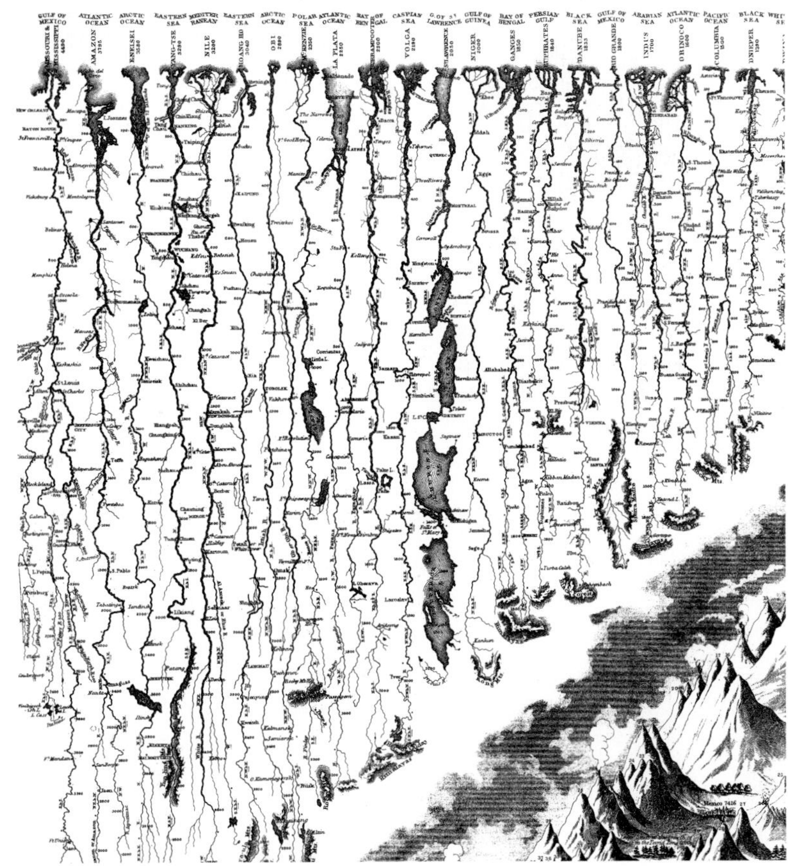

INTERPRETING THE VISUAL RECORD *This diagram showing the headwaters, tributaries, lakes, and deltas of major rivers was published in 1864. The rivers appear side by side.* ***Use a map of North America to identify the lake and river system in the diagram's center. This river is the heart of what functional region?*** TEKS

Surface Water As precipitation falls on continents and islands, it flows down hills and mountains toward the lowlands and coasts. The first and smallest streams from this runoff are called **headwaters**. As these headwaters join, they form larger streams, and farther downstream they eventually form rivers. Any smaller stream or river that flows into a larger stream or river is called a **tributary**. In the central United States the Arkansas, Missouri, and Ohio Rivers are all major tributaries of the Mississippi River. The whole region drained by a river and its tributaries is called a **watershed** or **drainage basin**. Thus, the Mississippi River watershed has hundreds of tributaries and a watershed of more than 1.2 million square miles.

On continents, surface water may collect in lakes. Many lakes, including the Great Lakes along the U.S.-Canada border, were carved by glaciers during the ice ages. Other lakes, like the Dead Sea in Israel and Jordan, lie in valleys created by continental rifts. Some lakes lie in volcanic craters, like Crater Lake in Oregon. Lakes provide water supplies, fish, and recreation opportunities. Because water heats and cools more slowly than land, large lakes can also reduce the severity of the climates on their shores. This moderating effect allows palm trees to grow on the shore of Lake Geneva in Switzerland. In fact, this effect is common on almost any lakeshore.

Some lakes are not connected to an ocean. These lakes, like the Great Salt Lake in Utah, may collect minerals from runoff. As water continually flows in and evaporates, more and more minerals build up in the lake water. Thus, the lake becomes more and more salty.

Surface water is also found in **estuaries**. An estuary forms where a river meets an inlet, or small arm, of the sea. In this semi-enclosed body of coastal water, seawater and freshwater mix. Chesapeake Bay on the Atlantic coast of the United States is a good example. Many estuaries provide good harbors, and most of the largest port cities in the United States lie on estuaries. Examples include New York, New Orleans, and Seattle. Estuaries are rich in shellfish and fish, so they must be carefully protected from pollution. Their usefulness as harbors often conflicts with their importance as shelters for animals.

Surface water is also found in **wetlands**. A wetland is any landscape that is covered with water for at least part of the year. They include bogs, coastal marshes, river bottomlands, and wooded swamps. The Everglades in Florida and the Okavango inland delta in southern Africa are two of the world's most famous wetlands. Wetlands support large populations of fish, shellfish, birds, and native plants. Many migrating birds depend on wetlands for food, water, and rest during their long journeys. Unfortunately, people have drained, paved, or filled many wetlands and used the land for farms, houses, and industrial sites. More than half the original wetlands in the United States have been destroyed. Preserving the remaining wetlands has become an important environmental issue.

The Everglades is still one of the world's great wetlands, although it is now about one half of its original size. The marshland once covered almost 9 million acres (3.6 million hectares).

Groundwater Water found below ground is called **groundwater**. When rainwater sinks into the ground, it is first stored in the soil. Plant roots reach down through the soil to take up this water. Water in the soil slowly moves downslope and deeper underground. The cracks and spaces in the rock immediately below the soil contain both air and water. However, farther down all the spaces inside the rock are filled with water. The level at which all the spaces are filled with water is called the **water table**. The level of the water table rises and falls with rainy and dry seasons.

In some areas, groundwater is being used up as the water is pumped out for both farms and cities. As a result, water tables may drop. In addition, the ground may settle or slump where it is no longer supported by water below. This process can damage buildings, roads, sewer lines, and other structures. In some places, big cracks open in Earth's surface. For example, these cracks have caused considerable damage in southern Arizona.

✓ **READING CHECK:** ***Physical Systems*** What are some common sources of water on and near land?

INTERPRETING THE VISUAL RECORD

The Okavango River empties into an area that receives little rainfall, forming a large swamp in what is otherwise a desert. The swamp is rich in vegetation, including reeds and water lilies. Examples of the animals that live there are storks, cranes, ducks, hippopotamuses, crocodiles, and lions. ***What kinds of plants and animals might live there if the Okavango River flowed to the sea?*** TEKS

INTERPRETING THE VISUAL RECORD

Floods can devastate human settlements, but they can also make human settlement possible. This satellite photo of the Nile River delta in northern Egypt shows how the river turns the desert into fertile farmland. For thousands of years, the Nile flooded along its length every year, bringing rich mud to the Egyptian fields. A dam now controls the Nile's waters. ***How do you think the Nile's flooding has affected the size of Egypt's population?*** TEKS

Floods

Floods occur when rivers carry more water than the stream channels can hold. They usually result from heavy rains or sudden snow melts. Floods erode land and destroy vegetation. They can also drown people and livestock. People who do not drown may die from starvation after the flood has destroyed their crops. Others may die from diseases caused by polluted drinking water. Flooding is natural, but human activity can make floods worse. For example, people increase surface runoff when they clear vegetation from land and cover the ground with pavement and buildings. Flooding increases if rainwater cannot sink quickly into the soil.

The damage flooding causes is often increased because, for many reasons, people choose to live next to rivers. Floodplains are fertile farmland. Cities and industries are typically located along rivers because of the easy water transport, waterpower, and abundant water supply. For thousands of years, governments have been trying to control floods with dams and artificial channels. These measures work to some degree, but flood disasters still occur.

READING CHECK: *Environment and Society* How do floods affect the environment and people?

TEKS Questions 2, 3, 4

Section 2 Review

Homework Practice Online
Keyword: SW3 HP4

Define desalinization, hydrologic cycle, headwaters, tributary, watershed, drainage basin, estuaries, wetlands, groundwater, water table

Reading for the Main Idea

1. *Physical Systems* How much of the water on Earth is liquid freshwater?
2. *Physical Systems* How and why do humans destroy wetlands?

Critical Thinking

3. **Evaluating** What might be some advantages and disadvantages of policies that limit economic development and water use in and around estuaries and wetlands?
4. **Drawing Inferences and Conclusions** How does the presence of buildings and roads make flooding worse?

Organizing What You Know

5. Create and label a pie graph showing the distribution of water on Earth.

Section 3 Natural Resources

READ TO DISCOVER

1. Why are soil and forests important resources?
2. What are the concerns about water quality and air quality?
3. What are some of the ways minerals are used?
4. What are the main energy resources, and how are they used?

WHY IT MATTERS

Preventing erosion of topsoil is a major concern in many countries, including the United States. Use CNNfyi.com or other **current events** sources to learn about soil conservation programs.

DEFINE

humus
leaching
contour plowing
soil exhaustion
crop rotation
irrigation
soil salinization
deforestation
reforestation
acid rain
aqueducts
aquifers
fossil water
ore
fossil fuels
petrochemicals
hydroelectric power
geothermal energy

Soil and Forests

Landforms and water are among Earth's most visible features. They are also essential natural resources. A resource is any physical material that makes up part of Earth and that people need and value. Some of Earth's natural resources are renewable, while others are nonrenewable. Renewable resources, such as soil and forests, are those that natural processes continuously replace. Nonrenewable resources are those that cannot be replaced naturally after they have been used.

Soil Building Soil is natural material that includes both rocky sediments and organic matter. The sediments can come from the rocks below the ground surface. They may also be transported to the site by streams and the wind. The organic material can be decayed plant and animal matter. Soil differs from place to place, and there are hundreds of soil types.

Thick soil makes the state of Mississippi a fertile farming area. The layer of soil in the picture was probably picked up elsewhere by the wind and deposited there.

Near the surface, bacteria, insects, and worms break down the plant and animal material into a mixture called **humus**. These creatures—along with larger animals like rodents—open up spaces in the soil. The spaces allow air and water into the soil. Weathered rock material, organic matter, gases, and water must all be present in the soil to support plant life.

One important factor in the composition of soil is the rock the soil particles come from—the parent rock. Minerals in the soil vary depending on the type of parent rock. For example, eroded sandstone produces a different type of soil than does eroded granite or limestone. Climate is another major factor in soil variation. Moisture, sunlight, temperature, wind, and plants and animals in the soil all influence rock weathering. Soils vary between landforms as well. For example, sloping ground, like the side of a ridge, tends to erode steadily. As a result, the soil there is usually shallow. In contrast, sediments generally build up on valley floors, creating deep soils.

Soil develops very slowly, taking a few or even hundreds of years. It forms distinct layers, called soil horizons. (See the diagram.) Most soils have three layers, called the A, B, and C horizons. The upper layer, the A horizon, is also called the topsoil. This horizon is rich in humus and has active populations of plants and small animals. This horizon also has the most plant roots. The B horizon, sometimes called the subsoil, is the middle layer. This horizon is mostly minerals, but the roots of large plants penetrate to these depths. The C horizon is composed of weathered rock fragments of the soil's parent material. The depths of all these horizons can vary greatly from place to place. Taken together, the soil horizons at any place are called the soil profile.

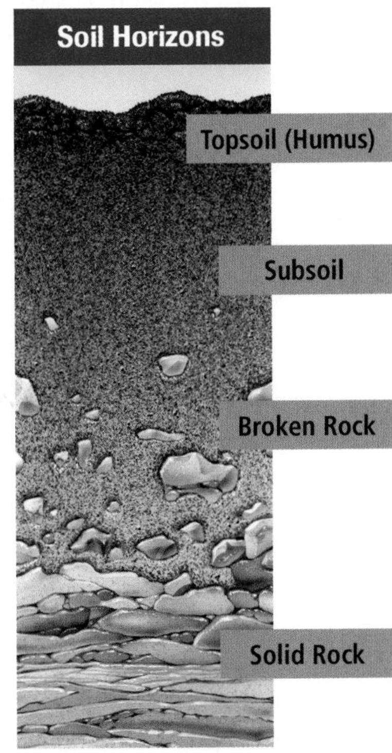

INTERPRETING THE DIAGRAM *The three layers of soil are the topsoil, subsoil, and broken rock.* ***What physical processes may have broken the rock beneath the subsoil?*** TEKS

Particularly important to soil profile development is the amount of rainwater moving through it. **Leaching** eventually moves the nutrients out of the reach of plants' roots. Leaching is the downward movement of minerals and humus in soils. Areas with abundant rain often have deep soils where many minerals have been leached downward. The result can be soil with low fertility even though the area's vegetation may be a rain forest. On the other hand, soils in dry areas are often shallow and leaching is limited. Soils here may contain abundant supplies of the minerals needed by plants.

Around the world, soil profiles vary according to climate and major vegetation types. The soils beneath semiarid grasslands in central Kansas are much the same as soils beneath semiarid grasslands in Ukraine. Similarly, the soils in the rainy tropics of South America resemble the soils in the rainy tropics of Africa. However, soil profiles also vary based on local environmental conditions. For example, soils next to swamps have abundant moisture and humus, no matter what the climate is like.

READING CHECK: *Physical Systems* Why do some dry areas have very fertile soil?

Sustaining Soil Resources Producing the world's food depends on soil. For this reason, preserving soil resources is important to all of us.

Soil erosion is a natural process. However, farming usually leads to increased erosion, removing soil faster than new soil can be formed. Maintaining A horizons is particularly important because this is where most plant nutrients are found. Farmers can control erosion by plowing less and by using **contour plowing**. Contour plowing works across the hill, rather than up and down the hill. This slows the movement of water down the slope, reducing erosion.

INTERPRETING THE VISUAL RECORD

Farmers are trying to preserve the soil of this field in southeastern Washington State by using contour plowing. ***What role might the angle of a field's slope play in contour plowing?*** TEKS

Crops draw certain nutrients out of the soil. If the same crop is planted year after year, a field may suffer nutrient loss. In extreme cases this loss can lead to **soil exhaustion**. This is a condition in which the soil becomes nearly useless for farming. The usual way to build up soil nutrients is with fertilizer. Some fertilizers are natural organic materials like manure and decayed plants. Chemical fertilizers are also used. Yields can also be maintained by **crop rotation**, or planting different crops in alternating years. Crop rotation gives a field time to replace naturally the nutrients used by each different crop.

People often look to the world's drylands to expand farming. Many dry regions can support farming if water is artificially supplied to the land—a process called **irrigation**. However, irrigation can lead to **soil salinization**, or salt buildup in the soil. Salt is destructive to nearly all crops. Small amounts of salt are often present in irrigation water. Evaporation of the water concentrates the salt on the surface of the soil. Irrigation can also cause salt to rise into the topsoil from the parent materials below. Soil salinization is a serious problem in many dry areas of the world, including the southwestern United States.

READING CHECK: *Environment and Society* How do farmers adapt to the need to avoid soil exhaustion?

Redwoods, the world's largest trees, were among Earth's first trees. They probably date from more than 100 million years ago. Once plentiful across North America, redwoods are now limited to California and Oregon in a narrow area near the Pacific Ocean.

Forests Forests protect soil from erosion, provide habitats for many different species, and yield useful products. In addition to wood, we get food, medicines, oils, and rubber from forests.

Deforestation is the destruction or loss of forests. The tropical rain forests of Africa, Asia, and Latin America are being cleared at a rapid rate today. Logging for wood, clearing for farmland and ranch land, and cutting wood for fuel are the main causes of deforestation. Deforestation is also a problem in North America.

Many countries are taking steps to protect their forests. For example, cutting trees has been outlawed in some areas. Many countries require that new trees be planted after an area has been cut. This replanting process is called **reforestation**. In many cases the young trees can be harvested after a few decades. Still, these new forests cannot fully replace the ecological diversity of the original old-growth forests.

READING CHECK: *Environment and Society* What policies and regulations have countries taken to prevent deforestation?

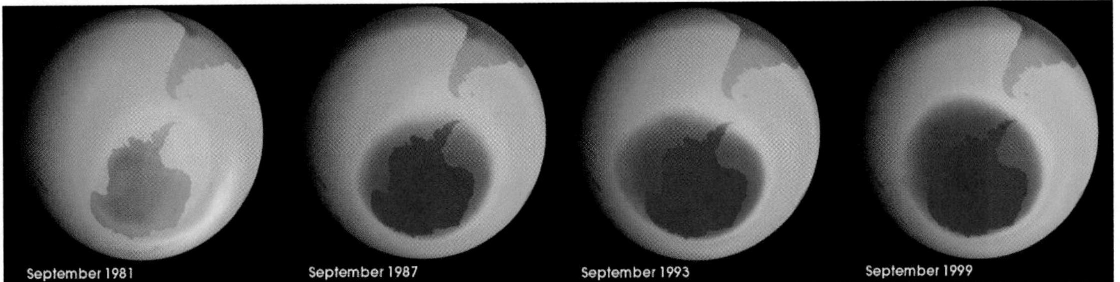

These satellite images show how the Southern Hemisphere's ozone layer has been damaged. The purple area is a hole in the ozone. During September 2000 the hole extended for the first time over a populated area—the city of Punta Arenas, Chile. People there were briefly exposed to high levels of ultraviolet radiation.

Air and Water

All living things on Earth depend on clean air and water. Fortunately, they are renewable resources. However, as the population grows and industry expands, our air and water supplies are at risk.

Air Humans, animals, and plants need the gases in air to survive. Pollution threatens our air supply. Automobiles and factories release smoke and fumes into the atmosphere. These chemicals build up in the air, particularly in large cities. The chemicals can interact with sunlight and create a mixture called smog. Some cities, such as Los Angeles and Mexico City, have terrible smog. At high enough levels, air pollution is dangerous to human health.

Some chemicals in air pollution form a mild acid when they combine with water vapor in the atmosphere. The acid can be similar in strength to vinegar. When it falls, this **acid rain** can damage trees and kill fish in lakes.

The effects of smog and acid rain are generally short-term. Air pollution also has the potential to cause long-term change in Earth's atmosphere. Certain kinds of pollution can damage a part of the atmosphere called the ozone layer. The ozone layer is important because it absorbs harmful ultraviolet radiation

Connecting to TECHNOLOGY

Irrigation

Irrigation water is delivered to farmland by many different methods. The oldest ways involve directing water into a field, usually from a small canal. This process can be as simple as flooding, in which water covers the entire surface of the field. Water can also be run into furrows—shallow trenches plowed between rows of crops. In either case, ditches are usually dug at the lower edge of the field. They allow the excess water to drain off the field. Proper drainage is very important for preventing soil salinization. If water is allowed to remain on a field and evaporate, salts are likely to concentrate in the soil.

More modern irrigation techniques use water more efficiently. Sprinkler irrigation is widely used in the United States. This method allows water flow to be controlled, from a mist to a heavy soaking. Sprinkler systems often use an automated device to travel over a field. The center-pivot system uses a long arm with sprinklers mounted on it. Anchored in the center of the field, the arm sweeps in a circle.

Drip irrigation, the most efficient method, applies water directly to the base of the plant with special "leaky" hoses. Drip irrigation is relatively expensive to set up and maintain. As a result, farmers generally use it only where water is very scarce. However, it is also used where a crop may be valuable enough to offset the high cost.

Analyzing Information How have different kinds of modern irrigation systems aided agriculture?

Sprinklers irrigate hay fields in Idaho.

The ancient Romans built this aqueduct in southern Spain. Rows of arches support the water channel at the top of the aqueduct. Some Roman aqueducts are still in use. Modern aqueducts are also crucial links in water supply systems. California's aqueduct system is the largest in the world.

from the Sun, protecting living things. As pollution damages the ozone layer, more solar radiation may cause an increase in skin cancer. You read about another long-term result of air pollution in the previous chapter. Some polluting gases may contribute to global warming.

Water In many parts of the world, maintaining a dependable supply of clean water is a major challenge. Cities and farms in dry regions often require water to be stored or transferred from other areas. For centuries, people have built complex systems of dams, canals, reservoirs, and **aqueducts**—artificial channels for transporting water.

In many dry areas, groundwater is the only dependable source of water. Thousands of years ago people learned to reach groundwater by digging wells. Today wells are drilled into the ground, and pumps bring the water to the surface. Wells are drilled to the depth of rock layers where groundwater is plentiful. These layers are called **aquifers**. Groundwater must be protected from pollution. This pollution is hard to remove.

Most groundwater is fed by rain. Because it can be preserved and managed like surface water, groundwater is usually a renewable resource. However, some groundwater, particularly in desert areas, was deposited long ago when the climate was wetter. This **fossil water** is not being replenished by rain. Today fossil water is being pumped to farmlands in many desert areas. These wells will run dry someday.

READING CHECK: ***Environment and Society*** How is pollution caused by humans affecting Earth's air and water?

Mining is dangerous! The deepest shafts of one South African gold mine reach 13,000 feet (3,962 m) below the surface. Temperatures in the mine would reach 140°F (60°C) if huge refrigeration units did not pump cool air from the surface.

Mineral Resources

Minerals are solid substances that come out of the ground. Examples include metals, rocks, and salt. Some minerals, such as the quartz in sand, are quite common. Others are hard to find—gold, for example. Over the centuries, people have developed countless uses for minerals.

Uses of Minerals Minerals are used in many processes and products, including construction, jewelry, and manufacturing. Rock, such as limestone, can be cut into blocks for building. Metals, from aluminum to zinc, are shaped into a vast range of products. Our daily lives are filled with metal objects, from airplanes to soft drink cans. Other important materials are made up of many different minerals. For example, sand that contains the mineral quartz is a key ingredient in glass.

Workers extract minerals from holes in the ground called mines. Some mines, like the gold mines in South Africa, tunnel deep underground. Other mines are open pits, some of which are miles across. Long ramps allow giant trucks to move in and out of the pit. The trucks move the **ore**—the mineral-bearing rock—to a processing plant where the valued mineral is removed.

Some minerals are truly rare. For example, gemstone-quality diamonds are minerals found in just a few countries. In addition, the deposits that contain diamonds are often only a few hundred yards across.

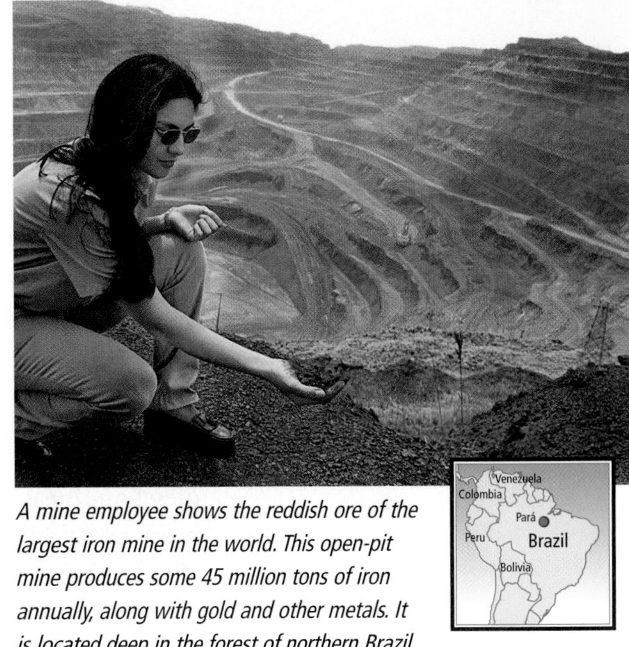

A mine employee shows the reddish ore of the largest iron mine in the world. This open-pit mine produces some 45 million tons of iron annually, along with gold and other metals. It is located deep in the forest of northern Brazil.

Mineral Recycling Today many products are made from recycled minerals. The advantages of recycling are obvious. For one thing, the more we recycle, the slower we will use up nonrenewable resources. For another, mines often cut great scars into the landscape. Processing plants belch dust and smoke and consume large amounts of energy. In contrast, using mineral products again can save money and spares damage to the environment. Recycling also saves space in landfills.

Recycling is not new. Precious metals like gold and silver have been reused over and over again through the centuries. However, increasing environmental awareness has encouraged more reuse of mineral products. Today aluminum, copper, and steel are often recycled.

READING CHECK: *Environment and Society* Why is recycling resources important?

Energy Resources

Some of the most important nonrenewable resources are sources of energy. These energy resources have become increasingly useful and valuable as people develop new technologies that need them for power. Energy resources include coal, natural gas, and petroleum. These substances are called **fossil fuels** because they were formed from the remains of ancient plants and animals. Coal comes from ancient swamps and bogs. Petroleum and natural gas are the remains of tiny plants and animals that lived in seas and lakes. Over millions of years, heat and pressure converted the dead plants and animals into oil and gas.

Another energy resource is uranium, a metallic element. Uranium is radioactive and provides the energy for nuclear power plants. Nuclear power does not pollute the air, but it creates waste material that remains dangerous for thousands of years. The problem of disposing of nuclear waste has limited the use of nuclear power.

Fossil Fuels Coal was the first fossil fuel to be used as a major energy source. It is a solid and is often found near the land surface. People have long used coal for heat. Coal also powered early steam engines and steel mills. Coal remains an important fuel for electric power generation. However, it has environmental disadvantages. Often it contains minerals that do not burn. This waste material either goes into the air as pollution or must be disposed of in landfills. Burning coal that contains sulfur is one of the causes of acid rain. However, coal has the advantage of being plentiful. At current rates of consumption, the world has enough coal for centuries to come.

Petroleum and natural gas come from wells drilled into deep sedimentary deposits. About 150 years ago, people began to use petroleum as fuel for lamps. The invention of the automobile later greatly expanded the market and need for petroleum, also called oil, as a fuel source. Today oil is made into gasoline, diesel and heating fuel, asphalt, and many other products. As oil has become more useful, it has become more valuable. Today its production and processing is a major industry. However, supplies are limited. Reducing the need for oil is important for our future.

Natural gas is usually odorless, but it is given an artificial smell so that leaks can be detected. Natural gas is particularly important for home and industrial heating. Both natural gas and oil are also used to generate electricity. They usually burn more cleanly than coal. Natural gas is growing in importance as a fuel. It is abundant, and supplies will certainly be available for decades to come. On the other hand, natural gas prices have been rising.

Fossil fuels are also important sources of chemicals. Coal has long been important in making dyes. Oil is the source of many important products called **petrochemicals**. Petrochemicals include the raw materials for many explosives, food additives, medicines, pesticides, and plastics. Having dependable sources of fossil fuels is a key factor for a country's economic success. Unfortunately, as

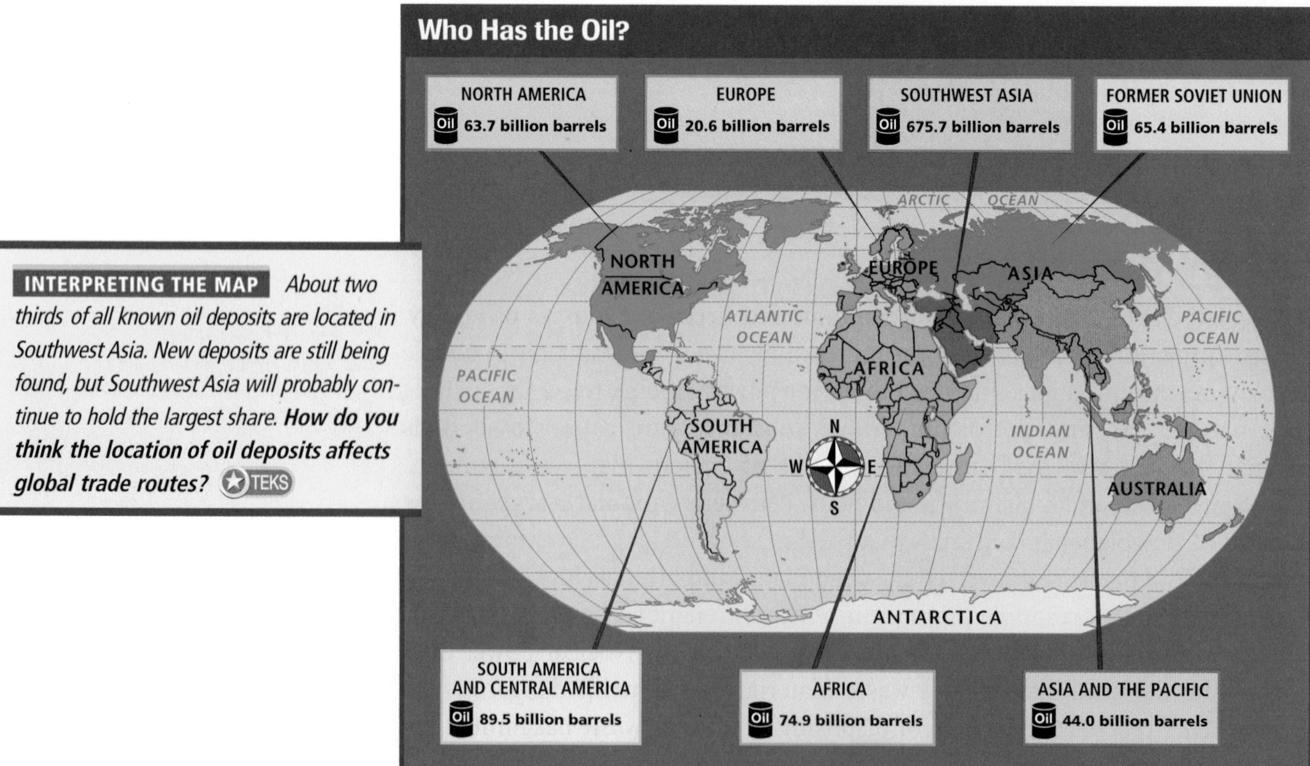

INTERPRETING THE MAP *About two thirds of all known oil deposits are located in Southwest Asia. New deposits are still being found, but Southwest Asia will probably continue to hold the largest share.* ***How do you think the location of oil deposits affects global trade routes?*** TEKS

fossil fuels are burned, they release carbon dioxide. Carbon dioxide is one of the gases, called greenhouse gases, that keep extra heat in the atmosphere.

TEKS **READING CHECK:** *Environment and Society* How are fossil fuels used?

FOCUS ON SCIENCE, TECHNOLOGY, AND SOCIETY

Renewable Energy Resources Renewable energy sources include waterpower, wind power, heat from Earth's interior, and solar energy. **Hydroelectric power**—electricity produced by moving water—is the most widely used type. Dams hold water back and force it to flow through narrow openings. As the water is pulled through these openings by gravity, it turns generators to produce electricity. Dams produce about 10 percent of the electricity used in the United States.

Wind has been used to propel ships for thousands of years. Windmills were also invented centuries ago. They were used to pump water out of wells or grind grain. Modern versions of windmills, called wind turbines, create electricity. Although wind contains great power, it may take thousands of wind turbines to equal the electricity of a conventional power plant. This fact limits the use of wind power.

The heat of Earth's interior, called **geothermal energy**, can also be used to generate electricity. Geothermal power plants, which capture this energy, are built in places with volcanoes or hot springs.

The energy of the Sun—solar energy—can also be used to create power. Special solar panels can absorb solar energy and convert it to electricity. Also, the Sun's rays can be concentrated and used to heat water or homes. Although solar energy has great promise, the technology needed to harness it is costly and complex. Breakthroughs in the future may make solar energy a major power source.

TEKS **READING CHECK:** *Environment and Society* What has been the effect of new technologies and new perceptions of fossil fuel resources?

INTERPRETING THE VISUAL RECORD *Some renewable energy sources can also be called flow resources. These resources must be used when and where they occur. They include running water, sunlight, and wind. For example, wind power can only be effective in places that have strong steady winds. Wind turbines, like these in southern California, harness wind to produce energy.* ***How might stricter regulations on the use of fossil fuels affect the use of flow resources?*** TEKS

TEKS Questions 1, 2, 3, 4, 5

Review

Define humus, leaching, contour plowing, soil exhaustion, crop rotation, irrigation, soil salinization, deforestation, reforestation, acid rain, aqueducts, aquifers, fossil water, ore, fossil fuels, petrochemicals, hydroelectric power, geothermal energy

Reading for the Main Idea

1. *Physical Systems* From which soil horizons do plants draw nutrients?
2. *Physical Systems* What are the two causes of soil salinization?

Critical Thinking

3. **Drawing Inferences and Conclusions** Are old-growth forests a renewable or nonrenewable resource? Why?
4. **Analyzing Information** How have changing perceptions of oil's usefulness made it more valuable over time?

Organizing What You Know

5. Copy the table below. Use it to list two short-term and two long-term threats to the atmosphere.

Short-term threats	Long-term threats

Geography for Life

World Ecosystems and Biomes

Plant and animal communities that cover large land areas are called biomes. Areas that share the same biome have similar ecosystems. These ecosystems display complex relationships among the area's climates, vegetation, soil, and geology. Some biomes even have the same names as some climate regions.

Different biomes also can be found in the oceans. In contrast to land biomes, marine biomes depend mainly on the depth at which the plants and animals live. For now, however, we will limit our discussion to biomes on land.

What's in a Biome?

A single biome can exist in places scattered around the world. Different species that live in areas with the same biome may have much in common. We can use rain forests as one case. The plants and animals in Africa's rain forests are different species than those of Brazil's rain forests. Monkeys of the Western Hemisphere, for example, are different from African monkeys. However, the two groups' appearance and behaviors are often similar because they have adapted to similar environments. Monkeys of both hemispheres spend much of their time in trees, because they find their main foods in trees.

These similarities are what make biomes such a useful way to classify large parts of the world. Geographers can make generalizations about factors like a biome's structure and seasonal growth patterns. Geographers can also generalize about ways people use the biome's plants and animals. In turn, this information can help people manage and conserve these environments.

World's Biomes

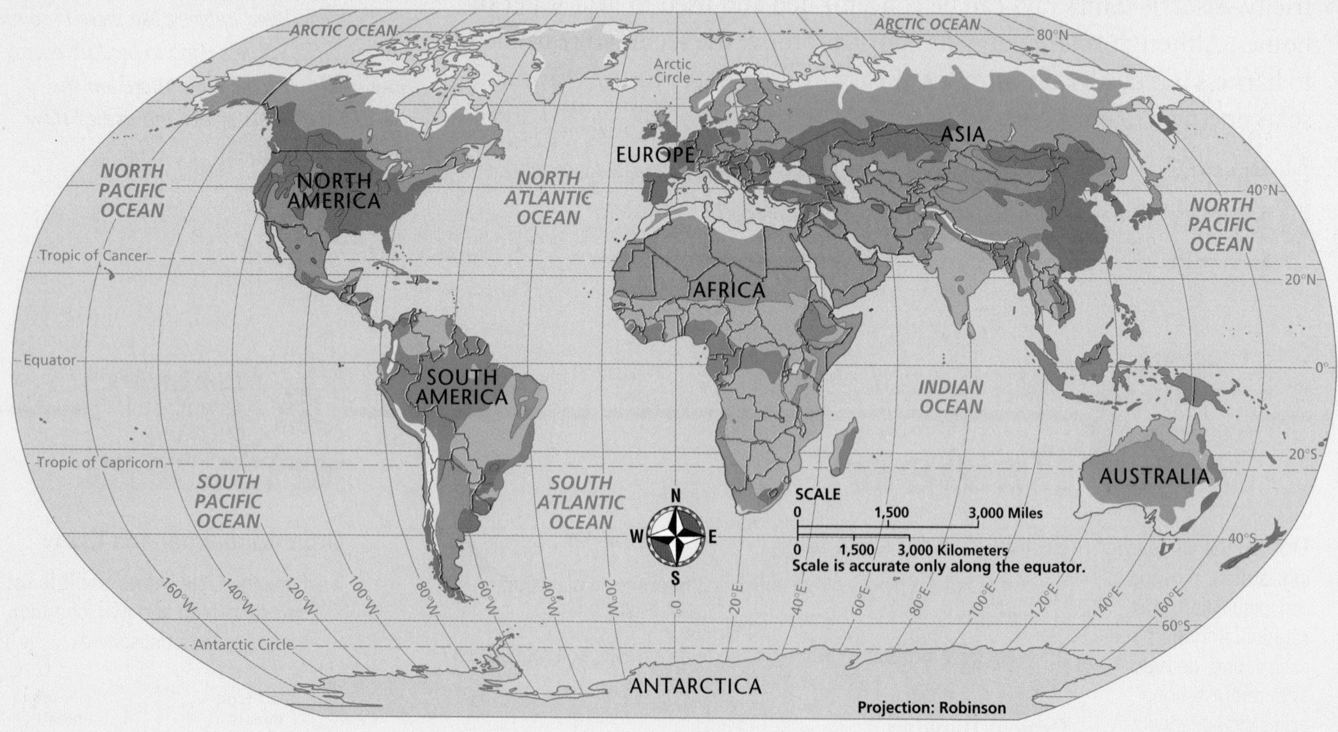

Distribution of Biomes

Latitude and the world climate zones have influenced the distribution of biomes. However, a biome results from the interaction of many factors. A sharp variation in one factor can result in a biome existing where you might not expect it. Perhaps the most dramatic examples are caused by variations in elevation. For example, the plants and animals that make up the tundra biome live in northern Alaska near the Arctic Circle. Similar life-forms can be found near the top of Kilimanjaro, a mountain on the equator in eastern Africa.

Variation in soil and geology can also determine a given place's biome type. The process of desertification provides a clear example. If animals overgraze a grassland area, wind or water can erode the topsoil. Without topsoil to support plant growth, the area can become a desert. This process can take place in a dramatically short time. On the other hand, fertile volcanic soil can support denser and more varied vegetation than thin nutrient-poor soil. Different types of fertile soil also support different plants. For example, an isolated pocket of sandy acidic soil supports a small forest of tall pines in Bastrop County, Texas. Surrounding the so-called Lost Pines are oak and pecan trees growing in different soils. Thus, two places with the same climate but different soils have different biomes.

Applying What You Know

1. **Drawing Inferences and Conclusions** Compare the map of world biomes to the map of world climate regions in Chapter 3. What connections do you see between the two distribution patterns?

2. **Analyzing Information** Describe the climate, soils, vegetation, and representative animals of the biome in which you live. What geological features do you think have influenced plant and animal life in your region? How?

Biome Legend

Biome	Climate	Soil	Vegetation	Typical Animals
Tropical rain forest	warm and rainy year-round	thin; poor in nutrients as a result of leaching	most diverse of all biomes; huge variety of trees, vines, and other plants	monkeys, lemurs, parrots, snakes, tree frogs, bats, pigs, small antelopes, tigers, jaguars, and leopards
Savanna	warm year-round; distinct rainy and dry seasons	generally poor in nutrients	tall grasses, some trees, and thorny shrubs	gazelles, rhinoceroses, giraffes, lions, hyenas, ostriches, crocodiles, and elephants
Semiarid and desert	dry; sunny and hot in tropical regions; wide temperature variations in middle latitudes	poor in organic matter but may have plentiful minerals	moisture-retaining plants such as cacti; shrubs and thorny trees	kangaroo rats, lizards, scorpions, snakes, birds, bats, and toads
Grassland	low rainfall; warm summers; cold winters	most fertile soils of all biomes	grasses; trees near water sources	lions; large grazing animals, including buffaloes, kangaroos and antelopes
Temperate forest	plentiful rainfall; warm summers; cold winters	very fertile; rich in organic nutrients	deciduous and evergreen trees; shrubs; herbs	deer, bears, badgers, squirrels, wolves, wild cats, red foxes, owls and many other birds
Mediterranean scrub forest	hot dry summers; cool wet winters	rocky; nutrient-poor soils	evergreen shrubs; scrubby trees; herbs	ground squirrels, deer, elk, mountain lions, coyotes, and wolves
Boreal forest	short warm summers; long cold winters	acidic soil	mosses, lichens, and coniferous trees	birds, rabbits, moose, elk, wolves, lynxes, and bears
Tundra	short cool summers; long cold winters	permafrost	mosses, lichens, sedges, and dwarf trees	rabbits, lemmings, reindeer, caribou, musk oxen, wolves, foxes, birds, and polar bears
Barren	cold year-round	thin; poor soils	few mosses and lichens in coastal areas	animals that depend on the sea including penguins, other birds, and seals

CHAPTER 4 Review

Building Vocabulary TEKS

On a separate sheet of paper, explain the following terms by using them correctly in sentences.

magma	wetlands
continental drift	groundwater
abyssal plains	humus
sediment	crop rotation
plateau	irrigation
delta	fossil water
headwaters	fossil fuels
watershed	

Locating Key Places 

On a separate sheet of paper, match the letters on the diagram of the hydrologic cycle with their correct labels.

evaporation	precipitation
condensation	groundwater
moist air	

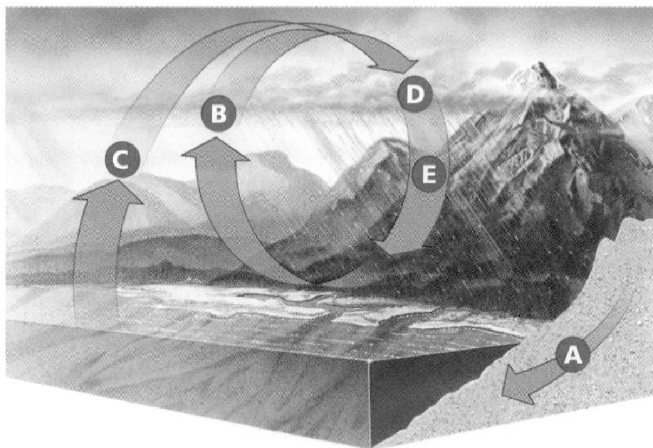

Understanding the Main Ideas TEKS

Section 1

1. *Physical Systems* What are the three types of movements possible at plate boundaries?
2. *Physical Systems* What are the two physical processes that wear down landforms on Earth's surface?

Section 2

3. *Physical Systems* What process makes some inland lakes salty?

Section 3

4. *Environment and Society* What are two short-term effects of air pollution?
5. *Environment and Society* What is a major drawback of using coal as an energy source?

Thinking Critically for TAKS TEKS

1. **Analyzing Information** Which would have a smoother shape, a young mountain chain or an old mountain chain? Why?
2. **Drawing Inferences and Conclusions** In an area where a rain forest has been cleared, why would you expect the soil to become exhausted quickly?
3. **Identifying Cause and Effect** What new technologies made fossil fuels more valuable over time? What effect have development and use of these resources had on the physical environment?

Using the Geographer's Tools TEKS

1. **Creating Maps** Review the maps of the United States in the atlas at the front of this textbook. Then sketch a map of the United States and shade the region drained by the Mississippi River system. In what ways do you think this and other river systems constitute a region?
2. **Analyzing Information** Sketch the soil horizons diagram. Label the A, B, and C horizons. Add arrows and symbols to the diagram to show the leaching process.
3. **Preparing Charts** Study the Plate Tectonics map. Create a chart listing each plate and the direction or directions of its movement, if any.

Writing for TAKS TEKS

Keep a journal listing all the natural resources you use during a single day. Then create a journal from the point of view of a person living 50, 100, or 200 years in the past. Remember that technology changes the way people use resources. When you are finished with your two journals, proofread them to make sure you have used standard grammar, spelling, sentence structure, and punctuation.

SKILL BUILDING

Geography for Life TEKS

Preparing Diagrams and Models

Physical Systems Use construction paper or modeling clay to create posters or three-dimensional models of the different types of plate boundaries. Show landforms that may result, such as mountain ranges.

Building Social Studies Skills

Interpreting Graphs

Study the line graph below. Then use the information from the line graph to help you answer the questions that follow. Mark your answers on a separate sheet of paper.

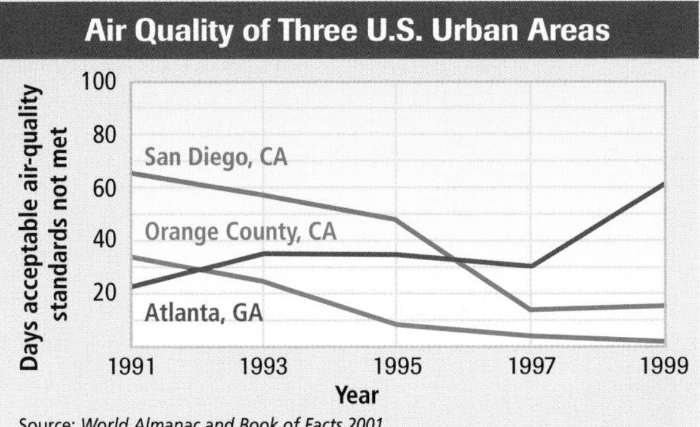

Source: *World Almanac and Book of Facts 2001*

1. How many days did Atlanta fail to meet acceptable air-quality standards in 1997?
   a. 31
   b. 61
   c. 14
   d. 3
2. Describe the overall pattern of air quality in Orange County and San Diego between 1991 and 1999. Then contrast those patterns with that of Atlanta during the same period.

Building Vocabulary

Answer the following questions on a separate sheet of paper.

3. *Magma* specifically means
   a. layer of Earth's interior between the crust and core.
   b. liquid rock.
   c. rock that contains minerals.
   d. Earth's solid center.
4. Many earthquakes have been caused by movement along the San Andreas Fault. In which sentence does *fault* have the same meaning as it does in the sentence above?
   a. It is partly the fault of the geologist that the exploration company has gone bankrupt.
   b. The fault in this circuit is preventing the light from coming on.
   c. The area of deforestation is a fault in an otherwise beautiful landscape.
   d. When layers of rock break and move, a fault results.

Alternative Assessment

PORTFOLIO ACTIVITY

Learning about Your Local Geography

Group Project: Field Work

As a group, learn about recycling in your community. If possible, tour your local recycling facility. Find out what materials are recycled and how they are gathered, transported, and processed. Interview workers at the recycling center or public officials to learn the costs and benefits of recycling to your community's economy and natural environment.

internet connect

Internet Activity: go.hrw.com
KEYWORD: SW3 GT4

Choose a topic about landforms, water, and natural resources to:

- create a newspaper on the causes and effects of earthquakes.
- understand the formation and uses of fossil fuels.
- analyze the different stages of the hydrologic cycle.

CHAPTER 5 Human Geography

People have shaped much of the world around us. Cities, farms, airports, and other features have been created to meet our ever-changing needs. The study of the world's people—how they live and how their activities vary from place to place—is called human geography.

The Shibuya district in Tokyo, Japan

Selling goods in Latacunga, Ecuador

Population Geography

READ TO DISCOVER

1. How do geographers study population?
2. What are some important trends in world population?

WHY IT MATTERS

People are always moving to new places. Use CNNfyi.com or other **current events** sources to find a news item about a group of people leaving an area for a new one.

DEFINE

demography
population density
birthrate
death rate
migration
emigrants
immigrants
push factors
pull factors
refugees

Studying Population

Is the population of your state growing or decreasing? Where do people choose to live? What is the average age of people in your state? These are the kinds of questions that population geographers ask. These geographers study the relationships between populations and their environments. They use maps, graphs, population pyramids, and a spatial perspective to study population patterns and trends. Population geography is closely related to **demography**—the statistical study of human populations. Statistics are information in number form. They help us learn about populations. Demographers collect statistics about populations and use them to forecast what future populations will be like. Such forecasts and demographic information can be very useful. For example, it can be used to decide where to build new schools or how to redraw political boundaries to reflect changing populations.

internet connect

GO TO: go.hrw.com
KEYWORD: SW3 CH5
FOR: Web sites about human geography

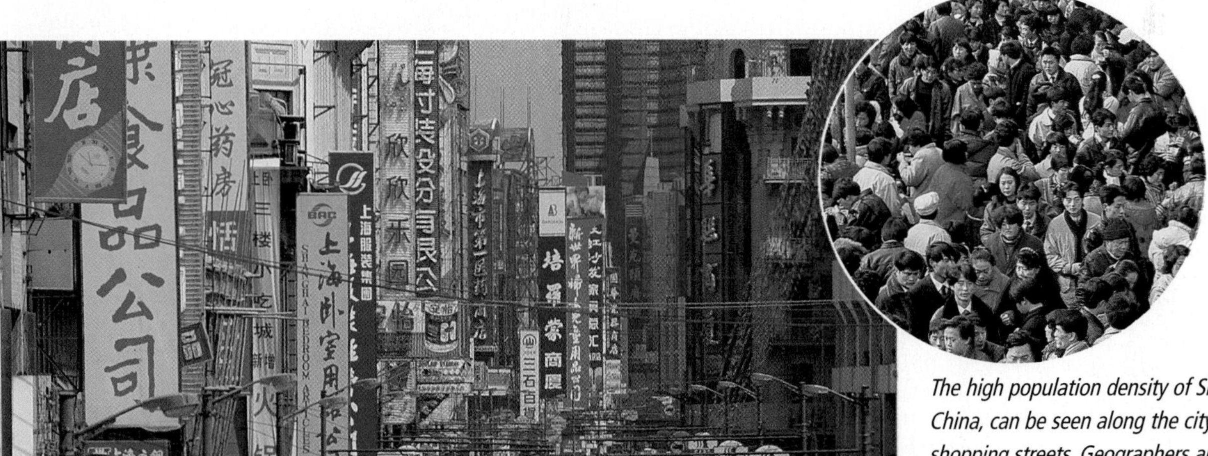

The high population density of Shanghai, China, can be seen along the city's bustling shopping streets. Geographers and demographers use statistics like population density to study the world's population.

Population Density in Selected U.S. Counties	
County, State	**People per Square Mile**
Loving, TX	0.17
Forest, WI	10
Blue Earth, MN	72
Honolulu, HI	1,441
San Francisco, CA	15,889
New York (Manhattan), NY	55,423

Source: *The World Almanac and Book of Facts 2001*

INTERPRETING THE CHART *Population densities in U.S. counties vary dramatically.* ***Based on the population data in this chart, how would you expect the human geography of Blue Earth County, Minnesota, to be different from that of New York (Manhattan), New York?*** TEKS

Population Density One important statistic geographers use is **population density**. This is the average number of people living in an area. It is usually expressed as persons per square mile or square kilometer. In this textbook you will find the population densities of the world's countries in the Fast Facts table in each unit.

Population densities around the world vary greatly. For example, Canada has just 8 persons per square mile. On the other hand, Bangladesh has 2,361 persons per square mile. Population density reflects the size of a country, the size of its population, and its environmental conditions. For example, Canada is a huge country. However, much of it is too cold to support large numbers of people. Most of Canada's 31.6 million people live along the country's southern border. Temperatures are warmer there. As a result, southern Canada is more densely populated than the rest of the country. Bangladesh is much smaller in area than Canada. Yet nearly all of Bangladesh is covered by rich farmland. Many of the country's roughly 131 million people are crowded onto that farmland. Within a country, smaller areas may have an even greater range of population densities. (See the chart of population density in selected U.S. counties.)

READING CHECK: ***Human Systems*** Why do population densities vary greatly among the world's countries?

World Population Density

Los Angeles
New York
Mexico City
São Paulo
Buenos Aires
Lagos
Karachi
Delhi
Mumbai (Bombay)
Dhaka
Kolkata (Calcutta)
Shanghai
Tokyo
Osaka
Jakarta

Persons per sq. mi.	Persons per sq km
520	200
260	100
130	50
25	10
3	1
0	0

● Metropolitan areas with more than 10 million inhabitants

SCALE
0 1,000 2,000 Miles
0 1,000 2,000 Kilometers
Projection: Robinson

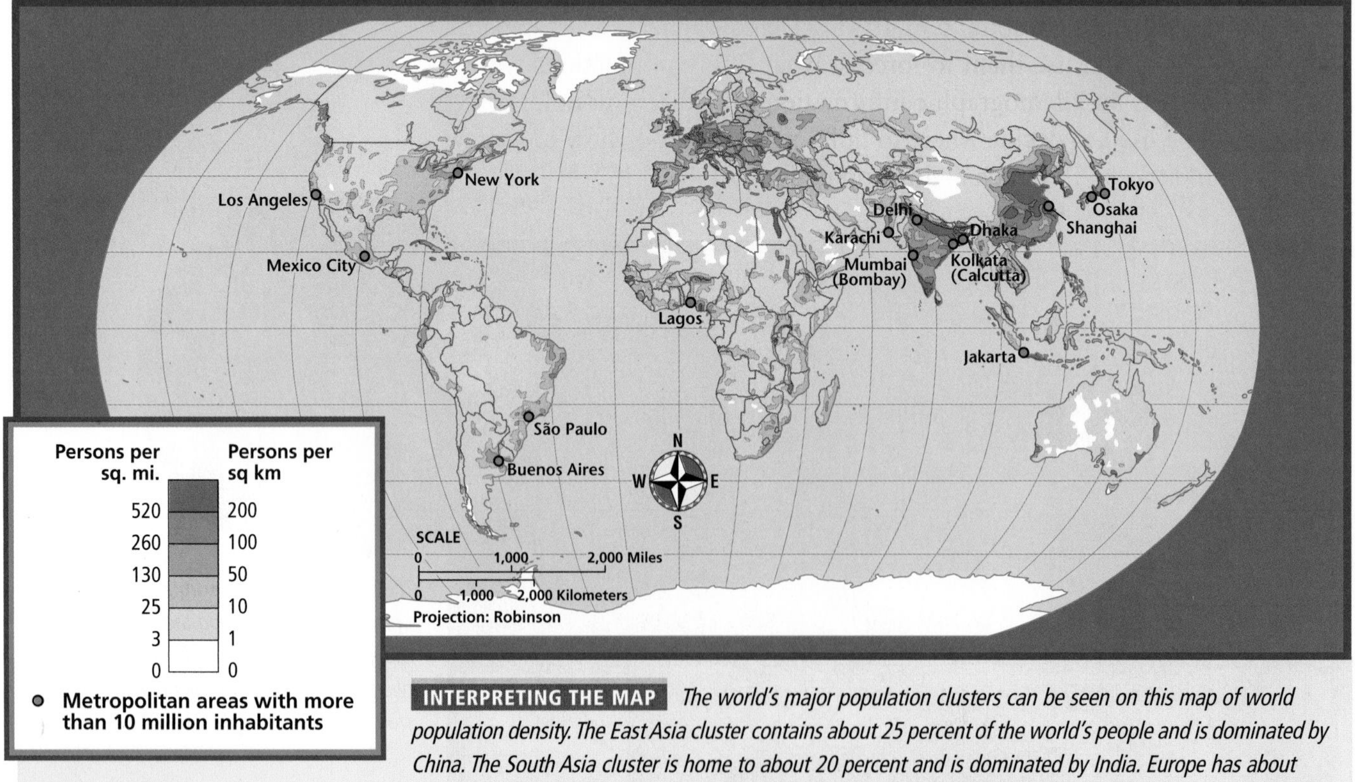

INTERPRETING THE MAP *The world's major population clusters can be seen on this map of world population density. The East Asia cluster contains about 25 percent of the world's people and is dominated by China. The South Asia cluster is home to about 20 percent and is dominated by India. Europe has about 12 percent of the world's people, while eastern North America has less than 5 percent.* ***How do you think the patterns on this map will change over the next 100 years? Why?*** TEKS

Nigeria is an example of a country with a high birthrate—there are about 40 births per 1,000 persons each year. The high birthrate, combined with a much lower death rate, has resulted in a young, rapidly growing society. About 44 percent of Nigeria's population is currently below the age of 15.

Population Distribution People are spread unevenly across Earth. Some places are crowded with people, while others are empty. Where do most of the world's people live? Why do people live in some places and not in others? About 90 percent of the world's people live in the Northern Hemisphere. About two thirds of these people live in the middle latitudes between 20° and 60° north. Many people there live in lowland areas, particularly along fertile river valleys near the edges of continents. Four areas of the world have great clusters of population. Those areas are East Asia, South Asia, Europe, and eastern North America. There are also smaller clusters, such as the Nile Valley in Egypt and the island of Java in Indonesia. (See the World Population Density map.)

Why is the world's population so unevenly distributed? The simple reason is that people tend to live in areas that are favorable for settlement. They also tend to avoid areas that are unfavorable. Therefore, places with mild climates, fertile soils, and an adequate supply of freshwater usually have more people than areas without these features. For the same reason, polar regions, deserts, and rugged mountains usually have few people.

READING CHECK: ***Human Systems*** Why is the world's population so unevenly distributed?

Population Change Geographers also use statistics to study how populations are changing. The number of people in any place is the result of three major factors. The first is how many people are born each year. This factor is expressed as the **birthrate**. The birthrate is the number of births each year for every 1,000 people living in a place. The second factor is how many people die each year. This is known as the **death rate**—the total number of deaths each year for every 1,000 people. The third factor is **migration**. Migration is the process of moving from one place to live in another. People who leave a country to live somewhere else are called **emigrants**. People who come to a new country to live are called **immigrants**. If large numbers of people are leaving or entering a country, it can have major effects on a population.

FOCUS ON GEOGRAPHY

Migration Migration is a common theme in human history. Scientists think that humans and their ancestors began migrating out of Africa between 1 and 2 million years ago, moving into Asia and Europe. They may have been looking for better hunting, or perhaps they just wanted to know what was over the horizon. The last continents discovered and settled by humans were Australia at least 40,000 years ago and, later, North and South America. To reach the Americas, people may have crossed a land bridge from Asia to Alaska that became available when sea levels fell during the last ice age. However, there is growing evidence that humans may have reached the Americas even earlier. Some may have traveled to the region in small boats along the coast.

Geographers study migration by analyzing **push factors** and **pull factors**. A push factor causes people to leave a location. A pull factor attracts people to a new location. Most people migrate for economic reasons. The push may be a lost job or lack of opportunity for promotion. The pull may be a better job or higher pay somewhere else.

Many other push and pull factors can cause people to move. For example, environmental conditions and hazards cause migration. Droughts and floods have often forced people to move. Over time, warm sunny climates have drawn people to Arizona, Florida, and Texas. People also flee political unrest and wars. People who have been forced to leave and cannot return to their homes are **refugees**. They often leave because they do not feel safe where they live or suffer discrimination. On the other hand, political or personal freedom can be a pull factor. For example, people have moved to find freedom in democratic countries like the United States and Canada. Finally, physical geography can affect the routes, flows, and destinations of migrants. For example, many early Americans migrated from the East Coast to the interior plains to farm. The physical

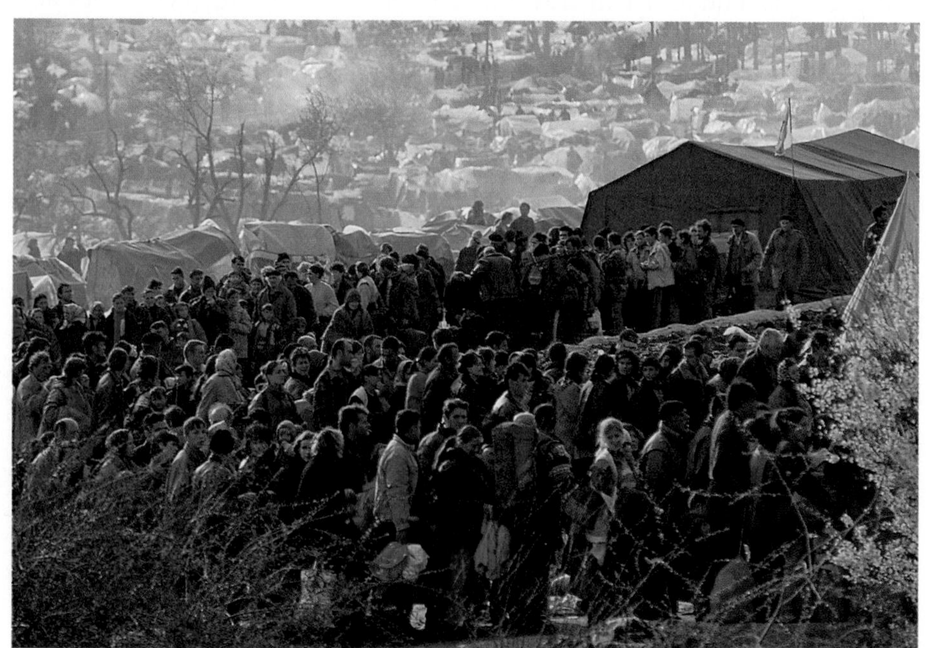

INTERPRETING THE VISUAL RECORD

In the late 1990s unrest in Kosovo, Yugoslavia, disrupted the lives of hundreds of thousands of ethnic Albanians and forced many people to flee the region. Here, refugees in neighboring Macedonia wait to be transported to nearby transition camps. ***How might such a sudden population increase strain the resources of Macedonia?***

geography of the United States, including the locations of the Appalachian Mountains and major rivers, greatly influenced these migration patterns.

READING CHECK: *Human Systems* What kinds of factors cause migration?

Natural Increase Geographers are also interested in the rate of natural population growth, or natural increase. This rate is based just on births and deaths—it does not take migration into account. You can find the rate by subtracting the birthrate from the death rate. This final number is expressed as a percentage. In the United States, the rate of natural increase is about 0.6 percent each year.

The rate of natural increase varies greatly in countries around the world. The highest rates are found in countries in Africa and Southwest Asia. Rates there are sometimes 3 percent or higher. The number of people living in those places is rising rapidly. Most countries in Central and South America and in Southeast Asia have more moderate rates of natural increase. The rates for these areas are somewhere between 1 and 3 percent. The lowest rates—less than 1 percent—are found in most European and North American countries. Australia, New Zealand, and Japan also have such low rates. Some countries, such as Italy and Russia, actually have negative growth rates. The number of people living in these countries is decreasing.

These percentages—1, 2, and 3—may sound small. However, they can add up to large population increases in a short period of time. For example, suppose the number of people living in a country grows at a rate of 3 percent. That country's population will double in only about 23 years! Geographers call the number of years needed to double a country's population its doubling time.

READING CHECK: *Human Systems* Where are the highest rates of natural increase found in the world? The lowest?

World Population Trends

The world's population has increased rapidly in the last 200 years. (See the graph of world population growth.) Earth is now home to more than 6 billion people. This number is increasing by nearly 80 million each year. That is almost 220,000 people a day! It is hardly surprising that worries about crowding and environmental problems are in the news.

In the year A.D. 1 the world's population was probably about 300 million. By 1600, it had doubled to 600 million. Then populations began to increase steadily. Farm technology and public sanitation improved, and the basics of modern medicine began. Cities grew larger. The world's population distribution was also changing as large population clusters grew in North America, South Asia, and East Asia. By about 1850, world population had doubled again to 1.2 billion. Since then, Earth's population growth has exploded. World population passed the 2 billion mark before 1930 and doubled again to 4 billion by 1975. In 2000 there were more than 6 billion people.

World Population Growth

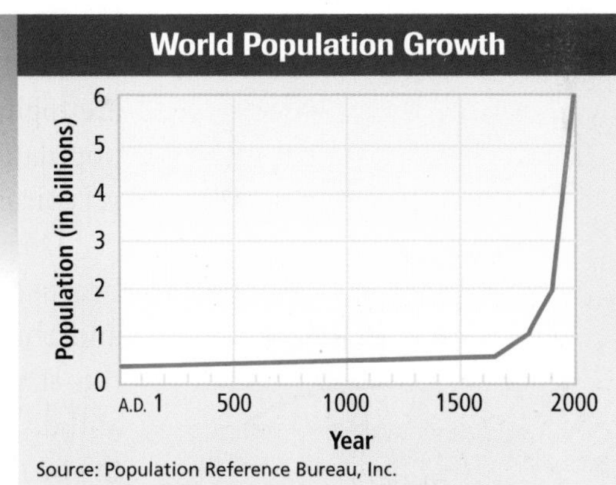

Source: Population Reference Bureau, Inc.

INTERPRETING THE GRAPH *This graph of world population growth shows the dramatic increase in the world's population over the last several hundred years. While population growth for most of human history was slow, it exploded as improved health, nutrition, sanitation, and medical care combined to lower death rates around the world.* ***What trends in past world population growth does this graph show? How has the world's population distribution changed during this time?*** TEKS

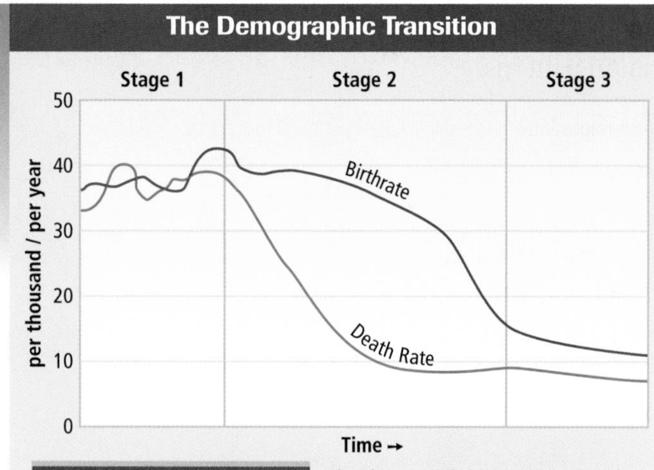

INTERPRETING THE GRAPH *The demographic transition model is based on changes that occurred in the populations of many Western European countries as they industrialized. New economic and social conditions shaped these countries' populations. As medical care and health improved, the death rate fell. Eventually, the birthrate also fell as modern urban societies developed.* ***During which stage in the model is population growth highest? How could you apply the demographic transition model to present geographic information about the countries of Western Europe?*** TEKS

Some people worry about overpopulation. Overpopulation is a situation in which the existing number of people is too large to be supported by available resources. However, there is much debate among geographers and other scientists about how many people the world can support. Also, birthrates in many countries have dropped in recent years. These changes will affect future population growth and resource needs around the world.

READING CHECK: ***Human Systems*** How many people are there in the world? How is this number changing each year?

The Demographic Transition The demographic transition is a model that shows how birthrates and death rates dropped in many Western countries as they developed modern economies and industries. Most of the world's richest and most technologically advanced countries have experienced a transition from high birthrates and death rates to low birthrates and death rates. In addition, many of the world's poorer countries are now in the middle of similar population changes. While many demographers think these countries will also experience a transition to lower birthrates and death rates, no one can predict exactly how this will happen.

During the first stage of the demographic transition, both birthrates and death rates are high. (See the graph of the demographic transition.) Parents have many children, but poor health conditions mean that many do not live to become adults. The infant mortality rate—the number of infants that do not survive their first year of life—is high. Since both the birthrate and death rate are high, the population neither grows nor decreases much. Instead, the total population remains relatively stable. This stage of the demographic transition is typical of countries that are mostly agricultural.

During the second stage, the death rate begins to fall. This happens partly because of improvements in medicine and health care, particularly for infants and children. As a result, the infant mortality rate drops significantly during this stage, and more children survive their early years. Better food production and distribution may also help lower the death rate. While the death rate falls, the birthrate remains high. Families grow as children and adults live longer. As a result, the total population grows. Also during this stage, economic improvements and the switch to more advanced farming technologies cause rural to urban migration and the rapid growth of cities. Near the end of the second stage, the birthrate begins to fall. As more people switch to life in a modern urban society, they marry later and have fewer children. As a result, population growth begins to slow again.

During the third stage, both birthrates and death rates are low. Thus, total population growth is low. All of the world's economically advanced countries have reached this final stage. These countries include the United States, Canada, Japan, Singapore, Australia, New Zealand, and nearly all European countries.

Between 1950 and 1998 the average life expectancy in the developing world rose from 40 to 63 years, and the world's population more than doubled.

These countries have all experienced a demographic transition from an agricultural society with high birthrates and death rates to an urban and industrial society with low birthrates and death rates.

READING CHECK: *Human Systems* How does the demographic transition explain trends in past world population growth?

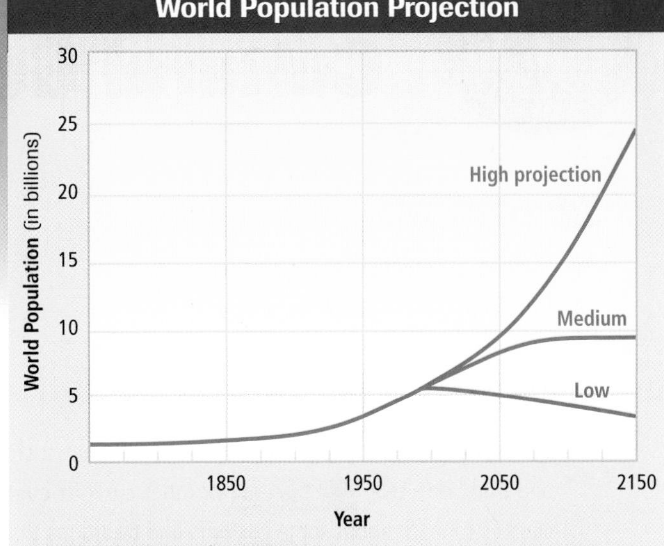

Source: United Nations Development Programme

INTERPRETING THE GRAPH *This graph shows three different population projections for the next 150 years: a low, medium, and high projection. The three projections are based on different assumptions about whether birthrates will increase or decrease in the future.* ***How different are these three projections? Which do you think most accurately predicts future population patterns? Why?*** TEKS

Future Populations How will the world's population change in the future? Will it continue to grow rapidly, or will lower birthrates lead to a slower rate of growth and a more stable population?

While we cannot be certain about how populations will change in the future, we can use demographic information to make population projections. Population projections are estimates of a future population's size, age, growth rate, or other characteristics based on current data. Forecasting future population growth is not easy. Population growth is tied to future birthrates, which cannot be known ahead of time. As a result, demographers often make several different projections. (See the graph.) In general, the farther into the future a projection is made, the less reliable it is.

Future world growth rates will depend on what happens in the countries that have high rates of natural increase. In some of these countries birthrates have been falling over the last 20 years. These countries seem to be moving toward the third stage of the demographic transition.

Regardless of how the world's population changes in the future, all countries will face population-related challenges. For example, caring for a growing number of children or older people will strain the resources of many countries. Maintaining public health, producing food, and protecting the environment will be important issues for populations around the world.

✓ **READING CHECK:** *Human Systems* Why is it so difficult to predict population growth?

TEKS Questions 1, 2, 3, 4, 5

Review

Define demography, population density, birthrate, death rate, migration, emigrants, immigrants, push factors, pull factors, refugees

Reading for the Main Idea

1. *Human Systems* What are some statistics that population geographers and demographers use to study populations?
2. *Places and Regions* Where are the four great clusters of world population?

Critical Thinking

3. **Making Generalizations and Predictions** How can the demographic transition model help geographers predict future population changes?
4. **Analyzing Information** What is one example of physical geography affecting the route and flow of migration?

Organizing What You Know

5. Create a chart like the one shown below. Use it to identify five push factors and five pull factors that can cause migration.

Push factors	Pull factors

Section 2 Cultural Geography

READ TO DISCOVER

1. How do geographers study culture?
2. How do cultures change over time?

DEFINE

culture
culture traits
culture region
ethnic groups
acculturation
innovation
diffusion
globalization
traditionalism
fundamentalism

WHY IT MATTERS

Our world's billions of people have many different customs and traditions. Use CNNfyi.com or other **current events** sources to learn about some customs and traditions that people have brought from other places to the United States.

Studying Culture

A key term for understanding human geography is **culture**. Culture includes all the features of a people's way of life. It is learned and passed down from parents to children through teaching, example, and imitation. Important parts of culture include a group's language, religion, architecture, clothing, economics, family life, food, and government. It also includes a people's beliefs, institutions, shared values, and technologies as well as its members' skills. What are some of the features of your culture?

Culture Traits Activities and behaviors that people often take part in are called **culture traits**. There are many kinds of culture traits. Some, such as learning to read and do math, are much the same around the world. This is true even though different cultures use different alphabets and symbols. Other culture traits vary from place to place. For example, most Americans eat with a knife, fork, and spoon. However, Chinese eat with chopsticks. Ethiopians eat

INTERPRETING THE VISUAL RECORD

The Amish are a religious group in North America with many distinct culture traits, such as simple clothing styles and the use of traditional farming techniques. These unique culture traits give Amish communities a distinctive cultural landscape. ***What elements of Amish culture can you see in this photograph?***

with their fingers or use bread to scoop their food. Each of these traits is considered correct where it is practiced.

Many culture traits are linked. For example, the Amish are a religious group in the United States and Canada. Traditional Amish beliefs favor a simple way of life and separation from many parts of modern society. People in Amish communities may not use automobiles, electricity, or telephones. Men are known for their simple dress, which includes broad-brimmed black hats and plain clothes. Women wear bonnets, long dresses, and shawls. Most Amish farmers do not use modern technology, such as tractors. Instead, they use horses to plow their fields. This example shows how the culture traits of religion, farming, and the use of technology can be linked.

READING CHECK: *Human Systems* How are Amish cultural traits linked, and how do they separate the Amish from the rest of society?

One important change in many cultures involves attitudes toward the role of women in society. In the 1900s women in many countries entered the workforce in large numbers, became politically active, and enjoyed more freedoms and independence. These cultural changes allowed many women to seek careers in traditionally male fields, such as steelmaking.

Culture Regions The world is made up of many different culture groups. Each group has its own way of life. This includes the way they use their land, the resources they depend on and value, and what religions they practice. It also includes settlement patterns, attitudes toward the role of women in society, forms of government, and other customs. These different ways of life generate distinctive cultural landscapes around the world.

An area in which people have many shared culture traits is called a **culture region**. An individual country may be a single culture region. For example, Japan has one dominant culture throughout the country. However, sometimes countries include many culture regions. For example, many countries in Africa include dozens of different **ethnic groups**. An ethnic group is a human population that shares a common culture or ancestry. African countries that are home to many different ethnic groups include Kenya, Nigeria, and South Africa.

Country borders sometimes divide culture regions and separate members of one ethnic group. For example, the Kurds are an ethnic and linguistic group in Asia. However, the borders of Iran, Iraq, and Turkey divide their traditional homeland. A culture region can also be made up of several countries. Australia and New Zealand make up one such culture region. The two countries share the same language and have similar traditions and systems of government.

READING CHECK: *Human Systems* What traits make a place's cultures distinctive?

Culture Change

Culture traits change through time. Sometimes, these changes are as simple as wearing new clothing styles. Ask some adults what clothing styles were popular when they were your age. Clearly, for many people clothing is more than just something that covers our bodies. It also reflects personal values, tastes, and style.

Often, changes in culture traits are even more complex than new fashions. For example, in the 1800s people in the United States walked or rode streetcars downtown to do their shopping. In the 1900s, buses and cars eventually replaced streetcars. Today modern expressways and subways take shoppers away

INTERPRETING THE VISUAL RECORD

Innovation and diffusion of ideas cause culture change in places like this area near Breves, Brazil. With satellite connections and other modern technologies, diffusion across the world is now possible in a matter of days. ***How might cultural patterns in this part of Brazil influence the process of diffusion?*** TEKS

from city centers to huge shopping malls in the suburbs. As transportation systems have changed, so have culture traits such as shopping patterns and personal mobility. Can you think of other examples of changing cultural traits in the United States?

Throughout history, general processes such as migration, war, and trade have caused cultures to change. These processes expose culture groups to new ways of life, including new languages, resources, and technologies. For example, when Spain colonized the Americas, Spanish settlers brought horses with them. Before Spanish settlement, American Indian cultures had not been exposed to these animals. Soon, however, they acquired horses, which became an important part of their culture. Many American Indian groups became skilled riders and used horses in hunting and in war.

The general processes of migration, war, and trade still cause culture change around the world. New culture traits are added as older ones fade from memory. Today modern communications and transportation systems have speeded culture change. Many of the world's young people seem to enjoy such fast change. They look forward to new technologies and ways of thinking.

When an individual or group adopts some of the traits of another culture, the process is called **acculturation**. Immigrants to the United States provide good examples. They often have to learn a new language and adopt a new way of life in this country. It is often difficult and stressful. However, through time, they often fully accept a new culture. When immigrant groups adopt all of the features of the main culture, it is called assimilation.

READING CHECK: ***Human Systems*** How do general processes such as migration, war, and trade cause culture change?

Innovation and Diffusion Two concepts help us understand how cultures change. The first is **innovation**—new ideas that a culture accepts. People are always thinking of new ways to do things. However, only new ideas that are useful will last. Some innovations happen just once and then spread throughout the world. For example, baseball was developed in the United States and later spread to the Caribbean and Asia. Other innovations are discovered in different places at different times. For example, ways for building boats developed independently among people all over the world.

The second important concept for understanding how cultures change is **diffusion**. Diffusion happens when an idea or innovation spreads from one person or group to another and is adopted. For example, jazz is an American form of music that took hold in New Orleans. Later, it spread to other parts of the United States and the world. Certain factors can aid or slow diffusion. For example, physical barriers like mountains and deserts can slow diffusion. Cultural similarities, such as shared languages, can aid diffusion from one group to another.

Types of Diffusion How does diffusion happen? Culture traits can spread to new places in several ways. Sometimes, information about a new idea or innovation spreads throughout a society. This is called expansion diffusion.

Culture traits also spread when people move to new places and take their culture with them. This is called relocation diffusion. That is how English, a European language, became the main language in Australia and New Zealand. Religions often spread through this type of diffusion. For example, Judaism diffused throughout the United States with European immigrants.

Culture traits sometimes spread from places of greater size and influence to smaller places. This is called hierarchical diffusion. For example, new fashions and music styles often begin in Los Angeles or New York. These large cities are centers of the entertainment industry. From these cities, the new styles spread to other large cities. Eventually, the latest fashions reach small towns.

READING CHECK: ***Human Systems*** What are some factors that might aid or slow diffusion?

Globalization Today communications networks like the Internet and satellite television deliver information constantly to people around the world. These innovations are spreading culture traits more quickly than at any other time in human history. As a result, a global set of culture traits is taking hold. For example, people around the world now eat the same kinds of fast-food and

INTERPRETING THE VISUAL RECORD

The spread of global businesses is an important part of the process of globalization. For example, American fast-food restaurants have diffused to Russia and Eastern Europe, Latin America, Asia, and other world regions. This fast-food restaurant is in Moscow. ***How might the spread of U.S.-based fast-food franchises such as this one be an example of cultural convergence?*** TEKS

wear the same types of jeans. Many also drive the same kinds of cars and enjoy listening to the same music. This process, in which connections around the world increase and cultures become more alike, is called **globalization**. Today it often has its roots in the United States. For example, American English, slang, popular culture, and businesses have spread to many countries and caused culture change. However, globalization also affects American culture. For example, Japanese electronics, German cars, and Italian fashions are all popular in the United States.

Globalization is an example of cultural convergence—different cultures blending together. This happens when the ideas, habits, and institutions of one culture come in contact with those of another culture. The spread of name-brand soft drinks throughout the world is an example. So is the popularity of Mexican food in the United States.

TEKS **READING CHECK:** ***Human Systems*** What are some examples of globalization and the spread of cultural traits?

INTERPRETING THE VISUAL RECORD

Geographers are fascinated by globalization and traditionalism. Our world is full of places that are modern and innovative and places that are traditional and historic. Here, bagpipers in Edinburgh, Scotland, play traditional Scottish music. The bagpipe has become a well-known symbol of Scotland. ***Besides bagpipes, what other examples of cultural traditions can you see in this photograph?***

Traditionalism The opposite of globalization is **traditionalism**. Traditionalism means following longtime practices and opposing many modern technologies and ideas. In some cases, increasing religious **fundamentalism** has been a reaction to the spread of modern culture. The word *fundamentalism* can describe any movement in which people believe in strictly following certain established principles or teachings. In fact, many people believe that old ways of doing things should not be changed. They argue that old traditions tie people to their community, religion, and ancestors.

Traditionalism contributes to cultural divergence—the process of cultures becoming separate and distinct. This process happens when one group protects its culture from outside influences and another group welcomes change. Places with traditional cultures preserve the past, and their landscapes change very little. For example, ethnic celebrations and festivals keep cultures separate and distinct. These festivals celebrate a culture's unique history, identity, and way of life.

READING CHECK: ***Human Systems*** Why do some favor traditionalism? Does traditionalism contribute to cultural convergence or cultural divergence?

Section 2 Review

TEKS Questions 1, 2, 3, 4, 5

Homework Practice Online

Keyword: SW3 HP5

Define

culture, culture traits, culture region, ethnic groups, acculturation, innovation, diffusion, globalization, traditionalism, fundamentalism

Reading for the Main Idea

1. ***Human Systems*** What do geographers study when they look at a region's culture?
2. ***Human Systems*** How does diffusion cause culture change?

Critical Thinking

3. **Supporting a Point of View** What do you think are the most important parts of your culture? Why?
4. **Contrasting** How might globalists and traditionalists view cultures, places, and regions differently?

Organizing What You Know

5. Copy the graphic organizer below. Use it to describe the three different types of diffusion. Then give two examples of each.

Expansion diffusion	Relocation diffusion	Hierarchical diffusion

Geography for Life

Geography and History

To understand our world, we must study its geography and its history at the same time. All historical events happen at a certain place. In turn, environmental conditions affect historical events.

Earth's landscapes and environmental conditions change over time. These changes in physical geography can determine the fates of entire civilizations. Most of these changes are slow and their effects very gradual. Other changes are more dramatic. For example, to the ancient Greeks, Ephesus was a key commercial city on the coast of what is now Turkey. Over time a river filled the city's harbor with sediment. When it was no longer useful as a port, Ephesus faded from importance.

Environmental changes such as climate shifts can also affect history. For example, a period of weather extremes began about A.D. 1300. In Europe heavy rains and early freezes caused crops to fail. In some places, many people starved to death. Because many survivors were poorly nourished, they were vulnerable to diseases, such as the plague. The spread of the plague, or Black Death, across Europe turned the 1300s into a century of disasters.

Physical features can also influence history and the distribution of culture groups. Consider why people have settled in certain places. You may live in the Mississippi River watershed in a city with a French name, such as St. Louis or Des Moines. Why did French explorers and settlers go there instead of farther east or west? The region's waterways give us the answer. The Great Lakes and the Mississippi River provided a route south and west from French territory in Canada. The Mississippi became a highway into the continent's heart for those who had crossed the Atlantic from France. Similarly, high mountains between Afghanistan and Pakistan seem like a huge wall in the path of travelers. However, the Khyber Pass cuts through this wall. In places this ravine is only 50 feet (15 m) wide. Yet over the centuries invading armies and migrating peoples have poured through the pass. Each group has left its stamp on the region's diverse cultures.

Just as physical features can make diffusion of peoples and ideas easier, they can also make it more difficult. For example, the Sahara forms a vast barrier across northern Africa. During the A.D. 600s, Arab armies brought Islam to the area north of the desert. However, the Sahara slowed the spread of Arab influence to the south. Other features of Africa's natural environments also limited European influences for a time there. The mosquitoes that carry malaria require a warm, rainy climate. Because the insects thrived in Africa's tropical forests, movement by Europeans into many areas was delayed. Many Africans who had lived in those regions for generations had some resistance to malaria. Europeans did not.

In turn, history affects both physical and human geography. For example, European sailors of the 1500s changed forever the human geography of the places they opened to settlement. An area's languages, religions, ethnic makeup, architecture, clothing, customs—all these would change. Humans have also changed physical geography. We have altered the land throughout our history.

Since ancient times, the Khyber Pass has been an important travel route between Afghanistan and Pakistan. Persians, Greeks, Mughals, Afghans, and other peoples have all passed through it.

Applying What You Know (TEKS)

1. **Summarizing** What are some ways in which geography affects history?
2. **Drawing Inferences and Conclusions** Look at a physical map of the world. How do you think certain deserts, mountain ranges, rivers, and other physical features affected migration patterns and the distribution of culture groups?

World Languages and Religions

READ TO DISCOVER

1. What is the geography of the world's languages?
2. What are the three main types of religions that geographers identify?

DEFINE

dialect
lingua franca
ethnic religions
animist religions
polytheism
universalizing religions
monotheism
missionaries
mosques
hajj

WHY IT MATTERS

Language and religion give character to places. They also are key causes of cultural differences. Use CNNfyi.com or other **current events** sources to learn about a conflict where religion is an important issue.

Geography of Languages

Language is one of the most important areas of study in geography. Language is important to culture because it is the main means of communication. For example, one generation passes customs and skills to the next mainly through language. Language is an important part of a culture's traditional celebrations, rituals, and ceremonies. Language also influences the routes and patterns of cultural diffusion and migration. For example, new information spreads more easily among places that have a common language. In addition, language differences can be a barrier to diffusion.

For geographers, one of the most basic facts about languages is that they have spatial characteristics. Languages are spoken in specific regions of the world and shape people's lives there. Patterns of speech also help make specific regions of the world distinctive. For example, Spanish is spoken throughout most of South America. The widespread use of Spanish in the region helps make South America distinct. However, most people in South America's largest country, Brazil, speak Portuguese. As a result, Brazil is a clear linguistic subregion within South America.

About 3,000–6,500 languages are spoken in the world today. Experts divide these languages into more than a dozen families. These families are groups of languages that experts believe have a common origin. (See the map of world language families.) Language families, in turn, are divided into language branches. For example, English is a language in the Germanic branch of the Indo-European family. Spanish and French are languages in the Romance branch of the same family. About 50 percent of the world's people speak an Indo-European language.

The language with the most speakers is Mandarin Chinese. (See the chart of principal languages of the world.) Mandarin belongs to the Sino-Tibetan language family. About 20 percent of the world's people speak a Sino-Tibetan language. Speakers of any one language might use a particular **dialect**. A dialect

Principal Languages of the World

Language	Speakers (in millions)
Mandarin	1,075
English	514
Hindi	496
Spanish	425
Russian	275
Arabic	256
Bengali	215
Portuguese	194
Malay-Indonesian	176
French	129

Source: *The World Almanac and Book of Facts 2000*

INTERPRETING THE CHART

Mandarin has by far the most speakers of any language in the world, while English is a distant second. ***Why do you think English has the second-highest number of speakers?***

World Language Families

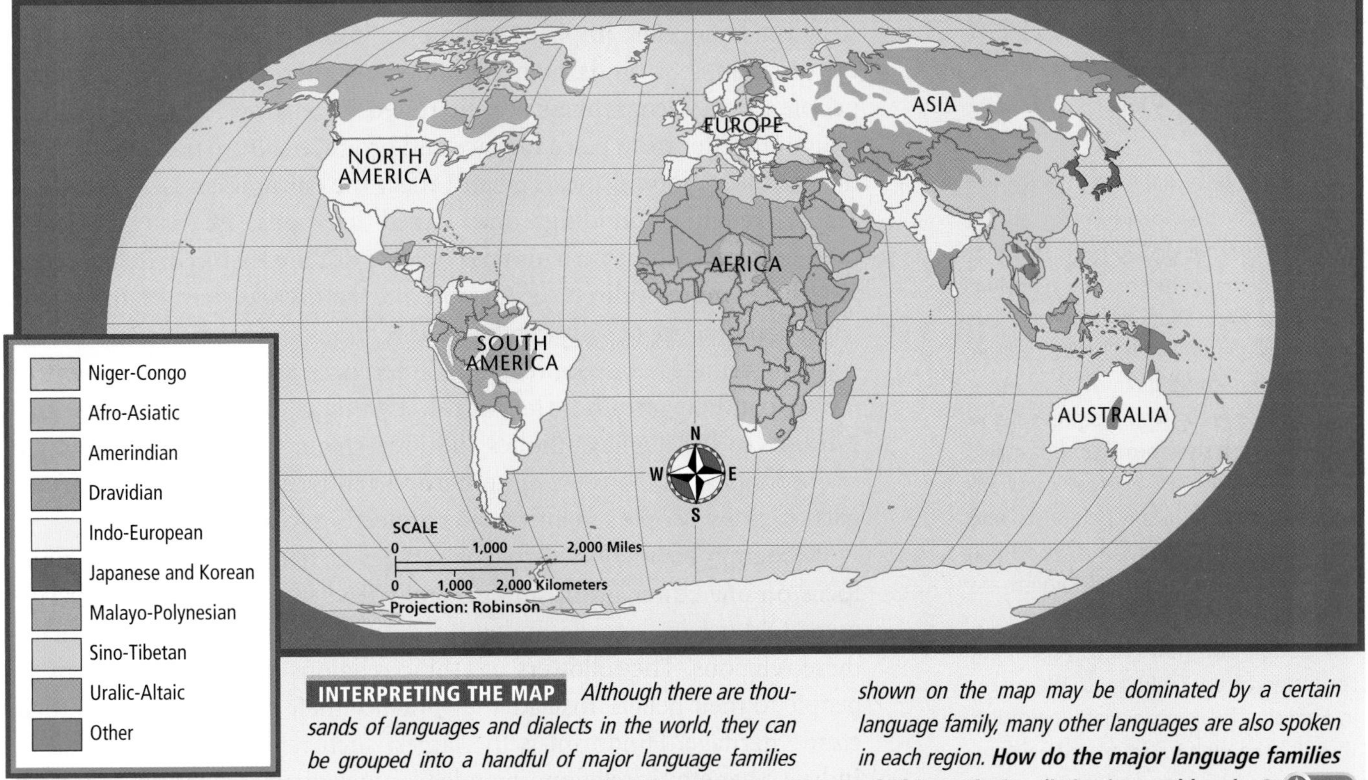

NORTH AMERICA
EUROPE
ASIA
AFRICA
SOUTH AMERICA
AUSTRALIA

- Niger-Congo
- Afro-Asiatic
- Amerindian
- Dravidian
- Indo-European
- Japanese and Korean
- Malayo-Polynesian
- Sino-Tibetan
- Uralic-Altaic
- Other

SCALE
0 1,000 2,000 Miles
0 1,000 2,000 Kilometers
Projection: Robinson

INTERPRETING THE MAP *Although there are thousands of languages and dialects in the world, they can be grouped into a handful of major language families thought to have a common origin. Although the areas shown on the map may be dominated by a certain language family, many other languages are also spoken in each region.* ***How do the major language families of Africa make it a distinctive world region?*** TEKS

is a regional variety of a language. For example, British English and American English are dialects of the same language. While different dialects can usually be understood by speakers of the same language, they contain distinctive words and pronunciation. These unique words and phrases can usually provide clues to where a speaker is from. Can you think of some dialects of English found in the United States?

To understand language patterns, geographers study where a language comes from and how it has spread. Languages spoken by very few people usually indicate remote populations whose people have not migrated. We can see examples of this in some mountain areas. For example, in the Caucasus Mountains many different languages are spoken by small groups of people. Some of these groups have been isolated in mountain valleys since ancient times.

English has become the most widespread language in the world. English traces its beginnings to Anglo-Saxon, which became distinct about 1,500 years ago on the island of Great Britain. By about 1500, it had developed into Modern English. The language spread rapidly when Britain began to colonize large areas of the world a few hundred years ago. Today English is the main language of globalization and the Internet. Millions of people now speak English as a second language. English words have also become part of other languages. For example, baseball is called *beisbol* in Spanish. The French refer to the weekend as *le weekend*. English has become a **lingua franca**—a language of trade and communication—for the whole world.

Some linguists believe that 90 percent of the languages spoken today may die out in the next 100 years.

READING CHECK: *Human Systems* How has English spread around the world?

Connecting to
TECHNOLOGY

English on the Internet

The Internet links millions of the world's computers. It developed out of efforts to make sure that U.S. authorities would be able to communicate after a nuclear war. Ever since then the Internet has been dominated by the United States and the English language. Today more than 80 percent of the Internet's home pages, or Web sites, are in English. Also, about 90 percent of Internet users are English-speakers. Many scientists and businesspeople around the world use the Internet—and therefore English—to communicate with each other.

People in non-English-speaking countries often build their Web sites in English. This is particularly helpful if they use the Web for international communication. In some countries, using English has become a status symbol. It suggests that a Web site's information may interest people around the world.

Analyzing In what way is the use of English on the Internet an example of cultural convergence? TEKS

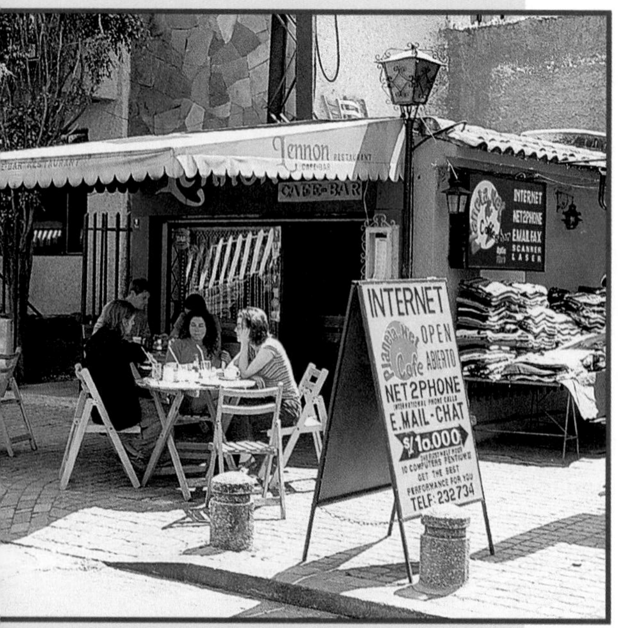

An Internet cafe in Ecuador

Geography of Religion

Religion is another important topic in human geography. It is a key culture trait that binds many societies together and gives meaning to people's lives. Geographers are interested in religion for many reasons. Religions differ from place to place and produce culture traits that can be mapped. Religions also can greatly affect the cultural landscape. For example, religious buildings and sacred locations are usually clearly marked places where certain types of behavior are required. Religion can also have more indirect effects on the landscape. For example, the Christian practice of burying the dead requires Christian societies to use land for cemeteries. Also, religious differences are a key component of many conflicts around the world. Conflicts between Hindus and Muslims in India and Catholics and Protestants in Northern Ireland are two examples. Perhaps even more importantly, religion is an important part of many people's identity and connection to a certain place.

Geographers identify three main types of religions. **Ethnic religions** focus on one ethnic group and generally have not spread into other cultures. Old beliefs, legends, and customs of different ethnic groups shape these religions. The followers of ethnic religions do not seek to convert people to their beliefs. Instead, they practice their religion as part of their ethnic heritage. Hinduism is the largest such religion. It is centered in India. Other ethnic religions include Confucianism and Taoism in China, Shintoism in Japan, and Judaism in Israel. (See the World Religions map.)

In **animist religions**, people believe in the presence of the spirits and forces of nature. **Polytheism**, or the belief in many gods, is an essential part of most of these religions. Animist religions are also often considered ethnic religions because particular peoples practice them. They are common in many traditional societies and likely were practiced long before more modern types of religions became common.

In contrast to ethnic religions, **universalizing religions** seek followers all over the world. They hope to appeal to people of many different cultures. More than half of the world's people follow these religions. Christianity and Islam are universalizing religions. Each is based on **monotheism**—the belief in one god. These religions also have **missionaries**, who help spread the religion. New converts are accepted through symbolic rituals and initiations. The ultimate goal of universalizing religions is to spread their beliefs to the entire world. Therefore, it is not surprising that universalizing religions are the most rapidly growing religions.

Religion provides a rhythm to daily life and shapes distinctive cultural patterns. In Islamic countries, for example, people stop to pray several times a day. When praying, they face toward Mecca, the spiritual center of Islam. Houses of worship called **mosques** are common. Important mosques are beautifully designed and decorated. Large crowds gather for prayers there, particularly on Friday at noon. Many people in Islamic countries also make a special religious journey to Mecca called a **hajj**. Islam requires its followers to make this journey at least once in their lifetime.

World Religions

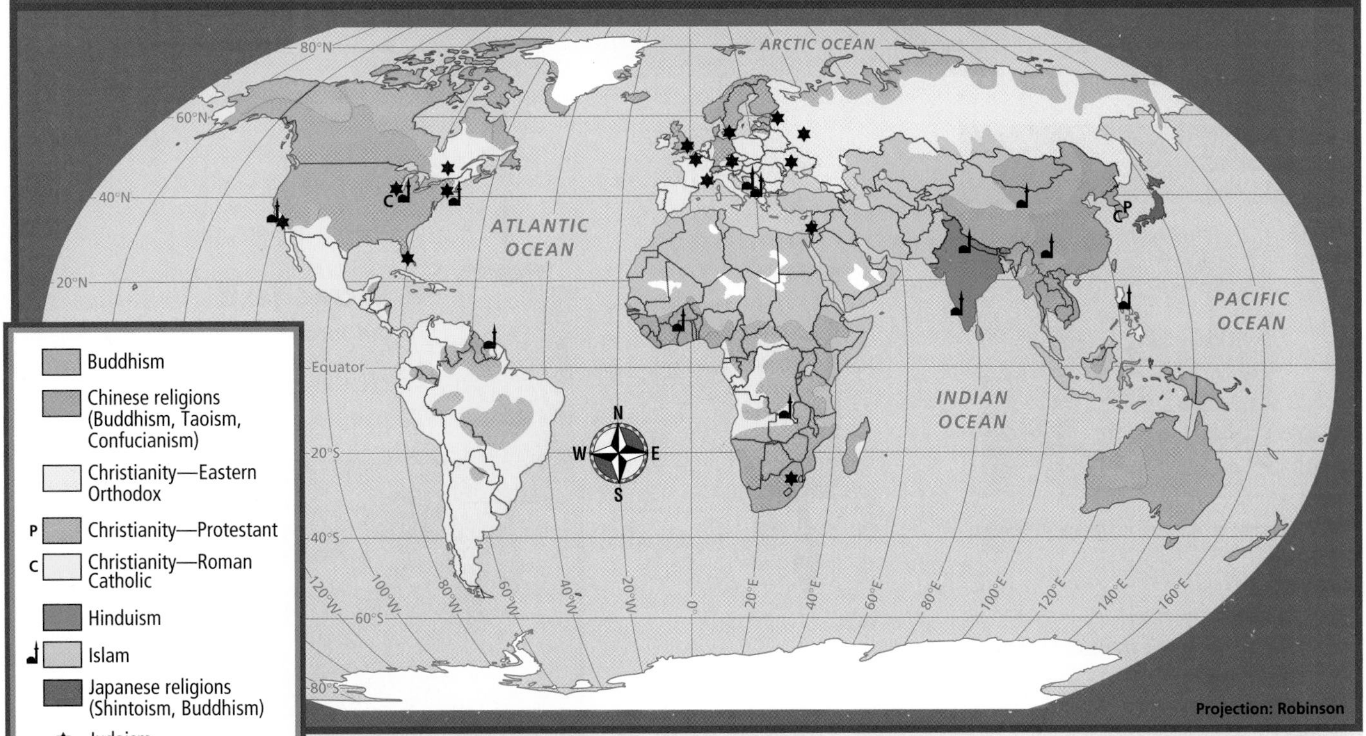

- Buddhism
- Chinese religions (Buddhism, Taoism, Confucianism)
- Christianity—Eastern Orthodox
- P Christianity—Protestant
- C Christianity—Roman Catholic
- Hinduism
- Islam
- Japanese religions (Shintoism, Buddhism)
- ✱ Judaism
- Local religions
- Uninhabited

ARCTIC OCEAN
ATLANTIC OCEAN
PACIFIC OCEAN
INDIAN OCEAN

Projection: Robinson

INTERPRETING THE MAP *The world's major religions each dominate certain regions of the world. For example, Buddhism is most widely practiced in Asia, while Hinduism dominates India. However, within each major world region, a huge number of smaller religions also exist. Also, a growing number of people are secular, or nonreligious.* ***Based on the map, which religions are practiced in the United States? In Africa? How widespread is Christianity? How widespread is Islam?***

While religion remains important around the world, in some places it has declined. A growing number of people in these places are secular, or nonreligious. Europe in particular has experienced declines in the number of people actively participating in religious activities, such as attending church each week. There are many possible reasons for this decline in religion. For example, the development of modern technologies and science have caused some people to question the beliefs of traditional religions.

READING CHECK: *Human Systems* What are the three main types of religions?

TEKS Questions 1, 4

Review

Homework Practice Online
Keyword: SW3 HP5

Define dialect, lingua franca, ethnic religions, animist religions, polytheism, universalizing religions, monotheism, missionaries, mosques, hajj

Reading for the Main Idea

1. *Human Systems* Why are communication and beliefs so important in human geography?
2. *Human Systems* About how many languages are spoken in the world today? Which language has the most speakers?
3. *Human Systems* What are some examples of ethnic religions?

Critical Thinking

4. **Analyzing** How has religion shaped the cultural landscape of your community?

Organizing What You Know

5. Copy the chart below. Use it to describe, compare, and contrast the three main types of religions.

Animist religions	Ethnic religions	Universalizing religions

World Religions

Abraham
c. 2000 B.C.

INTERPRETING BIOGRAPHY

According to biblical sources, Abraham was originally an inhabitant of Ur, where he was known as Abram. After establishing a covenant with God, he changed his name to Abraham, meaning, "father of many." This reflected a promise that Abraham would be the leader of many peoples.

Abraham is a sacred figure to three religions: Judaism, Christianity, and Islam. All three faiths trace their heritage to Abraham through the line of prophets descended from his sons. Some people consider Abraham the first historical figure to follow a faith with one supreme deity. ***How does Abraham's connection to modern religions reflect his name "father of many"?***

Sacred Text: ***the Torah***

Sacred Site: ***Jerusalem***

Sacred Symbol: ***Star of David***

Special Days

Passover, in spring; Rosh Ha-Shanah and Yom Kippur, in autumn; and Hanukkah, in late autumn or winter

Hindus consider it a sacred duty to bathe in the holy waters of the Ganges River. This ritual cleanses the bather's mind and spirit.

Sacred Sites:
the Ganges River, the city of Varanas

Sacred Texts:
The Vedas, Bhagavad Gita

Festival of Holi

Mahavira
c. 599–527 B.C.

INTERPRETING BIOGRAPHY

Over the years many religious leaders added to and expanded Hindu thought. One such person was Mahavira, also known as Vardhamana. He was born into a warrior clan in northeastern India. At the age of 30 he left his home and entered the forest to find spiritual fulfillment. He got rid of all his personal possessions. For more than 12 years, he wandered the countryside with nothing to his name and little contact with other people.

After he felt he had gained the answers to his questions about life, Mahavira began teaching others. He believed the key to enlightenment was to live apart from the material world as much as possible. Many early Hindus were influenced by his ideas. Eventually his beliefs became the basis of Jainism, a new religion. ***How did Mahavira influence Hindus?***

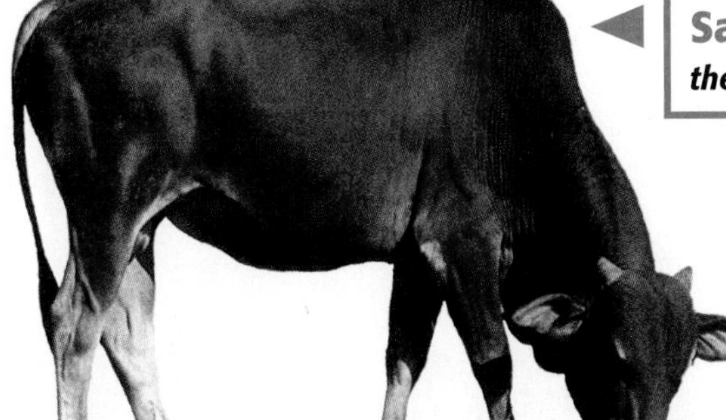

Sacred Creature:
the Cow

The cow is a particularly sacred animal in the Hindu faith in part because of the important role it has played in sustaining life.

Special Days

Festival of Holi, in spring; Diwali, or Deepavali (Festival of Lights) in autumn

Buddhism

Siddhartha Gautama
563 B.C.–483 B.C.

INTERPRETING BIOGRAPHY

Siddhartha Gautama was born the son of an Indian prince. At the age of 29, he left his palace and was shocked by the suffering he saw. As a result, he wondered about the great problems of life. "Why does suffering exist?" he asked. "What is the meaning of life and death?"

Gautama decided to spend the rest of his life seeking answers to his questions. In what is now called the Great Renunciation, he put aside all his possessions, left his family, and set out to search for truth. One day, while meditating under a bodhi tree near the town of Bodhgaya, Gautama realized the key to ending suffering. This led to the development of the Four Noble Truths and the Eightfold Path, which all Buddhists follow. After his experience under the bodhi tree, Gautama became known as the Buddha, or "enlightened one." ***How did Siddhartha Gautama reject his old life?***

Bodhi tree

Sacred Site: ***Bodhgaya***

Buddha Day festival

Sacred Objects: ***Statues of the Buddha***

Sacred Text: ***the Pali Canon***

Special Days

Buddha Day, celebrated at the full moon in May

Because Confucius emphasized the importance of education and learning, his followers celebrate his birthday as Teacher's Day.

Sacred Text:
The Analects

Sacred Symbol:
Yin-yang

Sacred Site:
Confucian Temple

Confucius
551 B.C.– 479 B.C.

INTERPRETING BIOGRAPHY

Westerners know K'ung Ch'iu as Confucius. He was born in the Chinese province of Lu. He spent much of his life tutoring and working in low-level government positions.

Confucius grew frustrated by other officials around him. In mid-life he left his government position and traveled the countryside promoting reform. After 13 years he returned to Lu to teach about the ideas he had formed during his travels.

Confucius had little to say about gods, the meaning of death, or the idea of life after death. For this reason some people do not consider Confucianism a religion, although the goal of Confucianism is to be in "good accord with the ways of heaven." Many followers of Confucianism practice his ideas as religion. ***Why do some people not consider Confucianism a religion?***

Special Days

Teacher's Day, in August or September

Christianity

Jesus
c. 6 B.C.– c. A.D. 30

INTERPRETING BIOGRAPHY

According to the Gospels, Jesus was born in Bethlehem, near Jerusalem, but grew up in Nazareth. He was a Jewish carpenter with a strong interest in religious matters. In time he began preaching. As he traveled through the villages of Judea, he assembled 12 disciples to help him preach.

Jesus often taught using parables, or stories intended to teach a moral lesson. His followers believe that Jesus was the Son of God and that he was resurrected after his death. ***Why do you think Jesus taught using parables?***

Sacred Text:
The Bible

Sacred Sites:
Bethlehem
Jerusalem

Sacred Symbol:
cross

A Christmas candlelight service

Special Days

Christmas, on December 25 (January 6 for some Orthodox churches); Easter, in the spring

▲ **Sacred Objects:**
prayer rugs

▲ **Sacred Text:**
Qur'an

Muhammad
C. A.D. 570 – A.D. 632

INTERPRETING BIOGRAPHY

In Islam, Muhammad is a messenger or prophet of God. Muhammad was born in Mecca (Makkah) and orphaned at an early age. He was from a respected but poor family. They belonged to a leading tribe of caravan merchants and keepers of Abraham's shrine and pilgrimage site, the Ka'bah.

Islam prohibits the use of images for Muhammad. The symbol above, which means "Muhammad is the Prophet of God," is often used in place of his picture. ***Why is a symbol used in place of Muhammad's image?***

Muslim woman praying during Ramadan

Thousands of Muslim pilgrims gather around the Ka'bah, in Mecca.

Sacred Sites:
Mecca (Makkah), Medina, Jerusalem

Special Days

Fast of Ramadan, during the entire first month of the Islamic year and Id al-fitr at the end of Ramadan; Id al-Adha at the end of the hajj

CHAPTER 5 Review

Building Vocabulary

On a separate sheet of paper, explain the following terms by using them correctly in sentences.

demography	diffusion
population density	dialect
migration	ethnic religions
culture	animist religions
culture traits	polytheism
ethnic groups	universalizing religions
acculturation	monotheism
innovation	

Locating Key Places

On a separate sheet of paper, match the letters on the map with their correct labels.

Hinduism	Christianity—Eastern Orthodox
Buddhism	Christianity—Protestant
Islam	Christianity—Roman Catholic

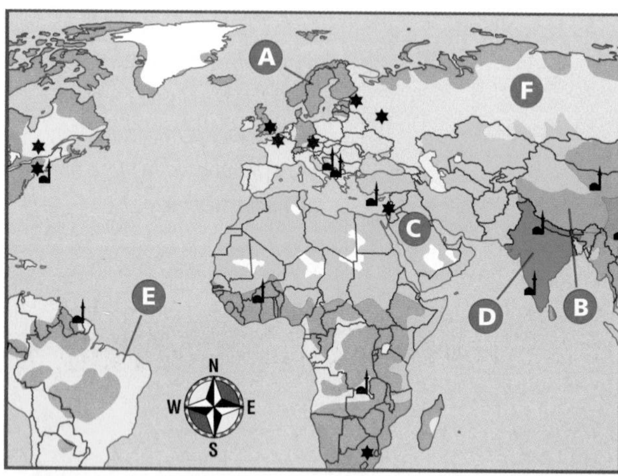

Understanding the Main Ideas

Section 1

1. *Human Systems* What do population geographers use to study population patterns and trends?
2. *Human Systems* What factors influence migration?

Section 2

3. *Human Systems* What are culture traits? What is one example of how culture traits vary from place to place?

Section 3

4. *Human Systems* Why is language fundamental to culture?
5. *Human Systems* Why is religion fundamental to culture?

Thinking Critically for TAKS

1. **Making Generalizations** How do you think innovations like computers, the Internet, and other modern telecommunications affect culture change today?
2. **Identifying Points of View** How might a people's culture influence the way they view other cultures, places, or regions?
3. **Identifying Cause and Effect** In what ways do you think religion shapes daily life and cultural patterns in this country as it does in Islamic countries? What are some examples?

Using the Geographer's Tools

1. **Interpreting Graphs** Study the World Population Growth graph in Section 1. How would you describe the world's population growth before about 1600? Afterward? What might this graph tell us about future population growth?
2. **Interpreting Maps** Study the World Religions map in Section 3. Which areas of the United States does the map identify as Roman Catholic? Why do you think this is so?
3. **Drawing Graphs** Use the information in the chart of Principal Languages of the World in Section 3 to draw a bar graph of the world's major languages. Which major world regions have many English-speakers?

Writing for TAKS

Imagine a typical day where you live. What activities are going on? How do these activities reflect local culture? What kinds of words and expressions do people use? What kinds of architecture, clothing, food, and music are common? Write a journal entry about the culture of your area. When you are finished with your journal entry, proofread it to make sure you have used standard grammar, spelling, sentence structure, and punctuation.

SKILL BUILDING

Geography for Life

Using Questionnaires

Human Systems Use a questionnaire to find out what other students in your class think would be a good innovation. First, design a short questionnaire that asks the questions you want the other students to answer. For example, you could ask: What innovation would you like to see developed in your lifetime? Why do you think this would be a good innovation? How would it change your life? When the questionnaires are completed, compare them. What did you learn?

Building Social Studies Skills

go.hrw.com TAKS PREP ONLINE keyword: SW3 T5

Interpreting Graphs

Study the bar graph below. Then use the information to help you answer the questions that follow.

1. Which country is a source of between 500,000 and 1,000,000 refugees?
   a. Somalia
   b. Iraq
   c. Afghanistan
   d. Sierra Leone
2. Which regions of the world appear to have the most difficult refugee problems? Support your answer.

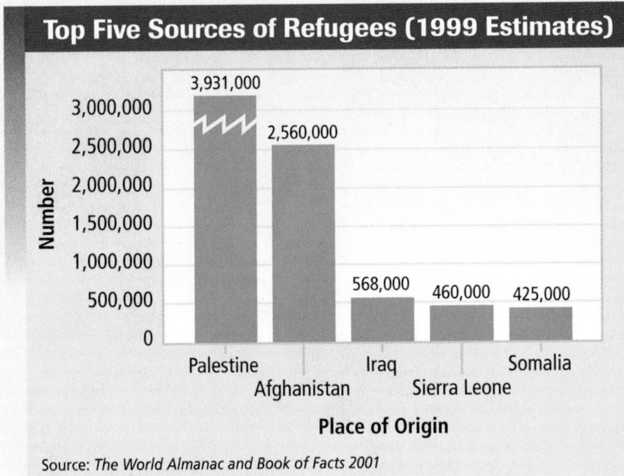

Source: *The World Almanac and Book of Facts 2001*

Analyzing Secondary Sources

Read the following passage and answer the questions on a separate sheet of paper.

> **"The demographic transition is a model that shows how birthrates and death rates dropped in many Western countries as they developed modern economies and industries. Most of the world's richest and most technologically advanced countries have experienced a transition from high birthrates and death rates to low birthrates and death rates. In addition, many of the world's poorer countries are now in the middle of similar population changes."**

3. Birthrates and death rates dropped in many Western countries as they developed what?
   a. widespread farming
   b. modern economies and industries
   c. resistance to diseases
   d. strong militaries
4. What do you think are examples of low birthrates and death rates? Of high birthrates and death rates?

Alternative Assessment

PORTFOLIO ACTIVITY

Learning about Your Local Geography TEKS

Group Project: Research

Plan, organize, and complete a research project about historical migration in your community. Check your local library to find information about the early settlement and history of your community. Talk to local historians to find out what they know about the settlement of your area. As a group, put together a short presentation on historical migration in your community that answers the following questions: When was your area settled? Where did the settlers come from? What brought them to your area? How did they influence the area's culture? How did physical geography affect the routes migrants took to reach your community?

internet connect

Internet Activity go.hrw.com
KEYWORD: SW3 GT5 TEKS

Choose a topic on human geography to:
- research and report on world population growth.
- examine push and pull factors that cause the migration of refugees.
- answer questions on demographic calculations and data from the United States Census Bureau.

CHAPTER 6 Human Systems

People affect Earth's geography. They use resources, earn a living, build communities, and compete for control of Earth's surface. All these activities shape the world around us.

Parliament building in Budapest, Hungary

The skyline of Dallas, Texas

Dogon women selling vegetables in Mali

Economic Geography

READ TO DISCOVER

1. What are the three main types of economic systems?
2. How are developed countries and developing countries different?

WHY IT MATTERS

The world's poorer countries are constantly struggling to improve their economies. Use CNNfyi.com or other **current events** sources to find news about a country facing challenges as it tries to develop its economy.

DEFINE

market economy
free enterprise
capitalism
command economy
communism
gross national product (GNP)
gross domestic product (GDP)
industrialization
literacy rate
developed countries
infrastructure
developing countries

Economic Systems

Economic geography deals with how people earn a living and use resources and with the links among economic activities. Economic geographers group money-making activities into four categories. These are primary (first order), secondary (second order), tertiary (third order), and quaternary (fourth order) activities. (See the chart.)

Primary Activities	Secondary Activities
• use natural resources directly • **location:** at the site of the natural resource being used • **examples:** wheat farming, iron mining	• use raw materials to produce or manufacture something new • **location:** close to the resource or close to the market for the finished product • **other factors affecting location:** labor, energy, and land costs • **examples:** processing wheat into flour, manufacturing steel
Tertiary Activities	**Quaternary Activities**
• provide services to people and businesses • **location:** usually near customers • **examples:** bakeries, car dealerships	• process and distribute information • **location:** anywhere • **other factors affecting location:** access to skilled workers, good transportation and communications systems, places with pleasant climates and a high quality of life • **examples:** plant-genetics research, automotive engineering

INTERPRETING THE CHART *The most basic economic activities, such as farming, extract resources directly from Earth and have locations that are closely tied to environmental conditions and the distribution of resources. More complex activities, such as software development, produce more sophisticated goods and have locations that are not as tied to environmental conditions.* ***What factors affect the location of the different types of economic activities?*** TEKS

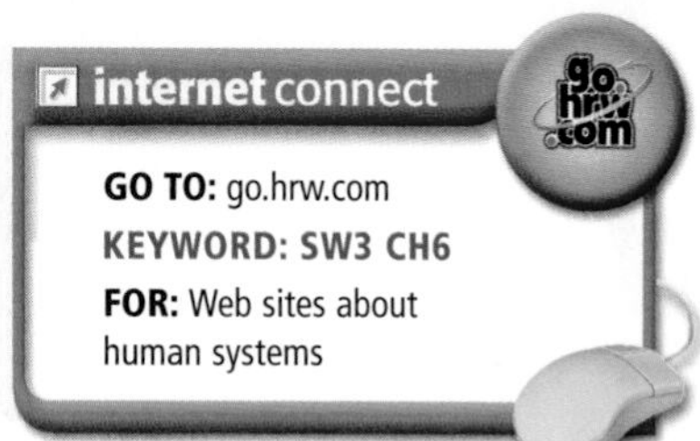

Economic Activities Economic activities that use natural resources directly are called primary activities. They include farming, fishing, forestry, herding, and mining. Primary activities provide the basic raw materials for industry. They are located at the site of the natural resource being used.

Secondary activities use raw materials to produce or manufacture something new. Examples include steelmaking, processing wheat into flour, and making lumber into plywood. Dairies take a raw material—fresh milk—and process it into products like cheese and ice cream. Often, a raw material goes through several stages before it becomes a finished product. Secondary activities are usually located close to the resource being used or close to the market for the finished product.

Tertiary activities provide services to people and businesses. For example, doctors, teachers, and dry cleaners provide personal and professional services. Store clerks, truck drivers, and restaurant staff provide retail and wholesale services. Tertiary activities are usually located near customers to serve them better. However, the use of computers and the Internet is dramatically changing the locations of many service industries. For example, more customers now shop online and buy products from Web sites.

In advanced economies many workers process and distribute information. These jobs are called quaternary economic activities. They require workers with specialized skills and knowledge. Jobs include research scientists, computer programmers, and government administrators. Quaternary activities are not tied directly to resources, environmental conditions, or access to markets. Increasingly, workers in this category can be located almost anywhere.

Economic Systems The four types of economic activities are found in different economic systems around the world. There are three main kinds of economic systems. Each uses resources and produces goods in a distinctive way. The most basic economic system is a traditional or subsistence economy. In this economy, people make goods for themselves and their families. There is

INTERPRETING THE VISUAL RECORD

Large shopping malls, such as this one in New York City's Trump Tower, are common in countries that have market economies and a free enterprise system. Shoppers here can find a wide range of stores and goods concentrated in one area. ***What types of goods would you expect to find in this shopping mall?***

little surplus or exchange of goods. As a result, there are few markets—places to buy and sell things. Traditional economies are found mostly in the world's poorer countries, particularly in rural areas.

A **market economy** is another type of economy. In a market economy, people freely choose what to buy and sell. Market economies are guided by the system of **free enterprise**. This system lets competition among businesses determine the price of products. The businesses' need to make profits drives their decisions. Businesses supply products and set prices to meet demand. Free enterprise is the basis of **capitalism**. In a capitalist system, businesses, industries, and resources are privately owned.

Market economies are characterized by specialization. Businesses and regions make the goods that they can sell for the highest profit. For example, the midwestern United States has ideal conditions for producing corn, hogs, and soybeans. As a result, farmers there produce these goods and export them around the world. Most of the world's rich countries have market economies. These countries include Australia, Japan, the United States, and most European countries. Goods made in these countries are traded around the world.

The third type of economy is one in which the government makes the major economic decisions. In a **command economy**, the government decides what to produce, where to make it, and what price to charge. Prices are not based on the market forces of supply and demand. For example, it may cost $1.00 to produce a loaf of bread. However, the government may set the price at 25 cents so that people can easily afford it. Communist countries have command economies. **Communism** is an economic and political system in which the government owns or controls almost all the means of production. Cuba and North Korea are communist countries.

TEKS **READING CHECK:** *Human Systems* What are the three main kinds of economic systems?

Economic Patterns, Resources, and Technology The creation and distribution of resources affect the locations of economic activities. Resources also affect the movement of products, capital, and people. How does this happen? The need for a resource draws businesses and workers to the place where it is found. Related businesses grow nearby. Businesses must then find ways to ship their products to markets.

Consider the California Gold Rush of 1849. People from all over the world went to California to mine for gold. As a result, San Francisco grew from a small town to a city of 25,000 people in

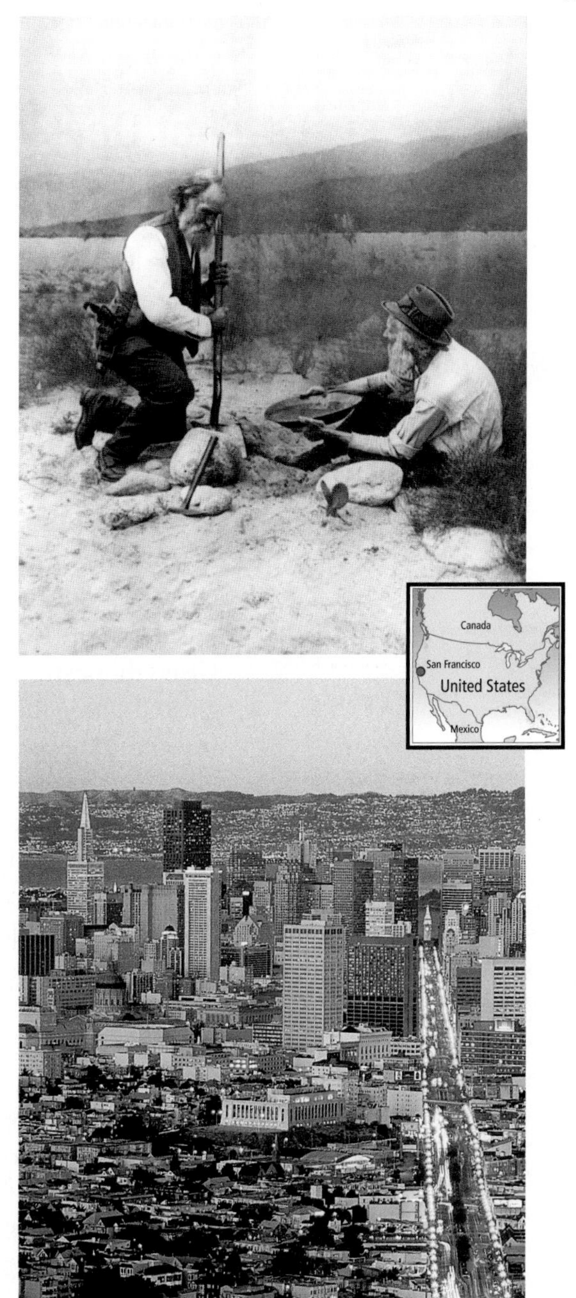

INTERPRETING THE VISUAL RECORD

The California Gold Rush of 1849 quickly transformed the city of San Francisco from a small quiet town to a city attracting thousands of immigrants from all over the world. As people and capital flocked to the city, a wide range of businesses sprang up to serve the miners, and real estate prices soared. ***How did the gold rush affect the movement of people and products to San Francisco?*** TEKS

just one year. The city also became an important financial and banking center. Banks and other businesses provided money to invest in the new mines. Transportation connections developed between gold-mining areas and ports.

Changes in technology, transportation, and communication also affect the location and patterns of economic activities. The development of refrigerated railroad cars, ships, and trucks is an example. The use of refrigerated transportation allowed farm goods to be shipped around the world without spoiling. As a result, the global production and trade of farm goods changed. For example, in the last 150 years, Argentina has become a major meat exporter. This would not have been possible without these changes in technology. Likewise, the development of computers and the Internet has affected the locations of many service and information industries. A growing number of people are now able to work at home and send their work around the world by e-mail.

TEKS **READING CHECK:** ***Human Systems*** How have changes in technology, transportation, and communication affected the locations and patterns of economic activities?

Level of Development

One of the most important topics for understanding world geography is understanding the level of development in countries around the world. Development refers to steady improvements in a country's economy and in people's quality of life.

Economic progress varies greatly among different countries and also within countries. Geographers use general measures of development to analyze such progress. You will find many of these measures in the tables at the beginning of Units 2 through 10.

Measures of Development One common measure of development is **gross national product (GNP)**. GNP is the total value of goods and services that a country produces in a year. GNP includes goods and services made by businesses owned by that country's citizens but located in foreign lands. **Gross domestic product (GDP)** includes only those goods and services created within the country. GDP becomes more useful when it is divided by the number of people living in a country. This gives us the per capita GDP, which can be used to compare income levels in different countries.

Another measure of development is the level of **industrialization.** Industrialization is the process by which manufacturing based on machine power becomes widespread in an area. In industrialized countries, many people work in manufacturing, service, and information industries. In other countries, most people work in primary economic activities, particularly farming. There are a number of additional measures of development. They include the average amount of energy people use. In addition, some measures look at the size and quality of a country's transportation and communications systems. For example, countries with more telephones per person tend to be more developed than countries with fewer telephones.

Literacy Rates

Country	Literacy Rate
China	82%
India	52%
Japan	99%
Mexico	90%
United States	97%

Source: Central Intelligence Agency, *2001 World Factbook*

INTERPRETING THE TABLE *Literacy rates reflect a country's standard of living. Along with other data, these rates can be used to compare levels of economic development in different countries.* ***Based on the information in this table, which countries do you think have the highest standards of living? Which country do you think has the lowest level of economic development?*** TEKS

Standard of Living A country's level of development, in turn, determines the standard of living of its people. Standard of living is measured by factors like amount of personal income, levels of education, and food consumption. **Literacy rate**—the percentage of people who can read and write—also reflects standard of living. Other measures of standard of living include quality of health care, technology level, and life expectancy—the average length of people's lives.

READING CHECK: *Human Systems* How is a country's standard of living measured?

Developed and Developing Countries Geographers organize the world's countries into two main groups. The richest countries are called **developed countries.** They have high levels of industrialization, and their people enjoy high standards of living. The world's developed countries include most countries in Europe, the United States, Canada, Japan, Australia, and others. Less than 25 percent of the world's people live in developed countries.

Developed countries share many features. (See the chart.) For example, each has a high per capita GDP. These countries also have high levels of education and good health care. Literacy rates are high. More literacy leads to more educated workers, who are more productive economically. Life expectancy is also high. Both birthrates and death rates are usually low. As a result, overall population growth is low.

Most people in developed countries live in cities and work in service or manufacturing industries. Few work in agriculture. The small number of farmers use advanced technology to produce large amounts of food. Developed countries also have good **infrastructure.** An infrastructure is a system of roads, ports, and other facilities needed by a modern economy. Developed countries have global market economies.

The world's poorer countries are called **developing countries** or less developed countries. These countries are less productive economically and have

Selected Countries' Statistics

		Per Capita GDP	Life Expectancy	Literacy Rate	Urban	TV Sets (per 1,000 persons)	Physicians
Developed Countries	Australia	$23,200	77, male; 83, female	100%	85%	524	1 per 389 persons
	Japan	$24,900	78, male; 84, female	99%	79%	682	1 per 522 persons
	United States	$36,200	74, male; 80, female	97%	76%	778	1 per 365 persons
Developing Countries	Afghanistan	$ 800	47, male; 45, female	31.5%	21%	4	1 per 6,690 persons
	Haiti	$ 1,800	48, male; 51, female	45%	35%	5	1 per 9,846 persons
	Mali	$ 850	46, male; 48, female	31%	29%	4	1 per 18,376 persons
Middle-Income Countries	Brazil	$ 6,500	59, male; 68, female	83%	81%	209	1 per 681 persons
	Mexico	$ 9,100	68, male; 75, female	90%	74%	251	1 per 613 persons
	Thailand	$ 6,700	65, male; 72, female	94%	21%	245	1 per 3,461 persons

Sources: Central Intelligence Agency, *The World Factbook 2001; The World Almanac and Book of Facts 2000, 2001;* U.S. Census

INTERPRETING THE TABLE *Demographic, economic, and social data for the world's countries varies greatly, as this table shows.* ***Based on the information in this table, what features do developed countries have in common? What features do developing countries have in common? How might literacy rates affect per capita GDP?*** TEKS

INTERPRETING THE VISUAL RECORD *These two photos from Australia and Afghanistan show some of the differences between developed and developing countries.* ***Based on these photos, how do you think daily life in these two places is different? What can you see in the photographs that might indicate the level of development in each place?*** TEKS

lower standards of living. This group includes most countries in Africa, Asia, Central and South America, and the Pacific Islands. Most of the world's people live in developing countries. These countries have low per capita GDPs. (See the chart.) In general, birthrates are high, and life expectancy is low. Grade schools are often available, but few people go to high school or college. Most people farm, and many homes do not have electricity. There are also not many computers, refrigerators, or televisions per person. There are few service businesses and manufacturing industries to provide jobs. Because rural areas offer few jobs, many people move to cities.

Between the world's richest and poorest countries are what some geographers call middle-income countries. Examples include Mexico, Brazil, Thailand, and Malaysia. They have features of both developed and developing countries. Their cities may be modern, but rural areas and small towns are often poor. Many of these countries have new industries, and many people are switching from rural life to city life. In many of these countries incomes are rising quickly. They may soon join the developed countries. However, some countries, like Argentina and South Africa, seem stuck in the middle-income category.

READING CHECK: *Places and Regions* What are some examples of developed and developing countries?

TEKS Questions 1, 2, 3, 4

Section 1 Review

Define market economy, free enterprise, capitalism, command economy, communism, gross national product (GNP), gross domestic product (GDP), industrialization, literacy rate, developed countries, infrastructure, developing countries

Reading for the Main Idea

1. *Human Systems* What are the main characteristics of traditional, market, and command economies?
2. *Places and Regions* What factors would you use to measure the level of economic development in a country?

Critical Thinking

3. **Analyzing Information** How might the discovery of valuable minerals nearby affect economic activities in your community?
4. **Analyzing Information** How do you think the development of new technologies such as refrigerated railroad cars, ships, and trucks can change people's perception of resources?

Organizing What You Know

5. Create a chart like the one shown below. Use it to list characteristics of developed and developing countries.

Developed countries	Developing countries

Urban and Rural Geography

READ TO DISCOVER

1. How have people used land throughout human history?
2. How does urban geography describe human settlements?
3. What are some of the ways people use land in rural areas?

DEFINE

domestication
urbanization
world cities
central business district (CBD)
edge cities
subsistence agriculture
shifting cultivation
pastoralism
market-oriented agriculture
agribusiness

WHY IT MATTERS

U.S. cities share many of the same challenges. Use CNNfyi.com or other **current events** sources to learn about a U.S. city that is trying to solve some of its problems.

Using the Land

Not all of the resources that humans need can be found in one place. Usually people must travel or trade to find everything they need. How people get resources from the land has greatly affected Earth's geography.

Origins of Domesticated Plants

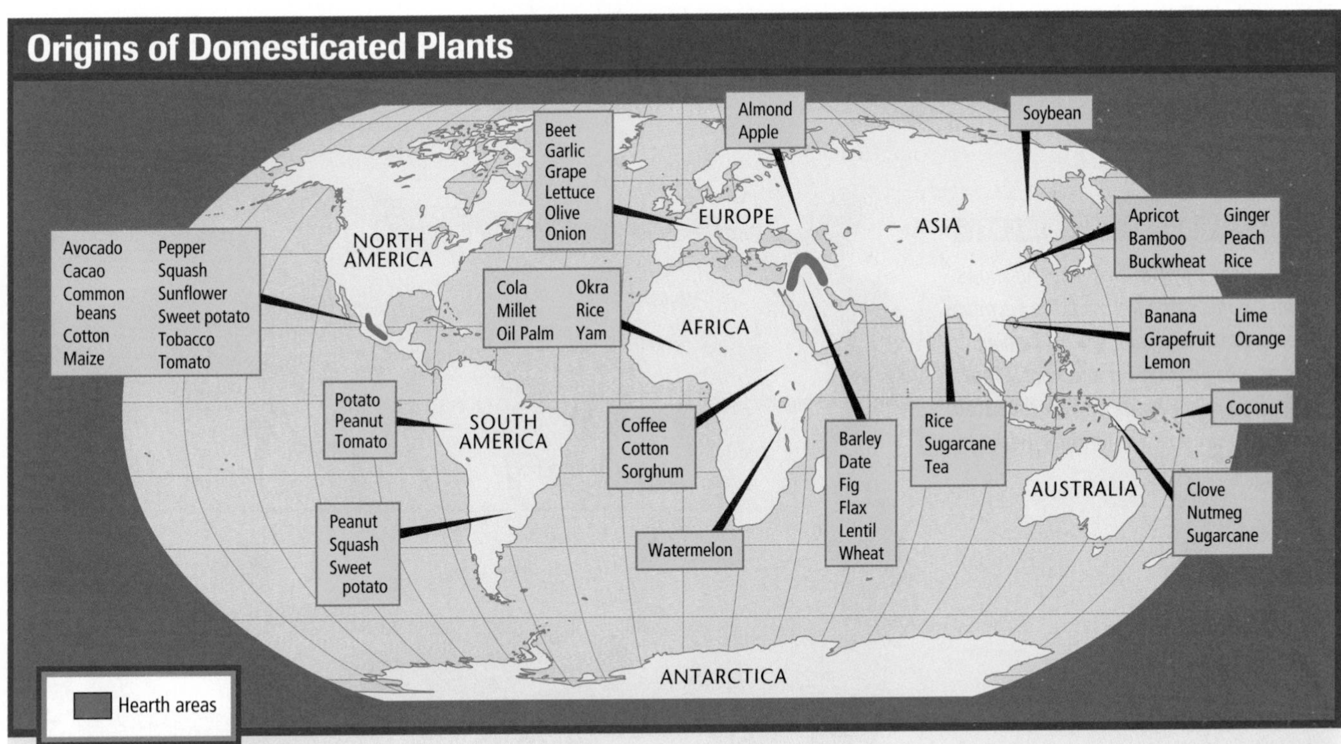

INTERPRETING THE MAP *This map shows the probable origins of many of the world's most important domesticated plants. It is a generalized map, as the exact origins of many domesticated plants are unknown, though their general area of origin is. Hearth areas were particularly important centers of plant domestication.* ***Which plants are native to Asia? To North America?***

Only about 150 of the world's 80,000 or so edible plants have been domesticated. Of these, just 20 provide about 90 percent of the world's food supply.

Hunting and gathering was the main way of life for most of human history. Often hunter-gatherers moved their camps with the seasons. They searched for different plants and animals throughout the year. Today few such societies are left. They remain mostly in environments that are too difficult for farming. One example is the Inuit. They hunt and fish along the Arctic shores of North America and Siberia.

Agriculture began about 10,000 years ago. It radically changed the way people saw the land around them and how they used it. Agriculture appeared when hunter-gatherers learned how to grow plants and tame animals for their own use. This innovation is called **domestication**. Many scientists think domestication first began in Southwest Asia. It also developed independently in other regions. (See the map of origins of domesticated plants.)

Learning how to raise animals and grow crops were two of the most important developments in human history. Farmers transformed the world's environments. They cleared land to plant crops like wheat and barley. They learned how to plow and irrigate land. Farming produced more food, so the same land could support more people. As a result, populations increased, and people could settle permanently in one place.

The first cities appeared in Southwest Asia more than 5,000 years ago. City life became possible when there was enough food so that some people did not have to farm. Instead, some of them worked as potters or weavers. Others became merchants and traders. Still others carried out governmental or religious tasks. Towns and cities began to grow. The growth in the proportion of people living in towns and cities is called **urbanization**. With the development of cities, population densities increased and communication became easier. As trade connected early cities, cultural diffusion increased.

READING CHECK: ***The Uses of Geography*** How did domestication change how people see their environment and use land?

INTERPRETING THE VISUAL RECORD

This ancient Egyptian wall painting shows domesticated cattle in Egypt. The Egyptians used cattle as draft animals for farming and as a source of meat and skins. ***What are some other kinds of domesticated animals that you are familiar with?***

General Areas of City Locations

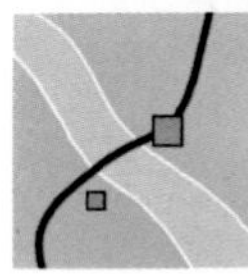
1 River crossing

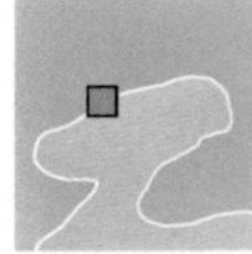
2 Natural harbor

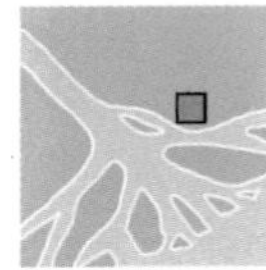
3 Head of delta

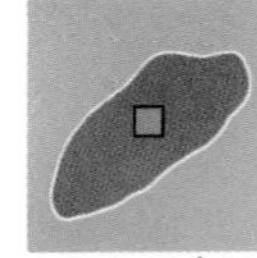
4 Defensive hilltop site

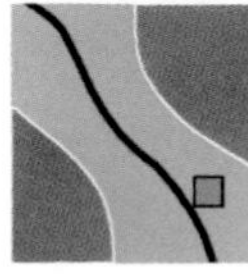
5 Defensive site controlling a pass

Urban Geography

Cities are centers of culture, trade, government, and ideas. Urban geography—the geography of cities—is an important subject. It includes the study of city locations, sizes, land use, and urban problems.

City Growth A number of important factors have influenced the site and growth of cities. One is location near key resources. Other factors include location along transportation and trade routes and at easily defended sites. (See the diagram.) Many of the world's greatest cities grew up where two or three of these factors were present. Once cities are established, continued access to other cities and resources allows them to grow and prosper.

Many cities are found near freshwater, which is a key resource. For example, many cities lie along rivers, particularly in dry regions. Cities are also found where waterfalls or rapids kept large boats from going farther upstream. Other cities are located near important mineral resources. For example, Johannesburg in South Africa grew near huge gold deposits. Houston, Texas, became an important city when oil deposits were found nearby.

Because resources are not distributed evenly, trade routes have always been vital to human societies. As a result, cities rose along these routes. For example, London and Philadelphia grew as great river ports near the ocean. Chicago expanded due to its railway connections near the shore of Lake Michigan. Singapore, in Southeast Asia, grew along a major shipping route.

Easily defended sites protect cities from invasion and allow them to prosper. For example, Jerusalem began on a hilltop that was easier to defend than surrounding lowlands. The first settlers in Paris lived on islands in the middle

INTERPRETING THE DIAGRAM

This diagram shows the types of locations that are likely to be selected as sites for villages, towns, or cities. Topography, access to resources and transportation routes, and other factors have long affected settlement locations. ***How might these locations allow cities to grow and prosper? Can you think of some examples of cities that have these types of locations?*** TEKS

Over the last 5,000 years, the world's population has gone from less than 1 percent urban to 45 percent urban.

of the Seine River for protection. How do you think resources, trade routes, and defensive sites influenced the location of towns and cities near you?

READING CHECK: *Places and Regions* What factors have influenced city location and growth?

Patterns in the Size and Distribution of Cities The world has many villages, fewer towns, still fewer cities, and a handful of giant urban areas. The larger places are, the fewer there are. Geographers call this pattern a hierarchy of urban places. Cities and settlements are different in size because they serve different functions and purposes.

Villages have only a few hundred people. Stores in villages usually sell basic goods, such as food or farm supplies. Towns are larger than villages. They may have a few thousand people. Towns have some stores that sell goods for daily life and others that sell items needed less often. These items may include books, cars, or furniture. Towns may also have local and area government buildings, like county courthouses.

Cities, with several thousand to a few million people, have more services. Cities have large shopping centers, government offices, and many businesses. They also have hospitals and perhaps museums and universities. Some cities grow into huge urban areas called **world cities**. World cities are the most important centers of economic power and wealth. Their economies are dominated by the headquarters of global banks and businesses like advertising and insurance. London, New York, and Tokyo are all world cities. Each is the financial, business, and government center for a huge area and is important to the global economy.

Central Place Theory

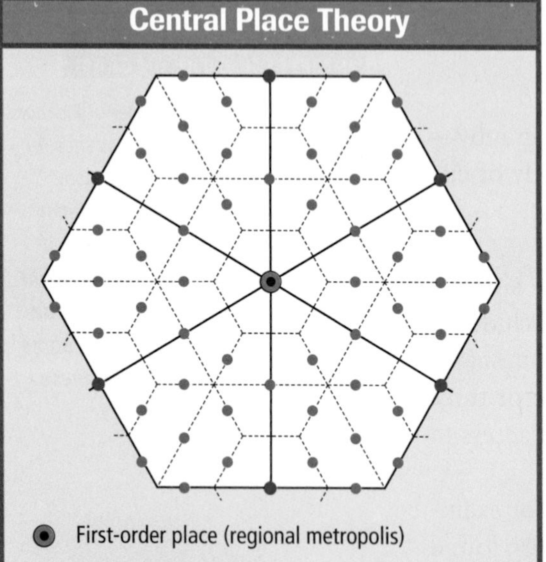

- First-order place (regional metropolis)
- Second-order place
- Third-order place
- Fourth-order place
- Main transport routes between regional metropolises

Source: *The Human Mosaic, Eighth Edition*, Terry G. Jordan

INTERPRETING THE MODEL *This model shows the distribution of different-sized urban places according to central place theory.* ***How can this model be applied to present information about the locations of urban places in the real world?*** TEKS

FOCUS ON GEOGRAPHY

Central Place Theory In the 1930s the German geographer Walter Christaller noticed that there was a regular pattern to the locations of different-sized urban places. Why, he wondered, was this so? Christaller developed a theory to explain the patterns in the size and location of cities. We call his idea central place theory.

Christaller thought that the best arrangement of different-sized cities serving different functions would take the shape of a hexagon. At the center of the hexagon would be a first-order place. Such a place would be the area's largest city and would have the most goods and services. Evenly spread around this city would be smaller towns ranked as second-order, third-order, or fourth-order, according to their size. These smaller towns would offer fewer goods and services. However, there would be more of them. According to Christaller, this pattern represented the most efficient arrangement of urban places.

However, in the real world cities and towns do not form a perfect hexagon. This is true because the most efficient arrangement is not the only factor involved. Environmental factors, such as mountains or rivers, affect settlement location. Cultural factors, such as political boundaries or trade routes, also play an important role.

Nevertheless, central place theory is useful because it does help explain certain geographic patterns. First, there are fewer large cities than small ones. Large cities are also generally farther apart. People usually find specialized goods and services in large cities but not in small towns. Finally, people in small towns often travel farther to find goods that they need less often. In short, the theory does not match the real world perfectly. However, it helps us understand how cities of different sizes are arranged and connected.

READING CHECK: ***Human Systems*** How does central place theory help explain patterns in the size and distribution of cities?

Urbanization Today Both the number and proportion of people living in cities has grown dramatically over the last century. Probably in the next 15 years, more than half of the world's people will live in cities. This growth results not only from the natural increase of populations but also from people leaving rural areas for cities.

In developed countries about 75 percent of the people live in cities. However, urban growth in the world's rich countries is now slow. There are several reasons for this trend. First, the populations of most rich countries are either stable or are increasing slowly. Also, many people prefer the peace and quiet of rural areas to fast-paced cities.

Less than half of the people in developing countries live in cities. However, these areas are rapidly being urbanized. As the populations of these countries grow, more of their people come to the cities looking for jobs. In fact, many of the world's largest cities today are found in developing countries.

READING CHECK: ***Places and Regions*** How do patterns of urbanization differ in developed and developing countries?

Urban Land Use Though they are all unique, cities around the world function in much the same way. They are home to businesses, government agencies, housing, industries, and religious and social groups. All these activities and groups are connected by different forms of transportation.

Most city centers are dominated by large stores, offices, and buildings. Such an area is called a **central business district (CBD)**. CBDs are transportation

INTERPRETING THE VISUAL RECORD

The city of São Paulo, Brazil, is an example of a huge, rapidly growing city in a middle-income country. Between 1975 and 2000 the city's population grew by an incredible 77 percent. Today São Paulo has more than 17 million people and is one of the five largest cities in the world. ***What do you think attracts immigrants from rural areas to cities like São Paulo?*** TEKS

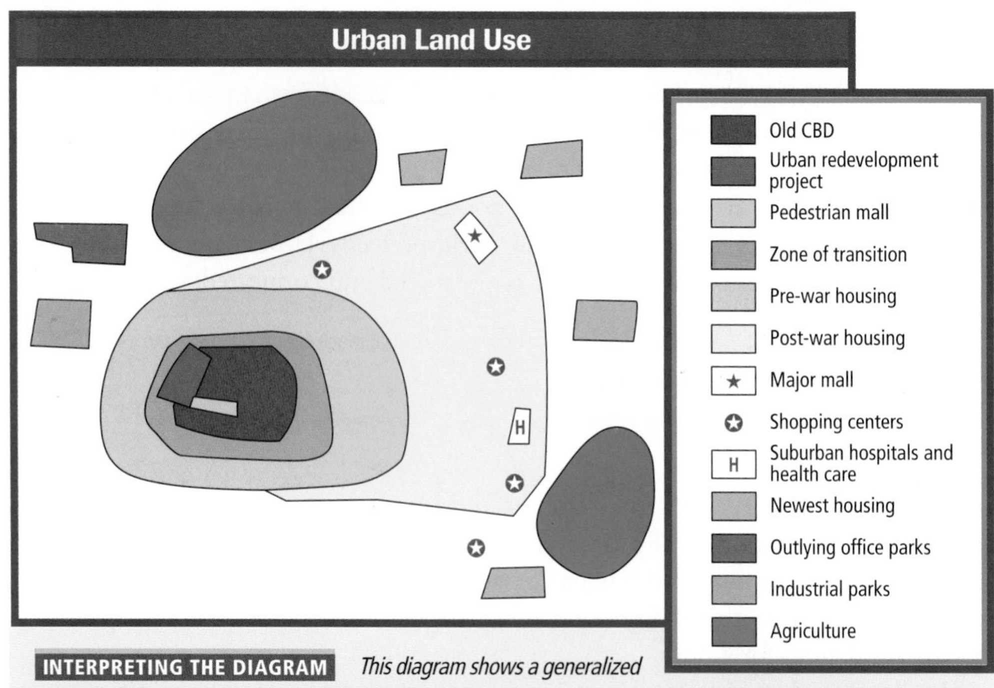

INTERPRETING THE DIAGRAM *This diagram shows a generalized pattern of land use typical in many large U.S. cities. Outlying suburbs and industries have grown as the traditional CBD declined.* ***Where are major malls and shopping centers located? How does the age of housing developments change as you travel away from the CBD? How might the growth of the city into new areas affect agriculture?***

hubs where roads, railroads and buses come together. Outside the CBD are factories and warehouses, which need larger amounts of land. Farther away from the city center, housing takes up most of the land. Small shopping strips with convenience stores line major streets, particularly at major road junctions. Still farther out are the suburbs, which usually have the newest houses. Some suburbs were once small towns that have been engulfed by a spreading city. Other suburbs are in areas that were once farmland. (See the Urban Land Use diagram.)

Modern cities are often ringed by one or more major highways. Along these roads, many new clusters of tall office buildings and big shopping centers have appeared. These clusters of large buildings away from the CBD are called **edge cities**. Increasingly the residents of the suburbs work and shop in edge cities. As a result, many people no longer need to visit the original CBD.

Traffic problems are common in large cities. Roads are clogged during rush hours, when large numbers of people travel to or from work. Traffic problems are particularly common in U.S. cities, such as Los Angeles (below), because Americans rely so heavily on cars as the main means of transportation. Cars transport few people per vehicle and require a lot of space to park.

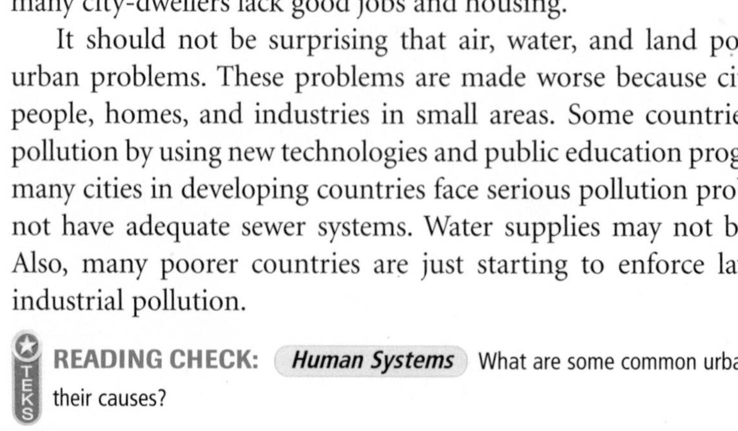

Urban Problems Most cities today face serious social, housing, transportation, or environmental problems. The key social problem in cities is often poverty. This is particularly true in developing countries. In those countries many city-dwellers lack good jobs and housing.

It should not be surprising that air, water, and land pollution are also urban problems. These problems are made worse because cities concentrate people, homes, and industries in small areas. Some countries have reduced pollution by using new technologies and public education programs. However, many cities in developing countries face serious pollution problems. Some do not have adequate sewer systems. Water supplies may not be safe to drink. Also, many poorer countries are just starting to enforce laws that control industrial pollution.

TEKS **READING CHECK:** ***Human Systems*** What are some common urban problems and their causes?

Rural Geography

Rural landscapes are found outside cities. Agriculture is the key economic activity in most rural areas. However, people who live there also work in forestry, mining, and recreation. Rural areas may have land set aside in national parks and wildlife preserves.

Subsistence Agriculture The kind of agriculture practiced most widely around the world is called **subsistence agriculture**. It is found in many parts of Africa, Asia, and Central and South America. In subsistence agriculture, food is produced by a family for its own needs. Anything extra—usually very little—may be sold for important supplies, like cooking oil, clothes, or fuel for a lamp. This kind of agriculture uses little if any machinery.

Subsistence agriculture takes different forms around the world. In many regions families farm the same fields for generations. In more difficult environments, such as tropical forests, **shifting cultivation** is common. In shifting cultivation, farmers clear trees or brush for planting. Sometimes they burn the debris. Then the fields are farmed for a few years, but fertility steadily decreases. The field is eventually abandoned, and then a new field is cleared. Crops such as maize (corn) in South America and manioc (cassava) in Africa are grown in this way.

Another type of subsistence agriculture is **pastoralism**—herding animals. Cattle, goats, horses, sheep, or other animals provide milk and meat for pastoralists. In addition, animal skins or hair are used for shelter and clothing. Pastoralists usually follow regular migration routes. In semiarid regions they migrate with their herds across open grasslands. Others move their animals up and down mountain slopes with the seasons. Pastoralists often trade animal products for grain or vegetables. A few animals may be sold in towns to buy other supplies. In pastoralist cultures, animals are more than just sources of food. Herds represent wealth and social prestige.

INTERPRETING THE VISUAL RECORD

Subsistence agriculture is characterized by low levels of technology. For example, this farmer in China is watering his crops without the help of modern farm machinery. ***How might the technology used in this photograph be different in a developed country with advanced farming technologies?*** TEKS

The von Thünen Model

Johan von Thünen (1783–1850) was a German scholar and farmer. He noticed regular patterns of land use in farming areas of southern Germany. Based on these patterns, von Thünen developed one of the first spatial models in geography. Von Thünen's model is used to explain the location of different types of agriculture in a market economy.

Von Thünen noted that land closest to urban areas has the highest value. Land values decrease as one moves away from the city. Therefore, the most intensive forms of agriculture—those that generate the most profit on the smallest plots of land—are located closer to urban areas. Here, activities such as dairy farming and market gardening are common. Less intensive agricultural activities—those that require large amounts of land to be profitable—take place farther away from cities. In these areas, activities such as ranching and grain farming are found.

Analyzing Information How does von Thünen's model explain the location of different types of agriculture in a market economy? TEKS

The von Thünen Model

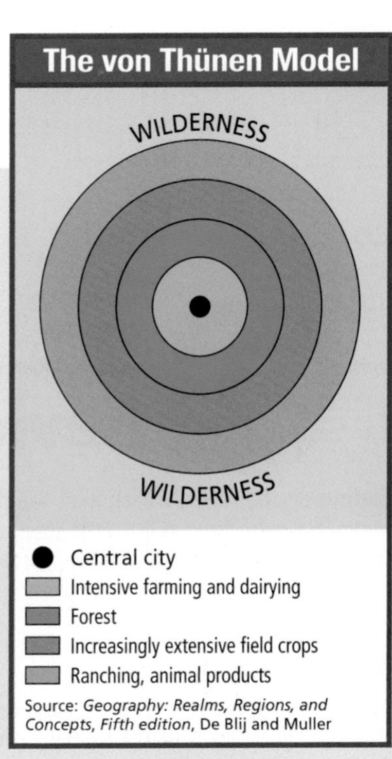

Source: *Geography: Realms, Regions, and Concepts, Fifth edition*, De Blij and Muller

INTERPRETING THE VISUAL RECORD

The application of advanced farming technologies to increase production is a key feature of market-oriented agriculture. Here, detailed soil fertility maps, GPS technology, and computers are used to apply the correct amount of fertilizer to each part of this farmer's fields. ***How might these technologies increase yields and reduce costs?*** TEKS

Market-Oriented Agriculture Another type of agriculture is **market-oriented agriculture**, or commercial agriculture. Under this system, farmers grow products to sell to consumers. They specialize in growing crops that other people want to buy and that they can sell at a profit. Often, crops are first sold to companies that process and package them. Then consumers buy the final products in stores. This kind agriculture is the most common type found in developed countries.

Scientific advances have made market-oriented agriculture very productive. Those advances include new animal breeds, fertilizers, pesticides, and plant varieties. In fact, commercial farmers depend on the latest technology to be successful. Their farms and ranches are often enormous in size. Yet because of technological advances, they require fewer and fewer workers. Some commercial farms are not owned by families at all. Instead, they are parts of large corporations. Advanced commercial farming is referred to as **agribusiness**. Agribusiness is the operation of specialized commercial farms for more efficiency and profits. It is characterized by huge farms with close links to other parts of the food industry.

READING CHECK: *Human Systems* What role has technology played in the development of market-oriented agriculture?

TEKS Questions 3, 4

Section 2 Review

Define domestication, urbanization, world cities, central business district (CBD), edge cities, subsistence agriculture, shifting cultivation, pastoralism, market-oriented agriculture, agribusiness

Reading for the Main Idea

1. *Human Systems* When do many scientists think agriculture began? Where and when did cities originate?
2. *Places and Regions* How is land use typically arranged in urban areas?

Critical Thinking

3. **Comparing** Why do you think market-oriented agriculture is more productive than subsistence agriculture?
4. **Analyzing Information** How might large-scale commercial farming be possible in areas with arid or semiarid climates? What environmental consequences might such farming have in dry regions?

Organizing What You Know

5. Using the information in this section, draw a sketch map showing land use in a typical urban area.

Geography for Life

Rural Settlement Forms

Settlements are purposely grouped and organized clusters of houses and buildings. Some rural settlements have only a few buildings grouped loosely together. We call these hamlets. Other settlements may be larger, such as villages. These settlements take various forms.

Perhaps you live in or have visited a rural community. In such places, homes and other buildings may be scattered and linked by a network of roads. Wooded areas or farmlands may separate these homes and other buildings from each other. On the other hand, a rural community might grow from a cluster of farmhouses separated from agricultural fields. Hamlets like these are found in many hilly rural areas throughout Asia and Europe. Some of these hamlets have a linear form. That is, buildings are clustered along features such as roads or streams. Sometimes hamlets are groups of buildings clustered at the intersection of several roads.

Larger villages may take forms similar to hamlets. Historically, houses in villages were clustered together for defensive reasons. People could defend themselves better against outsiders by grouping together. In fact, many old villages occupy easily defended sites, such as hilltops. Round villages and walled villages are often clustered together in this way. Many old cities around the world began behind strong defensive walls. Also, like hamlets, some villages are grouped along rivers, roads, streams, or other linear features. These places are called string villages. For example, in many lowland areas of western Europe, villages are located along waterways. Villages grouped around road intersections are called cluster villages.

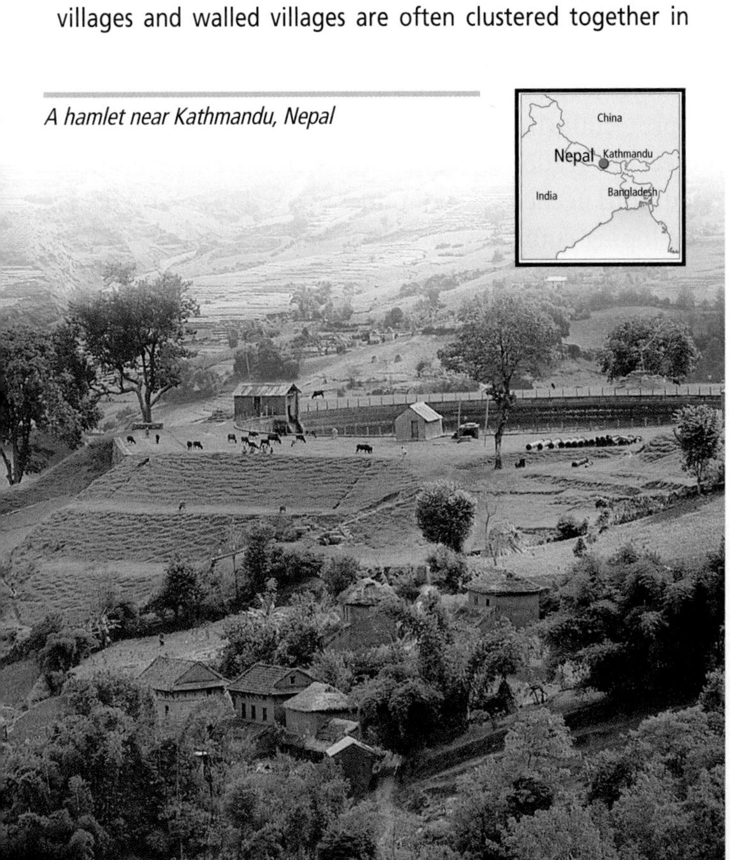

A hamlet near Kathmandu, Nepal

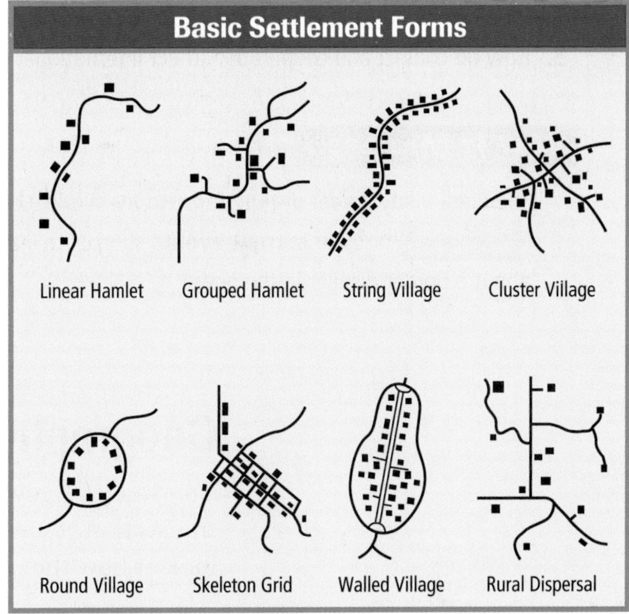

Human Geography, William Brown Publishers

Defense is not the only reason houses in a village are clustered together. Such settlement forms may also leave the best land available for farming. By occupying hilltops or rocky areas, more good farmland is available. For example, some Japanese villages are so tightly packed together that only narrow passages remain between houses. This pattern reflects the need to farm all the useful land. In the hilly parts of Europe, many villages are clustered on hillsides. This arrangement leaves the level land for farming.

Applying What You Know 

1. **Contrasting** How do string, or linear, settlement forms differ from cluster forms?
2. **Analyzing Information** Why do you think road intersections and rivers are common locations for hamlets and villages?

Section 3 Political Geography

READ TO DISCOVER

1. How are government and geography connected?
2. What are three main types of geographic boundaries?
3. How do conflict and cooperation affect international relations?

DEFINE

natural boundaries
cultural boundaries
geometric boundaries
nationalism
totalitarian governments
democracy
tariffs
quotas

WHY IT MATTERS

There are many border disputes around the world. Use CNNfyi.com or other **current events** sources to learn about a border dispute in the news.

Geography and Governments

The study of government and politics is an important part of geography. In this section you will learn how control of Earth's surface is divided. We will also look at how the culture of a place influences its government.

Government and Development There are about 200 countries in the world today. A country has an independent government, which has authority over territory within its borders. Governments are free to make their own laws and have their own leaders. Usually, governments negotiate and deal with each other in peace. We call this diplomacy. Governments interact with each other through trade agreements and international organizations. However, disputes between countries sometimes lead to conflict or war.

Good governments protect the lives and property of the people who live in a country. They also protect the freedoms and rights of their citizens. In doing so, they help ensure the conditions needed for economies to develop and for people to prosper. In many developing countries, governments are unstable. That is, they do not last long or have much authority. Corruption can also be a problem. Political leaders may use their power only to enrich themselves and their friends.

Officials sign the Treaty of Versailles in 1919, which ended World War I. International agreements and diplomacy through government representatives are basic features of the world's independent countries.

Cultural Beliefs and Government The cultural beliefs of different groups affect how governments are set up and operate. These beliefs influence government decisions and public policies. For example, a people's cultural beliefs might lead to laws that force businesses to close on special religious days. Cultural beliefs also affect the way citizens see their duties and responsibilities.

To see how these cultural influences can work, consider Israel, a country in Southwest Asia. After World War II, Israel was established as a homeland for Jews from anywhere in the world. Israel's role as a Jewish homeland has guided many of its government policies. For example, any Jew who wishes to become a citizen of Israel can do so. In fact, many Jews have moved there from around the world. Still, people in Israel debate the role of religion in government and society. About 20 percent of Israeli Jews strictly follow the beliefs of Judaism. These people are called Orthodox Jews. They tend to believe that religious values should play an important role in shaping government policy. Most other Israeli Jews seek to limit the role of religion in Israel. In fact, Israeli law guarantees religious freedom for all people, Jewish or not. About 20 percent of Israeli citizens are not Jews. Most of these people are Arabs who have very different cultural beliefs and practices. They have often opposed Israel's immigration, military, and foreign policies.

READING CHECK: *Human Systems* How can cultural beliefs influence citizenship practices and public policies?

Geographic Boundaries

Three main types of boundaries separate countries from each other. Boundaries that follow a feature of the landscape are called **natural boundaries**. Mountains make good natural boundaries. They are difficult to cross and are permanent markers. Rivers, on the other hand, are often troublesome boundaries. Many rivers are important transportation routes for more than one country. In addition, river channels may move. Other natural boundaries include deserts, lakes, and oceans.

Borders that are based on culture traits, such as religion or language, are called **cultural boundaries**. For example, the border between mostly Muslim Pakistan and mostly Hindu India was established largely along religious lines. The same is true on the island of Ireland. A border divides the island between mostly Protestant Northern Ireland and the mostly Roman Catholic Republic of Ireland. Many cultural boundaries are based on language. The boundary between Portugal and Spain is an example.

Boundaries that follow regular, geometric patterns are called **geometric boundaries**. These borders are usually straight lines drawn without regard to environmental or cultural patterns. Geometric boundaries are often based on lines of latitude or longitude. For example, the border between the United States and Canada lies mostly along the 49th parallel. In colonial Africa, European countries drew many geometric boundaries that are still used today. Some of these boundaries have led to problems because they divide the territory of different ethnic groups.

READING CHECK: *Human Systems* What are three different types of boundaries that geographers study?

The Rio Grande is a natural boundary between Mexico and the United States.

Conflict and Cooperation

The study of international relations is another important area of political geography. The field focuses on foreign policies and relations between the world's countries. Very few countries today are isolated. Instead, they are affected by events in both nearby and faraway places.

Political Conflicts Political conflicts, both within and among countries, are common. There are many reasons for this. Most people feel proud of their culture and country. Feeling pride and loyalty for one's country or culture group is called **nationalism**. These feelings are often expressed in special songs, symbols, and writings. Unfortunately, one group's pride can conflict with that of another. Competing feelings of nationalism have led to problems time and again. Conflicts can also result from other differing culture traits, such as religion. A modern example is Sri Lanka, where a Hindu minority has battled the Buddhist majority.

How a group of people should be governed has also been a source of conflict. One person or a small group of people governs some countries. They have full authority to make laws and decisions. In these **totalitarian governments** one person or a few people decide what is best for everyone. The people have little or no say in how their country is governed. Communist countries, such as North Korea and Cuba, have such governments. In other countries, all citizens have a voice in their government. These countries are based on **democracy**. This is a system in which the people decide who will govern. They choose their leaders by voting in free elections. Democratic governments value individual freedoms and human rights. The United States and many other countries have democratic governments.

Economic issues also lead to conflicts. For example, countries often establish **tariffs**—taxes on imports and exports. Countries also set **quotas**—limits to the amount of a product that can be imported. Tariffs and quotas usually help protect a country's industries from foreign competition. However, they can also cause trade disputes among countries.

Terrorism Some political conflicts involve the use of terrorism by one or more individuals or groups. Terrorism is the use of violence and fear as a political force. Terrorists act for many reasons. For example, some want independence for homelands that may be part of or under the control of another country. Others might work to further various political goals, such as different public or social policies. Terrorists do not usually act under the authority of a particular government. However, governments sometimes protect and even support the actions of terrorists who share similar political goals. Stopping terrorism is difficult, particularly in South America, Southwest Asia, and parts of Europe.

Even the United States has been directly affected by terrorism. One of the worst acts of terrorism in history happened on September 11, 2001. On that day, fundamentalist Muslim terrorists stunned the world by hijacking four commercial airliners. The terrorists crashed two of the planes into the twin towers of New York's World Trade Center, collapsing the buildings. The third plane hit the Pentagon, just outside Washington, D.C., and the fourth crashed in rural Pennsylvania. The attacks killed thousands of people. In the wake of these events, the U.S. government worked to enlist the cooperation of other

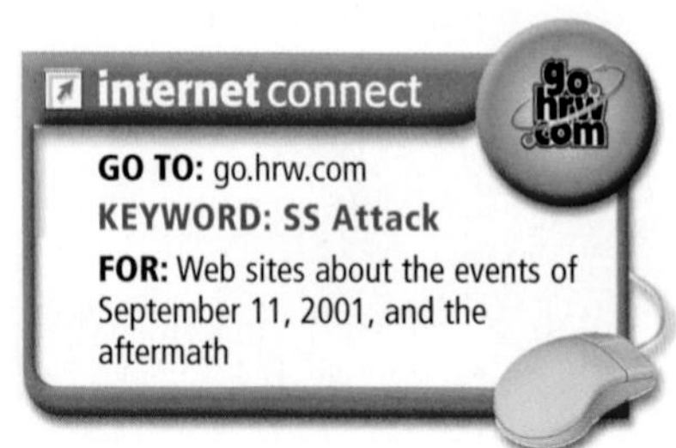

GO TO: go.hrw.com
KEYWORD: SS Attack
FOR: Web sites about the events of September 11, 2001, and the aftermath

countries in a global effort to defeat terrorism. This effort included cooperation in military, political, and security matters.

International Cooperation Countries often cooperate with each other. They do this for two main reasons—political benefits and economic benefits.

Political and military cooperation are most developed in the United Nations (UN). Nearly all of the world's countries are members of the UN. As a result, it is the most important international organization. In the UN, representatives of the world's countries can discuss international issues and voice their concerns. The UN's main goals are settling disputes between countries and trying to prevent wars. Sometimes, the UN sends peace-keeping military forces to warring regions. It also tries to solve global problems such as disease, hunger, and illiteracy.

Many countries benefit from economic cooperation and free trade. This cooperation can help countries produce goods at lower costs and reach larger markets. People can then buy those goods at lower prices. Economic cooperation can also end or reduce tariffs and quotas. For example, the World Trade Organization (WTO) works to make trade between countries fairer and easier. Most countries belong to the WTO. You will read about many other important economic organizations in the regional chapters throughout this textbook.

✓ **READING CHECK:** *Human Systems* What are the main reasons countries cooperate with each other?

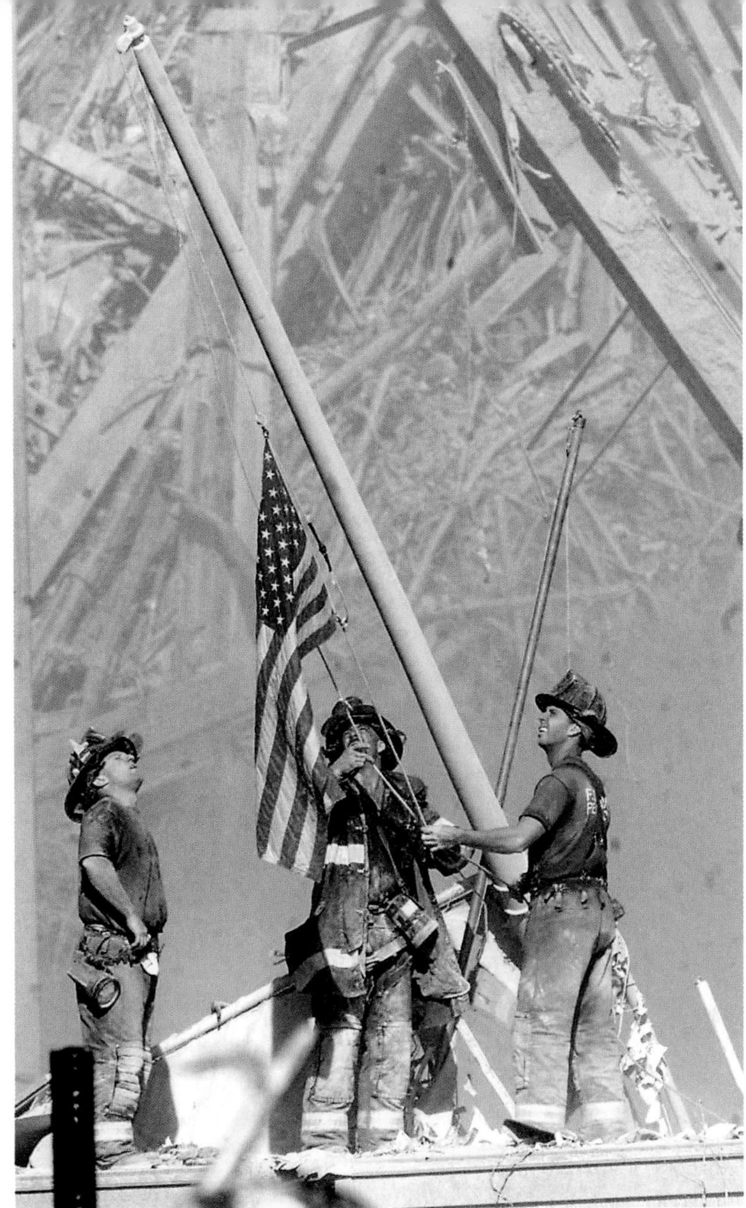

©2001 *The Record* (Bergen County, NJ), Thomas E. Franklin, Staff Photographer

INTERPRETING THE VISUAL RECORD *The terrorist attacks on the World Trade Center and the Pentagon in 2001 killed thousands of people. Rescue workers and firefighters, such as these, faced the grim task of searching for victims.* ***How might images like this one inspire people to work to defeat terrorism around the world?***

TEKS Questions 1, 2, 3, 4

Section 3 Review

Define natural boundaries, cultural boundaries, geometric boundaries, nationalism, totalitarian governments, democracy, tariffs, quotas

Reading for the Main Idea

1. *Human Systems* How do good governments promote economic development?
2. *Places and Regions* What physical and cultural factors might influence the ways a country's boundaries are established?
3. *Human Systems* How do organizations like the United Nations and the World Trade Organization promote international cooperation?

Critical Thinking

4. **Analyzing Information** What kinds of boundaries surround your state? What kinds surround the United States?

Organizing What You Know

5. Copy the chart below. Use it to describe the features of totalitarian and democratic forms of government. List examples of each.

Totalitarian	Democratic

Geography for Life

Technology and the Environment

People depend on their environment for clothing, food, fuel, and shelter. We also adapt to the environment and modify it to better suit our needs. Throughout history, major technological innovations have caused large-scale changes in the environment and in human societies.

Fire

One of the earliest examples of people using technology to change the environment is the use of fire. Early humans used fire to stay warm and to cook food. They also used fire to clear brush, improving grazing lands for the animals they hunted. Much later, people used fire to clear land for farming or raising livestock. By using fire, people changed the landscape over wide areas.

The Steam Engine

In the 1700s steam power became the main power source for industry. As well as powering factory machines, steam engines were used to power ships and trains. Steam-powered vehicles were very successful. They made travel faster, safer, and more reliable. As a result, people changed the way they thought about long-distance travel and trade. Soon railroad lines were built throughout Europe and North America, and then in countries around the world. Railroads helped unify huge countries like the United States, Canada, Russia, and India. Then steamships linked countries, and global trade increased.

As the use of steam engines increased, so did the demand for coal, their main fuel. As a result, steam engines indirectly led humans to modify the environment as they dug large coal mines around the world. In addition, burning coal caused air pollution, which damaged the environment. Industrial cities of the 1800s, such as London and Birmingham, had terrible smog.

Gasoline and Diesel Power

In the 1900s gasoline engines and then diesel engines began to replace coal-powered steam engines. Gasoline and diesel fuel were particularly suitable for cars and trucks. The rapid development of automobiles led people to modify the environment again with roads and highways. Webs of paved roads soon connected cities, towns, and rural areas. Today the United States alone has more than 2.3 million miles (3.7 million km) of paved roads.

In developed countries, gasoline and diesel machinery replaced human and animal power in agriculture. Patterns of food production and distribution around the world shifted as a result. Also, fewer farmers were needed to plow, plant, and harvest. These changes allowed more and more people to move from rural areas to the cities. This trend continues today as use of machines for agriculture is becoming more widespread in developing countries.

Electricity

With electricity, power could be transmitted over long distances. In addition, factories could be built farther from their power sources. Cities and roads could be lit more easily and safely. New electric machines also made many household jobs easier and faster, giving people more free time. In agriculture, electric motors powered farm machinery, water pumps, and machines used to process crops. Even fewer farmers were needed to feed the rest of the population. For example, in 1900 it took about four

The development of gasoline as a fuel for cars has led to environmental changes such as the expansion of roads and an increase in air pollution.

INTERPRETING THE VISUAL RECORD *This image of Earth's lights at night shows how people have used electricity to transform the environment. The brightest areas are generally the most urbanized, but not necessarily the most densely populated.* ***Which areas appear the brightest? Which areas are dark? What might this image indicate about the level of development in these regions?*** TEKS

farmers to feed 10 people. In the United States today, one farmer produces enough food to feed nearly 130 people.

The use of electric power has also caused direct environmental change. Coal and natural gas are among the resources used to generate electricity. Mining and drilling for these resources has altered, and even damaged, landscapes. Dams can also generate power. Building a dam affects the upstream environment by creating a lake where none had been. Downstream, flooding is reduced.

This dam on the Columbia River in Oregon shows how humans modify the environment to generate electricity.

Environmental Change

As these examples show, technology allows people to alter their environment in important ways. The rate of technological developments has greatly increased over the last several hundred years. Can you think of other technological innovations that have allowed societies to change their environment?

Applying What You Know

1. **Evaluating** How important are major technological innovations, such as fire, steam power, diesel, and electricity, that have been used to modify the environment?
2. **Comparing** How does environmental change caused by the development of gasoline power and diesel power compare to change caused by the steam engine?

CHAPTER 6 Review

Building Vocabulary TEKS

On a separate sheet of paper, explain the following terms by using them correctly in sentences.

- market economy
- command economy
- developed countries
- infrastructure
- developing countries
- domestication
- urbanization
- central business district (CBD)
- subsistence agriculture
- market-oriented agriculture
- natural boundaries
- cultural boundaries
- geometric boundaries
- nationalism
- democracy

Locating Key Places TEKS

On a separate sheet of paper, match the letters on the diagram with their correct labels.

- intensive farming and dairying
- forest
- wilderness
- field crops
- ranching, animal products
- central city

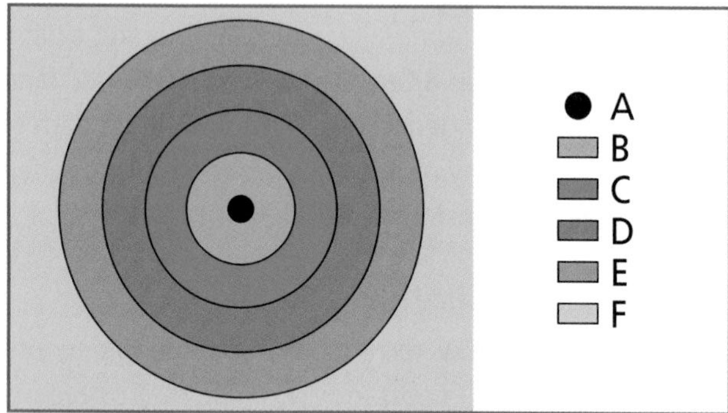

Understanding the Main Ideas TEKS

Section 1

1. *Human Systems* Which type of economic activity is located at the site of the resource being exploited? Which activity can be located almost anywhere?
2. *Human Systems* What are some measures of development?

Section 2

3. *The Uses of Geography* How did the development of early cities affect people's daily lives?
4. *Human Systems* How do people satisfy their basic needs with subsistence agriculture?

Section 3

5. *Human Systems* What are some of the ways countries interact with each other?

Thinking Critically for TAKS TEKS

1. **Evaluating Information** What factors do you think are most important in determining the level of economic development and standard of living of a country? Why?
2. **Summarizing** How have humans modified the physical environment since the development of agriculture?
3. **Analyzing** How have cultural beliefs influenced public policies and citizenship practices in your community?

Using the Geographer's Tools TEKS

1. **Interpreting Diagrams** Study the chart of economic activities in Section 1. How might changes in transportation affect the location of different economic activities?
2. **Interpreting Charts** Study the chart of developed and developing countries in Section 1. How might some of the different categories of information shown in the chart be related?
3. **Preparing Sketch Maps** Use the student atlas at the beginning of this book and the information in Section 3 to find a country that has both a natural boundary and a geometric boundary. Draw a sketch map of the country, showing what these boundaries are based on. Why do you think these boundaries were selected?

Writing for TAKS TEKS

Write a short article that summarizes how subsistence agriculture and market-oriented agriculture are different. What are the main differences between these two forms of agriculture? In general, how is technology used in each system? Where in the world is each type common? When you are finished with your article, proofread it to make sure you have used standard grammar, spelling, sentence structure, and punctuation.

SKILL BUILDING

Geography for Life TEKS

Gathering Field Data

Places and Regions Use a notepad and pencil to gather field data about urban land use in your area. Start in the downtown or CBD of your closest urban area. Take notes about what types of buildings you see, how high they are, and how they are used. Then walk in one direction away from the downtown area, taking more notes along the way. When you are done, try to answer the following questions: How is land used differently as one travels away from the downtown? Why do you think this is?

Building Social Studies Skills

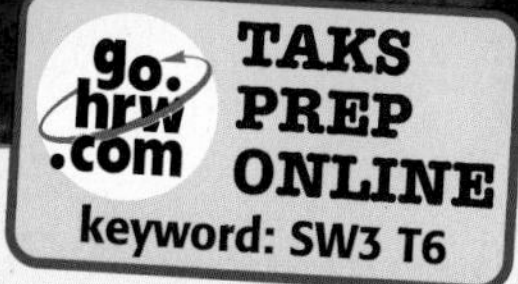

Interpreting Graphs

Study the line graph below. Then answer the questions that follow.

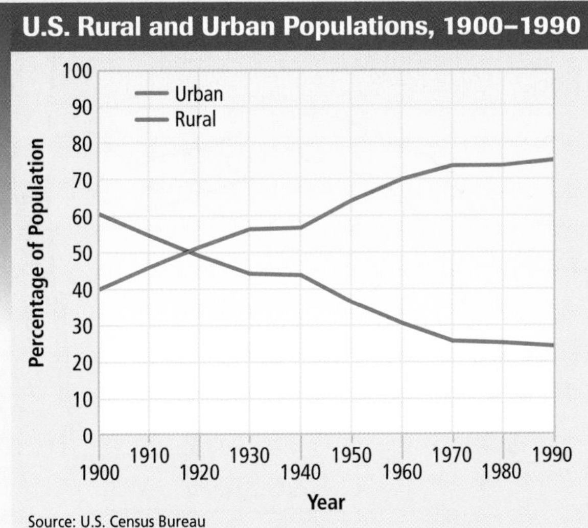

1. In 1990, how many more Americans were living in an urban environment than were living in a rural environment?
   a. 30 percent
   b. 40 percent
   c. 50 percent
   d. cannot be determined
2. During which three decades did the most dramatic increase in urban population occur?

Analyzing Secondary Sources

Read the following passage and answer the questions.

"The first cities appeared in Southwest Asia more than 5,000 years ago. City life became possible when there was enough food so that some people did not have to farm. Instead, some of them worked as potters or weavers. Others became merchants and traders. Still others carried out governmental or religious tasks. Towns and cities began to grow. The growth in the proportion of people living in towns and cities is called urbanization."

3. When did city life become possible?
   a. after people became potters and weavers
   b. more than 15,000 years ago
   c. when there was enough food that some people did not have to farm
   d. after trade connected early cities
4. Explain how the development of cities affected people's occupations.

Alternative Assessment

PORTFOLIO ACTIVITY

Learning about Your Local Geography TEKS

Group Project: Research

Plan, organize, and complete a research project on the economic activities in your community. Assign one person each to research one of four main kinds of economic activities that are found in your community. Check your local library or chamber of commerce to find information. Try to learn about the different jobs that people do and which activities employ the most people in your community. Then create a pie graph that estimates the percentages of people working in primary, secondary, tertiary, and quaternary activities. What did you learn from your research?

Internet Activity: go.hrw.com
KEYWORD: SW3 GT6 TEKS

Access the Internet through the HRW Go site to create a data profile of one country from each of the six populated continents. Your data profile should include information from the following categories: people, economy, environment, technology, and trade. Use information from the World Bank to create your data profile.

Skill Building for TAKS

WORKSHOP

Using Geographic Information Systems

As you read in Chapter 1, a geographic information system is a special computer system. A GIS stores, displays, and maps locations and their features. Geographers can use GIS software to examine relationships among our planet's physical and human features. The software helps researchers find patterns that may help solve complex problems. For example, geographers can use a GIS to identify areas that are favorable for growing crops. Map layers may show areas with ample water, rich soils, and favorable terrain for farming. Once a GIS merges those layers, good locations for farming become easier to identify. A GIS can also be helpful in urban planning. A planner might use a GIS to help plan an urban park or locate areas ideal for business, industry, or residential use. All of the information that a researcher needs to do such work can be translated into map layers. Combining those map layers gives the researcher a more complete image of the characteristics of the area under study. With the development of powerful personal computers, GIS technology has become more accessible. Insurance companies, farmers, local governments, health departments, and schools now use GIS technology to solve problems and to display spatial information.

Developing the Skill In this workshop you will practice making a simple GIS plan with graph paper and transparency overlays. However, a real GIS plan is a complex process. Such a plan requires special software and sets of data. The software is the tool that tells the computer what to do. The software must have a database program to manage the information, a spreadsheet program to process the data, and a programming language to write functions for the data. It must also have the ability to zoom in and out and to pan across the computer screen. For example, a researcher might choose to zoom in, or enlarge, a small area of a map. Doing so will let a researcher see more

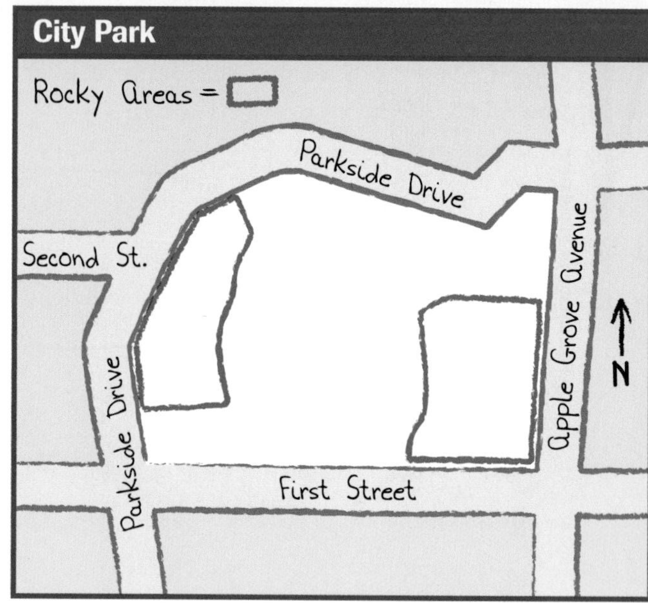

detail. At other times a researcher might want to zoom out, or view a larger area in less detail. Doing so will help the researcher see how a small place fits into surrounding areas.

The geographic data is the fuel for the whole process. For example, census data might tell a researcher the average personal income in different areas of a city. Data can also be gathered from field work. Consider how a park planner might use GIS to create or redesign a park. The planner will want to observe the terrain and other physical features of areas throughout the park. For example, that planner will want to identify areas with rocky terrain. (See the illustration.) He or she will also want to identify areas with existing vegetation, easy access to neighboring streets, and broad open spaces. That and other information can be displayed in map layers. By putting those layers together, the planner can decide which areas are best suited for gardens, parking, picnics, recreation, and other uses.

The most important part of GIS technology is the geographer or researcher using it. Because there is a risk of using too much data, the wrong data, or different scales of data, the geographer must be careful. For example, when using a GIS, it is very important that all the data provided is at the same scale and projection. In other words, each map layer should show the same area at the same scale.

Practicing the Skill

Now it is your turn to practice using a GIS. Work with a group of other students to design a new park in your community or redesign an existing park.

The tools you will need include graph paper, five transparency overlays, and transparency markers of different colors and width. You will also need a map of an existing park or another map of your community that you can use to find the best spot for a new park.

Create a base map for your park, using a black marker to show only park boundaries. You can then photocopy that base map onto your transparency overlays. Each overlay will become a layer in a detailed map of your park site.

Working with your group, study the park site you have chosen. Each group member should have a different task. Each will research the park site and create a map overlay based on his or her observations. Your group should create four map overlays in all:

- The first map layer should use one color to show the location of water sources, such as creeks, ponds, rivers, or streams.
- The second map layer should use one color to show areas with rocky terrain.
- The third map layer should use colors to mark areas with thick tree growth, scattered trees, or mostly grass.
- The fourth map layer should use one color to show contour lines. In this project, the contour lines do not have to be completely accurate. Simply use lines to identify flat, gently sloping, or steeply sloping areas. You might want to review the discussion of contour lines in Chapter 1. Because you could have many contour lines, you might want to analyze this map layer separately from the others.

When the map layers are completed, use an overhead projector to place the layers on top of each other. By doing this, the group should be able to identify areas that are best suited for gardens, picnics, parking, hiking trails, and other uses. In some cases you may need to use only one or two layers to identify areas suited for certain functions.

Then work with group members to develop a final "park plan." That plan should include a map of your park and a written summary. The summary should describe why different functions were assigned to specific areas within the park. Gather all of your materials into a formal plan folder and make a presentation to your class.

UNIT 2

The United States and Canada

The Statue of Liberty
Liberty Island, New York

CONNECTING TO

Literature

"Loo-Wit"* by Wendy Rose

Wendy Rose

(1948–) is a poet and an artist. Rose's poems reflect her Hopi heritage and focus on environmental themes. Before she began writing, she made up songs to express herself. Rose notes that "it has always been a tradition of my people [the Hopi of Arizona] to celebrate everything in life with song. For all occasions, both happy and sad, I sang my feelings."

The way they do
this old woman
no longer cares
what we think
but spits her black tobacco
any which way
stretching full length
from her bumpy bed.
Finally up
she sprinkles ash on the snow,
cold and rocky buttes
that promise nothing
but winter is going at last.
Centuries of cedar
have bound her to earth,
huckleberry ropes
lay prickly about her neck. . . .
Around her
machinery growls,
snarls and ploughs
great patches of her skin.
She crouches
in the north,
the source
of her trembling–
dawn appearing
with the shudder
of her slopes.

Blackberries unravel,
stones dislodge;
it's not as if
they weren't warned.

She was sleeping
but she heard the boot scrape,
the creaking floor;
felt the pull of the blanket
from her thin shoulder.
With one free hand
she finds her weapons
and raises them high;
clearing the twigs from her throat
she sings, she sings,
shaking the sky like a blanket
about her
Loo-wit sings and sings and sings!

* Loo-Wit *is a name that the Cowlitz tribe gave to Mount St. Helens, a volcano in Washington State. Mount St. Helens erupted violently in 1980 after being inactive for more than 120 years.*

Analyzing the Primary Source

1. **Supporting a Point of View** Why is this image of an elderly woman awakening from her sleep appropriate for a volcano such as Mount St. Helens?
2. **Analyzing** How does Loo-Wit warn of the impending eruption? Which of Loo-Wit's actions represents the eruption itself?

The World in Spatial Terms

The United States and Canada: Political

1. *Places and Regions* Which U.S. states share a border with Mexico? TEKS
2. *Environment and Society* Compare this map to the physical map. In which provinces is the Canadian Shield found?

Critical Thinking

3. **Comparing** How do the size of states in the eastern United States compare with those in the west? Why do you think this is? TEKS

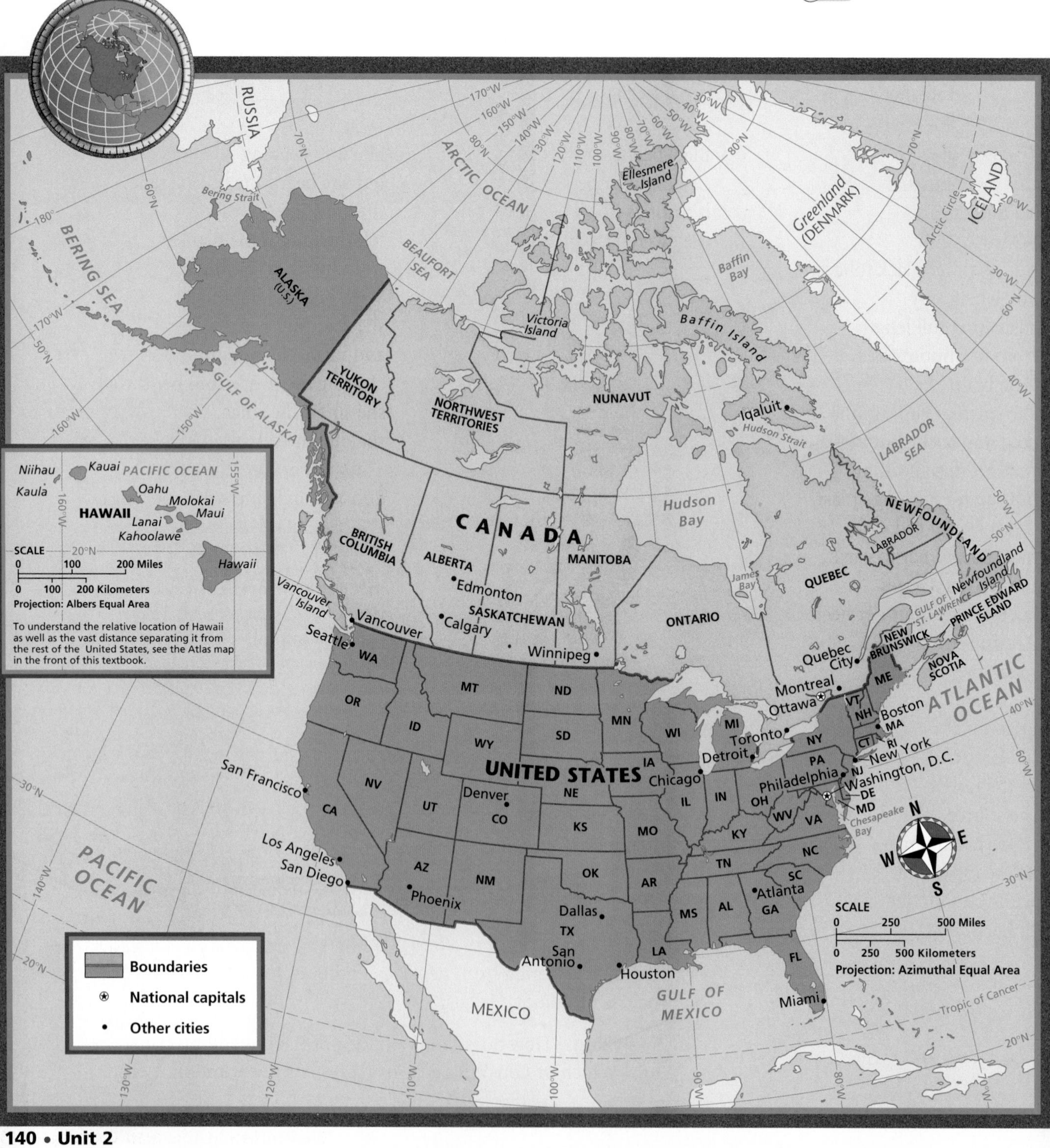

HAWAII

SCALE
0 100 200 Miles
0 100 200 Kilometers
Projection: Albers Equal Area

To understand the relative location of Hawaii as well as the vast distance separating it from the rest of the United States, see the Atlas map in the front of this textbook.

Boundaries
National capitals
Other cities

SCALE
0 250 500 Miles
0 250 500 Kilometers
Projection: Azimuthal Equal Area

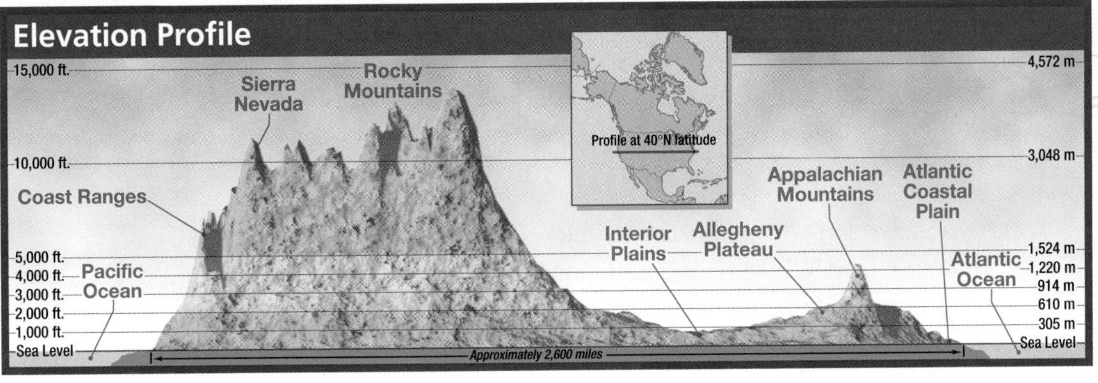

The United States and Canada: Physical

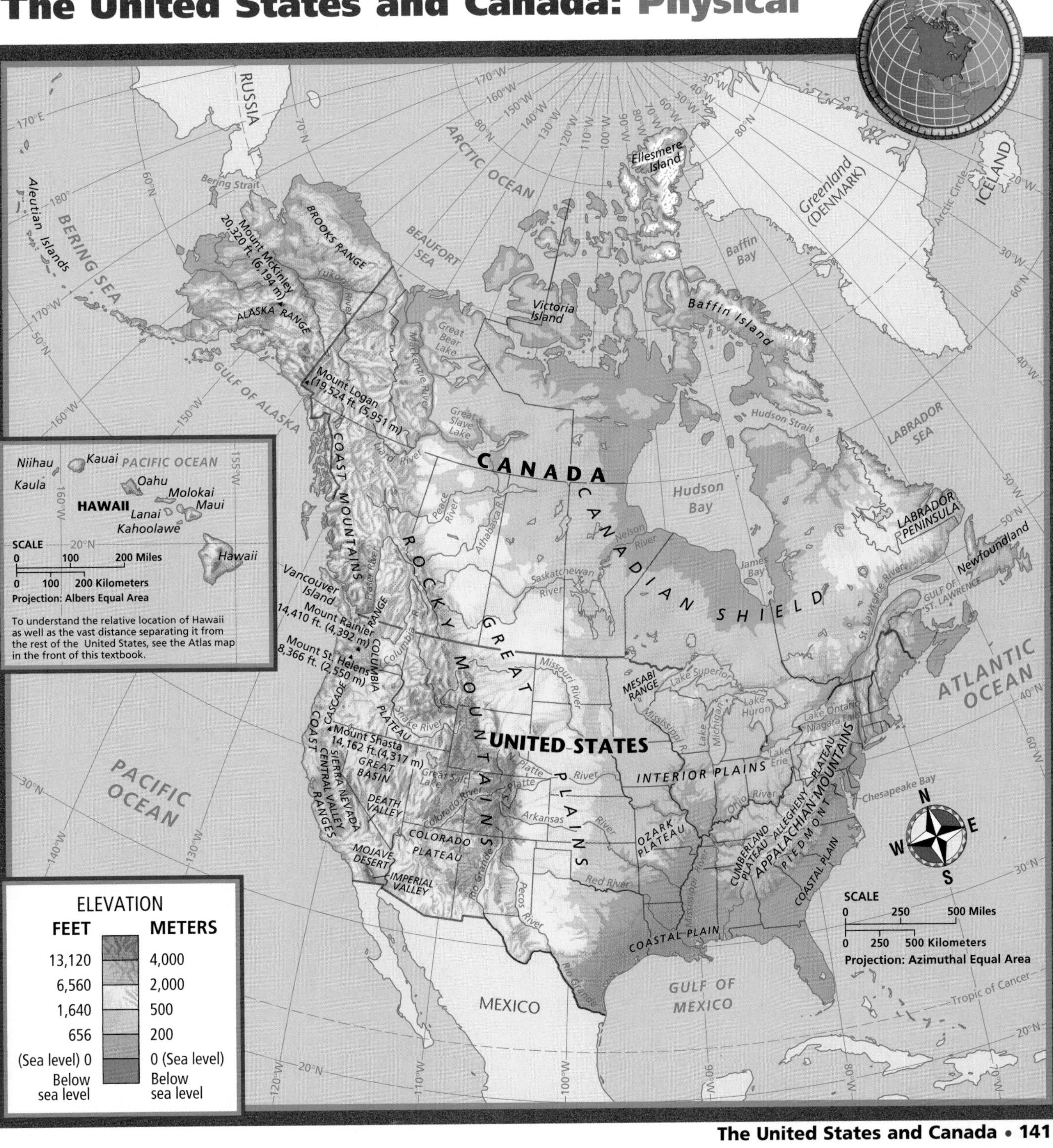

CANADA

UNITED STATES

RUSSIA

MEXICO

ARCTIC OCEAN

PACIFIC OCEAN

ATLANTIC OCEAN

GULF OF MEXICO

BERING SEA

BEAUFORT SEA

GULF OF ALASKA

LABRADOR SEA

Hudson Bay

Baffin Bay

Greenland (DENMARK)

ICELAND

Ellesmere Island

Victoria Island

Baffin Island

Aleutian Islands

Bering Strait

Hudson Strait

James Bay

LABRADOR PENINSULA

Newfoundland

GULF OF ST. LAWRENCE

Vancouver Island

BROOKS RANGE

ALASKA RANGE

Mount McKinley 20,320 ft. (6,194 m)

Mount Logan (19,524 ft. (5,951 m)

Mount Rainier 14,410 ft. (4,392 m)

Mount St. Helens 8,366 ft. (2,550 m)

Mount Shasta 14,162 ft. (4,317 m)

COAST MOUNTAINS

ROCKY MOUNTAINS

CANADIAN SHIELD

GREAT PLAINS

INTERIOR PLAINS

CASCADE RANGE

COAST RANGES

SIERRA NEVADA

CENTRAL VALLEY

GREAT BASIN

DEATH VALLEY

MOJAVE DESERT

IMPERIAL VALLEY

COLUMBIA PLATEAU

COLORADO PLATEAU

OZARK PLATEAU

MESABI RANGE

CUMBERLAND PLATEAU

ALLEGHENY PLATEAU

APPALACHIAN MOUNTAINS

PIEDMONT

COASTAL PLAIN

Great Bear Lake

Great Slave Lake

Great Salt Lake

Lake Superior

Lake Michigan

Lake Huron

Lake Erie

Lake Ontario

Niagara Falls

Chesapeake Bay

Yukon River

Mackenzie River

Liard River

Peace River

Athabasca R.

Saskatchewan River

Nelson River

Fraser River

Columbia River

Snake River

Missouri River

Mississippi River

Ohio River

Platte River

Arkansas River

Red River

Colorado River

Rio Grande

Pecos River

St. Lawrence River

Arctic Circle

Tropic of Cancer

HAWAII

PACIFIC OCEAN

Niihau

Kauai

Kaula

Oahu

Molokai

Maui

Lanai

Kahoolawe

Hawaii

SCALE

0 100 200 Miles

0 100 200 Kilometers

Projection: Albers Equal Area

To understand the relative location of Hawaii as well as the vast distance separating it from the rest of the United States, see the Atlas map in the front of this textbook.

SCALE

0 250 500 Miles

0 250 500 Kilometers

Projection: Azimuthal Equal Area

N E S W

ELEVATION

FEET	METERS
13,120	4,000
6,560	2,000
1,640	500
656	200
(Sea level) 0	0 (Sea level)
Below sea level	Below sea level

The United States and Canada: Climate

1. *Physical Systems* Compare this map to the physical map. Which major mountain ranges appear to influence the distribution of highland climate areas? TEKS
2. *Physical Systems* How do you think latitude influences the distribution of climates in Canada? TEKS

Critical Thinking

3. **Analyzing** Compare this map to the population map. How do you think climate has influenced the distribution of Canada's population? TEKS

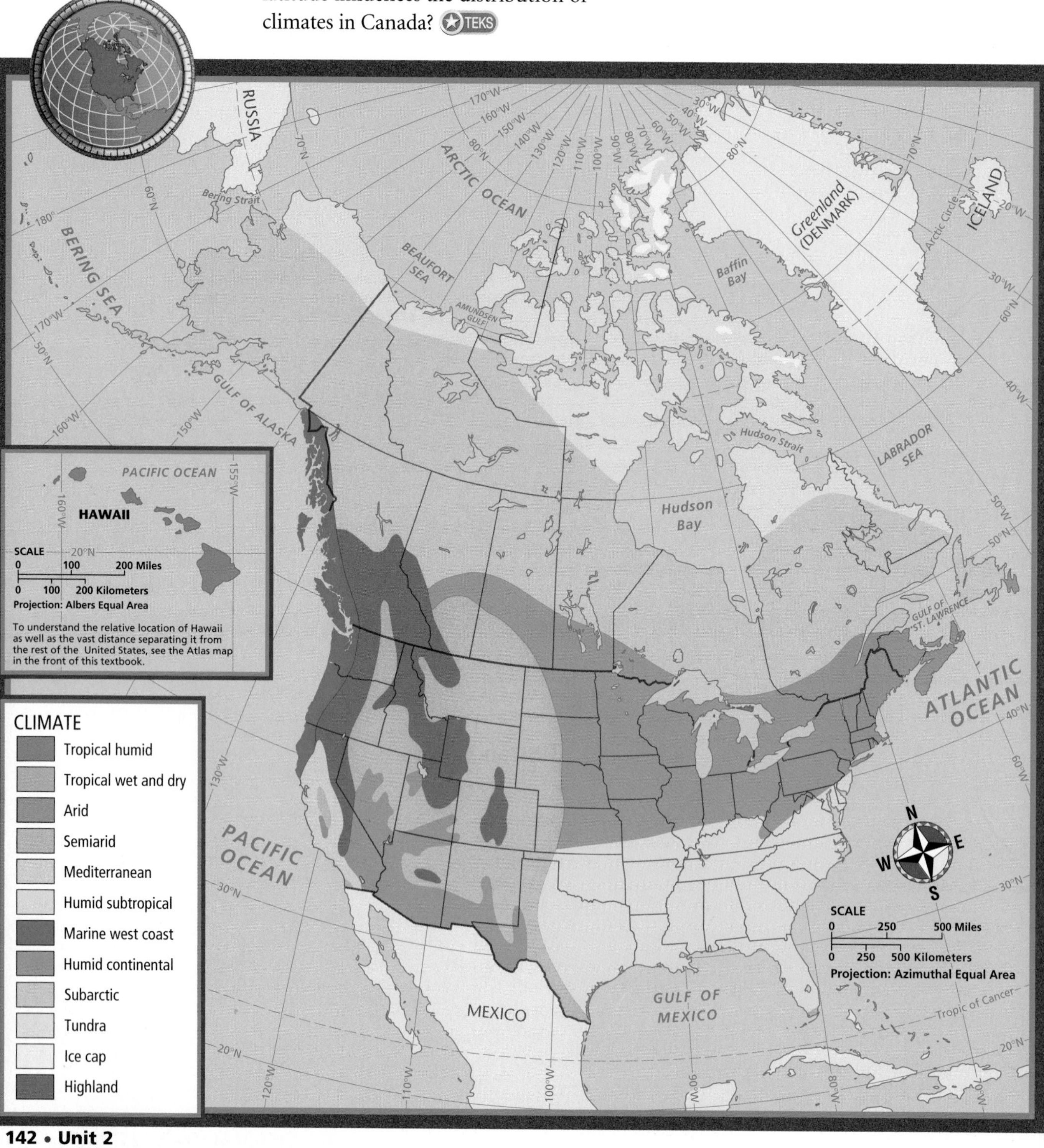

PACIFIC OCEAN

HAWAII

SCALE 0 100 200 Miles
0 100 200 Kilometers
Projection: Albers Equal Area

To understand the relative location of Hawaii as well as the vast distance separating it from the rest of the United States, see the Atlas map in the front of this textbook.

CLIMATE
- Tropical humid
- Tropical wet and dry
- Arid
- Semiarid
- Mediterranean
- Humid subtropical
- Marine west coast
- Humid continental
- Subarctic
- Tundra
- Ice cap
- Highland

SCALE 0 250 500 Miles
0 250 500 Kilometers
Projection: Azimuthal Equal Area

UNIT 2 ATLAS

The United States and Canada: Precipitation

1. *Physical Systems* Compare this map to the climate map. Which two climate regions receive the most precipitation? TEKS
2. *Places and Regions* Which region in the United States receives the lowest amount of precipitation? TEKS

Critical Thinking

3. **Making Generalizations** Compare this map to the climate map. How much precipitation falls in tundra climates? Why do you think these high-latitude climates have such low precipitation? TEKS

The United States and Canada: Population

1. **Human Systems** How are the population patterns of Canada and the United States different? TEKS
2. **Places and Regions** What are Canada's largest cities? TEKS

Critical Thinking

3. **Analyzing** Why do you think so many large U.S. cities are located on the coast? TEKS

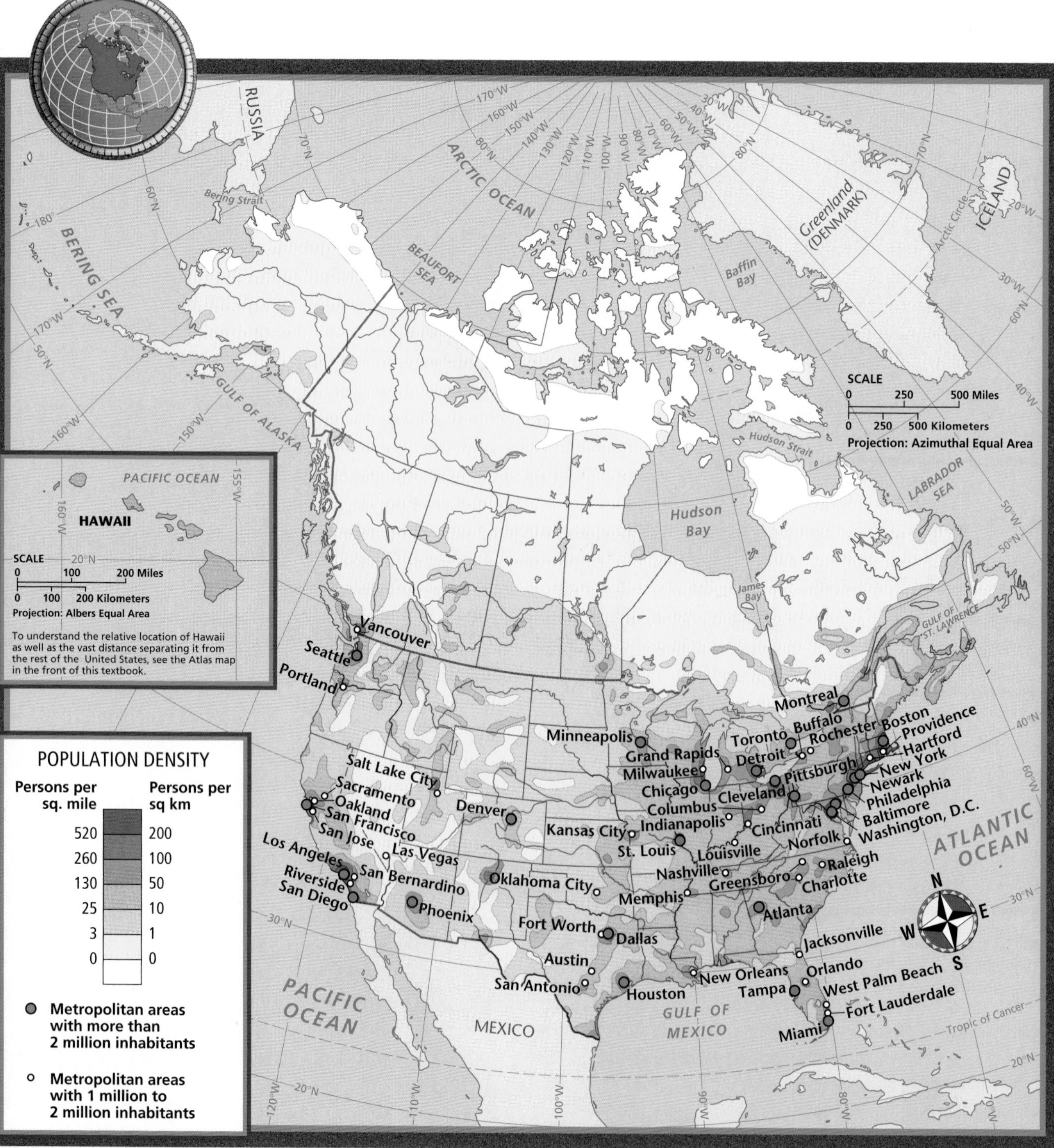

SCALE
0 250 500 Miles
0 250 500 Kilometers
Projection: Azimuthal Equal Area

HAWAII
PACIFIC OCEAN
SCALE
0 100 200 Miles
0 100 200 Kilometers
Projection: Albers Equal Area

To understand the relative location of Hawaii as well as the vast distance separating it from the rest of the United States, see the Atlas map in the front of this textbook.

POPULATION DENSITY

Persons per sq. mile	Persons per sq km
520	200
260	100
130	50
25	10
3	1
0	0

● Metropolitan areas with more than 2 million inhabitants

○ Metropolitan areas with 1 million to 2 million inhabitants

UNIT 2 ATLAS

The United States and Canada:
Land Use and Resources

1. *Environment and Society* Compare this map to the climate map. In which climates is timber production found?
2. *Places and Regions* Where in North America is dairy production an important economic activity? TEKS

Critical Thinking

3. **Analyzing** Compare this map to the physical map. In what physical environments is commercial farming found? Why do you think this is? TEKS

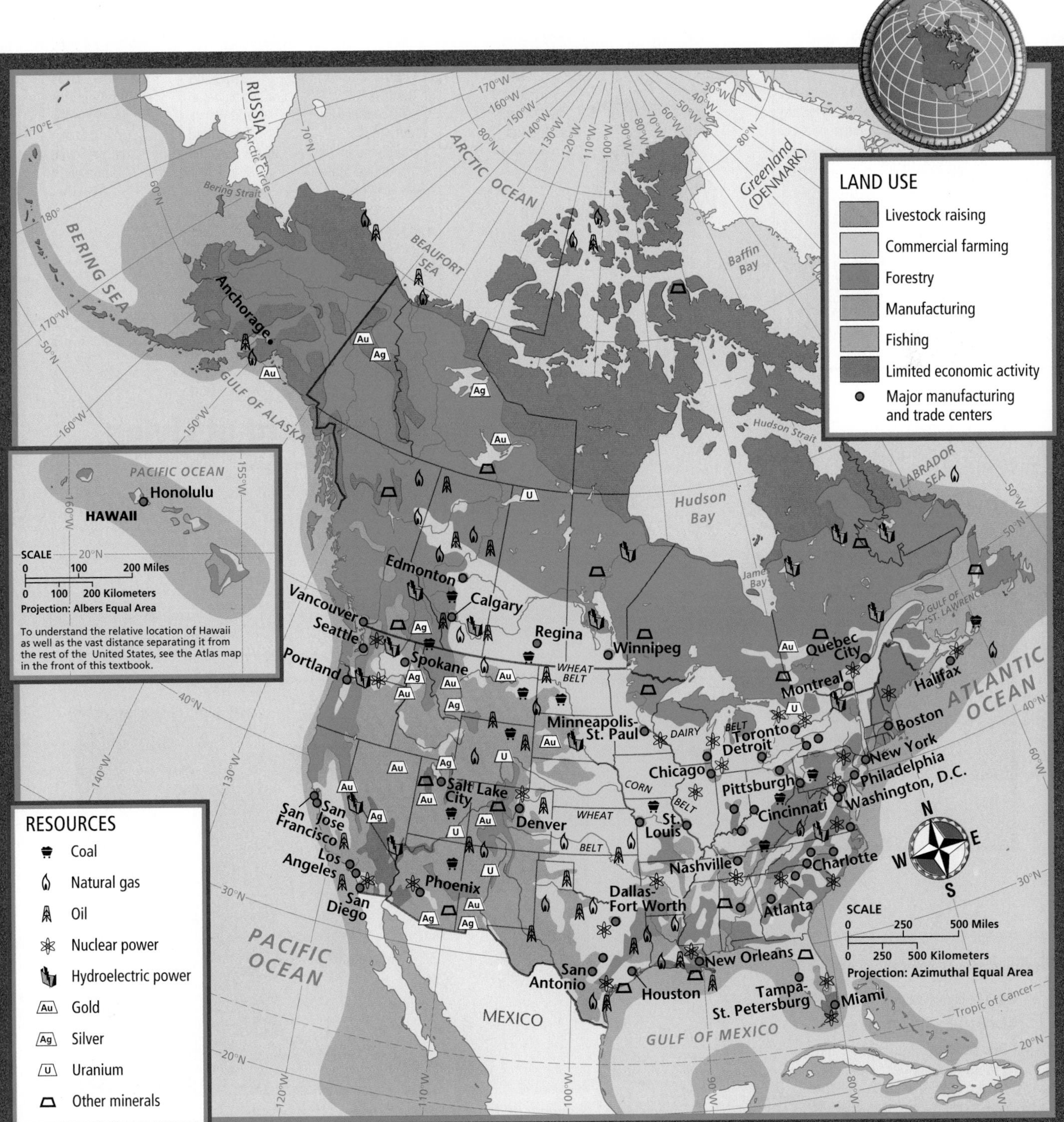

UNIT 2 ATLAS

Time Line: The United States and Canada

1500s–1700s
Europeans establish colonies in North America.

1763
Britain takes control of French Canada.

1776
The 13 American colonies declare independence.

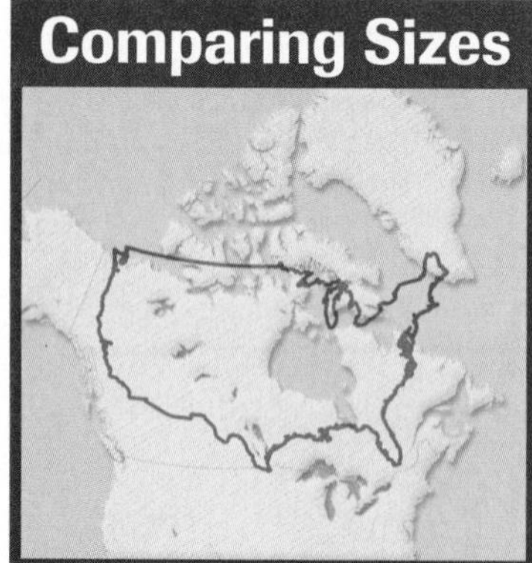

1800s
Americans and Canadians migrate westward.

1848
Gold is discovered in California.

1861–65
The Civil War takes place in the United States.

1867
The British government creates the self-governing Dominion of Canada.

1869
The first transcontinental railroad is completed in the United States.

1914–18
Canada fights in World War I. The United States enters the war in 1917.

1939–45
Canada fights in World War II. The United States joins the war in 1941.

1991
The Cold War between the United States and the Soviet Union ends.

2001
On September 11, terrorists attacked the World Trade Center and the Pentagon.

1500 · 1700 · 1800 · 1900 · 2000

The United States and Canada

Comparing Sizes

Comparing Standard of Living

COUNTRY	LIFE EXPECTANCY (in years)	INFANT MORTALITY (per 1,000 live births)	LITERACY RATE	DAILY CALORIC INTAKE (per person)
Canada	76, male 83, female	5	97%	3,093
United States	74, male 80, female	7	97%	3,603

Sources: Central Intelligence Agency, *2001 World Factbook; Britannica Book of the Year, 2000*

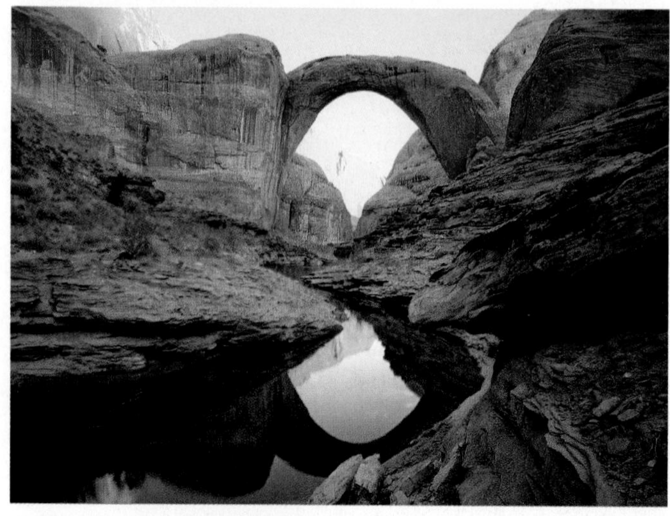

Rainbow Bridge, Lake Powell, Utah

internet connect

go.hrw.com

GO TO: go.hrw.com
KEYWORD: SW3 Almanac
FOR: Additional information and reference sources

Fast Facts: The United States and Canada

FLAG	COUNTRY Capital	POPULATION (in millions) POP. DENSITY	AREA	PER CAPITA GDP (in US $)	WORKFORCE STRUCTURE (largest categories)	ELECTRICITY CONSUMPTION (kilowatt hours per person)	TELEPHONE LINES (per person)
	**Canada** Ottawa	31.6 8/sq. mi.	3,851,788 sq. mi. 9,976,140 sq km	$ 24,800	74% services 15% manufacturing	15,744 kWh	0.59
	**United States** Washington, D.C.	281.4 76/sq. mi.	3,717,792 sq. mi. 9,629,091 sq km	$ 36,200	30% manage., prof. 29% tech., sales, admin.	12,260 kWh	0.69

Sources: Central Intelligence Agency, *2001 World Factbook; The World Almanac and Book of Facts, 2001; Britannica Book of the Year, 2000;* U.S. Census Bureau
The CIA calculates per capita GDP in terms of purchasing power parity. This formula equalizes the purchasing power of each country's currency.

The Grand Canyon National Park from the South Rim.

Natural Environments of North America

Historians think the Norse were the first Europeans to visit North America. The "Groenlendinga Saga" tells of these explorations, including Leif Eriksson's voyage to Vinland—believed to be what is now Newfoundland.

"They went ashore and looked about them. The weather was fine. There was dew on the grass, and the first thing they did was to get some of it on their hands and put it to their lips, and to them it seemed the sweetest thing they had ever tasted. Then they went back to their ship and sailed into the sound that lay between the island and the headland jutting out to the north.

. . . But they were so impatient to land that they could not bear to wait for the rising tide to float the ship; they ran ashore to a place where a river flowed out of a lake. . . . Then they decided to winter there, and built some large houses.

There was no lack of salmon in the river or the lake, bigger salmon than they had ever seen. The country seemed to them so kind that no winter fodder would be needed for livestock: there was never any frost all winter and the grass hardly withered at all.

In this country, night and day were of more even length than in either Greenland or Iceland: on the shortest day of the year, the sun was already up by 9 A.M. and did not set until after 3 P.M."

Prickly pear cactus in Arizona

Irises in Newfoundland, Canada

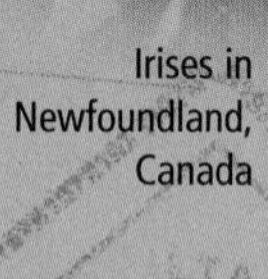

Black bear in Alaska

Physical Features

READ TO DISCOVER

1. What are the major landform regions in the United States and Canada?
2. What rivers and lakes are found in the region?

WHY IT MATTERS

Tourists from all over the world are drawn to the beautiful natural landscapes of the United States and Canada. Use CNNfyi.com or other **current events** sources to learn how important tourism is to the region's economy.

IDENTIFY

Continental Divide

DEFINE

barrier islands
piedmont
fall line
basins
hot spot

LOCATE

Piedmont
Appalachian Mountains

Locate, continued

Rocky Mountains
Mississippi River
Great Lakes
Great Plains
Canadian Shield
Hudson Bay
Cascade Range
Sierra Nevada
Coast Ranges
St. Lawrence River

Physical Regions of the United States and Canada

SCALE
0 500 1,000 Miles
0 500 1,000 Kilometers
Projection: Azimuthal Equal Area

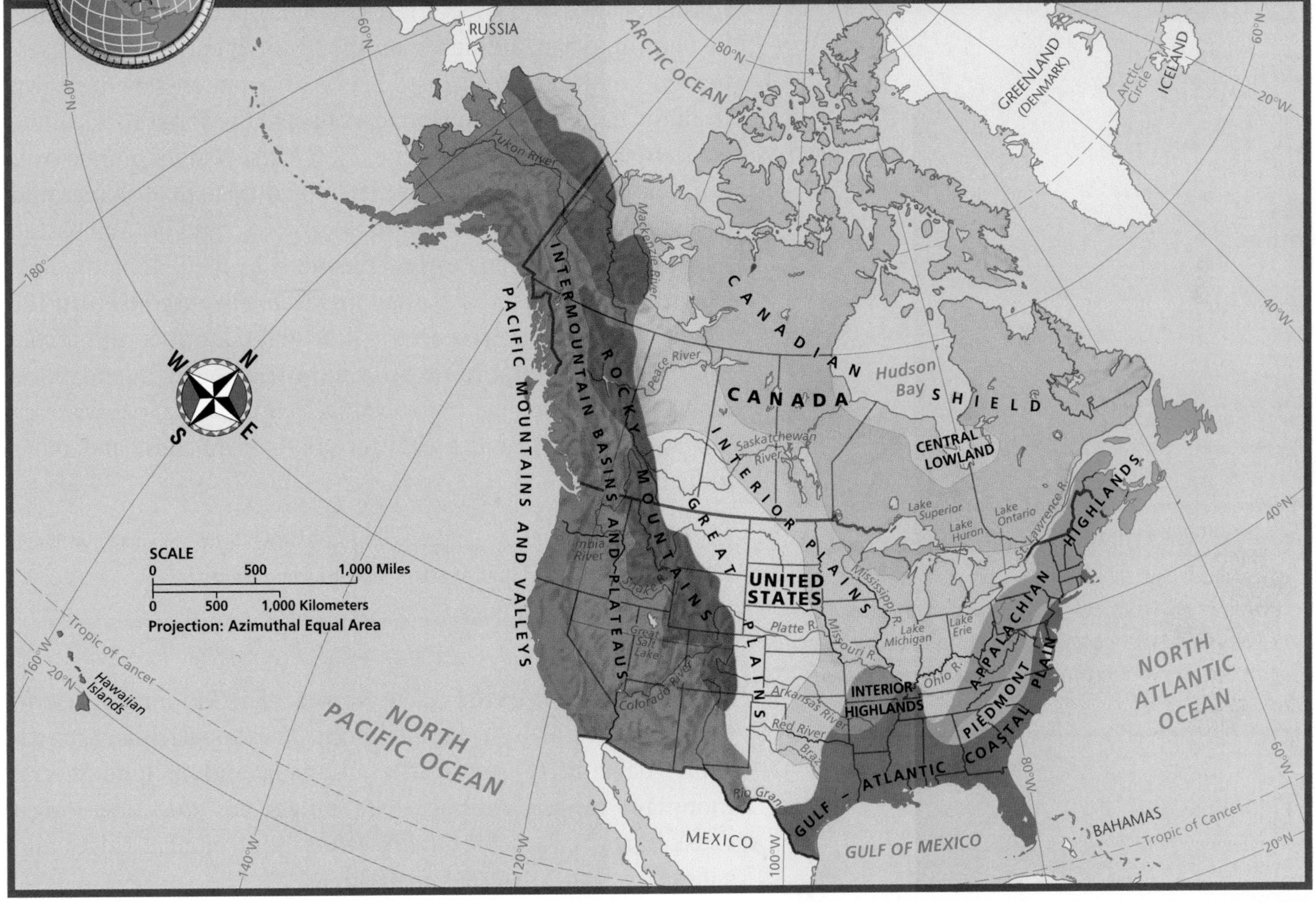

Landforms

The United States and Canada make up about 80 percent of the continent of North America. These two countries have some of the world's most spectacular scenery. Landforms range from vast plains to high mountains, plateaus, and volcanic islands. Most major landform regions stretch from north to south across both countries. (See the chapter map.) The landforms of the eastern half of the United States and Canada are older than those of the western half. Eastern mountains have been eroded, and rolling hills and flatlands cover most of the region. In contrast, the west has a younger landscape. There you will find steep mountains, active volcanoes, deep canyons, and high plateaus.

A long coastal plain stretches along the Atlantic Ocean and Gulf of Mexico from New England to Mexico. This low plain lies close to sea level and rises gradually inland. It is narrowest along the northern Atlantic coast but widens south of New York. Along some parts of the coastal plain, **barrier islands** have formed. Ocean waves and currents create these long narrow islands by depositing sand in shallow water.

Inland from the coastal plain lies the Piedmont, an upland region. A **piedmont** is an area at or near the foot of a mountain region. The Piedmont stretches from New Jersey to Alabama.

The Fall Line

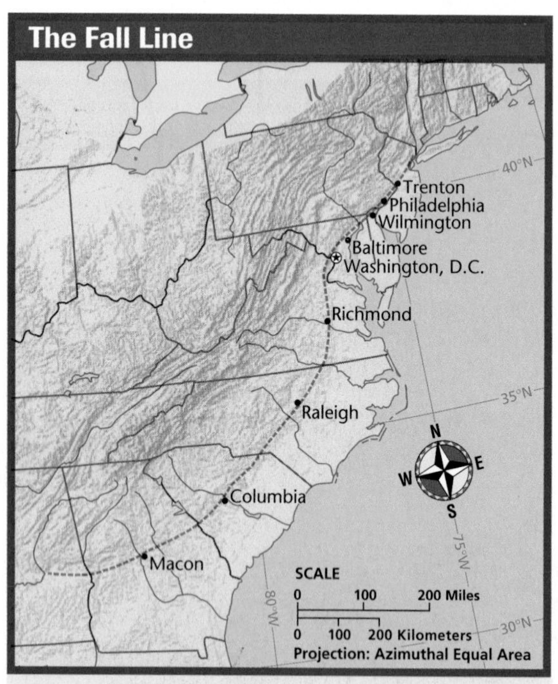

INTERPRETING THE MAP *Most cities on the fall line, including the ones shown on this map, mark the head of navigation on rivers. The head of navigation is the point that most ships are not able to sail past.* ***What economic activities do you think these cities were established to provide?*** TEKS

FOCUS ON GEOGRAPHY

The Fall Line The boundary between the Piedmont and the coastal plain is known as a **fall line**. This natural boundary has had an important influence on the historical geography of settlements in the eastern United States. River waters flowing down from the hard rock of the Piedmont reach the softer rocks of the coastal plain along the fall line. Here these waters plunge over rapids and waterfalls. Early settlers noted that small ships could easily reach the fall line from the ocean but could not sail past it. Partly as a result, many early settlements formed along the line. The tumbling waters of the fall line were also used to turn the waterwheels that powered early industries. Lumber and textile mills were two important early industries here. Inland ports along the fall line, like Philadelphia, Pennsylvania, became important transportation points for goods from these and other industries.

READING CHECK: ***Places and Regions*** Why were many settlements in the eastern United States founded along the fall line?

The East and Interior The Appalachian Mountains rise to the west and north of the Piedmont. The Appalachians stretch from Alabama to southeastern Canada and include several mountain ranges. Among these ranges are the Blue Ridge, Catskill, and Green Mountains. A series of parallel ridges and valleys form the eastern Appalachians.

The collision of eastern North America with Africa more than 300 million years ago created the Appalachians. Erosion has since lowered and smoothed the peaks of these mountains. The highest peaks only rise to more than 6,000 feet (1,829 m).

Between the Appalachians and the Rocky Mountains lie the vast interior plains. The Mississippi River and its many tributaries drain most of this region. Glaciers covered the northern interior plains, north of the Ohio and Missouri Rivers, during the last ice age. Today you will find thousands of lakes there, including the Great Lakes. This area has rolling hills, many river systems, and productive soils. The interior plains partly surround a highland region in Missouri, Arkansas, and Oklahoma. Like the Appalachians, these interior highlands are a region of old eroded uplands. They include the Ozark Plateau. Farther west are the Great Plains, a subregion of the interior plains. The Great Plains stretch from south-central Canada into Texas and Mexico and reach to the eastern edge of the Rocky Mountains. Elevations along this edge of the Great Plains reach more than 5,000 feet (1,524 m) above sea level.

North of the interior plains lies the Canadian Shield. This arc of ancient rocks covers nearly half of Canada. The Canadian Shield is centered on Hudson Bay. It stretches from the Arctic Ocean eastward to the Atlantic coast. This area has been thoroughly scraped by glaciers. This process left a rough rocky landscape with little soil for productive farmland.

The Canadian Shield contains some of the oldest rocks in the world. Some rocks there were formed at least 3.8 billion years ago.

The West The Rocky Mountains stretch from New Mexico to Canada. Many of the highest peaks reach more than 14,000 feet (4,267 m). The Rocky Mountains, or Rockies, are not a single range but several ranges. High plains and valleys separate these ranges. West of the Rockies lie the Cascade Range and the Sierra Nevada, two major mountain ranges located near the Pacific coast. The area between these ranges and the Rockies is called the intermountain region.

High plateaus with deep canyons, isolated mountain ranges, and desert **basins** make up most of the intermountain region. A basin is a lower area of land, generally surrounded by mountains. The Great Basin makes up a large area of the intermountain region in the United States. Most rivers there never reach the ocean. The Colorado River, which flows southward to the Gulf of California, is an exception. Farther west, at the edge of the Great Basin, is California's Death Valley. The lowest point in North America—at 282 feet (86 m) below sea level—is found there.

The Pacific coast region is made up of two major mountain systems and a series of valleys between these mountains. The Sierra Nevada and Cascade Range, or simply the Cascades, lie on the eastern edges of the Pacific coast region. The Sierra Nevada runs along California's eastern border. North of the Sierra Nevada, in northern California, Oregon, and Washington,

The Teton Range in Wyoming is one of the youngest mountain ranges of the Rockies. A combination of tectonic activity and ice age glaciation has created a spectacular landscape of rugged peaks and deep lakes in this area.

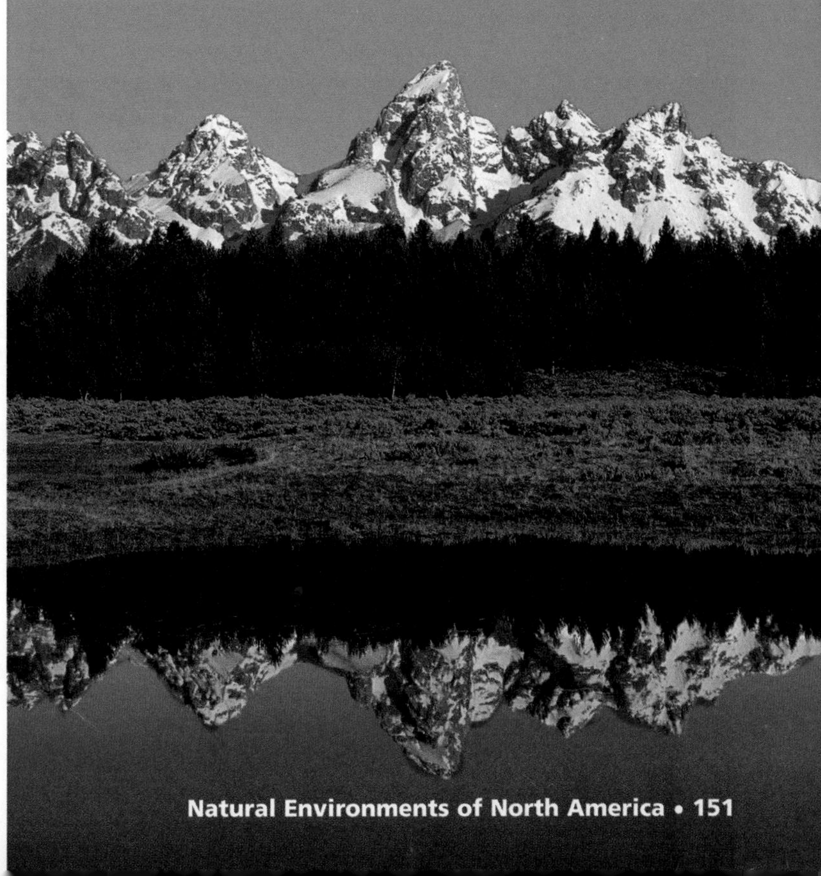

are the Cascades. A series of high volcanoes is found in this range. These volcanoes include Mount Rainier, Mount Hood, Mount Shasta, and Mount St. Helens. Along the Pacific Ocean the rugged Coast Ranges stretch from California to Canada. Between the Coast Ranges in the west and the Sierra Nevada and Cascades to the east are three fertile valleys. These are the Puget Sound lowland in Washington, the Willamette River valley in Oregon, and the Central Valley in California.

The rugged western United States is part of the Ring of Fire. The Ring of Fire is a tectonically active region around the edges of the Pacific. It has many active volcanoes and earthquake faults. On the eastern edge of this ring, the North American plate collides with the Pacific plate. All of the continental United States except some parts of California lies on the North American plate. Some areas of coastal California lie on the Pacific plate and are separated from the North American plate along the San Andreas Fault. Major earthquakes occur periodically along this fault.

The two westernmost U.S. states, Alaska and Hawaii, are also geologically active. The Hawaiian Islands are the tops of submerged volcanoes that rise from the ocean floor. They formed over a **hot spot**—a place where magma wells up to the surface from Earth's mantle. Alaska's southern coast is in a subduction zone, and powerful earthquakes sometimes strike there. The volcanic Aleutian (uh-LOO-shun) Islands extend into the Pacific from Alaska. North America's highest peak, Mount McKinley in the Alaska Range, reaches 20,320 feet (6,194 m) in elevation. Except for the Brooks Range, northern and interior Alaska have flat and hilly landscapes.

TEKS **READING CHECK:** ***Physical Systems*** How did glaciers affect the landscape of the Canadian Shield?

Connecting to THE ARTS

Ansel Adams

Ansel Adams (1902–84) is famous for his black-and-white photographs of America's beautiful natural landscapes. Growing up in California, Adams visited the Sierra Nevada and Yosemite National Park. He began photographing these places and many others as an adult. Adams published many books of his photographs of America's rugged mountains and national parks.

Ansel Adams not only loved the natural landscapes he photographed, he also tried to help protect them. From 1936 to 1973, Adams was director of the Sierra Club, a California-based conservation group. His work inspired the conservation of America's natural wonders. Through his beautiful photographs, Ansel Adams became one of the most well-known photographers of the 1900s.

Summarizing How were Ansel Adams's life and career tied to America's natural environments?

Yosemite National Park in California

Ansel Adams photographing the California coast

Bodies of Water

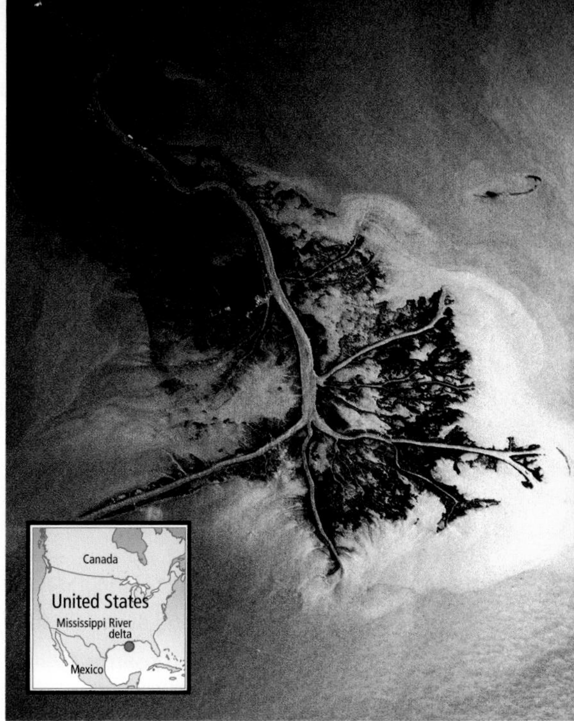

The crest of the Rockies marks the **Continental Divide**. This crest divides North America's major river systems into those flowing eastward and those flowing westward. To the east the Mississippi, Missouri, and Ohio Rivers make up the continent's major river system. This system drains most of the interior plains of the United States. It also provides an important network of waterways for trade and transportation. From its delta in southern Louisiana, the Mississippi deposits huge amounts of sediment into the Gulf of Mexico.

The second major river system in the interior plains is the St. Lawrence system. The St. Lawrence connects the Great Lakes to the Atlantic Ocean and drains most of southeastern Canada. In northwestern Canada, the Mackenzie River system also drains the interior plains and part of the Canadian Shield. Three large northern lakes—Lake Athabasca, Great Slave Lake, and Great Bear Lake—drain into the Mackenzie. Water from these lakes flows northward into the Arctic Ocean.

Long rivers like the Colorado, Columbia, Fraser, and Yukon flow west out of the Rockies. The Columbia and Fraser rivers flow into the Pacific Ocean. The Colorado flows southwestward into the Gulf of California. These rivers are important water sources for many people in the western United States and Canada. They are also used to produce hydroelectricity. The Yukon River flows across Alaska to the Bering Sea.

The United States and Canada have many lakes. In fact, North America has more large lakes than any other continent. Continental ice sheets created most of these lakes. During the ice ages, the ice sheets widened and deepened existing basins. Then as the last ice age ended and the ice melted, the basins filled with water. This process formed the Great Lakes and most of Canada's large lakes. In fact, much of the Canadian Shield is a waterlogged landscape covered with lakes and wetlands.

INTERPRETING THE VISUAL RECORD

This image taken from the space shuttle Challenger *shows the delta of the Mississippi River. The Mississippi deposits some 220 million tons of sediment into the Gulf of Mexico each year, forming what geographers call a "bird's foot delta." The shape of the delta is caused by the compaction and sinking of sediment.* ***How do you think this delta changes through time?*** TEKS

TEKS **READING CHECK:** *Physical Systems* How did glaciers create lakes in this region?

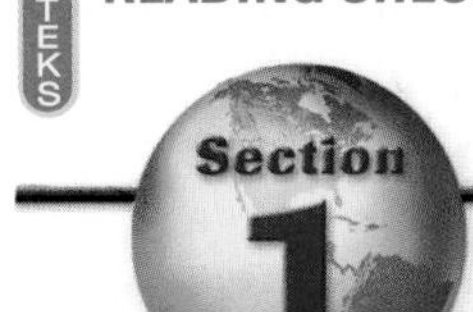

TEKS Questions 1, 3, 4, 5

Review

Homework Practice Online

Keyword: SW3 HP7

Identify Continental Divide

Define barrier islands, piedmont, fall line, basins, hot spot

Working with Sketch Maps On a map of the United States and Canada that you draw or that your teacher provides, label the Piedmont, Appalachian Mountains, Rocky Mountains, Mississippi River, Great Lakes, Great Plains, Canadian Shield, Hudson Bay, Cascade Range, Sierra Nevada, Coast Ranges, and St. Lawrence River. Where are the interior and coastal plains?

Reading for the Main Idea

1. *Places and Regions* How are the landforms of the eastern United States and Canada different from those of the western United States and Canada?
2. *Places and Regions* Where is the Great Basin located?
3. *Physical Systems* Which river drains most of the interior plains?

Critical Thinking

4. **Drawing Inferences and Conclusions** How do you think tectonic activity affects life in the western United States?

Organizing What You Know

5. Copy the graphic organizer below. Use it to describe the landforms and bodies of water in the interior plains, Canadian Shield, and Pacific coast region. What physical processes shaped these regions?

Interior plains	Canadian Shield	Pacific coast region

Section 2 Climates and Biomes

READ TO DISCOVER

1. Which climate types are found in the United States and Canada?
2. What are the major biomes of the region, and where are they found?

DEFINE

natural hazards
lichens

LOCATE

Gulf of Mexico
Hawaiian Islands

WHY IT MATTERS

Natural hazards in the United States and Canada destroy homes, ruin crops, and even take lives. Use CNNfyi.com or other **current events** sources to learn about how destructive natural hazards can be in North America.

Climates

The United States and Canada have a great variety of climates. (See the unit climate map.) For example, every climate type except an ice cap climate can be found in the United States. However, you will find an ice cap climate on some Arctic islands of far northern Canada.

Four major factors influence the distribution of climates in the United States. These factors are a middle-latitude location, prevailing winds, ocean currents, and high mountain ranges. Due to its more northerly location, Canada has mostly colder climates.

Use the unit climate map as we look at the distribution of climates across North America. The very southern tip of Florida has a tropical wet and dry climate. However, most of the southeastern quarter of the United States has a humid subtropical climate. This region stretches from the Atlantic coast to about 100° west longitude in western Texas and Oklahoma. Summers are hot

INTERPRETING THE VISUAL RECORD

Average winter temperatures in Toronto, Canada, are just below freezing. Although Lake Ontario does not completely freeze, the city's harbor is generally iced over from December to April. ***How might cycles of freezing and thawing affect this area's environment?*** TEKS

and humid, and winters are mild. Rainfall is distributed fairly evenly throughout the year, and thunderstorms are common. The warm waters of the Gulf of Mexico and the Gulf Stream influence this climate region. The Gulf Stream is a current in the Atlantic Ocean that moves warm tropical water northward along eastern North America.

The northeastern United States and parts of southern and southeastern Canada have a humid continental climate. This climate region stretches westward from the Atlantic coast to about 100° west longitude in Kansas. The region has four distinct seasons, including a warm humid summer and a cold snowy winter. The nearness of the Great Lakes and Atlantic Ocean moderates temperatures slightly and is a source of increased precipitation.

INTERPRETING THE VISUAL RECORD

The United States is hit by more tornadoes each year than any other country. Most occur between April and June in the central United States in an area known as Tornado Alley. ***How do you think hazardous environmental conditions such as tornadoes affect the natural environment?*** TEKS

West of 100° west longitude is the semiarid climate of the Great Plains. This climate supports vast grasslands and scattered trees. Cold air masses from the north meet warm moist air masses from the south over the Great Plains. Where these air masses come into contact with each other, violent storms can erupt. These storms can produce **natural hazards** like floods, hail, lightning, and tornadoes. Natural hazards are events in the physical environment that can destroy human life and property.

Because of its mountainous terrain, the intermountain area in the west has a variety of climates. Mountains block prevailing westerly winds that flow over the Pacific. This rain-shadow effect creates arid and semiarid climates on the leeward sides of mountains. For example, areas east of the Sierra Nevada and Cascades are dry. The Rockies have a highland climate. Temperatures and precipitation in the Rockies depend on elevation and local geography.

The Pacific coast region has two main climates, marine west coast and Mediterranean. The mild marine west coast climate dominates the coast from southeastern Alaska to northern California. These areas have cool wet winters and mild sunny summers. A Mediterranean climate is found in parts of southern and central California. This climate is known for its mild winters and long, sunny, dry summers.

Hawaii lies completely within the tropics. Because the Hawaiian Islands fall in the easterly trade wind belt, they are wetter on the windward eastern sides. These eastern areas have a tropical humid climate. Leeward, western slopes have a drier tropical wet and dry climate.

Far northern North America has a tundra climate. This extremely cold area stretches from northern Alaska across to Quebec and Newfoundland. Permafrost underlies much of the area. To the south, a subarctic climate is found. This large subarctic climate region covers most of Canada and Alaska.

READING CHECK: *Physical Systems* What main factors influence the distribution of climate types in the United States and Canada?

The world's largest living organism is a fungus in Oregon. This fungus stretches 3.5 miles (5.6 km) across and covers an area as big as 1,665 football fields. It is about 2,400 years old, lives underground, and spreads slowly from tree to tree.

Coast redwoods are the tallest trees in the world and can grow as high as 385 feet (117 m). Coast redwood forests are found only in a narrow belt along the Pacific Ocean in northern California and southern Oregon, where coastal fog is common. Water droplets from the fog collect on the trees' leaves and then drip down to the ground. This helps provide enough water for the huge trees to survive the dry summers.

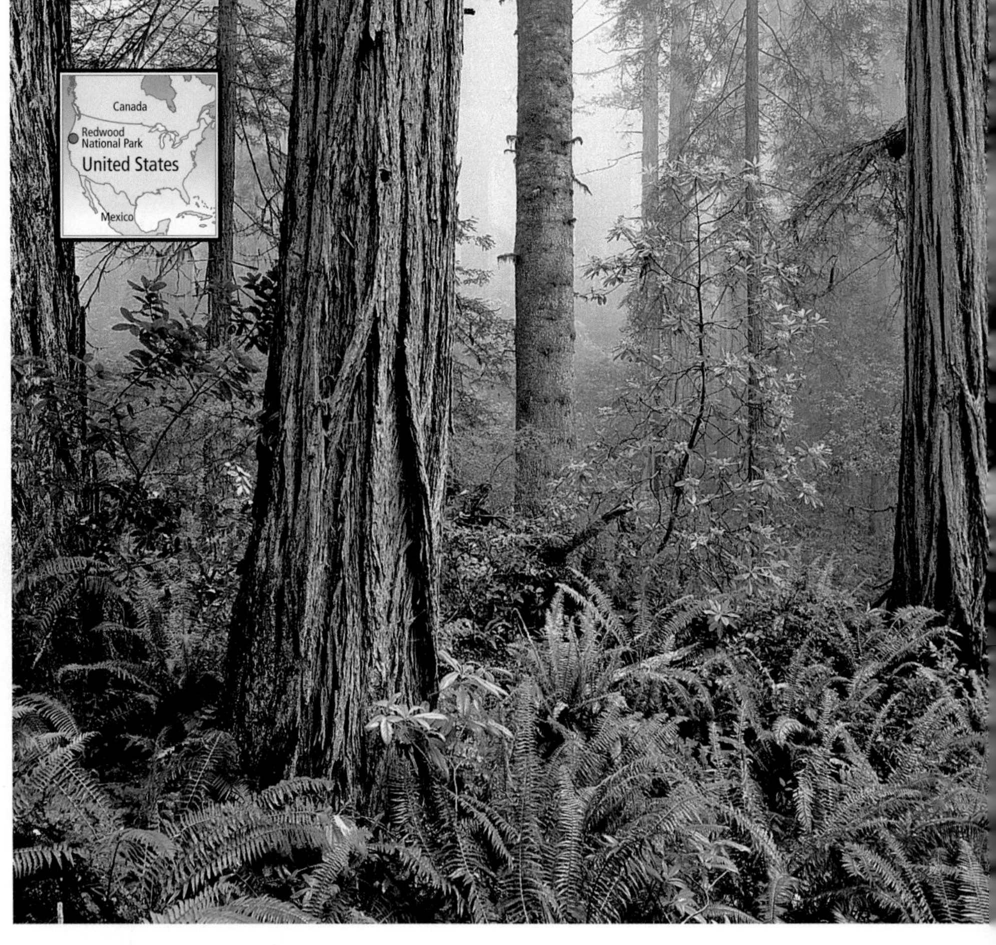

Plants and Animals

Climate patterns greatly influence North America's plant and animal life. In general, forests dominate humid areas, while grasslands or scrub cover more arid regions. However, human settlement has greatly altered the distribution of plants and animals in the region. For example, people have converted many forests and grasslands to farmland. This human activity has caused major disruptions in the natural vegetation and animal life.

The southeastern United States and much of the U.S. and Canadian west have a temperate forest biome. Different types of forests are found within these large areas. For example, mixed forests of hickory, oak, and walnut are common along the coastal plain in the southeastern United States. Deer, opossum, and raccoon are among the animals that live there. Temperate forests along the Pacific coast differ from those in the southeastern United States. Redwood trees form North America's densest and tallest forests along the coast of northern California. To the north in Oregon and Washington, Douglas fir trees become more common. Even farther north, in coastal southeastern Alaska, Sitka spruce trees are widespread. Wildlife along the Pacific coast includes black bears, eagles, hawks, and salmon.

Cacti like this cholla (CHOY-yuh) are a common sight in the Sonoran Desert of the southwestern United States. During infrequent rainstorms, wide root systems collect water, which is stored in the plant's stem. The absence of leaves and a waxy coating on the stem help prevent water loss through evaporation.

Much of the southwestern United States has a semiarid and desert biome. Creosote and mesquite bushes as well as many species of cacti cover open areas there. Coyotes, hawks, jackrabbits, and snakes live in this biome. A grassland biome stretches across the interior of North America. These grasslands, or prairies, once supported huge herds of bison. However, American settlers hunted the bison nearly to extinction. Over time farmers have also plowed under almost all of the original prairie. Today farmers use these grassland areas mainly for growing grains.

INTERPRETING THE VISUAL RECORD

Heavy rainfall and fertile volcanic soils help create a lush tropical environment in the Hawaiian Islands. More than 150 types of ferns and nearly 1,000 types of flowers grow there. Many of these plants do not grow anywhere else in the world. ***What geographic features do you think helped create Hawaii's unique vegetation patterns?*** TEKS

About half of Canada and Alaska have a boreal forest biome. This vast northern forest is one of the largest in the world. The main trees there are the spruce, fir, and pine. Great herds of caribou live in the forest during the winter. Deer, elk, moose, and wolves also inhabit the area. North of these vast forests is a treeless arctic tundra. Beneath the tundra surface is a layer of permafrost that can be up to 1,500 feet (460 m) deep. Tundra plants include grasses, small shrubs, mosses, and **lichens**—small plants that consist of algae and fungi.

Two smaller areas in the United States have tropical biomes. The tip of southern Florida has a savanna biome. The land there is swampy and covered with tall grasses. Palm trees and pine forests thrive in Florida. Hawaii has a tropical rainforest biome. Seeds carried by birds, ocean currents, and winds sprouted and grew in Hawaii's rich volcanic soils. As a result, a unique collection of plants and animals developed there. The remote location of the islands in the Pacific Ocean also influenced Hawaii's biogeography. Human activities have greatly altered Hawaii's natural ecosystems, however.

TEKS **READING CHECK:** *Places and Regions* What soil conditions will you find in the tundra biome of far northern Alaska and Canada? Which plants live there?

Section 2 Review

TEKS Working with Sketch Maps, Questions 1, 2, 3, 4, 5

go.hrw.com Homework Practice Online
Keyword: SW3 HP7

Define
natural hazards
lichens

Working with Sketch Maps
On the map you created in Section 1, label the Gulf of Mexico and the Hawaiian Islands. Which climate region do the warm waters of the Gulf of Mexico and the Gulf Stream influence?

Reading for the Main Idea

1. *Physical Systems* Why do the Great Plains have such violent weather?
2. *Places and Regions* How do climate patterns influence the distribution of vegetation in North America?
3. *Places and Regions* Why are the eastern sides of the Hawaiian Islands wetter than the western sides?

Critical Thinking

4. **Making Generalizations and Predictions** How might the tropical climates of Hawaii and southern Florida be important to agriculture in the United States?

Organizing What You Know

5. Draw a sketch map of North America. Use it to show the major biomes of the United States and Canada.

Section 3 Natural Resources

READ TO DISCOVER

1. What farming, forest, and water resources are found in the United States and Canada?
2. How rich is the region in energy and mineral resources?

DEFINE

alluvial soils
newsprint

LOCATE

Central Valley
Imperial Valley
Rio Grande valley
Colorado River
Grand Banks

WHY IT MATTERS

The many natural resources of the United States and Canada have helped make North America an economic powerhouse. Use CNNfyi.com or other **current events** sources to learn about the resources that make this region so strong economically.

Farming, Forests, and Water Resources

Abundant natural resources have helped make the United States and Canada very rich countries. The United States is an industrial giant and the world's leading agricultural producer. The country's diverse resources and strong economy help support a high standard of living. Canada also has a high standard of living, but its economy is much smaller. However, Canada is an important producer and exporter of key natural resources.

In both the United States and Canada, only about 3 percent of the population farms. However, both countries easily feed their people and have large food surpluses to export. Much of this success is a result of North America's large area and its wide variety of climates and soils. American and Canadian farmers in the Great Plains grow huge amounts of corn, soybeans, and wheat. They also raise cattle, hogs, and other livestock. In some places, fertile **alluvial soils**—soils deposited by streams or rivers—are particularly productive. These rich soils are found in the Mississippi Valley, California's Central and Imperial Valleys, and in the Rio Grande valley in Texas. With irrigation, these excellent soils support a wide range of fruits and vegetables.

INTERPRETING THE VISUAL RECORD

Much of the United States has productive farmland, such as this area in Oregon. ***How do you think the use of agricultural technology has changed the American landscape?*** TEKS

Forests are another important natural resource. Both the United States and Canada are leading producers and exporters of forest products. Canada's forests provide lumber and pulp for paper. Countries like Japan and the United States look to Canada for lumber, **newsprint**, and pulpwood. Newsprint is an inexpensive paper used mainly for newspapers. About one third of the United States is forested. The country's major commercial forests are located in the southeastern states and the Pacific Northwest. Early logging cleared large forest areas in the Northeast and the Midwest. Today much of the logging there takes place on private tree farms and in national forests.

As you have learned, North America has plentiful water resources. Water has been important to the economic development of both the United States and Canada. Many rivers, such as the Colorado, Fraser, and Tennessee, are used for irrigation and hydroelectricity. In fact, Canada and the United States are the world's two largest producers of hydroelectricity. North America's coastal waters also provide marine resources. Canada's fisheries are among the world's richest. However, overfishing and pollution have taken their toll, reducing fish catches in some areas. The Grand Banks area near Newfoundland is one of the most famous fishing areas in the world. Fisheries along the Atlantic coast are home to cod, haddock, lobster, and swordfish. Salmon is the main commercial fish on the Pacific coast. Fishers catch shrimp and shellfish in the Gulf of Mexico.

READING CHECK: ***Places and Regions*** What factors make the Central and Imperial Valleys and the Rio Grande valley productive farming regions?

A fishing crew brings in their catch from waters off the New England coast. New England's waters have long been a major fishing area in the United States.

Energy and Minerals

The United States and Canada are rich in energy and mineral resources. (See the map of resources of the United States and Canada.) The United States has about 25 percent of the world's coal reserves and is a major coal exporter. Most coal is mined in the Appalachians, Rockies, and interior plains. In Canada, coal is mined in Nova Scotia and in the western provinces of Saskatchewan, Alberta,

Resources of the United States and Canada

Oil (Petroleum)
Natural gas
Coal

SCALE 0 500 1,000 Miles
0 500 1,000 Kilometers
Projection: Azimuthal Equal Area

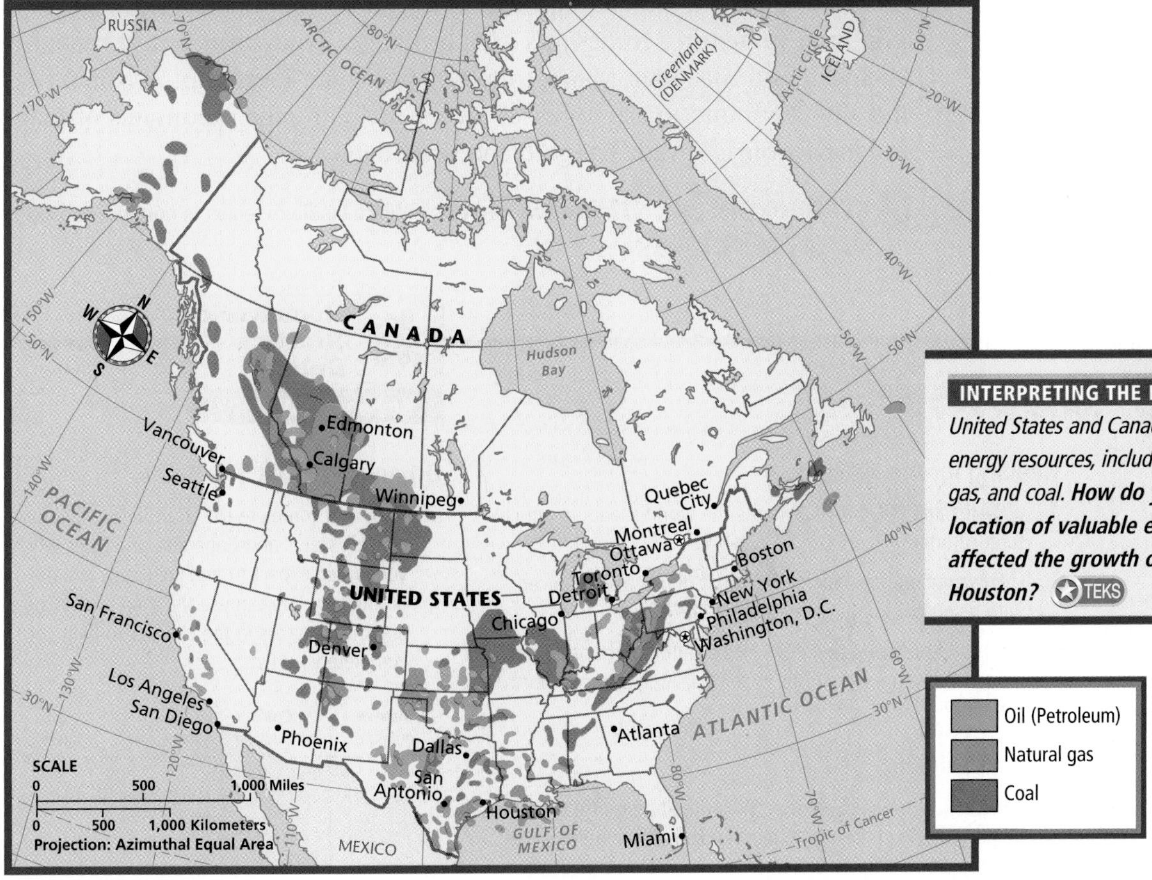

INTERPRETING THE MAP *Both the United States and Canada have abundant energy resources, including oil, natural gas, and coal.* ***How do you think the location of valuable energy resources affected the growth of cities such as Houston?*** TEKS

Oil wells dot the coastline of Huntington Beach, south of Los Angeles. California has been a leading producer and refiner of oil since the late 1800s, and petroleum is still the state's leading mineral product.

and British Columbia. These coal deposits are generally very thick and are located in unpopulated areas.

The United States is a major oil producer but uses much more oil than it produces. In fact, the United States has to import more than one half of the oil it needs. Most U.S. oil is produced on the Gulf Coast of Texas and Louisiana and in California and Alaska. These same areas are rich in natural gas, which is often found with oil. About 65 percent of Canada's oil and about 80 percent of its natural gas come from Alberta. Oil and gas deposits have been discovered off Canada's eastern and Arctic coasts as well.

The United States and Canada have a wide range of valuable mineral resources. The rocky Canadian Shield, once considered a wasteland, has many mineral deposits. Canada is a leading source of the world's nickel, zinc, and uranium. It is also a major producer of lead, copper, gold, and silver. Northern Canada even has diamond deposits. In the United States, valuable minerals are mined in the Appalachians, Rockies, and western mountain ranges. Iron has long been mined in Minnesota's Mesabi Range. Today iron is also mined in Michigan and Alabama. Major copper deposits are located in Arizona. Lead and zinc are found in a number of places, including the mountains of Idaho and in Missouri. Nevada has gold and silver deposits.

✓ **READING CHECK:** *Places and Regions* Which landform region in northern Canada has many mineral deposits?

TEKS Questions 3, 4, 5

Review

Define

alluvial soils
newsprint

Working with Sketch Maps

On the map you created in Section 2, label the Central Valley, Imperial Valley, Rio Grande valley, Colorado River, and Grand Banks. What ocean area near Canada is one of the most famous fishing grounds in the world?

Reading for the Main Idea

1. *Places and Regions* What is the world's leading agricultural country?
2. *Environment and Society* What goods are produced with resources from Canada's forests?
3. *Environment and Society* What human activities can influence the size of catches in some fishing areas in North America?

Critical Thinking

4. **Making Generalizations** Why might modern technology be particularly important to mining operations in the Canadian Shield?

Organizing What You Know

5. Copy and complete the chart below, identifying U.S. states and Canadian provinces where you would expect to find coal, gas, and oil production. You can use the map of coal, gas, and oil resources in the chapter to help you complete your lists.

State or province	Coal production	Gas production	Oil production

Environment and Society

Geography for Life

Wetlands in the United States

Wetlands are areas that are covered with water for at least part of the year. Many different kinds of wetlands exist. They include marshes, swamps, estuaries, and coastal areas affected by ocean tides. Wetlands in the United States range from the Arctic bogs of northern Alaska to The Everglades of Florida. In fact, wetlands are found in all 50 U.S. states.

Wetlands are important natural resources for many different reasons. They provide habitat for fish and other marine life, birds, mammals, and a great variety of plants. For example, wetlands are a source of food for migratory birds. In addition, wetlands provide important water resources. They hold back and then slowly release floodwaters and snowmelt. They are a source of groundwater and act as filters that cleanse water of pollutants. Coastal wetlands can prevent erosion and protect coastal areas from powerful waves and storms. In addition, wetlands provide recreation for many people. Boaters, birdwatchers, fishers, kayakers, photographers, and tourists all enjoy visiting wetland areas.

Despite their value as natural resources, most people viewed wetlands as bug-infested wastelands for much of American history. As a result, they did not think that conserving wetlands was particularly important. In fact, about half of the natural wetlands in the United States have been destroyed. Most were drained, filled, or paved over to make room for farmland or expanding urban areas. Canal construction for irrigation and flood control and the development of coastal highways have also taken their toll. California alone has lost more than 90 percent of its original wetlands. Most of this loss is the result of agricultural and urban growth.

Beginning in the 1970s and 1980s, however, American attitudes toward wetlands changed. Many people began to realize the value of protecting these diverse and useful environments. In 1986 the U.S. Congress passed the Emergency Wetlands Resources Act. This law required the U.S. Fish and Wildlife Service to study the country's remaining wetlands. The Fish and Wildlife Service reports its findings to Congress every 10 years.

According to the most recent Fish and Wildlife report, the United States has made much progress in slowing the loss of wetlands. From 1986 to 1997 the rate of wetland loss fell by 80 percent from the previous decade. Much of this decline was the result of new wetland policies and programs. These policies helped reduce the drainage and filling in of wetlands. They also helped restore wetlands. The country has not yet achieved the goal of reducing the rate of wetland loss to zero. However, Americans are now protecting and restoring more of these valuable environments.

Wetlands

SCALE
0 250 500 Miles
0 250 500 Kilometers
Projection: Albers Equal Area

N E S W

Mostly wetlands
Small wetlands, some seasonal

INTERPRETING THE MAP *Although wetlands are found in all 50 states, their distribution is uneven.* ***Which regions appear to have the highest concentration of wetlands? Why do you think that is?*** TEKS

Applying What You Know 

1. **Summarizing** Why are wetlands such important natural resources?
2. **Analyzing Information** What human activities reduced the amount of wetlands in the United States over time? What effects have conservation efforts had on the wetlands since the 1970s and 1980s? What long-term geographic and economic advantages do you think might come from these policies?

Building Vocabulary TEKS

On a separate sheet of paper, explain the following terms by using them correctly in sentences.

barrier islands	hot spot	alluvial soils
piedmont	Continental Divide	newsprint
fall line	natural hazards	
basins	lichens	

Locating Key Places TEKS

On a separate sheet of paper, match the letters on the map with their correct labels.

Piedmont	Great Lakes	Canadian Shield
Rocky Mountains	Great Plains	St. Lawrence River
Mississippi River		

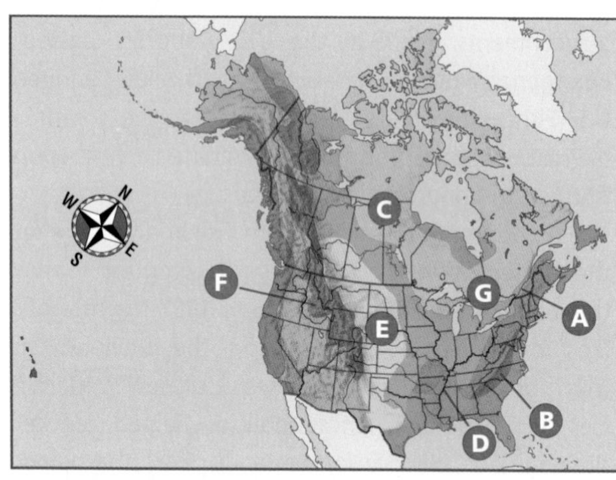

Understanding the Main Ideas TEKS

Section 1

1. *Physical Systems* What physical process forms barrier islands? Where are they found in North America?
2. *Places and Regions* What evidence of tectonic forces will you find in western areas of the United States and Canada?

Section 2

3. *Physical Systems* What can happen when different air masses come in contact with one another over the Great Plains?
4. *Places and Regions* What major factors influence climates in the United States? How does nearness to the Great Lakes and the Atlantic Ocean influence the humid continental climate region of the northeastern United States?

Section 3

5. *Places and Regions* About 25 percent of the world's reserves of which energy resource are found in the United States?

Thinking Critically for TAKS TEKS

1. **Drawing Inferences and Conclusions** Which mountain system do you think is older, the Appalachians or the Rockies? How might the physical geographic features of each range provide clues to their relative age?
2. **Analyzing Information** How have people changed the natural environment of the United States and Canada over time?
3. **Drawing Inferences and Conclusions** How do you think the locations of lakes and rivers in the United States and Canada affected the locations of settlements? Why?

Using the Geographer's Tools TEKS

1. **Analyzing Maps** Look at the map of coal, gas, and oil resources in Section 3. Which states and provinces appear to have the richest energy resources?
2. **Interpreting Climate Graphs** Study the climate graphs for Sacramento and Washington, D.C., in Chapter 1. Which city appears to have a wetter climate? Note that Sacramento is located in California's Central Valley. How might this location account for the difference in relative precipitation amounts for the two cities?
3. **Preparing Tables** Look at the map of the physical regions of the United States and Canada in Section 1. Create a table that lists these geographic regions and describes the climates and biomes of each.

Writing for TAKS TEKS

Choose one area of the United States and one area of Canada. Imagine that you went to both areas on a vacation. Write a letter to a friend comparing how people in each place depend on the natural resources there. When you are finished with your letter, proofread it to make sure you have used standard grammar, spelling, sentence structure, and punctuation.

SKILL BUILDING

Geography for Life TEKS

Creating Diagrams

Places and Regions Research the watershed, or drainage area, of the Mississippi River. Draw a diagram showing the Mississippi River and its main tributaries. Add other large rivers that flow into the main tributaries. Think of a way to visually represent how each tributary adds volume to the body of water it enters. Add other details, such as the length of the rivers or the areas they drain.

Building Social Studies Skills

Average Precipitation in Selected U.S. Cities

City	Average Annual Precipitation (in inches)
Albuquerque, NM	8.88
Boston, MA	41.51
Chicago, IL	35.82
Houston, TX	46.07
Miami, FL	55.91
Little Rock, AR	50.93
Reno, NV	7.53

Sources: *World Almanac and Book of Facts 2001,* National Climatic Data Center

Interpreting Tables

Study the table above. Then use the information to help you answer the questions that follow. Mark your answers on a separate sheet of paper.

1. Which city has the highest average annual precipitation?
   - a. Albuquerque, NM
   - b. Miami, FL
   - c. Little Rock, AR
   - d. Boston, MA
2. Which cities are probably located in an arid or semiarid climate region? How do you know that?

Building Vocabulary

To build your vocabulary skills, answer the following questions. Mark your answers on a separate sheet of paper.

3. After crossing the mountains, the pioneers entered a *basin.* In which of the following sentences does *basin* have the same meaning as it does in the sentence above?
   - a. Have you seen the sailing ships anchored in the basin?
   - b. The settlers had transported a large basin from Virginia to California.
   - c. Larry's mother washed the stained shirt in a basin.
   - d. Only one of the rivers in that basin reaches the ocean.
4. Earthquakes and volcanic eruptions are deadly *natural hazards.*
   Natural hazards are
   - a. the result of warfare between countries.
   - b. destructive phenomena in the physical environment.
   - c. dangerous obstacles along major roads and highways.
   - d. human activities that alter the natural environment.

Alternative Assessment

PORTFOLIO ACTIVITY

Learning about Your Local Geography TEKS

Group Project: Research

Plan, organize, and complete a research project about water resources. As a group, determine how your local area acquires, processes, and uses freshwater. Begin planning your project by brainstorming what you know about local water resources. How much rainfall does your area receive? What does the local watershed consist of? Who uses the most water? Where do they get that water? Plan how to find the answers to these and other questions. Divide the research and other tasks among group members. When you have completed your research, communicate your results through a report. Use maps, charts, or other visual aids to support your conclusions.

internet connect

Internet Activity: go.hrw.com

KEYWORD: SW3 GT7 TEKS

Access the Internet through the HRW Go site to examine the factors that have influenced city growth and economic development along the Great Lakes and St. Lawrence River. Then create a diagram or 3-D model that explains the role of transportation routes, types of economic activities, and resources of the area. Present your diagram or model to the class.

The United States

Trumpet player in New Orleans

The United States is home to more than 280 million people, roughly 5 percent of the world's population. Most of these people live in the 48 contiguous states between Canada and Mexico. The rest live in Alaska and Hawaii.

Hopi kachina from the southwestern United States

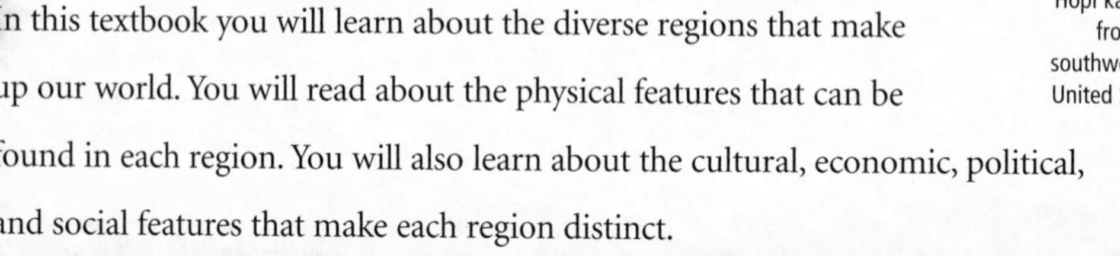

In this textbook you will learn about the diverse regions that make up our world. You will read about the physical features that can be found in each region. You will also learn about the cultural, economic, political, and social features that make each region distinct.

In addition, this textbook will introduce you to people from the world's many countries. Most of these individuals are students like you. They will tell you about their daily lives, activities, and cultures. Through their stories, you will learn something about different ways of life around the world.

We will begin our study of the world by looking at our own country, the United States. In this chapter you will study the history, culture, regions, and important geographic issues of the United States today. You can begin your study of the United States by looking at yourself and the students around you. What languages do you speak? Do you practice a religious faith? Do you live in a big city or in a small town? What are some values and beliefs that you think are important? Do you have a favorite sport? What do you like to eat? The answers to these and many other questions will help you understand what the geography of the United States is like.

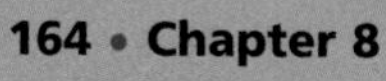

History and Culture

READ TO DISCOVER

1. What are some important events in the history of the United States?
2. What are some unique elements of American culture?

DEFINE

colonies
plantations
bilingual

LOCATE

Alaska
Florida
California
Mississippi River
Texas

WHY IT MATTERS

American popular culture is constantly spreading to other parts of the world, where it is often a major influence on other cultures. Use CNNfyi.com or other **current events** sources to find an article about the influence of American culture on faraway places.

Model T

The United States: Physical-Political

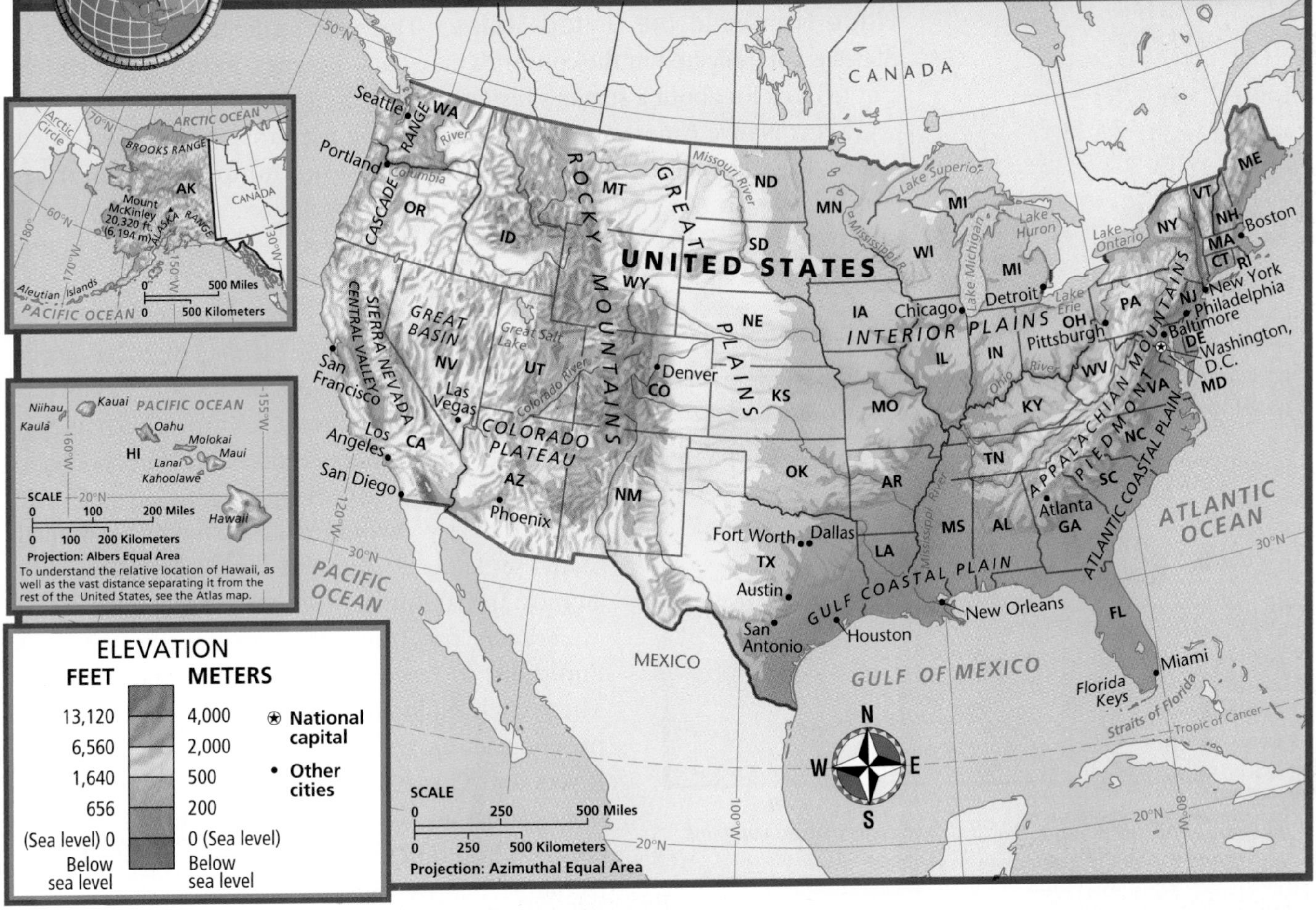

ELEVATION

FEET	METERS
13,120	4,000
6,560	2,000
1,640	500
656	200
(Sea level) 0	0 (Sea level)
Below sea level	Below sea level

⊛ National capital
• Other cities

SCALE: 0 – 250 – 500 Miles; 0 – 250 – 500 Kilometers
Projection: Azimuthal Equal Area

Hawaii inset — SCALE: 0 – 100 – 200 Miles; 0 – 100 – 200 Kilometers
Projection: Albers Equal Area
To understand the relative location of Hawaii, as well as the vast distance separating it from the rest of the United States, see the Atlas map.

The cliff dwellings of Mesa Verde National Park in Colorado were built between A.D. 1000 and 1300 by the Anasazi, the ancestors of the Pueblo Indians. The Anasazi grew crops on the mesa above these cliffs and in river valleys below. The cliff dwellings were abandoned about 700 years ago, possibly after severe droughts.

History

The ancestors of today's American Indians first settled North America at least 14,000 years ago. These early settlers probably came across an Ice-Age land bridge that linked Asia and Alaska. However, many researchers today think humans may have reached North America much earlier, possibly traveling by sea along the Alaskan coast. Eventually, these early settlers spread throughout the Americas. Over time many different cultures and languages emerged.

Spanish explorers reached the continent about 500 years ago. They claimed parts of what is now the United States from Florida to California. The Spanish were soon followed by the English and French. English colonists settled along the east coast of North America. The French explored the major river systems and Great Lakes in the interior of the continent.

FOCUS ON HISTORY

Place-Names in the United States The early settlement of the United States is still reflected in the country's many place-names. In fact, place-names can tell us a lot about a region's historical geography.

Many Native American place-names are still used throughout the United States. These names include the *Appalachians,* named for the Apalachee tribe. *Mississippi* is an Algonquian word that means "great river." A number of U.S. states, such as Alaska, Arizona, Kansas, Nebraska, and North and South Dakota, have names that originate from Indian words.

Spanish explorers arrived in the early 1500s. In the 1600s and 1700s Spanish settlers migrated north from Mexico into Texas, New Mexico, and California. They set up missions, towns, and forts such as San Antonio, El Paso, Albuquerque, Santa Fe, San Diego, Los Angeles, and San Francisco. Because of the Southwest's dry environment, the Spanish founded many settlements along rivers such as the Rio Grande. In California, they settled along the coast.

French fur trappers were among the first Europeans to reach the Great Lakes and Midwest. The French explored the interior of the country by traveling along the Mississippi and St. Lawrence Rivers and Great Lakes. As a result, many French place-names are found along these physical features today. These include New Orleans, St. Louis, Dubuque, and Detroit.

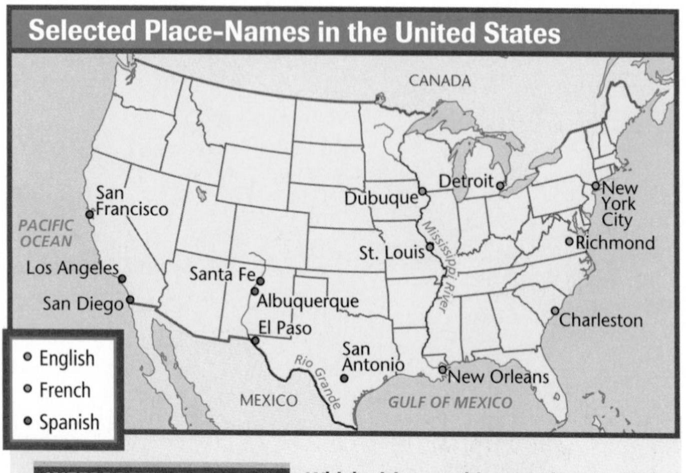

INTERPRETING THE MAP ***Which cities on this map do you think were named after European cities or to honor people?***

Many place-names in the east reflect that region's history as part of the British Empire until 1776. New York, Richmond, and Charleston are among the many English place-names there.

READING CHECK: *Environment and Society* What physical features and environmental conditions influenced early French and Spanish migration patterns and the distribution of European settlers?

Colonial History The British became the major influence on the early history of the United States. Beginning in the early 1600s they set up 13 **colonies** along North America's east coast. A colony is a territory controlled by people from a foreign land. Other early settlers included Dutch and Germans. In 1619 Europeans began bringing enslaved Africans to the colonies.

Overland travel was difficult in the early colonies. For a long time, water transportation was the colonists' main link to the outside world. In fact, nearly all the early colonial settlements were ports located on natural harbors or navigable rivers. New settlers migrated by sea to the growing coastal towns and inland trading posts on rivers.

The American colonies developed regional economies. In the southern colonies, the climate and soils were ideal for growing tobacco and cotton. Soils were less fertile and farming less productive in the northern colonies. The colonies in the north became centers for trade, shipbuilding, and fishing. Forests were an important resource at a time when ships were made from lumber. Eventually the colonists grew increasingly unhappy with British control over their economies. In 1776 they began a successful rebellion against colonial rule, which led to independence for the United States.

READING CHECK: *Environment and Society* How did physical geography affect migration patterns to colonial America?

Independence and Westward Expansion After independence, the United States set up a federal system of government. Under the U.S. federal system, power is divided between local, state, and national governments. Underlying all levels of government is the idea that ultimate power rests with the people. This idea and the U.S. model of democratic government have both diffused throughout the world.

During the 1800s many Americans and immigrants migrated westward in search of more and better farmland. By 1830, settlers had crossed the

Albert Bierstadt, *Emigrants Crossing the Plains,* 1867, oil on canvas, A011.IT: The National Cowboy Hall of Fame and Western Heritage Center, Oklahoma City, OK.

INTERPRETING THE VISUAL RECORD

This painting by the American artist Albert Bierstadt (1830–1902) shows a group of pioneers heading west across the Great Plains. In the 1800s some Americans thought the United States was destined to expand across the continent, an idea known as manifest destiny. ***What does Bierstadt's painting suggest about American perceptions of the West during the 1800s?*** TEKS

Territorial Expansion of the United States

INTERPRETING THE MAP

Within 100 years of independence, the borders of the United States had expanded to the Pacific Ocean. New land was added through negotiations, purchases, and wars. ***How did the United States gain most of its territory after 1783? Which was the largest single addition?*** TEKS

Mississippi River and settled as far south as Texas. Pioneers were soon settling on the Pacific coast. Many arrived after gold was discovered in California in 1848. However, few people settled in the deserts and mountains of the western United States or in the plains between the Mississippi River and Rocky Mountains. They called this grassland region the Great American Desert and believed it was too dry to support farming.

The boundaries of the United States shifted as settlement spread westward. By the mid-1800s the country stretched from the Atlantic to the Pacific coast. (See the map.) The government sold land cheaply or gave it away to encourage people to settle new areas.

As pioneers moved westward, they began to have bitter conflicts with American Indians. Many American Indians did not own land like the descendants of Europeans did. Instead, some Indians considered land a shared resource rather than someone's personal property. As settlers occupied and divided up land, they pushed American Indians farther west and onto reservations. Many died from warfare or from diseases carried by settlers.

INTERPRETING THE VISUAL RECORD

Once considered part of the Great American Desert, the southern Great Plains attracted many settlers in the 1800s. Some even described the area as a Garden of Eden. However, in the 1930s the area was hit by a severe drought made worse by poor farming practices. Strong winds carried away topsoil and caused huge dust storms. As a result, more than 350,000 people left the area, which became known as the Dust Bowl. ***How did changing perceptions of the Plains lead to shifts in settlement patterns?*** TEKS

Economic Development By 1830 the northeastern United States was industrializing. Industries and railroads spread. In the South, however, the economy was based on export crops like tobacco and cotton. Farmers grew these crops on **plantations**—large farms that produce one major crop. Many southern plantations used the labor of enslaved Africans. Economic differences between the North and South, and the South's insistence on maintaining slavery, eventually led to the Civil War. This war lasted from 1861 to 1865 and ended with the defeat of the southern, or Confederate, states. The federal government then moved to end slavery throughout the country.

After the Civil War, improvements in technology encouraged rapid westward migration. The transcontinental railroad was completed in 1869. This railroad made it much easier to move goods and people across the country.

Railroads also allowed major cities to develop far from navigable waterways. With new agricultural machinery, farms could produce more food using fewer people than ever before. Irrigation and better plows allowed farmers to grow crops in the Great American Desert. As a result, people's perceptions of the region changed, and they began to settle the area.

The development of industry attracted more people to the country's growing cities. Some came from rural areas. However, many were immigrants, mostly from Europe. Many European immigrants settled in the industrial cities of the Northeast. By 1920 more Americans lived in cities than in rural areas.

The 1900s In the 1900s the United States experienced major social, economic, and technological changes. The country fought in World War I in 1917 and 1918 and suffered through the Great Depression in the 1930s. U.S. forces also fought in World War II from 1941 to 1945. Since then, the United States has been one of the richest and most powerful countries in the world.

After World War II, the United States and the Soviet Union became rivals in the Cold War. Both countries built huge military forces and developed nuclear weapons. The two countries never formally went to war against each other. However, they supported different sides in small wars around the world. Since the collapse of the Soviet Union in 1991, the United States and Russia have had friendlier relations.

READING CHECK: *Human Systems* Why did many Americans and immigrants move westward during the 1800s?

Culture

Because of its long history of immigration, American culture includes traditions, foods, and beliefs from all over the world. In fact, the United States is one of the world's most culturally diverse countries.

People and Languages The diversity of the American people is perhaps their major characteristic. More than 99 percent of Americans are either immigrants or the descendants of immigrants. American Indians make up less than 1 percent of the population. Today immigrants from all over the world continue to arrive in the country.

Most Americans are of European descent. This group includes people whose ancestors came from Britain, Germany, France, and other European countries. About 12 percent of Americans trace their origins to Africa. Slightly more people identify themselves as Hispanic. Hispanics and Asians are the most rapidly growing parts of the population.

Since colonial times, English has been the main language of the United States. However, the United States has no official language. The second most widely spoken language in the country is Spanish. Spanish is particularly common along the U.S.-Mexico border. Many Spanish speakers are also **bilingual**, which means they are able to speak two languages. Hundreds of other languages are spoken in the United States, particularly in large cities.

Connecting to HISTORY

American Slang

American slang words have diverse and sometimes remote origins. Some slang words originate in a certain geographical area or social group and change meaning as they diffuse to other groups. For example, the word *funky* originated from British English, where it first meant "musty or foul-smelling." Over time people began to use the term to mean "filthy." Eventually the word crossed the Atlantic to the United States. African Americans adopted the word, which evolved a variety of new meanings. By the 1930s *funky* had come to mean "excellent" or "deeply satisfying." In the 1950s it was applied to a type of music. In the 1980s it acquired the meaning of "odd or strange." Today the word *funky* is used throughout the United States by many different ethnic and social groups and has a variety of meanings.

Preparing Maps Draw a sketch map tracing the diffusion of the word *funky* from Britain to the United States. How has the word been used over time? TEKS

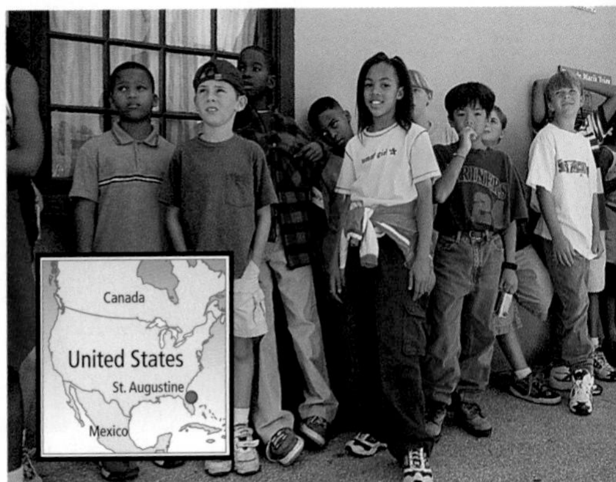

These schoolchildren in Florida illustrate the cultural and ethnic diversity of the United States. Immigration has long shaped the human geography of the country.

Diverse foods and cooking styles in the United States reflect regional and cultural differences. For example, the Cajuns of southern Louisiana are known for their spicy and flavorful dishes. Here, a restaurant owner in New Orleans shows off a dish of boiled crawfish, a local specialty.

Religion The cultural diversity of the United States can also be seen in its many religions. There are more than 1,200 religious groups in the country. Immigrants introduced many of these religions. Other religions were founded in the United States.

Christianity is the major religion in the country. European settlers brought it to North America. More than half of all Americans are Protestant Christians. Major Protestant groups include Baptists, Lutherans, and Methodists. About 25 percent of Americans are Roman Catholic Christians. Many American Catholics have Spanish, Italian, or Irish ancestors. Some 6 million Jewish Americans live in the United States—more than live in Israel. In recent years, immigration has increased the numbers of people who practice Buddhism, Hinduism, and Islam.

Settlement and Land Use The population of the United States is concentrated in the Northeast. (See the unit population map.) This pattern reflects the history of the country's settlement. Early European settlement was concentrated in the Northeast, and the first large U.S. cities were located there. Later, settlers moved westward.

Although the Northeast is still the most concentrated area of settlement, the country's population has been moving to the South and West. (See the map.) This shift in settlement reflects the decline of the country's old industrial region, once known as the Rust Belt. At the same time, warmer areas in the South and West, known as the Sun Belt, have attracted many people.

Settlement patterns also reflect land use in the country. For example, the most densely populated regions include the urban areas of the Northeast, Midwest, and Pacific coast. Less densely populated are the rich farmlands and ranch areas of the Midwest and West. The most sparsely populated areas are desert and mountain regions of the West.

READING CHECK: *Human Systems* How has America's long history of immigration shaped its culture?

Center of Population of the United States

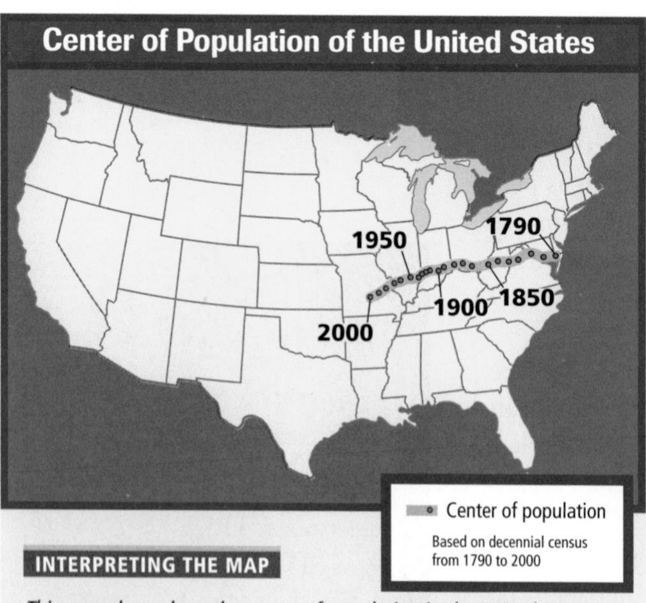

INTERPRETING THE MAP

This map shows how the center of population in the United States has changed throughout the country's history. ***What factors led to the changes in population distribution shown here? How do you think the distribution of the U.S. population will change in the future?*** TEKS

Education Education has been an important factor in the development of the U.S. economy. It has also helped create a high standard of living. Basic education is free and is required of all citizens. Nearly all children complete elementary school, and more than 80 percent graduate from high school. Many go on to study at colleges and universities.

The United States has one of the largest and best systems of higher education in the world. More than 3,000 colleges and universities are located in the country. Many, such as Harvard and the University of California at Berkeley, are world leaders in research and teaching. In fact, tens of thousands of students from all over the world come to study at universities in the United States.

READING CHECK: *Places and Regions* How does the U.S. education system make the country a distinctive region?

The Arts, Customs, and Traditions Americans have many different traditions and customs. In the arts, American writers, artists, musicians, filmmakers, and sculptors are internationally famous. The United States helped pioneer the development of motion pictures and still dominates the industry. American movies are shown all over the world. In architecture, the United States was the first country to build skyscrapers. This style of architecture has now diffused throughout the world. Well-known American writers and poets include Edgar Allan Poe, Mark Twain, Emily Dickinson, Ernest Hemingway, Maya Angelou, and many others.

The many ethnic and cultural groups that came to the United States brought their own musical styles. For example, Africans brought the rhythms of West African music. Europeans brought instruments and harmonies from their native lands. As many African Americans migrated to cities in the early 1900s, the musical traditions of Africans and Europeans blended together to form jazz. Jazz later diffused from cities such as New Orleans to the rest of the country. In fact, jazz is now popular around the world.

Other musical styles that originated in the United States include blues, country, rock 'n' roll, and rap. These styles have spread around the world through recordings, radio, and television. American music is a major influence on globalization and the diffusion of American culture.

Widely celebrated holidays in the United States include Christmas and Easter, which reflect America's Christian religious heritage. Many Americans celebrate other religious holidays. Americans also mark Independence Day, July 4, with fireworks and picnics. In addition to national celebrations, many towns and communities celebrate local historical events and personalities.

People in the United States enjoy watching and playing baseball, basketball, football, golf, and other sports. American athletes compete in international sporting events such as the Olympics. In fact, popular American sports such as baseball and basketball have diffused to other parts of the world and contributed to the spread of American culture.

Louis Armstrong (1901–71) was one of the most popular and influential jazz musicians of all time. Nicknamed Satchmo, he was a gifted trumpet player, singer, and bandleader. Armstrong's brilliant performances changed the way jazz was played and sung.

READING CHECK: *Human Systems* How did the distinctive cultural patterns of the United States lead to the innovation and diffusion of jazz?

TEKS Working with Sketch Maps, Questions 1, 2, 3, 4, 5

Define colonies, plantations, bilingual

Working with Sketch Maps On a map of the United States that you draw or that your teacher provides, label Alaska, Florida, California, Mississippi River, and Texas. Which state are the ancestors of Native Americans believed to have crossed into from Asia?

Reading for the Main Idea

1. *The Uses of Geography* How did migration over an Ice-Age land bridge affect the history and culture of the United States?
2. *Human Systems* How did American Indians and descendants of Europeans view land ownership differently?
3. *Environment and Society* How did irrigation and better plows change settlers' perceptions of the Great American Desert between the Mississippi River and the Rocky Mountains?

Critical Thinking

4. **Comparing** How do you think patterns of land use might lead to cultural differences between urban areas and rural farming areas?

Organizing What You Know

5. Copy the graphic organizer below. Use it to describe important cultural features that make the United States distinctive.

People and languages	
Religion	
Settlement and land use	
Education	
The arts, customs, and traditions	

The World in Spatial Terms

Geography for Life

A Small World after All?

Have you ever heard the expression, "the world is getting smaller"? Have you wondered what this expression means? As technology and transportation improve, it becomes easier and quicker to move people, products, and information to distant places. As a result, these places begin to seem closer because they become more accessible. However, these places are the same distance apart as they were before.

Many geographers are interested in the relationship between changes in technology, movement, and perception. They use the term *time-space convergence* to refer to the increasing nearness of places that happens as transportation and communication technologies improve. These improvements can steadily reduce the amount of time it takes to travel from place to place.

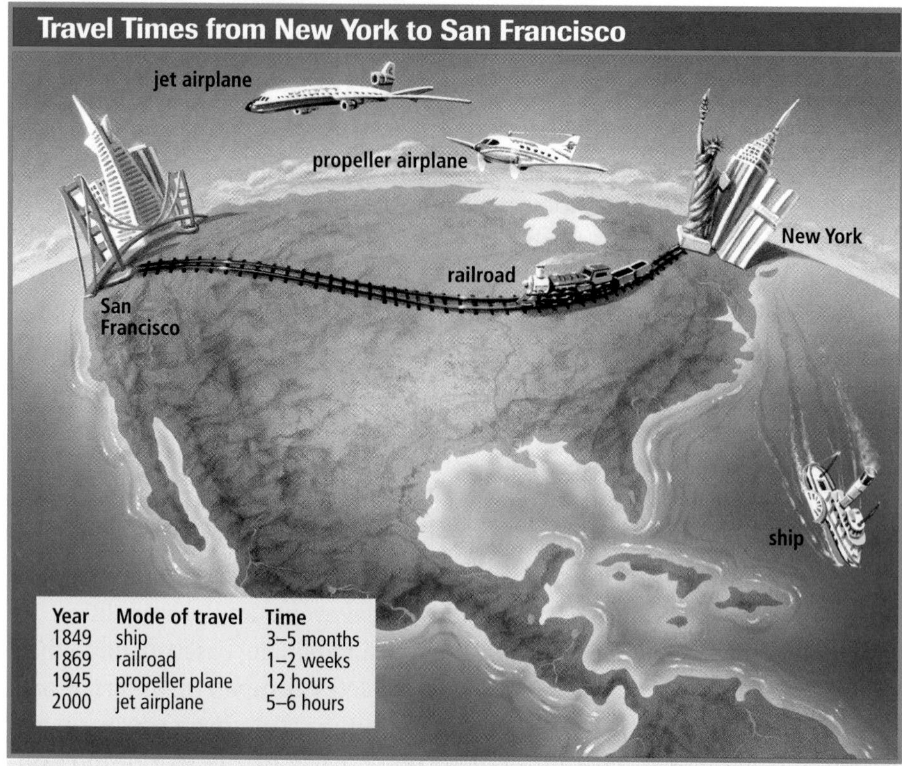

Year	Mode of travel	Time
1849	ship	3–5 months
1869	railroad	1–2 weeks
1945	propeller plane	12 hours
2000	jet airplane	5–6 hours

INTERPRETING THE DIAGRAM *As this diagram shows, changes in technology have dramatically reduced travel times between New York and San Francisco over time.* ***How have new forms of transportation altered travel routes? How do you think these changes in transportation technology affect the country's cultural geography?***

For example, imagine you lived in the mid-1800s and wanted to travel from New York to San Francisco. How long do you think it would take? What route would you follow? How difficult would the journey be? Most importantly, how would you perceive the journey? As you can see from the diagram, it took three to five months to travel from New York to San Francisco in 1849. Today, however, it is just a five or six hour flight. What was once a long, difficult ocean voyage is now a comfortable flight complete with dinner and a movie. In fact, some people now travel from coast to coast for weekend vacations or one-day business meetings.

Other modern technologies are also tying distant places closer together. For example, faxes, the Internet, and e-mail all help the free flow of information between some of the most isolated places. The same technologies also encourage the diffusion of culture traits and globalization. Together with modern travel, these technologies are helping make the world seem smaller and smaller.

Applying What You Know

1. **Summarizing** How can improvements in technology and transportation make the world seem smaller?
2. **Drawing Inferences and Conclusions** How do you think improvements in technology and transportation can change people's perceptions of geographic features? How might these changes lead to changes in human societies?

Section 2 Regions of the United States

READ TO DISCOVER

1. What is the economy of the Northeast like?
2. Why is the Midwest such an important farming area?
3. How is the geography of the South changing?
4. How have environmental conditions influenced the history of the West?

IDENTIFY

Megalopolis
Corn Belt
Dairy Belt
Wheat Belt
Silicon Valley

DEFINE

textiles
metropolitan area
arable
smog

LOCATE

Washington, D.C.
Boston
New York
Chicago
Detroit
St. Louis
Dallas
Houston
Miami
Los Angeles
Seattle
San Francisco

WHY IT MATTERS

The U.S. Census Bureau divides the United States into regions. Use CNNfyi.com or other **current events** sources to learn about how the Census Bureau organizes information about the regions of the United States.

The Northeast

The Northeast is the smallest and most densely populated region in the United States. It includes Maine, New Hampshire, Vermont, Massachusetts, Rhode Island, Connecticut, New York, New Jersey, Pennsylvania, West Virginia, Maryland, Delaware, and Washington, D.C. *D.C.* is the abbreviation for "District of Columbia." The Northeast is home to about a fifth of the country's population. Most of that population is concentrated in an urban corridor known as **Megalopolis**. A megalopolis is a group of cities that have grown into one large, built-up area. Boston, New York, Philadelphia, Baltimore, Washington, D.C., and their suburbs form the Megalopolis. A web of highway, rail, and air routes links these urban areas.

The Northeast is the political and financial center and most industrialized region of the United States. In fact, the country's first industries developed there. These early industries used running water from rivers to power machinery and produce **textiles**, or cloth products. A wide variety of manufacturing

INTERPRETING THE VISUAL RECORD

New York is located in the heart of Megalopolis. The term Megalopolis *was used in the 1960s by a French geographer to describe the huge urban area stretching from Boston to Washington, D.C. Today it is used to describe other massive urban areas around the world.* ***What other large urban area would you describe as a megalopolis?***

industries developed later. They developed in places favorable to certain industries, such as steelmaking. For example, consider Pittsburgh, Pennsylvania. Pittsburgh is located at the junction of two rivers and is near rich iron ore and coal deposits. This location helped make the city a major early center of the steel industry.

Good transportation connections were very important to industrial growth in the Northeast. The early growth of the region's cities was made possible by their good port sites, which allowed the movement of people and products. The completion of the Erie Canal in the 1820s provided access to the Great Lakes and the interior United States. Today a dense network of roads, railroads, and air routes crisscrosses the region and connects it to the rest of the country.

In the late 1900s manufacturing in the Northeast declined. Cheaper labor in other places forced many factories to close. Because of its declining older industries, the region became known as the Rust Belt. Meanwhile, economic growth and warm climates attracted people and industries to the South and West. The Northeast is still a major industrial area and has attracted newer industries. However, the region has lost the economic dominance it once enjoyed.

READING CHECK: ***The Uses of Geography*** What features helped Pittsburgh grow as a major steel center?

Cultures The Northeast is home to many different cultures. Until the mid-1960s most immigrants came from Europe. As a result, some cities in the region have large Greek, Irish, and Italian neighborhoods. In recent decades most immigrants have come from Latin America and Asia. These immigrants bring different cultures and ways of life with them, reshaping the region's cultural geography. For example, many Northeastern cities now host parades celebrating the Chinese New Year. Other neighborhoods are alive with the sounds of Latin American music and people speaking Spanish.

Cities The New York City area is the largest **metropolitan area** in the United States. A metropolitan area is a city and its surrounding built-up areas. More than 20 million people live in the New York area. Dutch colonists settled at the site of New York in the 1620s and called it New Amsterdam. The English renamed the city New York later after acquiring it from the Dutch. New York's excellent natural harbor at the mouth of the Hudson River makes it an ideal location for a port and trading center. Today New York is America's leading center of commerce, banking, advertising, fashion, and media.

Other cities in the Megalopolis have long been important commercial centers. Boston and Philadelphia date back to colonial times. Baltimore, on the western shore of Chesapeake Bay, was founded on the fall line. Baltimore became a major port because of its rail connections to interior coal mines, steel mills, and farming areas. Washington, D.C., is unique among the country's cities because it was planned and built to serve as the U.S. capital. Construction of the city began in the late 1700s.

Many farmers in the Northeast grow specialty crops that are shipped overnight to the region's large urban markets. In this photo, cranberries are being grown in southern New Jersey.

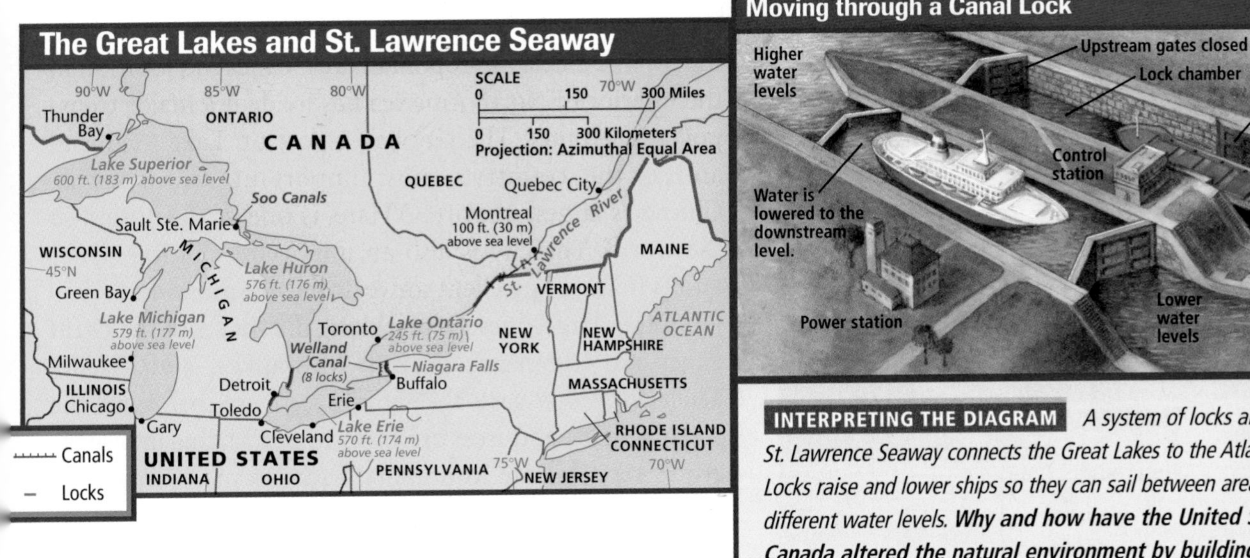

INTERPRETING THE DIAGRAM *A system of locks along the St. Lawrence Seaway connects the Great Lakes to the Atlantic Ocean. Locks raise and lower ships so they can sail between areas with different water levels.* ***Why and how have the United States and Canada altered the natural environment by building the St. Lawrence Seaway? How was technology used?*** TEKS

Government employment is the largest source of income today, followed by tourism.

TEKS **READING CHECK:** ***Places and Regions*** How is the character of the Northeast related to its cultural characteristics?

The Midwest

The Midwest is this country's major farming region and a leading producer of industrial goods. The region includes Ohio, Indiana, Illinois, Michigan, Wisconsin, Minnesota, Iowa, and Missouri. The combined population of these states is slightly smaller than that of the Northeast. The Great Lakes and Mississippi River link the Midwest's major cities to each other and to other regions. (See the Great Lakes and St. Lawrence Seaway map.) Railroads and highways are also major transportation connections.

Many settlers moved into the Midwest as transportation routes from the East Coast developed during the 1800s. Excellent farmland and growing cities attracted many of these settlers. In addition, between 1915 and 1930 hundreds of thousands of African Americans migrated to the region from the South. They came to find work in cities like Chicago and Detroit.

Agriculture The Midwest is one of the most productive farming regions in the world. Most of the region's land is **arable**, or fit for growing crops. In fact, some areas specialize in certain crops. One such region is the **Corn Belt**, which stretches from Nebraska to Ohio. Within the Corn Belt, Illinois and Iowa are the country's leading corn-producing states. Most of this corn is used to feed livestock such as beef cattle and hogs. The United States is also the world's major exporter of corn. Soybeans are another important Corn Belt crop. Soybeans are used to make margarine, vegetable oil, and bean curd (tofu).

The **Dairy Belt** is located north of the Corn Belt, where summers are cooler and soils are rockier and less fertile. The Dairy Belt includes Wisconsin and most of Minnesota and Michigan. Wisconsin—"America's Dairyland"—produces more butter and cheese than any other state. These dairy products are sold in the Midwest's cities and are shipped by refrigerated trucks to the rest of the country.

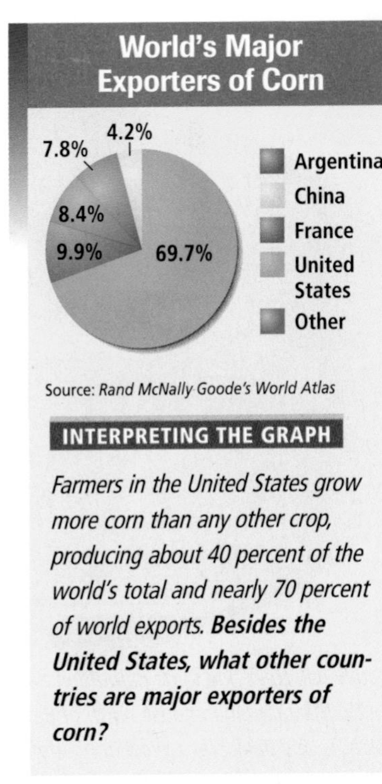

Source: *Rand McNally Goode's World Atlas*

INTERPRETING THE GRAPH

Farmers in the United States grow more corn than any other crop, producing about 40 percent of the world's total and nearly 70 percent of world exports. ***Besides the United States, what other countries are major exporters of corn?***

The Gateway Arch stands next to the Mississippi River in St. Louis. The arch is 630 feet (192 m) high and symbolizes the historical importance of St. Louis as the Gateway to the West.

Cities Chicago is the largest city in the Midwest and the third-largest U.S. metropolitan area. Chicago's location in the interior of North America has made it a major transportation center. The city has a port on Lake Michigan and is the country's most important railroad hub. Chicago's largest airport, O'Hare, is one of the busiest in the world. The city is also an important cultural center and is home to excellent universities and museums.

Other large cities located along the shores of the Great Lakes are Cleveland, Detroit, Milwaukee, and Toledo. Their locations gave the cities access to coal from the Appalachians and iron ore from upper Michigan, northern Wisconsin, and Minnesota. Each is a major manufacturing center. For example, Detroit has been the center for automobile manufacturing in the United States.

The Twin Cities of Minneapolis and St. Paul are located on the upper Mississippi River. They are major distribution centers for the agricultural products of the upper Midwest. St. Louis, Missouri, is located near the area where the Mississippi, Missouri, and Illinois Rivers flow together. The city began as a French fur-trading post in 1764. In the 1800s St. Louis was the center for pioneers heading west and was a major river port known as the Gateway to the West.

TEKS **READING CHECK:** ***Places and Regions*** How did access to resources help the development of cities like Cleveland, Detroit, Milwaukee, and Toledo?

The South

The South stretches in a great arc from Virginia to Texas. The region includes Virginia, North Carolina, South Carolina, Georgia, Florida, Alabama, Mississippi, Tennessee, Kentucky, Arkansas, Louisiana, and Texas. These states are home to a little more than 30 percent of the country's population, more than any other region.

Historically, the South was mainly rural and agricultural. The majority of the population lived on farms. Enslaved Africans worked on large cotton, rice, and tobacco plantations. The Civil War wrecked the South's economy. The South became the poorest region in the country. Its economy lacked the industries and railroads that were so important to the development of the Northeast and Midwest. As a result, many southerners migrated to northern cities in search of factory jobs.

The South has attracted new industries in recent decades. Since the 1960s many people have migrated back to the region. In addition, new immigrants have moved to the South from the Caribbean, Mexico, and other areas in Latin America. In parts of southern Florida, for example, Hispanics make up a majority of the population. Spanish is more widely spoken than English in some places.

Stately oak trees and large plantation homes are common in the South. This plantation house north of Baton Rouge, Louisiana, has columns and balconies typical of plantation architecture.

Economy Primary industries based on local raw materials are important in the South. For example, the region is a major source of lumber. In turn, forestry supports paper, pulp, and furniture industries. Farm products, particularly cotton and tobacco, are still important. Cotton provides raw material for a large textile industry. The textile industry is concentrated in the Piedmont of Georgia, the Carolinas, and Virginia. Southern states, particularly Texas and Louisiana, are major producers of mineral and energy resources.

In recent years, many new industries have developed in the South. These new industries take advantage of the region's lower wages, cheap land, and favorable laws and regulations. For example, high-tech and aerospace industries have been growing. Foreign automobile makers have also built factories here. The headquarters of several large banks are located in North Carolina. That state has also become an important center for the biotechnology industry.

The South's warm climate has made the region a popular tourist destination. For example, Orlando and other places in Florida rely heavily on tourism. The South attracts retirees from colder parts of the country. Some cities, such as Tampa, have become popular places for retired people to live.

Warm tropical weather and sandy beaches draw millions of people to Florida. Tourism is an important part of the state's economy.

Cities The Dallas–Fort Worth metropolitan area is the largest in the region. Dallas was founded in 1841 as a railroad shipping stop for cattle and cotton. Fort Worth was an early cattle-marketing center and is sometimes considered part of the West. Today it is a center for oil, grain, and aerospace industries. Houston and surrounding cities make up the second-largest metropolitan area in the South. Atlanta, Miami, and New Orleans are also transportation and commercial centers.

TEKS **READING CHECK:** ***Human Systems*** What has attracted new industries to the South in recent years?

The West

The West is the largest and most sparsely populated region of the United States. About a quarter of the country's population lives in this huge area. The interior West includes the Great Plains, Rocky Mountains, and an intermountain region west of the Rockies. States there include North Dakota, South Dakota, Nebraska, Kansas, Oklahoma, New Mexico, Colorado, Wyoming, Montana, Idaho, Utah, Nevada, and Arizona. Farther west are the Pacific states: California, Oregon, Washington, Alaska, and Hawaii.

Environmental conditions have played an important role in the history of the West. Much of the area consists of dry plateaus, deserts, and high mountains.

The San Juan Mountains, a range in the Rockies, rise dramatically in southern Colorado. The West has long been known for its spectacular scenery and natural landscapes.

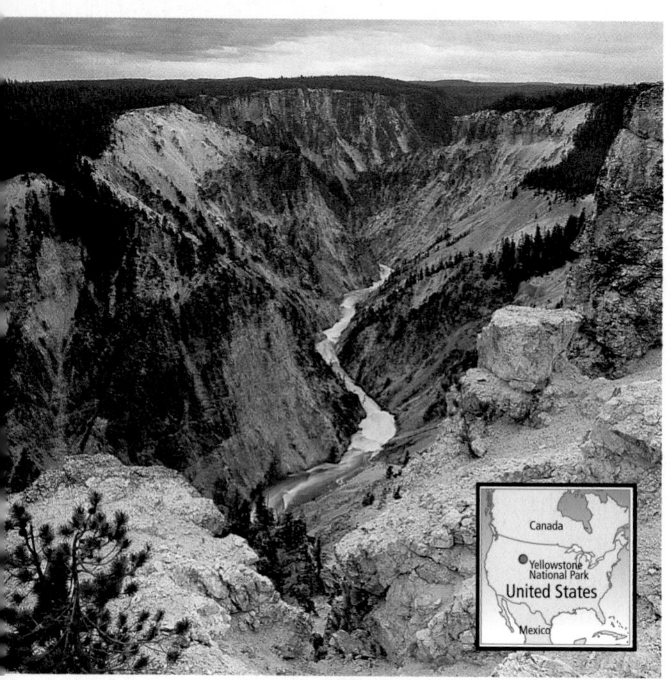

Early pioneers found the region difficult to travel across, inhospitable to live in, and almost impossible to farm. However, the opening of the transcontinental railroad in 1869 made travel much easier. As a result, many settlers moved into the region. In the 1900s aqueducts and irrigation systems opened up even more areas to settlement and farming.

The Interior West Historically, raising livestock has been a major economic activity in the interior West. In many areas, raising livestock is combined with wheat farming. The **Wheat Belt** stretches across the Dakotas, Montana, Nebraska, Kansas, Oklahoma, Colorado, and Texas. Irrigation water from the Ogallala Aquifer allows farmers to grow wheat and other crops. Overuse of this water, however, has lowered the water table of the aquifer. As a result, some farmers have begun to use more-efficient irrigation methods to try to preserve valuable groundwater.

INTERPRETING THE VISUAL RECORD

In 1872 the U.S. Congress established Yellowstone National Park. Yellowstone was the world's first national park. Since then, the national park concept has spread around the world. ***How might the diffusion of the national park concept have caused cultural change in the United States and elsewhere?*** TEKS

Mining is a key economic activity in the Rocky Mountains. Early prospectors struck large veins of gold and silver there. Today Arizona, New Mexico, and Utah are leading copper-producing states. Nevada is the leading gold-mining state. Lead and many other ores are also found in the interior West. Tourism is also important. Attractions include ski resorts like Aspen and Vail in Colorado and Taos in New Mexico. Each year millions of people visit the region's stunning national parks. These parks include Glacier in Montana, Grand Canyon in Arizona, and Grand Teton and Yellowstone in Wyoming.

READING CHECK: ***Environment and Society*** How did environmental conditions influence past migrations and settlement patterns in the West?

The Pacific States Most people in the West live in the Pacific states. California is home to some 34 million people, more than any other state. Today the Pacific Coast ranks second only to Megalopolis in economic importance.

Before World War II the economy of the Pacific states was based mostly on farming, forestry, and the film industry in Los Angeles. The growth of military bases in the region during World War II boosted the economy. This growth also drew migrants to the region. After the war, dams were built on the Columbia River to provide cheap hydroelectricity for aluminum smelters and other growing industries. Aircraft manufacturing in Seattle became Washington's largest industry.

In much of California, the warm Mediterranean climate allows a year-round growing season. However, rainfall is rare in the summer, which means crops must be irrigated. Building aqueducts to carry water from the mountains of northern California and the Colorado River to central and southern California created a boom in farming. Today agriculture uses about 80 percent of California's water supply. City residents must compete with farmers for scarce water resources.

In the late 1900s the development of computer technologies brought new industries to the Pacific states. **Silicon Valley**, located south of San Francisco, became the country's leading center of computer technology. Many software companies are located in the San Francisco Bay and Seattle areas.

Our Amazing Planet

The oldest known living tree is a bristlecone pine in Nevada that is thought to be about 4,900 years old.

The economies of Alaska and Hawaii depend heavily on their states' locations, natural resources, and scenery. The United States bought Alaska from Russia in 1867. It became a state in 1959. Alaska is the country's largest and least densely populated state. Its economy was initially based on fishing. However, after the discovery of large North Slope oil deposits, oil became Alaska's most valuable natural resource. Hawaii also became a state in 1959. Because of its strategic Pacific location, Hawaii is home to many military facilities. The state's tropical climate and fertile soils are used to grow crops like pineapples and sugarcane. Tourism is also a major industry.

INTERPRETING THE VISUAL RECORD

Located in northeastern Alaska, the Arctic National Wildlife Refuge is one of the most pristine ecosystems on Earth. However, there is growing pressure to open the area to oil exploration and production. Some people worry that oil production might cause environmental damage. Supporters of this development believe the environment can be protected at the same time. ***How do you think oil production might change this natural environment? Which point of view on oil production here do you agree with? Why?*** TEKS

Cities Los Angeles is the largest metropolitan area in the West and the second-largest in the country. The city began as a Spanish settlement in 1781. Railroad connections boosted the town's economy in the late 1800s. However, lack of water made further growth difficult. In 1913 a new aqueduct brought water from the slopes of the Sierra Nevada nearly 250 miles (400 km) away. As a result, the city grew rapidly. Major industries today include entertainment, oil refining, chemicals, and manufacturing. Because most of its growth has occurred during the automobile era, Los Angeles is a vast sprawling city. Automobile and factory exhaust creates **smog**, which often hangs over the city. Smog results from chemical reactions involving sunlight and pollutants from automobiles and industries.

Other big cities in the region include San Francisco, Seattle, San Diego, Phoenix, and Denver. San Francisco is located on an excellent deepwater harbor. It has long attracted immigrants and is one of the most diverse cities in the country. Seattle is an important port for trade with Asia. San Diego is also a port and is home to the most important naval base on the west coast.

READING CHECK: *Environment and Society* How have southern Californians modified their physical environment to make farming productive?

TEKS Working with Sketch Maps, Questions 1, 2, 3, 4, 5

Section 2 Review

Homework Practice Online

Keyword: SW3 HP8

Identify Megalopolis, Corn Belt, Dairy Belt, Wheat Belt, Silicon Valley

Define textiles, metropolitan area, arable, smog

Working with Sketch Maps

On the map you created in Section 1, label Washington, D.C., Boston, New York, Chicago, Detroit, St. Louis, Dallas, Houston, Miami, Los Angeles, Seattle, and San Francisco. Which cities make up Megalopolis? Which city is the largest metropolitan area in the West?

Reading for the Main Idea

1. *Places and Regions* How has urban growth and industrialization in the Northeast been tied to the region's good transportation connections?
2. *Environment and Society* What makes the Midwest such an important agricultural region?
3. *Places and Regions* How have new technologies affected the economies of Silicon Valley and the San Francisco Bay area in recent decades?

Critical Thinking

4. **Comparing** How do you think life for people in Megalopolis might be similar to life in rural parts of the United States? How do you think cultural or social attitudes might differ between people who live in such different places?

Organizing What You Know

5. Copy the chart below. Use it to describe some of the major political, economic, social, and cultural characteristics of the Northeast, Midwest, South, and West.

Northeast	South
Midwest	West

Section 3

Geographic Issues

READ TO DISCOVER

1. What are some important environmental issues in the United States?
2. What natural hazards affect the lives of Americans?
3. How are cities and population patterns in the United States changing?
4. How is the U.S. economy tied to other countries around the world?

WHY IT MATTERS

Environmental issues in the United States are often in the news and are the subject of much debate. Use CNNfyi.com or other **current events** sources to learn about environmental issues that are important today.

IDENTIFY

North American Free Trade Agreement (NAFTA)

DEFINE

gentrification
superpower
trade deficit

LOCATE

Gulf of Mexico
Columbia River
Colorado River

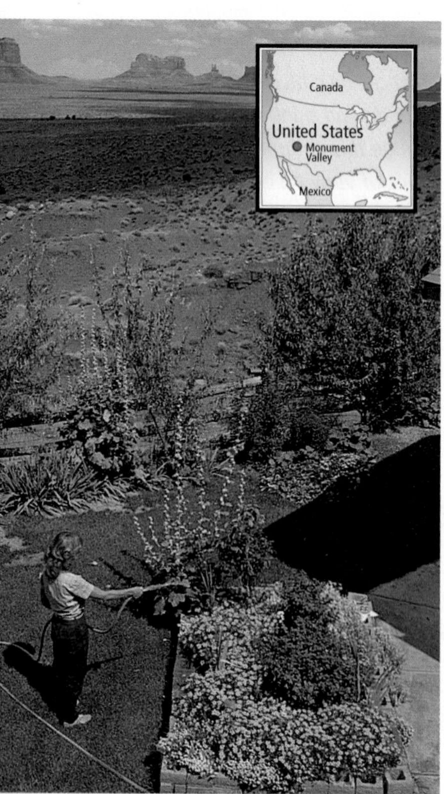

INTERPRETING THE VISUAL RECORD

Demand for water is high in parts of the western United States. However, state regulations often limit the amount of water available for personal use. ***How do you think regulations on water usage shape the geographic and economic character of western states?***  TEKS

Environmental Issues

The United States consumes more energy than any other country. As a result, the country produces huge amounts of waste, automobile exhaust, and other pollutants. A major challenge has been finding ways to reduce pollution and protect the environment.

Population growth and economic development have contributed to this challenge. For example, power plants and factories that burn coal and oil cause acid rain. This pollution kills trees and contaminates rivers and lakes. The problem has been particularly serious in parts of upstate New York and New England. Older factories in the Ohio River Valley produce pollution that reaches these areas. Laws restricting the emissions of pollutants reduced levels of acid rain in the 1990s. However, the problem has not been resolved.

The use of fertilizers has created problems in some places. For example, the Mississippi River carries fertilizers from farms to the Gulf of Mexico. These chemicals promote the growth of algae. Then bacteria that consume the algae use oxygen that marine life needs to live. This has created a "dead zone" in the waters off the Louisiana coast. Fertilizers used in Florida's sugarcane fields have also killed plants and wildlife in the Everglades.

In the West, dams on the Columbia and Snake Rivers produce hydroelectricity. However, they also block the path of migrating salmon. Dams are one of the main causes for a dramatic decline in the salmon population. A growing population and rapid economic development have also strained water resources in the West. Competition for the limited water resources of the Colorado River has been intense. Growing populations in both Arizona and California want the water. So much water is now drawn from the Colorado River that almost none reaches the Gulf of California.

TEKS **READING CHECK:** ***Environment and Society*** What are four examples of environmental challenges in the United States? What factors have contributed to these challenges?

Natural Hazards

Natural hazards present other challenges in regions around the country. For example, some large cities on the west coast are vulnerable to earthquakes. In fact, powerful earthquakes in recent years have caused serious damage in Los Angeles, San Francisco, and Seattle. Scientists predict that more large quakes will strike in the future. In areas where earthquakes are common, developers must follow strict building codes designed to limit earthquake damage. Many use building techniques and materials that can limit damage even to large structures.

Flooding threatens many parts of the United States. In some places, dams and levees have helped control the flow of rivers and have reduced flooding. However, heavy rains and snowmelt still cause major floods. Some engineers think that by controlling rivers too much, the potential for dangerous floods actually increases.

Tornadoes are threats in the Midwest and South, particularly in spring and summer. Hurricanes threaten areas along the east coast and Gulf of Mexico. Strong hurricane winds damage buildings and vegetation and reshape barrier islands. However, the greatest danger posed by hurricanes is flooding. High seas whipped up by a hurricane's winds can flood lowlying coastal areas and cause heavy damage. The barrier islands of Texas, Florida, and the Carolinas are particularly at risk. Governments enforce strict building codes in many areas. Seawalls and other devices also help limit damage. However, coastal buildings are still damaged and destroyed.

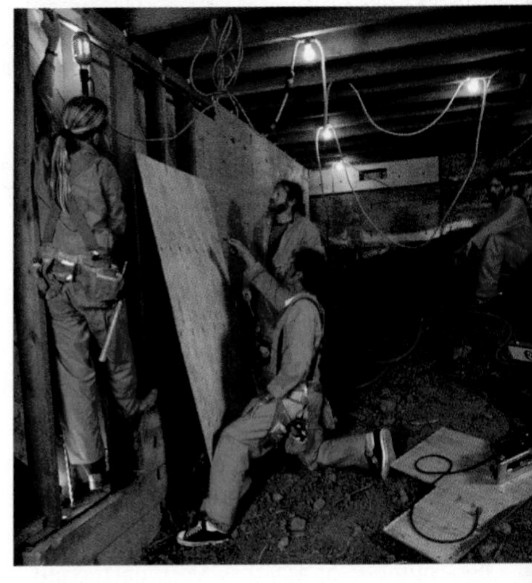

INTERPRETING THE VISUAL RECORD

People in Oakland, California, work to make a house safer in the event of an earthquake. Some engineering practices designed to make buildings more resistant to earthquakes include bolting homes to their foundations, strengthening supporting walls, and installing gas shut-off valves. ***How might these measures help houses resist earthquake damage?***

TEKS

✓ **READING CHECK:** ***Environment and Society*** What are some natural hazards that threaten parts of the United States?

Cities and Population

The residents of U.S. cities have seen major changes over the last 50 years. Many people have moved from inner cities to the suburbs. As a result, the populations of some large cities have dropped significantly, particularly in the Northeast. Businesses also began moving to suburban shopping malls and edge cities. Because of these shifts in population, many cities are unable to collect as much money from taxes. In fact, some cities find it difficult to provide basic services to poor and minority residents who remain.

In some urban areas wealthier people have been moving back to inner cities. They are often young professionals who buy run-down houses and restore them, a process known as **gentrification**. However, this process has its opponents. These opponents argue that gentrification increases property values and property taxes. These factors in turn push out low-income residents.

Immigration and changing population patterns have also greatly affected the United States. Historically, most immigrants to the country came from Europe and settled in the Northeast. However, most immigrants today come from Asia and Latin America. These immigrants settle in many different places. As a result, the ethnic and cultural composition of the country has changed significantly. This is the case particularly in the West and in parts of the South.

San Francisco's Chinatown is the largest Chinese community outside of Asia. Most Chinese there have maintained their language and customs.

Connecting to

GOVERNMENT

Voting Patterns and the Distribution of Political Power

In the U.S. system of government, the states play an important role. This is particularly true in presidential elections. When a state's voters cast ballots for president in November, they are really choosing members of the electoral college, or electors. The number of electors from each state is equal to its members in the U.S. House of Representatives and the U.S. Senate. To win the election, a presidential candidate must win at least 270 of the possible 538 electoral votes.

The size of a state's population determines how many seats that state gets in the U.S. House. As a result, states with the largest populations have the most seats. It follows, then, that they also have the most electoral votes. Therefore, changes in the population of each state affect the distribution of political power in the country. In fact, political power in the United States continues to shift as the population of the country moves westward and southward.

Look at the maps from the 1920 and 2000 presidential elections. These maps show the number of electoral votes each state had in each election. In 1920 several Northeast and Midwest states had the most electoral votes. This was because they had the largest populations. However, by 2000 California, Texas, and Florida each had more electoral votes than any other state but New York.

Analyzing What role do you think immigration might play in the distribution of political power in this country? How might these maps support your answer? TEKS

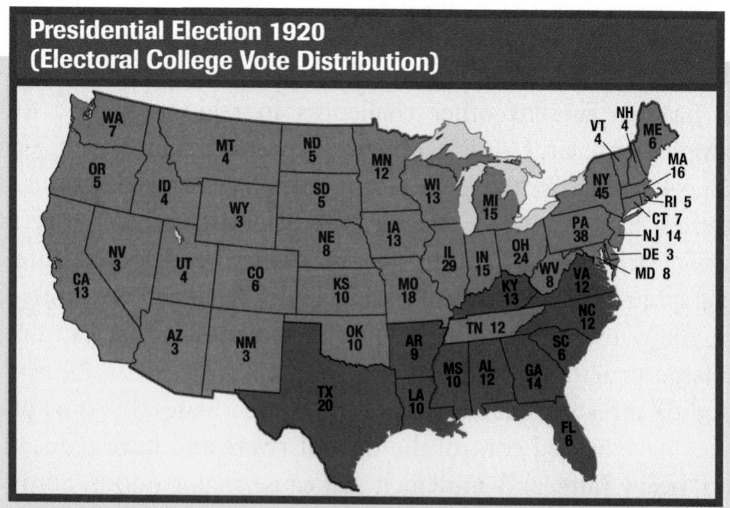

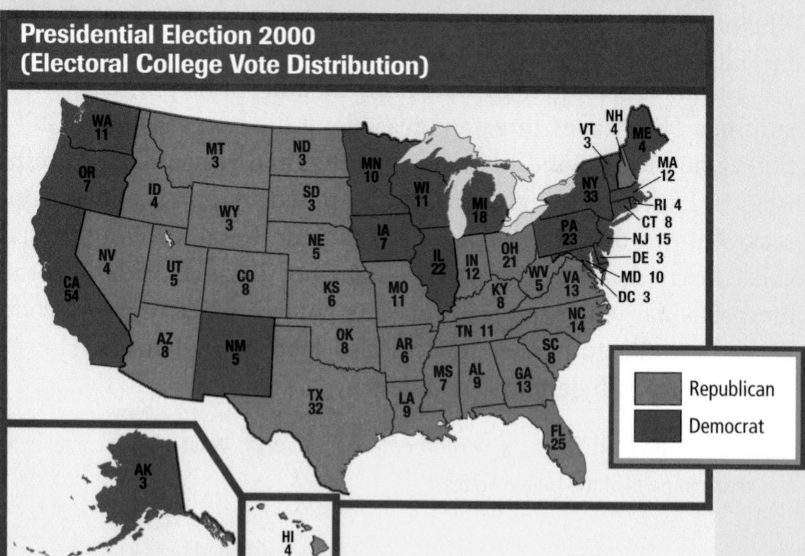

INTERPRETING THE MAPS *These maps show how voting patterns in the United States have changed over time. Democrats once dominated elections in the South. Republicans did better in the Northeast and West. Those voting patterns have changed somewhat. Shifts in migration patterns, political attitudes, and regional economics have all played roles in this change.* ***What does the 2000 map indicate about the relative strengths of the two major U.S. political parties today?*** TEKS

Changing population patterns have been a source of conflict in some places. Some people argue that the United States has historically been an English-speaking country, and they want it to remain so. They oppose bilingual education and think that new immigrants should be required to learn English. Also, some people think high levels of immigration keep wages down and make it harder for people to find jobs. Other people point to the long history of immigration to the United States. They say that immigration has strengthened the country and is needed to fill jobs.

READING CHECK: *Human Systems* How has the character of major U.S. cities been changing over the last 50 years?

The Port of Houston is one of the busiest in the United States. Billions of dollars worth of goods pass through it each year. These goods include automobiles, chemicals, grain, machinery, and oil.

Economy and Trade

When the Soviet Union collapsed in 1991 the United States became the world's only **superpower.** A superpower is a huge powerful country. The United States has the world's largest economy and is also the most powerful country politically and militarily.

However, the U.S. economy also relies on global trade. For example, the United States imports many raw materials and manufactured goods. In most years, the country has a **trade deficit**. A country has a trade deficit when the value of its exports is less than the value of its imports. In the past the government supported American industries by imposing tariffs on many imported goods. In recent years, however, the U.S. government has moved to lower barriers to free trade. For example, in 1992 the United States, Canada, and Mexico signed the **North American Free Trade Agreement (NAFTA)**. NAFTA eliminates many tariffs on products flowing between these countries. Since NAFTA took effect in 1994, trade between the three countries has boomed. Supporters argue that free trade will help American companies sell more products and create more jobs. Opponents say free trade allows American companies to move factories to countries with lower wages and business costs, causing unemployment in the United States.

The United States contains roughly 5 percent of the world's population but consumes 27 percent of the world's energy production.

READING CHECK: *Human Systems* How have shifts in U.S. government policies affected trade practices?

TEKS Working with Sketch Maps, Questions 1, 2, 3, 4, 5

Section 3 Review

Homework Practice Online
Keyword: SW3 HP8

Identify North American Free Trade Agreement (NAFTA)

Define gentrification, superpower, trade deficit

Working with Sketch Maps On the map you created in Section 2, label the Gulf of Mexico, Columbia River, and Colorado River. Which river is a major water source for California and Arizona?

Reading for the Main Idea

1. *Environment and Society* How has the use of fertilizers affected the environment along the Louisiana coast?
2. *Physical Systems* How do natural hazards such as hurricanes affect the environment?

Critical Thinking

3. **Drawing Inferences and Conclusions** How might this country's history of immigration influence immigration policies today?
4. **Identifying Points of View** How might people worried by the negative effects of gentrification in inner cities influence local public policies? In what ways might gentrification help people in inner-city neighborhoods?

Organizing What You Know

5. Copy the chart below. Use it to explain how technological innovations have helped Americans adapt to natural hazards.

Natural hazards	Innovations
Earthquakes	
Floods	
Hurricanes	

CASE STUDY

Creating Congressional Voting Districts

Human Systems Every 10 years the U.S. Census Bureau counts the population of the United States. The government then uses this data to create new election boundaries for the 435 U.S. congressional districts and state legislative districts. This process is called redistricting. Guiding the redistricting process is the "one person, one vote" principle. This principle requires that each congressmember or state legislator represent roughly the same number of people. As a result, states with large populations receive more members in the U.S. House of Representatives than states with fewer people. Also, more densely populated regions of a state get more representatives.

Gerrymandering

Even guided by the "one person, one vote" principle, redistricting officials still have many decisions to make. In fact, the redistricting process is a lesson in how the forces of conflict and cooperation play out in the U.S. political system. Members of rival political parties often work to draw district boundaries that are most favorable to their own party. Even moving district boundaries slightly can shift the political balance of power within the affected districts. Sometimes the process of redistricting creates some messy district maps. Boundaries might jut out here or there, creating districts with odd shapes. These odd shapes may be the result of a redistricting technique called gerrymandering.

Gerrymandering gets its name partly from that of Elbridge Gerry, a governor of Massachusetts in the early 1800s. In 1812 Gerry's administration enacted a law that divided his state into new senatorial districts. The new boundaries placed most of the supporters of Gerry's opponents into a few districts. This process helped ensure victory for members of Gerry's party in the remaining districts. It also created some rather oddly shaped districts. One district looked something like a salamander—or, as the governor's critics wryly noted, a "gerrymander." People now use the term *gerrymander* to describe the process of drawing district boundaries that unfairly favor a political party or group.

The term gerrymander *originated from this political cartoon. The cartoonist drew salamander features on a map of Boston-area voting districts to highlight their odd shapes. Since the districts' boundaries had been drawn by Governor Elbridge Gerry, the cartoon became known as the Gerry-mander.*

One Person, One Vote

Gerrymandering has not been the only way to unfairly favor a political party or group. During the first half of the 1900s, many states did not redraw electoral boundaries after each census. The boundaries did not move, but the people within them did. For example, many rural Americans migrated to cities, but the number of representatives from rural areas stayed the same. As a result, rural areas had relatively more political power than urban areas.

Some of the most extreme examples of political inequality between rural and urban areas occurred in the South. In 1960 just one quarter of Alabama residents elected a majority of the state's legislators. As a result, rural districts had more power than urban districts. For example, lawmakers in some urban districts of Alabama represented about 100,000 people each. In contrast, some rural legislators represented as few as 7,000 people. Tennessee's voting districts were also unfairly drawn. Populations in state senatorial districts ranged from 131,971 to 25,190. The most crowded representative district had 42,298 people. The smallest had only 2,340.

Over time, people in urban areas challenged the way these districts were created. They argued that the U.S. Constitution guaranteed equal representation in government—one person, one vote. Eventually, a series of U.S. Supreme Court cases forced states to redraw their electoral maps. Now all districts must contain about the same number of people.

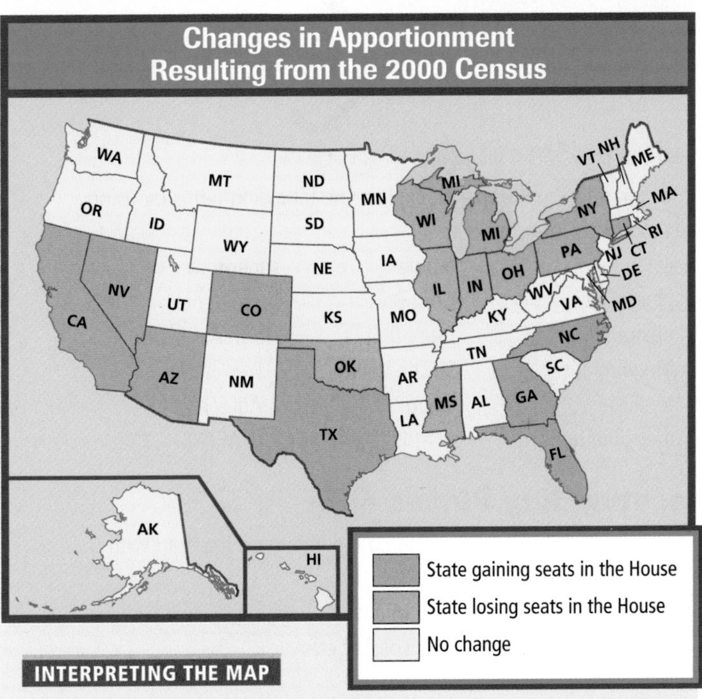

INTERPRETING THE MAP

Congressional districts are redrawn after each census to reflect changes in the distribution of the U.S. population. ***Based on the map, which regions of the United States do you think are gaining population? Which are losing population?*** TEKS

Redistricting Today

Other laws and court cases in the last half of the 1900s further shaped redistricting processes. Officials must now take care not to create districts that dilute the voting strength of racial and ethnic minority groups. These officials use census data and sophisticated software and mapping programs, such as a geographic information system (GIS). Some states created districts to ensure the election of more minority candidates. For example, after the 1990 census, Georgia officials stretched the 11th Congressional District along a narrow strip. Snaking across 260 miles (418 km) of the state, the new district linked African American communities in Atlanta and Savannah. As the planners had intended, voters in the new district elected an African American congressmember in 1994.

Critics call such redistricting practices gerrymandering. Some also argue that grouping many minority voters into a few districts dilutes their strength in other districts. In addition, the U.S. Supreme Court later ruled that districts cannot be based mainly on racial and ethnic characteristics. Courts still sometimes step in to resolve disputes because electoral maps can be very controversial.

As you can see, the redistricting process has become quite complicated. Still, despite the challenges, the new maps must be drawn. With Census Bureau data in hand, geographers and demographers play a role in congressional redistricting. By working with government officials and citizen groups, they help redraw our country's political geography every 10 years.

Georgia's 11th District, 1992

Athens, Atlanta, Augusta, Macon, Waynesboro, Columbus, Savannah, Albany, Brunswick

INTERPRETING THE MAP

In 1995 the U.S. Supreme Court declared Georgia's 11th District unconstitutional because they concluded that race was the driving force behind its boundaries. ***How do you think this conflict reshaped Georgia's political geography?*** TEKS

Applying What You Know

TEKS

1. **Summarizing** How do officials create congressional voting districts?
2. **Analyzing** How do you think the forces of conflict and cooperation influenced the creation of Georgia's 11th Congressional District?

Building Vocabulary

On a separate sheet of paper, explain the following terms by using them correctly in sentences.

colonies	arable	superpower
plantations	Corn Belt	trade deficit
bilingual	Dairy Belt	North American
Megalopolis	Silicon Valley	Free Trade
textiles	smog	Agreement
metropolitan area	gentrification	(NAFTA)

Locating Key Places

On a separate sheet of paper, match the letters on the map with their correct labels.

Alaska	Miami
Washington, D.C.	Los Angeles
New York	Seattle
Chicago	

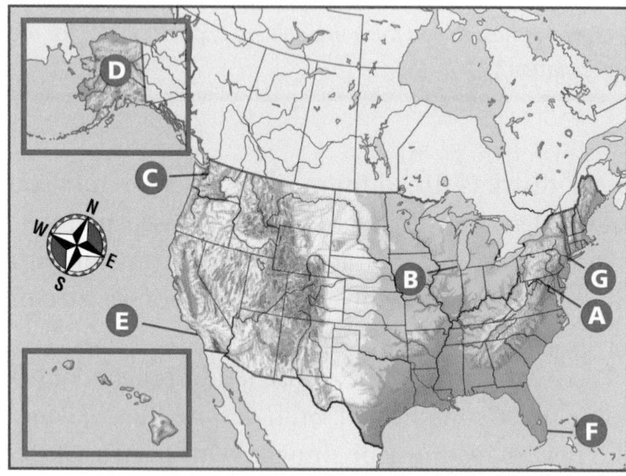

Understanding the Main Ideas

Section 1

1. *Environment and Society* How did changing perceptions of the Great American Desert lead to changes in society?
2. *Human Systems* How has immigration to the United States influenced the diffusion of religions and cultural change?

Section 2

3. *Environment and Society* How has competition for water resources from the Colorado River affected California and Arizona? How has the river itself been affected?
4. *Places and Regions* Which states are part of the Corn Belt, the Dairy Belt, and the Wheat Belt?

Section 3

5. *Places and Regions* When did the United States become the world's only superpower? Why?

Thinking Critically for TAKS

1. **Making Generalizations** Do you think a megalopolis is a formal, functional, or perceptual region? Why?
2. **Analyzing** How could the Northeast be described as an important economic, political, and cultural center?
3. **Identifying Points of View** What restrictions, if any, do you think should be placed on immigration to the United States? Why? Why might immigration be such a controversial issue?

Using the Geographer's Tools

1. **Analyzing Maps** Look at the Section 1 map showing place-names in the United States. How does the map reflect French, Spanish, and English settlement patterns?
2. **Analyzing Maps** Look at the Section 1 map showing U.S. expansion. During which century did the United States acquire most of its territory?
3. **Preparing Maps** Map the locations of different types of economic activities in your state. You can use the unit land use and resources map, the chapter, and other resources.

Writing for TAKS

Imagine that you are writing a script for a documentary about how life in Los Angeles has changed over time. Use the information in this chapter and in other sources to write a short introduction to your script. Be sure to answer the following questions: When was Los Angeles founded, and by whom? What resources have been important in the city's history and economic development? How big is the Los Angeles metropolitan area today? What challenges does the city face?

SKILL BUILDING

Geography for Life

Analyzing Primary Sources

Human Systems Use your local library to find firsthand accounts about the exploration, settlement, and westward expansion of the United States. What can these accounts teach us about American history and geography? What types of information do they contain? What types of information are they lacking? When might researchers want to study firsthand accounts?

Building Social Studies Skills

go.hrw.com TAKS PREP ONLINE
keyword: SW3 T8

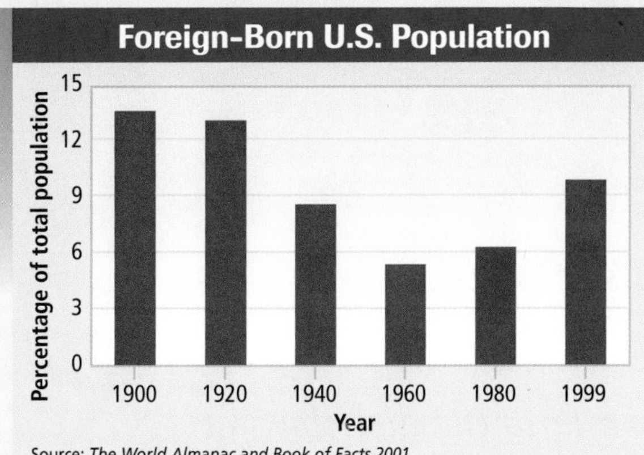

Source: *The World Almanac and Book of Facts 2001*

Interpreting Graphs

Study the bar graph above. Then use the information from the graph to help you answer the questions that follow.

1. In which year was the percentage of the U.S. population that was foreign-born highest?
   a. 1900
   b. 1920
   c. 1940
   d. 1999
2. Did immigration appear to increase or decrease in the middle part of the 1900s? How can you tell?

Analyzing Primary Sources

Read what historian William Cronon has to say about the Midwest. Then answer the questions that follow. Mark your answers on a separate sheet of paper.

> **"I came to know and care for a landscape that few who are not midwesterners ever call beautiful. Travelers, whether in the air or on the ground, usually see the Middle West less as a destination than as a place to pass through. Only after a long while does one appreciate that the very plainness of the countryside is its beauty. . . . When people speak . . . of the American heartland, this is one of the places they mean."**

3. According to Cronon, what is the source of the Midwest's beauty?
   a. its airports
   b. its travelers
   c. its plainness
   d. its cities
4. Why do you suppose Cronon thinks that travelers see the Midwest as simply a place to pass through?

Alternative Assessment

PORTFOLIO ACTIVITY

Learning about Your Local Geography TEKS

Individual Project: Research

People, places, and environments are connected and interdependent. For example, resources may attract people to a new place, and their activities may change it. Study your local town or community to learn about the relationship between its people and environment. What was the area like before it was settled? What is it like now? How have settlement and population growth affected the environment? How has economic development affected it?

internet connect

Internet Activity: go.hrw.com
KEYWORD: SW3 GT8 TEKS

Choose a topic on the United States to:

- take the GeoMap challenge to test your knowledge of U.S. geography.
- learn how people from different cultures celebrate holidays in the United States.
- create a brochure on national parks of the United States.

CHAPTER 9

Canada

Canada is the world's second-largest country. Only Russia is larger. In this chapter you will learn how the distribution of Canada's population is related to the country's cultural, economic, and political development.

Kwagiuth totem pole

Arctic wolf in northern Canada

Hello. My name is Bella Morris. I am Inuit, born and raised in the Canadian arctic. I grew up in a large extended family.

I spent my early childhood years traveling, fishing, and hunting on the Arctic tundra. In those early years I traveled by dog team, lived in an igloo, and watched my father and uncles hunt whales. I spent winter months listening to the legends and hunting stories of my people while learning to sew the warm fur clothing needed to survive in the Arctic.

When I was seven, I began my formal education at the residential school in Aklavik in the Mackenzie Delta, where I would live separated from my family for long months. Learning English was difficult, but when I finally mastered the language, school became much easier for me.

When I was 13, the other Inuit children and I were integrated into the public school system, where I met European (white) children for the first time. I did well in school, but I only went as far as ninth grade. After that, I either had to go home or train as a practical nurse or secretary, so I became a practical nurse. Later I went back to school, graduating with a bachelor's degree in education. Today I work to help bridge the cultural differences between the Inuit and Canadian Indian children and children of European and Asian descent.

History and Culture

READ TO DISCOVER

1. Which European countries played a role in Canada's early history?
2. What are some important features of Canadian culture?

DEFINE

provinces
hinterland

WHY IT MATTERS

Canada and the United States share the longest unguarded border in the world. Use CNNfyi.com or other **current events** sources to learn about the relationship between the two countries.

LOCATE

St. Lawrence River
Montreal
Quebec City
Toronto
Ottawa
Ottawa River
Whitehorse
Yellowknife
Windsor
Saguenay River
Vancouver

Canada: Physical-Political

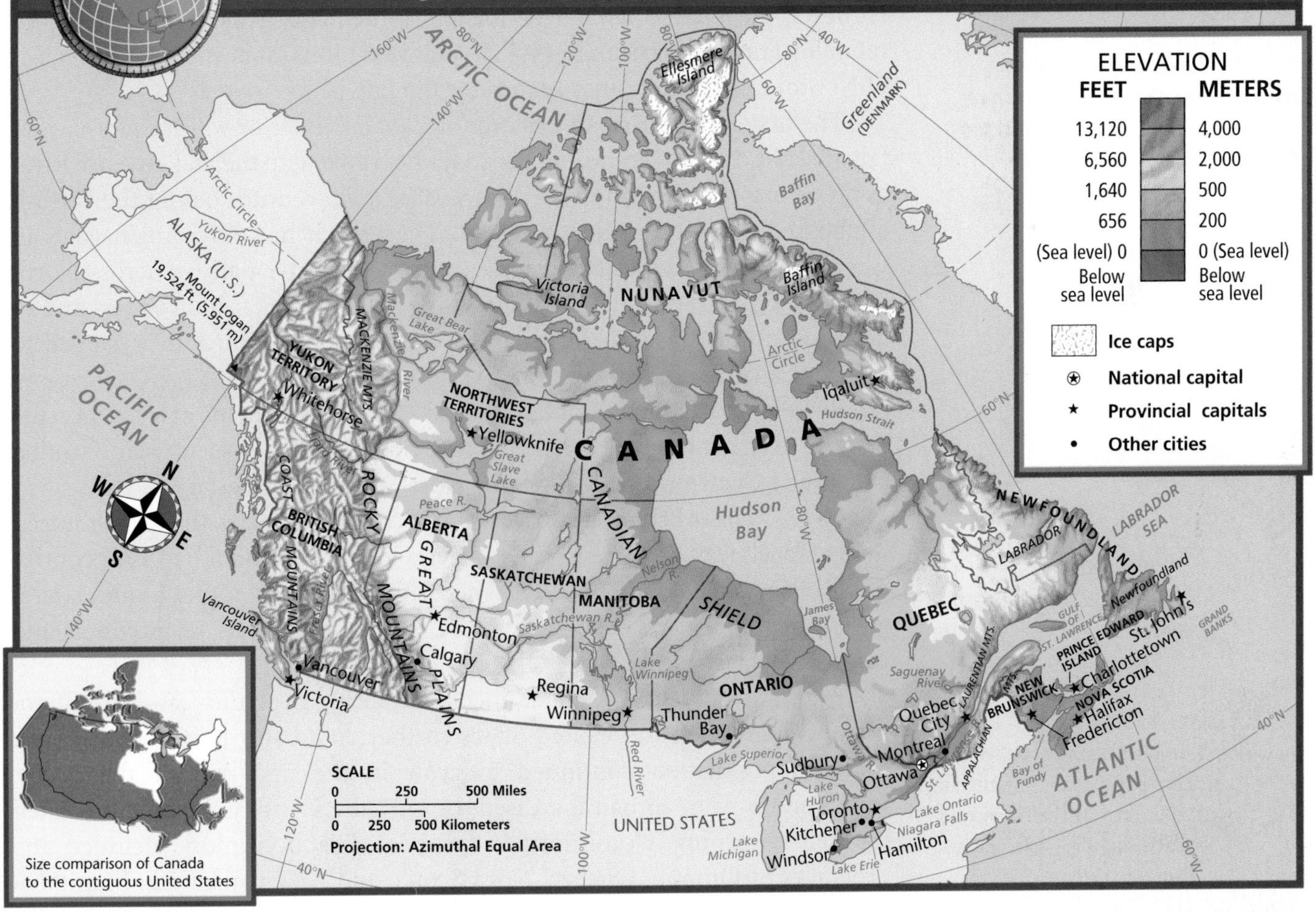

ELEVATION FEET	METERS
13,120	4,000
6,560	2,000
1,640	500
656	200
(Sea level) 0	0 (Sea level)
Below sea level	Below sea level

Ice caps
National capital
Provincial capitals
Other cities

SCALE: 0 250 500 Miles; 0 250 500 Kilometers
Projection: Azimuthal Equal Area

Size comparison of Canada to the contiguous United States

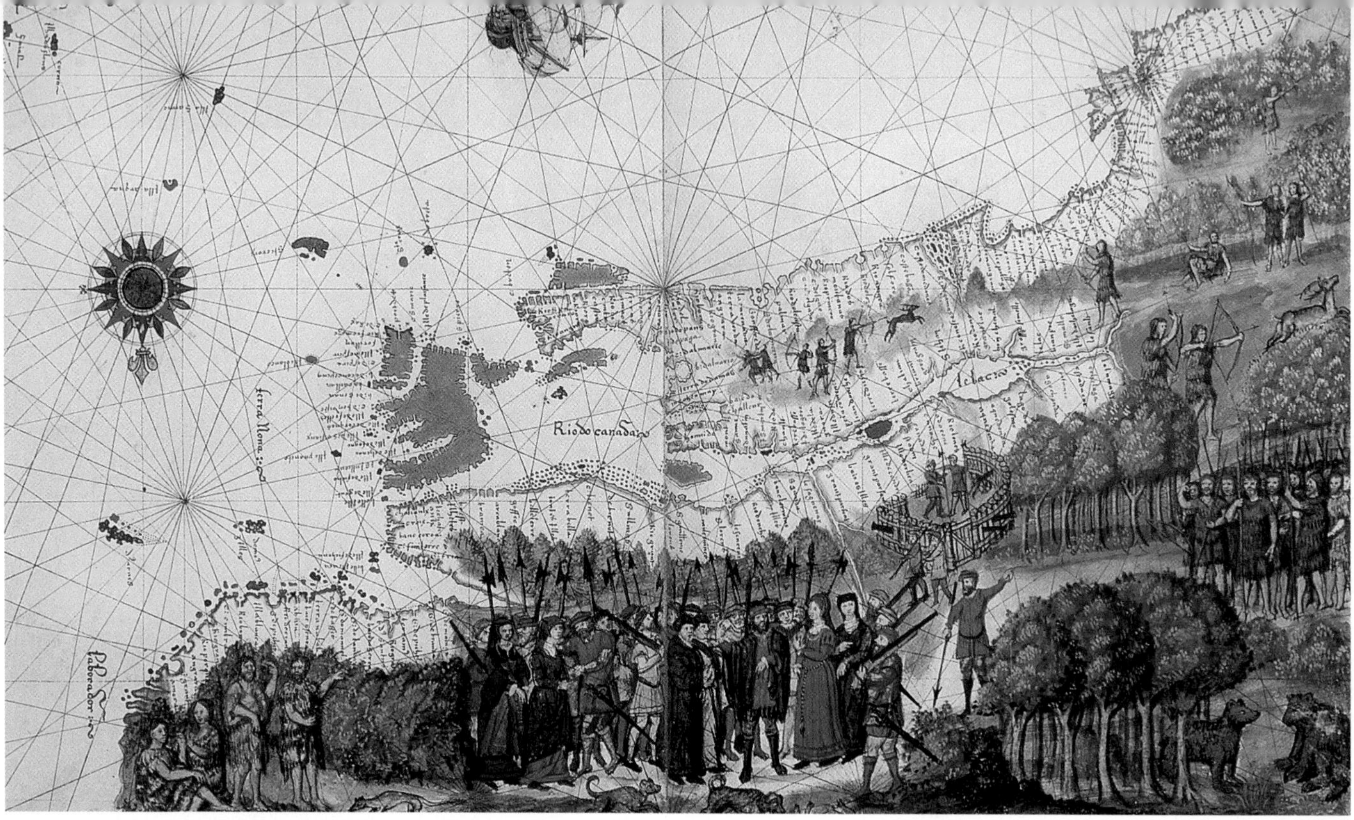

INTERPRETING THE VISUAL RECORD

This historical map shows the French explorer Jacques Cartier arriving in Canada in the early 1540s with a group of French colonists. Cartier's attempts to establish a permanent colony near what is now Quebec City failed, and he was forced to return to France. ***What European and Canadian Indian cultural features can you see in this illustration? What features of the region's natural environment are visible?***

In the 1960s archaeologists discovered the first clear evidence of Viking settlement in North America. At L'Anse aux Meadows on the northern tip of the island of Newfoundland, they found the remains of eight buildings used as Viking workshops and numerous artifacts, including nails, a brass ring, and a bronze cloak pin.

History

As in the United States, Native American societies were once found across Canada. The first Europeans to sail to Canada's eastern shores were Viking adventurers. They visited between A.D. 1000 and as late as the mid-1300s. However, the Vikings left no permanent settlements. More extensive exploration by Europeans began in 1497. In that year John Cabot explored the coasts of Newfoundland and other islands for the English.

The first great European explorer of Canada's interior was Jacques Cartier (zhahk kahr-TYAY) of France. In the 1530s he traveled up the St. Lawrence River as far as present-day Montreal. This was nearly a century before the English established colonies in New England. The French had three main goals in Canada. First, they wanted to find a northwest water passage across North America to Asia. Second, they wanted to exploit nearby fishing waters and to develop a trade for animal furs from North America. Third, they wanted to convert Canadian Indians to Roman Catholicism.

By 1608 the French established a permanent settlement at what became Quebec City on the St. Lawrence River. Soon, French settlers were farming along the St. Lawrence and in nearby Nova Scotia to the east.

In 1713 Great Britain took over Nova Scotia. Eventually, the British forced many French settlers there to leave. After a long war, Britain had won control of all of French Canada by 1763. The British organized Canada into several governmental districts called **provinces**. Today Canada has 10 provinces and three special territories.

British settlement in Canada increased during the American Revolution. Many colonists left the United States so they could stay under British rule. Canada's population continued to grow in the first half of the 1800s. Immigration from abroad increased. In 1867 the British government created the self-governing Dominion of Canada. The dominion included the provinces of Ontario, Quebec, Nova Scotia, and New Brunswick. Manitoba,

British Columbia, and Prince Edward Island joined them in the 1870s. Alberta and Saskatchewan did not become provinces until 1905. Newfoundland became part of Canada in 1949.

READING CHECK: *Human Systems* What European countries most influenced Canada's development?

internet connect

GO TO: go.hrw.com
KEYWORD: SW3 CH9
FOR: Web sites about Canada

Culture

More than 31 million people live in Canada. French and British culture have remained strong there, along with many influences from the United States. In addition, immigration has brought other Europeans and people from the Caribbean, Asia, and Africa. Canada's government encourages each group to hold on to its culture. As a result, Canada is a multicultural country.

People, Languages, and Religion About one fourth of all Canadians live in the province of Quebec. Quebec City is the provincial capital, and Montreal is Quebec's largest city. The province is the center of French-Canadian culture. In fact, more than 90 percent of Canadians who speak French as their first language live there. Most people in Quebec are Roman Catholic, which is the largest religion in Canada. (See the graph of religions in Canada.)

French Canadians in Quebec call themselves Quebecois (kay-buh-KWAH). They have worked to maintain cultural independence from the rest of Canada. This effort has influenced public policies and laws there. For example, official papers of the provincial government are written only in French. Also, signs on businesses and along Quebec's roads are in French as well. However, a minority in Quebec do not share the province's dominant French culture.

To the west, Ontario reflects British heritage much like Quebec symbolizes the French. British customs are still widespread in Toronto, Ontario's capital. French is seldom heard or seen on signs in the city. On the other hand, Ottawa (AH-tuh-wuh), which is in Ontario and is Canada's capital, is bilingual. English and French are commonly spoken in the city, which lies just across the Ottawa River from Quebec.

Many immigrants from the British Isles and southern and Eastern Europe settled provinces in the east and west. The residents of northern Canada include people who have left Canada's cities in the south. Many Inuit (once called Eskimos) and Canadian Indians also live there. Settlements such as Whitehorse and Yellowknife are classic frontier towns.

READING CHECK: *Places and Regions* In what way is Quebec culturally different from the rest of Canada?

Settlement and Land Use The St. Lawrence lowlands of southern Quebec and Ontario make up the most densely settled part of Canada. They are also Canada's most economically developed areas. There you will find a chain of cities that extends from Quebec City to Windsor, Ontario. This chain includes the cities of Montreal, Toronto, and Ottawa. Together all of these cities lead the country in wealth, industry, commerce, politics, and influence. This area forms the heartland of Canada. For this reason, we call Ontario and Quebec the Heartland Provinces.

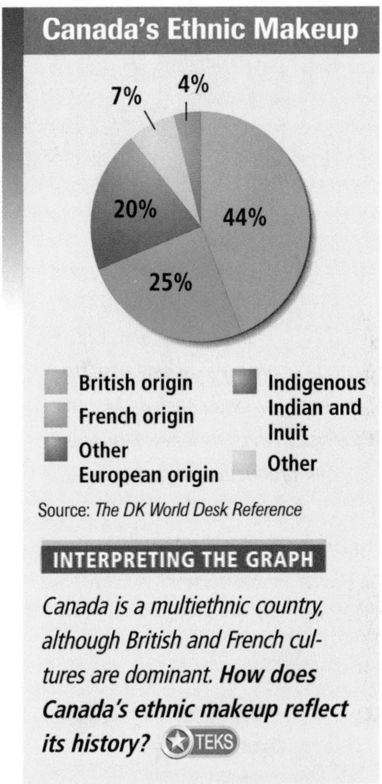

Source: *The DK World Desk Reference*

INTERPRETING THE GRAPH

Canada is a multiethnic country, although British and French cultures are dominant. ***How does Canada's ethnic makeup reflect its history?*** TEKS

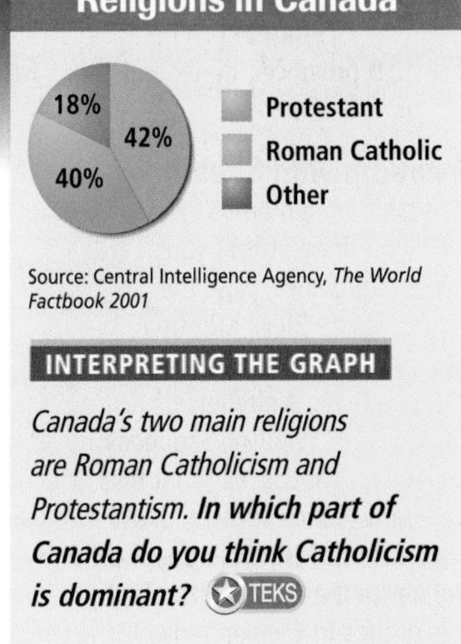

Source: Central Intelligence Agency, *The World Factbook 2001*

INTERPRETING THE GRAPH

Canada's two main religions are Roman Catholicism and Protestantism. ***In which part of Canada do you think Catholicism is dominant?*** TEKS

Canada's core area includes its largest cities in the St. Lawrence lowlands, such as Montreal (top). The Prairie Provinces, which include Manitoba (bottom), are known for their productive market-oriented agriculture.

Most people in the Heartland Provinces live in the south and east. In Quebec, for example, the only densely settled areas are in the St. Lawrence, Saguenay (sa-guh-NAY), and Ottawa River valleys. The forests and rocky uplands of the vast interior are nearly empty. Only isolated government centers, trading stations, and mining districts there have many people.

The Atlantic Provinces of the east are thinly populated. These coastal provinces are Newfoundland, New Brunswick, Nova Scotia, and Prince Edward Island. They form Canada's eastern **hinterland.** A hinterland is a region that lies far away from major population centers. Less than one twelfth of all Canadians live in the Atlantic Provinces.

In the west, settlement of most of Manitoba, Saskatchewan, and Alberta followed completion of the country's transcontinental railroad in 1885. Thousands of immigrant farmers rode the rails to their new homes in what are called the Prairie Provinces. The southern parts of these provinces were covered with prairie grasslands when European settlers arrived.

Today large farms stretch across the Prairie Provinces. However, few people are needed to work them because the level land encourages the large-scale use of farm machinery. As a result, Canada's fertile prairies remain only thinly settled. Still fewer people live farther north. The grasslands there give way to forests. Farther west is the Pacific Coast province of British Columbia. Vancouver, a major seaport, is the largest city there and the third-largest in Canada.

Forests, tundra, and rocky plains cover Canada's vast frigid north. Most of the land is underlaid by permafrost. Isolated towns and villages are scattered throughout this huge region of wilderness. It has only a few, usually gravel, highways.

READING CHECK: *Human Systems* What area of Canada is most densely settled?

TEKS Working with Sketch Maps, Questions 1, 2, 3, 4, 5

Section 1 Review

Define provinces, hinterland

Working with Sketch Maps On a map of Canada that you draw or that your teacher provides, label the St. Lawrence River, Montreal, Quebec City, Toronto, Ottawa, Ottawa River, Whitehorse, Yellowknife, Windsor, Saguenay River, and Vancouver. What river did early French settlers follow into Canada? How did this route influence the distribution of ethnic groups in Canada today?

Reading for the Main Idea

1. *Human Systems* How is Quebec's French culture reflected in the province's public policies? Give two examples.
2. *Human Systems* How is the fertile land of the Prairie Provinces used? How have technology and physical geography combined to affect the population there?

Critical Thinking

3. **Comparing** What characteristics do the provinces of Quebec and Ontario have in common? In what ways are they different?
4. **Making Generalizations and Predictions** Compare the geographies of northern Canada and southern Canada. In which region do you think people would generally need to be more self-sufficient? Explain your answer.

Organizing What You Know

5. Copy the chart below. Use it to list the provinces in each of Canada's regions. You can use the chart map to help you complete the chart. Do you think these are perceptual, formal, or functional regions? Why?

Atlantic Provinces		
Heartland Provinces		
Prairie Provinces		
Pacific Coast		
Canadian North		

Section 2

Canada Today

READ TO DISCOVER

1. What resources and activities drive Canada's economy?
2. What factors and processes have influenced the growth of Canada's cities?
3. How is Canada organized and governed?

DEFINE

parliament
consensus

LOCATE

Laurentian Mountains
Sudbury
Thunder Bay
Vancouver Island
Calgary
Edmonton
Winnipeg
Iqaluit
Baffin Island

WHY IT MATTERS

The United States imports more goods from Canada than from any other country. Use CNNfyi.com or other current events sources to learn more about U.S.-Canadian trade.

Economic Development

Canada today is a developed country with a market economy and high standard of living. Its most important trade partner is the United States. Both countries have good transportation systems and similar business practices. As a result, Canadian and American firms can easily do business together.

Over the last century Canada has shifted away from an agricultural economy. Today its economy is based mainly on manufacturing and service industries. Mining has also long been a major activity. In fact, no other country exports more minerals and metals than Canada. Agriculture remains important. Canada is a major exporter of farm goods, producing more than its small population needs. Where do you think those farm goods are produced? Where would you expect to find large manufacturing centers? Next we will look at how the economic geography varies across the country's regions.

INTERPRETING THE VISUAL RECORD

While Canada has a strong, modern economy based on manufacturing and services, economic development is still a challenge in some areas. For example, many small towns in eastern Canada, such as Prospect, Nova Scotia, have historically depended on primary activities such as fishing. These areas lag behind Canada's most economically developed areas. ***How would you describe the physical geography of this region?***

The Atlantic Provinces Life in this part of Canada is challenging. It has been the country's poorest region, with the lowest wages and the highest unemployment rate. In addition, long cold winters and thin rocky soils make farming difficult. Small farms that produce a variety of crops are found there. Crops do a little better in the milder climate of Nova Scotia.

Other economic activities have long depended on the resources of the sea and forests. However, the easy-to-reach old-growth forests are mostly gone. In addition, the Grand Banks area off Newfoundland—once one of the world's great commercial fishing grounds—has been overfished. As catches declined, unemployment increased. Today the government limits the number of fishing boats in the area. However, fishing is still important to the region's economy. With careful management, fish stocks may increase in the future.

The other natural resources of the Atlantic Provinces could help economic development. The mainland part of Newfoundland has important mineral deposits. Also, oil has been found offshore. Yet mining these resources provides few jobs, and other industries have been hard to develop. Why? The region's population is too small to provide a good home market. The major population and market centers in Quebec and Ontario are far away. With high unemployment, migration to the wealthier cities of Ontario and western Canada has been common. However, the region has finally made important progress in attracting new businesses in recent years.

Our Amazing Planet

Canada's Bay of Fundy has the highest tides in the world. They can be as high as 70 feet (21 m). These tides bring water from the North Atlantic Ocean into the narrow bay. The bore, or leading wave of the incoming water, can roar like a big truck as the tides rush in.

READING CHECK: ***Human Systems*** Why has industry been hard to develop in the Atlantic Provinces?

Quebec and Ontario As you read in Section 1, Quebec and Ontario make up Canada's heartland. Montreal is the industrial and financial center of Quebec. It also serves as a major port on the St. Lawrence Seaway. This is true even though much of the river is frozen for four months each year. Quebec City, located where the shores of the St. Lawrence River pinch together, is also a major port. Service industries are more important than manufacturing there. To the north, ski resorts dot the Laurentian (law-REN-chuhnz) Mountains.

Farming takes place in both southern Ontario and Quebec. However, manufacturing is the most important economic activity in Ontario. Southeastern Ontario is the chief manufacturing district of Canada. Toronto, Hamilton, Kitchener, Windsor, and other, smaller cities are located in this region. Factories there supply many of the needs of a modern industrial society and growing urban population.

Outside this part of Ontario, most of the province's cities are isolated. Some are service centers for remote mining districts. Sudbury, north of Lake Huron, developed around one of the world's largest deposits of nickel. Other

INTERPRETING THE VISUAL RECORD

Completed in 1959, the St. Lawrence Seaway connects the Great Lakes to the Atlantic Ocean through a combination of artificial and natural waterways. It allows ocean-going ships to reach ports as far as Lake Superior and is important economically for both Canada and the United States. ***How do you think the St. Lawrence Seaway affects the locations and patterns of economic activities in Canada?*** TEKS

cities have grown up at transportation junctions. The city of Thunder Bay on northwestern Lake Superior serves as the major port for wheat from the Prairie Provinces.

The Prairie Provinces Wheat is a major crop in the Prairie Provinces, and farmers there export most of it. Changes in the global wheat market and uncertain weather conditions can cause problems for these farmers. Once, as in the United States, the government guaranteed prices. This meant that farmers could count on making a profit from the sale of their wheat. However, the government is reducing aid, and the risks for individual farmers have increased. The results are ever-larger farms that use more modern technology and machinery. Those farms can grow more food with fewer workers.

Saskatchewan's economy is mostly agricultural. However, the province has other industries as well. For example, the province has the world's largest deposits of potash. Potash mining provides an important raw material that is used in fertilizers. Alberta's income is based mostly on fossil fuels, particularly oil. Rich oil fields are found there and in western Saskatchewan. The Rocky Mountains of southwestern Alberta are also a valuable natural resource. Their spectacular scenery attracts tourists from around the world.

READING CHECK: ***Places and Regions*** How have technology and reduced government aid changed farming patterns in the Prairie Provinces?

British Columbia The province of British Columbia, or "BC," stretches inland from the Pacific Coast. It is a land of mountains, plateaus, and fertile river valleys. British Columbia is rich in natural resources. Like the Pacific Northwest of the United States, much of the land is covered with forests of fir, spruce, and cedar trees. Income from forest products is substantial. In addition, salmon fishing and mining are important. Farmers use British Columbia's limited farmland mostly for growing fruits and vegetables as well as for dairying. Because of its location on Canada's west coast, BC trades with countries around the Pacific Rim. Japanese companies are important buyers of the province's forest products and minerals.

The Canadian North In the last 30 years, technology has helped make northern Canada less remote. Airplanes and satellite communications have tied the region more closely to the rest of the country. Now, in spite of its severe climate, the north has important promise for Canada's future. It is one of the modern world's great frontiers. Rich deposits of metals, diamonds, and fossil fuels have been discovered. In addition, supplies of freshwater there are huge. Years ago the indigenous peoples of this region lived by hunting and gathering. Some Inuit today still make a living this way. However, many now work for mining and construction companies, on military posts, or in the tourism industry.

READING CHECK: ***Physical Systems*** What effects have new technologies and discoveries of natural resources had on northern Canada?

Connecting to TECHNOLOGY

Inuit Igloos

An igloo is a traditional Inuit hunting shelter built from blocks of snow. The dome-shaped igloo remains comfortable in northern Canada's howling winter winds and subzero temperatures. How is that possible? The secret lies in turning the snow house into an ice house. The Inuit do this by heating the igloo's interior so that its inside walls begin to melt. The walls absorb the water until the snow blocks are soaked through. Then the heating is stopped, and cold outside air is allowed inside the igloo. The freezing air fuses the blocks and creates an airtight structure of solid ice.

Because ice insulates, the igloo traps warm air. Extreme cold is kept out. In fact, the temperature of the interior can be kept at nearly 55°F (about 13°C) without threatening the structure. A little water may run down the walls, but it freezes again.

Drawing Conclusions Building igloos from snow is one way humans have depended on and adapted to their environment. What traditional building materials might people in forested or warm treeless areas use? TEKS

Toronto's location on the northern shore of Lake Ontario provides access to Atlantic shipping and to major industrial centers in the United States.

Urban Environments

Canada's cities are generally well managed, clean, and safe. Toronto is Canada's largest city. It has a metropolitan population approaching 5 million. The city is also home to Canada's largest stock exchange, major banks, and insurance companies. Many other large Canadian companies are also located there. In addition, people there can visit great museums and other cultural institutions. Recent immigrants from Eastern Europe, the Caribbean, and China help make Toronto a multicultural city.

Founded on an island in the St. Lawrence River, Montreal has a population of about 3.5 million. It is Canada's second-largest city. Underground passageways and overhead glass tunnels connect many buildings in the city center. These structures protect people from the city's cold winter weather. Montreal's residents are proud of their subway, the Métro. It is patterned after the subway system in Paris.

Victoria, the capital of British Columbia, is located at the southeastern tip of Vancouver Island. The city is the home port for a large fishing fleet. Its old English charm also attracts many tourists every year. Nearby Vancouver, on the British Columbia mainland, is western Canada's most populous city. It has Canada's major ice-free harbor and is Canada's main Pacific port. A growing number of immigrants is adding to the metropolitan area's population of about 2 million. The city has also become a major center for movie and television productions.

Alberta has two rapidly growing cities, Calgary and Edmonton. Each is an important oil and agricultural center. Glass office towers stand out as striking structures on the Canadian prairie. However, Winnipeg, the capital of Manitoba, is the Prairie Provinces' chief city. All east-west rail traffic passes through Winnipeg. This makes the city an important collection and shipping point for the region's products.

READING CHECK: ***Places and Regions*** What factors have been important to the growth of major cities in the Prairie Provinces?

Government and Politics

Canada's ties to Great Britain have remained close. Britain's monarch is also Canada's monarch. Also like Britain, Canada is a democracy. It is governed by a prime minister and an elected **parliament**, or legislature. A minister, or premier, also heads each province's parliament. Provincial governments can levy taxes and set policies on issues such as education and civil rights. Canada has three northern territories spread across the Canadian Arctic and sub-Arctic: Yukon Territory, Northwest Territories, and Nunavut (NOO-na-voot). While they do not live in provinces, residents in the territories still have considerable control over local issues.

FOCUS ON GOVERNMENT

Nunavut Nunavut is Canada's newest territory. Canada's government created Nunavut out of the Northwest Territories in 1999. They created it to give the Inuit of the region a self-governing homeland. In fact, Nunavut's name is an Inuit word meaning "Our Land." This land covers about one fifth of Canada. However, it has less than 13 miles (21 km) of highways. Looked at another way, Nunavut is three times the size of Texas but has fewer than 30,000 people. Its people live in just 28 widely scattered communities.

To govern, Nunavut's leaders blend tradition with technology. Nunavut's 19-member elected assembly meets in Iqaluit (ee-KAH-loo-it), on southeastern Baffin Island. With some 4,200 people, it is by far the territory's largest town. Unlike most Canadian legislatures, where one political party is in control, Nunavut's assembly makes laws by **consensus**. Consensus means "general agreement." It is the traditional Inuit way of making decisions.

Nunavut's leaders also plan to use the Internet and e-mail to bring government to the widely scattered residents. These leaders plan to open community centers where people who do not have computers can access government offices. In addition, a number of government agencies are based in towns outside the capital. Scattering government agencies allows Nunavut's people to have better access to government jobs and services. Those jobs are considered important in Nunavut. This is because nearly as many people in Nunavut still hunt and fish for a living as work for wages.

READING CHECK: *Environment and Society*
In what ways do Nunavut's leaders blend tradition and technology to govern their territory?

INTERPRETING THE VISUAL RECORD

These children travel by boat to visit relatives in Nunavut, Canada's newest territory. The daily lives of many Inuit in northern Canada blend traditional activities with modern ones. ***How do you think diffusion of technology and changing trade patterns might cause cultural change among these Inuit children?*** TEKS

Section 2 Review

TEKS Working with Sketch Maps, Questions 2, 3, 4, 5

Homework Practice Online
Keyword: SW3 HP9

Define parliament, consensus

Working with Sketch Maps On the map that you created in Section 1, label Laurentian Mountains, Sudbury, Thunder Bay, Vancouver Island, Calgary, Edmonton, Winnipeg, Iqaluit, and Baffin Island. Where in Canada would you find ski and tourist resorts?

Reading for the Main Idea

1. *Physical Systems* What natural resources are important to Canada's economy?
2. *Places and Regions* What influenced the growth of important cities in Ontario and Quebec? Give some examples.

Critical Thinking

3. **Evaluating** What geographic and economic effects do you think government limits on fishing in the Grand Banks might have over time?
4. **Making Generalizations** Why did Canada create the territory of Nunavut? Why do you suppose the Inuit wanted this territory?

Organizing What You Know

5. Copy the chart. Use it to identify physical and human factors that define Canada's regions. You may want to review what you read about Canadian culture in Section 1. Note factors such as soils, climate, vegetation, language, trade, river systems, religion, economic activities, and natural resources.

Region	
Atlantic Provinces	
Ontario and Quebec	
Prairie Provinces	
British Columbia	
Canadian North	

Places and Regions

Geography for Life

Canadian Residential Preferences

Imagine that you could live anywhere in the United States or Canada. Where would you choose to live? Why? Would you pick a place near your home or a place far away? Are there some areas where you would not want to live? Why?

Many geographers are interested in how people perceive the world around them. These perceptions include people's thoughts about distant places. The knowledge and images that we have of different places are all a part of our mental maps—and all people have different mental maps. For example, think for a minute about the city of Montreal. What images and ideas come to your mind? Where did you get these ideas and images? Now suppose you had a friend who grew up in Montreal. How do you think his or her mental map would be different from yours? This example illustrates an important idea: people's culture and experience influence their opinions and perceptions of the world and its places.

In the 1970s and 1980s a cultural geographer named Herbert A. Whitney studied where a group of Canadians would most like to live. They could choose from places and areas in the United States and Canada. He surveyed people from all over Canada and asked them to indicate on a map which areas they liked and disliked. Most people surveyed were young, single college students. Whitney specifically chose this group because they tend to be the most mobile segment of society. He hoped that by analyzing their residential preferences, he could learn about possible future migration patterns.

Whitney's study produced some very interesting results. It showed that people in different parts of Canada had different preferences about where they wanted to live. For example, people in Quebec favored parts of the East Coast and southern United States more than people from Ontario and British Columbia. People in British Columbia had a strong preference for the western coasts of both countries.

However, people throughout Canada also shared some common perceptions. For example, most Canadians had a strong dislike of northern Canada as a place to live. In contrast, they had a strong like for the west. Also, Canadians generally had a strong like for areas near their homes and a dislike for distant areas. This last result might be expected. People usually have more knowledge of nearby areas and feel more comfortable there. For many people, distant places can seem different, strange, or even scary.

Applying What You Know

1. **Summarizing** What affects people's mental maps of distant places?
2. **Making Generalizations and Predictions** After studying the maps, where would you expect Canadians to migrate in the future?

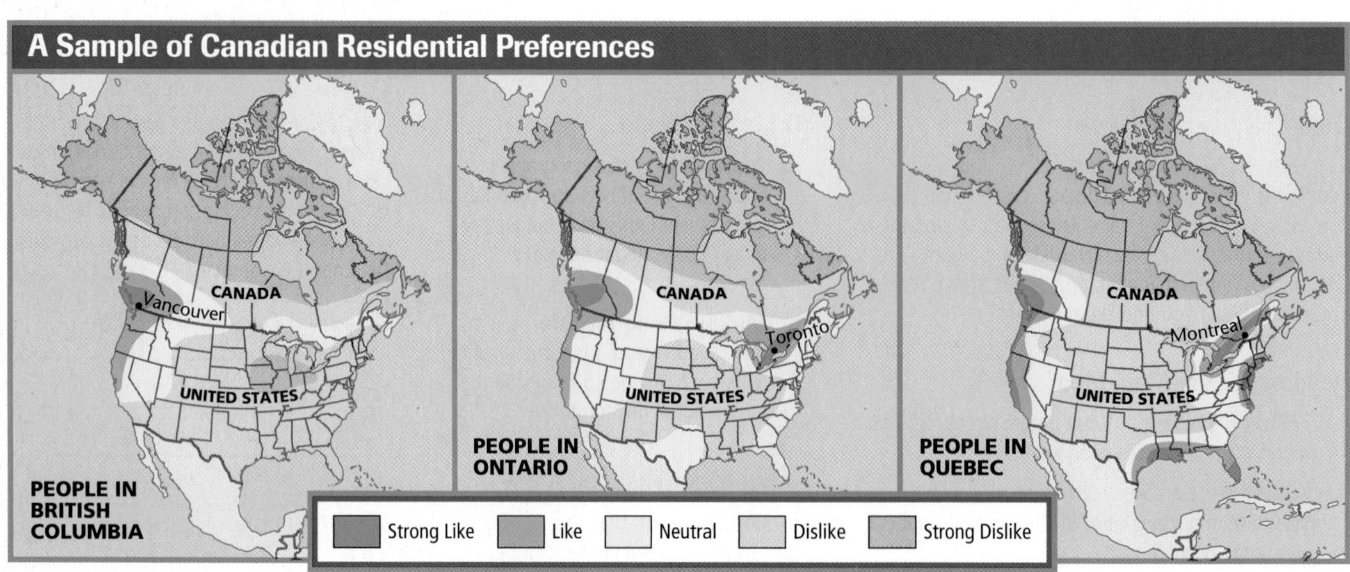

Section 3 Geographic Issues

READ TO DISCOVER

1. How does the United States influence Canada today?
2. How have geographic factors affected Canada's national unity?

DEFINE

regionalism
separatism

WHY IT MATTERS

If Canada split apart, the event could have important effects on the United States. Use CNNfyi.com or other **current events** sources to learn about the debate over Canada's political future.

Surveyors from the International Boundary Commission mark a section of the U.S.-Canada border.

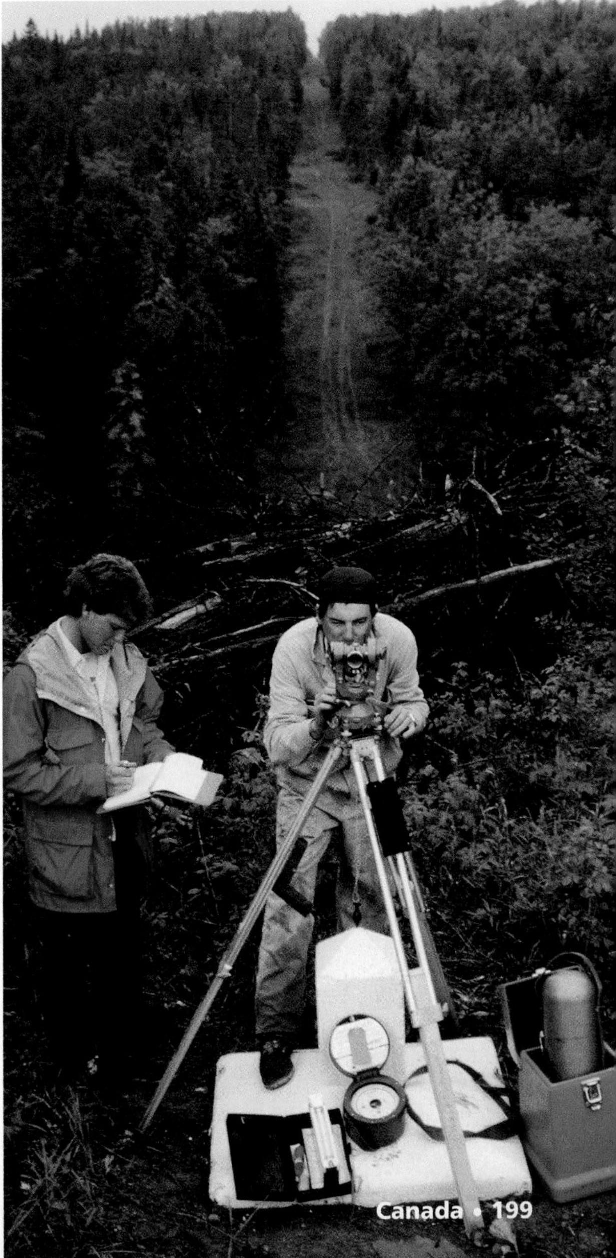

Our Northern Neighbor

The United States has always had much in common with its northern neighbor. English is the main language in the United States and most of Canada. The two countries also share similar histories as former European colonies. Many immigrants have made their new homes in these countries. In addition to such cultural and historical ties, the two countries have strong economic connections. Many companies based in the United States are giants in the Canadian economy. In addition, the U.S. and Canadian economies tend to rise and fall together. This relationship has become stronger since the signing of the North American Free Trade Agreement (NAFTA). NAFTA has resulted in increased trade and cooperation between the two countries.

In addition, Canada's cities in many ways seem more connected to U.S. cities than to each other. For example, Vancouver has common interests with Seattle. Farther east, Toronto has connections to many cities in the U.S. Midwest. Montreal has many business links to Boston.

The potential for cultural domination by the United States is a great concern for many Canadians. Canada's population is only about 31 million, and the great majority of Canadians live near the U.S. border. For this reason, the exchange of culture traits between the two countries can hardly be expected to be equal. With nine times the population of Canada, the United States often seems to overwhelm Canadians with its mass culture. For example, Canadians hear cultural, political, and economic information about the United States nearly every day. Yet many Americans seldom think about Canada.

Still, Canadians have a strong sense of political independence from the United States. Most Canadians would not want their country to join their neighbor to the south. However, Canadians today are debating the nature and unity of their country. This debate could help decide the future relations between Canada's provinces and the United States.

READING CHECK: *Human Systems* How has NAFTA affected the economy of Canada?

Regionalism and Separatism

The nature of the debate over a united Canada is linked to the country's physical and cultural geography. Canadians often show considerable **regionalism** when considering their country's important issues. Regionalism refers to the feeling of strong political and emotional loyalty to one's own region.

Canada's physical geography and isolated settlements help keep regionalism alive. For example, many Albertans believe their province shares too much of its income from the oil industry with the national government. British Columbians often feel separate from the rest of Canada, which lies far to the east across the Rocky Mountains and prairies. Many Canadians outside the Heartland Provinces believe the interests of Ontario and Quebec dominate in Ottawa, the national capital.

Even more critical to Canada's future than regionalism is **separatism.** This is the belief that certain parts of a country should be independent. Separatist feelings are strongest in Quebec. The separatist movement there has grown over the last 30–40 years, which has greatly affected Canada's government. Quebecois culture is specially protected under Canadian laws. In fact, civil law in Quebec is based on the French legal system rather than the English. Also, French is one of Canada's official languages, along with English. According to law, the children of French Canadians born outside of Quebec can choose to

INTERPRETING THE VISUAL RECORD

The landscape of Quebec City reflects the dominance of French culture in the region. ***What cultural features can you see in this photograph that help make French Canada a distinct region from the rest of the country?*** TEKS

have their children educated in French. Special immigration powers help Quebec to attract French-speaking newcomers. No other province has such status. Still, many French Canadians want Quebec to be an independent country. This issue has again and again threatened to break up Canada. In 1995, people in Quebec narrowly voted down a proposal that would have made their province an independent country. In fact, about 60 percent of the province's French-speaking citizens voted for it. People who pushed the proposal promise to bring the issue before voters again.

People in Ontario and Quebec have been arguing over language and culture for more than two centuries. Quebec even has what some people call "language police." These government officials make sure that signs are always in French. If English is also used, they make sure the French lettering is most prominent. Even Montreal has changed in recent years. The city once was an island of British-Canadian culture in Quebec. However, some corporations have left because they were worried about what an independent Quebec would mean for their businesses. Many have moved west to cities like Toronto and Calgary. Thousands of English-speaking residents have also left. Now Quebecois fill most important jobs in banking, education, insurance, and manufacturing.

The debate over Quebec has created the worst unity crisis in Canada's history. An independent Quebec would separate the Atlantic Provinces from the rest of Canada. Some Canadians worry that British Columbia, which has strong ties with countries around the Pacific, might also leave the union. Other provinces might demand their independence as well.

Still, Canadians seem to feel that these disputes are more like family squabbles than a national crisis. Canada remains peaceful because most Canadians believe everyone should have a chance to explain and debate their views. Most still support a united Canada rather than the idea of several small countries. Canadians continue to find strength in their diversity.

Street signs in French stand in front of a Catholic church in old Montreal. In Quebec both language and religion—two of the most important features of culture—are different from the rest of Canada and provide the basis for regionalism and separatism.

READING CHECK: ***The World in Spatial Terms*** How might the independence of Quebec affect the political geography of Canada?

TEKS Questions 1, 2, 3, 4, 5

Review

Define

regionalism, separatism

Reading for the Main Idea

1. ***Human Systems*** Why is Canada's culture so closely connected to that of the United States?
2. ***Places and Regions*** Why would an independent Quebec be a challenge for Canada?
3. ***Places and Regions*** For what reasons might regionalism in British Columbia be stronger than in most of the other provinces?

Critical Thinking

4. **Supporting a Point of View** Do you think that it has been a good policy to give Quebec such a special status in Canada? Explain your answer.

Organizing What You Know

5. Create a diagram like the one below and use it to identify the ways in which Canada and the United States are economically connected.

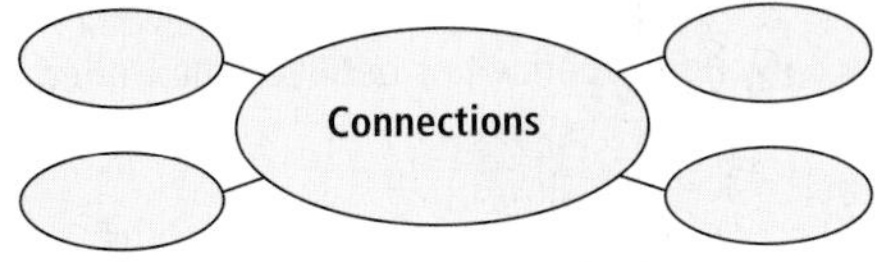

Review

Building Vocabulary TEKS

On a separate sheet of paper, explain the following terms by using them correctly in sentences.

provinces	parliament	regionalism
hinterland	consensus	separatism

Locating Key Places TEKS

On a separate sheet of paper, match the letters on the map with their correct labels.

Montreal	Ottawa	Winnipeg
Quebec City	Vancouver	Baffin Island
Toronto	Vancouver Island	

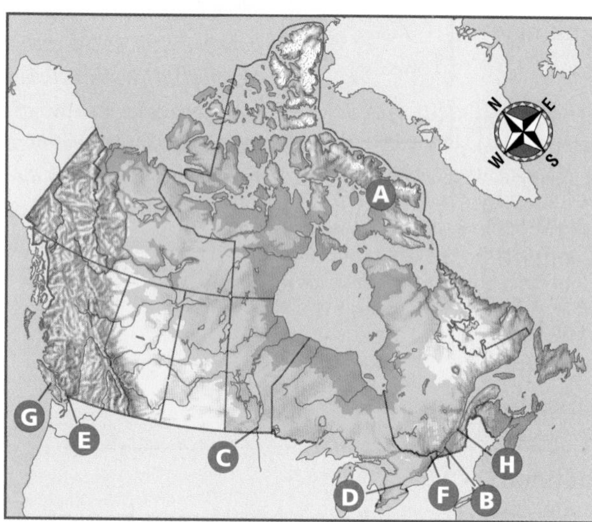

Understanding the Main Ideas 

Section 1

1. *Human Systems* What major groups have settled in Canada over time?
2. *Places and Regions* What characteristics make Ontario and Quebec Canada's heartland?

Section 2

3. *Places and Regions* How has Canada's economy changed over the last century?
4. *Human Systems* How is Canada's heritage reflected in its system of government? Why was Nunavut created?

Section 3

5. *Human Systems* What are some important cultural, historical, and economic ties between the United States and Canada?

Thinking Critically for TAKS TEKS

1. **Identifying Cause and Effect** How does Canada's cultural and physical geography contribute to regionalism?
2. **Drawing Inferences** Do you think major manufacturing centers will develop in northern Canada? Explain.
3. **Making Predictions** How might Canadians of French, English, and Inuit descent view their country differently?

Using the Geographer's Tools TEKS

1. **Analyzing Pie Graphs** Review the pie graph showing ethnic groups in Canada in Section 1. What are the origins of the largest ethnic group?
2. **Creating Pie Graphs** Use the unit Fast Facts table to create two pie graphs. One should show how the land area of North America is distributed between the United States and Canada, and the other how the continent's population is distributed. How does the population density of the United States compare to Canada?
3. **Creating Maps** Draw a map of what North America might look like if Canada were to dissolve. Indicate possible national boundaries, names of new countries, and likely capitals. What geographic and cultural features would link these countries?

Writing for TAKS TEKS

Review the discussions of economic development and standard of living in Chapter 6. Then analyze information from chapters in this unit and from the unit Fast Facts table to compare the level of development and standard of living in Canada and the United States. Write a short report that identifies the factors you have analyzed and the conclusions you have reached.

SKILL BUILDING

Geography for Life TEKS

Acquiring Geographic Information

Places and Regions Review the unit map of economic activities in Canada and the United States. Choose a Canadian province and investigate economic activities there. For example, what are the major items that the province manufactures? What crops or livestock are raised? Write a one-page summary of what you have learned. When you are finished with your summary, proofread it to make sure that you have used standard grammar, spelling, sentence structure, and punctuation.

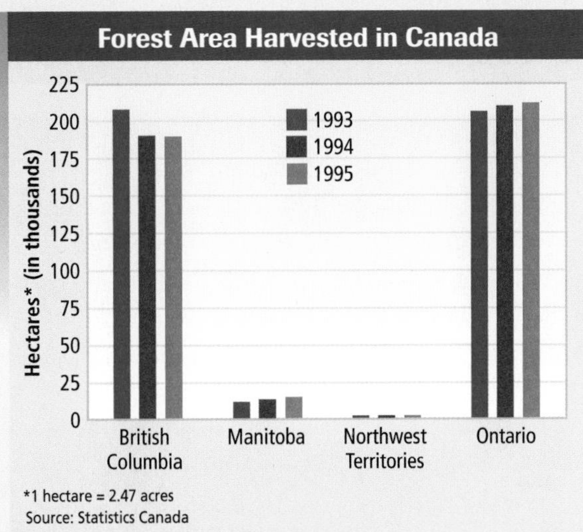

Interpreting Graphs

Use the graph above to answer the questions that follow.

1. Which province had the most forest area harvested in 1995?
   a. British Columbia
   b. Manitoba
   c. Northwest Territories
   d. Ontario
2. Why do you think the amount of forest area harvested in Manitoba and Northwest Territories is so small?

Analyzing Primary Sources

Writer Ian Darragh visited Prince Edward Island in the late 1990s and noted the Confederation Bridge, a new bridge to the mainland. Read his description, then answer the questions that follow.

> **"The eight-mile-long ribbon of concrete, which opened in June 1997, connects Prince Edward Island, Canada's smallest province, to the mainland for the first time in 5,000 years. . . . While people debate how the bridge may alter the environment of Northumberland Strait—from the swirl of the currents to the spring breakup of ice—it is already changing the ebb and flow of life on the island itself, a place long defined by close-knit communities and a slow-paced way of life."**

3. What features of the physical environment does the author suggest might be affected by the new bridge?
   a. a slow-paced way of life
   b. ribbons of concrete
   c. water currents and the spring breakup of ice
   d. the ebb and flow of life itself
4. In what ways does the author suggest that the bridge might affect the "ebb and flow of life on the island itself"?

Alternative Assessment

PORTFOLIO ACTIVITY

Learning about Your Local Geography

Group Project: Research

Plan, organize, and complete a research project with a partner about the cultural background of your community. From what countries or areas did early settlers in your community originate? What ethnic backgrounds do current residents have, and what languages are commonly spoken there? Working with your partner, create historical and current population maps of your community that illustrate your findings.

internet connect

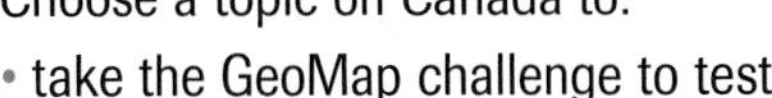

Choose a topic on Canada to:

- take the GeoMap challenge to test your knowledge of Canada's geography!
- research modern Quebec and create a newspaper.
- learn about writers, artists, and musicians of Canada.

New York

Human Systems New York has long been a beacon for people all over the world. Ever since the city's founding, waves of immigrants entering New York have helped shape the culture of the city as well as that of the United States. Today about 80 percent of all New Yorkers are of African, Irish, Italian, Jewish, or Puerto Rican descent. Many other ethnic groups and nationalities are also represented there. These others include immigrants from Africa, the Caribbean, China, Russia, other European countries, and Colombia as well as other Middle and South American countries. Scattered throughout the city are neighborhoods like Manhattan's Chinatown and Little Italy, which reflect unique cultural landscapes. In each neighborhood, visitors will find places of worship, restaurants and other businesses, and services that cater to resident ethnic groups.

A World City

New York's global ties have also helped make it a world financial capital. The city is home to major banks, insurance companies, and other financial institutions. More than 200 international banks from every major country in the world have offices in New York. In fact, of the 25 largest foreign branches of international banks in the United States in 2001, 20 had offices in the city. In addition, the New York Stock Exchange has attracted many of the world's largest firms that buy and sell stocks in companies. The Exchange is located on Wall Street, a place that has come to symbolize the country's financial power.

New York is also a major cultural center. It has many famous museums, galleries, theaters, and performance halls. Large media companies have offices in the city. Major American television networks are based there, and many movies and television shows seen around the world have been set in New York. The concentration of media in New York has helped make the city's concrete canyons and towering skyscrapers famous. In fact, for many people the concept of a "city" has been shaped by New York's striking skyline. In the same way that the Eiffel Tower has become a recognizable symbol of France and Paris, gleaming skyscrapers, many of which are found in New York, have become symbols of the United States.

These photos show the dramatic change in New York's skyline after the destruction of the twin towers of the World Trade Center on September 11, 2001. Today the Empire State Building is once again New York's tallest building.

September 11, 2001

For nearly three decades the twin towers of the World Trade Center dominated New York's skyline. These modern towers each soared higher than 1,360 feet. They joined the Statue of Liberty and the Empire State Building as among New York's most recognizable landmarks. It is perhaps the World Trade Center's status as a great symbol of modern capitalism and globalization that made it a target of attack in 2001. Fundamentalist Muslim terrorists flew two hijacked

New York police officers and firefighters worked to help survivors after the attack on the World Trade Center. Cities like New York must employ large numbers of police officers, firefighters, and other emergency personnel.

planes into the towers on September 11 of that year. Other hijacked planes crashed into the Pentagon outside Washington, D.C., and in rural Pennsylvania.

The destruction of the World Trade Center left thousands dead. New York mayor Rudolph Giuliani called the death toll "more than we can bear." Nonetheless, he vowed, "We're going to rebuild and rebuild stronger."

Looking Forward

New Yorkers and other Americans pulled together to rebuild, inspiring people around the world with their dignity and courage. People in some countries saw the acts of the terrorists as an attack on all of the world. Some declared with respect and unity that they too were Americans.

In New York, residents pondered their city's future. How had the attacks changed their city? One change was the loss of a sense of security many New Yorkers and other Americans had felt before the attacks. Americans had long felt protected from such horrible attacks by the vast ocean distances separating this country from the world's trouble spots. Commentators observed that the terrorist attacks changed life as New Yorkers and other Americans had known it. This realization refocused the efforts of Americans to defeat terrorism on a global scale.

Another change was a renewed sense of respect and admiration among New Yorkers for their city's fire and police departments. Hundreds of firefighters and police officers died when the World Trade Center towers collapsed. They had bravely rushed into the buildings in an attempt to rescue trapped victims.

Perhaps the most obvious change after the attack was New York's altered skyline. Gone were the gleaming towers of the World Trade Center, which had soared high above the city's other skyscrapers. Some people wanted the city quickly to rebuild the towers to replace much-needed office space. Many people felt that the site should include proper memorials to the thousands of victims of the terrorist attack. More symbolically, New Yorkers had to decide what any new complex would say about their city. "That's a real question for New York," said one scholar. "Whether the new design looks in or looks out: Will it be something for New Yorkers or something for New York's symbolism in the world?"

New York and other cities have places like parks and city squares for recreation and public gatherings. New York's Washington Square became a site for memorials to victims of the September 11, 2001, terrorist attacks on the World Trade Center and the Pentagon.

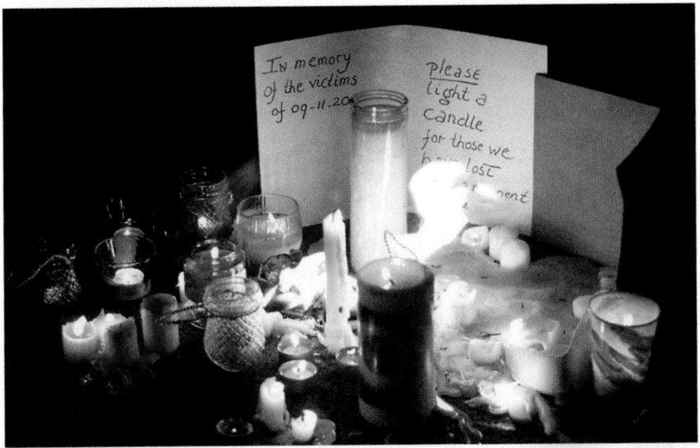

Applying What You Know 

1. **Summarizing** What are some economic, social, and cultural characteristics of New York?
2. **Making Generalizations** Why do you think some people outside the United States said they thought of themselves symbolically as Americans after the terrorist attacks on September 11, 2001?

Skill Building for TAKS

WORKSHOP 1

Using Mental Maps

As you have read, mental maps are maps that represent the mental image a person has of an area. They help us make sense of the world around us by organizing information we have about places. Those places range from our own homes to the whole world.

You will not find mental maps in a textbook. Instead, mental maps are images we have in our minds. These images include knowledge of features and spatial relationships in an area. They also include our perceptions and attitudes about that particular area. As such, mental maps of even the same places vary from person to person.

Each of us uses a variety of mental maps. You can probably think of many places about which you have a mental map. You likely have a mental map of the floor plan of your home. You probably also have a mental map of the neighborhood in which you live. You know where the homes of friends, certain streets, houses, schools, and places of business and worship are located. You can use your mental map to plan routes to each of those places. You can also use a mental map of routes you take when you travel outside your community.

Developing the Skill We can also use mental maps to organize how we think about our state, country, and world. For example, you have read that there are seven continents on Earth. Without reviewing a map of the globe, you might be able to sketch at least the locations of those continents on a map of the world. You might be able to do this even if you are not sure of the shape of each continent.

The sample sketch map on this page uses circles that show the general locations of the continents. Now close your eyes and visualize what you already know about these continents. As you think about the continents, consider the following questions:

- Can you place the name of each continent on your mental map?

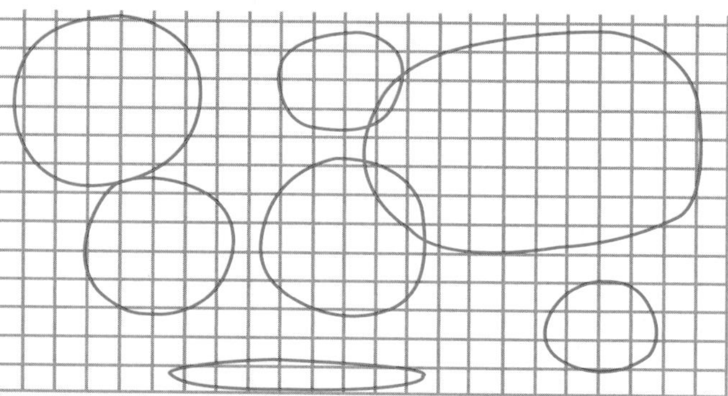

- What do you know about the shapes and relative sizes of each continent?
- Which continents lie in the northern half of the globe?
- Which continents are located south of the equator?
- How are the continents connected to or separated from each other?
- What oceans lie between the continents?

After you have thought about the continents, sketch their shapes and label them and the oceans on your own map of the world. Then compare your sketch map to the world map in the atlas at the front of this textbook. How does your sketch map differ from the atlas map? How is it similar? How did your perceptions of the world guide the way you sketched your map?

Practicing the Skill

1. Sketch the floor plan of your school from memory. Locate and label important places, including classrooms you use each day. Identify the fastest or easiest routes between your classrooms.
2. Sketch a map of your neighborhood from memory. On your map locate important places, landmarks, and the route you take to school.
3. Sketch a map of your state from memory. Include important cities, rivers, travel routes, and places you have visited or would like to see in the future.

WORKSHOP 2

Using Geographic Models

When geographers present ideas, they sometimes find it helpful to use models. In this context, *model* does not mean a small replica of something, like a model airplane. Rather, the models geographers use are more like the plans for different types of airplanes. You might look at these plans to construct an actual airplane. On the other hand, you might just study the plans to better understand how airplanes work.

Developing the Skill This textbook includes a number of models. Two of them are shown on this page and were discussed in Chapter 6. One model represents an idea called central place theory. It shows the ideal distribution of urban places of different sizes and the main transport routes that link these places. The model does not show how urban places are really distributed. Instead, it represents what one geographer thought was the best arrangement of such places, with each place serving a special function.

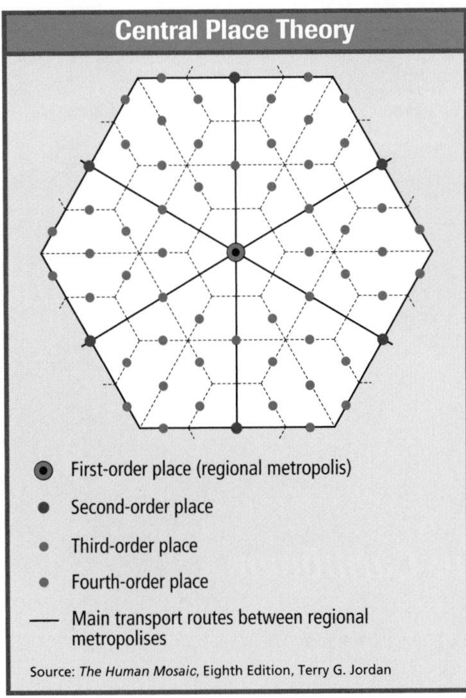

Source: *The Human Mosaic*, Eighth Edition, Terry G. Jordan

The von Thünen Model

WILDERNESS

WILDERNESS

- Central city
- Intensive farming and dairying
- Forest
- Increasingly extensive field crops
- Ranching, animal products

Source: *Geography: Realms, Regions, and Concepts*, Fifth Edition, De Blij and Muller

The second model represents patterns of land use in Germany observed by a geographer in the 1800s. Each circular band represents a different kind of land use. The distribution of land uses is related to land value and to distance from the central city. In short, this model is used to show which types of land use are most common nearest a city center and which are most common farther away.

The two models on this page have some common features. They both use color, although color may not be necessary if a model is not very complex. Both also have legends. The legends are similar to those you might find on a map. Each legend identifies symbols and colors that represent information in the model.

Practicing the Skill

1. Create a model for the ideal floor plan of a school. Your model should show how you think different areas of an ideal school might be arranged. For example, where should the cafeteria, classrooms, gymnasium, main offices, parking areas, and restrooms be located in relation to each other? Include a title, legend, any necessary labels, and a descriptive caption for your model.
2. Work with a group to study the layout of major cities around the world. Then work together to create a model that could be used to plan future cities. Your group should consider issues such as where to locate commercial and industrial areas, parks and recreational areas, and residential areas. Also consider the transportation network that should connect your model city's various areas. Draw your model on a posterboard. Then present and explain it to your class.

UNIT 3 Middle and South America

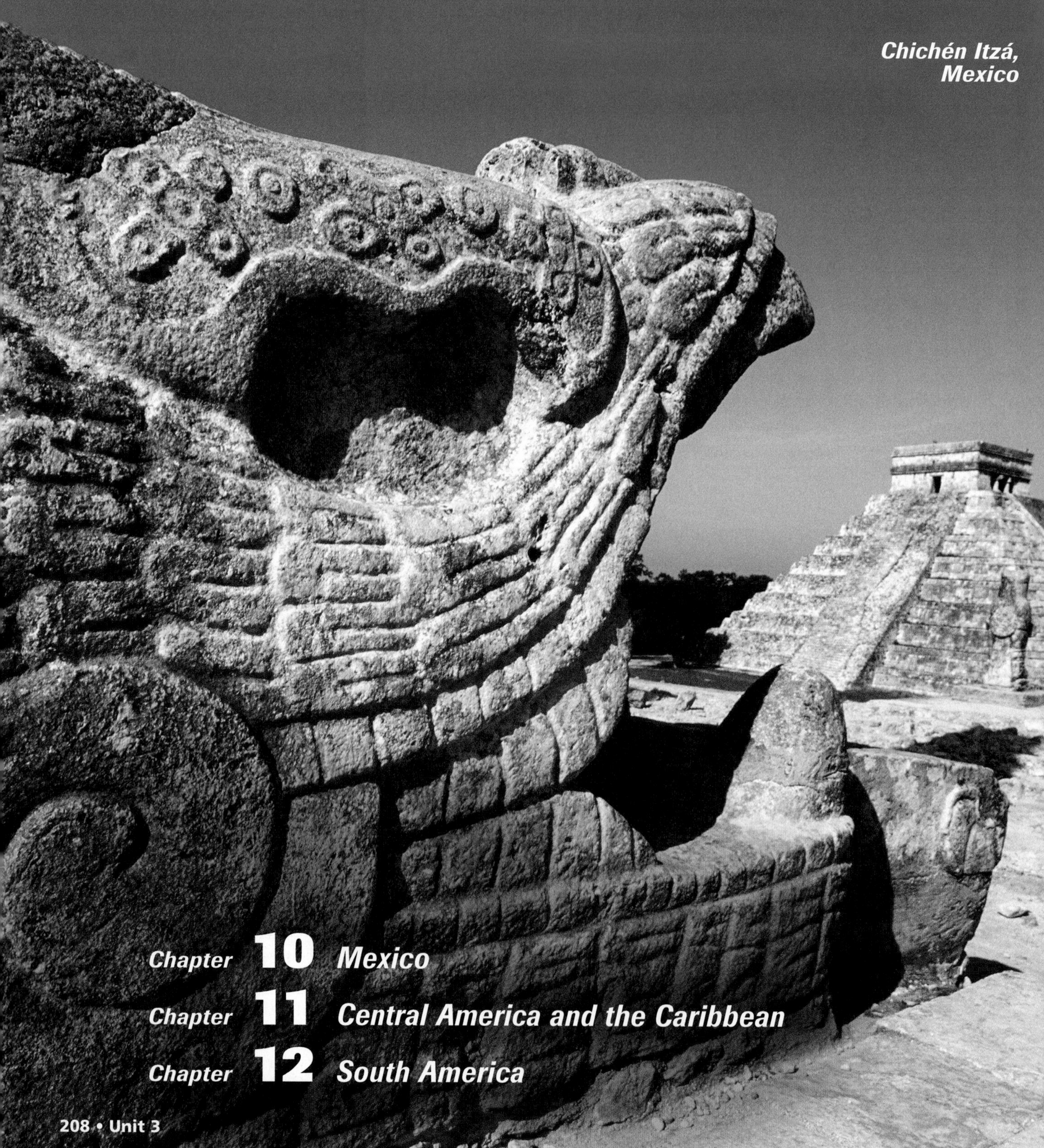

Chichén Itzá, Mexico

CONNECTING TO

Literature

The Temple of the Sun

from The Incas: The Royal Commentaries of the Inca *by Garcilaso de la Vega*

Garcilaso de la Vega
(1539–1616)—also known as El Inca—was the son of a Spanish conquistador and an Inca princess. Interested in Inca history, he spoke to many people in South America about the events they witnessed and took notes about what they said. He also learned Inca oral histories. In 1560 he went to Spain, where he eventually wrote his histories of Peru.

The Temple of the Sun was located on the site that today is occupied by the Church of San Dominique, and its walls, which are made of highly polished stone, still exist. . . .

The four walls were hung with plaques of gold, from top to bottom, and a likeness of the Sun topped the high altar. This likeness was made of a gold plaque. . . . There was no other idol in this temple, nor in any other, for the Sun was the only god of the Incas, whatever people may say. . . .

On either side of this Sun, were kept the numerous mummies of former Inca kings, which were so well preserved that they seemed to be alive. They were seated on their golden thrones resting on plaques of this same metal, and they looked directly at the visitor. Alone among them, Huaina Capac's body had assumed a peculiar pose, facing the Sun, as though from childhood, he had been its favorite son who deserved to be adored for his unusual virtues. . . .

The temple was decorated with five fountains. . . . Their pipes were of solid gold and their stone pillars were covered with either gold or silver, for the sacrifices were washed in these waters. I remember the last of these fountains which was used to water the garden of the convent that the Spaniards established on this sacred ground. One day it stopped working, to the great despair of the Indians who, not knowing where the water came from, were unable to repair it; and the garden dried up, in spite of their desire and their efforts to save it. This only shows how quickly the Indians lost their traditions, since, in the space of forty-two years, there was not one left who could say from whence came the waters that circulated throughout the temple of their god the Sun.

Analyzing the Primary Source

1. **Identifying Points of View** How does Garcilaso de la Vega portray the Inca?
2. **Evaluating Sources** Did Garcilaso de la Vega witness everything he wrote about in this passage? Give reasons for your answer.

UNIT 3 ATLAS

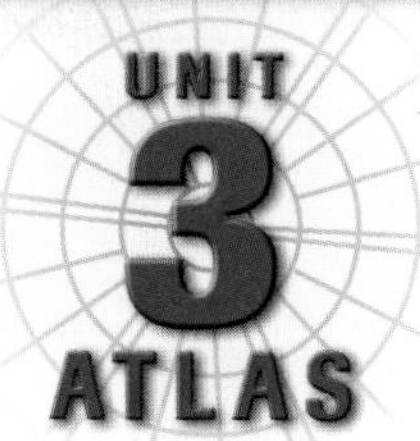

The World in Spatial Terms

Middle and South America: Political

1. *Places and Regions* Which South American country has two capitals?
2. *Places and Regions* What is the largest country in Middle America? In South America?

Critical Thinking

3. **Making Generalizations** How would you describe the shape of Chile? What problems might this shape create for Chile's government? TEKS

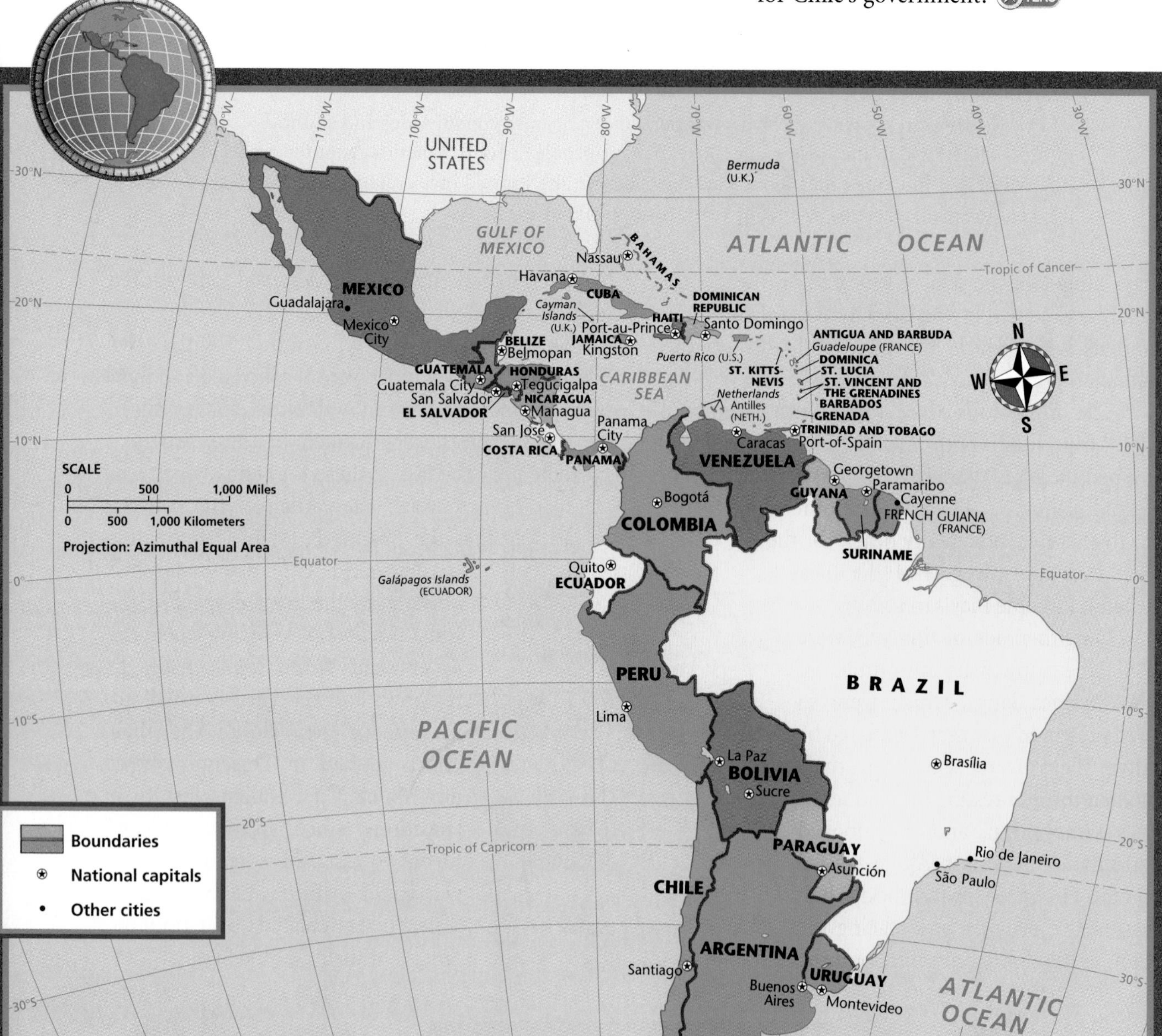

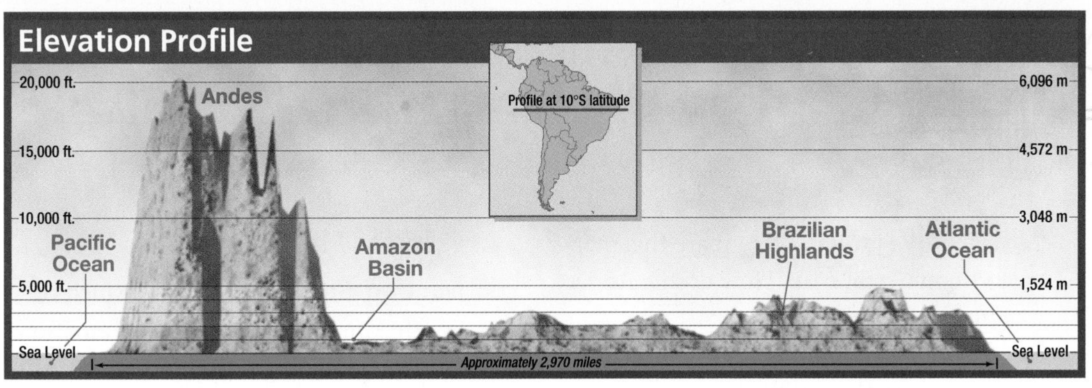

Middle and South America: Physical

UNITED STATES
Rio Grande
Rio Bravo
BAJA CALIFORNIA
SIERRA MADRE OCCIDENTAL
SIERRA MADRE ORIENTAL
GULF OF MEXICO
Bermuda (U.K.)
ATLANTIC OCEAN
Tropic of Cancer
BAHAMAS
GREATER ANTILLES
CUBA
YUCATÁN PENINSULA
Cayman Islands (U.K.)
JAMAICA
HAITI
Hispaniola
DOMINICAN REPUBLIC
LESSER ANTILLES
MEXICO
BELIZE
GUATEMALA
HONDURAS
EL SALVADOR
NICARAGUA
CARIBBEAN SEA
Lake Nicaragua
Panama Canal
COSTA RICA
ISTHMUS OF PANAMA
PANAMA
VENEZUELA
Lake Maracaibo
Orinoco River
LLANOS
Angel Falls
GUYANA
SURINAME
FRENCH GUIANA (FRANCE)
GUIANA HIGHLANDS
Cauca R.
Magdalena R.
COLOMBIA
Rio Negro
Equator
Galápagos Islands (ECUADOR)
ECUADOR
AMAZON BASIN
Amazon River
ANDES
PERU
BRAZIL
São Francisco R.
BRAZILIAN HIGHLANDS
PACIFIC OCEAN
Lake Titicaca
BOLIVIA
Lake Poopó
Paraguay River
Paraná R.
MATO GROSSO PLATEAU
ATACAMA DESERT
GRAN CHACO
Tropic of Capricorn
PARAGUAY
Iguazú Falls
Paraná
River
Uruguay
CHILE
Aconcagua 22,834 ft. (6,960 m)
PAMPAS
URUGUAY
Rio de la Plata
ATLANTIC OCEAN
ARGENTINA
PATAGONIA
Falkland Islands (U.K.)
TIERRA DEL FUEGO
CAPE HORN

N
E
S
W

SCALE
0 500 1,000 Miles
0 500 1,000 Kilometers
Projection: Azimuthal Equal Area

ELEVATION

FEET	METERS
13,120	4,000
6,560	2,000
1,640	500
656	200
(Sea level) 0	0 (Sea level)
Below sea level	Below sea level

Middle and South America: Climate

1. *Physical Systems* Compare this map to the physical map. How do the Andes appear to affect climates? TEKS
2. *Environment and Society* Compare this map to the political and population maps. Most of Brazil's population is concentrated in which climate regions?

Critical Thinking

3. **Making Generalizations** Much of South America's west coast has an arid climate. How might this climate be influenced by the cold Peru Current? TEKS

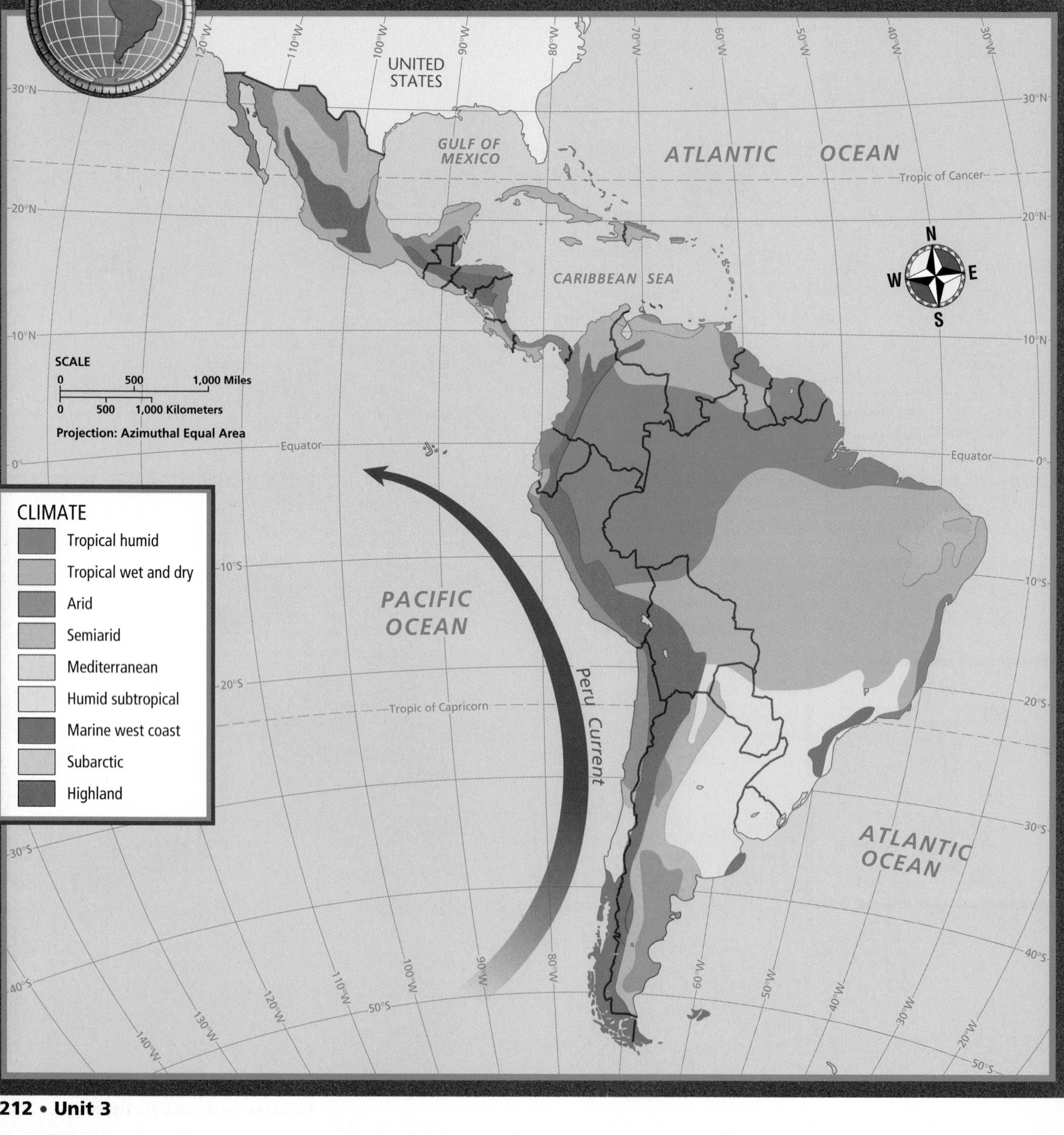

UNITED STATES
GULF OF MEXICO
ATLANTIC OCEAN
CARIBBEAN SEA
PACIFIC OCEAN
Peru Current
Tropic of Cancer
Equator
Tropic of Capricorn

SCALE
0 500 1,000 Miles
0 500 1,000 Kilometers
Projection: Azimuthal Equal Area

CLIMATE
- Tropical humid
- Tropical wet and dry
- Arid
- Semiarid
- Mediterranean
- Humid subtropical
- Marine west coast
- Subarctic
- Highland

Middle and South America: Precipitation

UNIT 3 ATLAS

1. *Places and Regions* Compare this map to the physical map. Which regions receive the most precipitation? TEKS
2. *Environment and Society* Compare this map to the land use map. How might precipitation patterns affect the location of hydroelectric power sites? TEKS

Critical Thinking

3. **Drawing Conclusions** Compare this map to the physical map. Which region of Argentina lies in a rain shadow? Based on the precipitation map, from which direction do you think the prevailing winds come? TEKS

UNIT 3 ATLAS

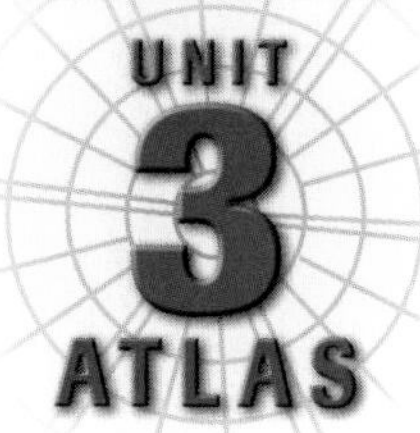

Middle and South America: Population

1. **Environment and Society** Compare this map to the physical map. How densely populated is the Amazon Basin? TEKS
2. **Places and Regions** Which region of South America is the most densely populated? TEKS

Critical Thinking

3. **Making Generalizations** Compare this map to the physical and climate maps. Which regions of Mexico are the least densely populated? How might climate affect population density there? TEKS

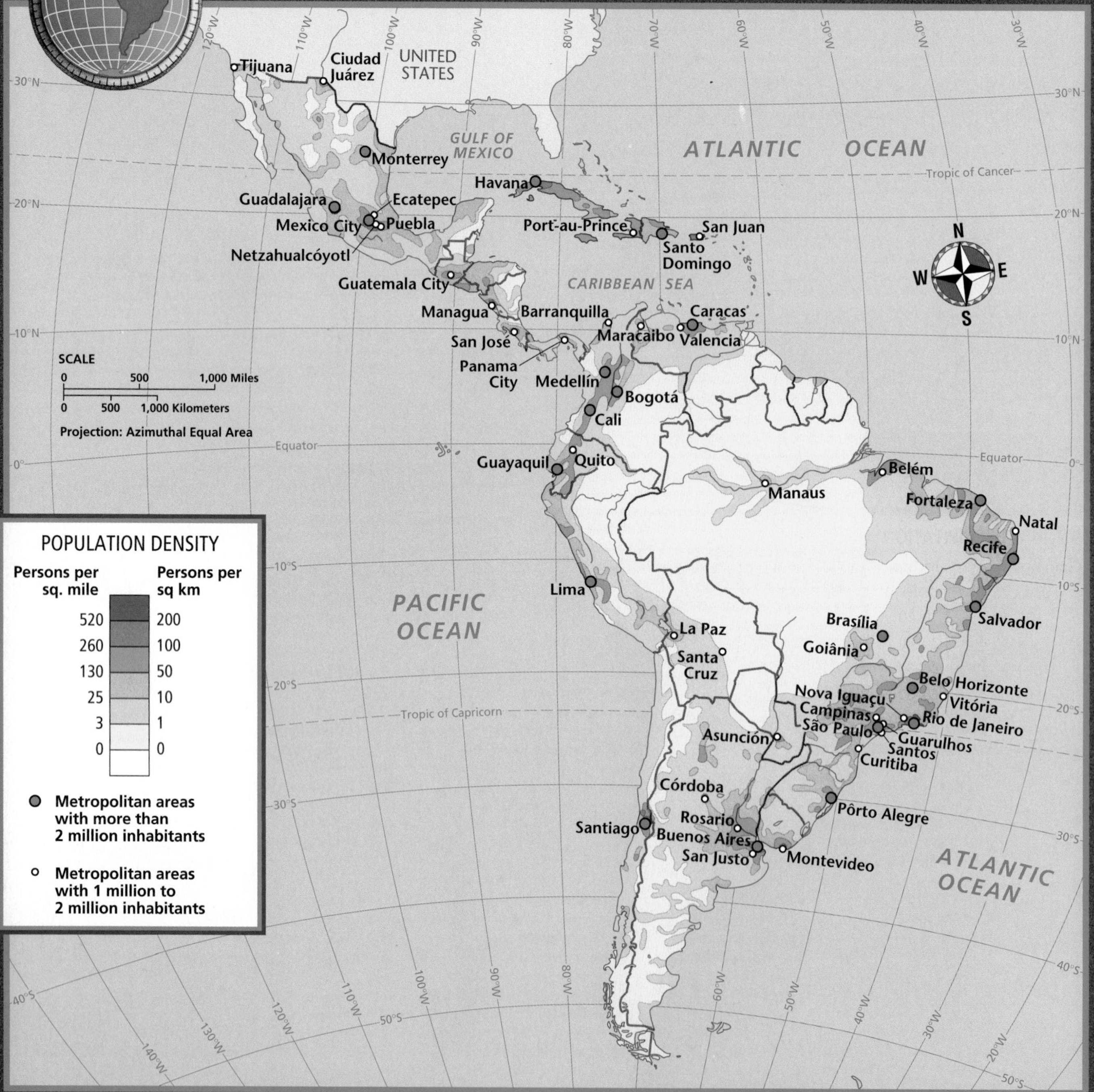

UNITED STATES
GULF OF MEXICO
ATLANTIC OCEAN
CARIBBEAN SEA
PACIFIC OCEAN
ATLANTIC OCEAN
Tropic of Cancer
Equator
Tropic of Capricorn
Tijuana
Ciudad Juárez
Monterrey
Guadalajara
Ecatepec
Mexico City
Puebla
Netzahualcóyotl
Guatemala City
Havana
Port-au-Prince
Santo Domingo
San Juan
Managua
San José
Panama City
Barranquilla
Maracaibo
Caracas
Valencia
Medellín
Bogotá
Cali
Quito
Guayaquil
Lima
La Paz
Santa Cruz
Manaus
Belém
Fortaleza
Natal
Recife
Salvador
Brasília
Goiânia
Belo Horizonte
Vitória
Nova Iguaçu
Campinas
São Paulo
Rio de Janeiro
Guarulhos
Santos
Curitiba
Asunción
Córdoba
Rosario
Santiago
Buenos Aires
San Justo
Montevideo
Pôrto Alegre
N
S
E
W
SCALE
0 500 1,000 Miles
0 500 1,000 Kilometers
Projection: Azimuthal Equal Area
POPULATION DENSITY
Persons per sq. mile
Persons per sq km
520 200
260 100
130 50
25 10
3 1
0 0
Metropolitan areas with more than 2 million inhabitants
Metropolitan areas with 1 million to 2 million inhabitants
30°N
20°N
10°N
0°
10°S
20°S
30°S
40°S
50°S
120°W
110°W
100°W
90°W
80°W
70°W
60°W
50°W
40°W
30°W
20°W
130°W
140°W

Middle and South America: Land Use and Resources

UNIT 3 ATLAS

1. *Environment and Society* Where in Middle and South America would plantation agriculture likely be found? TEKS
2. *Places and Regions* Compare this map to the political map. In which countries is oil production an important activity? TEKS

Critical Thinking

3. **Analyzing** Compare this map to the climate map. Which climate areas have limited economic activity? What are some economic activities that are found in these areas? TEKS

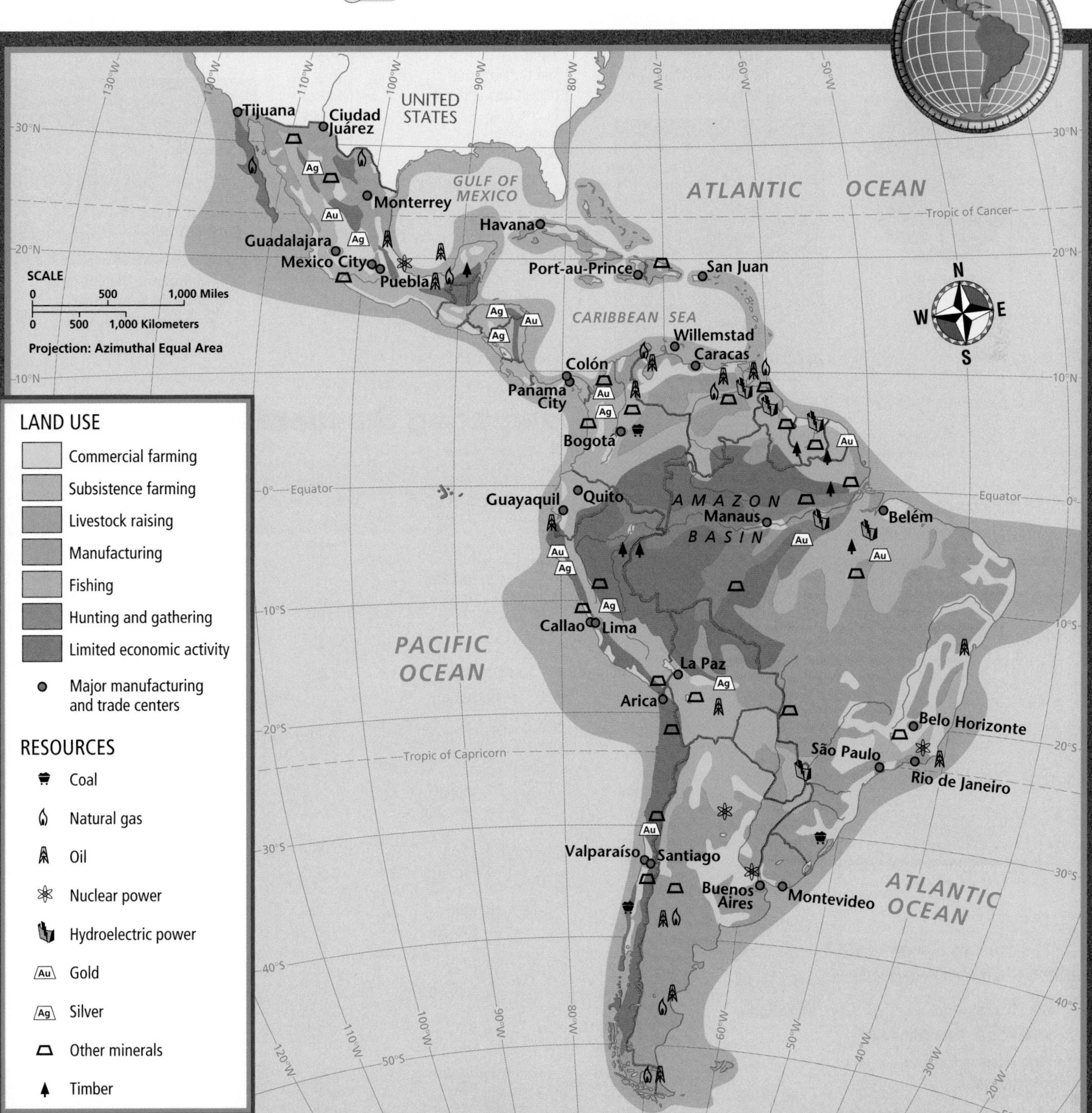

Time Line: Middle and South America

1494
Spain and Portugal sign a treaty to divide all lands to be explored.

1492

Columbus arrives in the Caribbean.

1810–1830
Spanish colonies in South America gain independence.

1821
Mexico wins independence from Spain.

1822
Brazil declares independence from Portugal.

1898
The United States takes Cuba and Puerto Rico from Spain during Spanish-American War.

1910–1917
The Mexican Revolution occurs.

1959
Fidel Castro comes to power in Cuba.

1981
Belize becomes the last Central American country to gain independence.

1992

Mexico, the United States, and Canada sign NAFTA.

1999
Panama takes control of the Panama Canal.

1450 | 1800 | 1900 | 2000

The United States and Middle and South America

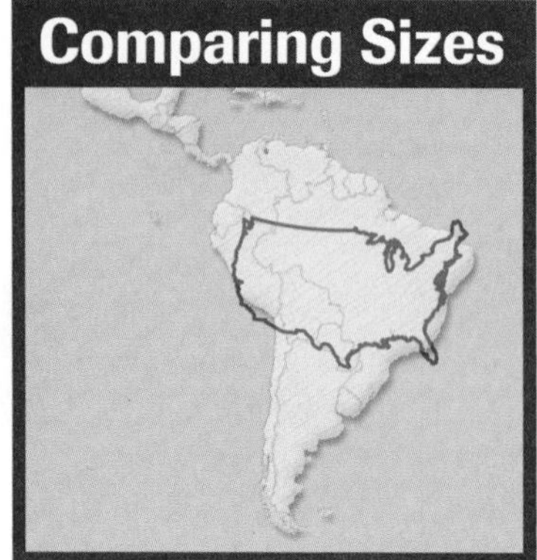

Comparing Standard of Living

COUNTRY	LIFE EXPECTANCY (in years)	INFANT MORTALITY (per 1,000 live births)	LITERACY RATE	DAILY CALORIC INTAKE (per person)
Argentina	72, male 79, female	18	96%	3,110
Brazil	59, male 68, female	40	83%	2,834
Colombia	67, male 75, female	24	91%	2,758
Costa Rica	73, male 79, female	11	95%	2,865
Cuba	74, male 79, female	7	96%	2,291
Ecuador	69, male 74, female	34	90%	2,436
Haiti	48, male 51, female	95	45%	Not available
Mexico	69, male 75, female	25	90%	3,136
Nicaragua	67, male 71, female	34	66%	2,311
United States	74, male 80, female	7	97%	3,603

Sources: Central Intelligence Agency, *2001 World Factbook; Britannica Book of the Year, 2000*

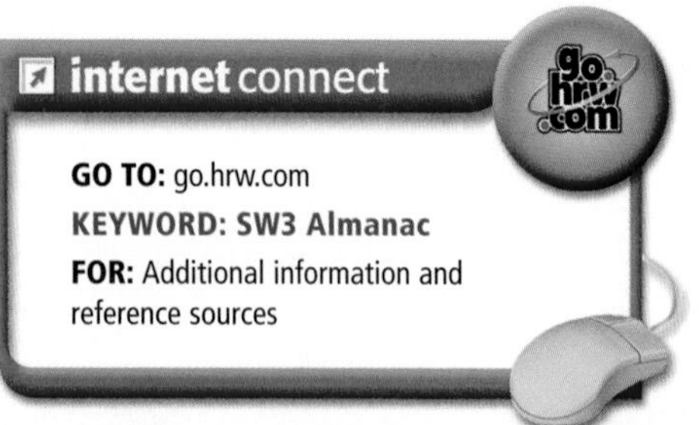

GO TO: go.hrw.com
KEYWORD: SW3 Almanac
FOR: Additional information and reference sources

Fast Facts: Middle and South America

FLAG	COUNTRY Capital	POPULATION (in millions) POP. DENSITY	AREA	PER CAPITA GDP (in US $)	WORKFORCE STRUCTURE (largest categories)	ELECTRICITY CONSUMPTION (kilowatt hours per person)	TELEPHONE LINES (per person)
	Antigua and Barbuda Saint John's	0.07 392/sq.mi.	171 sq. mi. 442 sq km	$ 8,200	82% comm., services 11% agriculture	1,320 kWh	0.42
	Argentina Buenos Aires	37 35/sq. mi.	1,068,296 sq. mi. 2,766,890 sq km	$ 12,900	24% services 21% manufacturing	2,063 kWh	0.20
	Bahamas Nassau	0.3 55/sq. mi.	5,382 sq. mi. 13,940 sq km	$ 15,000	50% other services 40% tourism	4,573 kWh	0.32
	Barbados Bridgetown	0.3 1,659/sq. mi.	166 sq. mi. 430 sq km	$ 14,500	75% services 15% industry	2,425 kWh	0.39
	Belize Belmopan	0.3 29/sq. mi.	8,867 sq. mi. 22,966 sq km	$ 3,200	38% agriculture 32% industry	672 kWh	0.12
	Bolivia La Paz, Sucre	8.3 20/sq. mi.	424,162 sq. mi. 1,098,580 sq km	$ 2,600	39% agriculture 14% public admin., services	406 kWh	0.04
	Brazil Brasília	174.5 53/sq. mi.	3,286,470 sq. mi. 8,511,965 sq km	$ 6,500	53% services 23% agriculture	2,027 kWh	0.10
	Chile Santiago	15.3 52/sq. mi.	292,258 sq. mi. 756,950 sq km	$ 10,100	59% services 27% industry	2,311 kWh	0.17
	Colombia Bogotá	40.3 92/sq. mi.	439,733 sq. mi. 1,138,910 sq km	$ 6,200	46% services 30% agriculture	1,005 kWh	0.13
	Costa Rica San José	3.8 191/sq. mi.	19,730 sq. mi. 51,100 sq km	$ 6,700	58% services 22% industry	1,405 kWh	0.12
	Cuba Havana	11.2 261/sq. mi.	42,803 sq. mi. 110,860 sq km	$ 1,700	51% services 25% agriculture	1,194 kWh	0.04
	Dominica Roseau	0.07 243/sq. mi.	291 sq. mi. 754 sq km	$ 4,000	40% agriculture 32% ind., commerce	815 kWh	0.27
	Dominican Republic Santo Domingo	8.6 456/sq. mi.	18,815 sq. mi. 48,730 sq km	$ 5,700	59% services, gov. 24% industry	790 kWh	0.08
	Ecuador Quito	13.2 120/sq. mi.	109,483 sq. mi. 283,560 sq km	$ 2,900	45% services 30% agriculture	712 kWh	0.07
	El Salvador San Salvador	6.2 768/sq. mi.	8,124 sq. mi. 21,040 sq km	$ 4,000	55% services 30% agriculture	583 kWh	0.06

Sources: Central Intelligence Agency, *2001 World Factbook; The World Almanac and Book of Facts, 2001; Britannica Book of the Year, 2000;* U.S. Census Bureau
The CIA calculates per capita GDP in terms of purchasing power parity. This formula equalizes the purchasing power of each country's currency.

Fast Facts: Middle and South America

FLAG	COUNTRY Capital	POPULATION (in millions) / POP. DENSITY	AREA	PER CAPITA GDP (in US $)	WORKFORCE STRUCTURE (largest categories)	ELECTRICITY CONSUMPTION (kilowatt hours per person)	TELEPHONE LINES (per person)
	Grenada Saint George's	0.09 / 681/sq. mi.	131 sq. mi. 340 sq km	$ 4,400	62% services 24% agriculture	1,251 kWh	0.30
	Guatemala Guatemala City	13.0 / 309/sq. mi.	42,042 sq. mi. 108,890 sq km	$ 3,700	50% agriculture 35% services	254 kWh	0.05
	Guyana Georgetown	0.7 / 8/sq. mi.	83,000 sq. mi. 214,970 sq km	$ 4,800	39% agric., forest, fish. 24% mining, manuf., const.	607 kWh	0.10
	Haiti Port-au-Prince	7.0 / 650/sq. mi.	10,714 sq. mi. 27,750 sq km	$ 1,800	66% agriculture 25% services	90 kWh	0.01
	Honduras Tegucigalpa	6.4 / 148/sq. mi.	43,278 sq. mi. 112,090 sq km	$ 2,700	50% services 29% agriculture	505 kWh	0.04
	Jamaica Kingston	2.7 / 628/sq. mi.	4,243 sq. mi. 10,990 sq km	$ 3,700	60% services 21% agriculture	2,278 kWh	0.13
	Mexico Mexico City	101.9 / 134/sq. mi.	761,602 sq. mi. 1,972,550 sq km	$ 9,100	56% services 24% industry	1,676 kWh	0.09
	Nicaragua Managua	4.9 / 98/sq. mi.	49,998 sq. mi. 129,494 sq km	$ 2,700	43% services 42% agriculture	461 kWh	0.03
	Panama Panama City	2.8 / 94/sq. mi.	30,193 sq. mi. 78,200 sq km	$ 6,000	61% services 21% agriculture	1,423 kWh	0.14
	Paraguay Asunción	5.7 / 37/sq. mi.	157,046 sq. mi. 406,750 sq km	$ 4,750	45% agriculture	334 kWh	0.05

Sea lion on the shore, Puerto Egas, Galápagos Islands

FLAG	COUNTRY Capital	POPULATION (in millions) / POP. DENSITY	AREA	PER CAPITA GDP (in US $)	WORKFORCE STRUCTURE (largest categories)	ELECTRICITY CONSUMPTION (kilowatt hours per person)	TELEPHONE LINES (per person)
	Peru Lima	27 / 55/sq. mi.	496,223 sq. mi. 1,285,220 sq km	$ 4,550	33% agriculture 28% services	639 kWh	0.05
	St. Kitts-Nevis Basseterre	0.04 / 384/sq. mi.	101 sq. mi. 261 sq km	$ 7,000	69% services 31% manufacturing	2,160 kWh	0.44
	St. Lucia Castries	0.16 / 662/sq. mi.	239 sq. mi. 620 sq km	$ 4,500	43% agriculture 39% services	647 kWh	0.23
	St. Vincent and the Grenadines Kingstown	0.12 / 773/sq. mi.	150 sq. mi. 389 sq km	$ 2,800	57% services 26% agriculture	658 kWh	0.18
	Suriname Paramaribo	0.4 / 7/sq. mi.	63,039 sq. mi. 163,270 sq km	$ 3,400	40% services 13% trade	4,150 kWh	0.15
	Trinidad and Tobago Port-of-Spain	1.2 / 591/sq. mi.	1,980 sq. mi. 5,128 sq km	$ 9,500	64% services 14% manuf., mining, quarrying	3,896 kWh	0.21
	Uruguay Montevideo	3.4 / 49/sq. mi.	68,039 sq. mi. 176,220 sq km	$ 9,300	33% services 25% government	1,753 kWh	0.25
	Venezuela Caracas	23.9 / 68/sq. mi.	352,143 sq. mi. 912,050 sq km	$ 6,200	64% services 23% industry	3,158 kWh	0.11
	United States Washington, D.C.	281.4 / 76/sq. mi.	3,717,792 sq.mi. 9,629,091 sq km	$ 36,200	30% manage., prof. 29% tech., sales, admin.	12,211 kWh	0.69

Machu Picchu, Peru

CHAPTER

10 Mexico

Mexico is part of a region we call Middle and South America. The region includes Central America and the Caribbean islands as well as South America. Mexico and the region's other countries share many cultural and historical features.

¡Hola! (Hello!) My name is Ellie. I live in a village, San Francisco Acatepec, near Puebla. Our house is made of adobe, or sun-dried clay bricks, and has a big yard. The ceilings are wood, with beams, and there are many arches.

Portrait of Maya ruler Pacal from Palenque, Chiapas State

We eat on a different schedule than you do in the United States. We eat a big breakfast, then at school we just have a snack. At lunch almost everybody goes to a vendor for *tortas,* which are like sandwiches in rolls. There are also little stands outside the school where we can buy fruit and raw vegetables and homemade potato chips. All come with a lot of very hot chile. We come home from school, and at about 3:00 P.M. we have a big *comida*—that is our dinner. Some of the things we really like are lentil soup, *chiles rellenos,* and fish. Tortillas go with everything. We almost never eat meat because my brother is a vegetarian. At about 8:00 P.M. we have *cena,* which is like a snack before bed.

Tarahumara woman selling baskets, Sonora State

If you came to visit me, I would show you the Zócalo—our town plaza, or central square. I could also take you to the pyramid at Cholula. It was the biggest thing ever built in the Americas before the Spaniards got here! There is a Spanish church at the very top. I'd also take you to see a lot of churches decorated with gold. If we had time, we'd go to the beach in Veracruz, a really pretty town about three hours away.

Natural Environments

READ TO DISCOVER

1. What are the main landforms of Mexico?
2. What climates, biomes, and natural resources does Mexico have?

DEFINE

isthmus
sinkholes

LOCATE

Mexican Plateau
Sierra Madre Oriental
Sierra Madre Occidental
Sierra Madre del Sur
Valley of Mexico
Mexico City
Orizaba
Isthmus of Tehuantepec
Yucatán Peninsula
Baja California

WHY IT MATTERS

Mexico shares a long border with the United States. As more industries have built factories along this border, many people have raised concerns about environmental damage. Use CNNfyi.com or other **current events** sources to learn about pollution along the border.

Mexico: Physical-Political

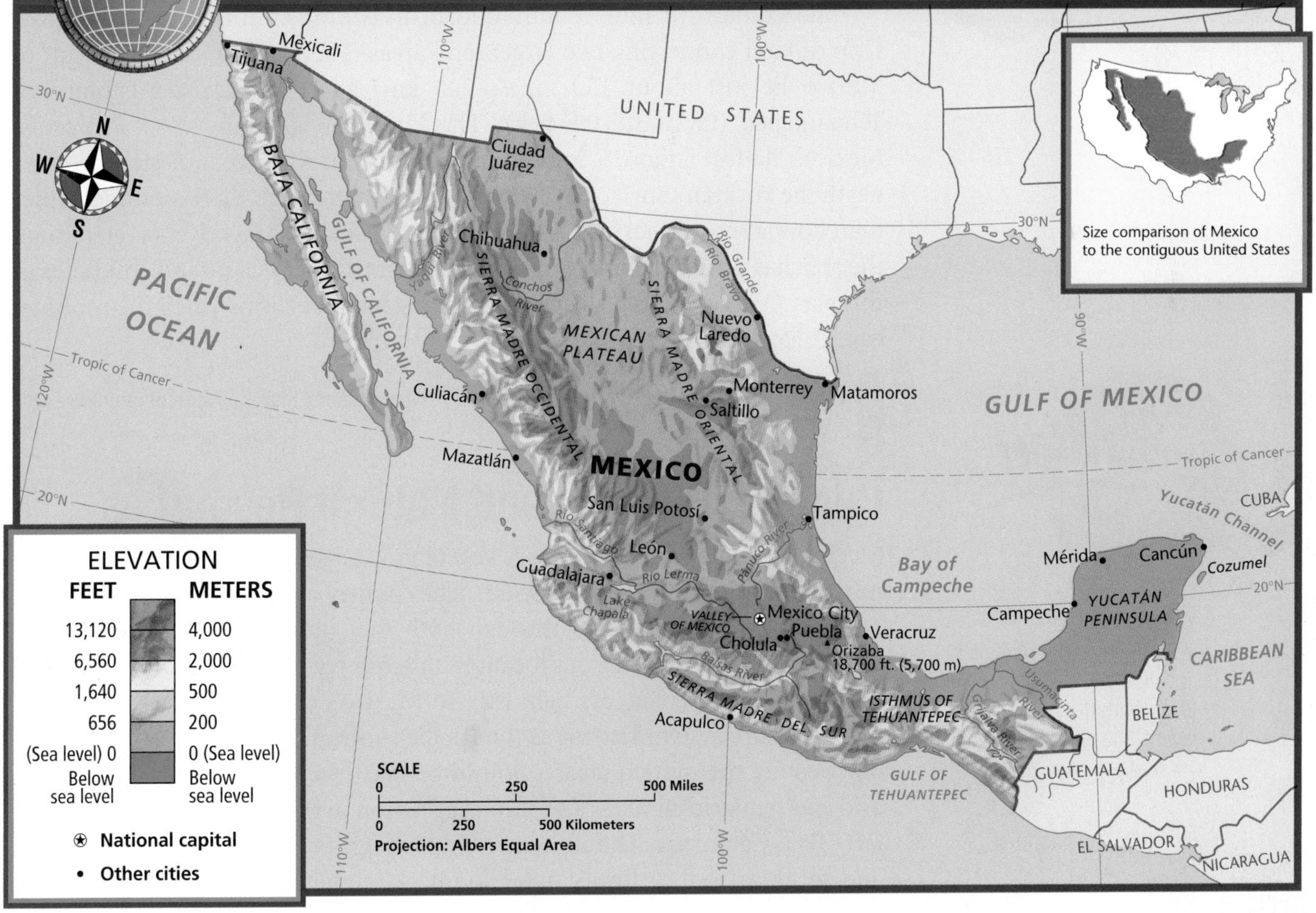

Size comparison of Mexico to the contiguous United States

ELEVATION FEET	METERS
13,120	4,000
6,560	2,000
1,640	500
656	200
(Sea level) 0	0 (Sea level)
Below sea level	Below sea level

⊛ National capital
• Other cities

SCALE: 0 – 250 – 500 Miles; 0 – 250 – 500 Kilometers
Projection: Albers Equal Area

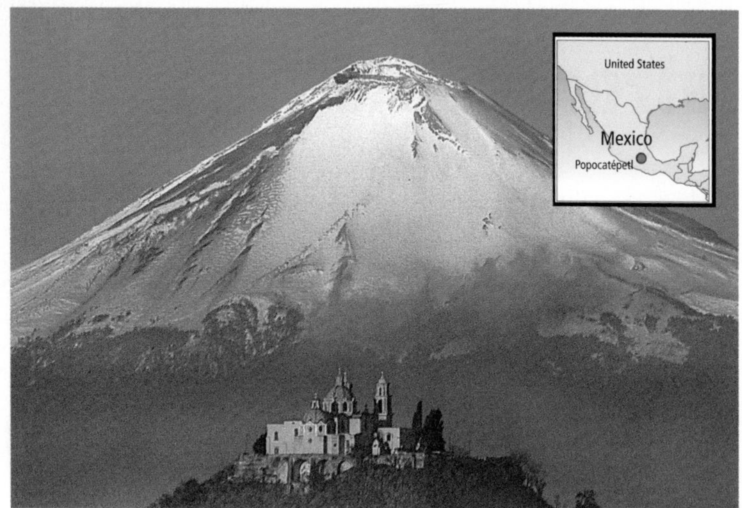

Popocatépetl, an active volcano, rises behind an ancient pyramid at Cholula and the church at its summit. More than 20 million people live within 50 miles of the volcano.

Landforms

Mexico is a large country, almost three times the size of Texas. Most of the country is made up of a rugged central plateau, called the Mexican Plateau. In places this plateau is as high as 9,000 feet (2,700 m). Three great mountain ranges border the Mexican Plateau. The Sierra Madre Oriental is in the east and the Sierra Madre Occidental in the west. Along the southern Pacific coast is the Sierra Madre del Sur. Coastal plains separate the mountain ranges from the sea. The eastern coastal plain is generally wider than the plain along the Pacific coast.

At the southern end of the Mexican Plateau lies the Valley of Mexico, where Mexico City is located. The floor of this broad valley is about 7,500 feet (2,280 m) above sea level. Until they get used to the altitude, visitors to the city may feel a shortness of breath and lack of energy due to the thin air. The mountains southeast of Mexico City include great volcanoes. The highest, Orizaba, soars to 18,700 feet (5,700 m). Because the area is tectonically active, earthquakes are also common.

In southern Mexico, the landforms become more complex, with many small mountain ranges, narrow valleys, and volcanoes. The rugged terrain makes overland travel difficult. Many villages are connected only by single-lane roads. Similar landscapes continue southward into Central America.

Mexico narrows in the south to form an **isthmus**. An isthmus is a narrow strip of land connecting two larger land areas. The Pacific Ocean and Gulf of Mexico lie just about 150 miles (240 km) apart at Mexico's Isthmus of Tehuantepec (tay-WAHN-tah-pek). The Yucatán (yoo-kah-TAHN) Peninsula, Mexico's flattest region, is located in southeastern Mexico. Limestone lies beneath the Yucatán's surface. Water tends to drain through this limestone, rather than flow across the surface. As a result, the Yucatán has few rivers even though the area has a humid climate. Erosion has created many caves and **sinkholes** in this area. A sinkhole is a steep-sided depression that forms when the roof of a cave collapses.

GO TO: go.hrw.com
KEYWORD: SW3 CH10
FOR: Web sites about Mexico

READING CHECK: *Physical Systems* What landform makes up most of central Mexico?

Climate, Biomes, and Natural Resources

Mexico's climate varies by region. Vegetation ranges from desert plants in the north to tropical forests in the south. The country also has a variety of natural resources. Mexico stretches through both the subtropical and tropical latitudes. Partly as a result of this location, the country has a range of tropical and dry climates. Three factors help explain Mexico's climates. First is a regional high-pressure system known as the Pacific subtropical high pressure cell. The dry weather this system creates dominates northwestern Mexico. The result is arid and semiarid climates and scrub vegetation in the north and west, particularly Baja California. The influence of this high-pressure system extends east and south as well. It limits the amount of rainfall over about two thirds of

Every fall, thousands of monarch butterflies migrate to the mountains of Mexico. Some of the butterflies travel more than 1,800 miles (2,900 km) to their winter home.

Mexico's land area. Therefore, dry grasslands and brush cover much of northern Mexico's plains.

The northeast tradewinds are the second factor affecting Mexico's climates. These winds bring humid air from the Gulf of Mexico and the Caribbean Sea. As a result, rains sweep into Mexico mostly from the east and southeast, particularly during the summer. In fact, the forested plains of southeastern Mexico have a tropical humid climate. The easterly winds also steer hurricanes toward Mexico's east coast. Hurricanes also strike the Pacific coast but are less common there.

Finally, climates vary dramatically with elevation. Mexico's highest levels of rainfall occur where the humid trade winds rise against the mountains of the southeast. This also leads to dry rain-shadow climates on the western sides of the mountains. Many people live in the mild environments of the mountain valleys. The valleys along Mexico's southern coast also have pleasant subtropical climates. The high elevations on the Mexican Plateau can have cool highland climates. During the winter, cold polar air sometimes flows southward across the Mexican Plateau. Snow can fall in the Sierra Madre Occidental. This cold air brings freezing temperatures as far south as Mexico City.

Mexico does not have many major rivers. Its largest rivers drain areas of central Mexico. These rivers provide the country with hydroelectric power and water for irrigation.

Mexico has many mineral resources. Centuries ago silver was the country's most valuable mineral product. Mexico remains the world's leading silver producer. The country also produces many other metals, including gold, iron, lead, and mercury.

Today petroleum is Mexico's most valuable natural resource. Mexico's great oil and natural gas fields lie along the Gulf of Mexico. Most of Mexico's oil is exported to the United States.

INTERPRETING THE VISUAL RECORD

Cuatro Ciénegas ("Four Marshes") is a unique area of sparkling water in the Chihuahua Desert of northern Mexico. These spring-fed pools shelter many species found nowhere else. Many other desert-dwelling species depend on the water. The horned lizard (inset photo) lives in the region's dry areas. ***Which mountain range do you think appears in the photo?***

TEKS **READING CHECK:** *Physical Systems* What three factors influence Mexico's climates? What is Mexico's most important natural resource?

Section 1 Review

TEKS Questions 1, 2, 3, 4, 5

Define
isthmus
sinkhole

Working with Sketch Maps
On a map of Mexico that you draw or that your teacher provides, label the Mexican Plateau, Sierra Madre Oriental, Sierra Madre Occidental, Sierra Madre del Sur, Valley of Mexico, Mexico City, Orizaba, Isthmus of Tehuantepec, Yucatán Peninsula, and Baja California. Where are most of Mexico's oil deposits located?

Reading for the Main Idea
1. *Physical Systems* What three mountain ranges border the Mexican Plateau?
2. *Physical Systems* Why does the Yucatán Peninsula have sinkholes?

Critical Thinking
3. **Analyzing Information** Why are the western sides of Mexico's mountains drier than the eastern sides?
4. **Drawing Inferences and Conclusions** What geographical factors do you think might benefit industrial growth in Mexico?

Organizing What You Know
5. Copy the following graphic organizer. Use it to describe Mexico's physical geography.

Landforms	Climate	Resources

2 History and Culture

READ TO DISCOVER

1. What were the cultures of Mexico like before the Spanish arrived?
2. How did Spanish control change Mexico?
3. What has Mexico's history been like since independence?

DEFINE

conquistadores
haciendas
plaza
mestizos
dictator

LOCATE

Acapulco
Cancún
Mazatlán

WHY IT MATTERS

For most of the 1900s, a single political party ruled Mexico. Recently Mexico's government has become more democratic. Use CNNfyi.com or other **current events** sources to find out about changes in Mexican politics.

Early Mexico

Mexico's early peoples belonged to many cultures, each with its own language. Some people were hunter-gatherers. Others were farmers. Their main crops included beans, corn, peppers, and squash. Farmers grew these crops together in the same plots. The bean plants climbed the corn stalks, while the squash and peppers grew between the corn plants. This combination maintained soil fertility and also provided a complete diet. This style of cultivation, called milpa, is still widespread today. In addition, corn tortillas and beans flavored with peppers are still bases of Mexican cooking.

Some of the American Indian peoples of what are now Mexico and Central America created highly complex and accomplished civilizations. These peoples included the Maya, Olmec, Toltec, and Zapotec. Many had large city centers with splendid avenues, plazas, and pyramids. The last of these civilizations was the Aztec. Skilled in warfare, the Aztec built an empire in what is now

INTERPRETING THE VISUAL RECORD

Palenque (puh-LENG-kay), in Chiapas State, was one of the main Maya cities. The building shown is Palenque's palace complex, which includes a four-story observatory. By watching the stars and planets, the Maya were able to devise a complex and accurate calendar. ***Examine the unit and chapter maps. How might the natural environment of the Chiapas area have affected Maya culture?*** TEKS

central and southern Mexico. Their splendid capital city, Tenochtitlán (tay-nawch-tee-TLAHN), occupied an island in a lake in the Valley of Mexico. When the Spaniards arrived, this city was one of the largest in the world. (See Cities & Settlements: Mexico City.)

READING CHECK: *Places and Regions* What were the main crops that farmers grew in ancient Mexico?

Our Amazing Planet

Before the Spanish conquest, the people of Mexico enriched their diets with protein-rich grasshoppers, larvae, locusts, and worms. Now these foods are popular in some Mexican restaurants.

The Colonial Period

In 1519 a band of Spanish adventurers landed on the eastern coast of Mexico. As they traveled inland, these **conquistadores** (kahn-kees-tuh-DAWR-ez), or conquerors, formed crucial military alliances with peoples who resented the Aztec. The Spaniards also had muskets and horses, which were unknown in the Americas at that time.

New diseases, such as smallpox, arrived along with the Spaniards. The American Indians had no resistance to these European diseases, which spread rapidly through their population, killing many. The high death rate weakened the Aztec Empire. This weakness, along with the Spaniards' other advantages, helped the small Spanish band capture the Aztec capital. With the city's fall, the Aztec Empire ended. The conquerors called their colony Nueva España, or New Spain. Colonists would build Mexico City on the ruins of Tenochtitlán.

Desire for gold and silver had been a major motive for Spain to colonize the Americas. The Spaniards expanded the existing mining operations. Gradually, agriculture became an important part of the colonial economy as well. The Indians had mostly owned and worked the land in groups. Lands they worked in common were called *ejidos* (e-HEE-thos). The Spanish organized these lands into **haciendas** (hah-see-EN-duhs). Haciendas were large estates usually owned by wealthy families but worked by many peasants, usually Indians.

Roman Catholic missionaries tried to convert the American Indians to Christianity. They established frontier outposts called missions. Towns often grew up around these churches. The open space, or **plaza**, in front of the church might become a center for the community market. The plaza is a common feature in towns throughout middle and South America, Spain, and other parts of southern Europe.

INTERPRETING THE VISUAL RECORD

This painting from 1579 shows the initial meeting in 1519 of Hernán Cortés, a Spanish conquistador, and Moctezuma, the Aztec ruler. Cortés took Moctezuma prisoner and conquered the Aztec lands. ***How can you tell from the picture that the relationship between the two peoples had changed over time?*** TEKS

Oaxacan families decorate graves on the Day of the Dead. On this holiday, which combines Christian and pre-Christian traditions, Mexicans remember their dead and celebrate the continuity of life.

Today most Mexicans are Roman Catholic. However, over time Christianity in Mexico has changed. In many cases pre-Christian beliefs and holidays combined with Christian beliefs. These distinctively Mexican traditions continue into the present.

Throughout Mexico's history the Spanish and American Indian cultures mixed. Early in the colonial period, most colonists were men, and marriage with American Indian women was common. Today the majority of Mexicans are **mestizos**, people of mixed European and Indian ancestry.

TEKS **READING CHECK:** ***Human Systems*** What factors helped the Spanish conquer the Aztec?

Mexico since Independence

The recent history of Mexico has been marked by rapid, sometimes violent, political change. Economic and cultural change there has been just as dramatic.

Independence and Revolution In 1810, Mexicans began to revolt against Spanish rule. Fighting continued until 1821, when Mexico finally won its independence. Although Spanish administrators returned to Spain, little really changed in Mexico. A few powerful families still controlled the economy and the government. In 1848 Mexico lost its northern territory from Texas to California following a war with the United States. Still, American and European investments during the late 1800s did lead to economic growth for Mexico. New mines were developed, and railroads were built. Modern industries grew in cities. Plantation agriculture expanded along the eastern coast. These plantations were large estates farmed by workers who lived on the property.

However, while a few Mexicans became rich, most remained poor. This economic inequality led to the Mexican Revolution, which lasted from 1910 to 1920. Following the revolution, the new government took the outward form of a democracy. The reality was much different. The president, who ruled much like a **dictator**, held almost all of the power. A dictator is a leader who rules with almost absolute authority. Mexico's government became much more involved in the national economy. It even began to directly control some industries. As a result, many foreign-owned businesses, particularly oil companies, were forced out of Mexico.

One result of the revolution was land reform. Large haciendas were broken up and given to peasant villages according to the old *ejido* system. Over time about half of Mexico's farmland became *ejidos*. While politically popular, the new land system was not very successful. The land given out was often poor and the plots small in size. Few farmers had the money to buy fertilizers and modern machinery. Many discouraged farmers moved to the cities. In 1992, *ejido* farmers won the right to sell their land, and many did. Some farmland has been combined into larger commercial farms. Some land near towns has become suburban housing tracts.

TEKS **READING CHECK:** ***Human Systems*** What kind of government resulted from the Mexican Revolution?

Connecting to ANTHROPOLOGY

Rites of Passage

Societies hold ceremonies, such as weddings, to mark the times when its members move from one stage of life to another. Anthropologists call these ceremonies rites of passage. In Mexico, the *quinceañera* (keen-se-ahn-YE-rah) celebrates a girl's fifteenth birthday. It also marks the end of her childhood. A *quinceañera* typically begins with a Catholic Mass and ends with a big fiesta. The event can be similar to a wedding in terms of preparation, special clothes, food, and costs. *Quinceañeras* may date back to an old Aztec ceremony that stressed to young women the importance of following society's rules.

Making Generalizations and Predictions How might *quinceañeras* contribute to a society's unity? 

Modern Mexico Since about 1990, Mexico has again opened its economy to foreign businesses. Mexico's factories turn out nearly all the products a large country needs. The country, once largely rural, is increasingly urban. In fact, three quarters of Mexicans now live in towns and cities. Many work in industrial and service jobs.

In 1992 Mexico joined Canada and the United States in signing the North American Free Trade Agreement, or NAFTA. This agreement lowered trade barriers between the three countries. Since the door to markets in the north opened wider, manufacturing in Mexico has expanded even more.

Tourism has become increasingly important to Mexico. Resort cities on the east and west coasts draw tourists from around the world. Popular spots include Acapulco, Cancún, and Mazatlán. Beach attractions, grand hotels, crafts, and restaurants are central to these cities' economies. Mexico City, old colonial towns, and dramatic scenery draw visitors also to the country's interior.

Daily life in Mexico is changing rapidly. Like their American neighbors, most Mexicans watch television and shop in modern stores. Traffic jams clog the cities during rush hours. Mexican family life is also changing. Families tend to be smaller now, with only two or three children. More and more women are working outside the home. Many more Mexicans are graduating from universities. Protestant churches are attracting new members, which challenges the traditional role of the Roman Catholic Church. Mexican politics have become more democratic than ever before. Even Mexican Indians, who for a long time had little influence on the government, are becoming participants in politics. While not fully economically developed, Mexico is a powerful partner in North American affairs.

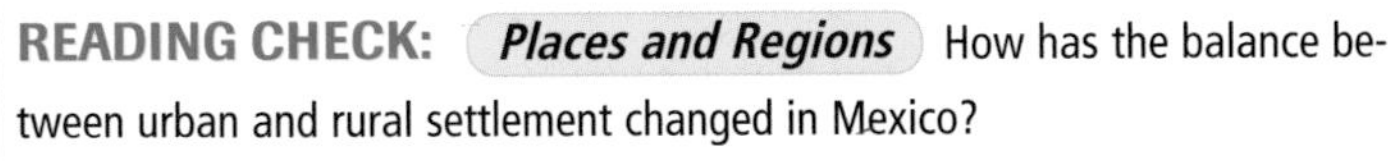

READING CHECK: *Places and Regions* How has the balance between urban and rural settlement changed in Mexico?

NAFTA Trade Flows

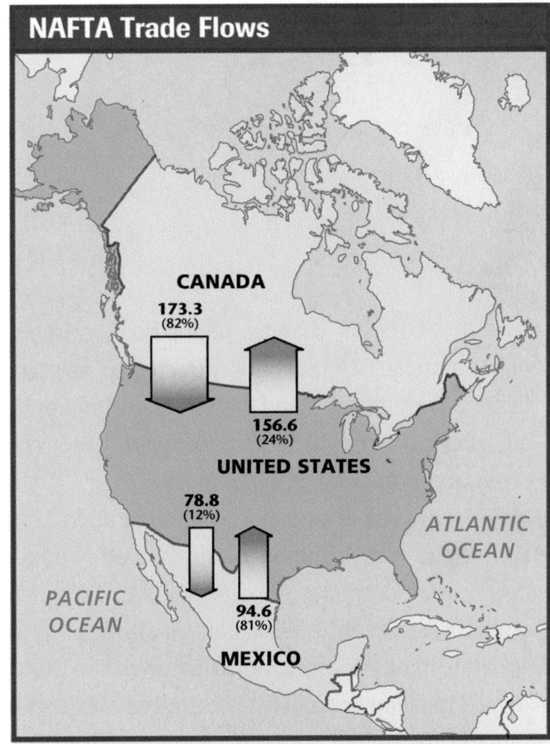

INTERPRETING THE MAP *This map shows trade among the United States and its NAFTA partners in billions of U.S. dollars. Numbers in parentheses show the percentage of each country's total exports. Trade among all three countries has greatly increased since NAFTA went into effect in 1994.* ***Which country does more business with the United States? Which country sends the smaller percentage of its exports to the United States?*** TEKS

TEKS Working with Sketch Maps, Questions 1, 2, 3, 4, 5

Section 2 Review

Homework Practice Online

Keyword: SW3 HP10

Define conquistadores, haciendas, plaza, mestizos, dictator

Working with Sketch Maps On the map you created for Section 1, label Acapulco, Cancún, and Mazatlán. Which city is on the Caribbean coast?

Reading for the Main Idea

1. *The Uses of Geography* What powerful empire ruled much of what is now Mexico at the time of Spanish arrival?
2. *Human Systems* What were the two most important economic activities of colonial Mexico?
3. *Human Systems* What agreement has lowered trade barriers between Mexico and the rest of North America? When was it signed?

Critical Thinking

4. **Analyzing Information** In what ways did Spanish colonization shape Mexican culture? How did precolonial beliefs shape the practice of Christianity in Mexico?

Organizing What You Know

5. Create a time line of Mexican history. Use it to list key events from the time of Spanish arrival to the present.

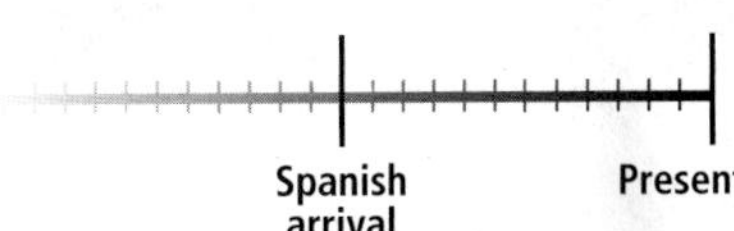

Geography for Life

Mexican Migration

Migration changes populations, cultures, and economic systems. Migrants are attracted to new areas by pull factors, such as jobs and better opportunities. They leave areas because of push factors, like war, changing economies, droughts, or poor living conditions.

Mexico has a long history of migration to the United States. The idea of going north for better opportunities is deeply rooted among Mexican youths, particularly in rural areas of west-central Mexico. In the 1870s large numbers of Mexican migrants went north to work. In the early 1900s American landowners actively recruited Mexicans to work on farms in the United States. However, the Great Depression in the 1930s temporarily slowed migration. Later, an agreement between the U.S. and Mexican governments encouraged further migration of temporary farmworkers from Mexico from 1942 to 1964. The Immigration Act of 1990 greatly increased the number of immigrants allowed to enter the United States. As a result immigration, both legal and illegal, increased.

More people emigrate from Mexico than from any other country. Some 7 million people now living in the United States were born in Mexico. The major cause of migration between the two countries is the difference in wages. Simply put, Mexican workers can earn more money in the United States than they can in Mexico. Many travel to the United States temporarily to work and send money to relatives in Mexico.

The Nogales border crossing connects Arizona and Sonora.

This migration has created a unique cultural landscape along the U.S.-Mexico border. Many Mexican Americans live along the border in Texas, New Mexico, Arizona, and California. Many border towns are linked economically, and border crossings between the two countries are among the busiest in the world. This benefits the economies of both countries. The border region has developed a unique culture that blends American and Mexican ways of life. In fact, many geographers recognize the U.S.-Mexico border area as a distinct region. This region is sometimes called the borderlands.

However, it is estimated that about 2 million Mexicans are living in the United States illegally. In recent years, the United States has stepped up its border patrols and tried to block illegal immigration. Many people caught by patrols and returned to Mexico will try to cross the border again. Many Mexican migrants suffer mistreatment from "guides" called coyotes whom they pay to lead them across the border. Some illegal immigrants try to enter the United States in remote desert areas. Unprepared for the harsh conditions, some have died.

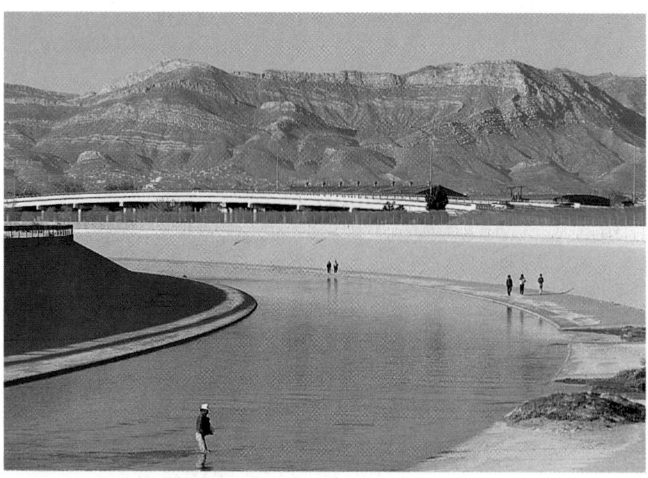

INTERPRETING THE VISUAL RECORD *Mexicans wade across the shallow Rio Grande in the El Paso, Texas, area. Here the river has often cut new channels, sometimes changing the country in control of hundreds of acres.* ***From what you can see in the photo, how have Mexico and the United States improved their ability to control their territory?*** TEKS

Applying What You Know

1. **Summarizing** How has migration from Mexico to the United States shaped the distribution of culture groups today?
2. **Drawing Inferences and Conclusions** What are some of the political, economic, social, and environmental factors that contribute to Mexican migration to the United States?

Section 3

Mexico Today

READ TO DISCOVER

1. What are the economic and cultural regions of Mexico?
2. What challenges face Mexico?

DEFINE

cash crops
maquiladoras

WHY IT MATTERS

Economic connections with Mexico are very important to the United States. Use CNNfyi.com or other **current events** sources to learn about recent economic developments in Mexico and how they affect this country.

LOCATE

Guadalajara
Campeche
Tampico
Veracruz
Monterrey
Tijuana
Ciudad Juárez

This cat figure from the state of Oaxaca is an example of Mexican folk art.

Mexico's Regions

Mexico is divided into 31 states and the capital district. Different parts of the country exhibit great geographical, economic, and cultural diversity. It is useful to divide the country into four regions for study.

Greater Mexico City Greater Mexico City is the cultural, economic, and political center of Mexico. This huge metropolis includes many smaller cities and may hold a fourth of Mexico's entire population. (See the graph on the next page.) It also generates much of the country's GDP.

Mexico City has monumental government buildings. It is also home to the country's largest university and greatest museums and theaters. The

The States of Mexico

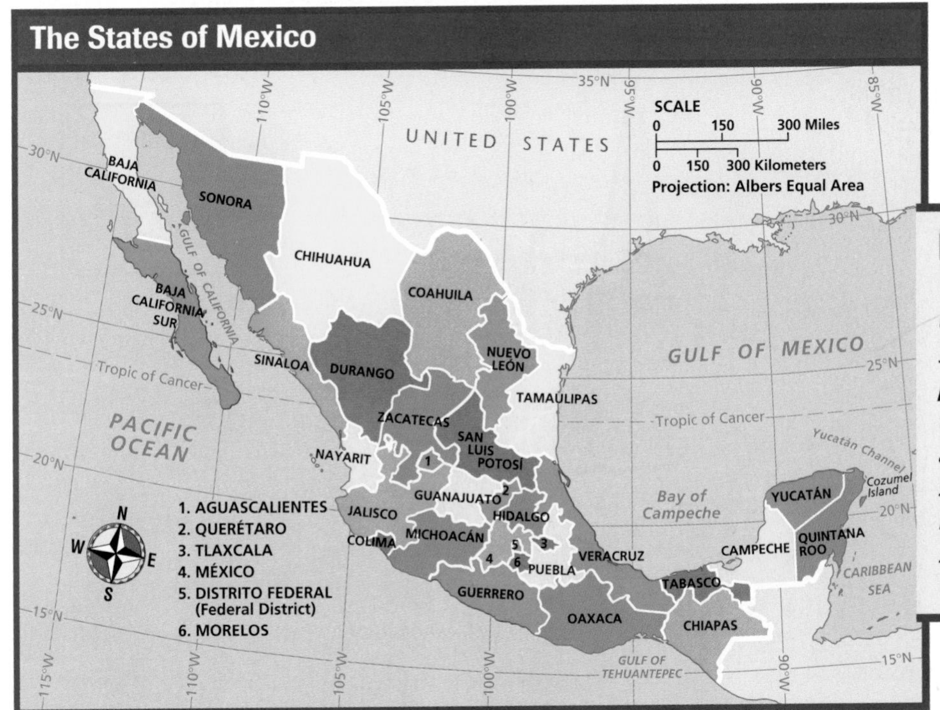

INTERPRETING THE MAP

The official name of Mexico is Estados Unidos Mexicanos, or United Mexican States. ***Compare this map to the physical map. What features seem to form natural borders for Sinaloa and Veracruz? Why might the smaller states be in south-central Mexico? What area has a political status similar to that of Washington, D.C.?*** TEKS

Mexico City's Population

Population (in millions)

25
20
15
10
5
0

1930 1960 1990

Year

Source: *Latin America and the Caribbean*, Blouet and Blouet, eds.

INTERPRETING THE GRAPH

Mexico City has experienced phenomenal growth. ***Approximately how many times larger was the 1990 population than the 1930 population? What do you predict for Mexico City's future growth?*** TEKS

headquarters of leading industries and banks, as well as the Mexican stock exchange, are located there. Millions of tourists come to Mexico City to see Aztec ruins and grand old colonial buildings. Major industries include cement, chemicals, construction, plastics, textiles, and tourism.

Mexico City's economic activity also draws many new residents. However, there are too few jobs. As a result, some people live in huge settlements of shacks built from waste wood and sheetmetal. Many people live without electricity, sewers, or a safe water supply.

Alongside this poverty, great wealth also exists in Mexico City. Exclusive boutiques and world-class restaurants attract tourists and upper-class residents. Districts with grand homes and luxury apartments extend for miles. At the city's center is the Zona Rosa, a stylish urban area.

Mexico City suffers from terrible air pollution, however. The city is located in a broad valley ringed by mountains. The mountains trap the pollution of thousands of factories and millions of cars. The government is trying to clear the air by reducing these emissions.

READING CHECK: ***Places and Regions*** What are some of Mexico City's leading industries?

Central Mexico Central Mexico stretches northwest of Mexico City and across the Mexican Plateau. Many cities there began as colonial mining or ranching centers. Mexico's second-largest city, Guadalajara, is located in this area. Guadalajara contains many buildings from the Spanish colonial period. Small towns with city squares and colonial churches are common.

Fertile valleys dot central Mexico. This was once colonial Mexico's great grain-producing region. Agriculture there is a mix of small family farms and medium-sized commercial farms. These farms grow a number of **cash crops**. Cash crops are crops grown for sale in a market. Trucks bringing fruits and vegetables from the area to the United States often jam border crossings. In recent years central Mexico has also been attracting new factories.

Construction of Guadalajara's immense cathedral began when the city was founded in 1542. It was not finished until the early 1700s. After being severely damaged by earthquakes, the old towers were replaced by the yellow-tiled spires that have become the city's symbol.

Gulf Lowlands and Southern Mexico Throughout much of Mexican history, the Gulf lowlands between the cities of Campeche and Tampico were lightly settled. People living there used the region's hot humid tropical forests and savannas for grazing or growing sugarcane. Now large forest areas have been cleared for commercial farming and ranching. The region also includes Veracruz, an important seaport and communications center.

In the 1990s armed rebels from Chiapas, in southern Mexico, called for a redistribution of wealth. In this photo, rebellion supporters carry an image of Emiliano Zapata, who supported land reform during the Mexican Revolution of 1910 to 1920.

Rich deposits of oil and natural gas have long been key to this region's economy. The area is booming because of these deposits. With the development of the oil industry have come oil refineries, pipelines, petrochemical complexes, port facilities, and fertilizer plants.

Southern Mexico includes the mountainous areas south of Mexico City and the plains of the Yucatán Peninsula. This is Mexico's poorest region. It has few cities and little industry. In addition, transportation and telephone service are poorly developed. Schools are inadequate. More and more migrants head north hoping for a better life, either in Mexico City or in the United States.

Southern Mexico is also Mexico's most traditional region. Village life there has changed little over the last hundred years. Subsistence agriculture is common, and handicrafts provide much of the cash income. Mexican Indians make up about half of this area's population. Many speak Indian languages. For example, many of Yucatán's rural people speak Mayan.

TEKS **READING CHECK:** ***Human Systems*** In what ways is southern Mexico the country's most traditional region?

Northern Mexico The large dry region of northern Mexico has become one of the most prosperous parts of the country. Much of the region's infrastructure is new and modern. The roads are good, and the telephones work well.

Monterrey is the great industrial city of the north. However, many other cities and towns have also industrialized and grown rapidly. Northern Mexico's factories and commercial farms draw migrant workers from all over Mexico. Cattle ranching, mining, and tourism are also important to northern Mexico's economy.

INTERPRETING THE VISUAL RECORD

Pinacate National Park is among the many scenic attractions in northern Mexico. The region contains both bustling cities and sparsely populated desert areas. ***From what you see in the photo, what are some of the physical processes that have affected northwestern Mexico in the past?*** TEKS

Signs in the small border town of Nuevo Progreso, Tamaulipas, advertise various services. Many tourists take day trips from South Padre Island or Brownsville, Texas, to Nuevo Progreso.

FOCUS ON GEOGRAPHY

Border Towns The part of Mexico along the border definitely "faces north" in more than one sense. Many businesses have links with the United States. In fact, American companies own many of the special factories, called **maquiladoras** (mah-kee-lah-DOHR-ahs), that lie along the Mexican side of the border. These factories employ hundreds of thousands of Mexicans. Workers there assemble products for export, mostly to the United States. These products range from auto parts to toys. In addition, irrigated farms provide fruits and vegetables to markets in the United States and Europe during winter.

The border region also has many cultural links to the United States. American music, television, and other forms of entertainment are popular. The Spanish spoken in this part of Mexico is full of English words. Pairs of towns that straddle the border are beginning to function more like single cities. For example, Tijuana is increasingly linked to San Diego, California. Every day thousands of people move back and forth between cities like Ciudad Juárez and El Paso, Texas.

READING CHECK: ***Places and Regions*** What links with the United States are found along Mexico's border region?

U.S.–Mexico Border Region

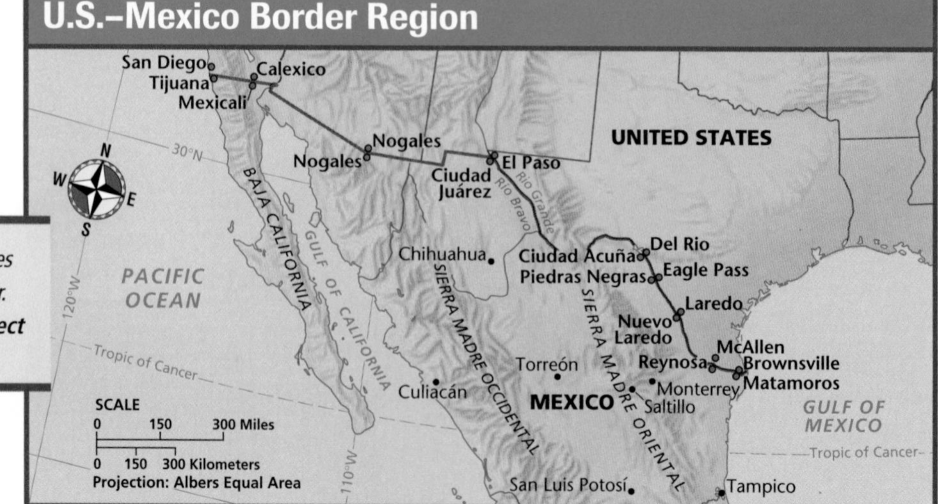

INTERPRETING THE MAP *Pairs of cities and towns lie along the U.S.-Mexico border.* ***How do some of the cities' names reflect their locations?***

Challenges for the Future

Mexico is changing rapidly. Mexican politics are becoming more democratic. To stimulate growth, the country's leaders are reducing government controls on the economy. NAFTA's supporters believe that the treaty will boost economic growth. The Mexican middle class continues to grow. However, problems remain.

Geographers see several interrelated challenges in Mexico. One such challenge is economic inequality. Many Mexicans are poor, and much of Mexico's wealth lies in the hands of a few rich people. In general, Mexican Indians have fewer economic opportunities than other citizens. Reducing poverty might help create greater political stability as well as improve the country's economy. It might also slow migration out of Mexico. Migration to the United States, both legal and illegal, is common. Many Mexicans have achieved success in the United States. Some send money to their families back in Mexico. However, migration also takes skilled workers away from Mexico's economy.

Another challenge is reducing crime, much of which results from widespread poverty. Mexico is a main route for smuggling drugs into the United States. Profits from the drug trade have even tempted some government officials, including police, to break the law. Curbing crime will be key for further economic and social progress.

Finally, improving the country's poor infrastructure presents another challenge. Many Mexican communities do not have clean water supplies or modern sewers. This can cause health problems. Many of Mexico's roads and railways are worn and out-of-date. Mexicans cannot sell their products if their goods cannot get to market. All these are problems that further economic progress may help solve.

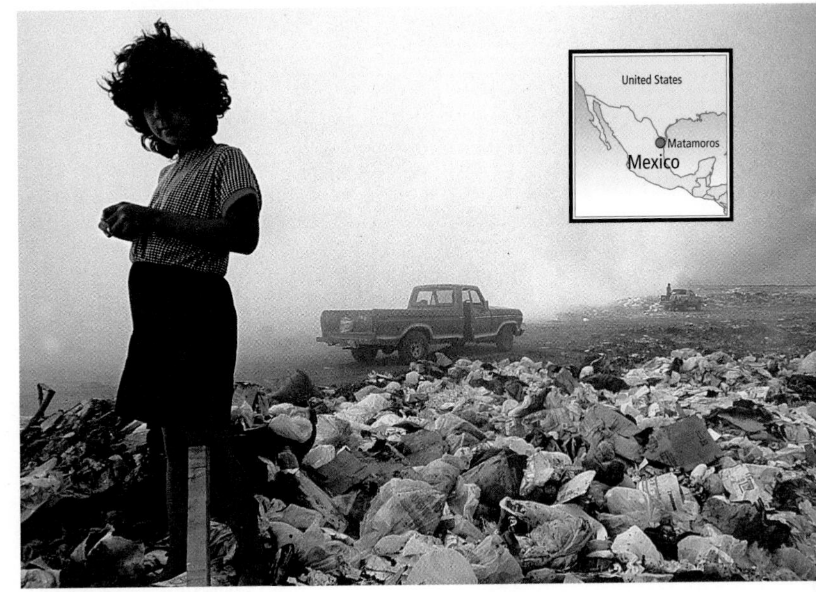

A young girl looks out over the Matamoros city dump. Several families routinely search the garbage for items they can sell or use. Hazardous wastes are often dumped there illegally. Garbage that is burned in the open sends thick smoke over the area. ***How might conditions at this dump affect the water supply?*** TEKS

TEKS **READING CHECK:** *Human Systems* What are some important challenges facing Mexico?

Section 3 Review

TEKS Working with Sketch Maps, Questions 1, 2, 3, 4, 5

go.hrw.com Homework Practice Online
Keyword: SW3 HP10

Define cash crops, maquiladoras

Working with Sketch Maps

On the map you created for Section 2, label Guadalajara, Campeche, Tampico, Veracruz, Monterrey, Tijuana, and Ciudad Juárez. Which cities are located at or near sea level? Which cities are located in mountainous areas or on the Mexican Plateau?

Reading for the Main Idea

1. *Environment and Society* What geographical features make Mexico City's pollution problems worse?
2. *Human Systems* Why is migration to the United States a disadvantage for the Mexican economy?

Critical Thinking

3. **Making Generalizations** In what ways is Mexico both a rich country and a poor country?
4. **Drawing Inferences** How might economic progress help increase political stability in Mexico?

Organizing What You Know

5. Copy the graphic organizer shown below. Use it to list the four economic and cultural regions of Mexico and their major economic activities.

Region	Economic activities
Greater Mexico City	
Central Mexico	
Gulf lowlands and southern Mexico	
Northern Mexico	

Mexico City

Environment and Society

Mexico City is one of the largest cities in the world and one of the oldest cities in the Americas. The city's rich history and tremendous growth have made it Mexico's cultural, economic, and political center as well. However, Mexico City also has serious environmental problems which the city's people are working to solve.

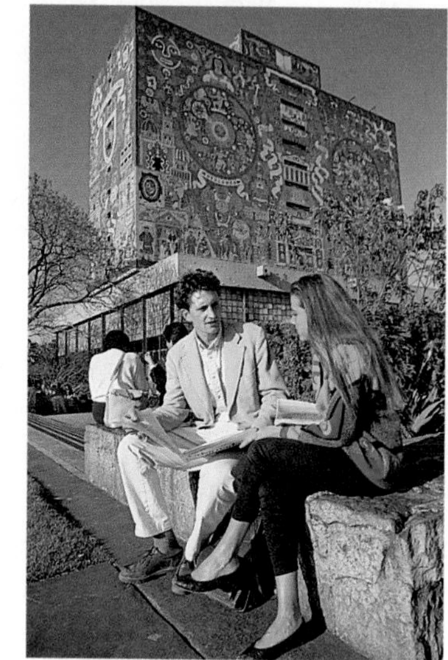

Mosaics decorate the library of the National Autonomous University of Mexico.

An Indian and European City

In the mid-1300s the Aztec founded a town called Tenochtitlán in central Mexico. According to legend, their god advised them to build where they saw an eagle eating a snake while perched on a cactus. This legend is now represented on Mexico's flag. When the Spanish arrived in 1519, Tenochtitlán was home to perhaps 300,000 people. It was larger than any city in Europe at that time. In 1521 the Spanish defeated the Aztec and destroyed their city. On its ruins a new Spanish settlement rose—what became known as the Ciudad de México, or Mexico City.

Today Mexico City is a lively modern city that is proud of its history and culture. One can visit more than 100 museums there. Modern skyscrapers tower over Aztec ruins and Spanish colonial buildings. Few other cities offer such cultural variety.

The historical center of Mexico City is its main square, the Zócalo, where the center of Tenochtitlán once stood. Today it is one of the world's largest public squares. Bordering the Zócalo are some of Mexico City's oldest and grandest buildings. The Metropolitan Cathedral dates to the mid-1500s and is the largest church in Latin America. In the 1700s and 1800s, the block-long National Palace was home to Mexico's rulers. Today it houses government offices. Inside, huge murals by Diego Rivera, one of Mexico's famous artists, trace Mexican history from Aztec times to the early 1900s. On the square's northeast corner are the remains of the Templo Mayor, the great Aztec ceremonial pyramid. Southwest of the Zócalo, Chapultepec Park is one of the world's largest urban parks. The area was once used by Aztec emperors.

Pockets of old historic buildings are scattered throughout Mexico City. Many mark former villages that have been absorbed by the city's rapid expansion. The modern city completely surrounds these old towns. A well-known modern area includes the National University in southern Mexico City. Many of its buildings are covered with murals. The main library is decorated with tile mosaics showing Mexico's history and scientific achievements.

Cars zoom around a traffic circle on Reforma Avenue. The monument honors those who fought for Mexico's independence from Spain.

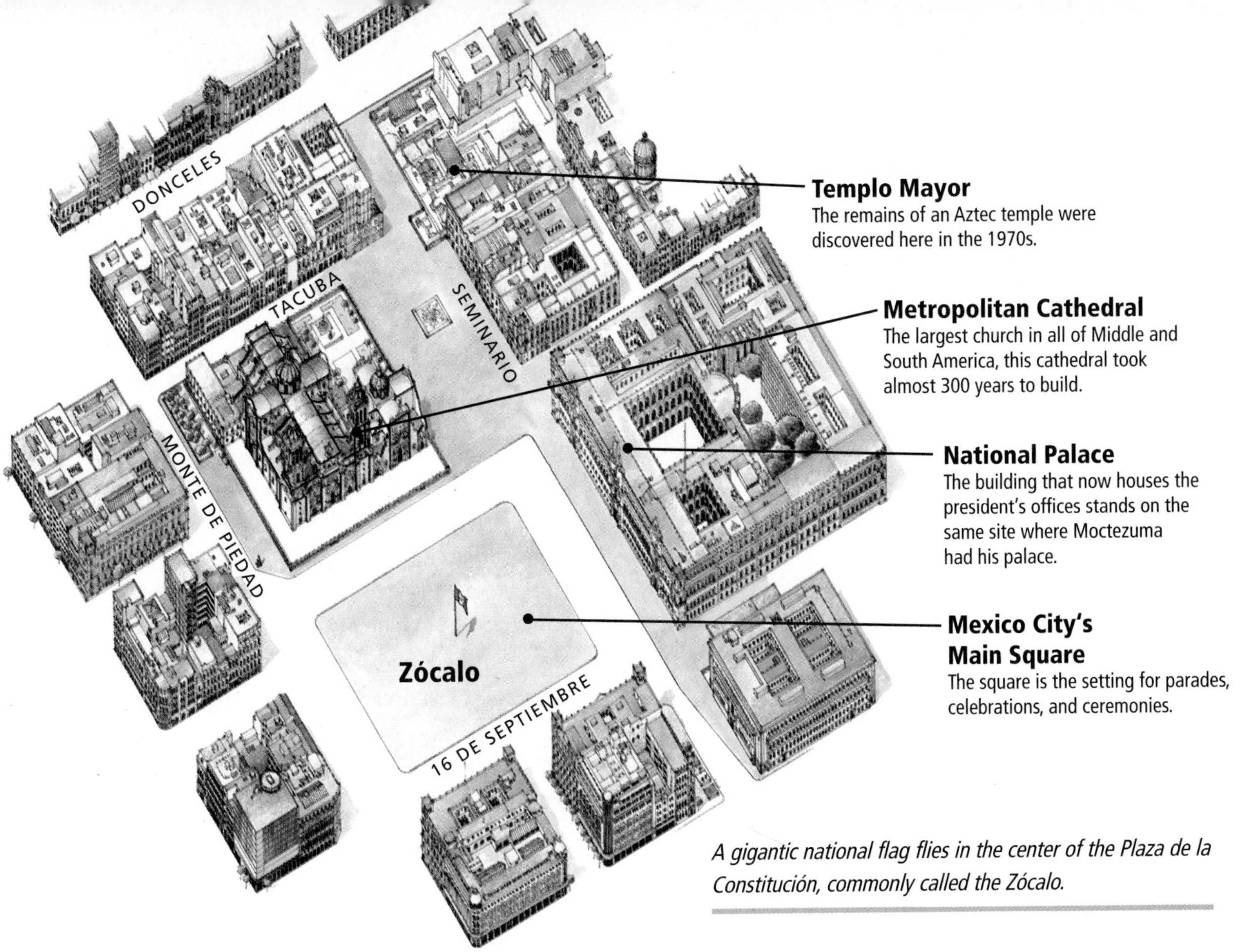

DONCELES
TACUBA
SEMINARIO
MONTE DE PIEDAD
16 DE SEPTIEMBRE
Zócalo

Templo Mayor
The remains of an Aztec temple were discovered here in the 1970s.

Metropolitan Cathedral
The largest church in all of Middle and South America, this cathedral took almost 300 years to build.

National Palace
The building that now houses the president's offices stands on the same site where Moctezuma had his palace.

Mexico City's Main Square
The square is the setting for parades, celebrations, and ceremonies.

A gigantic national flag flies in the center of the Plaza de la Constitución, commonly called the Zócalo.

Environmental Problems and Solutions

Some 18 to 25 million people live in the Mexico City metropolitan area today. By most accounts, Mexico City is now the second-largest urban area in the world, after Tokyo.

Mexico City's huge growing population has caused serious environmental problems including air pollution. Some 4.5 million motor vehicles and numerous factories release chemicals into the air. The resulting smog can cause breathing problems, sting eyes, and burn throats. As you read earlier, the city's location in a basin makes the situation worse. In addition, the city's industries and people use huge amounts of water. Heavy pumping from wells has lowered the water table under the city. This has caused the ground to settle and buildings in many areas to sink. Because the city lies on a dry lake bed the soils underneath are soft and unstable. In some parts of downtown, buildings have sunk as much as 29 feet (9 m) since 1900. The ground beneath the Metropolitan Cathedral has shifted so much that the building has cracked down the middle.

In 1985 a major earthquake damaged much of Mexico City. This experience raised awareness of the city's pressing environmental needs. As a result, work began in 1991 to save the Metropolitan Cathedral. Private companies have spent $300 million to repair other old buildings. Many have been turned into much-needed apartments.

Efforts were also launched to improve the city's air quality. Many factories were forced to close or cut production. In the late 1980s city officials developed a system to restrict the use of cars. Now cars with certain license plate numbers can only be driven on specific days. Since 1991 all new cars must have anti-pollution devices. Other laws ban nearly half of the city's older cars from the streets on "emergency days," when pollution levels are high. These measures seem to be working. In 1999 Mexico City experienced only 5 such days, compared to 95 of them in 1994.

Applying What You Know

1. **Summarizing** How has Mexico City changed since it was founded in the mid-1300s?

2. **Comparing** How do environmental changes in Mexico City compare to changes in large U.S. cities you have read about, such as Los Angeles?

CHAPTER 10 Review

Building Vocabulary TEKS

On a separate sheet of paper, explain the following terms by using them correctly in sentences.

isthmus	haciendas	dictator
sinkholes	plaza	cash crops
conquistadores	mestizos	maquiladoras

Locating Key Places TEKS

On a separate sheet of paper, match the letters on the map with their correct labels.

Sierra Madre Oriental	Baja California
Sierra Madre Occidental	Acapulco
Mexico City	Monterrey
Yucatán Peninsula	Ciudad Juárez

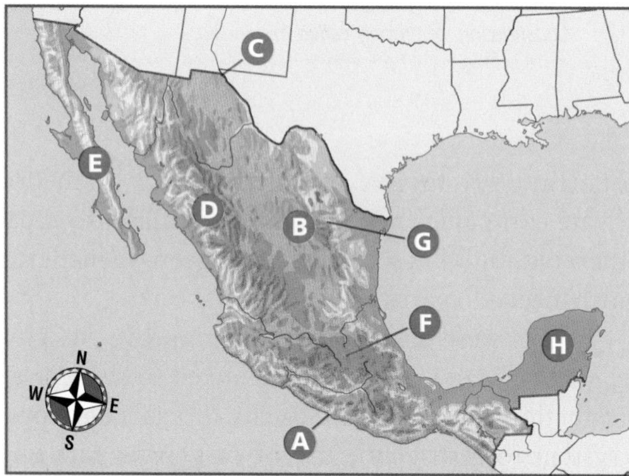

Understanding the Main Ideas 

Section 1

1. *Physical Systems* Why does northwestern Mexico have dry weather?

Section 2

2. *Human Systems* What peoples ruled Mexico before the arrival of Spaniards?
3. *Human Systems* Why is Roman Catholicism the most common religion in Mexico?

Section 3

4. *Places and Regions* In what ways is Greater Mexico City important to the entire country?
5. *Human Systems* How have Mexican politics changed in recent years?

Thinking Critically for TAKS TEKS

1. **Drawing Inferences and Conclusions** Review what you read about the resort cities on the east and west coasts of Mexico. How might "resort Mexico" be considered a functional region? How might it be a perceptual region?
2. **Comparing** How is Mexico City's status in Mexico similar to that of New York, Los Angeles, and Washington, D.C., in the United States? What problems might these cities share?
3. **Making Generalizations and Predictions** How might migration from Mexico shape cultural change in the United States?

Using the Geographer's Tools TEKS

1. **Analyzing Graphs** Look at the graph of Mexico City's population. If growth continued at the same pace, what do you think the city's population might be in 2020? What factors might affect the future growth of Greater Mexico City?
2. **Preparing Maps** Draw a sketch map of Mexico and its individual states. Then use the chapter map to locate and label the main cities. Use maps you find in library resources to check if you placed the cities correctly within the states. Revise your map if necessary.
3. **Preparing Maps** Create a map of the region near the border of Mexico and the United States. Use this textbook and other sources to identify Mexican and U.S. towns that lie across the border from each other. How might the economic and cultural features of Mexican border towns be different from cities farther south? What links would these border towns have with U.S. cities that other Mexican cities probably do not?

Writing for TAKS TEKS

Write the text of an imaginary interview with a Mexican farmer who grows one or more cash crops. Include the farmer's viewpoint on the importance of trade with the United States and Canada. How does lifting trade barriers affect this farmer?

SKILL BUILDING

Geography for Life

Illustrating Geographic Information

Human Systems Create a poster or sketch map illustrating the four regions of Mexico. Use pictures and captions to describe the main cultural, economic, and natural features of these regions. Include information about the relationships among these features.

Building Social Studies Skills

Interpreting Graphs

Use the information from the bar graph below to answer the questions that follow.

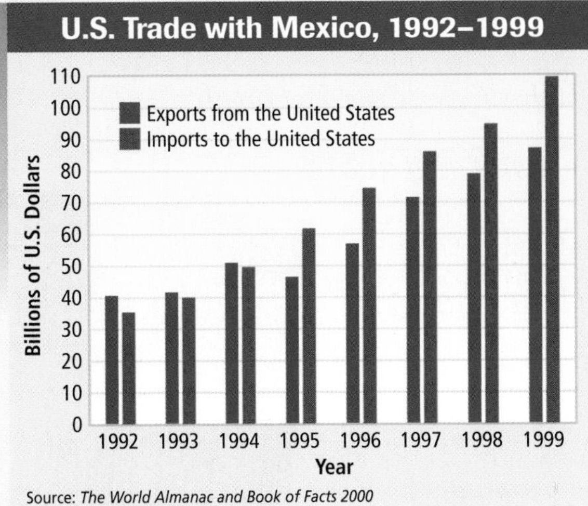

1. U.S. exports to Mexico increased every year except which one?
   a. 1993
   b. 1994
   c. 1995
   d. 1996
2. The North American Free Trade Agreement (NAFTA) took effect on January 1, 1994. What general changes in U.S. exports and imports can you see since then?

Analyzing Primary Sources

Read Michael Parfit's comments about Mexico. Then answer the questions that follow on a separate sheet of paper.

> **"Rough by reputation, exotic even in name, the Sierra Madre Occidental (or Mother Mountains of the West) remains one of the last untrammeled [untamed] wildernesses in Mexico. The northern portion of the 800-mile-long range is famed for its immense, mile-deep gorges—Mexico's answer to the Grand Canyon. Such daunting topography has proved a barrier to settlement for centuries—and a godsend to renegades from Geronimo to Pancho Villa."**

3. What makes part of the Sierra Madre Occidental "Mexico's answer to the Grand Canyon"?
   a. its 800-mile-long length
   b. mile-deep gorges
   c. its tall mountains
   d. a desert climate
4. Why do you think these mountains are both a barrier to settlement and a good place for people trying to hide from authorities?

Alternative Assessment

PORTFOLIO ACTIVITY

Learning about Your Local Geography

Individual Project: Field Work

You have read about efforts to improve air quality in Mexico City. Conduct research to learn about air pollution in your area. Contact your local air quality board, television station weather department, or a local environmental group. Request information on the air quality of your area. If possible, collect data going back several years. Graph the data. What patterns do you see over time?

internet connect

Internet Activity: go.hrw.com
KEYWORD: SW3 GT10 TEKS

Choose a topic on Mexico to:

- take the GeoMap challenge to test your knowledge of geography!
- create a brochure on Mexican holidays and culture.
- learn about the physical features and climate of Mexico's coastlines.

CHAPTER 11 Central America and the Caribbean

Central America and the Caribbean stretch across a large area of land and water between North and South America. Countries there share similar histories and face similar challenges today.

Mola design from Panama

Good morning. I am Aldayne, and I live in Kingston, Jamaica. My house is gray—the door is green and the windows are blue. It is all on one floor, with a yard where we grow roses.

I am a student at Kingley Preparation School, about five or six minutes from my house. I get up at 6:00 A.M. and have a peanut butter sandwich and tea—or sometimes I just have an apple and tea at school. When school is over, I go downtown to my mother's shop—she is a dressmaker. I hang out until 6:00 P.M. when we can go home together. I do my homework and sometimes do errands for her like buying buttons. By the time I get home, all my homework is done. While my mother makes dinner, I clean my shoes, get my clothes together, and get ready for the next day. My favorite dinners are rice and peas with spring chicken, or crispy fried chicken with chips (french fries).

We have three school terms a year starting in September. We go to school for three months and then have one month off. My favorite vacation month is summer when I can go to the country with my father. I go to the river, catch fish, and have cook-outs. When I grow up, I want to be a pilot.

Carnival in Trinidad

Natural Environments

READ TO DISCOVER

1. What physical processes have shaped the landforms of Central America and the Caribbean?
2. What is the climate like in Central America and the Caribbean?
3. What natural resources and environmental hazards are common in the region?

DEFINE

mangrove
bauxite

LOCATE

West Indies
Greater Antilles
Lesser Antilles
Cuba
Hispaniola
Jamaica
Puerto Rico
Bahamas
Martinique
Cayman Islands

WHY IT MATTERS

Central America and the Caribbean experience dangerous natural hazards. Use CNNfyi.com or other **current events** sources to learn about the effects of a recent earthquake, hurricane, or volcanic eruption in the region.

Central America and the Caribbean: Physical-Political

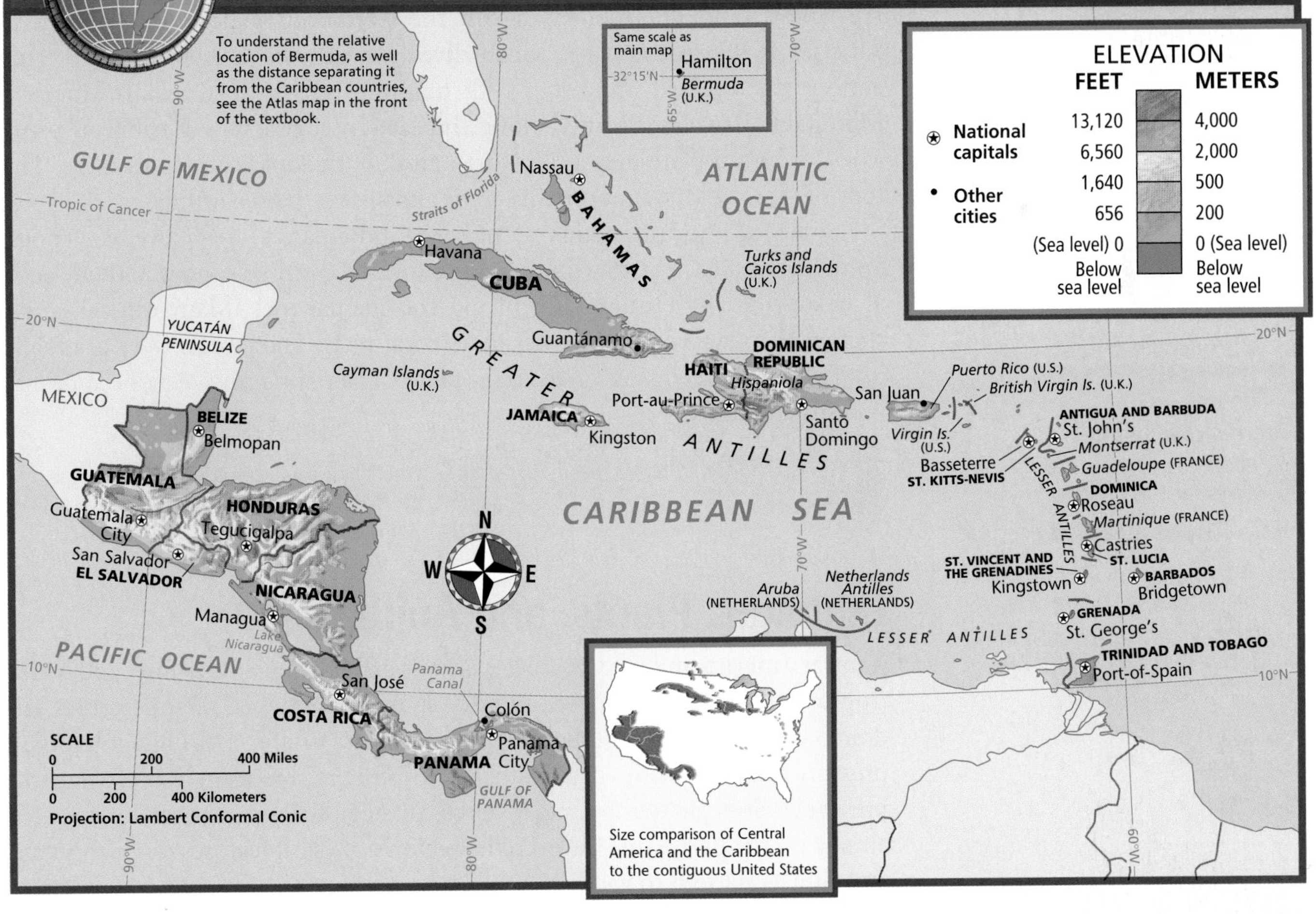

To understand the relative location of Bermuda, as well as the distance separating it from the Caribbean countries, see the Atlas map in the front of the textbook.

Same scale as main map
Hamilton
Bermuda (U.K.)
32°15'N
65°W

ELEVATION

FEET	METERS
13,120	4,000
6,560	2,000
1,640	500
656	200
(Sea level) 0	0 (Sea level)
Below sea level	Below sea level

⊛ National capitals
• Other cities

GULF OF MEXICO
Tropic of Cancer
Straits of Florida
ATLANTIC OCEAN
Nassau
BAHAMAS
Havana
CUBA
Turks and Caicos Islands (U.K.)
Guantánamo
GREATER ANTILLES
DOMINICAN REPUBLIC
HAITI
Hispaniola
Port-au-Prince
Santo Domingo
San Juan
Puerto Rico (U.S.)
British Virgin Is. (U.K.)
Virgin Is. (U.S.)
Cayman Islands (U.K.)
JAMAICA
Kingston
YUCATÁN PENINSULA
MEXICO
BELIZE
Belmopan
GUATEMALA
Guatemala City
HONDURAS
Tegucigalpa
San Salvador
EL SALVADOR
NICARAGUA
Managua
Lake Nicaragua
PACIFIC OCEAN
San José
COSTA RICA
Panama Canal
Colón
Panama City
PANAMA
GULF OF PANAMA
CARIBBEAN SEA
Aruba (NETHERLANDS)
Netherlands Antilles (NETHERLANDS)
LESSER ANTILLES
Basseterre
ST. KITTS-NEVIS
ANTIGUA AND BARBUDA
St. John's
Montserrat (U.K.)
Guadeloupe (FRANCE)
DOMINICA
Roseau
Martinique (FRANCE)
Castries
ST. LUCIA
ST. VINCENT AND THE GRENADINES
Kingstown
BARBADOS
Bridgetown
GRENADA
St. George's
TRINIDAD AND TOBAGO
Port-of-Spain
20°N
10°N
90°W
80°W
70°W
60°W
N
W
E
S

SCALE
0 200 400 Miles
0 200 400 Kilometers
Projection: Lambert Conformal Conic

Size comparison of Central America and the Caribbean to the contiguous United States

Landforms

Central America is an isthmus that links North and South America. The Pacific Ocean lies to the west. To the east is the Caribbean Sea and a group of islands called the West Indies. These islands extend in an arc from just south of Florida to Venezuela. The major island groups of the West Indies are the Greater Antilles and the Lesser Antilles.

The Greater Antilles include the large islands of Cuba, Hispaniola, Jamaica, and Puerto Rico. Northeast of Cuba are some 700 islands that make up the Bahamas. They lie entirely in the Atlantic Ocean. Hispaniola is divided between the countries of Haiti and the Dominican Republic. The Lesser Antilles include more than 20 small island countries and territories. Central America has seven countries. Beginning in the north, these countries are Belize, Guatemala, Honduras, El Salvador, Nicaragua, Costa Rica, and Panama.

Narrow coastal plains are found in much of Central America and the Caribbean. Rugged hills or mountains lie in the interior. The rugged terrain makes travel and communication difficult. The region has few long rivers.

INTERPRETING THE VISUAL RECORD

The beautiful twin peaks near Soufrière, St. Lucia, rise above the island's lush rain forests. ***What physical processes do you think created these peaks?*** TEKS

Tectonic processes have shaped the region. Central America, Jamaica, Hispaniola, and Puerto Rico lie on the Caribbean plate. West of Central America is the Cocos plate, which dives beneath the Caribbean plate. This action has created mountains throughout most of Central America. Mountains have also formed along the eastern edge of the Caribbean plate. The Caribbean plate meets the North and South American plates there. The tops of these volcanic mountains are islands in the Lesser Antilles.

Some of the mountains along plate boundaries are active volcanoes. Some islands, such as Martinique, were formed from volcanoes. Others, such as Hispaniola, have mountains of old continental rock. Many other islands, such as the Cayman Islands, began as coral reefs. These reefs were gradually uplifted by the collision of tectonic plates. Over time, the reefs became flat limestone islands.

internet connect

GO TO: go.hrw.com
KEYWORD: SW3 CH11
FOR: Web sites about Central America and the Caribbean

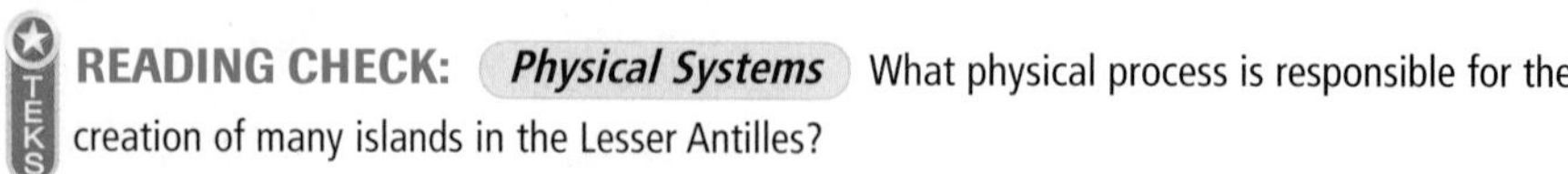
READING CHECK: ***Physical Systems*** What physical process is responsible for the creation of many islands in the Lesser Antilles?

Climates, Plants, and Animals

Central America and the Caribbean islands extend across the sunny and warm tropical latitudes. Tropical wet and dry climates are typical. Temperatures seldom vary more than 10°F between summer and winter. During winter, high pressure generally brings dry weather. A summer rainy season results when low pressure cells begin to move north across the region. Rain can then be expected almost every afternoon. However, the region's physical features cause this general climate pattern to vary.

In Central America, the climate zones follow the terrain. The Caribbean coast gets the full effect of moist trade winds. This results in a tropical humid climate with frequent rain, even in winter. Dense forests are common. The mountainous interior has a cooler highland climate. The eastern side of the mountains gets heavy rain as moist air rises and cools. However, mountain valleys in the west often lie in a rain shadow. These valleys and the western side of the mountains, along the Pacific coast, have drier climates. Scattered trees, shrubs, and tropical grasslands are found here. However, much of this vegetation has been cleared for plantations and ranches.

Throughout the Caribbean, elevation greatly affects climate. Islands with mountains and volcanoes often have heavy rain on their windward side. Rain shadows make other island locations somewhat drier. Lower islands do not produce any orographic effect. The rain that does fall quickly sinks into the limestone bedrock. As a result, the low islands have limited water resources. Bare ground and even cacti are common.

Thickets of **mangrove** trees dominate muddy tropical coastlines in Central America and all around the world. Mangroves are unusual because their roots grow in salt water. Shrimp and dozens of kinds of fish live among these roots. Marine life also thrives in shallow warm water and reefs off the Caribbean coast of Central America and the West Indies.

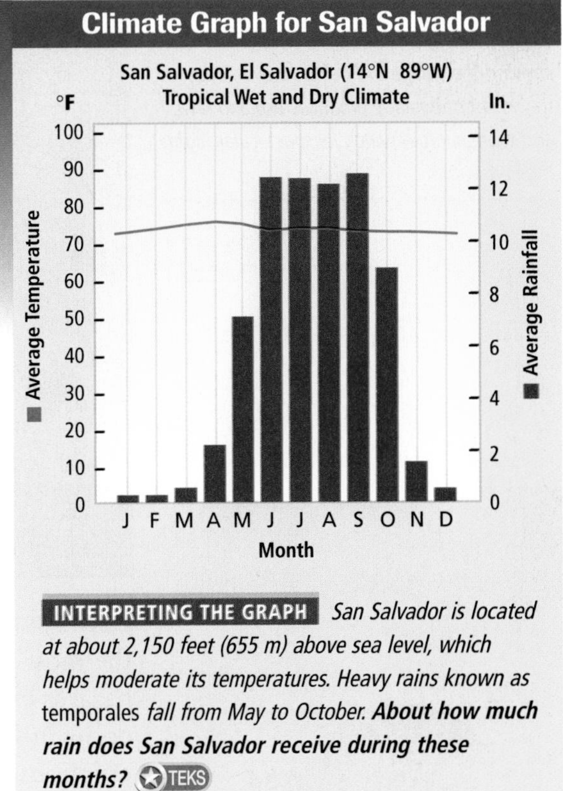

INTERPRETING THE GRAPH *San Salvador is located at about 2,150 feet (655 m) above sea level, which helps moderate its temperatures. Heavy rains known as* temporales *fall from May to October.* ***About how much rain does San Salvador receive during these months?*** TEKS

TEKS **READING CHECK:** ***Physical Systems*** How do mountains affect climates in the region?

Natural Resources and Environmental Hazards

The region's natural resources include its physical beauty. However, this beautiful environment is also known for its dangerous environmental hazards.

Natural Resources One of the region's greatest natural resources is its warm and sunny climate. That climate attracts millions of tourists from the United States, Canada, and Europe. Fertile soils, particularly in highland valleys, are also important natural resources. Rich fishing grounds lie along the coasts. They provide fish, lobsters, shrimp, and other seafood to coastal and island populations.

Mineral resources are found in Central America and in the continental rocks of the Greater Antilles. Small gold fields have been discovered throughout Central America. Jamaica has major deposits of **bauxite**, the ore from which aluminum is made. In addition, Cuba and the Dominican Republic produce nickel. Trinidad has oil.

Blue Hole, a circular limestone sinkhole, lies off the coast of Belize. Many divers come to explore this interesting formation.

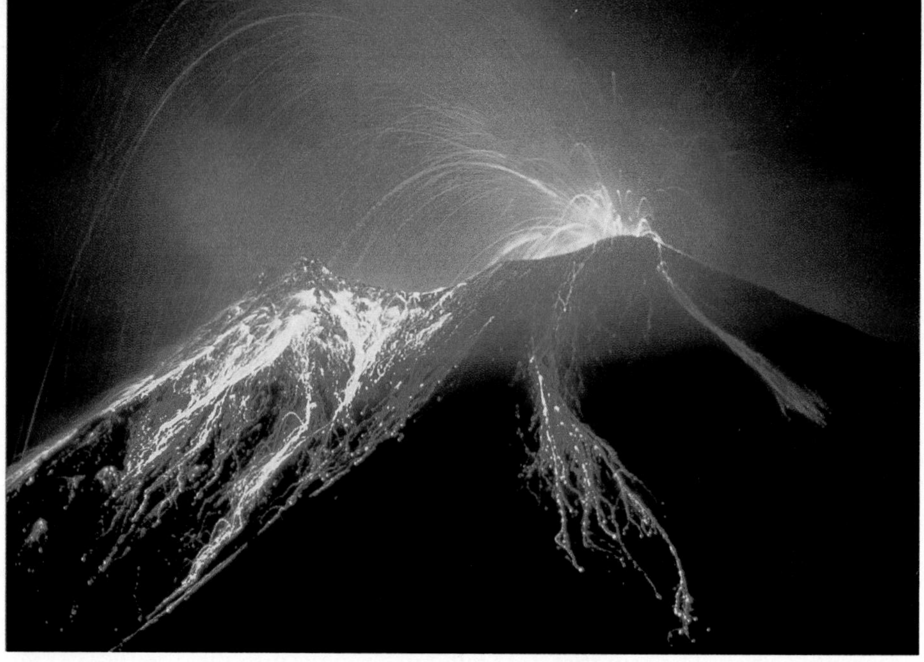

On July 29, 1968, Costa Rica's Arenal Volcano erupted, killing more than 70 people and destroying the village of Pueblo Nuevo. Today the volcano regularly produces ash and lava, and tourists often watch its spectacular nighttime eruptions.

Environmental Hazards Central America and the Caribbean have some of the most beautiful natural landscapes in the world. However, deadly natural hazards exist in the region as well. Earthquakes and volcanic eruptions often occur in Central America and in the Lesser Antilles. These earthquakes and eruptions have caused terrible destruction from time to time. For example, in 2001 two earthquakes just a month apart killed hundreds of people in El Salvador.

Hurricanes are another major hazard in the region. These tropical storms occur mostly in late summer. They bring destructive winds, heavy rains, and flooding. Damage has increased as populations have grown. This is true in the highlands, where farmers have cleared some slopes of forests. The heavy rains cause severe erosion and even mud slides in such areas. These mud slides can destroy homes, roads, and bridges. Crops and farm animals are also lost. In addition, it is difficult for aid to reach people needing food, shelter, and medical help.

In 1835 a powerful volcanic eruption in Nicaragua spread ash as far as Mexico City, more than 850 miles (1,400 km) away.

READING CHECK: *Environment and Society* Why does the clearing of forests make hurricanes more destructive?

TEKS Working with Sketch Maps, Questions 1, 2, 3, 4, 5

Review

Define mangrove, bauxite

Working with Sketch Maps On a map of Central America and the Caribbean that you draw or that your teacher provides, label the West Indies, Greater Antilles, Lesser Antilles, Cuba, Hispaniola, Jamaica, Puerto Rico, Bahamas, Martinique, and Cayman Islands. What parts of the region lie on the Caribbean Plate?

Reading for the Main Idea

1. *Physical Systems* What climate types are found in Central America? What are some factors that influence these climates?
2. *Places and Regions* How and why does the physical geography of Martinique differ from that of the Cayman Islands?

Critical Thinking

3. **Supporting a Point of View** Do you think that the region's climate should be considered a natural resource? Why or why not?
4. **Making Generalizations and Predictions** Where are volcanic eruptions and earthquakes *least* likely to occur: Central America, the Greater Antilles, or the Lesser Antilles? Why?

Organizing What You Know

5. Copy the chart below and use it to identify important landforms, climates, and natural resources in the region.

Landforms	Climates	Resources

Environment and Society

Geography for Life

Montserrat's Soufrière Volcano

You may have seen dramatic movies about volcanoes and other natural disasters that destroy entire cities. In many places around the world, such destruction is not a fictional threat. Take, for example, Montserrat (mawnt-ser-RAHT), a small island in the Lesser Antilles. Montserrat is a possession of Great Britain. On the southern part of the island sits a volcano known as Soufrière (soo-free-ER). For most of Montserrat's history, this volcano has been inactive. Islanders lived a quiet life, and tourism and agriculture provided the basis for the economy.

In 1995, however, the Soufrière Volcano began a series of major eruptions. The eruptions threw superheated clouds of ash, rock, and steam into the air. Debris rained down on the island, and piles of ash collected in many places. Lava and other material raced down the volcano at nearly 100 miles (160 km) an hour, mowing down trees.

As you might expect, the spectacular eruptions completely disrupted life on the island. Thousands of people had to flee their homes and move to the northern part of the island. In the evenings, they listened to radio updates on the volcano's daily activity. Hotels, restaurants, and other businesses closed their doors. The capital of Plymouth, located near the volcano, was abandoned and was later buried in mud and ash.

The government set up an exclusion zone on the southern half of the island. Authorities limited access to this zone to short trips. However, eventually some people stayed in the exclusion zone for longer time periods. That was a mistake. In 1997 an eruption killed at least 19 people. Eventually, some two thirds of the population fled the island. Most migrated to other Caribbean islands, Britain, or Canada.

The eruption of the Soufrière Volcano buried Montserrat's capital, Plymouth, in mud and ash.

Montserrat

INTERPRETING THE MAP *The eruption of the Soufrière Volcano forced the evacuation of more than half of Montserrat.* ***Why do you think the eruption of the volcano disrupted life on so much of the island?***

Applying What You Know

TEKS

1. **Summarizing** How have the eruptions of the Soufrière Volcano affected the physical and human geography of Montserrat?
2. **Problem Solving** What, if anything, do you think should be done to make Montserrat safe for its inhabitants?

Section 2

Central America

READ TO DISCOVER

1. How does Central America's history continue to shape the region today?
2. What economic, political, and social conditions exist in the region?

DEFINE

indigenous
mulattoes
cacao
ecotourism

LOCATE

Panama Canal
Colón
Panama City

WHY IT MATTERS

Much of the food that we eat and clothing that we wear comes from Central America. Use CNNfyi.com or other **current events** sources to learn about goods imported to the United States from Central America.

History and Culture

Together, the seven countries of Central America are only about three fourths the size of Texas. However, the combined population of these small countries is almost double that of Texas. Their colonial history continues to shape the culture of these countries today.

Colonization and Independence Nearly all of Central America's native, or **indigenous**, peoples were farmers when Spanish explorers arrived in the early

The Maya civilization included parts of present-day Guatemala, Belize, and Honduras. Tikal, in northern Guatemala, was a major Maya settlement that flourished between about A.D. *600 and 900.*

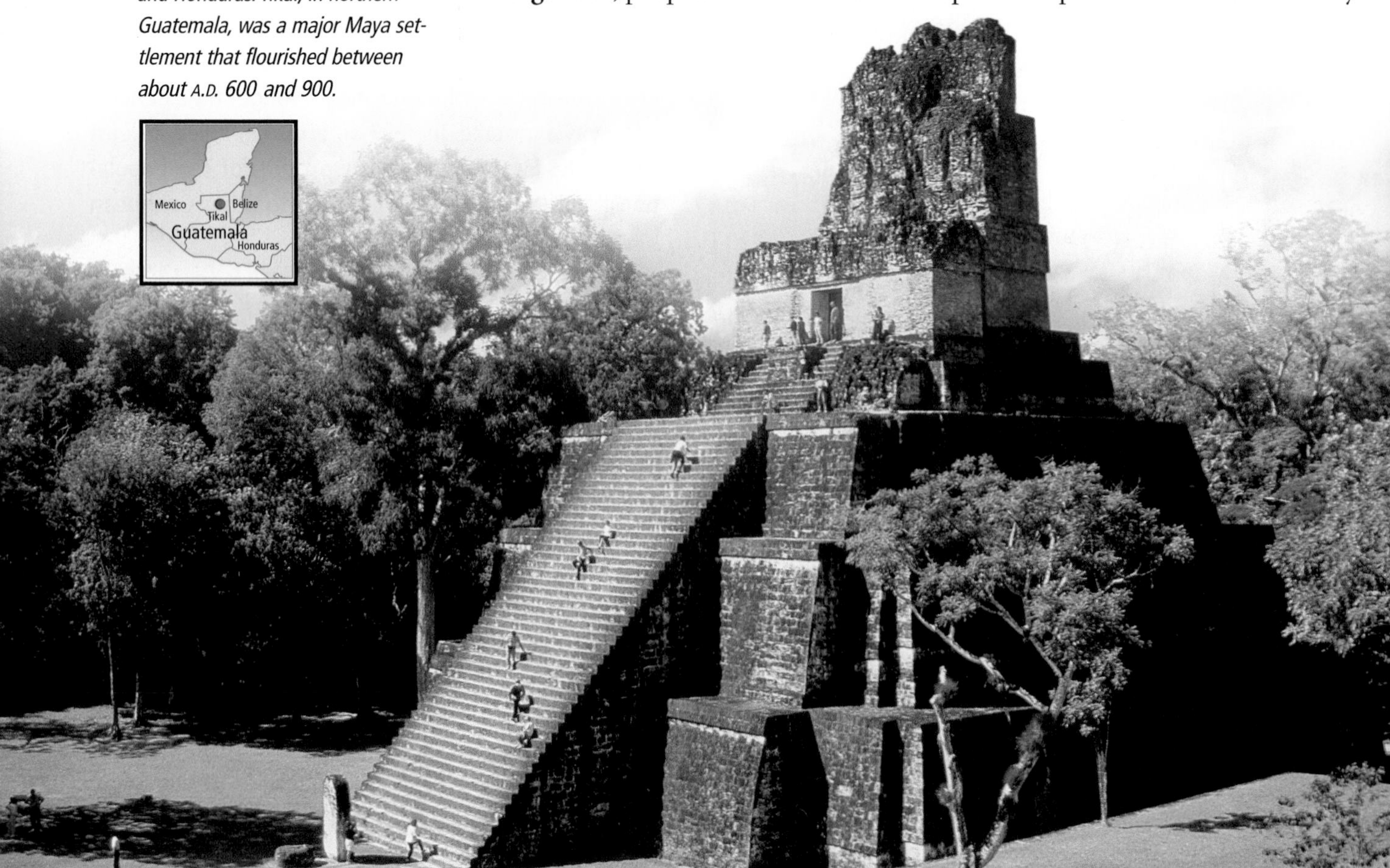

1500s. Over time, the Central American Indian population declined dramatically. One of the causes of this decline was the spread of European diseases. In addition, the Spaniards enslaved many Indians.

Early Spanish settlement spread from the Pacific coast into the highlands. Settlers built towns around central plazas and Roman Catholic churches. Climates along the Pacific were compatible with Spanish-style agriculture, and large estates developed. A few rich families owned the land, and most of the workers had no land rights. The Spaniards mostly ignored the Caribbean coast. British Honduras (what is now Belize) became the only non-Spanish colony in Central America. Eventually, Europeans brought enslaved Africans from the Caribbean islands to the area.

INTERPRETING THE VISUAL RECORD

Coffee is one of Guatemala's main export crops. It is grown mostly on large plantations in cool highland areas. ***To which countries do you think Guatemala exports coffee?***

Independence came to Spanish Central America in the 1820s. Little changed, however. Spanish officials left, but wealthy families continued to run the countries and their economies. Foreign companies, mainly from the United States and Great Britain, built railroads to cross the isthmus. Coffee plantations were founded along with railroads and became very important to the region. Bananas also became an important commercial crop. Large American firms controlled the banana business.

In the early 1900s the United States built the Panama Canal across central Panama. This canal has been an important economic resource. It allows ships to move from the Atlantic Ocean and Caribbean Sea to the Pacific Ocean. The United States controlled the canal until turning it over to Panama in 1999.

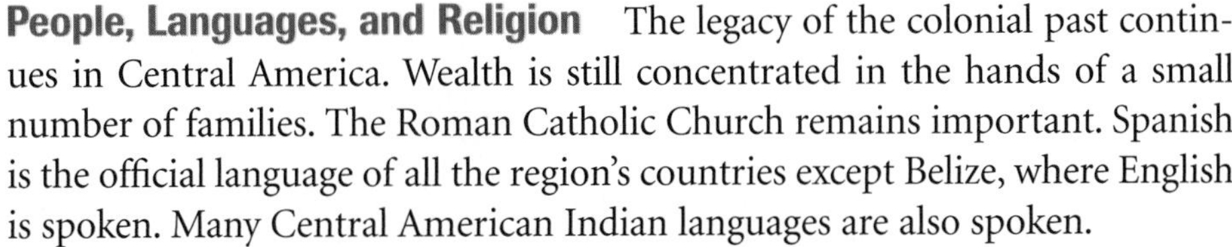

TEKS **READING CHECK:** ***Human Systems*** Which European country had the greatest influence on Central America's development?

People, Languages, and Religion The legacy of the colonial past continues in Central America. Wealth is still concentrated in the hands of a small number of families. The Roman Catholic Church remains important. Spanish is the official language of all the region's countries except Belize, where English is spoken. Many Central American Indian languages are also spoken.

The majority of Central Americans are mestizos. Some are **mulattoes**, people with both African and European ancestors. Small groups of Asians and Africans also live in the region. These groups are largely descended from laborers brought to the region to work on the plantations.

Most of the region's Central American Indians live in Guatemala. That country's population is almost evenly split between mestizos and Indians. The other countries have much smaller Indian populations. The populations of El Salvador, Honduras, Nicaragua, and Panama are overwhelmingly mestizo. Indians live on the Caribbean sides of Nicaragua and Panama. People of African descent are a significant group in Belize and Panama.

Costa Rica's people are mostly of Spanish descent. Few Central American Indians lived in the area when European colonists first arrived. As a result, the colonists did not have local labor to develop large estates. Instead, small family farms developed in this country.

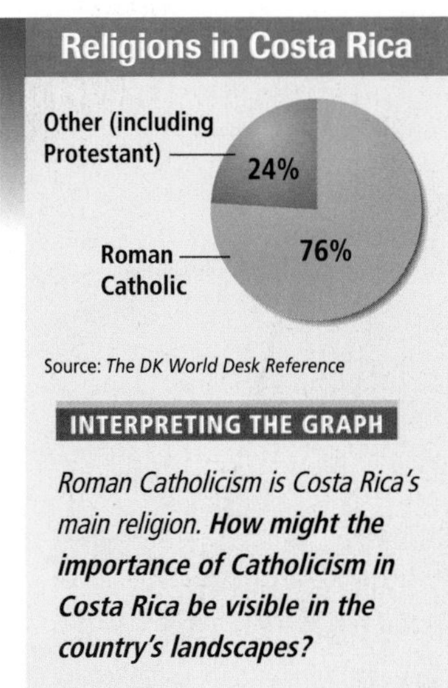

INTERPRETING THE GRAPH

Roman Catholicism is Costa Rica's main religion. ***How might the importance of Catholicism in Costa Rica be visible in the country's landscapes?***

Our Amazing Planet

The Belize Barrier Reef is the largest coral reef in the Western Hemisphere, extending for more than 180 miles (290 km) along the country's coast. The reef is home to manatees, crocodiles, sea turtles, and more than 500 species of fish.

Belize was the last to gain independence in 1981. People of African descent who speak English live along the coast. The inland forests include both Indians and Spanish-speaking settlers from Mexico and Guatemala.

READING CHECK: ***Human Systems*** What ethnic group has shaped the social character of Central America and makes up a majority of its population today?

Economic, Political, and Social Development

Central America continues to depend heavily on the export of coffee and bananas. Sugar, cotton, and **cacao** are also important commercial crops. Cacao is a type of tree from which we get cocoa beans. Those beans are then used to produce chocolate.

The influence of American and other foreign companies on these and other industries remains strong. In most countries, wealthy families have long had ties to foreign companies and to their own country's army. Such ties have often enabled these families to run their countries for their own benefit. However, population growth has strained this arrangement. As the people's need for land and demand for more political power have increased, unrest and violence have arisen. Many Central Americans have immigrated to the United States to escape these economic and political problems.

Reform and Development The need for land reform has been an important problem in the region. In El Salvador, for example, rich landowners control most of the land. They raise cash crops on large profitable estates. However, most Salvadorans are poor subsistence farmers. They survive by raising corn and beans on small farms. Similar inequalities have troubled

INTERPRETING THE VISUAL RECORD

Subsistence agriculture is common in the highlands of El Salvador. ***Based on the photo, what type of economic system do you think El Salvador has?*** TEKS

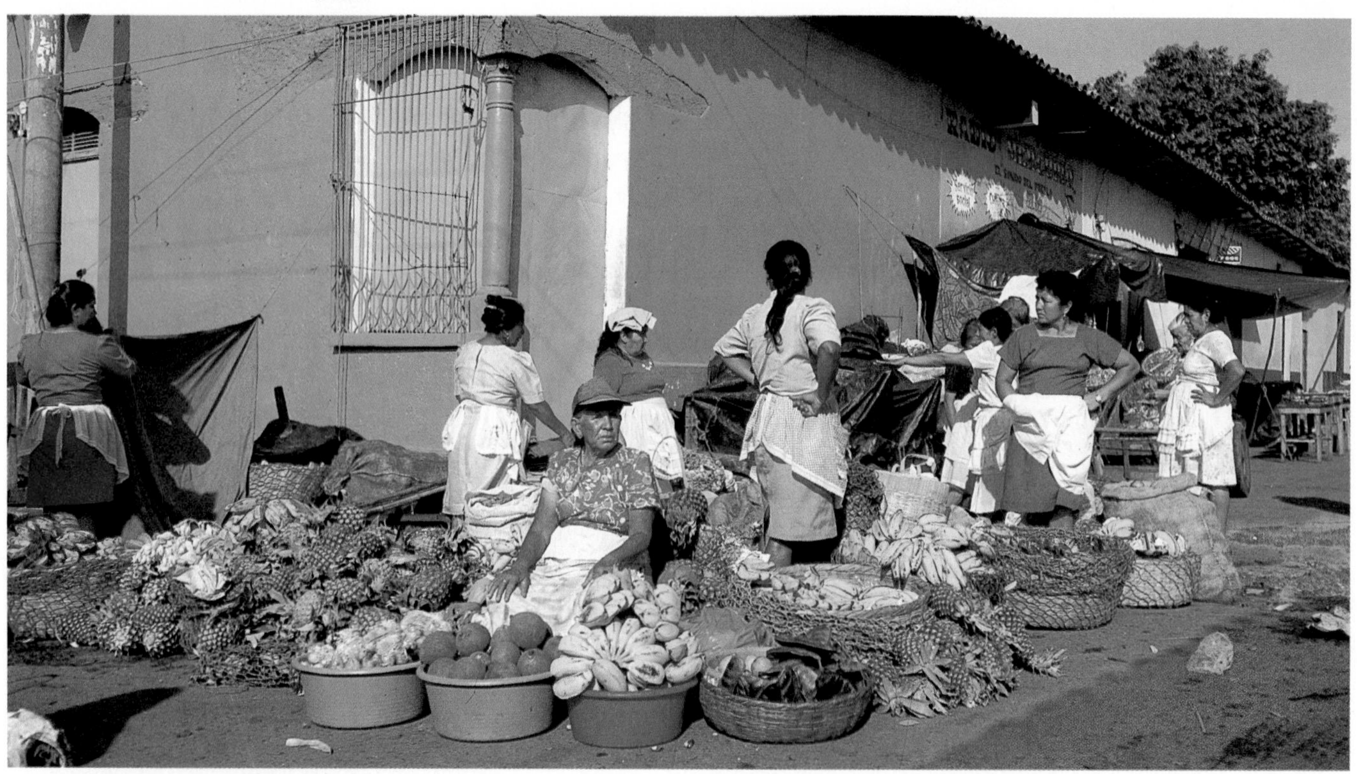

Nicaragua. As a result, both countries suffered through long periods of violence and civil war, particularly in the 1970s and 1980s. With these countries now at peace, their governments have a chance to build fairer societies. In fact, El Salvador and Nicaragua have both made important economic progress in recent years.

Guatemala, which has also suffered through long periods of violence and unrest, has attempted land reform. Market-oriented agriculture exists along the country's Pacific coast. Meanwhile, more than half of all Guatemalans live and farm in isolated highland villages. In nearby market towns, they sell what few goods they do not consume. Migrants to the country's northern lowland plains have found some useful land there. However, these migrants are burning forest areas, including national parkland, to clear land for farming.

INTERPRETING THE VISUAL RECORD

The Panama Canal, competed in 1914, is one of the most strategic artificial waterways in the world. ***How do you think the Panama Canal caused changes in world trade patterns?*** TEKS

To some, Panama seems like three different countries in one. In the east, toward South America, is a densely forested region with few people. In the middle lies a more prosperous area surrounding the Panama Canal. The cities at each end of the canal—Colón on the Caribbean and Panama City on the Pacific—are major industrial centers. The country's western areas are more rural, and both small farms and large plantations are common.

Coffee is particularly important to the economies of Honduras, Guatemala, Panama, and Costa Rica. Rugged mountains and valleys dominate Honduras. The rough terrain makes transportation and large-scale farming difficult. As a result, economists often regard Honduras as Central America's least-developed country. Costa Rica, on the other hand, has the highest standard of living in the region. Costa Rica's tradition of democracy, education, and political stability has recently attracted investments by foreign computer companies. Computer chips are now a major export. Costa Rica has also been a leader in developing tourism.

TEKS **READING CHECK:** ***Human Systems*** Which country has the highest standard of living in the region? What are some reasons for this?

A hiker in Costa Rica

FOCUS ON ECONOMICS

Ecotourism in Costa Rica One key to Costa Rica's successful tourism industry is the country's natural beauty. Would you pay to sit on a beautiful white-sand beach, where the rain forest extends nearly to the water's edge? Would you pay to walk in a misty mountain forest, surrounded by exotic birds? These are just two of many opportunities to get truly close to nature in Costa Rica.

INTERPRETING THE VISUAL RECORD

Visitors to Monteverde cloud forest in Costa Rica can explore the area's incredible diversity of plant and animal life by walking across suspension bridges located high in the rainforest canopy. ***How can ecotourism help promote economic development and environmental conservation at the same time?***

In the mid-1980s, Costa Rica found a way to make money and protect its natural environment at the same time. Its solution was **ecotourism**. This type of tourism focuses on guided travel through natural areas and on outdoor activities. It allows visitors to observe wildlife and learn about the environment. Today protected public or private nature preserves make up nearly 25 percent of Costa Rica's area. For a fee, people can tour these preserves with a naturalist who is an expert on the area's animals, plants, and physical geography.

Ecotourism has provided tremendous economic benefits to Costa Rica. Hotels and other businesses have sprung up around the preserves, providing jobs for the local people. The cost of tours also provides income for the economy. Today tourism is Costa Rica's leading industry. About two thirds of the country's vacationers are ecotourists.

TEKS **READING CHECK:** *Environment and Society* How is Costa Rica's natural beauty a valuable resource? What policies have helped the country protect this resource?

TEKS Working with Sketch Maps, Questions 1, 2, 3, 4, 5

Section 2 Review

go.hrw.com Homework Practice Online

Keyword: SW3 HP11

Define indigenous, mulattoes, cacao, ecotourism

Working with Sketch Maps On the map you created in Section 1, label the Panama Canal, Colón, and Panama City. Why has the Panama Canal been an important economic resource?

Reading for the Main Idea

1. *Human Systems* What evidence of Spanish colonization remains in Central America today?
2. *Environment and Society* How is the practice of agriculture similar in El Salvador, Guatemala, and Panama?

Critical Thinking

3. **Identifying Cause and Effect** Why do you think the region surrounding the Panama Canal is generally prosperous and industrial?
4. **Drawing Inferences and Conclusions** How do patterns of land use in Central America affect social, economic, and political conditions in the region?

Organizing What You Know

5. Copy the chart below. Use it to describe and compare the ethnic makeup of the Central American countries.

Belize	
Costa Rica	
El Salvador	
Guatemala	
Honduras	
Nicaragua	
Panama	

Section 3 The Caribbean

READ TO DISCOVER

1. What are some important events in the history of the Caribbean?
2. What cultural and population patterns are found in the region?
3. What activities support the economies of the Caribbean countries?

IDENTIFY

Santeria
Caricom

DEFINE

commonwealth
creole
voodoo

LOCATE

Santo Domingo
Havana
San Juan

WHY IT MATTERS

The Caribbean is a popular tourist destination for Americans. In addition, many people from the Caribbean have immigrated to the United States. Use CNNfyi.com or other **current events** sources to learn about emigration from the Caribbean or about tourism opportunities there.

History and Culture

Like Central America, much of the Caribbean's modern identity is linked to the period of European colonization. European settlers and the peoples they brought to the region shaped the cultures of the islands.

Colonization and Independence In 1492 Christopher Columbus landed on an island in the southern Bahamas. However, he thought that he had reached islands off the coast of Asia that Europeans called the Indies. Columbus was wrong, but the term West Indies stuck. People still use the term to describe the islands of the Caribbean region.

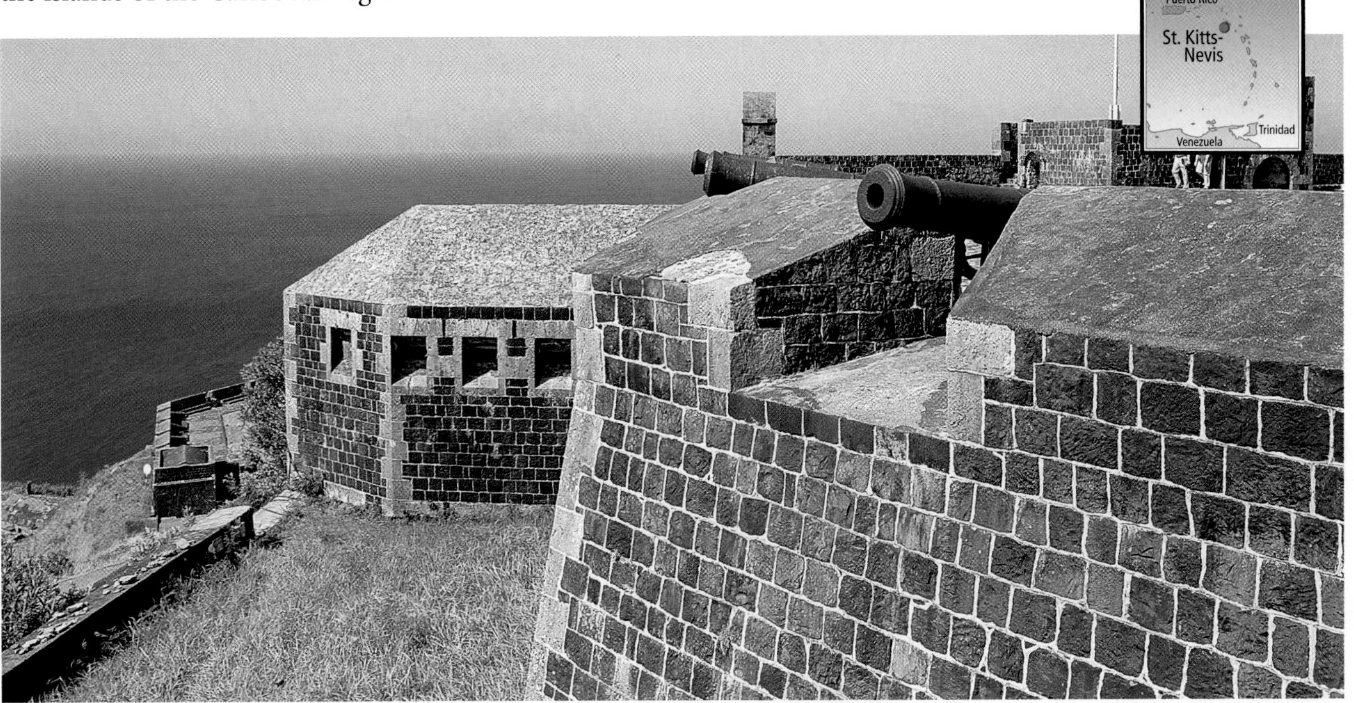

Brimstone Hill Fortress on the island of St. Kitts takes its name from the volcanic stone, known as brimstone, that was used to build its massive walls. The fortress was used by both British and French forces during the colonial period.

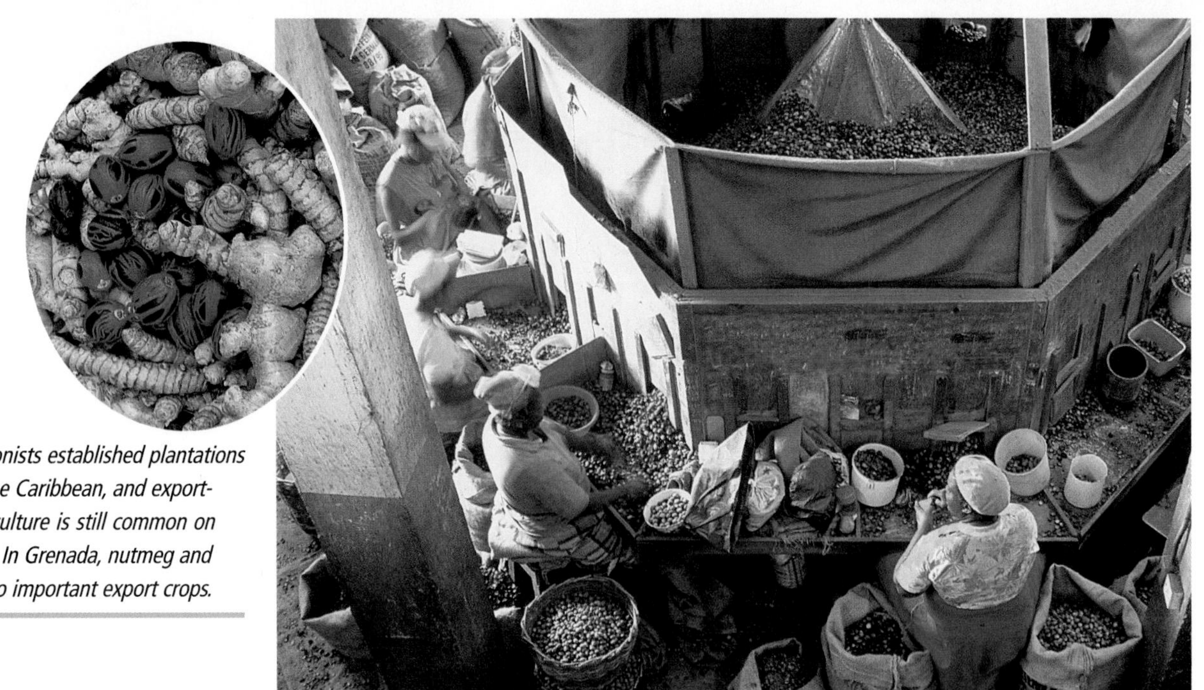

European colonists established plantations throughout the Caribbean, and export-oriented agriculture is still common on many islands. In Grenada, nutmeg and ginger are two important export crops.

The Spanish monarchs sponsored the voyage of Columbus, and the first wave of Europeans in the Caribbean was Spanish. They fanned out across the Greater Antilles looking for gold. The settlers brought bananas, citrus fruits, rice, sugarcane, and farm animals to the region.

When the Spanish found little gold, they lost interest in many of the small islands. However, the British, Dutch, and French competed for the islands. They reaped fabulous wealth from the sugar plantations they developed. Spain also developed sugarcane plantations in Cuba. To get workers for their farms, plantation owners turned to Africa. Over time, Europeans brought millions of enslaved Africans to the West Indies.

When slavery ended in the 1800s, former slaves could buy land if it was available. Some remained plantation workers. Others left the region, immigrating to Central America or the United States. Landowners then brought in laborers, mostly from South and East Asia, to work on the plantations.

Haiti won independence from France in 1804. The Dominican Republic gained independence from Spain by the mid-1800s. The United States took Cuba and Puerto Rico from Spain during the Spanish-American War in 1898. However, Cuba then became independent in 1902. Other Caribbean countries did not gain independence until the last half of the 1900s. Puerto Rico remains a **commonwealth** of the United States. A commonwealth is a self-governing territory associated with another country. Puerto Ricans are U.S. citizens. However, Puerto Rico has no voting representation in the U.S. Congress.

Indian, African, and European cultural traditions have combined to shape life throughout the Caribbean. In the Dominican Republic, for example, a blend of both Spanish and African cultures is found.

People, Languages, and Religions The Caribbean's population is largely descended from the Europeans and Africans who arrived during the colonial period. Most of the region's people are mulatto or otherwise of African descent. Nearly 40 percent of Cubans are of European descent. Haiti and Jamaica have the largest African populations in the region. Many descendants of Asian plantation workers live in Trinidad and Tobago. For example, more than a third of that country's people are East Indian. Small East Indian populations are also found in most other Caribbean countries. Jamaica and several smaller islands in the Lesser Antilles have small populations of Lebanese and Chinese as well. Most of the few Caribbean Indians who remain live in Dominica, St. Vincent and the Grenadines, Aruba, and the Netherlands Antilles.

The official language in each Caribbean country or territory depends on which European country colonized or controls it today. In most places, the main language is Spanish, English, French, or Dutch. Spanish is spoken in Cuba, the Dominican Republic, and Puerto Rico. English is also an official language in Puerto Rico. French and **creole** are official languages in Haiti. Creole is a blend of European, African, or Caribbean Indian languages. People on many other Caribbean islands often speak a creole language. In Aruba and the Netherlands Antilles, many people speak Papiamento. This creole language combines elements of Spanish, Dutch, and Portuguese.

Most people living on the Spanish- and French-speaking islands are Roman Catholic. Protestants make up the majority on the Dutch- and English-speaking islands. African traditions strongly influence religion in some places. In Haiti, for example, **voodoo** is important. Voodoo is a Haitian version of traditional African religious beliefs that are blended with elements of Christianity. Followers believe that good and bad spirits play an important role in daily life. Like voodoo, **Santeria** blends African traditions and Christian beliefs. It began in Cuba and has spread to other islands. In the 1800s Asian workers brought Hinduism and Islam to the islands.

Connecting to THE ARTS

Caribbean Music

Caribbean culture has contributed much to American music and dance. Calypso, a folk music from Trinidad, has ties to the music of Caribbean slaves of the early 1800s. It first became popular in the United States in the 1950s. More recently, reggae, salsa, and merengue have attracted a large American following. Reggae from Jamaica arrived in the 1970s. Salsa has its roots in the African rhythms and Spanish lyrics of Cuba and Puerto Rico. Merengue came to the United States from the Dominican Republic. It has long been part of rural folk music and dance traditions in that island country. Today people across the United States can find places to dance salsa and merengue.

Drawing Inferences and Conclusions What other part of the world has influenced the Caribbean music that has become popular in the United States?

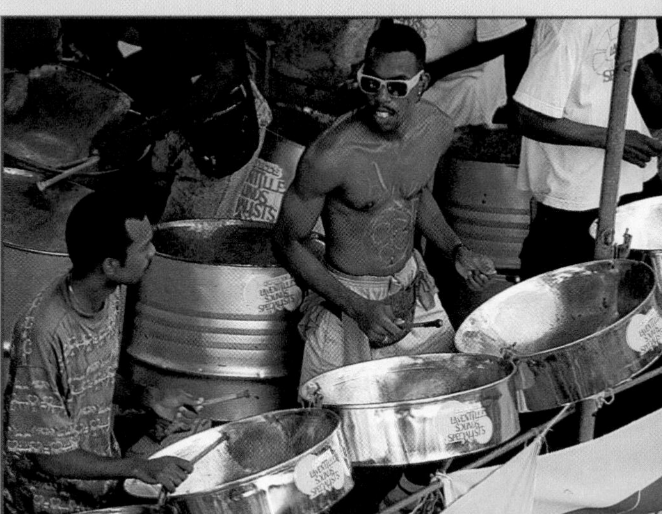

Steel drums, which originated in Trinidad, are a common element in Caribbean music. Steel drums are made from metal shipping drums and are played with rubber-tipped hammers.

Settlement and Land Use Since 1960 the Caribbean's total population has more than doubled. About 36 million people lived in the region in 2000. About 70 percent of the population lives in Cuba, Haiti, and the Dominican Republic. Santo Domingo, the Dominican Republic's capital, is the region's largest city. More than 3.5 million people live there. Havana, Cuba's capital, is home to more than 2.2 million people.

Population growth has created difficult problems for the region. The Caribbean islands have only one tenth the land area of Central America. Unemployment and underemployment are high. This has led to immigration to the United States, Canada, and Europe. Large Caribbean communities can

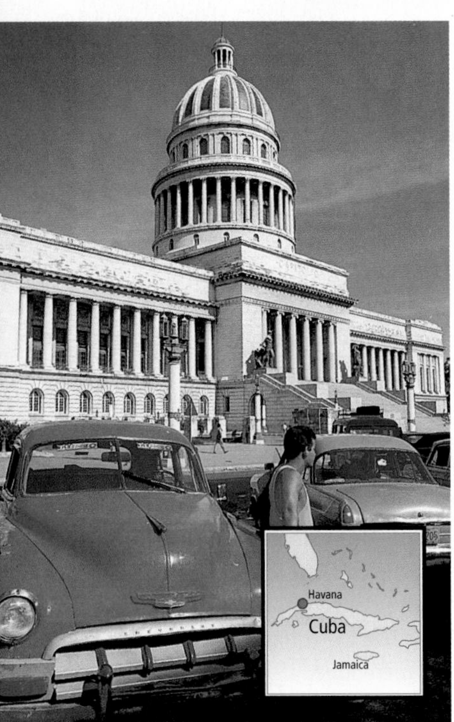

Since the 1960s the United States has banned trade with Cuba because it opposes Cuba's Communist government. As a result, most American cars in the country date from the 1950s.

be found in those places. For example, today more Puerto Ricans live in New York City than in San Juan, Puerto Rico's capital.

Another result of this population pressure has been increasing urbanization. People have had little choice except to move to towns and cities in search of new opportunities. Today more than half the people in the region live in towns and cities rather than farming communities. Partly because of this, most islands import much of their food.

READING CHECK: ***Places and Regions*** What geographic factors have contributed to emigration from the Caribbean?

Economic Development

Except for Cuba, most of the Caribbean countries and territories have market economies. Businesses and farms are privately owned, and their owners decide what and how much to produce. Cuba has a command economy. Its Communist government makes all the decisions about production.

Agriculture and Industry Despite efforts to encourage manufacturing, the region's economy has remained largely agricultural. Sugar, bananas, cacao, citrus fruits, and spices are the region's main exports. Jamaica produces nearly all of the world's supply of a seasoning called allspice. Most countries are trying to grow a wider range of crops. However, overall farm production and employment continue to decline. To promote industry and trade, countries in the region have formed an economic union called **Caricom**. Caricom is an abbreviation for the Caribbean Community and Common Market.

Cuba, the largest Caribbean island, is not a member of Caricom. In 1959 Fidel Castro came to power and soon after set up a Communist dictatorship there. The United States has banned trade with Cuba and restricted travel to the island. For years, Cuba concentrated on producing sugar on government-owned farms. Much of this sugar was sold to the Soviet Union. However, that country's collapse in the early 1990s severely hurt Cuba's economy. Recently, Cuba's government has emphasized the manufacture of farm machinery, steel, cement, clothing, food products, and consumer goods.

Cuba, Jamaica, Puerto Rico, and several countries in the Lesser Antilles have also developed important mining industries. Puerto Rico has the region's

INTERPRETING THE VISUAL RECORD

About three quarters of the land in Barbados is arable, and most of it is used to grow sugarcane. However, in recent years, the government has encouraged farmers to grow a wider range of crops. ***Based on the photo, how important do you think modern technology is for growing sugarcane in Barbados?*** TEKS

most industrialized economy. This is the result of Puerto Rico's easy access to the American market, low taxes, and special training programs for workers. In addition, wages in Puerto Rico are generally lower than in the United States. This means companies can produce goods less expensively there. Still, unemployment is high when compared to the United States.

INTERPRETING THE VISUAL RECORD

Tourism is a huge industry in island countries such as the Bahamas, where it supplies about 60 percent of the country's GDP and employs about 40 percent of the labor force. ***What problems might such a dependence on tourism cause?***

Tourism Many island leaders see tourism as the great hope for future economic growth. However, tourism also has its problems. On small islands, golf courses, resorts, and condominiums take land that could be used for farming or industry. Jobs in tourist industries are mostly seasonal, and pay is low. Furthermore, free-spending tourists often raise the cost of living on the islands. Plus, companies from developed countries build and operate most of the tourist facilities. As you might expect, then, most of the profits also go to these foreign companies. Still, tourism does bring needed income and reduces unemployment. Both of those benefits are important for island economies.

READING CHECK: ***Environment and Society*** What forms the basis of the economy in most of the Caribbean?

TEKS Working with Sketch Maps, Questions 1, 2, 3, 4, 5

Section 3 Review

Identify Santeria, Caricom

Define commonwealth, creole, voodoo

Working with Sketch Maps

On the map you created in Section 2, label the region's countries and Santo Domingo, Havana, and San Juan. Which city is the largest?

Reading for the Main Idea

1. ***Human Systems*** What factors help to explain the high rate of emigration from some Caribbean islands?
2. ***Human Systems*** How and why has the collapse of the Soviet Union affected the economy of Cuba?

Critical Thinking

3. **Drawing Inferences** Do you agree that emphasizing tourism is a good approach to economic growth in the Caribbean countries? Why or why not?
4. **Summarizing** How is the Caribbean's history reflected in its culture today?

Organizing What You Know

5. Copy the diagram. Use it to describe and then compare important cultural features of both the Caribbean islands and the countries of Central America. Where the circles overlap, identify cultural features that the regions share.

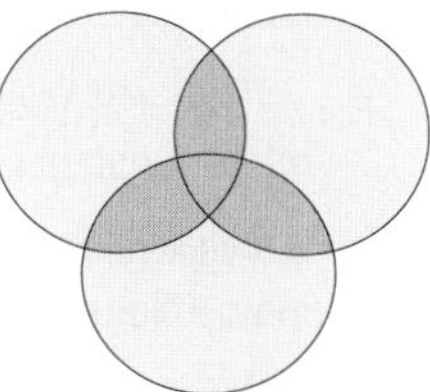

Review

Building Vocabulary TEKS

On a separate sheet of paper, explain the following terms by using them correctly in sentences.

mangrove	cacao	voodoo
bauxite	ecotourism	Santeria
indigenous	commonwealth	Caricom
mulattoes	creole	

Locating Key Places TEKS

On a separate sheet of paper, match the letters on the map with their correct labels.

Greater Antilles	Puerto Rico	Santo Domingo
Lesser Antilles	Bahamas	Havana
Cuba	Panama Canal	

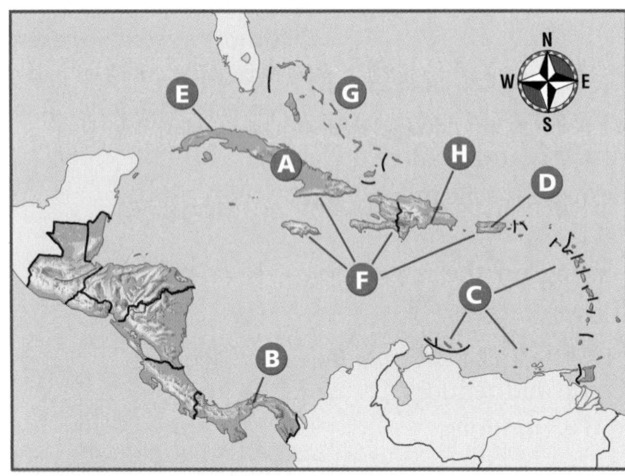

Understanding the Main Ideas TEKS

Section 1

1. *Physical Systems* What physical process has created the mountains of Central America? How do those mountains affect climates in Central America and the Caribbean islands?

Section 2

2. *Environment and Society* How are economics and politics linked in Central America?
3. *Human Systems* What economic activities are important in Central America? How is the economy of the region changing?

Section 3

4. *Places and Regions* How does Cuba differ economically from other Caribbean countries?
5. *Human Systems* What non-Caribbean countries have had the most influence in the region? Why?

Thinking Critically for TAKS TEKS

1. **Comparing** What similarities exist between the physical and human geography of Central America and the Caribbean?
2. **Identifying Cause and Effect** Why would the construction of railroads in Central America have encouraged the spread of plantations and market-oriented agriculture?
3. **Making Generalizations and Predictions** What might cultures and economies in Central America and Caribbean be like today if Europeans and North Americans had never become involved in the region?

Using the Geographer's Tools TEKS

1. **Analyzing Maps** Study the chapter physical-political map and the unit map showing land use and resources. What general conclusions can you draw about the location of plantation agriculture in the region?
2. **Analyzing Graphs** Review the climate graph for San Salvador in Section 1. Does the city have distinct wet and dry seasons? When do these occur? In which climate type would you expect to find this city?
3. **Creating Bar Graphs** Using the unit Fast Facts table, create bar graphs comparing the per capita GDPs of the countries in Central America and the Caribbean.

Writing for TAKS TEKS

Describe and compare the patterns of language and religion that distinguish Central America and the Caribbean as separate cultural regions. Write a short essay about how the two regions differ in these areas of culture. When you are finished with your essay, proofread it to make sure that you have used standard grammar, spelling, sentence structure, and punctuation.

SKILL BUILDING

Geography for Life TEKS

Organizing Geographic Information

Places and Regions Imagine that you are an official in charge of promoting tourism in a Central American or Caribbean country of your choice. Use the textbook, library, and Internet resources to collect information and materials for a tourist brochure. Then use maps, pictures, or other graphics and text information to identify physical and human geographical features that will attract tourists to your country.

Building Social Studies Skills

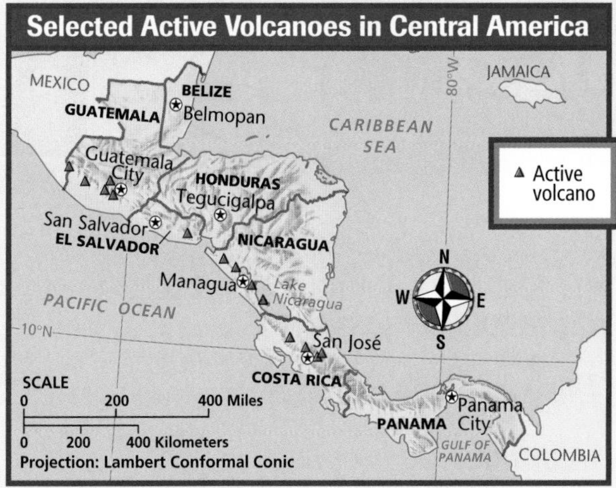

Interpreting Maps

Study the map above. Then answer the questions that follow.

1. Based on the map, which country does not have active volcanoes?
   a. Nicaragua
   b. Guatemala
   c. Costa Rica
   d. Belize
2. Which Central American countries would you expect to be most affected by earthquakes? Why?

Using Language

The following passage contains mistakes in grammar, punctuation, or usage. Read the passage and then answer the following questions. Mark your answers on a separate sheet of paper.

(1) European colonists established large plantations in Central America. (2) They grew crops like tobacco and sugarcane. (3) They forced the Central American Indians to work on plantations. (4) In gold mines elsewhere in the Americas some Indians were sent to work. (5) In addition many Africans were brought to the region as slaves.

3. In which sentence is a comma missing?
   a. 1
   b. 3
   c. 4
   d. 5
4. In which sentence does a prepositional phrase need to be moved so that it is closer to the verb it modifies?
   a. 1
   b. 2
   c. 4
   d. 5
5. Combine sentences 1 and 2 correctly.

Alternative Assessment

PORTFOLIO ACTIVITY

Learning about Your Local Geography TEKS

Group Project: Field Work

Plan, organize, and complete a research project about products from Central America and the Caribbean that are available in your community. Divide your community into zones. Have a group member survey stores in each zone to determine which products from Central America and the Caribbean—such as clothing, electronics, and food items—are sold there. Present your group's results on a sketch map of your community, showing the items that your group found in each location. Use the unit map showing land use and resources as a model for how to present your group's findings on your community sketch map.

internet connect

Internet Activity: go.hrw.com
KEYWORD: SW3 GT11

Choose a topic on Central America and the Caribbean to:

- create a postcard of an ecotour.
- visit the Panama Canal and learn about its history.
- take the GeoMap challenge to test your knowledge of Central American and Caribbean geography!

CHAPTER 12

South America

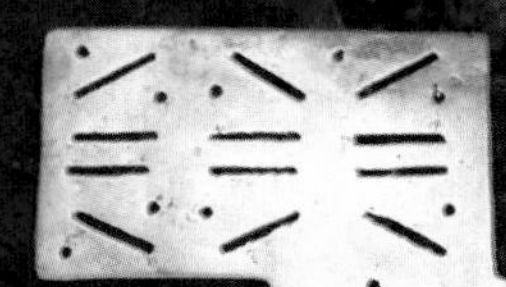

South America extends from the sunny beaches of the Caribbean Sea to the cold windswept shores of Cape Horn. The continent includes a marvelous variety of environments, peoples, plant life, and animals. Modern meets ancient in South America.

Gold chest ornament from Colombia

Head *llanero* on a Venezuela ranch

¡Hola! ¿Cómo estás? (Hello! How are you?) My name is Jorge, and I live in Armenia, a city in west-central Colombia. I live with my family and our big black dog, Rocca. Our house surrounds a courtyard where we grow flowers.

My father is a merchant and a farm owner. Our hacienda is in the country. About 60 people work for my father there, growing coffee, plantains, yucca, and fruits like strawberries and oranges. We also raise chickens and pigs. My dad has a fleet of trucks to carry our produce into the city for sale to groceries and restaurants.

I go to a big school. It has hundreds of kids, both boys and girls, in grades 1 through 12. I drive myself to school. You can drive in Colombia at 15 years old, and my dad let me have an old car. In school I am studying Spanish, French, and English, as well as science, math, and history. On many weekends we go to the farm and eat fresh fruit and swim in the large pool—the weather is warm all year round. In the summer, I like to go to the ocean at Cartagena and swim.

My favorite holiday is Christmas, when our whole family gathers. We have a big barbecue—a whole stuffed pig cooked in a large pit. One year, though, we had an earthquake after New Year's that lasted for 40 seconds. It was strange, because the earthquake moved like a wave. Some areas like the downtown were completely destroyed, while other areas, like my neighborhood, were not much affected.

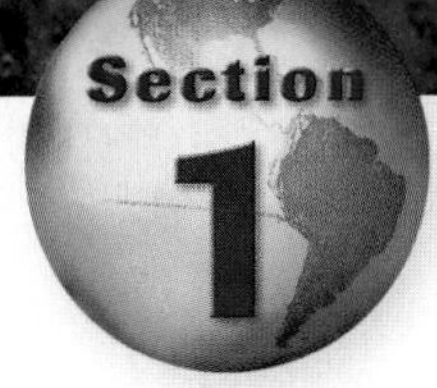

Natural Environments

READ TO DISCOVER

1. What are the major landforms and rivers of South America?
2. What climates, plants, and animals are found in South America?
3. What natural resources does the continent have?

WHY IT MATTERS

New businesses are developing products that use resources from South America's rain forests without harming the environment. Use CNNfyi.com or other **current events** sources to learn about these businesses and the products they sell.

IDENTIFY

Llanos
Gran Chaco
Pampas
El Niño
La Niña

DEFINE

tepuís
tree line
tar sands

LOCATE

Andes
Altiplano
Lake Titicaca
Guiana Highlands
Brazilian Highlands
Amazon River
Patagonia
Tierra del Fuego
Orinoco River
Paraná River
Río de la Plata
Atacama Desert
Lake Maracaibo

South America: Physical-Political

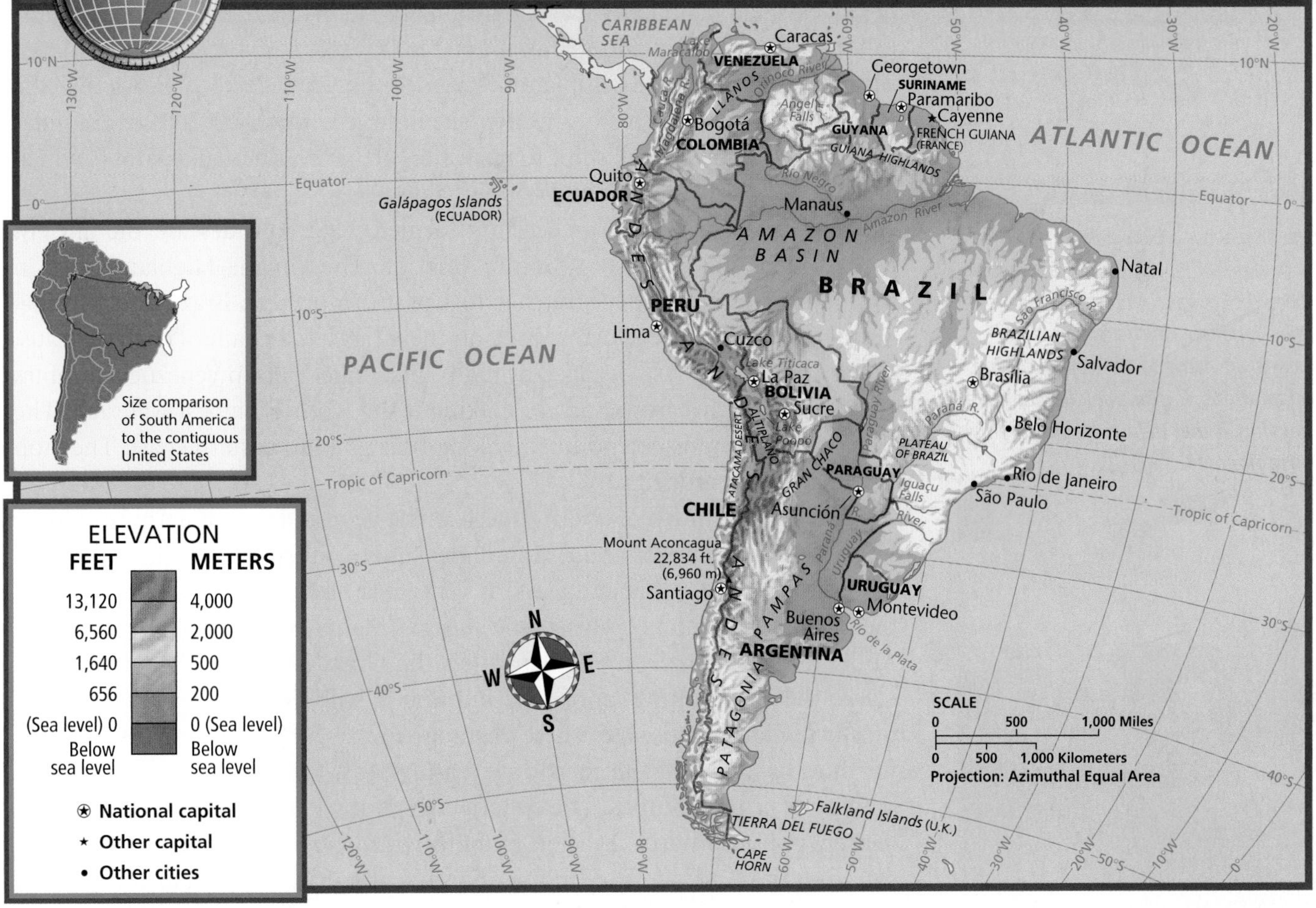

Size comparison of South America to the contiguous United States

ELEVATION

FEET	METERS
13,120	4,000
6,560	2,000
1,640	500
656	200
(Sea level) 0	0 (Sea level)
Below sea level	Below sea level

⊛ National capital
⋆ Other capital
• Other cities

SCALE
0 500 1,000 Miles
0 500 1,000 Kilometers
Projection: Azimuthal Equal Area

Landforms and Rivers

South America includes 12 countries and an overseas department of France—French Guiana. Brazil is the largest country. Colombia, Venezuela, Guyana, Suriname, and French Guiana lie along the continent's northern coasts. In the west, Ecuador, Peru, and Chile border the Pacific Ocean. Bolivia and Paraguay are cut off from the sea. Argentina and Uruguay lie along the Atlantic. The continent has bone-dry deserts, tropical rain forests, snowcapped mountains, high plateaus, and vast fertile grasslands.

GO TO: go.hrw.com
KEYWORD: SW3 CH12
FOR: Web sites about South America

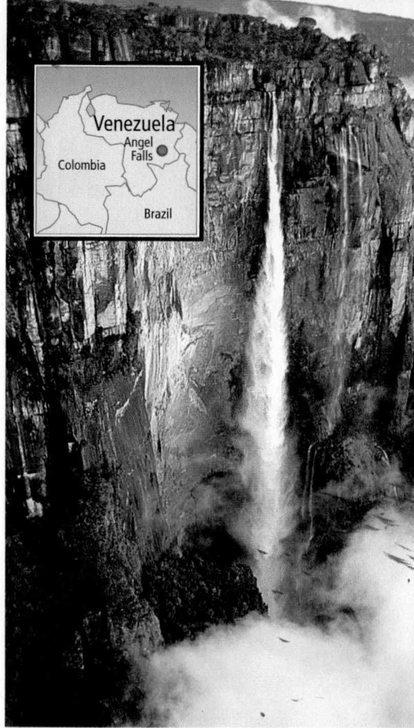

Angel Falls drops from the top of a tepuí. *The falls begin high above the clouds.*

Landforms South America's great mountain range, the Andes, extends along the continent's Pacific coast. The mountains stretch from the northern edge of the continent south to its tip. Mount Aconcagua, the highest peak in the range, rises to 22,834 feet (6,960 m). The collision of the Nazca and South American plates created the Andes. Continued tectonic activity causes volcanic eruptions and earthquakes.

In Peru and Bolivia the Andes divide into two great ranges. Between them lies an elevated plain known as the Altiplano. The name means "High Plateau" in Spanish. The Altiplano lies at about 12,000 feet (3,658 m) above sea level. More than 25 of the Altiplano's rivers drain into Lake Titicaca on the Peru-Bolivia border. This freshwater lake is large enough—3,200 square miles (8,288 sq km)—for small ships to navigate. Farther south is Lake Poopó, which is salty. Few people live along Poopó's banks.

INTERPRETING THE VISUAL RECORD

The Altiplano is a high plain area between two ranges of the Andes. Few trees grow there. The southern Altiplano gets little rainfall. Llamas, which are related to camels, are native to the Altiplano. ***What adaptations do you think llamas have developed that help them live in this environment?*** TEKS

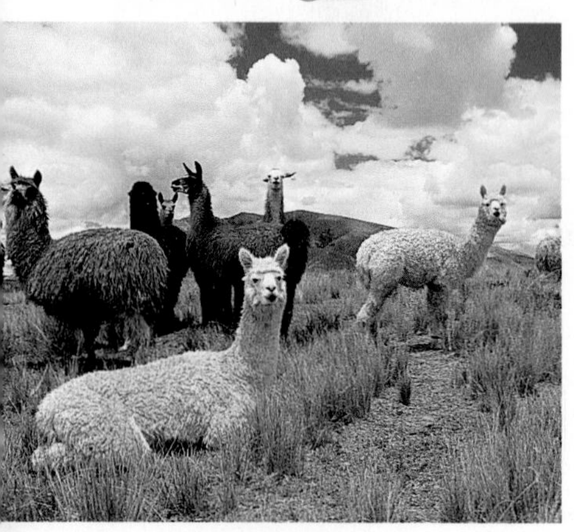

In contrast to the more recently formed rugged Andes are the ancient eroded highlands of eastern South America. The Guiana Highlands rise in southern Venezuela. They stretch across part of northern Brazil and Guyana, Suriname, and French Guiana. Erosion there has left a chain of high plateaus, called ***tepuís*** (tay-PWEEZ), edged by high cliffs. Angel Falls in Venezuela tumbles 3,212 feet (979 m) from a *tepuí,* making it the world's highest waterfall. The Brazilian Highlands extend inland along Brazil's southeastern coast. They too are ancient and eroded.

Plains cover much of South America. The largest plain is the Amazon River basin, which occupies about 2 million square miles (5.3 million sq km). Northeastern Colombia and western Venezuela have a large plains area called the **Llanos** (YAH-nohs). This name means "Plains" in Spanish. Between the Andes and the Brazilian Highlands lies the **Gran Chaco** (grahn CHAH-koh). *Chaco* means "hunting land," and hunting is still a popular pastime in this mostly semiarid landscape. These plains are so flat that water sometimes stands for months after the summer rainy season ends. Farther south are the wide grasslands of the **Pampas**. The eastern edge of the Pampas is Argentina's most densely populated area. Erosion by both wind and water have carried fertile soil from the Andes to the Pampas. South of the Pampas, semiarid Patagonia stretches all the way to Tierra del Fuego. Tierra del Fuego is an island divided between Argentina and Chile.

Rivers Only small rivers and streams flow west into the Pacific Ocean. However, three great river systems—the Amazon, Orinoco, and Paraná—drain the eastern part of the continent. The Amazon River is 4,000 miles (6,436 km) long. It is the world's largest river in volume, and no other river drains a larger area. So much water flows from the Amazon into the Atlantic Ocean that freshwater dilutes seawater more than 100 miles (161 km) from shore. Ocean-going ships can navigate up the Amazon for nearly 2,300 miles (3,700 km)—all the way to Iquitos (ee-KEE-tohs), Peru! The Orinoco River, which drains the western Guiana Highlands and the Llanos, also empties into the Atlantic Ocean. Several rivers together drain another large area far to the south. The longest of these rivers is the Paraná. The Paraná River flows into the Río de la Plata estuary between Argentina and Uruguay. The Paraná drains an area that includes the eastern slopes of the Andes and the highlands of eastern Brazil.

READING CHECK: ***Physical Systems*** What qualities make the Amazon River unique?

Our Amazing Planet

South America's giant anacondas, or water boas, are the longest snakes in the Western Hemisphere. They are also the heaviest snakes in the world. Anacondas can be as thick as a telephone pole, measure more than 30 feet (10 m) long, and weigh up to 300 pounds (135 kg).

Climates, Plants, and Animals

Because South America extends across more than 60 degrees of latitude, the continent has a variety of climate regions. The Amazon River basin is the world's largest tropical humid climate region. It also has the largest tropical rain forest in the world. More than 150 inches (380 cm) of rain fall there every year. Anacondas, bats, jaguars, monkeys, and countless other species live in the forest. Along the western edge of the basin, rain forests yield to the highland climates of the Andes. Environments of the central and northern Andes region can be divided into five zones, according to elevation. These zones range from hot and humid lands near sea level to frozen peaks high above the **tree line**.

Elevation Zones in the Andes

INTERPRETING THE DIAGRAM *Environments in the Andes change with elevation. Five different elevation zones are commonly recognized.* ***In which zone is the Altiplano located?***

INTERPRETING THE VISUAL RECORD

Dense rainforest vegetation grows along the Aguarico River in Ecuador. The toucan in the small photo is a rainforest resident. Birds and bats assist the survival of tropical rain forests. ***How might birds and bats play a role in reforestation?***

TEKS

A tree line is the line of elevation above which trees do not grow. (See the diagram.) Animals unique to South America have adapted to the Andes' harsh conditions. They include llamas, the related animals alpacas and vicuñas, and the Andean condor, a large vulture with a 10-foot (3 m) wingspan.

Many areas of South America have tropical wet and dry climates. They have wet summers and dry winters. The natural vegetation includes either dry forest or savannas, where a mixture of trees and grasses cover the plains. Southern South America has a variety of middle-latitude climates. Chile's central valley has a Mediterranean climate, with winter rains and summer droughts. Moist westerlies influence southern Chile. That area has a marine west coast climate. In southern Argentina the Andes create a rain shadow. As a result, Patagonia has semiarid and arid climates. Relatively few animals live in this area.

The driest region of South America is the Atacama Desert of northern Chile and southern Peru. A high pressure system and cool ocean currents bring dry weather to this area throughout the year. Although rain is extremely rare, fog and low clouds are common. They form when the cold Peru Current chills warmer air above the Pacific Ocean's surface. Cloud cover keeps air near the ground from being warmed by the Sun. This coastal area is one of the cloudiest—and driest—places on Earth. In fact, the area receives almost no sunshine for about six months of the year. Some people who live along this coast increase their water supply by "trapping" fog. Near the seashore, they set up plastic nets on which fog droplets condense. In this way a village can collect several thousand gallons of clean water per day.

About once or twice a decade, the dry Pacific coast is affected by an ocean and weather pattern called **El Niño** (ehl-NEEN-yoh). During an El Niño event, the eastern Pacific Ocean is warmer and the climate much wetter than normal. This pattern can alternate with **La Niña**, when Pacific waters are colder than normal. (See Geography for Life: El Niño.)

TEKS **READING CHECK:** ***Physical Systems*** How do mountains and elevation affect the climates of South America? How does the Peru Current affect weather patterns in Chile?

Natural Resources

South America has rich mineral deposits, fertile soils, and climates suitable for growing a range of crops. Many rivers, particularly in the Paraná and Amazon Basins, have been dammed to generate electricity and store water for irrigation. The rain forests provide rubber and timber. Many nuts and plants used for medicines come from the Amazon rain forest. Scientists hope other useful plants will be discovered there.

The mineral wealth that attracted Spaniards and Portuguese to the region centuries ago is still being developed. New gold and silver deposits have been found in Brazil and Colombia. Chile is the world's largest producer and exporter of copper. Brazil has enormous reserves of iron ore and bauxite—the main aluminum ore. Colombia has long been famous for its emeralds.

Several South American countries have petroleum deposits as well. The largest oil reserves are in Venezuela. The vast oil deposits surrounding Lake Maracaibo (mah-rah-KY-boh) have made Venezuela a leading oil-exporting country. In addition, oil deposits have been developed in Colombia and the upper Amazon Basin of Peru and Ecuador. More recent oil discoveries have been made off the coasts of Argentina, Brazil, and Chile.

Venezuela will have oil resources for years to come in the form of **tar sands**. Tar sands are rock or sand layers that contain oil. However, because the oil has to be cooked out of the rocks, production is expensive. Nonetheless, tar sands may become more important once the more easily pumped oil is gone.

✓ **READING CHECK:** *Physical Systems* What are the main mineral and energy resources of South America?

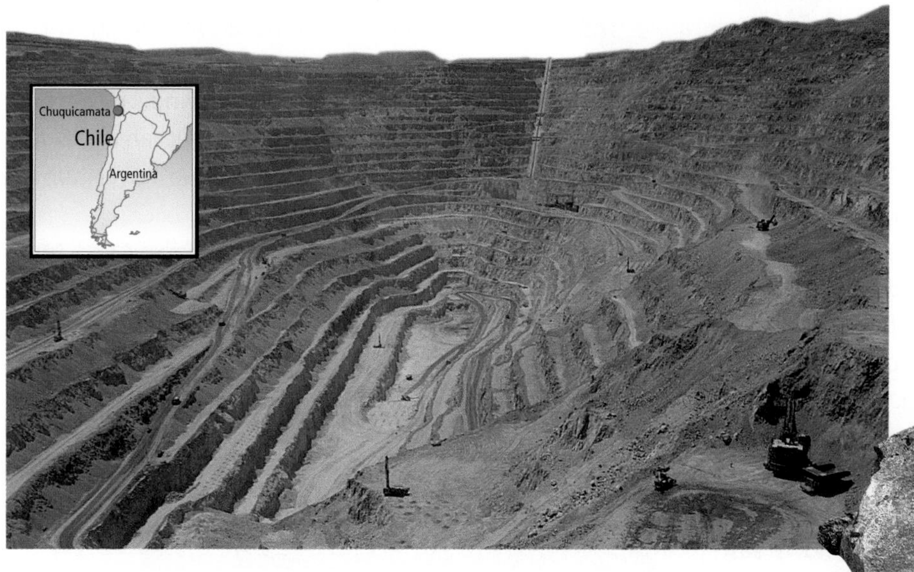

Chuquicamata mine, in the Atacama Desert, is the largest open-pit copper mine in the world. The ore sample shown displays copper's distinctive blue-green color.

TEKS Working with Sketch Maps, Questions 1, 2, 3, 4, 5

Section 1 Review

Homework Practice Online

Keyword: SW3 HP12

Identify Llanos, Gran Chaco, Pampas, El Niño, La Niña

Define *tepuís,* tree line, tar sands

Working with Sketch Maps

On a map of South America that you draw or that your teacher provides, label the Andes, Altiplano, Lake Titicaca, Guiana Highlands, Brazilian Highlands, Amazon River, Llanos, Gran Chaco, Pampas, Patagonia, Tierra del Fuego, Orinoco River, Paraná River, Atacama Desert, and Lake Maracaibo. In the margin of your map, explain why coastal Chile is both dry and cloudy.

Reading for the Main Idea

1. *Physical Systems* What is a basic difference between the Andes and the other highland regions of South America?
2. *Places and Regions* What are South America's three largest rivers, and what areas do they drain?

Critical Thinking

3. **Drawing Inferences** Look at the chapter map. To what physical conditions would you have to adjust if you moved from Rio de Janeiro to La Paz?
4. **Making Generalizations and Predictions** See the feature on the next page. How might El Niño and La Niña affect societies of the dry Pacific coast over time?

Organizing What You Know

5. Create a chart like the one below. Use it to identify the climates, vegetation, and animal life of the three landform regions listed.

	Climates	Vegetation	Animals
Amazon River basin			
Plains areas			
Andes			

Geography for Life

El Niño

Long ago, fishers noticed that once or twice a decade the normally cool waters off Peru's coast became warmer near Christmastime. Referring to the baby Jesus, they called this warming trend El Niño. *El Niño* means "The Boy" in Spanish. El Niño shows how a physical process that happens on one side of the planet can affect environments thousands of miles away.

This 1997 satellite image shows an El Niño developing in the eastern Pacific. The warmer water is in white.

Usually, the contrast in temperatures across the oceans helps create winds. Trade winds generally maintain a balance between warm western Pacific water and cool eastern Pacific water. Along the eastern Pacific, particularly off Ecuador and Peru, strong trade winds blow warm surface water westward. This allows colder water to rise and bring up nutrients from the depths, attracting fish.

When atmospheric pressure rises north of Australia, the winds calm. With the drop in wind, El Niño begins off the coast of South America. As easterly trade winds decrease, warm water in the western Pacific flows eastward. The warm water layer flows over cooler, nutrient-rich water. This blocks the normal upwelling of the cool water along North and South America. As a result, sea life suffers from a lack of nutrients. Fish that usually thrive off Ecuador and Peru head south in search of cooler waters and more food. Chilean fishers then catch more sardines. Meanwhile, in North America, the polar jet stream stays farther north over Canada. Less cold air moves into the upper United States, and the northern states enjoy a mild winter. During the 1997–98 El Niño, for example, people in northern states saved billions of dollars in heating costs. In addition, an El Niño event alters upper-level tropical wind patterns. States along the Atlantic Ocean and Gulf of Mexico benefit because fewer hurricanes strike land.

However, El Niño can also cause terrible destruction. Because oceans affect weather patterns, severe weather changes may accompany the change in ocean temperature. Droughts can occur in the Pacific Islands. Meanwhile, heavy rains may flood areas of coastal North and South America that are usually bone-dry.

Applying What You Know

1. **Summarizing** How does an El Niño event affect the environment?
2. **Drawing Conclusions** Why might people in areas affected by El Niño view it differently?

Normal and El Niño Ocean Conditions

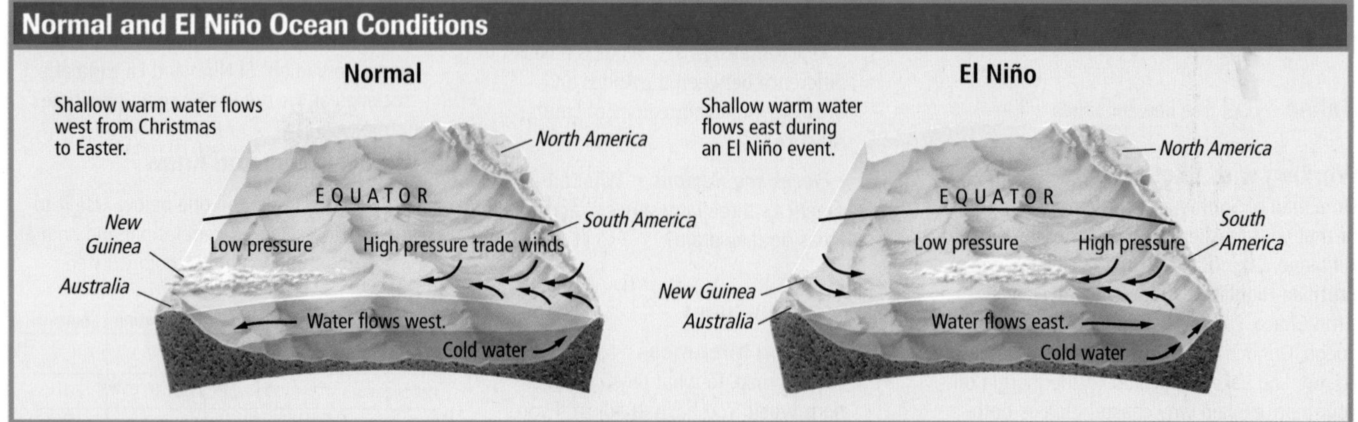

History and Culture

READ TO DISCOVER

1. What were some important events in the early history of South America?
2. How did the colonial era and independence affect South America?
3. What are some important features of South America's cultures?

WHY IT MATTERS

Some South American countries have experienced many changes in government since they became independent. Use CNNfyi.com or other **current events** sources to learn how political changes in one country can affect people elsewhere.

IDENTIFY

Chibcha
Inca

DEFINE

latifundia
buffer state
coup
manioc

LOCATE

Cuzco
Buenos Aires
Rio de Janeiro
Bogotá
La Paz
Quito

Early History

Most researchers believe people first entered South America from the north more than 12,000 years ago. They eventually inhabited even the continent's harshest environments. The first settlers were hunter-gatherers. Farming began in the region more than 5,000 years ago.

For several thousand years before Europeans arrived, kingdoms rose and fell in western South America. In the Colombian Andes, for example, the **Chibcha** ruled and developed gold-working skills. The **Inca**, however, founded South America's greatest early civilization. At its height the Inca Empire stretched from what is now Ecuador to central Chile. The Inca built paved roads and suspension bridges to connect their empire from the Pacific coast to the Amazon lowlands. Fine examples of Inca stone construction can still be seen in Cuzco, Peru, the Inca capital. Also common and still being used for farming are Inca terraced fields braced by stone walls.

The Mochica people who made this ceramic warrior lived in northern coastal Peru from about A.D. 100 to 700.

INTERPRETING THE VISUAL RECORD

Local women pose with their llamas before a wall at Sacsahuamán, an Inca site near Cuzco. Some of the stones weigh hundreds of tons. They were moved to the site without wheeled vehicles and are fitted closely together without mortar. ***What function might these walls have served?*** TEKS

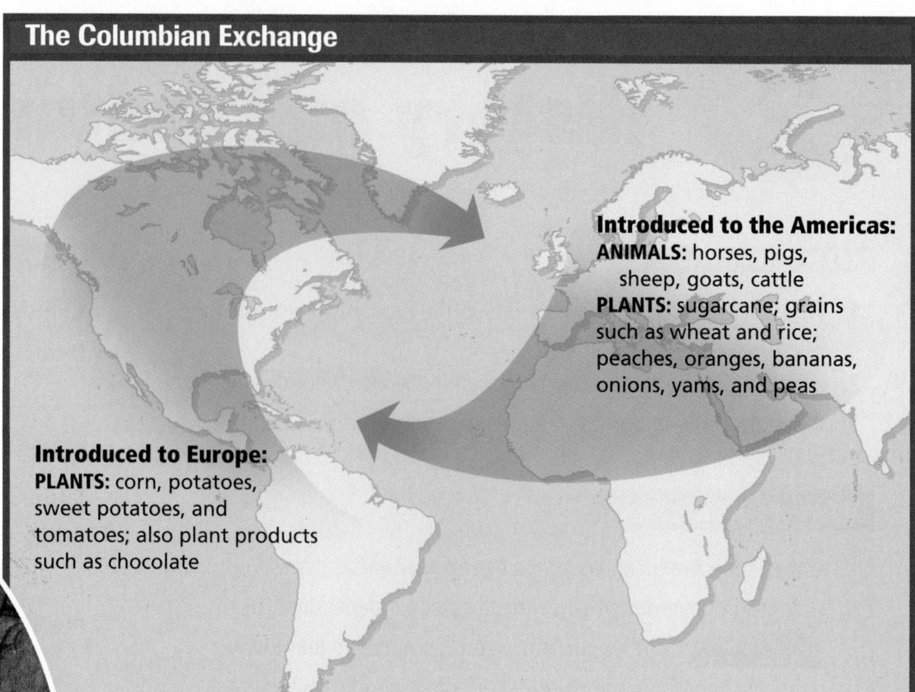

INTERPRETING THE VISUAL RECORD

Following the arrival of Columbus in the Americas, Europeans brought many plants and animals with them. In turn, they introduced American species to Europe. This process is called the Columbian Exchange. In the painting below, workers pick coffee beans on a South American plantation. Coffee was probably first grown in eastern Africa. ***How has this exchange of plants and animals affected the regions of contact?*** TEKS

When Spanish conquistador Francisco Pizarro heard about the Inca Empire he set out to conquer it. Unrest within the empire aided Pizarro's conquest in the 1530s. The Spanish looted and destroyed Inca buildings. They then built churches, government buildings, and Spanish-style plazas on the ruins. The Spanish also built a new capital city, Lima, near the coast.

TEKS **READING CHECK:** ***Human Systems*** How are Inca and early Spanish influences still evident in modern South America?

Spanish Settlement The Spanish focused their efforts at conquest on the western part of the continent. A 1494 treaty had divided South America between Spain and Portugal. Spain got the right to lands to the west of the treaty line and Portugal the lands to the east. The Spanish also focused on the west because the Inca, rich with gold and silver, lived there. The area was agriculturally productive and could provide the Spanish with a ready source of labor. In Spanish society owning land was a source of prestige, and the Spanish colonists soon took over South American Indian lands. They established a system of landed estates like they had known in Spain and forced the Indian peoples to work the land.

The colonists also introduced animals and agricultural products they had known in Europe. For example, they brought cattle, horses, sheep, sugarcane, and wheat. Over time, the colonists took American products like beans, chilies, corn, potatoes, and squash to Europe, Africa, and Asia.

Europeans also carried new diseases to South America. As in Mexico and Central America, these diseases killed millions of South American Indians. Europeans killed many others in the battles of conquest. Indians who labored in mines and on ranches and plantations often died from overwork. As a result, Indian populations fell sharply during the colonial period. Only a fraction of the original Indian population remained by the late 1500s.

Once the Spanish were established in Peru, their influence spread across the Altiplano into what is now Bolivia. There they expanded the Inca silver and gold mines. The Spanish colonized central Chile in the 1540s. Later they spread southeastward, herding cattle in what became Paraguay and Argentina. After a while, large estates called **latifundia** (lah-ti-FOOHN-dee-uh) spread across the Pampas. A South American Indian people known as the Guaraní already lived in the fertile lands of eastern Paraguay. The Spanish used the labor and food production skills of the Guaraní to further expand their settlement in the region.

Capoeira, a popular Brazilian folk dance, combines dancing, fighting, acrobatics, and music. It was possibly first developed as a martial art in the 1600s by escaped slaves who wanted to defend their freedom.

Portuguese Settlement Portuguese settlement began in the 1530s along the eastern coast of what is now Brazil. Portuguese nobles received royal land grants to set up large plantations. Their first important crop was brazilwood, which can be used to produce red and purple dye. However, sugarcane soon became the key crop. After the decline of the South American Indian population, colonists brought in enslaved Africans to work their estates. Cities in northeastern Brazil, like Natal and Salvador, remain from the sugar and slavery era. Meanwhile, the Portuguese spread cattle ranching inland. Their expansion southward and inland in the 1600s led to major mineral discoveries. This movement eventually led to the growth of cities like São Paulo and Rio de Janeiro.

TEKS **READING CHECK:** ***Human Systems*** How did the distribution of resources affect the location and patterns of movement of Spanish and Portuguese colonists?

Colonial Era and Independence

The Spanish colonies of South America gained their independence between 1810 and 1830. Wars in Europe had weakened Spain. Spanish authorities fled their colonies after a period of unrest and minor military battles. The independence of the United States had inspired colonial leaders. There the ideals of political freedom and cooperation between the colonies had led to the formation of a single large country. However, in South America several different countries formed after the Spanish left. The wealthy elites who ruled each country were oriented more toward Europe than toward each other. In addition, these countries tended to be isolated from each other. This was because the continent's size and rugged terrain made communication difficult.

Borders of the new countries mostly followed the divisions created during the colonial period. However, the frontier between Argentina and Brazil was not controlled well by either country. As a result, the new state of Uruguay was able to form. Uruguay is an example of a **buffer state**—a small country between two larger, more powerful countries.

Brazil followed a different path to independence. European wars forced the Portuguese royal family to flee to Rio de Janeiro, where they arrived in 1808. Later, after the king's return to Portugal, Brazil declared its independence in 1822.

The British and Dutch only recently granted their small colonies in the Guianas independence. British Guiana became

INTERPRETING THE VISUAL RECORD

Grand buildings rise on Independence Plaza in Montevideo, the capital of Uruguay. Spaniards founded the city in 1726 to balance Portuguese influence in the area. ***What can you infer about Montevideo's climate from the photo?*** TEKS

Connecting to

HISTORY

The Nazca Lines

One intriguing ancient site in South America is in the desert of southern Peru. There some 900 gigantic geometric and animal shapes are etched into the desert floor. Examples include a 1,000-foot (305 m) pelican and a 360-foot (110 m) monkey. One trapezoid covers 160,000 square yards (134,400 sq m). For the most part, these shapes can be seen clearly only from the air. The Nazca people, who lived in the area before the Inca, made the lines by moving surface stones aside to expose the lighter soil beneath. Although there are many theories, no one really knows why the Nazca Lines were drawn. One theory suggests that the lines and shapes were in some way related to water. Some researchers think they may have marked underground water sources. Many anthropologists, however, think the lines may have formed paths. They believe the Nazca walked these paths as part of rituals meant to ensure that there would be enough water for them and their lands.

Supporting a Point of View Do you think the anthropologists' theory about the Nazca Lines may be a logical one? Why or why not?

Guyana in 1966, and Dutch Guiana became Suriname in 1975. French Guiana remains a part of France.

In the end, the independence movements in South America did little to improve people's lives. Revolutions often changed the governments in the new countries. However, these revolutions only tended to replace one group of powerful families with another. Leadership was usually by one man, a dictator, who ruled with the support of wealthy colonial families. Meanwhile, life for most poor farmers, plantation workers, and city dwellers changed little. Moreover, there were few opportunities for improvement. Sometimes a group would take power by force. Such a change in government is called a **coup** (koo). Coups have been common throughout South American history. For example, Bolivia has experienced revolutions or military coups nearly 200 times since becoming independent in 1825.

READING CHECK: ***Human Systems*** Why has daily life not improved much over the years for many South Americans?

Culture

South American Indians, Europeans, Africans, and Asians have all played a part in the peopling of South America. Each culture has left its stamp on the continent's countries.

People and Languages Today the countries of South America vary widely in their ethnic makeup. For example, 97 percent of Argentines are of European ancestry, compared to only 7 percent of Ecuadorians. Bolivia's population has the largest percentage of South American Indians—55 percent. In other countries people of mixed ancestry called mestizos are in the majority. For example, 95 percent of Paraguayans are mestizos. Asians have also immigrated to the continent. In Guyana the descendants of workers from India make up about half of the population. Many Japanese have immigrated to Brazil and Peru.

The main European language spoken in each country reflects that country's colonial history. Therefore, people in most South American countries speak Spanish, but most Brazilians speak Portuguese. English, Dutch, and French are official languages in Guyana, Suriname, and French Guiana, respectively. In the Andes region, from 10 to 13 million people speak the Inca language, Quechua. Most Paraguayans speak Spanish and Guaraní. On Uruguay's border with Brazil, a mix of Portuguese and Spanish called Portunol is widely spoken.

Settlement Patterns To a large extent, densely populated areas of South America hug the coasts and reach only a few hundred miles inland. Many major cities—such as Buenos Aires, Lima, and Rio de Janeiro—are seaports. Some cities, such as Bogotá, La Paz, and Quito, lie in high Andean valleys. Much of the South American interior is thinly populated. Large areas, particularly in the

Amazon Basin, the Andes, the Guianas, and southern Argentina, have few people.

Religion and Traditions The Spanish and Portuguese colonists were Roman Catholics. Therefore, the majority of South Americans today are also Roman Catholic. South Asians and Indonesians in the Guianas have added Hindu temples and Islamic mosques to the landscape. Many indigenous peoples, such as those who live deep in the rain forest, follow their traditional religions.

Although South America is changing rapidly, traditional ways of life can still be found. Some rainforest peoples have had little contact with the outside world. They raise bananas, **manioc**, yams, and other crops. Manioc is a tropical plant with starchy roots. Some hunt with bows and arrows or blowguns and darts. As the Amazon Basin is developed, these people's lives will change. Also, although most ranchers use modern methods, some ranch hands still live much like the cowboys of the old American West. *Llaneros* work on the Venezuelan Llanos. Argentine cowhands herd cattle and horses on the Pampas. Some wear the traditional clothing that Argentina's gauchos, or cowboys, wore in the 1700s and 1800s. The gauchos also live on in Argentine literature and popular culture. For example, the poem *The Gaucho Martín Fierro* celebrates the independent life of the Argentine gaucho.

READING CHECK: *Places and Regions* What are two ways that Spanish and Portuguese heritage are expressed in South American culture?

Wearing a traditional derby hat, a Bolivian Indian woman sells folk medicines and good luck charms in a La Paz market. The Quechua and Aymara, Bolivia's two main ethnic groups, often combine local religious beliefs with Roman Catholicism.

TEKS Working with Sketch Maps, Questions 1, 2, 3, 4, 5

Section 2 Review

Identify Chibcha, Inca

Define latifundia, buffer state, coup, manioc

Working with Sketch Maps On the map you created in Section 1, label the countries of South America, Cuzco, Buenos Aires, Rio de Janeiro, Bogotá, La Paz, and Quito. Then shade in the areas that had been included in the Inca Empire.

Reading for the Main Idea

1. *Human Systems* What happened to the indigenous population following the arrival of Europeans in South America?
2. *Places and Regions* Where are the densely populated areas of South America? Which areas are thinly populated?

Critical Thinking

3. **Identifying Cause and Effect** How do you think the exchange of food products between the Old and New Worlds changed life in both places?
4. **Analyzing Information** What do the *llaneros* and rainforest peoples have in common?

Organizing What You Know

5. Create a time line like the one below. Use it to identify important periods and events in the history of South America.

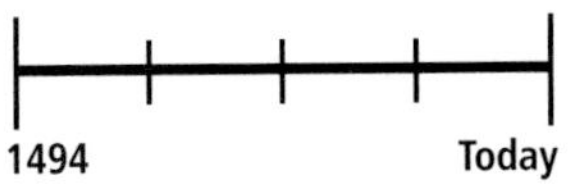

Section 3 South America Today

READ TO DISCOVER

1. What is the economy of South America like today?
2. What are South American cities like?
3. What issues and challenges face the people of South America?

IDENTIFY

Mercosur

DEFINE

minifundia
favelas
landlocked
terrorism

LOCATE

Manaus
Santiago
Lima

WHY IT MATTERS

Every day more people move from the South American countryside to cities that are already crowded. Use CNNfyi.com or other **current events** sources to learn about changes in settlement patterns in South America and around the world.

The Economy

Some South Americans enjoy a high standard of living, and the middle class is growing. However, all the continent's countries are considered developing or middle-income countries. Argentina, Brazil, Chile, Uruguay, and Venezuela have the strongest economies. See the unit's Fast Facts table for each country's per capita GDP. All of the countries have market economies.

Agriculture Agriculture in South America ranges from subsistence farming to huge commercial farms and ranches. When the large estates from the colonial period were broken up, small farms called ***minifundia*** (mi-ni-FOOHN-dee-uh), were created. Often these *minifundia* have poor land and are too small to be profitable. In most countries a few wealthy people own much of the best land. Inequality in land ownership is the basis of much poverty and unrest in South America.

Ranch workers herd horses on a flooded Argentine plain. From horseback the cowhands manage huge herds of cattle. Cattle are such a large factor in Argentina's economy that some residents eat beef at every meal. The rider pictured offers yerba maté, a tealike drink often served in a hollow gourd.

Market-oriented farming is most highly developed in two regions—Chile's central valley and the area bridging southern Brazil and northern Argentina. Brazil produces more coffee than any other country in the world. Colombia is the second-largest coffee producer. Colombia's newest industry is selling cut flowers. These flowers are flown every night to markets around the world. Farmers in Chile's central valley grow fresh fruits and vegetables during the South American summer. This farm produce sells well in the United States during our winter. Argentina specializes in producing and exporting wheat and beef. South American countries export a wide range of agricultural products, from cacao beans to potatoes to sunflower seeds.

Prospectors dig for gold in the mud of Brazil's rain forest.

FOCUS ON ECONOMICS

Amazon Basin Development Agriculture and mining have played important roles in the development of the Amazon Basin. Consider the history of Manaus, a major inland port of more than 1 million people. Manaus lies on the Río Negro about 1,000 miles (1,600 km) from the Atlantic Ocean. It began as a mission in 1669 and remained relatively isolated until the late 1800s. Then the demand for rubber—used for waterproofing and tires—soared. Brazil's rain forest produced large amounts of rubber. Manaus grew rich on the profits from harvesting and shipping the precious material. Great buildings, including a cathedral and an opera house, date from this period. When sources for cheaper rubber developed in Asia, Manaus declined in importance. By 1920 the boom was over.

However, other resources from the Amazon rain forests are being exploited today. For example, the hardwood trees themselves are valuable, and the land is in demand for ranching. In addition, oil exploration is expanding into the Amazon Basin. Bauxite, copper, gold, iron ore, manganese, and tin draw in miners. Manaus is again bustling, exporting the region's riches from its river port.

Development in the Amazon Basin has had a downside, however. About 17,000 square miles (44,000 sq km) of rain forest are cleared every year. This deforestation threatens the region's unique plant and animal life. Development also threatens the ways of life of Amazonian Indians who have long lived in the forested basin.

READING CHECK: ***Environment and Society*** How have the creation and distribution of resources affected development in the Amazon Basin?

Economic development of the Amazon Basin is a cause of deforestation.

Industry Most South American factories produce food items, consumer goods, or building materials for local markets. Larger countries also produce cars, trucks, and jet airplanes. Workers must assemble many export products, such as clothing and small appliances, by hand. Sometimes these industries import all the parts and raw materials needed for manufacturing, which limits profits. On the other hand, these industries provide jobs and training. This experience can lead to better jobs in more advanced industries.

Cooperation among the countries could lead to economic progress. For example, countries in the southern part of the continent have formed a trade

INTERPRETING THE VISUAL RECORD

Rio de Janeiro is famous worldwide for its pre-Lenten Carnaval. The days-long celebration enhances Rio's image as an exciting city. ***How might the dancers' costumes reflect Brazil's history?*** TEKS

organization called **Mercosur**. In Spanish, *Mercosur* stands for Southern Common Market. The purpose of Mercosur is to expand trade, improve transportation, and reduce tariffs among member countries. The full members are Argentina, Brazil, Paraguay, and Uruguay. The Andean countries have similar goals. However, each country's own interests often receive the most political support.

TEKS **READING CHECK:** ***Human Systems*** What is one problem that many export industries face?

Urban Environments

In most of South America's countries the leading cities are huge in comparison to the other cities. Large parts of each country's population live in these big cities. For example, one third of Chile's people now live in or around Santiago. Nearly one third of Peru's people live in Lima. These cities grow as people move there from the countryside looking for jobs.

Both push and pull factors are at work in the migration process. Rural poverty and limited good land push people away from their small farms. The prospect of a better job and more exciting life pulls them toward the cities. However, life in the city is often just as hard as it was in the village. Rural migrants have few of the skills needed for the modern workplace. As a result, few find good jobs. Urban poor people often live in the large slums that surround major cities. These areas have their own names in each country. In Venezuela they are called *ranchos,* in Chile *callampas,* which means "mushrooms." In Brazil they are called **favelas** (fah-VE-lahs). They are home to some 25 percent of Rio de Janeiro's people. In recent years the government has worked hard to improve living conditions in the favelas. However, crime and lack of basic services, such as sanitation and schools, still trouble the people who live there.

TEKS **READING CHECK:** ***Human Systems*** How do the leading cities of South America compare to the region's other cities?

Issues and Challenges

Overall, South American governments have become more democratic. Challenges remain, however. Many South Americans, both urban and rural, are poor. In some countries, such as Bolivia, high birthrates make development harder. A growing population, the need for resources, and concern for the environment create tensions throughout the region.

South America's environmental issues concern many people around the world. Of major importance are the great rain forests of the Amazon River basin. These forests have an incredible range of

Another environmental issue is protecting the unique animals of the Galápagos Islands, a small cluster of Pacific islands that belong to Ecuador. The marine iguana pictured belongs to a species found nowhere else. Oil spills and irresponsible tourism threaten the animals.

plant and animal species and hundreds of unique local ecosystems. They are also a vital source of oxygen for the whole planet. However, much of the Amazon forest may disappear within the next 100 years. As you have read, large parts of the forest are being cleared for farms and ranches. Other businesses harvest the forests' fine woods. Major mineral deposits attract prospectors and developers to the fragile forests. Other environmentally sensitive areas of the continent are also under threat.

Soil exhaustion is another environmental problem that threatens South America's future. This loss of soil nutrients has reduced the usefulness of large areas. In Brazil, for example, growing coffee—one of the country's major exports—reduces soil fertility. In other areas, overgrazing has stripped the land of plant life.

Political issues also cause conflict. Many South American countries have been involved in border disputes, often over areas with valuable resources. Following a war during the 1880s, Bolivia and Peru lost lands, and the mineral industries there, to Chile. This war left Bolivia **landlocked**. A landlocked country has no border on the ocean. Some of these border disputes still simmer. For example, Ecuador and Peru still dispute parts of their common border, as do Venezuela and Guyana.

Violence threatens the daily lives of many South Americans. **Terrorism**, or the use of fear and violence as a political force, is common. This is true today particularly in Colombia. Armed groups there often scare people away from voting places and control large parts of the country. Much of the violence is a result of the drug trade. Drug dealers have used illegal profits to support private armies and to buy off or assassinate judges and politicians. Violence has also recently troubled Bolivia and Peru.

TEKS **READING CHECK:** *Environment and Society* Why is the disappearance of the rain forest a global concern?

Derricks tap the oil of Lake Maracaibo, Venezuela. Before democratic reforms were approved, corrupt government officials skimmed off much of the oil wealth for themselves. Overdependence on oil still threatens Venezuela's stability. When oil prices fall, the whole economy suffers.

TEKS Working with Sketch Maps, Questions 2, 3, 4, 5

Section 3 Review

go.hrw.com Homework Practice Online
Keyword: SW3 HP12

Identify

Mercosur

Define

minifundia, favelas, landlocked, terrorism

Working with Sketch Maps

On the map you created in Section 2, label Manaus, Santiago, and Lima. Then shade in the areas where commercial farming is most developed.

Reading for the Main Idea

1. *Places and Regions* What are some agricultural specialties of Brazil, Colombia, Chile, and Argentina?
2. *Human Systems* How does Mercosur try to improve trade in South America?

Critical Thinking

3. **Identifying Cause and Effect** What push and pull factors are causing South American cities to grow? What is the result?
4. **Comparing** What challenges do many of South America's largest cities share?

Organizing What You Know

5. Create a graphic organizer like the one below. Use it to describe the main environmental and political issues facing South America. Also, describe the basic challenges facing the region as a whole.

Basic:	
Environmental: 1. 2.	Political: 1. 2.

CASE STUDY

Urban Planning in Brasília

Places and Regions Urban geographers are interested in how cities function. They are also concerned about the quality of life in cities. Working with elected officials, urban planners try to design cities that meet residents' needs. They try to plan for future urban growth and figure out how to provide resources to growing cities. Food, water, electricity, transportation networks, and waste disposal are just a few of urban residents' needs.

Brasília, Brazil's capital, illustrates some of the issues that urban planners face. Brazil constructed its new capital in the 1950s and 1960s. Designed by architects and urban planners, the city is located deep in Brazil's interior. Beginning in 1956, workers built an airstrip and flew in heavy construction machinery. They began laying out streets and pouring building foundations. By 1960 the main buildings were completed and the federal government began to move to the new capital. Today Brasília has a population of about 1.8 million.

The urban planners who designed Brasília wanted to create a new urban environment that would lead to a better society. Segregation was one of the main problems the city's planners hoped to avoid. Segregation is the separation of different economic and social groups into distinct neighborhoods. To prevent this, city planners designed many groups of identical six-story apartment buildings. They organized the buildings into giant "superblocks." Each had its own schools, shopping areas, and parks. The goal was to have everyone live in the same type of environment. If this happened, the planners thought the new residents would develop a more integrated and equal society. Differences in income and social status would be less likely to lead to social problems. Even Brasília's address system reflects this idea. For example, think about addresses along Rodeo Drive in Los Angeles or Park Avenue in New York City. Such addresses imply high social status or wealth. Instead of proper names, numbers and letters identify streets and buildings in Brasília. For example, a typical address is SQS 106-F-504. Such addresses offer no clues about the social or economic position of people who live there.

Despite the efforts of Brasília's planners, the city is segregated. The poor live mostly on the outer edges of the city. This happened because of the new city's isolated location. More than 100,000 workers relocated to the region between 1957 and 1960. Known as *candangos,* these workers lived in wooden shacks on the edge of the huge construction site. When the workers were no longer needed, officials declared the wooden shacks slums and ordered them torn down.

However, it soon became clear that the workers would stay. The planners had not anticipated the housing needs of these people. As a result, officials changed the original plan of the city to provide housing and services to *candangos* and their families. They created satellite settlements on the edges of the central

Four large statues known as the Four Evangelists stand at the entrance to Brasília's Metropolitan Cathedral. A bell tower is at the right.

Brasília

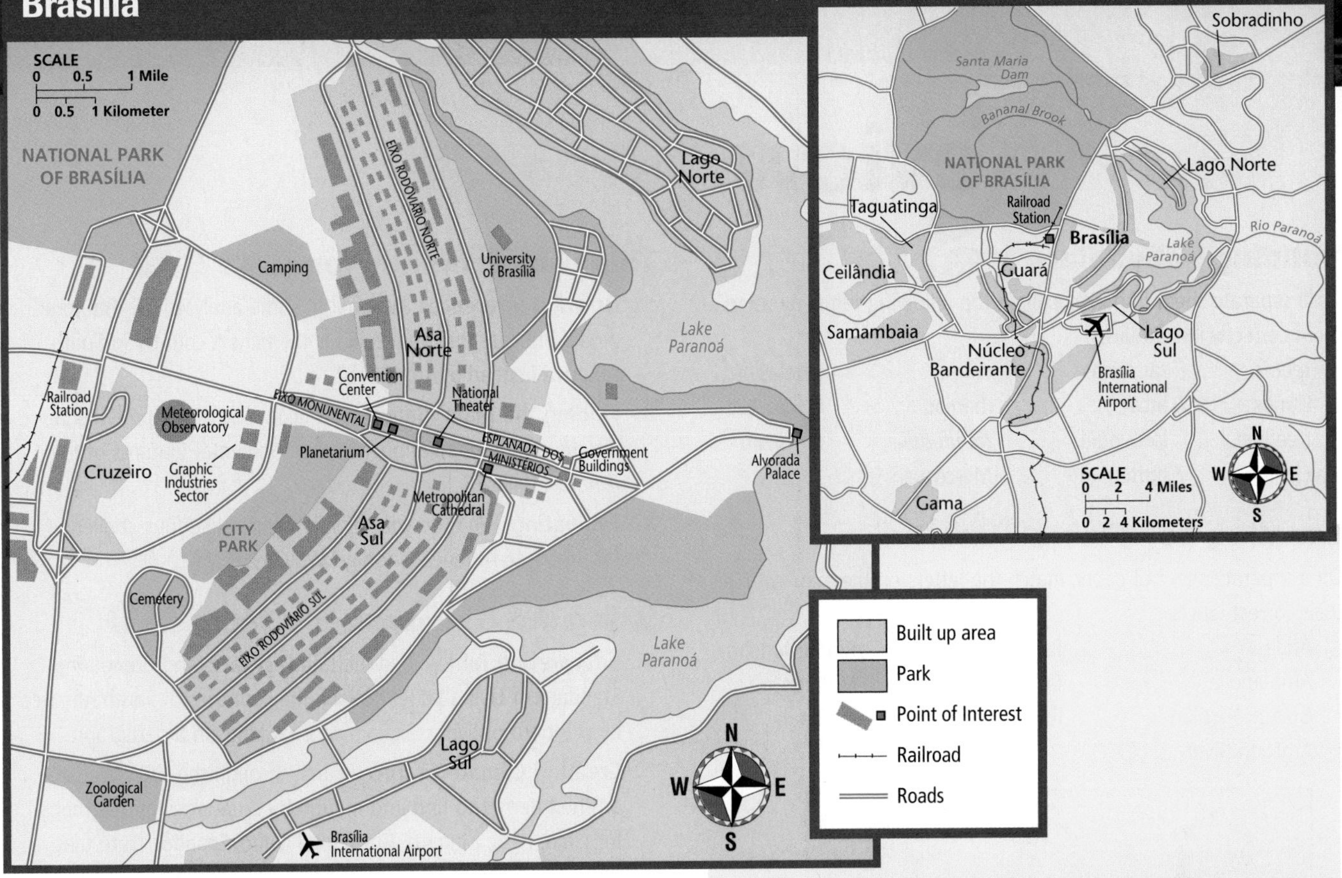

Built up area
Park
Point of Interest
Railroad
Roads

zone. These settlements have grown dramatically as rural Brazilians continue to move to Brasília in search of better jobs and pay. By the late 1980s about 75 percent of the urban area's population lived in the satellite towns.

These towns were not as strictly planned and organized as the central zone. Instead of the same apartment buildings, single-family houses are much more common. Some of Brasília's wealthy citizens have moved away from the central zone's strict building codes. On the city's outskirts, they are free to build big homes that display their wealth and social status. The satellite towns are known for the iron fences that surround homes to deter criminals. As you can see, Brasília has not developed as the integrated city originally planned. Instead, sprawling edge settlements surround the city's heart.

Applying What You Know

1. **Summarizing** How is Brasília different than planners had hoped?
2. **Comparing** How are political, economic, and social conditions interrelated in Brasília? How does this compare to other cities you have studied?

INTERPRETING THE MAP *The large map shows Brasília itself. The small one shows the city's surroundings. A neighborhood of superblocks appears in the photo.* ***How does Brasília's layout reflect its history? What are some of the satellite towns that appear on the small map?*** TEKS

CHAPTER 12 Review

Building Vocabulary

On a separate sheet of paper, explain the following terms by using them correctly in sentences.

tepuís	tar sands	coup	favelas
Llanos	Inca	manioc	landlocked
tree line	latifundia	*minifundia*	terrorism
El Niño	buffer state	Mercosur	

Locating Key Places

On a separate sheet of paper, match the letters on the map with their correct labels.

Andes	Patagonia	Rio de Janeiro
Altiplano	Cuzco	Manaus
Guiana Highlands	Buenos Aires	Santiago
Amazon River		

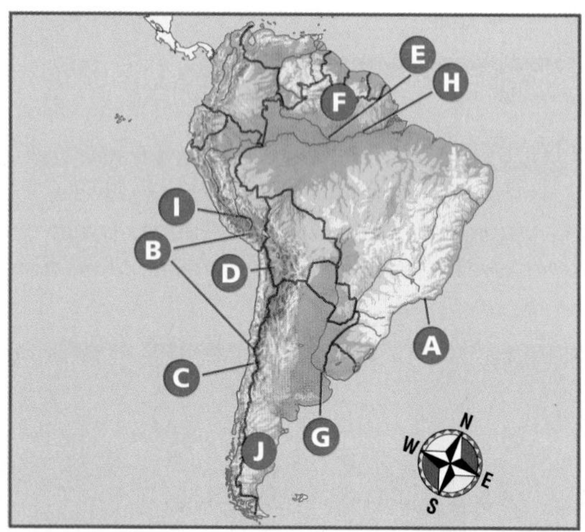

Understanding the Main Ideas

Section 1

1. *The World in Spatial Terms* What are the five climate zones of the Andes, and what vegetation is found in each?
2. *Environment and Society* What are some products that come from the rain forest?

Section 2

3. *Human Systems* What were some accomplishments of early South American peoples?
4. *Human Systems* How did Spanish and Portuguese settlement spread?

Section 3

5. *Environment and Society* What are three factors that create tensions in South America?

Thinking Critically for TAKS

1. **Drawing Inferences** Why might some analysts say that logic would indicate that Venezuela is the richest country in South America instead of Argentina?
2. **Finding the Main Idea** How is economic development shaping the urban, population, and environmental features of the Amazon Basin?
3. **Evaluating** What do you think is the most serious challenge facing South Americans today? Why?

Using the Geographer's Tools

1. **Interpreting Tables** Use the unit Fast Facts and Comparing Standard of Living tables to rank the countries of South America by standard of living. Explain your rankings in a paragraph.
2. **Creating Climate Graphs** Search through the HRW Website on the Internet to find information for creating climate graphs for Lima, Peru; Manaus, Brazil; and Quito, Ecuador. Note that these three places lie near the same latitudes. In which climate regions are these places located? What are the major climate features of each place? What factors do you think influence the climate in these places?
3. **Creating Maps** Create a map of South America that shows which countries were settled by the Spanish, Portuguese, British, Dutch, and French during the colonial era.

Writing for TAKS

Imagine that you are a subsistence farmer in Bolivia. Your cousin works on a large commercial farm in Chile's central valley. She has urged you to take an entry-level job there in farm management. Write a diary entry in which you reflect on your current life and how your life might change if you join her in Chile. When you are finished with your diary entry, proofread it to make sure you have used standard grammar, spelling, sentence structure, and punctuation.

SKILL BUILDING

Geography for Life

Comparing Maps

Environment and Society Search the Internet and library resources for aerial photographs and land use, resource, road, and vegetation maps of Brazil from different time periods. Use the photographs and maps to draw conclusions about the effects of mining, road construction, and timber harvesting on Brazil's rain forest.

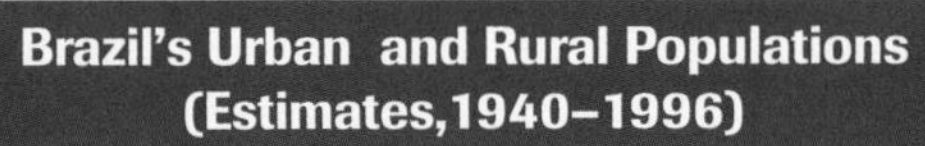

Building Social Studies Skills

go.hrw.com TAKS PREP ONLINE keyword: SW3 T12

Brazil's Urban and Rural Populations (Estimates, 1940–1996)

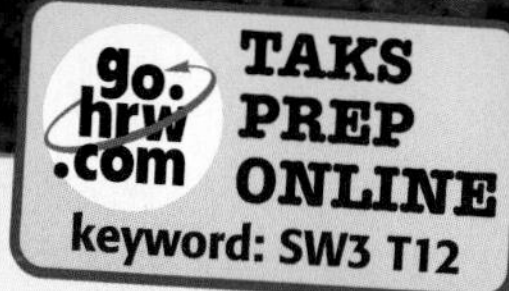

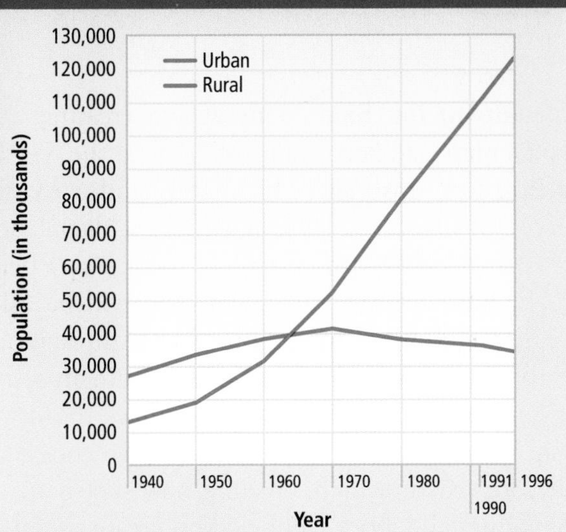

Interpreting Graphs

Use the line graph above to answer the questions that follow.

1. About when were Brazil's urban and rural populations closest in number?
   a. 1960
   b. 1964
   c. 1966
   d. 1970

2. From 1940 to 1996, the rate at which Brazil's urban population increased was very different from the rate at which its rural population decreased. Describe this difference.

Using Language

The following passage contains mistakes in grammar, punctuation, or usage. Read the passage and then answer the following questions.

"(1) Colombia is the most populous country of northern South America. (2) The national capital is Bogotá a city located high in the eastern Andes. (3) Most Colombians live in the fertile valleys and basins among the mountain ranges. (4) Rivers, such as the Cauca and Magdalena, flowing down from the Andes to the Caribbean and helping to connect settlements between the mountains and the coast."

3. In which sentence is a comma missing?
   a. 2
   b. 3
   c. 4
   d. none of them

4. Rewrite sentence 4 so that it is not a fragment.

Alternative Assessment

PORTFOLIO ACTIVITY

Learning about Your Local Geography TEKS

Group Project: Research

Indigenous people living in the Amazon rain forest are often in conflict with the loggers, miners, and ranchers who want to develop their land and its resources for profit. Plan, organize, and complete a research project on issues that have arisen in your state or region between longtime residents and business interests. Determine the relative success of conflict and cooperation in resolving those issues. Develop a hypothesis to describe what you find. Write a brief report and place it in your portfolio.

internet connect

go.hrw.com

Internet Activity: go.hrw.com
KEYWORD: SW3 GT12 TEKS

Choose a topic about South America to:

- trek through the Guiana Highlands and create a poster.
- create a brochure on Machu Picchu.
- learn about threats to the Amazon Basin.

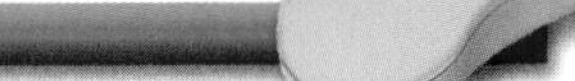

Skill Building for TAKS

WORKSHOP 1

Using Sketch Maps

As you have read, geographers use many kinds of maps to study our world. Sketch maps are simple kinds of maps. To create the simplest sketch map, you need only a pencil and a piece of paper. Depending on how complex you make the map, you might also use other tools. For example, you might use a compass, color pencils, or a ruler or other instrument with a straight edge.

Sketch maps are useful tools for communicating information in a visual way. They can show physical features, such as the locations of mountains, rivers, deserts, and oceans. They can also show human features, such as countries, cities, roads, and economic activities. Sketch maps can also show geographic distributions, such as natural resources, climate regions, vegetation patterns, and population density. Finally, sketch maps can show geographic relationships. For example, sketch maps might show relationships between climate and vegetation growth or between the locations of cities and nearby resources.

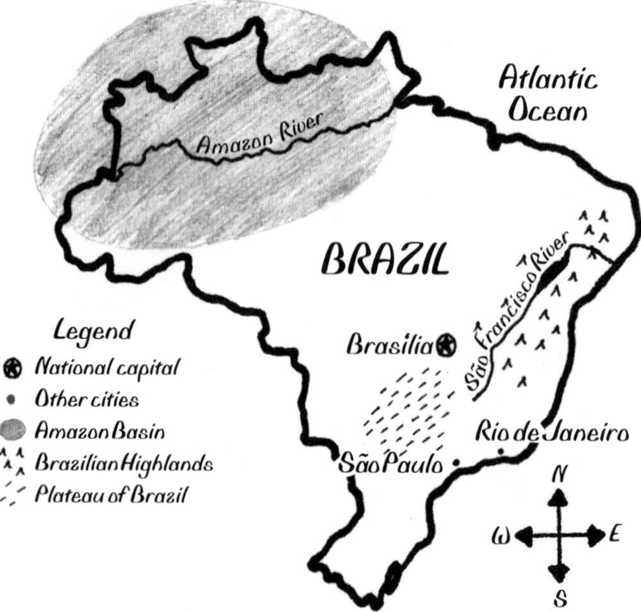

Developing the Skill One key to creating a useful sketch map is to keep it simple. Look at the sketch map of Brazil on this page. This map includes a variety of places and features, such as cities, highlands, and the Amazon River. If this map had too many labels, you might have trouble reading it. You will be asked to create sketch maps in section reviews throughout this textbook. If you are building a map from an activity in a previous section, make sure you have room in your map for new labels. If you do not have enough room on your existing map, create a new sketch map. The following guidelines will help you create useful sketch maps on your own:

- You can create sketch maps from a model, such as another map, or from a mental map.
- You can use a pencil to trace a model map's outline onto a thin sheet of paper. As an alternative, you might create your own freehand sketch map.
- Do not crowd your sketch map. Identify places that are important for your purpose. Leave out places that are not relevant or required.
- Assign symbols to represent certain kinds of places. For example, small dots or circles might represent cities. Shaded areas or special marks might show highland areas. Create a legend that identifies each symbol you use.
- Make sure the information in your map is clear. You might want to point out geographic features, distributions, and relationships in a caption.

Practicing the Skill

1. Study the political and physical maps of South America at the beginning of this unit. Then create a sketch map of the continent, labeling each country, each national capital, and at least three major rivers.
2. Locate a map of your county. Then create a sketch map that shows important towns, roads, and rivers there.

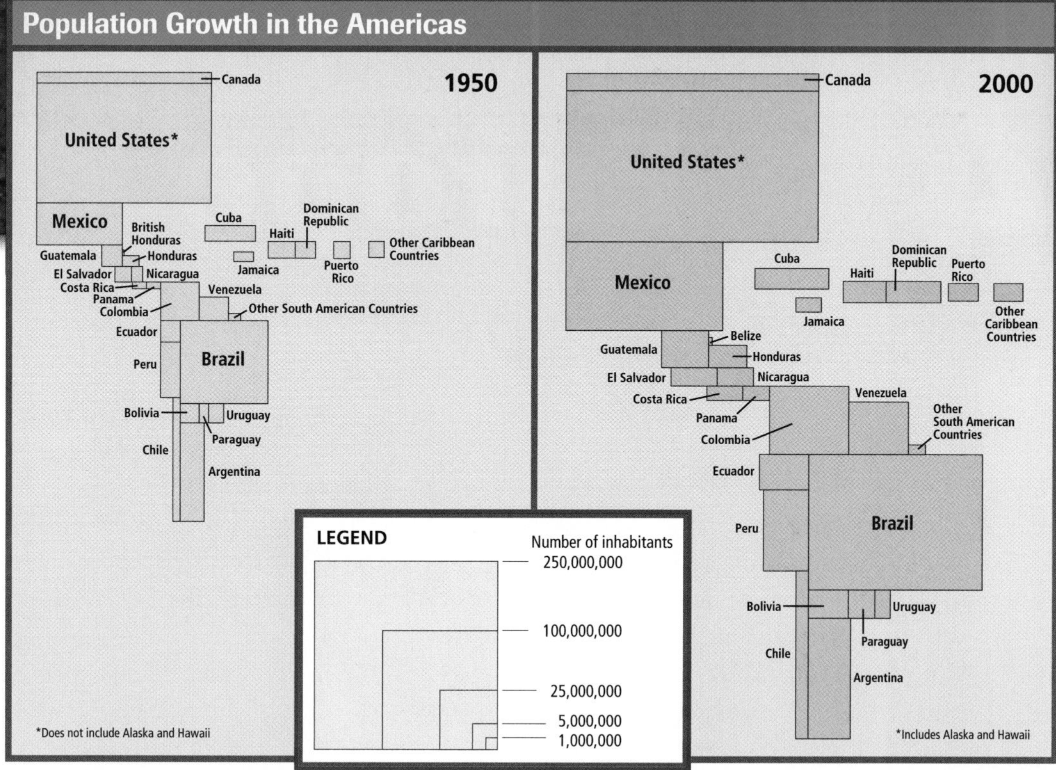

WORKSHOP 2

Using Cartograms

We are accustomed to looking at maps on which countries and states have been drawn in proportion to their geographic area. Yet it is possible to create maps that have been drawn so that sizes are proportional to something other than geographic area. Such maps are called cartograms. Actually, a cartogram is a cross between a map and a graph. Cartograms let geographers display statistical information visually. The size of countries or states can be created in proportion to many measures, such as population or GDP. For example, a cartogram showing all of the U.S. states in proportion to their population would look quite different from a map showing states by geographic area.

Developing the Skill Using a cartogram is much like reading a map. First, read the title and legend to identify the kind of information presented on the cartogram. The cartogram on this page compares the populations of countries in North and South America in 1950 and 2000. The legend shows how the size of a country changes depending on how many people live in that country. The largest countries on these cartograms are the United States, Brazil, and Mexico. Therefore, you know that those three countries are also the most populous countries in the Americas.

Practicing the Skill

You can make cartograms by hand or by using special computer software. Follow the instructions below to make a cartogram of South America by hand. You will need the following materials: color pencils or markers, scissors, an almanac or other database, graph paper, and a map of South America.

1. Find the total GDP for each country in South America.
2. Determine a scale for sizing each country by GDP. For example, you might use one square unit of area per $10 billion or $100 billion.
3. Cut out one piece of graph or grid paper for each South American country. Each piece should be cut in proportion to a country's GDP. For example, a country with a GDP of $10 billion should be shown as 10 times smaller than a country with a GDP of $100 billion.
4. Secure the pieces on paper or poster board. Each shape should be cut and pasted to reflect a country's approximate location in South America. You might color each piece so that it is easier to distinguish between countries.
5. Create a legend and title.
6. Write a short caption describing your cartogram and summarizing what it shows. Then present your cartogram to the rest of the class.

UNIT

4 Europe

Notre Dame Cathedral, Paris

CONNECTING TO

Literature

From "Pericles' Funeral Oration" in

Thucydides' History of the Peloponnesian War

translated by Benjamin Jowett

Thucydides' *History of the Peloponnesian War* is still read for the insights it provides into the conduct of war. In this excerpt Pericles is speaking at the funeral of those Athenians who have died in the war.

Our form of government does not enter into rivalry with the institutions of others. We do not copy our neighbors, but are an example to them. It is true that we are called a democracy, for the administration is in the hands of the many and not of the few. But while the law secures equal justice to all alike in their private disputes, the claim of excellence is also recognized; and when a citizen is in any way distinguished, he is preferred to the public service, not as a matter of privilege, but as the reward of merit. Neither is poverty a bar, but a man may benefit his country whatever be the obscurity of his condition. There is no exclusiveness in our public life, and in our private intercourse we are not suspicious of one another, nor angry with our neighbor if he does what he likes; we do not put on sour looks at him which, though harmless, are not pleasant. While we are thus unconstrained in our private intercourse, a spirit of reverence pervades our public acts. . . .

. . . I would have you day by day fix your eyes upon the greatness of Athens, until you become filled with the love of her; and when you are impressed by the spectacle of her glory, reflect that this empire has been acquired by men who knew their duty and had the courage to do it, who in the hour of conflict had the fear of dishonor always present to them, and who, if ever they failed in an enterprise, would not allow their virtues to be lost to their country, but freely gave their lives to her as the fairest offering which they could present at her feast. The sacrifice which they collectively made was individually repaid to them; for they received again each one for himself a praise which grows not old. . . . Make them your examples.

Analyzing the Primary Source

1. **Analyzing Information** How does Pericles appeal to Athenians?
2. **Supporting a Point of View** How do you think Athenian ideals are reflected in American democracy?

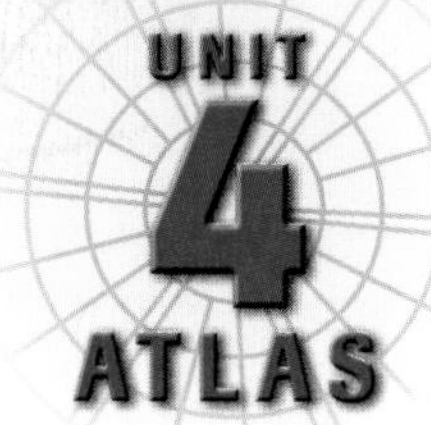

The World in Spatial Terms

UNIT 4 ATLAS

Europe: Political

1. **Places and Regions** Which European countries are island countries?
2. **Places and Regions** Compare this map to the physical map. Which two countries occupy the Scandinavian Peninsula?

Critical Thinking

3. **Analyzing Information** Compare this map to the physical map. Where in Europe do political boundaries coincide with physical barriers? TEKS

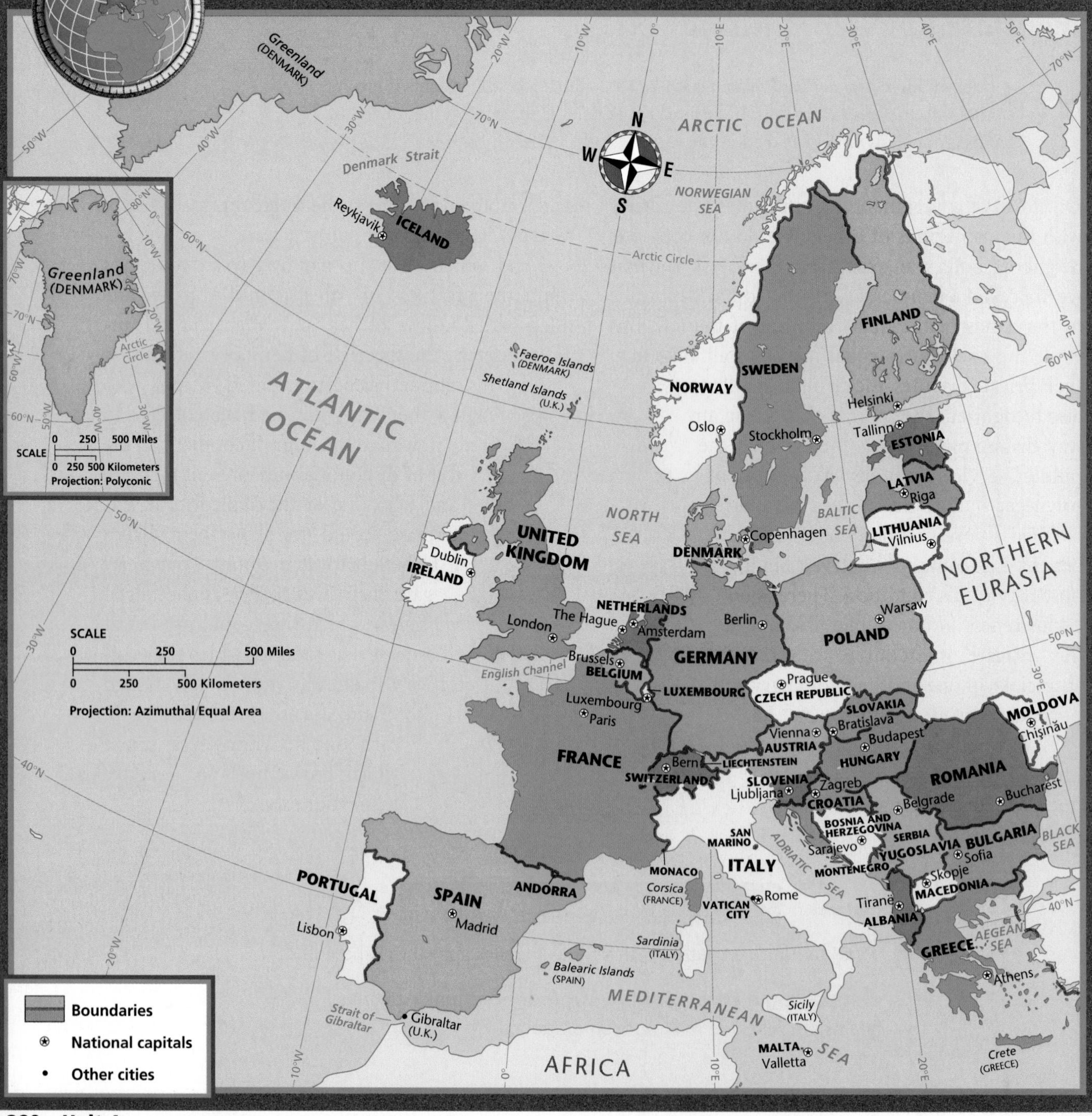

Europe: Physical

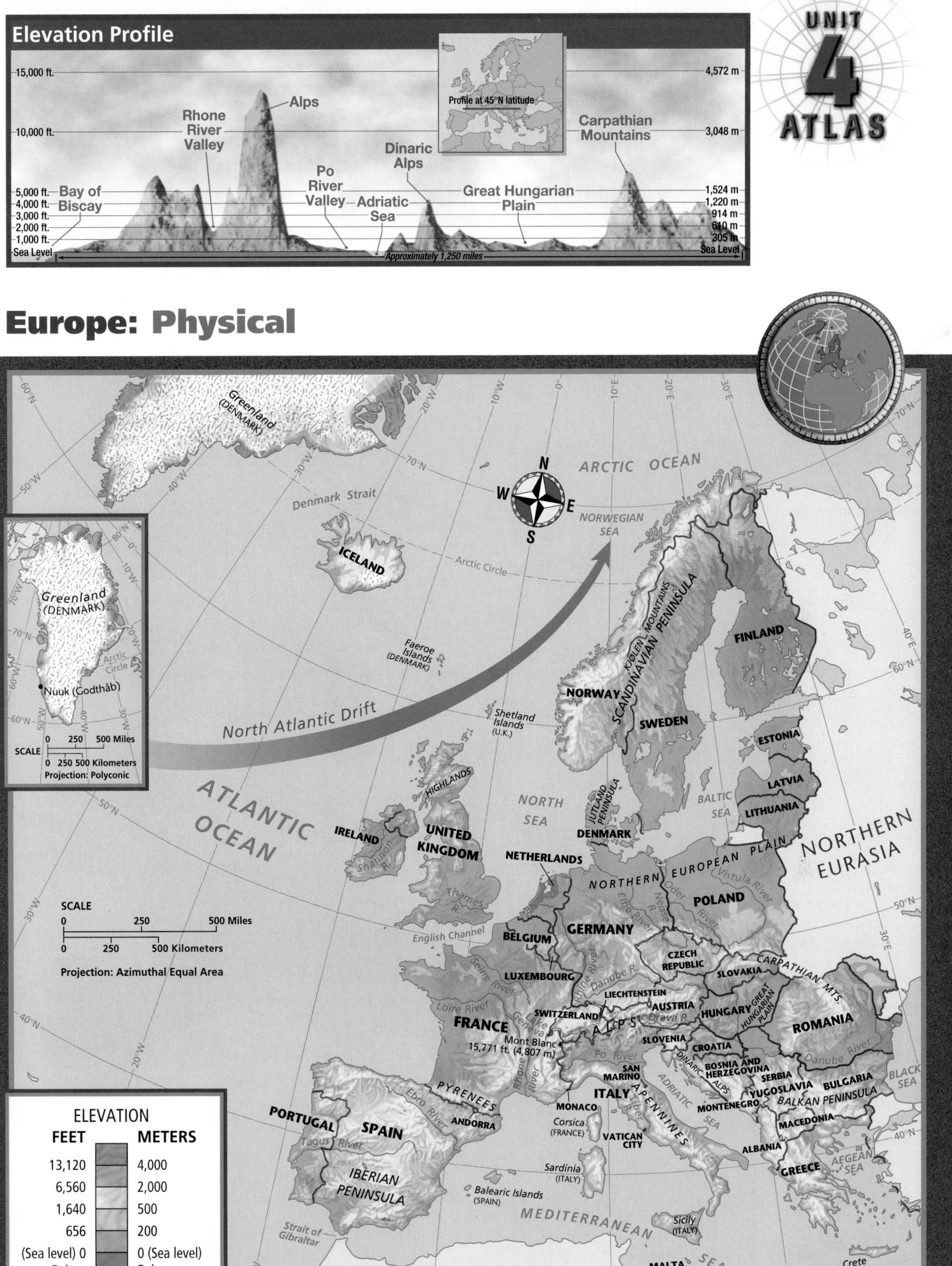

Europe: Climate

1. **Physical Systems** Which climates cover large parts of Europe? TEKS
2. **Physical Systems** Which climate types are not found in Europe? Why do you think this is so? TEKS

Critical Thinking

3. **Comparing** How is the climate of western Europe different from that of eastern Europe? How might the positions of these two areas on the continent of Europe influence these differences? TEKS

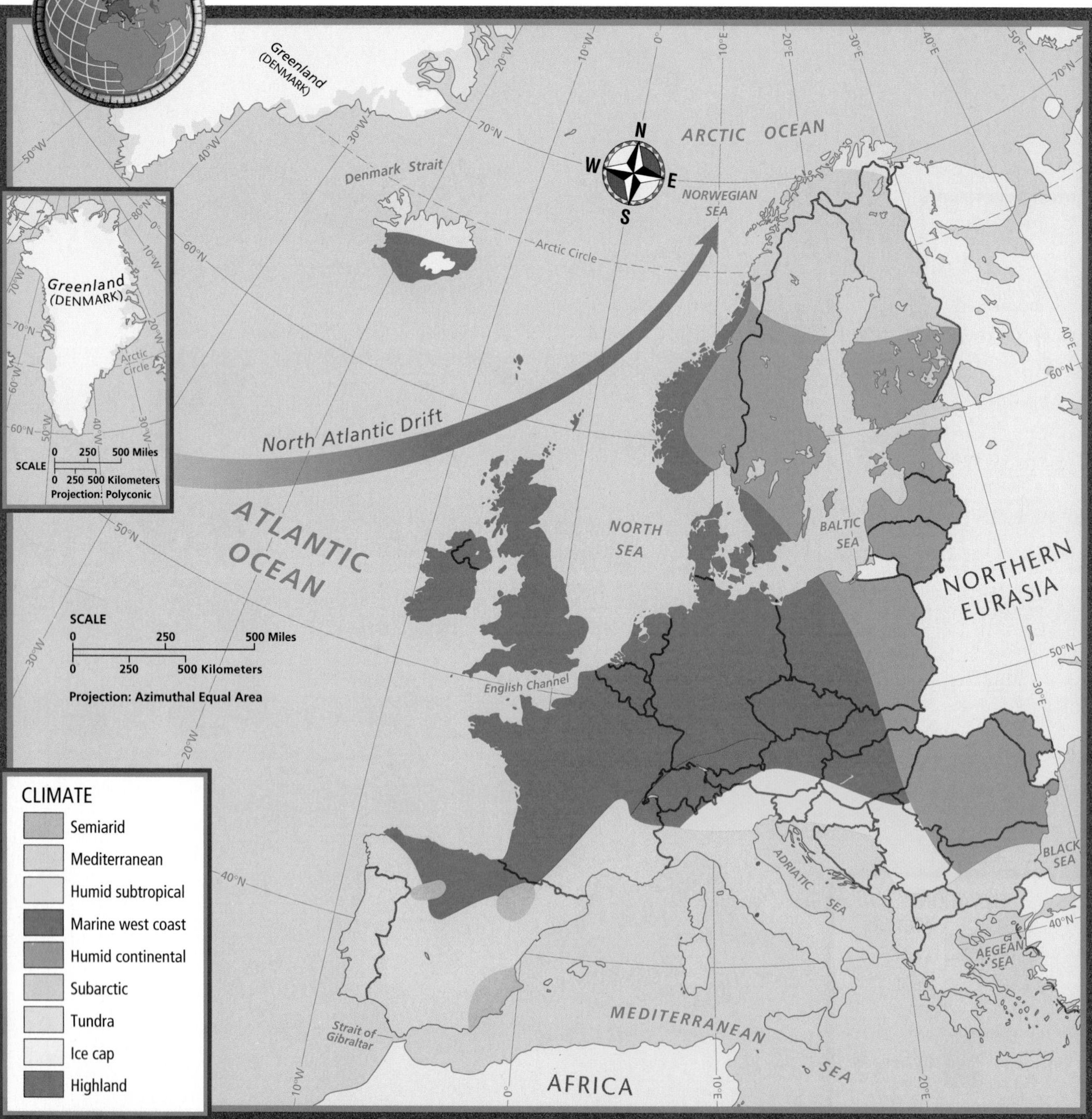

Greenland (DENMARK)

ARCTIC OCEAN

NORWEGIAN SEA

Denmark Strait

Arctic Circle

North Atlantic Drift

ATLANTIC OCEAN

NORTH SEA

BALTIC SEA

NORTHERN EURASIA

English Channel

ADRIATIC SEA

BLACK SEA

AEGEAN SEA

Strait of Gibraltar

MEDITERRANEAN SEA

AFRICA

SCALE 0 250 500 Miles; 0 250 500 Kilometers

Projection: Azimuthal Equal Area

Inset: Greenland (DENMARK). SCALE 0 250 500 Miles; 0 250 500 Kilometers. Projection: Polyconic

CLIMATE
- Semiarid
- Mediterranean
- Humid subtropical
- Marine west coast
- Humid continental
- Subarctic
- Tundra
- Ice cap
- Highland

Europe: Precipitation

1. *Physical Systems* Compare this map to the climate map. How much precipitation falls in Spain's semiarid regions? TEKS

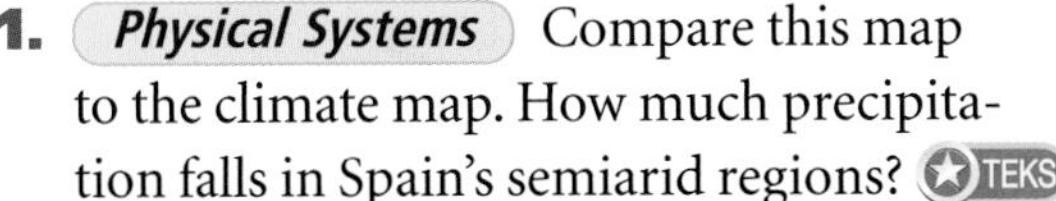

2. *Places and Regions* Compare this map to the political map. How much precipitation do most areas in France and Germany receive? TEKS

Critical Thinking

3. **Analyzing Information** Which areas receive the most precipitation? Why do you think this is so? TEKS

ANNUAL PRECIPITATION

Centimeters	Inches
Under 25	Under 10
25–50	10–20
50–100	20–40
100–150	40–60
150–200	60–80
Over 200	Over 80

Europe: Population

1. *Environment and Society* Compare this map to the physical map. What major lowland region in Europe has a high population density?
2. *Places and Regions* Which country in Europe has the lowest average population density? TEKS

Critical Thinking

3. **Drawing Inferences** Compare this map to the physical and climate maps. Why do you think Europe's average population density is so high? TEKS

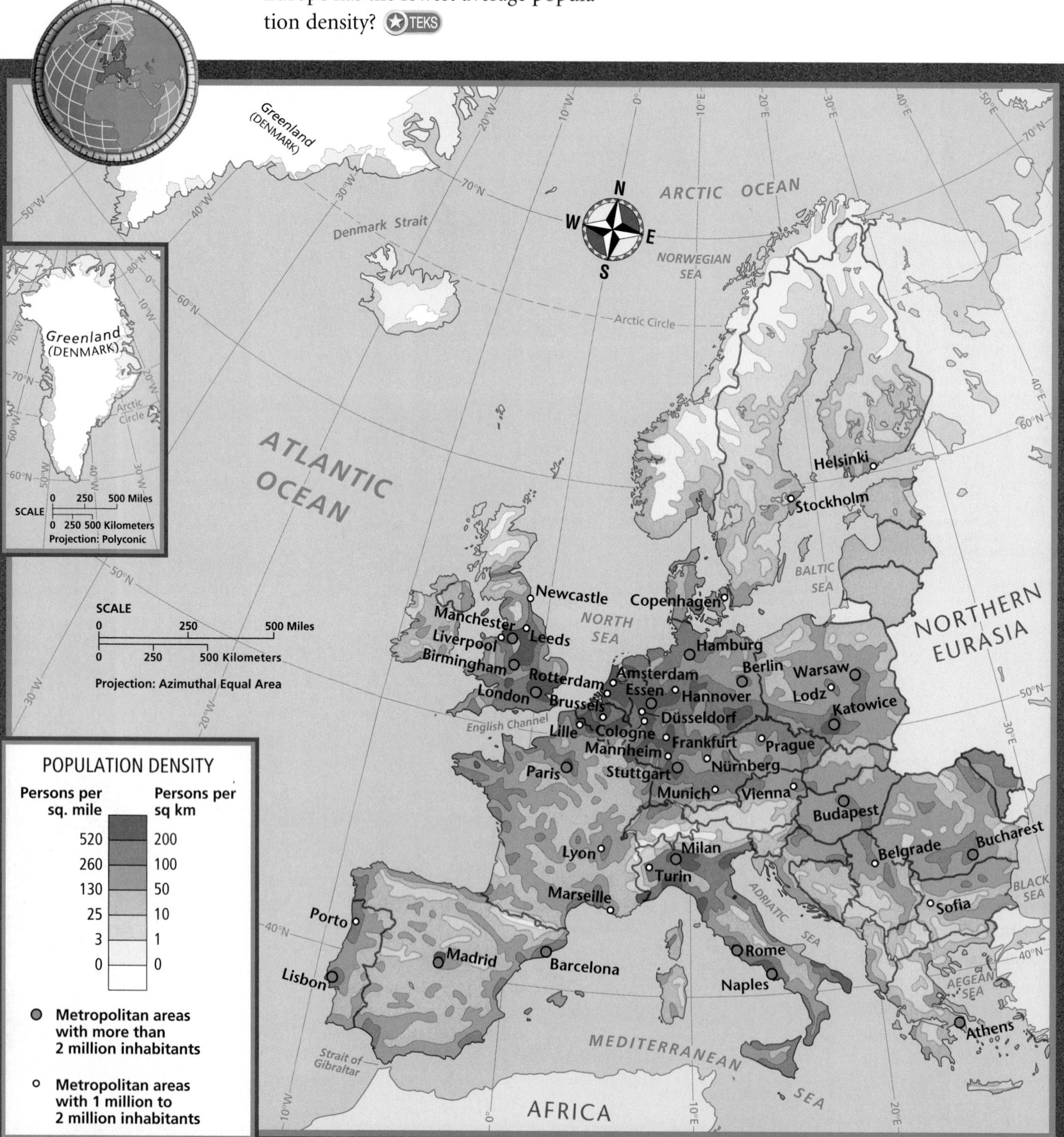

SCALE
0 250 500 Miles
0 250 500 Kilometers
Projection: Azimuthal Equal Area

Greenland (DENMARK)
SCALE 0 250 500 Miles
0 250 500 Kilometers
Projection: Polyconic

POPULATION DENSITY

Persons per sq. mile	Persons per sq km
520	200
260	100
130	50
25	10
3	1
0	0

● Metropolitan areas with more than 2 million inhabitants

○ Metropolitan areas with 1 million to 2 million inhabitants

UNIT 4 ATLAS

Europe: Land Use and Resources

1. *Places and Regions* What are the main economic resources of the North Sea?
2. *Human Systems* What is the most widespread type of land use in Europe? TEKS

Critical Thinking

3. **Drawing Inferences** Is subsistence agriculture found in Europe? What does this indicate about Europe's level of economic development? TEKS

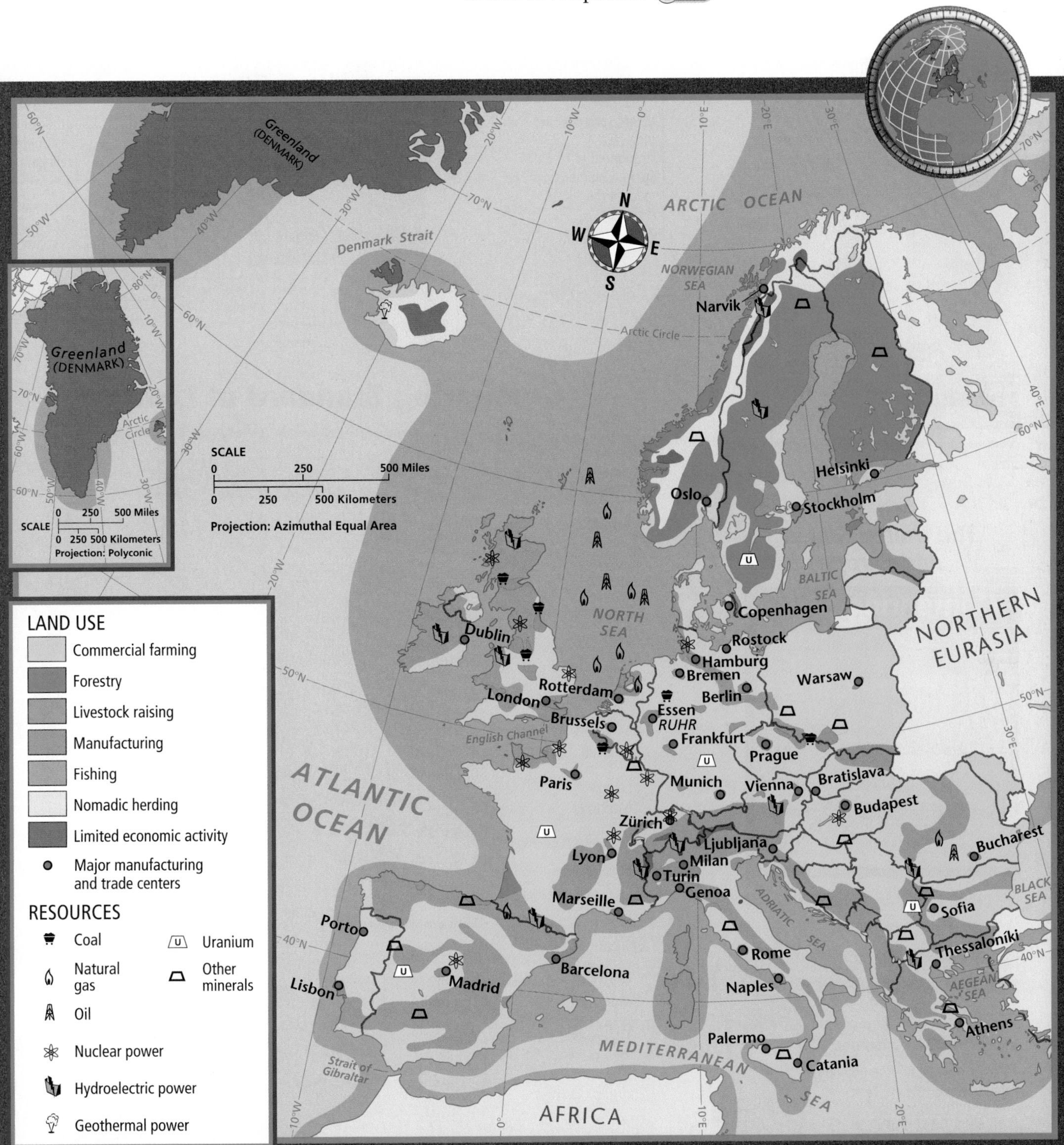

Time Line: Europe

1000

1066
William of Normandy conquers England.

1300

1300s–1500s
The Renaissance renews interest in learning across Europe.

1347
The Black Death begins its sweep across Europe.

1400

1492
Moors surrender their last stronghold in Spain. Columbus reaches the Americas.

1800

1800s
The steam engine revolutionizes transportation across Britain and Europe.

1900

1914
World War I erupts.

1939
Germany invades Poland, sparking World War II.

1957
The European Economic Community, or EEC, is created. Later, it evolves into the European Union.

1990
East and West Germany are reunited.

1991
Soviet Union collapses.

The United States and Europe

Comparing Sizes

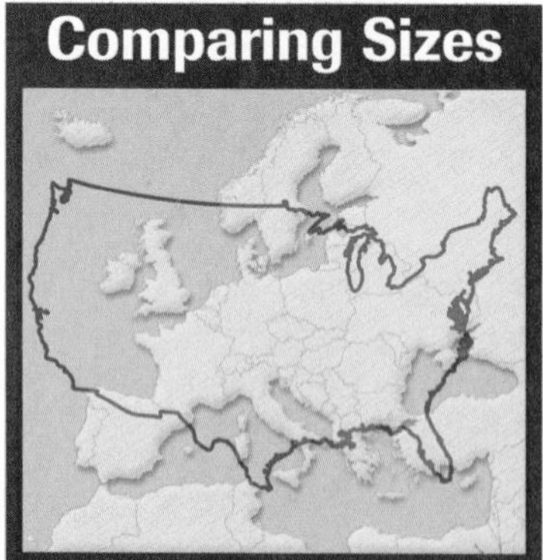

internet connect

GO TO: go.hrw.com
KEYWORD: SW3 Almanac
FOR: Additional information and reference sources

Comparing Standard of Living

COUNTRY	LIFE EXPECTANCY (in years)	INFANT MORTALITY (per 1,000 live births)	LITERACY RATE	DAILY CALORIC INTAKE (per person)
Albania	69, male 75, female	40	93%	2,324
France	75, male 83, female	4	99%	3,588
Germany	74, male 81, female	5	99%	3,265
Greece	76, male 81, female	6	95%	3,561
Ireland	74, male 80, female	6	98%	3,638
Moldova	60, male 69, female	43	96%	Not available
Poland	69, male 78, female	9	99%	3,307
Spain	75, male 83, female	5	97%	3,338
United Kingdom	75, male 81, female	6	99%	3,149
Yugoslavia	71, male 77, female	17	93%	3,396
United States	74, male 80, female	7	97%	3,603

Sources: Central Intelligence Agency, *2001 World Factbook; Britannica Book of the Year, 2000*

Fast Facts: Europe

FLAG	COUNTRY Capital	POPULATION (in millions) / POP. DENSITY	AREA	PER CAPITA GDP (in US $)	WORKFORCE STRUCTURE (largest categories)	ELECTRICITY CONSUMPTION (kilowatt hours per person)	TELEPHONE LINES (per person)
	Albania Tiranë	3.5 / 316/sq. mi.	11,100 sq. mi. 28,748 sq km	$ 3,000	50% industry, services 50% agriculture	1,532 kWh	0.02
	Andorra Andorra la Vella	0.07 / 374/sq. mi.	181 sq. mi. 468 sq km	$ 18,000	78% services 21% industry	Not available	0.49
	Austria Vienna	8.2 / 252/sq. mi.	32,378 sq. mi. 83,858 sq km	$ 25,000	68% services 29% industry, crafts	6,531 kWh	0.49
	Belgium Brussels	10.3 / 871/sq. mi.	11,780 sq. mi. 30,510 sq km	$ 25,300	73% services 25% industry	7,319 kWh	0.46
	Bosnia and Herzegovina Sarajevo	3.9 / 199/sq. mi.	19,741 sq. mi. 51,129 sq km	$ 1,700	51% manufacturing 15% services	684 kWh	0.08
	Bulgaria Sofia	7.7 / 180/sq. mi.	42,822 sq. mi. 110,910 sq km	$ 6,200	43% services 31% industry	4,305 kWh	0.42
	Croatia Zagreb	4.3 / 199/sq. mi.	21,831 sq. mi. 56,542 sq km	$ 5,800	Not available	3,148 kWh	0.34
	Czech Republic Prague	10.3 / 337/sq. mi.	30,450 sq. mi. 78,866 sq km	$ 12,900	55% services 40% industry	5,154 kWh	0.38
	Denmark Copenhagen	5.4 / 322/sq. mi.	16,639 sq. mi. 43,094 sq km	$ 25,500	79% services 17% industry	6,149 kWh	0.89
	Estonia Tallinn	1.4 / 82/sq. mi.	17,462 sq. mi. 45,226 sq km	$ 10,000	69% services 20% industry	4,782 kWh	0.33
	Finland Helsinki	5.2 / 40/sq. mi.	130,127 sq. mi. 337,030 sq km	$ 22,900	32% public service 22% industry	15,768 kWh	0.55
	France Paris	59.6 / 282/sq. mi.	211,208 sq. mi. 547,030 sq km	$ 24,400	71% services 25% industry	6,696 kWh	0.59
	Germany Berlin	83.0 / 602/sq. mi.	137,846 sq. mi. 357,021 sq km	$ 23,400	64% services 33% industry	5,964 kWh	0.54
	Greece Athens	10.6 / 209/sq. mi.	50,942 sq. mi. 131,940 sq km	$ 17,200	59% services 21% industry	4,080 kWh	0.51
	Hungary Budapest	10.1 / 281/sq. mi.	35,919 sq. mi. 93,030 sq km	$ 11,200	65% services 27% industry	3,486 kWh	0.31
	Iceland Reykjavík	0.3 / 7/sq. mi.	39,768 sq. mi 103,000 sq km	$ 24,800	60% services 13% manufacturing	23,655 kWh	0.60
	Ireland Dublin	3.8 / 142/sq. mi.	27,135 sq. mi. 70,280 sq km	$ 21,600	64% services 28% industry	4,794 kWh	0.41

Sources: Central Intelligence Agency, *2001 World Factbook; The World Almanac and Book of Facts, 2001; Britannica Book of the Year, 2000;* U.S. Census Bureau
The CIA calculates per capita GDP in terms of purchasing power parity. This formula equalizes the purchasing power of each country's currency.

Fast Facts: Europe

FLAG	COUNTRY Capital	POPULATION (in millions) / POP. DENSITY	AREA	PER CAPITA GDP (in US $)	WORKFORCE STRUCTURE (largest categories)	ELECTRICITY CONSUMPTION (kilowatt hours per person)	TELEPHONE LINES (per person)
	Italy Rome	57.7 / 496/sq. mi.	116,305 sq. mi. 301,230 sq km	$ 22,100	62% services 33% industry	4,722 kWh	0.43
	Latvia Riga	2.4 / 96/sq. mi.	24,938 sq. mi. 64,589 sq km	$ 7,200	65% services 25% industry	1,809 kWh	0.31
	Liechtenstein Vaduz	0.03 / 525/sq. mi.	62 sq. mi. 160 sq km	$ 23,000	53% services 45% ind., trade, const.	Not available	0.61
	Lithuania Vilnius	3.6 / 143/sq. mi.	25,174 sq. mi. 65,200 sq km	$ 7,300	50% services 30% industry	2,719 kWh	0.29
	Luxembourg Luxembourg	0.4 / 444/sq. mi	998 sq. mi. 2,586 sq km	$ 36,400	83% services 14% industry	13,881 kWh	0.71
	Macedonia Skopje	2 / 209/sq. mi.	9,781 sq. mi. 25,333 sq km	$ 4,400	23% agriculture 18% mining, manuf.	2,928 kWh	0.20
	Malta Valletta	0.4 / 3,234/sq. mi.	122 sq. mi. 316 sq km	$ 14,300	71% services 24% industry	3,888 kWh	0.47
	Moldova Chişinău	4.4 / 339/sq. mi.	13,067 sq.mi. 33,843 sq km	$ 2,500	40% agriculture 14% industry	1,304 kWh	0.14
	Monaco Monaco	0.03 / 42,456/sq. mi.	0.75 sq. mi. 1.95 sq km	$ 27,000	22% manufacturing 22% public admin.	Not available	0.97
	Netherlands Amsterdam, The Hague	16 / 997/sq. mi.	16,033 sq. mi. 41,526 sq km	$ 24,400	73% services 23% industry	6,117 kWh	0.57
	Norway Oslo	4.5 / 36/sq. mi.	125,181 sq. mi. 324,220 sq km	$ 27,700	74% services 22% industry	24,602 kWh	0.61
	Poland Warsaw	38.6 / 320/sq. mi.	120,728 sq. mi. 312,685 sq km	$ 8,500	50% services 28% agriculture	3,106 kWh	0.21
	Portugal Lisbon	10.1 / 282/sq. mi.	35,672 sq. mi. 92,391 sq km	$ 15,800	60% services 30% industry	3,767 kWh	0.53
	Romania Bucharest	22.4 / 244/sq. mi.	91,699 sq. mi. 237,500 sq km	$ 5,900	40% agriculture 35% services	2,002 kWh	0.17
	San Marino San Marino	0.03 / 1,158/sq. mi.	23.6 sq. mi. 61.2 sq km	$ 32,000	60% services 38% industry	Not available	0.66
	Slovakia Bratislava	5.4 / 287/sq. mi.	18,859 sq. mi. 48,845 sq km	$ 10,200	46% services 29% industry	3,965 kWh	0.36
	Slovenia Ljubljana	2 / 247/sq. mi.	7,820 sq. mi. 20,253 sq km	$ 12,000	38% manufacturing 15% trade	5,193 kWh	0.37

FLAG	COUNTRY Capital	POPULATION (in millions) POP. DENSITY	AREA	PER CAPITA GDP (in US $)	WORKFORCE STRUCTURE (largest categories)	ELECTRICITY CONSUMPTION (kilowatt hours per person)	TELEPHONE LINES (per person)
	Spain Madrid	40.0 205/sq. mi.	194,896 sq. mi. 504,782 sq km	$ 18,000	64% services 28% manuf., mining, const.	4,735 kWh	0.43
	Sweden Stockholm	8.9 51/sq. mi.	173,731 sq. mi. 449,964 sq km	$ 22,200	74% services 24% industry	14,515 kWh	0.68
	Switzerland Bern	7.3 457/sq. mi.	15,942 sq. mi. 41,290 sq km	$ 28,600	69% services 26% industry	7,121 kWh	0.66
	United Kingdom London	59.6 631/sq. mi.	94,525 sq. mi. 244,820 sq km	$ 22,800	80% services 19% industry	5,583 kWh	0.58
	Vatican City Vatican City	0.0009 5,235/sq. mi.	0.17 sq. mi. 0.44 sq km	Not available	Not available	Not available	Not available
	Yugoslavia Belgrade	10.7 270/sq. mi.	39,517 sq. mi. 102,350 sq km	$ 2,300	41% industry 35% services	3,091 kWh	0.19
	United States Washington, D.C.	281.4 76/sq. mi.	3,717,792 sq. mi. 9,629,091 sq km	$ 36,200	30% manage., prof. 29% tech., sales, admin.	12,211 kWh	0.69

The French Alps

CHAPTER 13 Natural Environments of Europe

In this chapter you will learn about Europe's climates, resources, and physical features, including its great rivers. Here British writer Jan Morris (1926–) reflects on the importance of the Rhine River to Europe's economy.

Sunflower, Spain

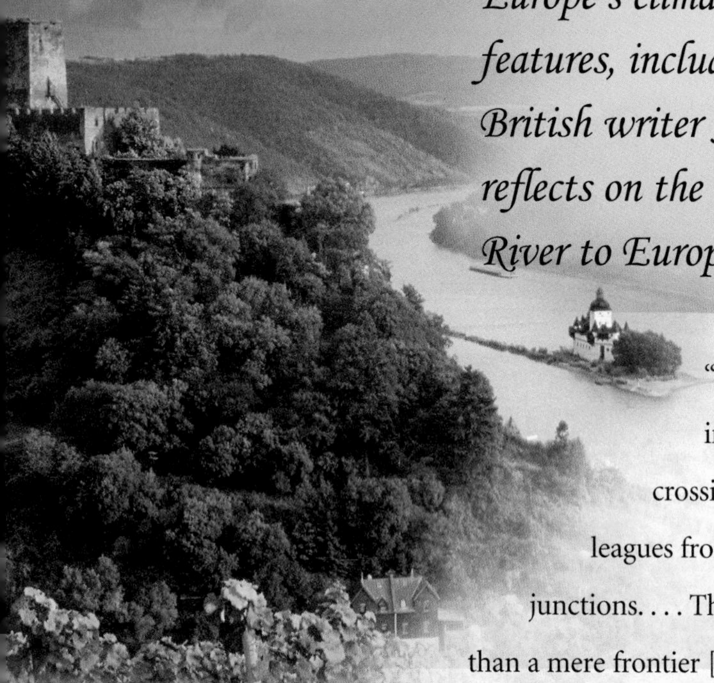

"The boat people of the inland waterways form an inner community of Europe, forever on the move, crossing the old frontiers constantly and meeting colleagues from all over the continent at the big river ports and junctions. . . . The supreme European river is the Rhine—far more than a mere frontier [border], . . . but a majestic communication.

Rüdesheim in Germany is . . . one of the best (or worst) places to gauge the importance of the river and its valley as a conductor of traffic. . . . There is seldom a silent moment on the Rhine at Rüdesheim, scarcely a moment without the plod, hurtle or judder of the river's purpose. The Rhine is the busiest of all waterways. As a highway it begins at Konstanz, on the frontier between Switzerland and Germany, where a large zero on a riverside board tells the barge-captain that he has 1,165 kilometers [700 miles] to go to the North Sea. By the time he gets to Rotterdam he will have passed beneath some 150 bridges, sailed along the littorals [shores] of six nations and helped to define a continent. The Rhine, said Thomas Carlyle [a Scottish writer of the 1800s], was his 'first idea of a world river,' and a world river it is, because the goods it carries across Europe to the sea are distributed across all earth's oceans."

Rhine River, Germany

Highland cow, Scotland

1 Physical Features

READ TO DISCOVER

1. What are Europe's major landform regions?
2. What are the region's main rivers and bodies of water?

WHY IT MATTERS

Europe's Rhine and Danube Rivers suffer from major pollution problems. Use CNNfyi.com or other **current events** sources to learn about river pollution and what can be done to clean up polluted rivers.

DEFINE

fjords
polders
dikes
navigable

LOCATE

Ural Mountains
Mediterranean Sea
Scandinavian Peninsula
Iberian Peninsula
Italian Peninsula

Locate, continued

Balkan Peninsula
Northern European Plain
Alps
Carpathian Mountains
Pyrenees
Black Sea
Bosporus
North Sea
Rhine River
Danube River

Natural Environments of Europe

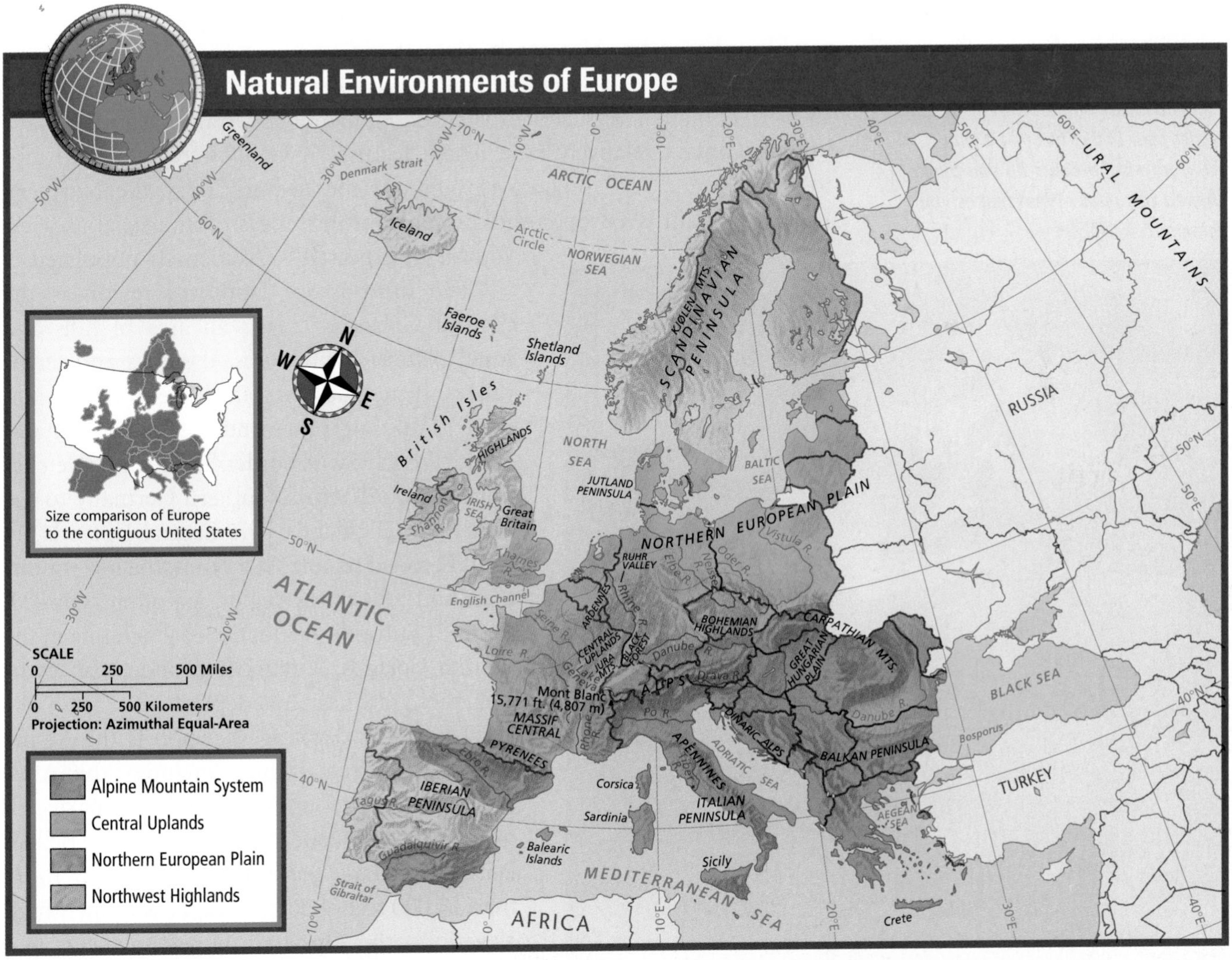

Size comparison of Europe to the contiguous United States

SCALE
0 250 500 Miles
0 250 500 Kilometers
Projection: Azimuthal Equal-Area

- Alpine Mountain System
- Central Uplands
- Northern European Plain
- Northwest Highlands

Our Amazing Planet

In 1783 a volcano in Iceland erupted continuously for 10 months, devastating large areas of the island. Poisonous gases released during the eruption killed about 75 percent of Iceland's livestock. Many people died from the resulting famine.

Landforms

Europe stretches from the Atlantic Ocean to the Ural Mountains and from the Arctic Ocean to the Mediterranean Sea. In this unit we will study the part of Europe that lies generally west of Russia, Belarus, and Ukraine. Compared to the other continents, Europe is small. However, within Europe's small area is a complex variety of landforms, islands, and peninsulas. Major islands include Great Britain, Ireland, and Iceland. Major peninsulas include the Scandinavian, Iberian, Italian, and Balkan Peninsulas.

Europe can be divided into four major landform regions. These regions are the Northwest Highlands, the Northern European Plain, the Central Uplands, and the Alpine mountain system. (See the map.) The Northwest Highlands is an ancient eroded region of rugged hills and low mountains. In the north it includes the hills of Ireland and England, the Scottish Highlands, and the mountains of Scandinavia. Northwestern France and some of the Iberian Peninsula are also part of the Northwest Highlands. During the last ice age, glaciers scoured the landscapes of Scandinavia and much of the British Isles. Glaciers also carved **fjords** (fee-AWRDZ) along Norway's coast. Fjords are narrow deep inlets of the sea set between high rocky cliffs. When the ice melted, the retreating glaciers left behind thin soils and thousands of lakes.

To the south lies the Northern European Plain. This broad coastal plain stretches from France's Atlantic coast all the way to the Urals. Most of the plain is less than 500 feet (152 m) above sea level. Many rivers flow across it before reaching the ocean. As a result, river towns and port cities have developed there. For example, large cities like Paris and Berlin are located in this region. In fact, its many rivers, short distances, and smooth terrain have long made human contact relatively easy. These features have allowed culture groups to travel, trade, and migrate throughout the region. Today the Northern European Plain is Europe's most important farming and industrial area. As you might expect, it is also densely populated.

INTERPRETING THE VISUAL RECORD

The Matterhorn, on the Switzerland-Italy border, rises 14,691 feet (4,478 m). ***What physical process do you think shaped this peak? What makes you think so?*** TEKS

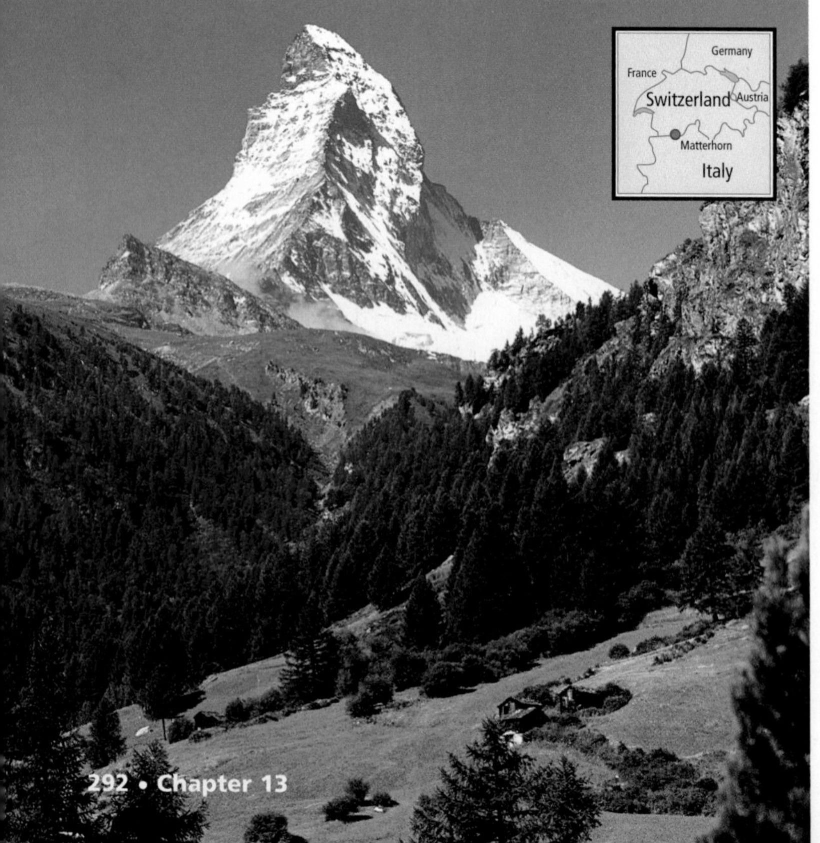

The third major landform region is the Central Uplands. This is an area of hills and small plateaus, with forested slopes and fertile valleys. It includes the Massif Central (ma-SEEF sahn-TRAHL) of France and the Jura Mountains on the French-Swiss border. The region stretches northeastward across southern Germany to the Bohemian Highlands. The Central Uplands are an old eroded region. As a result, the low mountains and hills in the region are often rounded. Many of Europe's productive coal fields lie in the Central Uplands. A number of industrial towns and cities grew near coal deposits.

The last and youngest region is the Alpine mountain system, which includes the Alps, Europe's major mountain range. The Alps stretch from France's Mediterranean coast to the Balkan Peninsula. Many peaks reach heights of more than 14,000 feet (4,268 m). Because of their high elevations, the Alps have large snowfields and glaciers. Avalanches are fairly common in winter. Although the Alps are high mountains, historically they have not been a serious barrier to human interaction. People have crossed the Alps through mountain passes for thousands of years to trade and travel. Other ranges in the Alpine system include the Carpathian (kahr-PAY-thee-uhn) Mountains in Eastern Europe and the Apennines (A-puh-nynz) in Italy. The Pyrenees (PIR-uh-neez) of France and Spain are also part of this system.

Beginning about 65 million years ago, tectonic processes formed the mountains of the Alpine system. At that time, the African plate began pushing against the Eurasian plate. This caused the mountains to rise. Tectonic activity continues today. A subduction zone off the coasts of southern Italy and Greece still creates powerful earthquakes and volcanoes. Because it lies on a tectonic plate boundary, Iceland also experiences volcanic eruptions and earthquakes.

internet connect

GO TO: go.hrw.com
KEYWORD: SW3 CH13
FOR: Web sites about the natural environments of Europe

TEKS **READING CHECK:** *Physical Systems* How have physical processes affected the shapes of mountains and hills in the Central Uplands?

Connecting to TECHNOLOGY

Polders

The Dutch have long used technology to shape their natural environment. For hundreds of years, they have been "creating" land by reclaiming it from the sea. Lands reclaimed from the sea are called **polders**.

To create polders, the Dutch built earthen walls called **dikes** along the shoreline. Then they used windmills to pump out the seawater behind the dikes. The Dutch used the drained lands for farming or for housing. By using polders to grow crops and raise livestock, the Dutch greatly increased the amount of available farmland. In fact, the Netherlands is an exporter of agricultural goods. The Dutch export products such as flowers, grains, potatoes, and sugar beets, particularly to other European countries.

Today more than 25 percent of the Netherlands lies below sea level. A national system of dams, dikes, and floodgates holds back the sea, and water is constantly pumped out. This system ranks as one of the wonders of the modern world. The largest dike, 19 miles (31 km) long and 100 yards (91 m) thick, closes off a large inlet. Completed in 1932, this dike allowed for the creation of four huge polders. Farms and cities have sprung up on these lands.

Comparing In what other areas of the world, or against what environmental hazards, might Dutch techniques for creating polders be useful? TEKS

Waddenzee (saltwater)
IJsselmeer Dam 1927–32
Wieringermeer 1927–30
IJsselmeer (freshwater)
Northeast Polder 1937–50
East Flevoland 1950–57
South Flevoland 1959–68
Amsterdam
— Dike
← Pumping station
Polder area

Creating a Polder

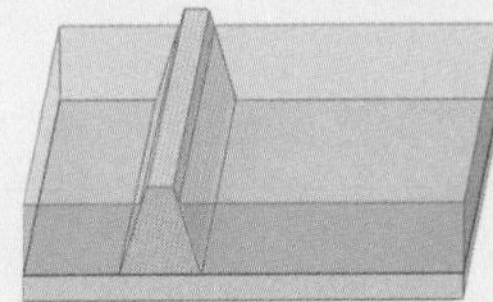

Year 1
Dikes are built around the area to be reclaimed. Pumps and canals drain the water.

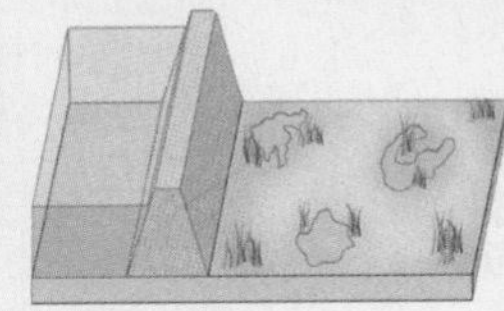

Years 2–3
The water level falls. Seeds blow into the area and salt-tolerant plants grow.

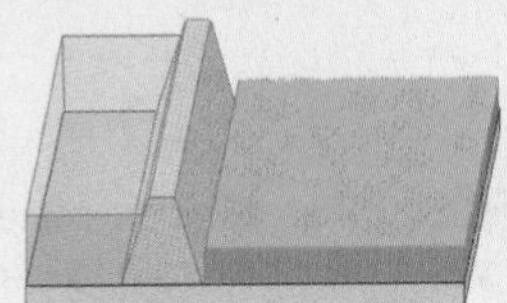

Years 4–6
Reeds are planted over a net of woven twigs. The reeds draw up more water.

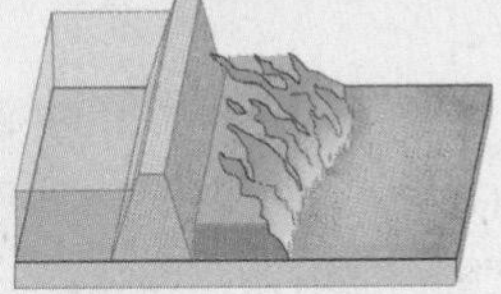

Year 7
The reeds are burned. Heavy plows turn their roots and the ash into the soil.

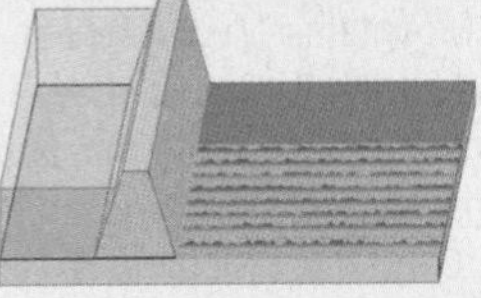

The land is ready for crops. Within 15 years the polder looks like it has been farmed forever.

In Romania, an artificial canal connects the Danube River to the Black Sea. The canal allows ships to avoid the marshy Danube Delta and shortens voyages by many miles. Locks, shown above, lift ships going upstream to the river's level and lower ships going downstream to sea level. By using rivers and canals, ships can now travel all the way from the Black Sea to the North Sea.

Water

Europe is nearly surrounded by water. To the south lies the Mediterranean Sea. It is connected to the Black Sea by the narrow Bosporus (BAHS-puh-ruhs). Geographers consider the Bosporus a boundary between Europe and Asia. The Arctic Ocean, North Sea, and Baltic Sea wash the shores of northern Europe. The shallow North Sea has long been important for trade and fishing. The smaller Baltic Sea freezes over during the winter months. To the west of Europe lies the North Atlantic Ocean. For centuries, European explorers, fishers, and merchants have traveled the waters of the Atlantic.

Europe's long, irregular coastline has hundreds of good natural harbors. These harbors are generally located near the mouths of **navigable** rivers, making Europe ideally situated for trade by sea. A navigable river is one that is deep enough and wide enough for shipping. Canals connect many rivers in Europe. For example, France's Canal du Midi lets boats and barges travel between the Atlantic Ocean and the Mediterranean Sea. Many interior towns and cities across Europe have access to the sea through canals and rivers.

The Rhine and Danube stand out among Europe's most important rivers. Many cities and industrial areas line their banks, and barges carry goods along their courses. The Rhine rises in the Swiss Alps. It then flows northwestward through Germany and the Netherlands before entering the North Sea. The Danube begins in the uplands of southern Germany. It flows eastward through nine countries in central and eastern Europe. It empties into the Black Sea. Unfortunately, large amounts of pollution enter the ocean from these and other rivers. Cleaning up and controlling pollution in Europe's rivers is a major environmental challenge.

READING CHECK: *Environment and Society* How do Europe's interior towns and cities have access to the sea?

TEKS Questions 1, 2, 3, 4, 5

Section 1 Review

Define fjords, polders, dikes, navigable

Working with Sketch Maps

On a map of Europe that you draw or that your teacher provides, label the Ural Mountains, Mediterranean Sea, Scandinavian Peninsula, Iberian Peninsula, Italian Peninsula, Balkan Peninsula, Northern European Plain, Alps, Carpathian Mountains, Pyrenees, Black Sea, Bosporus, North Sea, Rhine River, and Danube River. Which river rises in the Swiss Alps, flows through the Netherlands, and empties into the North Sea?

Reading for the Main Idea

1. *Physical Systems* How did continental ice sheets shape the landscapes of the Northwest Highlands?
2. *Environment and Society* How has Europe's natural environment made human contact relatively easy?

Critical Thinking

3. **Making Generalizations** What are some physical features that probably shaped migration routes in Europe? How do you think they did so?
4. **Analyzing Information** Considering what you know about Europe's natural environments, where would you expect to find many of its largest and most important cities and settlements?

Organizing What You Know

5. Copy the chart shown below. Use it to describe Europe's four major landform regions: the Northwest Highlands, Northern European Plain, Central Uplands, and the Alpine mountain system.

Northwest Highlands	
Northern European Plain	
Central Uplands	
Alpine mountain system	

Environment and Society

Geography for Life

A Peninsula of Peninsulas

Look at a map of Europe. You will notice that the continent is actually a large peninsula made up of many smaller peninsulas. You might also notice Europe's jagged outline. What do you think created these features? How might they have affected the region's history?

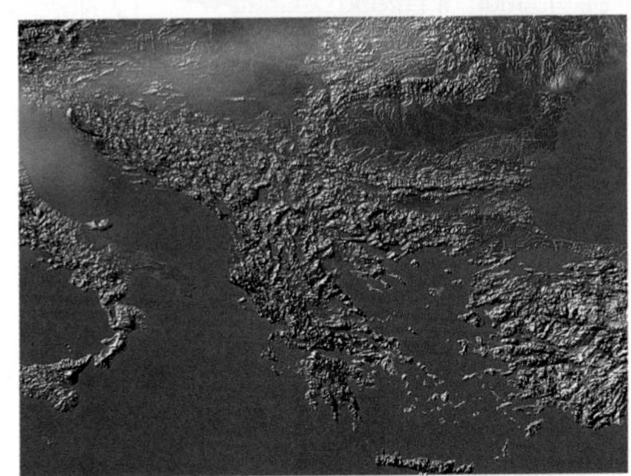

This computer-enhanced satellite image shows the Balkan and Italian Peninsulas clearly.

Much of Europe's present-day coastline has taken shape since the end of the last ice age about 10,000 years ago. Since that time, sea levels have risen, flooding lowlands and changing Europe's shoreline. For example, in northern Europe the Baltic Sea formed from the melting Scandinavian ice sheet. The North Sea and Irish Sea also took their present form after the last ice age. Rising sea levels flooded the mouths of many rivers. This process formed estuaries that are now deep ocean ports.

Párga, Greece, overlooks a picturesque harbor. Like that of many towns on Europe's peninsulas, Párga's harborside location has played a major role in its history and economy. Tourists also enjoy the town's setting.

Geographers have long noticed the remarkable influence of the sea on Europe. The sea greatly affects climate and rainfall patterns. Warm ocean currents bring mild temperatures and rainfall to much of western and northern Europe. In some places, these effects are felt far inland.

Europe's peninsular geography and rugged coastline have also influenced its history. Harbors along rocky shores have long offered protection for ships. Since ancient times, the joining of land and water has provided opportunities for exploration, fishing, sea trade, and political and military power. Early peoples like the Phoenicians, Greeks, and Vikings sailed and explored Europe's intricate coastline. In fact, Europe has long been a place of contact between peoples and cultures. From the Italian Peninsula, which juts into the Mediterranean Sea, the Romans ruled a vast empire. Later European culture groups turned to the sea for global colonial and economic power. Spanish and Portuguese explorers sailed around the world in the 1500s, setting up trading posts and colonies. The British, Dutch, French, and other Europeans followed. For example, in the 1700s and 1800s Great Britain used the seas to become the world's dominant colonial and sea power.

Applying What You Know

1. **Summarizing** How has Europe's peninsular geography influenced its history?
2. **Making Generalizations** How do you think Europe's peninsular geography has affected the locations of its cities and settlements?

Section 2 Climates and Biomes

READ TO DISCOVER

1. How do ocean currents affect the distribution of Europe's climates?
2. Which biomes are found in this region?

IDENTIFY

North Atlantic Drift

LOCATE

British Isles

WHY IT MATTERS

Warm ocean currents moderate Europe's climate. Use CNNfyi.com or other **current events** sources to learn about how the circulation of ocean waters can affect climate and weather.

Climates

Europe has three major climate types: marine west coast, humid continental, and Mediterranean. (See the unit climate map.) The marine west coast climate is found throughout most of northern and western Europe. This climate region includes southern Iceland and the British Isles. It also stretches across northern continental Europe from northern Spain into Poland and Slovakia. Frequent Atlantic storms bring clouds and rain. Rainfall averages between 20 and 80 inches (51 and 203 cm) a year. (See the unit precipitation map.) Snow and frosts can occur in winter. Temperatures are mostly mild, and cloudy, drizzly, and foggy days are common.

INTERPRETING THE VISUAL RECORD

Ireland is one of the European countries that have a marine west coast climate. This hillside is in a part of Ireland where rainfall is very heavy—more than 60 inches (150 cm) per year. ***What is the connection between Ireland's climate and its nickname—the Emerald Isle?*** TEKS

Areas from interior Norway and Sweden south to the Black Sea have a humid continental climate. This climate has four distinct seasons, including a cold snowy winter and mild to cool humid summer. Winters are severe in parts of this climate region. Periodic summer droughts affect Hungary and Romania.

High mountains, particularly the Alps, separate these first two climates from Europe's third major climate region to the south. Most of southern Europe has a Mediterranean climate. This region gets between 10 and 30 inches (25 and 76 cm) of rainfall a year. Most rainfall comes during the mild winter. Occasional North Atlantic storms pushed by the westerly winds bring rain at that time. Long, dry, and sunny summers are typical.

Smaller climate regions are found in other parts of Europe. For example, a subarctic climate stretches across northern Norway, Sweden, and Finland. The northernmost parts of these countries, along with interior and northern Iceland, have a tundra climate. A small humid subtropical climate region is located south and southeast of the Alps. This area stretches from Italy's Po Valley into the Balkans. In parts of Spain, high mountains block moist ocean air from reaching farther inland. A semiarid climate is found there.

Compared to world regions of similar latitude, much of Europe enjoys mild climates. Winter temperatures are particularly mild for such high latitudes. These mild temperatures are caused by the moderating influence of the **North Atlantic Drift**.

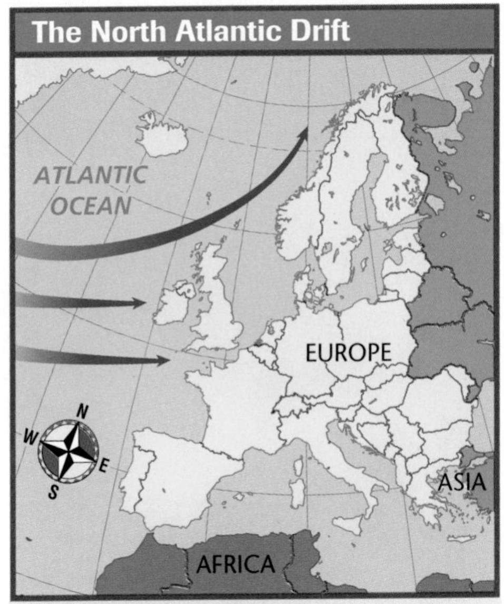

INTERPRETING THE MAP *Northern Europe's temperatures are relatively mild, thanks to the North Atlantic Drift.* ***What are some of the coastal countries affected by the current? Use the unit atlas to find the large island west of Iceland that is far from this ocean current and, as a result, is much colder.*** TEKS

FOCUS ON GEOGRAPHY

The North Atlantic Drift The North Atlantic Drift is a warm ocean current. It originates off the coast of North America and is fed by the warm tropical waters of the Gulf Stream. (See the map.) The North Atlantic Drift warms the air above it. Then prevailing westerly winds carry this warm moist air across much of northwestern Europe. The winds bring mild temperatures and rain. These conditions allow farmers to grow crops as far north as Sweden and Iceland. Also, the warm waters keep seaports in Norway and at Murmansk, Russia, free of ice.

READING CHECK: *Physical Systems* How do the North Atlantic Drift and prevailing westerly winds affect Europe's climates?

Plants and Animals

Human activities have affected Europe's plants and wildlife severely. For thousands of years, people there have hunted animals and cleared forests for timber and farmland. The growth of towns, cities, and roads has also changed the natural environment. Some waterways have been polluted. As a result, many species have become extinct from loss of habitat. Some other creatures, such as bears, lynx,

White storks build their nests on a rooftop in Spain.

INTERPRETING THE VISUAL RECORD *A member of the Sami people of northern Scandinavia works with his reindeer herd. Reindeer are known as caribou in North America. Their wide hooves allow them to walk more easily on snow.* ***What food sources do you think the tundra environment provides for the reindeer during the winter?*** TEKS

wolves, and wild horses, survive mainly in areas where they are protected. Despite these changes, however, Europe can be divided into four major biomes. These biomes include temperate forest, Mediterranean scrub forest, boreal forest, and tundra.

Most of Europe lies within a temperate forest biome. Trees such as ash, beech, maple, and oak are common. Badgers, deer, and a variety of birds live in this environment. Today fields and towns occupy much of the land. Only remnants of the dense forests that once covered much of the landscape remain in this region.

You will find a Mediterranean scrub forest biome in some drier areas in southern Europe. Small trees, shrubs, and drought-resistant plants are typical of the region. Animals such as wild boars and wild sheep still roam remote Mediterranean mountain areas.

Large parts of northern and central Europe lie within the boreal forest biome. These northern forests make up most of Europe's remaining woodlands. Finland, Norway, and Sweden all have large evergreen forests. Trees here, such as pine, spruce, and fir, provide most of Europe's timber for building and papermaking. However, logging and other human activities have greatly reduced the area's animal life.

Far northern Europe has a tundra biome. This biome includes much of Iceland and northern Scandinavia. The land in this cold treeless area is frozen most of the year. During the short Arctic summer, the tundra thaws, and many swamps and marshes form. Millions of migratory birds visit during the summer. Reindeer and foxes are among the tundra's mammals.

READING CHECK: *Physical Systems* In which biome would you find trees such as ash, beech, maple, and oak?

TEKS Questions 1, 2, 3, 4, 5

Section 2 Review

Identify

North Atlantic Drift

Working with Sketch Maps

On the map you created in Section 1, label the British Isles. Which climate dominates the British Isles?

Reading for the Main Idea

1. *Places and Regions* Where in Europe is a marine west coast climate found?
2. *Physical Systems* How do the Alps affect the distribution of climates in Europe?
3. *Environment and Society* What human activities have affected Europe's plants and wildlife?

Critical Thinking

4. **Drawing Inferences and Conclusions** What advantages might forestry industries in Norway, Sweden, and Finland have over forestry operations in other European countries?

Organizing What You Know

5. Create web diagrams like the one below to describe the climates of Europe. Use as many circles as you need to provide important details about each climate.

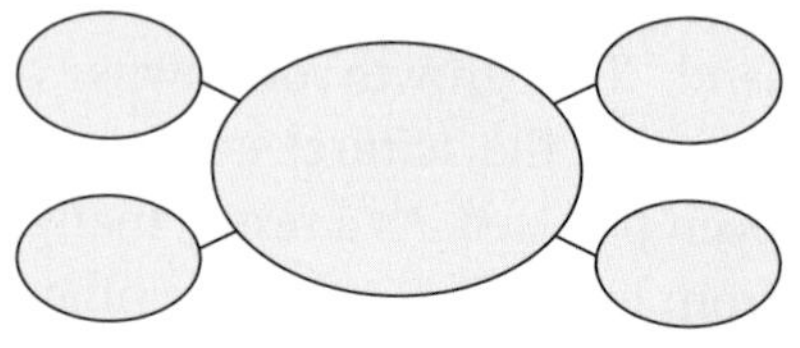

Natural Resources

READ TO DISCOVER

1. Where are Europe's forest, soil, and fishery resources located?
2. What energy and mineral resources are found in this region, and where are they located?

DEFINE

loess

LOCATE

Po Valley
Guadalquivir River

WHY IT MATTERS

Overfishing and pollution have affected fishing industries in Europe and around the world. Use CNNfyi.com or other **current events** sources to learn about the challenges facing the world's fishing industries.

Forests, Soils, and Fisheries

Most of Europe's original forests were cleared centuries ago. For example, clearing or overgrazing in ancient times destroyed nearly all of the Mediterranean area's original oak woodlands. Only a scrub-plant community covering the hillsides remains. Air pollution and acid rain have destroyed many more trees throughout the continent. Large areas of timber-producing forests exist only in limited areas, such as Sweden and Finland. As a result, most European countries must now import lumber. However, most European countries also have reforestation and forest protection programs.

Europeans have made good agricultural use of their soils. In fact, more than half of Europe's land area is used for farming. Some of the best soils are found in the Northern European Plain. Farmers grow a variety of grains there and raise cattle and hogs. Some of these soils developed from **loess**—fine-grained, windblown soil that is very fertile. Such soils can keep their fertility for many years. In southern Europe, alluvial soils are particularly productive. River valleys, such as Italy's Po (POH) Valley and Spain's Guadalquivir (gwah-dahl-kee-VEER) River valley, are major farming centers. With the help of irrigation, these soils produce a wide range of crops. Farmers grow lemons, oranges, and many different vegetables. Farmers in southern Europe also raise goats, hogs, and sheep.

Autumn tints grapevines growing in southwestern Germany. Partly because they can grow in many different soils, grapes are grown in several European countries.

Lavender, which is used in perfumes and soaps, grows in a region of southern France called Provence. Fertile soil and a temperate climate make farming possible throughout the country. Each region has its specialties. France's agricultural bounty makes it one of the world's largest exporters of farm products.

Europe produces many crops, such as large amounts of grapes, olives, potatoes, and wheat. Efficient methods and modern technology have made Europe's crop yields among the highest in the world. Farmers use chemical fertilizers to enrich the soil. They also rotate crops to maintain fertility. Modern machinery is used in planting and harvesting. However, some areas lag behind in farm production. This is the case in Europe's formerly communist countries. Farming technology is often outdated there.

Throughout history, fishing has been an important part of the European economy. Fishing villages dot Europe's coasts, and fishing boats can be found in all waters bordering the continent. Europe's best fisheries are located in the North Atlantic and Arctic Oceans and in the North Sea. Coastal waters, particularly where the warm North Atlantic Drift mixes with cold polar waters, are excellent fishing grounds. Iceland, Norway, Spain, and Denmark are major fishing countries. However, overfishing and coastal pollution threaten the future of the fishing industry in the Mediterranean and North Atlantic.

READING CHECK: ***Places and Regions*** Where are some of the best farming soils in Europe found?

Energy and Minerals

To meet its current industrial and energy needs, Europe must rely heavily on mineral imports. Europe's technologically advanced economies lack sufficient supplies of critical natural resources, such as oil, iron, and other metals.

Europe does have large deposits of coal, however. Some countries, such as Germany, Britain, and Poland, have mined coal for hundreds of years. (See the unit land use and resources map.) In fact, Germany's Ruhr coal field is one of the world's largest. During Europe's industrial era, coal and iron were essential to the creation of the steel and manufacturing industries. However, in the 1900s oil replaced coal as the most important energy source. Europe became increasingly dependent on imported oil. Today oil and gas from Southwest Asia, Russia, and Africa power the economies of most European countries.

INTERPRETING THE VISUAL RECORD

Workers cut marble from a quarry near Carrara, Italy. For centuries, sculptors have favored the white stone. ***How might changes in technology have affected the marble industry of Carrara?*** TEKS

Europe's main oil and natural gas deposits lie beneath the North Sea. These deposits were discovered in the early 1960s. They have greatly benefited the economies of Norway and Britain. Both countries are now energy exporters. The Netherlands also produces and exports natural gas.

Hydroelectricity is another important energy source. It is produced in mountainous Norway, Sweden, and Switzerland. France produces ocean tidal power and solar power. Iceland uses geothermal energy to heat homes and generate electricity. Nuclear power also supplies energy, particularly in France, Belgium, Bulgaria, and Sweden. However, many Europeans are worried about the long-term safety of nuclear power.

Other mineral resources in Europe include iron ore, uranium, lead, and zinc. Sweden and France both have large deposits of iron ore in upland regions. France also has sizable uranium deposits. Lead, zinc, and other metals are found in Spain and southern Europe. Marble, a stone used for building and sculpting, has long been mined in parts of southern Europe, such as in Carrara, Italy.

✓ **READING CHECK:** ***Places and Regions*** Where are Europe's main oil and natural gas fields?

The world's largest deposits of amber are found along the shores of the Baltic Sea. Some deposits of this fossilized tree resin date back to perhaps 60 million years ago. The preserved bodies of ancient insects have been found in some deposits. The yellowish translucent amber is usually made into jewelry.

Section 3 Review

TEKS Questions 1, 2, 3, 4, 5

Homework Practice Online

Keyword: SW3 HP13

Define loess

Working with Sketch Maps On the map you created in Section 2, label the Po Valley and Guadalquivir River. Which of these areas is a major agricultural region in Spain?

Reading for the Main Idea

1. ***Places and Regions*** What types of soils are found in river valleys such as Italy's Po Valley? How useful are these soils for agriculture?
2. ***Physical Systems*** How does modern farming technology affect European agriculture?
3. ***The Uses of the Geography*** How did the growing importance of oil as an energy source in the 1900s affect Europe?

Critical Thinking

4. **Problem Solving** How might Europeans lessen their dependence on imported oil?

Organizing What You Know

5. Copy the chart shown below. Use it to list Europe's main energy resources and where they are found. Add as many rows as you need.

Energy source	Location

Building Vocabulary

On a separate sheet of paper, explain the following terms by using them correctly in sentences.

fjords
polders
dikes
navigable
North Atlantic Drift
loess

Locating Key Places

On a separate sheet of paper, match the letters on the map with their correct labels.

Ural Mountains
Mediterranean Sea
Scandinavian Peninsula
Iberian Peninsula
Italian Peninsula
Northern European Plain
Alps
Pyrenees
North Sea
Rhine River

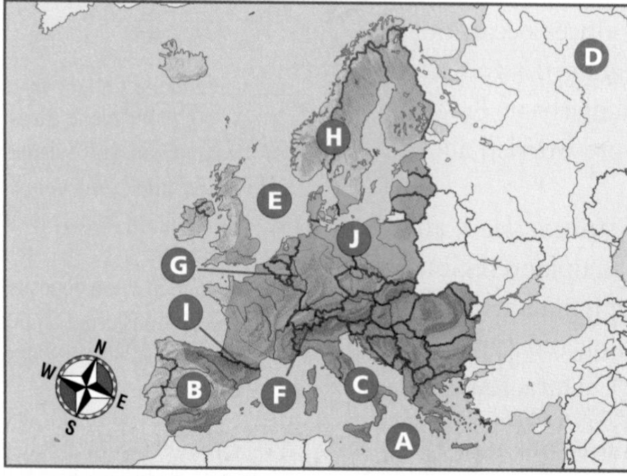

Understanding the Main Ideas

Section 1

1. *Physical Systems* How have tectonic processes shaped the physical environment of Europe?
2. *Places and Regions* Where can you find glacially scoured landscapes in Europe? What are these landscapes like?

Section 2

3. *Places and Regions* What are Europe's three major climate types?
4. *Places and Regions* Which biome do large parts of northern and central Europe have? How have people affected it?

Section 3

5. *Environment and Society* Where are Europe's best fisheries located?

Thinking Critically for TAKS

1. **Supporting a Point of View** What potential dangers do you see with creating polders that are below sea level?
2. **Summarizing** Why does Europe have such mild climates compared to other world regions of similar latitude?
3. **Comparing** How does the environment of the Northern European Plain compare with the environment of the Great Plains in North America?

Using the Geographer's Tools

1. **Analyzing Maps** Study the unit physical and political maps. What area of Europe is generally below sea level?
2. **Creating Climate Graphs** Go to the HRW Web site on the Internet to find statistics you need to create climate graphs. Then create climate graphs for London, Rome, and Berlin. Describe the climate patterns you see in your graphs.
3. **Preparing Diagrams** Use the information in Section 2 to prepare a diagram showing how the westerlies and North Atlantic Drift bring mild temperatures to northwestern Europe. You might want to use arrows to show the general direction of wind flow. How would this pattern be different if a high mountain range bordered the coast of western and northern Europe?

Writing for TAKS

Write a report about tectonic activity in Europe. Explain why, how, and where tectonic activity occurs in Europe. How does this tectonic activity affect the continent? When you are finished with your report, proofread it to make sure you have used standard grammar, spelling, sentence structure, and punctuation.

SKILL BUILDING

Geography for Life

Using Research Skills

Environment and Society Plan a research project to study the decline of commercial fishing in Europe. First, determine a goal for your research or a question that you want answered. For example, you might want to know how the amount of fish caught in Europe has changed over the last 20 years. Then find information that relates to your question, such as statistics or articles about the fishing industry. When you have gathered enough information, write a short report that answers your research question and suggests reasons for the decline of commercial fishing in the region. When you are done, you might want to suggest other possible areas for future research.

Building Social Studies Skills

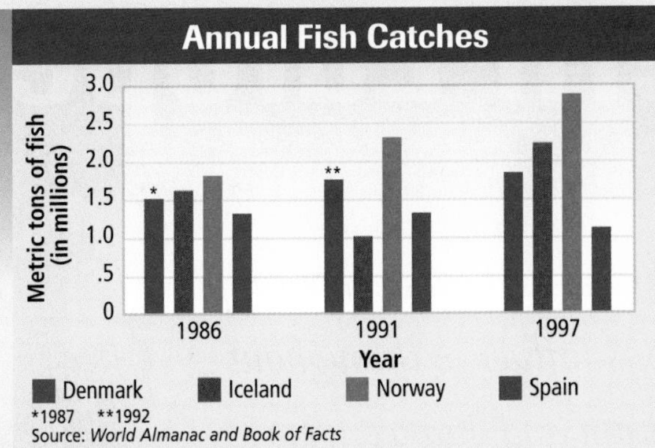

Interpreting Graphs

Study the graph above. Then use information from the graph to help you answer the questions that follow.

1. From 1986 to 1991 and from 1991 to 1997 which country had the most dramatic increase in its fish catch?
   a. Denmark
   b. Iceland
   c. Norway
   d. Spain

2. A comparison of the 1991 and 1997 fish catches of the four countries shows that only one country had a clear decrease. Which country was it? Support your answer.

Interpreting Secondary Sources

Read the following passage and answer the questions that follow. Mark your answers on a separate sheet of paper.

"To create polders, the Dutch built earthen walls called dikes along the shoreline. Then they used windmills to pump out the seawater behind the dikes. The Dutch used the drained lands for farming or for housing. By using polders to grow crops and raise livestock, the Dutch greatly increased the amount of available farmland. In fact, the Netherlands is an exporter of agricultural goods. The Dutch export products such as flowers, grains, potatoes, and sugar beets, particularly to other European countries."

3. One step that is part of the process of creating polders is
   a. flooding valleys by opening river dams.
   b. building dikes along the shoreline.
   c. using windmills to generate electricity.
   d. building homes that stand above local water levels.

4. How have the polders helped the Netherlands become an exporter of agricultural goods?

Alternative Assessment

PORTFOLIO ACTIVITY

Learning about Your Local Geography TEKS

Group Project: Field Work

Plan, organize, and complete a research project with a partner that compares plant and animal life in your area with those in Europe. First, study the plants and animals in your area by doing field work. Work together to observe your area's wildlife. You may want to make drawings of what you see. Then use a library to find information about the plants and animals that you observed. What kind of biome do you live in? What plants and animals are common to that biome? Finally, compare your biome to the biomes of Europe. Does the same biome exist in Europe? If so, where? How are the plants and animals of your area similar to and different from those in Europe?

internet connect

Internet Activity: go.hrw.com
KEYWORD: SW3 GT13 TEKS

go.hrw.com

Choose a topic on the natural environments of Europe to:

- describe the environmental impact of oil drilling in the North Sea.
- understand how technological innovations affect the maintenance of polders.
- create a poster about fjords along the Norwegian coast.

CHAPTER 14 Northern and Western Europe

The countries of northern and western Europe have had a tremendous influence throughout the world. Great Britain and France were among the greatest colonial powers. They spread their languages, educational, and political systems worldwide.

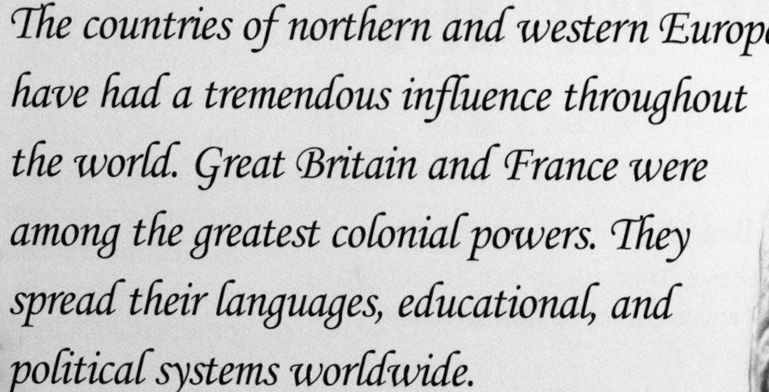

Gathering tulips in the Netherlands

Walrus-ivory chess pieces from Scotland

Hi. My name is Lars, and I live in Tromsø, in far northern Norway. I live in a new house in the middle of town. We are so close to the sea that we eat fish almost every day. For breakfast I have a slice of bread with a sweet goat cheese, smoked salmon, shrimp, or cod liver spread. I take two more of these sandwiches for lunch at school. For dinner we have salmon or cod or another fish about four or five times a week. The other days we like to eat pizza and hamburgers.

In the winter everyone skis to school along trails that look like snowy streets. We live above the Arctic Circle, and it is completely dark in the winter. Because of this, the trails must have streetlights. The Sun does not shine at all in Tromsø from the end of November to January 20 or so. On January 20, when the Sun appears again for just a few minutes, we have a big celebration for Sun Day.

In the summer, the Sun never sets, so there is no nighttime at all. This is my favorite time of year. We have a huge bonfire and a cookout on June 21. But it still can be very cold. Last year the temperature in the summer was usually about 6° or 7°C (about 43° or 44°F).

The British Isles

READ TO DISCOVER

1. How has history affected the culture of the British Isles?
2. Why are the cultures of Ireland and the United Kingdom similar?
3. How has the British economy changed over the last 200 years?
4. What issue has caused tension in Northern Ireland?

DEFINE

sequent occupance
famine
constitutional monarchy
nationalized

LOCATE

British Isles
England
Wales
Scotland
Northern Ireland

Locate, continued

Dublin
London
Thames River
Glasgow
Edinburgh
Birmingham
Belfast

WHY IT MATTERS

Since the 1960s many British musicians have helped to shape popular music in the United States. Use CNNfyi.com or other **current events** sources to learn about how modern British music has influenced world culture.

Northern and Western Europe: Physical-Political

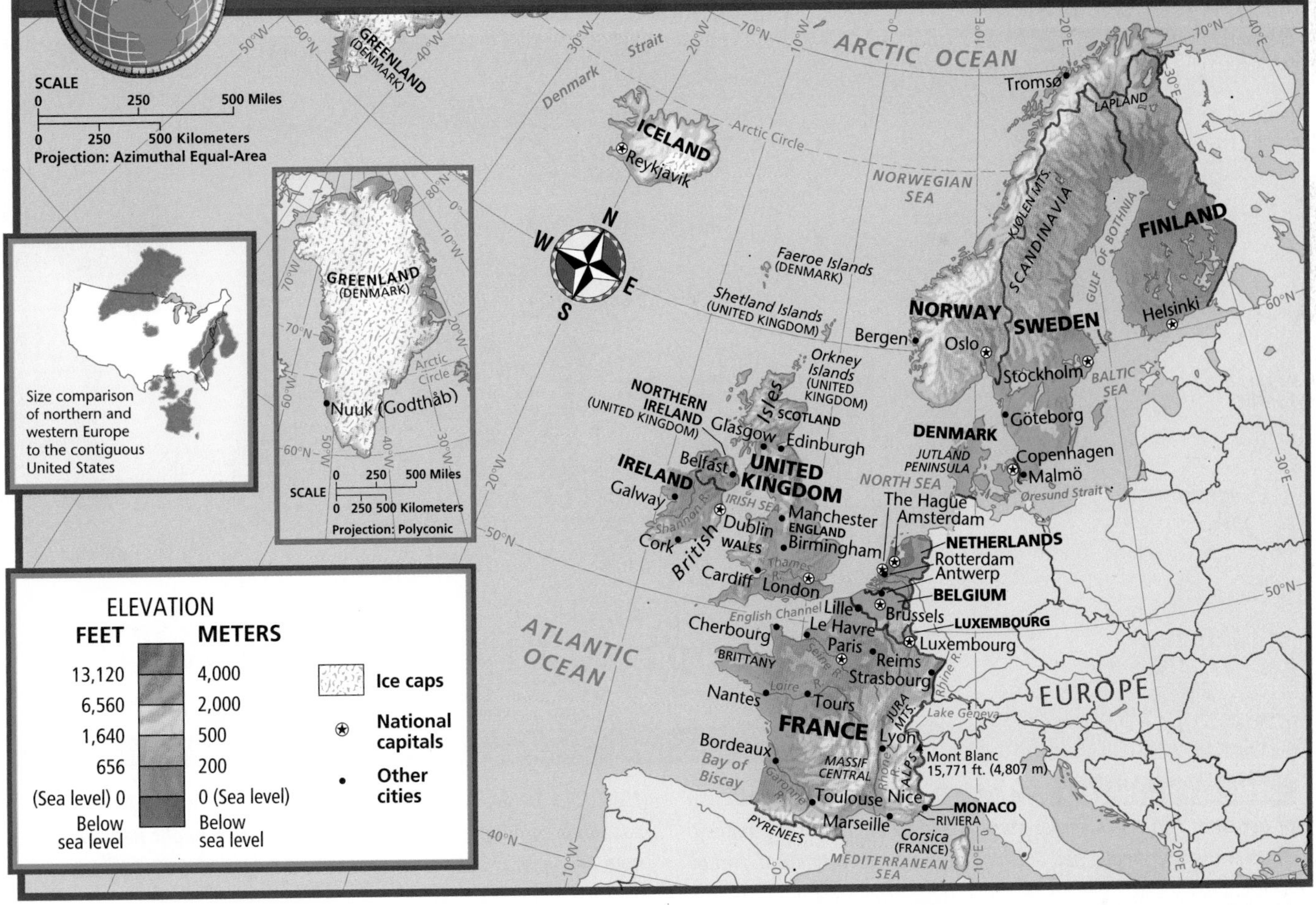

SCALE
0 250 500 Miles
0 250 500 Kilometers
Projection: Azimuthal Equal-Area

Size comparison of northern and western Europe to the contiguous United States

GREENLAND (DENMARK)
SCALE 0 250 500 Miles
0 250 500 Kilometers
Projection: Polyconic

ELEVATION

FEET	METERS
13,120	4,000
6,560	2,000
1,640	500
656	200
(Sea level) 0	0 (Sea level)
Below sea level	Below sea level

Ice caps
National capitals
Other cities

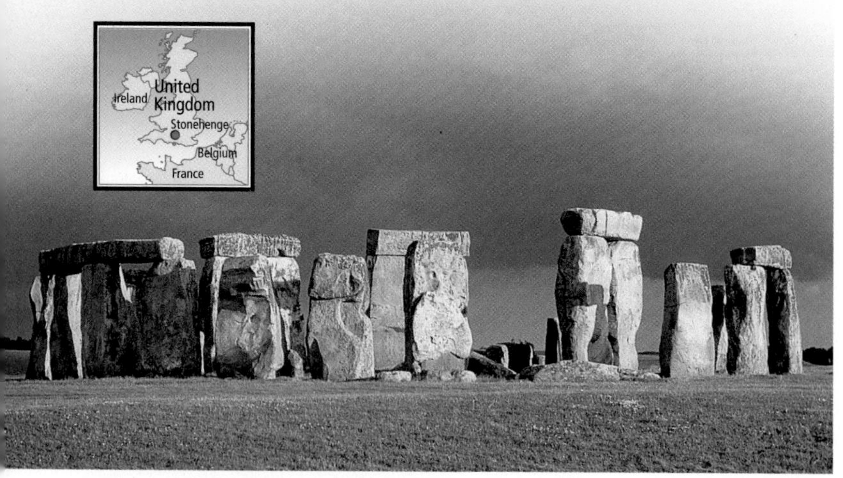

History

The British Isles are made up of two independent countries—the Republic of Ireland and the United Kingdom of Great Britain and Northern Ireland. The Republic of Ireland occupies all but the northern part of the island of Ireland. The United Kingdom is often referred to as just Great Britain or even Britain. It includes four political regions—England, Wales, Scotland, and Northern Ireland. Great Britain can be divided into two physical regions, lowland Britain and highland Britain. Most of England, except the far north, is part of lowland Britain. Northern England, Wales, and Scotland make up highland Britain.

INTERPRETING THE VISUAL RECORD

Stonehenge, an ancient complex of massive stone circles in England, was built beginning about 3100 B.C. Modern archaeologists are unsure exactly why Stonehenge was built. ***How is Stonehenge an example of sequent occupance?*** TEKS

Lowland Britain has developed a complex cultural geography. About 5,000 years ago, the earliest settlers left their mark with monuments like Stonehenge. Later the Celts—the ancestors of the Scots, Welsh, and Irish—occupied the island. Then the Romans came and built fortified towns. Later, Angles and Saxons, two Germanic tribes, came and drove the Celtic peoples to highland Britain. Vikings from Scandinavia raided the coastal areas and also built settlements. In 1066 William of Normandy conquered England. Normandy is now part of France. Each of these peoples left an imprint on lowland Britain. This process of settlement by successive groups of people, each group creating a distinctive cultural landscape, is called **sequent occupance**.

The British Empire, 1920

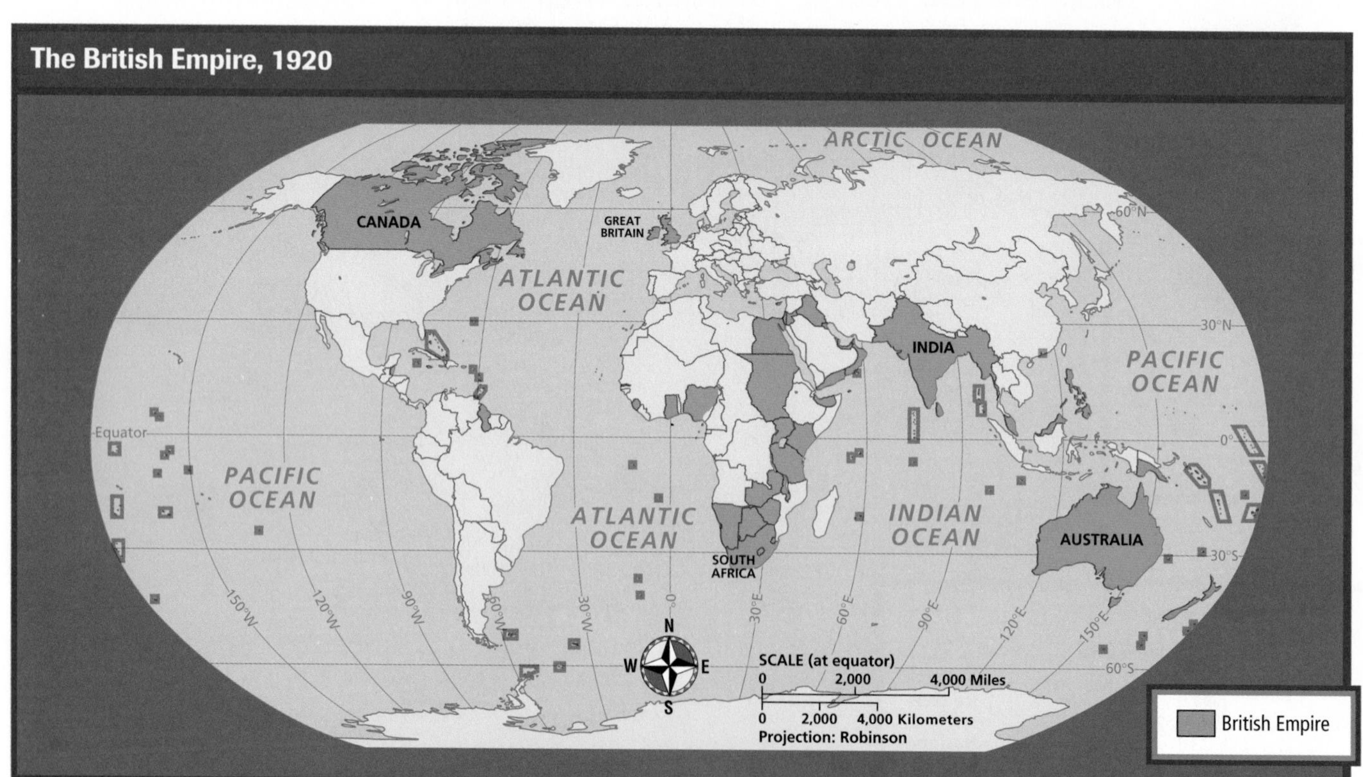

INTERPRETING THE MAP *The British Empire was a worldwide system of territories and dependencies administered by the British government. Areas within the empire had different levels of self-government, and by 1920, colonies such as Canada, South Africa, and Australia largely managed their own affairs.* ***What geographic factors might help explain how Britain was able to control such a large empire?*** TEKS

The British Empire In the 1600s and 1700s, British explorers and settlers founded colonies around the world. By 1801 England had brought Ireland, Scotland, and Wales into the United Kingdom. The surrounding ocean helped protect this kingdom. The British also built a powerful navy to take further advantage of the sea. During the 1800s more than one fourth of the world's land was ruled by the British Empire. The empire's colonies provided raw materials for British industries. The colonies also served as markets for finished goods. The empire spread the English language, Christianity, British law, sports, and other British customs around the globe.

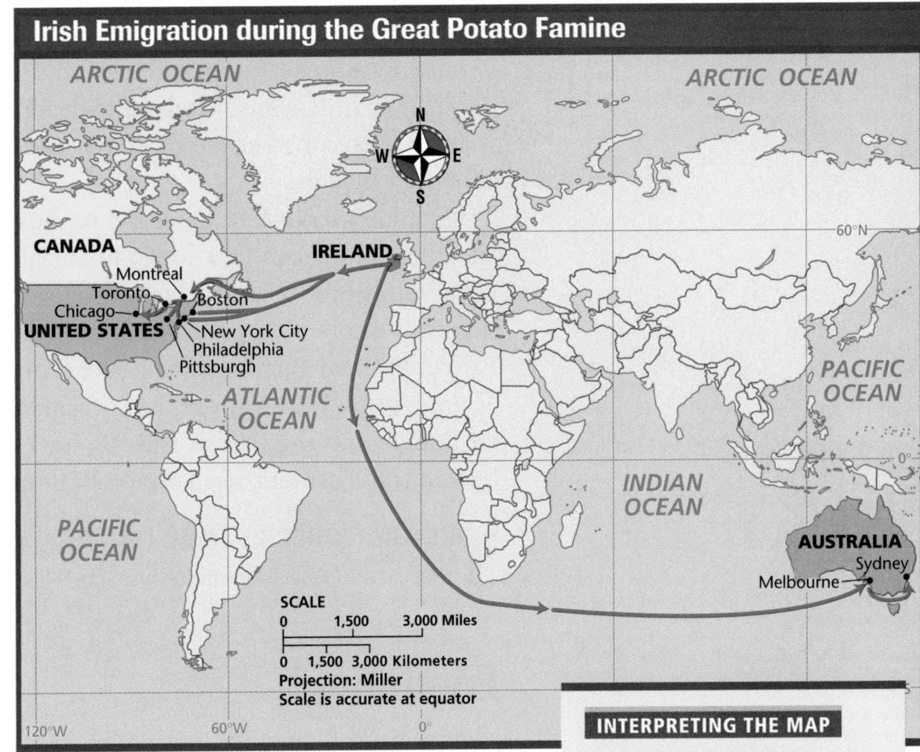

INTERPRETING THE MAP

Ireland's Great Potato Famine dramatically changed the country's population pattern as more than 1 million Irish immigrated to North America, Britain, and Australia. ***What geographic factors, both physical and cultural, might explain the migration patterns shown on the map?*** TEKS

The colonies that became the United States were part of the British Empire when they declared independence in 1776. Over time, most parts of the empire gained independence. Most former colonies became members of the Commonwealth of Nations. They still meet to discuss economic, business, and scientific matters of common concern.

Ireland did not win independence from the British until 1921. Before independence, life had long been hard for many Irish. Then in the mid-1800s Ireland suffered from a potato **famine**. A famine is a widespread shortage of food that may lead to severe hunger and starvation. About 1 million Irish died when the potato crop failed for several years in a row. The famine, poverty, and a lack of economic opportunities led many Irish to immigrate to other countries. Many migrated to the United States. (See the map.)

TEKS **READING CHECK:** *Environment and Society* Why did many people emigrate from Ireland during the mid-1800s?

Culture

Because of their shared history, Ireland and Great Britain share many cultural features. Social life is often centered around local eateries. Sports such as soccer, rugby, and cricket are popular. In addition, English is the main language of both countries. However, a small number of Irish also speak Irish Gaelic, and some Scots speak Scottish Gaelic.

The countries also differ in important ways. Both Ireland and Britain are democracies. However, their governments are organized differently. Ireland is a republic, and the president is the head of state. Britain is a **constitutional monarchy**. That is, a king or queen is the head of state, but a parliament led by a prime minister serves as the lawmaking branch of government.

GO TO: go.hrw.com
KEYWORD: SW3 CH14
FOR: Web sites about northern and western Europe

Another important difference is religion. The vast majority of people in Ireland are Roman Catholic. However, Protestants make up a majority of the population in Northern Ireland and the rest of the United Kingdom. As you will read, tensions between Catholics and Protestants have led to violence in Northern Ireland.

Dublin is Ireland's capital and its most populous city. London is the largest city, the cultural center, a world financial center, and the capital of the United Kingdom. Both cities are home to government buildings and famous landmarks. Visitors to London can see historic buildings like the Houses of Parliament and Buckingham Palace. For centuries, London's location on the Thames (TEMZ) River made that city an ideal port for trade between continental Europe and the British Isles. Today Heathrow Airport, one of the world's busiest travel centers, ties London to thousands of cities around the globe.

READING CHECK: ***Human Systems*** What are some cultural features that Britain and Ireland share? What is one major difference?

Economy

The Industrial Revolution began in Britain. By the 1700s the country had developed coal and iron mining and a large labor force. Britain also built a good transportation network that used rivers and canals. By the early 1800s the British had built the world's first railroads. Later in the century, London built the first subway system. (See Geography for Life: The London Underground and Mass Transit.) All of these features aided industrial development.

Industrial Rise and Decline Britain's early industries included iron and steel, shipbuilding, and textiles. Innovations such as spinning machines and steam power revolutionized how fabrics were produced. Wool and cotton from British colonies as well as from the United States supplied the textile industry. Industrial growth spread from London and central England to southern Wales, Scotland, and Northern Ireland. Cities like Glasgow and Edinburgh became industrial centers. Trade of raw and finished products between Britain and its colonies further aided development.

INTERPRETING THE VISUAL RECORD

This painting from the late 1700s shows a steam engine being used to dig a coal mine and represents the beginnings of the Industrial Revolution in England. ***How are people in this painting using technology to modify the physical environment?*** TEKS

Throughout much of the 1700s and 1800s, Britain dominated global trade. (See Case Study: Global Trade.) By 1900, however, the British had lost their dominance to foreign competition. By the mid-1900s Britain's coal mines and traditional industries were in rapid decline. British industries suffered because many were inefficient and because their products were not in demand. In the years after World War II, the United Kingdom **nationalized** many industries to try to stop the decline. Nationalized industries are those that are owned and operated by the government. They are protected from domestic competition.

Changing Fortunes Today Britain has returned most industry to private ownership, and the British economy is strong. Many early industrial cities, such as Glasgow and Birmingham, have benefited from urban renewal. They have also attracted high-tech industries. Although the coal industry has declined, oil and gas wells in the North Sea have helped the economy. In addition, much of Britain's labor force works in service industries rather than manufacturing. Tourism is also an important industry.

The Irish economy traditionally was based on farming. However, Ireland now has one of Europe's most rapidly developing economies. Low taxes and a well-educated workforce have attracted foreign companies. They use the country as a door to other European markets. The main industries are now banking, computers, electronics, and food processing. Immigrants from other countries are moving to Ireland to get jobs. This situation is quite different from the past, when Irish emigrated from their poor country.

READING CHECK: *Human Systems* How has Ireland's economy changed over time?

Orangemen parade in Portadown, Northern Ireland. The Orangemen are members of the Orange Society, a Protestant organization formed in 1795 to try to maintain Protestant control in Northern Ireland. In recent years, Orangemen parades in the region have led to street violence between Protestants and Catholics.

Issues and Challenges

One of the greatest challenges facing the people of the British Isles is violence in Northern Ireland. The Irish call the problems there "the troubles." Most of the people in Northern Ireland are descendents of Protestant English and Scottish settlers. A large minority are Irish Catholic. Many Catholics believe that union with the Republic of Ireland would protect them from discrimination in employment, housing, and government. However, Protestants want to stay part of the mostly Protestant United Kingdom.

The disagreement between the two groups has led to violence, particularly in the city of Belfast. British troops have tried to keep the peace. However, terrorist groups from both sides have killed thousands of people. A 1998 agreement created a shared government between Northern Ireland, the United Kingdom, and the Republic of Ireland. However, the future of this arrangement is not clear.

READING CHECK: *Human Systems* How do Roman Catholics and Protestants in Northern Ireland view their region differently?

Section 1 Review

TEKS Questions 1, 2, 3, 4, 5

Homework Practice Online
go.hrw.com
Keyword: SW3 HP14

Define

sequent occupance, famine, constitutional monarchy, nationalized

Working with Sketch Maps

On a map of northern and western Europe that you draw or that your teacher provides, label the United Kingdom, Republic of Ireland, British Isles, England, Wales, Scotland, Northern Ireland, Dublin, London, Thames River, Glasgow, Edinburgh, Birmingham, and Belfast. In the margin of your map, identify the capital of the United Kingdom.

Reading for the Main Idea

1. *Human Systems* How has London's location affected its growth?
2. *Human Systems* How have cultural differences led to the division of Ireland?
3. *Human Systems* How did Great Britain's history as a naval power contribute to the diffusion of cultural traits? How did innovation in Britain spur the Industrial Revolution?

Critical Thinking

4. **Analyzing Information** What geographical factor has influenced Britain's power to control territory around the world? How do you think that factor has influenced Britain's role in foreign affairs?

Organizing What You Know

5. Create a chart like the one below. Use it to list differences between Britain's old industrial economy and its modern economy. Write a paragraph describing how Britain's economy has changed over the past 200 years.

Britain's old industrial economy	Britain's modern economy

Geography for Life

The London Underground and Mass Transit

Greater London is home to about 7 million people. The city is also home to the world's first underground rail system. Plans for the Underground were part of an improvement plan of the mid-1800s. Construction began in 1860. Workers dug trenches along streets and built brick walls to support their sides. Then these trenches were covered with brick arches, and the roads were restored above them. London's clay soils made construction of the underground railways easier. The clay was easy to excavate and provided the raw material to make bricks for tunnel walls.

In 1863 the first subway line opened, using steam locomotives. These trains burned coal, producing unpleasant fumes. However, despite the pollution, the Underground was successful from the very beginning. During its first year, the line carried 9.5 million passengers! In 1890, electric trains began to replace steam engines.

Expansion of the Underground, which Londoners call the tube, continued over the years. Improved tunneling techniques were developed after World War I. These techniques made possible a rapid expansion of the underground network. As the system spread out from central London, large areas of rural land became prime locations for new housing developments. As a result, the expansion of the Underground contributed to the development of London's modern suburbs. Now 253 miles (408 km) of track connect 275 stations. Each year, passengers log more than 920 million journeys on the Underground.

London's transit system served as an example for others. Why are such mass-transit systems important? Cities originally developed as places where people and resources were located close together. As cities grew and places became widely separated, improved transportation networks became more important.

However, planning for transportation needs has not always kept up with urban growth. Traffic jams on major highways and roads are common. Heavy traffic increases air pollution. Parking presents other problems. Building parking lots and garages takes up valuable land. Mass-transit systems, including buses, subways, and surface trains, can help solve these problems. Cities with efficient mass-transit systems usually run more smoothly than those in which people must depend mainly on cars. Mass-transit networks help make many of the world's big cities more livable. Examples include Mexico City, Moscow, New York, and Paris.

Applying What You Know 

1. **Summarizing** How has the development of mass-transit systems made big cities more livable?
2. **Drawing Inferences and Conclusions** How big do you think cities have to be to build and operate underground rail systems? What types of mass transit would you expect to find in smaller cities?

INTERPRETING THE VISUAL RECORD *London's mass-transit system includes the Underground and a citywide system of buses, which provide connections to the many Underground stations.* ***Why do you think London's mass-transit system is so effective at moving large numbers of people?***

Ireland
United Kingdom
London
Belgium
France

France

READ TO DISCOVER

1. What is French culture like?
2. What are some of the main industries in France?
3. What challenges does France face today?

DEFINE

primate city

LOCATE

Paris
Seine River
Lyon
Lille
Marseille
Alps
French Riviera
Corsica

WHY IT MATTERS

Many Europeans have been concerned about the spread of mad cow disease. Recently, the disease was found in French livestock. Use CNNfyi.com or other **current events** sources to learn about this and other dangerous livestock diseases.

History and Culture

France is one of Europe's largest and most influential countries. Like the United Kingdom, France's culture shows the imprint of successive waves of migrants. Some of the peoples that have shaped French culture include the Gauls, Romans, Franks, and Vikings.

The mistral, a powerful wind that blows from the Alps across southern France, can reach speeds of up to 100 miles per hour (161 kmh).

French Society France has a strong cultural identity unified by language and religion. Although some people also speak regional dialects and languages, most speak French. About 90 percent of France's population is Roman

France has produced many world-famous artists, including Pierre-Auguste Renoir (1841–1919). Renoir's Le Moulin de la Galette, *a scene showing life in Paris, is considered a masterpiece of impressionism. Impressionism is a style of painting that developed mainly in France in the late 1800s. It attempts to show what one's first impression of a scene is.*

The Eiffel Tower is a landmark and symbol of Paris. It was built from 1887 to 1889 to celebrate the 100-year anniversary of the French Revolution. At 984 feet (300 m), the Eiffel Tower was the tallest human-made structure in the world until the completion of New York City's Chrysler Building in 1930.

Catholic. The French government spends money to promote French culture and language.

France has had a long and friendly relationship with the United States. However, today some French people worry about the influence of American culture in Europe. They see the spread of American fast food and media, such as movies, as a threat to their own culture. Some French dislike the fact that English words are creeping into French. They think the United States is responsible because English is becoming the global language of business and technology. To counter this trend, a 366-year-old government agency guards the French language from foreign influences. For example, it has declared that e-mail must be called *courrier électronique* (KOOH-ree-ay ay-lek-trohn-EEK). Some French are also concerned that American corporations are buying a growing number of French businesses.

Cities Paris is the capital and **primate city** of France. A primate city is one that ranks first and dominates a country in terms of population and economy. Paris is also one of Europe's largest and most important cities. About 11 million people live in the metropolitan area. The city was founded more than 2,000 years ago on an island in the middle of the Seine (SAYN) River. Today Paris is France's center for banking, business, communications, education, government, and transportation. The city is also a center for fashion, French culture, and tourism. Important regional cities include Lyon (LYOHN), Lille (LEEL), and the Mediterranean seaport of Marseille (mar-SAY).

READING CHECK: ***Human Systems*** How has France reacted to the influences of American culture?

French Waterways

In the 1800s canals and rivers were a popular and economical way to move goods. Later the waterways fell from favor with the introduction of railroads and long-distance trucking. However, today there is a new interest in using waterways for commerce. France is working to increase trade on its rivers and canals. It has Europe's longest system of canals deep enough to move commercial barges. By moving more goods by water, the government hopes to reduce traffic on its roads.

To accomplish this goal, the government lifted many old regulations. River channels have also been deepened, and locks have been removed. These changes have helped speed up travel on the waterways. As a result, more businesses are shipping their goods by water.

Why is such shipping important? Barges are more than twice as energy-efficient as trains and more than five times as efficient as trucks. The increased energy efficiency is good for the economy as well as for the environment. In addition, boats are not as noisy as trains or trucks. This means that they can operate at night in urban areas without bothering people. Boats also provide a safer way to carry dangerous goods.

Analyzing How has the French perception of their water resources changed? What has been the result of this changed perception? TEKS

A canal bridge over the Loire River

INTERPRETING THE VISUAL RECORD

A hiker views mountain scenery near Chamonix in the French Alps. France's physical and cultural landscapes attract tourists from around the world and help make tourism one of the country's leading industries. ***What type of physical feature is shown in this photo? What physical processes created it?*** TEKS

Economy

France has a highly diversified, developed economy. Its workers are some of the most productive in the world. This is true even though they have the shortest workweek—35 hours—and some of the longest vacations—one month—of any workers in the industrial countries.

The French are famous for fashion design. They also produce perfumes, cosmetics, jewelry, glassware, and furniture. Tourism is also important. Millions of people visit Paris, ski in the French Alps, and enjoy the famous Mediterranean coast known as the French Riviera.

Farming remains an important part of the economy. France is second only to the United States in agricultural exports. The French produce high-quality food products and a great variety of agricultural produce. Farmers provide wheat, sugar beets, olives, grapes, and dairy products. France is the world's leading wine producer in both variety and export income.

France's early industries were centered in the northeast near large deposits of coal and iron ore. Those heavy industries are now in decline. However, the country's high-tech industries are developing rapidly. The south is a growing center for aviation, communications industries, and space technology.

READING CHECK: ***Environment and Society*** Where were early French industries centered? What natural resources were found there?

Issues and Challenges

One of the major issues facing France today is the government's powerful influence over the economy. The French economy is both highly taxed and highly regulated, and many industries are government-owned. Many argue that these controls hurt innovation and creativity. They also make it hard for private businesses to grow. This situation is changing as state-owned businesses and industries are turned over to private owners. This process is called privatization. However, as privatization occurs, many government workers are faced with unemployment.

France's Exports

machinery and transportation equipment; chemicals; iron and steel products; agricultural products; textiles and clothing

Source: Central Intelligence Agency, *The World Factbook 2000*

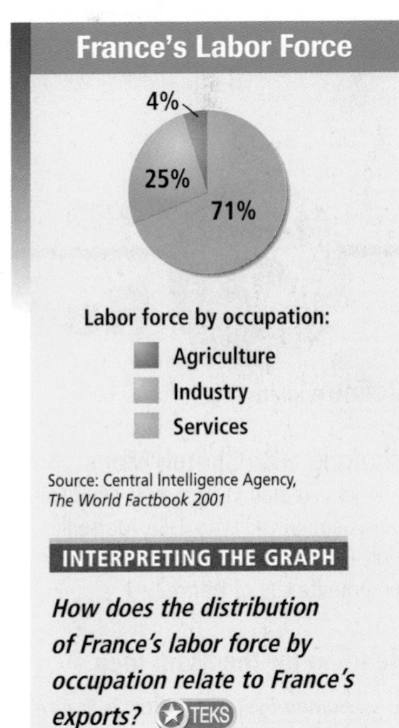

INTERPRETING THE GRAPH

How does the distribution of France's labor force by occupation relate to France's exports? TEKS

Corsica is known for its beautiful Mediterranean landscape, which features rugged mountains and dramatic coastal cliffs. However, economically the island is less developed than much of France, which has led many Corsicans to migrate to the mainland for jobs.

Since the early 1900s many Algerians and Moroccans have been migrating to France in search of jobs. In fact, North Africans now form the largest immigrant group in the country. Immigrants have also come from former French Indochina—Vietnam, Cambodia, and Laos—and from former French colonies in West Africa. Many immigrants live in poorer sections of the major French cities. France's many immigrant communities have helped create distinctive urban landscapes. Many neighborhoods where immigrants live feature non-European restaurants and shops. Bringing these immigrants into French society is another important challenge.

After the British, the French maintained the second-largest colonial empire in the world. France has tried to maintain ties with some of those former colonies, particularly in Africa. The French also have overseas territories that are departments of France. In other words, they are considered part of France. One such department is French Guiana, in South America. Other French territories are mainly islands in the Caribbean, the South Pacific, and the Indian Ocean. They include Guadeloupe, Tahiti, and Réunion. Recent independence movements on the South Pacific island of New Caledonia have led to violence. Violence has also occurred on the large and rugged Mediterranean island of Corsica. Corsica has been part of France since 1768. However, many Corsicans consider themselves culturally distinct from France and want their island to become independent.

TEKS **READING CHECK:** ***Environment and Society*** Why have many Moroccans, Algerians, and other groups of people been immigrating to France?

TEKS Questions 1, 2, 4, 5

Section 2 Review

Homework Practice Online

Keyword: SW3 HP14

Define primate city

Working with Sketch Maps On the map you created in Section 1, label France, Paris, Seine River, Lyon, Lille, Marseille, Alps, French Riviera, and Corsica. What is the primate city of France?

Reading for the Main Idea

1. ***Human Systems*** How is France's culture important to its economy?
2. ***Human Systems*** How are public policies and decision making influenced by French cultural beliefs?
3. ***Human Systems*** Where does France still have overseas possessions?

Critical Thinking

4. **Making Generalizations** Why do you think some French workers at government-owned companies might face losing their jobs as their companies are turned over to private ownership?

Organizing What You Know

5. Construct a word web like the one below. Use it to identify sources of American influence on French culture.

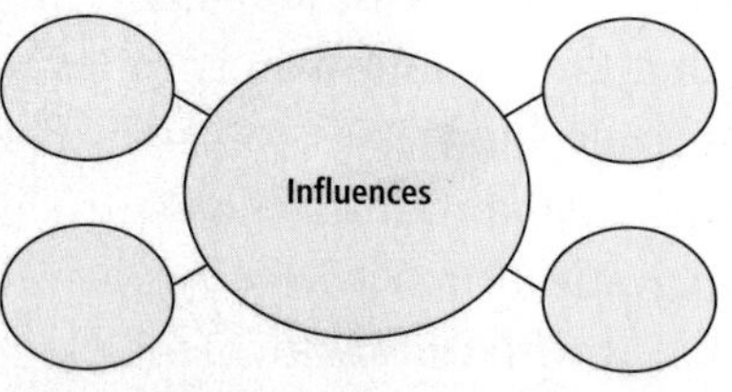

Section 3 The Benelux Countries

READ TO DISCOVER

1. What historical ties do the Benelux countries share?
2. What are the cities and economies of the Benelux countries like?

WHY IT MATTERS

In 1999 a new currency called the euro was introduced in Europe. Use CNNfyi.com or other **current events** sources to learn about how and why the euro was created.

IDENTIFY

European Union

DEFINE

cosmopolitan

LOCATE

North Sea
Flanders
Wallonia
Brussels
Antwerp
Amsterdam
Rotterdam
The Hague

History and Culture

Belgium, the Netherlands, and Luxembourg make up the Benelux countries. For many years, Belgium and Luxembourg were part of the Netherlands. Because of their position between three powerful countries—France, Germany, and the United Kingdom—all three of the Benelux countries have been fought over by foreign powers. After World War II, the Benelux countries established a political and economic union. Their early economic association planted the seed that eventually led to the creation of the **European Union**.

Vianden Castle in Luxembourg was built between the A.D. 1000s and 1300s.

FOCUS ON GOVERNMENT

The European Union The European Union, or EU, is an organization of countries interested in increasing economic and political cooperation between its members. The EU was established on November 1, 1993. However, its origins date back to the 1950s.

In the past, many European leaders tried to unite the continent politically and economically. They failed because they used force rather than cooperation to bring different countries together. After World War II, some European countries began forming alliances based on mutual aid rather than military strength. These alliances tried to tie members more closely together economically and politically. One way of doing this was to eliminate trade barriers among members of the same alliance.

Belgium, the Netherlands, Luxembourg, France, Italy, and West Germany formed such an organization in 1957. They called it the European Economic Community, or EEC. Later the name was shortened to just the EC. Over time, the EC grew, joined with other organizations, and became the EU.

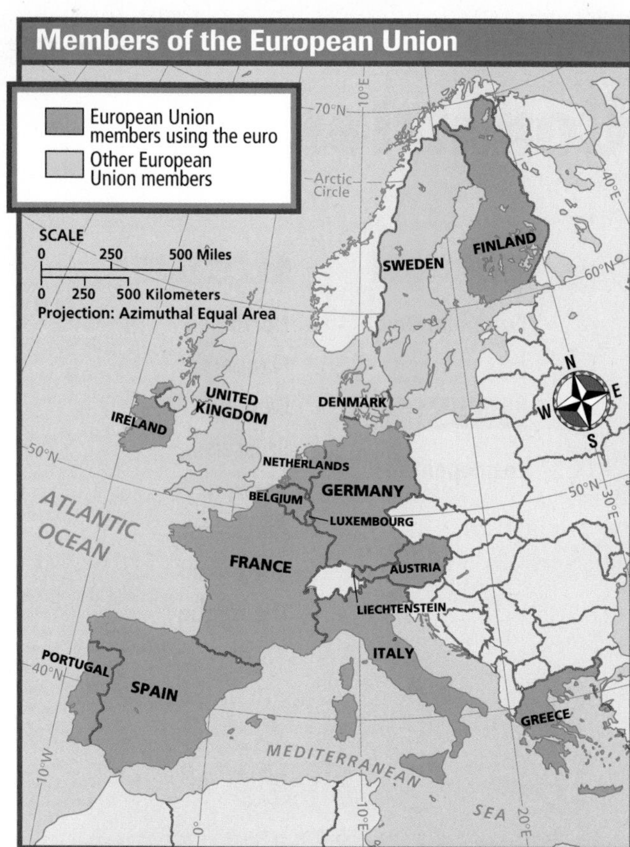

INTERPRETING THE MAP *The 15 EU countries set common tariffs for goods imported from nonmembers and have allowed the free movement of goods, labor, and capital among member-countries, creating one of the most important free trade organizations in the world.* ***How has the EU affected boundaries and political divisions within Europe?*** TEKS

Today the EU includes 15 countries. It has increased cooperation among members in the areas of trade, law-making, and social issues. The EU also introduced a common currency, the euro. Most, but not all, EU countries adopted the euro in 2002. (See the map.)

In February 2002 the EU held a constitutional convention to address issues such as common defense and taxes and how an enlarged EU will work. In December 2002 the EU voted to admit 10 new members—mostly from Eastern Europe. If all goes as planned, these 10 countries will become members on May 1, 2004. The EU must also decide whether, and how, to expand membership to other countries—such as Turkey—that want to join.

READING CHECK: *Human Systems* What role does the EU play in Europe today?

Land Reclamation The name *Netherlands* means "low lands." Large areas of this country are below sea level. In fact, early in the country's history, much of the land in the Netherlands was coastal marshes and wetlands. However, people in the region have long worked to reclaim land from the sea. Today farms, towns, and industrial centers are located on polders below sea level. The dike and polder system has been very successful. However, the Dutch—the people of the Netherlands—worry about floods during severe North Sea storms. Rising global sea levels might also become a problem.

Language Dutch is spoken in the Netherlands and in northern Belgium. The dialect of Dutch spoken in northern Belgium is also called Flemish. About 60 percent of Belgians speak Dutch. Many Belgians also speak French. In fact, the country is divided into two cultural regions. The northern coastal region is known as Flanders. The French-speaking Belgians in the southern portion of Belgium are known as Walloons, and the region is known as Wallonia. The people in Belgium generally view themselves as either Flemish or Walloon rather than Belgian. Luxembourg, to the south, has three official languages: German, French, and Luxembourgian. Luxembourgian is a language related to German and Dutch.

✓ **READING CHECK:** *Human Systems* What ties exist among the Benelux countries?

Urban and Economic Environments

The headquarters for the EU is in Brussels, the capital of Belgium. The city's central location in Europe and good transportation connections make it an ideal headquarters for the EU. Brussels is also the headquarters for the North Atlantic Treaty Organization (NATO) and many international corporations. As you might expect, Brussels is one of Europe's most **cosmopolitan** cities. A cosmopolitan city is one that is characterized by many foreign influences. The port of Antwerp is Belgium's second-most-important city. The Belgian economy is based on industry, agriculture, and services for international business. The country is also known for diamond cutting, quality carpets, and chocolate.

The Dutch economy is known for agriculture, particularly dairy products and flowers. For example, Dutch cheese and tulips are world famous. The Dutch economy today is very diversified, and exports are important. The economy is also one of the best performers in the EU. Natural gas deposits are found in the coastal and offshore region of the North Sea. However, the Netherlands is very dependent on imported oil.

The most urbanized and industrialized area in the Netherlands is known as the Randstad, or "Ring City." Here you will find the largest cities—including Amsterdam, the capital, and Rotterdam, one of the world's busiest seaports. These cities are strung together in a crescent shape. The Dutch parliament and International Court of Justice are located in The Hague. The Dutch population is well educated, productive, and supported by expensive government social programs.

Luxembourg is a forested and hilly country between Belgium, France, and Germany. It is the smallest member of the European Union, but it has the highest per capita GDP in the world—$36,400. Luxembourg has long been a steel producer. However, today international banking is most important. The small country has a constitutional monarchy.

INTERPRETING THE VISUAL RECORD

The Netherlands is an important exporter of flowers and is famous for its brightly colored tulip fields. Tulips are one of the most popular garden flowers in the world, and almost 4,000 varieties have been developed. ***Based on the photo, what environmental factors might make the Netherlands ideally suited to using modern technology in agriculture?*** TEKS

TEKS **READING CHECK:** ***Environment and Society*** Why is Brussels an ideal location for the headquarters of many European businesses?

TEKS Questions 1, 2, 3, 4, 5

Section 3 Review

Keyword: SW3 HP14

Identify European Union

Define cosmopolitan

Working with Sketch Maps On the map you created in Section 2, label the three Benelux countries, North Sea, Flanders, Wallonia, Brussels, Antwerp, Amsterdam, Rotterdam, and The Hague. Where is the headquarters of the EU?

Reading for the Main Idea

1. ***The World in Spatial Terms*** How has the location of the Benelux countries influenced their history?
2. ***Human Systems*** What large European organization promotes cooperation among members in the areas of trade, lawmaking, and social issues? What are two important issues debated by members today?

Critical Thinking

3. **Making Generalizations** Why would rising sea levels be a concern in the Netherlands?
4. **Comparing and Contrasting** How do you think the Randstad in the Netherlands compares to the megalopolis of the northeastern United States? What similar political, economic, social, and environmental features would you expect to find in the two regions?

Organizing What You Know

5. Copy the chart. Use it to list the languages spoken in each of the Benelux countries.

Countries	Languages
Belgium	
The Netherlands	
Luxembourg	

4 Scandinavia

READ TO DISCOVER

1. How are the cultures of Scandinavia similar to and different from each other?
2. What does the economy of this region rely on?
3. In what areas do most people in Scandinavia live?

DEFINE

uninhabitable
geysers
socialism

WHY IT MATTERS

Many houses in Iceland use geothermal heat as an energy source. Use CNNfyi.com or other **current events** sources to learn about other alternative energy sources.

LOCATE

Lapland
Copenhagen
Oslo
Helsinki
Stockholm
Greenland
Faeroe Islands
Reykjavik

Norway is sometimes called the Land of the Midnight Sun because the Sun does not set in the northern parts of the country for about one month every summer.

History and Culture

Five countries make up Scandinavia—Norway, Sweden, Denmark, Finland, and Iceland. These countries are the northernmost countries in Europe. In the past the region was known for the fierce Viking sailors and warriors who raided the shores of Europe. However, today the countries of Scandinavia are known for their modern economies and high standards of living.

Scandinavians share many cultural traits. For example, almost all Scandinavians are Protestant Lutheran. Except for Finnish, Scandinavian languages are closely related. Finnish belongs to the same language family as Hungarian and Estonian. However, speakers of Danish, Swedish, and Norwegian can generally understand one another. Also, all of the Scandinavian countries have democratic governments. These cultural similarities help make Scandinavia a clear cultural region.

INTERPRETING THE VISUAL RECORD

Viking longships, like this one from Norway, featured many technological innovations. They were lighter, faster, easier to sail, and more durable than other ships of their time. ***How might Viking improvements in ship design have allowed the diffusion of Viking culture?*** TEKS

Lapland stretches across northern Norway, Sweden, and Finland. This region is mainly tundra and is populated by the Lapps—or Sami, as they call themselves. The Sami probably originated in central Asia. Their economy was traditionally based on reindeer herding, but today most Sami earn a living from tourism. Despite this, many Sami have maintained some of their traditional culture.

INTERPRETING THE VISUAL RECORD

The Scandinavian countries all have a primate city. For example, Copenhagen is Denmark's capital and cultural, artistic, and economic center. ***What are the primate cities of the other Scandinavian countries?***

Settlement Patterns Most Scandinavians live in the southern parts of their countries, where climates are warmer. For example, most Norwegians live in coastal plains areas or along narrow fjords. More than half of the population lives along the southeastern coast of the country. Most of the ports on that coast remain free of ice all year.

Scandinavian countries have healthy and well-educated populations with long life spans and low birthrates. In general, their populations are growing slowly and are heavily urban. About 85 percent of Sweden's population is found in urban areas. More than one fourth of Danes and Norwegians live in or around their respective capital cities of Copenhagen and Oslo. Most Finns live near Helsinki, the country's capital and leading seaport. Stockholm is Sweden's capital and largest city.

TEKS **READING CHECK:** ***Human Systems*** Which country's language is not related to the languages of the other Scandinavian countries?

Greenland and Iceland Greenland is not really green. About 85 percent of it is covered by a thick ice cap. The icy interior of the island is **uninhabitable**. An uninhabitable region is one that cannot support human life and settlements. Only Greenland's rocky coastline is fit for human habitation. Most Greenlanders live along the southwestern coast. In the 900s, the Vikings founded settlements on Greenland's coast. Early Viking settlers tried to attract others by describing the island as a place with plenty of vegetation and a mild climate. This description drew new settlers to Greenland. However, these early settlements died out in the 1400s. Scandinavian settlers did not colonize the island again until the 1700s. Today Greenland is a self-governing territory of Denmark. Denmark also governs the Faeroe (FAHR-oh) Islands in the North Atlantic Ocean.

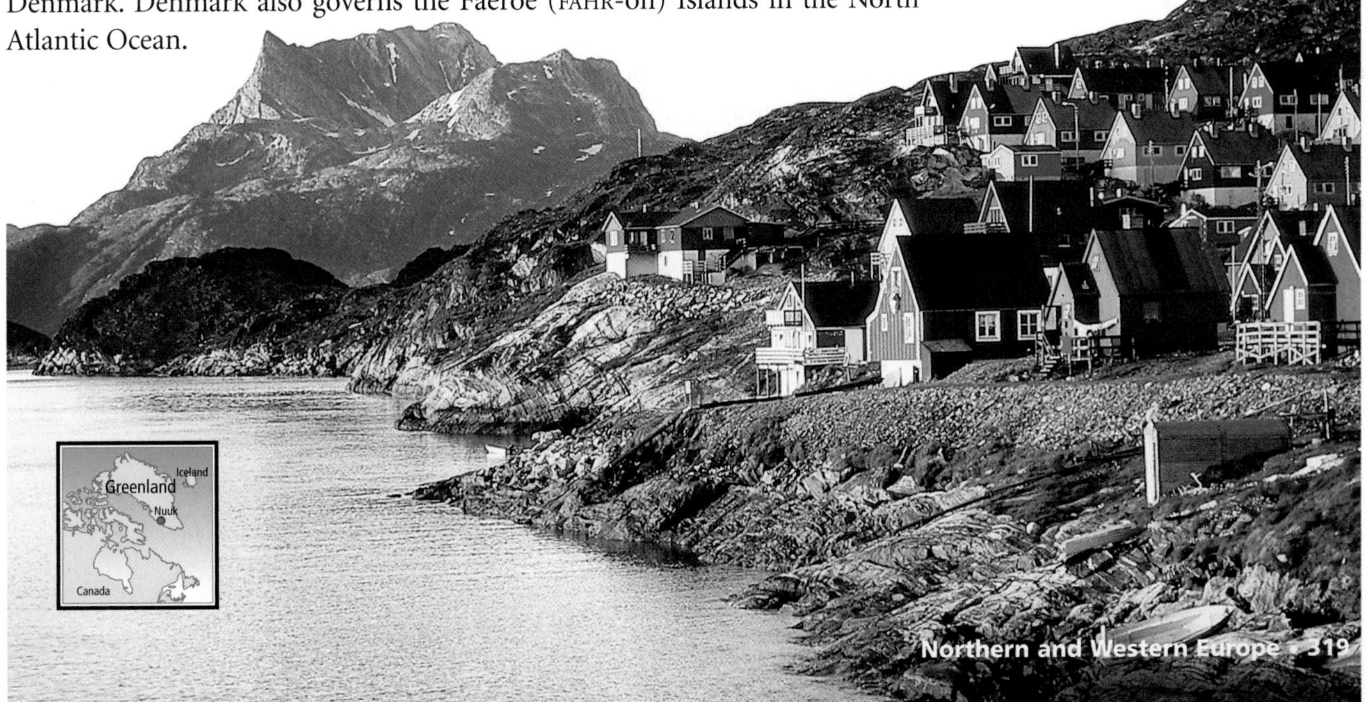

INTERPRETING THE VISUAL RECORD

Greenland's capital, Nuuk, is located on the island's western coast, where the warm West Greenland Current helps moderate temperatures. ***How do you think this ocean current has influenced settlement patterns in Greenland?*** TEKS

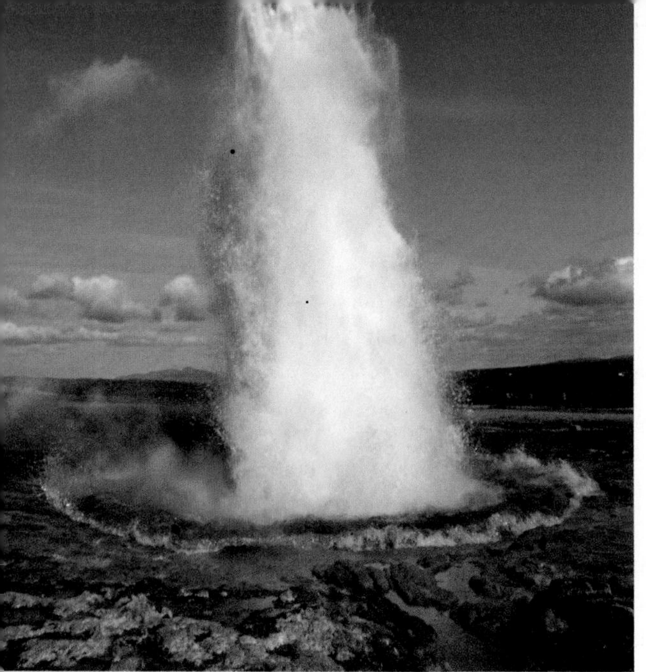

Volcanic activity in Iceland generates many geysers. Some can shoot water as high as 1,640 feet (500 m).

In Iceland, which is greener than its name implies, all the people live along the narrow coastal plains. That is because the island's interior of ice-covered lava rock is also uninhabitable. Most Icelanders live in or near the capital Reykjavik (RAY-kyah-veek). The country is a member of NATO but has not shown an interest in joining the EU.

Iceland has tremendous geothermal energy because of its location on a mid-ocean ridge where volcanic activity is common. Underground water rises and steams as **geysers** in many locations on the island. A geyser is a hot spring that shoots water into the air. The hot water is used to heat homes and vegetable greenhouses. The island also has hydropower potential. In the future, Iceland may be able to export geothermal energy and hydroelectricity to Europe across an underwater cable.

READING CHECK: ***Environment and Society*** Where will you find the human populations of Greenland and Iceland?

Economy

The Scandinavian countries all have high standards of living. For example, Denmark is one of the EU's most prosperous countries and has one of the highest per capita GDPs in Europe. High-tech industries and export-oriented economies maintain the high standards of living.

Economic Development Finland has been transformed from an exporter of natural resources to a manufacturing country. Finland produces and exports high-tech goods. Its products include advanced telecommunications equipment, cellular phones, and computer software. The Swedes produce a variety of high-tech and high-value products. These goods include automobiles, cellular phones, aircraft, and industrial robots.

In addition to manufacturing, commercial agriculture—particularly the dairy and meatpacking industries—is important to the Danish economy. The paper- and wood-products industries are well developed in Sweden, where forests cover half the country.

Fishing is also important in Scandinavia, particularly in Iceland and Norway. Norway has a large commercial fishing fleet. However, it is offshore oil and gas from the North Sea that makes Norway a rich country. Most of Norway's oil profits are invested for the future. The country also has hydroelectric plants that produce a surplus of electricity, which is exported.

Fishing has long been important to Norway's economy. In fact, Norway has one of the largest commercial fishing industries in the world and exports more fish than any other European country.

Economic Change During much of the last half of the 1900s, Sweden's economy was a mix of capitalism and **socialism**. Socialism is an economic system in which the government owns and controls the means of producing goods. Most of Sweden's industries remained privately owned. However, the government controlled some businesses, and it levied high taxes. These high taxes still pay for a large system of government welfare and services. For example, the government pays for almost all the educational, medical, and childcare needs of its citizens. For example, all residents of Sweden are covered by national health insurance. Compared to many other countries, health conditions in Sweden are very good. The government also pays for programs to help parents raise their children. For example, parents can share up to one year of paid time off from work before their child reaches the age of eight. They also receive tax-free payments to help pay for the costs of raising children.

By the late 1990s about 60 percent of Swedes relied on the government for work or welfare payments. The Danes also have a well-developed welfare system. High taxes in Scandinavia pay for environmental protection and support sports and the arts.

Many Swedish economists blame a costly welfare system for the economic problems the country has been experiencing since 1991. In recent years Sweden's government has tried to lessen its influence over the economy. Still, many Swedes do not like the idea of cutting back the welfare system that they have. As a result, the government has preserved many of its expensive social programs.

TEKS **READING CHECK:** *Human Systems* Why are taxes in Scandinavian countries like Sweden so high?

INTERPRETING THE VISUAL RECORD *The Øresund bridge, which connects Copenhagen, Denmark, with Malmö, Sweden, opened in 2000. The massive bridge cost more than $2 billion and provides road and rail connections between the Scandinavian Peninsula and the rest of Europe.* ***How might this bridge affect the locations and patterns of economic activities in Scandinavia?*** TEKS

TEKS Questions 1, 2, 4

Review

Define
uninhabitable, geysers, socialism

Working with Sketch Maps On the map you created in Section 3, label the Scandinavian countries, Lapland, Copenhagen, Oslo, Helsinki, Stockholm, Greenland, Faeroe Islands, and Reykjavik. Which large island is part of Denmark?

Reading for the Main Idea

1. *Environment and Society* How did early Viking settlers try to shape perceptions of Greenland to draw others there? How accurate were their descriptions of the island?
2. *Human Systems* What are the standards of living and the economies of Scandinavian countries like?
3. *Human Systems* What is the major source of Norway's wealth?

Critical Thinking

4. **Making Generalizations** What might be one advantage and one disadvantage of the social welfare system found in many Scandinavian countries?

Organizing What You Know

5. Create a chart like the one below. U it to list important economic activi ties and energy resources in the Scandinavian countries.

Countries	Economic a and energy

CASE STUDY

Global Trade

Human Systems Have you ever thought about how far away some of the things you buy originate? Many of the products we buy are transported great distances before they reach the cash register. This flow of goods is part of a global system of trade. For example, raw materials like cotton, iron, oil, or wood are often harvested or mined far away from the factories that transform them into finished goods. Products like blue jeans, cars, coffee tables, and toys may then travel even greater distances from the factory to the store.

The roots of modern global trade go back some 500 years. At that time, long-distance trading networks began to develop as European countries explored and colonized the Americas and Asia. These countries were aided by improvements in ship design and navigation equipment. By the 1540s Portugal had established a chain of trading posts all the way to Japan. Portuguese ships brought valuable spices from Asia to buyers in Europe. At the same time, Spain set up its own trading networks. Spain mined gold and silver in Mexico, Bolivia, and Peru. These valuable minerals were then shipped back to Spain to enrich the royal treasury. Other ships brought tobacco from North America and chocolate from Mexico.

In the 1600s and 1700s Dutch, English, and French traders competed with Spanish and Portuguese merchants to control global trade. This competition led to a rapid growth of new colonies and trading patterns. Europeans explored distant parts of the world and discovered new foods and drinks. These new, exotic goods began to pour into Europe. For example, coffee, originally from Ethiopia, was introduced to Europe through Yemen. The first coffeehouse in London was set up in 1652. By the late 1600s coffeehouses were becoming popular in Boston, New York, and Philadelphia.

Other goods also appeared on European markets. In 1669 an English trading company made its first shipment of Chinese tea to London from Java. Dutch traders brought spices such as cloves and nutmeg from the East Indies. Later, they also traded cinnamon, coffee, jewels, and pepper. They also participated in the slave trade. In 1624 the Dutch set up a port on Manhattan and developed a fur-trading business with American Indians.

Over time, European merchants and government officials developed new economic systems to meet consumer demands for all of these new products. Specifically, they created the plantation system. This system produced large amounts of agricultural goods in the tropical and subtropical regions of the globe. The plantation system typically relies on a large labor force to raise a single crop on a large tract of land. After the crop is harvested, it is exported to distant markets. Large regions of Asia, Africa, and the Americas developed into colonial plantations. The plantations produced coffee, cotton, spices, sugarcane, tea, and many other goods.

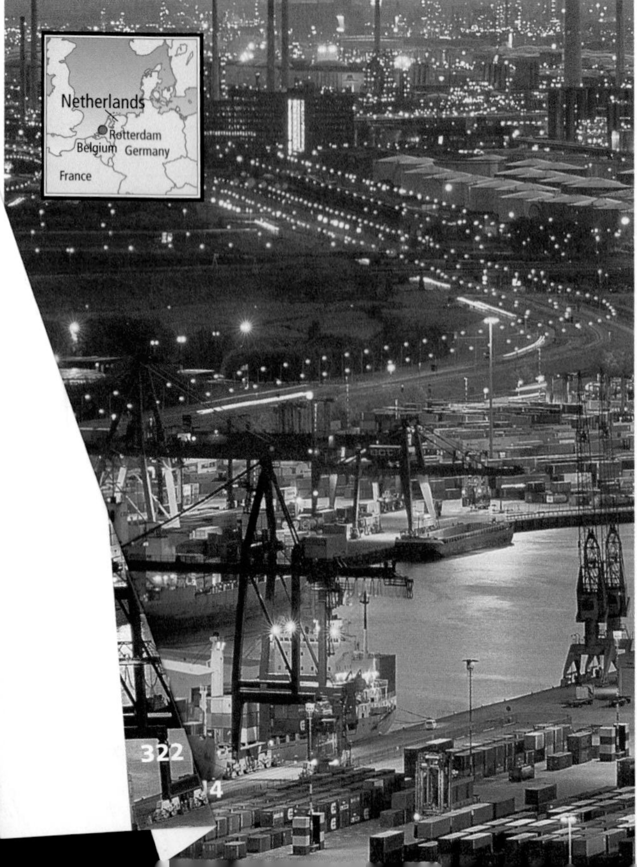

Rotterdam is one of the busiest ports in the world, and its history is closely tied to the development of global trade. This port near the mouth of the Rhine River became internationally important in the 1600s when Dutch traders began importing goods to Europe from the East Indies. Today Rotterdam handles a huge range of goods, including oil and petroleum products, grains, and many other goods.

Trade Routes in the Middle Ages

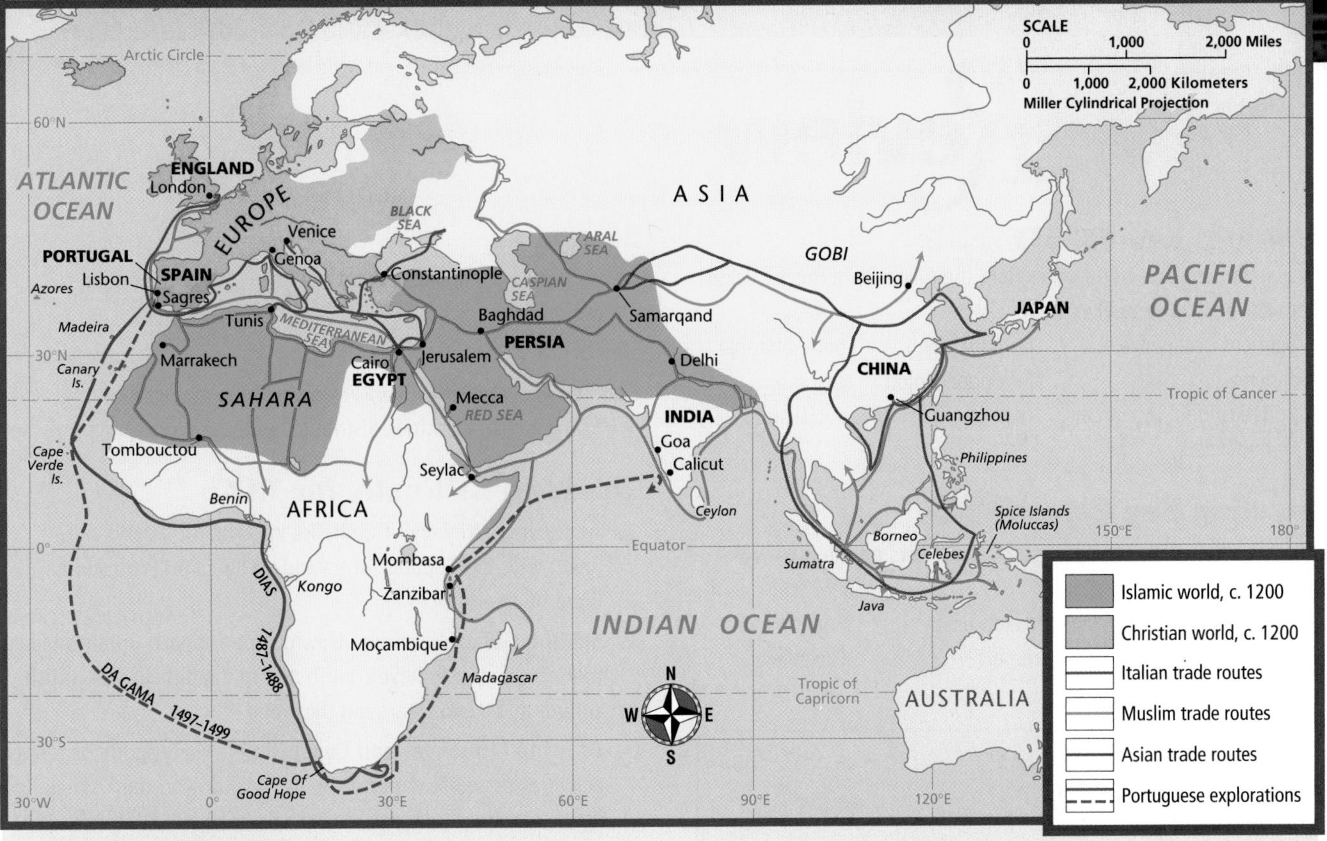

INTERPRETING THE MAP *In the Middle Ages, extensive trade routes had already been developed in Europe, Africa, and Asia. However, Portuguese and then Spanish explorations beginning in the late 1400s altered these patterns.* ***How did global trade patterns change by the 1700s? Why did they change, and what were the implications of these changes?*** TEKS

In the late 1700s the Industrial Revolution had a major influence on global trade. The development of steam engines created a new way to transport goods across large distances—the railroad. A strong navy and control of the seas were no longer the only ways to develop global trading links. The vast resources of interior Asia, Africa, and the Americas became more accessible and began to appear on the global market. Also, the iron and steel needed to build railroads were suddenly in great demand. Germany's Ruhr Valley had a good supply of iron and steel and soon became a major manufacturing area. The invention of the cotton gin in 1793 also greatly affected global trade. Much of the southern United States developed a plantation economy based on cotton. By 1861 the United States grew more than 80 percent of the world's cotton. More than half of the cotton was shipped to Manchester, England, where it was made into cloth.

Today major trading routes crisscross the entire globe. Many common items that we use every day have done more traveling than most people do in a lifetime. A car may be made of German steel, Saudi Arabian plastic, and British glass. New technologies constantly reshape global trade patterns. For example, computers and electronic trading now allow people to shop on the Internet. They can order things easily from around the world. Global trade has also changed the way many companies do business. Companies can buy resources from distant places, locate factories in many different countries, and sell their products around the world. As a result, global trade is now a major force behind globalization. Expanding trade networks allow the same products to become familiar all over the world.

Applying What You Know

1. **Summarizing** How have global trade patterns changed since the 1500s? How have these patterns affected life around the world?

2. **Making Generalizations** Suppose that European countries had not developed huge colonial empires. What other factors may have influenced global trade patterns over time?

Building Vocabulary (TEKS)

On a separate sheet of paper, explain the following terms by using them correctly in sentences.

sequent occupance	primate city	uninhabitable
famine	European Union	geysers
constitutional monarchy	cosmopolitan	socialism
nationalized		

Locating Key Places (TEKS)

On a separate sheet of paper, match the letters on the map with their correct labels.

Northern Ireland	Luxembourg	Copenhagen
London	Brussels	Greenland
Seine River		

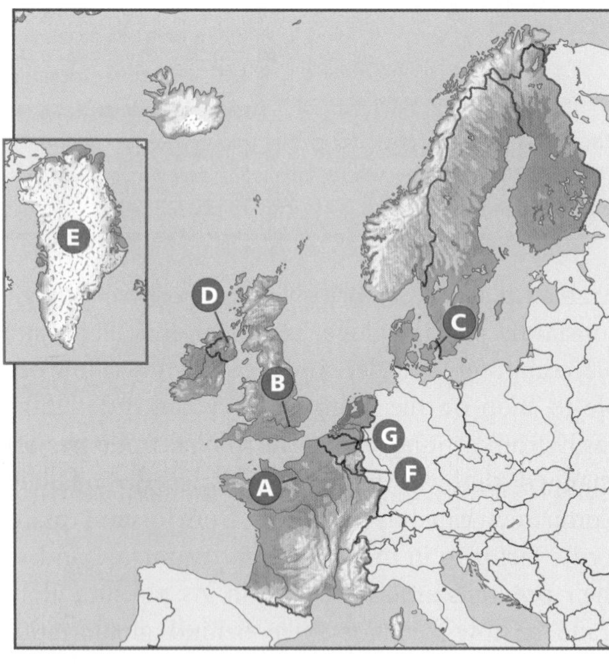

Understanding the Main Ideas (TEKS)

Section 1

1. *Human Systems* How did Britain control a vast empire? How is it tied to former colonies today?
2. *Human Systems* How has Ireland's economy changed in recent years? How have these changes influenced migration?

Section 2

3. *Places and Regions* What features make Paris the primate city of France?

Section 3

4. *Places and Regions* What are Belgium's two language regions?

Section 4

5. *Environment and Society* How is the population of Scandinavian countries distributed with regard to cities?

Thinking Critically for TAKS (TEKS)

1. **Analyzing** Why might Catholics in Northern Ireland want union with the Republic of Ireland? Why might Protestants there be against it?
2. **Making Generalizations** How do you suppose Britain's industrial economy may have contributed to the diffusion of British culture and customs around the world?
3. **Drawing Inferences and Conclusions** Many sports began as activities essential to everyday life. How do you think skiing has been essential to daily life in Scandinavia?

Using the Geographer's Tools (TEKS)

1. **Analyzing Maps** Review the map of the British Empire in Section 1. Then list the continents on which you would expect to find people who speak English today. Explain.
2. **Analyzing Maps** Review the map of the European Union in Section 3. Which countries do you think the EU might expand to in the future? Why?
3. **Creating Bar Graphs** Use statistics from the unit Fast Facts table to construct a bar graph comparing the population density of countries in northern and western Europe. Which country is the most densely populated?

Writing for TAKS (TEKS)

How is the European Union similar to and different from the United States? Do you think the EU countries will unite? Write a short report explaining your point of view.

SKILL BUILDING

Geography for Life (TEKS)

Observing the Weather

Places and Regions How is your local climate influenced by the wind? Set up a wind sock and record the direction and force of the wind for a week. Also, record temperature and precipitation. Graph your information and explain the information shown.

Building Social Studies Skills

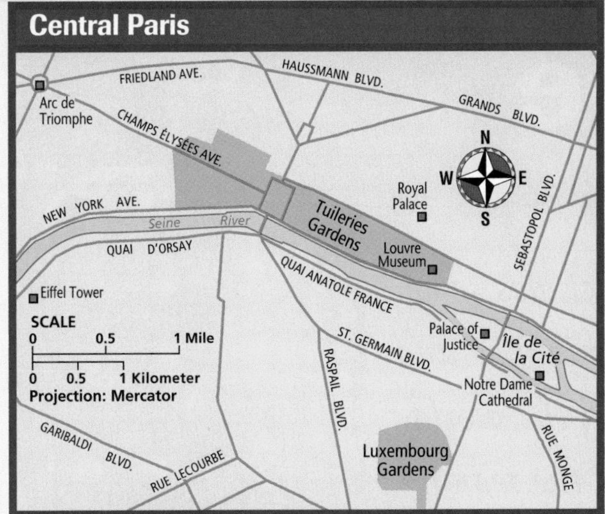

Interpreting Tourist Maps

Study the tourist map above. Then answer the questions that follow.

1. Which point of interest is located south of the Seine?
   a. Notre Dame Cathedral
   b. Eiffel Tower
   c. Arc de Triomphe
   d. Louvre Museum

2. Suppose you wanted to take a walking tour along the Seine. What nearby sites of interest would you be able to visit along the tour?

Analyzing Primary Sources

Read the following description of London by Simon Worrall and then answer the questions.

"The whole world lives in London. Walk down Oxford Street and you will see Indians and Colombians, Bangladeshis and Ethiopians, Pakistanis and Russians, Melanesians and Malaysians. Fifty nationalities with communities of more than 5,000 make their home in the city, and on any given day 300 languages are spoken. It is estimated that by 2010 the population will be almost 30 percent ethnic minorities, the majority born in the U.K. [United Kingdom]."

3. According to the author, in 2010 the population of London will
   a. be made up of mostly immigrants.
   b. have very few ethnic minorities.
   c. be decreasing as immigration slows.
   d. be almost 30 percent ethnic minorities.

4. What point is the author trying to make when he says "the whole world lives in London"? What details does he provide to support this point?

Alternative Assessment

PORTFOLIO ACTIVITY

Learning about Your Local Geography TEKS

Group Project: Research

In Sweden, government programs take care of the educational and medical needs of Swedish citizens. In your own area, who is responsible for paying the costs of people's educational and medical needs? With your group, research how each set of needs is addressed. Present your information in a chart.

internet connect

Internet Activity: go.hrw.com
KEYWORD: SW3 GT14 TEKS

Choose a topic on northern and western Europe to:

- learn the history of skiing in Norway.
- compare and contrast major cities in the region.
- research daily life in the region.

CHAPTER 15
Central Europe

Crown, Holy Roman Empire

Central Europe includes some of the most industrialized and richest countries in the world. Other countries in the region are slowly recovering from decades of Communist rule.

Shepherd with alpenhorn, Switzerland

Grüss dich! (Hello!) My name is Lizzi (LEE-zee), and I live in the village of Deutenhausen. Deutenhausen is near Munich in Bavaria, in southern Germany. I am in the eighth grade at the gymnasium (high school). I live on a farm with my three older sisters, my parents, and my grandmother. My parents are farmers and also own a restaurant. In the summer, I make sure the cows have enough water. I also help my parents in the restaurant by chopping vegetables for the salads. When I grow up, I hope to become a doctor and work in an emergency room.

In the morning I drink warm milk fresh from our cows and eat fresh bread baked in the restaurant. Our kitchen is a huge room where everyone hangs out. At about 7:30 A.M. each day, I take the bus to school in Weilheim, which is about 2 miles (3 km) away. My favorite subject is art. I also study German, geography, Earth science, English, and Latin. Next year I will start classical Greek.

When school is over at 12:30 P.M., I go home to have lunch with my grandmother. Afterwards, I play with my friends outdoors, even though it often rains. We splash in the creek, race our bikes, and climb up to the church steeple to hear the bells ring.

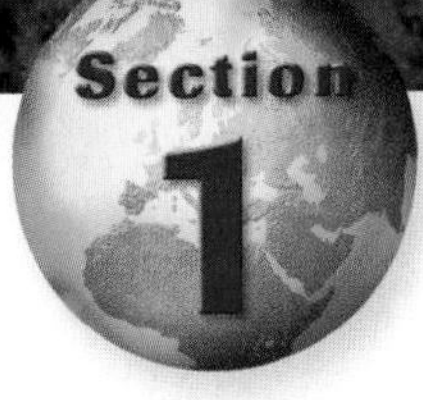

Germany

READ TO DISCOVER

1. What are some key events in the history of Germany?
2. What are some features of German culture?
3. What is Germany's economy like?
4. What issues and challenges does Germany face today?

DEFINE

alliances
balance of power

LOCATE

Berlin
Bavaria
Ruhr Valley

Ivory carving from Dresden, Germany

WHY IT MATTERS

The German government once invited foreign workers to the country in order to solve a labor shortage. Now some Germans are taking jobs in other countries. Use CNNfyi.com or other **current events** sources to learn about labor shortages and how countries deal with them.

Central Europe: Physical-Political

SCALE
0 100 200 300 Miles
0 100 200 300 Kilometers
Projection: Azimuthal Equal Area

ELEVATION

FEET	METERS
13,120	4,000
6,560	2,000
1,640	500
656	200
(Sea level) 0	0 (Sea level)
Below sea level	Below sea level

⊛ National capital
• Other cities

Size comparison of Central Europe to the contiguous United States

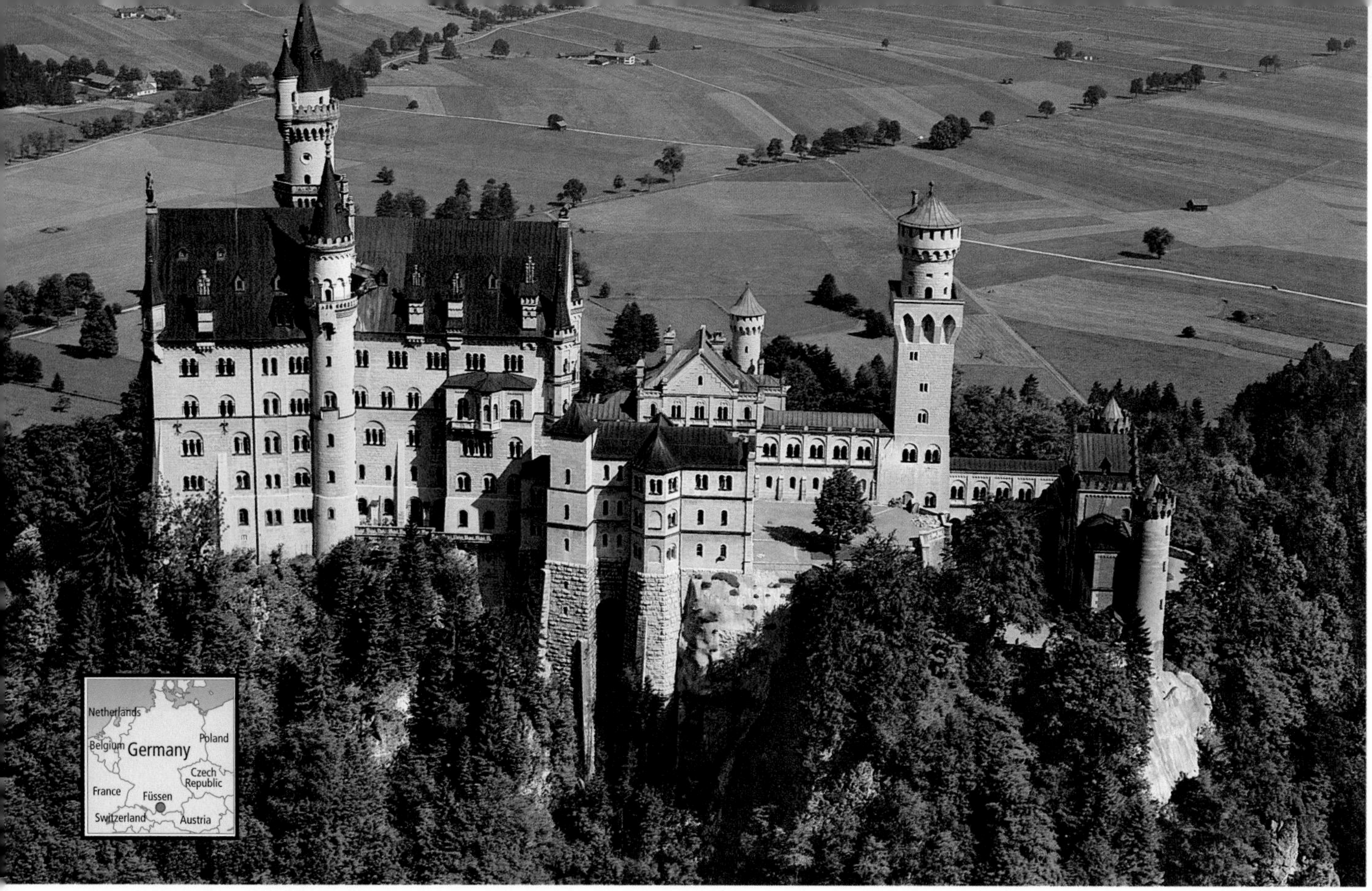

King Louis II of Bavaria had this castle, Neuschwanstein, built in the mid-1800s. It is a fanciful version of a medieval German castle. Louis spent his family fortune and part of the country's treasury to pay for the castle. It is now a major source of income for Bavaria's tourism industry. Every year, approximately 1 million tourists visit Neuschwanstein.

History

From its location in the heart of Europe, Germany has helped shape the continent's history. Many Germanic tribes fought against the Roman Empire. During the A.D. 700s a ruler called Charlemagne united several German kingdoms. Later the region broke into hundreds of small states, each with its own ruler. The German states became part of a loose confederation called the Holy Roman Empire. By the 1300s, about 100 northern German towns formed a trading group known as the Hanseatic League. This group dominated trade in the Baltic region. By the 1700s, a number of powers controlled or strongly influenced the German states. Among these powers were the German state of Prussia and the Habsburg Empire, which later became the Austro-Hungarian Empire.

Prussia led the movement to create a single German country. Northern and southern German states united in 1871. From 1890 to 1914, Germany prospered and became a great industrial and military power. Germany's army and navy were among the strongest in Europe.

The World Wars The rapid rise of German power worried other European countries, particularly France, Great Britain, and Russia. As a result, many European countries formed military **alliances**. An alliance is an agreement between countries to support one another against enemies. Countries that are joined in an alliance are called allies.

These alliances helped maintain a **balance of power** in the region for some time. A balance of power exists when countries or alliances have such equal levels of strength that war is prevented. World War I erupted in 1914 partly because the balance was upset. Britain, France, Russia, and later the United States

internet connect

GO TO: go.hrw.com
KEYWORD: SW3 CH15
FOR: Web sites about Central Europe

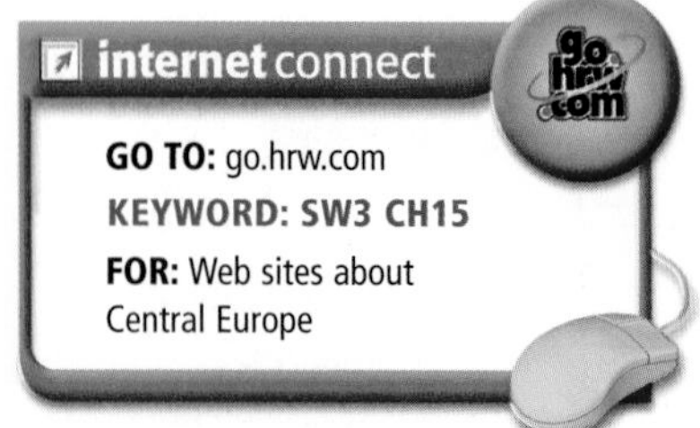

joined forces against Germany. Germany was allied with the Austro-Hungarian Empire, Bulgaria, and the Ottoman Empire. After World War I ended in 1918, Germany had to accept harsh peace terms imposed by the victors. Germany's economy also collapsed in the 1920s. Food shortages, high inflation, and high unemployment caused severe hardships.

Germany's economic and political problems helped bring the Nazi Party to power in 1933. Adolf Hitler was the Nazi leader. Under Hitler, Germany rebuilt its military and allied itself with Italy and Japan. In 1939 Germany invaded Poland, sparking World War II. Fighting soon involved most of the European continent and later much of the world. The United States, Britain, the Soviet Union, and other allies finally defeated Germany in 1945. Some 50 million people had lost their lives. Germany and much of Europe lay in ruins.

Our Amazing Planet

From 1961 to 1989, Berlin was split into a communist East and a capitalist West by a heavily guarded concrete wall. During this time, about 5,000 people escaped over, under, or through the wall into West Berlin.

Division and Reunification The Allied victors of World War II divided Germany. Soviet troops occupied eastern Germany as well as most of Eastern Europe. British, French, and U.S. troops occupied western Germany. Over time, two countries emerged from this division, East Germany and West Germany. Communist governments ruled East Germany as they did other Eastern European countries. West Germany became a democracy. West Germany also rebuilt rapidly with U.S. aid and soon became a global economic power. However, the economy of East Germany lagged. In 1990, following the collapse of communism, East and West Germany reunited.

TEKS **READING CHECK:** ***The Uses of Geography*** How did World Wars I and II affect Germany?

Culture

Today Germany has a democratic system of government. Berlin is the capital. (See Cities & Settlements: Berlin.) The country is divided into 16 states, or *Länder,* which vary in size and population. Bavaria, in the south, is the largest German state in area.

A traditional carnival celebration winds through the streets of Mainz. The city serves as the capital of Rhineland-Palatinate, one of the Länder *of southwestern Germany. Mainz is famous as the home of Johannes Gutenberg, the inventor of movable type printing.*

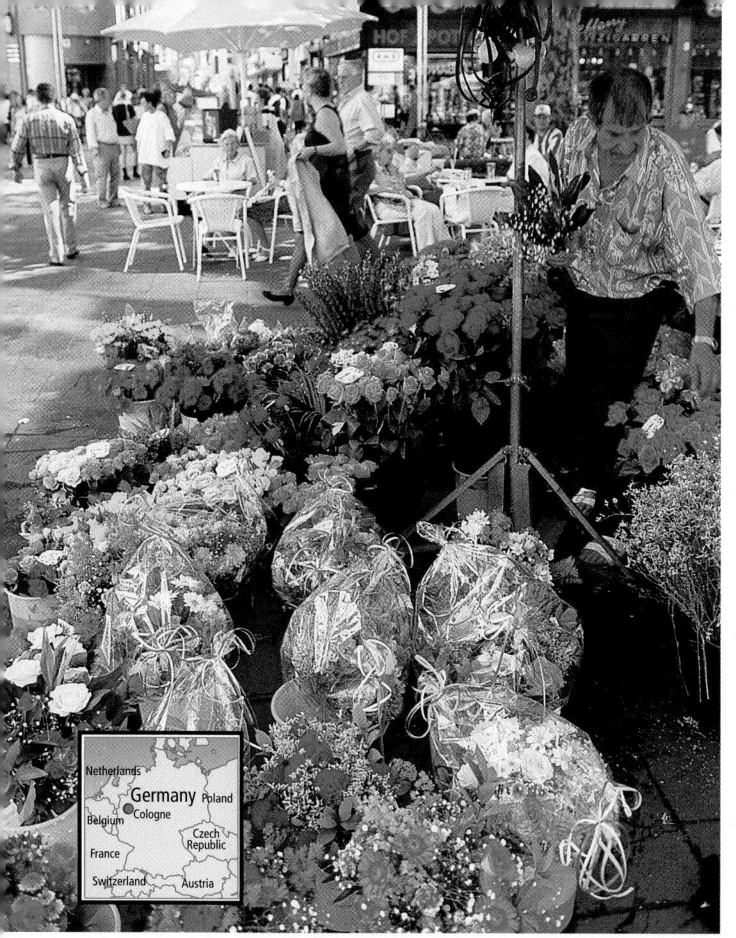

German is the dominant language, although it has several regional dialects. Novels, plays, and poetry written in German have enriched world literature. German composers and artists have also created great works. About a third of Germans are Roman Catholic, and a larger number are Protestant. Southern and western areas are more Catholic than northern and eastern regions. Many Germans do not attend religious services of any kind. German food features pork, sausages, veal, and cheeses. Rich pastries are popular desserts.

In recent years concern over Germany's environment has grown. Many people worry about the effects of air and water pollution and acid rain. "Green parties" are well established in government. These parties have helped pass laws to protect the country's natural environment. As a result, Germany now has some of the strictest environmental laws in the world. Interest in the environment also goes beyond legislation. For example, many Germans spend their vacations hiking, camping, or volunteering with environmental organizations.

READING CHECK: ***Environment and Society*** How is concern for the environment part of Germany's political culture?

INTERPRETING THE VISUAL RECORD

Flowers are grown as a cash crop near western German cities. ***How may strict environmental laws have affected the marketability of Germany's crops?*** TEKS

Economy

Germany is an economic powerhouse. In fact, the country's GDP is the fourth-largest in the world, behind the United States, Japan, and China. However, Germany's per capita GDP is much higher than that of China, which has a far larger population.

Germany is one of the most prominent members of the European Union (EU). Most of Germany's trade is with other EU members. In addition, German has become a widely used language for business in Central Europe. German businesspeople are major investors in other Central European countries. The United States and Japan are other major trading partners.

The German economy is diverse. Businesses manufacture machinery, automobiles, electronics, and medical equipment. Chemicals, steel, and high-tech computer equipment are also important products.

Coal, iron ore, and other minerals helped make Germany an industrial power. In fact, the Ruhr Valley in western Germany is a major industrial center. Industries there developed around huge coal deposits. Today the Ruhr Valley is an almost continuous belt of cities and industries.

Almost half of the country's land is available for agriculture. German agriculture is efficient. Thus, farmers make up less than 3 percent of the population. Grains, potatoes, and sugar beets are major crops.

Nuclear power has provided about a third of Germany's electricity. However, in 2001 the government decided to gradually close all the nuclear power plants. Germany imports almost all of its oil.

READING CHECK: ***Places and Regions*** How productive is the German economy?

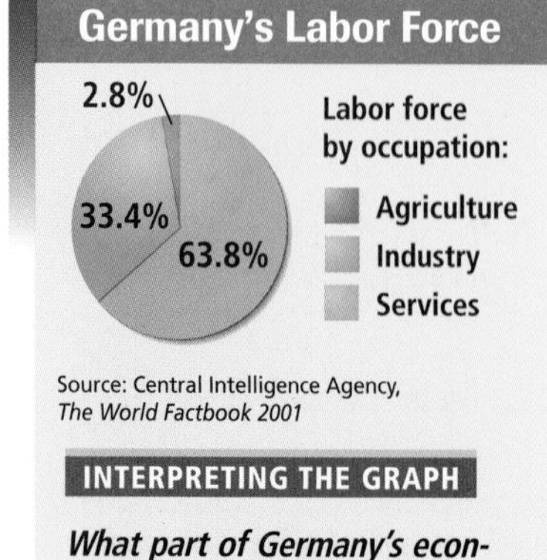

Source: Central Intelligence Agency, *The World Factbook 2001*

INTERPRETING THE GRAPH

What part of Germany's economy has the largest number of employees? How does Germany's labor-force distribution compare to that of most developing countries? TEKS

Issues and Challenges

One important issue in Germany today is the country's changing population. (See Geography for Life: Germany's Aging Population.) Low birthrates, longer life expectancies, and a large number of immigrants are all changing Germany's population. The largest group of immigrants are from Turkey. They have migrated to Germany to work in its growing industries. Most live clustered together in the neighborhoods of big cities. Their Islamic religion, Turkish language, and distinct culture add to their isolation in Germany. In some cases, prejudice and violence against Turks and other groups have been a problem. Many other immigrants to Germany are ethnic Germans from the former Soviet Union.

Since 1990 Germany has tried to bring the standard of living in the east up to that of the west. This effort has been difficult and costly. Many inefficient factories in the east were closed down. As a result, unemployment soared. With unemployment at almost 20 percent in eastern Germany, some Germans are becoming migrant workers in other European countries. Eastern Germany also suffered from heavy pollution during the Communist era, and the cleanup of the environment is just beginning.

Some easterners, or *Ossis*, feel that they are treated as second-class citizens. They resent that westerners, or *Wessis*, have a higher standard of living. Some *Ossis* also miss the lower costs of living and guaranteed jobs and housing that they had under communism. Even citizenship practices sometimes differ between *Ossis* and *Wessis*, a result of Germany's long period of division. For example, May Day, or May 1, was a major holiday in East Germany during the Communist era. It was a day to honor workers. Today some *Ossis* still organize parades to celebrate May Day. *Ossis* do this even though it is not an official holiday in the reunified Germany.

INTERPRETING THE VISUAL RECORD *More than 2 million people of Turkish descent live in Germany today. Some are victims of persecution and discrimination, even though many were born in Germany. Only recently have new laws offered immigrant Turks full German citizenship.* ***How have cultural beliefs shaped the political opportunities available to German Turks?*** TEKS

READING CHECK: ***Places and Regions*** How did different economic and political systems in East and West Germany influence economic, social, and environmental differences between the two regions after reunification?

TEKS Working with Sketch Maps, Questions 1, 3, 4, 5

Section 1 Review

Homework Practice Online

Keyword: SW3 HP15

Define alliances, balance of power

Working with Sketch Maps On a map of Central Europe that you draw or that your teacher provides, label Germany, Berlin, Bavaria, and the Ruhr Valley. In the margin of your map, identify the capital of Germany.

Reading for the Main Idea

1. ***Human Systems*** How was Germany divided after World War II?
2. ***Human Systems*** What role does Germany play in European economies?
3. ***Environment and Society*** What is the Ruhr Valley? Around what natural resource did its industries and cities grow?

Critical Thinking

4. **Comparing** In what ways might people in eastern Germany have mixed views about the effects of reunification on their lives?

Organizing What You Know

5. Copy the time line below. Use it to identify important periods and events in Germany's history after 1871.

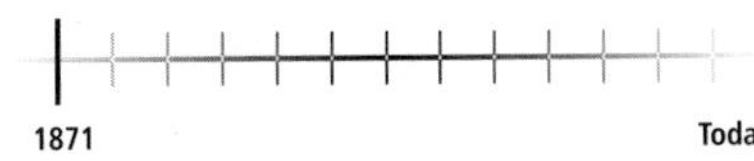

Berlin

Human Systems For much of its long history, Berlin seemed to be on the road to greatness. In the mid-1900s, though, a series of events intervened to block its path. Today, barely a half century after the city's near-destruction and later division, Berlin is once again a vibrant city. It still does not enjoy the status of London, Paris, or Rome. However, Berlin at last seems ready to rejoin the ranks of the great European cities.

A History of Triumph and Tragedy

Berlin was founded in the early 1200s as a trading village. Although it became the capital of a small independent state called Brandenburg, the town grew slowly. In 1670 its population was only 12,000. By then, Brandenburg had merged with a neighboring state to form the powerful kingdom of Prussia. In the early 1700s Prussia's king made Berlin his capital. The king and later rulers turned the town into a great city. By 1750 Berlin had become a thriving commercial center and home to some 100,000 people. In the late 1800s Prussia united the region's other states to form Germany. Berlin became the new country's capital. By 1880 the city's population had reached 1.3 million.

While it was Prussia's capital, Berlin became a center for the arts, education, and literature. Among the city's residents in the 1800s were political philosopher Karl Marx and the composer Felix Mendelssohn. Over time French, Jewish, Polish, and Russian immigrants flocked to the city. Despite political and economic problems following Germany's defeat in World War I, Berlin continued to flower. The city remained a cultural center during the 1920s. In the 1930s, however, Adolf Hitler came to power and led Germany into World War II. Bombs and invading armies then destroyed much of Berlin.

After the war Britain, France, the United States, and the Soviet Union divided Berlin and the rest of Germany into zones. The British, French, and American zones in Berlin soon merged into what was called West Berlin. The Soviet zone became East Berlin. All of Berlin was located deep within Soviet-controlled East Germany. In 1961 the Communists built a high concrete wall around West Berlin to keep East Germans from fleeing to the west. The wall did not come down until 1989, as both East Germany and the Soviet Union began to collapse. In 1990 Berlin, like all of Germany, was officially reunited.

A Future of Cautious Hope

Most of the Berlin Wall is gone now. Small sections still stand as a monument to Berliners' struggle for unity and freedom. In some ways, however, the city remains divided. For example, when the Communists controlled East Berlin they seized houses owned by West Berliners and turned them into public housing. After reunification, the former owners began to reclaim their property. Many East Berliners have been forced to give up homes where they lived for years. In addition, many of East Berlin's inefficient government-run businesses closed. Some 275,000 easterners lost their jobs. The east has seen a great deal of commercial construction since reunification. However, East Berliners

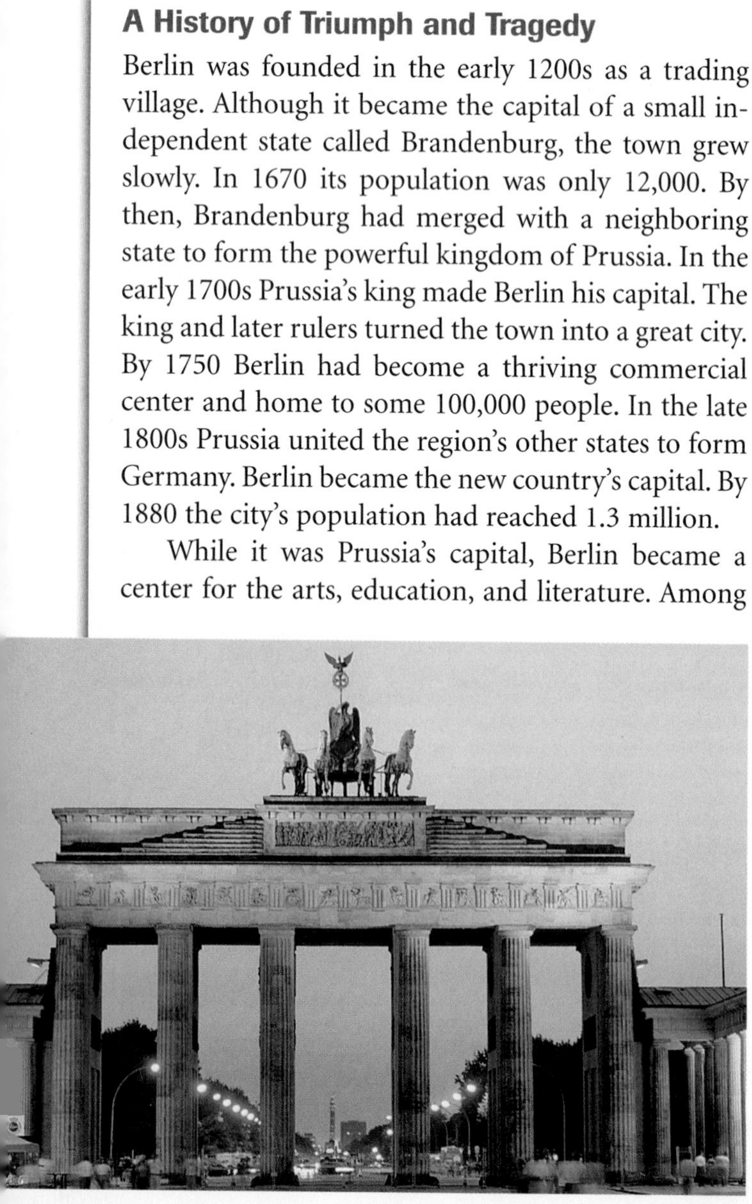

In 1989 the Brandenburg Gate was opened to traffic for the first time in 28 years. The gate had been blocked by the Berlin Wall.

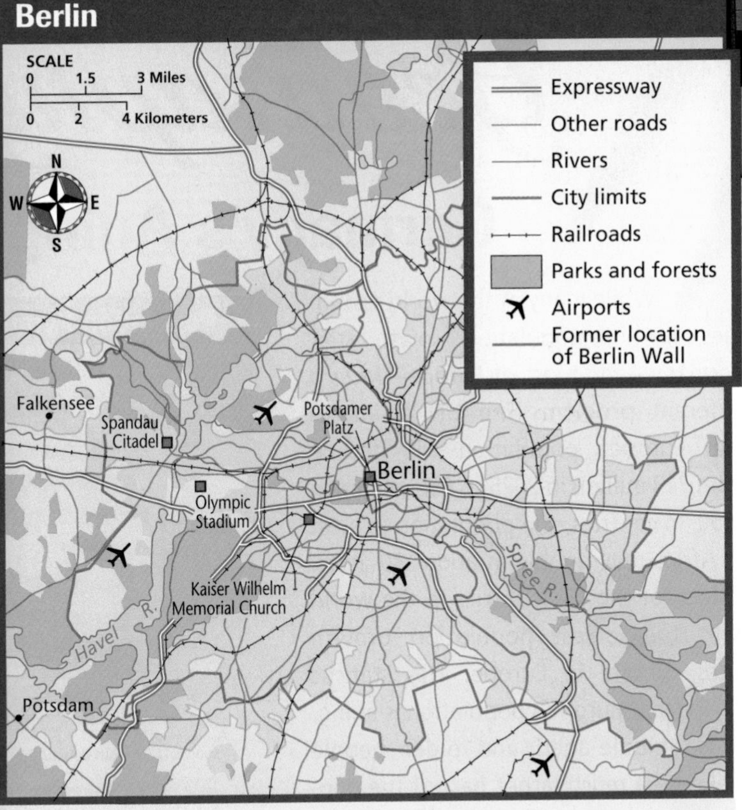

Berlin

SCALE 0 1.5 3 Miles; 0 2 4 Kilometers

Expressway
Other roads
Rivers
City limits
Railroads
Parks and forests
Airports
Former location of Berlin Wall

Falkensee
Spandau Citadel
Potsdamer Platz
Berlin
Olympic Stadium
Kaiser Wilhelm Memorial Church
Havel R.
Spree R.
Potsdam

INTERPRETING THE MAP *Potsdamer Platz was the cultural and economic heart of Berlin until it was destroyed in World War II. Developments like the Sony Center, seen at left, help to re-create the Platz's prestige.* ***How might the location of Potsdamer Platz allow it to again develop into a major economic center?***

resent the fact that westerners have many of the jobs created by Berlin's recent growth. For their part, West Berliners resent the attention, money, and development focused on East Berlin. Berliners have a phrase to describe the tension—*die Mauer im Kopf,* or "the wall in the head." It refers to the psychological and emotional barrier that still lingers in the city.

Yet if concrete remains a reminder of Berlin's recent past, it is also a symbol of the city's future. At one point in the late 1990s, some 1,200 construction cranes dotted the Berlin skyline. In 1991 Potsdamer Platz was a huge vacant field just east of the wall in downtown Berlin. Today it is packed with workers, shoppers, and entertainment-seekers. This gleaming new city center houses offices for international corporations, apartments, and general office buildings. Eastern Berlin needs more construction. More than 100,000 new businesses were started in Berlin during the 1990s. In addition, thousands of new emigrants from Eastern Europe and the former Soviet Union have flooded the city. In the first five years after reunification, Berlin's population jumped from 3 million to 3.5 million.

Today western Berlin is energetic and prosperous. Meanwhile, the eastern half of the city is shedding its Communist past and looking toward the future. Life in the former East Berlin is no longer dreary. Many of its beautiful old buildings have been restored as art galleries and cafés. Residents dress in the latest styles. Cultural divisions between east and west continue to lessen. The golden era of the 1920s is returning to the city. The "new" Berlin is again one of the liveliest cities in the world.

Applying What You Know

1. **Summarizing** Why are there still some divisions among Berliners even though the Berlin Wall has been torn down?

2. **Contrasting** What were the urban environments of East and West Berlin like before the wall came down? What are they like today?

Geography for Life

Germany's Aging Population

While many of the world's countries face high population growth rates, Germany—and most of Europe—faces a different problem. The UN estimates that Europe's population will decline from 713 million in 2025 to 658 million in 2050. In contrast, the population in most of Africa may double in about 30 years. In fact, of the 45 countries in the world with the lowest population growth rates, 30 are in Europe. This statistic means that Europe's population is aging. Thus, middle-aged and older people make up a much larger part of the population, and relatively few young people live there.

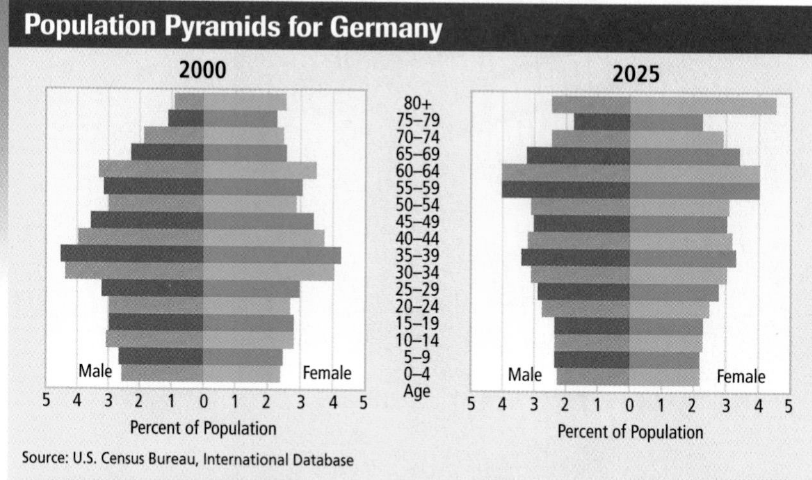

INTERPRETING THE GRAPH *In 2025 what percentage of Germany's population will be 80 and older? What trend do you see at the bottom of the 2025 pyramid?*

The total fertility rate—the average number of children a woman has during her lifetime—that is needed to replace a population is 2.1. However, in 2000, Germany's total fertility rate was about 1.38. Therefore, fewer children are born in Germany than are needed to replace the population. This means that young people are making up a smaller percentage of Germany's population. At this rate the population will gradually decline.

Germany's population is also aging because people are living longer. Good health care helps people survive illnesses and accidents. Now 16 percent of the German population is aged 65 or older. This figure will probably rise to more than 23 percent by 2025. Such a trend will increase demand for health care and government pensions. Pensions are regular payments to retired people. As more and more older people retire, young people will carry the increasing burden of supporting the retirees. How will Germany accommodate this change?

Germany could increase its birthrate or its worker productivity. It could decrease pension benefits or increase immigration. Immigrants are part of German society. When the economy was booming in the 1950s, the country needed workers. People from Turkey and other countries responded to the German government's invitation to become "guest workers." When this campaign ended in 1973, most Germans assumed the workers would return to their homelands. Yet many Turks stayed. Nearly 2 million Turks lived in Germany by the mid-1990s.

Since 1990, ethnic Germans and others from the former Soviet Union and Eastern Europe have also moved to Germany. Their return could help slow the aging of Germany's population. However, many Germans worry about their country's ability to absorb so many immigrants. Furthermore, the unemployment rate is currently high—as much as 20 percent in some areas. As a result, the government now discourages ethnic Germans from returning.

Applying What You Know

TEKS

1. **Analyzing** How is the continued aging of Germany's population shown on the population pyramids? What do you think Germany's population pyramid will look like in 2050? What do you predict for Germany's future population growth?

2. **Evaluating** Compare Germany's situation to Oman's. Oman's population growth rate is 3.46 percent. Thus, Oman has a young population that will double in about 20 years. What might be the consequences of Oman's young population? Create a chart listing the advantages and disadvantages of Germany's and Oman's population structure.

2 The Alpine Countries

READ TO DISCOVER

1. What are some important features of Austria's history, culture, and economy?
2. What are the political, cultural, and economic features of Switzerland?

DEFINE

confederation
cantons
neutral
multilingual

WHY IT MATTERS

The headquarters of the Red Cross and Red Crescent (the name Red Crescent is used in Muslim countries) is located in Geneva, Switzerland. Use CNNfyi.com or other **current events** sources to learn more about this international humanitarian organization.

LOCATE

Vienna
Danube River
Geneva
Basel
Bern
Zürich

Austria

Austria and Switzerland are both located in the Alps, the most mountainous region of Central Europe. Germanic culture has deeply influenced these two Alpine countries. The areas of both countries were settled by Germanic tribes after the fall of the Roman Empire. Like Germany, Austria became part of the old Holy Roman Empire. From the 1400s onward the Holy Roman emperor was a member of the Habsburgs, a powerful family of German nobles. Many different ethnic groups lived within the empire. Each had its own language, local government, and legal system. The Holy Roman Empire was united only by its allegiance to the emperor and for the defense of the Roman Catholic Church.

The Austrian Empire, under Habsburg control, eventually replaced the Holy Roman Empire. At the height of their power, the Habsburgs ruled Spain, the Netherlands, Germany, Italy, and parts of eastern Europe. The Austrian Empire was a major power in Europe in the 1800s. In 1867 the Austrians agreed to share political power with the Hungarians. The Austrian Empire became the Austro-Hungarian Empire. That empire collapsed at the end of World War I. Hungary and other parts of the empire won independence. In addition, the Habsburgs lost power in Austria, which then became a democratic republic. Germany took over Austria shortly before World War II, uniting the two countries. After the war, the Allies occupied Austria. The country became independent again in 1955.

Today Austria is a country about the same size as South Carolina. Nearly all Austrians speak German, and more than

Austria's Habsburg emperors lived in elegant palaces like Schloss Belvedere in Vienna, seen here. The schloss, *which means "palace," now houses a museum and a botanical garden.*

75 percent are Roman Catholic. Vienna, Austria's capital and largest city, is located on the banks of the Danube River. Vienna was the political and cultural capital of Central Europe during Habsburg rule. Historic palaces, churches, and performance halls beautify the city. In addition, many great artists and composers once lived there. Among the most famous were Wolfgang Amadeus Mozart and Ludwig van Beethoven.

Austria is a member of the EU and has a diverse economy. Austrian industries include steel, machinery, and chemicals. Forestry and hydropower are also important. Austria is famous for its high-quality wood, glass, textile, and ceramic handicrafts. The scenic Alps and the country's cultural attractions draw many tourists. Austrian ski resorts are world-famous.

After World War II, Austria kept up trade relations with many countries in Eastern Europe. These ties strengthened after the Soviet Union collapsed in the early 1990s. Today many American and Western European companies base their Eastern European operations in Austria.

The Austrian government encourages the country's farmers to maintain traditional rural customs. Agriculture itself is not a major factor in the Austrian economy, but the preservation of rural ways helps draw millions of tourists each year.

TEKS **READING CHECK:** ***Human Systems*** Why is Vienna an important city culturally and historically?

Switzerland

The history of Austria and Switzerland began to diverge in the late 1200s. At that time Swiss states began to form alliances to protect themselves against invading Austrian armies. Switzerland became independent of Habsburg rule in the 1600s. Today Switzerland is a **confederation**, or a group of states joined together for a common purpose. The country is made up of 26 **cantons**, or states. Each canton has self-government for all issues not reserved for the federal government. The federal government controls national policies, such as defense, international relations, and social programs.

Since Switzerland was formed, it has generally been a **neutral** country. A neutral country is one that does not take sides in international conflicts or alliances. In fact, Switzerland has not been involved in any recent wars. To preserve their neutrality, the Swiss have even resisted joining international organizations. For example, Switzerland is not a member of the EU. The country does participate in international affairs, however. In fact, many world and regional organizations have their headquarters in the Swiss city of Geneva. Geneva also hosts many international conferences.

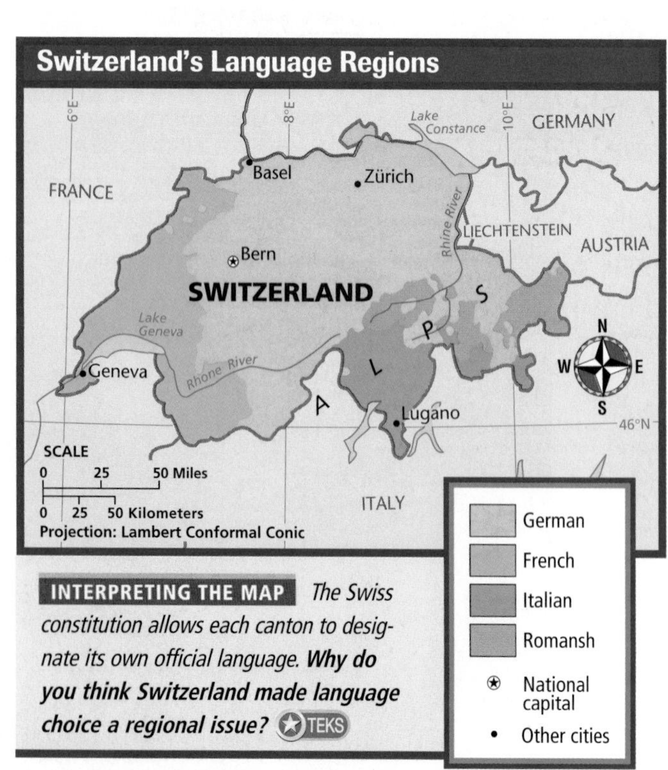

INTERPRETING THE MAP *The Swiss constitution allows each canton to designate its own official language.* ***Why do you think Switzerland made language choice a regional issue?*** TEKS

Switzerland has four major languages: German, French, Italian, and Romansh. (See the map of Switzerland's language regions.) However, many Swiss are **multilingual**, meaning they speak several languages. English is rapidly becoming the country's fifth language. About 46 percent of the Swiss are Roman Catholic, and 40 percent are Protestant. As you can see, Switzerland is a culturally diverse country. However, it is one of the world's most stable countries.

INTERPRETING THE VISUAL RECORD

The Swiss Alps are a favorite destination for people from around the world. Tourists are drawn to mountain resorts like this one near Bern. ***How can you tell from the photo that Alpine architecture has been adapted to the area's climate?*** TEKS

Switzerland's largest cities are Basel, Bern, Geneva, and Zürich. Zürich, the largest, is a leading world-banking center. Basel is a transportation center on the Rhine River in northwestern Switzerland. Bern, the capital, is centrally located between the country's German-speaking and French-speaking populations.

Switzerland has one of the world's highest standards of living. Immigrant workers make up about one fourth of the population and are vital to Swiss business. International banking and insurance are important segments of the economy. Switzerland also produces chemicals, pharmaceuticals, watches, and some farm goods. Most of the country's farm output comes from dairy products. Swiss cheese and chocolate are world-famous. This scenic mountain country also attracts crowds of tourists. Timber production once played a role in the economy. However, air pollution has damaged the woodlands. In fact, pollution has harmed more than 35 percent of the country's forests. The Swiss government now limits tree cutting.

READING CHECK: *Human Systems* How has Switzerland's traditional foreign policy shaped the country?

TEKS Working with Sketch Maps, Questions 2, 3, 4, 5

Section 2 Review

Homework Practice Online
Keyword: SW3 HP15

Define confederation, cantons, neutral, multilingual

Working with Sketch Maps On the map that you created in Section 1, label Austria, Switzerland, Vienna, the Danube River, Geneva, Basel, Bern, and Zürich. Which city is a leading world-banking center?

Reading for the Main Idea

1. *Places and Regions* What are some of Austria's main economic products and industries?
2. *Environment and Society* What are Switzerland's largest cities? In which language region is the capital located?
3. *Environment and Society* How has air pollution affected Switzerland's forests?

Critical Thinking

4. **Drawing Inferences and Conclusions** How do you think Switzerland's system of government has helped it maintain unity despite the country's cultural divisions?

Organizing What You Know

5. Create a chart like the one shown below. Use it to compare the cultures and economies of Austria and Switzerland.

	Austria	Switzerland
Culture		
Economy		

Section 3 Poland and the Baltics

READ TO DISCOVER

1. What is the history of Poland and the Baltic countries?
2. What are the urban environments and economy of Poland like today?
3. What influences have shaped culture in the Baltic countries?

DEFINE

exclave
ghetto

LOCATE

Kaliningrad
Warsaw
Vistula River
Kraków
Gdańsk

WHY IT MATTERS

In 1999 Poland became one of three former Soviet-bloc states to join NATO. Use CNNfyi.com or other current events sources to learn more about NATO's relationship with former communist countries in Europe.

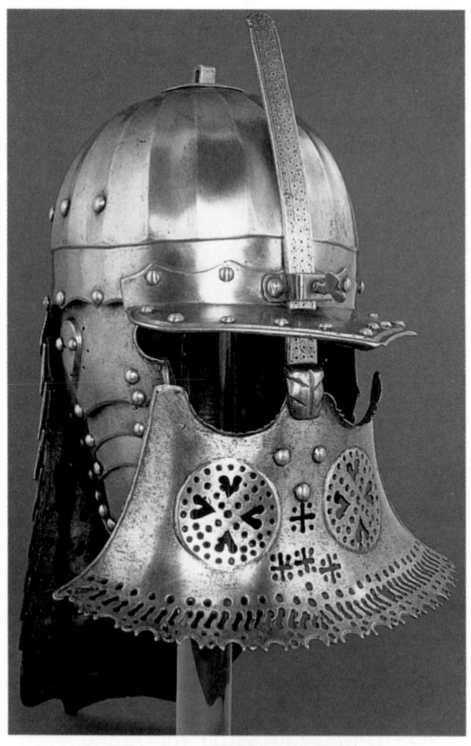

In 1386 the monarchs of Poland and Lithuania united their countries. This union, the Commonwealth of Two Nations, became a powerful force in European affairs. After about 1550, however, costly wars and poor leadership weakened the Commonwealth. Russia took it over in the mid-1700s. This Polish helmet dates from about 1640.

History

Poland gets its name from a Slavic people who moved into the area long ago. Their name, *Polanie,* came from a Slavic word meaning "plain" or "field." In fact, Poland's landscape is filled with plains and rolling hills.

Poland is the largest of the European countries that once made up what was called the Soviet bloc. These countries were allied with the Soviet Union from shortly after World War II until the early 1990s. Poland had also been under Russian control during part of the 1700s and 1800s. During that period, Austria and Prussia (and later Germany) also occupied areas that make up Poland today. Poland became independent after World War I. During and after World War II, however, the Soviet Union occupied the country. A Communist government then ruled the country for more than 40 years.

The Baltic countries also gained independence from Russia after World War I. These countries are Estonia, Latvia, and Lithuania. They stretch northward from Poland to the Gulf of Finland. The Soviet Union took over these countries during World War II. However, they regained their independence in 1991.

While most people have embraced the move toward democracy and capitalism, some people in the Baltics and Poland favor a return to Communist rule. Russian cultural influences linger in the Baltic countries. In fact, some of the countries have large Russian minority populations. Also, Russia still controls Kaliningrad (kuh-LEE-nin-grat). The city, along with its surrounding territory, is an **exclave.** An exclave is an area separated from the rest of a country by the territory of other countries.

TEKS **READING CHECK:** *Human Systems* What role have foreign countries played in the history of Poland and the Baltic countries?

FOCUS ON HISTORY

Kaliningrad The Russian exclave of Kaliningrad is slightly larger than the state of Connecticut. The rest of Russia lies more than 200 miles (322 km) away, across Lithuania and Belarus. Lithuania borders the city on the north and east. Poland lies to the south.

Founded by German knights in the 1200s, the city and surrounding area were once part of Germany. The Germans called the city Königsberg (KOOH-niks-berk), or "King's City." Königsberg eventually became the seat of the Prussian government. During World War II, it became a staging area for German attacks against the Soviet Union. When the Soviets defeated Germany, they took control of the area.

The Soviets wanted to remove as much of Königsberg's German culture as possible. They renamed the area Kaliningrad after a Soviet leader. They also destroyed historical sites. Finally, the Soviets forced the German residents to leave. Some went to Germany, while the Soviets sent others to prison camps. Ethnic Russians and other Slavs then moved into the abandoned homes. Kaliningrad became a key base for the Soviet navy. The city is still Russia's only Baltic port that is free of ice all year.

After the collapse of the Soviet Union, Kaliningrad began to build a market economy. A skilled workforce and the port's access to richer European markets may contribute to that goal. In addition, both Russian and foreign companies doing business in Kaliningrad receive special tax breaks that increase profits. Still, some Russian leaders value the city's status as a military base and want to isolate it from foreign influences. Their efforts might slow Kaliningrad's economic growth. In short, the city's future may be determined as much by outsiders as by the people who live there.

Kaliningrad

SCALE 0 50 100 Miles; 0 50 100 Kilometers
Projection: Conic Equidistant

Baltic Sea
Kaliningrad
LITHUANIA
Neman River
Pregolya River
KALININGRAD (RUSSIA)
POLAND
To Russia
To Russia
To Russia
20°E
22°E
54°N
Roads
Railroad

INTERPRETING THE MAP *Kaliningrad is Russia's only port on the Baltic Sea that can be used all year.* ***Why would Kaliningrad be of strategic military importance to Russia?***

READING CHECK: ***Human Systems*** How did Kaliningrad become a Russian city?

Poland

Nearly all of Poland's people are ethnic Polish and speak Polish, a Slavic language. In addition, the population is overwhelmingly Roman Catholic.

Warsaw is Poland's capital and its transportation hub. The city lies along the Vistula River. Evidence of early settlement on the city's site dates back more than 1,000 years. Warsaw became the capital of the kingdom of Poland in the late 1500s. In the early 1900s Warsaw had the largest urban Jewish population in the world. When the Germans took over Warsaw during their invasion of Poland, they forced the Jews to live in a **ghetto**. A ghetto is a section of a city where a minority group is forced to live. The Jews defied the Germans in the Warsaw Ghetto Uprising of 1943. More than 60,000 Jews died as a result. World War II devastated

The Solidarity Party, a labor union that became a political party, led Poland's struggle to break away from communism. Poland was the first country to leave the Soviet bloc. Here an old Soviet monument bears red paint thrown by anticommunist protesters.

Poles play a game of street chess. The people of Poland are proud of their country's diverse cultural heritage. Poland has strong traditions of art, literature, and music. Since the 1950s, film has also become a prominent means of expression in Poland. Polish movie directors have achieved worldwide fame for their work.

the city. After the war, the Poles rebuilt Warsaw. Today more than 2.2 million people live there.

Farther south along the Vistula lies Kraków (KRAH-kow). The beautiful medieval city has a university, monuments, and museums. Poland's main seaport is Gdańsk (guh-DAHNSK), on the Baltic coast. Gdańsk has been a shipbuilding city since the 1500s.

Poland's economy has made progress since the end of the Communist era. Many successful Polish companies have emerged. In addition, Poland has attracted foreign investment. Auto and glass manufacturing has grown. Still, the traditional coal and steel industries are lagging. Poland's economic future took an upturn in 1997 when the country adopted a new constitution. The constitution committed the country to a free-market economy and to turning over many government-controlled companies to private ownership. More economic progress is needed before the country can join the EU.

Much of Poland's farming activity takes place in productive soils created by thick deposits of loess. Cereals, potatoes, and sugar beets are the main crops. Still, farmers have suffered as the country has moved from communism to capitalism. Many do not have work. Others have moved to the cities to look for jobs.

READING CHECK: *Human Systems* How has the operation of Poland's economy changed since the end of the Communist era?

Poland's Exports

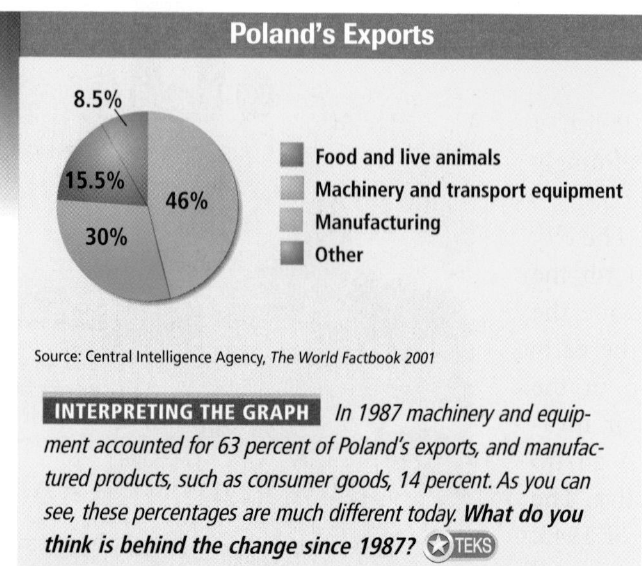

Source: Central Intelligence Agency, *The World Factbook 2001*

INTERPRETING THE GRAPH *In 1987 machinery and equipment accounted for 63 percent of Poland's exports, and manufactured products, such as consumer goods, 14 percent. As you can see, these percentages are much different today.* ***What do you think is behind the change since 1987?*** TEKS

The Baltic Countries

During the Middle Ages, two groups of people lived in what are now the Baltic countries. The Balts occupied modern Latvia and Lithuania. Finns from Scandinavia made up the other group. They settled in Estonia. Lithuania remained an independent country for many years, but Latvia and Estonia did not. First Vikings and then German knights called the Teutonic Order invaded and conquered these countries. The Teutonic Knights brought a strong German element into Baltic society. They also helped spread Christianity to the Latvians and the Estonians. Lithuania did not become Christian until later.

INTERPRETING THE VISUAL RECORD

Riga, the capital and largest city of Latvia, is a port city on the Baltic Sea. The city's architecture and society reflect the influence of many different cultures. Germany, Poland, Russia, and Sweden have all helped to shape Latvian society. ***How has the city's location influenced Riga's cultural development?*** TEKS

Historically, the Baltic Sea was one of the busiest trade routes in northern Europe. People who met in Baltic ports exchanged both goods and information. The influence of this meeting of cultures can still be seen in the Baltic countries. For example, the Estonian language is related to Finnish. Also, like the Finns, almost all Estonians are Lutheran. Latvia has ties to Sweden, a result of a long history of trade between the two countries. Lithuania, on the other hand, is closer culturally to Poland and the Roman Catholic Church. Most Lithuanian folk festivals are tied to church holidays. Folk music is an important part of these festivals. Many of the instruments used are similar to Polish musical instruments. One example is the *cymbaly,* a type of percussion instrument. Also, Russian minorities in each of the Baltic countries keep Russian cultural traditions alive.

Estonia, Latvia, and Lithuania share some challenges. They are trying to rebuild their economies after years of Soviet rule. Their citizens are also cleaning up environmental pollution from the Soviet era. Small populations and limited natural resources make trade essential to all three countries.

TEKS **READING CHECK:** ***Human Systems*** Why is trade essential to the Baltic countries?

TEKS Working with Sketch Maps, Questions 1, 2, 3, 4, 5

Review

Homework Practice Online

Keyword: SW3 HP15

Define exclave, ghetto

Working with Sketch Maps On the map that you created in Section 2, label Poland, Estonia, Latvia, Lithuania, Kaliningrad, Warsaw, the Vistula River, Kraków, and Gdańsk. In which city did Jews living in the ghetto rise up against the Germans?

Reading for the Main Idea

1. ***Human Systems*** What countries once controlled areas that are now part of Poland?
2. ***Human Systems*** How are Poland's three main cities related to the country's politics, sea trade, and history?
3. ***Environment and Society*** What are two challenges facing the Baltic countries?

Critical Thinking

4. **Making Generalizations** Why might many people in Poland and the Baltic countries want closer ties to Western Europe than to Russia?

Organizing What You Know

5. Draw a map illustrating cultural and historical ties between the countries discussed in this section and other European countries. Note some specific cultural connections, such as language and religion.

4 The Czech Republic, Slovakia, and Hungary

READ TO DISCOVER

1. What are some similarities and differences in the histories of the Czech Republic, Slovakia, and Hungary?
2. What are the Czech Republic and Slovakia like today?
3. How has the fall of communism affected Hungary?

DEFINE

complementary region

LOCATE

Prague
Bratislava
Budapest

Wood carvings from Hungary

WHY IT MATTERS

Czechoslovakia split peacefully into two countries in 1993. Use CNNfyi.com or other **current events** sources to learn about other countries that have split apart in recent years.

History

The Czech Republic, Slovakia, and Hungary lie south of Germany and Poland. Slavic peoples have long lived in Slovakia and the Czech Republic. About 90 percent of Hungary's people belong to a non-Slavic ethnic group—the Magyars.

The Czech Republic, Slovakia, and Hungary were once part of the Austro-Hungarian Empire. They gained independence after World War I. The Czech Republic and Slovakia formed one country, Czechoslovakia. The union of the two parts of Czechoslovakia had economic advantages. Czech lands had mineral resources and industries. Slovakia was mostly agricultural. Together they formed a **complementary region**. The combining of two areas with different activities or strengths, each of which benefits the other, forms a complementary region. At the time, Czechoslovakia was one of the world's 10 most industrialized countries.

Germany occupied Czechoslovakia during World War II. The Soviet Union then occupied Czechoslovakia and Hungary at the end of World War II and set up Communist governments there. The Soviets invaded Hungary in 1956 and Czechoslovakia in 1968 to keep control of both countries. Soviet control finally ended in the early 1990s. In 1993 the Czechs and Slovaks decided to separate peacefully into two countries.

READING CHECK: *Human Systems* In what way did the Czech Republic and Slovakia form a complementary region?

INTERPRETING THE VISUAL RECORD

Medieval castles and ruins are scattered across Slovakia's landscape. The fortress shown below dates back to the 1200s. ***Why were castles like this one built on hilltops?*** TEKS

Austro-Hungarian Empire (before World War I)

SCALE 0 200 400 Miles
0 200 400 Kilometers
Projection: Azimuthal Equal Area

GERMANY
RUSSIA
Prague
L'viv
Danube R.
Vienna
Budapest
ROMANIA
ITALY
Belgrade
BULGARIA
SERBIA
ALBANIA
GREECE

Austrian Empire
Kingdom of Hungary

Hungary and Czechoslovakia (after World War I)

Hungary, Slovakia, and the Czech Republic (1993)

INTERPRETING THE MAP *Political boundaries in Central Europe have changed significantly in the last century. For example, Czechoslovakia was created after World War I from part of the defeated Austro-Hungarian Empire. In 1993 the Czech Republic and Slovakia separated from each other. This division into two countries was not a violent event.* ***What cultural factors do you think led to the creation of these states as separate countries?*** TEKS

The Czech Republic and Slovakia

The Czech Republic is made up of the regions of Bohemia and Moravia. About 40 percent of Czechs are Roman Catholic. About the same percentage are not religious. The country's capital and largest city is Prague (PRAHG). Prague is located on seven hills along the Vltava (VUHL-tuh-vuh) River. This historic city was founded more than 1,100 years ago. It is a cultural, university, and tourist center with a rich architectural heritage. The urban region is also the center of the country's major industries.

American culture has increasingly influenced Prague in recent years. With the end of the Communist era, trade and cultural links with the United States grew. A community of American businesspeople and students also began to grow over time. American English-language schools have popped up across the city. In addition, Prague cinemas show the latest releases from Hollywood. Even American fast food is easy to find in the central city.

The Czech Republic is a hilly region with good supplies of coal, iron ore, and uranium. The Czechs became well known for the production of fine steel and glass products. Farms grow mainly cereals and sugar beets. The country's economy had one of the strongest and most stable economies of the former Soviet-bloc countries. However, political and economic problems in the late 1990s slowed progress. Part of the problem was that the government still had too much influence over the economy. Still, the Czech Republic has attracted foreign investment and tourism. The country continues to move toward a market economy. It wants EU membership and has already joined NATO.

Slovakia is the poorer eastern half of the two former regions of Czechoslovakia. The move to a capitalist system there has been hard. Unemployment is high. The capital and largest city is

Prague, the capital of the Czech Republic, has become a popular destination for tourists. Historical attractions like the Charles Bridge on the Vltava River are gathering spots for visitors and residents alike.

Connecting to

HISTORY

A Shifting Region

Geographically, the countries discussed in this chapter lie in the middle of Europe, which stretches eastward to the Ural Mountains. However, throughout history, people's perceptions about just what makes up Central Europe have changed.

For example, in the early 1900s Germany was considered part of Central Europe. Then Germany was divided into East Germany and West Germany from the end of World War II until 1990. Communist East Germany, Poland, Czechoslovakia, Hungary, Romania, and Bulgaria were tied to the Soviet Union. These so-called Soviet-bloc countries made up a region known as Eastern Europe. (See the map.)

Today Germany is reunited. Poland, Hungary, and other former Soviet-bloc countries have strong ties to the West. Once again, many people consider these countries part of Central Europe. To the east, the former Soviet Union has split into several countries. Perhaps these countries will build stronger ties with the West. Then the boundaries of Central Europe may march farther east.

Making Generalizations In what ways is Central Europe a perceptual region? In what ways might it be a functional or formal region? TEKS

Budapest

Expressway

Primary road

Point of interest

INTERPRETING THE MAP

Budapest, the capital of Hungary, benefits from international commerce. Many foreign investors have chosen the city as a base from which to ship their products around the world. ***Why do you think Budapest is a suitable location for these enterprises?***

Bratislava. It is located on the Danube River and on a major railway junction. This location makes the city ideally located for the trade of both goods and ideas. Although Bratislava lies in the far west of Slovakia, it is the country's educational and cultural center. Institutions from Bratislava stage ballets, concerts, operas, and plays all across the country. Slovaks also have a strong folk culture, which is still evident in Slovakia's art and music.

READING CHECK: ***Environment and Society*** What factor has made Bratislava central to Slovakia's culture and economy?

Hungary

The Hungarians speak Magyar, rather than a Slavic language. Magyar is related to Finnish and has its origins in Central Asia. About two thirds of Hungary's population is Roman Catholic.

Hungarians have a rich history of folk and music traditions. Hungarian music is heavily influenced by Roma, or Gypsy, rhythms. To Hungarians, the delivery of a song is as important as the melody and lyrics. Therefore, many Hungarian musicians are gifted actors and dancers as well.

Budapest is Hungary's major city and capital. It is made up of Buda and Pest, which lie across the Danube from each other. The two communities joined in 1873 to form Budapest. Almost one fourth of the country's population lives in or near the capital. Most businesses and industries are also located there. Its location along highways, rail lines, and the Danube help make the city the national transportation center.

Migration from rural areas to Budapest slowed in the 1990s. The slowdown came as rural areas experienced improvements in water, sewage, and other services. Also, many people have been moving out of Budapest to live in the suburbs. The suburbs have fewer problems with crowded housing and pollution. Because of these factors, Budapest's population declined a little in the 1990s.

The origins of goulash, a traditional Hungarian stew based on meat, onions, and paprika, date from the A.D. 800s. Then the dish was a cooked mixture that was dried in the sunshine so it could be packed in bags made from sheep's stomachs.

Hungary is located on a broad agricultural plain in the central Danube Basin. Farming still plays a major role in the country's economy. Potatoes, sugar beets, and wheat are the main crops. Farmers also raise livestock, particularly cattle and hogs. Almost half of Hungary's population lives in small farming villages and towns.

The Communist government began to allow some private ownership of businesses in the 1960s. Over time, private businesses helped the economy grow. Still, after the end of Communist rule in the early 1990s, the rapid move to a market economy was hard. Today, however, most of the country's businesses are privately owned. In recent years Hungary has attracted new industries, foreign investment, and tourists. In fact, it has one of the strongest economies in the region today. Hungary recently joined NATO and seeks EU membership.

The Great Hungarian Plain accounts for approximately half of Hungary's area. Most of Hungary's farms are located on this fertile plain.

READING CHECK: *Environment and Society* What factors affected the growth of Budapest in the 1990s?

TEKS Working with Sketch Maps, Questions 2, 3, 5

Section 4 Review

Define complementary region

Working with Sketch Maps On the map that you created in Section 3, label the Czech Republic, Slovakia, Hungary, Prague, Bratislava, and Budapest. What are the capital cities of the Czech Republic, Slovakia, and Hungary?

Reading for the Main Idea

1. *Human Systems* What are the main ethnic backgrounds of the people in the Czech Republic, Slovakia, and Hungary?
2. *Human Systems* What American cultural influences grew in Prague after the end of the Communist era?
3. *Human Systems* What was Hungary's economy like in the 1960s? How strong is it today?

Critical Thinking

4. **Making Generalizations** Why do you think American and other Western companies might want to expand into former Communist-ruled countries like Hungary and the Czech Republic?

Organizing What You Know

5. Create a chart like the one below. Then use it to compare political, economic, and cultural features of Prague, Bratislava, and Budapest.

Prague	Bratislava	Budapest

Building Vocabulary

On a separate sheet of paper, explain the following terms by using them correctly in sentences.

alliances	cantons	exclave
balance of power	neutral	ghetto
confederation	multilingual	complementary region

Locating Key Places

On a separate sheet of paper, match the letters on the map with their correct labels.

Berlin	Vienna	Geneva	Prague
Ruhr Valley	Danube River	Warsaw	Budapest

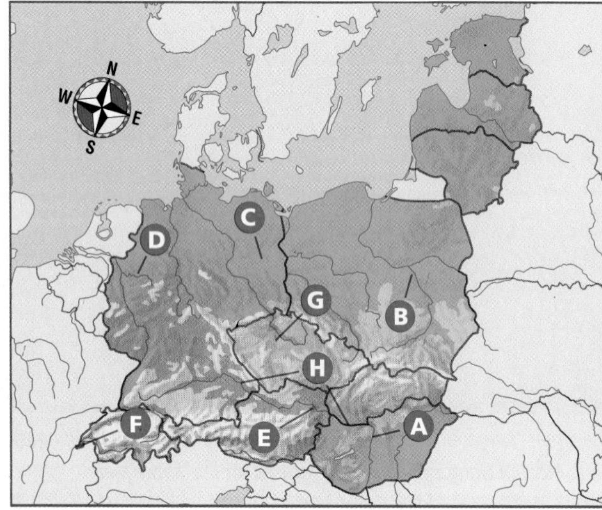

Understanding the Main Ideas

Section 1

1. Human Systems How did World Wars I and II shape the history of Germany?
2. Human Systems What problems has eastern Germany faced since reunification?

Section 2

3. Human Systems Why have many American and Western European companies chosen Austria as a base for their European operations?

Section 3

4. Environment and Society Why is loess important to Poland's economy?

Section 4

5. Human Systems How does Hungary's language differ from the languages of neighboring countries?

Thinking Critically for TAKS

1. **Analyzing** How might differing viewpoints affect public policies regarding economic development in eastern Germany?
2. **Analyzing** What role have rivers played in the early development of Central Europe's major cities?
3. **Comparing and Contrasting** Which of Central Europe's former communist countries, other than the former East Germany, do you think may offer the brightest future for its people? Why?

Using the Geographer's Tools

1. **Analyzing Maps** Review the map of Swiss language regions. What are the major languages spoken in the country? In which region are three of the country's largest cities located? Which cities are these?
2. **Analyzing Tables** Use the unit Fast Facts and Comparing Standard of Living tables to rank countries in Central Europe by economic development and standard of living. How may the political histories of the countries have influenced their development and standard of living?
3. **Creating Maps** Research historical information about Central Europe since 1900. Then create a political map of the region, using colors or other tools to identify former communist countries that once were part of the Austro-Hungarian Empire, Germany, or Russia. Note that the territory of some countries was controlled by more than one of those three powers.

Writing for TAKS

People in the former communist countries of Europe are eager to learn about ways of life in other countries. Write a letter to a real or imaginary teenager in one of those European countries. Note similarities and differences you might expect to see between life in this country and in your pen pal's country.

SKILL BUILDING

Geography for Life

Documenting Environmental Change

Environment and Society Environmental pollution is a legacy of the Communist era in Central Europe. What caused this pollution? Which areas are particularly polluted, and why? How might the pollution be cleaned up? Investigate these questions and write a script for a brief documentary film presenting your findings. Include a script, illustrations, maps, and other visual materials.

Building Social Studies Skills

Interpreting Graphs

Study the pie graphs below. Then use the information from the graphs to help you answer the questions that follow. Mark your answers on a separate sheet of paper.

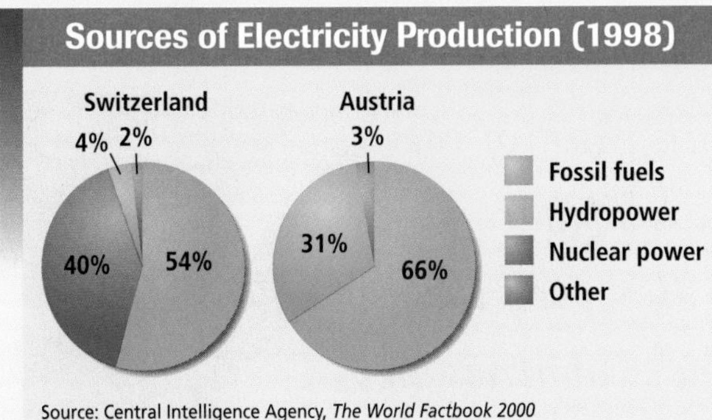

1. More than 90 percent of Switzerland's electricity is produced by
   - **a.** hydropower alone.
   - **b.** fossil fuels and nuclear power.
   - **c.** nuclear power alone.
   - **d.** hydropower and nuclear power.
2. What is the single largest source of electricity in both Austria and Switzerland? What clues might this answer give you about the physical geography of the two countries?

Building Vocabulary

To build your vocabulary skills, answer the following questions. Mark your answers on a separate sheet of paper.

3. Few countries have remained *neutral* for long periods of time. In which sentence does *neutral* have the same meaning as it does in the sentence above?
   - **a.** The solution that resulted from the experiment was neutral.
   - **b.** The dining area should be painted in a neutral color.
   - **c.** The counselor has remained neutral throughout the dispute.
   - **d.** The car rolled down the hill because it was left in neutral.
4. *Alliance* means the same as
   - **a.** a largely self-governing state within a country.
   - **b.** a group of states with opposing views on world issues.
   - **c.** an agreement between countries for support against enemies.
   - **d.** an agreement between countries to establish new borders for overseas colonies.

Alternative Assessment

PORTFOLIO ACTIVITY

Learning about Your Local Geography TEKS

Individual Project: Research

Throughout history many peoples, empires, rulers, invasions, and wars have affected the countries of Central Europe. Use library and Internet resources to research the history of your state. Note which countries or peoples have controlled the area over time. Explain how your state's history may have shaped cultural features there today. Write a short report about your findings. Then proofread it to make sure you have used standard grammar, spelling, sentence structure, and punctuation.

internet connect

KEYWORD: SW3 GT15 TEKS

Access the Internet through the HRW Go site to research the change in global trade patterns of three Central European countries. Then create a database that shows those changes. Use the Holt Grapher to represent your information in graph form. Finally, write a hypothesis to explain the changes that occurred in each country's trade and note the implications of these changes.

CHAPTER 16
Southern Europe and the Balkans

Southern Europe is made up of three large peninsulas. Most of the countries there share a similar physical geography. However, the cultures of the region's countries are very different.

Red-figure pelike (wine container) from ancient Greece

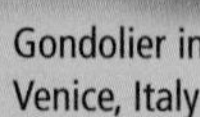

Gondolier in Venice, Italy

Aupa! (Basque for "What's up?") My name is Nagore Perez España, and I am 17 years old. I live in Bilbao, which is the biggest city in the Basque Country in northern Spain. My parents and I live on the seventh floor of an apartment block that is 14 stories tall. Our balcony overlooks a mountain and a road. Our apartment block is near other apartment blocks just like ours, and there is a big square in between them. We have always played there since we were kids.

I attend the Elorrieta Institute, the local public high school, which is about a 15-minute walk from my apartment. I am in the equivalent of your senior year of high school. I have not decided what to do for a career, but I'm thinking of doing something related to the tourism industry. In Spain there are good possibilities of finding jobs in tourism.

I go skiing with my parents about five times a year and sometimes with my friends as well. In the summer, or when the weather is good, we go climbing mountains and also to the beach. On Saturdays I sometimes eat lunch out with my parents. We prefer the local dishes rather than hamburgers, pizzas, or hot dogs. Other people eat those things for an occasional change of pace. On Sundays I always eat lunch with my parents at my grandmother's. She cooks such good food!

The Iberian Peninsula

READ TO DISCOVER

1. How have past events affected Spain?
2. How is Portugal both similar to and different from Spain?

DEFINE

autonomy
cork

WHY IT MATTERS

Spain's tourist industry is one of the largest in the world. Use CNNfyi.com or other **current events** sources to learn about historical and cultural sites in Spain or neighboring Portugal.

LOCATE

Madrid
Balearic Islands
Strait of Gibraltar
Barcelona
Bay of Biscay
Lisbon
Porto

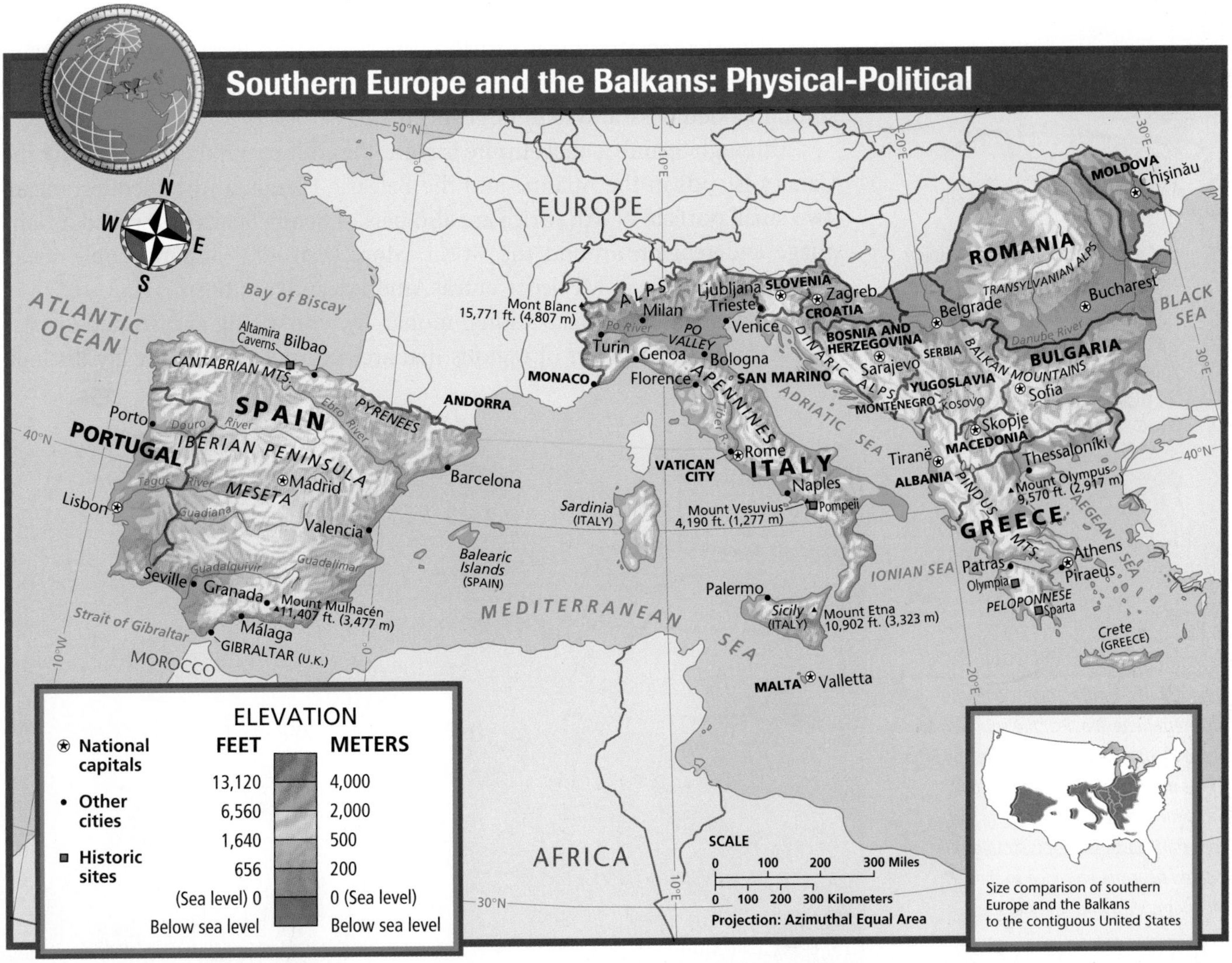

Southern Europe and the Balkans: Physical-Political

ELEVATION

FEET	METERS
13,120	4,000
6,560	2,000
1,640	500
656	200
(Sea level) 0	0 (Sea level)
Below sea level	Below sea level

⊛ National capitals
• Other cities
▪ Historic sites

SCALE
0 100 200 300 Miles
0 100 200 300 Kilometers
Projection: Azimuthal Equal Area

Size comparison of southern Europe and the Balkans to the contiguous United States

Spain

Spain and Portugal share the Iberian Peninsula, or Iberia. Spain, southern Europe's largest country in area, covers about 85 percent of Iberia. Much of Spain is surrounded by water. At one time, this feature helped make Spain a great seafaring country.

History and Government The early history of Spain is similar to that of most of the Mediterranean region. Various Mediterranean peoples, including the Romans, have ruled the area. In the A.D. 700s an Arabic people called the Moors invaded Iberia from North Africa. The Moors brought the Islamic religion, new irrigation techniques, and new crops to Iberia. They also built universities and brought many crafts and trades. However, Christian rulers eventually forced the Moors out of Spain. In 1492 the Moors surrendered their last stronghold there—Granada in southern Spain.

During the 1500s Spain used its strong navy to build a worldwide empire. At its peak, the Spanish Empire included most of Central and South America. Spain also ruled what is now the southwestern United States. In addition, the empire included small colonies in Africa and islands in the Caribbean and Pacific. However, Spain had lost almost all its overseas empire by the end of the 1800s. During the 1800s the country was also shaken by wars. Spaniards who wanted a monarchy fought those who wanted a more democratic government. The struggle for power continued in the 1900s and led to a terrible civil war in 1936. The democratic forces lost the war. A dictator, Francisco Franco, then ruled Spain from 1939 to 1975. After Franco's death Spain quickly made a transition to a democratic system of government.

Although Spain's global empire is gone, the country retains control over the Canary Islands in the Atlantic and the Balearic Islands in the Mediterranean. Two small ports in North Africa are also part of Spain. Spain has spread its language and religion around the world. More than 400 million people speak Spanish. Many live in Mexico, Central America, and South America.

Today Spain is a constitutional monarchy with a king and elected legislature. Spain has a number of regions that are culturally or historically distinct.

Our Amazing Planet

Some 14,000 years ago, people painted bison, a deer, and horses on a cave ceiling near what is now Altamira, Spain. Today the paintings are treasured as some of the world's greatest artworks.

The Rock of Gibraltar towers over the Mediterranean Sea near the southern tip of the Iberian Peninsula. Captured by British forces from Spain in the early 1700s, Gibraltar is now a British colony and important naval base. It is strategically located near the entrance to the Mediterranean from the Atlantic Ocean.

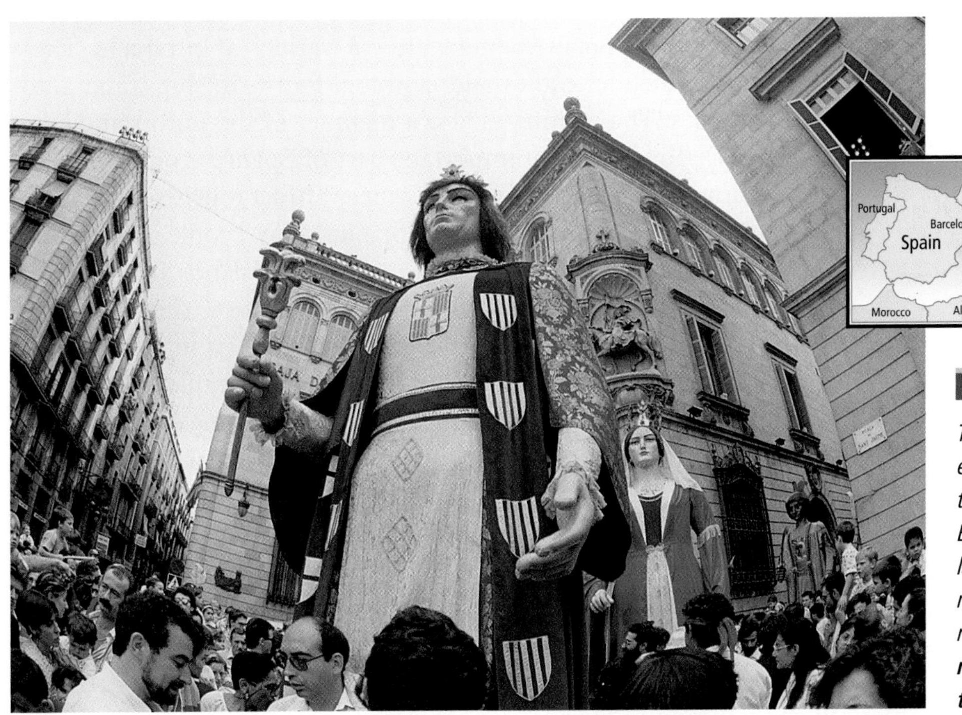

INTERPRETING THE VISUAL RECORD

Thousands of festivals take place across Spain each year and are important events in local towns and communities. Many festivals in Barcelona feature gegants, *or "giants." These large figures are actually men on stilts underneath elaborate costumes. The* gegants *often represent biblical or historical figures.* ***Why might festivals like this one be important to a culture's identity?***

In the past, some of these regions were independent kingdoms. After Franco's death in 1975, some wished to be independent again. To prevent this, Spain's central government gave the country's 17 regions different levels of **autonomy**, or self-government. These regional governments make decisions about health and social programs, urban planning, education, and other local issues. The central government still controls policies for the whole country, such as foreign relations and national defense.

TEKS **READING CHECK:** ***Human Systems*** How did Spain spread the Spanish language and Roman Catholicism around the world?

People and Culture Nearly all people in Spain today are Roman Catholic. About 75 percent speak Castilian Spanish. Castilian is a dialect from the Castile region around Madrid, the country's capital. Spain's other languages include Basque and Catalan, which are both spoken in regions near the Pyrenees.

Spain's villages and cities have many open spaces for people to meet. The plaza is a common feature in Spanish towns. A plaza is a square surrounded by public buildings, such as a church, a marketplace, and government offices. These squares serve as social gathering places, particularly on warm evenings and weekends. Plazas are also found throughout Central and South America and much of southern Europe.

Moorish influences can still be seen in Spain. For example, Arabic architectural styles are common in many towns, particularly in southern Spain. These styles include horseshoe-shaped arches and geometric decorations. Many natural features and settlements with Moorish place-names still dot the Spanish landscape. (See Geography for Life: Arabic Place-Names in Iberia.)

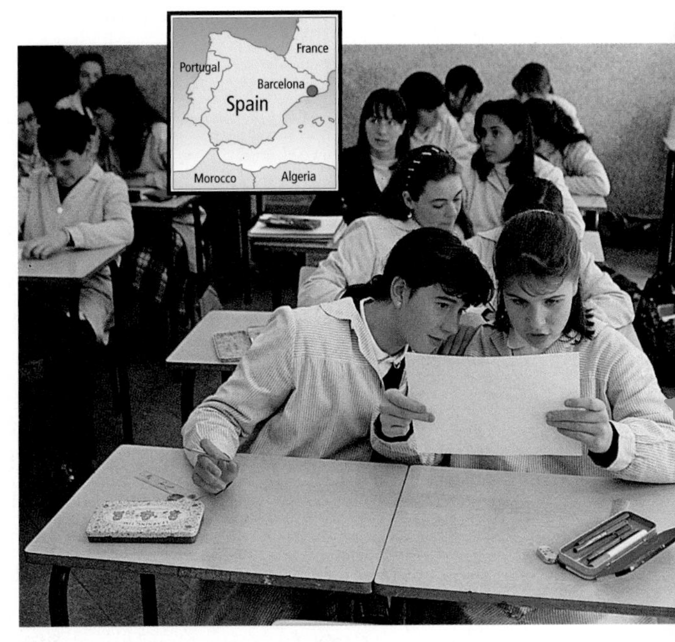

INTERPRETING THE VISUAL RECORD *In mostly Roman Catholic Spain the church has long played an important role in the educational system. For example, many students attend Catholic schools. However, in recent decades the government has increased its control over Spain's private religious schools.* ***How does this classroom compare to yours? Do you think parochial schools are as common in your community?*** TEKS

Major Exports of Spain and Portugal	
Spain	**Portugal**
machinery, motor vehicles, foodstuffs, other consumer goods	clothing and footwear, machinery, chemicals, cork and paper products, hides

Source: Central Intelligence Agency, *The World Factbook 2000*

INTERPRETING THE CHART *Spain's most important exports are motor vehicles and a wide range of foodstuffs, such as fresh fruits, olive oil, vegetables, and nuts. Portugal's leading exports include clothing, footwear, and cork.* ***What does the information in this chart suggest about the level of development in Spain and Portugal?*** TEKS

Economy Spain manufactures a variety of products. These goods include textiles and clothing, footwear, ships, and automobiles. The country is also a member of the European Union (EU). Tourism is an important part of Spain's economy. Warm sunny weather and beautiful scenery attract tourists to areas like the Costa del Sol in southern Spain. Many people also visit the Pyrenees and the Balearic (ba-lee-AR-ik) Islands in the Mediterranean Sea. Famous historical and cultural sites are found throughout the country. Although tourism is a major source of income, it has also caused problems. For example, tourism has brought more traffic, pollution, and overbuilding to scenic areas.

Agriculture also plays a major role in Spain's economy. The country is a leading producer of olive oil and wine. Farmers also grow many crops that the Moors brought to the region, such as citrus fruits. The area around Valencia is particularly famous for producing oranges. Other crops, such as corn, potatoes, and tomatoes, were first imported from Spain's American colonies. Spain shipped these crops to many places in its empire. Now crops that were originally American are common in Europe and other parts of the world.

Issues and Challenges Spain's economy has been growing rapidly over the last several decades. This growth further improved when the country joined the EC in 1986. Continued economic development is one of Spain's goals. However, the country still has one of Western Europe's highest unemployment rates.

Immigration has also become an issue. Morocco lies across the Strait of Gibraltar just about 8 miles (13 km) from Spain. Many North Africans cross the strait to find work in Spain and other European countries. They usually move to the big cities, such as Barcelona, Spain's main port. However, many of these immigrants do not find jobs.

Spain also faces political challenges. As you read earlier, Spain's regions have a certain amount of autonomy. Several of those regions used to have active independence movements. However, since the regions received more autonomy, some of those movements have quieted. However, the Basque still work to make their region a separate country. Catalonia borders France and the Mediterranean Sea. The Basque Country lies along the Bay of Biscay and the Pyrenees in northern Spain. The Basques have ancient origins. Their ancestors were among the earliest settlers of Europe. The ancient Basque language seems to be unrelated to any other European language. Some Basques have turned to violence in an effort to win independence from Spain. One group, known as ETA, has killed many government officials and others. Non-Basques in the region, as well as many Basques, oppose ETA and their violent struggle for Basque independence.

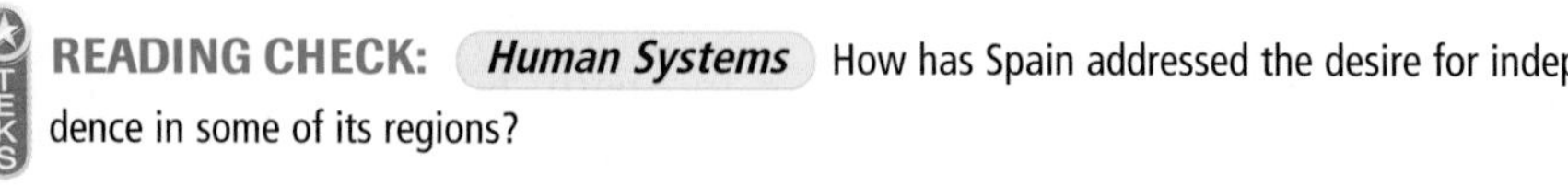
TEKS **READING CHECK:** ***Human Systems*** How has Spain addressed the desire for independence in some of its regions?

Portugal

Portugal lies in the western part of Iberia and faces the Atlantic. Both Portugal and Spain have many cultural similarities.

History and Culture Much of Portugal's history closely mirrors Spain's. Portugal, too, was under the control of Rome and then the Moors. Like Spanish, the Portuguese language developed from Latin, the language of the Romans. Portuguese also contains many words from Arabic. These words became part of the language when Moors ruled the region. After the Moors were driven from Iberia, the Portuguese built a powerful colonial empire. That empire included parts of Africa, Asia, and South America. Portugal's former colonies include Angola, Brazil, Mozambique, and part of Timor, an island in Southeast Asia. The Atlantic islands of Madeira and the Azores are all that remain of the old empire.

Economy Like Spain, after losing its empire Portugal entered a period of economic decline and limited personal freedoms. Now Portugal's government is democratic, and its economy, helped by membership in the EU, is growing. The country's new freeways, car factories, and high-speed trains reflect this growth. However, even the moderate increase in prosperity has drawn immigrants to Portugal, particularly from North Africa. They crowd into Portugal's cities, which include Lisbon and Porto. Lisbon is the country's largest city and capital. Porto, the second-largest city, is a major seaport in the north.

As in Spain, tourism is important to Portugal. The country is also the world's leading **cork** producer. Cork is a bark that is stripped from the trunks of cork oaks. These trees grow in southern Europe and North Africa. Some types of insulation and flooring are made from cork. However, most cork is used to seal wine and other bottles. Portugal, along with other Mediterranean countries, is a major producer and exporter of wine.

Much of Portugal is covered with cork oak trees, which provide the basis for the country's cork industry. Some trees can provide cork for more than 100 years.

READING CHECK: *Human Systems* How are the histories of Spain and Portugal similar?

TEKS Questions 1, 2, 3, 4, 5

Section 1 Review

Homework Practice Online

Keyword: SW3 HP16

Define
autonomy, cork

Working with Sketch Maps
On a map of southern Europe and the Balkans that you draw or that your teacher provides, label Spain, Portugal, Madrid, Balearic Islands, Strait of Gibraltar, Barcelona, Bay of Biscay, Lisbon, and Porto. Use the description in the text to shade in the Basque Country.

Reading for the Main Idea
1. *Places and Regions* How did Roman and Moorish rule in Spain and Portugal influence the countries' modern cultural landscapes?
2. *Places and Regions* How has Spain influenced the diffusion of foods between the Americas and Europe?
3. *Human Systems* How is immigration affecting urbanization in Spain and Portugal?

Critical Thinking
4. **Analyzing Information** Why has crossing the Strait of Gibraltar been an important migration route?

Organizing What You Know
5. Copy the chart below. Use it to show how economic development, urbanization, and environmental change are shaping Spain.

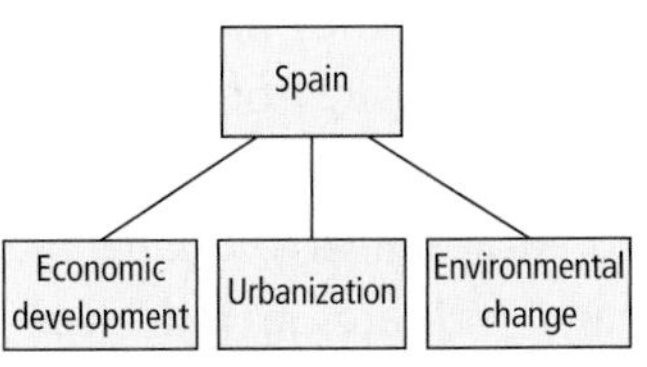

Geography for Life

Arabic Place-Names in Iberia

Settlements and the patterns they form on Earth's surface can provide a historical record to modern researchers. One of the patterns that tells a story is the arrangement of place-names, also known as toponyms. Place-names can provide clues about certain ethnic groups that live in a region or about invasions that have happened there. They may also indicate other information, such as the physical conditions at the time the name was first used.

Many place-names consist of two parts—generic and specific. The generic part refers to features found in many places, like mountains, rivers, or valleys. The specific part refers to something unique to a certain place. For example, consider Battle Creek, a city in Michigan. *Battle* refers to a specific event that happened in 1824. *Creek* is the generic term and refers to a common physical feature.

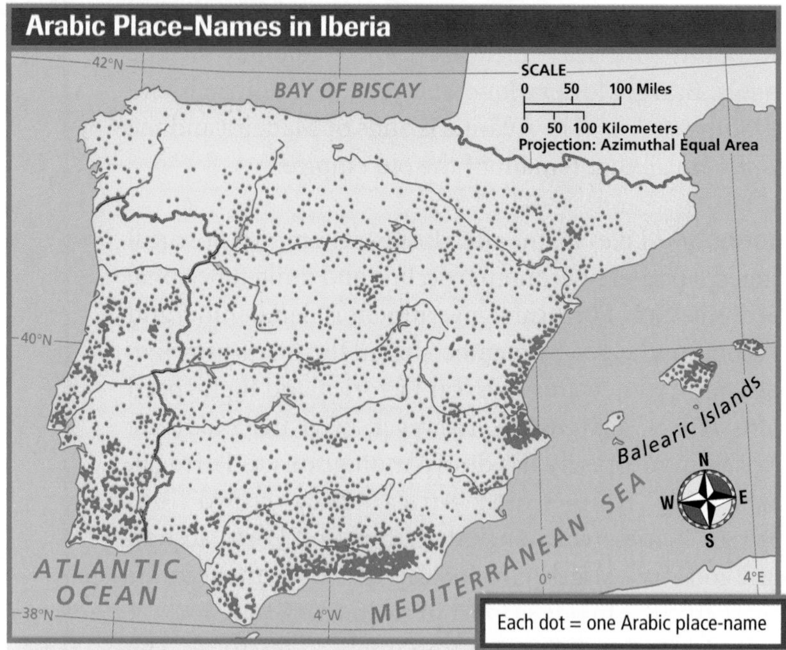

INTERPRETING THE MAP *Arabic place-names are found throughout Iberia but are most common in coastal areas of southern and eastern Spain and southern Portugal.* ***What might be some reasons that Arabic place-names are most densely concentrated in these areas?***

The study of place-names can also help us learn about places where different groups of people lived before recorded history. For example, some European place-names date back hundreds or even thousands of years. In many places, few written records remain from these eras.

The Iberian Peninsula provides a good illustration of how place-names can be studied to learn about the past. Moors from North Africa invaded the region in the A.D. 700s and ruled for about seven centuries. They brought the Arabic language with them to Spain and Portugal. Although the Moors were eventually driven out by speakers of Romance languages, their language survives in many Iberian place-names. An example is the prefix *guada-*, which derives from the Arabic word *wadi*, meaning "river" or "stream." The prefix *guada-* appears in the names of some major Iberian rivers. *Guadalquivir*, from the Arab name *Wadi al-Kabir,* means "the great river" because *al-Kabir* means "the most great." Similarly, *Guadalimar* comes from *Wadi al-Ahmar* and indicates a red river. Note that these terms also contain generic *(guada-)* and specific *(-quivir, -imar)* parts. Other Arabic place-names in Iberia include *Madrid,* which comes from the Arabic name *Medshrid* and relates to the abundant wood supply in the region that the Moors used for building.

The many Arabic names in Spain and Portugal indicate the Moors' cultural imprint on the region. In what other ways do you think the Moors influenced Spanish and Portuguese cultures?

Applying What You Know

1. **Summarizing** How do place-names help geographers learn about the past?
2. **Analyzing Maps** Examine the map. What might the map indicate about the direction from which the Moors invaded? What does it tell you about the extent of their empire?

Section 2 The Italian Peninsula

READ TO DISCOVER

1. How has Italy's history affected its culture?
2. What is Italy like today?

WHY IT MATTERS

Italian Americans are a large ethnic group in the United States. Use CNNfyi.com or other **current events** sources to learn about the history and culture of Italian American communities in this country.

IDENTIFY

Renaissance

DEFINE

microstates

LOCATE

Sicily
Sardinia
Alps
Rome

Locate, continued

Florence
Genoa
Venice
Milan
Turin
Po River
Bologna
Trieste
Naples

History and Culture

Italy occupies the boot-shaped peninsula that stretches southward from the middle of Europe into the Mediterranean Sea. Italy also includes two large islands, Sicily and Sardinia. To the north, Italy is separated from the rest of Europe by the Alps. Despite this barrier, Italy has influenced European culture for more than 2,000 years. Italians have created some of the world's most beloved architecture, literature, music, painting, and sculpture.

History The Etruscans created one of the earliest civilizations to occupy the Italian Peninsula. Later, the Romans set up a republic in central Italy about 500 B.C. Over time the Romans built a huge empire. This empire stretched across much of Europe, North Africa, and Southwest Asia. The city of Rome lay

In A.D. 79 Mount Vesuvius near present-day Naples erupted and completely buried three Roman towns in volcanic ash. In Pompeii, the largest of the towns, archaeologists have found the remains of more than 2,000 victims.

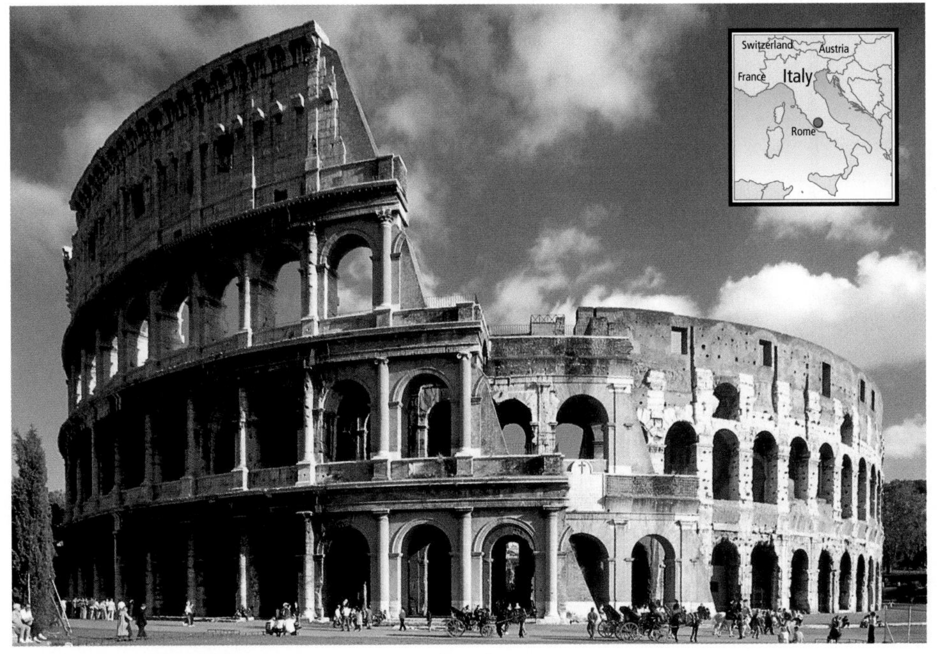

INTERPRETING THE VISUAL RECORD

The Colosseum in Rome is just one example of the many engineering marvels built by the Romans. Officially dedicated in A.D. 80, the Colosseum was used to stage battles between soldiers, slaves, and wild animals and could hold some 50,000 spectators. ***How do you think the Colosseum influenced the architecture of modern sporting arenas today?*** TEKS

Connecting to HISTORY

The Plague

A terrible plague, which Europeans came to call the Black Death, swept across Europe in the mid-1300s. Trading ships brought it from Asia to Italian and other southern European ports in 1347. Infected fleas living on rats spread one form, bubonic plague. A second form, pneumonic plague, could also be spread by infected people. Together the two forms spread across almost all of Europe. (See the map.)

A few areas were spared, but the plague killed as many as 30 million people. That total would have been about a third of Europe's population. The many deaths disrupted people's ways of life. Workers, in short supply as so many died, demanded higher wages. Tensions between upper and lower classes increased. In addition, the power and influence of the Roman Catholic Church declined. Some people—terrified by the spreading death—turned to other beliefs and practices for comfort and protection.

Drawing Inferences and Conclusions How might modern technologies allow disease to spread even more rapidly today?

INTERPRETING THE MAP ***How did trade routes influence the diffusion of the plague? What effects did the plague have on the regions into which it spread?*** TEKS

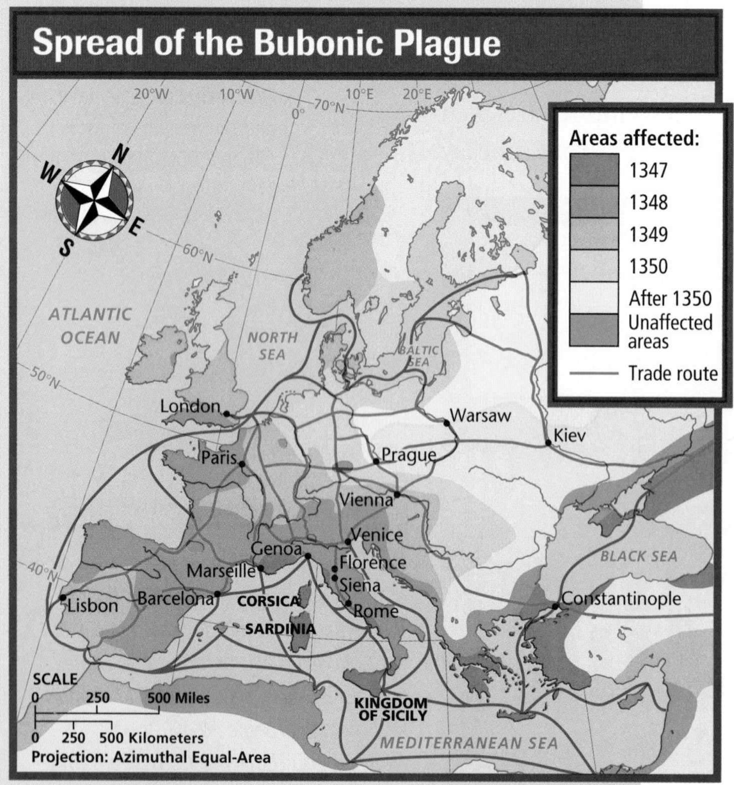

at its center. It was one of the first cities to have more than 1 million people. Roman culture, including language, laws, architecture, and urban lifestyles, diffused throughout the empire. The Romans also helped spread Christianity throughout Europe. The Western Roman Empire collapsed in the A.D. 400s. Italy remained a mix of separate states and cities for long afterward.

The influences of Roman law, literature, and language can still be seen in many European countries. For example, all the Romance languages—including Catalan, French, Italian, Portuguese, Romanian, and Spanish—are derived from Rome's Latin language. Rome is also the headquarters of the Roman Catholic Church. The seat of the church is at Vatican City, an independent country in the heart of Rome.

During the Middle Ages many Italian cities grew rich from trade. These great cities included Florence, Genoa, and Venice. During this time, northern Italy was one of the wealthiest and most culturally advanced regions in Europe. Trade not only brought wealth to Italy but also new ideas. Almost 1,000 years after the fall of the Roman Empire, Italy became the center of the **Renaissance** [re-nuh-SAHNS]. *Renaissance* is a French word meaning "rebirth." It describes the renewed interest in learning that spread throughout Europe from the 1300s to the 1500s. This time was particularly important for the development of architecture, painting, and sculpture. Some of the world's most famous artists, such as Leonardo da Vinci, Michelangelo, and Raphael, worked during this time. Today millions of people visit Florence, Rome, and Venice to see the great art and architecture of the Renaissance.

Italy did not become a united country until 1861. It fought on the side of the Allies in World War I. In the early 1920s a dictator named Benito Mussolini took control of Italy's government. He formed an alliance with Germany, and the two countries were allies during World War II. However, Mussolini was finally overthrown in 1943, and Italy was later controlled by Allied forces. Since World War II, Italy's economy and industries have grown tremendously. Today the country is a member of NATO and the EU.

TEKS **READING CHECK:** ***Human Systems*** What role did Italy play in the Renaissance, and how does this affect its economy today?

FOCUS ON GOVERNMENT

Europe's Microstates On a mountain slope in west-central Italy may be Europe's oldest country. The people of San Marino trace their country's history back to the A.D. 300s. At that time a small group of Christians came to the area. There they found a safe place to live and practice their religion. Mountains and wise choices of allies have isolated and protected the country since then.

Today San Marino is surrounded by Italy. It is one of the world's smallest countries—just 24 square miles (61 sq km) in area. In Europe just Monaco in southern France and Vatican City in Rome are smaller. These countries are so small that we call them **microstates**. A number of other tiny countries are also found in Europe. Andorra lies in the Pyrenees between Spain and France. Liechtenstein (LIKT-uhn-shtyn) is located in the Alps between Austria and Switzerland. Malta is a small island country between Italy and Africa.

These countries have survived for a variety of reasons, such as physical isolation and international treaties. They have different kinds of governments, although all but Vatican City have elected legislatures. Vatican City is all that is left of the old Papal States. The Papal States occupied much of central Italy from A.D. 754 to 1870. The pope heads Vatican City's government as well as the Roman Catholic Church. Church officials elect the pope. He then rules with absolute authority for the rest of his life.

Tourism and trade are important to the economies of most of

Independent since 1929, Vatican City is the world's smallest country at just 109 acres (44 ha). About 870 people live in Vatican City, home to extensive museums, famous works of art, banks, a post office, and a radio station.

the microstates. Low taxes have attracted foreign citizens and businesses. Malta is Europe's only island microstate. It has limited freshwater and other natural resources. However, many tourists visit the country, which has some of the oldest stone temples in the world.

READING CHECK: *Human Systems* How have Europe's microstates managed to survive until modern times?

People and Culture Most people in Italy are Roman Catholic and speak Italian. Some people in the northern part of the country also speak French, German, or Slovene.

Much of modern Italian culture was first developed during the Renaissance. Italian food may be the most famous part of this culture. Delicious sauces, pastas, sausages, and pastries can be found cooking in almost every Italian home. Many of those foods come from recipes first used by Italian chefs in the 1400s. More modern Italian foods, such as pizza, have also become popular around the world.

Italian daily life is similar to that in other Mediterranean countries. The main meal is in the middle of the day. Afterwards some people rest a while before returning to work. Italians often spend evenings with their friends and family, eating and discussing the latest news.

Central Italy is the country's political and cultural center. There you will find Rome, Italy's capital and largest city. The city has spread out along the banks of the Tiber River. Ruins of ancient Roman buildings still stand there. Among the most famous are the Colosseum and the Forum.

The endurance of Rome's historical buildings stands in contrast to the endurance of its many governments. In fact, Italy has had more than 50 governments since the end of World War II. Some have lasted just a few months. They have been so unstable partly because many political parties are represented in the country's parliament. Since none has a majority, parties must join to form what are called coalition governments. These temporary alliances usually do not last long. Still, Italy has a strong democracy, and its people have many freedoms. These advantages have helped the country make great economic progress in the last half century.

READING CHECK: *Human Systems* Why has Italy had many different governments since World War II?

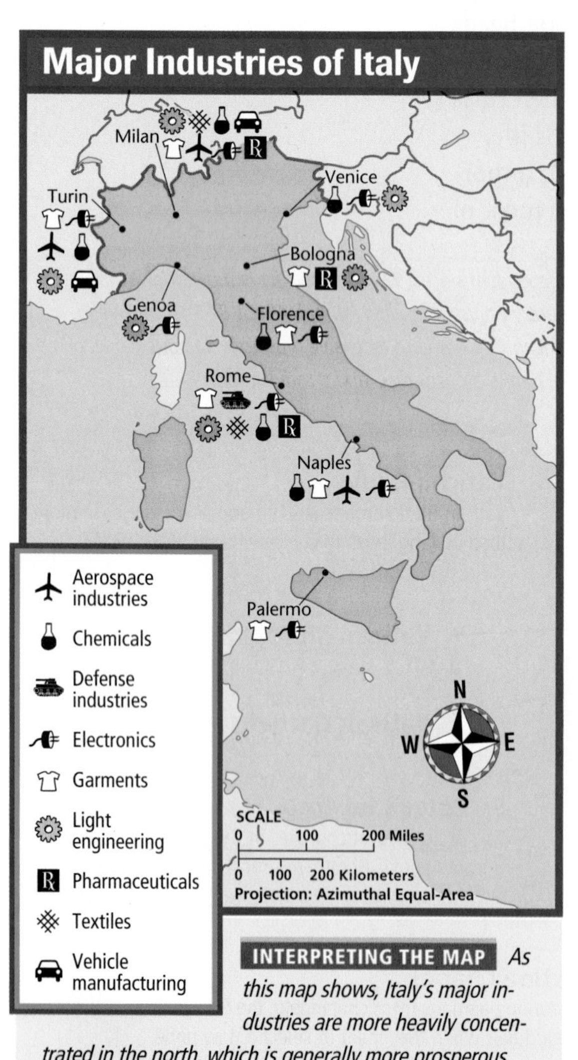

INTERPRETING THE MAP *As this map shows, Italy's major industries are more heavily concentrated in the north, which is generally more prosperous and developed than the south.* ***Based on the map, which cities seem to be Italy's most important industrial centers?*** TEKS

Italy Today

Italy is a modern developed country. Its GDP is similar to that of France and Great Britain. Italy produces agricultural and manufactured products that are sold around the world. The most famous of these products are probably Italian automobiles, designer clothes, and fine food.

North and South Italy has two main economic regions. The north is rich and industrial. The south is poorer and more agricultural. The south is known as the Mezzogiorno (MET-soh-gee-OR-noh), which means "midday" and refers to the area's bright sunshine. The dividing line between the two regions lies just south of Rome.

In the north are the large industrial cities of Milan, Genoa, and Turin and the rich farmlands of the Po River valley. Fertile soils make the Po Valley the "breadbasket" of Italy. To the east and south of the valley are other rapidly developing cities. These cities include Bologna (boh-LOH-nyah), Florence, Trieste (tree-ES-tay), and Venice.

Southern Italy, Sicily, and Sardinia are drier and poorer. They have high poverty and unemployment rates. The south produces farm products such as olives, citrus fruits, and grapes. However, it lags far behind the north in developing a modern economy. Soil erosion and deforestation have long troubled the area. The Italian government and the EU give aid to this region. Still, the economy there has not advanced significantly. Naples is southern Italy's largest city.

Italy's Population, 2000

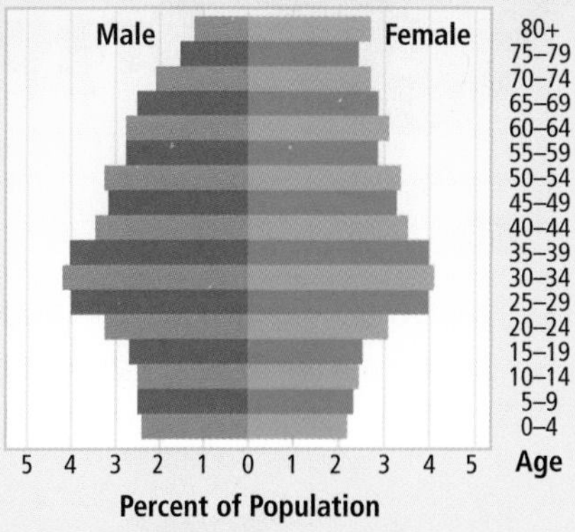

Male | Female

Age: 80+, 75–79, 70–74, 65–69, 60–64, 55–59, 50–54, 45–49, 40–44, 35–39, 30–34, 25–29, 20–24, 15–19, 10–14, 5–9, 0–4

5 4 3 2 1 0 1 2 3 4 5

Percent of Population

Source: U.S. Census Bureau, International Data Base

INTERPRETING THE GRAPH *In 2000 Italy's rate of natural increase was –0.1 percent, and the country's population was expected to shrink in the coming years.* ***Based on the graph, in about how many years will Italy face the challenge of supporting large numbers of retirees?*** TEKS

Issues and Challenges Developing southern Italy's economy remains a challenge. Another issue is the country's aging population. Italy has one of the lowest birthrates in the world. As a result, Italy's population is becoming older. This means that there are fewer young workers to replace older workers as they retire. Most of Italy's population growth is from immigration.

Pollution threatens not only Italy's future but also its past. Heavy traffic, smog, and wear and tear have damaged many historical monuments. Neglect has also taken its toll on some of Italy's cultural sites. The government has recognized the problem and is taking steps to protect and preserve important monuments. For example, Rome now limits the number of trucks that are allowed to drive through its historic center.

READING CHECK: ***Places and Regions*** What are three important challenges facing Italy?

TEKS Questions 1, 2, 3, 4, 5

Review

go.hrw.com Homework Practice Online

Keyword: SW3 HP16

Identify

Renaissance

Define

microstates

Working with Sketch Maps

On the map that you created in Section 1, label Italy, Sicily, Sardinia, Alps, Rome, Florence, Genoa, Venice, Milan, Turin, Po River, Bologna, Trieste, and Naples. Where is the seat of the Roman Catholic Church?

Reading for the Main Idea

1. *Human Systems* What great empire once ruled the Mediterranean from Italy? How is its influence seen in European culture today?
2. *Environment and Society* What importance does the Po River valley have to Italy?

Critical Thinking

3. **Comparing** How are northern Italy and southern Italy different economically?
4. **Identifying Cause and Effect** How might trade among Italian cities during the Middle Ages have spurred the Renaissance?

Organizing What You Know

5. Create a time line like the one shown below. On your time line, list important years, periods, and events in the history of Italy.

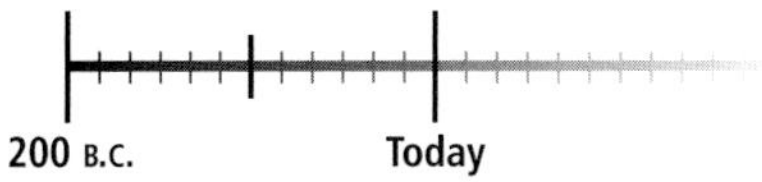

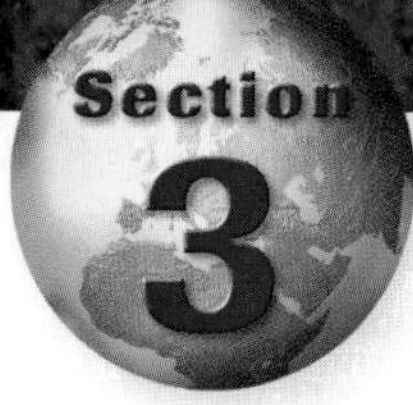

Greece and the Balkan Peninsula

READ TO DISCOVER

1. How did Greece develop into a modern country?
2. Why are the western Balkans politically unstable?
3. What changes are occurring in the eastern Balkans?

DEFINE

city-states
enclaves

LOCATE

Crete
Mount Olympus
Athens
Kosovo
Belgrade
Sarajevo
Bucharest

WHY IT MATTERS

The Balkan Peninsula is one of Europe's most volatile regions. In recent decades it has been the scene of much ethnic conflict. Use CNNfyi.com or other **current events** sources to learn about recent efforts toward peace in the region.

Greece

At the southern tip of the Balkan Peninsula lies Greece. The country is made up of many peninsulas, islands, and rugged mountains. The largest Greek island is Crete. The highest peak in Greece is Mount Olympus. It has an elevation of 9,570 feet (2,917 m).

History Greece was the site of one of Europe's earliest and most advanced civilizations. Civilization there can be traced back more than 2,500 years. Long ago many Greeks lived in a number of powerful **city-states.** A city-state is a self-governing city and its surrounding area. Each city was independent from the others. The people of those cities made great contributions to the arts, government, philosophy, science, and sports. Those contributions have influenced much of Europe and other places. For example, the ancient Greeks developed early systems of democratic government.

INTERPRETING THE VISUAL RECORD

The Parthenon in Athens is one of the finest examples of Greek architecture. Built in the 400s B.C. as a temple to the Greek goddess Athena, the building was later used as a Christian church and as a mosque. ***What features of the Parthenon's architecture do you think might have influenced the design of important buildings in this century?***

Over time, Greece fell under the control of outside invaders. The Persians invaded Greece but were defeated. Later, Romans and Ottoman Turks dominated Greece. It was not until 1829 that Greece became an independent country. After World War II the country slipped into a bloody civil war. A series of elected governments followed, but military leaders took over the government in 1967. In 1974 the country returned to democratic government.

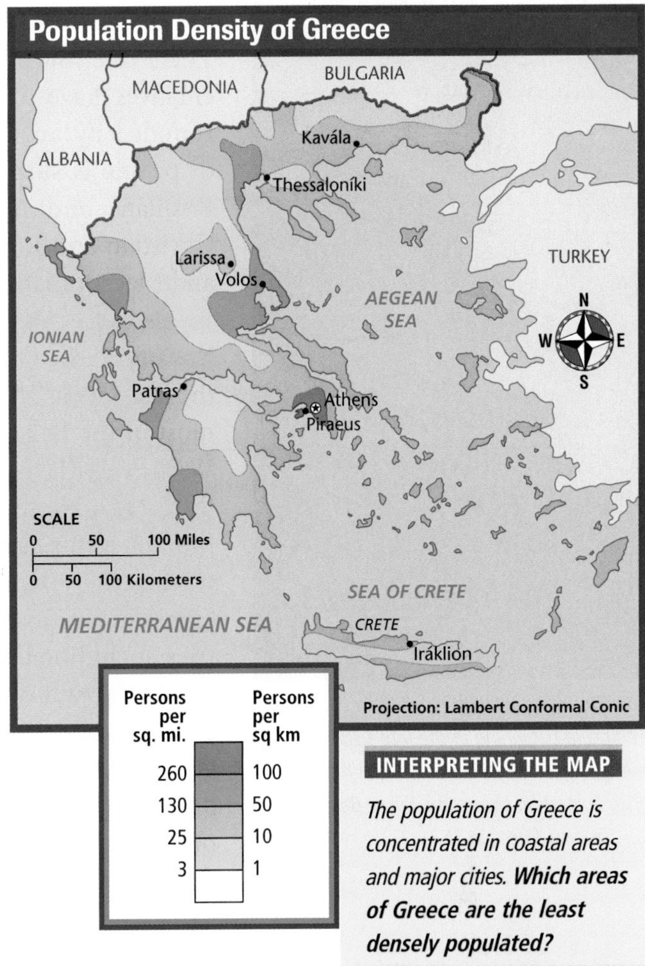

INTERPRETING THE MAP

The population of Greece is concentrated in coastal areas and major cities. ***Which areas of Greece are the least densely populated?***

People and Culture Greece's long history of foreign rule influenced its culture. These influences are particularly evident in Greek cooking. Common foods include Turkish dishes such as baklava, a honey-based pastry. However, the Greeks kept their language. In addition, about 98 percent of the population is Greek Orthodox Christian.

Many social changes are taking place in Greece today. Once poor and agricultural, Greece is becoming an industrialized country. In fact, Greece's economy is becoming more like those of northern and central Europe. With an increase in wealth, education and opportunities for women have also increased. Unlike those of earlier generations, modern Greek women seek education and jobs outside the home. In fact, women hold important positions in the country's government, private industries, and universities.

Economy, Issues, and Challenges Greece has made much economic progress since joining the EC in 1981. However, the country remains relatively poor by European standards. Still, many illegal immigrants enter Greece to fill low-paying jobs. Most are from Albania.

Greece's lack of population growth is much like the situation in Italy and Spain. The country is also urbanizing rapidly. Today about 30 percent of Greeks live in the capital and largest city, Athens. That city's growth has created terrible smog, traffic, and pollution. This smog and other pollution threatens the health of Greeks as well as the ancient monuments that draw tourists. As a result, the government of Athens has passed regulations designed to ease pollution problems.

READING CHECK: ***Human Systems*** How has economic change in recent years affected women in Greece?

The Western Balkans

Albania and what was once Yugoslavia make up the mountainous western Balkans. Bosnia and Herzegovina (also called just Bosnia), Croatia, Macedonia, and Slovenia are former Yugoslav republics. Serbia and Montenegro remain united and continue to use the name *Yugoslavia.*

The Ottomans invaded the Balkan Peninsula beginning in the late 1300s and eventually controlled much of the region. This illustration shows the fall of Belgrade to Ottoman forces in 1521.

This area has one of the most diverse human populations in Europe. There one finds a complicated mix of languages and religions. Many ethnic **enclaves** have formed there. An enclave is an area that is completely surrounded by another region. So, ethnic enclaves are areas where one group of people is surrounded by one or more other ethnic groups. For example, Bosnians and Albanians are Muslims surrounded by Eastern Orthodox Christian and Roman Catholic peoples. Those peoples include Serbs, Croats, and Macedonians.

History The Ottoman Turks, who were Muslim, invaded southern Europe in the late 1300s. They conquered most of the Balkan Peninsula. By World War I, most of the Balkans had won independence from the Ottomans. After World War I, Yugoslavia was formed to unite the various Slavic peoples who lived in the area. These peoples included the Bosnians, Croats, Macedonians, Montenegrins, Serbs, and Slovenes. A monarchy governed and held the country together. Albania, which had claimed independence in 1912, remained separate. However, many Albanian people had moved to neighboring countries, including the Kosovo region of Yugoslavia.

At the end of World War II, the Soviet Union occupied all of the Balkan countries but Greece. The occupied countries became communist. Yugoslavia's government was led by the dictator Tito. He kept the country united until his death in 1980. Ten years later communism crumbled in Yugoslavia and the rest of Eastern Europe. Yugoslavia then began to split apart. Slovenia and Croatia declared independence in 1991 and fighting soon broke out among Slovenes, Croats, and Serbs.

Bosnia and Herzegovina

Areas of Ethnic Majority, 1991

Areas of Ethnic Control, 1995

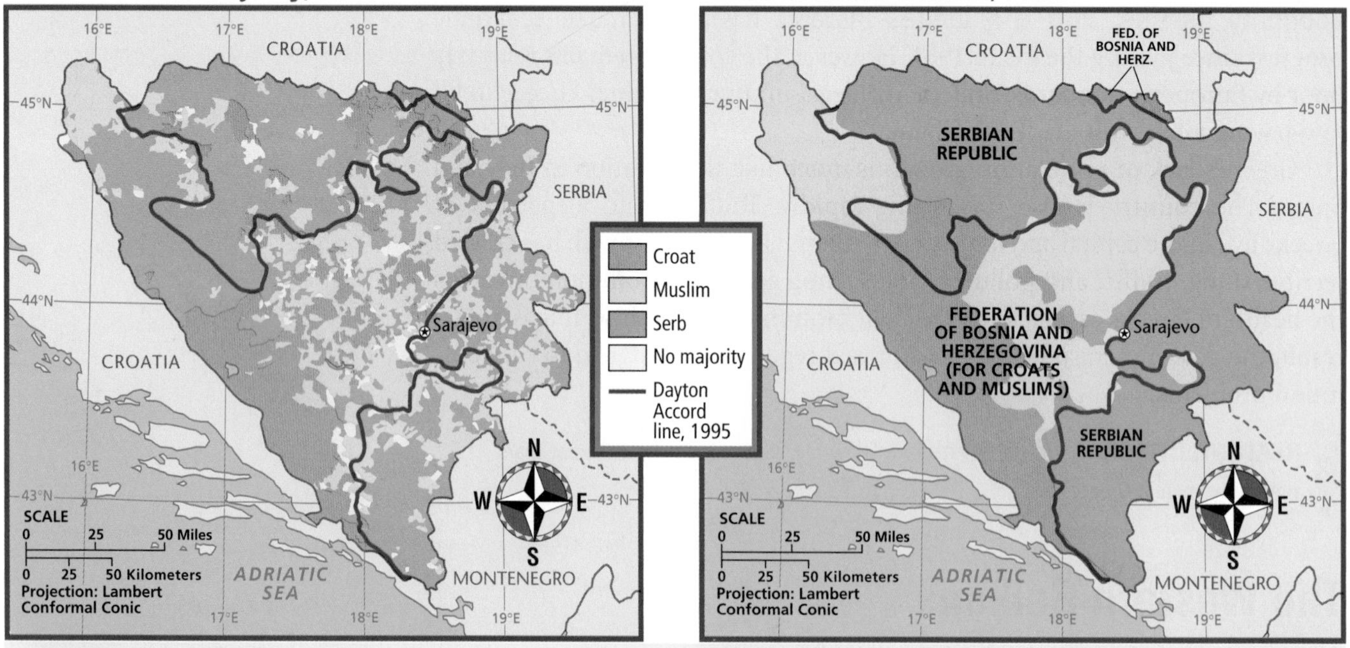

Croat
Muslim
Serb
No majority
Dayton Accord line, 1995

INTERPRETING THE MAP *Before the war in Bosnia the region's major ethnic groups were intermixed (left). After the war, however, the country's ethnic map had changed dramatically (right). In 1995 the Dayton Accord established a line between a Serbian-controlled area and an area controlled by Croats and Muslims.* ***How has conflict influenced the control of different regions in the country? How might these areas of control have influenced the creation of the Dayton Accord line?*** TEKS

The city of Mostar was heavily damaged during the war in Bosnia. Before the war, the city's population included many Croats, Muslims, and Serbs. However, today there are no Serbs in the city, and the Croat and Muslim populations are geographically and politically divided.

The fighting was brief in Slovenia, which borders Austria, Hungary, and Italy. However, fighting in Croatia slid into bloody civil war. Serbia sent weapons and supplies to the Serb rebels in Croatia. When Bosnia declared independence in 1992, civil war broke out there too. Bosnian Muslims, Serbs, and Croats all fought to control the country or parts of it.

Serbs in Croatia and Bosnia refused to accept separation from the rest of Yugoslavia. They did not want to live in countries controlled by people of other ethnic groups. The government in Serbia supported them. However, when the terrible fighting ended in late 1995, Croatia and Bosnia had won independence. Keeping the peace has not been easy.

In 1997 periodic fighting between ethnic Albanians and Serbs in Kosovo got much worse. Tensions between the mostly Muslim Albanians and Christian Orthodox Serbs stretch back centuries. The United States and its NATO allies intervened to stop the fighting in 1999. Kosovo's future is still uncertain.

During the war in Kosovo, many ethnic Albanians fled to Albania. These refugees were forced to leave for their own safety. The resulting population increase further strained the limited resources of Albania, Europe's poorest country. Today Albania continues to adjust from a long period of isolation under its old Communist government. That government lost power after elections in 1992.

People and Culture As you can see, culture in the western Balkans is a complex mix of religions, ethnic backgrounds, and political views. For example, the Serbs and Macedonians are Slavic. They practice an Orthodox Christian religion. The Croatians are also Slavic, but most are Roman Catholic. Most Albanians are Muslim. All of these peoples wish to preserve their particular heritage.

Belgrade is the largest city in the region. It and Sarajevo (sar-uh-YAY-voh)—the once beautiful capital of Bosnia—have been heavily damaged by war. Many cities and towns across the region must be rebuilt. Also, many people who fled Croatia and Bosnia during the fighting want to return to their old homes. However, it is not clear whether this will be possible.

Economy, Issues, and Challenges Slovenia has rapidly built up trade with the EU. It has also attracted foreign businesses and tourists. Coal, oil, minerals, and an electronics industry support its growing economy. The country also promotes tourism, which it hopes to develop further. However, other countries in the area have had major problems. War, corruption, and a lack of modernized industries have left them with weak economies.

Continuing unrest makes this area's economic future uncertain. The United States and other NATO countries have peacekeeping troops there. Recent government changes in Yugoslavia may also help. Many observers hope the new government will focus on solving economic problems and calming ethnic tensions.

TEKS **READING CHECK:** ***Human Systems*** What characterizes the cultural geography of the western Balkans?

The Eastern Balkans

Bulgaria, Romania, and Moldova lie in the eastern Balkans. Those countries have not experienced the ethnic fighting seen in the western Balkans. Therefore, they are more politically stable than their neighbors. However, as in other formerly communist countries of Eastern Europe, economies are underdeveloped, and standards of living are lower than in other regions of Europe.

History and Culture The countries of the eastern Balkans have been controlled by the Roman, Byzantine, and Ottoman Empires. The region gained independence from the weakened Ottoman Empire during the late 1800s and early 1900s. After World War II the area came under the control of the Soviet Union. Communist control ended in the early 1990s.

Bulgarians are a Slavic people and share some culture traits with other Slavs. Like Russians, for example, Bulgarians use the Cyrillic alphabet. Most people in Bulgaria follow the Bulgarian Orthodox religion. In Romania about 7 percent of the population is ethnic Hungarian, but the majority is ethnic Romanian. They speak a language derived from Latin. Most Romanians follow the Romanian Orthodox religion. Most of Moldova's people are closely related to Romanians and are Eastern Orthodox Christians.

Many people in Romania live in rural areas and suffer from a lack of access to modern technology. In fact, nearly half of the country's population is rural. In general, farmers work their fields without the help of modern farm machinery. Also, there is a lack of many consumer goods, which means that people rely on homemade goods for some of their needs.

INTERPRETING THE VISUAL RECORD

The huge Danube River delta in Romania has many strips of land called grinduri *that farmers use to grow crops. Reeds that grow in the region's shallow waters are also used to make paper and fibers for textiles.* ***What challenges might farmers in this region face?***

Economy, Issues, and Challenges The collapse of communism brought great changes to the eastern Balkans. Moldova, Bulgaria, and Romania are now working to adjust to democracy and a free-market economy. Moldova is a small densely populated country with rich soils and a mild climate. The struggling economy is based on producing fruit, grains, and wine. About one fourth of Bulgaria's labor force works in agriculture. However, the government has made efforts to attract modern industries. Romania is trying to expand an economy based on agriculture, oil, coal, and low-technology industries. Bucharest is the country's capital and major industrial center.

Most people in the eastern Balkans have living standards well below those of other Europeans. Many people of these countries lack health care and safe water, particularly in the villages. A lack of housing in urban areas is also a problem. Many people are leaving to find work in other countries. As a result, Bulgaria's population is decreasing.

READING CHECK: ***Human Systems*** How have the economies and politics of the eastern Balkans changed in recent years?

Review

TEKS Questions 1, 2, 3, 4, 5

Define

city-states
enclaves

Working with Sketch Maps

On the map that you created in Section 2, label Greece, the Balkan countries, Crete, Mount Olympus, Athens, Kosovo, Belgrade, Sarajevo, and Bucharest. In the margin of your map, identify the capital of Greece.

Reading for the Main Idea

1. ***Human Systems*** How did ancient Greece contribute to modern government?
2. ***Places and Regions*** How has cultural conflict shaped the political boundaries of the western Balkans?
3. ***Human Systems*** How are the economies of the eastern Balkan countries changing?

Critical Thinking

4. **Identifying Points of View** Why do you think ethnic Serbs living in Croatia and Bosnia and Herzegovina opposed independence from Yugoslavia? Why do you think Croats and Bosnians may have wanted independence even at the risk of war?

Organizing What You Know

5. Copy the flowchart below and use it to trace events in Yugoslavia since the death of Tito in 1980.

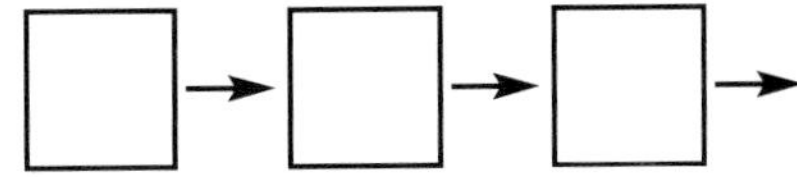

Building Vocabulary (TEKS)

On a separate sheet of paper, explain the following terms by using them correctly in sentences.

autonomy	microstates
cork	city-states
Renaissance	enclaves

Locating Key Places (TEKS)

On a separate sheet of paper, match the letters on the map with their correct labels.

Madrid	Sicily
Balearic Islands	Kosovo
Strait of Gibraltar	Bucharest
Rome	Athens

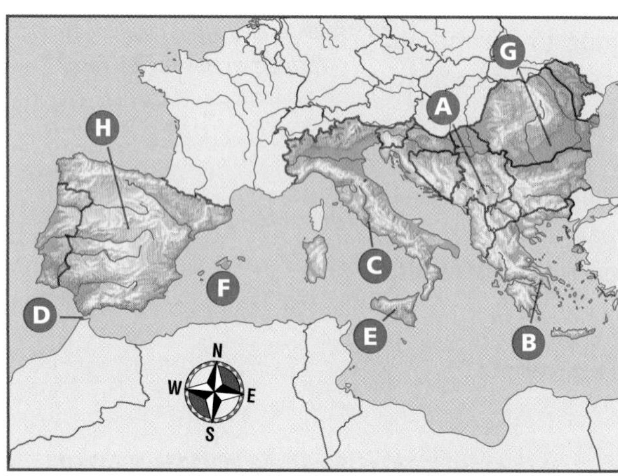

Understanding the Main Ideas (TEKS)

Section 1

1. *Human Systems* How have events in Spain's history influenced other areas of the world?
2. *Human Systems* How does Portugal's history mirror Spain's?

Section 2

3. *Places and Regions* Which of Italy's two major regions is the richest and most industrialized? Why?

Section 3

4. *Human Systems* How has urbanization contributed to environmental problems in Greece?
5. *Human Systems* What factors have shaped the boundaries of Yugoslavia?

Thinking Critically for TAKS (TEKS)

1. **Making Generalizations** Why might some people in the Basque Country and Catalonia see themselves more as Basque and Catalan than Spanish? How might this affect their view of the rest of Spain?
2. **Drawing Inferences and Conclusions** How might northern Italy's location near central Europe have affected its economic growth?
3. **Analyzing Information** What factors have limited economic development and improvement in the standard of living in the Balkan countries?

Using the Geographer's Tools (TEKS)

1. **Analyzing a Population Pyramid** Study the population pyramid for Italy. What features tell you that Italy's population is decreasing?
2. **Creating Maps** Use the unit atlas to create a map of Europe. Identify which countries make up the Iberian, Italian, and Balkan Peninsulas.
3. **Creating Bar Graphs** Use statistics from the unit Fast Facts table to construct a bar graph comparing per capita GDP in the countries of southern Europe and the Balkans.

Writing for TAKS (TEKS)

Compare patterns of culture in Spain and Canada. How has each country been affected by cultural differences among its people? How have those differences affected political issues, boundaries, public policies, points of view, and events in each? You might want to do further research by using secondary sources of information like magazines. Write a short report about your findings. When you are finished with your report, proofread it to make sure you have used standard grammar, spelling, sentence structure, and punctuation.

Creating an Elevation Profile

Places and Regions Create an elevation profile of Italy. The profile should show the elevation of features along a line from west to east across the peninsula, just north of Rome. Identify important points along the line. You can use the chapter map as well as atlas maps from your classroom and library to prepare your diagram.

Building Social Studies Skills

Interpreting Graphs

Study the graph below. Then answer the questions that follow.

1. Which country is projected to lose the most people by 2050?
   a. Italy
   b. Spain
   c. Greece
   d. Portugal
2. Describe the projected general population trend in the region between now and 2050.

Building Vocabulary

To build your vocabulary skills, answer the following questions.

3. *Autonomy* means
   a. economic self-sufficiency.
   b. a large cooperative group.
   c. self-government.
   d. a territory or area of influence.
4. An *enclave* is
   a. a very small country.
   b. an extreme shortage of food.
   c. a legal restriction against the movement of freight.
   d. an area completely surrounded by another region.

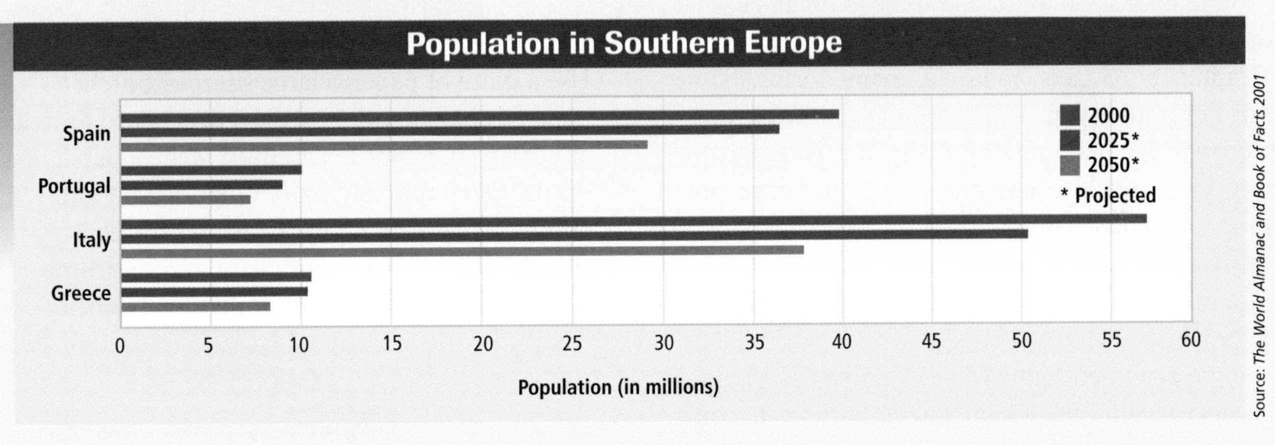

Alternative Assessment

PORTFOLIO ACTIVITY

Learning about Your Local Geography TEKS

Group Project: Field Work

With a group, plan, organize, and complete a research project about features in your community that might be developed into popular tourist attractions. Identify scenic locations, historical sites, cultural events, and geographic features—such as an appropriate climate—that might draw tourists. Prepare a questionnaire to survey area residents about the attractions in your community. Present your findings in a tourism-guide brochure for your community.

internet connect

go.hrw.com

Internet Activity: go.hrw.com
KEYWORD: SW3 GT16 TEKS

Choose a topic on southern Europe and the Balkans to:

- create a postcard of the islands and peninsulas of the Mediterranean.
- research conflict in the Balkans.
- understand how geographic factors and global trade influenced the development of pizza as a common food.

Skill Building for TAKS

WORKSHOP 1

Using Graphic Organizers

Some people say a picture is worth a thousand words. A graphic organizer creates a picture of information. It allows you to "see" patterns and relationships. We use graphic organizers every day. For example, calendars and daily planners are types of graphic organizers. The section reviews throughout this textbook include a question that asks you to use a graphic to organize what you have learned.

Graphic organizers come in many forms. Time lines, for example, identify important events in historical sequence. Charts and tables help you gather, compare, and analyze information. Another kind of graphic organizer is an idea web, which you will learn about in this workshop.

Graphic organizers can be used for many purposes. Some organizers help you brainstorm ideas, problems, and solutions. Others allow you to identify connections or tell stories. Still others help you communicate complex ideas and develop ideas for writing.

Idea Web

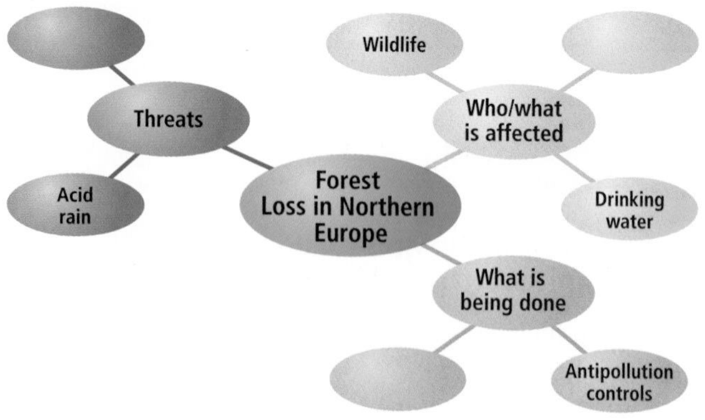

Developing the Skill Idea webs, also called word webs, are very useful graphic organizers. These organizers are maps that show how different categories of information relate to each other. They provide a structure for ideas and facts. They can also help you brainstorm and organize your writing tasks.

A major topic or concept is located at the web center. Supporting details radiate outward from the center. The example provides a main topic in the center: forest loss in northern Europe. The subtopics, or details, are provided in bubbles that grow from the main topic. Just as when you brainstorm and one thought leads to another, each detail can have its own bubbles. The details become more and more specific as the web grows. The web can grow until you run out of either ideas or space.

Practicing the Skill

Use a piece of paper, a large sheet of butcher paper, a chalkboard, or a dry-erase board to make an idea web. Suppose you must write a paper on the geography of Spain. What do you know about Spain's geography? Consider topics like the physical environment, demographics, and trade. Build your idea web individually or in a small group.

WORKSHOP 2

Using Databases and Analyzing Statistics

Statistics you can use in your study of geography are often organized into databases. Databases condense and organize information. They can help you answer important geographic questions. In addition, databases can help you understand geographic relationships.

A telephone book is a database. The books you keep on a shelf at home make up a database of some of the things you have read. Other examples of databases include statistical tables (like the one shown on the next page), almanacs, and CD–ROMs. Even maps are databases. Maps organize and show information about a variety of features. For example, a precipitation map shows average annual precipitation amounts for the country or region under study.

Comparing Statistics: The Former Yugoslav Republics

Country	Population (in millions)	Per Capita GDP (in U.S. $)	Electricity Consumption (kilowatt hours per person)	Telephone Lines (per person)
Bosnia and Herzegovina	3.9	$1,720	538 kWh	0.10
Croatia	4.3	$5,100	3,042 kWh	0.36
Macedonia	2	$1,050	3,036 kWh	0.23
Slovenia	2	$10,300	5,531 kWh	0.39
Yugoslavia (Serbia/Montenegro)	10.7	$2,300	3,390 kWh	0.21

Source: Central Intelligence Agency, *2000 World Factbook; The World Almanac and Book of Facts 2001*

As you work with databases, you will probably note that statistics can vary from one source to another. For example, the U.S. Central Intelligence Agency's World Factbook and some other database might differ in their population totals or economic figures for a country. The sources may have used different formulas for calculating data, such as economic statistics. In addition, they may have collected the statistics at different times in a year. One source might offer more recent information than others. As a result, it is often a good idea to check two or three sources for information you need.

Developing the Skill A number of key terms are important for learning how to create and use databases for organizing statistics and other information. A *file* stores a database on a computer. A short file name for the example on this page might be Statistics. *Fields* are names for categories of information in a database. The example above has five fields: Country, Population, Per Capita GDP, Electricity Consumption, and Telephone Lines. *Records* group fields that are all related to the database's topic or idea. All of the fields in the example are a record of statistics about the former Yugoslav republics.

To create a database on a computer, you need a database management system. Such a system is part of special software developed for record keeping. It provides instructions on how to organize and name your fields. Following are some key points to remember when creating and using any database:

- Read the headings and labels to determine the kinds of information included in the database. Are the figures provided in metric or in standard measurement? Do the numbers represent a fraction of a larger number? If you were looking at economic data for various European countries, one source may use GNP and another GDP. It would be important to understand the differences between such data and how those differences affect the research you are conducting.
- Be sure to identify the original source for the data. When was the information collected? How was the information collected?
- Always look up any unfamiliar terms so you are sure to understand exactly what the data represents.

Practicing the Skill

Use a variety of historical and statistical databases to create a profile of population change in Europe. Collect data on Austria, Germany, Italy, and other European countries with negative or slow population growth rates. You will need to create your own databases to organize your information. You can use a software database program or create a database by hand. For example, you might create a table that tracks changes in population from 1800 to what is projected in 2050 for the countries you choose to study. As an alternative, you could create maps that use color to show population growth rates over time. For example, you might shade countries with the slowest growth rates red and the highest growth rates blue. Include a file name and identify fields you use.

UNIT 5 Russia and Northern Eurasia

St. Basil's Cathedral, Moscow

CONNECTING TO *Literature*

THE WORLD I LEFT BEHIND

by Luba Brezhneva

Luba Brezhneva

(1943–), the niece of Soviet leader Leonid Brezhnev, was born in Russia and lived there until 1990. In *The World I Left Behind* she tells the story of her life in the former Soviet Union. Here Breshneva writes about when she and her husband, Mischa, moved from Moscow to live in the rural village of Korovino. She also provides some insights into the culture of rural Russia.

The bath, with its steam and heat, was a true haven for body and soul. . . . Yakov Maksimovich [Brezhneva's father] and his son, Ilya, would whip each other with hot birch twigs (to open the pores) and raise volleys of steam by sprinkling herbal broth on the stove. Climbing in turn to the top bunk, where it was hottest—"so the heat will enter the bones"—they stayed in the steam room till they were woozy and as red as lobsters. Then they would run to the pond—'to cast off the heat.' And then they would begin again. Finally came an unhurried conversation as they lay on benches in the dressing room. "You've gotten mighty thin, Ilya," Yakov once said as he examined his son's slim body. "The city is eating you up."

On Saturdays, while the men heated the stove, brought water, and bathed, the women would be in the house hastily finishing their domestic chores, baking pies, and cleaning. The Brezhnev men used the bathhouse before the women. . . .

Saturdays also meant scrubbing the wooden floor of the house, a time-consuming ritual that began in the morning. It would first be swept, then doused with hot water and ashes. While still warm, it would be scraped with a knife until the wood glowed a tawny yellow brown. Then it would be rinsed. Finally patchwork rugs would be laid down. The furniture was washed in the same way, since it was made of unpainted wood.

Illness in the family also meant it was time to heat the bath. The patient was steamed in the bath and whipped with birch twigs, then smeared with goat fat and forced to drink a hot broth of bitter herbs. Finally he or she was wrapped in an old sheepskin coat and laid out on the stove to sleep, assured that "you'll be fit as a fiddle by morning."

Analyzing the Primary Source

1. **Analyzing Information** What kind of climate do you think is typical in Korovino? What clues do you find in the passage about how much life has changed in the village over time?

2. **Summarizing** According to Brezhneva, what is life like for women in rural Russia?

The World in Spatial Terms

Russia and Northern Eurasia: Political

1. *Places and Regions* Compare this map to the physical map. Which countries in the region are largely mountainous?
2. *Places and Regions* Compare this map to the climate map. Which countries in the region have large areas with an arid climate? TEKS

Critical Thinking

3. **Making Generalizations** What do the names Kazakhstan, Kyrgyzstan, Tajikistan, Turkmenistan, and Uzbekistan all have in common? How do you think the names of these countries might relate to the ethnic groups that live there?

SCALE
0 500 1000 Miles
0 500 1000 Kilometers
Projection: Two-Point Equidistant

Boundaries
National capitals
Other cities

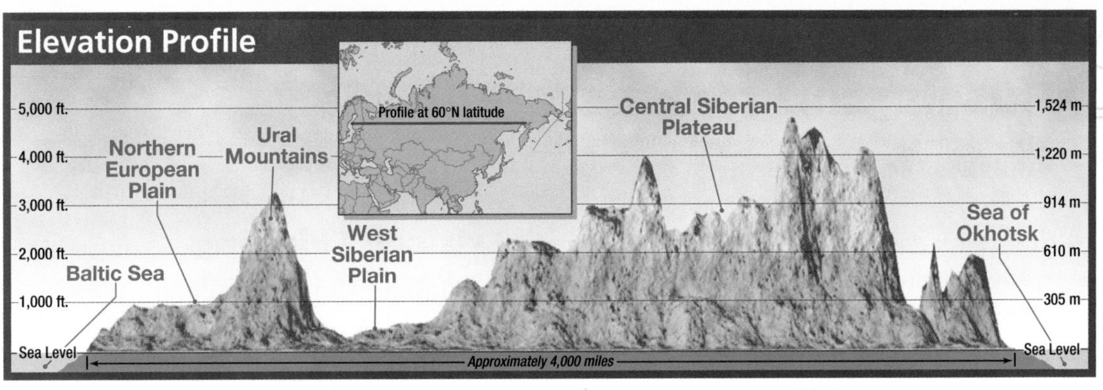

Russia and Northern Eurasia: Physical

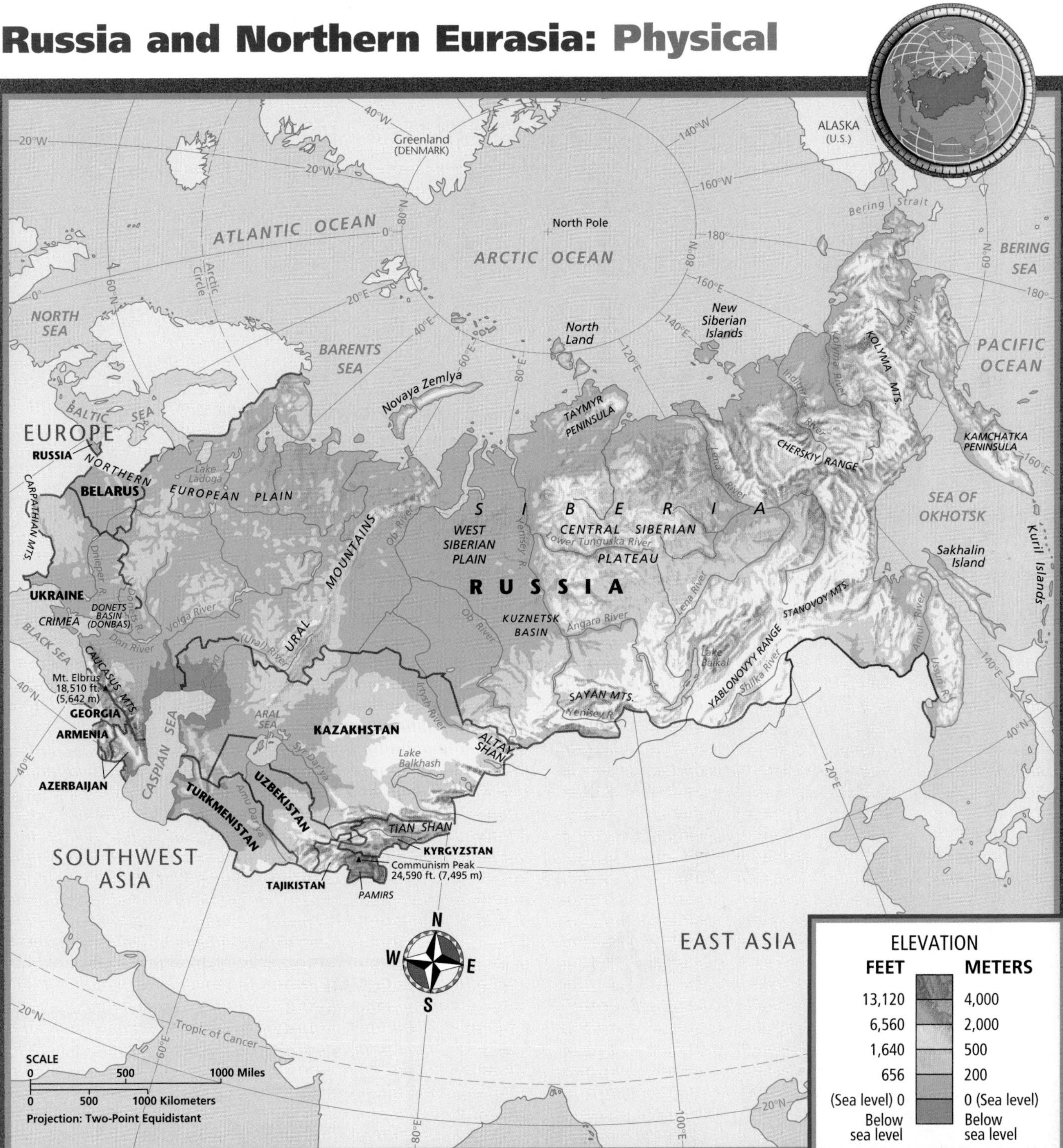

ATLANTIC OCEAN
ARCTIC OCEAN
North Pole
Greenland (DENMARK)
ALASKA (U.S.)
Bering Strait
BERING SEA
PACIFIC OCEAN
NORTH SEA
BARENTS SEA
BALTIC SEA
Arctic Circle
Novaya Zemlya
North Land
New Siberian Islands
TAYMYR PENINSULA
KOLYMA MTS.
Kolyma River
Anadyr R.
Indigirka River
CHERSKIY RANGE
KAMCHATKA PENINSULA
SEA OF OKHOTSK
Kuril Islands
Sakhalin Island
EUROPE
RUSSIA
BELARUS
NORTHERN EUROPEAN PLAIN
Lake Ladoga
CARPATHIAN MTS.
Dnieper R.
UKRAINE
DONETS BASIN (DONBAS)
Donets R.
CRIMEA
BLACK SEA
Don River
Volga River
CAUCASUS MTS.
Mt. Elbrus 18,510 ft. (5,642 m)
GEORGIA
ARMENIA
AZERBAIJAN
CASPIAN SEA
URAL MOUNTAINS
(Ural) River
Ob River
WEST SIBERIAN PLAIN
Yenisey R.
S I B E R I A
CENTRAL SIBERIAN PLATEAU
Lower Tunguska River
Lena River
R U S S I A
KUZNETSK BASIN
Angara River
Lake Baikal
SAYAN MTS.
Yenisey R.
STANOVOY MTS.
YABLONOVYY RANGE
Shilka River
Amur River
Ussuri R.
ARAL SEA
KAZAKHSTAN
Irtysh River
ALTAY SHAN
Lake Balkhash
Syr Darya
Amu Darya
UZBEKISTAN
TURKMENISTAN
TIAN SHAN
KYRGYZSTAN
Communism Peak 24,590 ft. (7,495 m)
TAJIKISTAN
PAMIRS
SOUTHWEST ASIA
EAST ASIA
Tropic of Cancer
N
W
E
S

SCALE
0 500 1000 Miles
0 500 1000 Kilometers
Projection: Two-Point Equidistant

ELEVATION	
FEET	METERS
13,120	4,000
6,560	2,000
1,640	500
656	200
(Sea level) 0	0 (Sea level)
Below sea level	Below sea level

Russia and Northern Eurasia: Climate

1. **Places and Regions** Compare this map to the political map. What are Russia's major climates? TEKS
2. **The World in Spatial Terms** Compare this map to the physical map. What climate types are found near the Black Sea? TEKS

Critical Thinking

3. **Making Generalizations** Why do you think cold climates cover so much of Russia? TEKS

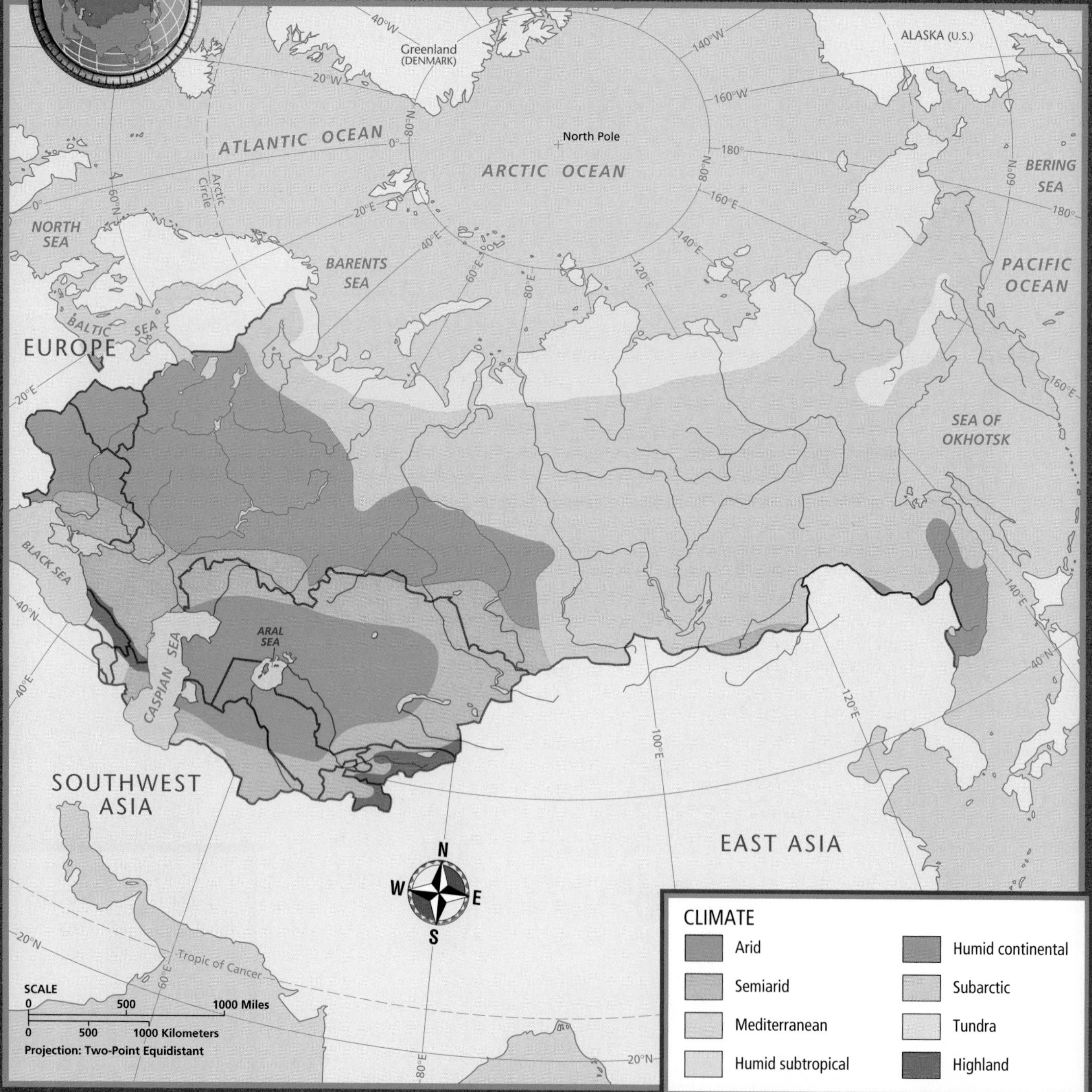

CLIMATE

- Arid
- Semiarid
- Mediterranean
- Humid subtropical
- Humid continental
- Subarctic
- Tundra
- Highland

SCALE: 0–500–1000 Miles; 0–500–1000 Kilometers

Projection: Two-Point Equidistant

Russia and Northern Eurasia: Precipitation

1. *Physical Systems* How are precipitation patterns different near the Black Sea and Caspian Sea? TEKS
2. *Environment and Society* Compare this map to the political and population maps. How are population density and precipitation patterns in Russia related? TEKS

Critical Thinking

3. **Making Generalizations** Why might the eastern part of the region receive little precipitation? TEKS

Centimeters	Inches
Under 25	Under 10
25–50	10–20
50–100	20–40
100–150	40–60
150–200	60–80
Over 200	Over 80

Russia and Northern Eurasia: Population

1. **Places and Regions** Which Russian metropolitan areas have more than 2 million inhabitants? TEKS
2. **Environment and Society** Compare this map to the climate map. Which climate types do the most densely populated areas in the region have? TEKS

Critical Thinking

3. **Analyzing Information** Compare this map to the physical map. How are rivers and lakes in Siberia related to population density? TEKS

POPULATION DENSITY

Persons per sq. mile	Persons per sq km
520	200
260	100
130	50
25	10
3	1
0	0

● Metropolitan areas with more than 2 million inhabitants

○ Metropolitan areas with 1 million to 2 million inhabitants

SCALE
0 500 1000 Miles
0 500 1000 Kilometers
Projection: Two-Point Equidistant

Russia and Northern Eurasia: Land Use and Resources

UNIT 5 ATLAS

1. *Environment and Society* Compare this map to the climate map. What is the main economic activity in areas with tundra and arid climates? TEKS
2. *Environment and Society* How have Russians adapted Siberian rivers to produce energy? TEKS

Critical Thinking

3. **Analyzing Information** Compare this map to the climate map. How do you think farmers have adapted to environmental conditions in arid areas? How can you tell? TEKS

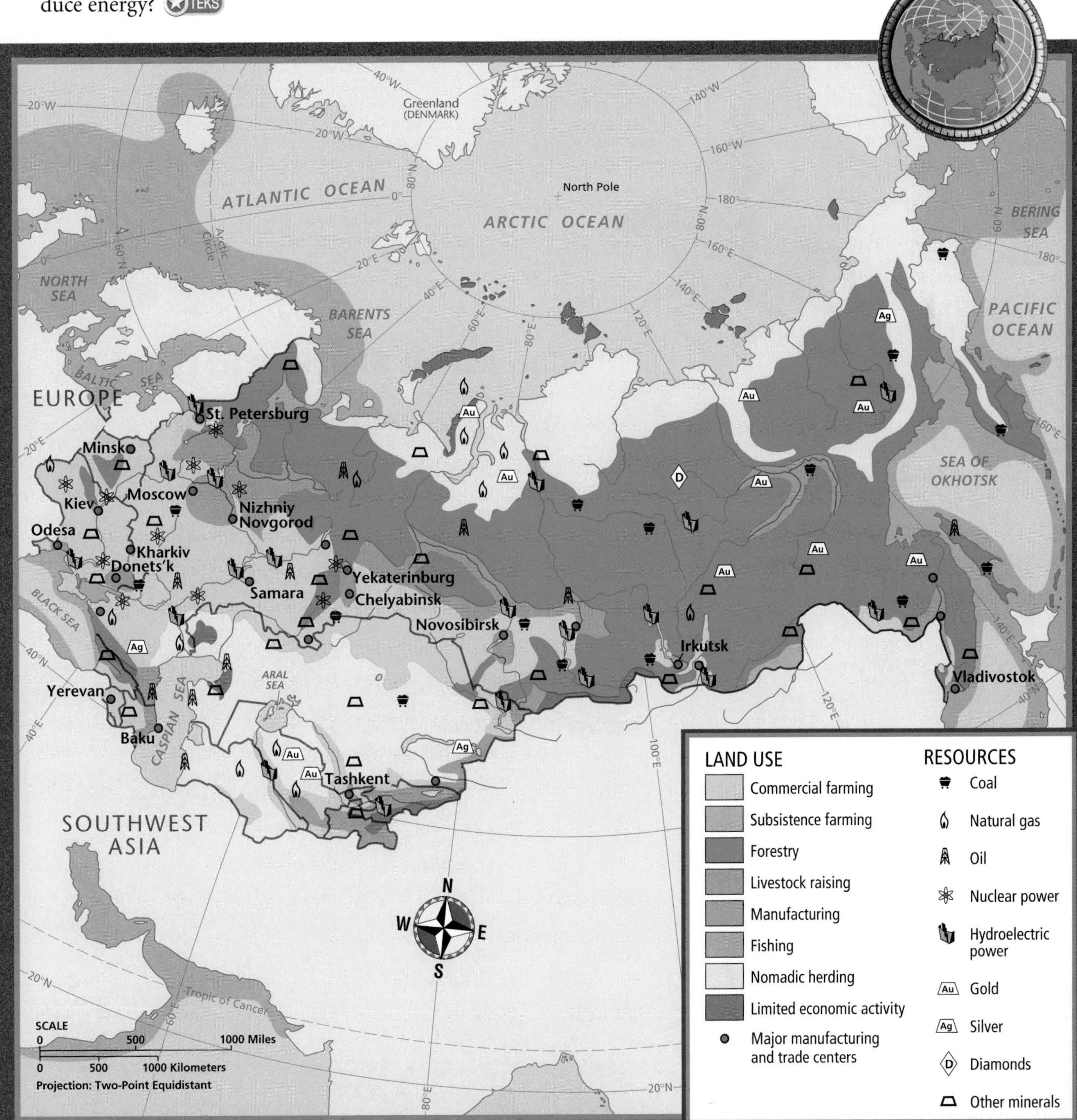

Time Line: Russia and Northern Eurasia

330 B.C. Alexander the Great invades Central Asia.

A.D. 600s Turkic-speaking peoples establish kingdoms in Central Asia.

700s Arabic speakers invade Central Asia, bringing Islam with them.

800s The city of Kiev becomes an important center for trade between the Mediterranean and Baltic Sea areas.

1218 Mongols, led by Genghis Khan, begin a long period of rule over Central Asia.

1547 Ivan IV crowns himself czar of all Russia.

1682–1725 Peter the Great takes over lands along the Baltic Sea and expands Russian control into what are now Belarus and Ukraine.

late 1700s Catherine the Great rules Russia.

1800s Russians spread into the Caucasus and Central Asia.

1917 The Bolsheviks overthrow the Russian government in what becomes known as the Russian Revolution.

1941 Germany invades the Soviet Union during World War II.

1991 The Soviet Union collapses.

330 B.C. | A.D. 600 | 700 | 800 | 1200 | 1300 | 1400 | 1500 | 1600 | 1700 | 1800 | 1900

The United States and Russia and Northern Eurasia

Comparing Sizes

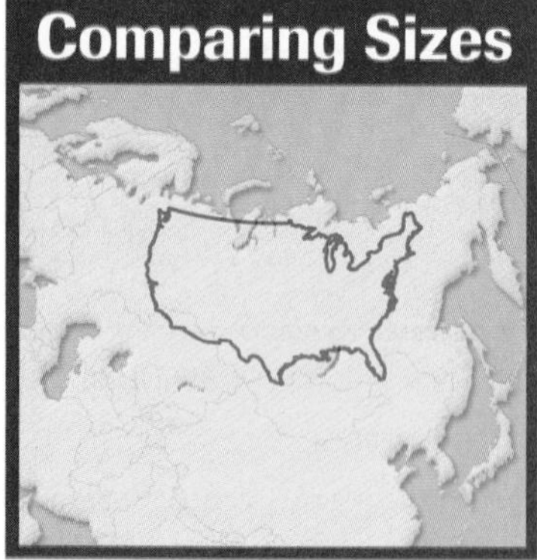

Comparing Standard of Living

COUNTRY	LIFE EXPECTANCY (in years)	INFANT MORTALITY (per 1,000 live births)	LITERACY RATE	DAILY CALORIC INTAKE (per person)
Armenia	62, male 71, female	41	99%	Not available
Azerbaijan	59, male 67, female	83	97%	Not available
Georgia	61, male 68, female	52	99%	Not available
Kazakhstan	58, male 69, female	59	98%	Not available
Kyrgyzstan	59, male 68, female	77	97%	Not available
Russia	62, male 73, female	20	98%	2,926
Tajikistan	61, male 67, female	116	98%	Not available
Turkmenistan	57, male 65, female	73	98%	Not available
Ukraine	60, male 72, female	21	98%	Not available
Uzbekistan	60, male 68, female	72	99%	Not available
United States	74, male 80, female	7	97%	3,603

Sources: Central Intelligence Agency, *2001 World Factbook; Britannica Book of the Year, 2000*

internet connect

GO TO: go.hrw.com
KEYWORD: SW3 Almanac
FOR: Additional information and reference sources

Fast Facts: Russia and Northern Eurasia

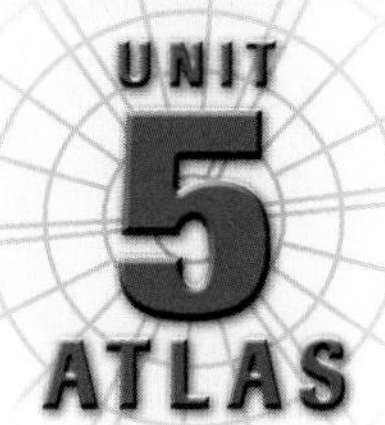

FLAG	COUNTRY Capital	POPULATION (in millions) POP. DENSITY	AREA	PER CAPITA GDP (in US $)	WORKFORCE STRUCTURE (largest categories)	ELECTRICITY CONSUMPTION (kilowatt hours per person)	TELEPHONE LINES (per person)
	Armenia Yerevan	3.3 290/sq. mi.	11,506 sq. mi. 29,800 sq km	$ 3,000	55% agriculture 25% services	1,859 kWh	0.17
	Azerbaijan Baku	7.8 232/sq. mi.	33,436 sq. mi. 86,600 sq km	$ 3,000	53% services 32% agric., forestry	1,986 kWh	0.09
	Belarus Minsk	10.4 129/sq. mi.	80,154 sq. mi. 207,600 sq km	$ 7,500	41% services 40% industry, const.	2,671 kWh	0.22
	Georgia Tbilisi	5.0 185/sq. mi.	26,911 sq. mi. 69,700 sq km	$ 4,600	40% services 40% agriculture	1,426 kWh	0.12
	Kazakhstan Astana	16.7 16/sq. mi.	1,049,150 sq. mi. 2,717,300 sq km	$ 5,000	50% services 27% industry	2,638 kWh	0.11
	Kyrgyzstan Bishkek	4.8 62/sq. mi.	76,641 sq. mi. 198,500 sq km	$ 2,700	55% agriculture 30% services	2,154 kWh	0.07
	Russia Moscow	145.5 22/sq. mi.	6,592,735 sq. mi. 17,075,200 sq km	$ 7,700	55% services 30% industry	5,006 kWh	0.21
	Tajikistan Dushanbe	6.6 119/sq. mi.	55,251 sq. mi. 143,100 sq km	$ 1,140	50% agriculture 30% services	2,239 kWh	0.06
	Turkmenistan Ashgabat	4.6 24/sq. mi.	188,455 sq. mi. 488,100 sq km	$ 4,300	44% agriculture 37% services	1,039 kWh	0.08
	Ukraine Kiev	48.8 209/sq. mi.	233,089 sq. mi. 603,700 sq km	$ 3,850	44% services 32% industry	3,008 kWh	0.19
	Uzbekistan Tashkent	25.2 146/sq. mi.	172,741 sq. mi. 447,400 sq km	$ 2,400	44% agriculture 36% services	1,727 kWh	0.08
	United States Washington, D.C.	281.4 76/sq. mi.	3,717,792 sq. mi. 9,629,091 sq km	$ 36,200	30% manage., prof. 29% tech., sales, admin.	12,260 kWh	0.69

Sources: Central Intelligence Agency, *2001 World Factbook; The World Almanac and Book of Facts, 2001; Britannica Book of the Year, 2000;* U.S. Census Bureau
The CIA calculates per capita GDP in terms of purchasing power parity. This formula equalizes the purchasing power of each country's currency.

Traditional wedding, Tashkent, Uzbekistan

CHAPTER 17

Russia, Ukraine, and Belarus

Russia, Ukraine, and Belarus formed the core of the old Russian and recent Soviet empires. In this chapter you will read how the region's location, resources, and leaders have made it the focus of major events.

Easter egg designed by Peter Carl Fabergé

Russian folk dancers

Privyet! (Hi!) My name is Polina, and I am 17. I live in an apartment in the center of Moscow with my parents and my basset hound, Marquise. Our apartment has two rooms. Every day except Sunday I wake up at 7:00 A.M., have some bread and cheese with tea, and take the subway to school.

I attend State School 637 and will graduate this spring. A few years ago we had to choose whether to study science or humanities. I chose humanities. My favorite subjects are history, literature, and English—my history teacher is great! During the day we have five or six classes with a 15-minute break between each one. During the breaks, I often eat a *pirozhki,* a small meat pie, at the school snack bar. At 2:00 P.M. I go home for lunch (meat, potatoes, and a salad of cooked vegetables and mayonnaise) and a nap. When I wake up, I go out with my friends to a park. On Sundays my friends and I cheer our favorite soccer team, Lokomotiv. This year, we took the train with other fans to matches in the Russian cities of St. Petersburg and Yaroslavl' and in Belarus.

In July I will take the entrance exams for university. I hope to win a place in law school, but if my exam results aren't good enough, I will go to night school. The best students get a free place in a state university just on the exam results, but other students have to pay tuition.

Section 1 Natural Environments

READ TO DISCOVER

1. What landforms and rivers are found in Russia, Ukraine, and Belarus?
2. What factors influence the region's climates and vegetation?
3. What natural resources does the region have?

WHY IT MATTERS

Russia has large coal deposits, but burning coal causes air pollution. Use CNNfyi.com or other **current events** sources to learn about industries that use coal and efforts to clean up the pollution they cause.

IDENTIFY

Eurasia

DEFINE

icebreakers
taiga

LOCATE

Baltic Sea
Black Sea
Ural Mountains

Locate, continued

Caucasus Mountains
Caspian Sea
Northern European Plain
Crimean Peninsula
Volga River
West Siberian Plain
Central Siberian Plateau
Kamchatka Peninsula
Lake Baikal
Murmansk
Sakhalin Island

Russia, Ukraine, and Belarus: Physical-Political

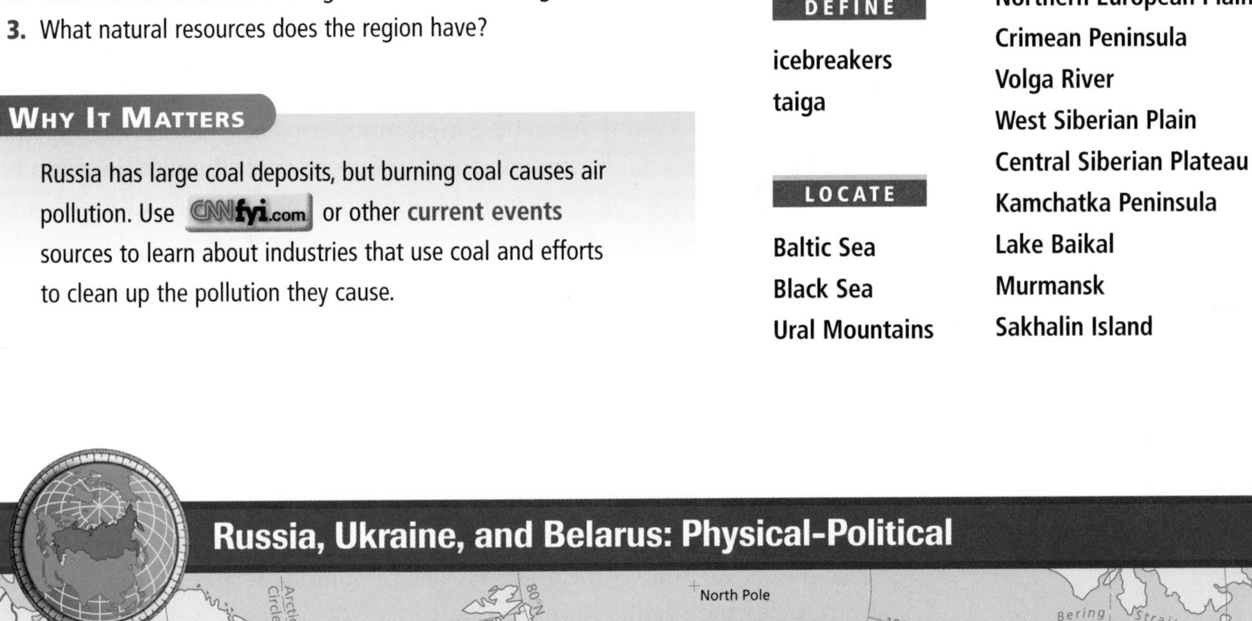

internet connect

GO TO: go.hrw.com
KEYWORD: SW3 CH17
FOR: Web sites about Russia, Ukraine, and Belarus

Landforms and Rivers

Together Russia, Ukraine, and Belarus cover about 12 percent of the world's land area. Russia alone extends more than 6,000 miles (9,600 km) from east to west. The huge country stretches across **Eurasia** from the Baltic Sea and Black Sea to the Pacific Ocean. *Eurasia* is the name given to Europe and Asia when they are considered one landmass or continent. Russia is the world's largest country in area. No other country shares borders with more countries. Much of northern Russia lies above the Arctic Circle.

The Ural Mountains divide the region. Areas west of the Urals—including Ukraine and Belarus—are part of Europe. Those to the east lie in Asia. The part of Russia that is east of the Urals is known as Siberia. The region's remaining three countries—Armenia, Azerbaijan, and Georgia—are in the Caucasus Mountains. These high mountains lie between the Black Sea and the Caspian Sea. The highest point in Europe is on Russia's southern border with Georgia in the Caucasus Mountains. There Mount Elbrus soars to 18,510 feet (5,642 m). An active tectonic zone, the Caucasus region suffers from severe earthquakes.

Ice-age glaciers and long-term erosion shaped the broad plains that are the region's major landforms. Much of the European area shares the Northern European Plain with countries farther west. Thus, the European areas have low elevations. In fact, Belarus has no point over 1,135 feet (346 m) above sea level. Southern Belarus and northwestern Ukraine contain the Pripet Marshes. These marshes make up the largest swamp in Europe. Ukraine's highest point is located where the Carpathian Mountains cross the country's western borders. Peaks on the Crimean Peninsula in southeastern Ukraine, a popular tourist area, are slightly lower.

Russia's Ural Mountains are more like high rolling hills. For this reason, road and rail crossings there need no major tunnels. West of the Urals, the gently rolling terrain of the Volga River basin dominates the heart of Russia. East of the Urals is the thinly populated West Siberian Plain. The Ob River

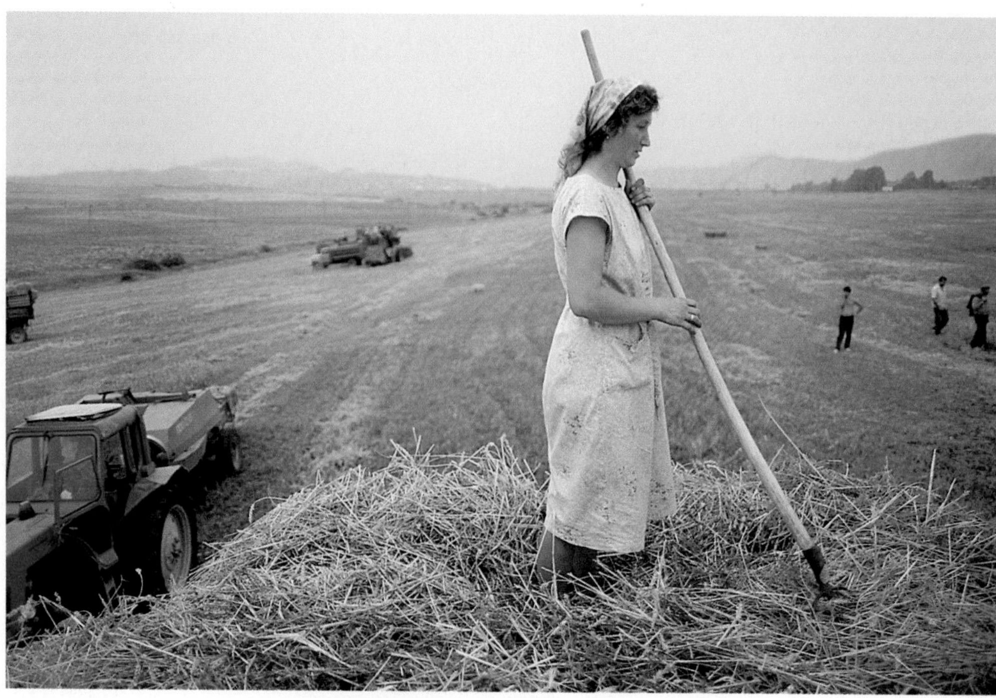

Plains like the Ukrainian farmland pictured here spread for vast distances across Eurasia.

creates a huge swamp area there. In the Russian Far East, beyond the Central Siberian Plateau, are high snowy ranges. Among these are the active volcanoes of the Kamchatka (kuhm-CHAHT-kuh) Peninsula.

The Dnieper, Don, and Volga are three of the largest south-flowing rivers in the region. These important shipping channels also supply water for hydroelectric projects and cities. The major Siberian rivers, such as the Ob, Yenisey, and Lena, flow northward to the Arctic Ocean from mountains in the south. One of the Yenisey's tributaries, the Angara River, flows through southern Siberia from Lake Baikal. Sometimes called the Jewel of Siberia for its beauty, Lake Baikal is the deepest lake in the world. It holds about one fifth of the world's freshwater!

Few creatures can live more than about 400 feet (120 m) below the Black Sea's surface. Too little oxygen and too much hydrogen sulfide create an environment that is poisonous to most life forms, including most bacteria.

TEKS **READING CHECK:** ***Physical Systems*** What factors shaped the region's main landform type?

Climates and Vegetation

Russians sometimes joke that winter lasts for 12 months and then summer begins. As you can see on a map, much of the region is in the same latitudes as northern Canada and Alaska. The weather can be harsh. However, the region offers a wealth of resources to those who can brave the elements.

Climates Much of the country lies in the humid continental, subarctic, and tundra climate regions. During the year's five coldest months, rivers and canals throughout the region freeze. In these cold climates a polluted icy fog often hangs over cities during winter. Created by fumes and smoke from cities, this fog is trapped over the cities by the cold air. In the region's northern areas permafrost is widespread and deep. When the permafrost's surface layer melts in summer, buildings tilt, highways buckle, and railroad tracks slip sideways.

Harsh conditions prevail in the area's eastern two thirds. Any ocean winds that might bring moisture and moderate temperatures cannot reach far inland. As a result, parts of the interior are very dry. Siberia's severe winters often bring temperatures below −40°F (−40°C). At one of the coldest places outside of Antarctica, Verkhoyansk in Siberia, the thermometer has reached −90°F (−68°C).

The region's European third has the mildest climates. In addition, the soils there are better for agriculture and human settlement. Moisture from the Atlantic Ocean far to the west brings winter snow and summer rain. In the Russian Far East, coastal areas receive rain-bearing winds from the Pacific Ocean.

The cold climate and small amount of warm coastline reduce Russia's access to the sea. The Arctic Ocean can freeze all

Although in much of Siberia snowfall is relatively light, the cold temperatures ensure that the snow stays on the ground for months. The village of Ust'-Anzas, in southern Siberia, lies under a blanket of snow. The sign on the building tells travelers that inside they can buy tickets on Aeroflot, Russia's national airline.

the way south to Russia's northern shores. Ship and barge traffic there requires using **icebreakers**. These are ships that can break up ice in frozen waterways. However, warm waters of the North Atlantic Drift reach around northern Norway to northwestern Russia. There you will find Murmansk, Russia's only large ice-free Arctic port.

Vegetation Differences in climate cause plant life to vary from north to south. Tundra vegetation grows along the northern coast. Low shrubs, mosses, and wildflowers are common there.

To the south is the **taiga**, a forest of mainly evergreen trees that covers half of Russia. Fir, larch, pine, and spruce are common. Farming is limited there because of the short growing season, acidic soils, and permafrost.

Farther south, in Belarus and in European Russia, you will find mixed deciduous-coniferous forest. This type of forest also grows along the coast of the Sea of Japan in the Russian Far East.

Still farther south is the drier grassland known as the steppe. Rich soil called *chernozem* (Russian for "black earth") has built up on the steppe. The grassland, long used for grazing, was plowed under by the 1800s. It has become one of the world's major grain-producing areas. In the past, people of the steppe fleeing invaders found safety in the taiga farther north. These landscapes are often featured in Russian literature.

TEKS **READING CHECK:** ***Places and Regions*** Which vegetation area allows grain production on a large scale and why?

INTERPRETING THE VISUAL RECORD

The taiga forms a wide band across northern Russia from its western borders to its Pacific coast far to the east. It provides a wealth of forest resources. ***How is taiga vegetation different from forests farther south?*** TEKS

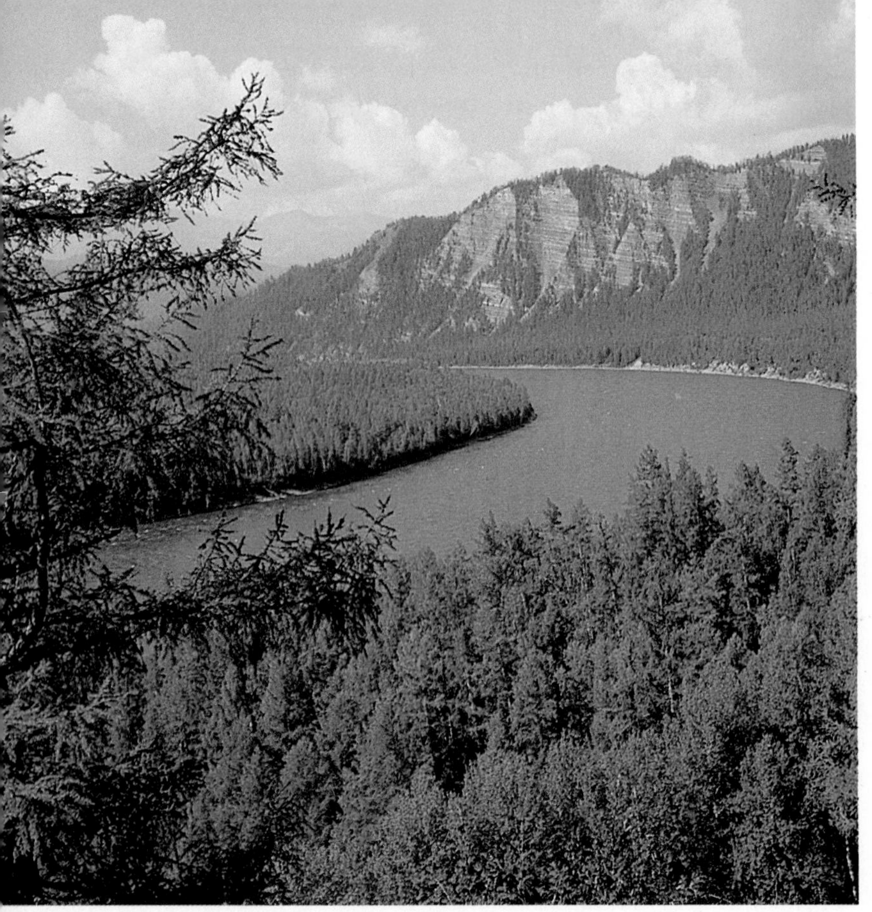

Natural Resources

Russia's forest, energy, and mineral resources are among the richest in the world. Yet much of this wealth was wasted because the government pushed production over conservation. Some of the remaining resources are in remote areas or are of low quality.

The taiga provides wood for building products and paper pulp. Steady logging west of the Ural Mountains has cleared many areas. However, in Siberia the taiga can provide forest resources for a long time to come. Eastern Siberia also has gold and diamond mines.

Coal, hydroelectricity, natural gas, and oil are the region's main energy resources. Huge oil reserves in the Caspian Sea area are being tapped by all the countries around the sea. Oil and gas fields between the Volga River and the Ural Mountains have been crucial to the region's development. They helped the Volga River basin become Russia's industrial heartland. Large reserves east of the Urals in the Ob River basin now supply most of Russia's oil and gas. The world's largest network of pipelines carries fuel from that area to Moscow, St. Petersburg, and for export to Europe. Sakhalin (sah-kah-LEEN) Island and the Kamchatka Peninsula also have energy resources. Russia's first geothermal power station is in Kamchatka. Geothermal water is put to other uses too, such as heating greenhouses and fish farms.

Russia and Ukraine have many large coal mines. Those coal reserves could last for centuries. The region is also rich in metals, such as copper, gold, iron ore, manganese, nickel, and platinum.

READING CHECK: *Environment and Society* Why is it hard for Russia to profit from some of its natural resources?

INTERPRETING THE VISUAL RECORD

Limited access to the ocean restricts the region's fishing industry. Here fishers net sturgeon for their eggs where the Volga River flows into the Caspian Sea. Served fresh as caviar, the sturgeon's eggs are an expensive delicacy. Sturgeon are now threatened with extinction. ***How might the Caspian Sea's location affect the price of caviar?*** TEKS

Section 1 Review

TEKS Working with Sketch Maps, Questions 1, 2, 3, 4, 5

Homework Practice Online

Keyword: SW3 HP17

Identify Eurasia

Define icebreakers, taiga

Working with Sketch Maps On a map of Russia, Ukraine, Belarus, and the Caucasus that you draw or that your teacher provides, label the Baltic Sea, Black Sea, Ural Mountains, Caucasus Mountains, Caspian Sea, Northern European Plain, Crimean Peninsula, Volga River, West Siberian Plain, Central Siberian Plateau, Kamchatka Peninsula, Lake Baikal, and Sakhalin Island. In the margin, identify Russia's only ice-free port.

Reading for the Main Idea

1. *Physical Systems* Why are earthquakes common in the Caucasus Mountains?
2. *Environment and Society* What are three of the European region's major south-flowing rivers? What are three functions that they serve?
3. *Places and Regions* Why is western Russia wetter than most of Siberia?

Critical Thinking

4. *Environment and Society* Why might developing Siberia's resources be difficult?

Organizing What You Know

5. Create a chart like the one shown below. Use it to describe the tundra, taiga, and steppe regions. Refer to this unit's climate and precipitation maps for more information.

	Climate	Soil conditions	Vegetation
Tundra			
Taiga			
Steppe			

Section 2 History and Culture

READ TO DISCOVER

1. What are some major events in the growth of the Russian Empire?
2. How did the Soviet Union develop, and what was life like for its citizens?
3. What are some features of the region's culture?

WHY IT MATTERS

Since the fall of the Soviet Union, tensions between ethnic Russians and other groups have grown. Use CNNfyi.com or other **current events** sources to learn about ethnic conflict in Russia.

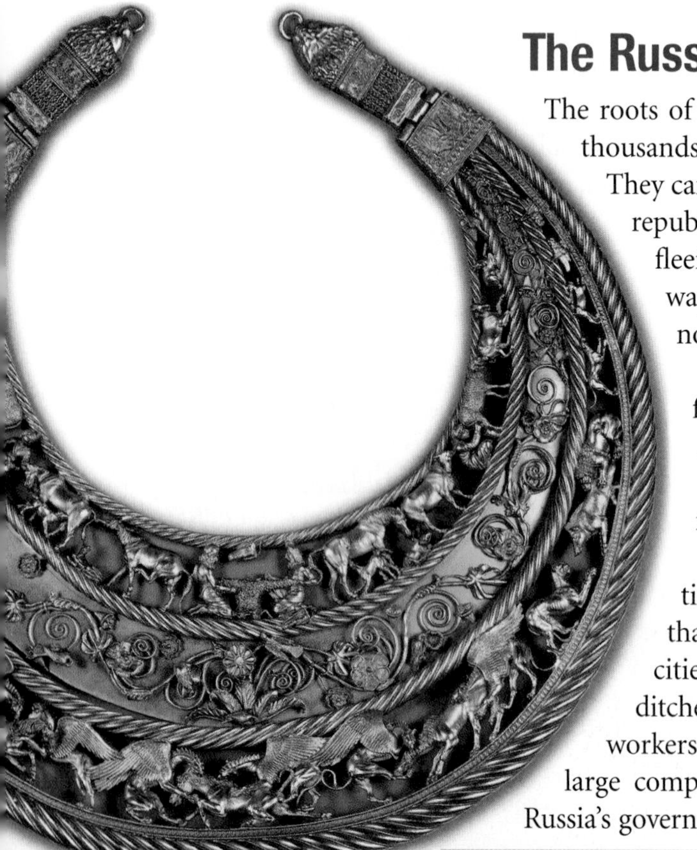

IDENTIFY

Slavs
Rus
Cossacks
Bolsheviks

DEFINE

czar
serfs
abdicate
soviets
autarky
gulag
shatter belt

LOCATE

Kiev
Moscow
Sea of Okhotsk
St. Petersburg
Amur River
Minsk

The Russian Empire

The roots of the Russian Empire lie in the grassy plains of the south. For thousands of years, people moved across the steppe, usually east to west. They came from what are now Mongolia, China, and the Central Asian republics. Bringing their herds with them, these peoples were often fleeing droughts and wars. Each wave of newcomers brought new ways of life to the region. The main people to settle in what are now Russia, Ukraine, and Belarus were the **Slavs**.

In the A.D. 800s the city of Kiev became an important center for trade between the Mediterranean and Baltic Sea areas. Among Kiev's early leaders were Scandinavian traders called **Rus** (ROOS). The name *Russia* comes from this word, which also referred to Slavic peoples in the region.

Merchants also traveled into the forests farther north. Over time these merchants founded new towns. Some of the towns that were located on high banks where rivers joined grew into cities. Moscow is an example. A prince ordered workers to dig ditches and build dirt walls on the site of an older settlement. The workers topped the dirt walls with a wooden wall. This fort became a large compound called the Kremlin. Its walls would eventually shelter Russia's government buildings, churches, and palaces.

INTERPRETING THE VISUAL RECORD *This intricately detailed gold necklace was made by the Scythians—one of the early peoples who moved across the Eurasian steppe. They flourished from the 700s to the 300s B.C. The Scythians were known for their skill in warfare and on horseback.* ***Why do you think metalworking and jewelry making were valued art forms for a people such as the Scythians?***

Over time Christianity increasingly influenced the region. By the 1100s Eastern, or Orthodox, Christianity had become the main religion of Kiev. In 1240, Mongol invaders from Central Asia destroyed Kiev. They made the region the western outpost of their growing empire. For the common people, though, life went on much as it had before.

In Russian, Ivan IV was called Grozny, which means "Awe Inspiring." He was indeed terrible at times, as he brutally suppressed the noble class—or boyars—and lashed out at enemies. Yet during Ivan's reign, the empire expanded, and printing was introduced to Russia.

Conquest and Expansion While the Mongols remained in power, several states emerged. The strongest was Muscovy, north of Kiev. Its chief city was Moscow. In the late 1400s Ivan III, the prince of Moscow, won control over parts of Russia from the Mongols.

In 1547 Ivan IV, who became known as Ivan the Terrible, crowned himself **czar** (ZAHR) of all Russia. The word *czar,* or *tsar,* comes from the Latin word *caesar* and means "emperor." Under Ivan IV the Russian Empire stretched from north of Kiev to the Arctic Ocean and east to the Urals.

Gradually Russian fur trappers, hunters, and pioneers migrated eastward into Siberia. By 1637, explorers reached the Pacific coast at the Sea of Okhotsk (uh-KAWTSK). **Cossacks**, a hardy people from the southern steppe frontiers, played an important role in the eastward expansion.

Russia gained some European territory under Czar Peter the Great, who ruled from 1682 to 1725. He took over lands along the Baltic Sea. He also expanded Russian control in what are now Belarus and Ukraine. Peter had St. Petersburg built for his capital. (See Cities & Settlements: St. Petersburg.) Catherine the Great ruled Russia during the late 1700s. She took the northern side of the Black Sea and encouraged settlers to move to the Volga region. This expansion brought many non-Russian peoples within the Russian Empire.

In the 1800s Russians spread into the Caucasus and Central Asia. Much of the population there was Muslim. For a brief time, Russia controlled what is now Alaska. There was even a Russian fort and farming settlement in California. By 1860 Russia had taken much of the Amur (ah-MOOR) River region that had been claimed by China. After Russia lost a war with Japan (1904–05), the country retreated to its current borders with China and North Korea.

End of an Empire Russia started to industrialize by the late 1800s, but it remained largely a country of poor peasant farmers. These farmers, called **serfs**, worked for a lord. Serfs were bound to the land, which means they could not leave the lord's land permanently without his permission. The serfs were freed in the 1860s, but rural poverty did not end. Soon life got worse for many Russians. Poor harvests led to food shortages. There was also an economic depression. By the start of World War I in 1914, the foundations of Russian society were on shaky ground. Russia suffered huge losses in the war, and social and economic problems worsened. Finally, the czar was forced to **abdicate**, or resign, in early 1917. A republic was set up but had little success. In the fall of 1917, a small group called the **Bolsheviks** overthrew the government, an event known as the Russian Revolution.The czar and his family were killed.

TEKS **READING CHECK:** ***The Uses of Geography*** What are some factors that led to the fall of the czar?

INTERPRETING THE VISUAL RECORD

Vladimir Lenin gives a speech during the Russian Revolution. Lenin believed in what he called "the dictatorship of the proletariat." Proletariat *means "working class."* ***How would Lenin's plan for government contrast with what you have already learned about dictators or dictatorships?*** TEKS

The Soviet Union

The Bolsheviks, led by Vladimir Lenin, wanted to remake Russia using the ideas of German philosopher Karl Marx. Marx thought that the people of the working classes were victims of capitalism. Like Marx, Lenin thought the solution was communism. Under communism, the workers were to elect governing local bodies called **soviets** to pass laws and make decisions. The Russian Empire was renamed the Union of Soviet Socialist Republics (USSR), also known as the Soviet Union. The Soviet Union eventually included 15 republics, each based largely on ethnic territories.

These workers at a collective farm share a meal. Collective farms, or kolkhozy *in Russian, were made up of many small holdings grouped into a single unit for joint operation under government supervision. Peasants were forced to join* kolkhozy.

Life in the Soviet Union The Soviet Union soon became a one-party, totalitarian state led at first by Lenin. After Lenin's death in 1924, Joseph Stalin took power. Stalin's brutal rule lasted until 1953. Both Lenin and Stalin tried to promote a single Soviet culture. They had names of cities and streets changed to honor communism's heroes. In addition, because it was the language of the political leadership, Russian spread to non-Russian ethnic groups.

Soviet economic planners set up a command economy. They also followed a policy of **autarky** (AW-tahr-kee). Under this system a country tries to produce all the goods that it needs. Trade with capitalist nations was very limited. Without competition, however, efficiency and product quality often fell. Production of consumer goods and services lagged far behind that of the United States and Western Europe.

The government ran large state farms, but agriculture faced constant problems. Food production was often low on the state farms. Millions of peasants died of starvation or in prison during the forced change to the new farming methods. Small private plots, which families worked in their spare time, produced about one fourth of the country's food.

Personal freedoms were strictly limited. People who disagreed with Communist leaders could be jailed. Under Stalin, millions were sent to terrible labor camps. Many of those camps were in the far north, both east and west of the Urals. This network of labor camps was called the **gulag**. Soviet leaders also tried to stop religious worship. They believed that religion would lessen people's loyalty to the state. Many Christian, Jewish, and Muslim houses of worship were closed or destroyed.

Yet the Soviet government did have some successes in education and health care. For example, by the 1980s some 90 percent of the people could read and write. Many people, including women, became doctors. In fact, basic health care was free and widely available. Most able workers had jobs.

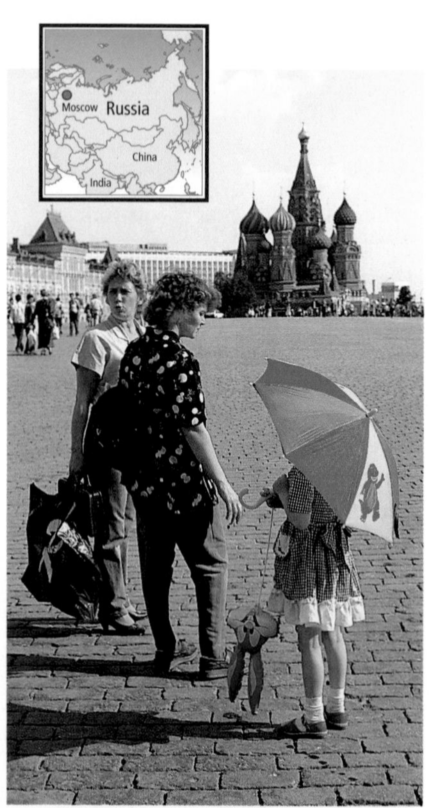

INTERPRETING THE VISUAL RECORD

A child and her mother walk across Red Square in Moscow. ***What clues in the photograph may indicate that regional trade patterns have changed since the Soviet era?*** TEKS

A New Beginning Finally, the government allowed some economic and political changes in the 1980s. However, the Soviet Union began to fall apart in 1990 and collapsed at the end of 1991. Each of the 15 former Soviet republics became independent. The new countries kept the same boundaries as the old republics even when they divided ethnic groups.

Life changed quickly for the people of the former Soviet Union. Today citizens can finally choose among candidates in elections. News from around the world now flows more freely. Religious freedoms have also expanded. In addition, communism is being replaced by capitalism. Shoppers can buy new consumer products. American fast food companies have opened restaurants there. In the new market economies, many businesses that had been owned by the government are in private hands. In many ways, however, the rapid change has caused severe hardships. You will read more about these problems in Section 3.

READING CHECK: *Human Systems* What are some ways that life has changed since the Soviet Union collapsed?

Among the peoples experiencing change are the Khanty of the Ob River basin. They are trying to protect their land from damage done by the oil industry.

Culture

Russia, Ukraine, and Belarus share a strong sense of cultural identity. There are many similarities in language, religion, and customs. However, there is great cultural diversity within Russia. In fact, Russia has at least 60 different ethnic groups.

People and Languages Language is an important source of national identity in the region. At least 85 percent of Russians are Slavs and speak Slavic languages. The region's Slavic languages are written in the Cyrillic alphabet, which was developed from an ancient Greek script. More than 95 percent of Ukrainians and about 98 percent of Belarusians are also Slavic. In fact, the great majority of Eurasia's more than 300 million Slavs live in these three countries.

As the Russian Empire grew, it pulled in many non-Slavic peoples. During the Soviet era, the lands where these non-Slavic peoples lived became special republics within the country. Russia has 21 of these republics today. Members of non-Slavic groups there speak different languages. Many books have been published in these languages. However, in most of the republics, Russian speakers are in the majority. In some of the republics, the non-Slavic languages are disappearing.

FOCUS ON HISTORY

The Peoples of the Caucasus Republics Some of Russia's ethnic republics are located in the Caucasus region in the south. Also in the Caucasus are the former Soviet republics of Armenia, Azerbaijan, and Georgia. Those three republics are independent countries today. The ethnic republics and countries in the Caucasus lie in a band of land that separates the Black and Caspian Seas. This area is made up of the high rugged Caucasus Mountains. The region's different ethnic groups developed within the hundreds of small isolated valleys in this mountain range.

The Caucasus is also what geographers call a **shatter belt**. It is a zone of frequent boundary changes and conflicts. Often shatter belts are located between major powers. Throughout history, peoples from the south—Turks, Persians, Arabs—and the north—Russians, Mongols, Tatars—have fought over the Caucasus. Ethnic tensions still trouble the region.

Republics of the Russian Federation and Ethnic Composition

Republic	Ethnic Composition
Adygea	Russian
Alania	Ossetian
Bashkortostan	Russian
Buryatia	Russian
Chechnya	Chechen
Chuvashia	Chuvash
Dagestan	Avar
Gorno-Altay	Russian
Ingushetia	Ingush
Kabardino-Balkaria	Kabard
Kalmykia	Kalmyk
Karachay-Cherkessia	Russian
Karelia	Russian
Khakassia	Russian
Komi	Russian
Mari El	Russian
Mordvinia	Russian
Sakha	Russian
Tatarstan	Tatar
Tyva	Tyva
Udmurtia	Russian

Source: Centre for Russian Studies

INTERPRETING THE CHART

The chart lists the ethnic group that makes up the majority in each of the republics of the Russian Federation. Note that the Russian Federation is the country's formal name.

Georgia reached a golden age during the reign (1184–1212) of Queen Tamara. When she rallied her troops before going into battle, the soldiers cheered their "king" Tamara. At the time, there was no word for *queen* in the Georgian language.

The region's physical geography and history have shaped its cultural geography. For example, many different languages are spoken in the Caucasus today. The three main languages are very different from each other. In Azerbaijan a Turkic dialect is most common. Except for some words borrowed from Persian, Armenian seems to be unrelated to any other living language. Georgian is one of few members of the South Caucasian language family. Some people believe the language might be related to northern Spain's Basque tongue, the origins of which are mysterious.

Religions here are as diverse as languages. The Armenian Christian Church is very old. A majority of Georgians belong to an Eastern Orthodox Church that is independent from the Russian Orthodox Church. Most Azerbaijanis and some people in Russia's southern republics, like Chechnya, are Muslim. Near the northeast edge of the Caucasus are Mongolian Kalmyks, whose faith is similar to Tibetan Buddhism.

READING CHECK: ***Environment and Society*** How have the physical geography and history of the Caucasus affected the region's cultural diversity?

Settlement Just 25 percent of Russia lies in Europe. However, 80 percent of its population lives there. Russians east of the Urals are concentrated in a southern corridor of transportation routes, warmer weather, and steppe environments.

Russia, Ukraine, and Belarus all have many large cities. More than two thirds of the population lives in cities. More than 9 million people live in and around Moscow, the region's most populous urban area. More than 5 million live in St. Petersburg to the northwest. Kiev, Ukraine's capital, and Minsk, the capital of Belarus, are also among Europe's larger cities.

All three countries are losing population. Many people have emigrated. Also, the death rate is higher than the birthrate. Poor health-related behaviors, including heavy smoking and alcohol abuse, are some key reasons for this trend. In addition, the collapse of the Soviet Union plunged many people into poverty. The old health care system also fell apart, cutting people off from medical care. No other part of the world has seen such population losses.

Population Pyramid, 2000

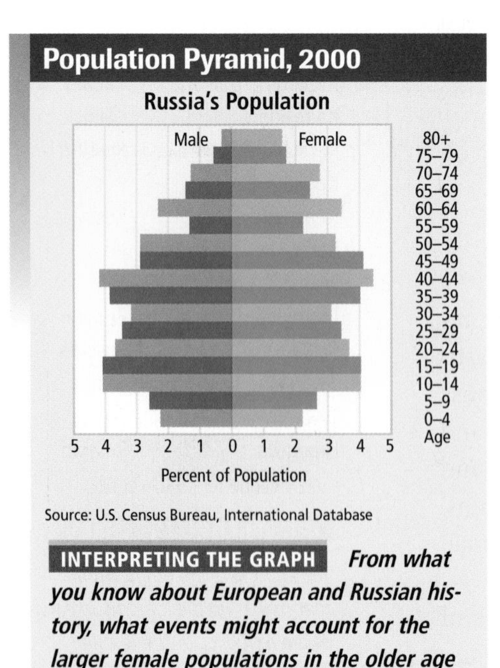

Source: U.S. Census Bureau, International Database

INTERPRETING THE GRAPH ***From what you know about European and Russian history, what events might account for the larger female populations in the older age ranges? How can you tell that Russia's population is declining?*** TEKS

Religion and Education Even after years of Communist rule, almost every city and village in the region has a prominent Christian church. The main religion is Eastern Orthodox Christianity. Church architecture often features an onion-shaped dome. Only the parts of Belarus and Ukraine bordering Roman Catholic Poland and Slovakia have many Catholic churches. Protestant churches are rare but increasing due to recent missionary activity. Also increasing is the number of Islamic mosques in the larger cities and in the Volga and Caucasus areas. Muslim minorities are common there.

Russia, Ukraine, and Belarus have inherited from the Soviet Union an emphasis on education and on scientific and technical training. They stress not just engineering but also the arts, humanities, and foreign languages. The best schools are in the national capitals of Kiev, Minsk, and Moscow.

Food, Traditions, and Customs Food reveals the influences of cold climates. Over much of the north, hardy grains grown are barley, oats, rye, and wheat. Dark rye bread and barley soup are common foods. Small buckwheat pancakes called blini are served with sour cream. Cold-weather vegetables like beets, cabbage, and potatoes go into borscht. Borscht is a traditional soup that sometimes has meat.

Belarusians, Russians, and Ukrainians often drink tea. This preference comes from centuries of ties to tea-growing areas nearby in the Caucasus and Central Asia. Fruit juices and spices are often added.

Environmental differences between the forested north and the open steppes of the south are seen in rural architecture. The people of the forested north used wood to fashion their cottages, churches, and other buildings. Elaborate wood carving decorates the front of homes and other buildings. On the steppe, where there is much less wood, people often built sod homes. Roofs and walls were blocks of grassy turf. Sometimes people dug the buildings partly into the ground. This helped keep buildings cooler in summer and warmer in winter. However, today most city people of all regions live in large apartment houses.

Outside of the big cities are country cottages, called dachas (DAH-chuhs). Members of the urban middle class spend weekends and holidays in these homes. Keeping a dacha was also a way to escape being spied upon during the Soviet era.

Dachas range from mere sheds to quaint cottages, like this one, and ornate palaces. They offer relaxation, relief from city pollution, and a place to raise vegetables. Much of Russia's food is grown on dacha land.

READING CHECK: *Human Systems* What are some traditions that have survived changes in government?

TEKS Working with Sketch Maps, Questions 1, 2, 3, 4, 5

Section 2 Review

Identify
Slavs, Rus, Cossacks, Bolsheviks

Define czar, serfs, abdicate, soviets, autarky, gulag, shatter belt

Working with Sketch Maps On the map you created in Section 1, label Russia, Ukraine, Belarus, Kiev, Moscow, Sea of Okhotsk, St. Petersburg, Amur River, and Minsk. In the margin of your map, name the two U.S. states where Russia held land during the 1800s.

Reading for the Main Idea

1. *The World in Spatial Terms* What lands were added to the Russian Empire under Ivan IV, Peter the Great, and Catherine the Great?
2. *Human Systems* What subjects do the region's educational systems emphasize? What other subjects do students learn?

Critical Thinking

3. **Analyzing** Why does Russia have many non-Slavic peoples? What factors account for the cultural diversity of the Caucasus?
4. **Making Generalizations and Predictions** Using data and graphics from the chapter and the unit population map, describe the population characteristics of modern Russia. Do you think Russia's population will grow or decline in the near future? Why?

Organizing What You Know

5. Create a graphic organizer like the one shown below. Use it to provide information about the major religions of the region.

Eastern Orthodoxy	Roman Catholicism	Protestantism	Islam

The Uses of Geography

Geography for Life

Mapping Napoléon's Russian Disaster

Maps, diagrams, and graphs are widely used to display geographic information. The geographer must choose the best way to represent the relevant information. No single map can show everything. However, certain kinds of maps can show a remarkable amount.

Consider the approach that a French engineer took in creating a famous historical map. In 1861 Charles Joseph Minard illustrated French emperor Napoléon's 1812 invasion of Russia. On the eve of the invasion, Napoléon dominated much of Europe. However, his campaign in Russia was a disaster.

Minard's design tells a sad tale of death and misery. He used a shaded band to illustrate the changing size of Napoléon's army. Look at the left edge of the map, which shows the Polish-Russian border near the Neman River. Minard drew a thick band representing the 422,000 soldiers who swept into Russia across the river. He narrowed the band's width to show how battle losses gradually shrank the army's size as it marched eastward. When he reached Moscow, Napoléon led just 100,000 troops. After the people of Moscow burned the city, the army turned around in October to return to France. Minard drew a darker, lower band to illustrate the retreat. A temperature scale across the bottom of the map is tied to the darker band. As you can see, bitter cold weather took a terrible toll on the army. Many soldiers froze or starved to death. After crossing the Berezina River, the army straggled back to Poland. Only about 10,000 soldiers survived the complete journey.

This illustration shows the army's size, its location on certain dates, its route, and temperatures the soldiers faced. Displaying a series of events that occurred over a vast space and several months is not easy. We might call Minard a storyteller as well as an engineer and cartographer.

Applying What You Know

1. **Summarizing** What features and information did Minard show on his map?
2. **Drawing Inferences and Conclusions** What do you think is shown by the extra arms that branch off of the main bands?

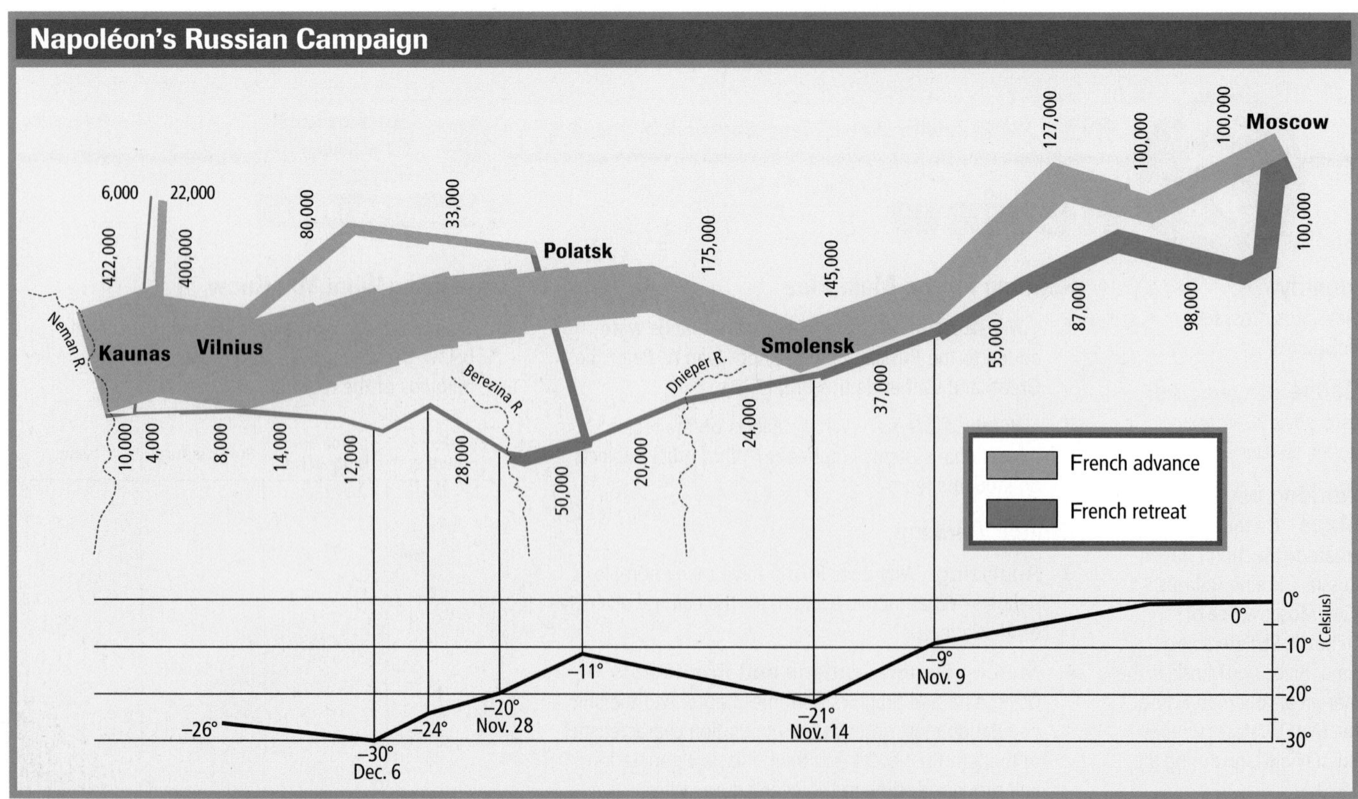

The Region Today

READ TO DISCOVER

1. How have the economies of areas within the region developed?
2. What challenges does the region face?

DEFINE

light industry
heavy industry
smelters

WHY IT MATTERS

Any major economic and political changes in the world's largest country are of worldwide interest. Use CNNfyi.com or other **current events** sources to learn about the latest developments in Russia.

LOCATE

Vladivostok
Khabarovsk
Kuril Islands
Chernobyl

Economic Development

Belarus, Russia, and Ukraine are changing their economies to compete in new markets. The countries are working to develop **light industry**. Light industry focuses on the production of consumer goods, such as clothing or housewares. **Heavy industry**, which usually involves manufacturing based on metals, is becoming less important. Cities are becoming more like those in richer countries. New shopping centers, stores, and sidewalk stalls are opening. Paint and better maintenance brighten old apartment houses. Single-family houses, even some luxury homes, are being built.

The Moscow Region Moscow, with its huge Kremlin, has symbolized Russia for centuries. The city became the home of the Russian Orthodox Church in the 1300s and Russia's capital in the 1400s. Most Russians have looked to Moscow as their country's heart and soul. This was true even while St. Petersburg was the capital from 1712 to 1918.

Today greater Moscow is Russia's most important economic region. It is the national center of communications, culture, education, finance, politics, and transportation. More than 70 institutions of higher learning are there. As a result, Moscow's economic advantages are many. Roads, rails, and air routes link the capital to all points in Russia. The city's location also gives its businesses access to raw materials and labor.

The economic region around Moscow stretches for many miles in all directions. Millions of Russians live and work within the area's network of transportation routes and job sites. Among the transportation links is the world's busiest subway. The area also has electrified railroads and a major beltway.

INTERPRETING THE VISUAL RECORD

Two women relax near their salon at GUM, a Moscow shopping mall of more than 150 stores that receives some 300,000 visitors per day. GUM stands for Gosudarstvenny Universalny Magazine, *or "State Department Store."* ***How does GUM compare to your community's shopping centers?*** TEKS

Ballet dancers perform Pyotr Tchaikovsky's Swan Lake *at the Mariinsky Theater in St. Petersburg. Tchaikovsky was one of Russia's many great composers.*

The St. Petersburg Region Moscow reflects Russia's old values and traditions. In contrast, St. Petersburg represents the country's desire for Western ideas and practices. Located on the Gulf of Finland, it has been called the Venice of the North for its many canals. St. Petersburg has good transportation facilities. The city's location also eases trade and transportation links with other European cities. Major products include chemicals, machinery, ships, and textiles. Many cultural attractions and universities draw tourists and high-tech industries.

The Volga and Urals Regions Heavy industry lines Russia's Volga River and the Ural Mountains. Hydroelectricity is abundant there. Dams that produce power have also turned the Volga into a chain of lakes. Refineries and petrochemical plants process oil and gas. Russia's largest car and truck factories are in the area.

Nearly every important mineral except coal and oil has been discovered in the Urals. These resources laid the base for industrial development. Copper and iron **smelters**, factories that process metal ores, are still important.

Siberia For centuries, Russians saw Siberia as a frontier treasure chest of furs, gold, and lumber. However, opening this cold harsh region has been difficult. Now Siberian settlement, farming, and industry mostly follow the Trans-Siberian Railroad. The building of the railway started in 1891. It eventually connected Moscow to Vladivostok (vla-duh-vuh-STAHK) on the Sea of Japan. At about 5,800 miles (9,330 km), it is the longest single rail line in the world. Workers completed a more direct railway, called the Baikal-Amur Mainline (BAM), across eastern Siberia in 1989. Permafrost and other difficult conditions made building these lines a great feat.

Lumbering, mining, and oil production are Siberia's most important industries. Because wages are higher in Siberia, some Russians move there to work. Still, large areas of Siberia have few people or none at all.

INTERPRETING THE VISUAL RECORD

Workers lay a pipeline that will transport natural gas westward from Siberia. ***How do you think these workers adapt to Siberia's environment in order to do their jobs?*** TEKS

The Russian Far East Russia has a long coastline on the Pacific Ocean. There, in the Russian Far East, much land remains heavily forested. Summer weather is mild enough for farming in the Amur River valley. Khabarovsk (kuh-BAHR-uhfsk), the main inland city, has factories that process forest and mineral resources. Vladivostok is a naval base and the chief seaport and fishing center.

Sakhalin Island, with its oil and mineral resources, lies off the eastern coast of Siberia in the Sea of Okhotsk. The Kuril (KYOOHR-eel) Islands, which are important for commercial fishing, are farther east. Russia took the islands from Japan at the end of World War II. Japan claims that four of them should be returned. If the two countries settle the issue, Japan may invest more in the Russian Far East and Siberia.

Ukraine and Kiev Kiev is Ukraine's capital. Sheltered by high bluffs in the Dnieper River valley, it is an attractive city. About 10 percent of Ukraine's population lives there. The city also has a large share of the country's economic activity. Like Moscow, Kiev is centrally located in a region rich with agricultural, energy, industrial, and human resources. Kiev's winning soccer team, Dynamo, is an important symbol for the city.

Wheat, sunflowers (for cooking oil), and sugar beets are common crops in Ukraine. The country exports a wide variety of fruits, vegetables, and animal products. Ukraine's heavy industry is based on coal, iron, manganese, and other metals. These resources led to concentration of metalworking in the Donets Basin and along the Dnieper River. Ukraine's moderate climate, access to expanding markets, and resources may help it attract new investment over time.

INTERPRETING THE VISUAL RECORD *Kiev is one of the oldest cities in Europe.* ***What characteristic of the region's housing patterns is visible in the photo?***

Connecting to ECONOMICS

Trading on the Russian stock market

The Russian Stock Market

Stock markets allow businesses to grow by using other people's money. In turn, when investors buy shares in businesses they get the chance to make a profit. The New York Stock Exchange began operating in 1792. In contrast, Russia's stock market organized in 1994. For the first time, Russian citizens could buy shares in businesses that had previously been run by the government. More than 70 percent of the Russian economy was in private hands by 1995. By 1997 investment in Russian stocks by both banks and individuals was booming. However, a year later overestimation of businesses' worth, scandals, and swindles caused stock prices to fall. Many investors' profits were wiped out. Buying stock on the Russian exchange is still risky. Most stocks are cheap, but buyers can easily lose their money. On the other hand, those willing to do their homework and take risks can reap big rewards.

Making Generalizations and Predictions How would building a stable stock market contribute to Russia's efforts to build a strong market economy? TEKS

Belarus and Minsk Belarus has few mineral resources and generally poor soil. As a result, the country has relied on its educated labor force to build its economy. The remaining forests support wood products industries. Peat is still used as a fuel, even though burning it causes air pollution. Minsk, the capital, has many of the country's industries. Its outdated motor vehicle and consumer-goods plants are left over from the Soviet era.

READING CHECK: *Places and Regions* What economic advantages do some of these areas have?

Issues and Challenges

Belarus, Russia, and Ukraine face serious challenges as they move from command to market economies and democracy. Holding free elections was an early and fairly easy step. Much harder is creating the social and economic structures that support peace and prosperity.

Political and Economic Challenges Tension between supporters and opponents of reform and among ethnic groups has grown. Unemployment and crime have increased. The gap between rich and poor is also growing. Public health care has declined. Many older, unemployed, and ill people find that the safety net the old Soviet government provided is gone. Still, Russians have experienced relatively peaceful changes in government after free elections.

Placing business in private hands has had mixed results. A few people have become rich, but some did so through unfair means. Many of the newly rich do not pay their taxes. Some have turned to crime to protect their wealth and power. In addition, members of the new middle class do not feel secure. Many of them fear that the government may again take over homes and businesses.

Many economists argue that several features of the region's economies need reform. For example, factories and transportation systems need to be repaired and modernized. Corrupt officials and managers should be replaced. Also, more businesses must switch to making better goods that people around the world really want to buy. Rules that limit movement of people, money, and goods should be changed. Courts that should be able to force payment of debts, but cannot, need to be strengthened.

INTERPRETING THE VISUAL RECORD

Homeless poor people have created this tent city near Red Square in Moscow. ***How might this woman's views on Russia's market economy compare to the opinions of a trader on Russia's stock market?*** TEKS

Geographical Challenges The Soviet Union was committed to developing local economies in remote places. This policy is less important today. People are moving from their homes in Siberia and other distant areas back to the European heartland. Some observers fear that whole industrial and mining districts will be emptied.

The Soviet history of environmental pollution created another serious challenge. In its rush to make the country an economic power, the Soviet

Smelters in the Murmansk area have contributed to high pollution levels. According to reports, acid rain has killed all forests within a 12 mile (20 km) radius of Monchegorsk, the town pictured.

government paid little attention to environmental issues. As a result, huge areas are ruined by pollution. Today the region's governments have little money to repair damage or require environmental safeguards. Therefore, these problems will remain for some time.

Perhaps the worst example of environmental damage in the former Soviet Union is in Ukraine. In 1986 a disastrous accident happened at the nuclear power plant at Chernobyl, north of Kiev. The Soviet government tried to cover up the story but failed. Radiation from explosions and fires contaminated millions of acres of forest and farmland. It spread as far away as Sweden and France. People cannot return to the immediate area for many years to come.

Finding solutions to these environmental problems and other challenges will be difficult. However, they are not impossible to overcome. The future of the countries that once belonged to the Soviet Union is not necessarily a prisoner to the past.

READING CHECK: *Human Systems* What political, economic, social, and environmental challenges do people in the region face today?

TEKS Working with Sketch Maps, Questions 1, 2, 3, 4, 5

Section 3 Review

Homework Practice Online

Keyword: SW3 HP17

Define light industry, heavy industry, smelters

Working with Sketch Maps

On the map you created in Section 2, label Vladivostok, Khabarovsk, Kuril Islands, Donets Basin, and Chernobyl. Circle the area at the center of a dispute with Japan.

Reading for the Main Idea

1. *Human Systems* What basic change in emphasis is occurring in Russian industry?
2. *Human Systems* Why might Ukraine attract new investment?

Critical Thinking

3. **Identifying Cause and Effect** How did Soviet economic policies affect the region's environment? What are some examples?
4. **Drawing Inferences and Conclusions** Why might Russia's unstable political system delay economic progress?

Organizing What You Know

5. Copy the graphic organizer below. Use it to identify factors that could fuel each subregion's economic growth. One is started for you.

Moscow Region	
St. Petersburg Region	good transportation
Volga and Urals Region	
Siberia	
Russian Far East	
Ukraine and Kiev	
Belarus and Minsk	

St. Petersburg

Places and Regions It has been called Russia's Window on the West and the Venice of the North. It has appeared on maps as St. Petersburg, Petrograd, and Leningrad. However, by any name, St. Petersburg is a beautiful and important city. In fact, in the 1990s the city was officially recognized as the cultural capital of Russia. It was also the political capital for a long period of Russia's history. Now St. Petersburg is regaining its reputation as one of the world's great historical cities.

More than 200 pounds of gold cover the dome of St. Isaac's Cathedral. Like many other St. Petersburg monuments, it is now a museum.

A City Born of War

Russia was an isolated and poorly developed country for much of its early history. In the late 1600s the Russian czar Peter I wanted a seaport through which trade and the latest European ideas could enter Russia. However, Sweden controlled the Baltic Sea to the west, while the Turks controlled the Black Sea to the south. In 1703 the Russians drove the Swedes from the Baltic's eastern shore. There, where the Neva River empties into the sea, the czar—Peter the Great—founded a new city. He modeled his city, which carries his name, after London and Amsterdam. He hired French and Italian architects to design it. Peter himself laid the foundation stones for the city's fortress on May 27 of 1703. This is the city's official founding date.

St. Petersburg's success was ensured in 1712 when it replaced Moscow as Russia's capital. Peter ordered the country's nobles to move to the new city. Many built grand homes there. Over the next 200 years St. Petersburg developed into Russia's chief port and industrial center. It also became a center for art, literature, and music. The culture that developed there was both European and Russian.

In 1914 the city's name was changed to Petrograd—the Russian form of *St. Petersburg*. In 1917 the Russian Revolution broke out, and Petrograd was a center of revolutionary activity. The new Communist rulers then moved the capital back to Moscow. In 1924, after the Soviet leader Lenin died, the city was renamed Leningrad. It was not called St. Petersburg again until 1991.

World War II caused heavy damage in Leningrad. German troops surrounded the city for 872 days and shelled it constantly. However, Leningrad never surrendered and the siege was finally broken. When the war ended, people began restoring the city's old buildings to their original splendor. This expensive and painstaking work continues today, more than half a century later.

Environmental Challenges

Besides the destruction of war, St. Petersburg has had to deal with a difficult physical environment. The city is built on more than 40 islands. These islands are created by the Neva River delta and by smaller rivers that flow into the Neva near its mouth. Because of the many water channels that course through St. Petersburg, special construction methods have been required to keep buildings from sinking. St. Isaac's Cathedral, for example, rests on 10,000 upright tree trunks driven into the ground in the early 1800s.

St. Petersburg's location exposes it to threats from the sea. In fall and early winter, storms and strong winds move across the Baltic Sea from the west. These storms drive seawater upstream at the mouth of the

One of the 1,057 rooms and halls of the Hermitage

The Winter Palace's grand facade

Interior of the palace's winter garden as it appeared in 1840

The Winter Palace was built during the reign of Peter's daughter Elizabeth I in the mid-1700s. It was home to the country's rulers until the Russian Revolution in 1917. Today the palace is part of the State Hermitage Museum, which houses the art amassed by Russia's rulers.

Neva River. Since the downtown area is just a few feet above sea level, flooding often occurs. In fact, St. Petersburg has suffered more than 270 major floods during its 300-year history.

St. Petersburg, which is home to some 5 million people, is located quite far north for a major city. Winters are long and cold, with daytime highs averaging about 23°F (–5°C). Because the city is so far north, winter daylight hours are short—about six hours each day. For about three weeks in June and July, the sky does not get completely dark. From about 11:00 P.M. until 3:00 A.M. the evening twilight merges into dawn. These are St. Petersburg's famous "white nights." Many celebrations and cultural events take place during this time of the year. For example, the Stars of the White Nights Festival brings music lovers from all over the world to St. Petersburg.

Russia's Cultural Capital

Compared to other places in Russia, St. Petersburg fared well during the more than 70 years of Communist rule. The government poured large sums of money into arts and culture. Under government sponsorship, the city's Kirov Ballet became one of the world's great dance companies. Government funds were also used to rebuild historic palaces and to maintain the city's many spectacular museums. Despite the Communists' limits on free expression, a secret community of artists and musicians developed and thrived.

Today music and art are alive all over St. Petersburg. The city's best-known attraction, the State Hermitage Museum, is home to one of the world's greatest art collections. Moscow may be Russia's political capital. However, the people of St. Petersburg are determined that their city will be the capital of Russian history and culture.

Applying What You Know

1. **Summarizing** How does St. Petersburg differ from most other major world cities?
2. **Analyzing Information** How has St. Petersburg's history as a cultural and political capital influenced its economy?

Building Vocabulary

On a separate sheet of paper, explain the following terms by using them correctly in sentences.

Eurasia	czar	Bolsheviks	gulag
icebreakers	serfs	soviets	shatter belt
taiga	abdicate	autarky	light industry
Slavs			

Locating Key Places

On a separate sheet of paper, match the letters on the map with their correct labels.

Ural Mountains	Kiev
Caucasus Mountains	Moscow
Crimean Peninsula	St. Petersburg
Volga River	Vladivostok
Central Siberian Plateau	

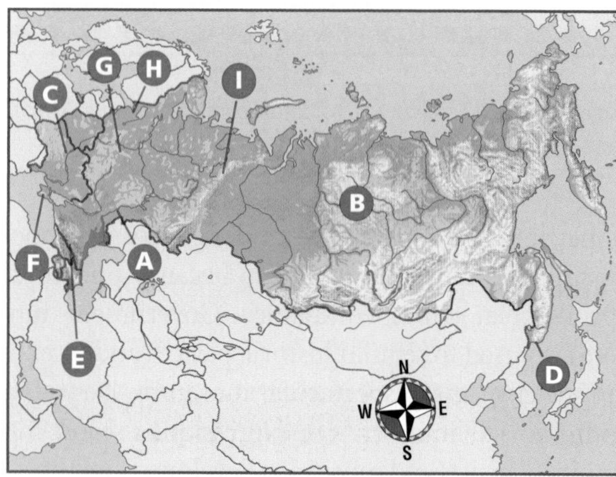

Understanding the Main Ideas

Section 1

1. *Physical Systems* What are the main physical characteristics of the two huge areas west and east of the Ural Mountains?
2. *Physical Systems* What are some resources that Russia has in large quantities?

Section 2

3. *The Uses of Geography* Across what physical region did early migrants come to Russia and its neighbors? How did those newcomers shape the region's culture?
4. *Human Systems* Where do most of the region's people live?

Section 3

5. *Environment and Society* What environmental problems remain from the Soviet era?

Thinking Critically for TAKS

1. **Analyzing Information** Review the unit population map. What geographical, political, and economic factors might explain the population distribution in Siberia and the Russian Far East?
2. **Drawing Inferences and Conclusions** Why do you think crime is a big problem for Russia today, even though it was not during the Soviet era?
3. **Contrasting** How does Russia's market economy differ from the command economy of the Soviet Union?

Using the Geographer's Tools

1. **Analyzing Graphs** Study the population pyramids for Russia. Have the greatest changes been for men or women? Can you predict a trend in age distribution for the future? Why or why not?
2. **Creating Tables** Design a table that identifies important political, economic, social, and cultural characteristics of the major regions discussed in Section 3.
3. **Preparing Maps** Draw an outline map of Russia, Ukraine, and Belarus. Use three different colors to mark the extent of the tundra, taiga, and steppe regions. Create a symbol for a major product of each climate region and place it correctly.

Writing for TAKS

As you have read, most people in the region live in large apartment houses. Make a list of questions you would ask people there about life in their city. How do you suppose life in such a city might be different from life in your community? What factors might account for these differences? Write a paragraph about your answers to these questions. When you are finished with your paragraph, proofread it to make sure you have used standard grammar, spelling, sentence structure, and punctuation.

SKILL BUILDING

Geography for Life

Evaluating Primary and Secondary Sources

Human Systems Ethnic conflicts have caused problems in the Caucasus region. Use the Internet and other resources to find articles from newspapers, journals, and other sources about those issues and events in the region. Compare the articles for background information provided, amount of detail, and bias. Create a chart to show your findings.

Building Social Studies Skills

Interpreting Graphs

Study the pie graph below. Then use the information from the graph to help you answer the questions that follow.

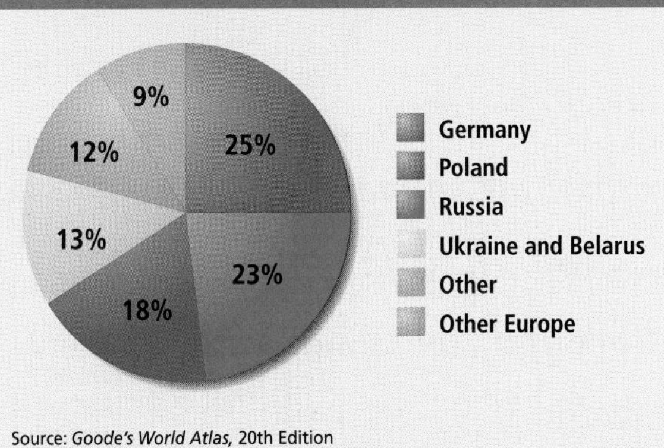

1. Which two countries together accounted for nearly half the world's rye production from 1994 through 1996?
   a. Germany and Belarus
   b. Russia and Belarus
   c. Poland and Russia
   d. Poland and Germany
2. In which hemisphere and on what continent is most of the world's rye produced?

Using Language

The following passage contains mistakes in grammar, punctuation, or usage. Read the passage and then answer the following questions.

"[1] Agriculture is important to Ukraines economy. [2] The country is the world's largest producer of sugar beets. [3] Its food-processing industry makes sugar, from the sugar beets. [4] Farmers also grow fruits, potatoes, vegetables, and wheat. [5] Grain is maked into flour for baked goods and pasta."

3. Which sentence contains an error in the punctuation of a possessive form of a noun?
   a. 1
   b. 2
   c. 3
   d. All are correct.
4. Which sentence contains an error in the use of commas?
   a. 3
   b. 4
   c. 5
   d. All are correct.
5. Rewrite sentence 5 to correct an error in verb tense.

Alternative Assessment

PORTFOLIO ACTIVITY

Learning about Your Local Geography TEKS

Group Project: Research

To build the railways across Siberia, Russian engineers had to deal with vast distances and harsh climate conditions. Highway and railway construction was also difficult in some parts of the United States. Plan, organize, and complete a group research project about how the physical geography of your state and local area affected highway or rail construction. Which bodies of water had to be crossed? Did mountains or deserts present problems? How did climate affect the workers? Prepare a short illustrated report about the answers you find to these geographic questions.

internet connect

Internet Activity: go.hrw.com
KEYWORD: SW3 GT17 TEKS

Choose a topic about Russia, Ukraine, or Belarus to:

- research the rise and fall of the Soviet Union.
- create a poster of a journey through the Caucasus.
- learn the causes and results of the Chernobyl disaster.

CHAPTER 18

Central Asia

Until recently Kazakhstan, Kyrgyzstan, Tajikistan, Turkmenistan, and Uzbekistan were unfamiliar to many people in other countries. Now these lands are being discovered by investors and tourists and are changing rapidly after decades of Soviet rule.

Silver and gold wolf's head from a battle flag, Kazakhstan

Man riding a donkey, Uzbekistan

Salam! (Hi!) My name is Leila. I live in Turkmenistan. Here we go to school six days a week and study 18 different subjects. There is no choice of courses. We just switched from writing with the Cyrillic alphabet to writing with the Latin alphabet.

After school we go home and have lunch. Then my sister and I divide the chores. My brothers have outdoor chores, like tending our fruit and vegetable garden or taking the sheep out to graze along the canal. The sheep are kept in a shed in back. Many of our neighbors keep cows. We grow cabbage, spinach, carrots, beets, turnips, parsley, dill, pomegranates, and grapes.

It takes a long time to cook dinner because everything is made from scratch. The oven is outside in the courtyard. For fuel we use the woody stems of cotton plants. The smoke has a wonderful smell. To keep the mosquitoes away, we scatter the seeds of desert plants on a hot dish to create smoke. We wash our hands before dinner and sit on the floor around low tables to eat with our hands, three or four people to one dish.

Every year in October, each student from seventh grade through university goes to the country to pick cotton, because hand-picked cotton is more valuable. We work for eight hours every day and have to pick at least 50 kilograms (110 lbs) a day. We get paid about 10 cents per kilo. This is the only paid job students can have.

Natural Environments

READ TO DISCOVER

1. What are the major landforms and rivers of Central Asia?
2. What climates, biomes, and natural resources does the region have?

DEFINE

zinc

LOCATE

Altay Shan
Tian Shan
Pamirs
Kopet-Dag
Caspian Sea
Aral Sea
Amu Dar'ya
Syr Dar'ya
Irtysh River
Lake Balkhash
Issyk-Kul
Kara-Kum
Kyzyl Kum

WHY IT MATTERS

Some of the largest oil reserves in the world are in the Caspian Sea region of Central Asia. Use CNNfyi.com or other **current events** sources to learn how the oil industry affects governments, economies, and people there.

Central Asia: Physical-Political

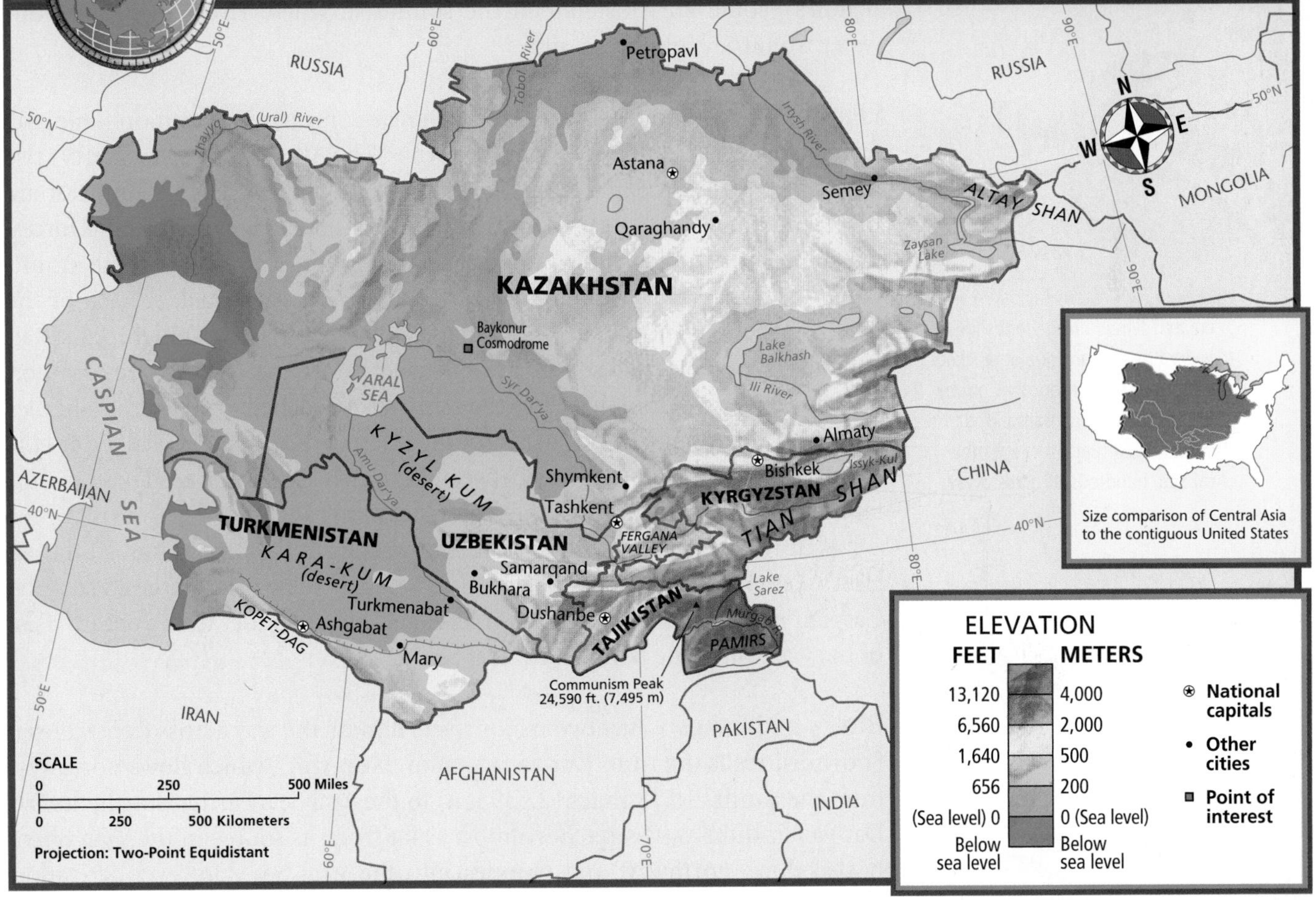

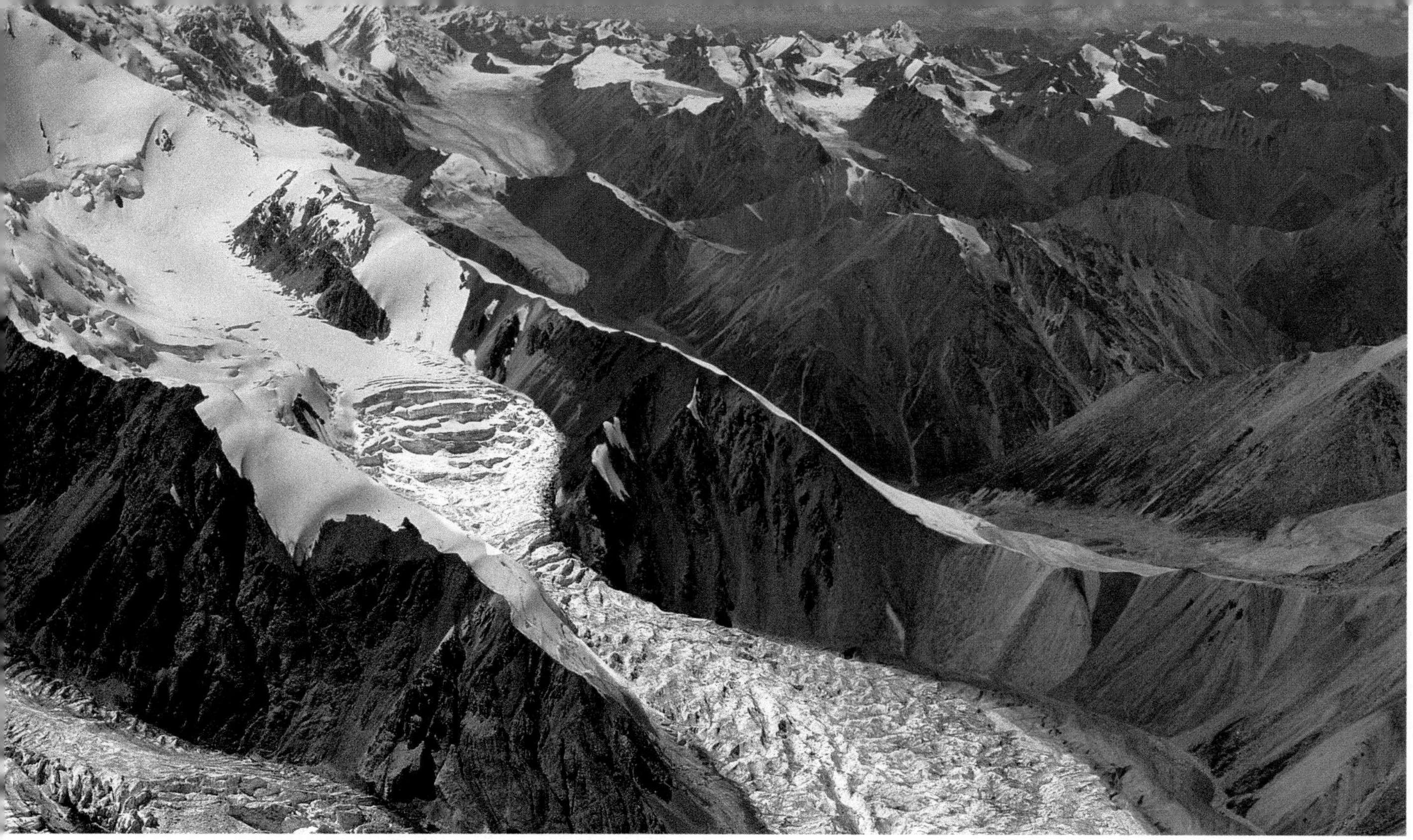

The Tian Shan range contains steep ridges and deep valleys. Glaciers like this one fill many of the valleys.

Landforms, Rivers, and Lakes

The five countries of Central Asia are all landlocked. Semiarid grasslands are found in the north. To the east, plateaus rise above barren deserts. Most of the region's people are clustered in the southeast, where rivers bring precious water from the high mountains.

Lake Sarez in eastern Tajikistan was created when a large landslide triggered by an earthquake blocked the Murgab River. If another quake breaks the natural dam, floodwaters could reach as far as the Aral Sea, hundreds of miles away.

Landforms Central Asia is a land of great contrasts in elevation—ranging from below sea level to lofty mountain peaks. The Altay Shan (al-TY SHAHN) rise in the far northeast. *Shan* means "mountain" in Chinese. In the southeast are the Tian Shan (TYEN SHAHN) and Pamirs (puh-MIRZ) ranges. Tectonic forces, which push the Indian subcontinent into the rest of Asia, created these mountains. Tectonic activity sometimes causes disastrous earthquakes. Each of the five countries has mountains more than 10,000 feet (3,050 m) high. Tajikistan contains the region's tallest peak at 24,590 feet (7,495 m). Glaciers are common in Central Asia's mountains. Tajikistan's massive Fedchenko Glacier is 44 miles (71 km) long. Along Central Asia's southwestern rim, the Kopet-Dag (koh-PET-DAHG) form a rugged boundary between Turkmenistan and Iran. These mountains are lower and drier than the region's eastern ranges.

Plateaus and plains stretch north and west from the mountains. At the region's western edge lies the Caspian Sea, which is the world's largest lake. The Caspian is 92 feet (28 m) below sea level and has no outlet to the ocean. East of the Caspian is the landlocked Aral Sea.

Rivers and Lakes Just two major rivers flow all the way across Central Asia. Snowmelt feeds the Amu Dar'ya (AH-moo DAHR-yuh), which flows northwest from the Pamirs 1,578 miles (2,539 km) to the Aral Sea. Farther north, the Syr Dar'ya (sir duhr-YAH) stretches almost as far from its source in the Tian Shan. It also flows northwest and empties into the Aral Sea. However, irrigation drains much of the water from these rivers. For example, a canal flows from the Amu Dar'ya across most of southern Turkmenistan. By the time they reach the Aral Sea they are little more than trickles. (See Geography for Life: The Shrinking Aral Sea.) The Syr Dar'ya and its tributaries provide water for the densely populated Fergana Valley.

The Irtysh (ir-TUHSH) River of eastern Kazakhstan flows northward into Russia. There it joins the Ob River, which drains into the Arctic Ocean. The Irtysh also provides water for crops, reservoirs, hydropower stations, and industrial cities.

Central Asia has some interesting lakes. Shallow Lake Balkhash has freshwater where the Ili River and other streams enter it. However, the lake is salty at its eastern end. Issyk-Kul (is-sik-KUHL) never freezes over even though it lies about a mile above sea level in the Tian Shan. The lake's warm water moderates the area's otherwise cold climate. (See Focus on Geography: Issyk-Kul and Tourism.)

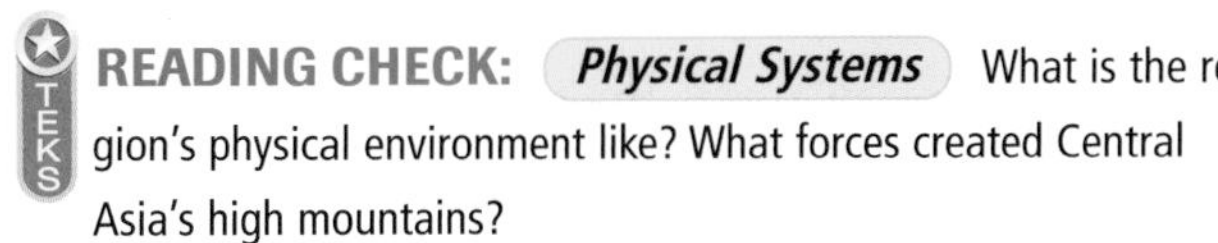

READING CHECK: ***Physical Systems*** What is the region's physical environment like? What forces created Central Asia's high mountains?

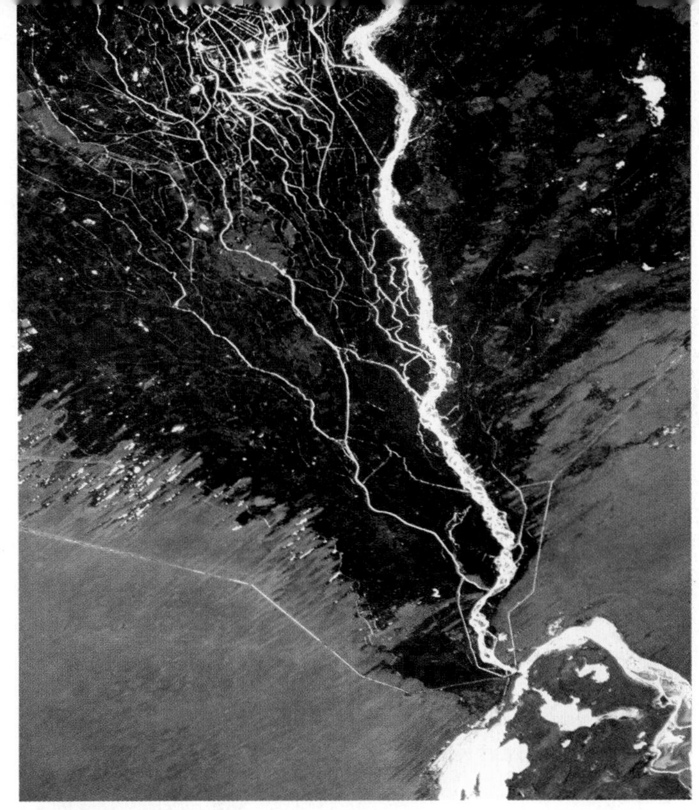

INTERPRETING THE VISUAL RECORD ***What are the straight narrow lines in this aerial view of the Amu Dar'ya? How might government planners use this photo?*** TEKS

Climates, Biomes, and Natural Resources

Most places in Central Asia have harsh climates. A location in the heart of Asia means the region is far removed from oceanic influences. Also, high mountains form a barrier to warm moist winds from the Indian Ocean and create a rain shadow. As a result, precipitation totals are low, and there are extreme seasonal temperature ranges. Summer temperatures can rise to 115°F (46°C). Winter lows have dropped to –36°F (–38°C).

Climates As you can see on the unit climate map, Central Asia has mostly semiarid and arid climates. The region has two large deserts—the Kara-Kum ("Black Sand") and Kyzyl Kum ("Red Sand"). One of the world's greatest tracts of drifting sand dunes is found in the Kara-Kum. Stony ground is more typical of the Kyzyl Kum.

Southern Turkmenistan has a small area with a Mediterranean climate. With its sunny skies and rugged surrounding scenery, the city of Ashgabat was a center of Soviet filmmaking. Some areas in the foothills of the Tian Shan also have mild weather. For example, Almaty, Kazakhstan, gets about 23 inches (58 cm) of rain each year. Spring and fall are pleasant there.

Plant and Animal Life The region's highest peaks are too cold, dry, and windy for vegetation. However,

INTERPRETING THE VISUAL RECORD

A Bactrian camel carries a rider across this desert in Central Asia. ***How did the domestication of the camel allow people to adapt to desert life?*** TEKS

The number of snow leopards is decreasing quickly. Corruption and the lack of effective government controls make protecting the big cats very difficult.

deciduous forests grow at middle elevations. In fact, Central Asia is known for its many walnut trees. Evergreen trees grow at higher elevations. Many alluvial fans, foothills, and river deltas support grasses and shrubs. Animals that live in the mountains include deer, pheasants, and wild boar. One of the world's most beautiful and endangered big cats, the snow leopard, also finds shelter in the high southeastern ranges.

Like many arid places, southern Central Asia has glorious vegetation in the spring. Desert shrubs and grasses bloom briefly before the searing summer heat begins. A tree unique to Central Asia, the saxaul, is one of the few large plants found in the desert. The tree's dense, heavy wood burns like charcoal. Desert peoples have used it as firewood for thousands of years. Desert wildlife includes antelope, wildcats, and wolves. Domesticated camels, goats, and sheep also graze on desert grasses.

TEKS **READING CHECK:** *Physical Systems* How is the mountain vegetation of Central Asia different from its desert vegetation?

Natural Resources In this arid region, water is the most precious resource. Although rainfall is low, snowmelt from the eastern ranges flows into rivers. Large dams along these rivers generate hydroelectricity.

Energy and mineral resources abound in Central Asia. Coal deposits are common on Kazakhstan's eastern plateau. The richest oil fields are in the west. The area around the Caspian Sea is a particularly important source of oil. Huge oil reserves there will last for many years. (See Case Study: Pipelines or Pipe Dreams?) Large gas reserves lie below Turkmenistan's desert basin.

Kazakhstan has most of the region's mines, and it exports copper, iron, lead, and nickel. Another export is **zinc**, an element important in metal processing and other industries. Uzbekistan is a major gold producer.

✓ **READING CHECK:** *Physical Systems* Where are the region's main energy resources found?

Section 1 Review

TEKS Questions 1, 2, 3, 4, 5

Define zinc

Working with Sketch Maps On a map of Central Asia that you draw or that your teacher provides, label the Altay Shan, Tian Shan, Pamirs, Kopet-Dag, Caspian Sea, Aral Sea, Amu Dar'ya, Syr Dar'ya, Irtysh River, Lake Balkhash, Issyk-Kul, Kara-Kum, and Kyzyl Kum. Which body of water is the world's largest lake?

Reading for the Main Idea

1. *Physical Systems* Where are the region's mountain ranges? How do they affect precipitation patterns in Central Asia?
2. *Physical Systems* What climates are found in the region? Where would you find a mild climate with reliable precipitation?

Critical Thinking

3. **Making Generalizations and Predictions** How might the location of resources affect Central Asia's population density in the future?
4. **Drawing Inferences and Conclusions** How might Central Asia's physical geography limit foreign trade for the region's countries?

Organizing What You Know

5. Create a chart like the one below. Use Section 1 and the physical-political map to complete it. Add boxes for the region's remaining countries.

Country	Landforms	Bodies of water
Kazakhstan		

Environment and Society

Geography for Life

The Shrinking Aral Sea

People often change the environment in the process of using natural resources. Sometimes when humans alter nature, they bring prosperity to some areas while creating crises in others.

Perhaps nowhere else is this process more visible than at the Aral Sea. The Aral Sea, which is really a salt lake, lies between Kazakhstan and Uzbekistan. The sea has shrunk by about 60 percent since 1960. It once covered about 26,000 square miles (68,000 sq km) and was the world's fourth-largest lake. Today the sea covers only about 11,000 square miles (28,000 sq km). The former port city of Mŭynoq now lies 30 miles (48 km) from the Aral Sea's shore. The sea's level has dropped 48 feet (16 m). As a result, the sea has become saltier, and few fish can survive in it. Moreover, some 10,000 square miles (26,000 sq km) of former seafloor have become a desert of sand and salt.

What happened? Cotton farming is largely responsible for the sea's drying. The Soviet government wanted to establish a profitable cotton industry. To water the cotton fields, the Soviets built a network of irrigation canals. This irrigation system still diverts large amounts of water from the Syr Dar'ya and Amu Dar'ya, the only rivers that flow into the Aral Sea. Very little river water reaches the shrinking sea. In fact, more water evaporates from the sea than flows into it. In addition, the sea and the surrounding land are polluted with agricultural chemicals. Wind spreads the chemicals, along with the sand and salt, ruining cropland and damaging the health of area residents.

Experts from many countries are trying to find ways to repair environmental damage in the Aral Sea region. Little hope remains for the sea itself. The Aral Sea will probably soon be just a cluster of small salty lakes.

Applying What You Know 

1. **Evaluating** What geographic and economic impact have past and present water policies had on the Aral Sea and the surrounding area?
2. **Making Generalizations and Predictions** Use the map, graph, and text to infer how the shrinking of the Aral Sea will affect the surrounding area. How might settlement patterns, population distribution, economic conditions, and political conditions in the area be affected?

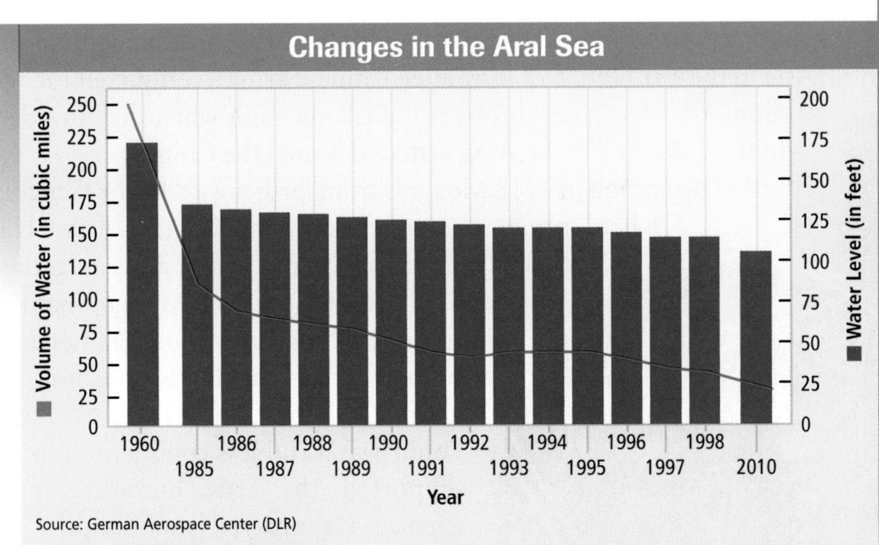

Source: German Aerospace Center (DLR)

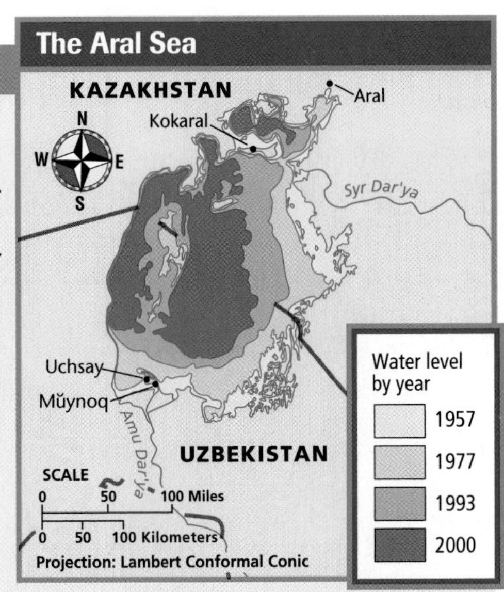

INTERPRETING THE GRAPH *Both this graph and the map beside it demonstrate the drying of the Aral Sea.* ***What relationship can you see between the surface area of the sea and its depth? How might changes in water level affect the rate at which the sea's volume decreases?***

Section 2 History and Culture

READ TO DISCOVER

1. How have various cultures and invaders affected the region's history?
2. What are some features of Central Asian cultures?

DEFINE

caravans
monoculture
nomads
transhumance
yurts

LOCATE

Samarqand
Fergana Valley

A warrior's armor from Kazakhstan

WHY IT MATTERS

Central Asians are rediscovering their cultural heritage after the Soviet government suppressed that heritage for more than 70 years. Use CNNfyi.com or other **current events** sources to learn about Central Asian heroes, holidays, legends, and literature.

History

Humans have lived in Central Asia for thousands of years. In fact, many migrant peoples and invaders have left their imprint on the region's history and cultures. For a while, parts of the area belonged to the Persian Empire. Alexander the Great brought Greek influences when he invaded in the 300s B.C. Merchants pursuing profits instead of conquest also came through Central Asia. Great **caravans** passed through as they brought silk and other luxury goods from China along routes called the Silk Road. A caravan is a group of people traveling together for protection. Other traders would eventually sell these products in Europe and the Mediterranean region.

Armies and Empires Turkic-speaking peoples established kingdoms in Central Asia in the A.D. 600s. Not long after, Chinese armies conquered the region. Arabic speakers brought the Islamic faith when they invaded in the 700s. In 751 the Arab armies defeated the Chinese.

Beginning in 1218 Mongols from farther east, led by Genghis Khan, began a 200-year rule of Central Asia. The Mongols destroyed many cities and irrigation systems. A Turkic-speaking Mongol named Timur (tee-MOOR), also known as Tamerlane, rose to power in the 1300s and built an empire. It lasted from 1370 to 1405. Timur was a ruthless conqueror, but he supported the arts, literature, and science. He ordered the building of beautiful gardens, mosques, and palaces at his capital, Samarqand. After Timur's death, his empire broke up into city-states. In the late 1400s Europeans began

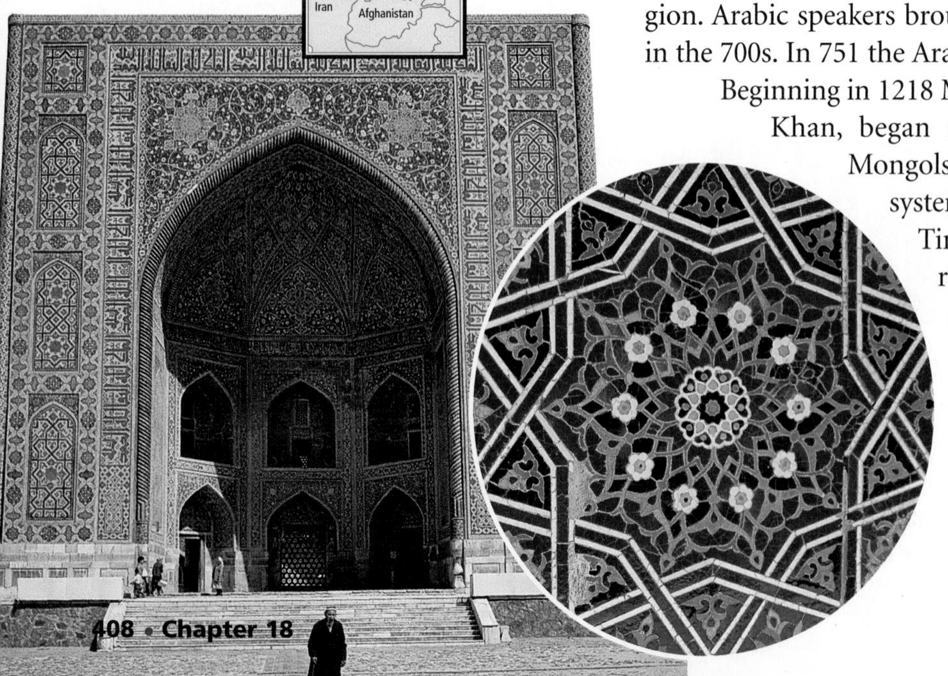

Built in Samarqand in the mid-1600s, the Tilla-Kari served both as a mosque and as a madrasa, *or Islamic school. The complex's exterior walls are decorated with geometric designs worked in ceramic tiles, like the detail in the small photo. Some of Samarqand's historic buildings are decorated with gold and marble.*

to sail to East Asia, avoiding the Silk Road. As a result, trade through Central Asia declined, and the region became isolated.

Russian and Soviet Rule Under the czars, Russia conquered and colonized Central Asia in the mid-1800s. The conquest began in the north and moved south. Following the soldiers were settlers who irrigated and farmed the desert. The Russians also built railroads throughout the region. Railroads helped the Russians create a stronger military presence between India, which was under British control, and Russia. The railroads also offered better access to the region's resources. Expanded cotton and oil production followed.

A teenager skates past a statue of Vladimir Ilich Lenin in Bishkek. The statue is left over from Kyrgyzstan's Soviet era. Lenin was a leader of the Russian Revolution.

Resistance to Russian rule grew during the 1910s. However, the Soviet government, which took power in Russia after the Russian Revolution in 1917, eventually crushed the resistance. In an effort to weaken that resistance, the Soviets drew new political boundaries that separated people and resources. The Central Asian countries were called republics, but they were fully under Soviet control. Russian migration to the region increased, bringing in several million people.

The Soviets built huge irrigation projects, which helped the Soviet Union become a leading cotton producer. In short, the government built a cotton **monoculture**—the cultivation of a single crop—in the region. People who had moved with their herds were forced to settle on large government-owned farms. During World War II, factories and more Russians moved east into Central Asia.

As the Soviet Union collapsed in 1991, the Central Asian republics declared their independence. Troublesome international boundaries remained from the Soviet era. For example, the fertile Fergana Valley had been divided among three countries. Those countries had to share the main rivers and irrigation canals. Today just southwest of the Fergana Valley, Uzbekistan enclaves lie in Kyrgyzstan. Enclaves of Tajikistan lie in Kyrgyzstan and Uzbekistan. The complex boundaries are now difficult to control. As a result, the governments cannot stop smugglers who ship illegal drugs through the area. The problem is most severe for Uzbekistan's government.

READING CHECK: ***Human Systems*** What are some ways that Russian and Soviet rule changed Central Asia?

Culture

The five Central Asian countries lie in a region sometimes called Turkistan. The suffix *-stan* means "place" or "land" in Turkish. Turkic languages and ethnic groups have long dominated the area. As a result, many people once called the whole region Turkistan. Turkic heritage is still strong in the region today.

Irrigated farming was the traditional way of life in the region's southern areas. Herding was traditional in the north. A large segment of the north's population was made up of **nomads.** Nomads are people who move often from place to place. The region's nomads moved herds from mountain pastures in the summer to lowland pastures in the winter. This practice is called **transhumance**.

Unique homes, called **yurts**, made moving with the herds possible. A yurt is a movable round house of wool felt mats. The mats are placed over a wood frame. Today these homes remain a symbol of the region's nomadic heritage. Even people in cities may put up yurts for weddings and funerals.

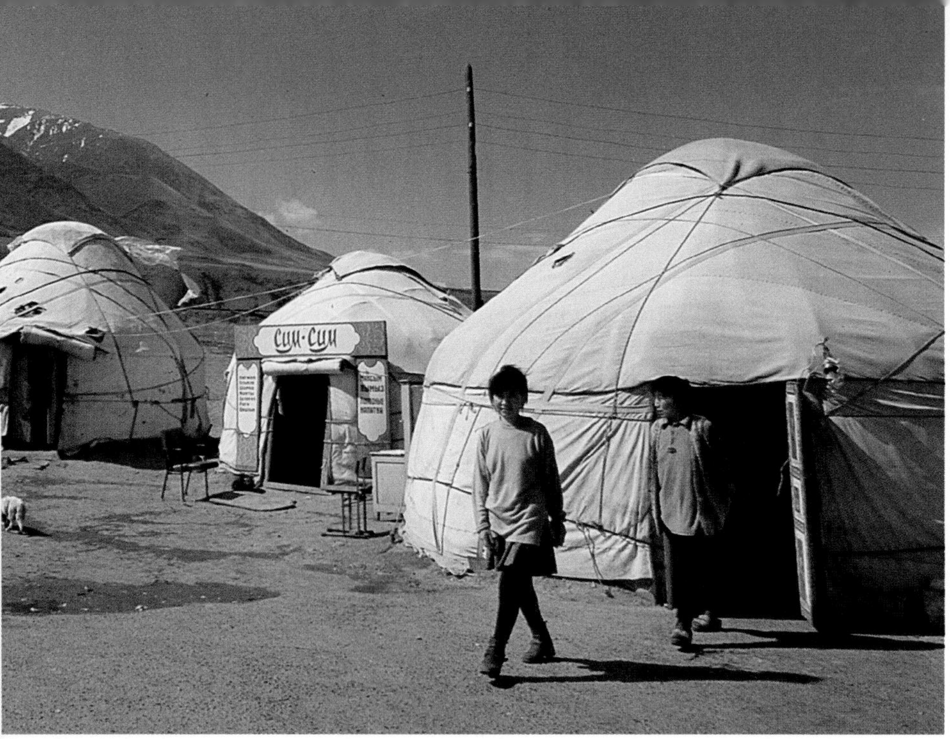

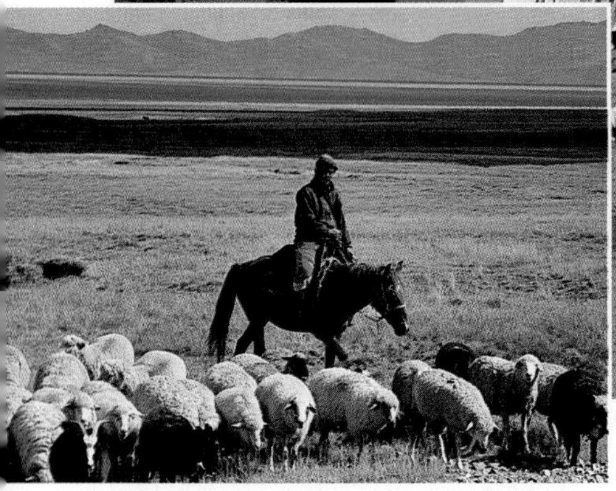

INTERPRETING THE VISUAL RECORD

Historically, most people of Central Asia have been farmers or nomadic herders like the Tajik man on the left. Many, however, have begun to settle in permanent communities and to seek new occupations. On the right, hungry Kyrgyz travelers can choose from several roadside restaurants housed in yurts. ***How does this picture suggest a blending of traditional and modern cultures?***

People, Languages, and Religion Almost two thirds of Central Asians speak a Turkic language. Many people identify themselves as Turkic. In fact, about 30 percent of the world's 125 million speakers of Turkic languages live in Central Asia. Four major Turkic languages are represented in the region. However, Tajikistan's language belongs to a different group and is related to the main language of Iran.

Russian is the main language and ethnic identity for a sizable minority of Central Asians. In fact, Russian is still an official language in some of the countries there. However, Russian language and culture in general are losing status. For example, the Latin alphabet is replacing the Cyrillic alphabet, in which Russian is written. In addition, millions of people have returned to Russia since 1991.

Traders and conquerors brought many religions to Central Asia. Islam is the main religion. Most of the region's Christians are ethnic Russians and belong to the Russian Orthodox Church. During the Soviet era, the government tried to get rid of all religions. It closed more than 35,000 mosques, churches, and Islamic schools. Many of these buildings were abandoned or destroyed over time. Since 1991 the remaining buildings have reopened. They are now powerful symbols of ethnic pride. The beautiful mosques and Islamic schools in Samarqand and other cities are major tourist attractions. Exquisite turquoise tile mosaics decorate many of them.

Source: Central Intelligence Agency, *The World Factbook 2001*

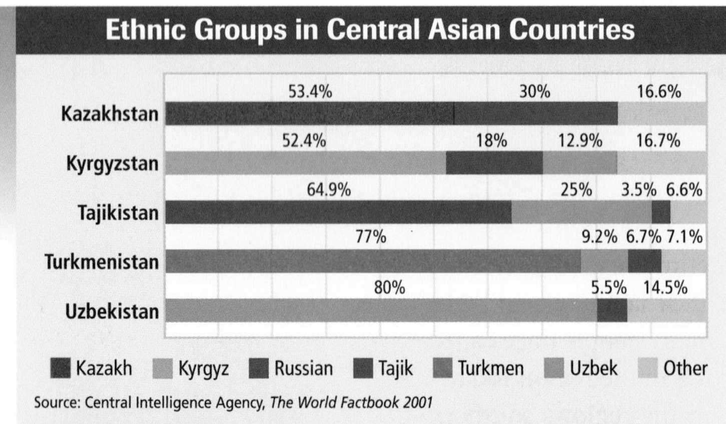

INTERPRETING THE GRAPH *Except for the Tajiks, who are more connected to the settled culture of Iran, the region's other major ethnic groups are all Turkic peoples.* ***What role do you think the distribution of ethnic groups played in shaping the borders of Central Asia?*** TEKS

Traditions and Education Textiles are among Central Asia's best-known art forms. Felt yurts, distinctive wool hats, and silk fabrics are examples. Silk textiles are produced in both homes and factories. Mulberry trees growing along roads, fields, and canals provide the leaves on which silkworms feed.

Sheep, goats, and other animals grow long hair that is excellent for weaving into carpets. Central Asian carpets, often made by Turkmen nomads, are famous. Red colors and geometric designs are common on most of the carpets. Factory-made copies from Europe and declining Silk Road trade caused production of carpets to fall. Now the region's traditional weaving crafts are being revived because the tourist trade is growing.

Distinctive headgear is part of traditional dress in Central Asia. The sheepskin hats of the Turkmen (top) contrast with the embroidered felt caps worn by Uzbeks (bottom).

Like other cultural crossroads, Central Asia has varied food traditions. Noodles probably came from China. The region imports tea from China today. Tea is very popular, and teahouses are a common sight in many areas. From the nomad culture came meat dishes and dairy products, some made with soured milk. Grilled meats such as lamb are featured in many meals. The farming way of life provides bread and rice dishes. Fruits and vegetables are plentiful. Apples, apricots, lemons, pumpkins, and watermelons are among the many fruits grown locally.

Communism limited personal freedoms, but the Soviet system did stress education and health care. State-run schools replaced education in Islamic schools. The results of widely available public services have continued in the post-Soviet period. Literacy rates are well above world averages. In addition, life expectancy in most Central Asian countries is longer than the world average, particularly for women.

TEKS **READING CHECK:** *Human Systems* In what ways did education in Central Asia change during the Soviet era?

TEKS Questions 1, 2, 3, 4, 5

Review

Define caravans, monoculture, nomads, transhumance, yurts

Working with Sketch Maps On the map you created for Section 1, label the countries of Central Asia, Samarqand, and the Fergana Valley. Which three countries have complex boundaries in the Fergana Valley region?

Reading for the Main Idea

1. *Human Systems* Which invaders destroyed Central Asian cities and irrigation systems? Which conqueror beautified Samarqand?
2. *Places and Regions* What are some cultural features that make Central Asia distinctive?

Critical Thinking

3. **Drawing Inferences and Conclusions** How might the region's cultural geography be different today if the Arab armies had not defeated the Chinese in A.D. 751?
4. **Analyzing Information** Why might Turkistan be considered a perceptual region?

Organizing What You Know

5. Create a graphic organizer like the one below. Use it to identify the many invasions of Central Asia. Add more boxes if necessary. Then use more arrows or other symbols to describe connections between the groups.

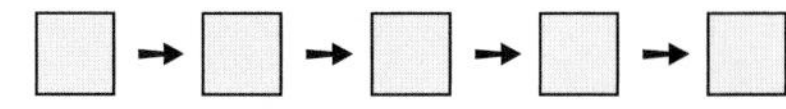

The Region Today

READ TO DISCOVER

1. How has the economy of Central Asia changed over time?
2. What are the region's cities like?
3. What issues must Central Asia face to improve its economy?

DEFINE

dryland farming

LOCATE

Bukhara
Tashkent
Bishkek
Almaty
Astana

WHY IT MATTERS

Many people in Kazakhstan suffer from ailments blamed on nuclear testing during the Soviet era. Use CNNfyi.com or other **current events** sources to learn about these and other health problems in Central Asia.

Economic Changes

Agriculture remains important to Central Asia. Traditional herders raise camels, cattle, goats, horses, and sheep. Crops include cotton, fruits, barley, rice, tobacco, and vegetables. Uzbekistan is the world's third-largest exporter of cotton. To retain the soil's nutrients, farmers are now adding other crops to the region's cotton monoculture. Kyrgyzstan's fertile soils in the Tian Shan foothills allow a mix of **dryland farming** and irrigated crops. Dryland farming relies on rainfall instead of irrigation. Turkmenistan depends on the world's longest irrigation channel to water grain and cotton fields in the country's southern areas.

Mining and industry offer a chance for future wealth. Huge reserves of oil, gas, and minerals await development. However, having to use outdated equipment slows industrial growth. Corruption, poor transportation links, and a lack of cash for investment also hurt development. Moreover, many skilled Russian workers and managers are leaving the region.

INTERPRETING THE VISUAL RECORD

Workers pick cotton in a field outside Bukhara, Uzbekistan. During the harvest season, students and government workers from across the country are brought in to help with the picking. ***Why might farmers be forced to rely on student labor to harvest the cotton crop?*** TEKS

The need for more foreign trade is another factor limiting development. Russia is not the reliable customer that the Soviet Union once was. Other markets are far away. Even products for local markets must often cross international borders.

READING CHECK: *Places and Regions* What factors limit the Central Asian countries' ability to develop their mines and industries?

Urban Environments

Central Asia has relatively few big cities because, throughout its history, most of the people have been nomads or farmers. As a result, few of the cities have many

Connecting to
TECHNOLOGY

Baykonur Cosmodrome

The space race between the Soviet Union and the United States began in 1957. In that year the Soviets launched the first artificial satellite *Sputnik* into space. *Sputnik* blasted off from the Baykonur Cosmodrome in what is now Kazakhstan. The world's first manned orbital flight and the flight of the first woman in space also began at Baykonur. It is still Russia's most important location for launching space vehicles. However, the noise and activity have not completely changed the surrounding environment. In fact, wild camels and horses roam the site!

For decades, the Soviet Union tried to keep Baykonur's location, and even its existence, a secret. Today a detailed map of the complex is available on the Internet.

Analyzing Maps Study the chapter physical-political map. Why do you think the Soviet Union established the Baykonur Cosmodrome in that location? TEKS

In 1963 Valentina Tereshkova became the first woman in space. The letters CCCP on her helmet stand for "Union of Soviet Socialist Republics" in Russian.

old features. Those that lie along ancient trade routes, such as Bukhara and Samarqand, still have colorful markets and blue-tiled mosques. Tashkent, the region's largest city, has an Old Town of mud-brick houses and narrow streets. However, most of Tashkent is a Soviet-era city of plain apartment buildings, factories, and offices. Smaller cities, such as Bishkek, also offer few cultural or historical sites. They do serve as starting points for tourists exploring the mountains and other scenic areas. As tourists discover the Central Asian countries, some new banks, restaurants, and shops are opening in the cities.

In 1997 Kazakhstan moved its capital from Almaty to Astana. Government leaders wanted to rebuild the city into a glorious capital, but funds are not available. On the other hand, the new location is closer to Russia and Europe. This new location might strengthen links between Kazakhstan and Russia and help Kazakhstan keep its Russian population.

READING CHECK: ***Human Systems*** What caused the old cities of Central Asia to grow? What evidence remains from these earlier times?

Issues and Challenges

The Central Asian countries must overcome tremendous challenges to ensure economic and social progress. Oil and gas deposits may bring in plenty of cash someday, but all the countries still face an uncertain future. Many of the people are poor and have few opportunities to improve their lives.

The region's location creates a basic problem. As you have read, the countries are landlocked. As a result, they are cut off from major

The bazaar in Andizhan, Uzbekistan, recalls the lively trade of the Fergana Valley city's history as a stop on the Silk Road. However, unemployment in the Fergana Valley is high. In some areas, up to half of the young men do not have jobs.

Different types of environmental challenges face Central Asia. Left: A slag heap contrasts with the snowy slope behind it. Slag is a waste product created when metal ores are refined. It is sometimes used to build roads or to make fertilizer. However, much of Central Asia's slag is contaminated with radiation or poisonous substances. Right: Stranded ships provide testimony of the Aral Sea's receding shoreline.

global trade routes. Even the Persian Gulf and the Black Sea are hundreds of miles away. Kazakhstan stretches the width of Central Asia, separating the other countries from their longtime economic partner, Russia. Shipping goods great distances for sale to other countries adds costs. Because they cost more than products from nearby countries, these goods find few buyers.

The need for water in this dry land is at the heart of many problems. The region's major rivers cross international borders. Water sources are in Tajikistan and Kyrgyzstan, while many water consumers are in Uzbekistan and Turkmenistan. Upstream users want to dam the rivers for hydroelectricity. Downstream users want the water for irrigation. Water has also been wasted. For example, Turkmenistan's canal lacks a concrete lining, so leaks are common. These leaks create swamps, high water tables, and salty soils.

Political Problems Changing from command to market economies has not been easy. The change to private enterprise is most advanced in Kazakhstan and Kyrgyzstan. It is slower in Uzbekistan and Turkmenistan. Corruption and the lack of democracy are major obstacles to economic growth. Tajikistan is recovering from years of civil war and still depends heavily on foreign aid from Russia, Uzbekistan, and international agencies.

Ethnic conflict also threatens the region. Various groups have committed violent acts. For example, in 1989 riots erupted among Uzbeks and others in the Fergana Valley. In 1999 an Islamic rebel group kidnapped foreign travelers. The region remains tense.

About 90 percent of the world's caviar—salted fish eggs—comes from sturgeon in the Caspian Sea. Sturgeon can grow up to 14 feet (4 m) long, weigh 1,300 pounds (590 kg), and carry 200 pounds (91 kg) of eggs worth up to $250,000 in the United States.

Environmental Issues Soviet agricultural, industrial, and military practices damaged Central Asia's land and water. For example, testing of biological weapons contaminated the Aral Sea region. In addition, beginning in 1949, the Soviet Union tested hundreds of nuclear bombs in an isolated area of northeastern Kazakhstan. Officials did not tell residents of the dangers from radiation. In fact, the testing exposed about 1.5 million people to radiation. Now birth defects, cancer, and other ailments plague the area's people.

Overuse of chemicals to increase crop yields has made some farmlands useless. In addition, places where uranium was mined and processed are now toxic. Money to clean up these sites is lacking. Many of the Russian experts and workers who know how to do the job are leaving the area.

TEKS **READING CHECK:** ***Environment and Society*** How did various Soviet practices affect the region's environment?

FOCUS ON GEOGRAPHY

Issyk-Kul and Tourism Although some areas in Central Asia have been polluted, others are clean and beautiful. For example, Kyrgyzstan's Issyk-Kul is an environmental and cultural treasure.

A gorgeous mountain setting surrounds the big lake, which covers 2,355 square miles (6,099 sq km). The lake is quite deep, at 2,303 feet (702 m), and it has a surface elevation of 5,279 feet (1,609 m). The lake never freezes, partly because of underwater hot springs. In fact, *Issyk-Kul* means "Hot Lake." Many streams tumble down to Issyk-Kul and its beaches. The clear blue water is slightly salty but pleasant for swimming. Fish are abundant. Because the lake's warmth moderates the climate, the surrounding area is a productive agricultural region. Apples are a local specialty.

Early civilizations and later conquerors seem to have valued the lake region. They left burial mounds, carvings in rock faces, and other artifacts. Some of these artifacts are up to 2,500 years old.

The Soviets built more than 100 resorts, spas, and similar establishments along the lake's shores. Most visitors were government officials because the Soviet navy tested torpedoes and other weapons in the area. Today tourists can rent converted navy boats for short trips. Local tourism prospers, as Kyrgyzstan's major cities are not far away. Hotels and travel facilities are springing up at the lake's east end. Many tour organizers now arrange hikes to the nearby mountains. A preserve the size of Switzerland has been proposed to protect Issyk-Kul's unique qualities.

READING CHECK: *Environment and Society* How has the development of tourism affected the character of Issyk-Kul?

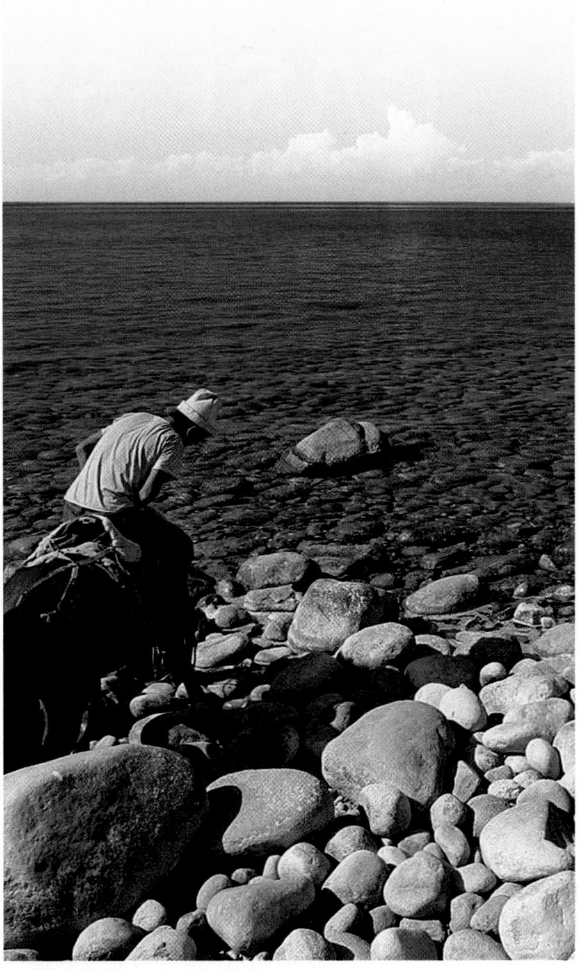

INTERPRETING THE VISUAL RECORD *A rider explores a rocky area of Issyk-Kul's shoreline. The area's temperate climate provides a habitat for a wide range of wildlife. Some 40 mammal species and 200 bird species live there.* ***How do you think Issyk-Kul's wildlife compares to animal populations farther from the lake?*** TEKS

TEKS Questions 1, 2, 3, 4, 5

Review

Define

dryland farming

Working with Sketch Maps

On the map you created for Section 2, label Bukhara, Tashkent, Bishkek, Almaty, and Astana. Where is the world's longest irrigation channel?

Reading for the Main Idea

1. *Environment and Society* How are farmers in Uzbekistan improving the soil?
2. *Places and Regions* What advantages does the new location for Kazakhstan's capital present? What problem remains?

Critical Thinking

3. **Solving Problems** How might international cooperation help the region's countries overcome the obstacles to trade that are posed by their location?
4. **Comparing Points of View** How do different countries in the region want to use water from rivers that cross their borders? How do you think these differing points of view make it difficult for countries in the region to develop policies regarding water use?

Organizing What You Know

5. Create an idea web to describe the geographic, political, and environmental challenges facing Central Asia.

CASE STUDY

Pipelines or Pipe Dreams?

Environment and Society Azerbaijan, Kazakhstan, and Turkmenistan sit astride some of the largest oil reserves in the world. Countries inside and out of the region are working to unlock the potential of this rich natural resource. However, they have faced a number of geographic and political hurdles.

Estimates of Caspian oil reserves vary widely, from 40 billion barrels to 200 billion barrels. Even the lower amount is more than the proven oil reserves in the United States and the North Sea combined. The higher amount could be double that of Kuwait's proven oil reserves of some 100 billion barrels. Kuwait is one of Southwest Asia's richest oil-producing countries. The Caspian region might also hold huge amounts of natural gas, another valuable energy resource.

Outside powers have long tried to control the oil along the Caspian. For example, German armies tried but failed to overrun the oil fields of the nearby Caucasus during World War II. Since the collapse of the Soviet Union, oil exploration in the region has expanded. In addition, the contest to further develop the oil resources there has heated up.

An employee walks near an oil-pumping station south of Turkmenbashy, Turkmenistan. Oil fields in this area produce some 7,500 barrels of oil per day.

The main problem in developing these fields has not been how to drill for the oil. The problem has more to do with getting the oil to global markets. Remember that the Caspian Sea is really a lake and has no natural outlet to the world's oceans. The only way to get the oil out of the region is to use overland routes to seaports in other countries. Oil companies and interested countries have proposed a variety of routes, including pipeline routes, that could transport the oil to these ports. Routes through Iran, Russia, and across Georgia and Turkey have been among the proposals. Even pipelines to China and Pakistan have been suggested, but they probably would be very expensive to build. However, the main factor is not cost but who will control the pipeline, and thus the oil.

On the surface, one would think that choosing a route should not pose a problem. Simply build a pipeline across the shortest distance to the nearest port, right? Things are rarely that easy, though, particularly in this part of the world. Many proposed routes cross through the Caucasus region, a shatter belt. As you

Alternative Pipeline Routes for Central Asian Oil

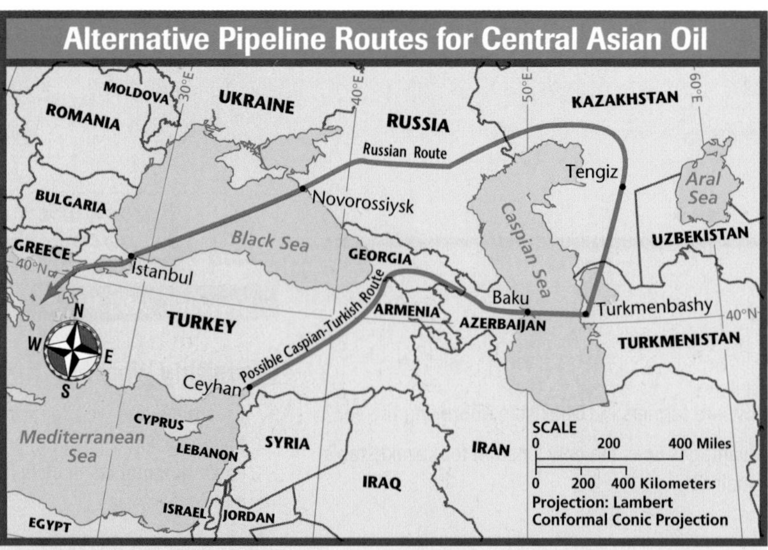

INTERPRETING THE MAP *Turkey is concerned about proposed pipeline routes. Because it lies between Central Asia and the Mediterranean Sea, Turkey will be involved with oil transport along either of the routes shown.* ***How would the choice of each port affect the movement of capital, goods, and people through Turkey? Which route does the Turkish government favor?*** TEKS

Tankers carry about 1.7 million barrels of oil through the Bosporus every day. Turkey's government fears a major accident in the strait and favors an alternate oil-transport route.

have read, a shatter belt is a zone of frequent boundary changes and conflicts. Often these changes are caused by the area's position between major powers. In fact, Iran, Russia, and Turkey have long competed for influence in the Caucasus and Central Asia. Some Russians see it as an area over which they have had, and should still have, influence.

The struggle for influence in the region has been particularly important in the debate over pipelines. Many of the companies that want to develop the Caspian oil fields are American or European. For political reasons, these companies and Western governments do not want to build oil pipelines across Iran or Russia. On the other hand, Russian leaders worry that Western countries—particularly the United States—already have too much influence in the region. They think moving the oil through Russian pipelines would limit those outside influences.

These competing fears and desires have led to competing pipeline plans. For example, Russia wanted oil companies to use a new pipeline across southern Russia. The pipeline could transport the oil to tankers at a Russian seaport on the Black Sea. (See the map.) The Russian plan faced opposition, however. Critics worried the Russians might use the pipeline to influence the region's affairs. They also worried about the route tankers would have to take from the Russian port. Tankers on that route would pass through the narrow Bosporus, the strait that connects the Black Sea to the Mediterranean Sea. This fact worried Turkey, which straddles the Bosporus. The Turks pointed out that the amount of maritime traffic through the narrow strait would more than double. This situation could create long traffic jams on the sea lane. Safety was also a concern because millions of people live on or near the Bosporus. An explosion or oil spill could be devastating in this densely populated area.

Turkey, its NATO ally the United States, and some oil companies campaigned for another route. Under their plan, tankers would ship the oil across the Caspian Sea to Baku in Azerbaijan. A new pipeline would then carry the oil across Georgia to a Turkish port on the Mediterranean Sea. (See the map.) However, this plan also presented challenges. One was the cost—it would be the most expensive route. Another challenge was that the pipeline would go through a territory that has experienced ethnic and religious conflict in recent years. War or other political violence could stop the movement of oil along the route. Most important, Russia, which still holds the most power in the region and is able to influence Georgia, opposed the route.

For Azerbaijan, Kazakhstan, and Turkmenistan to market their product, compromises will have to be made on many levels with the Russians, the oil companies, and regional neighbors. Economic and political cooperation among the parties involved will be necessary for the Central Asian republics to benefit from their oil resources.

Applying What You Know

1. **Summarizing** What are two of the pipeline routes that have been proposed for transporting Central Asian oil? How would each of these routes affect the ability of neighboring countries to influence Central Asian affairs?

2. **Analyzing Information** What are the drawbacks associated with each of the two pipeline routes discussed at length here?

Building Vocabulary

On a separate sheet of paper, explain the following terms by using them correctly in sentences.

zinc	nomads	yurts
caravans	transhumance	dryland farming
monoculture		

Locating Key Places

On a separate sheet of paper, match the letters on the map with their correct labels.

Tian Shan	Syr Dar'ya	Samarqand
Aral Sea	Lake Balkhash	Tashkent
Amu Dar'ya	Kyzyl Kum	Astana

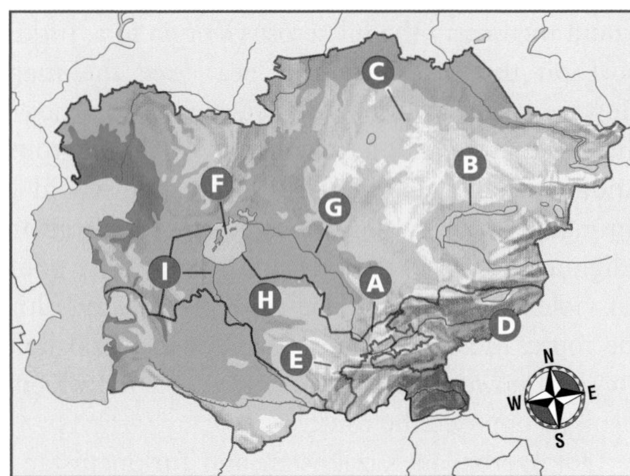

Understanding the Main Ideas

Section 1

1. *Human Systems* Why is herding common in the region's northern areas and farming more common in the south?

Section 2

2. *The World in Spatial Terms* How did boundary changes during the Soviet era affect the control that Central Asian governments have over their territory?
3. *Human Systems* What are the main languages and religions in Central Asia? Which language is losing status?

Section 3

4. *Environment and Society* What resources of Central Asia offer the possibility of wealth in the future?
5. *Places and Regions* How does Issyk-Kul contrast with some other areas in Central Asia?

Thinking Critically for TAKS

1. **Problem Solving** If you were in charge of developing tourism in Central Asia, how would you go about doing your job?
2. **Identifying Points of View** How do you think Central Asia's indigenous peoples might view Russian culture and ethnic Russians who live in their countries today?
3. **Analyzing Information** In what ways can traditional ways of life still be seen in the region?

Using the Geographer's Tools

1. **Using Statistics** Use the unit Fast Facts and Comparing Standard of Living tables to rank the countries of Russia and Northern Eurasia by economic development and standard of living. What political features might help or hurt the region's development?
2. **Using Databases** Examine the graph of ethnic groups in Section 2. Then use the Internet or library resources to find databases that provide similar figures for a year or years before 1991. In what ways have the percentages changed? What do you suppose accounts for the changes?
3. **Creating Graphs** Use the unit atlas to create bar graphs of the workforce structure and per capita GDPs of the Central Asian countries. How does the workforce structure of these countries seem to affect wealth?

Writing for TAKS

Imagine that you are a Russian teen living in Kazakhstan. Your parents are thinking about moving the family back to Russia. Write a diary entry in which you discuss reasons why you want to stay in Kazakhstan or why you want to leave. In your entry, analyze the economic and social effects on your family of each possibility. When you are finished with your entry, proofread it to make sure you have used standard grammar, spelling, sentence structure, and punctuation.

SKILL BUILDING

Geography for Life

Analyzing Statistics

Environment and Society Review the statistical and other data in Geography for Life: The Shrinking Aral Sea. How have physical processes and human activity affected settlement patterns and population distribution in the Aral Sea region? How have economic conditions suffered? What political changes might occur? How has the area's resource base lost value?

Building Social Studies Skills

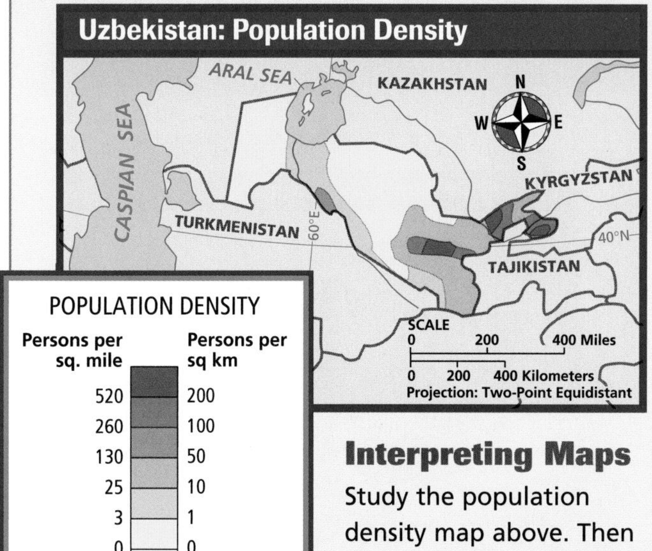

Interpreting Maps

Study the population density map above. Then use the information from the map to help you answer the questions that follow. Mark your answers on a separate sheet of paper.

1. Uzbekistan's population
   - **a.** is least dense in the west and in central areas.
   - **b.** is most dense in the far west.
   - **c.** is evenly scattered across the country.
   - **d.** is so small that few areas are densely populated.

2. Where would you expect to find Uzbekistan's largest cities? Why?

Using Language

The following passage contains mistakes in capitalization, grammar, and punctuation. Read the passage. Then answer the following questions. Mark your answers on a separate sheet of paper.

(1) About a third of Kazakhstan's people are Ethnic Russians. (2) Kazakh and Russian—both official languages.
(3) Many ethnic Kazakh's grow up speaking Russian at home and they have to learn Kazakh in school today.

3. Rewrite line 2 to make it a complete sentence.
4. Which sentence contains incorrect capitalization?
   - **a.** 1
   - **b.** 2
   - **c.** 3
   - **d.** None of them
5. One sentence contains an error in the use of an apostrophe. It also has a missing comma. Rewrite the sentence correctly.

Alternative Assessment

PORTFOLIO ACTIVITY

Learning about Your Local Geography TEKS

Group Project: Research

You have read about the importance of the Baykonur Cosmodrome in Kazakhstan. Plan, organize, and complete a group research project about a major scientific facility in your state or area. Contact officials there to create a list of some of the projects being completed at the facility. Then have group members research the national or international significance of the projects. Use the information you find to create a brochure about the facility.

internet connect

Internet Activity: go.hrw.com
KEYWORD: SW3 GT18 TEKS

Choose a topic on Central Asia to:

- create a poster that analyzes the increasing use of technology in agriculture and its consequences for the Aral Sea.
- research the Silk Road and its history in global trade.
- follow nomads in Kazakhstan.

Skill Building for TAKS

WORKSHOP 1

Using Graphs

Graphs paint a clear picture of numerical data and the relationships among the data. Bar and line graphs are common graphs. Bar graphs can show how the value of a certain item changes over time, or they can compare the values of different items. For example, the bar graph on this page shows the changing water level in the Aral Sea over time. A line graph can also show changes over time. The example shows changes in the volume of water in the Aral Sea over time.

A third type of graph, a pie graph, is particularly useful when considering parts of a whole. Such a graph looks like a pie cut into slices. The complete pie, or circle, represents the whole, and each wedge represents a share of the whole. These shares are expressed in percentages. The pie graph on this page shows the world's largest rye producers. The wedges are drawn in proportion to each country's share of world rye production.

Developing the Skill Creating graphs is relatively easy. First, determine which kind of graph is best for showing the information you want to communicate. Second, give the graph a descriptive title. Third, identify the values, or numbers, you want to show.

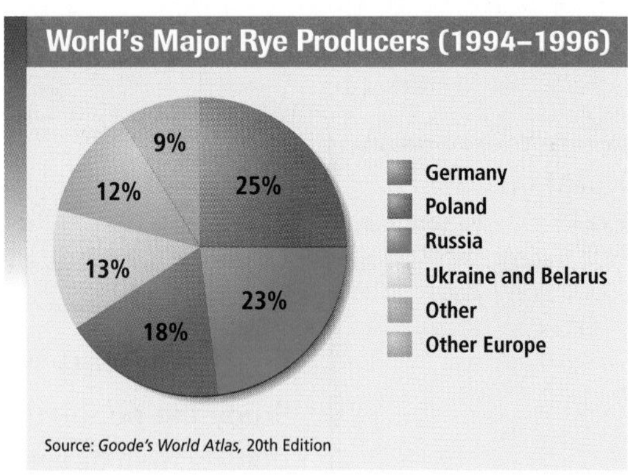

Bar graphs and line graphs include a side axis and a bottom axis. As you can see in the example below, values are labeled along the side axis. The years studied for volume and water level are placed along the bottom axis. The bars are then drawn so that their tops are aligned with the appropriate value along the side axis. The line graph identifies visible or invisible points that are also aligned to values along the side axis. You can also label values at the top of each bar or at dots along a line graph.

Pie graphs include a circle that represents the whole value of something. To create an effective pie graph, determine how many wedges will clearly fit into the circle. You can also include labels showing the percentage that each wedge represents. A legend, like the one above, can match colors to wedges they represent—in this case, particular countries that grow rye.

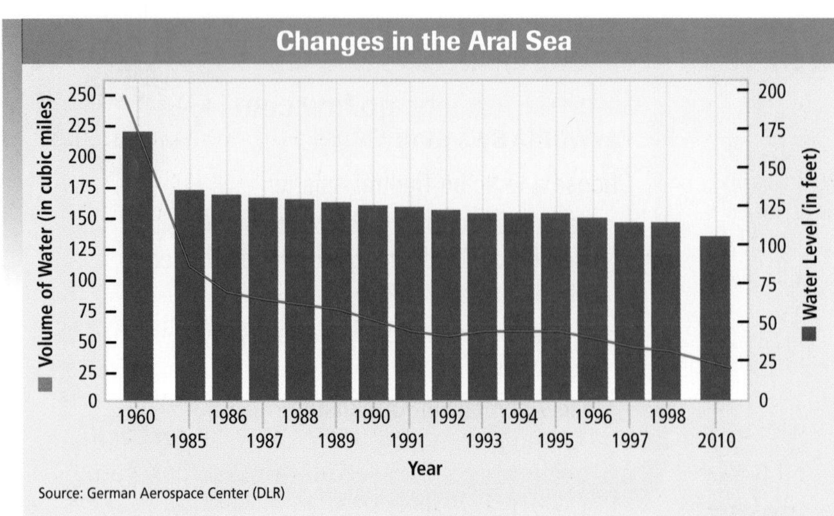

Practicing the Skill 

1. Create a pie graph showing the combined populations of Belarus, Russia, and Ukraine. Each wedge in the graph should represent one country.
2. Create a line graph showing the population of Russia from 1992 to the present.
3. Create a bar graph that compares the per capita GDP of each country you have studied throughout this unit.

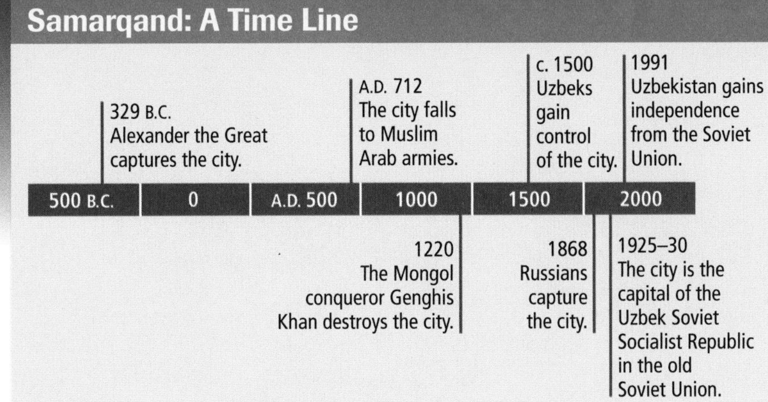

WORKSHOP 2

Using Time Lines

Time lines are visual tools that trace a chronology of events over a period of time. Time lines have a variety of uses. They can show family histories, town histories, and dates of important inventions. Time lines can also serve as records of important events in the history of a place or region. For example, the time line above shows important events in the history of Samarqand, an ancient city in Uzbekistan. Some time lines, such as those at the beginning of each regional unit in this textbook, are horizontal. However, time lines may also be vertical. The key point to remember is that a time line must progress from one point in time to another. Dates along a time line, such as the one on this page, identify key points and events.

Developing the Skill A variety of resources can help you create a time line. Newspapers, almanacs, the Internet, and encyclopedias are great resources for identifying dates and events. For example, those resources could help you create a time line of events in your state's history. You can also use biographical sources to create simple birth-to-death time lines for important historical figures. Newspapers allow you to follow events of local, national, and world news stories as they have unfolded. From this information you can create a time line for both recent and past events.

Time lines are usually read from left to right, but they can be drawn in any direction. Some time lines, for example, may read from top to bottom or bottom to top. In any case, be sure to label the events with their dates so the reader can follow the time line you create. Your labels might also include brief descriptions and illustrations. Also, space the dates and their events in proportion to the amount of real time between them. For example, if you identify important events in 1950, 1960, and 2000, you should place more space between 1960 and 2000. The extra space represents the longer elapsed time between those dates.

Remember that the years B.C. are counted backward to 1 B.C. In other words, an event that happened in 50 B.C. occurred more recently than an event in 100 B.C. Years A.D. are counted from 1 forward. If a particular date is approximated, you may place the abbreviation *c.* before it. That abbreviation stands for the Latin word *circa*, meaning "approximately." You might also include beginning and ending dates in your descriptions of events that stretch over a period of two or more years. For example, the time line on this page shows the years in which the city of Samarqand served as Uzbekistan's capital. Other examples include wars, important eras, and events that occurred repeatedly over a series of years.

Practicing the Skill

1. Study the time line above. Which dates are B.C.? Which dates are A.D.?
2. Create a time line of your life. Begin with your birth and include important events along the way. If you like, predict important events that may happen later in your life. You might include the year you graduate from high school, the year you graduate from college, and the year you hope to have reached a certain career goal.
3. Create a time line for your home state. What are the most important historical events in your state's history? You can illustrate your time line with sketches and pictures.

UNIT 6 Southwest Asia

Blue Mosque, İstanbul

CONNECTING TO

Literature

FROM THE EPIC OF GILGAMESH *translated by N. K. Sandars*

Gilgamesh is the hero of this ancient story that was popular all over Southwest Asia. In this passage Utnapishtim (oot-nuh-peesh-tuhm), whom the gods have given everlasting life, tells Gilgamesh about surviving a great flood.

In those days . . . the people multiplied, the world bellowed like a wild bull, and the great god was aroused by the clamor.[1] Enlil (en-LIL) heard the clamor and he said to the gods in council, "The uproar of mankind is intolerable[2] and sleep is no longer possible by reason of the babel."[3] So the gods agreed to exterminate[4] mankind. Enlil did this, but Ea (AY-uh) because of his oath warned me in a dream. . . . "Tear down your house, I say, and build a boat. . . . Then take up into the boat the seed of all living creatures."

Utnapishtim does as he is told. He builds a boat, fills it with supplies, his family, and animals. Then terrible rains come and flood Earth.

When the seventh day dawned the storm from the south subsided, the sea grew calm, the flood was stilled; I looked at the face of the world and there was silence, all mankind was turned to clay. . . . I opened a hatch and the light fell on my face. Then I bowed low, I sat down and I wept . . . for on every side was the waste of water. I looked for land in vain, but fourteen leagues distant there appeared a mountain, and there the boat grounded; on the mountain of Nisir the boat held fast. . . . When the seventh day dawned I loosed a dove and let her go. She flew away, but finding no resting-place she returned. . . . I loosed a raven, she saw that the waters had retreated, she ate, she flew around, she cawed, and she did not come back. Then I threw everything open to the four winds. I made a sacrifice and poured out a libation.[5]

[1]**clamor:** *noise*
[2]**intolerable:** *not bearable*
[3]**babel:** *confusing noise*
[4]**exterminate:** *kill off*
[5]**libation:** *poured as an offering to a god*

Analyzing the Primary Source

1. **Summarizing** Why did the god bring the flood?
2. **Analyzing** Why does Utnapishtim cry?

The World in Spatial Terms

Southwest Asia

Political

1. *Places and Regions* Which country in the region is landlocked?
2. *Places and Regions* Which country in the region has coastlines on both the Red Sea and Persian Gulf?

Critical Thinking

3. **Analyzing Information** Compare this map to the physical map. What geographic factors do you think allow Iraq to control its territory? TEKS

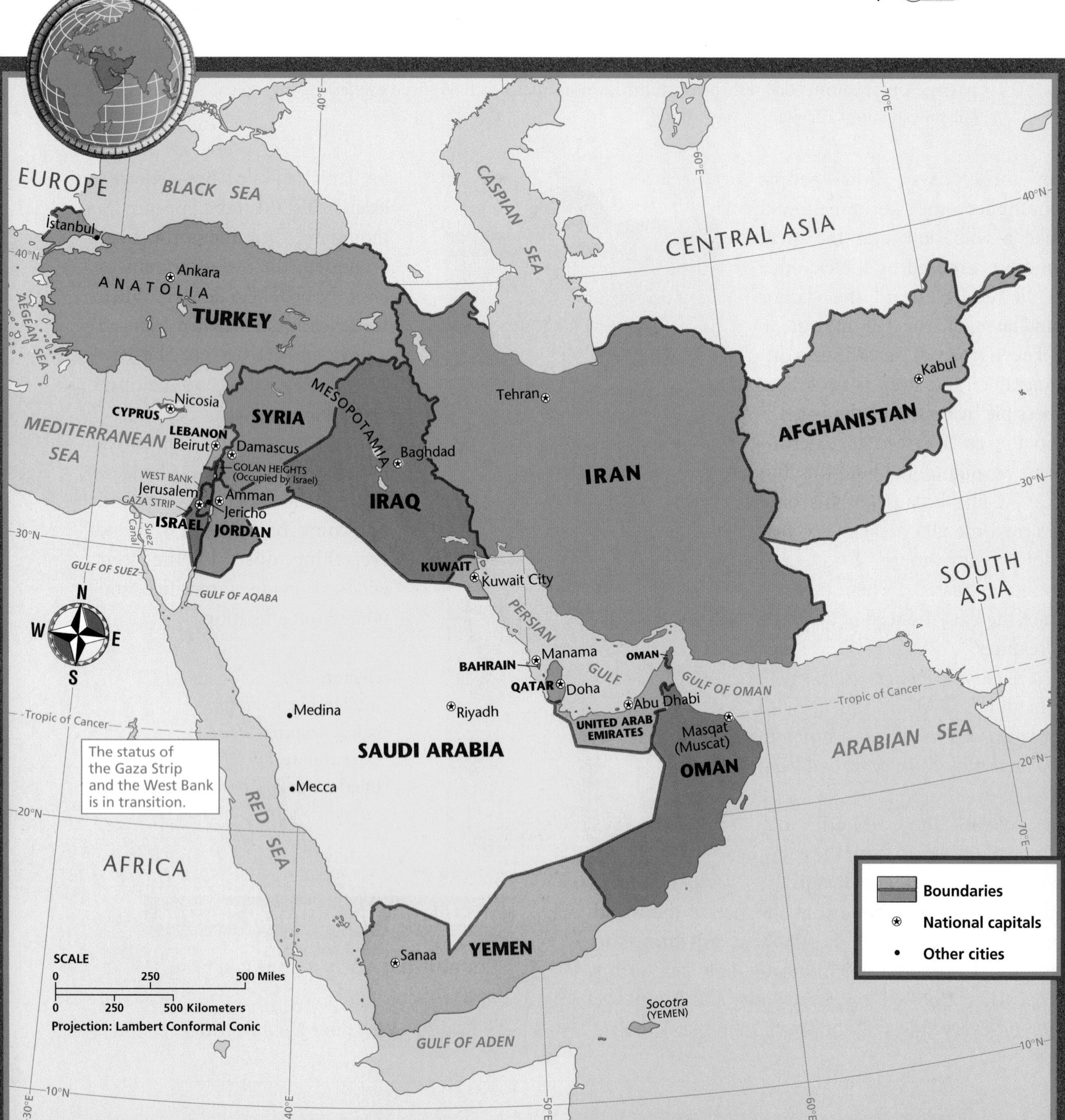

The status of the Gaza Strip and the West Bank is in transition.

SCALE
0 250 500 Miles
0 250 500 Kilometers
Projection: Lambert Conformal Conic

Boundaries
National capitals
Other cities

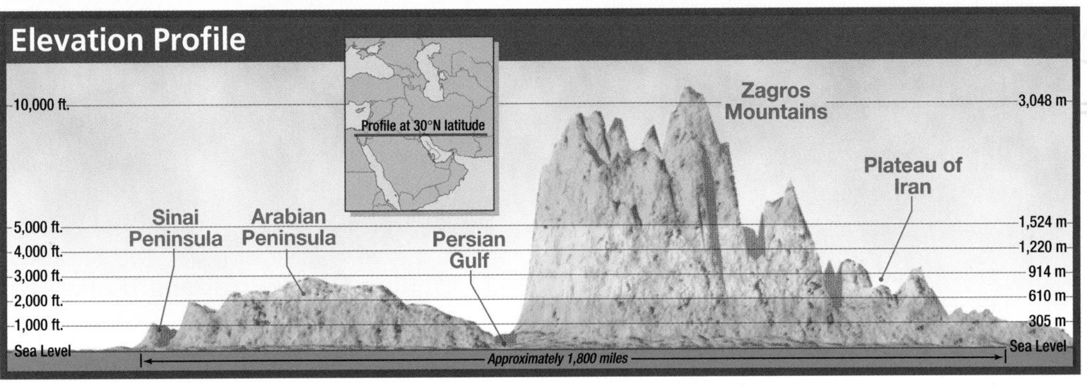

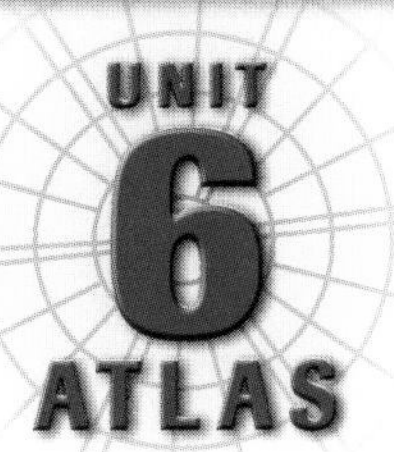

Southwest Asia: Physical

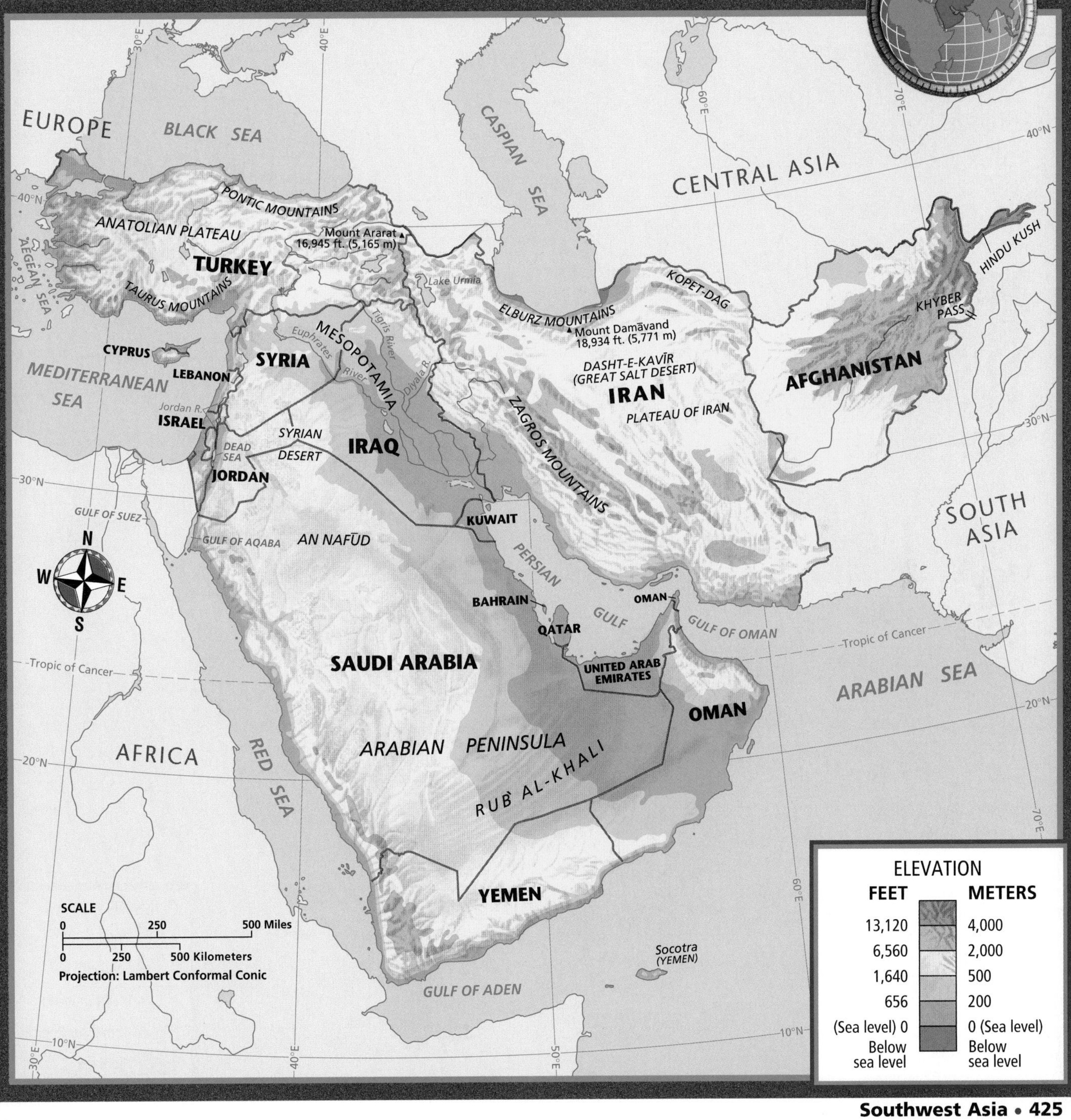

EUROPE
BLACK SEA
CASPIAN SEA
CENTRAL ASIA
PONTIC MOUNTAINS
ANATOLIAN PLATEAU
Mount Ararat 16,945 ft. (5,165 m)
TURKEY
AEGEAN SEA
TAURUS MOUNTAINS
Lake Urmia
KOPET-DAG
HINDU KUSH
ELBURZ MOUNTAINS
Mount Damāvand 18,934 ft. (5,771 m)
KHYBER PASS
CYPRUS
SYRIA
MESOPOTAMIA
Euphrates River
Tigris River
Diyala R.
DASHT-E-KAVĪR (GREAT SALT DESERT)
IRAN
AFGHANISTAN
MEDITERRANEAN SEA
LEBANON
Jordan R.
ISRAEL
PLATEAU OF IRAN
ZAGROS MOUNTAINS
SYRIAN DESERT
IRAQ
DEAD SEA
JORDAN
GULF OF SUEZ
KUWAIT
SOUTH ASIA
GULF OF AQABA
AN NAFŪD
PERSIAN GULF
BAHRAIN
OMAN
QATAR
GULF OF OMAN
Tropic of Cancer
SAUDI ARABIA
UNITED ARAB EMIRATES
ARABIAN SEA
OMAN
AFRICA
RED SEA
ARABIAN PENINSULA
RUB' AL-KHALI
YEMEN
Socotra (YEMEN)
GULF OF ADEN
40°N
30°N
20°N
10°N
30°E
40°E
50°E
60°E
70°E

SCALE
0 250 500 Miles
0 250 500 Kilometers
Projection: Lambert Conformal Conic

ELEVATION		
FEET		METERS
13,120		4,000
6,560		2,000
1,640		500
656		200
(Sea level) 0		0 (Sea level)
Below sea level		Below sea level

Southwest Asia: Climate

1. *Physical Systems* What is the most widespread climate type in Southwest Asia? TEKS
2. *Physical Systems* Compare this map to the physical map. What do you think the environment in most of the Arabian Peninsula is like? TEKS

Critical Thinking

3. **Making Generalizations** Compare this map to the physical map. What are some areas where mountain barriers might be a factor affecting climate? TEKS

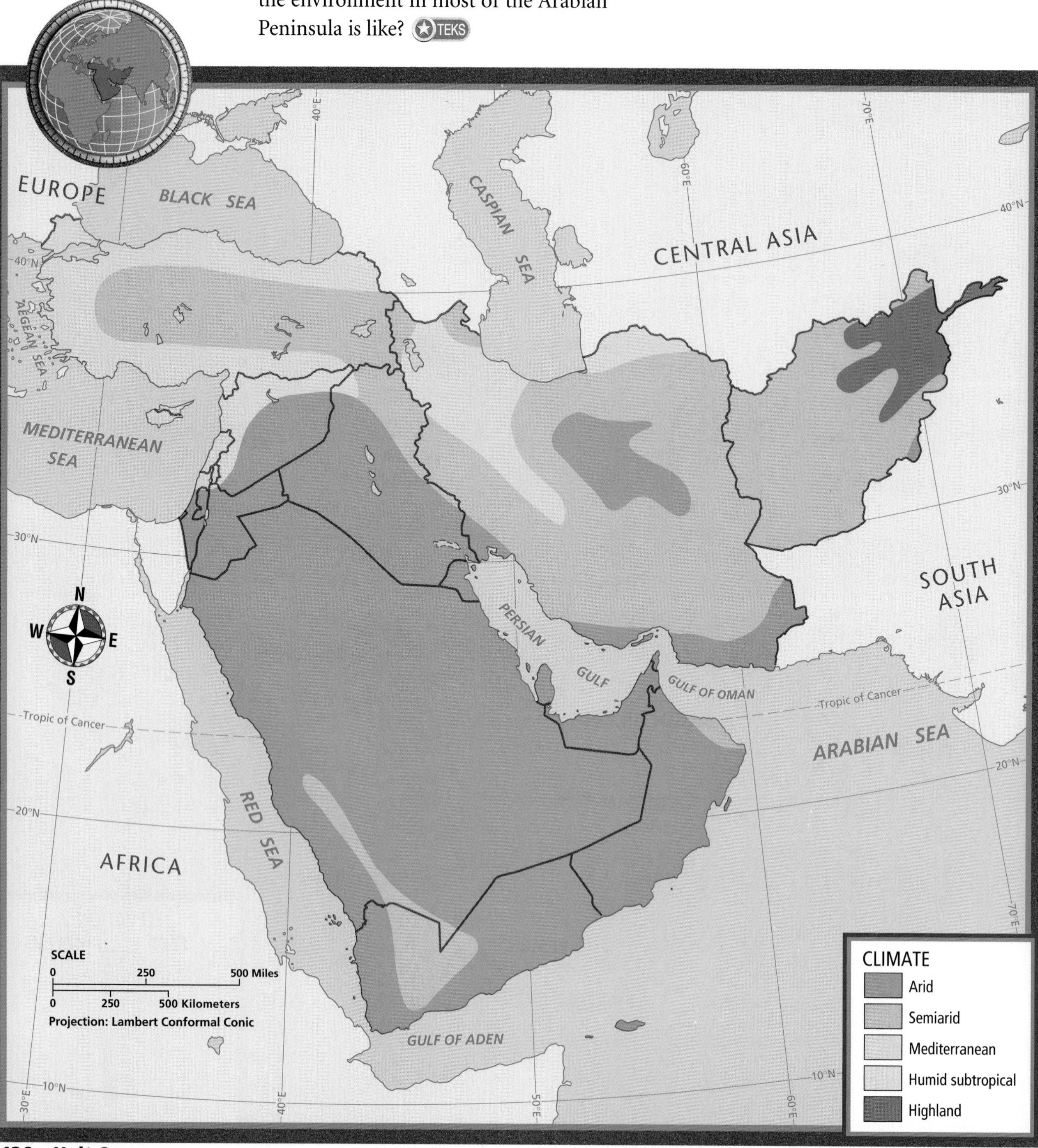

EUROPE
BLACK SEA
CASPIAN SEA
CENTRAL ASIA
AEGEAN SEA
MEDITERRANEAN SEA
PERSIAN GULF
GULF OF OMAN
SOUTH ASIA
ARABIAN SEA
RED SEA
AFRICA
GULF OF ADEN
Tropic of Cancer
40°N
30°N
20°N
10°N
30°E
40°E
50°E
60°E
70°E

SCALE
0 250 500 Miles
0 250 500 Kilometers
Projection: Lambert Conformal Conic

CLIMATE
- Arid
- Semiarid
- Mediterranean
- Humid subtropical
- Highland

UNIT 6 ATLAS

Southwest Asia: Precipitation

1. *Places and Regions* How much precipitation do most areas on the Arabian Peninsula receive? TEKS
2. *Places and Regions* Which country on the Arabian Peninsula receives the most precipitation? TEKS

Critical Thinking

3. **Drawing Conclusions** Compare this map to the physical map. Based on the distribution of precipitation in the region, which area do you think is known as the Fertile Crescent? TEKS

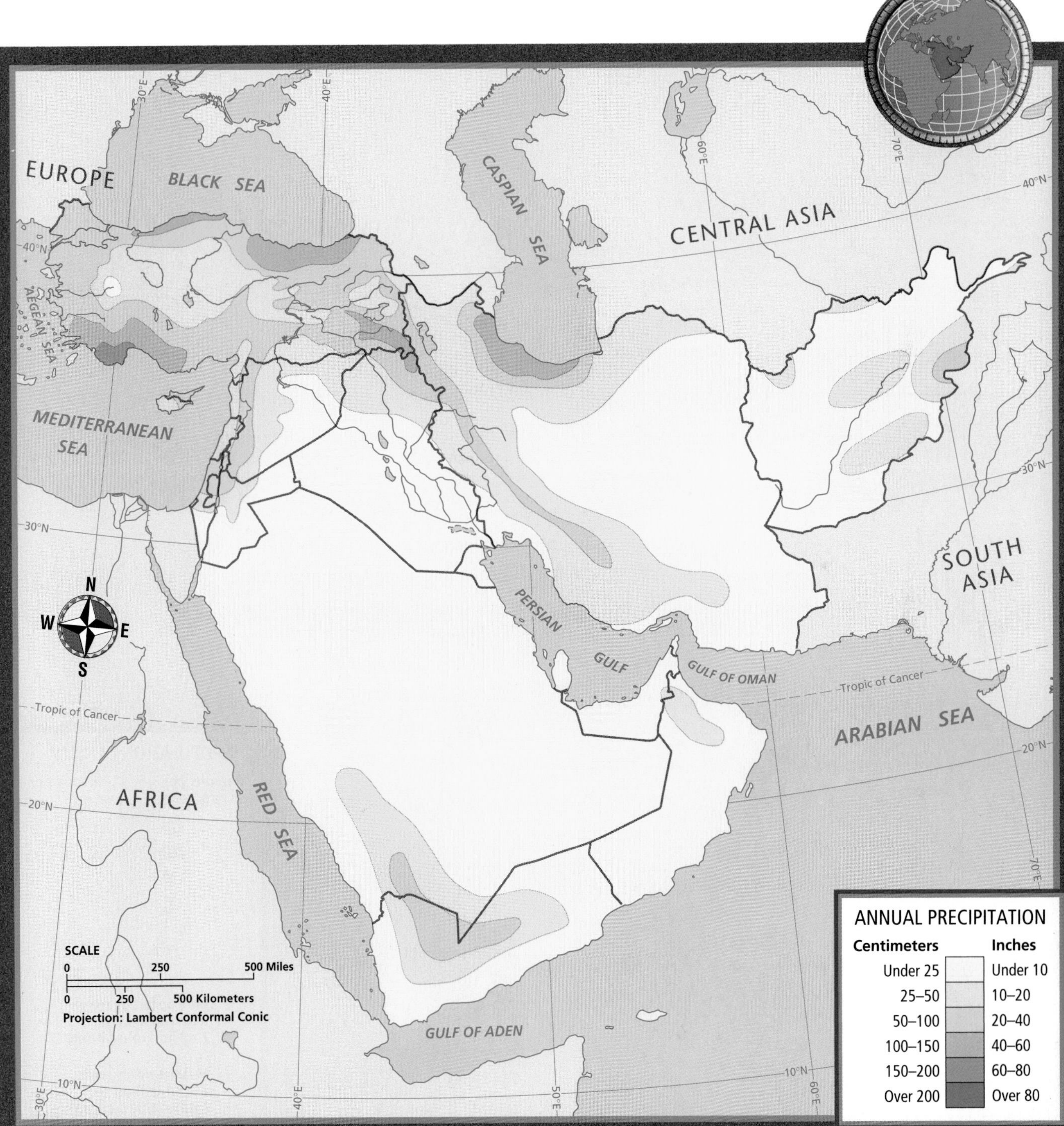

Southwest Asia: Population

1. **Environment and Society** Which country in the region appears to have the highest overall population density? TEKS
2. **Places and Regions** Compare this map to the physical and land use maps. How do you think İstanbul's location has influenced the city's growth? TEKS

Critical Thinking

3. **Analyzing Information** Compare this map to the physical map. Why do you think many coastal areas have higher population densities than areas in the interior?

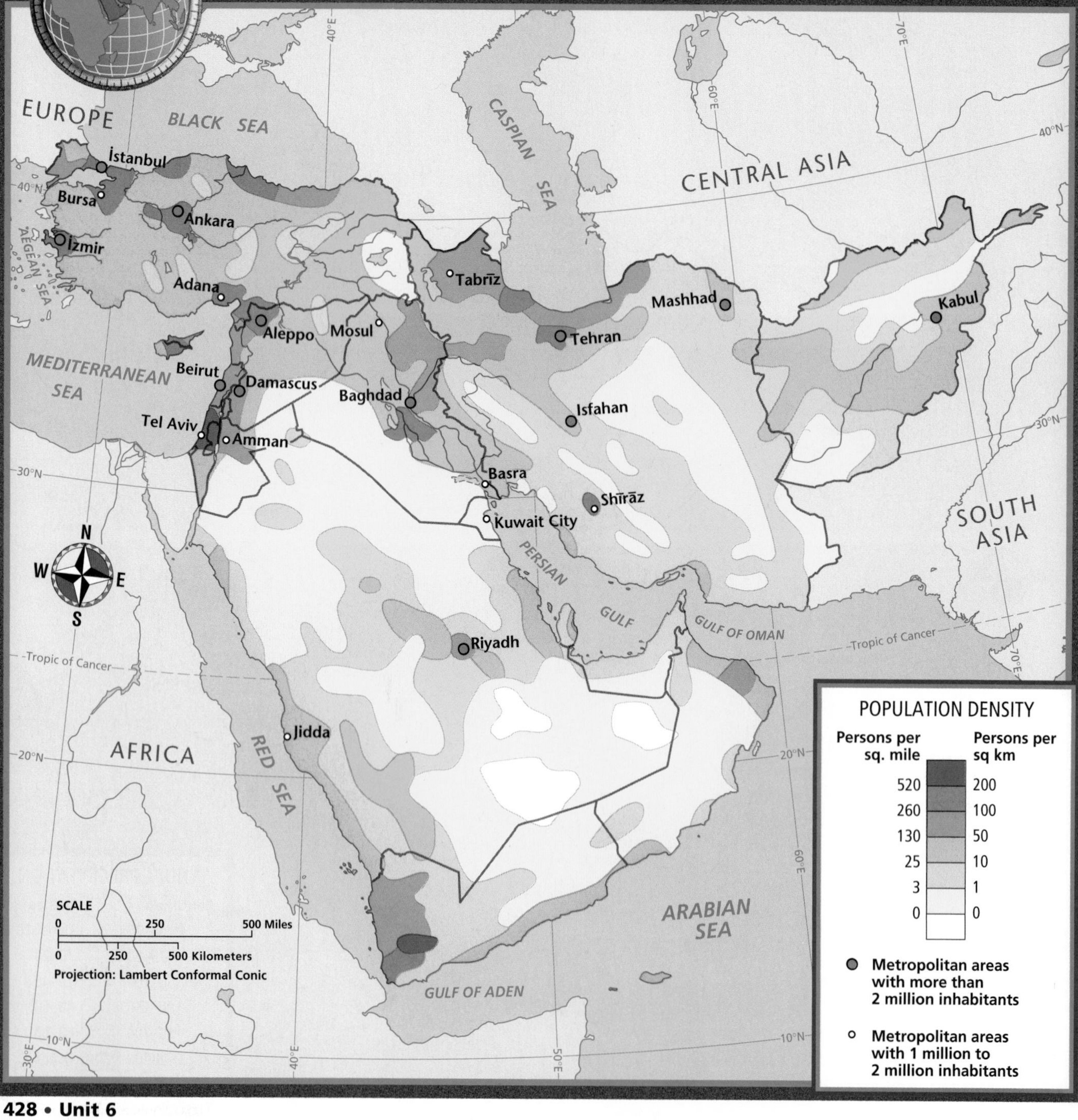

POPULATION DENSITY

Persons per sq. mile	Persons per sq km
520	200
260	100
130	50
25	10
3	1
0	0

● Metropolitan areas with more than 2 million inhabitants

○ Metropolitan areas with 1 million to 2 million inhabitants

SCALE: 0 – 250 – 500 Miles; 0 – 250 – 500 Kilometers

Projection: Lambert Conformal Conic

Southwest Asia:
Land Use and Resources

UNIT 6 ATLAS

1. *Places and Regions* Where do oil and natural gas seem to be the most concentrated in Southwest Asia?
2. *Environment and Society* Compare this map to the population map. How are land use and population distribution in Israel related?

Critical Thinking

3. **Analyzing Information** Compare this map to the climate map. How do you think farmers have adapted to arid and semiarid climates in the region? What clues to your answer do you find in the maps? TEKS

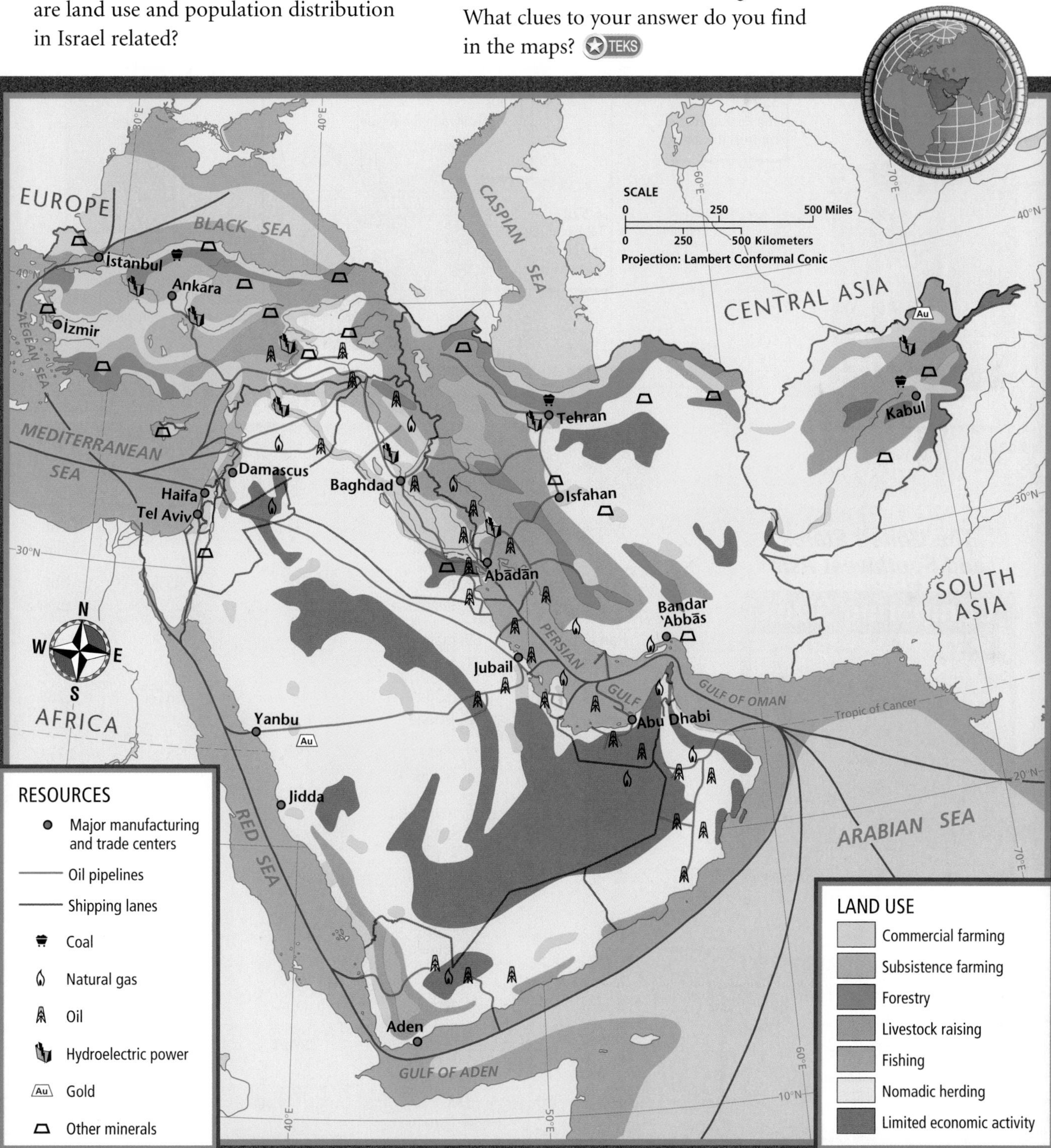

Time Line: Southwest Asia

3500 B.C.
Sumerians live in Mesopotamia.

c. 1000 B.C.
Hebrews establish a kingdom between the Jordan River and the Mediterranean Sea.

550 B.C.
The Persian Empire controls the region.

A.D. 400s
The Roman Empire and its control of the region crumble.

c. 570
Muhammad, the Prophet of Islam, is born.

1200s
Mongols invade the region.

1453
Muslim Ottomans take control of previously Christian Constantinople.

1918
The Ottoman Empire is defeated in World War I and collapses four years later.

1948
Israel becomes an independent country.

2001
Efforts to establish a lasting Arab-Israeli peace continue.

3500 B.C. | 1000 B.C. | A.D. 400 | 1200 | 1400 | 1900

Comparing Standard of Living

COUNTRY	LIFE EXPECTANCY (in years)	INFANT MORTALITY (per 1,000 live births)	LITERACY RATE	DAILY CALORIC INTAKE (per person)
Afghanistan	47, male 45, female	147	32%	Not available
Bahrain	71, male 76, female	20	85%	Not available
Iran	69, male 71, female	29	72%	2,995
Iraq	66, male 68, female	60	58%	2,268
Israel	77, male 81, female	8	95%	3,271
Kuwait	75, male 77, female	11	79%	3,160
Saudi Arabia	66, male 70, female	51	63%	2,746
Syria	68, male 70, female	34	71%	3,296
Turkey	69, male 74, female	47	85%	3,593
Yemen	58, male 62, female	69	38%	2,025
United States	74, male 80, female	7	97%	3,603

Sources: Central Intelligence Agency, *2001 World Factbook; Britannica Book of the Year, 2000*

The United States and Southwest Asia

Comparing Sizes

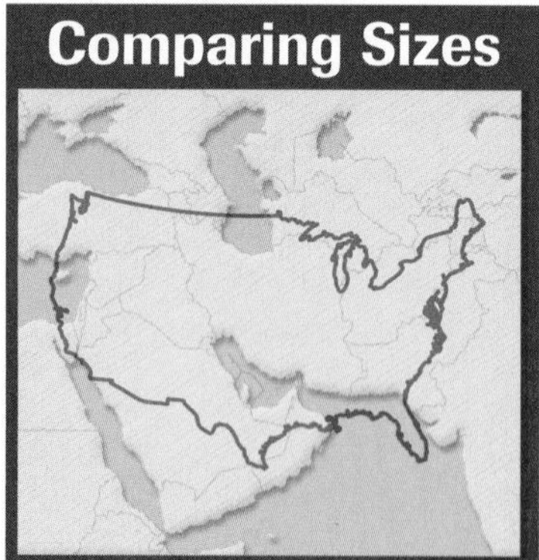

internet connect

GO TO: go.hrw.com
KEYWORD: SW3 Almanac
FOR: Additional information and reference sources

Fast Facts: Southwest Asia

FLAG	COUNTRY Capital	POPULATION (in millions) / POP. DENSITY	AREA	PER CAPITA GDP (in US $)	WORKFORCE STRUCTURE (largest categories)	ELECTRICITY CONSUMPTION (kilowatt hours per person)	TELEPHONE LINES (per person)
	Afghanistan Kabul	26.8 / 107/sq. mi.	250,000 sq. mi. 647,500 sq km	$ 800	70% agriculture 15% industry 15% services	18 kWh	0.001
	Bahrain Manama	0.6 / 2,700/sq. mi.	239 sq. mi. 620 sq km	$ 15,900	79% industry, commerce, services	8,913 kWh	0.24
	Cyprus Nicosia	0.8 / 214/sq. mi.	3,571 sq. mi. 9,250 sq km	$ 16,000 (Gr.) $ 5,300 (Tu.)	73% services (Gr.) 56% services (Tu.)	Not available	0.64
	Iran Tehran	66.1 / 104/sq. mi.	636,293 sq. mi. 1,648,000 sq km	$ 6,300	42% services 33% agriculture	1,449 kWh	0.10
	Iraq Baghdad	23.3 / 138/sq. mi.	168,753 sq. mi. 437,072 sq km	$ 2,500	52% services 12% agriculture	1,173 kWh	0.03
	Israel Jerusalem	5.9 / 741/sq. mi.	8,019 sq. mi. 20,770 sq km	$ 18,900	31% public service 20% manufacturing	5,372 kWh	0.47
	Jordan Amman	5.2 / 151/sq. mi.	35,637 sq. mi. 92,300 sq km	$ 3,500	52% services 11% industry	1,280 kWh	0.08
	Kuwait Kuwait City	2.0 / 297/sq. mi.	6,880 sq. mi. 17,820 sq km	$ 15,000	50% gov., social svs. 40% services	14,377 kWh	0.20
	Lebanon Beirut	3.6 / 904/sq. mi.	4,015 sq. mi. 10,400 sq km	$ 5,000	62% services 31% industry	2,167 kWh	0.19
	Oman Musqat	2.6 / 32/sq. mi.	82,031 sq. mi. 212,460 sq km	$ 7,700	24% public admin. 15% construction	3,061 kWh	0.08
	Qatar Doha	0.8 / 174/sq. mi.	4,416 sq. mi. 11,437 sq km	$ 20,300	51% services 22% construction	10,882 kWh	0.18
	Saudi Arabia Riyadh	22.8 / 30/sq. mi.	756,981 sq. mi. 1,960,582 sq km	$ 10,500	63% services 25% industry	4,904 kWh	0.14
	Syria Damascus	16.7 / 234/sq. mi.	71,498 sq. mi. 185,180 sq km	$ 3,100	40% agriculture 40% services	997 kWh	0.08
	Turkey Ankara	66.5 / 221/sq. mi.	301,382 sq. mi. 780,580 sq km	$ 6,800	38% agriculture 38% services	1,797 kWh	0.29
	United Arab Emirates Abu Dhabi	2.4 / 75/sq. mi.	32,000 sq. mi. 82,880 sq km	$ 22,800	60% services 32% industry	14,177 kWh	0.38
	Yemen Sanaa	18.1 / 89/sq. mi.	203,849 sq. mi. 527,970 sq km	$ 820	71% agriculture 13% trade	123 kWh	0.02
	United States Washington, D.C.	281.4 / 76/sq. mi.	3,717,792 sq. mi. 9,629,091 sq km	$ 36,200	30% manage., prof. 29% tech., sales, admin.	12,211 kWh	0.69

Sources: Central Intelligence Agency, *2001 World Factbook; The World Almanac and Book of Facts, 2001; Britannica Book of the Year, 2000;* U.S. Census Bureau
The CIA calculates per capita GDP in terms of purchasing power parity. This formula equalizes the purchasing power of each country's currency.

CHAPTER 19

The Persian Gulf and Interior

The human geography of the Persian Gulf and interior Southwest Asia is changing as the modern world mixes with the old. In this chapter you will read about changes affecting the region's cultures and economic development.

Illustrated manuscript from ancient Persia

Date harvest in Bahrain

Greetings from Iran! My name is Mitra. I live with my family on the top floor of a house in Tehran, the capital. The roof of our house is flat, and in nice weather we sit up there and look out at the mountains. In the summer we sleep there too.

My favorite holiday is No Ruz, our New Year's festival, in the spring. We get two weeks off from school. We go see all our relatives, who give us clean new money from the bank. We visit the older people first, like my mother's family in Tehran, and then they visit us. The second week we may go to Isfahan, where my father is from, and see my grandmother. Lots of sweets and pastries are served at these visits. At dinner we may have special New Year's dishes. We also set up a table with a cloth with seven food items on it that begin with the *s* sound in our language. This spread is called the cloth of seven dishes. The foods on the table symbolize rebirth, health, happiness, prosperity, joy, patience, and beauty. For example, *serkeh* is vinegar and represents age and patience. You might compare this with your putting up a Christmas tree.

My mother's family is from the north, near the Caspian Sea. During our vacation, we go to her family's summer house. I like to go to the beach and swim. There is one beach for boys and another for girls. I also love to go to the Tehran bazaars with my mother. Food shops, fruit markets, gold shops—they are all mixed together. The shop owners pull out the best things to show us. We bargain for everything.

1 Natural Environments

READ TO DISCOVER

1. What landforms and rivers can be found in the Persian Gulf area and the interior of Southwest Asia?
2. How does the region's physical geography affect its climates and biomes?
3. What natural resources does the region have?

WHY IT MATTERS

Huge oil deposits lie in the Persian Gulf region. How the countries there manage this resource affects the U.S. economy. Use CNNfyi.com or other **current events** sources to learn about the relationships the United States has with oil-rich countries in this region.

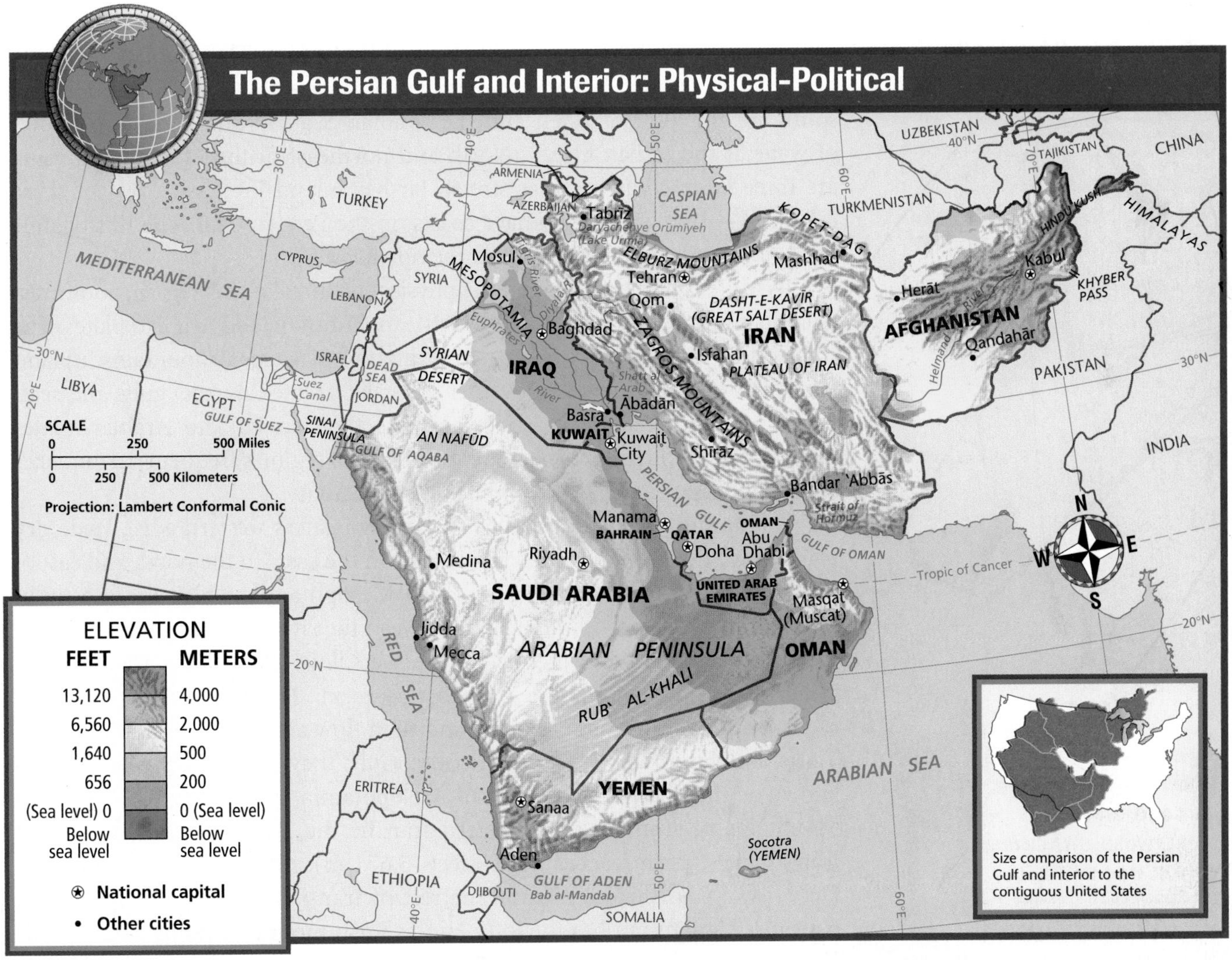

DEFINE

exotic rivers
oasis

LOCATE

Persian Gulf
Arabian Peninsula
Red Sea
Gulf of Aden
Arabian Sea
Mesopotamia

Locate, continued

Tigris River
Euphrates River
Shatt al Arab
Zagros Mountains
Elburz Mountains
Kopet-Dag
Hindu Kush
Rub' al-Khali
An Nafūd
Caspian Sea

The Persian Gulf and Interior: Physical-Political

SCALE
0 250 500 Miles
0 250 500 Kilometers
Projection: Lambert Conformal Conic

ELEVATION

FEET	METERS
13,120	4,000
6,560	2,000
1,640	500
656	200
(Sea level) 0	0 (Sea level)
Below sea level	Below sea level

⊛ National capital
• Other cities

Size comparison of the Persian Gulf and interior to the contiguous United States

INTERPRETING THE VISUAL RECORD

Eastern Afghanistan lies in the shadow of the rugged Hindu Kush mountain range. Crossing these mountains is difficult, as only a few major passes cut through the range. ***How do landforms like the Hindu Kush divide a continent into regions?*** TEKS

Landforms and Rivers

The region formed by the Persian Gulf and interior Southwest Asia includes Saudi Arabia and the smaller countries of the Arabian Peninsula. These countries are Bahrain, Kuwait, Oman, Qatar, the United Arab Emirates, and Yemen. This region is also often referred to as the Middle East. The Arabian Peninsula lies between the Red Sea and the Persian Gulf. (See the chapter map.) To the south are the Gulf of Aden and the Arabian Sea. Beyond these two bodies of water is the Indian Ocean. North and northeast of the Arabian Peninsula are three large countries that stretch farther inland into Asia. Two of these countries—Iraq and Iran—have coasts on the Persian Gulf. The third country, landlocked Afghanistan, lies to the northeast.

Tectonic forces have shaped the physical features of this region. Southwest Asia sits near the intersection of the African, Eurasian, and Arabian plates. The collision of these plates has created a mixture of rugged mountains, upland plateaus, and valleys. Plate movement has also created narrow gulfs and seas, which are bordered by coastal plains. As the African and Arabian plates move apart, the Red Sea is becoming wider. The region's frequent earthquakes are reminders of Earth's continuing tectonic activity.

Mountains stretch along the Arabian Peninsula's western edge. Wide dry plains slope down toward the Persian Gulf in the east. To the north and east of the Arabian Peninsula is a region called Mesopotamia, which lies mostly in Iraq. In Greek *Mesopotamia* means "between the rivers." In fact, Mesopotamia is a wide plain through which two great rivers flow. These rivers are the Tigris (TY-gruhs) and the Euphrates (yooh-FRAY-teez). They are **exotic rivers**, or rivers that begin in humid regions and then flow across dry areas. Before the Tigris and Euphrates reach the Persian Gulf, they join in a single channel known as the Shatt al Arab. East of Mesopotamia are the Zagros (ZA-gruhs) Mountains of Iran. In northwestern Iran lie the Elburz Mountains, where Iran's highest peak reaches 18,605 feet (5,671 m). Another range, the Kopet-Dag, rises in northeastern Iran. The rest of Iran is mainly made up of high plateaus. Afghanistan has the lofty Hindu Kush mountain range. Some peaks

GO TO: go.hrw.com
KEYWORD: SW3 CH19
FOR: Web sites about the Persian Gulf and interior

there rise higher than 24,000 feet (7,300 m). The Hindu Kush range is the western extension of the world's highest mountains, the Himalayas.

READING CHECK: *Physical Systems* What physical process is shaping the Red Sea?

The Rub' al-Khali is the world's largest region of active sand dunes—dunes that migrate over time. The English name for this region is the Great Sandy Desert.

Climates, Plants, and Animals

Hot and dry climates dominate the region. Rains come mostly during the winter when the westerly winds of the middle latitudes bring occasional cyclonic storms. The southern interior is a mostly uninhabited desert called the Rub' al-Khali (ROOB ahl-KAH-lee), or "Empty Quarter." Farther north lies the An Nafūd (ahn nah-FOOD), a desert of reddish sand. At its widest point, this desert stretches to 140 miles (225 km).

The region's mountains provide water to the valleys below. An orographic effect produces these more humid climates. The region's wettest climate is in Iran in a narrow zone along the southern shore of the Caspian Sea. Here winds blow southward over the water and pick up moisture. As the air rises along the Elburz Mountains, rain falls. The Zagros Mountains and the mountains of Yemen also have more humid climates.

The lowlands of Saudi Arabia along the Persian Gulf are among the hottest places in the world. Because of subtropical high pressure with clear skies and little shade, the summer daytime temperatures often climb above 114°F (46°C). Summer rains are almost unknown here, but nearness to the sea keeps the humidity high. However, inland areas are very dry. These areas experience rapid cooling at night because the air holds such a small amount of moisture. In fact, after sunset desert temperatures can drop some 30 degrees in just a few hours. Elevation also influences temperatures, so the region's highlands are much cooler than lowlands in general. Both Iran and Saudi Arabia have mountain resorts where people can escape the summer heat. Winter skiing is popular in Iran's Elburz Mountains.

Shrubs and grasses cover the region's wide dry plains. Trees are common only in mountain regions and the usually dry streambeds. The highest plains are grasslands. In the driest areas the ground is bare rock and sand. In some places the soil is so salty that no plants can grow. Nearly all the region's plants have adapted to survive long periods without rain. Roots either grow deep or spread out to capture as much water as possible. Many plants have developed leaves or stems that allow them to store moisture.

Hunting by humans and competition from domestic animals have made life difficult for the region's larger wild animals. Gazelles and wild goats were common a few centuries ago. Hyenas, leopards, lions, and tigers as well as herds of wild camels and donkeys once roamed the region. Most large wild mammals are now rare or restricted to a few game reserves. Today all the camels and donkeys are domesticated. Reptiles, including lizards and several poisonous snakes, are still common.

READING CHECK: *Physical Systems* What kinds of climates dominate the region? What produces humid climates in the region?

One of the driest areas in the world, the Rub' al-Khali receives an average of less than 4 inches (10 cm) of rainfall each year. Few water sources are found in this desert, but oil reserves have been discovered beneath its sands.

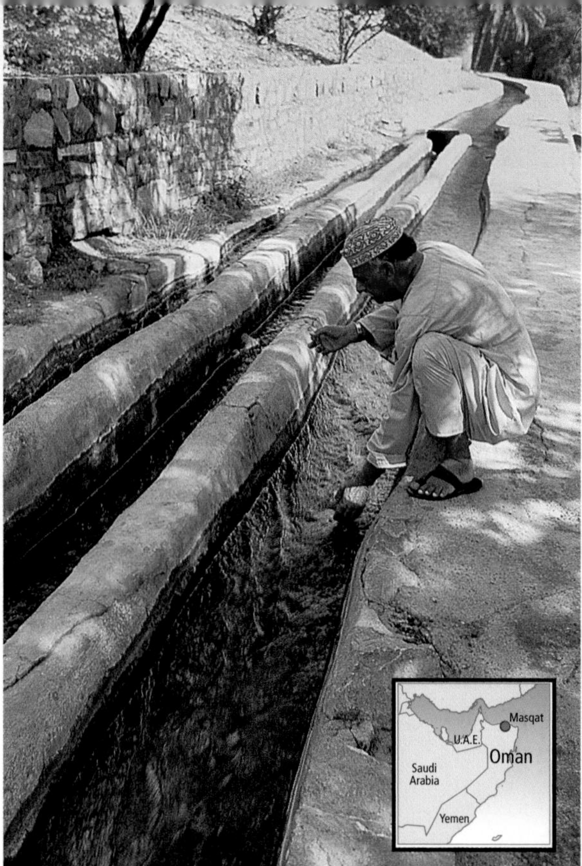

INTERPRETING THE VISUAL RECORD *An Omani man collects water from a local* falaj, *or aqueduct. These channels are dug to carry water from desert springs to farms and villages. Many of the* falaj *systems in use today were built more than 1,000 years ago.* ***How do you think irrigation systems like this one have changed the landscape of dry areas in Oman?*** TEKS

Natural Resources

The region's two most important natural resources are oil and water. Oil is plentiful, but water is not. The Tigris and Euphrates Rivers are the main sources of water in Iraq. Canals lead away from these rivers, bringing precious water to the surrounding dry lands. In the high plateaus and mountains of northern Iran, farmers depend on rain for agriculture. Farmers in most other places must irrigate their fields.

Surface water is rare in the desert areas. It can be found only at an **oasis**, an area where a spring bubbles to the surface. People have made many of these springs into productive wells. Wells may also reach water below a dry river bed. Deep wells may tap into fossil water, groundwater that is not being replaced by rainfall. Fossil water is not a renewable resource. Desalinization of seawater provides another source of freshwater. In general, however, only wealthy countries get freshwater this way. This is because the process uses large amounts of power and is expensive. Saudi Arabia produces more desalinized water than any other country.

Oil is the region's most valuable natural resource. The oil reserves along the Persian Gulf are the largest in the world. Iraq, Oman, and Yemen also have important oil deposits. However, the region's countries have few other resources for developing industry. Only Iran has the potential to develop an industrial economy from its own reserves of metallic ores.

TEKS **READING CHECK:** *Places and Regions* How can technology expand the region's supply of freshwater?

TEKS Questions 1, 3, 4, 5

Review

Define

exotic rivers
oasis

Working with Sketch Maps

On a map of Southwest Asia that you draw or that your teacher provides, label the Persian Gulf, Arabian Peninsula, Red Sea, Gulf of Aden, Arabian Sea, Mesopotamia, Tigris River, Euphrates River, Shatt al Arab, Zagros Mountains, Elburz Mountains, Kopet-Dag, Hindu Kush, Rub' al-Khali, An Nafūd, and Caspian Sea. What do the names *Rub' al-Khali* and *Mesopotamia* mean?

Reading for the Main Idea

1. *Physical Systems* What are four physical features that influence what plants can grow in different places in the region?
2. *Environment and Society* What are the region's two most precious natural resources?

Critical Thinking

3. **Making Generalizations and Predictions** Which landforms in the region seem most favorable for human settlement? Which seem the least favorable? Why?
4. **Identifying Cause and Effect** Why do parts of Saudi Arabia have high temperature variations?

Organizing What You Know

5. Copy the chart below. Use it to list major physical features that can be found in the region's largest countries. Use the information in Section 1 and the chapter map.

Feature	Afghanistan	Iran	Iraq	Saudi Arabia
Coastal plains				
Interior plains				
Major rivers				
Plateaus				
Mountains				

Section 2 History and Culture

READ TO DISCOVER

1. How have peoples, empires, and Islam affected the history of the Persian Gulf area and interior Southwest Asia?
2. What are the major features of the region's cultures?

WHY IT MATTERS

Many Americans trace their heritage to Southwest Asia. Use CNNfyi.com or other **current events** sources to learn about recent immigrants from Southwest Asia and their cultures.

IDENTIFY

Sunni
Shi'ism

DEFINE

dynasty
imams

LOCATE

Mecca
Medina

From Empires to Independence

The world's first civilizations developed in the area known as the Fertile Crescent. This arc of productive land extends northward from the Persian Gulf and through the plains of the Tigris and Euphrates Rivers. It continues to Asia Minor and the Mediterranean coast. Asia Minor is the Asian part of what is now Turkey. Many of the plants and animals found on farms throughout the world today may have been first domesticated in the Fertile Crescent region. By about 3000 B.C. a people called the Sumerians built the world's first known cities in southern Mesopotamia. These cities depended on nearby irrigated fields of wheat and barley. City merchants traded goods from throughout the region.

The ruins of Persepolis reflect the former glory of the Persian Empire, which once stretched from Egypt to Afghanistan. King Darius I built Persepolis in about 500 B.C. as the capital of his empire.

Connecting to MATH

Arabic Numerals

The symbols that most of the modern world uses to write numbers are known as Arabic numerals. Just 10 symbols are needed to write a number of any size. These symbols—0 through 9—probably evolved from letters in the Arabic alphabet. The mathematical system itself, in which a symbol's value depends on its place in the number, actually began in India. Arab traders brought the system back to Southwest Asia and later spread its use into Europe. The Arabic system proved much simpler than the old Roman system it gradually replaced. For example, CCCXXXIII in Roman numerals became 333 using Arabic numerals. This simplification of the number system aided the growth of science and commerce. Muslim scholars made important advances in art, astronomy, literature, medicine, and mathematics.

Identifying Cause and Effect How do you think Arabic numerals may have aided the growth of science and commerce? TEKS

The rich resources of Mesopotamia attracted invaders again and again. Akkadians conquered the Sumerian cities in about 2350 B.C. and created the region's first real empire. That empire extended to the Zagros Mountains and the Mediterranean coast. The islands in the Persian Gulf became major trading centers. Eventually, these trade centers linked the merchants of Mesopotamia and India.

Over time cities grew and declined in the region. Again and again invaders from the west, east, and north battled the peoples of the plains and set up new empires. About 550 B.C. an empire developed in Persia, where Iran is today. The Persians conquered both Mesopotamia and Asia Minor. Theirs was one of the largest, richest, and most powerful empires in world history. Later, the Greeks and then the Romans controlled much of the region for a time.

The Rise of Islam The prophet Muhammad, who lived in the region from about A.D. 570 to 632, established Islam. Islam is one of the world's most widely practiced religions. Muhammad was born in Mecca, a city in the western part of the Arabian Peninsula. At about the age of 40, a religious experience changed his life. He reported that a messenger of God, the angel Gabriel, told him to preach the word of God. The word *God* in Arabic is *Allah*. Muhammad spread Allah's message to his followers, called Muslims. Muslims are people who practice Islam. A holy book called the Qur'an (Koran) contains what Muslims believe to be Allah's message to Muhammad.

Muhammad established a Muslim community centered at Medina. By the time Muhammad died, Islam had spread to most of the Arabian Peninsula. After his death, Arab armies carried Islam as far west as Morocco and Spain in a little more than a century. Over time Islam spread to Central Asia, India, and Southeast Asia along land and sea trade routes. Muslims in all these places now face Mecca, Islam's holiest city, when they pray.

Gaining Independence Empires continued to rise and fall in Southwest Asia until modern times. In the 1200s the Mongols swept out of Central Asia to conquer what are now Afghanistan, Iraq, and Iran. Rulers called the Safavids (sah-FAH-weedz) came to power in Iran in the early 1500s. As their power expanded eastward, the Safavids seized Afghanistan from the Muslim rulers of India. Historians consider the rule of Safavids, which lasted more than 200 years, a golden age of Persian culture. Literature, architecture, and the arts flourished. Persian carpets, ceramics, and textiles became renowned. The Safavid **dynasty**, or line of hereditary rulers, ended in the mid-1700s. The expanding British and Russian Empires tried to control Iran and Afghanistan in the 1800s. During the 1900s both became independent countries.

The history of the western part of the region followed much the same pattern. In the 1500s the Ottoman Turks conquered Mesopotamia and the east and west coasts of the Arabian Peninsula. They held much of this area until the early 1900s, when the British took over. Iraq and Saudi Arabia emerged as independent countries in 1932. Kuwait, Bahrain, Qatar, the United Arab Emirates, and Yemen did not become independent from Britain until the 1960s and 1970s.

READING CHECK: *Human Systems* What are some of the peoples and empires that have controlled parts of the region?

Culture

Islam is the unifying cultural feature of the region. However, the region's long history of changing empires and migrations has resulted in the presence of many ethnic groups today.

People and Languages Most people in the Persian Gulf and interior Southwest Asia are Arabs, and Arabic is the dominant language. The spread of Islam encouraged the spread of the Arabic language over time. To read the Qur'an, Muslims had to learn Arabic, which also became the common language of scholarship and trade. Trade routes have long connected the distant parts of the Islamic world. Eventually, all Arabic speakers became known as Arabs. Today Arabic place-names are found in Spain and Morocco, Central Asia, and India. This diffusion of Arabic words is a result of migration, trade, and the spread of Islam.

Like Saudi Arabia, Iraq is an Arab country. In addition, there are more than a million Arabs in southern Iran. However, non-Arab ethnic groups are also numerous in the region. For example, the Kurds, who live in the borderlands of Iran, Iraq, Syria, and Turkey, are Muslims but not Arabs. The Kurds have never had their own country. Their desire for self-rule is a source of political unrest in some countries, including Iraq.

Cultural diversity is even more complex in Iran and Afghanistan. Most of Iran's people are Persians who speak Farsi. The Kurds, Baloch, Bakhtiari, and Hazara also speak languages related to Farsi. However, Persians dominate Iran and hold most of the important positions in Iranian society. A number of other ethnic groups speak Turkic languages. Turkmen communities are found in northeastern Iran. In the northwest are the Azeri people, Iran's largest group after the Persians. The Qashqai people of the southern Zagros Mountains also speak a Turkic language.

In Afghanistan the Pashtun make up the largest ethnic group. The name *Pashtun* really refers to a number of tribes that speak the Pashtu language. They are closely related to the Tajiks to the north as well as to small tribes in eastern Iran. A number of other ethnic groups also live in Afghanistan. Yet people's loyalties often rest more with their clan and family than with their ethnic group.

READING CHECK: *Human Systems* What processes aided the diffusion of Arabic culture?

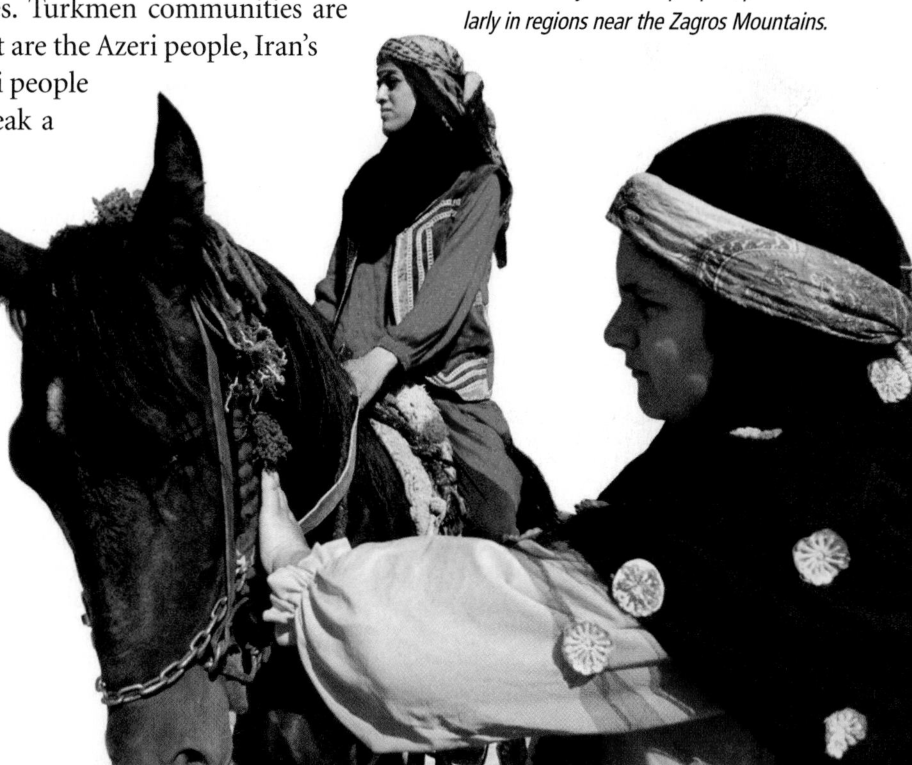

Horse races in Iran between different nomadic groups are popular events. Iran is home to many nomadic peoples, particularly in regions near the Zagros Mountains.

Each year some 2 million Muslims from around the world travel to Mecca, Islam's holiest city. All Muslims are expected to make this sacred journey, called the hajj, at least once during their lifetime. The pilgrims gather around the Ka'bah, a sacred shrine, for several days of prayer. Historically the experience of the hajj has served as a common bond among all Muslims, including followers of Sunni Islam and Shi'ism. However, overcrowding, political disputes, and religious differences have also led to tense confrontations and violence inside the holy city.

FOCUS ON CULTURE

Religion and Society Islam has split into two main branches and many different groups over the centuries. The two main branches are **Sunni** Islam and **Shi'ism**. Their differences center around who can become religious leaders, called **imams**, in Muslim society. Sunni groups choose their imams, who serve mainly as leaders of prayer. The Shia—those who practice Shi'ism—allow only descendants of the prophet Muhammad's family to become imams. The Shia also rely on imams to interpret the Qur'an and other sacred texts containing teachings that govern personal conduct.

Today about 90 percent of Muslims are Sunnis, and 10 percent are Shia. Sunnis are found everywhere Islam has spread. Because they make up the majority of Muslims, Sunnis are sometimes called orthodox Muslims. Shi'ism is concentrated in Iran, southern Iraq, Yemen, Syria, and Lebanon. Shia imams are particularly important in Iran. In that country they have considerable political as well as religious authority.

Both Sunnis and Shia share many of the same basic beliefs and practices of Islam. For example, members of both groups are expected to make a religious pilgrimage to Mecca. However, throughout history there have also been conflicts between the two groups. Some conflicts have been caused by the persecution of one group by the other.

READING CHECK: *Human Systems* How does the role of a Sunni imam differ from that of a Shia imam?

TEKS Questions 1, 2, 4, 5

Section 2 Review

Homework Practice Online
Keyword: SW3 HP19

Identify Sunni, Shi'ism

Define dynasty, imams

Working with Sketch Maps On the map you created in Section 1, label the countries of the Persian Gulf and interior Southwest Asia, Mecca, and Medina. Then use the description in Section 2 to shade in the Fertile Crescent.

Reading for the Main Idea

1. *Human Systems* On what did the early cities of the Fertile Crescent depend for their growth? In what way is farming today connected with the early history of the Fertile Crescent region?
2. *Human Systems* Why will you find the Arabic language, Arabic place-names and Islam in places outside of Southwest Asia today?
3. *Places and Regions* Which ethnic group is most widely spread throughout the region? Which groups dominate Iran and Afghanistan?

Critical Thinking

4. **Making Generalizations** What major cultural feature is common throughout the region? What other cultural features could you use to further divide the Persian Gulf and interior Southwest Asia into more regions?

Organizing What You Know

5. Create a diagram like the one shown below. Use it to identify the various peoples who conquered or controlled the region. Add more boxes as needed.

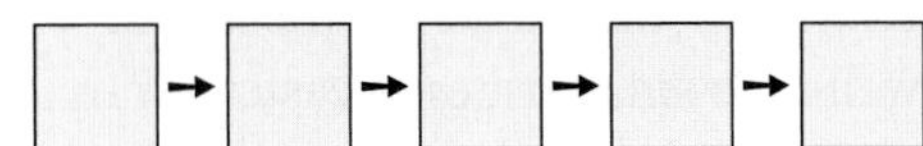

The Region Today

READ TO DISCOVER

1. On what activities do the region's economies depend?
2. What are the region's cities like?
3. What are some important issues in the region today?

WHY IT MATTERS

U.S. military forces were involved in the Persian Gulf War in 1991 and remain in the region today. Use CNNfyi.com or other **current events** sources to learn about recent military actions in the Persian Gulf region.

IDENTIFY

Bedouins
OPEC

DEFINE

ayatollahs
theocracy

LOCATE

Tehran
Baghdad
Kabul
Riyadh
Strait of Hormuz

Economic Development

Oil and gas production is central to the economies of the countries along the Persian Gulf. Saudi Arabia alone produces some 8.25 million barrels of oil a day. Oil wealth has helped modernize the economies of the region. However, many people continue to follow traditional rural ways of life and culture. Arid climates and a rugged landscape make farming difficult. In addition, farmers find fertile soils mainly in the river valleys, on high plateaus, and at a few oases. Partly as a result of this, all the region's countries must import food. Many of the more humid areas have been overgrazed, leading to soil erosion. People have cut down most of the mountain forests that once existed. Some countries in the region are now trying to conserve soil and native plants.

Most farmers practice subsistence agriculture, producing only enough to support their families. They sell any surplus at local markets. Barley and wheat are the most common grains grown in the area. Farmers may also raise livestock—mostly sheep, goats, and some cattle. Because Islam forbids eating pork, pigs are not raised. Commercial farms are typically found near the large cities. They provide fruits and vegetables for city markets. Where irrigation is available, farmers grow citrus fruits, dates, grapes, nuts, and olives for sale. Specialized commercial farms, such as modern dairies and chicken farms, are becoming more common in the oil-rich countries. Saudi Arabia even grows flowers for export to Europe.

Nomadic herders, some of whom are known as **Bedouins**, live in outlying dry lands. They tend to move their camels, goats, or sheep in regular routes as the seasons change. These herders trade their animals, animal products, and handicrafts in towns. More and more Bedouins are leaving this traditional way of life and taking jobs in cities and on the new modern farms.

INTERPRETING THE VISUAL RECORD

Although Bedouins make up only a small percentage of Iraq's population, they move with their herds across much of the country. However, recent laws have restricted the lands that are available to Bedouins for grazing. ***How do you think these laws are changing Bedouin culture?***

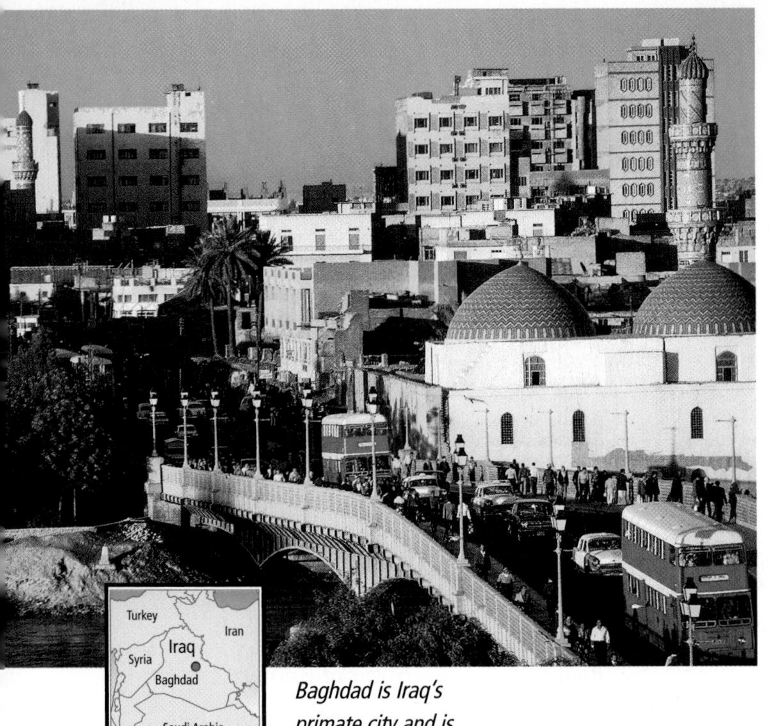

Baghdad is Iraq's primate city and is home to almost one third of its population. It was also one of the leading cities of ancient Mesopotamia, a cradle of early civilization. Modern Baghdad is a city of contrasts, featuring ancient mosques, bazaars, cafes, high-rise apartments, luxury hotels, and traffic congestion.

In most countries people proudly maintain traditional crafts. Artisans use local materials in products sold around the world. The region's beautiful and valuable wool rugs have been famous for centuries. Rugs are still made by hand, using traditional designs. Each community has its own distinctive designs.

Modern manufacturing in the region is limited. It focuses on building materials, food products, and household supplies. The region imports most of its electrical appliances as well as cars and trucks. The only large industries involve oil refining and related chemical manufacturing. However, these industries are highly automated and provide few jobs.

READING CHECK: ***Environment and Society*** How do some of the region's people maintain their traditional ways of life?

Urban Environments

Among the largest cities in the region are the national capitals. These cities include Tehran, Iran; Baghdad, Iraq; Kabul, Afghanistan; and Riyadh, Saudi Arabia. As elsewhere in the world's developing regions, many people have migrated to cities from rural farms and villages. They go there looking for jobs. These people often build housing on the fringes of cities. Sometimes they bring farm animals, which then live in urban neighborhoods.

Many of the region's cities are ancient. In the older sections, life goes on today much as it did centuries ago. Old-style buildings are still in use, and many are one or two stories tall. The narrow twisting streets reflect the fact that they predate the use of cars and trucks. The stalls in a central marketplace, called a bazaar, are covered to protect them from sunlight and rain. Merchants display piles of cloth, household utensils, rugs, and spices. Craft items used to be clustered in separate sections of the bazaar. Now buyers find goods of many types, including imports, mixed together throughout the market. A neighborhood mosque is usually nearby.

In contrast, the newer sections of cities have modern buildings and air-conditioned shopping malls. The avenues are wide and clogged with cars. People live in high-rise apartment buildings. With fast-food outlets and gas stations, some of these neighborhoods look like those in the West.

Governments, Issues, and Challenges

The region's politics and concerns for the future center around three basic themes. One theme is the use of oil wealth. Another is the desire of some to preserve the authority of traditional leaders. A third theme involves the role of Islam in a modernizing world. Individual countries emphasize these themes differently.

Oil Wealth and Power Saudi Arabia's oil wealth gives the country a special position in world affairs. Saudi Arabia's huge reserves make it the world's largest oil exporter. It is a key member of the Organization of Petroleum Exporting Countries, or **OPEC**. This organization influences oil prices by controlling supply. By reducing or increasing oil production, it can affect the economies of countries around the world. Saudi Arabia's role as caretaker of Islam's holiest city, Mecca, adds to its influence. Muslims travel to this city from all over the world. A conservative monarchy whose power is absolute rules Saudi Arabia from the capital, Riyadh. Defending Arab traditions helps the monarchy maintain control. Economic success, mostly due to the oil industry, has also limited opposition to the Saudi monarchy within the country.

The al-Ghawar field in Saudi Arabia is the biggest oil field in the world. It is more than 150 miles (240 km) long and contains some 82 billion barrels of oil.

In contrast to Saudi Arabia, Iran's politics have been unstable in recent decades. Rebellions by minority ethnic groups have been frequent. In 1979 a revolution toppled Iran's monarchy. A government dominated by **ayatollahs** came to power in Tehran. Ayatollahs are religious leaders of the highest authority among Shia Muslims. Today Iran is a **theocracy**, a country governed by religious law. Many religious leaders view Western ideas as a threat to public morality. As a result, some have tried to isolate the country from Western influences. However, many Iranians are seeking more personal freedoms.

Countries around the world pay attention to Iran's political situation. The huge tankers that ship oil from the Persian Gulf must pass through the Strait of Hormuz. This waterway narrows to just 35 miles (56 km) off Iran's southern coast. Iran could cut off a large portion of the world's oil supply by blocking the strait.

Iraq is ruled by a dictator, Saddam Hussein, who has used the country's oil revenues to build a large military. Iraq invaded Iran in 1980, seeking to gain control of the oil-rich region along the Shatt al Arab. The bloody war ended nearly a decade later, with neither country the winner. Then in 1990 Iraq invaded Kuwait to capture its rich oil fields. Iraq's leader claimed Kuwait had historically been Iraqi territory. The invasion threatened the stability of the region. A group of countries led by the United States drove back Iraqi forces in 1991. This conflict became known as the Persian Gulf War. Since the war, Iraq's relations with other countries have been strained.

TEKS **READING CHECK:** *Places and Regions* What geographic factors have affected Iraq's international relations in recent decades?

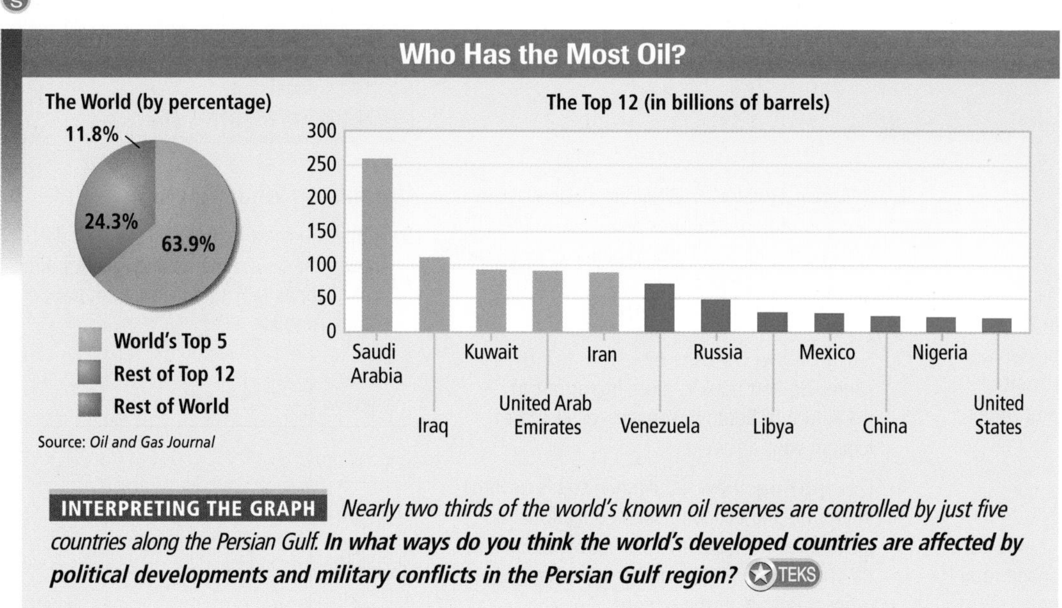

INTERPRETING THE GRAPH *Nearly two thirds of the world's known oil reserves are controlled by just five countries along the Persian Gulf.* ***In what ways do you think the world's developed countries are affected by political developments and military conflicts in the Persian Gulf region?*** TEKS

Saudi Arabia's oil economy produces huge incomes for the country's royal family, government, and businesspeople. The country's elites use some money to buy imported luxury goods, such as automobiles. In recent decades, the modernization of the country has greatly altered many traditional ways of life.

Islam, Society, and Change In Afghanistan, ethnic and political rivalries have long plagued the country. A group called the Taliban came to power in the 1990s. They were driven by an extreme version of Sunni Islam. The Taliban established strict laws governing Afghans' lives.

After the September 11 terrorist attacks, U.S. officials focused on the aid the terrorists received from the Taliban government. U.S. and allied forces attacked terrorist camps and Taliban military targets. The Taliban regime soon collapsed. A *loya jirga,* a traditional council, with representatives from all ethnic groups was held in June 2002. It elected Hamid Karzai as the country's transitional president.

Freed from the Taliban, Afghans experienced new liberties. Some of the most significant changes came in the lives of women. They were able to attend school and work outside their homes—rights denied by the Taliban.

In other countries in the region, women have more educational and economic opportunities. For example, some women work outside the home. Still, Islamic traditions encourage women to value roles as wives and mothers. Partly as a result, population growth rates are often high. Large families are common. Education for all segments of the population, including women, has increased. However, many school systems are inadequate, and many well-educated people cannot find jobs.

Oil, the influence of foreign cultures, and communications technology are changing the region. Over time these changes will be greater than any development since the spread of Islam. For example, much of the region's oil is available for export. Modern society runs on oil. Homes, power stations, and transportation depend on oil or oil products. Oil is also a basic raw material for products like fertilizers, industrial chemicals, and plastics. The importance of oil in global trade has given the oil-rich countries much economic and political power.

TEKS **READING CHECK:** *Human Systems* What three themes are reflected in the region's politics and concerns for the future?

Section 3 Review

TEKS Questions 1, 2, 3, 4, 5

Homework Practice Online
Keyword: SW3 HP19

Identify Bedouins, OPEC

Define ayatollahs, theocracy

Working with Sketch Maps On the map you created in Section 2, label Tehran, Baghdad, Kabul, Riyadh, and the Strait of Hormuz. Why is the Strait of Hormuz important to the global oil trade?

Reading for the Main Idea

1. *Environment and Society* In what ways does the water supply shape rural economic activity in the region?
2. *Human Systems* What are some ways in which traditional cultures and ways of life have been retained in the region's cities?

Critical Thinking

3. **Supporting a Point of View** Which resource do you think is more important to the future of Southwest Asia—oil or water? Explain your answer.
4. **Comparing** How has Afghanistan's government affected economic and educational opportunities for women? How do opportunities for women there compare to opportunities in other countries in the region?

Organizing What You Know

5. Copy the chart below and use it to describe the governments and recent political histories of Afghanistan, Iran, Iraq, and Saudi Arabia.

Afghanistan	
Iran	
Iraq	
Saudi Arabia	

Places and Regions

Geography for Life

The Ecological Trilogy

The relationships between living things and their environments are important to geographers. For example, geographers study the ways pollution affects ecosystems. They also study how the overuse of natural resources like rain forests and water affect the organisms that depend on them. The study of how living things interact with and depend on each other and the environment is called ecology. Geographers are also interested in human ecology—the ways human beings interact with and depend on the environment and each other. Southwest Asia is just one region in which geographers have studied human ecology.

In the 1960s a geographer developed a model he called the ecological trilogy. This model described how the three main ways of life in Southwest Asia depended on each other. Those three ways are life as a villager, as a pastoral nomad, and as a town- or city-dweller.

At the base of the trilogy were the peasant farmers who lived in the region's small villages. They served the city people and nomads by growing basic food crops like wheat and barley. Although they may not have wanted to do so, the villagers also provided the city with soldiers, tax money, and workers. On the other hand, cities and towns offered villages technological improvements, educational opportunities, and other services.

The three parts of the ecological trilogy are shown in the scenes below. On the left, a farmer tends crops outside his village. In the center, nomadic herders take their flock to market. On the right, people live and work in a modern urban environment in Tehran.

Nomads also had an important role in the trilogy. Nomads supplied villagers with animal products like cheese, meat, milk, and wool. They also provided desert herbs and medicines. City people supplied nomads with cooking utensils and factory-made clothing. However, city people and nomads interacted less than villagers and nomads.

New patterns of living in Southwest Asia have changed the relationships within the trilogy. In recent decades, government doctors, teachers, improved roads, cellular phones, and television have come to villages and nomads. Many villagers, particularly young people, move from rural areas to the city. When they arrive, the villagers change the culture of the city. Changing ways of life for nomads are also affecting this trilogy. As nomads settle in more permanent communities, the number of people who follow their herds declines. As the cities grow, they expand outward to engulf villages.

Although changes have occurred, each of the three parts of the ecological trilogy still supports the others. In addition, no matter where they may live now, individuals still identify themselves as city people, nomads, or villagers.

Applying What You Know

TEKS

1. **Summarizing** Which part of the trilogy serves as a base for the model? Why?
2. **Supporting a Point of View** How might the ecological trilogy of Southwest Asia differ from a similar model for the United States?

Building Vocabulary

On a separate sheet of paper, explain the following terms by using them correctly in sentences.

exotic rivers	Shi'ism	OPEC
oasis	imams	ayatollahs
dynasty	Bedouins	theocracy
Sunni		

Locating Key Places

On a separate sheet of paper, match the letters on the map with their correct labels.

Persian Gulf	Shatt al Arab	Mecca
Mesopotamia	Elburz Mountains	Baghdad
Tigris River	Rub' al-Khali	Riyadh

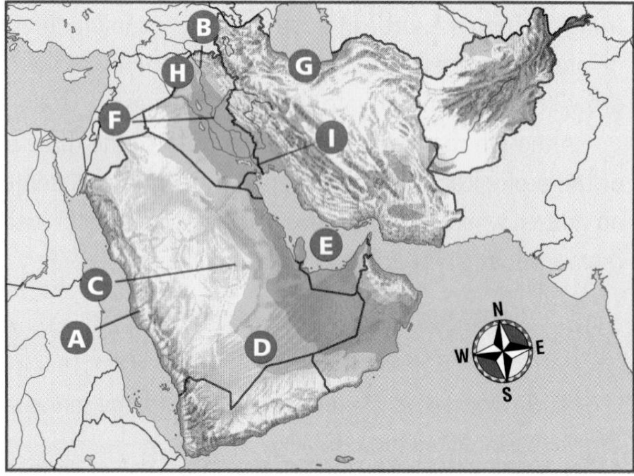

Understanding the Main Ideas

Section 1

1. *Physical Systems* In what ways has the plant life of the region adapted to the conditions there?

Section 2

2. *Environment and Society* Why has Mesopotamia been such an attractive target for invasion throughout history?
3. *Places and Regions* What is the major language spoken in Iran? In Iraq, Afghanistan, and Saudi Arabia?

Section 3

4. *Human Systems* What changes are taking place in the traditional rural economy and culture of the region?
5. *Human Systems* What are two important factors accounting for Saudi Arabia's influence in world affairs today?

Thinking Critically for TAKS

1. **Evaluating** What policies would you recommend to governments working to protect water resources in the region? Why would those policies not be practical for some countries? What geographic and economic effects might these policies have?
2. **Drawing Inferences and Conclusions** Why might stability in the Persian Gulf region be important to developed countries?
3. **Comparing and Contrasting** How do views about Islamic teachings and the role of Islam in society vary among governments in the region? How do cultural beliefs influence public policies regarding women in Afghanistan?

Using the Geographer's Tools

1. **Analyzing Maps** Review the unit maps to study the climates and land use and resources of Iraq. In which climate region do you see large areas of commercial farming? What is the probable source of water for commercial farming there?
2. **Analyzing Statistics** Use the unit Fast Facts table to rank the countries of the region by GDP, from highest to lowest. Then prepare a second list, ranking them by literacy rate, from highest to lowest. Compare the two rankings. What does this comparison reveal about the relationship between educational level and a country's level of economic development?
3. **Creating a Population Pyramid** Use the HRW Web site to find statistics about Iran's population. Use those statistics to create a population pyramid that describes the population characteristics of Iran. Then predict future growth trends there.

Writing for TAKS

Imagine that you are part of a Bedouin family that is about to move to a city in Saudi Arabia. Write a short journal entry describing how your way of life is about to change. After you have completed your journal entry, proofread it to make sure that you have used standard grammar, spelling, sentence structure, and punctuation.

SKILL BUILDING

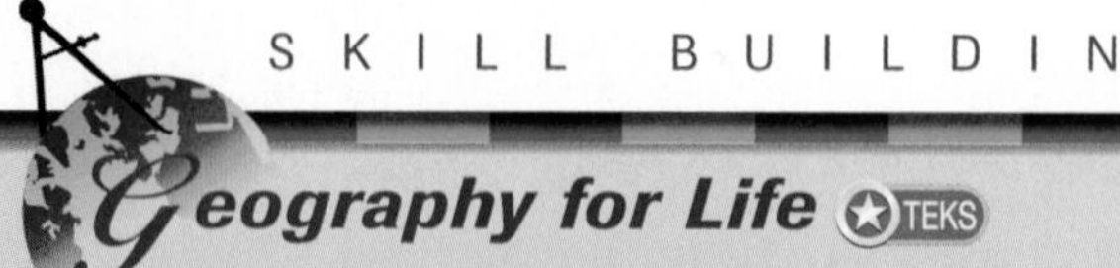

Geography for Life

Organizing Geographic Information

Human Systems Use Internet or library resources to identify the largest sources of oil imports into the United States. Also investigate how the amount of imported oil compares to the total produced from wells in the United States. Then plan, design, and create a graph that shows your findings.

go.hrw.com TAKS PREP ONLINE keyword: SW3 T19

Building Social Studies Skills

Major Earthquakes in Iran (1978–1997)

Date	Location	Estimated number of deaths
Sept. 16, 1978	Northeastern Iran	15,000
June 11, 1981	Southern Iran	3,000
July 28, 1981	Southern Iran	1,500
June 20, 1990	Western Iran	40,000+
Feb. 28, 1997	Northwestern Iran	1,000+
May 10, 1997	Northern Iran	1,560

Source: *World Almanac and Book of Facts 2001*

Interpreting Charts

Study the chart above. Then answer the questions that follow.

1. Which region of Iran suffered the most major earthquakes during the period?
   a. northern Iran
   b. southern Iran
   c. southwestern Iran
   d. southeastern Iran
2. On what dates did Iran suffer its three deadliest earthquakes during this period? List the earthquakes from deadliest to least deadly.

Building Vocabulary

To build your vocabulary skills, answer the following questions. Mark your answers on a separate sheet of paper.

3. *Theocracy* means the same as
   a. a country governed by religious law.
   b. a population that shares a common cultural background.
   c. the policy of gaining control over territory outside a country.
   d. a movement that stresses the strict following of basic traditional principles.
4. *Fossil water* means the same as
   a. water surrounded by the bones of ancient dinosaurs.
   b. Persian Gulf water processed into freshwater.
   c. groundwater not being replaced by rainfall.
   d. lake water supplied by streams and old rivers.

Alternative Assessment

PORTFOLIO ACTIVITY

Learning about Your Local Geography

Group Project: Field Work

The handmade rugs of Southwest Asia are famous throughout the world. Plan, organize, and complete a research project with a partner to discover what craft industries are important in your community. Conduct a survey to determine what craft items are produced there. Where are these items sold? Is your community widely known for any crafts? Present your findings in a report. Include illustrations of some of the community's craft products. When you have completed your report, proofread it to make sure that you have used standard grammar, spelling, sentence structure, and punctuation.

internet connect

Internet Activity: go.hrw.com
KEYWORD: SW3 GT19 TEKS

Choose a topic on the Persian Gulf and interior to:

- report on Islamic culture.
- research oil productivity and transfer your information into a graph or chart.
- understand the technology of desalinization.

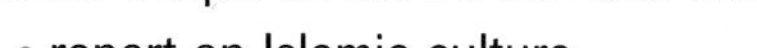

CHAPTER 20 The Eastern Mediterranean

Complex relationships among geography, history, and religion are typical in the eastern Mediterranean region. All of the region's countries have been involved in conflicts both long ago and in recent years.

Pitcher from Syria

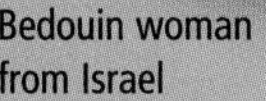

Bedouin woman from Israel

Türkiye'den selamlar! (Greetings from Turkey!) My name is Adalet, and I am in the tenth grade at a private school in Ankara, the capital of Turkey. I live in a high-rise apartment outside the city with my mom and dad. We have a view of the city, the distant mountains, and, of course, the parking lot. In the summers, I like to stay with my grandparents in their summer house on the Aegean Sea, near the ancient Greek and Roman ruins at Ephesus. I go to the beach there with my friends and stay until sundown. Back at the house, my grandpa and grandma love to cook delicious Turkish food like *kofte,* a spicy ground lamb dish.

On school days I get up at 8:00 A.M., put on my school uniform, and eat cornflakes or bread and cheese. The parents of boys and girls in my neighborhood have hired a bus to take us to school. It is a big school that goes from first grade through high school with about 800 students per grade. I am studying biology, physics, algebra, geometry, history, Turkish, and English. The science and math classes (and the English class, of course) are taught in English, the others in Turkish. At lunch my friends and I go to a little food stand nearby and eat hot dogs or grilled lamb on a skewer with bread, tomatoes, onions, and peppers. To drink, we have *ayran,* a drink made with yogurt.

Natural Environments

READ TO DISCOVER

1. What landforms and rivers are found in the eastern Mediterranean region?
2. What climates, biomes, and natural resources does the region have?

DEFINE

potash
magnesium

LOCATE

Anatolia
Dardanelles
Bosporus
Sea of Marmara
Pontic Mountains
Taurus Mountains
Jordan River
Dead Sea
Tigris River
Euphrates River
Syrian Desert
Negev

WHY IT MATTERS

In 2000 a new reservoir threatened to submerge important historical sites in Turkey. Use CNNfyi.com or other **current events** sources to learn about the many ways that dams and reservoirs affect places around the world.

The Eastern Mediterranean: Physical-Political

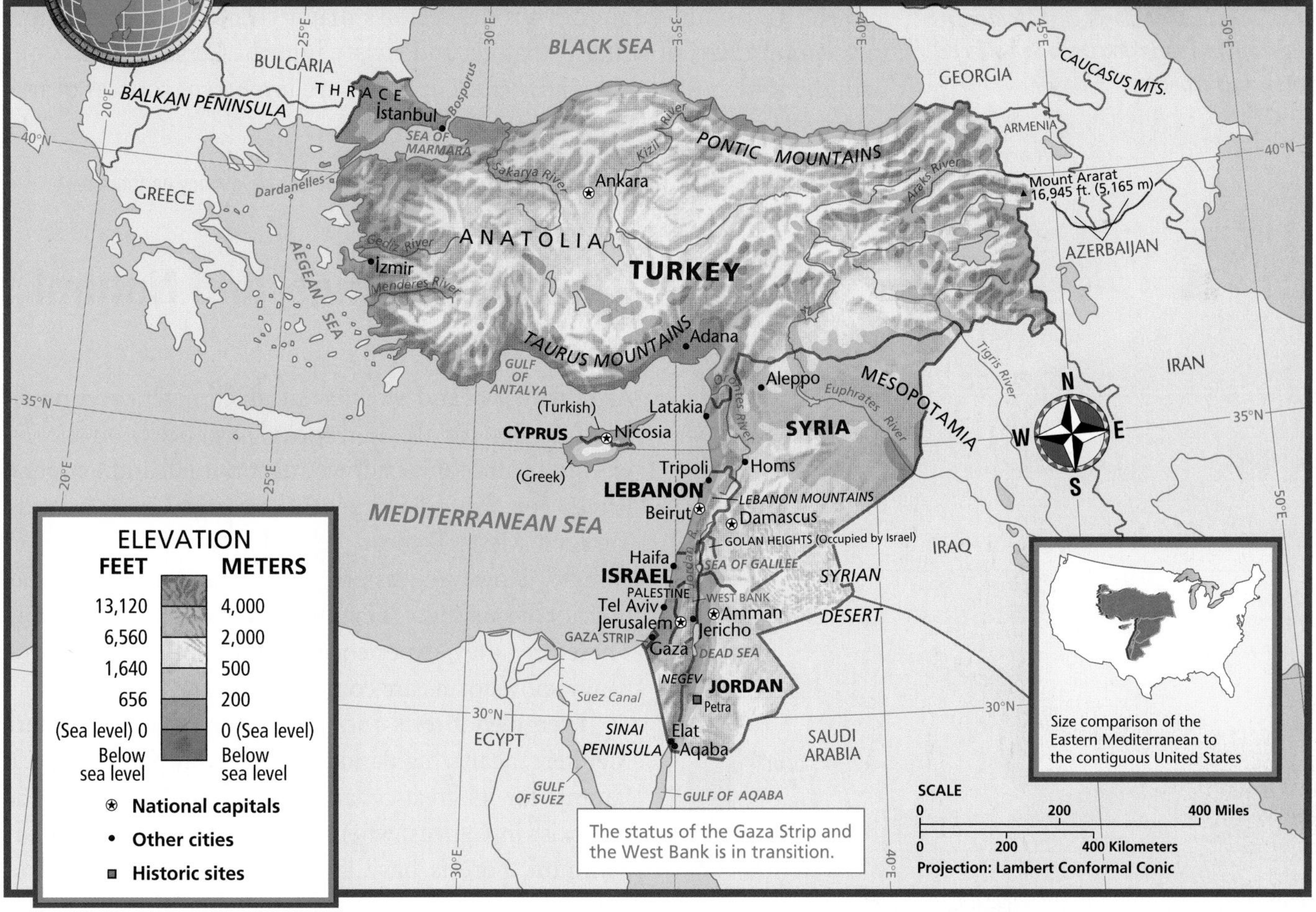

ELEVATION

FEET	METERS
13,120	4,000
6,560	2,000
1,640	500
656	200
(Sea level) 0	0 (Sea level)
Below sea level	Below sea level

⊛ National capitals
• Other cities
▪ Historic sites

Size comparison of the Eastern Mediterranean to the contiguous United States

SCALE
0 200 400 Miles
0 200 400 Kilometers
Projection: Lambert Conformal Conic

The status of the Gaza Strip and the West Bank is in transition.

GO TO: go.hrw.com
KEYWORD: SW3 CH20
FOR: For Web sites about the eastern Mediterranean

Landforms and Rivers

The eastern Mediterranean region is part of an area often called the Middle East. It consists of six countries. Israel, Jordan, Lebanon, Syria, and Turkey are on the mainland. Cyprus is an island in the Mediterranean Sea.

The region lies on two continents. A small part of Turkey is on Europe's Balkan Peninsula. Most of Turkey is in Asia, an area known as Anatolia (a-nuh-TOH-lee-uh). Three narrow connected bodies of water—the Dardanelles (dahrd-uhn-ELZ), the Bosporus, and the Sea of Marmara (MAHR-muh-ruh)—separate Europe from Asia.

European Turkey has plains and hills. A narrow coastal plain rims the western edge of the Asian part of Turkey. Two mountain systems run from east to west across Anatolia. They include the Pontic Mountains in the north and the Taurus Mountains in the south. Between them lies the Anatolian Plateau, which has many peaks and valleys of its own. Turkey's geology includes many faults and folds. Along with continued mountain building, these geological features also cause devastating earthquakes.

The coastal plain continues south from Turkey along the coasts of Syria, Lebanon, and Israel. Farther inland one finds plateaus, hills, and valleys. A rift valley extends northward from Africa into Syria. Hills rise on both sides of the rift. Between the rift's ridges the Jordan River flows south into the Dead Sea. This unusual sea lies 1,312 feet (400 m) below sea level. Its shore is the lowest land on Earth's surface. The Dead Sea was once part of the Mediterranean. Today it no longer has an outlet to the ocean. The sea is so salty that all swimmers can easily float in it.

The Tigris and Euphrates are the region's major rivers. Both rivers begin in the mountains of eastern Anatolia and empty into the Persian Gulf. Along with the Jordan River, the Tigris and Euphrates are important sources of irrigation water.

READING CHECK: ***Physical Systems*** Why does Turkey have many earthquakes? TEKS

INTERPRETING THE VISUAL RECORD

Top: A tourist floats in the Dead Sea. Bottom: The Dead Sea lies in a desert valley. The shoreline has receded nearly a mile in the last 40 years. Only 10 percent of the Jordan River's water flows into the sea. Growing populations and drought conditions place heavy demands on the Jordan. ***How might a shrinking Dead Sea affect the area's economy?*** TEKS

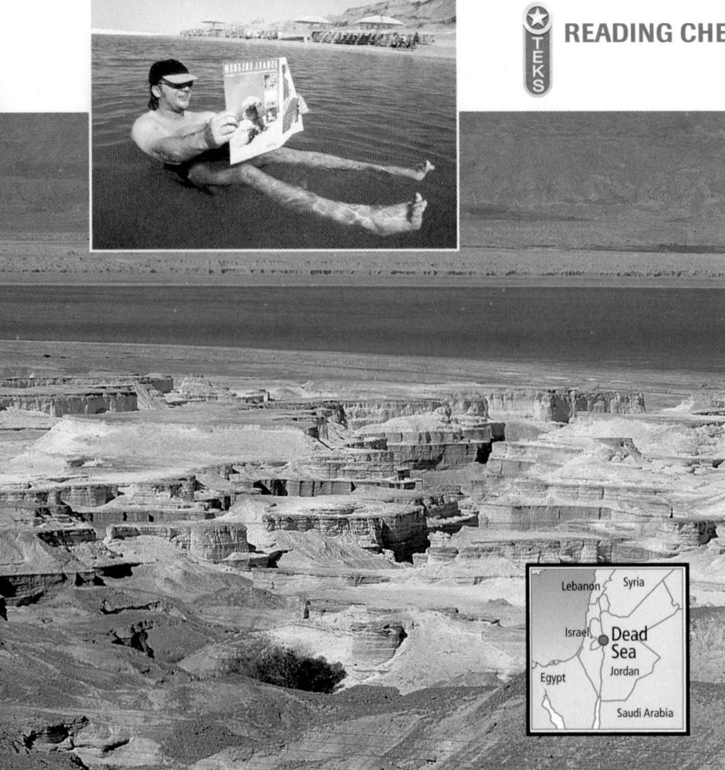

Climates, Biomes, and Natural Resources

Arid, semiarid, and Mediterranean climates cover nearly all of the eastern Mediterranean. Distance from the sea, elevation, and rain shadows affect rainfall and temperatures. For example, like the nearby Greek islands, Turkey's Mediterranean coast has a mild and sunny climate. However, eastern Turkey and the Anatolian interior lie inland. They are also at higher elevations. Therefore, they experience bitterly cold winters, and heavy snowstorms are common.

Evergreen forests once covered much of the eastern Mediterranean's highlands. Lebanon, in particular, was famous for its great cedar trees. Areas of forested land still exist in Cyprus, where cedar, cypress, and pine are common. Forests have largely disappeared elsewhere,

Forests of magnificent cedar trees were once common in Lebanon. In fact, the cedar is pictured on Lebanon's flag. However, people began cutting down the big trees long ago. Now few remnants of the great forests remain, but reforestation efforts are underway.

however. They are victims of centuries of farming, herding, shipbuilding, and firewood collecting that began more than 2,000 years ago.

In semiard areas like Anatolia, plant and animal life can be diverse. Farther south, the Syrian Desert covers much of both Jordan and Syria. Another desert, the Negev (NE-gev), lies in southern Israel. In desert biome areas, plant and animal life is scarce. Yet some hardy species, such as jackals, lizards, and snakes, are able to survive. People also live in the desert. Small groups of nomads move their flocks of sheep and goats with the seasons.

Valuable mineral and other natural resources can be found throughout the region. Turkey has supplies of coal, copper, and iron ore. Some oil and natural gas deposits are distributed across the region. The Dead Sea also provides Israel and Jordan with certain minerals, including **potash** and **magnesium**. Potash is used to process wool and to make fertilizers, glass, and soft soaps. Magnesium is a light metal that is valuable in certain industries, such as aerospace.

In the Cappadocia region of central Turkey, volcanic rock has eroded into an unusual landscape of columns, cones, pillars, and towers. People carved into the rock to create shelters. Some of the caves are now homes, stores, and churches.

READING CHECK: *Places and Regions* Why do eastern Turkey and the Anatolian interior have cold winters?

Section 1 Review

TEKS Working with Sketch Maps, Questions 1, 2, 3, 4

Define

potash
magnesium

Working with Sketch Maps

On a map of the eastern Mediterranean that you draw or that your teacher provides, label Anatolia, Dardanelles, Bosporus, Sea of Marmara, Pontic Mountains, Taurus Mountains, Jordan River, Dead Sea, Tigris River, Euphrates River, Syrian Desert, and the Negev. What parts of the region would you expect to be densely populated?

Reading for the Main Idea

1. *Places and Regions* What are the main landforms of Turkey? Which two major rivers begin in Turkey?
2. *Physical Systems* What has happened to the region's evergreen forests?
3. *Physical Systems* What factors affect rainfall and temperatures in the region?

Critical Thinking

4. **Analyzing Information** Why might the Dead Sea be considered a natural resource?

Organizing What You Know

5. Create a chart that lists the countries in the region and the different landforms found in each. Refer back to Section 1 and the chapter's physical-political map.

Country	Landforms

CASE STUDY

Palestine–Setting Boundaries

Human Systems Perhaps one of the most persistent themes in human history has been the struggle over territory. Today the world is divided into almost 200 countries. As you read in Chapter 6, a variety of boundaries separate these countries. Some are natural boundaries, such as mountains and rivers. Others are cultural boundaries. The border between mostly Hindu India and mostly Muslim Pakistan is an example of a cultural boundary. Finally, some boundaries are geometric, such as lines of latitude.

Sometimes the competing desires of two or more peoples to control a piece of land have been settled peacefully. However, such disputes have also led to war. The reasons for this are many. They are often cultural, reflecting the strong ties a particular people have to a place. They can also be economic, reflecting the desire for the natural resources found in an area. Both of these factors have played a part in the long Arab-Israeli conflict.

An Israeli soldier patrols a Jewish settlement in Hebron. The West Bank town has been the scene of many clashes between Palestinians and Jews.

Conflict and Cooperation

Palestine is an old Greek name for the eastern edge of the Mediterranean. It includes the modern country of Israel, the West Bank, and the Gaza Strip. For Jews and Arabs there alike, Palestine is the land of their ancestors. It is also home to many historical and cultural sites held sacred by millions of people. Many of the most important sites are located in Jerusalem. (See Cities & Settlements: Jerusalem.)

A Palestinian woman pumps water from an aquifer. Control of water resources is a primary source of conflict in the West Bank.

As you have read, hundreds of thousands of Arabs left Palestine after Israel declared independence in 1948. Most settled in the West Bank, Gaza Strip, and neighboring Arab countries such as Jordan. Threatened by surrounding Arab armies, Israel captured the West Bank and Gaza Strip during the so-called Six-Day War in 1967. Many Israelis thought that having control over these territories was vital to their country's security. Arabs demanded that the Israelis withdraw. The dispute over the territories grew more heated and violent over time.

In the 1990s Israel's government and Arab Palestinian leaders tried to end their long dispute through negotiation. They discussed many issues. Among them was how to draw a boundary between Israel and any new Palestinian state in the occupied territories. The two sides made significant progress. In fact, over time Israel returned parts of the West Bank and Gaza Strip to Palestinian control. However, the talks between Israelis and Palestinians bogged down in 2000. Later that year, violence again erupted, taking many lives.

INTERPRETING THE MAPS *The debate over new boundaries between Israel and a possible new Palestinian country involves a number of issues. Control of water resources in the dry region is one of them. Aquifers extend across the region. New boundaries will determine who controls the water stored in these aquifers. The West Bank would become part of the new country of Palestine. The detailed map of the West Bank shows current and planned Israeli settlements as well as areas of Palestinian control.* ***Why do you think establishing stable borders around Israeli settlements might be difficult? How might Israelis and Palestinians compromise over control of the region's aquifers?*** TEKS

Working toward Compromise

Many issues divided Israelis and Palestinians when the talks bogged down in 2000. Three primary issues involved Jewish settlements in the occupied territories, the status of Jerusalem, and access to water.

The settlement issue is a result of Israel's government allowing Jewish settlers to set up communities in the occupied territories. Today these settlements are scattered between Palestinian-controlled lands in the West Bank and Gaza Strip. The two sides cannot agree on what to do with those settlements.

Another issue is control of Jerusalem. The city had been divided between Israel and Jordan from 1948 to 1967. It was reunited under Israeli rule during the 1967 war. Jerusalem is a difficult issue because Jewish, Muslim, and Christian religious sites are found there. In addition, both Israelis and Palestinians want the city as their capital.

Finally, water is an issue. This dry region has limited water supplies. Aquifers are a crucial water source. One large aquifer straddles the dividing line between Israel and the West Bank. Mountain rainfall, mostly from the West Bank, supplies much of the aquifer's water. However, much of that stored water lies on the Israeli side of the aquifer. Agreement over access to the aquifer and the use of its water has been difficult to reach. In addition, Israel wanted to control land along the Jordan River, another important water source. This land would also act as a buffer zone between the West Bank and Jordan, an Arab country.

To the north the Sea of Galilee presented a similar problem. The Sea of Galilee lies along the old border between Israel and Syria. However, in the 1967 war Israel captured a Syrian area that included part of the freshwater sea's shore. In addition, springs in the area replenish the sea. Syria has demanded that Israel withdraw from this land before agreeing to a final peace treaty. However, Israelis fear that doing so would limit their supply of water from the sea.

These issues have long divided Israelis and their Arab neighbors. Until they are resolved, new boundaries dividing their countries will remain in question.

Applying What You Know 

1. **Summarizing** What are some of the issues that make agreement over new boundaries between Israel and a Palestinian state difficult to reach?
2. **Comparing** Why might both Israelis and Palestinians want Jerusalem as their capital? What kind of compromise do you think might settle this issue?

History and Culture

READ TO DISCOVER

1. How have various peoples and empires influenced the eastern Mediterranean?
2. How did the modern state of Israel develop?
3. What are the peoples and cultures of the region like?

WHY IT MATTERS

Many people go to Israel to visit places important to Christianity, Islam, or Judaism. Use CNNfyi.com or other **current events** sources to learn about those sites.

IDENTIFY

Zionism
Holocaust
Kurds

DEFINE

sultans
mandates

LOCATE

İstanbul
Ankara
Jerusalem
Nicosia

The Phoenicians were an early seafaring people from what is now Lebanon. Although their homeland was small, they traded as far as Spain and may even have sailed around Africa.

The Rise and Fall of Empires

Some of history's earliest civilizations developed in the eastern Mediterranean region. Farming supported the growth of permanent settlements as far back as 8000 B.C. Over time, the Egyptians, the Hittites of Asia Minor, the Persians, and others ruled all or parts of the region. In about 1000 B.C. a people called the Hebrews set up a kingdom between the Jordan River and the Mediterranean Sea. That area has been known as Palestine and Israel at different times. The Hebrews practiced Judaism, which is the dominant religion in Israel today.

The Roman and Byzantine Empires The Romans conquered the eastern Mediterranean between 200 B.C. and A.D. 106. During this time, most Jews were exiled from Palestine because they resisted Roman authority. In the first century A.D., followers of Christianity began to spread their beliefs from the eastern Mediterranean throughout the Roman Empire. By the late 300s Christianity was the empire's official religion.

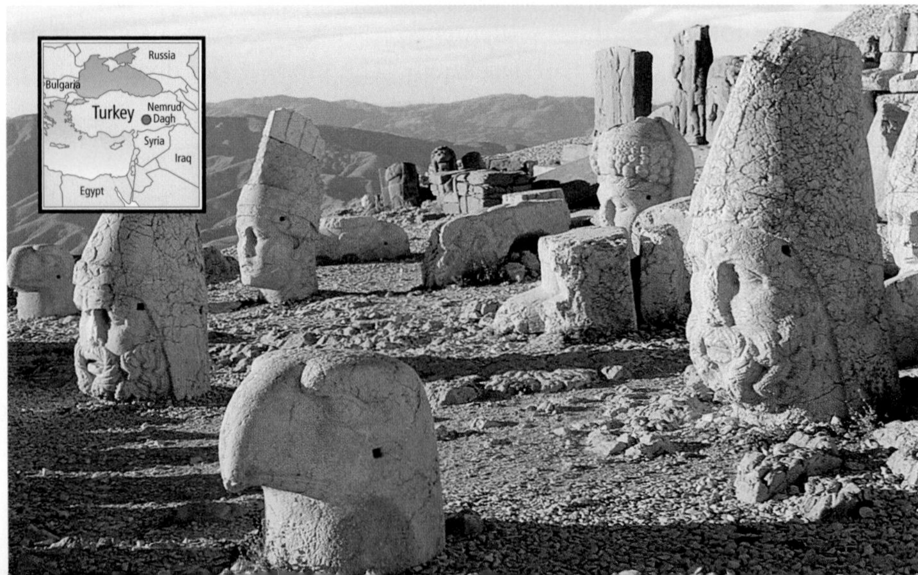

INTERPRETING THE VISUAL RECORD

These gigantic heads at Nemrud Dagh, now in eastern Turkey, are more than 2,000 years old. They lie in what was the kingdom of Commagene, where Greek, Persian, and Roman influences mixed. ***How do you think physical geography contributed to the combination of cultural influences at Commagene?*** TEKS

During the A.D. 400s the Western Roman Empire crumbled. The eastern part of the empire survived, however, and came to be known as the Byzantine (BIZ-uhn-teen) Empire. Christianity also became divided. The Eastern Orthodox Church broke away from the Roman Catholic Church. Constantinople—now İstanbul—was both the center of the Orthodox Church and the capital of the Byzantine Empire. It became a city of great beauty, power, and wealth. The Byzantine Empire ruled most of the eastern Mediterranean as well as areas far beyond the region. However, other peoples soon invaded, shrinking the empire.

Ottoman sultans lived in the Topkapi Palace from when it was built in the 1400s until the 1800s. It is a huge complex of courtyards, gates, and rooms. The Topkapi Palace Museum is now one of İstanbul's main tourist attractions.

The Arabs and Islam In the A.D. 600s Arab Muslim armies swept north out of the Arabian desert. They rapidly established an empire in southwestern Asia and northern Africa. The Byzantines lost much territory as a result, including the area that is now Israel.

Turkic Muslims captured Jerusalem from the Arabs in 1077. When Turks threatened Constantinople, the pope called for Christians to go to war against the Muslims. This series of wars was known as the Crusades. Between 1095 and the late 1200s, crusader armies from all over Europe invaded the region. For a while, the Crusaders held Jerusalem and some cities in Syria. However, Muslim armies later forced them out.

The Ottoman Empire In the 1300s the Ottoman Turks established another Muslim empire in the region. The Ottoman rulers were called **sultans**. They took Constantinople in 1453 and made it the Ottoman capital. By the 1600s the Ottoman Empire included most of Southwest Asia. The empire also stretched into parts of eastern Europe and most of North Africa. However, political struggles, corruption, and rivalries with other countries slowly weakened the empire. Over time ethnic minorities within the empire's borders also began to push for independence.

During World War I the Ottoman Empire fought on the losing side. After the war, a general named Mustafa Kemal (later known as Atatürk) took over the government. He created the Republic of Turkey and established its capital at Ankara. Beyond Turkey, the former Ottoman territories became **mandates** of Great Britain and France. These mandates were territories placed under another country's control. They were to become independent eventually. The British and French mandates included Iraq, Syria, Lebanon, and Palestine. After World War II these mandates gained independence. However, as you will see, a major dispute arose in Palestine.

READING CHECK: *Human Systems* How have different empires shaped the eastern Mediterranean region?

Palestine and Modern Israel

In the late 1800s European Jews began a movement called **Zionism**. Zionism called for Jews to set up their own country or homeland in Palestine, which was then under Ottoman rule. That land had traditionally been important to Jewish culture, history, and religion. After World War I thousands of Jews moved to the area, which became a British mandate in 1920. Later, many Jews who were fleeing persecution in Europe immigrated to Palestine. In fact, during World War II, Germany's Nazis murdered millions of Jews in what became known as the **Holocaust**. As a result, Jewish immigration to Palestine increased during the 1930s and 1940s. Arabs already living in Palestine soon felt threatened and became angry at the growing Jewish presence there. Today Jews make up about 80 percent of Israel's population. Most of the rest are Arab.

In 1947 the United Nations voted to divide Palestine into Jewish and Arab states. When the British withdrew from Palestine the next year, the Jewish leadership declared itself the independent state of Israel. Arab armies from Egypt, Iraq, Jordan (then called Transjordan), Lebanon, and Syria then invaded Israel. Israel pushed the Arab forces back and won more land. Many Palestinian Arabs fled to Jordan, Lebanon, and other Arab countries. At the same time, Jordan and Egypt took over the remaining Arab portions of Palestine. More wars between Arabs and Israelis occurred in 1956, 1967, 1973, and 1982. During the 1967 war, Israel gained Arab land west of the Jordan River, called the West Bank. It also took the Gaza Strip, a small piece of land on the Mediterranean coast. The Palestine Liberation Organization (PLO) became a force in the region around this time. Groups that wanted to establish an independent Palestinian state formed the PLO in 1964 to coordinate their efforts. Violent conflict between Palestinians and Israelis has continued despite efforts toward peace.

Jewish settlement in the occupied lands has slowed those efforts. Thousands of Jews have moved into the West Bank and Gaza Strip. Jewish settlement has also been an issue in the Golan Heights. Israel took that hilly region from Syria in the 1967 war. Many Israelis do not want to give up these areas. They think the occupied lands are important to the security of Israel. From the Golan, Israeli troops can guard the Jordan River, a crucial water

INTERPRETING THE VISUAL RECORD

Israeli soldiers patrol the West Bank. ***From the photo, how can you tell that some of the West Bank's residents have adopted aspects of Western culture, while others maintain traditional ways?*** TEKS

source. The Golan also blocks Syrian access to northern Israel. In addition, the West Bank separates Israel from Jordan and Arab countries to the east. Resolving the problem of the occupied lands remains an important foreign policy issue. (For more information on these issues, see Case Study: Palestine—Setting Boundaries.)

READING CHECK: ***Human Systems*** Why did many Jews immigrate to Palestine during the 1930s and 1940s?

Culture

Language in this region is often an important part of an ethnic group's identity. The many languages spoken there reflect the area's many cultural influences. Arabic, Hebrew, and Turkish are the most common languages. Britain and France introduced English and French when they ruled the area in the 1900s. Many Jews immigrated to Israel from the former Soviet Union in the 1990s, bringing the Russian language with them. The **Kurds**, who live mainly in southeastern Turkey and neighboring countries, make up Turkey's largest minority. They have their own Kurdish language.

INTERPRETING THE VISUAL RECORD

Seeing Hebrew used in public daily life would have amazed Jews a hundred years ago. For centuries the language was used only in religious and scholarly contexts. ***What evidence of cultural convergence do you see on this street sign?*** TEKS

FOCUS ON CULTURE

Reviving a Language Hebrew was spoken in the region more than 2,000 years ago. Hebrew is the language of the Jewish Bible, or Torah. In addition, it has always been the language of Jewish prayer and religious ceremonies. Scholars of Jewish law and literature also used Hebrew. The language was not used by most Jews in daily life, however. The rebirth of Hebrew began among Jewish immigrants to what is now Israel. They developed a more modern form of the language. When Israel became independent in 1948, Hebrew became an official language. New immigrants to Israel start on the road to citizenship by learning Hebrew. Arabic is the official language for Israel's Arab minority.

READING CHECK: ***Human Systems*** How do Israel's two official languages reflect the region's distinctive history?

In İstanbul, this Muslim mystic, or dervish, seeks spiritual perfection by whirling.

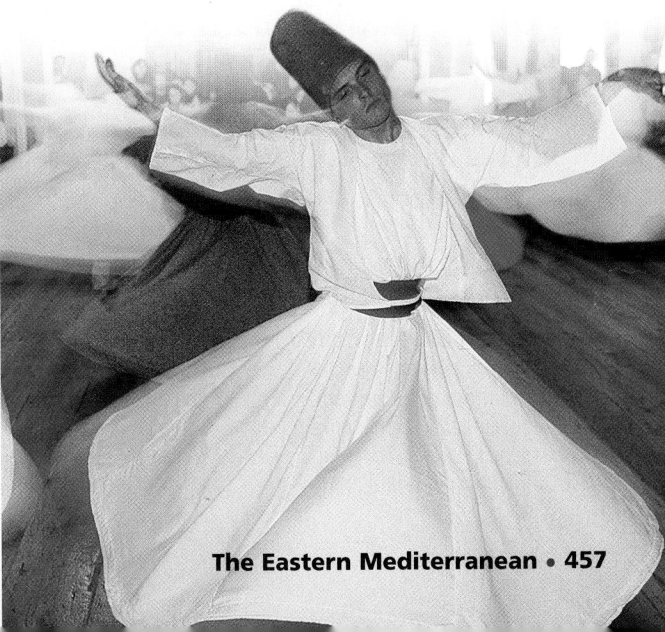

Religion Most people in the eastern Mediterranean are Muslim, Jewish, or Christian. All of the faiths are monotheistic, or based on the belief in one God. Judaism was the first of the three to develop. Its roots date back to about 2600 B.C. Today most Jews in the eastern Mediterranean live in Israel.

Christianity developed out of Judaism and spread during the Roman era. Today there are significant Christian minorities in most countries of the region, particularly in Lebanon and Syria. Muslims form the majority in all countries of the eastern Mediterranean except Israel.

These nomadic herders are spending the winter in the mountains of Lebanon. The nomadic way of life is becoming less common. Many nomads are moving to towns to look for work or settling on government-owned land.

Settlement People in the eastern Mediterranean often live in communities of similar cultural backgrounds. In fact, most of the major cities have different sections that were historically occupied by particular ethnic or religious groups. For example, Jerusalem was divided into Armenian, Christian, Jewish, and Muslim quarters. (See Cities & Settlements: Jerusalem.)

A division has occurred in modern times on Cyprus. A line runs all the way across the island. Greek Cypriots live south of the line, while Turkish Cypriots live north of it. (See Geography for Life: Cyprus—A Divided Island.)

Traditions and Customs Members of the same ethnic group may follow different religions. For example, Christian Arabs share the Arabic language and literature of Islamic Arabs. Yet they maintain their separate religious identity. In contrast, members of the same religion may belong to various ethnic groups. For example, the region's Muslims—whether Arabs, Kurds, or Turks—share many traditions. Turks and Kurds, while Islamic, are distinct from Arabs in their language and culture.

Cultural divisions also separate Turkey's urban and rural populations. Middle-class Turks tend to share the lifestyle and attitudes of middle-class Europeans. However, most rural Turks have more traditional views, such as those concerning the role of women. They prefer that women work as homemakers.

Jewish religious law influences Israel's traditions and customs. For example, because Saturday is the weekly holy day in Judaism, most Israeli businesses are closed on that day. Although most Israelis practice Judaism, Israel is also multiethnic because Jews have emigrated there from all over the world. Many of these immigrants have come from Germany, Russia, and other European countries. Others are from North Africa and Southwest Asia. Israel also has a large number of Jews from Ethiopia.

READING CHECK: *Human Systems* How does religion shape cultural patterns in the region?

Review

TEKS Questions 1, 2, 3, 4, 5

go.hrw.com Homework Practice Online
Keyword: SW3 HP20

Identify
Zionism
Holocaust
Kurds

Define
sultans
mandates

Working with Sketch Maps
On the map you created in Section 1, label the countries of the eastern Mediterranean and İstanbul, Ankara, Jerusalem, and Nicosia. Which of these four cities is in both Europe and Asia?

Reading for the Main Idea
1. *Human Systems* What are some cultural features that Christian Arabs share with Muslim Arabs?
2. *Human Systems* Why is Israel's population so multiethnic?

Critical Thinking
3. **Drawing Inferences** Why do many Israelis oppose giving up the occupied lands and their Jewish settlements there? What might Palestinians and other Arabs offer in return for those lands?
4. **Identifying Cause and Effect** Why do many people in the eastern Mediterranean speak Arabic?

Organizing What You Know
5. Create a time line like the one shown below. Enlarge the section between A.D. 1 and Today. On your time line, list important years, periods, and events in the history of the eastern Mediterranean.

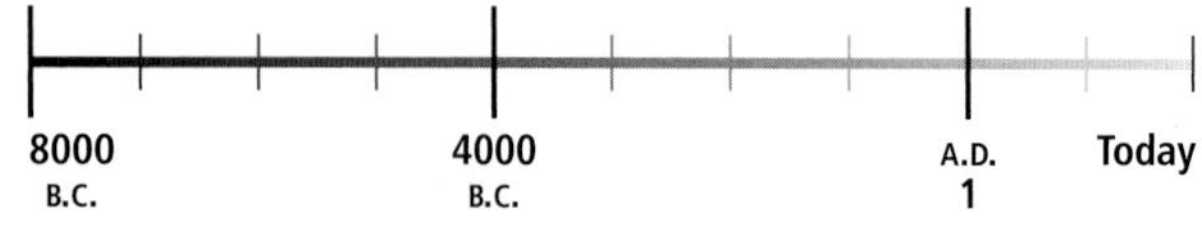

Human Systems

Geography for Life

Cyprus—A Divided Island

Cyprus lies in the Mediterranean Sea near Turkey. About 78 percent of its 763,000 people speak Greek and share a Greek heritage. Turkish Cypriots make up 18 percent of the population. Conflict between these groups marks the island's modern history.

Greeks have lived on Cyprus for at least 3,000 years. Turks later took over the island in the 1500s. Cyprus then came under British control in 1878. In 1960 Cyprus became an independent republic. Not long afterward, Turkish and Greek Cypriots clashed over sharing power in the new government. In addition, the Turkish Cypriots feared that the Greek Cypriots would join Cyprus to Greece. Fighting soon broke out. In 1964 the United Nations sent a peacekeeping force to Cyprus. In 1974 Turkey invaded and took control of the northern 40 percent of the island. Refugees—both Greek and Turkish—fled to the south. In 1983 Turkish Cypriots declared the northern territory an independent country—the Turkish Republic of Northern Cyprus (TRNC). No other country besides Turkey recognizes the TRNC. All other governments recognize Cyprus as a single country with a Greek Cypriot government.

Cyprus remains a divided island. A buffer zone called the Green Line separates Turkish Cyprus from Greek Cyprus. Nicosia, the capital of both Greek and Turkish Cyprus, lies along this line in the island's center.

Thousands of Turkish troops occupy northern Cyprus. Most countries refuse to trade with that region. As a result, the north must depend on aid from Turkey. However, the south has received help from Great Britain, Greece, the United States, and the United Nations. This aid has made construction of new businesses, housing, port facilities, and roads possible. Funds from other countries also allowed southern Cyprus to expand tourist facilities at its beaches and mountain resorts. The Greek state of Cyprus is also on a fast track for membership in the European Union. Greek Cypriots hope to unify the island and resettle Greeks in the north. Northern Turkish Cypriots, with backing from Turkey, still want two separate states.

Applying What You Know 

1. **Summarizing** What are two reasons why Greek Cypriots might feel that the Turks should give up their claims to the island?
2. **Identifying Points of View** How do Greek Cypriot and Turkish Cypriot views of the island differ?

Cyprus

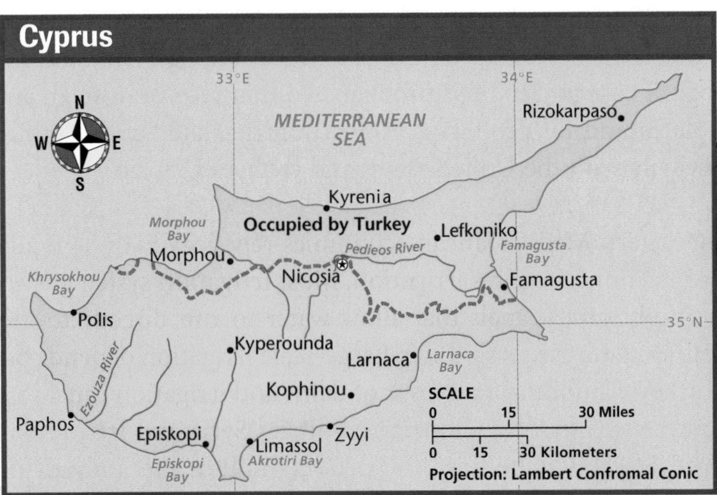

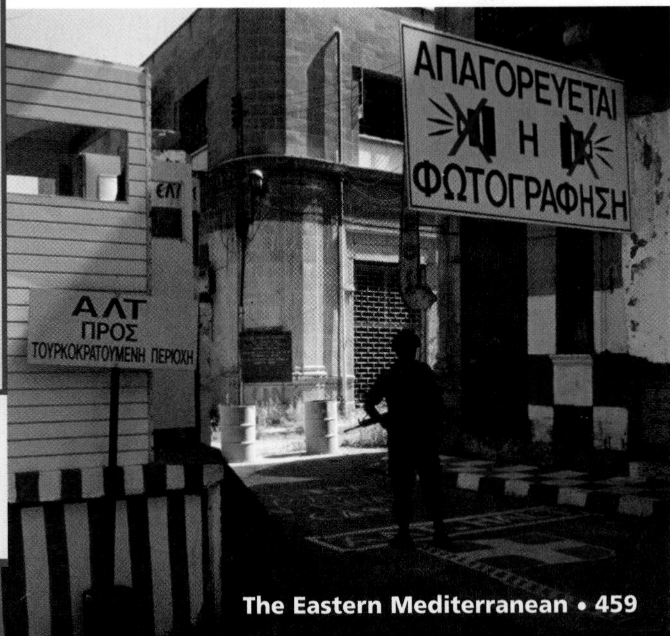

INTERPRETING THE VISUAL RECORD *In the photo, a member of the Cyprus National Guard patrols the Green Line in Nicosia. The blue and white paint echoes the colors of the Greek flag.* ***Why might both the Turkish and Greek Cypriots want Nicosia for their capital?*** TEKS

Section 3 The Region Today

READ TO DISCOVER

1. On what activities do the eastern Mediterranean economies rely?
2. What are the cities of the region like?
3. What challenges do the people of the region face?

DEFINE

souks
secular

LOCATE

Tel Aviv
Haifa
Damascus
Beirut

WHY IT MATTERS

Recent developments have led to important changes in the West Bank and Gaza Strip. Use CNNfyi.com or other **current events** sources to learn about the latest news from those areas.

A Turkish carpet seller displays his wares. Turkey's national government plays a major role in basic industries such as mining, but the textile and clothing industries are almost entirely in private hands.

Economic Development

Many factors have slowed the economic development of the eastern Mediterranean. For example, the land itself has caused problems for Turkey. Earthquakes there have interrupted its economic growth. Political problems are more widespread. Hostile relations between Israel and its neighbors have prevented the creation of normal economic links throughout the region. Also, hundreds of thousands of Palestinians live as refugees in other countries. Many live in Jordan and Lebanon. The resulting population increase has strained those countries' resources. At the same time, Israel has encouraged and absorbed waves of Jewish immigrants from other parts of the world. Lebanon and Cyprus have suffered devastating civil wars.

The economies of Syria and Jordan are underdeveloped and suffer from high unemployment. Lack of resources, a weak educational system, and an outdated technological base add to their problems. Jordan receives foreign aid from the United States and oil-producing Arab countries like Saudi Arabia. Israel also trades heavily with the United States and receives U.S. aid.

Agriculture The eastern Mediterranean economies rely heavily on agriculture. In many places farming requires irrigation. Most irrigation systems used in the region are simply small canals that allow water to run directly to the fields. Some of these systems use fossil water. Large-scale irrigation depends on the major rivers. Turkey is building a network of dams and irrigation canals on the Tigris and Euphrates Rivers. Neighboring countries Syria and Iraq resent these projects, however. This is because the projects reduce the amount of water downstream that is available to them.

Farmlands in Turkey's Mediterranean coastal valleys produce most of that country's crops. Livestock raising is common in Anatolia. Despite a lack of water, agriculture is productive in Israel as well. Careful irrigation makes growing fruits and vegetables possible in semiarid parts of the country.

Industry Israel is the most technologically advanced country in the eastern Mediterranean. The country has built much of its economy on high-tech industries. Diamond cutting is another major Israeli industry. Most diamond-cutting factories can be found in the Tel Aviv area. Polished diamonds are the country's leading export. Israel also has important chemical industries centered in the port city of Haifa. Turkey's industry is the second-most developed in the region. Textiles are Turkey's leading industry. The country's industrial sector includes modern urban factories and small-scale industries in rural areas.

Tourism Tourism is a major industry for the eastern Mediterranean. In Cyprus, tourists flock to beaches along the island's southern coast. Tourists come to Turkey to see Greek and Roman ruins, Ottoman palaces, and busy carpet markets. Many who travel to Israel want to visit religious sites, such as those in Bethlehem and Jerusalem, or float in the salty Dead Sea. However, the region's continuing unrest makes the tourist industry fragile. Israel's economy can handle a drop in tourism because it is well diversified. Other countries like Jordan, however, suffer when tourism slows.

✓ **READING CHECK:** ***Human Systems*** How do the economies of Israel and Turkey differ from those of Jordan and Syria?

Israeli workers cut and polish diamonds. Israel's diamond industry is one of the largest in the world and accounts for about 25 percent of the country's export earnings.

Urban Environments

The urban population of the eastern Mediterranean is growing rapidly. The region's high birth rates and migration from rural areas have fueled this growth. Many cities are crowded. Housing tends to be small and cramped, and traffic congestion and smog get worse every year.

Many of the cities in the eastern Mediterranean are ancient. In fact, Damascus, Syria, is probably the oldest city in the world. These old cities' centers usually consist of twisting narrow streets where traditional craftspeople sell their wares. Open-air markets called **souks** are also common. In many of

The Gaza Strip is a small, crowded piece of coastal land that has practically no resources. More than a million Palestinians live there.

Connecting to HISTORY

Petra

Petra, a desert city built about 2,000 years ago by an Arab people, is Jordan's main tourist attraction. It features huge buildings cut directly into the cliffs. How could this city of 30,000 people thrive in the desert? Protecting the water supply was crucial. Petra's builders installed canals, terraces, and hundreds of underground storage tanks to collect and store the area's scant rainfall. Dams in the nearby hills caught the water from flash floods after rare downpours. This water conservation system is the hidden marvel of Petra.

When the caravans were routed away from Petra, the city began to decline. Earthquakes did further damage. Neglecting the water storage system, which filled with sand, probably sealed the city's fate.

Problem Solving How might further study of Petra's water system affect public policy in Jordan? TEKS

Top: A water channel at Petra
Bottom: Petra's Royal Tombs

these markets, shops that sell the same items are clustered together. For example, vendors on Jerusalem's Christian Quarter Road specialize in religious souvenirs and Palestinian textiles. Typically, the newer parts of the city surround the old center. Most service-oriented business and government offices are in the newer areas.

War has damaged some cities, including Nicosia, Cyprus, and Beirut, Lebanon. In its better days Beirut had been called the Paris of the Middle East for its culture, glamour, and scenery. In recent years, Lebanon's government has worked hard to rebuild Beirut's center, making it a thriving and modern commercial zone. People there hope this rebuilding will help the country's economy recover from years of warfare.

TEKS **READING CHECK:** ***Human Systems*** How are the centers of eastern Mediterranean cities often different from the areas that surround them?

Issues and Challenges

The countries of the eastern Mediterranean face many issues and challenges. The most pressing of these are political ones. However, social and environmental problems also threaten stability.

One long-term challenge is the status of the West Bank and Gaza Strip. In the early 1990s the Israeli government began talks with the PLO. They agreed that the Palestinians would get control of parts of the West Bank and most of the Gaza Strip. However, the process has not been completed. Demonstrations, riots, and violence continue, and the future of the peace process remains uncertain.

Other ethnic and religious conflicts trouble other parts of the region. In Turkey, Armenian and Kurd minorities complain of unfair treatment by the government. Turkey's Kurds identify strongly with Kurds living in Iran, Iraq, and Syria. Some Kurds have fought for independence from Turkey. Thousands of people have been killed as a result

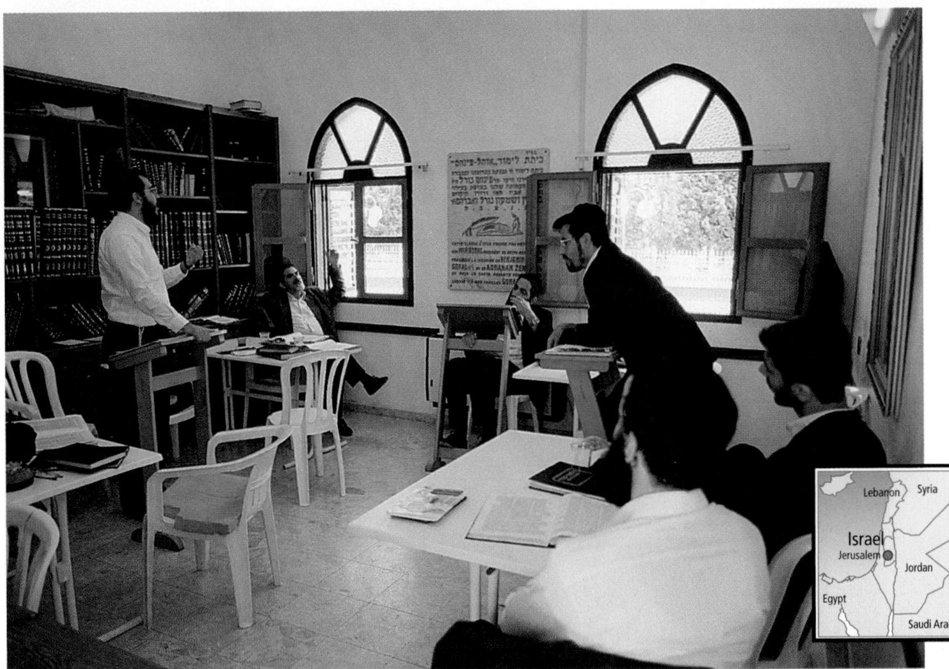

INTERPRETING THE VISUAL RECORD

Students debate issues related to religious law at a Jerusalem yeshiva—a school for the study of Jewish traditions. The role of religion in public life is a source of conflict between Jews and Arabs and among Jews. For example, many ultra-Orthodox Jews refuse to celebrate Israel's independence day because they feel the country is too secular. ***What are some aspects of religion in Israel that make the country distinctive?*** TEKS

of this struggle. In addition, Islamic fundamentalists in Turkey have objected to the country's **secular**, or nonreligious, political system. That criticism has turned violent at times. On the other side of the issue, the Turkish government has been criticized for limiting the religious freedom of Muslims.

As you read earlier, United Nations troops have tried to keep the peace between ethnic Greeks and ethnic Turks in Cyprus. In Lebanon a civil war among several different religious militias broke out in 1975. Christians belonging to various factions fought several Muslim groups. Fighting there lasted for 15 years. Jordan is divided between "original" Jordanians and Palestinian refugees. These Palestinians now make up a slight majority of Jordan's population. Many still live in refugee camps, where health care and other social services are poor.

The eastern Mediterranean's main environmental concern is a lack of water. Droughts, a growing population, and pollution all threaten the limited freshwater of the region. Also, overgrazing has damaged semiarid grasslands in parts of Jordan and Syria.

READING CHECK: ***Human Systems*** Why might some Kurds want Turkey to change its boundaries?

TEKS Questions 1, 3, 4

Section 3 Review

Homework Practice Online

Keyword: SW3 HP20

Define souks, secular

Working with Sketch Maps On the map you created in Section 2, label Tel Aviv, Haifa, Damascus, and Beirut. Which city is rebuilding after a civil war?

Reading for the Main Idea

1. ***Environment and Society*** Why do some other countries resent Turkey's proposed dam and irrigation system?
2. ***Human Systems*** What factors have fueled the rapid growth of cities in the region?
3. ***Environment and Society*** What problems threaten the region's scarce freshwater supplies?

Critical Thinking

4. **Analyzing Information** In what ways does the relationship between Israel and the Palestinians affect the rest of the region?

Organizing What You Know

5. Create a word web in which you describe Israel's agriculture, cities, industry, and the challenges it faces.

Jerusalem

Human Systems Jerusalem is one of the world's oldest cities and is deeply sacred for three major world religions: Christianity, Judaism, and Islam. Christians honor the city as the place where Jesus preached, died, and rose again. The city has long been a center of culture and faith for Jews. Muslims believe that from the site Muhammad departed on a spiritual journey through the skies.

During Jerusalem's 5,000 year history, many groups have fought to control the city. In fact, it has been captured and occupied more than 36 times over the centuries. Even today, control of Jerusalem is central to disputes between the Israelis and Palestinians.

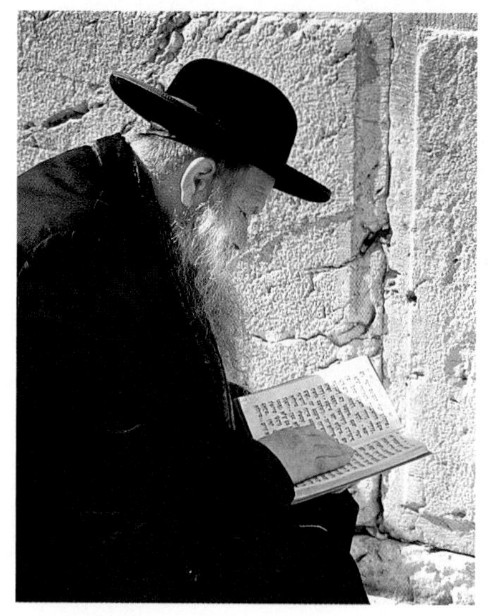

A Jewish man prays at Jerusalem's Western Wall, also called the Wailing Wall.

An Ancient and Sacred City

About 1000 B.C. King David ruled the Israelites from the city of Hebron. However, the king wanted a more centrally located capital from which he could further unite the tribes of Israel. This desire led him to seize the fortress of Zion from the Jebusites and found a new city, Jerusalem, there. When the king moved the sacred Ark of the Covenant to Jerusalem, the city became the religious center of Israel. David's son, King Solomon, expanded the city and built the Temple Mount, into which the Ark was placed.

In 586 B.C. the Babylonians captured Jerusalem and destroyed Solomon's temple. Although it was later rebuilt, the Romans destroyed it again in A.D. 70. All that remains of the temple today is a high wall on the western edge of the Temple Mount. This Western Wall—sometimes called the Wailing Wall—is the holiest place in Judaism. Each day people go to the wall to pray. They place slips of paper with prayers written on them between the narrow cracks in the wall.

The Church of the Holy Sepulchre, the most sacred shrine of Christianity, is also located in Jerusalem. Completed about A.D. 335 by the Roman emperor Constantine, it occupies the site where Christians believe Jesus was crucified, buried, and resurrected. Thousands of Christian pilgrims visit the church each year and retrace "the Way of the Cross" through Jerusalem's narrow streets.

Arab Muslims captured Jerusalem in the early A.D. 600s. In the late 600s they built a mosque on the Temple Mount to house the rock from which they believe Muhammad ascended to the heavens. Known as the Dome of the Rock, it is one of Islam's holiest sites. Some Jews believe the rock may also be the foundation stone on which the Ark of the Covenant rested in its ancient temple. Some even want to rebuild the temple. However, that could not be done without destroying the Dome of the Rock and other sacred Islamic sites on the Temple Mount. Conflicts like this over space sacred to different religions are common in Jerusalem.

The Modern City

Jerusalem is located on hills overlooking the Jordan River valley. In the heart of Jerusalem lies the Old City, the most ancient part of the city. Walls surround the Old City, which is separated into four quarters. These are the Armenian Quarter, Christian Quarter, Jewish Quarter, and Muslim Quarter. Each quarter contains religious buildings and houses for the members of its community.

In the 1860s the first settlements were built outside the walls of the Old City. Today more than

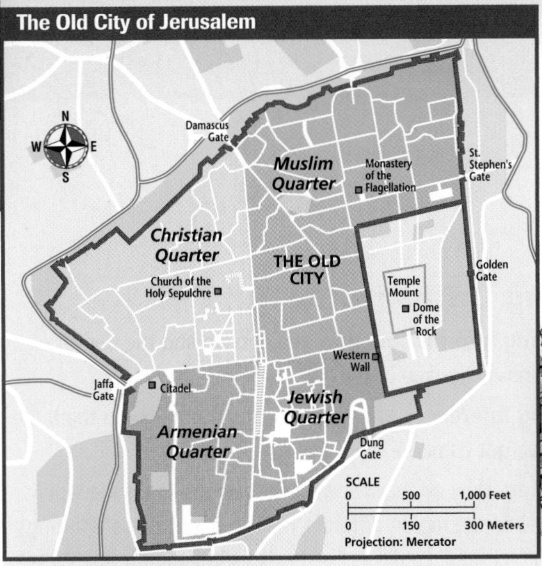
The Old City of Jerusalem
N
W
E
S
Damascus Gate
Muslim Quarter
Monastery of the Flagellation
St. Stephen's Gate
Christian Quarter
THE OLD CITY
Golden Gate
Church of the Holy Sepulchre
Temple Mount
Dome of the Rock
Western Wall
Jaffa Gate
Citadel
Jewish Quarter
Armenian Quarter
Dung Gate
SCALE
0 500 1,000 Feet
0 150 300 Meters
Projection: Mercator

INTERPRETING THE MAP *The Old City of Jerusalem is divided into four quarters. Part of the Muslim Quarter is pictured at the right.* ***How is the city's plan related to its political and historical characteristics?*** TEKS

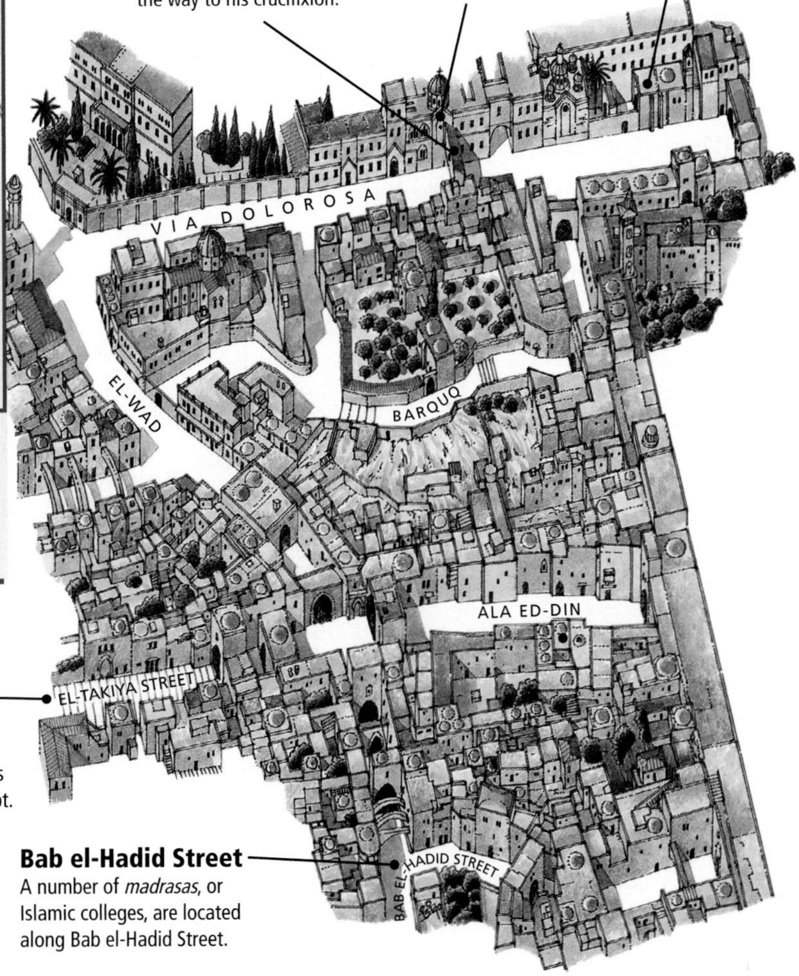
Ecce Homo Arch
The arch spans the Via Dolorosa, a street that follows the path Christians believe Jesus took on the way to his crucifixion.
Convent of the Sisters of Zion
Monastery of the Flagellation
VIA DOLOROSA
EL-WAD
BARQUQ
ALA ED-DIN
EL-TAKIYA STREET
El-Takiya Street
Examples of Mameluke architecture can be found along this narrow stepped street. The Mamelukes were Islamic rulers who once controlled the region from Egypt.
BAB EL-HADID STREET
Bab el-Hadid Street
A number of *madrasas*, or Islamic colleges, are located along Bab el-Hadid Street.

600,000 people live in Jerusalem, most outside its ancient walls. Many of Jerusalem's newer neighborhoods lie to the west and north. Commercial and government centers are also located west of the Old City.

About one third of Jerusalem's residents are Palestinian Arabs. Most live in East Jerusalem. The remaining two thirds of the population are mostly Jews, and many live in West Jerusalem. Neighborhood divisions throughout the city are often based on religious practices and preferences. For example, Arab East Jerusalem includes both Muslim and Christian neighborhoods. In West Jerusalem some communities are organized according to how strictly the residents practice Judaism.

When the modern state of Israel was established in 1948, Jerusalem was divided into Israeli and Jordanian sectors. Israel captured the entire city during the Six-Day War in 1967. Since then, Israel's government has controlled all of Jerusalem, including mostly Arab East Jerusalem. This has become a major political issue as Israelis and Palestinians struggle to make peace. Both sides consider Jerusalem a sacred city and want to control it. So far, negotiators have not been able to resolve this difficult issue.

Applying What You Know

TEKS

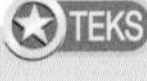

1. **Summarizing** Why have Jews and Muslims sought control of Jerusalem for so long?
2. **Comparing and Contrasting** Recall what you learned about Mecca in Chapter 19. What are some similarities and differences in the religious importance of Mecca and Jerusalem? What other places have you studied that are also important religious centers?

Building Vocabulary TEKS

On a separate sheet of paper, explain the following terms by using them correctly in sentences.

potash	mandates	Kurds
magnesium	Zionism	souks
sultans	Holocaust	secular

Locating Key Places TEKS

On a separate sheet of paper, match the letters on the map with their correct labels.

Dead Sea	İstanbul	Beirut
Euphrates River	Jerusalem	

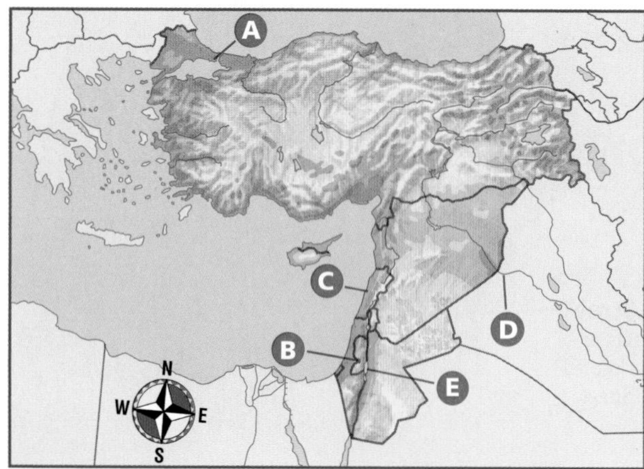

Understanding the Main Ideas TEKS

Section 1

1. *Physical Systems* What physical feature extends north from Africa to create the Jordan River valley and the Dead Sea?
2. *Places and Regions* How has the region's vegetation changed over time?

Section 2

3. *Human Systems* Why did many Jews immigrate to the region that later became Israel?
4. *Places and Regions* What are the three main religions of the eastern Mediterranean? What major feature do these religions share?

Section 3

5. *Places and Regions* What two countries are recovering from many years of ethnic and religious conflict?

Thinking Critically for TAKS TEKS

1. **Analyzing Information** How are Zionism and the revival of the Hebrew language linked?
2. **Drawing Inferences and Conclusions** How do you think the Dead Sea got its name?
3. **Analyzing** Why do many Jewish Israelis and Arabs view political issues in the region differently? How might these differing views affect public policies like Jewish settlement in the West Bank and Gaza Strip?

Using the Geographer's Tools TEKS

1. **Analyzing Maps** Review the chapter physical-political map. What areas could become part of a new Palestinian state? What problems might the geography of this new state pose for a Palestinian government?
2. **Analyzing Statistics** Review the unit Comparing Standards of Living and Fast Facts tables. Which country or countries appear to be the most developed and have the highest standard of living? What statistics have led you to this conclusion?
3. **Preparing Graphs** Go to **go.hrw.com** on the HRW Web site to find the information you will need to prepare a climate graph for one of the region's cities. Write a caption describing the city's climate and the factors affecting its climate.

Writing for TAKS TEKS

Review the information about the conflict between Israelis and Palestinians. Compare the Israeli and Palestinian points of view. Then find out how other countries view the conflict. Write a short report about your findings. You might want to create diagrams illustrating some of the key points in the conflict.

SKILL BUILDING

Geography for Life TEKS

Analyzing Geographic Information

Environment and Society Use Internet or library resources to investigate several of the eastern Mediterranean's non-European languages. Then answer the following questions: How similar are these languages to each other? How and when were they brought into the region? Where are they spoken today? Then draw a map showing where the languages are spoken and compare your map to the chapter physical-political map. Draw inferences about what geographic and political factors can create a language boundary. Write a paragraph describing your findings.

Building Social Studies Skills

Interpreting Charts

Study the chart below. Then use the information from the chart to help you answer the questions that follow. Mark your answers on a separate sheet of paper.

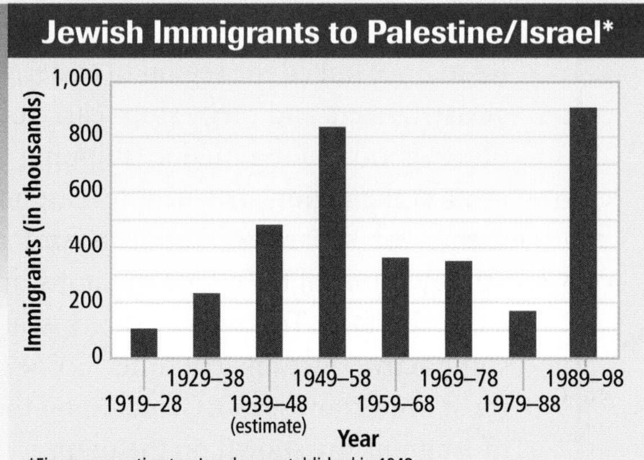

*Figures are estimates. Israel was established in 1948.
Source: American-Israeli Cooperative Enterprise

1. After Israel gained independence in 1948, the fewest number of Jewish immigrants to the area arrived between
   a. 1919 and 1928.
   b. 1929 and 1938.
   c. 1969 and 1978.
   d. 1979 and 1988.
2. What general pattern do you see in immigration to Palestine and Israel over time?

Interpreting Secondary Sources

Read the following passage and answer the questions that follow.

"In the 1300s the Ottoman Turks established another Muslim empire in the region. They took Constantinople in 1453 and made it the Ottoman capital. By the 1600s the Ottoman Empire included most of Southwest Asia. The empire also stretched into parts of eastern Europe and most of North Africa. However, political problems and rivalries with other countries slowly weakened the empire. Over time ethnic minorities within the empire's borders also began to push for independence."

3. What event happened in 1453?
   a. The Republic of Turkey was created.
   b. The Ottoman Turks created an empire.
   c. The Ottoman Empire was defeated.
   d. The Ottomans captured Constantinople and made it their empire's capital.
4. What are three reasons that the Ottoman Empire weakened?

Alternative Assessment

PORTFOLIO ACTIVITY

Learning about Your Local Geography TEKS

Group Project: Research

You have read about how cultural beliefs can influence public policies and decisions. Plan, organize, and complete a group research project about the various ethnic or cultural groups that live in your area. You might use library or Internet resources to begin your research. Then interview community government leaders to learn how cultural beliefs have influenced local policies. Finally, prepare a short report discussing your findings.

internet connect

Internet Activity: go.hrw.com
KEYWORD: SW3 GT20 TEKS

Choose a topic on the eastern Mediterranean to:

- research ancient and modern Jerusalem and create a newspaper.
- send a postcard from the Dead Sea and learn about its cultural value and its value as a natural resource.
- understand the sources of conflict and efforts to relieve it in Cyprus.

Skill Building for TAKS

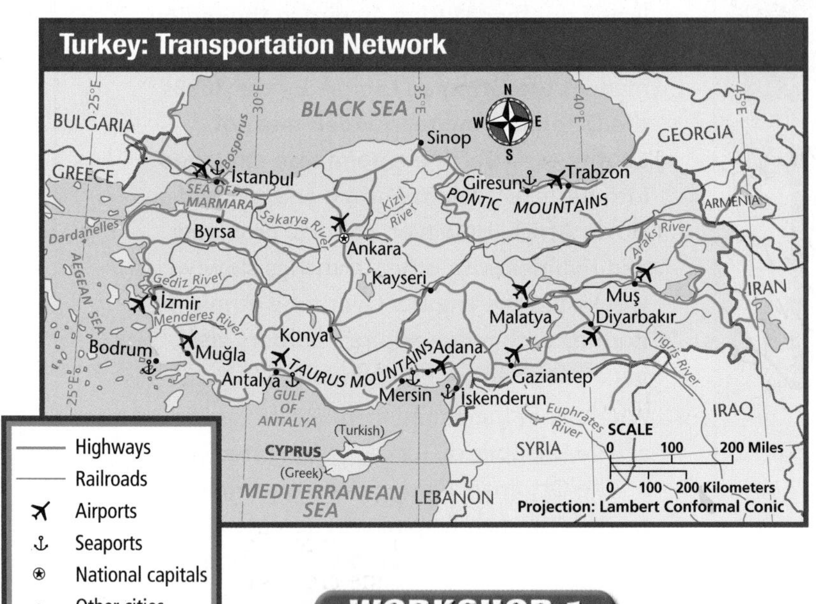

WORKSHOP 1

Using Transportation Maps

Some maps help us do more than study physical features or locations. They can also identify important transportation routes. Road maps, for example, include roads and highways with numbers and symbols that identify each route. Some transportation maps, such as the example on this page, are more complex. Such maps include a web of railways, major roads and highways, and important seaports and airports. All of those features link different places.

Developing the Skill Transportation maps are not difficult to understand. A map legend shows the kinds of transportation routes featured on the map. Common symbols include those for major roads, railways, airports, and seaports. Sometimes highways and railways are labeled on these kinds of maps. Distances along roads and railways may also be marked in miles, kilometers, or both. If not, a distance scale helps you estimate the distance between places shown on the map. Mapmakers may also show other cities or well-known cultural features to serve as landmarks on the map.

The map on this page shows Turkey's transportation network. Turkey and the rest of Southwest Asia have long been located along major transportation and trade routes. For example, the ancient Silk Road once linked the region to eastern Asia. Today tankers and cargo ships carry oil and other goods along major shipping lanes around the region. The Bosporus, a narrow strait that separates European and Asian Turkey, is one of the region's major shipping lanes. In fact, Turkey has many seaports, some active since ancient times. Some of the most important seaports, including the city of İstanbul, are shown on the map.

Turkey also has many major airports. As you can see, the country's land transportation network includes railways, roads, and highways. Coal, other minerals, and grain are transported along Turkish railways for hundreds of miles to ports or markets. A complex road network also links towns throughout the country. Increasing automobile and truck traffic led to the construction of a bridge across the Bosporus, which was completed in 1973. This bridge, the Trans-European Motorway, has become important for trade between Europe and Southwest Asia.

Practicing the Skill

1. What transportation features link Ankara to the rest of Turkey?
2. Identify major Turkish seaports shown on the map.
3. Prepare a transportation map of your state. Identify important railways, roads, highways, airports, and seaports.

WORKSHOP 2

Analyzing Aerial Photographs

Aerial photographs are important tools geographers use to view Earth's surface. Satellites and other spacecraft orbiting the planet provide many aerial images.

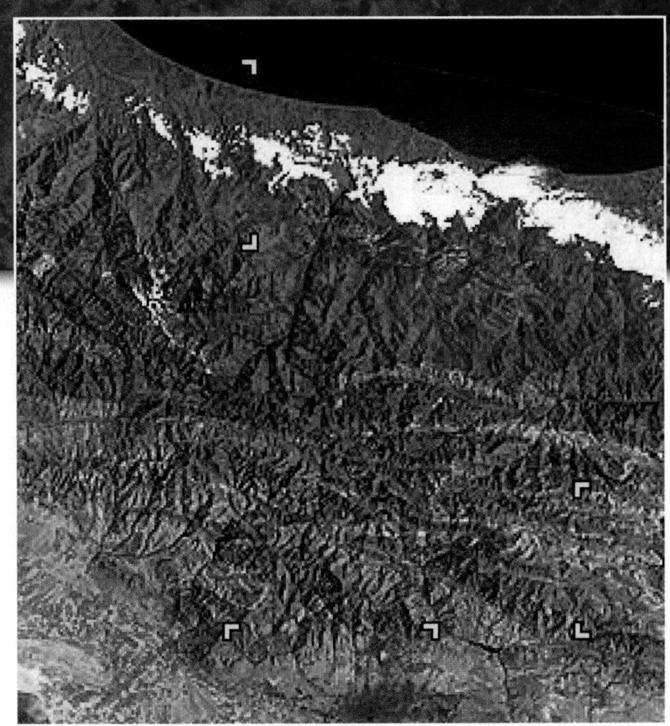

They use a variety of sensors to produce these images. These sensors include microwave detectors and scanners that measure and record electromagnetic radiation. Satellites send the data they gather to receiving stations on the ground. The data can then be converted into an image resembling a photograph.

The technology used in satellite imagery has links to World War II. Military researchers developed a special kind of color film during that war. It is known as false-color infrared film. The film captures the electromagnetic radiation that objects reflect. When this film is used, objects do not appear in their natural colors. The leaves of plants reflect very well. The healthier the vegetation is, the brighter the reflection, which appears in red on the film. Human structures like buildings, roads, and parking lots appear blue. Water appears black, unless it contains sediment or a lot of plant life.

Aerial photographs have been important to researchers. For example, geologists can use false-color aerial photographs to locate faults and other physical features on Earth's surface. The water that collects in such features encourages vegetation growth. This growth may be thicker than what is found in the surrounding area. As a result, areas along these physical features may show up in red, the color in which vegetation appears. In addition, the film clearly shows differences between vegetation and human features. As a result, it is useful in urban mapping.

Developing the Skill The following checklist can help you analyze satellite and aerial photographs. It will help you identify features in such images:

✓ Look at the shape of objects. Does the shape appear natural or human-made?

✓ Study the image's tone, or the lightness or darkness of its objects. Tone creates contrast and helps you distinguish between features.

✓ Analyze color, which fades with increasing distance. Colors also identify features.

✓ Try to locate a feature in the area that has a standard size, such as a football field. Compare other features to the one you have selected.

✓ Identify patterns, such as the repeated forms of sand dunes. Note that shadows can hide a feature or show the profile of its shape.

✓ Note the site and situation of a particular image. What is the relationship of the object to surrounding features? Look for links.

✓ Review the context of the image. Get as much information as possible about when and how the image was created.

Practicing the Skill

Use the checklist discussed above to analyze the aerial photograph on this page. The photograph shows an area of the Elburz Mountains in northern Iran. The city of Tehran (not shown) lies to the south, and the Caspian Sea is to the north. Create and complete a chart for your checklist. Then answer the following questions:

1. What do you suppose the white objects in the photograph represent?
2. What effect do the mountains appear to have on weather and climate patterns in the area?
3. Where is the thickest vegetation in the region? What might account for these vegetation patterns?

UNIT 7 Africa

The Great Sphinx and pyramids, Giza, Egypt

CONNECTING TO

Literature

"African Song" *by Richard Rive*

Richard Rive
(1930–89) was a high school English teacher in South Africa before becoming a writer. His short stories and novels focus on the lives of black Africans under apartheid, a system of racial segregation that ended only recently. The selection below is from his short story "African Song." The story is about a young man named Muti, who attends a meeting to protest apartheid. It reflects the pride that black Africans take in their ancient roots.

And then everyone was standing and Muti watched fascinatedly as the people sang; but still he sat because he had no pass. And what Muti knew must happen was happening because the blue uniforms were coming nearer. . . . And still the people sang. . . .

And as they [the protesters] sang there was a deep calm.

And this is what they sang.
Nkosi Sikelel' Afrika . . . . .

which means God bless Africa. God bless the sun-scorched Karoo and the green Valley of a Thousand Hills. . . . God bless this Africa of heat and cold, and laughter and tears, and deep joy and bitter sorrow, God bless this Africa of blue skies and brown veld [the open grazing land], and black and white and love and hatred, and friend and enemy. . . .

God bless this Africa, this Africa which is part of us. God protect this Africa. God have mercy upon Africa.

And still they sang.

Maluphakonyisw' Upshondo Lwayo. . . . .

which is lift up our descendents. And Muti thought of himself and wondered if he were a better man than his father, and his father's father and the many before him. For he felt like the Great Bird that flies higher and higher till it is a brother to the sun and can see the land even before the White man came. . . .

But Muti did not understand. Where were the cities and the towns and the villages? And the buildings and the shops? And where were the ones who lived in the cities and the towns? And the White ones and the Black ones? . . . And when he searched even further for his own people, he found them at last, and then his heart burst with pride. For he saw proud warriors with plumes of ostrich feathers, and shaking armlets which clicked as they raised the hands. And these warriors were huge ones and proud, and lifted high the legs and stamped upon the earth so that the ground shook. For they danced the dance of the young men and it was a vigorous dance and required much strength. And they were fearsome to behold.

Analyzing the Primary Source

1. **Summarizing** What image does Rive create of Africa's physical geography?
2. **Finding the Main Idea** What message does Rive's writing convey about Africa's origins and the roles of black and white Africans in the continent's history and development?

The World in Spatial Terms

Africa: Political

1. *Places and Regions* Where are the capitals of Algeria, Tunisia, and Libya located?
2. *The World in Spatial Terms* How many landlocked countries are there in Africa? How many island countries are there?

Critical Thinking

3. **Analyzing Information** Compare this map to the physical map. Why do you think Namibia's northeastern boundary is shaped the way it is?

Boundaries
National capitals
Other cities

SCALE
0 500 1,000 Miles
0 500 1,000 Kilometers
Projection: Azimuthal Equal Area

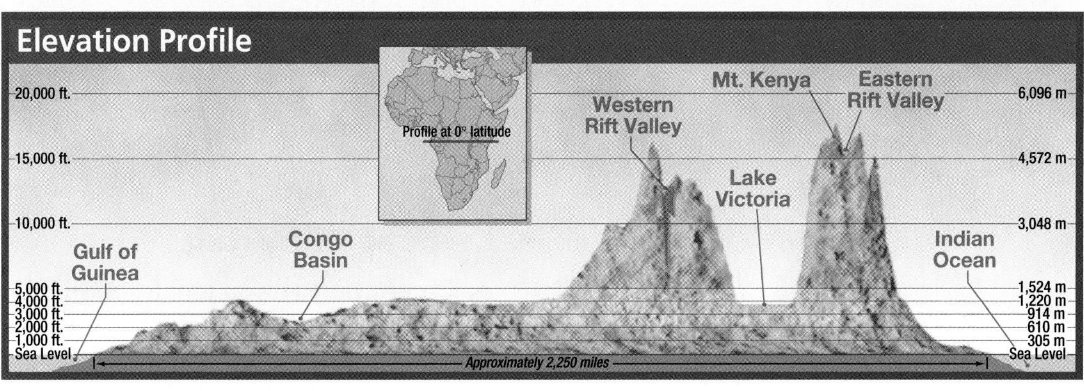

Africa: Physical

EUROPE
SOUTHWEST ASIA
MEDITERRANEAN SEA
ATLANTIC OCEAN
INDIAN OCEAN
Strait of Gibraltar
Canary Islands (SPAIN)
ATLAS MOUNTAINS
MOROCCO
TUNISIA
ALGERIA
LIBYA
EGYPT
WESTERN SAHARA (Occupied by Morocco)
SAHARA
AHAGGAR MOUNTAINS
LIBYAN DESERT
Suez Canal
ISTHMUS OF SUEZ
GULF OF SUEZ
ARABIAN DESERT
Aswān High Dam
NUBIAN DESERT
RED SEA
Nile River
Blue Nile
White Nile
Tropic of Cancer
MAURITANIA
MALI
NIGER
CHAD
SUDAN
ERITREA
DJIBOUTI
GULF OF ADEN
ETHIOPIA
ETHIOPIAN HIGHLANDS
SOMALIA
SAHEL
CAPE VERDE
Senegal River
Niger River
SENEGAL
GAMBIA
GUINEA-BISSAU
FOUTA DJALLON
GUINEA
SIERRA LEONE
LIBERIA
CÔTE D'IVOIRE
BURKINA FASO
GHANA
TOGO
BENIN
NIGERIA
Lake Chad
Benue River
CAMEROON
CENTRAL AFRICAN REPUBLIC
Ubangi R.
Bomu R.
Uele River
GULF OF GUINEA
EQUATORIAL GUINEA
SÃO TOMÉ AND PRÍNCIPE
GABON
REPUBLIC OF THE CONGO
Congo River
CONGO BASIN
DEMOCRATIC REPUBLIC OF THE CONGO
RWANDA
BURUNDI
UGANDA
KENYA
GREAT RIFT VALLEY
Lake Victoria
Kilimanjaro 19,341 ft. (5,895 m)
SERENGETI PLAIN
TANZANIA
Lake Tanganyika
Zanzibar
SEYCHELLES
Equator
CABINDA (ANGOLA)
Kasai River
KATANGA PLATEAU
Lake Malawi (Nyasa)
COMOROS
MALAWI
ANGOLA
ZAMBIA
Victoria Falls
Zambezi R.
Mozambique Channel
MADAGASCAR
MAURITIUS
MOZAMBIQUE
ZIMBABWE
NAMIBIA
NAMIB DESERT
BOTSWANA
KALAHARI DESERT
Tropic of Capricorn
SWAZILAND
LESOTHO
Orange R.
DRAKENSBERG
SOUTH AFRICA
N W E S

ELEVATION

FEET	METERS
13,120	4,000
6,560	2,000
1,640	500
656	200
(Sea level) 0	0 (Sea level)
Below sea level	Below sea level

SCALE
0 500 1,000 Miles
0 500 1,000 Kilometers
Projection: Azimuthal Equal Area

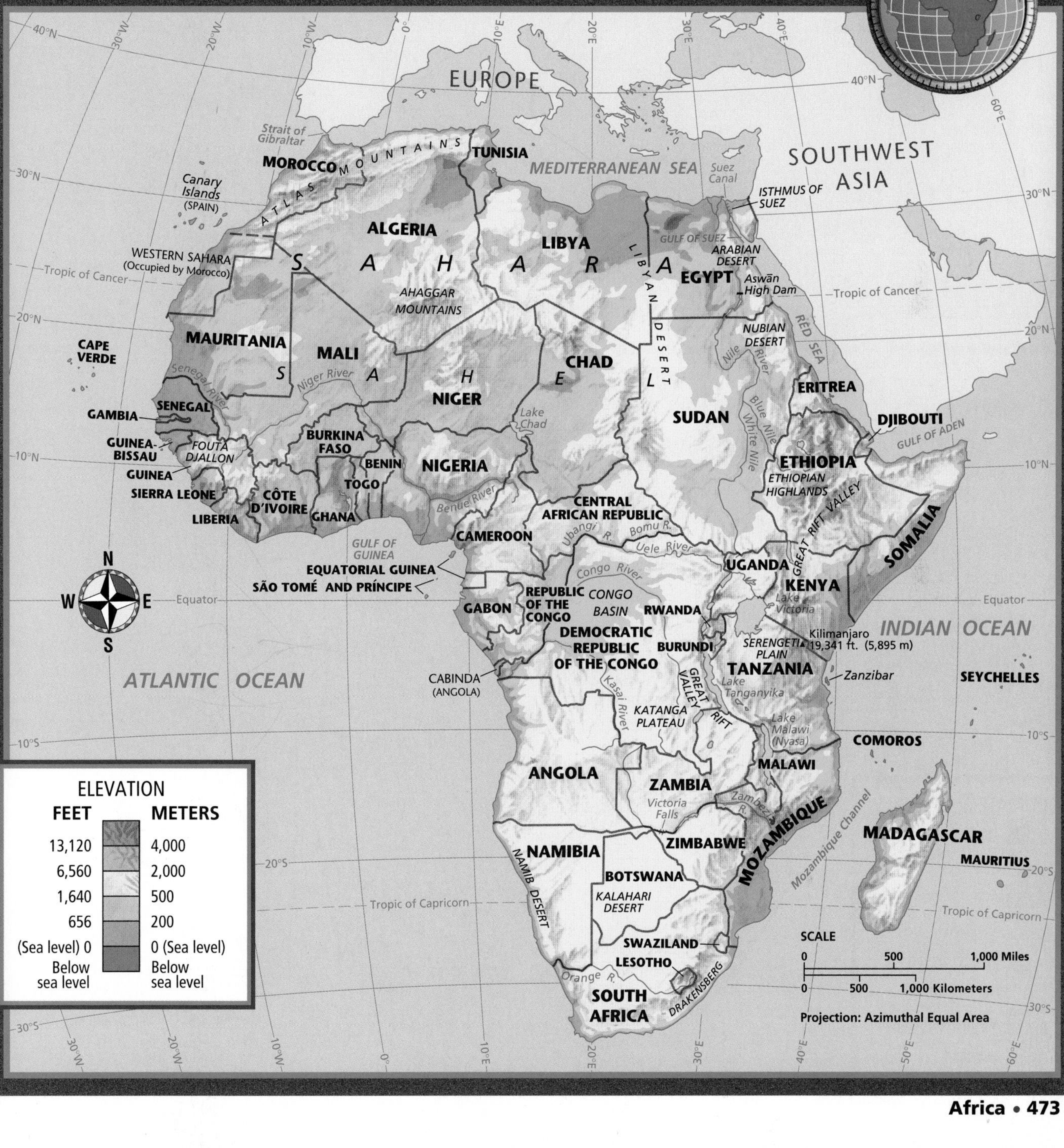

UNIT 7 ATLAS

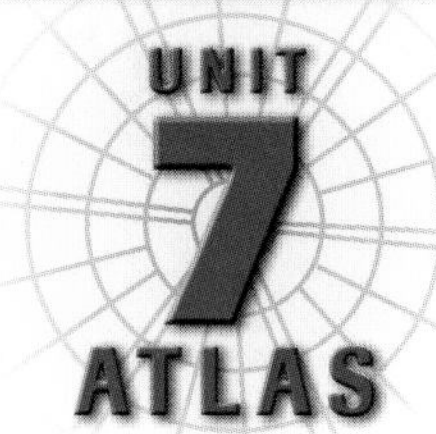

Africa: Climate

1. **Places and Regions** Compare this map to the physical map. Which country has a large area of highland climate? TEKS
2. **Places and Regions** Compare this map to the political map. Which countries have a Mediterranean climate? TEKS

Critical Thinking

3. **Comparing** If you traveled north from the equator, which climate regions would you pass through? Which would you pass through if you traveled south from the equator? TEKS

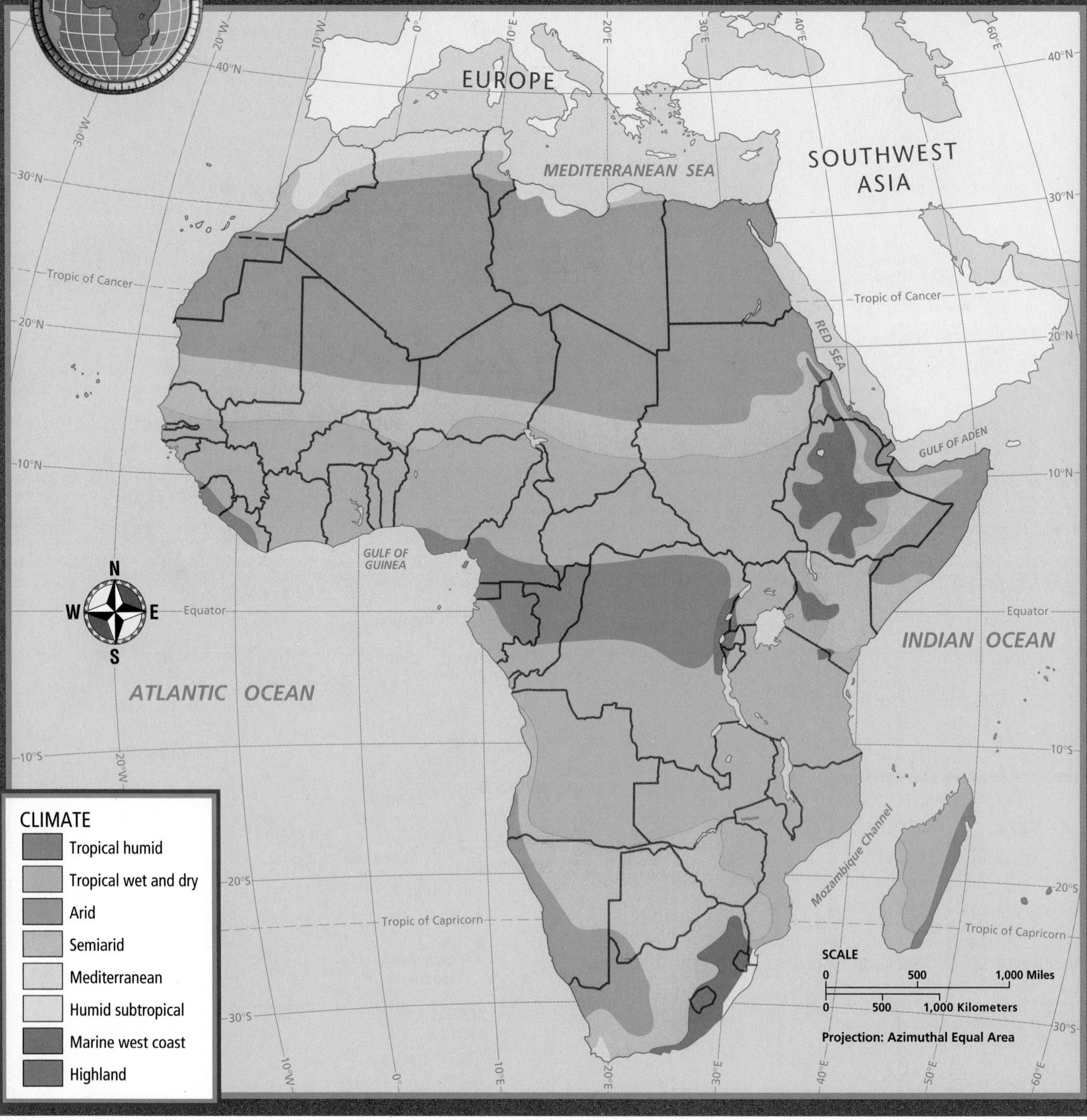

EUROPE
MEDITERRANEAN SEA
SOUTHWEST ASIA
RED SEA
GULF OF ADEN
GULF OF GUINEA
ATLANTIC OCEAN
INDIAN OCEAN
Mozambique Channel
Tropic of Cancer
Equator
Tropic of Capricorn
40°N
30°N
20°N
10°N
10°S
20°S
30°S
30°W
20°W
10°W
0°
10°E
20°E
30°E
40°E
50°E
60°E
N
S
E
W
CLIMATE
Tropical humid
Tropical wet and dry
Arid
Semiarid
Mediterranean
Humid subtropical
Marine west coast
Highland
SCALE
0
500
1,000 Miles
0
500
1,000 Kilometers
Projection: Azimuthal Equal Area

Africa: Precipitation

1. *Physical Systems* Compare this map to the physical map. Which mountain range in northern Africa receives the most precipitation? TEKS
2. *Physical Systems* Between which latitudes do the highest amounts of precipitation fall? TEKS

Critical Thinking

3. **Analyzing Information** Compare this map to the climate and population maps. How do you think people have adapted to living in arid and semiarid regions? TEKS

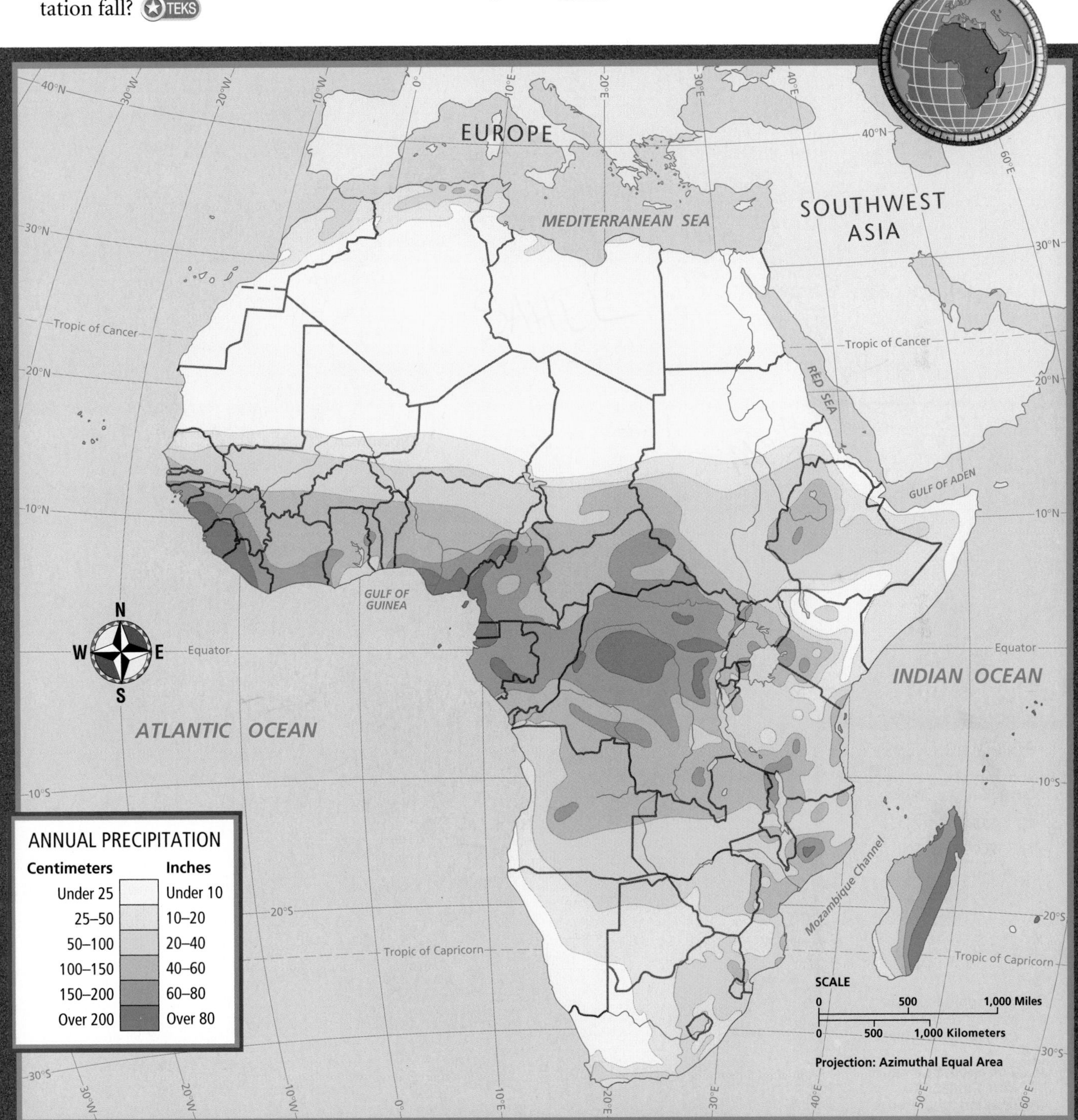

Africa: Population

1. *Places and Regions* Compare this map to the physical map. Which country in North Africa has an area with more than 520 persons per square mile (200 per sq km)? Along what physical feature are these people concentrated? TEKS
2. *Places and Regions* Compare this map to the climate map. Why do you think much of North Africa and parts of southern Africa have such low population densities? TEKS

Critical Thinking

3. **Making Generalizations** Based on the map, do you think most African countries have largely rural or largely urban populations? Why? TEKS

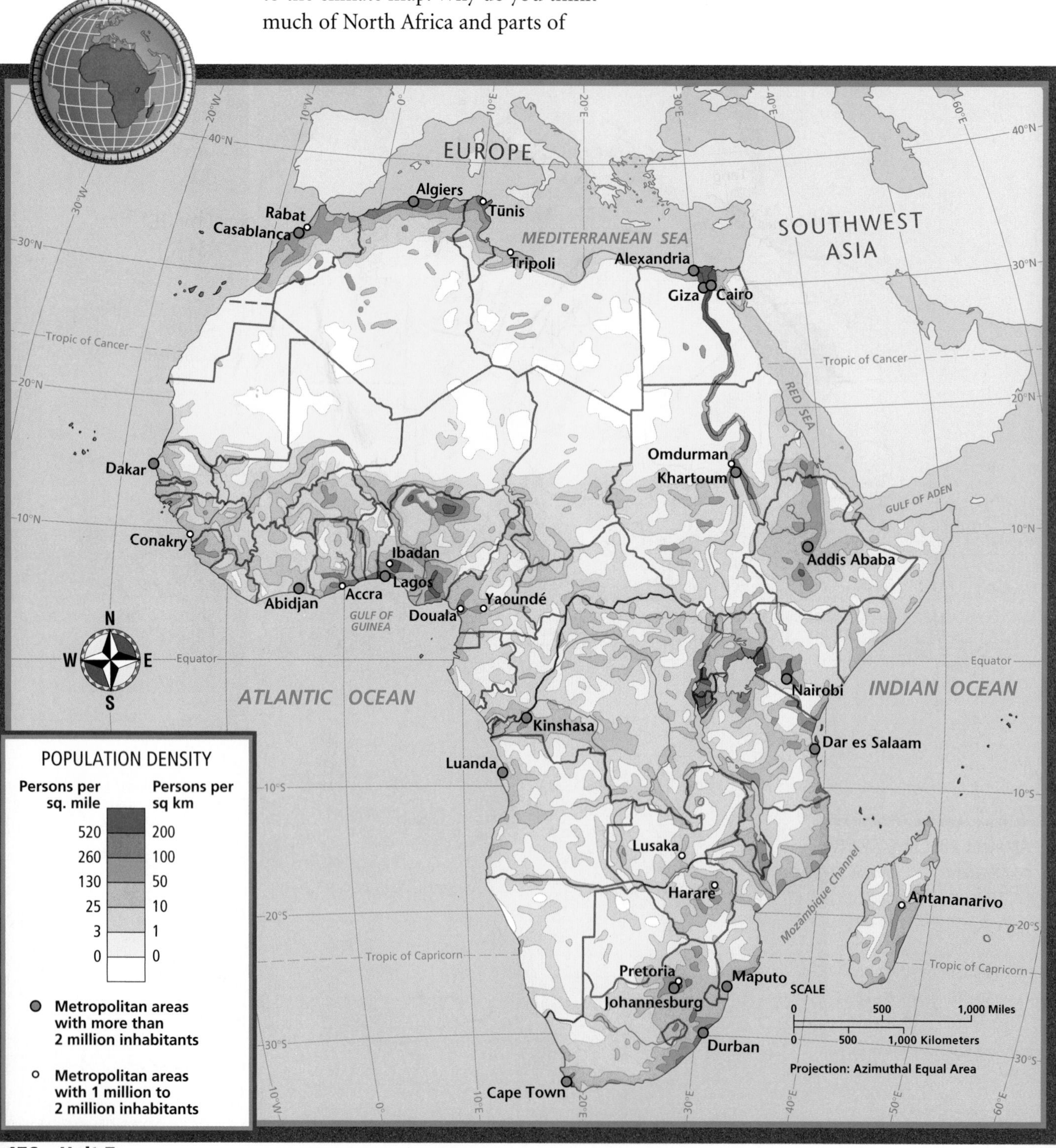

EUROPE
SOUTHWEST ASIA
MEDITERRANEAN SEA
RED SEA
GULF OF ADEN
ATLANTIC OCEAN
INDIAN OCEAN
GULF OF GUINEA
Mozambique Channel
Algiers
Tūnis
Rabat
Casablanca
Tripoli
Alexandria
Giza
Cairo
Omdurman
Khartoum
Addis Ababa
Dakar
Conakry
Ibadan
Lagos
Accra
Abidjan
Douala
Yaoundé
Nairobi
Kinshasa
Dar es Salaam
Luanda
Lusaka
Harare
Antananarivo
Pretoria
Johannesburg
Maputo
Durban
Cape Town
Tropic of Cancer
Equator
Tropic of Capricorn

POPULATION DENSITY

Persons per sq. mile	Persons per sq km
520	200
260	100
130	50
25	10
3	1
0	0

● Metropolitan areas with more than 2 million inhabitants

○ Metropolitan areas with 1 million to 2 million inhabitants

SCALE
0 500 1,000 Miles
0 500 1,000 Kilometers
Projection: Azimuthal Equal Area

UNIT 7 ATLAS

Africa: Land Use and Resources

1. **Environment and Society** Compare this map to the climate map. How do farmers appear to have adapted to environmental conditions in arid and semiarid areas? TEKS
2. **Human Systems** What are some countries that appear to have mostly traditional economies? TEKS

Critical Thinking

3. **Drawing Inferences** How widespread is subsistence farming in Africa? What might this indicate about Africa's level of development? TEKS

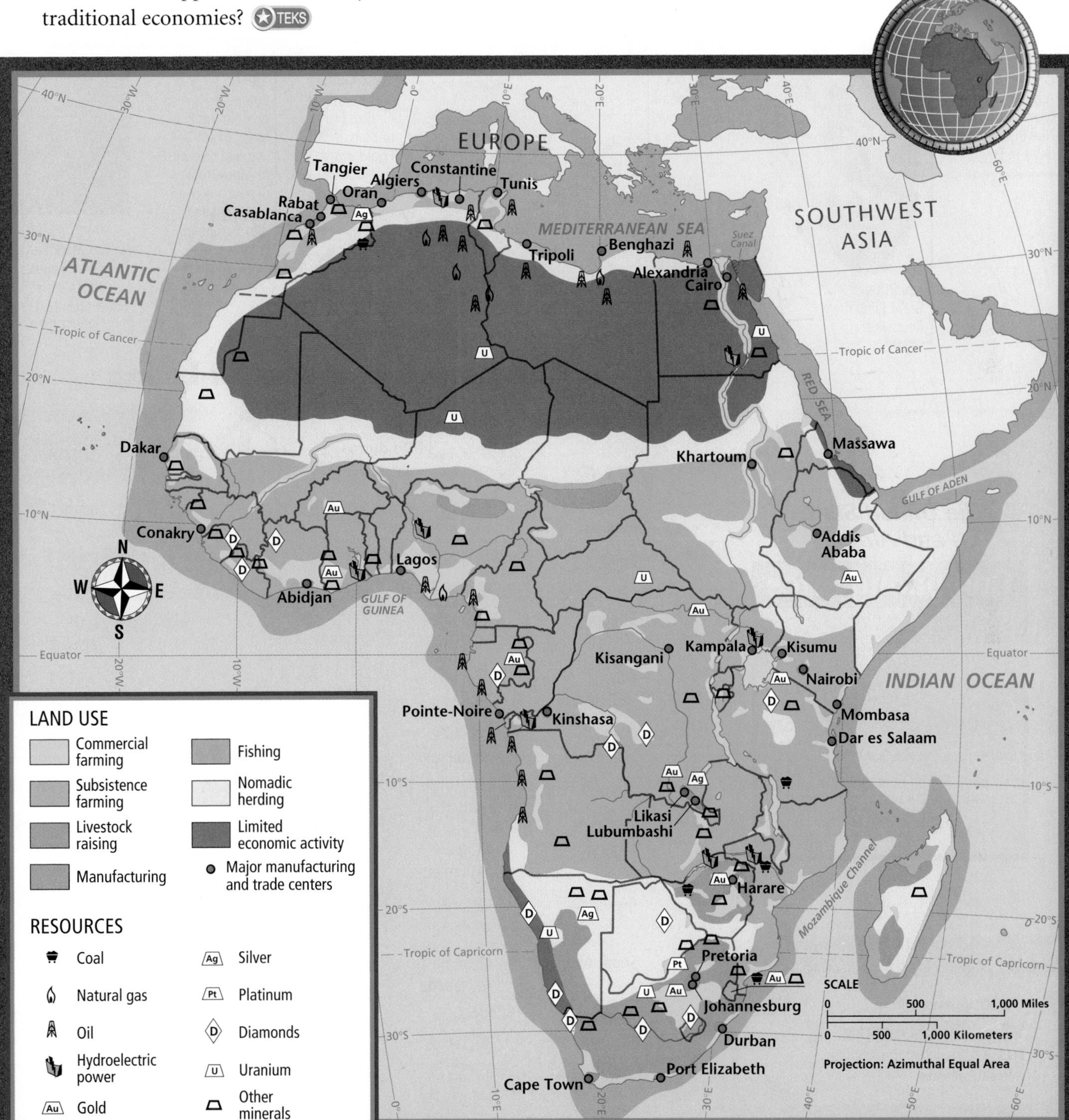

Time Line: Africa

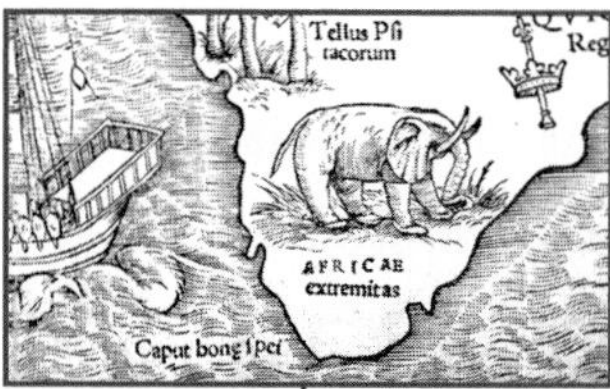

c. 3000 B.C. A great civilization grows along the Nile River and its delta in Egypt.

A.D. 200s Great Zimbabwe flourishes.

A.D. 600s Arab armies from Southwest Asia move across North Africa.

mid 1400s European explorers arrive in West Africa by sea.

late 1400s Portuguese sailors begin exploring the coast of southern Africa.

late 1800s The discovery of diamonds and gold draws people from all over the world to southern Africa.

1884 European powers meet in Berlin to try to settle colonial disputes in Africa. Most of Africa is eventually divided into European colonies.

1899–1902 The Boers and the British fight for control of South Africa's mineral wealth in the Boer War.

1950s and 1960s Most African countries gain their independence.

1994 Nelson Mandela is elected South Africa's first black president.

3000 B.C. | A.D. 600 | 1400 | 1500 | 1800 | 1900

Comparing Standard of Living

COUNTRY	LIFE EXPECTANCY (in years)	INFANT MORTALITY (per 1,000 live births)	LITERACY RATE	DAILY CALORIC INTAKE (per person)
Congo, Democratic Republic of the	47, male 51, female	100	77%	1,879
Egypt	62, male 66, female	60	51%	3,327
Ethiopia	44, male 46, female	100	36%	Not available
Kenya	47, male 48, female	68	78%	1,991
Libya	74, male 78, female	29	76%	3,126
Morocco	67, male 72, female	48	44%	3,157
Nigeria	51, male 51, female	73	57%	2,508
Rwanda	38, male 40, female	119	48%	Not available
South Africa	48, male 49, female	60	82%	2,890
Zimbabwe	39, male 36, female	63	85%	1,965
United States	74, male 80, female	7	97%	3,603

Sources: Central Intelligence Agency, *2001 World Factbook; Britannica Book of the Year, 2000*

The United States and Africa

Comparing Sizes

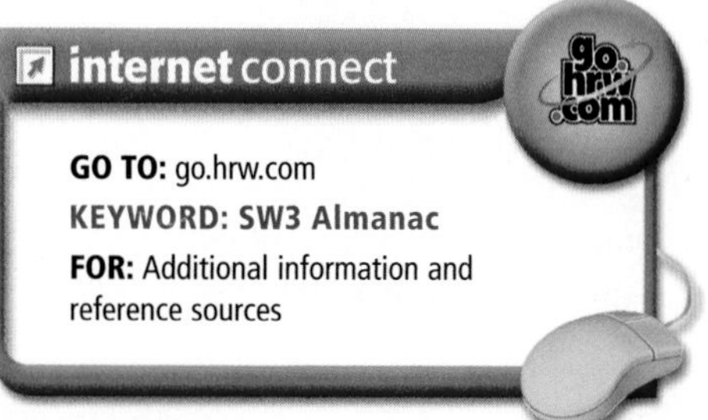

GO TO: go.hrw.com
KEYWORD: SW3 Almanac
FOR: Additional information and reference sources

Fast Facts: Africa

FLAG	COUNTRY Capital	POPULATION (in millions) POP. DENSITY	AREA	PER CAPITA GDP (in US $)	WORKFORCE STRUCTURE (largest categories)	ELECTRICITY CONSUMPTION (kilowatt hours per person)	TELEPHONE LINES (per person)
	Algeria Algiers	31.7 35/sq. mi	919,590 sq. mi. 2,381,740 sq km	$ 5,500	29% government 25% agriculture	681 kWh	0.07
	Angola Luanda	10.4 22/sq. mi.	481,351 sq. mi. 1,246,700 sq km	$ 1,000	85% agriculture 15% industry, serv.	132 kWh	0.006
	Benin Porto-Novo	6.6 152/sq. mi.	43,483 sq. mi. 112,620 sq km	$ 1,030	55% agriculture 21% trade	77 kWh	0.005
	Botswana Gaborone	1.6 7/sq. mi.	231,803 sq. mi. 600,370 sq km	$ 6,600	24% services 22% agriculture	956 kWh	0.05
	Burkina Faso Ouagadougou	12.3 116/sq. mi.	105,869 sq. mi. 274,200 sq km	$ 1,000	90% agriculture	22 kWh	0.003
	Burundi Bujumbura	6.2 579/sq. mi.	10,745 sq. mi. 27,830 sq km	$ 720	93% agriculture 4% government	26 kWh	0.003
	Cameroon Yaoundé	15.8 86/sq. mi.	183,567 sq. mi. 475,440 sq km	$ 1,700	70% agriculture 13% ind., commerce	204 kWh	0.005
	Cape Verde Praia	0.4 260/sq. mi.	1,557 sq. mi. 4,033 sq km	$ 1,700	25% agriculture 19% construction	92 kWh	0.11
	Central African Republic Bangui	3.6 15/sq. mi.	240,534 sq. mi. 622,984 sq km	$ 1,700	74% agriculture 8% trade	27 kWh	0.003
	Chad N'Djamena	8.7 18/sq. mi.	495,752 sq. mi. 1,284,000 sq km	$ 1,000	85% agriculture	10 kWh	0.0008
	Comoros Moroni	0.6 711/sq. mi.	838 sq. mi. 2,170 sq km	$ 720	80% agriculture	27 kWh	0.01
	Congo, Democ. Republic of the Kinshasa	53.6 59/sq. mi.	905,563 sq. mi. 2,345,410 sq km	$ 600	65% agriculture 19% services	85 kWh	0.0004
	Congo, Republic of the Brazzaville	2.9 22/sq. mi.	132,046 sq. mi. 342,000 sq km	$ 1,100	52% agriculture, forestry, fishing	141 kWh	0.008
	Côte d'Ivoire Yamoussoukro	16.4 132/sq. mi.	124,502 sq. mi. 322,460 sq km	$ 1,600	51% agriculture 12% manuf., mining	194 kWh	0.01
	Djibouti Djibouti	0.5 54/sq. mi.	8,494 sq. mi. 22,000 sq km	$ 1,300	75% agriculture 14% services	363 kWh	0.02
	Egypt Cairo	69.5 180/sq. mi.	386,660 sq. mi. 1,001,450 sq km	$ 3,600	49% services 29% agriculture	865 kWh	0.06
	Equatorial Guinea Malabo	0.5 45/sq. mi.	10,830 sq. mi. 28,051 sq km	$ 2,000	58% agriculture 8% services	40 kWh	0.008

Sources: Central Intelligence Agency, *2001 World Factbook; The World Almanac and Book of Facts, 2001; Britannica Book of the Year, 2000;* U.S. Census Bureau
The CIA calculates per capita GDP in terms of purchasing power parity. This formula equalizes the purchasing power of each country's currency.

Fast Facts: Africa

FLAG	COUNTRY Capital	POPULATION (in millions) POP. DENSITY	AREA	PER CAPITA GDP (in US $)	WORKFORCE STRUCTURE (largest categories)	ELECTRICITY CONSUMPTION (kilowatt hours per person)	TELEPHONE LINES (per person)
	Eritrea Asmara	4.3 92/sq. mi.	46,842 sq. mi. 121,320 sq km	$ 710	80% agriculture 20% ind., services	36 kWh	0.005
	Ethiopia Addis Ababa	65.9 151/sq. mi.	435,184 sq. mi. 1,127,127 sq km	$ 600	80% agriculture 12% govt., services	23 kWh	0.002
	Gabon Libreville	1.2 12/sq. mi.	103,346 sq. mi. 267,667 sq km	$ 6,300	60% agriculture 25% services, gov.	777 kWh	0.03
	Gambia Banjul	1.4 323/sq. mi.	4,363 sq. mi. 11,300 sq km	$ 1,100	75% agriculture 19% ind., comm., serv.	49 kWh	0.02
	Ghana Accra	19.9 216/sq. mi.	92,100 sq. mi. 238,540 sq km	$ 1,900	60% agriculture 25% services	280 kWh	0.01
	Guinea Conakry	7.6 80/sq. mi.	94,925 sq. mi. 245,857 sq km	$ 1,300	80% agriculture 20% ind., services	92 kWh	0.003
	Guinea-Bissau Bissau	1.3 94/sq. mi.	13,946 sq. mi. 36,120 sq km	$ 850	78% agriculture	39 kWh	0.006
	Kenya Nairobi	30.8 137/sq. mi.	224,961 sq. mi. 582,650 sq km	$ 1,500	75–80% agriculture	132 kWh	0.009
	Lesotho Maseru	2.2 186/sq. mi.	11,720 sq. mi. 30,355 sq km	$ 2,400	86% subsistence agriculture	25 kWh	0.009
	Liberia Monrovia	3.2 75/sq. mi.	43,000 sq. mi. 111,370 sq km	$ 1,100	70% agriculture 22% services	125 kWh	0.002
	Libya Tripoli	5.2 8/sq. mi.	679,358 sq. mi. 1,759,540 sq km	$ 8,900	54% services, gov. 29% industry	3,354 kWh	0.07
	Madagascar Antananarivo	16.0 71/sq. mi.	226,656 sq. mi. 587,040 sq km	$ 800	76% agriculture 12% manufacturing	47 kWh	0.003
	Malawi Lilongwe	10.5 231/sq. mi.	45,745 sq. mi. 118,480 sq km	$ 900	86% agriculture	90 kWh	0.004
	Mali Bamako	11.0 23/sq. mi.	478,764 sq. mi. 1,240,000 sq km	$ 850	80% agriculture, fishing	38 kWh	0.002
	Mauritania Nouakchott	2.7 7/sq. mi.	397,953 sq. mi. 1,030,700 sq km	$ 2,000	47% agriculture 39% services	51 kWh	0.009
	Mauritius Port Louis	1.2 1,657/sq. mi.	718 sq. mi. 1,860 sq km	$ 10,400	36% const., industry 24% services	985 kWh	0.19
	Morocco Rabat	30.6 178/sq. mi.	172,413 sq. mi. 446,550 sq km	$ 3,500	50% agriculture 35% services	439 kWh	0.05

FLAG	COUNTRY Capital	POPULATION (in millions) POP. DENSITY	AREA	PER CAPITA GDP (in US $)	WORKFORCE STRUCTURE (largest categories)	ELECTRICITY CONSUMPTION (kilowatt hours per person)	TELEPHONE LINES (per person)
	Mozambique Maputo	19.4 63 sq. mi.	309,494 sq. mi. 801,590 sq km	$ 1,000	81% agriculture 13% services	16 kWh	0.003
	Namibia Windhoek	1.8 6/sq. mi.	318,694 sq. mi. 825,418 sq km	$ 4,300	47% agriculture 33% services	1,084 kWh	0.06
	Niger Niamey	10.4 21/sq. mi.	489,189 sq. mi. 1,267,000 sq km	$ 1,000	90% agriculture 6% ind., commerce	39 kWh	0.002
	Nigeria Abuja	126.6 355/sq. mi.	356,667 sq. mi. 923,768 sq km	$ 950	70% agriculture 20% services	137 kWh	0.004
	Rwanda Kigali	7.3 719/sq. mi.	10,169 sq. mi. 26,338 sq km	$ 900	90% agriculture	26 kWh	0.002
	São Tomé and Príncipe São Tomé	0.2 428/sq. mi.	386 sq. mi. 1,001 sq km	$ 1,100	38% agriculture 23% services	96 kWh	0.02
	Senegal Dakar	10.3 136/sq. mi.	75,749 sq. mi. 196,190 sq km	$ 1,600	60% agriculture	115 kWh	0.01
	Seychelles Victoria	0.08 453/sq. mi.	176 sq. mi. 455 sq km	$ 7,700	71% services 19% industry	1,867 kWh	0.25
	Sierra Leone Freetown	5.4 196/sq. mi.	27,699 sq. mi. 71,740 sq km	$ 510	62% agriculture 20% services	41 kWh	0.003
	Somalia Mogadishu	7.5 30/sq. mi.	246,199 sq. mi. 637,657 sq km	$ 600	71% nomadic agric. 29% industry, serv.	32 kWh	Not available
	South Africa Pretoria	43.6 93/sq. mi.	471,008 sq. mi. 1,219,912 sq km	$ 8,500	45% services 30% agriculture	3,955 kWh	0.12
	Sudan Khartoum	36.1 37/sq. mi.	967,493 sq. mi. 2,505,810 sq km	$ 1,000	80% agriculture 10% ind., commerce	45 kWh	0.01
	Swaziland Mbabane	1.1 165/sq. mi.	6,704 sq. mi. 17,363 sq km	$ 4,000	70% private sector 30% public sector	179 kWh	0.03
	Tanzania Dar es Salaam	36.2 99/sq. mi.	364,898 sq. mi. 945,087 sq km	$ 710	80% agriculture 20% ind., commerce	59 kWh	0.004
	Togo Lomé	5.2 235/sq. mi.	21,925 sq. mi. 56,785 sq km	$ 1,500	65% agriculture 30% services	99 kWh	0.005
	Tunisia Tunis	9.7 154/sq. mi.	63,170 sq. mi. 163,610 sq km	$ 6,500	55% services 23% industry	894 kWh	0.07
	Uganda Kampala	24.0 263/sq. mi.	91,135 sq. mi. 236,040 sq km	$ 1,100	82% agriculture 13% services	44 kWh	0.002
	Zambia Lusaka	9.8 34/sq. mi.	290,584 sq. mi. 752,614 sq km	$ 880	85% agriculture 9% services	607 kWh	0.008
	Zimbabwe Harare	11.4 75/sq. mi.	150,803 sq. mi. 390,580 sq km	$ 2,500	66% agriculture 24% services	611 kWh	0.02
	United States Washington, D.C.	281.4 76/sq. mi.	3,717,792 sq. mi. 9,629,091 sq km	$ 36,200	30% manage., prof. 29% tech., sales, admin.	12,211 kWh	0.69

CHAPTER 21 North Africa

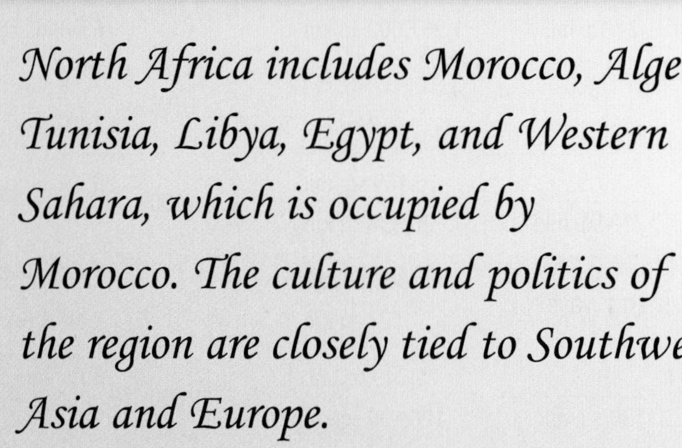

North Africa includes Morocco, Algeria, Tunisia, Libya, Egypt, and Western Sahara, which is occupied by Morocco. The culture and politics of the region are closely tied to Southwest Asia and Europe.

Pendant from Algeria

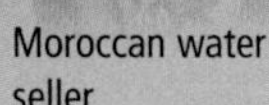

Moroccan water seller

Ahlan! (Hi!) My name is Shaimaa, and I am 18. I live with my mother and my little sister in an apartment about one hour from downtown Cairo. I am in my third year of high school.

Every day but Friday, I get up at 7:00 A.M., drink a glass of milk, and then meet my friends. We travel to school together on the metro (mass transit). My school is an all-girls school. I am in the humanities and social sciences track, so I am studying philosophy, psychology, Arabic, English, and history, my favorite. We all have religious education in school also—I study Islam with the other Muslim girls while the Christian girls meet with their religious teacher. At about 3:00 P.M., I get home from school, eat a big lunch of chicken and vegetables, and sleep for a couple of hours. When I wake up, I have a lot of homework—about six hours' worth. At 10:00 P.M., we have another small meal of cheese, yogurt, or beans before bedtime.

On Fridays I go to movies with my girlfriends and walk along the Nile River, or I stay home and listen to music. Sometimes I take a taxi to the beach with my family. For big holidays, we go to the mosque very early in the morning. Then we go to my grandmother's house with my 12 aunts and uncles and 16 cousins! After Ramadan (an Islamic holy month), we put on our new clothes and go around to all my relatives' houses, where we get presents of new money.

1 Natural Environments

READ TO DISCOVER

1. What landform regions are found in North Africa?
2. What factors influence the region's climates?
3. What natural resources does the region have?

DEFINE

erg
reg
depressions
wadis

LOCATE

Atlantic Ocean
Red Sea
Mediterranean Sea
Atlas Mountains
Sahara
Qattara Depression
Nile River
Strait of Gibraltar

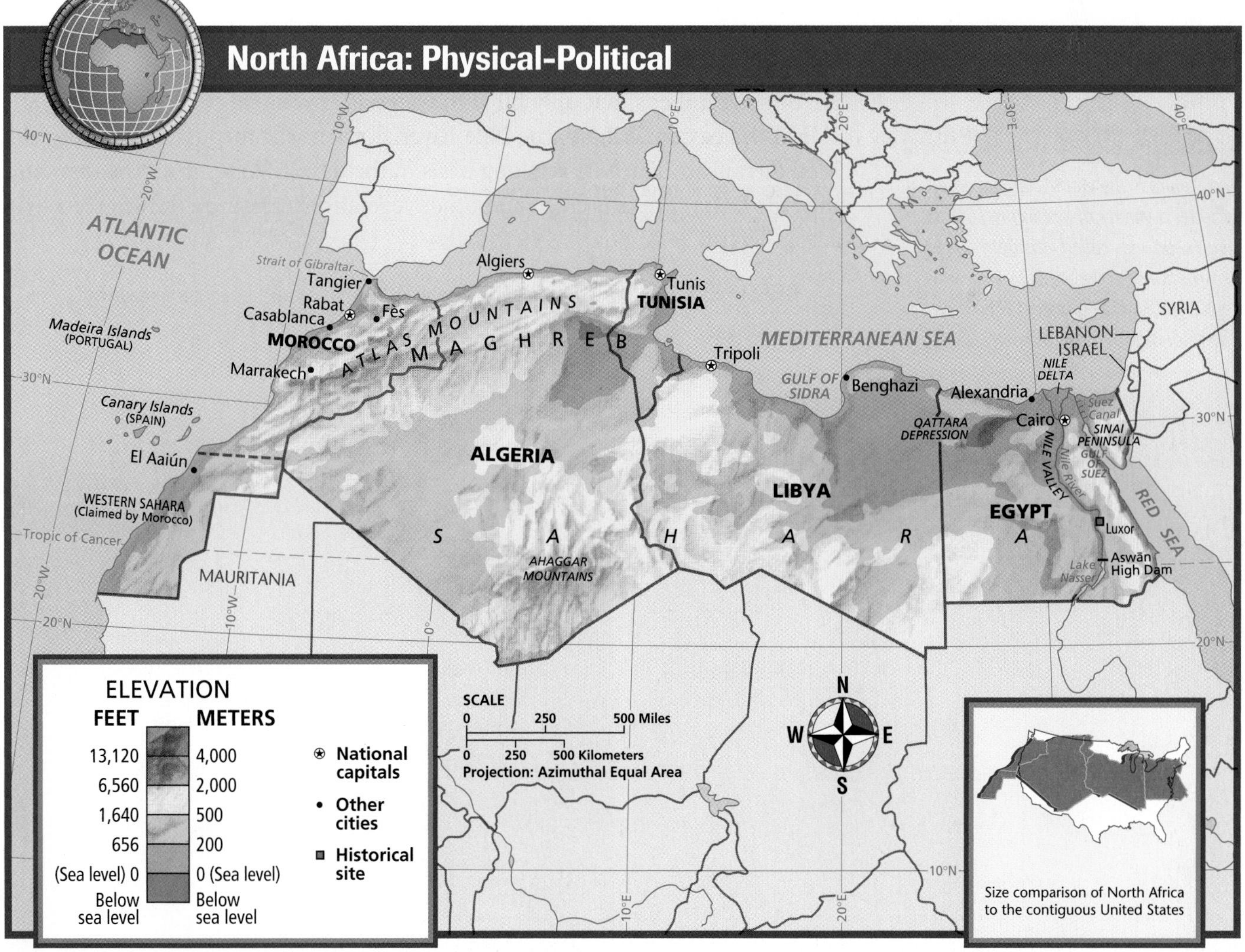

WHY IT MATTERS

North Africa is part of the world's largest desert region. Use CNNfyi.com or other **current events** sources to learn about the cultures, landscapes, and ways of life in very dry places.

North Africa: Physical-Political

Landforms

North Africa stretches from the Atlantic Ocean to the Red Sea. On the north it is bordered by the Mediterranean Sea. Coastal plains are the main landforms where North Africa meets the Mediterranean Sea and the Atlantic Ocean. From Morocco to Tunisia, the coastal plains quickly give way to the Atlas Mountains. These mountains run parallel to the Atlantic and Mediterranean coasts from northern Tunisia to the Atlantic Ocean. The Sahara is the world's largest desert. It lies south of the Atlas Mountains and coastal plains. The Sahara extends across all of North Africa. This vast desert acts as a natural barrier between North Africa and the rest of the African continent.

Over thousands of years, the Sahara has undergone cycles of wet and dry periods. At times there has been enough rain to support grasslands and abundant animal life. However, most of the Sahara today is a barren expanse of rock and sand. The great desert covers about 3.5 million square miles (9 million sq km) of land—roughly the size of the entire United States. Because there are few plants, wind and rain can erode the land easily. As a result, bare rock surfaces are common. The basins below these rocky ridges fill with eroded sediment. Some basins are covered with high, shifting sand dunes that create a sea of sand, called an **erg**. In other areas, wind blows the sand and dust away, leaving a gravel-covered plain. This type of eroded landform is called a **reg**.

The Sahara also has large low areas called **depressions**. In Egypt the Qattara Depression, 440 feet (134 m) below sea level, is a wilderness of quicksand and salt marsh. Other depressions have large dry lakebeds where water briefly collects during rare rainstorms. Rainwater also carves out **wadis**, which are dry streambeds that only fill with water after rain falls.

In the eastern Sahara, the Nile River flows north through Egypt into the Mediterranean. The Nile is a long oasis in the desert. Water from the river and the Nile Delta supports crops and other vegetation, creating a fertile green strip across Egypt.

READING CHECK: ***Physical Systems*** What has characterized the climatic history of the Sahara?

INTERPRETING THE VISUAL RECORD

The Sahara is the world's largest desert. It includes a variety of landforms, such as giant sand seas called ergs (left) and extensive gravel-covered plains known as regs (right). ***What physical forces create these distinctive landforms?*** 

The thorny argan tree is found mainly in Morocco. Goats climb these trees to eat the olive-like fruit that they produce.

Climates, Plants, and Animals

Most of North Africa's vegetation and wildlife are restricted to the areas with a Mediterranean climate. In the Sahara, plants and animals must be very hardy to survive. Temperatures there can climb above 130°F (54°C) in the summer. They can drop to below freezing in the winter. As in other desert areas, the daily range of temperatures is also great. Daytime temperatures are very hot, but at night, temperatures cool dramatically.

Climates North Africa's Mediterranean climate areas are found along the coast. Warm dry summers and mild rainy winters are common there. Some areas with a semiarid climate lie between coastal areas and the Sahara.

An arid climate covers most of North Africa. A subtropical high-pressure system keeps the region very dry. The system creates a wide band of dry lands, of which the Sahara is a part, across Asia and Africa. A rain shadow caused by the Atlas Mountains also contributes to the region's dry climate. A hot dry wind called the harmattan (har-muh-TAN) often sweeps southward across the Sahara. In the Saharan ergs, the winds can cause violent sandstorms that block out sunlight for days. Sometimes hot desert winds blow Saharan dust across the Atlantic Ocean to the Caribbean and the southern United States.

Plants and Animals Vegetation in the Sahara is limited to grasses and short shrubs. Few trees can grow in such dry regions. Plants and animals are concentrated around oases, where there are sources of water. Date palms are found around oases. Insects, snails, and small reptiles are common, along with a few larger animals such as gazelles and hyenas.

Morocco, in particular, has a great diversity of plant and animal life. The country is close to Europe, separated by just 8 miles (13 km) across the Strait of Gibraltar. As a result, many European as well as African species of plants and animals are found in Morocco. Also, migratory birds pass through Morocco every year as they fly between Europe and Africa. The country also has forests in the mountains.

The highest temperature ever recorded on Earth was 136°F (58°C) on the northern edge of the Sahara in Libya.

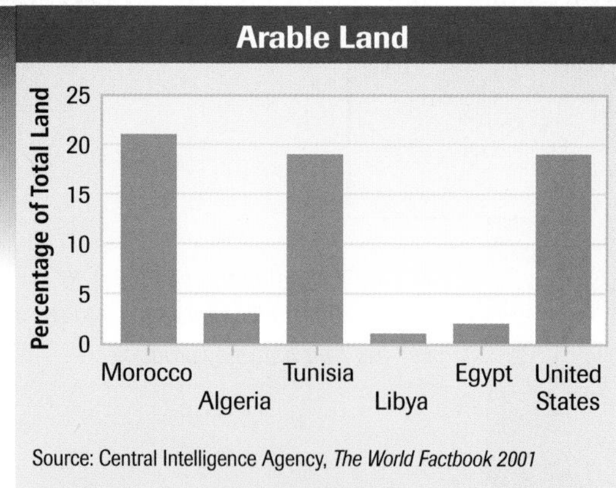

Source: Central Intelligence Agency, *The World Factbook 2001*

INTERPRETING THE GRAPH *Morocco and Tunisia have the highest percentage of arable land in North Africa, while Algeria, Libya, and Egypt have a far lower percentage.* ***Based on the graph, what can you infer about environmental conditions in the different countries of North Africa?*** TEKS

Egypt also has abundant bird life. More than 300 species of birds are found in the Nile Valley and Nile Delta. These include the egret, flamingo, golden oriole, heron, pelican, and stork. The bird life in Egypt benefits from the great variety of plant life along the Nile River. However, outside of the Nile Valley and Delta, Egypt is almost all desert, with limited plant and animal life. This is true of most of Algeria and Libya as well. Tunisia's milder climates support forests, grasslands, and many kinds of animals.

READING CHECK: ***Places and Regions*** What two factors create the Sahara's dry climate?

Natural Resources

Oil and natural gas are North Africa's most valuable natural resources. Oil is found in every country in North Africa, but Libya has the largest reserves. Other important resources include iron ore, lead, phosphates, and zinc.

There are rich fishing grounds off Morocco's Atlantic coast. The main catch is sardines. Fishers there catch more than 350,000 tons of sardines every year. In fact, Morocco is the world's largest exporter of this type of fish.

Although much of North Africa is desert, rain or irrigation makes farming possible in areas with good soil. The region produces many crops, including grapes, olives, dates, grains, and vegetables. River water irrigates fields all along the Nile and in the Nile Delta. These farmlands have helped make Egypt an important cotton producer. Rice is also an important crop in the Nile Valley. Farmers in the desert oases of Algeria, Libya, and Egypt grow crops such as date palms.

READING CHECK: ***Environment and Society*** Where do farmers in Egypt get water for growing crops?

Review

TEKS Questions 1, 2, 3, 4, 5

Homework Practice Online
Keyword: SW3 HP21

Define erg, reg, depressions, wadis

Working with Sketch Maps

On a map of North Africa that you draw or that your teacher provides, label the countries of the region and Atlantic Ocean, Red Sea, Mediterranean Sea, Atlas Mountains, Sahara, Qattara Depression, Nile River, and Strait of Gibraltar. In the margin of your map, identify the shortest water crossing between Africa and Europe.

Reading for the Main Idea

1. ***Physical Systems*** How is an erg formed? How is a reg formed?
2. ***Places and Regions*** Where are the region's wettest climates located?
3. ***Physical Systems*** What is one reason that Morocco's plant and animal life is so diverse?

Critical Thinking

4. **Analyzing Information** In what way might the Mediterranean Sea be considered an important natural resource in the region?

Organizing What You Know

5. Create a chart like the one shown below. Use the chart to list and describe the region's climates, plants and animals, and natural resources.

Climates	Plants and animals	Natural resources

History and Culture

READ TO DISCOVER

1. What peoples have settled in and ruled North Africa?
2. What are the people and culture of the region like today?

WHY IT MATTERS

Archaeological discoveries in Egypt and other parts of North Africa tell us much about early human history. Use CNNfyi.com or other **current events** sources to learn about recent archaeological research in the region.

IDENTIFY

Berbers

DEFINE

silt
pharaohs
hieroglyphs

LOCATE

Alexandria
Cairo
Marrakech
Maghreb
Suez Canal
Sinai Peninsula
Casablanca
Fès

History

The first people in North Africa were hunter-gatherers. They lived in areas where the climates were best. By 4000 B.C., much of North Africa had become desert. The human population became concentrated along the Mediterranean coast, desert oases, wadis, and the Nile River. Every year, the Nile would flood. These floods spread **silt**, which is fertile finely ground soil, over the river's banks.

Early Peoples Beginning about 3000 B.C., a great civilization grew along the Nile River and its delta in Egypt. A series of kingdoms arose, ruled by monarchs called **pharaohs**. These rulers were considered gods and had complete power over the Egyptian people. The Egyptians built great pyramids and other monuments that still stand today. They developed a writing system that used pictures and symbols called **hieroglyphs**. Egyptian astronomers created a 365-day calendar. They also learned to predict the annual floods of the Nile. Every year, when Sirius—the brightest star in the sky—appeared above the horizon at sunrise, the flood would soon follow. Because the Egyptians depended on the Nile's floods for farming, this information was key to their survival. From 1570 B.C. to 1085 B.C., Egypt expanded its power into the area that is now Syria, Israel, and Libya.

Later, as Egyptian power weakened, foreigners began to control much of North Africa. Those foreigners included the Phoenicians, Greeks, and Romans. The

This ancient Egyptian illustration from the Book of the Dead shows Horus, the falcon-god, introducing an Egyptian to the presence of Osiris, god of the underworld. The Book of the Dead was a collection of texts, spells, and formulas placed in tombs to help the dead in the afterlife. It contains many outstanding examples of ancient Egyptian art.

In Greek mythology, the giant Atlas held the world on his shoulders somewhere near the western end of North Africa. The Atlas Mountains are named for him. In the 1500s, collections of maps or charts came to be called atlases because mapmakers included images of Atlas holding Earth in them.

Phoenicians were sailors and traders from what is now Lebanon. They set up many Mediterranean trading colonies such as Carthage, which was founded in about 800 B.C. in modern-day Tunisia. Alexander the Great, at the head of a Greek army, founded the city of Alexandria in Egypt in 332 B.C. The Roman Empire became a great power in North Africa after it destroyed Carthage in 146 B.C. After the Roman Empire crumbled in the A.D. 400s, a Germanic tribe called the Vandals moved south through Spain into Africa. They set up a kingdom in what is now Libya. In the A.D. 500s the Byzantine Empire, which had been the Eastern Roman Empire, recaptured most of North Africa.

The Arabs and Islam Byzantine rule over North Africa was short-lived. In the 600s Arab armies from Southwest Asia swept across North Africa. They reached Africa's Atlantic coast and conquered Morocco by the early 700s. Arab armies also crossed into Iberia in the year 711. Most people in North Africa became Muslims. In addition, Arabic became the main language of the area.

Under Arab rule, Cairo and other North African cities became great centers of Islamic culture and education. Cities like Marrakech in Morocco became centers of trade between central and western Africa, Europe, and Arabia. Such cities grew rich trading gold, ivory, and spices as well as slaves. However, in the 1500s outsiders again invaded North Africa. The Ottomans—Muslims based in what is now Turkey—took control first of Egypt and then of Libya, Tunisia, and Algeria.

TEKS **READING CHECK:** ***Human Systems*** What religion and language did the Arabs bring to North Africa?

Colonialism The Ottoman Empire ruled much of North Africa until the late 1800s. Earlier in that century, western European powers had begun to take over parts of the region. Beginning in the 1830s, France moved to control the Maghreb (West), including Tunisia, Algeria, and part of Morocco. Spain took

INTERPRETING THE VISUAL RECORD

French colonists in Algeria found that the region was perfect for vineyards, and Algerian wine became a major export under French rule. However, it is mainly produced for export because most Algerians are Muslim and are forbidden to drink alcohol. ***How did cultural patterns influence the diffusion of crops to Algeria and affect the country's cultural landscapes?*** TEKS

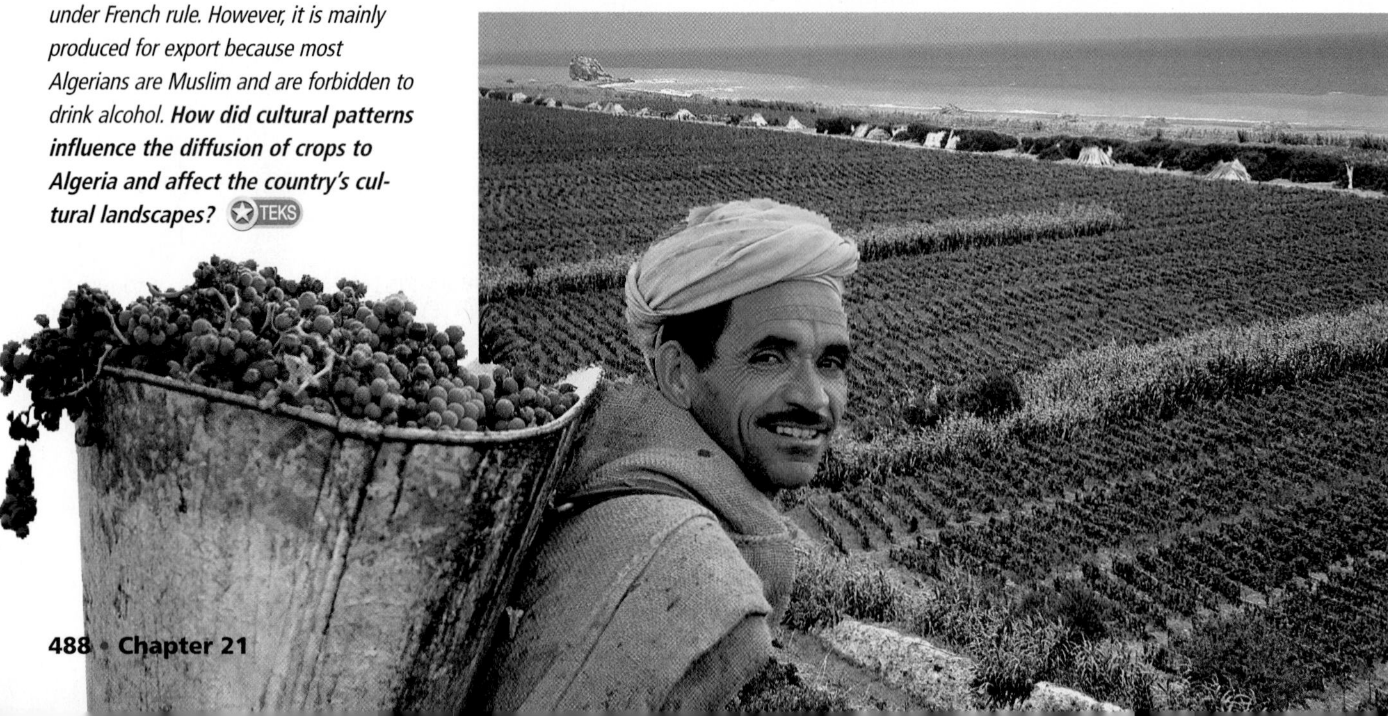

control of northern Morocco. Thousands of French, Spaniards, and Italians settled in North Africa in the decades that followed. More than 100 years later, the large number of French living in Algeria would complicate that country's struggle for independence.

In 1882 Great Britain took over Egypt. Britain wanted control of the Suez Canal, which connects the Mediterranean with the Red Sea. The canal was an important trade link between Europe and Britain's colony of India. Italy completed the European conquest of North Africa by taking Libya from the Ottoman Empire in 1912.

Independence North Africans resented European rule. They did not have the same rights as the European settlers. Over time North Africans worked to win their independence. In 1922 Egypt gained limited independence from Great Britain. However, independence efforts across the region became stronger after World War II ended in 1945.

In 1952 a group of Egyptian military officers led a revolution that brought complete independence from Britain. France granted Tunisia and Morocco their independence in 1956. Yet the French fought a bloody war to hold on to Algeria. When Algeria finally won independence in 1962, most of the French population left.

Libya became an independent kingdom in 1951. In 1969, military officers led by Mu'ammar Gadhafi overthrew the monarchy. Gadhafi declared the country a socialist republic and adopted anti-Western policies.

READING CHECK: *Human Systems* How did independence affect Algeria's human geography?

Culture

The countries of North Africa share a similar history and Muslim culture. Still, there is a great deal of variation among them.

People and Languages Nearly all of the people of North Africa consider themselves Arab or Arab-Berber. The **Berbers** are a cultural group that lived in North Africa long before waves of Arab armies crossed the continent. Also, small groups of desert nomads called Bedouins live along the Sinai Peninsula in Egypt.

Arabic is the official language of every country in North Africa. However, the people of each country speak their own version of Arabic. In some rural areas Berber dialects are also common. Because of the influence of colonization, many people also speak European languages. French is still widely used in Algeria, Morocco, and Tunisia. Italian is spoken in Libya, and English is used in Egypt.

Settlement and Land Use Most North Africans live along the Mediterranean coast or in the foothills of the Atlas Mountains. An exception to this pattern is found in Egypt. About 99 percent of that country's 69 million people live in the Nile Valley and Delta. Together, those populated areas make up only 3 percent of Egypt's land. Cairo, Egypt's

Connecting to THE ARTS

Naguib Mahfouz

In 1988 Egyptian writer Naguib Mahfouz became the first Arab to receive the Nobel Prize for Literature. Mahfouz was born in Cairo in 1911. He has written some 40 novels and short story collections as well as many plays and screenplays.

Mahfouz's first books were set in ancient Egypt. Later works dealt with contemporary subjects. In *The Cairo Trilogy,* for example, the story of a large middle-class Egyptian family connects to colonial and modern political history.

Mahfouz has often written about social issues, such as the status of women and political prisoners. *Children of the Alley* features characters that symbolize Adam, Moses, and other religious figures. Some people criticized the book, saying it was disrespectful to Islam. Nevertheless, Mahfouz is one of the region's most respected writers.

Evaluating If you were to write a novel about North Africa, which parts of its history or culture would you use as background? Why would you choose those topics? TEKS

capital, lies in the Nile Delta. With a population of more than 10 million, Cairo is the largest urban area in North Africa. Egypt also has the largest population of any Arab country.

Urban overcrowding is a problem across all of North Africa. People from the countryside are pouring into cities in hopes of finding work and a better life. Casablanca, the largest city in Morocco, must absorb about 30,000 new migrants every year.

READING CHECK: ***Human Systems*** In what areas are most of the population of North Africa found?

FOCUS ON CULTURE

The Medina Most old Arab cities in North Africa developed within the protective walls of a Casbah, or fort. As the population of the city grew, the buildings within the city's walls were built higher and closer together. Space was limited, and streets were as narrow as possible, often twisting at odd angles. The high walls and narrow streets also created shade that kept people and buildings cool in the hot climate.

When colonial governments took over North Africa, they built European-style cities around the old Arab city, or medina. However, people did not abandon the medina for the wide boulevards and spacious air-conditioned buildings of newer areas. Many medinas in North Africa remain lively places where people live and go for social interaction, shopping, and prayer. One of the most famous medinas in the world is in Fès, Morocco. There, tens of thousands of people crowd into a square mile of densely packed buildings.

READING CHECK: ***Human Systems*** What are the medinas of old Arab cities of the region like?

INTERPRETING THE VISUAL RECORD

The medina in Tunis was built during the A.D. *600s and is the cultural and historical focus of the city. During the colonial era, the French built a new area around the medina known as the* ville nouvelle, *or "new city."* ***What architectural features can you see in the photo that are typically found in medinas?***

Religion Most North Africans are Muslim, except for very small Christian and Jewish minorities. Islam plays a major role in North African life. For example, the five daily prayers punctuate life and mark the time for appointments. In addition, Fridays are special days when Muslims meet in mosques for prayer. In many cities across North Africa, businesses close on Thursday and Friday before opening again on Saturday. Businesses also close early for religious holidays, such as *Id al-Adha.* During this holiday, Muslim families sacrifice a sheep in honor of the willingness of Abraham to sacrifice his own son to prove his devotion to Allah. Islamic holidays are celebrated according to a lunar calendar. As a result, holidays shift over the years, being celebrated earlier each year according to the Western calendar.

INTERPRETING THE VISUAL RECORD

Cafés are popular places for men to socialize throughout North Africa. There, they play chess or dominoes, talk, or simply watch people passing by on the street. Most women socialize only at home. ***How are the social customs of cafés in North Africa different from those of cafés in this country?*** TEKS

Traditions and Customs Many North Africans wear traditional clothing. While there are many regional variations, in general North African clothing is long and loose. Such styles are ideal for the region's hot climates. Men and women often wear caftans and hooded robes made with a variety of fabrics. In Egypt the caftan and the *gallibiya,* or pants and a long shirtlike garment, are popular. Many women dress according to Muslim tradition. Their clothing covers all of the body except the face and hands.

When people greet each other in the street, they often shake hands and then touch their hand to their heart. If they are family or very close friends, they will kiss each other on the cheek. The number and pattern of the kisses vary from country to country.

Celebrations such as marriages are very important to North Africans because the family is central in Arab culture. Weddings can last for several days. Except for the last day of the wedding, the women's and men's celebrations are held separately.

TEKS **READING CHECK:** ***Human Systems*** How and why are traditional clothes of different North African countries generally similar?

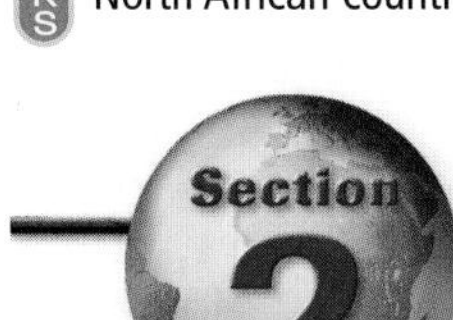

TEKS Questions 1, 2, 3, 4, 5

Section 2 Review

go.hrw.com **Homework Practice Online**

Keyword: SW3 HP21

Identify Berbers

Define silt, pharaohs, hieroglyphs

Working with Sketch Maps On the map you created in Section 1, label Alexandria, Cairo, Marrakech, Maghreb, Suez Canal, Sinai Peninsula, Casablanca, and Fès. Which city was founded by Alexander the Great?

Reading for the Main Idea

1. ***Human Systems*** Who were some early peoples that ruled over areas of North Africa?
2. ***Human Systems*** Why was control of the Suez Canal important to the British Empire?
3. ***Places and Regions*** What is the main ethnic group, language, and religion in the region?

Critical Thinking

4. **Identifying Cause and Effect** Why do many people in Algeria, Morocco, and Tunisia speak French as a second language today?

Organizing What You Know

5. Create a time line like the one shown below. On your time line, list important years, periods, and events in the history of North Africa.

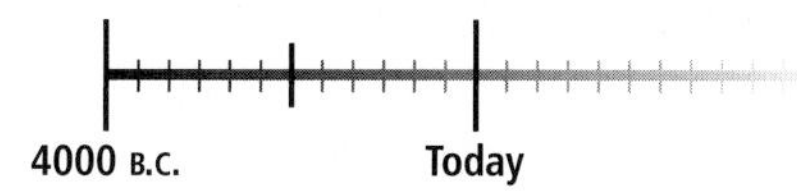

Section 3 The Region Today

READ TO DISCOVER

1. What are economies and cities of North Africa like?
2. What challenges do the people there face?

DEFINE

free port
fellahin

LOCATE

Tangier
Luxor
Tripoli
Tunis
Algiers
Aswān High Dam

WHY IT MATTERS

The politics of North Africa today are strongly influenced by Islamic fundamentalism. Use CNNfyi.com or other **current events** sources to learn about how this movement affects countries in other parts of the world.

Economic and Urban Environments

The countries of North Africa face issues typical of developing countries. These issues include controlling government spending, keeping inflation down, and reducing restrictions on businesses and foreign trade. Morocco has tried to address the last of these issues by giving the city of Tangier status as a **free port**. A free port is one where almost no taxes are placed on the goods unloaded there from other places.

Morocco and the other North African countries have taken other steps to strengthen economic ties with Europe. For example, in the late 1990s Tunisia entered an association agreement with the European Union. This agreement aims to improve trade between Tunisia and EU countries.

INTERPRETING THE VISUAL RECORD

Economic development and rapid population growth affect life in large North African cities such as Cairo. The city's famous pyramids, visible in the background, are increasingly hemmed in by the sprawl of Africa's largest city. ***Based on the photo, what environmental and urban problems do you think residents of Cairo face?*** TEKS

Economic Activities Oil and natural gas production form the backbone of the Libyan and Algerian economies. Egypt and Tunisia also have significant oil industries. When oil prices are high, these countries benefit. When prices are low, their economies weaken. As a result, they are trying to diversify their economies so they do not rely so much on oil and gas.

Agriculture is also a very important part of this region's economy, despite the dry climates. In Egypt millions of **fellahin**, or peasant farmers, work the fertile land along the Nile Valley. The fellahin make up about 40 percent of the Egyptian work force. The only country in North Africa that does not have a strong farming sector is Libya, which must import about 75 percent of its food.

Another important economic activity in North Africa is tourism. It is particularly important in Egypt, Morocco, and Tunisia. All three of these countries, and Algeria as well, have worked hard to attract tourism. However, in Algeria tourism almost entirely collapsed after political violence began in 1992. Tourism in Egypt also suffered a blow in 1997 when terrorists killed a group of foreign tourists at the ancient ruins of Luxor. Since then, tourism has slowly improved in Egypt but remains low in Algeria.

As you can see, the region has rich resources and productive industries. However, there still are not enough jobs. Rapid population growth makes this problem worse. In Libya and Algeria unemployment is as high as 30 percent. Many skilled and educated North Africans leave the region. They often find better jobs in Europe or in oil-rich Arab countries in Southwest Asia.

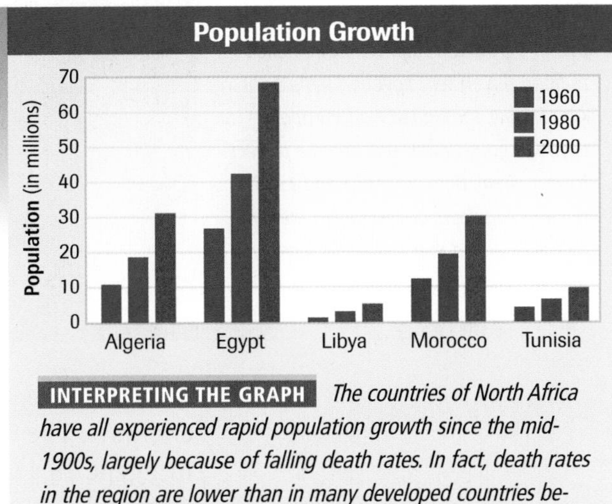

INTERPRETING THE GRAPH *The countries of North Africa have all experienced rapid population growth since the mid-1900s, largely because of falling death rates. In fact, death rates in the region are lower than in many developed countries because of the high percentage of young people.* ***How do you think the populations of these countries will change in the future? Why?*** TEKS

Urban Environments Among the largest cities in North Africa are Casablanca, in Morocco, and the capitals of the region's countries. Those capitals include Cairo, Egypt; Tripoli, Libya; Tunis, Tunisia; and Algiers, Algeria. These cities have a mix of modern and traditional buildings. Many large cities in North Africa are becoming more crowded as the region's population grows. At the same time, large numbers of people continue to migrate from the countryside to the cities. Thus, these large cities have grown outward with a ring of crowded slums surrounding the older core.

In cities such as Cairo, there is not enough housing. People crowd into slums, even setting up tents on rooftops or in rowboats along the Nile. Communities have even developed in cemeteries, where people convert tombs into bedrooms and kitchens.

READING CHECK: ***Human Systems*** How has rapid population growth been a problem for the region's economies and cities?

Issues and Challenges

The countries of North Africa share many of the same issues and challenges. Among these challenges are poverty and political unrest. The region also has some of the same political and social problems that face its Arab neighbors in Southwest Asia.

Political Issues Islam has long been an important factor in North African politics. Today many Islamic fundamentalists believe that government should be based strictly on the laws of Islam. For the most part, the region's governments want to limit the role of Islam. Sometimes this leads to violent clashes. For example, in Algeria in 1992, an Islamic party seemed set to win an election and take power. To keep this from happening, the government canceled the election and suspended parliament. Violence broke out, and the country was soon involved in a civil war. Since then thousands have died despite some progress in efforts to end the fighting.

In the 1950s, when the planned construction of the Aswān High Dam threatened to flood the ancient Egyptian temples of Abu Simbel, the United Nations and Egyptian government sponsored a project to save them. Between 1963 and 1968 a team of engineers disassembled the temples and moved them to a new location that was more than 200 feet (60 m) higher, thus saving them from destruction.

The United States has been involved in the politics of North Africa in several ways. The United States gives large amounts of military and economic aid to Egypt. In Libya, Mu'ammar Gadhafi has pursued anti-American policies. Gadhafi has seen himself as a leader of all Arabs and a defender of Arab causes. This has caused clashes between the Libyan and U.S. governments.

READING CHECK: ***Human Systems*** What religion plays an important role in the region's politics? Why?

The Nile River is the world's longest river. It begins in eastern Africa and flows northward for 4,160 miles (6,693 km) to the Mediterranean Sea.

Environmental Challenges Environmental issues in North Africa include desertification, pollution from oil refining, and polluted water supplies. In Egypt the environmental health of the Nile is currently a major concern. Construction of the Aswān High Dam across the upper Nile River was begun in 1960. Once completed, the dam became a major source of hydroelectric power. The reserves of water stored behind the dam provide water for crops year-round. This water also lets farmers open up new land for farming. These have been important benefits.

However, the dam has also stopped the annual flooding of the Nile. Before the dam was built, fields along the river were renewed every year by new deposits of silt. Now Egyptian farmers must buy fertilizer to keep the soil productive. The fertilizers have polluted the Nile and crippled the Egyptian fishing industry. Also, without the silt, the Nile Delta is slowly shrinking, and the Mediterranean coast is suffering from severe erosion.

READING CHECK: ***Environment and Society*** How has the heavy use of fertilizers affected the Nile?

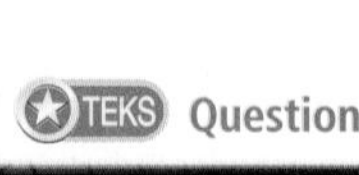

TEKS Questions 1, 2, 3, 4, 5

Section 3 Review

Define free port, fellahin

Working with Sketch Maps On the map you created in Section 2, label Tangier, Luxor, Tripoli, Tunis, Algiers, and Aswān High Dam. What is Egypt's source of hydroelectric power?

Reading for the Main Idea

1. ***Human Systems*** What problems face the petroleum industry in North Africa?
2. ***Human Systems*** What helps make tourism an important industry in North Africa?
3. ***Human Systems*** What are some issues facing cities in North Africa?

Critical Thinking

4. **Analyzing Information** How might the physical geography of the region contribute to the overcrowding of its cities?

Organizing What You Know

5. Create a word web like the one below. Use it to describe the positive and negative effects of building the Aswān High Dam for Egypt.

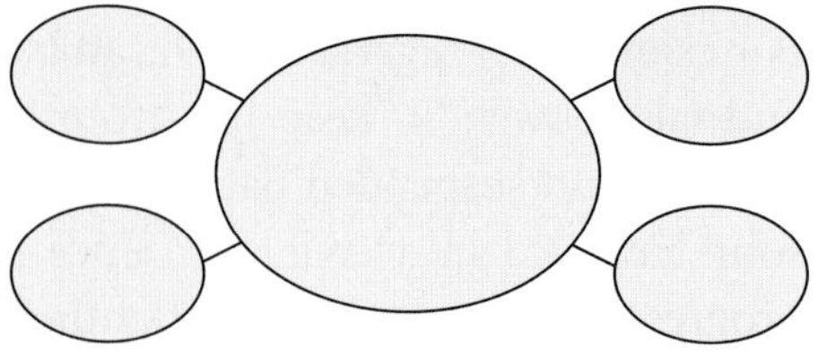

Geography for Life

Western Sahara

The phrase *Western Sahara* refers to more than a part of Africa's great desert. It is also a political unit—the only one in Africa that is not independent.

Western Sahara lies on Africa's northwest Atlantic coast. It shares borders with Algeria, Mauritania, and Morocco. Western Sahara was a Spanish colony from 1884 to 1976. When Spain withdrew from the area, no one was left as the clear authority. Three groups wanted to control Western Sahara. Both Morocco and Mauritania claimed the area. However, the Saharawis, the region's indigenous nomadic people, wanted self-rule. The Polisario—a nationalist organization—represented the Saharawis and opposed occupation by the other countries. Morocco moved some of its citizens into the western part of the disputed land. In return, the Polisario proclaimed independence. Following a change within its own government, Mauritania pulled out of Western Sahara in 1979.

During the 1980s Morocco built a sand wall along the border between Western Sahara and Mauritania. It stretches from Algeria all the way to the Atlantic Ocean. (See the map.) Morocco wanted the wall to keep Polisario troops from entering the western portion of the territory and southern Morocco. In 1988 Morocco and the Polisario accepted a United Nations peace plan, which was followed in 1991 by a cease-fire. In the year of the cease-fire, the UN also established a settlement. As part of the plan, the UN was to poll the people of Western Sahara for their views. They would be asked if they wanted official union with Morocco or independence. However, it was difficult to compile a list of eligible voters. In addition, the Moroccans continued to fortify the sand wall they had constructed. They also denied UN monitors easy travel through occupied Western Sahara. While governments quarrel over their land, many Saharawis wait in refugee camps in Algeria.

Morocco and the Polisario still disagree. The main dispute involves identifying who should be allowed to vote, which further delays the election. Morocco takes advantage of the delays by moving even more people to the region. Morocco also continues to exploit Western Sahara's mineral resources. Meanwhile, the UN tries to maintain the cease-fire and avoid armed conflict.

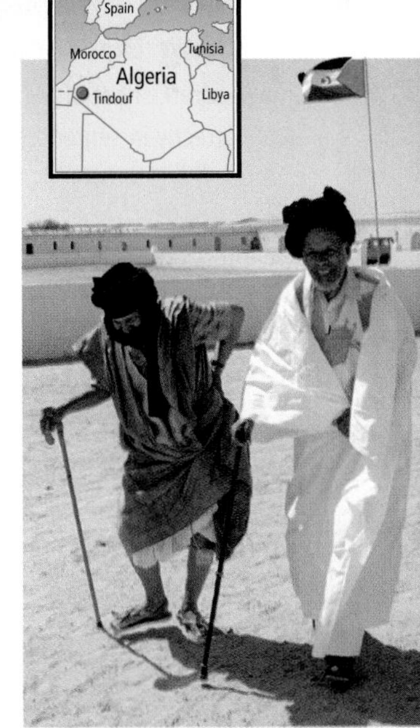

Two Saharawi leaders arrive for a meeting near Tindouf, Algeria, headquarters of the Polisario Front and home to thousands of refugees from Western Sahara.

Western Sahara

Phosphate mine
Rich fishing
Moroccan fortified wall

15°W
10°W
25°N
Canary Islands (SPAIN)
MOROCCO
ALGERIA
MAURITANIA
ATLANTIC OCEAN
El Aaiún (Laayoune)
Al Mahbas
Smara
Boukra
Tfaritiy
Boujdour
Galtat Zemmour
Ad Dakhla
Awsard
Aghwinit
Bir Gandouz
Techla
N S E W

SCALE
0 100 200 Miles
0 100 200 Kilometers
Projection: Lambert Conformal Conic

INTERPRETING THE MAP *The Moroccan defensive wall built in the 1980s stretches across much of Western Sahara, which Morocco claims.* ***Based on the map, about how long is the wall? Where is it in relation to the settlements in the region?***

Applying What You Know

1. **Summarizing** Why is there a conflict between the Saharawi people and Morocco?
2. **Analyzing Maps** Examine the map. How does the distribution of resources affect which part of Western Sahara Morocco most wants to control? How is Morocco exercising this control?

Building Vocabulary

On a separate sheet of paper, explain the following terms by using them correctly in sentences.

erg	wadis	hieroglyphs
reg	silt	free port
depressions	pharaohs	fellahin

Locating Key Places

On a separate sheet of paper, match the letters on the map with their correct labels.

Atlas Mountains	Sinai Peninsula
Nile River	Fès
Strait of Gibraltar	Aswān High Dam
Suez Canal	

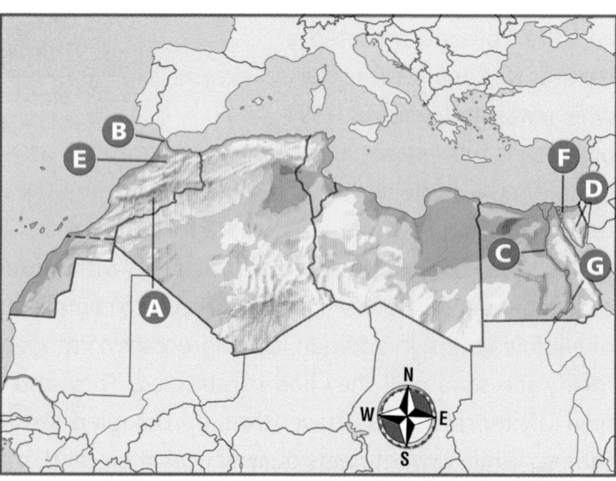

Understanding the Main Ideas

Section 1

1. *Physical Systems* How have wind and water shaped important physical features of the Sahara?
2. *Places and Regions* Why are plants and animals in Egypt limited mostly to areas along the Nile Valley and Delta?

Section 2

3. *Human Systems* Which ethnic group makes up the majority of North Africa's population? When did this ethnic group first move into North Africa?
4. *Places and Regions* What is the main religion of North Africa? What are two minority religions there?

Section 3

5. *Places and Regions* Why have many people in the region left to find work elsewhere? What has made the problem worse?

Thinking Critically for TAKS

1. **Drawing Inferences** Look at a world map. In what ways do you think the Suez Canal has influenced trade routes and patterns since it was built? Why might changes in these trade patterns have been important to European countries?
2. **Analyzing** In what ways does Islam influence the political, economic, social, and cultural characteristics of the region?
3. **Drawing Inferences and Conclusions** What do you think would be the effects of removing the Aswān High Dam?

Using the Geographer's Tools

1. **Analyzing Photographs** Look at the photograph of the medina in Section 2. How does it compare to the layout of your community's oldest area? What activities do you find in each?
2. **Analyzing Statistics** Review the Fast Facts and Comparing Standard of Living tables at the beginning of the unit. Then rank the levels of economic development and standard of living of the region's countries, from highest to lowest. Write a paragraph explaining your rankings. Note which statistics you believe were most important in your ranking.
3. **Preparing Maps** Create a map of the region that shows important agricultural areas. Shade the countries that have oil. Finally, label the Aswān High Dam and identify its importance as an energy resource.

Writing for TAKS

Review what you read in Unit 6 about the Tigris and Euphrates Rivers in Southwest Asia. Then write a report comparing the importance of major rivers to the history and economies of Southwest Asia and North Africa. What role did the rivers play in the development of early civilizations in each region? Why? When you are finished with your report, proofread it to make sure you have used standard grammar, spelling, sentence structure, and punctuation.

SKILL BUILDING

Analyzing Geographic Information

Environment and Society Find some recipes from North Africa and determine whether the dishes are similar to Southwest Asian, European, other African cuisine, or a mixture. Then compare the ingredients with typical crops and farm animals from the region. Which products are used most often? Write a paragraph describing your findings.

Building Social Studies Skills

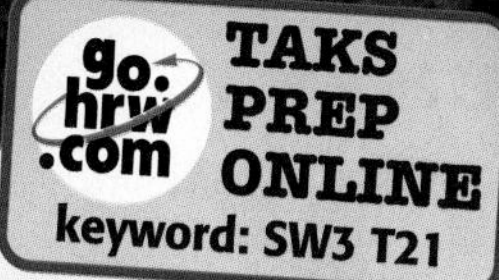

Interpreting Satellite Images

Study this satellite image of the Nile Delta in Egypt. Cultivated areas are shown in red. Then use the information from the image to help you answer the questions that follow.

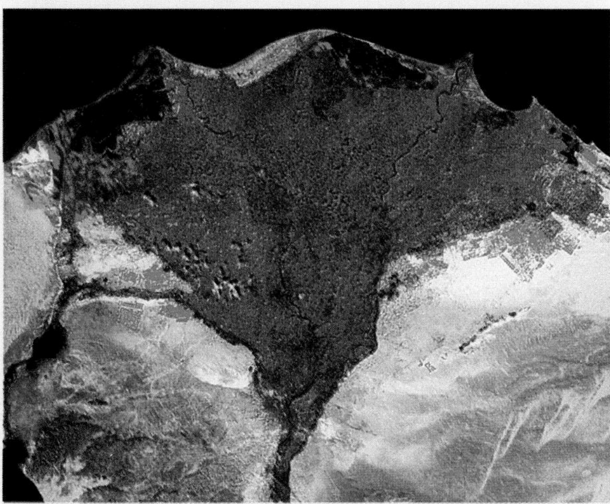

1. According to this satellite image, the Nile Delta is
   a. not cultivated.
   b. only partly cultivated.
   c. highly cultivated.
   d. mostly desert.

2. Which areas in the satellite image would you expect to be densely populated? Why?

Building Vocabulary

To build your vocabulary skills, answer the following questions. Mark your answers on a separate sheet of paper.

3. *Silt* is
   a. fertile finely ground soil.
   b. a fine fabric.
   c. a sea of sand.
   d. a type of veil.

4. The caravan of traders experienced many difficulties while crossing a *depression.*
   In which sentence does depression have the same meaning as it does in the sentence above?
   a. A tropical depression is beginning to form in the gulf.
   b. A salt marsh is located at the southern edge of the depression.
   c. The depression in the cotton market took the merchants by surprise.
   d. By the third day of the sandstorm, a slight depression had set in among the tourists.

Alternative Assessment

PORTFOLIO ACTIVITY

Learning about Your Local Geography

Group Project: Research

Is there an abundant supply of drinking water where you live? Does your community have to carefully protect its water resources? Work with a partner to research the location of the water resources available to your community. Find out about the quality of the water and what you can do as an individual to protect your water resources from pollution. Interview officials from your city government about any water management plans they have for the future. Create a map that illustrates some of the information you gathered. Finally, prepare a short report comparing the water resources of your state and those of North Africa.

internet connect

Internet Activity: go.hrw.com
KEYWORD: SW3 GT21 TEKS

Choose a topic on North Africa to:

- learn about ergs, regs, and other desert features and send a postcard from the Sahara.
- create a brochure on the design, construction, and function of the pyramids of ancient Egypt.
- take the GeoMap challenge and test your knowledge of North Africa's geography!

CHAPTER 22 West and Central Africa

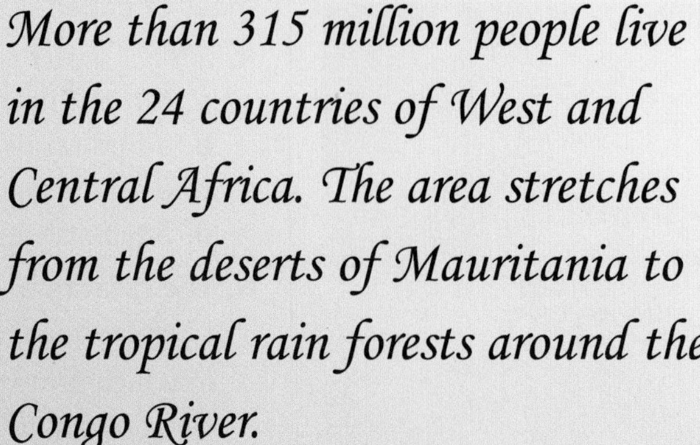

More than 315 million people live in the 24 countries of West and Central Africa. The area stretches from the deserts of Mauritania to the tropical rain forests around the Congo River.

Mask from Côte d'Ivoire

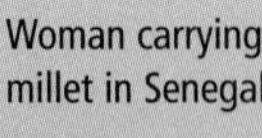

Woman carrying millet in Senegal

Ina kwanna! (Good morning!) My name is Ousseina. I have an identical twin sister named Hassana. We live in Niamey, the capital of Niger, with our grandmother, parents, and relatives. Every morning my sister and I eat breakfast with my grandmother. My favorite breakfast is millet porridge (a boiled grain) with yogurt and sugar.

When my sister and I were very small, we went to school to learn the Qur'an and the five daily prayers. We learned how to read and write in Arabic. We had to sit neatly in rows on beautiful mats and repeat verses and prayers after our religious teacher, the Malam. Now we go to a different school. In school we study literature, science, geography, history, and math. Some of our friends speak Hausa like us, but others speak a different language of Niger. School lasts from 8:00 A.M. to noon, with a 30-minute break for a snack. If we are hungry we buy mangoes, sugarcane, guavas, or bananas from street vendors. When we are thirsty, we drink baobab fruit juice or lemon and ginger juice. After school we help our grandmother pound spices for dinner, sweep the room, and wash the dishes. Sometimes after we do our homework we go to a friend's house to watch TV.

Natural Environments

READ TO DISCOVER

1. What are West and Central Africa's main landforms and rivers?
2. Which climates and biomes are found here?
3. What are some of the region's important resources?

IDENTIFY

Sahel

DEFINE

desertification

LOCATE

El Djouf
Niger River
Lake Chad
Congo Basin
Congo River

WHY IT MATTERS

Deserts are expanding in semiarid environments in West and Central Africa and around the world. Use CNNfyi.com or other **current events** sources to learn about the growth of desert areas.

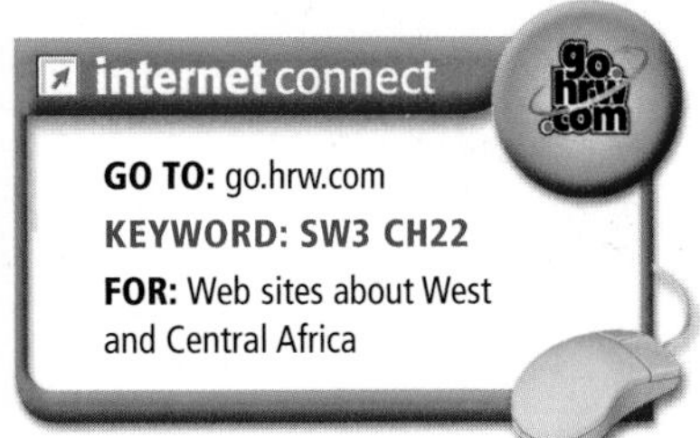

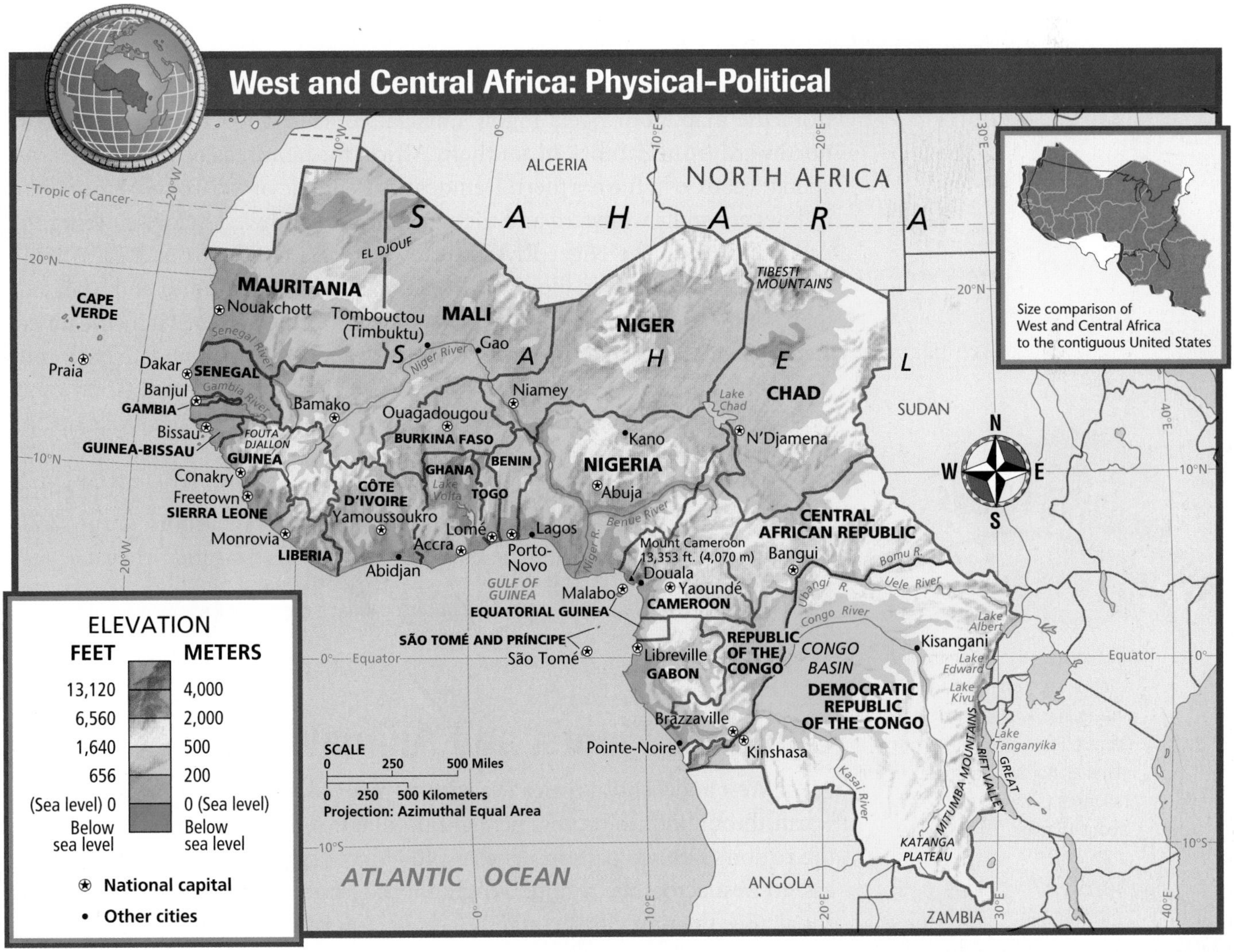

West and Central Africa: Physical-Political

Size comparison of West and Central Africa to the contiguous United States

ELEVATION

FEET	METERS
13,120	4,000
6,560	2,000
1,640	500
656	200
(Sea level) 0	0 (Sea level)
Below sea level	Below sea level

⊛ National capital
• Other cities

SCALE
0 250 500 Miles
0 250 500 Kilometers
Projection: Azimuthal Equal Area

INTERPRETING THE VISUAL RECORD

Thickets of coastal mangroves, such as these in the Gambia River delta, made it very difficult for early European explorers to reach land in parts of West and Central Africa. Often these explorers had to anchor their ships offshore and use small boats to reach the mainland. ***How might the physical environment of this region have influenced the locations of coastal settlements?*** TEKS

Landforms and Rivers

Plains and low hills make up most of the landscapes of West and Central Africa. There are also a few highland areas and broad depressions. (See the chapter map.) In the west is El Djouf (JOOF), a desert region in eastern Mauritania and western Mali near the Niger (NY-juhr) River. Lake Chad, on the border of Cameroon, Chad, Nigeria, and Niger, lies in the middle of a depression. To the south is the Congo Basin, a huge, wet tropical lowland in Central Africa. The Congo River drains this region.

The Congo and the Niger are two of Africa's major rivers. As you can see from the map, both rivers follow unusual courses. The Congo's waters flow northward from Zambia, in southern Africa, for hundreds of miles. The river changes course in the northern Democratic Republic of the Congo. From there it flows generally westward and then toward the southwest before entering the Atlantic Ocean. The Niger River's headwaters are not far from West Africa's Atlantic coast. However, the river flows northeast across Guinea and Mali and then southeast through Niger and Nigeria before entering the Gulf of Guinea. Geographers think that the upper courses of these great rivers date back millions of years. At that time, Africa was part of the supercontinent Gondwana. Both rivers probably flowed into large inland lakes. As Gondwana broke up, the lakes drained, and each river cut a channel to the sea.

A low coastal plain runs along the Atlantic shoreline of West and Central Africa. Most of the coastline is straight, with few natural harbors. Large sandbars and mangrove trees line the coasts of Ghana, Nigeria, and Senegal.

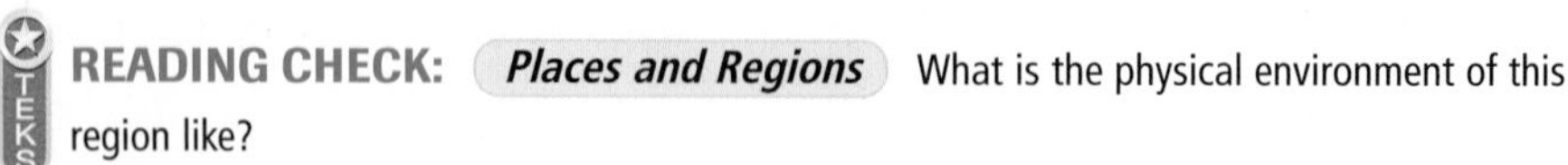

READING CHECK: ***Places and Regions*** What is the physical environment of this region like?

The tiny country of Gambia lies in an area confined to the navigable river valley of the Gambia River. Gambia extends inland along the river for about 200 miles (322 km) and is only about 15 miles (24 km) wide.

Climates, Plants, and Animals

All of West and Central Africa lies within the tropics. Therefore, most areas are warm throughout the year. There are no major mountain ranges to break up the region's climate pattern. As a result, climate regions form bands that run east to west across the region. Geographers describe this pattern of climates as zonal. (See the unit climate map.)

Arid Environments In the north, areas farthest from the equator have an arid climate. Here, the Sahara extends into northern Niger, Mali, and Mauritania. Along the southern edge of the Sahara is a region of semiarid climate called the **Sahel** (sah-HEL). The vegetation in this area includes scattered trees, shrubs, and grasses. The Sahel extends from Senegal and Mauritania in the west to Sudan in eastern Africa.

FOCUS ON GEOGRAPHY

The Sahel The Sahel only receives about 4 to 8 inches (10 to 20 cm) of rainfall each year. This small amount of rain may vary greatly from year to year. In some years rain is plentiful. In other years there may be almost no rain at all in some areas. When there are several years of below average rainfall in a row, major droughts occur. In the past long droughts in the Sahel have caused widespread famines.

Most people in the Sahel are subsistence farmers. They grow crops like peanuts and grains or raise cattle and goats. In recent years the number of people and livestock has risen. The increase has put pressure on the natural environment, particularly in times of drought. People cut down trees for firewood and clear land for crops. Livestock eat grasses, leaving the soil exposed. When it rains, valuable soil is washed away. This erosion leaves the land barren and makes farming even more difficult.

The combination of droughts and a growing population in the Sahel have caused **desertification**. This means desert conditions have spread into semiarid or marginal areas. In fact, the Sahara is slowly expanding southward.

READING CHECK: ***Physical Systems*** How have droughts and population growth affected the environment in the Sahel?

INTERPRETING THE VISUAL RECORD

Despite difficult environmental conditions, farmers in Africa's Sahel region attempt to grow crops where they can. Here in Mali, farmers plant their crops under shade trees to protect them from the Sun. However, rainfall is undependable, and the region often suffers from devastating droughts. A long period of drought in the late 1960s and 1970s killed some 100,000 people. ***How might environmental conditions in the Sahel influence migration and population patterns?*** TEKS

Tropical Environments South of the Sahel is a zone of tropical wet and dry climate. Northeast winds from the Sahara bring hot, dry, dusty conditions in winter months. Summer winds blow in the opposite direction—from the ocean—and bring rain. Open grasslands with shrubs and small trees are common. Elephants, giraffes, zebras, and other large animals once roamed freely. However, growing human populations and the conversion of grassland into farmland have led to a serious decline in wildlife populations.

The climate zone closest to the equator is tropical humid. Rain falls year-round here, and temperatures rarely drop below 65°F (18°C). Most of Africa's dense tropical rain forests are in this part of Central Africa. Tall trees form a complete canopy in some areas. Canopies are formed by the uppermost layer of the trees, where the limbs spread out. Leaves then block sunlight from reaching plant life on the ground below. This results in areas of open ground underneath the trees. The rain forests are home to a huge number of birds and insects. Monkeys, chimpanzees, and endangered gorillas also live there.

INTERPRETING THE VISUAL RECORD

Many of Africa's tropical rain forests are being cut down, which leads to the extinction of plants and animals. Rapid deforestation is a serious problem in Nigeria, where some 12 percent of the country is forested. Nigeria's large and rapidly growing population strains its forest resources as people clear trees for farmland, fuelwood, and timber bound for export. ***How might deforestation in Nigeria lead to economic development and environmental change?*** TEKS

Natural Resources

West and Central Africa have a wide variety of natural resources. Tropical timber, good soils for farming, and many minerals are found here. (See the unit Land Use and Resources map.) Some countries are world leaders in the production and export of certain farm or mining products.

The most valuable energy resource in the region is oil. Nigeria is Africa's largest oil producer. Smaller but significant oil reserves are found in other countries, such as Gabon and Cameroon. The Democratic Republic of the Congo is rich in minerals, including copper, diamonds, and cobalt. However, political problems and poor transportation systems have kept these resources from being fully developed.

West Africa is the world's major source of cacao, or cocoa beans. The tree that bears cacao is native to tropical areas of South America. Europeans brought it to West and Central Africa during the Colonial era. Cocoa beans are used to make chocolate. Côte d'Ivoire (koht-dee-VWAHR) is the world's leading producer. Ghana, Nigeria, and Cameroon are also important growers. Coffee, coconuts, and peanuts are also among the region's main exports. These crops grow well in the region's tropical environments.

READING CHECK: *Places and Regions* From what world region did cacao come? How did it get to Africa and what role does it play in the region's economy?

TEKS Working with Sketch Maps, Questions 1, 2, 4, 5

Section 1 Review

Identify Sahel

Define desertification

Working with Sketch Maps

On a map of West and Central Africa that you draw or that your teacher provides, label the countries of the region, El Djouf, Niger River, Lake Chad, Congo Basin, and Congo River. Which river flows through Nigeria before entering the Gulf of Guinea?

Reading for the Main Idea

1. *The Uses of Geography* How do geographers explain the irregular course of the Congo and Niger Rivers?
2. *Physical Systems* What factors affect the distribution of climates in this region?
3. *Places and Regions* What are some major resources in West and Central Africa?

Critical Thinking

4. **Problem Solving** What could be done to try to halt desertification in this region?

Organizing What You Know

5. Copy the graphic organizer below. Use it to compare and contrast the arid and tropical environments found in this region.

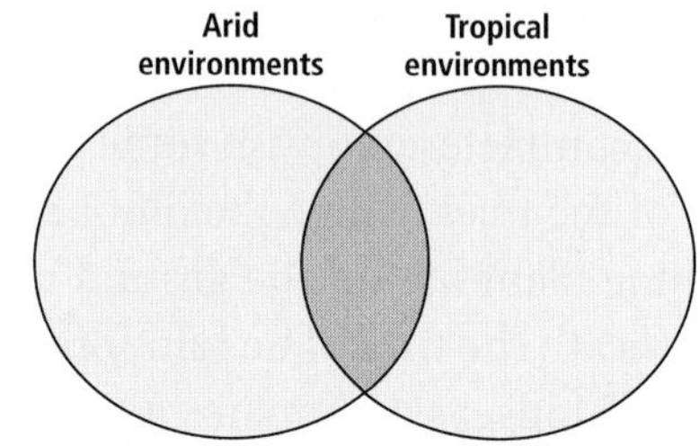

2 History and Culture

READ TO DISCOVER

1. What are the main eras in the history of West and Central Africa?
2. What are some features of this region's cultures?

DEFINE

staple
millet
sorghum

WHY IT MATTERS

Many plantation crops grown in West and Central Africa during colonial times, such as cacao, are still important today. Use CNNfyi.com or other **current events** sources to learn about important crops in this region.

LOCATE

Tombouctou

Beautiful gold objects, such as this staff from Ghana decorated with an alligator, have long attracted traders to West and Central Africa.

History

Great kingdoms once ruled large areas of West and Central Africa. One of the earliest was Ghana in the A.D. 800s. (See the map.) Ghana was a trading state. It exported products like gold and cloth to North Africa. The kingdom also participated in the slave trade. Traders brought to Ghana a variety of products,

Empires of Africa

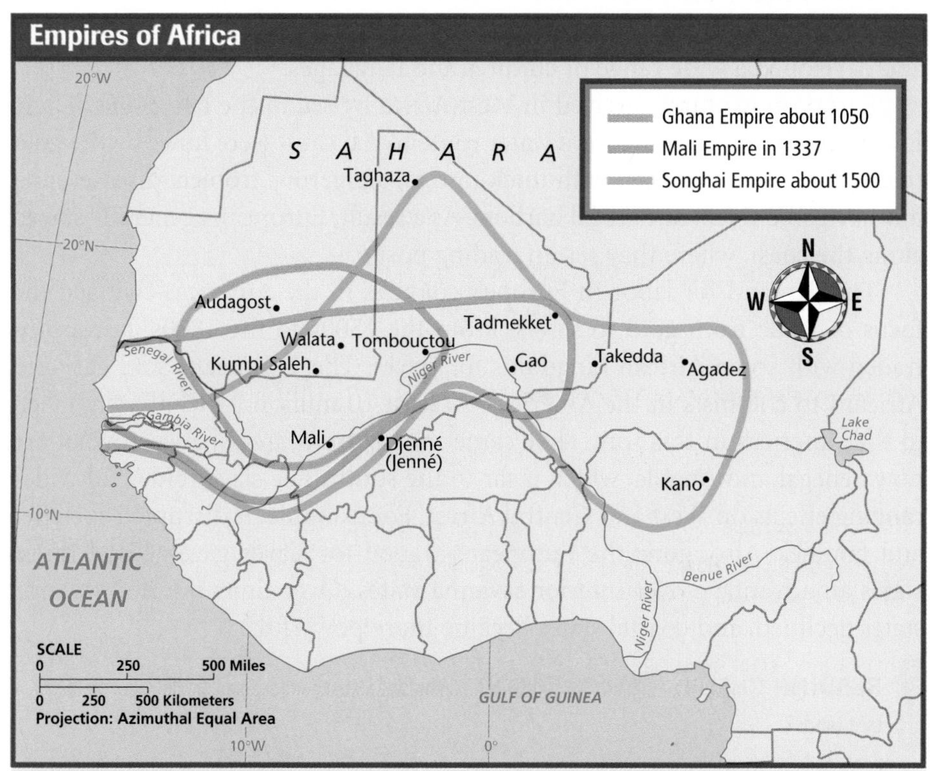

INTERPRETING THE MAP *Ghana, Mali, and Songhai were all great trading empires in West Africa that grew rich from trade between tropical areas to the south and arid regions to the north.* ***Which areas of West Africa did all three empires control?***

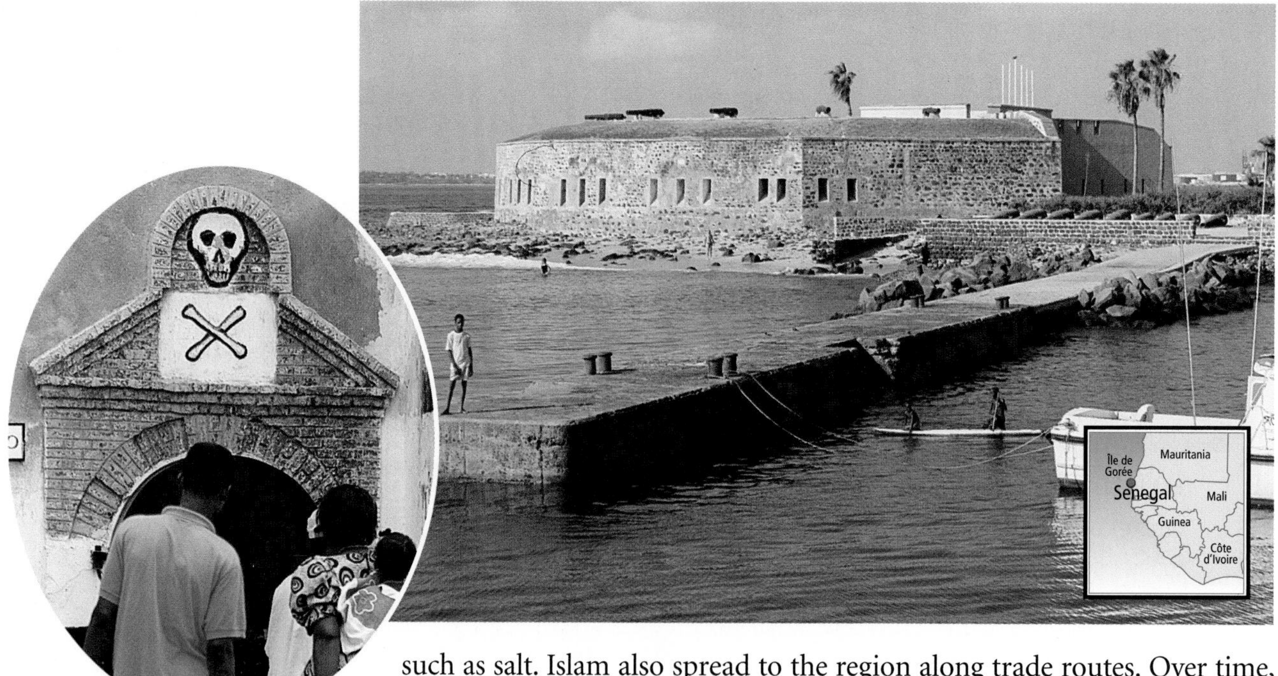

INTERPRETING THE VISUAL RECORD

The slave trade greatly affected the human geography of West and Central Africa. It altered settlement patterns, economic systems, and cultural landscapes, and caused widespread human suffering. Slaves were often held captive in coastal forts and prisons in places such as Ghana (left) and Senegal (right) before they were taken to the Americas on ships. ***How do you think the slave trade in West and Central Africa shaped the distribution of culture groups around the world?*** TEKS

such as salt. Islam also spread to the region along trade routes. Over time, rulers of many West African kingdoms became Muslim.

Many empires rose and fell here over the centuries. For example, Mali replaced Ghana as the most powerful kingdom during the 1200s. The city of Tombouctou (tohn-book-TOO) (also called Timbuktu) became an important center of trade and education during this time.

To the south lived the people of the tropical rain forests. Forest peoples traded less with distant lands. This was partly because the rain forest provided them with many different resources. Also, the dense vegetation of the forest made it difficult to travel long distances. Forest peoples lived in small groups and developed a wide range of cultures and languages.

European explorers arrived in West Africa by sea in the late 1400s. These Europeans were searching for a water route to Asia and were lured by the gold trade. They found an area with thick forests, dangerous tropical diseases, and few navigable rivers or natural harbors. As a result, Europeans generally stayed along the coast, where they set up trading posts.

The demand for labor in Europe's colonies in the Americas changed the focus of trade from gold to slaves. From the 1500s to the 1800s, Europeans traded with some African kingdoms for slaves. The Europeans sold enslaved Africans to colonists in the Americas. At least 10 million Africans were taken to the Americas in this way. They came mostly from areas between what are now Senegal and Angola, which is far to the south. The slave trade had wide-ranging effects on West and Central Africa. For example, it disrupted societies and families. Also, guns the Europeans traded for slaves gave coastal forest states an advantage over interior savanna states. Over time, interior savanna states declined, and coastal states became more powerful.

READING CHECK: ***Human Systems*** Why did forest peoples not trade more with distant lands?

The Colonial Era By the mid-1800s the slave trade was coming to an end. Europe was industrializing, and countries there wanted minerals and tropical farm products they could not produce at home. The climates and soils of West and Central Africa were good for growing many of these products, such as cocoa, peanuts, and rubber. As a result, many European countries sought political control over African territories.

During the colonial period many West and Central Africans quit subsistence farming and started working for wages. Some worked on plantations. Others moved to cities. Unlike most precolonial settlements, these new colonial cities were located along the coast. These locations offered better connections to Europe and the Americas.

The Postcolonial Era Africa's colonial era lasted less than 100 years. In 1957 Ghana became independent. Other countries soon followed. By 1976 all African countries in this region were independent.

Although the colonial era was relatively short, it had a major effect on West and Central Africa. Before the colonial era most people in the region were subsistence farmers. Afterward, many worked in the new commercial economy. Local economies had been based on trading gold, salt, and ivory. They now depended on the export of minerals and farm products. Modern medicine and infrastructure improved many people's lives, particularly in cities. However, people also faced new and difficult problems. For example, many earned low wages or were unemployed. Also, rival ethnic groups had to share power in newly independent countries. This caused serious political rivalries.

TEKS **READING CHECK:** ***The Uses of Geography*** How did the colonial era affect the economies of countries in West and Central Africa?

The Colonial Scramble

In 1884, European powers met in Berlin to settle colonial disputes in Africa. The conference set up rules designed to govern how outside powers took control of African lands. In a short time European countries had divided most of Africa among themselves. The French, British, and Belgians took most of West and Central Africa. Portugal, Spain, and Germany also claimed colonies in the region. (See the map.)

The effects of the European scramble for colonies can still be seen in Africa. Borders there today are largely those drawn by Europeans. When creating these borders, the colonial Europeans ignored Africa's existing states, ethnic groups, and natural environments. Some colonies included many different peoples. These peoples spoke many different languages and dialects. New borders also blocked traditional migration and trade routes. These borders created political problems, particularly when African colonies won independence. They also left a number of African countries landlocked. These problems remain a difficult challenge facing Africans today.

Summarizing How did the scramble for European colonies shape the borders and countries of modern Africa? TEKS

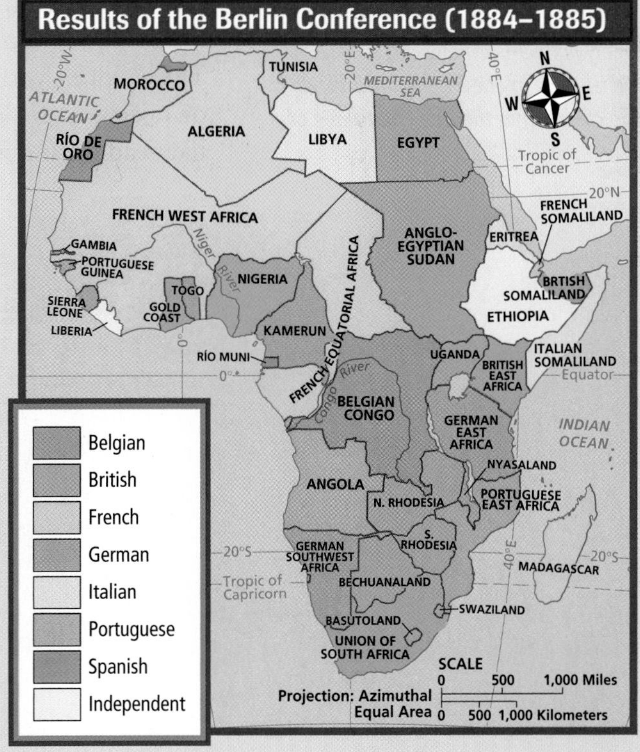

Nigeria has more than 250 ethnic groups, each with its own language. Linguists think that most Nigerian languages have been spoken in about the same places for thousands of years.

Culture

West and Central African societies are very diverse. Like the rest of Africa, peoples here reflect three major cultural influences. Those influences are traditional African cultures, Islam, and European culture. These influences have been called Africa's triple heritage.

People, Languages, and Settlement Most of the languages spoken here belong to the Niger-Congo language family. Within this family there are hundreds of different languages. Arabic is also spoken in northern areas. During the colonial era, English and French became lingua francas in much of Africa. These languages are still widely used in the region today.

Most West and Central Africans live in rural areas or small villages and rely mainly on farming. However, the populations growing the fastest are those in the cities. In most countries, the largest city is the capital. Except in landlocked countries, capitals are generally located on the coast. Most were set up as ports and government centers during the colonial era.

Religion and Education Islam is the main religion in the Sahel. However, many Christians live to the south, between the Sahel and the Atlantic coast. Many people also practice traditional African religions. These people believe that spirits—particularly the spirits of their ancestors—play an important part in their lives. People seek advice and help from spirits in times of trouble or sickness. Ancestral spirits are also honored in ceremonies.

Literacy rates are generally low throughout West and Central Africa. Only a small percentage of people in most countries finish high school. Very few have the chance to go to college. The main obstacle to education is poverty. In many poor families, parents need their children to work. Older children often take care of their younger brothers and sisters so their parents can work. At home, parents and elders provide what education they can for their children. The children learn about family and group traditions and about growing crops or raising animals. Many children do not have the opportunity to learn skills like reading and writing.

INTERPRETING THE VISUAL RECORD

Most people in West and Central Africa live in rural areas and practice subsistence agriculture. Nomadic lifestyles are found in dry northern environments such as the Sahel. ***Why might a nomadic lifestyle be common in the arid conditions of the Sahel?*** TEKS

West and Central Africans have a rich tradition of dance. Here in Cameroon, Juju dancers dressed as forest animals perform a dance about the destruction of the forest.

Food, Traditions, and Customs Most people in the region produce their own food. Some grow only a few **staple** crops. Staple crops are the main food crops of a region. Such crops in West and Central Africa include cassava, corn, and yams. **Millet** and **sorghum** are also staple crops here. These two grains are resistant to drought, growing well in dry areas like the Sahel. In the cities many people eat foods introduced from other countries. For example, bread has become an important staple in some places. However, the wheat needed to make bread does not grow easily here. As a result, wheat and other introduced food crops must be imported.

Customs and traditions differ among the region's many ethnic groups. For example, the lifestyles of Muslims in the north differ greatly from those of rain forest peoples in the south. Generally, societies are based on extended families. Families consist of several households, including the head of the family, his wife or wives, their unmarried children, and their married daughters and their husbands and children. Members of the extended family work together to support the household and take care of older people and young children.

READING CHECK: *Places and Regions* How is Africa's triple heritage reflected in the culture of the region?

TEKS Questions 1, 2, 3, 4, 5

Review

Keyword: SW3 HP22

Define staple, millet, sorghum

Working with Sketch Maps

On the map you created in Section 1, label Tombouctou. On which river is Tombouctou located?

Reading for the Main Idea

1. *Human Systems* What did kingdoms in West and Central Africa trade with people in North Africa?
2. *Places and Regions* Where will you find most capital cities in the region's countries that are not landlocked? Why?
3. *Human Systems* What are the most common religions practiced here?

Critical Thinking

4. **Drawing Inferences and Conclusions** In what ways do you think the physical environment influenced cultural patterns and the distribution of culture groups in West and Central Africa?

Organizing What You Know

5. Copy the time line below. Use it to list major events and periods in West and Central Africa's early history, colonial era, and postcolonial era.

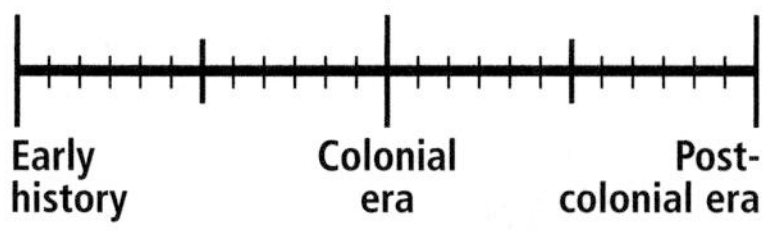

Human Systems

Geography for Life

Nigeria's Ethnic Diversity

Some countries, such as France and Japan, have a dominant culture and a clear national identity. In contrast, a national identity is hard to find in Nigeria, Africa's most populous country. Nigeria was granted full independence from Great Britain in 1960. After enduring years of harsh and corrupt military rule, Nigerians adopted a new constitution in 1999. That same year, an elected civilian government came to power. Although Nigeria's government is becoming more democratic, ethnic conflicts are still a major challenge.

To understand the country's politics, we must study Nigeria's diverse people and cultures. Nigeria has 36 states and more than 250 ethnic groups. Each group has unique customs, languages, and traditions. A strong Muslim presence dominates the north. The Hausa and Fulani together are the largest Muslim groups. Other major ethnic groups of the north include the Kanuri. Christian influence is strong in the south. The Yoruba people dominate the southwest. About half of them are Christian, but half are Muslim. The mainly Roman Catholic Igbo (Ibo) are the largest ethnic group in the southeast. The Ibibio and other peoples also live there in large numbers.

Religions in Nigeria

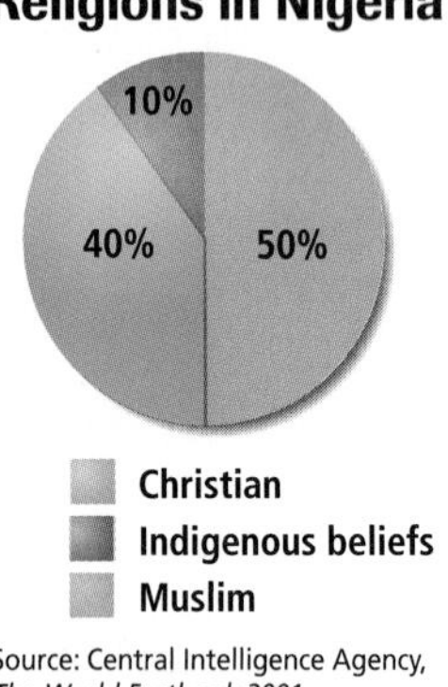

Source: Central Intelligence Agency, *The World Factbook 2001*

Some ethnic groups in the country want regional independence and a larger share of local wealth. Yet Nigeria's leaders fear that granting these requests would divide the country. Religious conflict has also appeared. For example, some northern states want to apply Islamic law to criminal offenses. Islamic law, called sharia, has long been used as the basis of family law in northern Nigeria. Sharia bans alcohol and allows severe punishments, such as cutting off a hand, for certain crimes. While officials say sharia would apply only to Muslims, many Christians are concerned.

Since military rule ended, hundreds of Nigerians have died in ethnic and religious fighting, particularly in the cities. Many more people have fled the violence by returning to their ethnic or religious homelands.

Nigeria's Major Ethnic Groups

The previous military government might have stepped in to end the conflicts. Some observers say the military could use the ethnic clashes as an excuse to retake power. Others say that the army is now too weak and that most Nigerians would fight to defend democracy. Because no easy solution to the conflicts exists, Nigerians are finding they must learn to live with their differences.

Applying What You Know

1. **Summarizing** Why is Nigeria unstable even though its military government has been replaced by a democratically elected government?
2. **Making Predictions** Do you think Nigeria will eventually split into several countries? Why or why not? How might democracy help Nigeria's diverse peoples live with their differences?

Section 3 The Region Today

READ TO DISCOVER

1. How economically developed are West and Central African countries?
2. What major challenges do the countries face today?

DEFINE

dual economies

LOCATE

Lagos
Kinshasa
Abidjan
Accra
Douala

WHY IT MATTERS

Most countries in West and Central Africa have very high population growth rates. Use CNNfyi.com or other **current events** sources to learn about the rapidly growing populations of these African countries.

Level of Development

West and Central Africa is a region of developing countries. On average, people here earn less and live shorter lives than people in other parts of the world. They also have lower levels of education. (See the unit Fast Facts table.) Some countries are better off than others. For example, Gabon is one of the richest countries in Africa because of its oil reserves. In contrast, landlocked Mali is among the poorest countries in the world. It lies in the Sahel region and has few resources.

The countries of West and Central Africa have **dual economies**. In a dual economy, some goods are produced for export to wealthy countries. Meanwhile, another part of the economy produces goods and services for local people. For example, cash crops like rubber and cocoa or minerals like diamonds and bauxite might be produced and exported. On the other hand, subsistence farmers produce food for their own use. Street vendors and local markets sell clothing, food, and services to passersby.

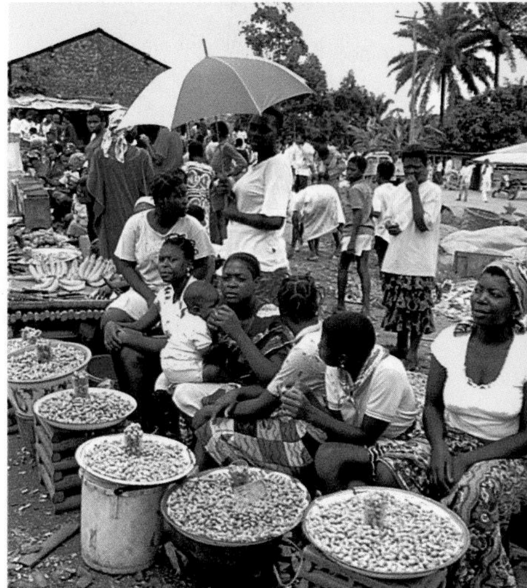

In a dual economy, goods like rubber are produced for export (above), while other goods are produced for local people (below).

Agriculture In the Sahel grasslands, farmers raise cattle and goats and move their herds in search of grazing lands. In tropical rain forest areas, cassava, millet, and yams are all staples. Farmers there have long planted several different crops in a single field. This kind of farming works well in tropical environments. If one crop is damaged by disease, other crops provide enough food for people to survive. Because different crops mature at different times, farmers do not have to harvest all their crops at once or keep food in storage.

The development of market economies under colonialism affected traditional farming. Plantations and ranches made it hard for herders to move their animals around to different grazing lands. Thus, herders have been forced to stay in one place. This leads to overgrazing and soil erosion. In turn, herders cannot keep as much livestock as in the past. As a result, they either have to migrate to more fertile areas or to the cities.

READING CHECK: *Human Systems* How has the development of market economies affected agriculture in the region?

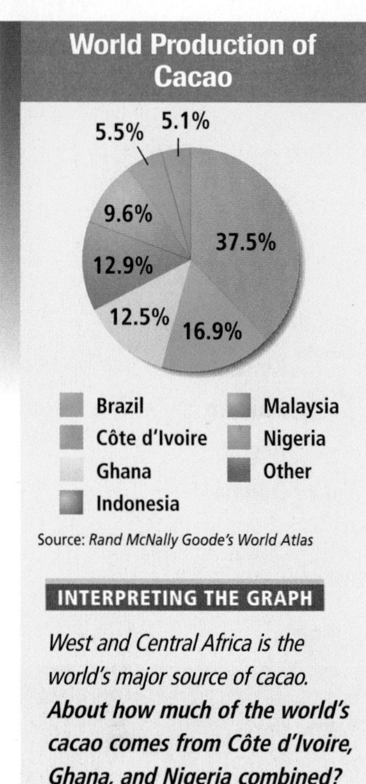

Source: *Rand McNally Goode's World Atlas*

INTERPRETING THE GRAPH

West and Central Africa is the world's major source of cacao. ***About how much of the world's cacao comes from Côte d'Ivoire, Ghana, and Nigeria combined?***

Economic Activities and Global Trade Most of the region's countries export primary rather than secondary goods. For example, Côte d'Ivoire exports cocoa beans, but the manufacture of chocolate often takes place in the developed countries. Similarly, Guinea exports bauxite, but the manufacture of aluminum takes place elsewhere.

Many countries in the region depend heavily on only a few main exports. This practice has two major disadvantages. First, it makes economies vulnerable to changes in the price of their main exports. For example, about 95 percent of Nigeria's export earnings come from selling oil. When the price of oil drops, the Nigerian economy suffers. Likewise, if the price of cocoa falls, the economy of Côte d'Ivoire can be hurt. Second, the export of primary goods is less profitable than the export of manufactured goods. For example, it is more profitable to sell peanuts processed into peanut butter and oil than to sell raw peanuts. Manufacturing does take place in some areas. However, most West and Central African countries do not have adequate facilities to process their primary products. As a result, they miss out on much of the wealth their raw materials create.

Cities In the early 1960s there were very few large cities in West and Central Africa. Only a handful of cities had populations greater than 300,000. After independence, however, cities grew very rapidly. For example, Lagos, Nigeria, grew from 760,000 people in 1960 to 4.5 million by 1980. Today some 13 million people live there. In the Democratic Republic of the Congo, Kinshasa, which had a population of 450,000 in 1960, has about 5 million people today. Other large cities include Abidjan in Côte d'Ivoire, Accra in Ghana, and Douala in Cameroon.

This rapid urban population growth has caused housing shortages. Many people live in crowded shantytowns without electricity or running water. In sharp contrast to the poor shantytowns are more prosperous downtown areas. These areas often look similar to the downtown areas of European cities. They

INTERPRETING THE DIAGRAM

Cacao beans are processed into liquid paste, which is then used to make cocoa powder, chocolate, and other products. ***The production of chocolate is what type of economic activity?***

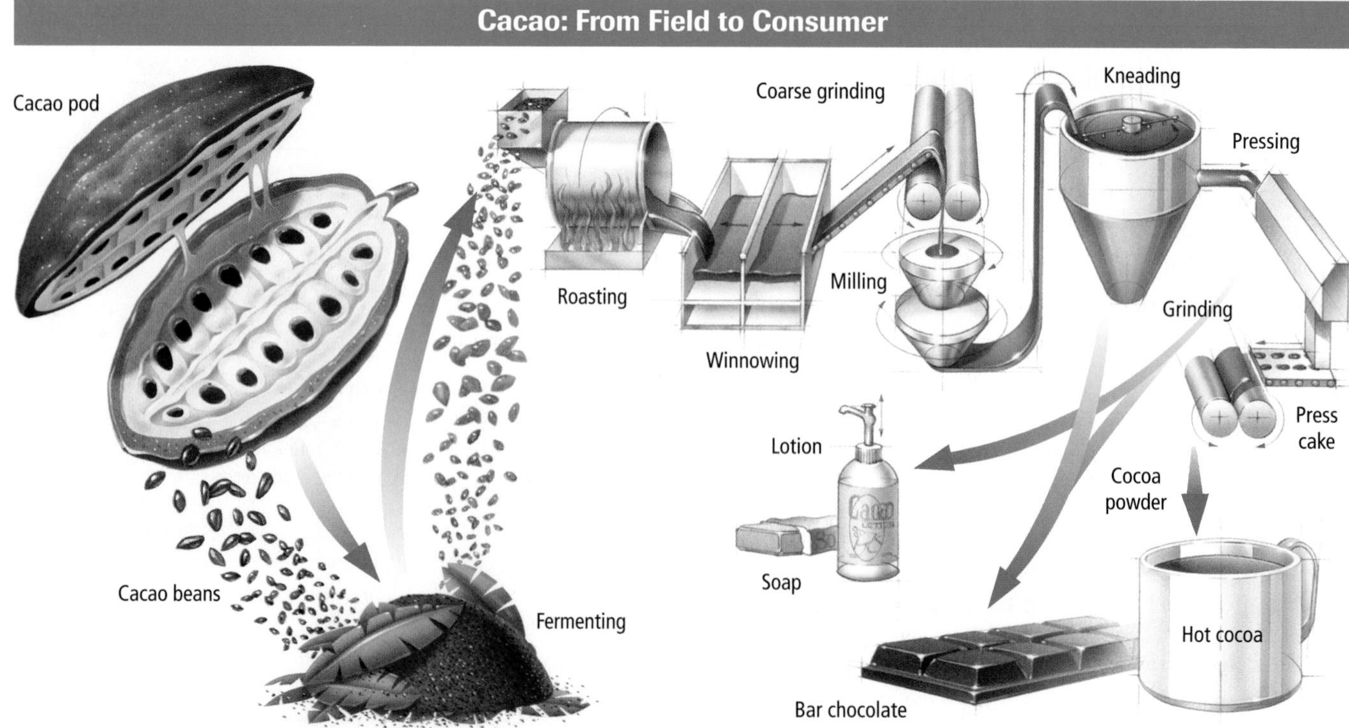

have busy roads full of cars and buses. Tall buildings dominate the central city. Neon signs and billboards advertise international products like soft drinks and electronics.

✓ **READING CHECK:** *Places and Regions* What types of goods are West and Central Africa's most important exports?

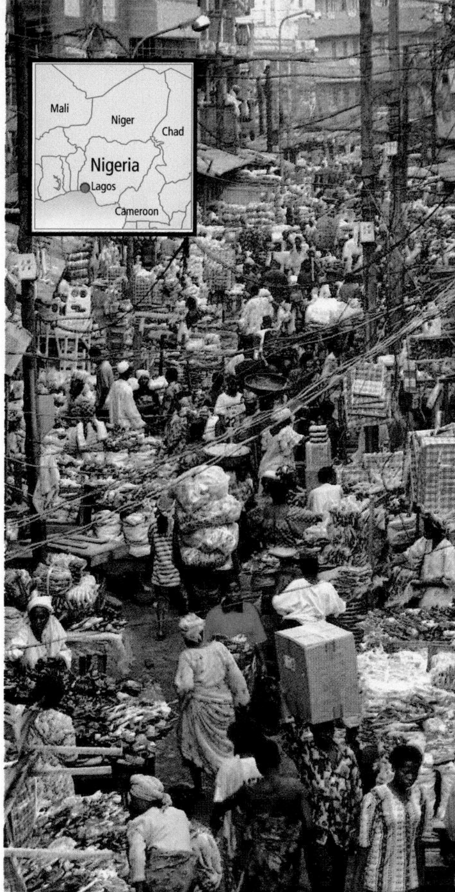

With about 13 million people, Lagos, Nigeria, is one of the 15 largest cities in the world. As in many cities in developing countries, the population of Lagos is growing rapidly.

Issues and Challenges

The countries of West and Central Africa face many challenges. Economic development is probably the most important challenge. Issues like population growth, health care, political problems, and protecting the environment all affect development.

The region's population is growing rapidly. This rapid growth has caused many problems. Since independence, agricultural production has not kept pace with population growth. In some places, food production has even declined. Food shortages and malnutrition have become more common.

Recently, many countries have suffered wars and conflicts. In the past 10 years alone, civil wars have been fought in the Democratic Republic of the Congo, Liberia, and Sierra Leone. Thousands of people have been killed in these civil wars.

Destruction of the natural environment is a serious problem. Lumber companies harvest tropical rain forests for timber. Grasslands are cleared for farming. Such clearing has led to the extinction of some plants and animals. In the Sahel, desertification has degraded the land. As a result, people have migrated southward in search of farmland and food.

One of the most serious problems facing the region is disease. Malaria has long been a problem. In addition, HIV—the virus that causes AIDS—has spread rapidly. There is no cure for HIV infection, and treatments are very expensive. Poor, malnourished, and poorly educated people are particularly vulnerable to the disease.

TEKS **READING CHECK:** *Places and Regions* What challenges does the region face?

TEKS Questions 1, 3, 4, 5

Review

Define dual economies

Working with Sketch Maps

On the map you created in Section 2, label Lagos, Kinshasa, Abidjan, Accra, and Douala. Which city has grown to about 13 million people today?

Reading for the Main Idea

1. *Places and Regions* What level of development do all the countries share?
2. *Human Systems* What disadvantages do West and Central Africa's countries face when they depend on just a few exports?

Critical Thinking

3. **Supporting a Point of View** Who do you think benefits the most from a dual economy?
4. **Analyzing** How is economic development in West and Central Africa tied to population growth, health care, political problems, and environmental change? How might rapid urbanization make some of these problems worse?

Organizing What You Know

5. Copy the chart below. Use it to list some major challenges the countries of West and Central Africa face. In addition, suggest some possible solutions to these challenges.

Challenges	Possible solutions

CITIES & SETTLEMENTS

The Tuareg: A Nomadic Way of Life

Human Systems How hard would it be to give up a way of life that your family has treasured for generations? This is the question facing many Tuareg (TWAH-reg), a nomadic people of North and West Africa. For more than 1,000 years they have raised camels, goats, and sheep in the Sahara. However, climate and politics are threatening to end the Tuareg's traditional way of life.

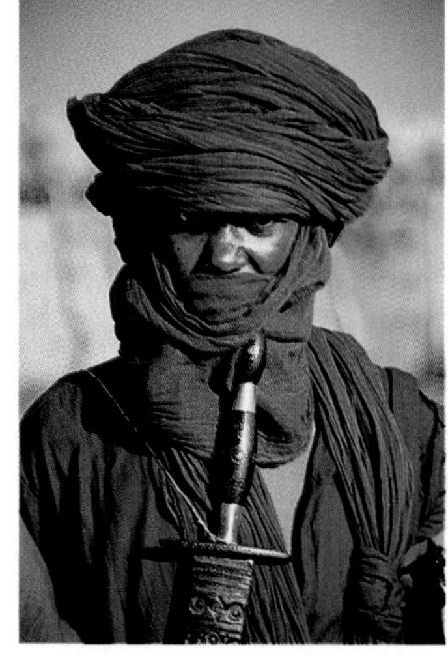
Tuareg men wear blue veils in the presence of women and strangers, although this practice is less common in cities.

The "Blue Men" of the Sahara

As many as 1 million Tuareg live in Africa. Countries with Tuareg populations include Algeria, Burkina Faso, Libya, Mali, and Niger. Over the centuries, many have settled in towns in the Sahel. However, many others still live in the desert. These Tuareg graze their herds on sparse desert plants. When the plants in one area are gone, they move to a new area. Generally, the Tuareg spend only about two weeks in one place. "My father was a nomad, his father was a nomad, I am a nomad," one herder explains. "This is the life that we know. We like it." Fiercely independent, the Tuareg call their ancient way of life *adima,* meaning "far from town."

The Tuareg's independent spirit can also be seen in the arrangement of their desert camps. Tuareg travel together in small groups of relatives and friends. In camp, however, group members live apart. Each family's tent is several hundred yards away from the others. Visiting is common, but Tuareg generally do not share food or care for another family's livestock. Their diet consists mainly of fruits and grains people get through trade or from other Tuareg who farm in oases. Goat milk and cheese provide protein. On special occasions, a sheep or goat is slaughtered to provide meat.

Tents are made from goat skins that are sewn together and stretched over a rectangular frame. Inside, family members sleep on carpets or mats. A sheepskin blanket provides warmth on chilly desert nights. Family tents belongs to Tuareg women. When a woman marries, she receives a tent made by her female relatives.

Tuareg men wear cloth veils wrapped around the face and head. Because these veils are traditionally dyed blue, Tuareg are sometimes called the Blue Men. The veils help protect against windblown desert dust. Long robes are also practical in the desert. They keep sweat from evaporating too quickly, which helps protect against dehydration.

Class divisions are important in Tuareg society. A family's position is passed down from father to son. Many Tuareg oasis dwellers are members of the servant classes. Their role is to provide food and other items needed by the upper-class herders of the desert.

These Tuareg boys are making toy camels. Camels have long been an important part of the Tuareg's nomadic way of life.

The campsites of Tuareg nomads have tents and animals such as donkeys and chickens.

Living as a nomad is a sign of high status among the Tuareg and is their preferred way of life.

A Changing Way of Life

In recent years it has become harder for the Tuareg to maintain their nomadic lifestyle. Some governments in the region have limited movement across their borders because of warfare and unrest. As a result, some Tuareg cannot travel to much-needed grazing lands.

Environmental changes have also disrupted the Tuareg's way of life. For example, rainfall in the Sahara has decreased in recent decades. Since the 1960s two major droughts have dried up already scarce water resources and reduced grazing areas. The lack of water has nearly wiped out the herds of many Tuareg. Some nomads have been forced to settle near towns. They survive by gardening and selling crafts and camel rides to tourists. However, for many Tuareg the lure of their traditional life never dies. "Each time I earn a little money I buy a goat or a sheep," says one former nomad. "I save up so that I can have enough animals to return to the desert."

Still, others fear that the old ways are gone forever. This fear has weakened the Tuareg's resistance to government education for their children. For generations the Tuareg viewed schools as a government program designed to limit their movement. Now, however, many believe that education will free their children from a way of life they fear has no future. Some nomadic groups even let government teachers travel with them in the desert. Schools have also been set up at water sources in the desert. Even so, many Tuareg children still refuse to accept that they will not live as their parents and grandparents did. "I like taking care of camels," one 15-year-old insists. "I don't know the world. The world is where I am."

Applying What You Know

1. **Summarizing** How is the Tuareg's traditional nomadic life changing? What is causing these changes?
2. **Analyzing Information** Why do you think some Tuareg are determined to maintain their traditional way of life?

Building Vocabulary TEKS

On a separate sheet of paper, explain the following terms by using them correctly in sentences.

desertification · sorghum · staple · dual economies · millet

Locating Key Places TEKS

On a separate sheet of paper, match the letters on the map with their correct labels.

El Djouf · Tombouctou · Niger River · Lagos · Congo Basin · Kinshasa · Congo River · Abidjan

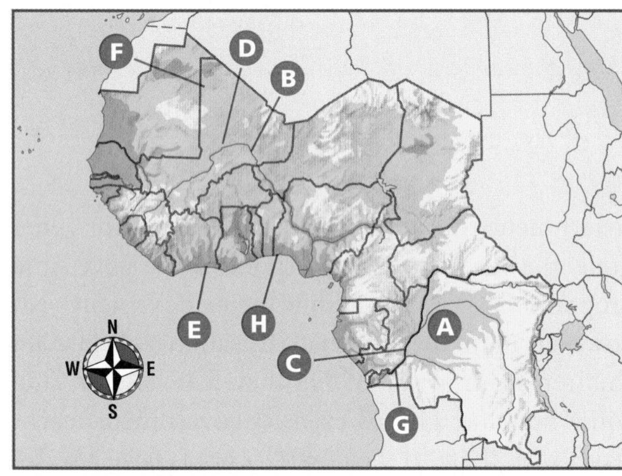

Understanding the Main Ideas TEKS

Section 1

1. Places and Regions In which latitudes are the countries of West and Central Africa located? How does this affect the region's climates?
2. Physical Systems What is the environment of the Sahel like?

Section 2

3. The Uses of Geography How did Islam spread into West and Central Africa?
4. Places and Regions In which part of West and Central Africa is Arabic spoken?

Section 3

5. Places and Regions How has rapid population growth affected cities in West and Central Africa?

Thinking Critically for TAKS TEKS

1. **Supporting a Point of View** What evidence do you think geographers may have used to support their theory that the Congo and Niger Rivers once flowed into large inland lakes? What tools might geographers use to study this theory?
2. **Making Generalizations** How might being landlocked affect economic development, settlement, and population distribution in some countries in the region? How might it influence political conditions in a country?
3. **Analyzing Information** How did the colonial history of West and Central Africa affect cultural, economic, and political characteristics of the countries there?

Using the Geographer's Tools

1. **Analyzing Maps** Look at the map of Empires of Africa in Section 2. How are the names of these empires related to countries in the region today?
2. **Interpreting Diagrams** Analyze the cacao diagram in Section 3. Why do you think the manufacture of chocolate from cacao takes place in countries outside this region?
3. **Preparing Tables** Use the pie graph of world cacao production in Section 3 to create a table that ranks the leading producers of cacao in Africa from highest to lowest. In which part of Africa are these countries found?

Writing for TAKS TEKS

Imagine that you are a researcher studying how farmers in the Sahel might adapt best to the difficult climate conditions. Compare techniques that farmers use in other semiarid regions of the world. Which might be useful in the Sahel? Why? Write a short report for a research journal explaining your suggestions. When you are finished with your report, proofread it to make sure you have used standard grammar, spelling, sentence structure, and punctuation.

SKILL BUILDING

Geography for Life TEKS

Analyzing Data

Places and Regions Use the unit Comparing Standard of Living and Fast Facts table to study the life expectancies, literacy rates, and per capita GDPs of countries in West and Central Africa. What patterns do you see? Is there a wide range of figures from country to country? What do you think are reasons for this?

Building Social Studies Skills

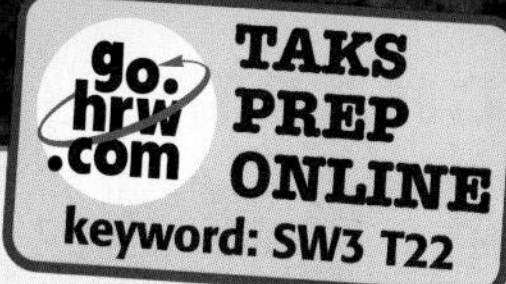

Interpreting Maps

Use the transportation map below to answer the questions that follow.

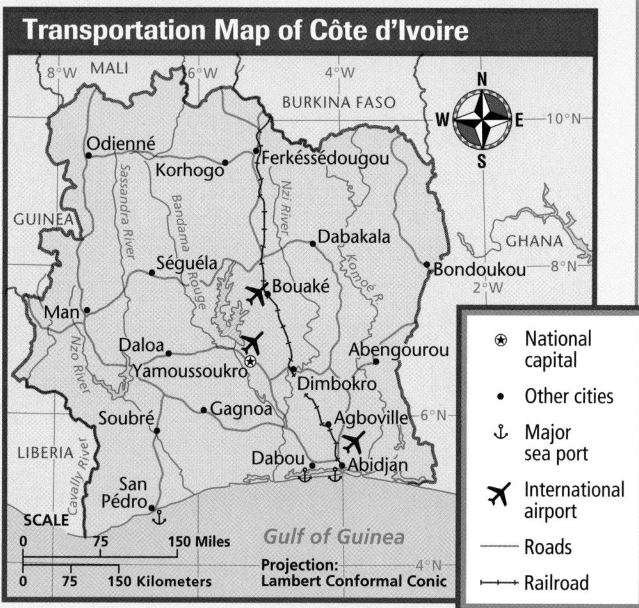

1. According to the map, what transportation facilities does the country's capital have?
   a. airport, railroad, roads
   b. seaport, railroad, roads
   c. airport and seaport only
   d. road and airport only

2. Which city do you think may be the country's busiest transportation center? Why?

Building Vocabulary

To build your vocabulary skills, answer the following questions.

3. *Millet* is
   a. a type of hammer used to build homes in the region.
   b. an army unit stationed only in capitals of the region.
   c. a drought-resistant grain grown in the Sahel.
   d. a state in Nigeria.

4. Some people in the region grow only a few *staple* crops, such as yams or corn. In which sentence does *staple* have the same meaning as it does in the sentence above?
   a. The staple of this cotton is not fine enough.
   b. Milk is a staple in the production of chocolate.
   c. Please staple together the reports on crops grown in Central Africa.
   d. The starch from cassava, a staple in West Africa, is used to make bread.

Alternative Assessment

PORTFOLIO ACTIVITY

Learning about Your Local Geography

Individual Project: Research

Research imported food crops in your area. First, look at what major food crops are produced in your area. Then list common foods that must be imported from other places. A trip to a local grocery store may help you figure out what these imports are. Often, the produce sections in grocery stores give the origin of imported foods. Once you have your list of imported foods, research where these items are produced. Do they come from other regions, states, or countries? Are some foods imported from great distances? Why do you think these foods are imported and not grown locally?

internet connect

Internet Activity: go.hrw.com

Access the Internet through the HRW Go site to research the impact of political and ethnic conflict, economic development, and human migration on the Congo Basin. Then write a report in which you describe how the above factors have affected the people as well as the ecosystem in the region. Use standard grammar, spelling, sentence structure, and punctuation.

CHAPTER 23

East Africa

East Africa includes Africa's largest country, Sudan, and small ones—Burundi, Rwanda, and Uganda. Kenya and Tanzania lie on the Indian Ocean. Ethiopia, Eritrea, Somalia, and Djibouti occupy the Horn of Africa.

Pyramids at Meroë, Sudan

Masai people, Kenya

Indemin adderu! (Good morning.) My name is Tsiyon. I live in Addis Ababa, the capital of Ethiopia, but far from the city center. I live with my parents and my older brothers. My parents sleep in the main house, while my brothers and I sleep in a smaller house near the kitchen, which is in a separate building. My father works for the game department doing research on the wild animals of our country.

In the mornings I ride to school with my oldest brother Undemageni, who works in a garage. I attend Freyhewat Number Two Junior Secondary School, where I study science, English, math, sports, and Amharic, Ethiopia's official language. Someday I want to be a doctor.

At noon, I eat a lunch of *injera* (a spongy sourdough pancake) and stew. Most days it is a meat stew, but on Wednesdays and Fridays, when Christians here fast, it is only vegetables. When I get home, after I help my mother sweep the house, I do my homework. On Saturdays, I go to church with my mother.

My favorite holidays are Christmas and New Year's. On our New Year's Day we have new clothes and a big feast. Because here in Ethiopia we use an old calendar system, our December 25 is around your January 3, and our new year begins on your September 11.

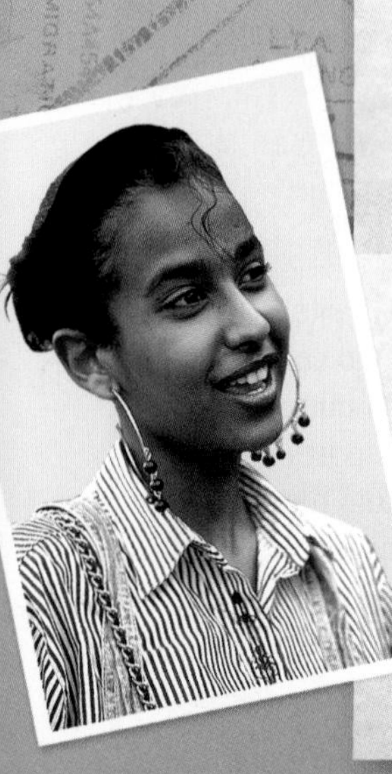

Natural Environments

READ TO DISCOVER

1. What landforms, rivers, and lakes are found in East Africa, and what physical processes have shaped the land?
2. Why does East Africa have a variety of climates and biomes, and on what natural resources does the region depend?

DEFINE

tsetse fly

LOCATE

Great Rift Valley
Lake Malawi
Lake Tanganyika
Lake Victoria
Serengeti Plain
Nile River
Blue Nile
White Nile
Sudd

Why It Matters

Illegal hunting, called poaching, and habitat destruction threaten the wildlife of East Africa. Use CNNfyi.com or other **current events** sources to learn about efforts to protect these magnificent animals.

East Africa: Physical-Political

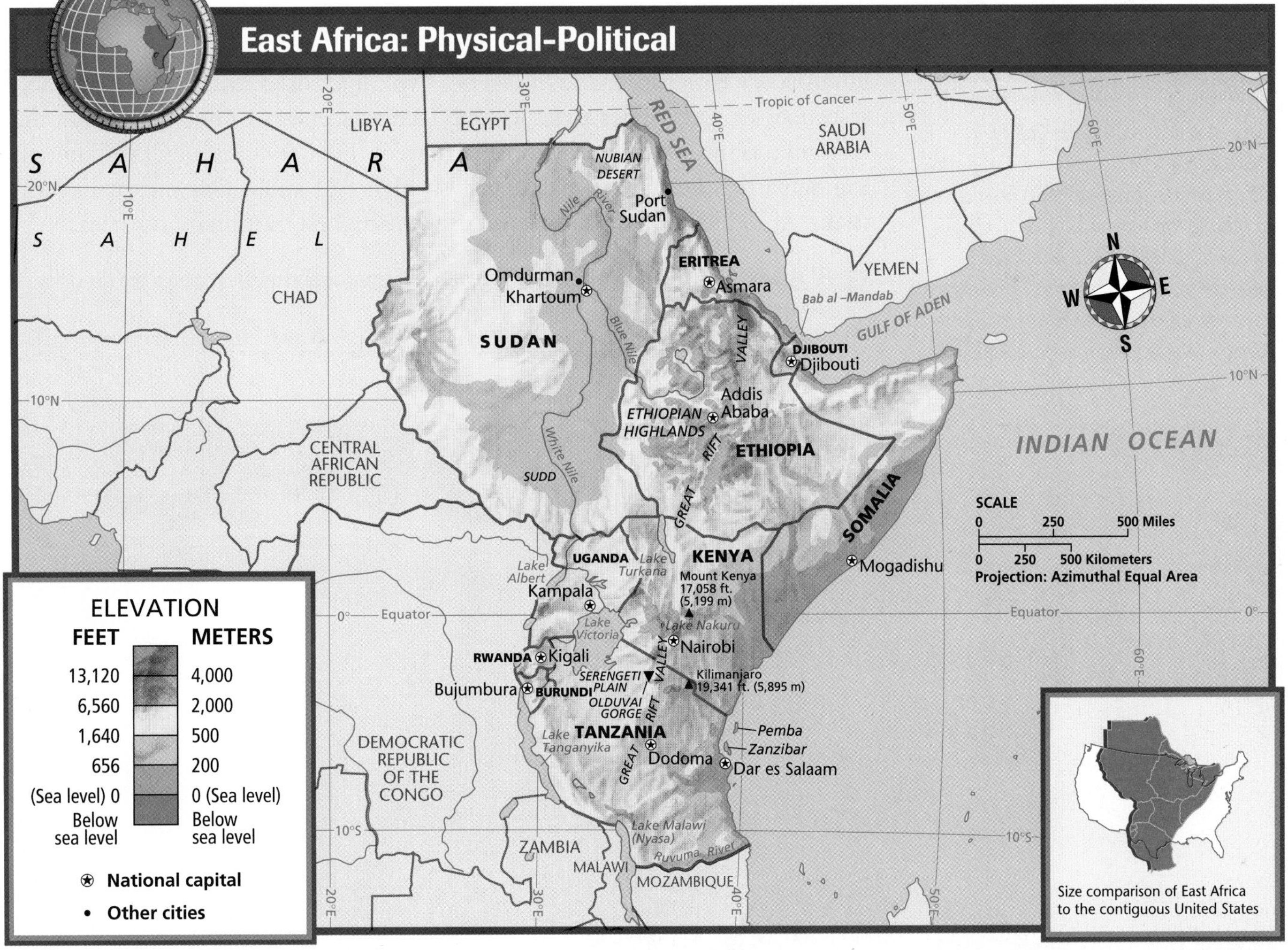

ELEVATION

FEET	METERS
13,120	4,000
6,560	2,000
1,640	500
656	200
(Sea level) 0	0 (Sea level)
Below sea level	Below sea level

⊛ National capital
• Other cities

SCALE
0 250 500 Miles
0 250 500 Kilometers
Projection: Azimuthal Equal Area

Size comparison of East Africa to the contiguous United States

Landforms and Water

Tectonic processes have played an important role in shaping the physical landscape of East Africa. Forces beneath Earth's surface have lifted the region, cracking it apart. This resulted in the formation of two rift valleys. Mountains and plateaus lie along these rifts.

FOCUS ON GEOGRAPHY

The Rifts The Great Rift Valley is a series of geological faults. These faults run from the Jordan Valley in Southwest Asia all the way to Mozambique in southern Africa. In East Africa the rift system is referred to as the Western and Eastern Rift Valleys. The Western Rift Valley begins in the south near Lake Malawi. It continues northward through the valleys of Lake Tanganyika and three smaller lakes. The Western Rift then disappears in southern Sudan. Nearby rainy highlands feed this valley's great lakes. Most of the rift system's lakes are very deep. In fact, some of the lake floors are below sea level. In contrast, the continent's largest lake, Lake Victoria, is shallow. Lake Victoria fills a depression on the high plateau between the Western and Eastern Rifts.

The Eastern Rift Valley begins in Mozambique and continues northward across the Serengeti Plain. In Kenya it passes through Lake Turkana before crossing Ethiopia. At its northern end the Eastern Rift stretches to the floors of the Red Sea and the Gulf of Aden and into Southwest Asia.

Volcanoes have erupted within and near both rifts. The highlands of Ethiopia are therefore made of layers of volcanic rock. Together, the rifts and volcanoes of East Africa produce spectacular scenery. Kilimanjaro, near the Tanzania-Kenya border, is the most famous of the rift's volcanoes. Even though Kilimanjaro is near the equator, it is so high that snow always caps its twin peaks. At 19,341 feet (5,895 meters), it is the highest mountain in Africa.

TEKS **READING CHECK:** ***Physical Systems*** What physical processes created the rift valleys of East Africa?

INTERPRETING THE VISUAL RECORD

Europeans first reached the Great Rift Valley in 1848. Until then, some Europeans did not believe that there could be snow so close to the equator, even on a mountain as tall as Kilimanjaro, which is pictured below. ***How might this discovery have affected Europeans' ideas about Africa's landscapes in general?***  TEKS

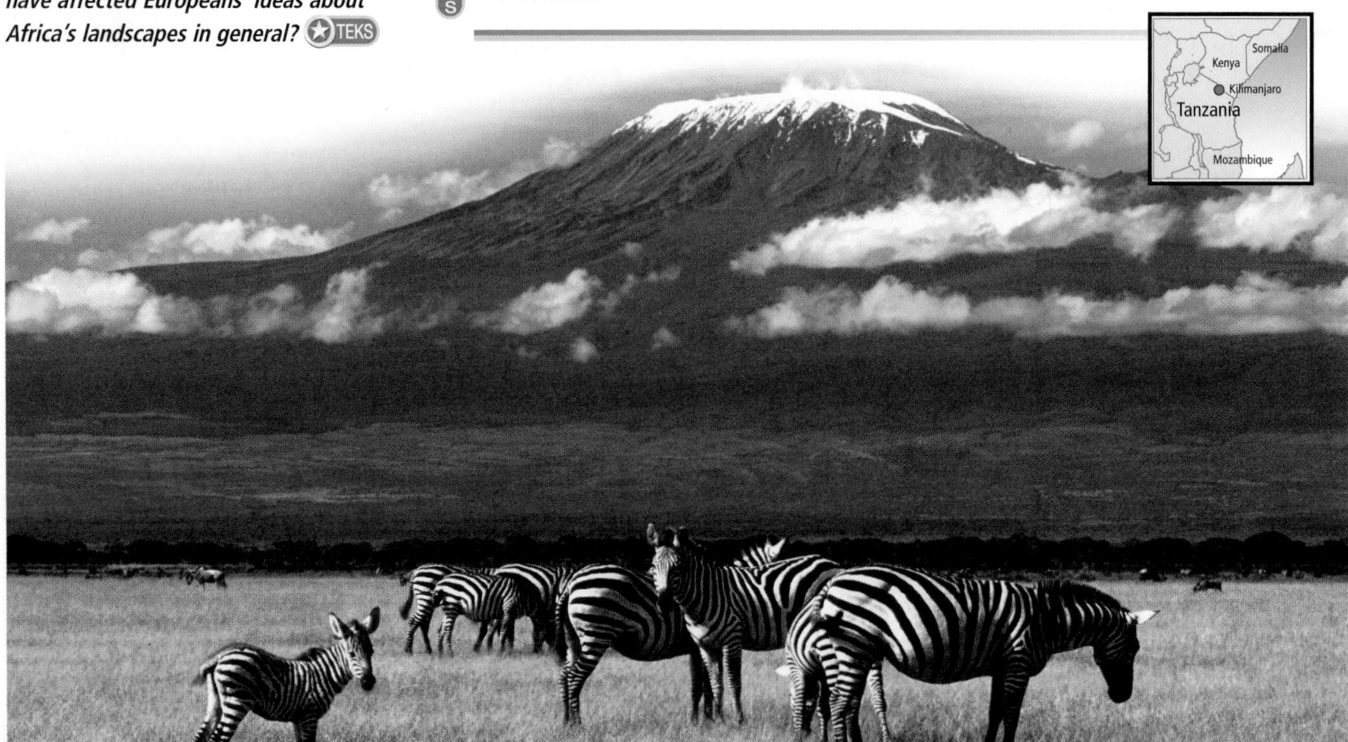

East Africa's major river, the Nile, flows northward through Sudan. Ancient Egypt depended on the Nile, but the Egyptians did not know where the river began. During the 1800s European explorers searched for the Nile's source. They found that the Nile's headwaters are in two different areas. The Blue Nile, which begins in the highlands of northern Ethiopia, provides most of the Nile's water. The waters that form the White Nile drain from Lake Victoria and through Lake Albert. Farther north, the White Nile almost ends in wetlands called the Sudd, in southern Sudan. The area's high temperatures cause about half of the White Nile's water to evaporate. The Blue Nile and the White Nile join in northern Sudan.

Climates, Biomes, and Natural Resources

Latitude and variations in elevation influence the climates of East Africa. Along the equator, distinct wet and dry seasons alternate. As a result, the vegetation on the high plains is a mixture of savannas and forests. Mountain slopes receive heavy rainfall, and the region's forests grow there. The highlands of Kenya and Uganda have a pleasant springlike climate year-round. British colonists found the climate comfortable and settled in these highlands.

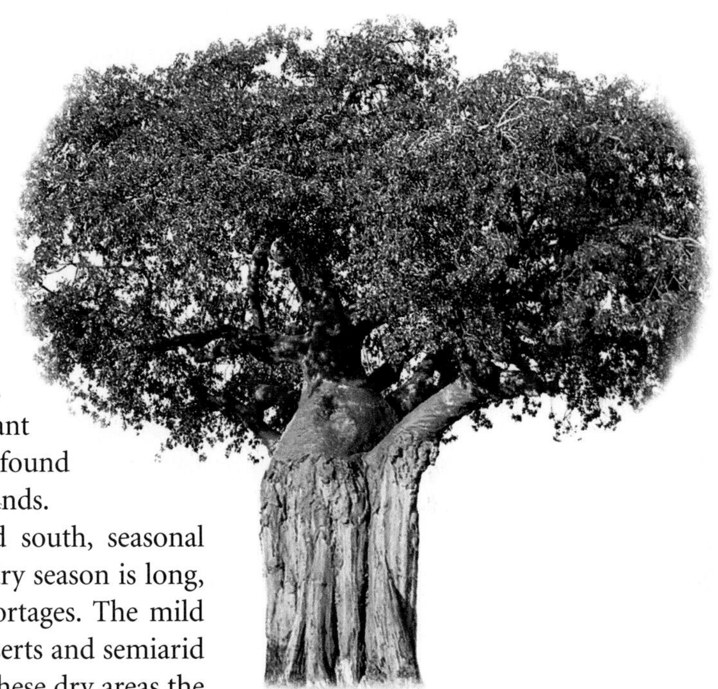

Baobab (BOW-bab) trees are one of the few types of trees that grow on the African savanna. They can grow as large as 30 feet (9 m) in diameter and as high as 60 feet (18 m). The trunks store water for the tree during droughts. People sometimes hollow out the trunks to use as temporary shelters.

Farther from the equator, to the north and south, seasonal droughts are more common. In areas where the dry season is long, trees are shorter and better adapted to water shortages. The mild moist Ethiopian plateau stands high above the deserts and semiarid areas of Sudan, Somalia, and northern Kenya. In these dry areas the vegetation is limited to thorny shrubs and tough grasses. Northern Sudan reaches into the Sahara. The Nile River forms oases through the desert's bare rocks and shifting sand.

Weather is often unpredictable in East Africa. Droughts have affected southern Sudan and the Horn of Africa several times in recent decades. As the grass dies, cattle also die, and people who depend on their livestock begin to starve. People and their animals then migrate to areas that still have some vegetation. This results in overgrazing and desertification. Too much rain also causes problems. After unusually heavy rains, locust populations may increase. Then these big grasshoppers swarm, devouring the plant life in their path. Again, people and animals go hungry. These natural events sometimes play a role in the region's social and political conflicts.

Animals You may have seen television programs about the wildlife of East Africa's Serengeti Plain. Oddly, the giraffes, lions, wildebeests, zebras, and other animals owe their survival partly to a pest—the **tsetse fly**. Tsetse flies carry a human disease called sleeping sickness. While many native animals are immune, the tsetse fly can spread a deadly disease to livestock. As a result, the area has few farmers or herders, leaving large wild animal populations undisturbed. Many areas with the best views, most wild animals, and fewest people have become national parks. By using modern pesticides, the savannas

In 1954 a swarm of locusts covering an area of about 77 square miles (200 sq km) invaded Kenya. Scientists estimated that there were 10 billion locusts in the swarm.

might be made safe for grazing and farming. However, doing so may also create pressure to open the parks to people who want to graze their herds there.

Natural Resources In contrast to some other regions in Africa, East Africa in general is not rich in energy or mineral resources. Sudan began producing oil only recently. Tanzania produces gems, including sapphires and diamonds. Small gold deposits can be found along the rifts. The region's soils are also not very productive. Soils in the dry lands often have too much salt or lime to be fertile. In humid areas, the soil may be too sticky or hard to work easily. Fertile soil and a humid climate are found only in the highlands above the rifts. This rich soil helps explain why the small countries of Rwanda and Burundi can support dense populations. (See the unit population map.)

East Africa's scenery can be considered a valuable natural resource. Sparkling clean beaches stretch along the Indian Ocean. Resorts in the highlands offer both wildlife viewing and hiking through the forests. The savannas' great parks offer tourists remarkable encounters with wildlife. Expanding tourism is an economic goal of many of the region's countries.

READING CHECK: *Places and Regions* How does the climate of Ethiopia's highlands compare to the climate of northern Sudan? Where is heavy rainfall most common?

INTERPRETING THE VISUAL RECORD

A miner searches for tsavorite, a type of green garnet found, so far, only in Kenya and Tanzania. This rare gemstone was discovered in 1968. ***How might the discovery of new resources create both benefits and problems for a country?*** TEKS

Section 1 Review

TEKS Questions 1, 2, 3, 4, 5

Keyword: SW3 HP23

Define
tsetse fly

Working with Sketch Maps
On a map of East Africa that you draw or that your teacher provides, label the Great Rift Valley, Lake Malawi, Lake Tanganyika, Lake Victoria, Serengeti Plain, Nile River, Blue Nile, White Nile, and the Sudd. Which rivers join to form the Nile River?

Reading for the Main Idea
1. *Physical Systems* What landforms have tectonic forces created in East Africa?
2. *Physical Systems* What factors influence climates in the region?

Critical Thinking
3. **Analyzing Information** How does drought affect the region's environment?
4. **Analyzing Information** How is the region's scenery an important economic resource?

Organizing What You Know
5. Create a chart of the main climate regions of East Africa and the vegetation found in those regions. Use information from the unit climate map and Section 1 to complete your chart.

Climate regions	Vegetation

Section 2 History and Culture

READ TO DISCOVER

1. What were some important developments in East Africa's early history?
2. How did European exploration and colonization affect the region?
3. What are the peoples and cultures of East Africa like today?

IDENTIFY

Swahili

DEFINE

ivory
sisal

LOCATE

Olduvai Gorge
Nairobi
Kampala

WHY IT MATTERS

The native languages of East Africa changed through contact with Arab traders. Use CNNfyi.com or other **current events** sources to learn how trade is changing other languages.

Early History

The world's oldest archaeological artifacts have been found in East Africa. The evidence they provide tells us much about how the human race developed. Olduvai Gorge in Tanzania is one of the most important archaeological sites in the world. Some of the earliest human remains have been found there.

First Civilizations Little evidence remains of East Africa's earliest cultures. Instead of keeping written records, most African peoples kept oral histories. These oral histories were stories of events and families that people memorized and passed from one generation to the next. Unfortunately, much of this information has been lost over time.

This beadwork cap was made by Masai artisans of East Africa.

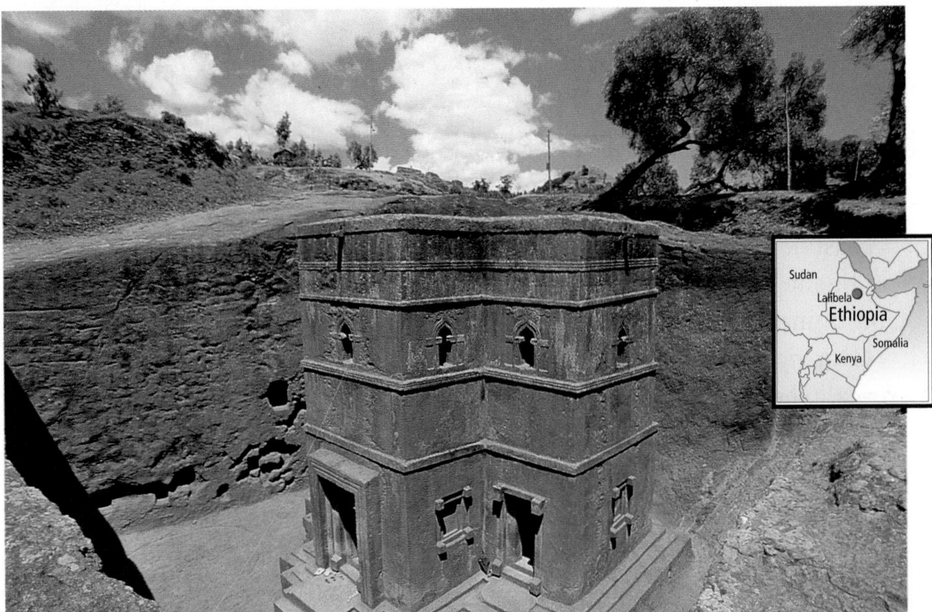

Ethiopia's ancient history is unique in the region. Ethiopians had adopted Christianity several centuries before European missionaries arrived. This church at Lalībela dates from more than 800 years ago. It was carved from solid rock entirely below ground level.

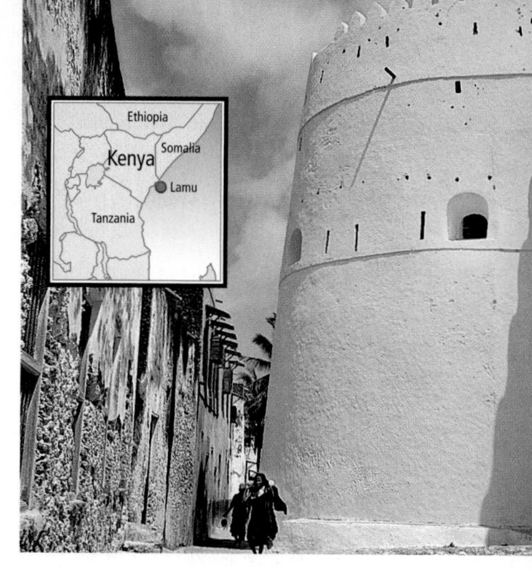

INTERPRETING THE VISUAL RECORD

In the 1800s Arabs built the fort shown on the island of Lamu. They wanted to defend the island against other merchants who hoped to control the area's ports and trade. ***Why do you think so many countries were interested in controlling trade through East Africa?***

We do know about a few early civilizations in the region, however. For example, one of the wealthiest and most powerful kingdoms was Kush. Kush controlled much of the middle Nile River valley, in what is now Sudan, and briefly ruled Egypt. Like the Egyptians, the people of Kush built pyramids. The world's largest cluster of pyramids is at Meroë (MER-oh-wee), the capital of Kush. By A.D. 350, Kush had been conquered by nearby Aksum (AHK-soom). Aksum, in the highlands of what is now Ethiopia, was a city of traders and merchants. For a time Aksum controlled western Arabia. In the A.D. 300s Aksum's kings adopted Christianity.

Arab Connections Arab traders began sailing southward along Africa's Indian Ocean coast about 1,500 years ago. They established ports for trading the gold and **ivory** that Africans brought from the interior. Ivory is the cream-colored material that makes up the tusk of an elephant. It is used in making jewelry and handicrafts. At the ports these goods were traded for products from as far away as India. Slave trading also took place there. The **Swahili** (swah-HEE-lee) language developed during this period. Swahili's grammar comes from original languages of the African coast. Over time, many Arabic words were added to the language. Swahili is now a common language of the region. It is spoken as far west as the Congo.

READING CHECK: ***Human Systems*** How did the spread of Arab influences affect language in East Africa?

European Influence

In the 1500s the Portuguese built the first European forts on the coast of East Africa. However, disease and rough terrain made travel inland difficult. As a result, for about 300 years European contact with East Africans took place mostly on the coast.

In the mid-1800s, European and American explorers, missionaries, and traders began to move inland. Sir Henry Morton Stanley, a journalist, explored the region during the 1870s. After traveling inland from the coast he followed the Western Rift Valley. In time, Stanley crossed the entire continent. Many other explorers fanned out across the region in search of precious minerals and ivory.

INTERPRETING THE VISUAL RECORD

Ceremonies like this Christian wedding in Kenya often blend elements of traditional cultures and of European influences. ***What evidence of European influence can you see in the photo?*** TEKS

During the late 1800s European powers scrambled to claim territory in Africa. They drew colonial boundaries without giving thought to Africa's human or physical geography. Some colonial boundaries divided ethnic groups or grouped traditional enemies. Some even limited access to water.

The Europeans quickly established colonies, although Ethiopia remained largely free of colonial rule. Soon products from African mines and plantations supplied European economies. Cash crops included coffee, cotton, tea, and **sisal** (SY-suhl). Sisal is a strong durable plant fiber used to make rope and

twine. However, during the colonial era most East Africans still practiced traditional subsistence agriculture.

The Europeans built cities, hospitals, ports, roads, and schools in the areas where there were useful natural resources for export. European settlers began to farm the region's fertile highlands. Most of East Africa's modern capital cities sprang up during this time. For example, both Nairobi and Kampala served as early railway stations. The colonizers also provided a small number of Africans with a European education. Many of these Africans later led independence movements. Most countries of East Africa gained their independence during the 1950s and 1960s. In 1977 Djibouti became the last colony to win independence. Eritrea became a separate country when it broke away from Ethiopia in 1993.

INTERPRETING THE VISUAL RECORD

Women celebrate after Eritrea's independence is declared in 1993. Eritrea's constitution praises women as influential and heroic in the country's struggle for freedom. ***What might this statement imply about social and economic opportunities for women in Eritrea?***

READING CHECK: ***Human Systems*** What were some pull factors that drew Europeans to East Africa?

Culture

East Africa includes several hundred ethnic groups. Their cultures have given the world a rich heritage of architecture, art, folk tales, and music. The groups can be organized into three categories according to language. The Nilotic peoples live in the Nile River area on the plains of Sudan. Most Nilotic peoples are herders. Several Nilotic peoples migrated southward into the highlands a few centuries ago. Among these were the Masai and Tutsi. The second group is made up of Cushitic-speakers. Their lands run from the Red Sea coast through the Horn of Africa. This group includes the Amhara of the Ethiopian highlands and the Somali of the coast. The third group, Bantu-speakers, live farther south. They include the Kikuyu of Kenya and the Hutu of Rwanda. Bantus also moved into southern Africa.

People who follow Arab traditions live mostly along the Indian Ocean coast. Africans of South Asian descent also live in the region. During the colonial period, their ancestors came to work as merchants and craftspeople.

The Dinka of Sudan are among the world's tallest people. Manute Bol, a Dinka almost 7 feet, 7 inches tall, played in the National Basketball Association. Many Dinka women are also more than six feet tall.

Religions of East Africa

Country	Christian	Muslim	Traditional Beliefs	Other
Burundi	60%	1%	39%	
Djibouti		94%		6%
Eritrea	45%	45%		10%
Ethiopia	37%	43%	17%	3%
Kenya	66%	6%	26%	2%
Rwanda	45%		50%	5%
Somalia		100%		
Sudan	5%	70%	20%	5%
Tanzania*	26%	31%	42%	1%
Uganda	66%	16%	18%	

*except Zanzibar, which is 99% Muslim

Source: *The DK Geography of the World*

Religion and family traditions are important aspects of daily life for East Africans. (See the chart.) Religions vary both within and among ethnic groups. However, most of the cultures honor ancestors. Many people believe the spirits of ancestors are strong forces in daily life.

Traditional religions are animist. Followers of animist religions believe the natural world contains spirits—in animals, mountains, trees, and waters. Many Africans also combine ancient forms of worship with later religions. Christianity came to Ethiopia more than 1,500 years ago. Missionaries during the 1800s and 1900s also spread Christianity. Arabs brought Islam to the region several centuries ago as well. New mosques throughout the region indicate Islam's continued growth. Rural communities are often dominated by one religion or the other. However, the ceremonies and holidays of many religions help make life in the cities exciting.

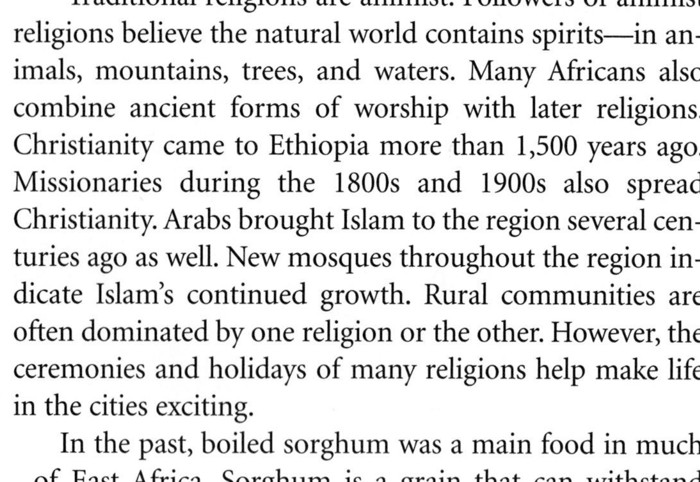

An Ethiopian woman makes injera. *Diners scoop up spicy vegetable or meat stews laid on top of the* injera *with pieces of the bread.*

In the past, boiled sorghum was a main food in much of East Africa. Sorghum is a grain that can withstand drought. Sometimes the boiled grain was mixed with roasted beef or lamb. Other basic foods were sour milk and animal blood. Roots, berries, and game added to the diet. These were called bush foods because they were gathered in the wild. However, many new foods began appearing during the colonial era. Cornmeal, potatoes, rice, and wheat bread are now common. American fast food is widely available in the cities. Your favorite soft drink might be for sale in even the smallest village. Ethiopia and Eritrea have a unique food tradition based on a grain called teff. A large flat rubbery bread called *injera* is made from teff flour. *Injera* is served as large platter-shaped loaves.

TEKS **READING CHECK:** *Human Systems* What are the main religions of East Africa?

TEKS Questions 1, 2, 3, 4, 5

Review

Identify

Swahili

Define

ivory, sisal

Working with Sketch Maps

On the map you created in Section 1, label the countries of East Africa, the Olduvai Gorge, Nairobi, and Kampala. What is the significance of the Olduvai Gorge?

Reading for the Main Idea

1. *Places and Regions* Where did Europeans build cities? What factor caused Nairobi and Kampala to grow?
2. *Human Systems* What are the three main language groups of East Africa? What influence did Arabic-speakers have on language in the region?

Critical Thinking

3. **Analyzing Information** Why might it be said that the colonial period created both benefits and problems for East Africa?
4. **Drawing Inferences and Conclusions** How might the availability of new foods influence health and culture in the region?

Organizing What You Know

5. Create a time line like the one shown below. Use it to identify and describe important events and periods in East Africa's history.

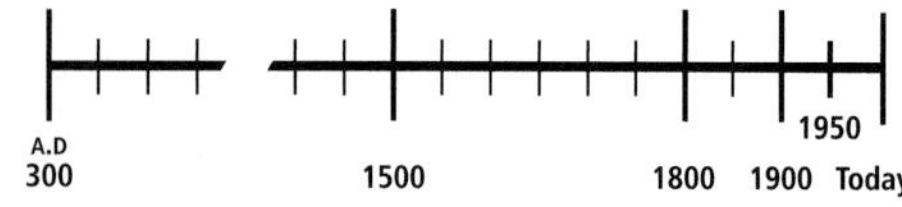

Section 3 The Region Today

READ TO DISCOVER

1. What roles do agriculture, industry, trade, and tourism play in the economies of East Africa?
2. What are the region's cities like?
3. What issues and challenges do East Africans face?

DEFINE

gum arabic
genocide

LOCATE

Zanzibar
Dar es Salaam
Mombasa
Djibouti
Addis Ababa
Khartoum
Omdurman

WHY IT MATTERS

East Africa remains a politically unstable region. Use CNNfyi.com or other **current events** sources to learn about the most recent efforts to bring lasting peace and prosperity to East Africa.

Economic Development

Farming, trade, and industry all contribute to East Africa's economies. Traditional economies based on small-scale subsistence agriculture are common in rural areas. Manufacturing and global trade play minor roles.

Some of the region's poorest people earn a small income by gathering wild plant products. In Ethiopia this includes picking coffee beans from wild coffee trees. In Sudan people gather **gum arabic**, the sap of acacia trees. Gum arabic is a sticky substance that binds the ingredients of many candies and medicines.

Kenyan coffee growers prepare beans for weighing. A majority of the Kenyan farmers who grow coffee work small plots of land. Many of these growers work together in cooperative societies to process and sell their crops.

Agriculture Farming and herding form the basis of East Africa's economies. Humid highlands, such as those in Ethiopia, have many small subsistence farms. Cattle, goats, and sheep graze the dry lowland plains and plateaus. Farming there is possible only on irrigated land near oases.

In the region's cultures, women are often the primary farmers. Men take care of the livestock. Depending on the climate, the important food crops are beans, corn, rice, sorghum, and wheat. Farmers grow an increasing number of cash crops. These include coffee, cotton, sugarcane, and tea. Areas along the Indian Ocean coast produce their own distinctive crops, such as cloves and coconuts. Zanzibar, an island off Tanzania's coast, is a major producer of cloves.

A few large commercial farms and plantations can be found in East Africa. These farms have modern technology like tractors and trucks as well as modern seeds and fertilizers. Although few in number, these commercial farms produce crops for export and food for the region's cities.

TEKS **READING CHECK:** *Environment and Society* What are some of the region's agricultural products?

INTERPRETING THE VISUAL RECORD

Hydroelectric power provided by Owen Falls Dam has enabled Jinja to become Uganda's main industrial center. More than 99 percent of Uganda's electricity comes from hydropower. Several other East African countries get more than 90 percent of their electricity from hydropower. ***What river does the dam control? Use the small inset map and the physical-political map to find out.*** TEKS

Industry and Economic Change All of the countries in the region have developing economies. Raw materials make up most of the exports. In contrast, East Africa imports many manufactured goods. The region's major ports include Dar es Salaam (dahr-es-sah-LAHM), Tanzania; Mombasa, Kenya; and the city of Djibouti (ji-BOO-tee). Manufacturing in the region centers around basic consumer goods, processed food, and building materials.

Over time the region's countries have had a mix of command and market economies. Beginning in the 1970s, for example, Ethiopia's leaders worked to create a command economy. The central government took over all of the country's agricultural land and urban rental property. It also took over banks, insurance companies, and many other businesses. However, the economy suffered throughout this period. To make matters worse, in 1984 a terrible drought struck. Famine and starvation became widespread. In the 1990s a new government began economic reforms. Ethiopia is slowly developing a market-oriented economy. Still, it remains one of the world's poorest countries.

Kenya has the highest per capita GDP in the region. However, progress has been slow, partly because of the government's mismanagement of the economy. Rapid population growth has also put pressure on the economy. (See Geography for Life: Population Growth in Kenya.)

Tourism has great potential for economic growth in the region. Fascinating animals, cool highlands, snowcapped mountains, clean beaches, and cultural events all draw tourists from developed countries. Resorts, restaurants, safari lodges, and taxi companies provide many jobs. Tourists also buy traditional arts and crafts as souvenirs. Political violence has hurt the tourist industry in some countries, however. Therefore, maintaining stability in the region is important to the industry. In addition, many people argue that only by preserving the environment will the countries of East Africa continue to draw a steady stream of visitors.

READING CHECK: ***Human Systems*** How did Ethiopia's government try to create a command economy beginning in the 1970s?

Urban Environments

Addis Ababa (AHD-dis AH-bah-bah), Ethiopia's capital, is the region's largest city. Regional organizations, such as the Organization of African Unity, have their headquarters there. A railroad through Djibouti connects Addis Ababa to the Indian Ocean coast. Nairobi, Kenya, is the most important commercial center in East Africa. The city's manufactured goods range from cement to soap. A national park—where gazelles, lions, zebras, and other animals live—lies within Nairobi's city limits. Dar es Salaam is a transportation hub. Tanzania's government is scheduled to move inland from that coastal city to Dodoma by 2005. Khartoum, which is Sudan's capital, and Omdurman (ahm-duhr-MAN) are the largest cities in Sudan. They lie across the Nile from each other.

East Africa's rapidly growing cities have superhighways and glittering skyscrapers. However, run-down buildings and slums often surround these symbols of economic progress. The number of people moving to the cities from the countryside is larger than the number of available jobs. Many of the newcomers have high hopes but possess only farming skills and little education. The unemployed and those with temporary jobs live in the slums that ring many cities. Providing better housing is a pressing issue for governments. However, providing good government under these conditions is difficult too. Political unrest sometimes is a problem. High crime rates contribute to that unrest.

READING CHECK: *Human Systems* Why are the region's cities growing rapidly?

Issues and Challenges

Populations in East Africa have risen dramatically during the past 30 years. A rapidly growing population underlies many of the problems that face the region's governments and people. Overpopulation contributes to widespread poverty. In many places there is not enough food to go around. Trying to grow food where the land cannot support farming can hurt the environment. Health services and educational opportunities are also spread thin. (See Geography for Life: Population Growth in Kenya.)

Ethnic conflicts, between countries and between groups within countries, have also presented problems. Often central to these conflicts are struggles over land and fair distribution of government aid and jobs. For example, in Ethiopia the Amhara have long been the dominant ethnic group. However, the Oromo and the Tigre are now demanding what they believe is their share of power and influence. Similarly, the Baganda, whose traditional lands include areas along Lake Victoria, dominate Uganda. Complaints from the peoples of northern Uganda have led to repeated unrest. For decades, Sudan has been torn by war between the Islamic government and the animist and Christian peoples of the south. The government has tried to force Islam and Arabic ways of life on the people of southern Sudan, who want more autonomy.

Ethnic hatreds have even led to **genocide**. Genocide is the intentional destruction of a people. The worst case happened in Rwanda in 1994.

Connecting to SCIENCE

Endangered Cheetahs

Among the many animals that attract tourists to East Africa is the cheetah. However, loss of habitat and illegal hunting threaten these graceful cats. They are also threatened by their own genes. Because of their reduced numbers, Africa's cheetahs have been able to mate only with their close relatives. Today cheetahs are inbred. As a result, the cheetahs in some populations share nearly all the same genes. This lack of genetic diversity leaves cheetahs vulnerable to disease and early death. Breeding programs now aim to restore the health of the species by mixing genetic material of captive and wild cheetah populations.

Problem Solving How might the tourist industry help save the cheetah? TEKS

Residents of a Nairobi slum carry firewood to their homes. East Africa is a poor region. Just under half of Kenya's population lives in poverty.

There the more numerous Hutu tried to wipe out the Tutsi. Armed bands killed hundreds of thousands of people.

Violence has also been a problem in Somalia. There, however, ethnic conflict has not been the issue. In Somalia most of the people are ethnic Somali and Muslim. The country has often had no central government of any kind. Instead, different clans have fought over grazing rights and the control of port cities such as Mogadishu. In addition, Somalia has tried to annex nearby parts of Ethiopia by military invasions. Relations with Ethiopia remain uneasy.

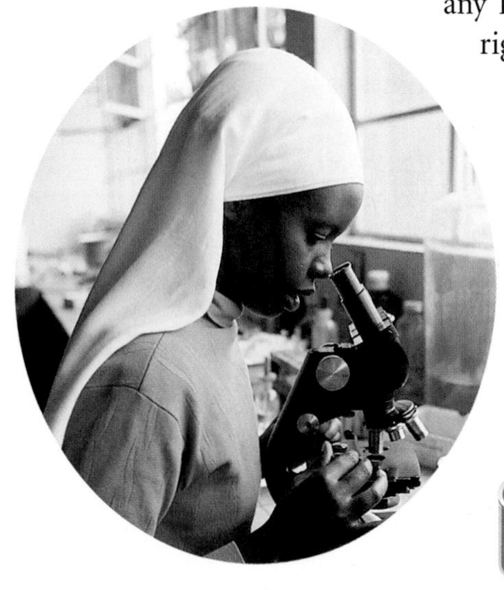

Diseases such as AIDS, cholera, and malaria kill thousands of East Africans each year. Healthcare workers like this Rwandan woman are working to help the sick, but with limited resources.

As you can see, the leaders of the East African countries face many challenges. These challenges include building stable countries in spite of ethnic fighting. Promoting economic progress, protecting the environment, and providing for the health and educational needs of growing populations are also important.

TEKS **READING CHECK:** *Human Systems* How have political and economic tensions contributed to problems in the region?

TEKS Working with Sketch Maps, Questions 1, 2, 3, 4, 5

Section 3 Review

Define

gum arabic, genocide

Working with Sketch Maps

On the map of East Africa that you created in Section 2, label Zanzibar, Dar es Salaam, Mombasa, Djibouti, Addis Ababa, Khartoum, and Omdurman. Which of East Africa's cities has the most people?

Reading for the Main Idea

1. *Human Systems* What are the most common economic activities in East Africa? How do some of the poorest people survive?
2. *Places and Regions* What are some features that many of the region's cities have in common?

Critical Thinking

3. **Categorizing** Are the countries of East Africa developed or developing? On which factors do you base your answer?
4. **Analyzing Information** How do you think overpopulation contributes to political unrest?

Organizing What You Know

5. Create a sketch map on which you identify the places where some conflicts have occurred in recent decades in the region. Use one color or symbol for ethnic conflicts and another for conflicts not related to ethnic identity.

Geography for Life

Population Growth in Kenya

If the current growth rate continues, the world's population will grow by almost a billion people every 12 years! Countries where big families bring high status and ensure care for elders later in life have the highest growth rates. In addition, farming families often want many children so they can work in the fields. All these factors apply to Kenya. In 1988 Kenya had a population growth rate of 4.2 percent—the highest any country has ever recorded. Kenya's growth rate soared because more children were born, fewer children died, and modern medicine kept more people alive longer.

After Kenya won independence in 1963, the country's economy expanded quickly. However, by the 1980s the economy could no longer keep pace with the population. Too many resources went just to supply the needs of the huge population. Little was left over for development or investment.

In addition, Kenya's population explosion created conflict between people and their environment. Arable land, clean water, firewood—all are in short supply. About 75 percent of Kenyans live on 10 percent of the land. Some farmers have moved onto land not suitable for farming. Efforts to farm these lands have led to environmental damage and the destruction of wildlife habitat. Meanwhile, conservationists argue that protecting wildlife serves a long-term goal of supporting the tourist industry. Kenyans who live in urban areas have other troubles. In the cities they compete for a limited number of jobs.

Kenya's government has tried to address these problems. For example, Kenya was the first African country south of the Sahara to adopt an official policy on population. Government officials realized that only by controlling population growth would the country prosper.

Fortunately, in recent decades the average family size in Kenya has dropped, from 8 children to 5.3 children. Women's healthcare has also improved. Few women have access to family planning services, however. Women also have few economic or educational opportunities. In countries where women have access to education they soon get the chance to work outside the home. In those countries the birthrate drops, and standards of living rise. The long-term solution to Kenya's high birthrate may lie in increasing opportunities for Kenyan women.

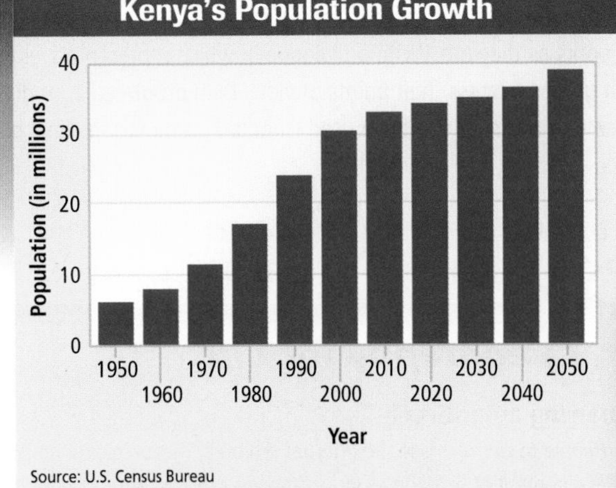

Source: U.S. Census Bureau

INTERPRETING THE GRAPH *Between 1970 and 1980 the number of Kenyan women who received an education, like those at the right, more than tripled.* ***How does the graph reflect a connection between education and slowing population growth?*** TEKS

Applying What You Know TEKS

1. **Summarizing** How does a high population growth rate limit Kenya's progress?
2. **Making Predictions** Develop and defend a hypothesis about how many people might live in Kenya 100 years from now. What factors should you take into account? What are some factors that might make the figure much higher or lower?

Building Vocabulary

On a separate sheet of paper, explain the following terms by using them correctly in sentences.

tsetse fly	sisal
ivory	gum arabic
Swahili	genocide

Locating Key Places

On a separate sheet of paper, match the letters on the map with their correct labels.

Great Rift Valley	Blue Nile	Dar es Salaam
Lake Victoria	White Nile	Addis Ababa
Serengeti Plain	Nairobi	

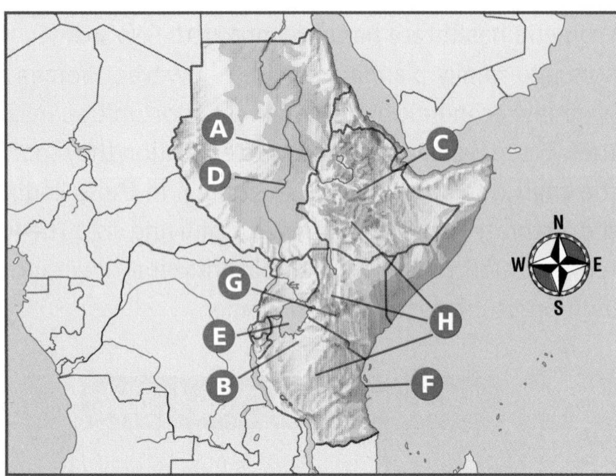

Understanding the Main Ideas

Section 1

1. *Physical Systems* How does elevation affect climate in East Africa?

Section 2

2. *Places and Regions* What are two ways that Arab traders influenced East African languages and religion?
3. *Human Systems* How have other cultures changed the traditional religions and diet of East Africa?

Section 3

4. *Human Systems* What activities form the basis of the region's economy?
5. *Places and Regions* Why does tourism hold great economic potential for the region?

Thinking Critically for TAKS

1. **Drawing Inferences and Conclusions** How may East Africa's natural environments have affected the routes, flows, and destinations of African and European migration and settlement?
2. **Comparing** As you have read, humid highland areas in the region have many farms. Dry plains are used for grazing livestock. Where do you find farming and livestock grazing in the United States and other countries you have studied?
3. **Analyzing Information** How is rapid population growth shaping economic, urban, and environmental change in the region?

Using the Geographer's Tools

1. **Using Maps and Tables** Review the unit maps and the population density figures in the unit Fast Facts table. How does physical geography affect population density in East Africa?
2. **Creating Pie Charts** Convert the information in the Section 2 religions chart into pie charts. Note that you will need to make a design adjustment for Tanzania.
3. **Preparing Maps** Create an outline map of East Africa and surrounding regions. Then after reviewing information in Chapter 4, shade areas on the African and Arabian tectonic plates and note the direction of their movement. Finally, note landforms and other features created by tectonic activity.

Writing for TAKS

Some farmers and herders want to clear the tsetse flies from the animal parks and use the land for agriculture. Others want to keep the parks as they are. Write a dialogue in which representatives of both groups express their points of view. Then proofread your dialogue to make sure you have used standard grammar, spelling, sentence structure, and punctuation.

SKILL BUILDING

Geography for Life

Drawing a Bar Graph

Human Systems Use the Internet or library resources to find the average number of people per doctor in each East African country and in the United States. Display your findings in a bar graph. Let the vertical axis represent the number of people. Place the countries along the horizontal axis at the bottom. Write a paragraph in which you draw conclusions about health care and the standard of living in East Africa.

Building Social Studies Skills

Interpreting Maps

Use the information from the map above to answer the questions that follow.

1. In Ethiopia, the Blue Nile passes through which climates?
   a. semiarid; highland
   b. arid; semiarid
   c. arid; tropical wet and dry
   d. tropical wet and dry; highland
2. Which area of Ethiopia would you expect to be the least densely populated? Why?

Analyzing Primary Sources

Read the following description of Ethiopia's Blue Nile, written by Virginia Morell during an expedition. Then answer the questions.

> **"With the river roaring through its canyon a good half mile below us, we trekked past clusters of round, thatch-roofed homes and fields of teff [a grain] edged with low stone walls and clumps of daisies. On both sides of the gorge the land rose in broad-shouldered, terraced mountains, each flat bit of land quilted with a patchwork of fields that shimmered green and gold in the sun. In many fields small groups of men, women, and children [were] weeding each row by hand."**

3. What kinds of homes did the author see along the gorge?
   a. homes made from teff
   b. homes made from low stone walls
   c. thatched-roof homes
   d. homes with terraces
4. What features of the local physical geography does the passage describe?

Alternative Assessment

PORTFOLIO ACTIVITY

Learning about Your Local Geography TEKS

Group Project: Interviewing

As you have read, much of Africa's precolonial history was passed down orally. Plan, organize, and complete a group research project on oral history in your community. Interview longtime residents about how they recall a certain local weather event, such as a drought or tornado. Create a list of questions that you will ask each person. Include questions about how the person experienced the event and what may have caused it. Also ask how the event affected the interviewee's feelings for or opinions about his or her community. Videotape the interviews and show them to the class. As a group, discuss how the interviews are similar or different.

internet connect

Internet Activity: go.hrw.com
KEYWORD: SW3 GT23 TEKS

Choose a topic on East Africa to:

- research United Nations efforts to reduce ethnic conflict in the region.
- learn about the region's many ethnic groups.
- create a brochure on the climate and wildlife of Kilimanjaro.

CHAPTER 24 Southern Africa

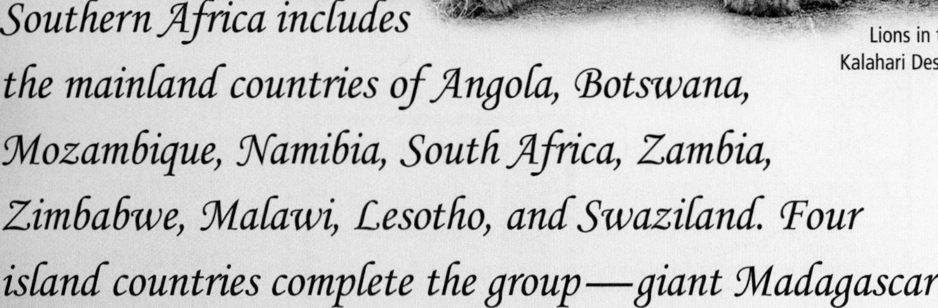
Lions in the Kalahari Desert

Southern Africa includes the mainland countries of Angola, Botswana, Mozambique, Namibia, South Africa, Zambia, Zimbabwe, Malawi, Lesotho, and Swaziland. Four island countries complete the group—giant Madagascar and tiny Comoros, Mauritius, and Seychelles.

Woman in Lesotho

Abatsu! (Good morning!) First of all I would like to ask about you—if you are fine, then I am fine also. My name is Kha//an[1] and I am a San person from Namibia. I am a student in the ninth grade at the Tsumkwe[2] Secondary School. We do not have a house to sleep in, only a small shelter to store our things. We live near the gate in the fence at M'Kata. If you turn east when you get there, you will see our blankets under the tree. When we need shade or shelter, we just go and sit under the tree.

I live with my elder brother. My mother and father died when I was five years old, and my brother took care of me. We do not have any animals. Instead, the government helps us by giving us food rations.

At school I live in a hostel. For our main meal at noon we have maize meal, tea, and milk. I always attend classes as well as evening study periods. My favorite subjects are English, history, physical science, and mathematics.

Aside from my school days, my favorite times are when my brother and I do small jobs to earn some money. When I grow up, I want to be a doctor. I will have to work very hard. I read all the books I can get and ask the teachers to explain things to me. In my free time, I read newspapers and try to understand the words I do not know. My favorite game is soccer, which we play on Wednesday afternoons.

[1] The "//" is a click made by clucking the tongue at the sides of the mouth.
[2] The *k* in Tsumkwe is another kind of click. It is made by placing the tongue on the roof of the mouth and bringing it down with a "pop."

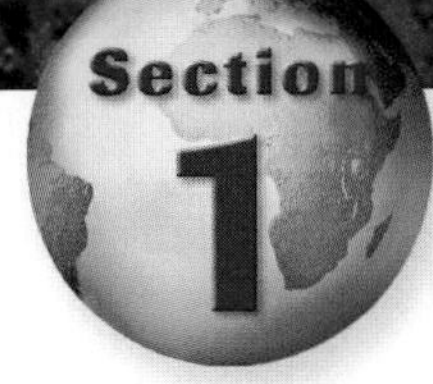

Natural Environments

READ TO DISCOVER

1. What are the main landforms and rivers of southern Africa?
2. What climates, biomes, and natural resources are found in the region?

DEFINE

escarpment
biodiversity
veld

WHY IT MATTERS

The countries of southern Africa struggle to balance wildlife conservation with people's need for land and resources. Use CNNfyi.com or other **current events** sources to learn about wildlife conservation in the region.

LOCATE

Drakensberg
Orange River
Limpopo River
Zambezi River
Namib Desert
Kalahari Desert
Okavango Swamps

Southern Africa: Physical-Political

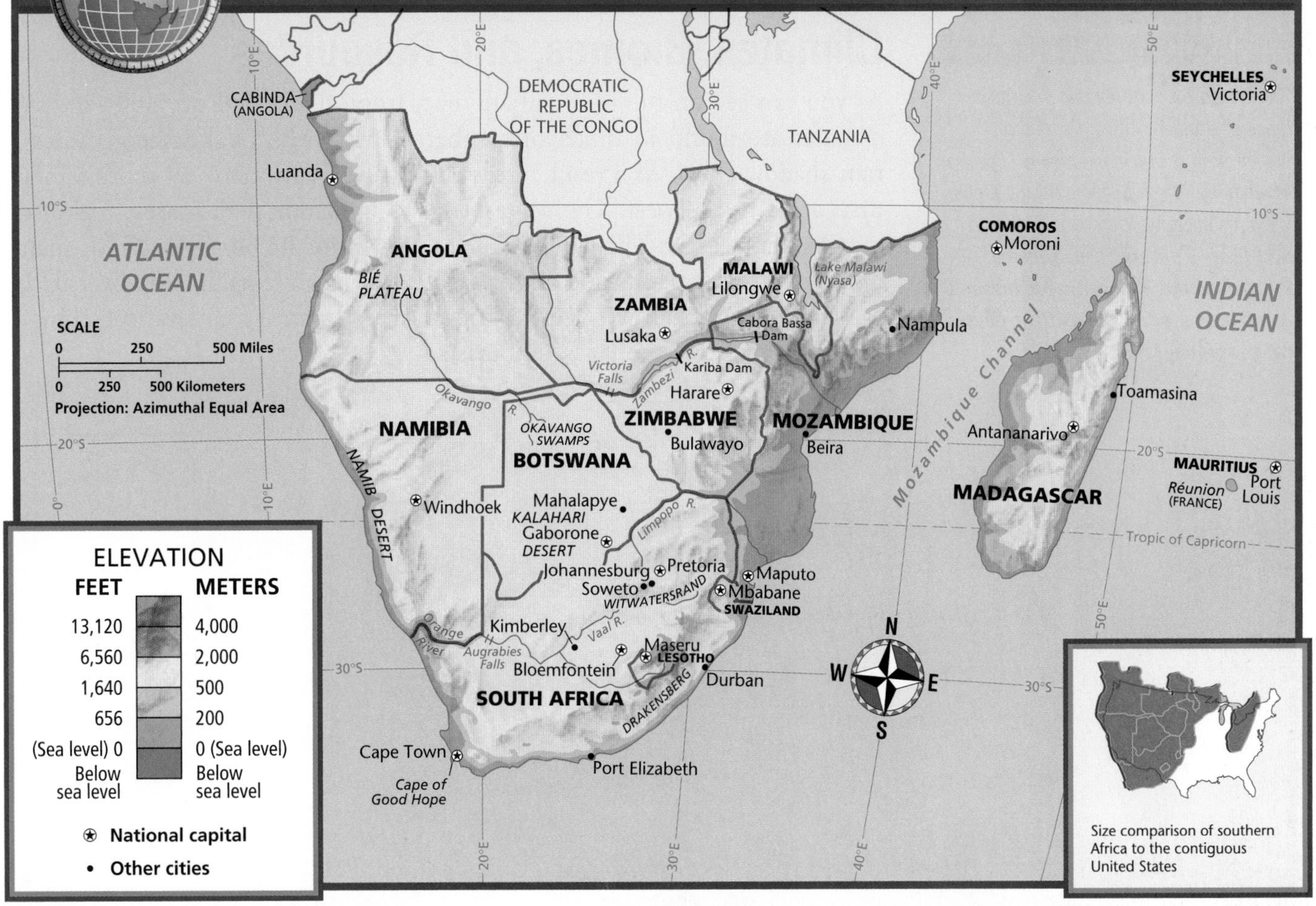

SCALE
0 250 500 Miles
0 250 500 Kilometers
Projection: Azimuthal Equal Area

ELEVATION
FEET | METERS
13,120 | 4,000
6,560 | 2,000
1,640 | 500
656 | 200
(Sea level) 0 | 0 (Sea level)
Below sea level | Below sea level

⊛ National capital
• Other cities

ATLANTIC OCEAN
INDIAN OCEAN
CABINDA (ANGOLA)
DEMOCRATIC REPUBLIC OF THE CONGO
TANZANIA
SEYCHELLES
Victoria
COMOROS
Moroni
Luanda
ANGOLA
BIÉ PLATEAU
ZAMBIA
MALAWI
Lilongwe
Lake Malawi (Nyasa)
Lusaka
Cabora Bassa Dam
Nampula
Mozambique Channel
Victoria Falls
Zambezi R.
Kariba Dam
Harare
Toamasina
Okavango R.
OKAVANGO SWAMPS
ZIMBABWE
MOZAMBIQUE
Antananarivo
NAMIBIA
BOTSWANA
Bulawayo
Beira
MAURITIUS
Port Louis
Réunion (FRANCE)
MADAGASCAR
NAMIB DESERT
Windhoek
Mahalapye
KALAHARI DESERT
Gaborone
Limpopo R.
Tropic of Capricorn
Johannesburg
Pretoria
Soweto
WITWATERSRAND
Maputo
Mbabane
SWAZILAND
Kimberley
Vaal R.
Orange River
Augrabies Falls
Maseru
LESOTHO
Bloemfontein
Durban
SOUTH AFRICA
DRAKENSBERG
Cape Town
Cape of Good Hope
Port Elizabeth
N E S W
10°S 20°S 30°S 0° 10°E 20°E 30°E 40°E 50°E

Size comparison of southern Africa to the contiguous United States

Our Amazing Planet

One of the world's great waterfalls is located on the Orange River. At flood time, Augrabies (oh-KRAH-bees) Falls is several miles wide, and 19 separate waterfalls flow into a ravine 11 miles (18 km) long.

Landforms and Water

Along southern Africa's coastline is a narrow coastal plain. This plain is less than 100 miles (160 km) wide in most places. Farther inland, a high plateau reaches more than 4,000 feet (1,220 m) above sea level. Most of the region lies on this plateau. Between the coastal plain and the plateau is an **escarpment**, a steep face at the edge of a plateau or other raised area. In South Africa the Drakensberg (DRAH-kuhnz-buhrk) range forms part of the escarpment. These mountains include a peak 11,425 feet (3,482 m) high.

Several major rivers flow across southern Africa. The Orange River starts in the Drakensberg and flows west across South Africa. It is the only major river in the region that drains into the Atlantic Ocean. Dams on the Orange River produce hydroelectricity. They also allow irrigation and economic development in central South Africa. The Limpopo, which is sometimes called the Crocodile River, drains into the Indian Ocean. The Zambezi River is another major source of hydroelectric power. It is more than a mile wide above Victoria Falls, where it drops 355 feet (108 m).

As southern Africa's rivers flow from the high interior plateau to the low-lying coastal plain, they form waterfalls and rapids. In addition, sandbars partially block the mouths of the Limpopo and Zambezi Rivers. As a result, large ships cannot sail upriver to the interior.

READING CHECK: ***Places and Regions*** What are the three main landform regions of southern Africa? What are three major rivers in the region?

Climates, Biomes, and Resources

As you can see on the unit climate map, tropical wet and dry and semiarid climates are found in much of southern Africa. The Drakensberg causes a rain-shadow effect. As a result, areas east of the escarpment get more rainfall than areas to the west. The wettest area is the tropical rainforest region of eastern Madagascar. This large island is known for its **biodiversity**, or many different types of plants and animals. (See Geography for Life: Biodiversity in Madagascar.) Cape Town, South Africa, has a pleasant Mediterranean climate that helps make it popular with tourists.

INTERPRETING THE VISUAL RECORD

The Drakensberg is the main mountain range of southern Africa. The name Drakensberg *is Dutch and means "Dragon Mountains." The Zulu name for the range is* Kwathlamba, *which means "Piled-Up Rocks" or "Barrier of Pointed Spears."* ***What does the Zulu name for these mountains suggest about their physical geography?***

Because a cold current flows off the Atlantic coast of southern Africa, the evaporation rate there is low. The dry air influences the Namib (NAH-mib) Desert, which has some of the world's highest sand dunes. Some parts of the Namib get as little as 0.5 inches (13 mm) of rain per year. Plants get water from dew and fog rather than rain. Beetles, lizards, and snakes live in the Namib, but relatively few mammals can survive there.

Moving inland, rainfall gradually increases as the Namib gives way to the Kalahari Desert. Because it is not as dry as the Namib, plant life in the Kalahari ranges from grasses to palm trees. Many kinds of animals, including antelope and elephants, live there. The Okavango Swamps of northern Botswana are particularly rich in plant and animal life. A river empties into a maze of channels and shallow lakes within the area.

Along its eastern edge, the Kalahari merges into the grassland or **veld** (VELT) of South Africa. The grassy, high plains in central South Africa north and west of Lesotho make up what is called the highveld. The middleveld, a savanna region of short trees and bushes, is found at lower elevations. Still another of South Africa's regions is the lowveld, which includes dry tall-tree savannas as well as forests at the base of the Drakensberg.

Southern Africa has many valuable energy and mineral resources. For example, Angola has petroleum reserves. South Africa and Zimbabwe (zim-BAH-bway) have enormous coal deposits. Most of the region's electricity is generated by hydropower or by burning coal. Gold and platinum are among South Africa's many metal ore deposits. South Africa, Botswana, and Namibia have productive diamond mines. Zambia's rich copper deposits make it a major producer of that metal. In fact, miners have worked Zambia's iron and copper deposits for more than 1,500 years.

READING CHECK: *Physical Systems* How does the Drakensberg affect climate patterns in southern Africa?

INTERPRETING THE VISUAL RECORD *Chimanimani National Park in eastern Zimbabwe is located in a savanna region.* ***What can you see in this photo that shows this is a savanna biome?***

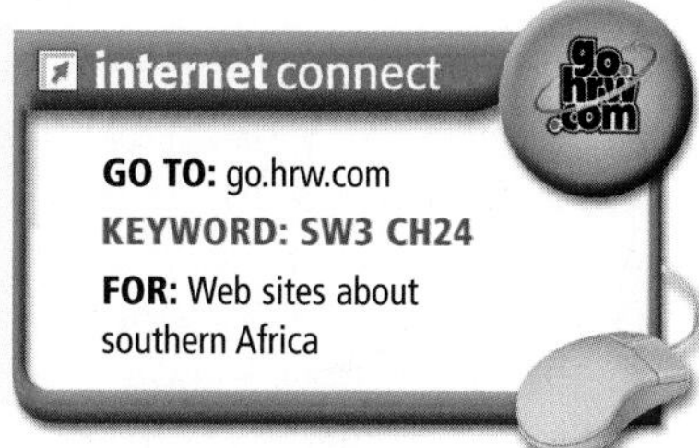

TEKS Working with Sketch Maps, Questions 1, 2, 3, 4, 5

Section 1 Review

Define

escarpment
biodiversity
veld

Working with Sketch Maps

On a map of southern Africa that you draw or that your teacher provides, label the countries of the region and the Drakensberg, Orange River, Limpopo River, Zambezi River, Namib Desert, Kalahari Desert, and Okavango Swamps. What famous feature makes the Zambezi River distinctive?

Reading for the Main Idea

1. *Places and Regions* Which landform region covers the largest area of southern Africa?
2. *Physical Systems* How does the cold ocean current off the Atlantic coast influence southern Africa's climates?

Critical Thinking

3. **Finding the Main Idea** What are the two deserts in this region, and how are their environments different?
4. **Drawing Inferences and Conclusions** Do you think southern Africa's rivers are vital links between the region's interior and global trade? Why or why not?

Organizing What You Know

5. Create a graphic organizer like the one below. Use it to describe the landforms, climates, and resources of southern Africa. Use the unit atlas and the information in Section 1.

Landforms	Climates	Resources

Physical Systems

Geography for Life

Biodiversity in Madagascar

You may have heard people call various places a "Garden of Eden." The term often refers to a place of great beauty and wonder. If wonder is the defining feature, then Madagascar might be something of a Garden of Eden.

This big island off the southeast coast of Africa is home to a wide range of unique plants and animals. Some 80 percent of the country's species live nowhere else on Earth. Because Madagascar has been an island for millions of years, these species have developed in isolation. In addition, Madagascar has a wide range of environments. The east coast has a tropical rain forest. Farther west are grasslands and scattered trees. A desert lies at the island's southern tip. Madagascar's isolation and its many environments contribute to its remarkable biodiversity.

Madagascar has more than 70 species of songbirds and parrots. There are about 800 butterfly species. More than 10,000 varieties of plants exist, from 1,000 orchid species to 6 species of baobab trees. Scientists often discover previously unknown plants on the island.

Ring-tailed lemur

Geographers can use Madagascar as a laboratory for studying the ways that species interact. For example, consider the small monkeylike mammals called lemurs. Lemurs live in trees and are active mainly at night. They are native to Madagascar and are also found on the nearby islands of Comoros. Lemurs of different species may be as small as mice or as big as medium-sized dogs. They eat mainly leaves and fruit. Therefore, lemurs depend on the forests. In an interesting twist, the forests also depend on the lemurs. Why is this so? When lemurs eat fruit from the trees, they also eat the seeds. Because the seeds do not digest, the lemurs spread them far from the parent tree in their droppings. This is the main way that many trees spread into new areas.

Today, however, Madagascar's natural environments are seriously threatened. Loss of habitat from deforestation is the biggest problem. Only some 10 percent of Madagascar's forests remain. Much of the damage comes from slash-and-burn farming. As the population grows, people clear more land for crops. Many conservation organizations now call Madagascar their top priority for preserving biodiversity. The country's government is working closely with these groups. Officials encourage citizens to take pride in the island's biodiversity. Ecotourism, better farming practices, and protected national parks may offer hope for Madagascar's environment.

Native Vegetation Zones of Madagascar

INTERPRETING THE MAP *Madagascar's government has passed laws to protect the country's native plants and animals.* ***How do you think the efforts of conservation groups can influence the government's environmental policies?*** TEKS

Applying What You Know

1. **Summarizing** Why does Madagascar have so many different plants and animals?
2. **Drawing Inferences and Conclusions** How might Madagascar preserve its biodiversity?

Section 2 History and Culture

READ TO DISCOVER

1. What are some important events in the history of southern Africa?
2. What are the region's cultures like?

WHY IT MATTERS

South Africa's past racial policies caused years of international isolation for the country's government. Use CNNfyi.com or other **current events** sources to learn about South Africa's international relations today.

IDENTIFY

Afrikaners
African National Congress

DEFINE

apartheid
sanctions

LOCATE

Great Zimbabwe
Cape of Good Hope
Luanda
Maputo
Cape Town
Durban
Kimberley
Johannesburg
Pretoria
Harare

History

The first inhabitants of southern Africa were hunter-gatherers and animal herders. Bantu-speaking peoples, the ancestors of most modern southern Africans, began migrating to the region around A.D. 100. Most early Bantu peoples were farmers who raised crops such as beans and sorghum. They also herded cattle, goats, and sheep. Unlike the people who already lived in the region, the Bantu knew how to make iron tools.

The Bantu peoples established several powerful kingdoms. Great Zimbabwe was the center of a wealthy farming and cattle-raising community. Farther south, the Sotho established a kingdom in the 1800s in what is now Lesotho. The realm of the Zulu lay east of the escarpment.

Portuguese sailors began exploring the southern African coast in the late 1400s. They reached the coast of what is now Angola in 1483. Soon Portuguese traders and merchants, eager to trade for spices in Asia, ventured farther. To get to Asia, they had to sail around the southern tip of Africa and then cross the Indian Ocean. The journey was long and difficult. Ships had to stop along the way for supplies. Therefore, the Portuguese set up small bases along the southern African coast. However, dense vegetation, the threat of disease, and the lack of navigable rivers discouraged them from moving very far inland.

READING CHECK: *Environment and Society* What factors discouraged the movement of early Portuguese settlers into the interior of southern Africa?

A San man drinks water from a hollowed ostrich egg. The San, who live in the Kalahari Desert in Botswana, Namibia, and South Africa, were historically nomadic hunter-gatherers. However, most have now settled in villages.

INTERPRETING THE VISUAL RECORD

Tulbagh, northeast of Cape Town in South Africa, was founded in 1699 by Dutch farmers. ***How does the architecture in this photo reflect European influence?*** TEKS

Boer soldiers pose with their weapons during the Boer War. During the war, well-armed Boer soldiers held off a much larger British army for nearly three years.

The Colonial Period In 1652 the Dutch set up a small settlement at the Cape of Good Hope. The Mediterranean climate at the Cape made it a good place to farm. Other Europeans, including French and Germans, joined the Dutch farmers. These European settlers were known as Boers (BOHRZ), which means "farmers" in Dutch. They thought of Africa as home and called themselves **Afrikaners** (a-fri-KAH-nuhrz). Over time, they developed a new language, called Afrikaans (a-fri-KAHNS). It combined elements of Dutch with African words. Speakers of German, French, and English also influenced the development of the language. Afrikaans even included words learned from Asians who had been brought to the Cape as slaves and laborers.

In the early 1800s, Great Britain took over the Cape area. In response, many Afrikaners moved inland. They wanted to be free from British rule and govern themselves. After fierce battles with Bantu-speaking peoples, the Afrikaners set up two independent republics in the interior.

During the late 1800s, the discovery of diamonds and gold drew people to southern Africa from all over the world. Both the Boers and the British wanted to control the region's mineral wealth. This led to the Boer War of 1899 to 1902. In the end, all of what is now South Africa came under British control. Britain then granted the Union of South Africa independence in 1910.

Other European countries claimed different parts of the region. The Portuguese kept control of Angola and Mozambique (moh-zahm-BEEK). The territory of South-West Africa (now known as Namibia) was a German colony. After Germany's defeat in World War I, South Africa took control of Namibia. Britain took Bechuanaland (now Botswana), Basutoland (Lesotho), Swaziland, and Northern and Southern Rhodesia (now Zambia and Zimbabwe).

READING CHECK: ***Human Systems*** Why did the Afrikaners move into the interior of southern Africa?

Independence Most of southern Africa remained under colonial rule until the late 1900s. During the 1960s many independence movements grew. Britain granted independence to most of its remaining colonies. The Portuguese colonies—Angola and Mozambique—did not win independence until the 1970s. Africans in Rhodesia had also been fighting to end white-minority rule. Finally, Southern Rhodesia won independence in 1980. The country changed its name to Zimbabwe. Northern Rhodesia became Zambia.

Conflicts continued long after independence in some areas. In Mozambique, a civil war between the Communist government and rebels lasted until the 1990s. In 1975 a Communist government took power in Angola. Rebels supported by the United States fought against the government, which was backed by the Soviet Union and Cuba.

Although the other countries in the region had cast off rule by white minorities, Afrikaners continued to control South Africa. Whites held political power, owned most of the land, and controlled the economy. However, black Africans made up most of the population. Since 1948, black South Africans had been denied political rights under a system of laws called **apartheid** (uh-PAHR-tayt). The term means "separateness." These laws forced black South Africans to live in separate areas and use separate facilities from whites. People of mixed race, called Coloureds, and Asian immigrants made up other South African ethnic groups. Their status was slightly higher than that of black South Africans.

Queen Nzinga is one of Angola's heroes. During the 1600s she led fighters into battle against Portuguese slave traders who were pushing farther into the interior.

Ending Apartheid Apartheid made South Africa an outcast in the world community. Eventually, some countries began to place economic **sanctions** on South Africa. Sanctions are penalties intended to force a country to change its policies. For example, a group of countries may refuse to provide economic aid to another country. Political organizations inside South Africa also pushed for an end to apartheid. The best known of these groups was the **African National Congress** (ANC), founded in 1912.

In 1990 the South African government finally began to change its policies. In that year it released the leader of the ANC, Nelson Mandela, from prison. The government began to get rid of the apartheid system. In 1994 South Africa held its first elections open to all citizens. Nelson Mandela was then elected South Africa's first black president.

FOCUS ON HISTORY

The Legacy of Apartheid With the election of Nelson Mandela, South Africa entered a new era. According to new laws, all citizens must be treated equally. However, the country still faces serious challenges. Providing economic opportunities, education, and health care to the entire population has proven difficult. Poverty remains a problem, and on average, white South Africans are still much wealthier than black South Africans. In addition, crime has increased in the cities. One reason is that young people who had worked against apartheid are now frustrated with the slow pace of change. Poverty combined with frustration has led to violence. Finally, divisions among black ethnic groups have caused new problems. The long-term stability of the country may depend on whether the government can improve people's standard of living.

READING CHECK: *Human Systems* What challenges has the end of apartheid left for South Africa?

Nelson Mandela was imprisoned by South Africa's government from 1962 to 1990 for his antiapartheid activities. After his release, Mandela assisted with South Africa's transition to a democratic system of government. He won the Nobel Peace Prize in 1993 and became the country's first black president in 1994.

Swazi men dance in a traditional harvest festival. Music and dance help bring a sense of unity to Swazi communities.

Culture

Southern Africa's cultural mix reflects its history. Over time, many peoples have settled in the region. Some aspects of their traditions have remained distinct. In other ways they have combined in a process of acculturation. Today African traditions are strongest in rural areas and small towns. In the cities, many people have adopted American and European customs.

People, Languages, and Religion Bantu languages of the Niger-Congo family are widely spoken in southern Africa. Two of these Bantu groups are Sotho and Nguni. Sotho speakers, including Tswana and Basuto peoples, live mainly in the interior. Nguni speakers, including the Zulu and Xhosa peoples, live closer to the coasts. The white population, concentrated in South Africa, speaks mainly English and Afrikaans. Smaller white populations live in Zimbabwe and Namibia and speak mostly English.

Because of the diversity of languages in each country, governments often rely on European languages. For example, English is commonly used in Zambia and Zimbabwe. Portuguese is the official language in Angola and Mozambique.

Before the arrival of Europeans, most southern Africans practiced traditional religions. These religions still have many followers. These followers may believe that ancestors and the spirits of the dead have divine powers. Diseases and misfortune may be explained as the work of spirits.

Europeans brought Christianity to southern Africa. Roman Catholicism is common in areas that the Portuguese colonized. The Dutch Reformed Church is the largest Christian denomination in South Africa. Today millions of Africans belong to Christian churches. Many others belong to churches that blend the teachings of Christianity with traditional African religious practices. Islam is also practiced. Some Muslims are descended from enslaved people who came from Southeast Asia in the 1600s and 1700s. South Asians brought Hinduism to southern Africa.

Language Families of Africa

	Afro-Asiatic	Niger-Congo	Khoisan	Nilo-Saharan	Malayo-Polynesian
English	**Arabic**	**Swahili**	**Nama**	**Kanuri**	**Malagasy**
mother	umm	mama mzazi	//gûs	yâ	reny
child	walad	mtoto	/gôaï	táda	zaza
head	rā`s	kichwa	tanas	kəlâ	loha
water	mā`	maji	/gami	njî	rano
tree	šajarah	mti	heis	kəská	hazo
house	bayt	nyumba	omi	fáto	trano
red	aḫmar	-ekundu	/awa	cimê	mena
eat	akala	la	‡û	búkin	homana
go/walk	ḏahaba/mašā	enda	!gû	lengîn	mamìndra

INTERPRETING THE CHART *Like the rest of Africa, southern Africa has an incredible diversity of languages grouped into several main language families. For example, San is a Khoisan language, while Zulu is a Bantu language of the Niger-Congo family.* ***How might language differences affect cultural diffusion in the region?***

Women buy and sell food in a marketplace in Moroni, the capital of Comoros. Some 33 percent of the country's population lives in urban areas.

Settlement and Land Use The wetter eastern part of the region has long been more densely populated than the drier western part. Some small rural villages have a traditional settlement pattern. In the center of the village is a pen, called a kraal (KRAWL), where cattle are kept at night. Around the pen, villagers build small houses with wooden poles, clay, and grass roofs.

Europeans founded most of the region's cities. The Portuguese established Luanda in Angola and Maputo in Mozambique. These cities were important seaports and administrative centers. The Dutch established Cape Town in what is now South Africa. Another South African city, Durban, was a major British port. As Europeans moved into the interior, they also set up mining towns and administrative centers. For example, South Africa's Kimberley and Johannesburg started as mining camps. Pretoria, South Africa, and Harare, Zimbabwe, began as government centers. Today many southern African cities are large. However, most southern Africans still live in small villages.

TEKS **READING CHECK:** *Human Systems* How did colonization influence the development of religion in the region?

Connecting to ANTHROPOLOGY

Marriage Customs

Anthropologists often study how marriages are arranged. Some societies in South Africa allow polygamy. Polygamy is a practice in which a person may have more than one spouse. However, under the old apartheid government, additional wives and their children had no official status as their husband's legal family. As a result, they also had no legal protection.

Under the black-majority government, laws have changed, and women benefit. South Africa's 1998 constitution recognizes traditional marriages as legal. Now, if a man wants to divorce any of his wives he must take the case to court. Before the laws changed, the husbands' families could end a marriage, which put women and children at risk financially.

Summarizing How have cultural beliefs influenced public policy in South Africa? TEKS

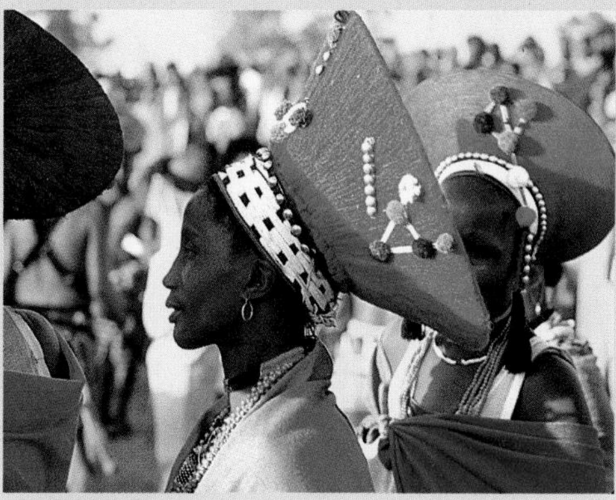

A Zulu wedding in South Africa

TEKS Questions 1, 2, 3, 4, 5

Section 2 Review

Homework Practice Online
Keyword: SW3 HP24

Identify Afrikaners, African National Congress

Define apartheid, sanctions

Working with Sketch Maps On the map you created in Section 1, label Great Zimbabwe, Cape of Good Hope, Luanda, Maputo, Cape Town, Durban, Kimberley, Johannesburg, Pretoria, and Harare. Which of these places provided the name for a modern country?

Reading for the Main Idea

1. *Places and Regions* Which three European countries had colonies in southern Africa?
2. *Places and Regions* How were the cities of southern Africa established?

Critical Thinking

3. **Drawing Inferences and Conclusions** Why are European languages still used in several countries of southern Africa?
4. **Identifying Points of View** Why do you think nonwhite South Africans were so unwilling to accept the system of apartheid? Why do you think many whites wanted the system to continue? How did public policies change in the 1990s?

Organizing What You Know

5. Create an idea web in which you describe, compare, and contrast the languages, religions, land-use practices, and customs of southern Africa.

3 The Region Today

READ TO DISCOVER

1. What are the main economic activities in southern Africa?
2. What are the region's cities like?
3. What challenges face the people of southern Africa?

DEFINE

informal sector

LOCATE

Cabinda
Soweto

WHY IT MATTERS

Many musicians in southern Africa combine traditional African musical styles with American jazz and rock 'n' roll. Use CNNfyi.com or other **current events** sources to learn about contemporary southern African music.

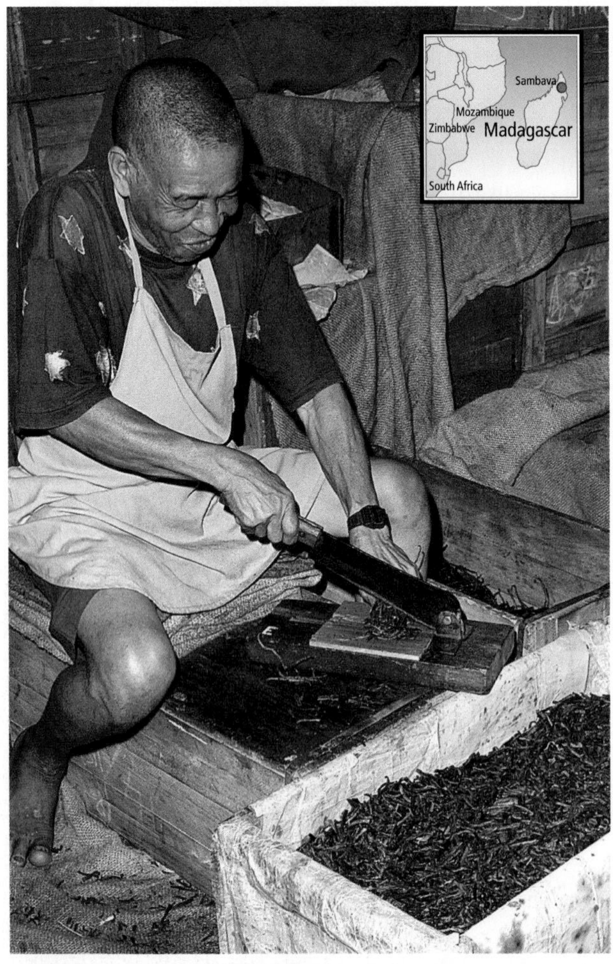

Vanilla is a major export of Madagascar and Comoros. Here, a worker in Madagascar prepares vanilla beans for processing. First the beans are dried and crushed. Then alcohol is used to draw out the vanilla flavor.

Southern African Economies

All the countries in southern Africa are classified as developing countries. However, South Africa is sometimes considered a middle-income country. It has the most developed economy in Africa. South Africa's market economy includes agriculture, manufacturing, and mining industries. The country's economy is much larger than all of the other economies in the region combined. In contrast, Mozambique is one of the poorest countries in the world. Its mainly traditional economy relies on farming.

Agriculture Farming, whether market-oriented or subsistence, is the most common economic activity in southern Africa. Most farmers practice subsistence agriculture. They depend on their crops and livestock for their survival. Farmers may also sell some of their produce in local markets. In some parts of the region, market-oriented agriculture is important. These farms are generally large and rely on modern machinery. Commercial farmers either sell their products in cities or export them. In Zimbabwe, commercial farms produce the country's most important export crop, tobacco. Angola grows coffee, while Madagascar exports vanilla. South Africans grow corn, fruits, and wheat on modern, mechanized farms.

READING CHECK: ***Human Systems*** How does market-oriented agriculture differ from traditional subsistence agriculture in southern Africa?

Business and Industry Minerals and oil are increasingly important to the economies of southern Africa. The largest mineral exporter in the region is South Africa. It produces more gold than any other country. Some mines are as deep as 13,000 feet (3,962 m). The country also exports many other metals. Angola pumps oil from deposits located offshore and in the small exclave of Cabinda. An exclave is a part of a country that is separated by the territory of another country. Cabinda lies on the Atlantic coast north of the Congo River. It is separated from the rest of Angola by the Democratic Republic of the Congo.

Botswana once had a small market economy that relied heavily on beef exports. Then diamonds were discovered in the late 1960s. Today Botswana is one of the world's largest producers of diamonds. The country's economy is also one of the fastest growing in Africa.

Being dependent on exports of a few primary products such as minerals can be risky. For example, copper is Zambia's most important export mineral. However, the price of copper on the world market goes up and down. When copper prices fall, Zambia's entire economy is hurt. On the other hand, South Africa has a more diversified economy. It exports a variety of agricultural, industrial, and mineral products. As a result, the country is less affected by world price changes of certain products.

Southern Africa's Major Imports (in selected countries)

Country	Imports
Angola	machinery and electrical equipment, vehicles and spare parts, medicines, food, textiles, military goods
Comoros	rice and other foodstuffs, consumer goods, petroleum products, cement, transport equipment
Mozambique	machinery and equipment, mineral products, chemicals, metals, foodstuffs, textiles
Namibia	foodstuffs, petroleum products and fuel, machinery and equipment, chemicals
South Africa	machinery, foodstuffs and equipment, chemicals, petroleum products, scientific instruments
Swaziland	motor vehicles, machinery, transport equipment, foodstuffs, petroleum products, chemicals
Zambia	machinery, transportation equipment, fuels, petroleum products, electricity, fertilizer, foodstuffs, clothing

Source: Central Intelligence Agency, *The World Factbook 2001*

INTERPRETING THE CHART *Southern Africa is dependent on imports to meet many of its basic needs.* ***What does this trade pattern suggest about the economic development of these countries?*** TEKS

Visitors to southern African cities often notice the many small businesses operating on street corners and empty lots. Some even operate in bus stations. Women sell fruit and cooked food from roadside stands. Small children sell souvenirs to tourists. Men use portable equipment to fix cars along the roadside. These businesses are part of the **informal sector** of the region's economy. This sector is made up of people who do not work for formal businesses. These people may not have set hours, employment benefits, or even contracts. In addition, their income is not taxed. The informal sector includes self-employed people and small family-owned businesses. People in such businesses usually work long hours, and their incomes are generally low. This is the only way that many poor people, particularly those in cities, can make a living.

Parts of southern Africa have become popular tourist destinations. On the mainland, wildlife attracts thousands of visitors each year. Tourists travel to the wild-game parks of Botswana, Namibia, South Africa, and Zimbabwe to see African animals in their natural habitats. The tropical island countries of Comoros, Mauritius, and Seychelles attract visitors to their beaches and coral reefs. Partly as a result, the average income of the people of Mauritius is now among the highest in the region.

TEKS **READING CHECK:** ***Human Systems*** In what ways do small businesses characterize the economies of the region?

The clear waters of the Seychelles attract divers from around the world. Tourism accounts for more than half of the country's GDP.

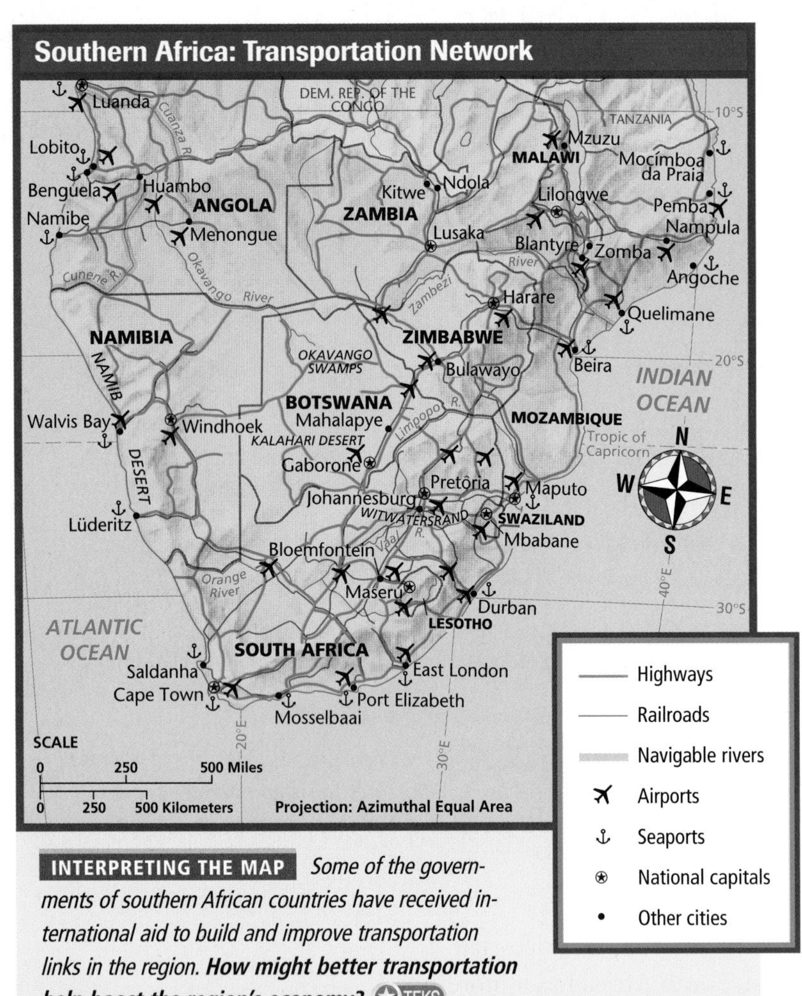

INTERPRETING THE MAP *Some of the governments of southern African countries have received international aid to build and improve transportation links in the region.* ***How might better transportation help boost the region's economy?*** TEKS

Urban Environments

Southern African cities are places of great contrasts. In the suburbs, wealthy businesspeople, foreigners, and government officials may live in large, comfortable houses. Downtown areas have tall buildings and well-stocked stores. In the poorer areas, many people are crowded into small homes. Others live in shanties—rough houses made of scrap wood, metal sheeting, or mud bricks. Although these shantytowns may look chaotic to outsiders, many are well-organized communities. In some shantytowns, elected councils run schools. They also set rules and present residents' concerns to the government.

Greater Johannesburg is the largest urban area in southern Africa. It is home to some 4 million people and is in the center of South Africa's industrial heartland.

During the apartheid era, South Africa created residential areas for nonwhite workers. These settlements, called townships, kept the nonwhites separate from the whites. Soweto, which lies a few miles from Johannesburg, is the largest of these settlements. Soweto's name comes from the first letters of the name *South Western Townships.* It covers some 40 square miles (100 sq km). Soweto has a few grand homes. However, most people live in small houses, shacks, and apartments built as dormitories for migrant workers. Soweto was at the forefront of the struggle against apartheid in South Africa. Many of its residents struggled against the government for the equality of all South Africans.

READING CHECK: ***Human Systems*** How did access to nearby Johannesburg contribute to the creation and growth of Soweto?

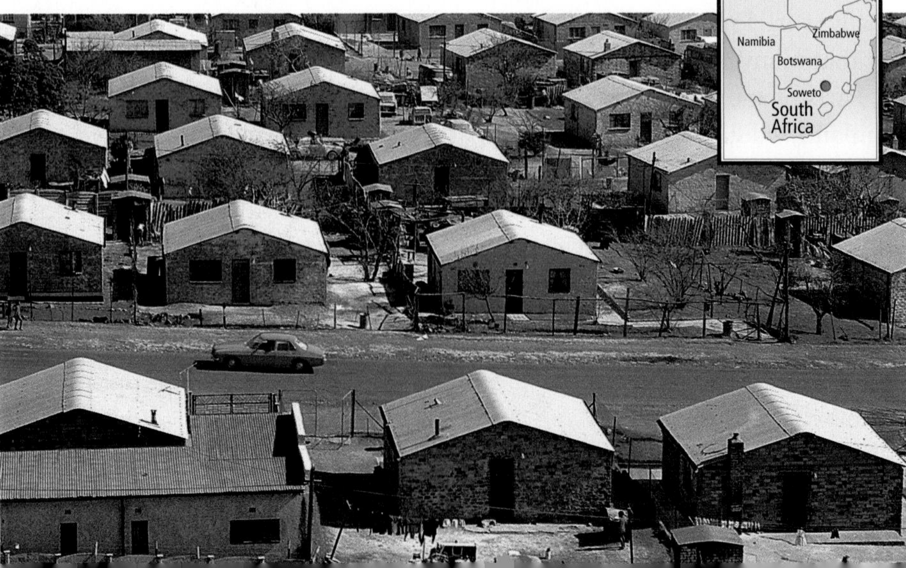

The township of Soweto is home to people from South Africa's many different indigenous groups.

Challenges

Poverty is the most serious problem facing southern Africa. Many people cannot afford to eat a balanced diet. As a result, they are more likely to get sick. Young children are particularly vulnerable. Moreover, many of southern Africa's poor are unemployed.

The region's cities are growing rapidly, mainly because people migrate from rural areas looking for work. The region's high birthrates also contribute to the rapid growth of cities and lead to a lack of suitable housing. In addition to housing shortages, the crowded cities face serious environmental problems. Smog from cars' exhaust fumes and smoke from coal and wood fires add to urban pollution problems.

Other environmental threats include the droughts and floods that often strike the region. In 1999, for example, major floods in Mozambique left some 200,000 people stranded. Even worse flooding in 2000 displaced more than 1 million people and devastated the country's economy. In addition, Madagascar's rain forests are being cut down, and soil erosion is increasing. (See Geography for Life: Biodiversity in Madagascar.)

Disease is also a major problem in southern Africa. In some countries, more than a quarter of the population is infected with HIV, the virus that causes AIDS. For example, in Botswana some 36 percent of adults between the ages of 15 and 49 are infected with HIV. Because of HIV/AIDS, the average life expectancy of southern Africans is falling. Southern African governments are trying to educate their people about the disease in the hope of slowing its spread. (See Case Study: The Geography of Disease.) The region's governments have also called on Western drug companies to make expensive medicines more affordable for southern Africans.

INTERPRETING THE VISUAL RECORD *The populations of the region's cities are growing rapidly. For example, Antananarivo, Madagascar, grew from a population of about 800,000 in 1990 to 1.5 million in 2000.* ***How might increased crowding compound the region's existing problems?***

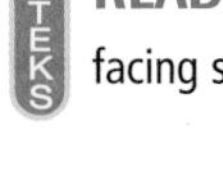

READING CHECK: *Human Systems* What are the main challenges facing southern Africa? How are they interrelated?

TEKS Questions 1, 2, 3, 4, 5

Review

Define informal sector

Working with Sketch Maps On the map you created in Section 2, label Cabinda and Soweto. Why were townships like Soweto created?

Reading for the Main Idea

1. *Places and Regions* What makes South Africa the most economically developed country in the region?
2. *Human Systems* What are some of the main cash crops grown on commercial farms in southern Africa?

Critical Thinking

3. **Finding the Main Idea** Why might governments in the region want to pursue policies that encourage the development of a wide range of resources?
4. **Analyzing Information** How might the countries of southern Africa protect and develop their tourist industries?

Organizing What You Know

5. Copy the following graphic organizer. Use it to identify the causes of some challenges facing countries in southern Africa.

Challenge	Cause
Rapid urban growth	
Poor nutrition	
Pollution	
Environmental damage	

CASE STUDY

The Geography of Disease

Environment and Society One constant throughout human history has been the threat of disease. Many societies have suffered from epidemics—the outbreak and rapid spread of a disease. Medical geographers and other health experts track the diffusion of such epidemics.

A number of factors influence the diffusion of disease. One factor is the interaction of humans with their environment. For example, consider the creation of Lake Volta in the early 1960s. The West African country of Ghana created the lake by building a huge dam across the Volta River. This dam produces hydroelectricity, and the lake provides irrigation water for crops. However, many people who settled along the lakeshore became infected with a deadly parasite. Experts tracked the source of the parasite to snails that lived along the shore.

Location is another important factor in the spread of disease. Certain climates, for example, may encourage the growth of infectious microbes and the insects that carry them. The unsanitary living conditions often found in developing countries also make it easier for diseases to spread. Worse still, poor countries often do not have the health-care resources to fight epidemics. Another factor in the spread of disease is migration. War refugees and other migrants can carry infections with them.

Medical geographers see some of these factors at work in Africa today. Two of the most difficult health threats there are malaria and Acquired Immune Deficiency Syndrome, or AIDS.

Malaria

Malaria is not a new disease. Researchers have uncovered records of ancient infections as far back as the 400s B.C. In fact, some researchers believe a malaria epidemic may have played a role in the fall of the Roman Empire. As early as the late 1400s, Europeans may have unknowingly carried the disease to the Americas.

Location and climate play important roles in the spread of malaria. Much of Africa lies in the tropics. Mosquitoes and other insects thrive in the humid tropical and subtropical climates there. One kind of mosquito that lives there carries the parasite that causes malaria. These mosquitoes are also found in

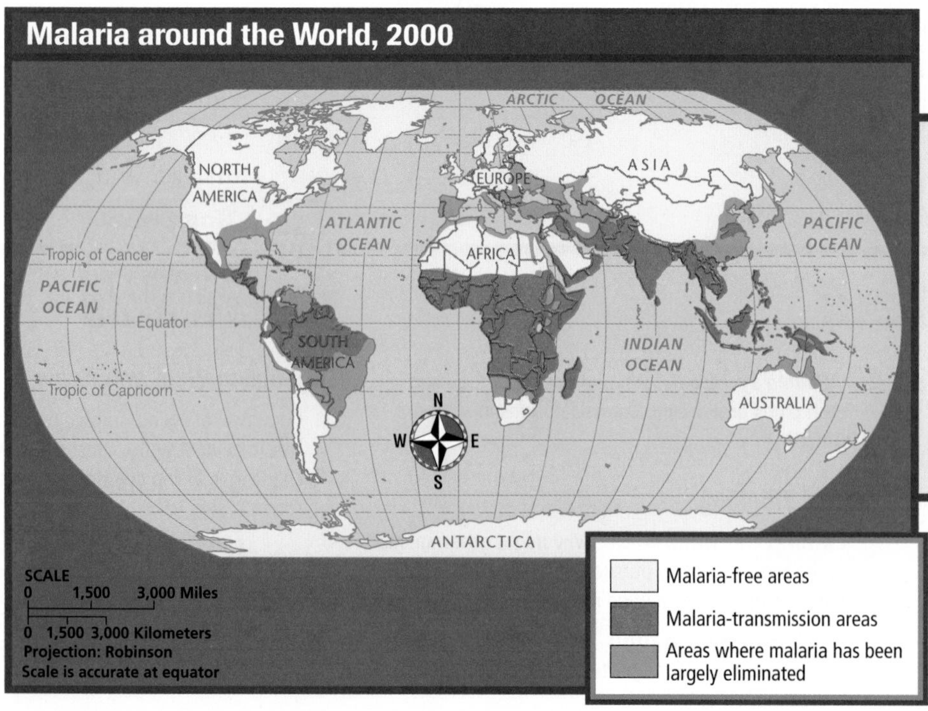

INTERPRETING THE MAP

More cases of malaria are reported each year in Africa than anywhere else in the world. Yet, as this map shows, malaria is virtually unknown in the northern and far southern parts of the continent. ***What does this map suggest about the habitat of the mosquito that spreads malaria?*** TEKS

other tropical and subtropical regions around the world. (See the map.)

Experts believe there are 300 to 500 million new malaria cases worldwide each year, mostly in Africa. The annual death toll from the disease in Africa alone may be about 1 million. In fact, as many as 30 percent of all hospital admissions in Africa may be for malaria. Patients who survive can develop immunity to the disease. In the meantime, however, workers lose time at their jobs, and children miss school, often for a week or more. The death of a working family member can cause even more financial problems.

The economic costs of the disease are great. Some experts point to the benefits that might have resulted if malaria had been eliminated 35 years ago. They say that the total GDP of Africa south of the Sahara might be more than 30 percent greater than it is today.

AIDS

The movement of people has been an important factor in the diffusion of HIV, the virus that causes AIDS. HIV may infect more than 36 million people around the world today. The vast majority of those infections have occurred in Africa. (See the graph.)

Researchers still debate the origin of AIDS. However, among the first identified cases of the disease were those around the lake district of the Great Rift Valley in eastern Africa. Many researchers believe HIV probably first spread from there along trade and transportation routes. Over time, modern means of transportation, particularly air travel, helped spread HIV around the world.

AIDS has hit eastern and southern Africa particularly hard. In South Africa, for example, one in five adults is infected with HIV. In Botswana the figure is more than one in three. As a result, life expectancy has fallen dramatically. As with malaria, the economic costs of AIDS are staggering. These costs could multiply with the rapid spread of the disease in other regions, particularly Asia.

Finding Solutions

The costs of treating malaria, AIDS, and other deadly diseases are often more than developing countries can pay. In addition, some diseases, such as AIDS, have no cure. Sometimes drugs may extend life, but they are very expensive. Some experts say that richer countries should play a bigger role in providing resources for the battle against these diseases. Education and prevention programs also help.

One success story is the battle against smallpox. As recently as 1967, smallpox was a dreaded disease that caused as many as 2 million deaths each year. However, world health officials started a massive prevention program. This program included widespread vaccination and other efforts. The last natural case of smallpox occurred in Somalia in 1977. In coming years medical geographers and other experts hope to add other diseases, including malaria and AIDS, to the list of success stories.

HIV/AIDS around the World

Percentage of world HIV/AIDS cases

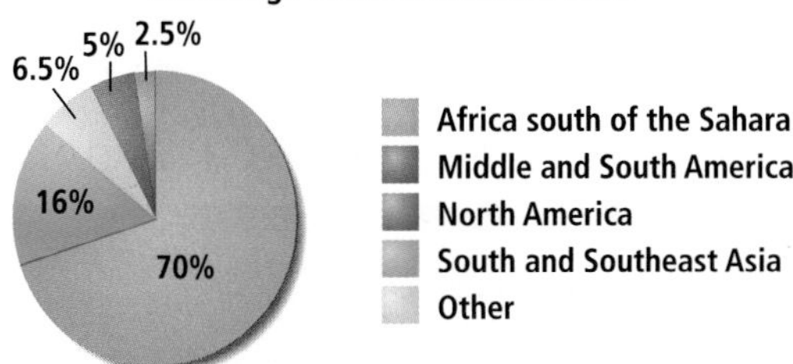

Source: UNAIDS/World Health Organization, 2000

Applying What You Know

TEKS

1. **Problem Solving** What are some factors that contribute to the diffusion of disease today? How do you think international efforts might help slow or stop the spread of diseases in developing regions such as Africa?

2. **Analyzing** How have AIDS and malaria affected the economic development of Africa south of the Sahara? What demographic effects has AIDS had on southern Africa?

CHAPTER 24 Review

Building Vocabulary TEKS

On a separate sheet of paper, explain the following terms by using them correctly in sentences.

escarpment	apartheid
biodiversity	sanctions
veld	African National Congress
Afrikaners	informal sector

Locating Key Places TEKS

On a separate sheet of paper, match the letters on the map with their correct labels.

Drakensberg	Namib Desert	Johannesburg
Orange River	Cape of Good Hope	Harare
Zambezi River	Luanda	Cabinda

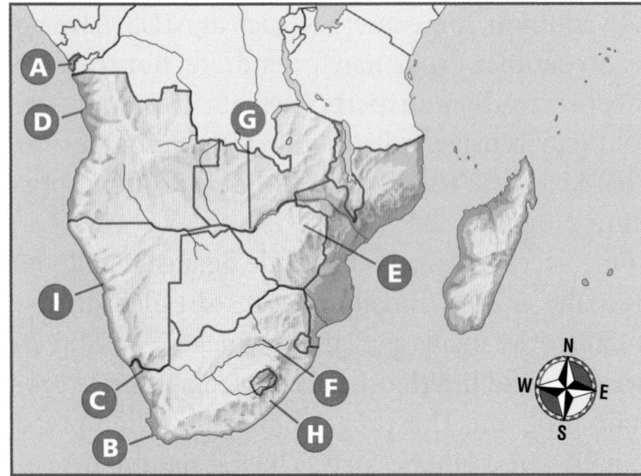

Understanding the Main Ideas 

Section 1

1. *Physical Systems* What are two reasons the western part of southern Africa is drier than eastern parts?

Section 2

2. *Places and Regions* Which Europeans settled in what is now South Africa? How did these peoples interact with each other?
3. *Human Systems* What was the apartheid system? When did it end?

Section 3

4. *Environment and Society* What are some features that attract tourists to southern Africa?
5. *Human Systems* Why has life expectancy in southern Africa dropped in recent years? How do you think such changes might affect future population growth in the region?

Thinking Critically for TAKS TEKS

1. **Analyzing** Why do you think European countries were so intent on colonizing all of southern Africa?
2. **Comparing** How do you think life in large cities of southern Africa compares to life in large cities in other developing countries? What political, economic, social, and environmental factors might account for similarities and differences?
3. **Comparing** In what world regions and kinds of societies do you think informal economic activities provide the most opportunities for women and religious minorities?

Using the Geographer's Tools TEKS

1. **Analyzing Maps** Examine the map of transportation routes in Section 3. Which country seems to have the largest area not served by major transportation links? How do the region's landlocked countries seem to get their goods to seaports?

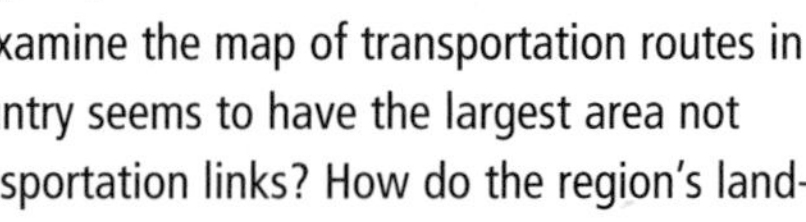

2. **Creating Maps** Use the unit atlas to create a map of Africa. Label each country. Then use a different color to shade the regions of North Africa, West and Central Africa, East Africa, and southern Africa.
3. **Creating Maps** Use the climate and land use maps to create a series of maps on tissue paper or other transparent material. One map may show arid and semiarid climate regions, while the other shows wetter climate regions. Then create maps for each type of land use common in the region. Using your maps as if they were layers in a geographic information system, make generalizations about the location of climate regions and particular economic activities.

Writing for TAKS TEKS

Imagine that you have moved from a small subsistence farm to a big city to find work. Will you work in industry or the informal sector? If you find work, how will your standard of living change? Write a letter to your family in which you compare your old way of life with life in the city.

SKILL BUILDING

Geography for Life TEKS

Summarizing Geographic Information

Human Systems Create an idea web in which you analyze South Africa's political, economic, social, and cultural characteristics. Note historical factors that are important in your analysis.

Building Social Studies Skills

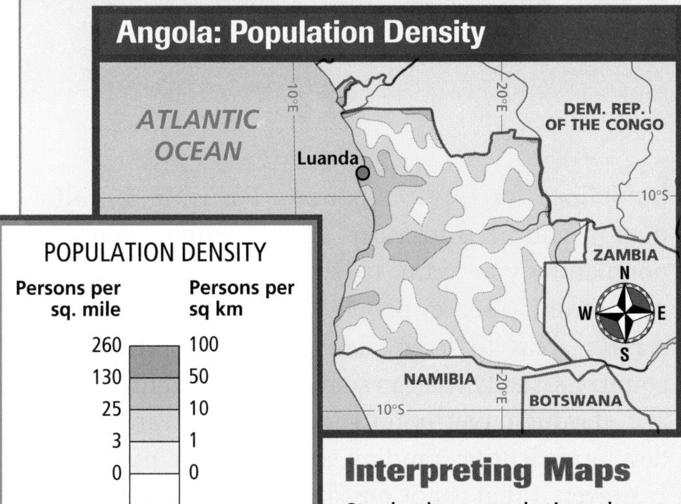

Interpreting Maps

Study the population density map above. Then answer the questions that follow.

1. Based on the map, which of the following statements is accurate?
   a. Luanda is located in one of Angola's least densely populated regions.
   b. Luanda is located in an area with between 25 and 130 people per square mile.
   c. Land along the coast is uninhabited.
   d. Angola has no cities with more than 1 million people.

2. Which areas of Angola are most densely populated?

Using Language

The following passage contains mistakes in grammar, punctuation, and usage. Read the passage and then answer the following questions. Mark your answers on a separate sheet of paper.

(1) Botswana is a large, landlocked and semiarid country. (2) Thanks to mineral resources and stable political conditions. (3) Botswana is one of Africa's success stories. (4) Cattle ranching and mining of copper and diamonds is the principal economic activities.

3. Which sentence contains an error in subject-verb agreement?
   a. 1
   b. 2
   c. 3
   d. 4

4. Combine word groups 2 and 3 to form one correct sentence.

Alternative Assessment

PORTFOLIO ACTIVITY

Learning about Your Local Geography TEKS

Individual Project: Research

You have read how mineral resources contribute to southern African economies. What mineral resources are mined in or near your area? Conduct research to learn where each resource is found, how it is processed, and how it is used. Create maps or graphic organizers to organize and present your findings.

internet connect

Internet Activity: go.hrw.com
KEYWORD: SW3 GT24

Choose a topic on southern Africa to:
- learn about the Namib Desert.
- investigate the history and legacy of apartheid.
- create a brochure of the animal and plant life on a South African safari.

Skill Building for TAKS

WORKSHOP

Using Questionnaires and Field Interviews

A common instrument for gathering research data is the questionnaire. Questionnaires are documents that ask individuals to provide certain information about themselves. They have a variety of uses. For example, they can provide feedback from customers, identify special needs people in an area might have, or collect statistical information for a census.

Questionnaires can be sent thousands of miles away to people the researcher may never even see. However, researchers can also conduct field interviews, in which they talk to people face-to-face. A field researcher might want to know, for example, how people in an area use local natural resources and how government policies have affected their use of those resources. Questionnaires would not let the researcher probe for additional details he or she might need to know.

Developing the Skill Questionnaires and field interviews can be used to answer geographic questions and infer geographic relationships. For example, the U.S. Census Bureau uses questionnaires to track population growth in the United States. Answers from the questionnaires also help researchers learn how population is distributed across the country. They also reveal age, race, income, and other characteristics of a population. On this page you can see part of a questionnaire that the Census Bureau used in 2000.

Questionnaires can ask many types of questions. Some census questions ask respondents to write answers, such as names. Other questions ask respondents to check boxes next to the answers given on the form. Still other questions might match numbers or letters to answers. For example, if there are four possible answers, the answers may be coded 1, 2, 3, and 4. Answers coded in this way can help researchers create databases from the information they gather.

Guidelines for Questionnaires Focus on the information you need to learn. Then carefully choose the population you want to survey. For example, you might want to survey people in certain age groups or in certain areas to learn about living conditions in a country. Usually you can survey only a certain percentage of your target population. Often researchers will develop scientific random samples of a population. However, if that is not possible, try to question a sample that is representative of the population. You can use the following checklist to create a questionnaire:

- ✓ Decide what information you want to gather. Make sure your questions are designed to help you reach a specific objective.

2000 U.S. Census Questionnaire (excerpt)

4. What is this person's age and what is this person's date of birth? *Print numbers in boxes.*

Age on April 1, 2000 | Month | Day | Year of Birth

→ **NOTE: Please answer BOTH Questions 5 and 6.**

5. Is this person Spanish/Hispanic/Latino? *Mark ☒ the **"No"** box if **not** Spanish/Hispanic/Latino.*

- ☐ **No,** not Spanish/Hispanic/Latino
- ☐ Yes, Puerto Rican
- ☐ Yes, Mexican, Mexican Am., Chicano
- ☐ Yes, Cuban
- ☐ Yes, other Spanish/Hispanic/Latino—*Print group.*

6. What is this person's race? ***Mark ☒ one or more races** to indicate what this person considers himself/herself to be.*

- ☐ White
- ☐ Black, African Am., or Negro
- ☐ American Indian or Alaska Native—*Print name of enrolled or principal tribe.*
- ☐ Asian Indian
- ☐ Japanese
- ☐ Native Hawaiian
- ☐ Chinese
- ☐ Korean
- ☐ Guamanian or Chamorro
- ☐ Filipino
- ☐ Vietnamese
- ☐ Samoan
- ☐ Other Asian—*Print race.*
- ☐ Other Pacific Islander—*Print race.*

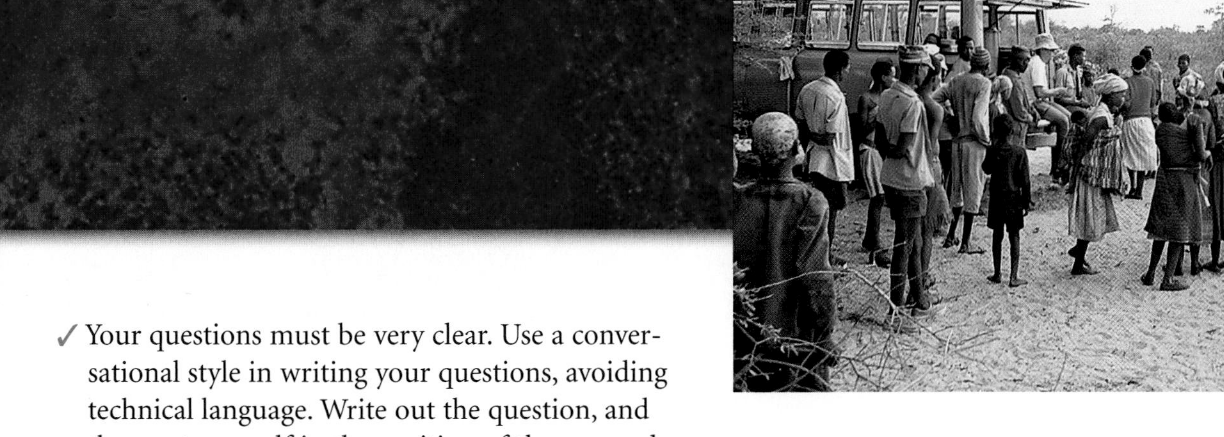

- ✓ Your questions must be very clear. Use a conversational style in writing your questions, avoiding technical language. Write out the question, and then put yourself in the position of the respondent. Is the question clear? Would you know how to respond to it?
- ✓ Ask only those questions that will provide the information you need.
- ✓ Structure the survey so that the questions follow a logical order.
- ✓ Whenever possible, use questions that require respondents to choose a specific answer. Examples include multiple-choice questions, yes-or-no-questions, and questions that ask respondents to rank items in order. Respondents will better understand the purpose of your questions, and the time it takes to complete the questionnaire will be shortened.
- ✓ Avoid bias in writing questions. Make sure respondents of different backgrounds can answer your questions. Also, do not ask leading questions. That is, the questions should not encourage people to answer in a certain way.
- ✓ Test the questionnaire on a small group of people. Ask each person if he or she easily understands and can answer the questions.
- ✓ In order to organize and analyze the responses, you may need to code the questions. Assign a value—a number or letter—to each question's response. Then use the codes to create a database of questions and answers.

Guidelines for Field Interviews Before you begin conducting field interviews, you need to do some background research. Study your topic and the people or place you are researching. Then follow these guidelines to conduct a professional field interview:

- ✓ Plan a general outline for the interview and script some questions. Break the general topic into more specific parts. Then design focused questions that are not overly broad.
- ✓ Use a variety of questions, but avoid yes-or-no questions and questions that can be answered with a single word. Unlike with a questionnaire, you want people to explain their answers.
- ✓ Set up the interview in advance and send the list of questions you will ask. Doing so will help people better prepare for the interview and provide more-detailed answers.
- ✓ Ask for permission to record the interview.
- ✓ Keep questions brief and clear.
- ✓ Follow your plan and have a copy of the questions for your respondents in case they do not have their own copy.
- ✓ Be a good listener and try to determine the meaning of the responses. Ask follow-up questions that draw out additional details about previous answers.
- ✓ After the interview, submit a copy of the interview to the respondents. Ask respondents to check that their answers have been properly recorded. As a courtesy, you might send your final report to your respondents for their review.

Practicing the Skill

Work with a group of three or four partners to research either the ways people in your community try to conserve natural resources at home or attitudes toward local traffic issues and transportation networks. Use a questionnaire and field interviews to gather information about your topic. Create questions with other group members, then assign certain members to manage the distribution of questionnaires and conduct field interviews. Other members should organize the questions and answers into databases. Finally, analyze the information you gather and develop a report for your class.

UNIT 8 South Asia

Taj Mahal, Agra, India

CONNECTING TO Literature

"Hundred Questions" *from the* Mahabharata

translated by R. K. Narayan

The "Hundred Questions" comes from Book 2 of the *Mahabharata,* an important Hindu religious text. In its present form, the *Mahabharata* is thought to have been composed sometime between the 300s B.C. and the A.D. 300s. Central to the *Mahabharata* is the Hindu concept of dharma or sacred duty. Performing these duties contributes to the universal order. In this excerpt a yaksha (forest divinity or nature spirit) challenges Yudhistira to answer questions. Yudhistira's brothers failed to heed the yaksha's challenge and died.

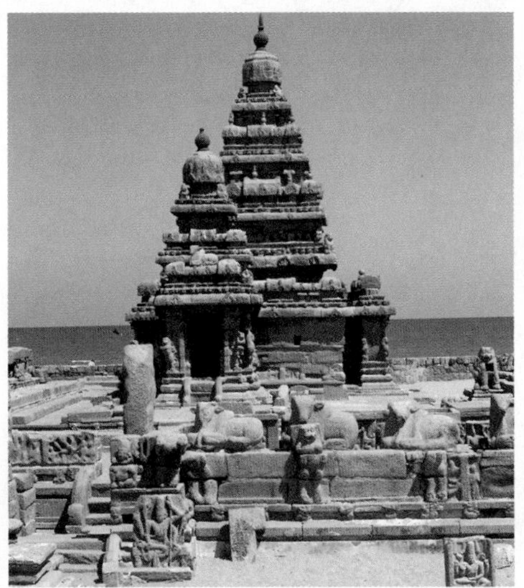

Yudhistira said humbly, "What god are you to have vanquished these invincible brothers of mine, gifted and endowed with inordinate strength and courage?" . . .

At this request he saw an immense figure materializing beside the lake, towering over the surroundings. "I am a yaksha. . . . If you wish to live, don't drink this water before you answer my questions." . . .

To . . . questions on renunciation, Yudhistira gave the answers: "Pride, if renounced, makes one agreeable; anger, if renounced, brings no regret; desire, if renounced, will make one rich; avarice, if renounced, brings one happiness. True tranquility is of the heart. . . . Mercy may be defined as wishing happiness to all creatures. . . . Ignorance is not knowing one's duties. . . . Wickedness consists in speaking ill of others." . . .

There were a hundred or more questions in all. Finally, the yaksha said, "Answer four more questions, and you may find your brothers—at least one of them—revived. . . . Who is really happy?"

"One who has scanty means but is free from debt; he is truly a happy man."

"What is the greatest wonder?"

"Day after day and hour after hour, people die, and corpses are carried along, yet the onlookers never realize that they are also to die one day, but think they will live for ever. This is the greatest wonder of the world."

"What is the Path?"

"The Path is what the great ones have trod. When one looks for it, one will not find it by study of scriptures or arguments, which are contradictory and conflicting." . . .

The yaksha said, "You have indeed pleased me with your humility and the judiciousness of your answers. Now let all your brothers rise up and join you."

Analyzing the Primary Source 

1. **Summarizing** What does the yaksha tell Yudhistira he must do?
2. **Drawing Inferences** What do you think is meant by the Path? What features of other religions with which you are familiar might be similar to the Path?

The World in Spatial Terms

South Asia: Political

1. *Places and Regions* Which country is by far the largest in South Asia?
2. *Places and Regions* Compare this map to the physical map. Which country is almost surrounded by India and dominated by the Ganges Delta? TEKS

Critical Thinking

3. **Analyzing Information** Compare this map to the physical map. Which part of Pakistan would you expect to have the highest population density? TEKS

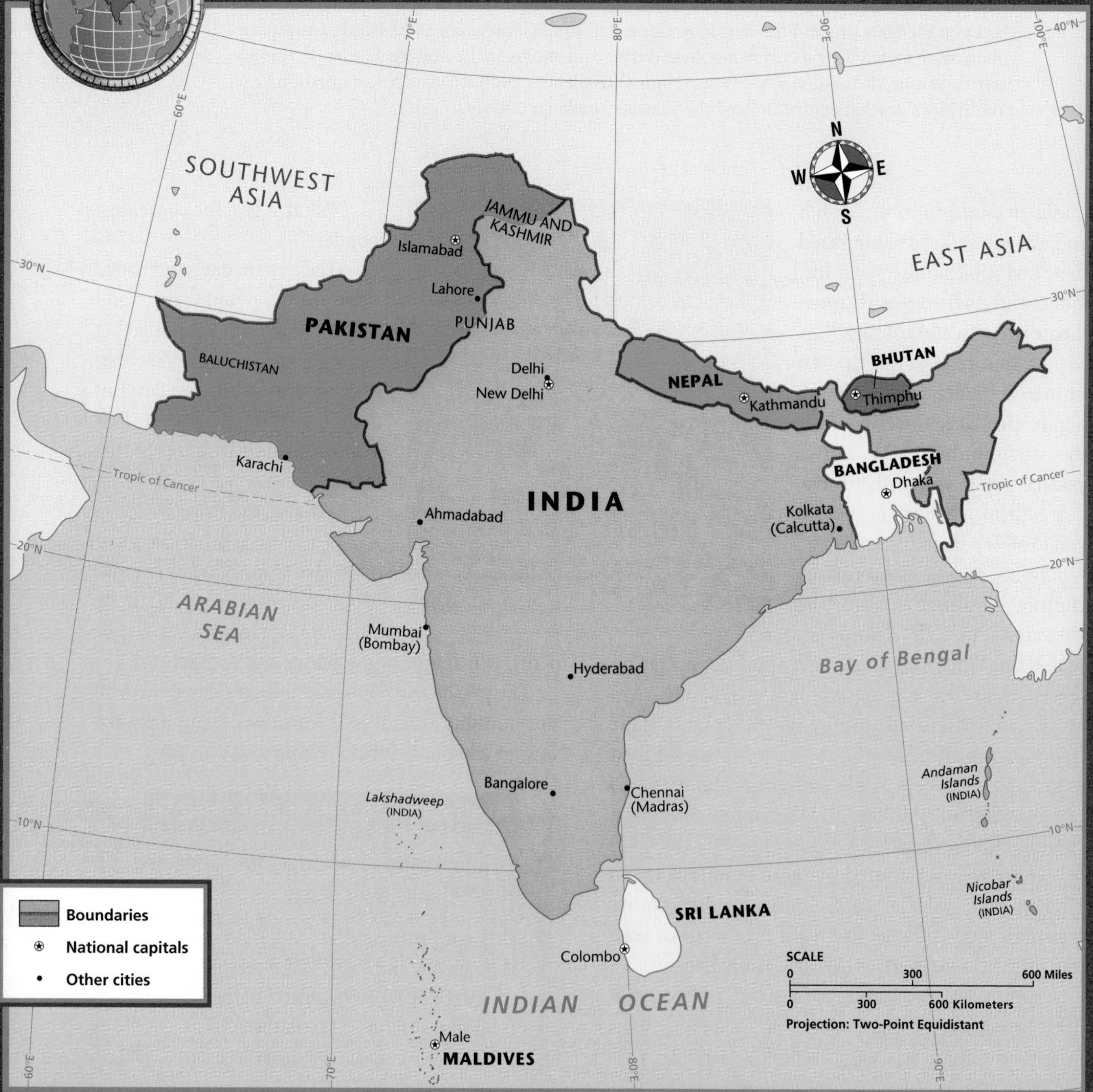

SOUTHWEST ASIA
EAST ASIA
N
S
E
W
JAMMU AND KASHMIR
Islamabad
Lahore
PUNJAB
PAKISTAN
BALUCHISTAN
Karachi
Delhi
New Delhi
NEPAL
Kathmandu
BHUTAN
Thimphu
BANGLADESH
Dhaka
INDIA
Ahmadabad
Kolkata (Calcutta)
Tropic of Cancer
ARABIAN SEA
Mumbai (Bombay)
Hyderabad
Bay of Bengal
Bangalore
Chennai (Madras)
Lakshadweep (INDIA)
Andaman Islands (INDIA)
Nicobar Islands (INDIA)
SRI LANKA
Colombo
INDIAN OCEAN
Male
MALDIVES
40°N
30°N
20°N
10°N
60°E
70°E
80°E
90°E
100°E

Boundaries
National capitals
Other cities

SCALE
0 300 600 Miles
0 300 600 Kilometers
Projection: Two-Point Equidistant

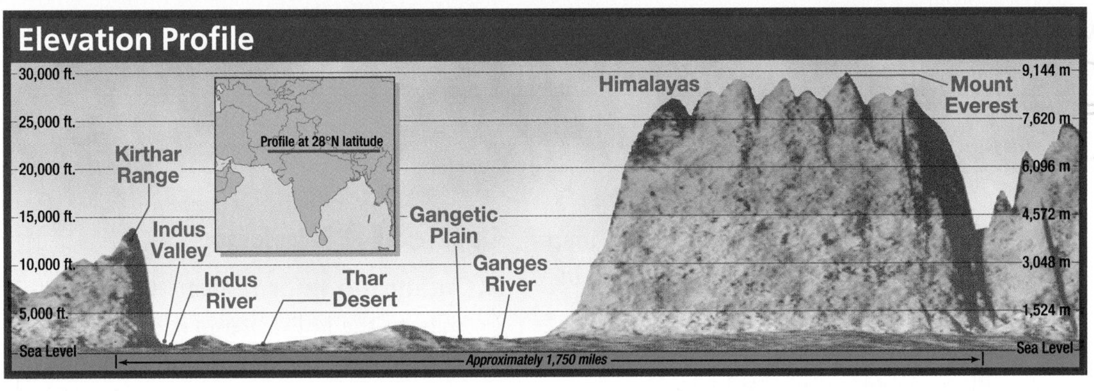

South Asia: Physical

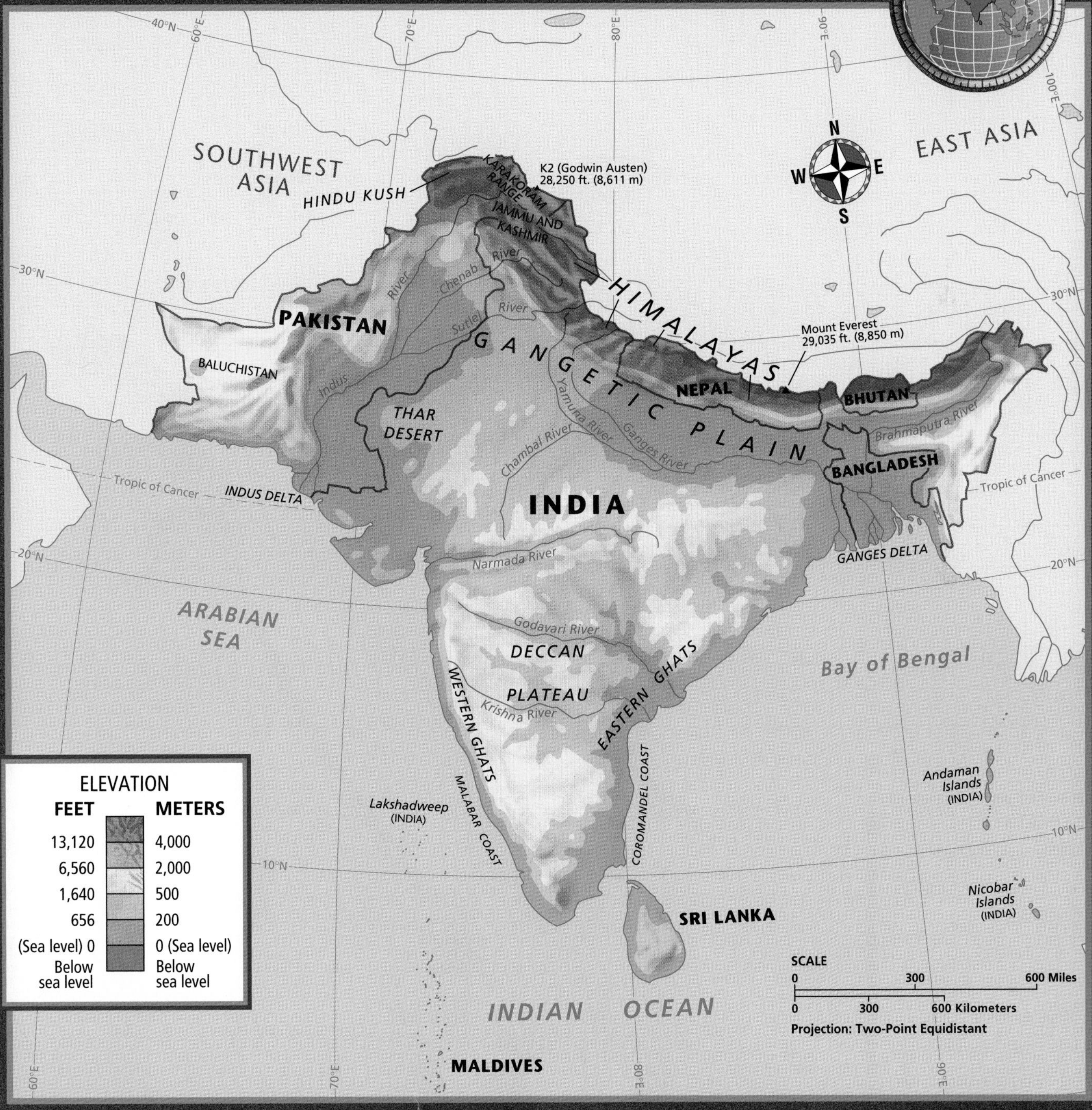

SOUTHWEST ASIA
EAST ASIA
HINDU KUSH
KARAKORAM RANGE
K2 (Godwin Austen) 28,250 ft. (8,611 m)
JAMMU AND KASHMIR
HIMALAYAS
Mount Everest 29,035 ft. (8,850 m)
PAKISTAN
BALUCHISTAN
Indus River
Chenab River
Sutlej River
GANGETIC PLAIN
NEPAL
BHUTAN
THAR DESERT
Yamuna River
Ganges River
Chambal River
Brahmaputra River
BANGLADESH
Tropic of Cancer
INDUS DELTA
INDIA
GANGES DELTA
Narmada River
ARABIAN SEA
Godavari River
DECCAN PLATEAU
Krishna River
WESTERN GHATS
EASTERN GHATS
Bay of Bengal
MALABAR COAST
COROMANDEL COAST
Lakshadweep (INDIA)
Andaman Islands (INDIA)
Nicobar Islands (INDIA)
SRI LANKA
INDIAN OCEAN
MALDIVES
40°N
30°N
20°N
10°N
60°E
70°E
80°E
90°E
100°E
N
S
E
W

ELEVATION

FEET	METERS
13,120	4,000
6,560	2,000
1,640	500
656	200
(Sea level) 0	0 (Sea level)
Below sea level	Below sea level

SCALE
0 300 600 Miles
0 300 600 Kilometers
Projection: Two-Point Equidistant

South Asia:
Climate

1. *Places and Regions* Compare this map to the physical map. Which landform in southern India is dominated by a semi-arid climate? TEKS
2. *Places and Regions* Which climate type dominates Nepal and Bangladesh? TEKS

Critical Thinking

3. **Analyzing Information** Compare this map to the political and precipitation maps. How do you think the wet monsoon airflow affects life in Bangladesh? TEKS

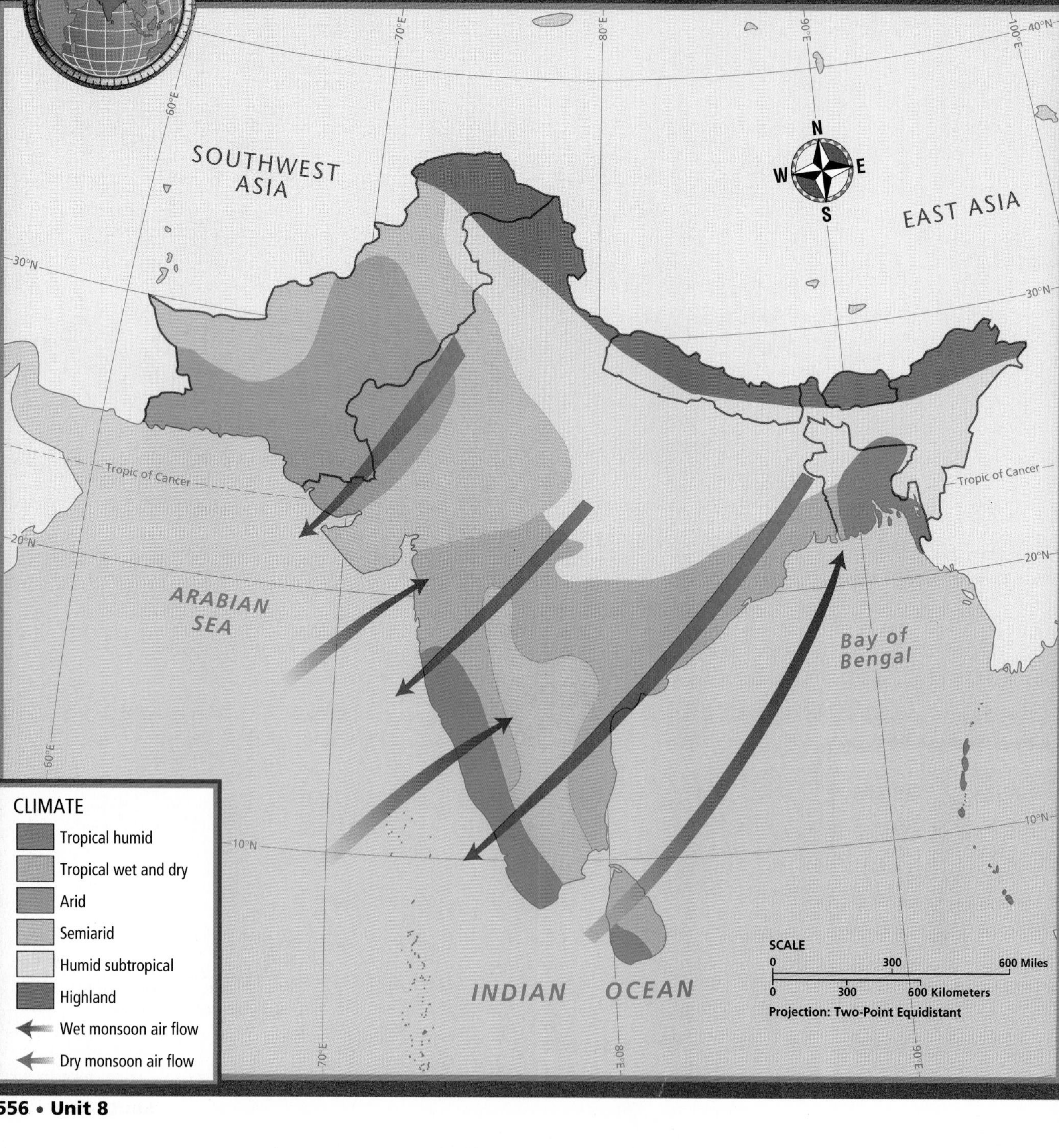

SOUTHWEST ASIA
EAST ASIA
ARABIAN SEA
Bay of Bengal
INDIAN OCEAN
Tropic of Cancer
40°N
30°N
20°N
10°N
60°E
70°E
80°E
90°E
100°E
N
S
E
W

CLIMATE
Tropical humid
Tropical wet and dry
Arid
Semiarid
Humid subtropical
Highland
Wet monsoon air flow
Dry monsoon air flow

SCALE
0 300 600 Miles
0 300 600 Kilometers
Projection: Two-Point Equidistant

South Asia: Precipitation

1. **Physical Systems** Compare this map to the physical and political maps. Which mountains in southern India cause a rain-shadow effect? TEKS
2. **Environment and Society** Compare this map to the climate and political maps. How do you think the wet monsoon airflow affects the amount of precipitation in Bangladesh? TEKS

Critical Thinking

3. **Analyzing Information** Compare this map to the physical map. How do you think the Himalayas affect the distribution of precipitation in South Asia? TEKS

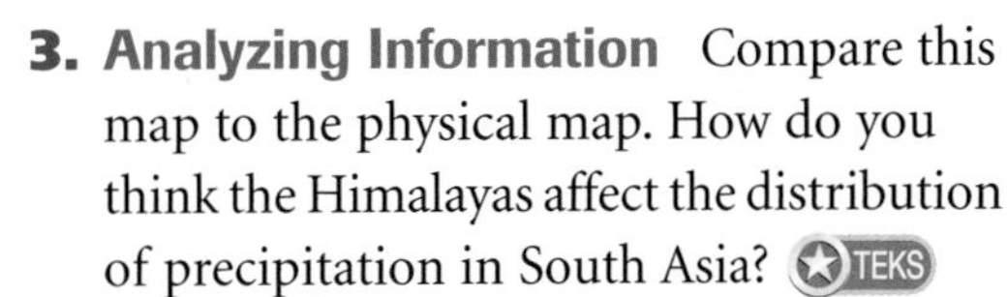

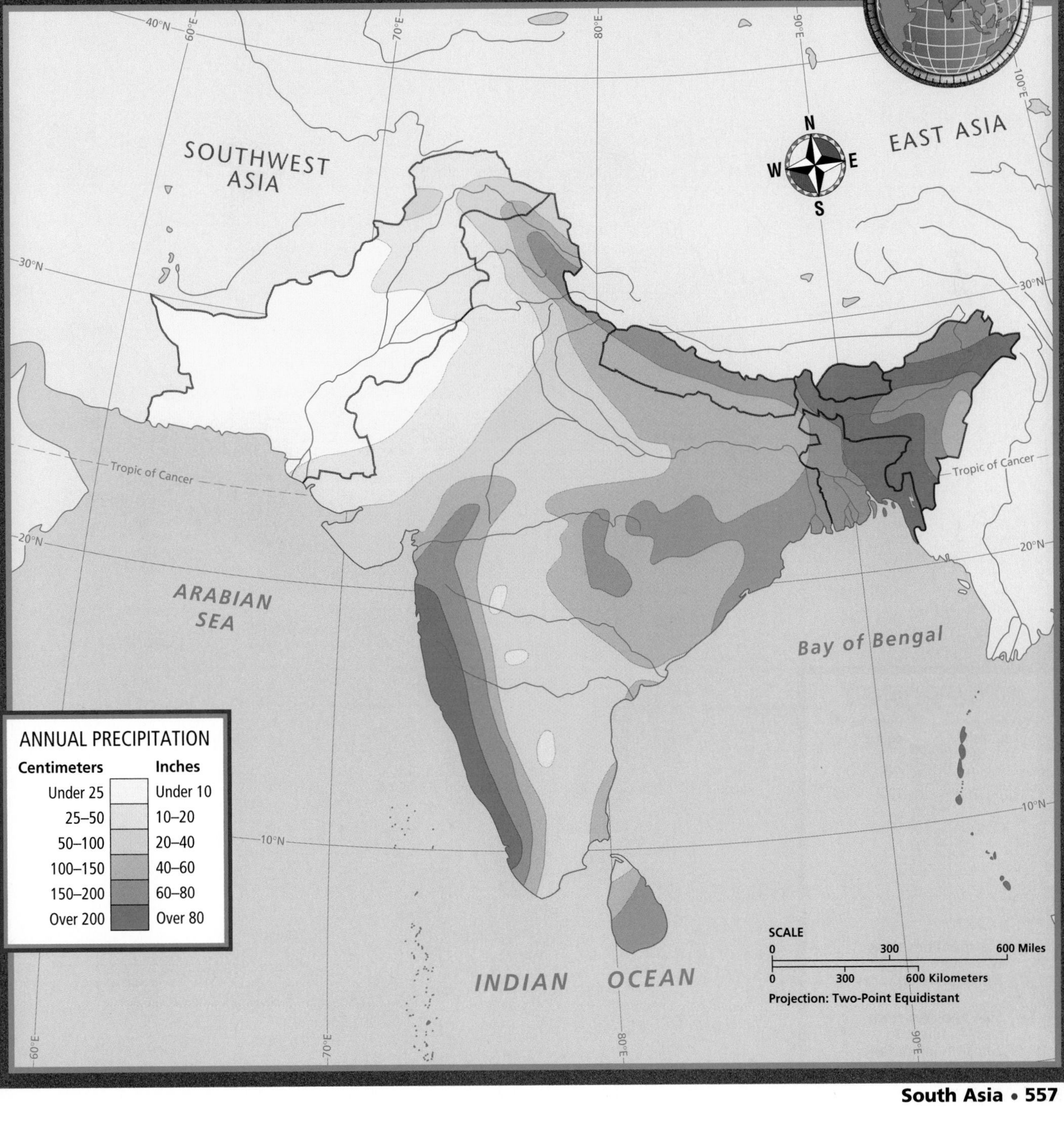

SOUTHWEST ASIA

EAST ASIA

ARABIAN SEA

Bay of Bengal

INDIAN OCEAN

Tropic of Cancer

40°N 30°N 20°N 10°N

60°E 70°E 80°E 90°E 100°E

N S E W

ANNUAL PRECIPITATION	
Centimeters	**Inches**
Under 25	Under 10
25–50	10–20
50–100	20–40
100–150	40–60
150–200	60–80
Over 200	Over 80

SCALE
0 300 600 Miles
0 300 600 Kilometers
Projection: Two-Point Equidistant

South Asia: Population

1. **Environment and Society** Compare this map to the physical map. Which two major rivers flow through densely populated northern areas? TEKS
2. **Places and Regions** Which country in the region has the highest overall population density? TEKS

Critical Thinking

3. **Analyzing Information** Compare this map to the physical, climate, and land use maps. What do population density patterns suggest about the ways South Asians have adapted to their physical environments? TEKS

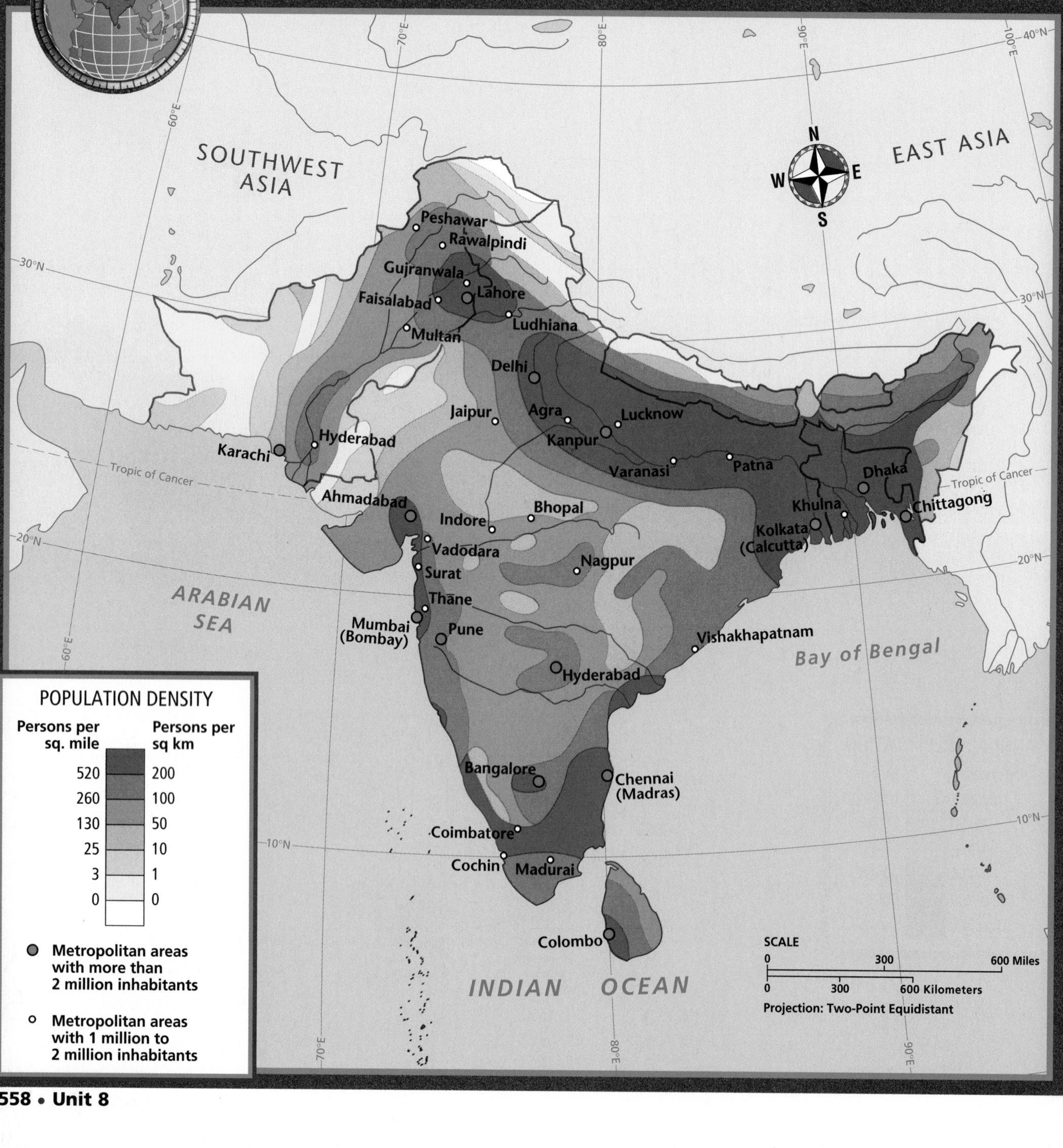

SOUTHWEST ASIA

EAST ASIA

ARABIAN SEA

Bay of Bengal

INDIAN OCEAN

Tropic of Cancer

Peshawar, Rawalpindi, Gujranwala, Lahore, Faisalabad, Multan, Ludhiana, Delhi, Jaipur, Agra, Lucknow, Kanpur, Varanasi, Patna, Karachi, Hyderabad, Ahmadabad, Indore, Bhopal, Dhaka, Khulna, Chittagong, Kolkata (Calcutta), Vadodara, Surat, Nagpur, Thane, Mumbai (Bombay), Pune, Vishakhapatnam, Hyderabad, Bangalore, Chennai (Madras), Coimbatore, Cochin, Madurai, Colombo

POPULATION DENSITY

Persons per sq. mile	Persons per sq km
520	200
260	100
130	50
25	10
3	1
0	0

Metropolitan areas with more than 2 million inhabitants

Metropolitan areas with 1 million to 2 million inhabitants

SCALE

0 300 600 Miles

0 300 600 Kilometers

Projection: Two-Point Equidistant

UNIT 8 ATLAS

South Asia:
Land Use and Resources

1. *Environment and Society* Compare this map to the political map. What is the dominant type of land use in coastal India? TEKS

2. *Places and Regions* Which part of India seems to have the highest concentration of minerals?

Critical Thinking

3. **Analyzing Information** Compare this map to the physical and political maps. How is land use in Nepal related to elevation? TEKS

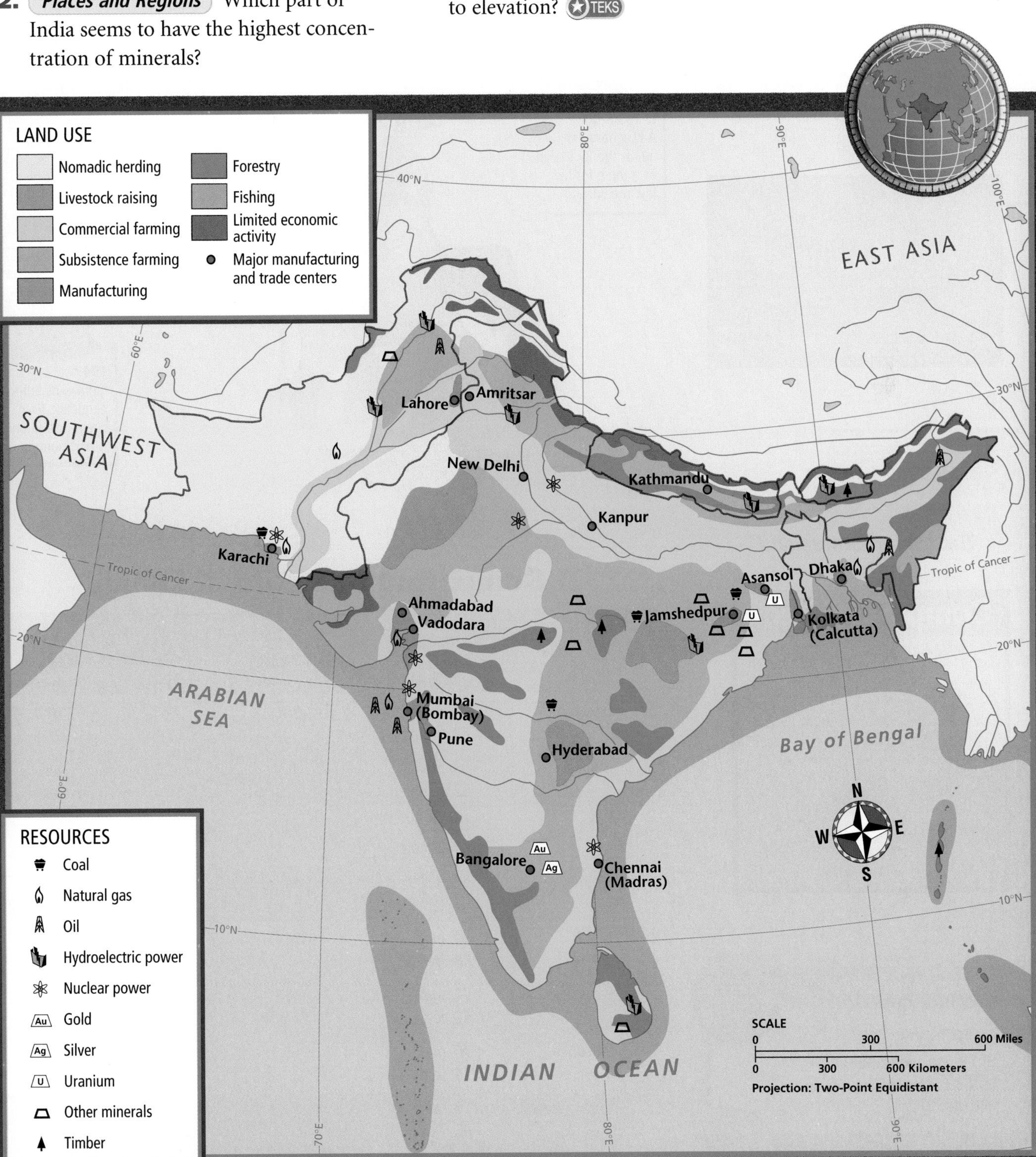

Time Line: South Asia

2500 B.C.	400 B.C.	1600 A.D.	1700	1800	1900

c. 2500 B.C.
Indus Valley civilization flourishes.

400s B.C.
Siddhārtha Gautama founds Buddhism.

1600 A.D.
The British establish the East India Company.

c. 1630–50
A Mughal ruler builds the Taj Mahal as a tomb for his favorite wife.

1858
The British government takes control of India.

1920
Mohandas K. Gandhi begins an independence movement of nonviolent disobedience in India.

1947
India wins independence from Great Britain and is partitioned.

1971
Bangladesh becomes independent.

The United States and South Asia

Comparing Sizes

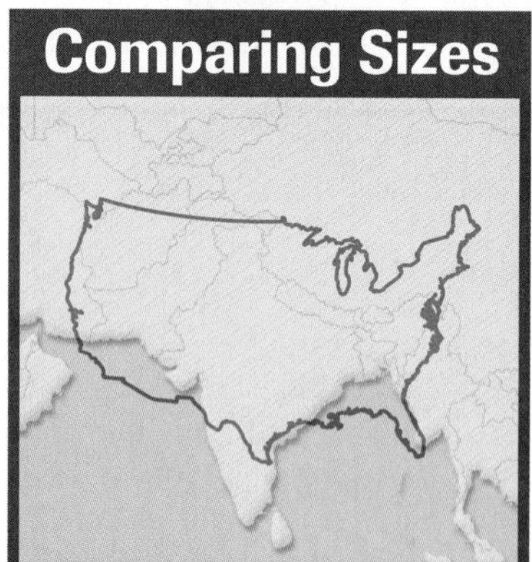

Comparing Standard of Living

COUNTRY	LIFE EXPECTANCY (in years)	INFANT MORTALITY (per 1,000 live births)	LITERACY RATE	DAILY CALORIC INTAKE (per person)
Bangladesh	61, male 60, female	70	56%	2,017
Bhutan	53, male 52, female	109	42%	Not available
India	62, male 64, female	63	52%	2,388
Maldives	61, male 64, female	64	93%	2,485
Nepal	59, male 58, female	74	28%	2,367
Pakistan	61, male 62, female	81	43%	2,475
Sri Lanka	70, male 75, female	16	90%	2,334
United States	74, male 80, female	7	97%	3,603

Sources: Central Intelligence Agency, *2001 World Factbook; Britannica Book of the Year, 2000*

internet connect

GO TO: go.hrw.com
KEYWORD: SW3 Almanac
FOR: Additional information and reference sources

Fast Facts: South Asia

FLAG	COUNTRY Capital	POPULATION (in millions) POP. DENSITY	AREA	PER CAPITA GDP (in US $)	WORKFORCE STRUCTURE (largest categories)	ELECTRICITY CONSUMPTION (kilowatt hours per person)	TELEPHONE LINES (per person)
	**Bangladesh** Dhaka	131.3 2,361/sq. mi.	55,598 sq. mi. 144,000 sq km	$ 1,570	63% agriculture 26% services	85 kWh	0.004
	**Bhutan** Thimphu	2.0 113/sq. mi.	18,147 sq. mi. 47,000 sq km	$ 1,100	93% agriculture 5% services	93 kWh	0.003
	**India** New Delhi	1,030 811/sq. mi.	1,269,338 sq. mi. 3,287,590 sq km	$ 2,200	67% agriculture 18% services	412 kWh	0.03
	**Maldives** Male	0.3 2,679/sq. mi.	116 sq. mi. 300 sq km	$ 2,000	60% services 22% agriculture	302 kWh	0.07
	**Nepal** Kathmandu	25.3 465/sq. mi.	54,363 sq. mi. 140,800 sq km	$ 1,360	81% agriculture 16% services	52 kWh	0.009
	**Pakistan** Islamabad	144.6 466/sq. mi.	310,401 sq. mi. 803,940 sq km	$ 2,000	44% agriculture 17% industry	399 kWh	0.02
	**Sri Lanka** Colombo	19.4 766/sq. mi.	25,332 sq. mi. 65,610 sq km	$ 3,250	45% services 38% agriculture	289 kWh	0.03
	**United States** Washington, D.C.	281.4 76/sq. mi.	3,717,792 sq. mi. 9,629,091 sq km	$ 36,200	30% manage., prof. 29% tech., sales, admin.	12,211 kWh	0.69

Sources: Central Intelligence Agency, *2001 World Factbook; The World Almanac and Book of Facts, 2001; Britannica Book of the Year, 2000;* U.S. Census Bureau
The CIA calculates per capita GDP in terms of purchasing power parity. This formula equalizes the purchasing power of each country's currency.

Bengal tiger

CHAPTER 25

India

About one out of every six people in the world lives in India, which is home to more than a billion people. This huge population includes people from many different ethnic and religious groups.

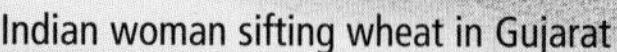
Indian woman sifting wheat in Gujarat

Royal portrait from Rajasthan

Hi! My name is Rojo, and I am 16. I live in Vaduthala, a small town outside the city of Cochin in the state of Kerala, southern India. My street is near a lake that is surrounded by coconut and banana trees.

When I'm relaxing at home, I like to read Indian entertainment magazines and watch television. My favorite program is about Hanuman, the monkey-god. The living room is my favorite place in the house because the ceiling fan keeps the mosquitoes away and keeps it cool. When my mom makes fish curry, she usually goes outside to the cooking terrace. That way, the whole house doesn't smell like fish. She grinds coconut there too.

I am a senior in high school. I wear a school uniform with navy blue pants and a grey shirt. Students in other schools have different uniforms, so it is very colorful in the morning when we all walk to school. We don't like walking in the summer monsoon season because we get wet. My favorite subject is math. I am studying a computer language based on math. At school we speak English. At home I speak Malayalam, the most common language of Kerala.

On Sundays we go to the Catholic church in the next town. Then we come back to the house and have a big lunch, including chicken curry, fish curry, sweet bread, eggs, and a vegetable dish. At Christmas they decorate the church with colored lights.

Natural Environments

READ TO DISCOVER

1. What are the major landform regions and rivers of India?
2. Which climate types and resources does India have?

WHY IT MATTERS

Summer monsoon rains in India can cause floods that ruin water supplies and wash away crops, homes, and livestock. Use CNNfyi.com or other **current events** sources to learn about the dangers posed by floods.

DEFINE

subcontinent

LOCATE

Himalayas
Ganges River
Gangetic Plain
Deccan Plateau
Western Ghats
Eastern Ghats
Narmada River
Godavari River
Krishna River
Brahmaputra River
Thar Desert

India: Physical-Political

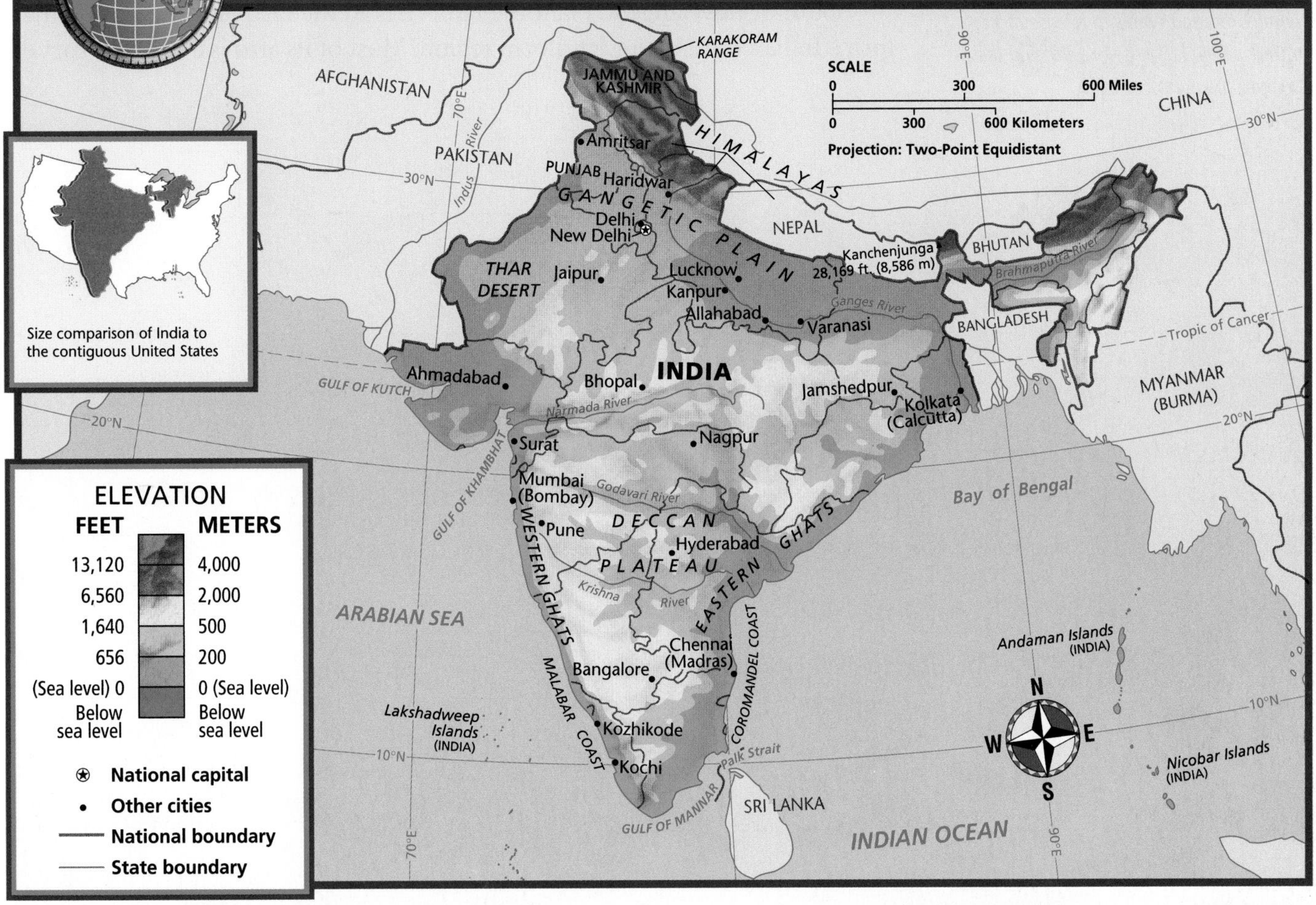

KARAKORAM RANGE
AFGHANISTAN
JAMMU AND KASHMIR
SCALE
0 300 600 Miles
0 300 600 Kilometers
Projection: Two-Point Equidistant
CHINA
PAKISTAN
Indus River
Amritsar
PUNJAB
Haridwar
HIMALAYAS
GANGETIC PLAIN
Delhi
New Delhi
NEPAL
Kanchenjunga 28,169 ft. (8,586 m)
BHUTAN
Brahmaputra River
THAR DESERT
Jaipur
Lucknow
Kanpur
Allahabad
Varanasi
Ganges River
BANGLADESH
Tropic of Cancer
Size comparison of India to the contiguous United States
GULF OF KUTCH
Ahmadabad
Bhopal
INDIA
Narmada River
Jamshedpur
Kolkata (Calcutta)
MYANMAR (BURMA)
GULF OF KHAMBHAT
Surat
Nagpur
Mumbai (Bombay)
Godavari River
Bay of Bengal
Pune
DECCAN PLATEAU
Hyderabad
WESTERN GHATS
EASTERN GHATS
Krishna River
ARABIAN SEA
COROMANDEL COAST
Chennai (Madras)
Andaman Islands (INDIA)
Bangalore
MALABAR COAST
Lakshadweep Islands (INDIA)
Kozhikode
Kochi
Palk Strait
GULF OF MANNAR
SRI LANKA
Nicobar Islands (INDIA)
INDIAN OCEAN
N S E W
30°N 20°N 10°N 70°E 90°E 100°E

ELEVATION

FEET	METERS
13,120	4,000
6,560	2,000
1,640	500
656	200
(Sea level) 0	0 (Sea level)
Below sea level	Below sea level

⊛ National capital
• Other cities
National boundary
State boundary

internet connect

GO TO: go.hrw.com
KEYWORD: SW3 CH25
FOR: Web sites about India

Physical Features

The world's highest mountains separate India and its neighbors from the rest of Asia. Together these countries form the Indian **subcontinent**. A subcontinent is a very large landmass that is smaller than a continent. In addition to this large landmass, India also includes two island territories located in the Indian Ocean.

Tectonic forces have shaped the Indian subcontinent. The subcontinent was once part of the supercontinent of Gondwana but then broke away. It slowly drifted northward. About 50 million years ago it pushed into Asia, forcing up a mountain system called the Himalayas (hi-muh-LAY-uhz). This mountain-building process still goes on today, causing severe earthquakes.

Landforms The Himalayas are one of India's main landform regions. They contain Kanchenjunga (kuhn-chun-JUHN-guh), which is the third-highest mountain in the world and India's highest point. At lower elevations there are farms, forests, and steep gorges. The Kashmir Valley, between two arms of the mountains, is famous for its beauty. The great Ganges (GAN-jeez) River begins high in the Himalayas' melting snow and glaciers.

South of the Himalayas is the Gangetic (gan-JE-tik) Plain. This low flat region stretches about 1,500 miles (2,415 km) across northeastern India. Its rich soil was brought down from the mountains by the Ganges River, forming the world's largest alluvial plain. In some places this alluvial layer is more than 25,000 feet (7,620 m) thick. For thousands of years, people have settled on the Gangetic Plain to farm.

South of the Gangetic Plain the land rises to the Deccan, a peninsula that forms India's third major landform region. Most of its area is a plateau. Part of

The Himalayas form a barrier between the alluvial plains of northern India and the Plateau of Tibet to the north. They are the highest mountains in the world, with more than 110 peaks rising to elevations of 24,000 feet (7,325 m) or more above sea level. The Himalayas affect weather patterns in India by blocking cold air from moving south during winter and by forcing summer monsoon rains to release most of their precipitation before crossing the range to the north.

the Deccan Plateau is very old and eroded. The newer part is made up of lava layers. Where the volcanic rocks have weathered and irrigation is possible, there are fertile grain fields. Two low mountain ranges, the Western Ghats (GAWTS) and the Eastern Ghats, form the plateau's edges. Narrow plains lie on both the Arabian Sea and Bay of Bengal coasts.

Great Rivers Several important rivers cut across the Deccan Plateau. The Narmada (nuhr-MUH-duh) River flows to the west. Hundreds of dams are being built on the Narmada. Although the reservoirs created by the dams store much-needed water, the projects have also displaced thousands of people. The Godavari (go-DAH-vuh-ree) and Krishna (KRISH-nuh) Rivers flow to the east. An irrigation and canal system linking the deltas of these two rivers has made the region a productive rice-growing area.

Like the Ganges, the Brahmaputra River begins high in the Himalayas. It flows eastward through Tibet and then turns to the south, flowing through eastern India. In Bangladesh the Brahmaputra joins the Ganges to form a huge delta.

READING CHECK: *Places and Regions* In what landform regions are India's major rivers located?

INTERPRETING THE VISUAL RECORD *Which of India's three main landform regions can you see in this satellite image?*

Climates and Natural Resources

As you can see from the Unit 8 climate map, India has six climate types. They range from a highland climate in the Himalayas to a tropical humid climate along the southwest coast.

The Thar (TAHR) Desert, also known as the Great Indian Desert, lies in the northwest. It extends into neighboring Pakistan. The prevailing dry monsoon winds and a high-pressure zone keep the region arid and hot. Compared to the densely populated Gangetic Plain, the Thar Desert has few residents.

India's plant life tends to reflect rainfall amounts. Vegetation ranges from scrubby trees in the Thar Desert to evergreen forests in the western Himalayas. The country's animal life is varied. Crocodiles, deer, elephants, mongooses, monkeys, tigers, more than 1,200 bird species, and nearly 400 snake species live in India.

The Indian cobra has a toxic bite. Cobras kill several thousand people each year, partly because they often visit houses at night to catch rats.

READING CHECK: *Places and Regions* What six climate types are found in India?

The Monsoons India's monsoons strongly influence the country's climates. The wet summer monsoon usually begins about June. A large low-pressure

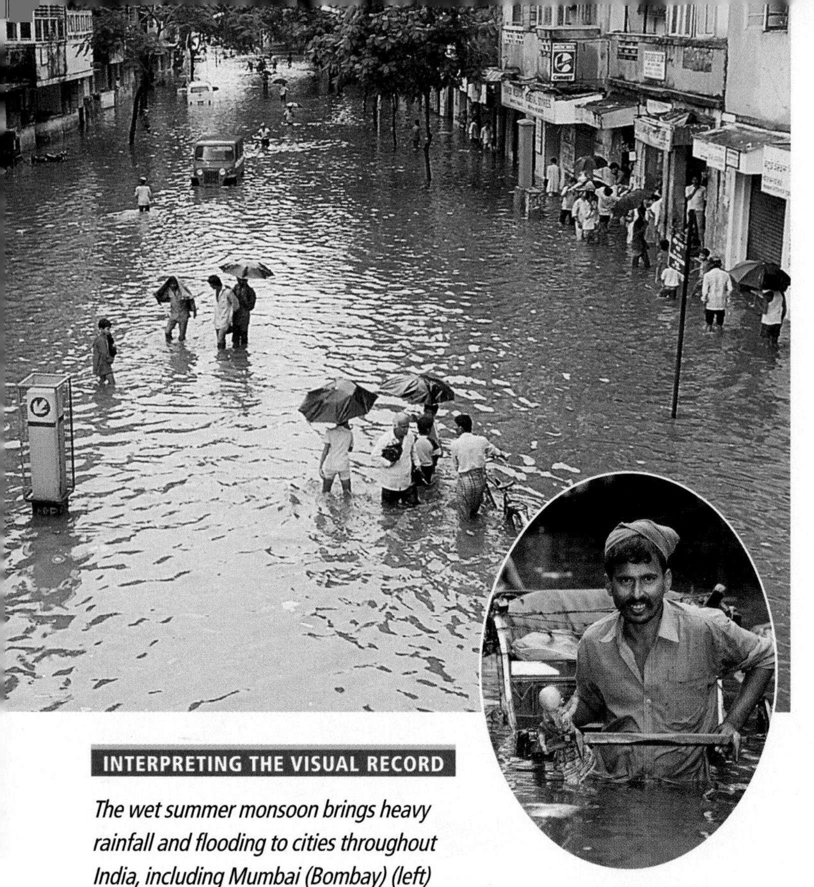

INTERPRETING THE VISUAL RECORD

The wet summer monsoon brings heavy rainfall and flooding to cities throughout India, including Mumbai (Bombay) (left) and Kolkata (Calcutta) (right). While Mumbai's average annual rainfall is some 71 inches (180 cm), about 24 inches (61 cm) fall in the month of July. ***How do you think heavy monsoon rains and floods affect the environment?*** 

area over interior Asia pulls moist air inland from the Indian Ocean. As the moist air flows inland, it brings rain. The heaviest rains fall where the monsoon meets the Western Ghats and the foothills of the Himalayas. Some areas receive more than 400 inches (1,016 cm) of rain per year.

The winter monsoon lasts from November through March. During those months cold dry winds blow from Asia's interior into India. As the winds drop down from the mountains they warm and become even drier. As a result, much of India has a warm dry winter.

Unusually heavy monsoon rains can cause terrible floods. If the monsoon is late or the rains stop early, crops may die and people may go hungry. Sometimes, within the same season one part of India can suffer drought while another has floods.

READING CHECK: *Physical Systems* What causes India's monsoons?

Natural Resources India's soils and rivers are its most important natural resources. About 56 percent of the country is arable, in contrast to only 19 percent in the United States.

India also has many mineral resources, including the world's fourth-largest coal reserves. There are large deposits of iron ore and bauxite, an aluminum ore, much of which is exported. India has some petroleum deposits but must import oil to meet its needs. The country's uranium mines support several nuclear power plants. Rivers supply hydroelectric power.

READING CHECK: *Environment and Society* Why are India's soils and rivers called its most valuable natural resources?

Section 1 Review

TEKS Working with Sketch Maps, Questions 1, 2, 4, 5

Define

subcontinent

Working with Sketch Maps

On a map of India that you draw or that your teacher provides, label the Himalayas, Ganges River, Gangetic Plain, Deccan Plateau, Western Ghats, Eastern Ghats, Narmada River, Godavari River, Krishna River, Brahmaputra River, and Thar Desert. In the margin of your map, identify the world's largest alluvial plain.

Reading for the Main Idea

1. *Physical Systems* What physical process created the Himalayas? What physical process created the Gangetic Plain?
2. *Places and Regions* The distribution of India's plant life reflects what feature of the country's climate? Where will you find scrubby trees? Evergreen forests?
3. *Environment and Society* What are some of India's important mineral resources?

Critical Thinking

4. **Drawing Inferences and Conclusions** Why might Indian farmers look forward to the summer monsoon with both hope and fear?

Organizing What You Know

5. Create a chart like the one below. Use it to name the rivers and climates in India's three main landform regions. Use the unit maps to help you.

Landform region	Rivers	Climates

Places and Regions

Geography for Life

The Holy Ganges

The Ganges River begins high in the Himalayas as melting glacial ice. It then flows more than 1,500 miles (2,505 km) southeastward across India's northern plains. Millions of people depend on the Ganges. They use the river as a source of drinking water and fish. Farmers use it to irrigate their crops. The Ganges also provides the country with a trade and transportation route.

To many people, the Ganges is much more than a river. It is Hinduism's holy river. The Ganges is part of India's folklore, history, and mythology. Indians call it the goddess Ganga. Every day you can find people bathing in the river's water, drinking it, and standing in it while they pray. Hindus believe that just touching the water can wash away their sins. People with various ailments come from near and far seeking the healing powers of the water. Perhaps more than 1 million people enter the Ganges somewhere along its course each day. The cities of Allahabad, Haridwar, and Varanasi are considered particularly sacred bathing sites. Temples line the banks of the Ganges in these cities. Wide stone staircases called ghats lead down to the water.

More than 1 million pilgrims visit Varanasi, India, each year to bathe in the holy Ganges.

Garbage and pollution are a common sight in the waters of the Ganges today.

Tradition says that the Ganges is pure and that nothing can pollute its water. Yet huge amounts of sewage flow into the river, making it one of the world's most polluted waterways. Waste from dozens of cities and thousands of villages is dumped into the river. Runoff from farms carries soil full of farm chemicals into it. Hundreds of factories also dump industrial waste into the Ganges. People even come to die in the river. As a result of these pollutants, many who enter the Ganges get sick.

In spite of these dangers, devout Hindus readily drink and bathe in the Ganges' waters. The government has proposed programs to reduce the river's pollution. While some Indians support a cleanup, others insist that the Ganges remains pure. Because the government lacks wide public support for its programs, it has done little to clean up the Ganges.

Applying What You Know 

1. **Summarizing** Why is the Ganges River holy to Hindus? How might non-Hindus perceive it?
2. **Supporting a Point of View** Construct an argument for or against the need to clean up the Ganges River from the perspective of either a traditional Hindu or an environmentalist.

History and Culture

READ TO DISCOVER

1. What were the major events and empires of India's early history?
2. How did European contact affect India?
3. What religions are practiced in India?
4. What are some other features of India's culture?

WHY IT MATTERS

Hinduism is the world's oldest major religion. Use CNNfyi.com or other **current events** sources to learn about important Hindu beliefs and practices.

IDENTIFY

Sanskrit
Dalits
Sikhism
Jainism

DEFINE

pantheon
sepoys
boycott
partition
reincarnation
dharma
karma
caste

LOCATE

Delhi

The Ellora Caves in western India were carved from cliffs between about 200 B.C. and A.D. 1000. They served as temples and monasteries for followers of three of India's religions: Buddhism, Hinduism, and Jainism.

Early Indian Civilizations

The first highly developed civilization in the Indian subcontinent developed in the Indus River valley. Named after a modern town near the site of one of its ancient cities, this culture is known as the Harappan civilization. It was centered in what is now Pakistan but also extended into India. The Harappan way of life was based on farming and trade. For India's history we look first at people who came through the western mountain passes to invade the fertile north.

The Aryans By about 1500 B.C. a warlike seminomadic people were moving into northern India from central Asia. They called themselves Aryans. The Aryans spread their influence gradually throughout northern India. In turn, Aryan culture changed as it mixed with cultures already in the area. The Dravidians, farming peoples of the interior, were slowly pushed south. They later developed advanced kingdoms in the Deccan.

The Aryans brought an Indo-European language to the subcontinent. They spoke an early form of **Sanskrit**. Hindi, the major language of modern India, developed from Sanskrit, which is still used for some religious ceremonies. The name *Himalaya* is an example of a Sanskrit word. It means "Home of Snows."

Aryan religion included many of the basic ideas that became part of Hinduism, such as a large **pantheon**. A pantheon is all the gods of a religion. The Aryans also introduced a strict system of social classes. These and other concepts developed into Hinduism, India's main religion today.

During the Aryan period, many kingdoms rose in the Gangetic Plain. These kingdoms were supported by farming the plain's rich soils. Also during the Aryan period Siddhārtha Gautama, who later became known as the Buddha, taught the concepts of Buddhism.

Islamic Empires About A.D. 1000, Muslim armies began attacking northwestern India. In the early 1200s a Muslim kingdom was founded at Delhi. In 1398 Timur, a conqueror from the interior of Asia, invaded India and sacked Delhi.

A descendant of Timur and Genghis Khan nicknamed Bābur—"the Tiger"—invaded India in the 1520s. He took over most of northern India and founded the Mughal (MOO-guhl) Empire. Bābur was not only a brilliant general but also a gifted poet. After an unstable period Bābur's grandson Akbar reunited the Mughal Empire and expanded it into central India. Akbar was an effective ruler as well as a successful conqueror. He allowed the peoples he conquered to practice their own religions. Although the ruling Mughals were Muslims, most of the region's people continued to practice Hinduism.

The Gangetic Plain's fertile land and large population helped the Mughal Empire grow rich. During this period Shāh Jahān built the world-famous Taj Mahal. Many other grand monuments remain from the Mughal Empire.

Religious tolerance ended with the rule of Shāh Jahān's son Aurangzeb. He placed heavy taxes on Hindus and destroyed many temples. When Aurangzeb died in 1707, he left a weakened empire. Europeans took the opportunity to expand their influence.

TEKS **READING CHECK:** *Places and Regions* How did invasions alter India's early cultures?

Connecting to HISTORY

Early Empires

Almost all of India was first united by the Mauryan (MOW-ree-uhn) Empire. It was founded in about 320 B.C. Aśoka, the most famous Mauryan emperor, gave up war after he saw the suffering he had caused. He adopted Buddhism and furthered its spread across India. Aśoka's thoughts and accounts of his deeds were carved on rocks and pillars placed throughout the empire.

India entered a golden age under the Gupta Empire. From A.D. 320 to about 500, architecture, art, literature, mathematics, and medicine all flourished. Gupta universities drew students from as far away as Java, in southeastern Asia, to study philosophy and other subjects. Art, learning, and trade continued to prosper after the Gupta period came to an end.

Analyzing Information
In what ways did the Mauryan and Gupta Empires contribute to the early development of India?

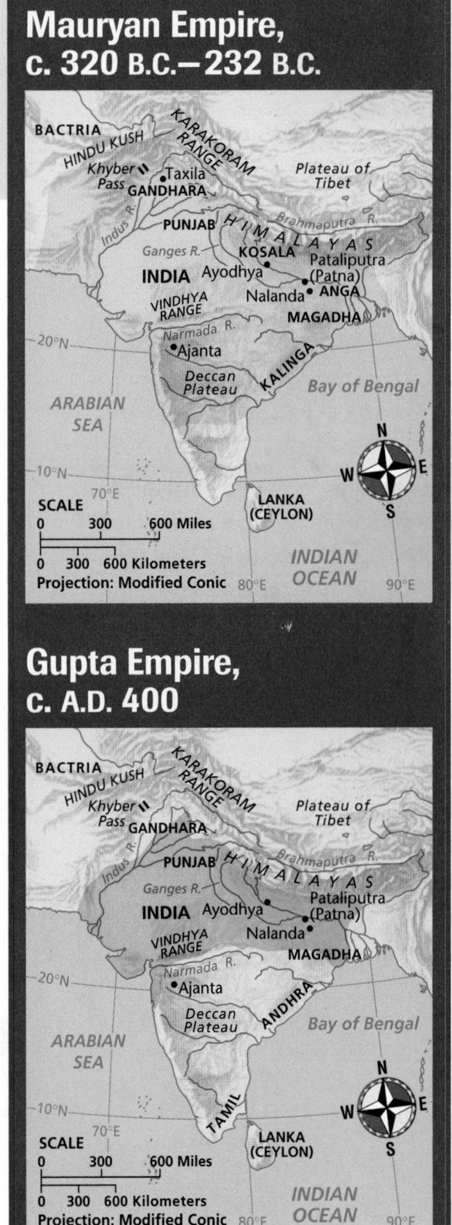

European Influence

Europeans arrived in India in the late 1490s to trade, expand their empires, and spread Christianity. The Portuguese came first, followed by the Dutch, French, and British. Powerful trading companies, such as Great Britain's East India Company, built successful businesses in the region. Indian cotton became particularly important as Britain's growing textile industry needed tons of the raw fiber. Soon Britain and France became rivals in India.

British Rule In the 1700s the British defeated the French. By the mid-1800s Britain controlled about half of the subcontinent. The East India Company controlled India for the British government.

However, foreign rule angered many Indians. In 1857 a rebellion broke out among the **sepoys**, the Indian troops under the command of

British officers. The sepoys killed the officers and their families. When the revolt spread across northern India, British troops rushed to the conflict. Both sides reported vicious acts of cruelty. In the end, the British crushed the revolt. The mutiny convinced the British government to end the East India Company's role and to rule India directly beginning in 1858.

India, which at that time included present-day Pakistan, Bangladesh, and Sri Lanka, became a British colony. It was called the "jewel in the crown" of the British Empire. Indian products, such as cotton, jewels, and tea, flowed to Britain. India was also a market for merchandise from British factories. To ensure an efficient flow of goods, the British organized the construction of railroads, roads, and ports. In addition, they brought the English language and English systems of education, law, and government.

Mahatma Gandhi led India's independence movement and became internationally famous for his doctrine of nonviolent protest and civil disobedience. Gandhi also worked to unite India's Hindus and Muslims, end discrimination, and promote education and economic development for India's poor.

Independence The British did not treat Indians as equals, and many Indians wanted independence. Others simply wanted fair treatment. In 1885, educated middle-class Indians organized the Indian National Congress. At first, the Congress asked only for more rights, such as a larger share of government jobs. The British refused this request.

Demands for independence increased in the early 1900s. A young lawyer named Mohandas K. Gandhi led the independence movement. His followers called him the Mahatma, or "Great Soul." Gandhi believed that nonviolent noncooperation was the best way to bring about change and achieve independence. Gandhi led peaceful protest marches and urged Indians to **boycott**, or refuse to buy, British goods, particularly cloth. When he was thrown into jail, Gandhi protested by going on hunger strikes.

Gandhi's efforts were effective. After World War II the British government granted India its independence. Britain and the Indian National Congress wanted India to become one country. However, India's Muslims demanded a separate state. Hostility between Hindus and Muslims grew.

To avoid civil war, in 1947 the British government divided the colony into two parts, India and Pakistan. This division is called the **partition** of India. Pakistan, which included what is now Bangladesh, was mostly Muslim. India was mostly Hindu. The region of Kashmir was divided between the two countries. However, the new boundaries left large numbers of Hindus in Pakistan and many Muslims in India. Panic broke out as some 16 million people fled to the country where their religion held a majority. Perhaps as many as 1 million people died in riots and massacres. Even Gandhi fell prey to the violence. In 1948 he was shot and killed by a Hindu extremist. Religious tension is still a pressing issue for India.

Today India is the world's most populous democracy. A large percentage of the people vote. There are 28 states, each with its own legislature, and seven territories. India's government is based on British models. It includes a multiparty parliament and a prime minister.

The Gateway of India in Mumbai (Bombay) was built to commemorate the 1911 visit of King George V and Queen Mary. On February 28, 1948, the last British troops to leave the newly independent country of India set sail from the Gateway of India.

READING CHECK: *Human Systems* Why did the British want control of India, and how did they lose control?

Religion

According to its constitution, India's government is officially secular. That is, it does not support any one religion, and all faiths are equal before the law. At the same time, religion is a powerful force in Indian society. A large majority of the people take part in religious activity on a daily basis. Some Indian political parties are based on religion.

The statue of Bahubali in the state of Karnataka is an important pilgrimage site for followers of Jainism, one of India's indigenous religions. Every 12 to 14 years a "head anointing ceremony" is performed on the statue as milk, yogurt, butter, saffron, gold coins, sandalwood, and other items are poured over its head. Jain pilgrims believe these objects acquire a powerful spiritual energy from the statue.

Hinduism About 80 percent of India's people are Hindu. Hinduism is an ancient religion and the largest ethnic religion in the world. There are many gods and goddesses in Hinduism. Although its pantheon is large, Hinduism teaches that all gods and all living beings are part of a single spirit.

Three Hindu beliefs, **reincarnation**, **dharma**, and **karma**, are closely related. Followers of Hinduism believe that the soul is reborn again and again in different forms. This process is called reincarnation. The importance of doing one's duty according to one's station in life is called dharma. Karma is the positive or negative force caused by a person's actions. Hindus believe that people who fulfill their dharma earn good karma and may be reborn as persons of higher status. Those with bad karma may be reborn with lower status or as animals or insects.

The **caste** system is another practice central to Hinduism. A caste is a group of people who are born into a certain position in society. The Aryans developed this system thousands of years ago. It assigned people to one of four major classes, or castes, according to their occupations. The Brahmins were the highest caste. These priests and intellectuals were the only people who could read and write. Below the Brahmins were the Kshatriyas (kuh-SHA-tree-uhz), or warriors. Next came the Vaisyas (VYSH-yuhz), the traders and merchants, and then the Sudras, the farmers and laborers. There are now thousands of subcastes.

Below these four classes were the **Dalits**, which means "the oppressed." They were not viewed as part of any caste. They held jobs that higher classes saw as unclean, such as removing dead animals from the street. Dalits were forbidden from having contact with Indians of any caste.

Under the caste system, a person is born into a caste and cannot move into another. Some Indians believe that taking a job of another caste or marrying into another caste can be punished by rebirth at a lower social level. In this way, belief in reincarnation, dharma, and karma helps maintain the caste system.

The Indian constitution abolished the caste system. It declared that the poor treatment of Dalits was illegal. Still, the caste system remains a part of Hindu life, particularly in villages. Today some Dalits are educated and have good jobs. Most live in poverty, however. Violence against the Dalits still occurs.

Islam and Other Religions Islam is the largest minority religion in India, although estimates of its followers vary. Some experts think that India has the second-largest Muslim population in the world—second only to Indonesia. Perhaps 11 to 14 percent of the population is Muslim. Most Muslims live in northern India. Their presence there reflects northern India's history as the heart of former Muslim empires.

Christianity arrived in India long before the Portuguese landed there. Accounts from the A.D. 500s describe well-established Christian churches along the west coast. Today about 20 million Indians are Christian. Although Buddhism first developed in India and flourished before the Mughal period, its importance faded under later rulers. Less than 1 percent of the population is Buddhist today.

Two other religions that began in India are **Sikhism** and **Jainism**. Sikhism combines the Muslim belief in one God with the Hindu beliefs of reincarnation and karma. Because it preaches the equality of all people, Sikhism rejects the caste system. Jainism emphasizes a strict moral code based on preserving life.

TEKS **READING CHECK:** ***Human Systems*** How does the role of religion in India's constitution contrast with its role in Indian society?

India's Culture

Ways of life vary widely among India's people. The population includes many different ethnic groups. The majority of the population—72 percent—are descended from the Aryans. About one quarter of the people are Dravidians, most of whom live in the south. Smaller groups account for the rest. India's more than 1,000 languages and dialects reflect its many ethnic groups. More than 20 of these languages have at least 1 million speakers. The national language and the main language for about 30 percent of the people is Hindi. English is the most widely used language in commerce and politics. India is unique among the world's countries because it gives another 15 languages official status. Bengali, Marathi, Tamil, Telugu, and Urdu are examples. Sanskrit is also an official language.

Clothing India's tropical climates demand cool clothing. Many Indian women wear traditional clothing styles. The sari is a rectangle of cloth, usually about 18 feet (5.5 m) long, that is wrapped around the body.

INTERPRETING THE VISUAL RECORD

Saris are made from cotton, silk, or synthetic fabrics and often come in bright colors. Dressy versions may be richly embroidered in gold thread or trimmed with sequins and can be wrapped and draped in many ways. Women can express differences in status, age, occupation, region, or religion by how they wear their saris. ***How might saris help make India a distinctive region?***

A tight-fitting blouse called a *choli* is worn underneath. Also popular are pajama-like pants worn under a tunic.

In the cities, Indian men wear typical Western clothing more often than women. However, a villager may be more comfortable in a garment such as the *lungi.* Like the sari, a *lungi* is a length of fabric wrapped around the waist. Loose-fitting pants are also popular for men. Some Indian men, particularly Sikhs, wear turbans. Sikhs wear turbans as a public symbol of their religious beliefs.

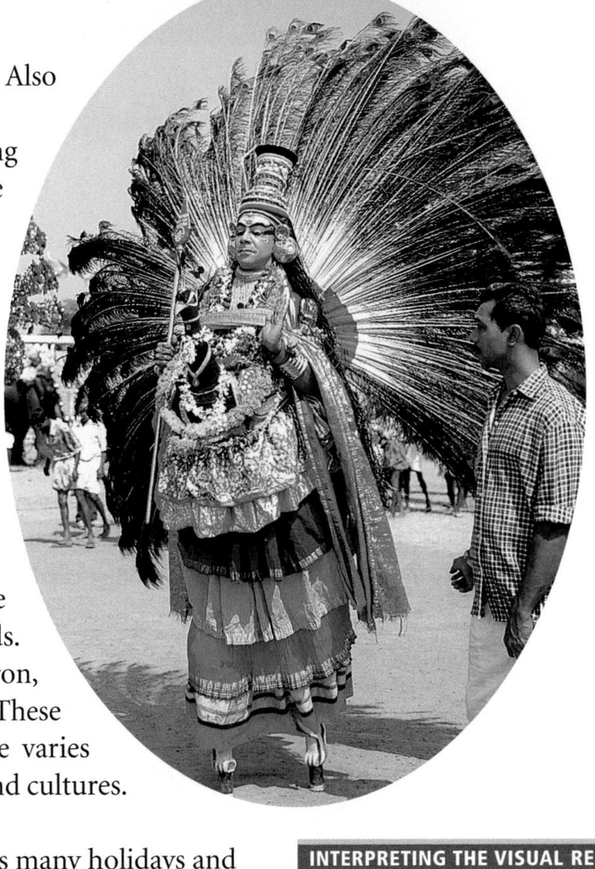

Food Many Americans are familiar with a packaged Indian spice mixture called curry powder. It flavors a gravy-based dish that is served over rice. However, real Indian curry powder does not come from a jar but is ground fresh every day. Cooks are proud of the unique mixtures they create from up to 20 herbs, spices, and seeds. Cardamom, chilies, cinnamon, cumin, mace, nutmeg, saffron, sesame seeds, and turmeric are among those often used. These spice mixtures are used in many recipes. Indian cuisine varies widely, reflecting the country's different climates, crops, and cultures.

Festivals With its wealth of ancient traditions, India has many holidays and festivals. Each religion has its special days, and other holidays mark political events. Many festivals are annual; some occur every few years.

Holi is observed mainly in northern and central India and celebrates spring and the triumph of good over evil. It is one of the most lively festivals. Everyone takes to the streets with colored water, powdered dyes, and water balloons. They smear, sprinkle, or splatter each other with color. Many other activities may be part of Holi, from wrestling tournaments to giving clothes to new brides. Pongal is a joyful three-day harvest festival of southern India. It features a parade of cattle decorated with beads, bells, and flowers.

INTERPRETING THE VISUAL RECORD

A dancer performs a peacock dance during a religious festival in Kerala in southern India. Religious festivals are celebrated throughout India and are an important part of many people's lives. All of India's major religions celebrate festivals during the year. Some festivals celebrate the seasons, phases of the Moon, or harvests. Others celebrate important dates in the lives of religious leaders. ***How does this festival compare to festivals that you have seen?*** TEKS

READING CHECK: ***Places and Regions*** Why does India have so many languages and such a wide range of customs?

TEKS Working with Sketch Maps, Questions 1, 2, 3, 4, 5

Section 2 Review

Identify
Sanskrit, Dalits, Sikhism, Jainism

Define
pantheon, sepoys, boycott, partition, reincarnation, dharma, karma, caste

Working with Sketch Maps
On the map you created in Section 1, label Delhi. Which conqueror destroyed Delhi?

Reading for the Main Idea
1. ***Human Systems*** What are the origins of Hindi and Hinduism in India?
2. ***Human Systems*** What were some results of the partition of India?
3. ***Human Systems*** What are the two most widely used languages in India? How many minor languages and dialects are spoken there?

Critical Thinking
4. **Identifying Cause and Effect** How did growing global trade beginning in the 1400s affect India over time?

Organizing What You Know
5. Create a graphic organizer like the one below. Use it to identify and compare key features of India's main religions.

Hinduism	Islam	Sikhism	Jainism

3 India Today

READ TO DISCOVER

1. What are the main features of India's economy?
2. How does life in India's villages compare to life in its cities?
3. What challenges does India face today?

DEFINE

cottage industries
jute

LOCATE

Kolkata (Calcutta)
Bangalore
Mumbai (Bombay)
Chennai (Madras)
Varanasi

WHY IT MATTERS

Rapid population growth, such as India is experiencing, affects many countries. Use CNNfyi.com or other **current events** sources to learn about the problems caused by growing populations.

India's Economy

India's economy is extremely varied and is expanding rapidly. It includes many different ways of making a living, from subsistence farming to the most advanced computer technology.

Agriculture Farming is the basis of India's economy. It contributes 25 percent of the country's gross domestic product (GDP). Farms cover about half of India's total area. Some of the major crops are rice, wheat, tea, sugarcane, and sorghum. Half of the world's mangoes come from India. No country grows more peanuts, sesame seeds, or tea than India. Only China grows more rice.

Half of India's farms are less than 2.5 acres (1 ha). A majority of farmers own their land. However, as land is divided among a family's sons, the farms become smaller and thus less profitable.

In recent decades, India has become almost self-sufficient in food-grain crops. Yet if the summer monsoon rains are late, inadequate, or too heavy, some people may still go hungry. The government constantly explores ways to make the country's agriculture less dependent on the summer monsoon.

INTERPRETING THE VISUAL RECORD

A pepper harvest in northern India creates a colorful scene. Native to the Americas, peppers were first introduced to Europe by the Spanish and later spread throughout Asia. Spicy peppers are now an important part of Indian cuisine. ***How did the diffusion of peppers from the Americas affect the landscape in this photo?*** TEKS

FOCUS ON SCIENCE, TECHNOLOGY, AND SOCIETY

The Green Revolution The government has made many efforts to increase food production for India's rapidly growing population. The need to grow more food was highlighted in 1943 when a disastrous famine struck eastern India. An estimated 4 million people starved to death. Although food distribution problems played a role in the famine, there was also a clear shortage of food. Government officials knew they had to increase the country's agricultural production. Their efforts, and the partial success achieved between 1967 and 1978, became known as the Green Revolution.

The Green Revolution consisted of three main elements. These were increasing the amount of cultivated land, harvesting two crops per year from existing farmland, and increasing yields with genetically improved seeds. Expanding the amount of farmland had been going on for centuries. Growing two crops each year was more difficult to accomplish. Since the monsoon rains ordinarily made only one crop possible, officials decided to use irrigation to create an artificial second rainy season. They built large and small irrigation projects to trap rainwater from the natural monsoon. Scientists also developed new strains of corn, millet, rice, and wheat that produced more grain per acre. In the areas where they were used, the new types increased yields dramatically.

Within a few decades the Green Revolution changed India from a country plagued by starvation to one that could usually feed its people. However, the Green Revolution had a downside. The irrigation projects, which required huge dams, displaced people and disrupted the environment. The new grain varieties used more dangerous pesticides and fertilizers. Many farmers who could not afford irrigation or chemicals got lower yields than they had from older varieties. Moreover, the new techniques did not work in areas where water was scarce or where farmers had no money for investment.

READING CHECK: ***Environment and Society*** How did the Green Revolution affect food production and the environment in India?

INTERPRETING THE VISUAL RECORD

About two thirds of all Indians are farmers. While many do not have access to advanced farming technologies, others have benefited greatly from the innovations of the Green Revolution. For example, the use of high-yielding varieties of seeds, particularly for wheat, has led to dramatic increases in some areas. ***How is this farmer using animals? How might the use of more advanced technologies change the way this person farms?***

Industry Although India is largely agricultural, its industrial production ranks about tenth among the world's countries. When its GDP is distributed among its billion-plus population, however, the per capita GDP is very low. As a result, most geographers and economists describe India as a developing country.

Millions of Indians make a living by working at home in small-scale industries called **cottage industries**. Many of these skilled workers are women. They weave silk fabrics, carve wooden statues, and make silver and gold lace, as well as many other beautiful handicrafts.

India also has large-scale commercial manufacturing. The country's many cotton and woolen mills produce textiles, the leading export. Factories in Kolkata (Calcutta) process **jute**, a plant fiber used to make burlap and rope. Factories use the steel from India's own steel mills to make durable goods such as diesel engines and cars.

INTERPRETING THE VISUAL RECORD

Many Indian women work in cottage industries. Here, women in Rajasthan dye fabrics that they will sell in local markets. ***How do you think cottage industries help people meet their basic needs?*** TEKS

Industry requires power. To increase its industrial production, India must increase its power supply. The country already imports large amounts of oil. India's rivers could supply additional hydropower, but dams can create new problems. To attract more foreign investment, India must strengthen its infrastructure. Airlines, communications systems, railroads, ports, and roads must be improved. These systems are needed to operate factories, move materials and products, and help people do their jobs efficiently. In India most of these systems lag behind what is available in other countries.

The government has been successful in attracting high-tech businesses. For example, Bangalore, in southern India, has so many computer companies that some have compared it to the Silicon Valley of the United States. (See Cities & Settlements: Bangalore.) Many high-tech and service industry workers belong to a growing middle class.

TEKS **READING CHECK:** ***Human Systems*** What are some ways in which India's market economy is improving?

Cities and Villages

India has more than half a million villages, where more than 70 percent of the people live. Most of these villages have fewer than 1,000 residents. At the same time, India has enormous cities. Two of them—Mumbai (Bombay) and Kolkata—rank fifth and eighth among the world's largest cities. More Indians are moving into the cities to look for work, but finding none. In fact, India's big cities are growing about twice as fast as its small towns.

City Life Crowding, noise, smog, and traffic are part of daily life in India's cities. In general, city dwellers wear European-style clothing and read

English-language newspapers. Most work in factories and offices. Higher education is available in the cities, and women work alongside men in professional and factory jobs.

A small number of rich businesspeople and landowners live in the cities. The working middle class makes up a larger part of urban society. Most live in small apartments and can afford many manufactured goods. However, most of the urban population lives in poverty. Many are homeless and live in the streets. Many others live in shacks built from scraps of cardboard, metal, and wood. They have no clean water or sanitation. Each day they must look for work.

Mumbai has about 18 million residents. It is the center of India's huge movie industry. So many films are produced there that the city has been nicknamed Bollywood. Kolkata has more than 12 million residents and is one of the world's busiest seaports. Although it has a large homeless population, Kolkata is a vibrant place that promotes itself as the City of Joy. Delhi is the hub of northern India. The city's old section contains beautiful Mughal buildings. The new part of the vast city—New Delhi—is India's capital. The East India Company founded Chennai (Madras) on the southeast coast. It is now a major rail and industrial center. Varanasi, on the Ganges River, is a holy city for Hindus.

Our Amazing Planet

In Mumbai (Bombay), some 20,000 people known as *dhobis* make a living washing clothes. The *dhobi* identifies the laundry's owner with marks most people cannot see. Sometimes police identify bodies by tracing *dhobi* marks on the clothing.

Village Life For millions of India's villagers, life goes on as it has for generations before them. Only recently have paved roads, electricity, or telephone services reached many villages. Some villages can only be reached on foot. Sanitation is still poor, and many people do not have clean drinking water. Medical services are very limited. More than half of the rural population cannot read or write.

Farm families tend to be large because children have to help in the fields. In addition, both Hindu and Muslim cultures value large families. In general, sons are valued more highly than daughters. They will also take care of their parents when they are old. Therefore, many rural couples continue to have children until they have at least two boys. Men dominate village society. When a woman marries, she becomes a member of her husband's family.

READING CHECK: ***Human Systems*** Why do many people leave the villages for the cities? What do they find there?

Challenges

India remains a country of contrasts. While millions struggle with extreme poverty and inequality, a few have wealth and privilege. Life for many Indians has improved in recent decades, but the country faces several issues that may cloud its future.

Population and Poverty Many of India's problems are connected to rapid population growth.

Kolkata (Calcutta), former capital of British India from 1773 to 1912, is one of India's largest cities. It has a very high population density of about 85,500 persons per square mile, and overcrowding is a major problem. Yet, Kolkata continues to attract new immigrants, particularly from nearby rural areas of eastern India.

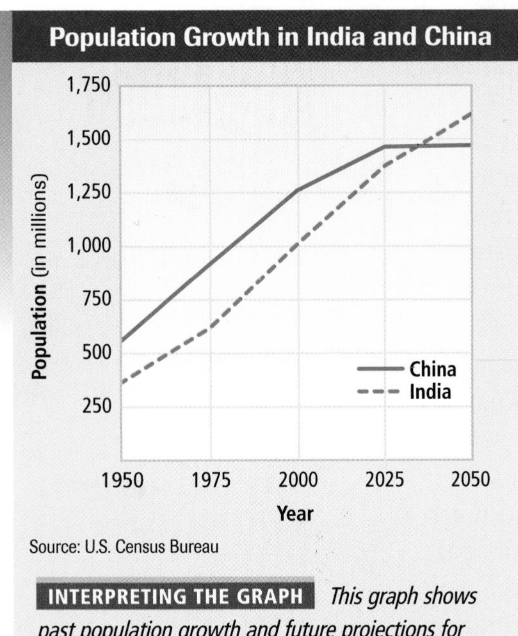

INTERPRETING THE GRAPH *This graph shows past population growth and future projections for the world's two most populous countries.* ***Which country has a larger population today? Which country is projected to have a larger population in 2050?*** TEKS

Because of improved sanitation and health care, the death rate has dropped much faster than the birthrate. The result is a high population growth rate. In addition, more than a third of the people are 14 or younger. Therefore, as this group enters the child-bearing years the population will continue to grow even if the birthrate drops. Government programs encouraging smaller families have had limited success.

The large population often overwhelms services, from communication to transportation. Millions of Indians in both rural and urban areas are still dreadfully poor. Many people never attend school, and this lack of education limits their opportunities.

Environmental Concerns A major challenge for India is to clean up and protect its environment. As much as 70 percent of India's surface water may be polluted. Poorly regulated industries and inadequate sewer systems are partly to blame.

India's large population has strained the country's land resources. Deforestation, overgrazing, and soil erosion have caused serious environmental damage. In fact, half of India's forests have been cut down since the country became independent. The government does, however, support conservation. It has created 80 national parks, 440 wildlife sanctuaries, and 23 tiger reserves. Once on the brink of extinction, India's tigers now number in the thousands. Their future is not secure, however. Illegal trade in tiger skins and bones continues.

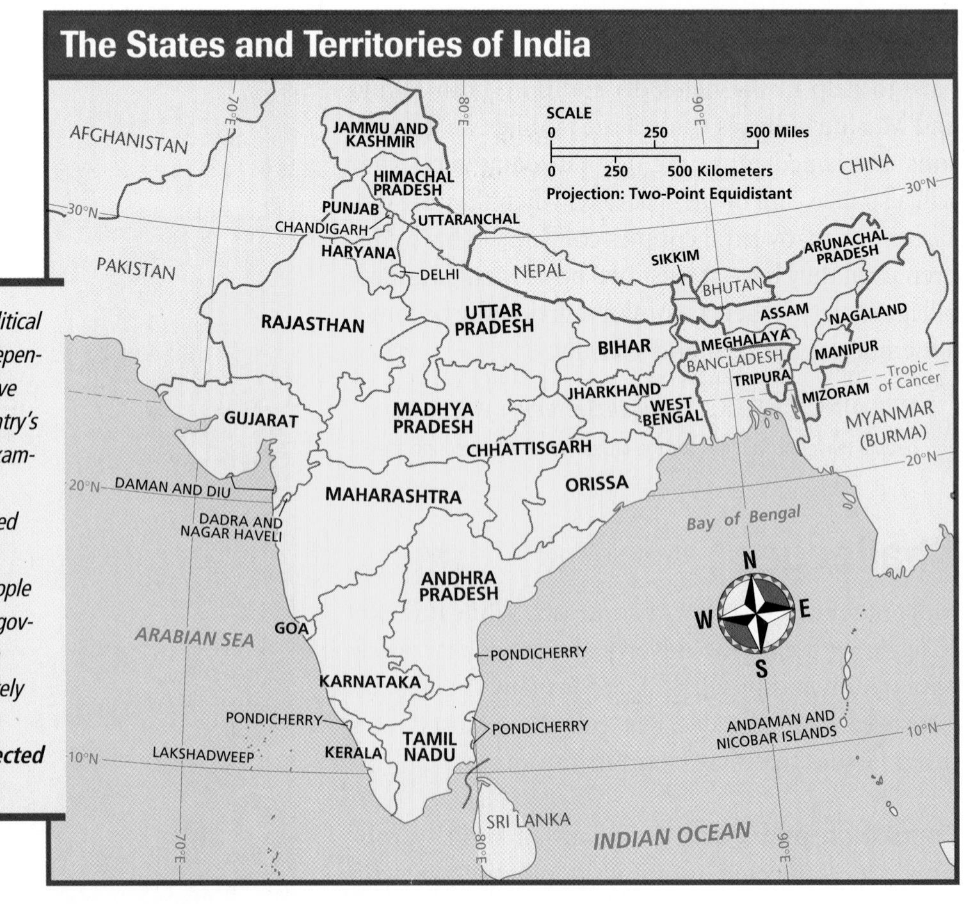

INTERPRETING THE MAP *India's political geography has changed greatly since independence in 1947. Over time, India's states have been reorganized to better reflect the country's major language and culture regions. For example, in 1960 the state of Bombay was divided into Gujarat and Maharashtra based mainly on language differences. In 1961 Nagaland became a state for the Naga people after an armed struggle with the national government. Three new states were created in 2000, and India's internal borders seem likely to change further in the future.* ***How have forces of conflict and cooperation affected India's political geography?*** TEKS

Political Issues India faces challenges from neighboring countries and its own citizens. Ever since the partition, India and Pakistan have clashed over the region of Kashmir in the far northwest. At the time of partition, most people in Kashmir were Muslims. However, unrest in the region led Kashmir's ruler to seek a union with India. This decision led to an armed struggle between India and Pakistan over control of the disputed region. Today Pakistan controls the northern and western areas of Kashmir. India controls southern and southeastern Kashmir as part of the state of Jammu and Kashmir. A line of control separates the two areas of Kashmir. This line was agreed to in 1972 after several clashes between Indian and Pakistani troops. Many Muslim Kashmiris want the entire region to be part of Pakistan. Their movement and the government's efforts to crush it have resulted in the deaths of about 10,000 civilians since 1989. Because both India and Pakistan have nuclear weapons, this tension creates concern around the world. The two countries also disagree over how much water India should take from the upper Indus River. India has border disputes with China as well. China had never accepted the boundaries of northeastern Kashmir. In the 1950s Chinese forces entered Ladakh, a region in Kashmir. This led to border clashes between India and China. Today China controls the northeastern part of Ladakh. However, India does not recognize Chinese claims in the region.

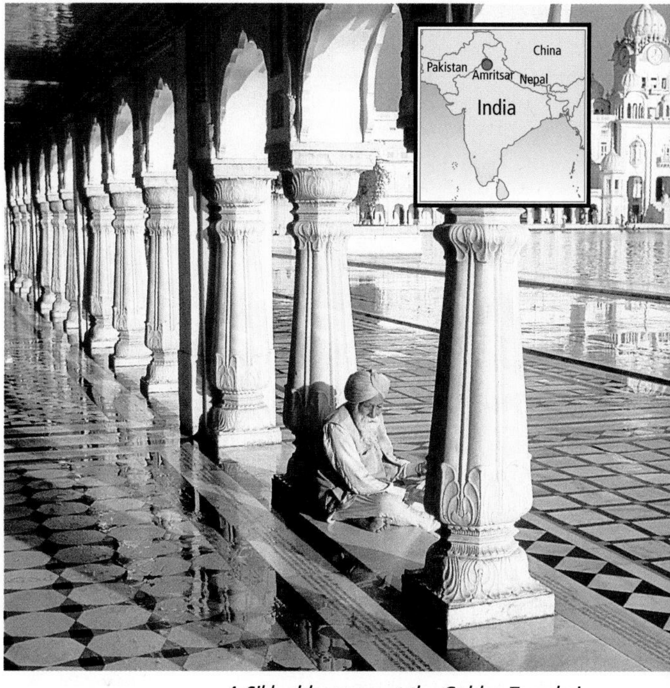

A Sikh elder prays at the Golden Temple in Amritsar, India. The Golden Temple is the spiritual center for the world's Sikhs and has figured prominently in their struggles with India's government. In 1984, Indian troops attacked Sikh extremists who were using the temple as a fortress and refuge.

Conflicts between India's Hindus and Muslims sometimes erupt into violence. Leaders from each group have accused the government of favoring the other. Sikhs have also demanded more political power in Punjab. In the early 1980s some Sikh extremists began to call for an independent country of their own called Khalistan, or "Land of the Pure." They began a campaign of terrorism against India's government and people. Ethnic groups in other parts of India have also demanded more autonomy, with violence the frequent result.

TEKS **READING CHECK:** *Environment and Society* What issue underlies several of India's challenges?

Section 3 Review

TEKS Working with Sketch Maps, Questions 1, 2, 3, 4, 5

Define cottage industries, jute

Working with Sketch Maps

On the map you created in Section 2, label Kolkata (Calcutta), Bangalore, Mumbai (Bombay), Chennai (Madras), and Varanasi. Why is Mumbai sometimes called Bollywood?

Reading for the Main Idea

1. *Environment and Society* In what ways was the Green Revolution unsuccessful?
2. *Human Systems* What types of goods are produced in India's cottage industries? In its commercial industries?
3. *Human Systems* Why do geographers consider India a developing country even though the country's industrial production is among the highest in the world?

Critical Thinking

4. **Analyzing Information** Why is control of Kashmir disputed? Why do you think this conflict has lasted for more than 50 years?

Organizing What You Know

5. Copy the graphic organizer shown below. Fill in each oval with one of the major challenges facing India. Then draw arrows among the ovals to show how the issues are related. Write sentences or phrases near the arrows to describe the connections.

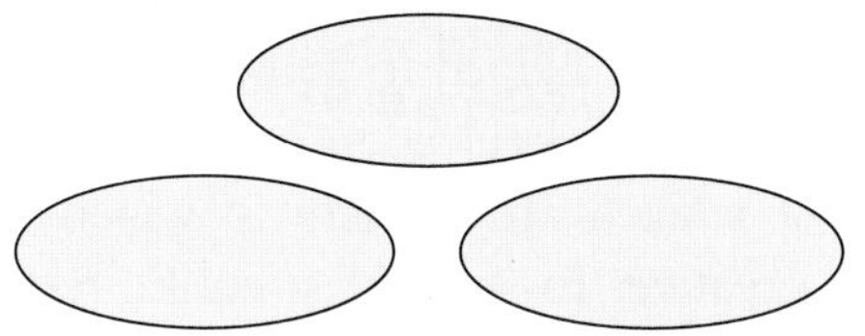

CITIES & SETTLEMENTS

Bangalore

Places and Regions Located on the Deccan Peninsula, Bangalore is India's sixth-largest city. With nearly 5 million people, it is one of the fastest-growing cities in Asia. This tremendous growth has brought much change to the city. Bangalore's development illustrates some of the forces that shape a city and its culture. It also shows how a rapidly changing urban environment can affect people's lives.

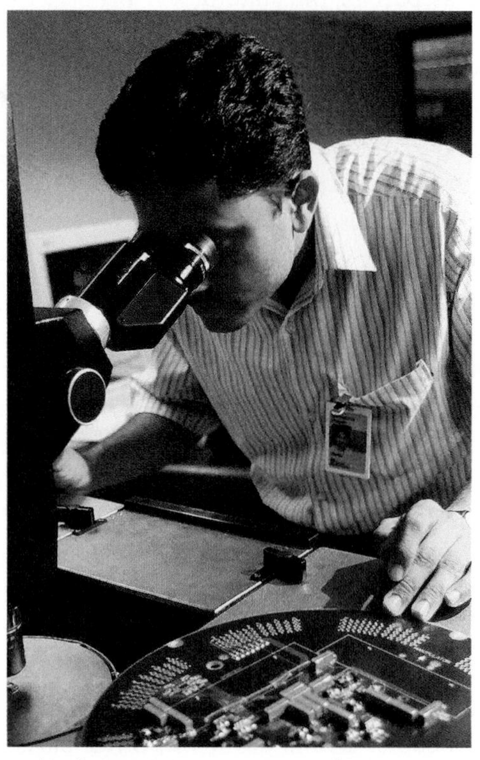

A chip designer inspects a computer chip in Bangalore. The city is home to hundreds of technology companies employing an estimated 75,000 people.

Garden City to Silicon City

Bangalore may have the most pleasant climate of any Indian city. Humidity is low, and summer temperatures average in the 80s. Even in winter it seldom gets colder than 60°F (16°C). Rainfall is also moderate, averaging less than 40 inches (102 cm) per year. These conditions inspired Bangalore's former Indian and British rulers to create many gardens, lakes, and parks throughout the city. Trees were planted along many streets to shade and beautify them. Because it had such a lovely environment, Bangalore was often called the Garden City. It became a popular vacation and retirement spot for wealthy Indians. It was also a headquarters for British officials during part of the colonial era. Many of these wealthy Indians and British officials built grand mansions in the city.

In the 1950s and 1960s Bangalore's pleasant environment helped to make it India's scientific research center. Indian leaders believed the city would provide a relaxed atmosphere in which scientists could be creative. To support scientific research in Bangalore, India's government invested heavily in the city. As a result, Bangalore now has three universities, 14 engineering colleges, 47 technical schools, and many research institutes. These include India's most advanced military and space research centers and several private industrial training institutes. These developments brought a large number of highly educated people to Bangalore. Eventually, they helped transform the Garden City into an Indian version of Silicon Valley in the United States.

Booming Bangalore

Because a large percentage of India's population does not use computers, the country is generally not thought of as a high-tech leader. For this reason, you may be surprised to know that India is second only to the United States in computer software exports. Most of this software comes from Bangalore. At first, the city's major export was not software. It was people. In the 1980s American companies began to computerize their operations on a wide scale. Yet there were not enough American software engineers and programmers to fill all the jobs in those fields. As a result, many of Bangalore's computer science graduates left to take high-paying jobs in the United States. However, two developments soon turned Bangalore into a boomtown. American companies found that computer science salaries in India were one fourth the size of those elsewhere. In addition, improved telecommunications made it easier for American companies to contract with Indian software firms.

Bangalore's booming high-tech economy has created an increase in traffic congestion and air pollution.

They could also set up their own computer operations in India. Soon more than 100 Indian, American, and multinational computer firms sprang up in Bangalore.

A Changing City

Bangalore changed quickly. Some of the grand old homes were torn down to make room for new high-rise office buildings. Others were gutted and filled with cubicles to become offices. Still others were divided into apartments for the flood of people who came to fill new software jobs. Thousands of cars and motorcycles soon clogged once quiet streets where horse-drawn carriages had traveled as late as the 1970s. The air became more polluted. Bangalore's explosive growth put a strain on its other systems as well. An overloaded electrical system began causing power outages, sometimes several times a day. In addition, the demand for housing and office space caused living and business costs to skyrocket.

Some of the high-tech companies have left Bangalore. Many are now located in so-called technology parks that have sprung up in nearby suburbs. These parks have their own water, power, and communication systems. Some also provide housing, schools, stores, and recreational facilities. Company employees thus have little need to go into the city. In fact, many buildings within Bangalore stand empty. In some ways, the Garden City's rapid growth has wilted. The city government is looking at ways to revive the city's glory.

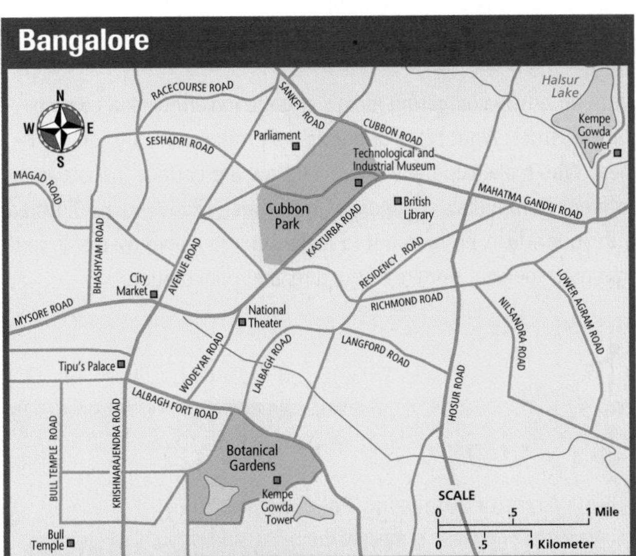

INTERPRETING THE MAP *The heart of Bangalore includes important governmental and cultural buildings, as well as parks, gardens, and lakes.* ***What features on the map do you see that might have inspired people to call Bangalore the Garden City?***

Applying What You Know 

1. **Comparing** How does Bangalore's experience compare to cities or communities in your region that have grown rapidly? How have the political, economic, social, and environmental relationships in those places changed?
2. **Problem Solving** What actions might Bangalore's government take to revive the city?

CHAPTER 25 Review

Building Vocabulary TEKS

On a separate sheet of paper, explain the following terms by using them correctly in sentences.

subcontinent	partition	caste
pantheon	reincarnation	cottage industries
sepoys	dharma	jute
boycott	karma	

Locating Key Places TEKS

On a separate sheet of paper, match the letters on the map with their correct labels.

Himalayas	Western Ghats	Brahmaputra River
Ganges River	Eastern Ghats	Thar Desert
Gangetic Plain	Narmada River	Kolkata (Calcutta)

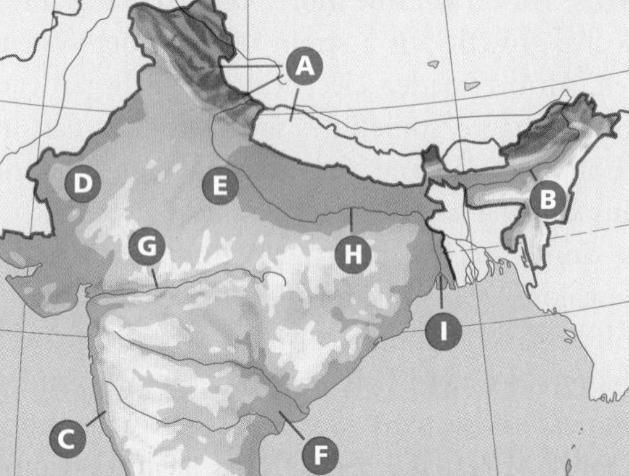

Understanding the Main Ideas TEKS

Section 1

1. *Places and Regions* What geographic features, such as river systems and climates, characterize India's three major landform regions?
2. *Physical Systems* At what time of the year do monsoons bring rainfall to India, and where are the rains the heaviest?

Section 2

3. *Human Systems* In what ways have invasions, trade, and outside control affected Indian society and culture?
4. *Human Systems* What is India's main religion? Which other religions are found there?

Section 3

5. *Places and Regions* What conflicts does India have with Pakistan?

Thinking Critically for TAKS TEKS

1. **Drawing Inferences and Conclusions** How does the timing of the summer monsoons relate to the tilt of Earth on its axis?
2. **Comparing and Contrasting** How does Indian village life compare to city life?
3. **Making Generalizations and Predictions** How might rapid population growth make it hard for India to develop its economy?

Using the Geographer's Tools TEKS

1. **Analyzing Maps** Study the map of India's states and territories in Section 3. What are some of India's largest states? What are some of the smallest? What might this indicate about the distribution of political power in India?
2. **Analyzing Graphs** Study the graph of population growth in India and China in Section 3. What was the combined population of India and China in 2000?
3. **Preparing Graphs** Create a pyramid graph of India's four major castes, using the information in Section 2. At the top of the pyramid, list India's highest caste. At the bottom, list the lowest caste. Which group is below the lowest caste?

Writing for TAKS TEKS

Write an article comparing India's cottage industries and commercial industries. What types of goods are produced in cottage industries? Who makes up the workforce? How are cottage industries different from India's commercial industries? When you are finished with your article, proofread it to make sure you have used standard grammar, spelling, sentence structure, and punctuation.

SKILL BUILDING

Geography for Life TEKS

Skills Activity: Analyzing Political Instability

Places and Regions Review the discussion of political issues in Section 3. Which areas in India are politically unstable? How is religion a factor in the country's internal and external conflicts? How does political instability in India compare to political instability in other areas of the world that you have studied? Generally, what are some of the forces that affect political stability?

Building Social Studies Skills

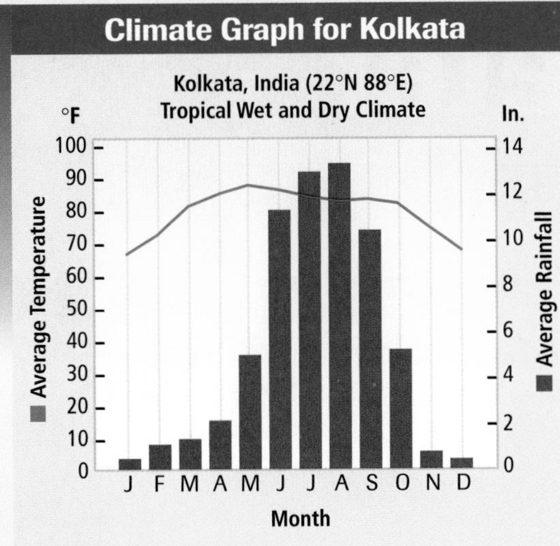

Interpreting Graphs

Use the climate graph to answer the following questions.

1. During which months is the average rainfall above 12 inches?
   a. June, July
   b. April, May
   c. August, September
   d. July, August
2. Which three months would be the least favorable to explore Kolkata by car? Why?

Analyzing Primary Sources

Read the following excerpt from an article that Geoffrey C. Ward wrote after visiting India during the 50-year anniversary of its independence from Britain. Then answer the following questions.

> **"For all its newfound modernism India remains steeped in religion. The pious cacophony [discordant sounds] I hear from my window each dawn attests to that. First comes chanting from a Sikh gurdwara [temple], which is soon partly drowned out by the sound of temple bells and the voice of a priest offering prayers from a Hindu temple dedicated to Siva. Then, louder than the rest, comes the wobbly tenor of a Muslim muezzin [crier], proclaiming the greatness of Allah from a mosque."**

3. Which religion is not mentioned in the excerpt?
   a. Islam
   b. Judaism
   c. Hinduism
   d. Sikhism
4. In what ways does the writer imply that the past and traditions are still important to modern India?

Alternative Assessment

PORTFOLIO ACTIVITY

Learning about Your Local Geography TEKS

Individual Project: Research

Plan, organize, and complete a research project on population growth in your community. In your report answer the following questions: What is the history of population growth in your community during the last 50 years? How many people live in your community today? Is the population increasing or decreasing? What are the reasons for the trend? What do you think will happen in the next 20 years? How has population change affected life in your community?

internet connect

Internet Activity: go.hrw.com
KEYWORD: SW3 GT25 TEKS

Choose an activity on India to:

- explore the regions of India.
- take the India GeoMap Challenge!
- learn about economic and social issues in India today.

CHAPTER 26

The Indian Perimeter

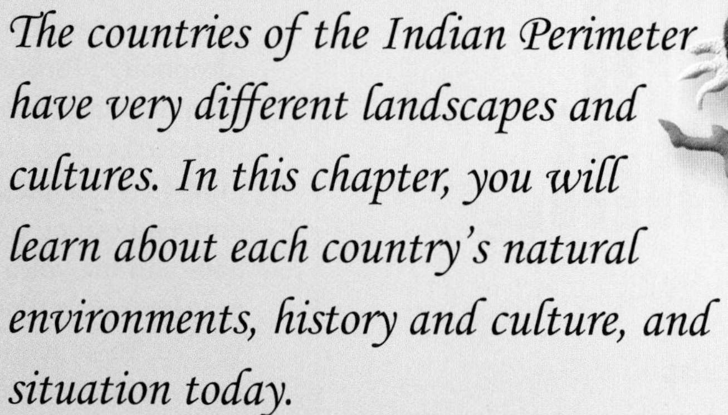

The countries of the Indian Perimeter have very different landscapes and cultures. In this chapter, you will learn about each country's natural environments, history and culture, and situation today.

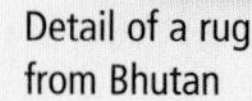

Detail of a rug from Bhutan

Bragpa man in Bhutan

A Saleem Ale Hem (God's peace be upon you)! I am Rehan. I am an only child and live with my parents in Karachi, a big sprawling city like Los Angeles. On one side is the sea, and on the other is the desert. I love Pakistan. It is a poor country, but it is very beautiful.

I really like to watch Indian movies. Even though relations between India and Pakistan are tense, Indian movies are very popular. Movies open at the same time in India and Pakistan. We do not make movies here.

I attend a boys' private school styled after the British public school system. The school is 157 years old. After school I play tennis or cricket before going home. Our house is connected to my grandparents' house, and we usually eat our meals all together. Dinner is almost always rice with goat meat or beef, along with lentils or vegetables. We never eat pork because we are Muslim and it is forbidden.

Next year, I am going to go to America with my mother. My parents want me to go to a world-class university. Also, Pakistan is not safe. Even in big cities like Karachi, the crime that upper-class people fear most is kidnapping. My father is a doctor. He will have to stay in Pakistan to take care of his parents and keep up his medical practice. My father has to care for his parents because both his brothers are living in the United States.

Natural Environments

READ TO DISCOVER

1. What are the main physical features of the Indian Perimeter?
2. What types of climates, plants, and animals are found there?
3. What natural resources do countries in the region have?

WHY IT MATTERS

Bangladesh is often hit by devastating tropical storms, which can kill people and destroy homes and crops. Use CNNfyi.com or other **current events** sources to learn about the effects of such storms.

DEFINE

storm surge

LOCATE

Indus River
Baluchistan
Thar Desert
Mount Everest
Tarai
Palk Strait
Chittagong Hill Tracts

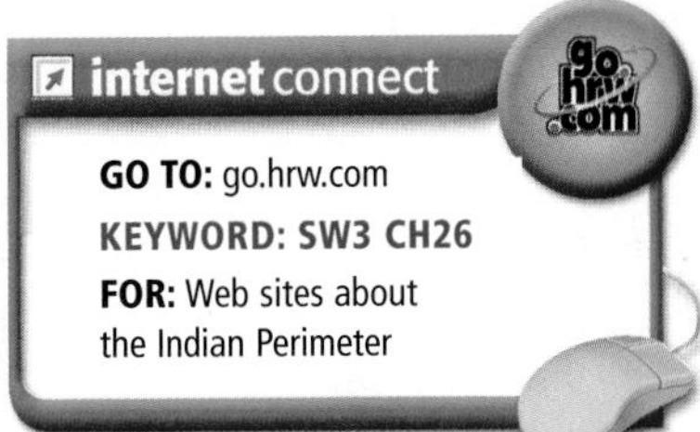

The Indian Perimeter: Physical-Political

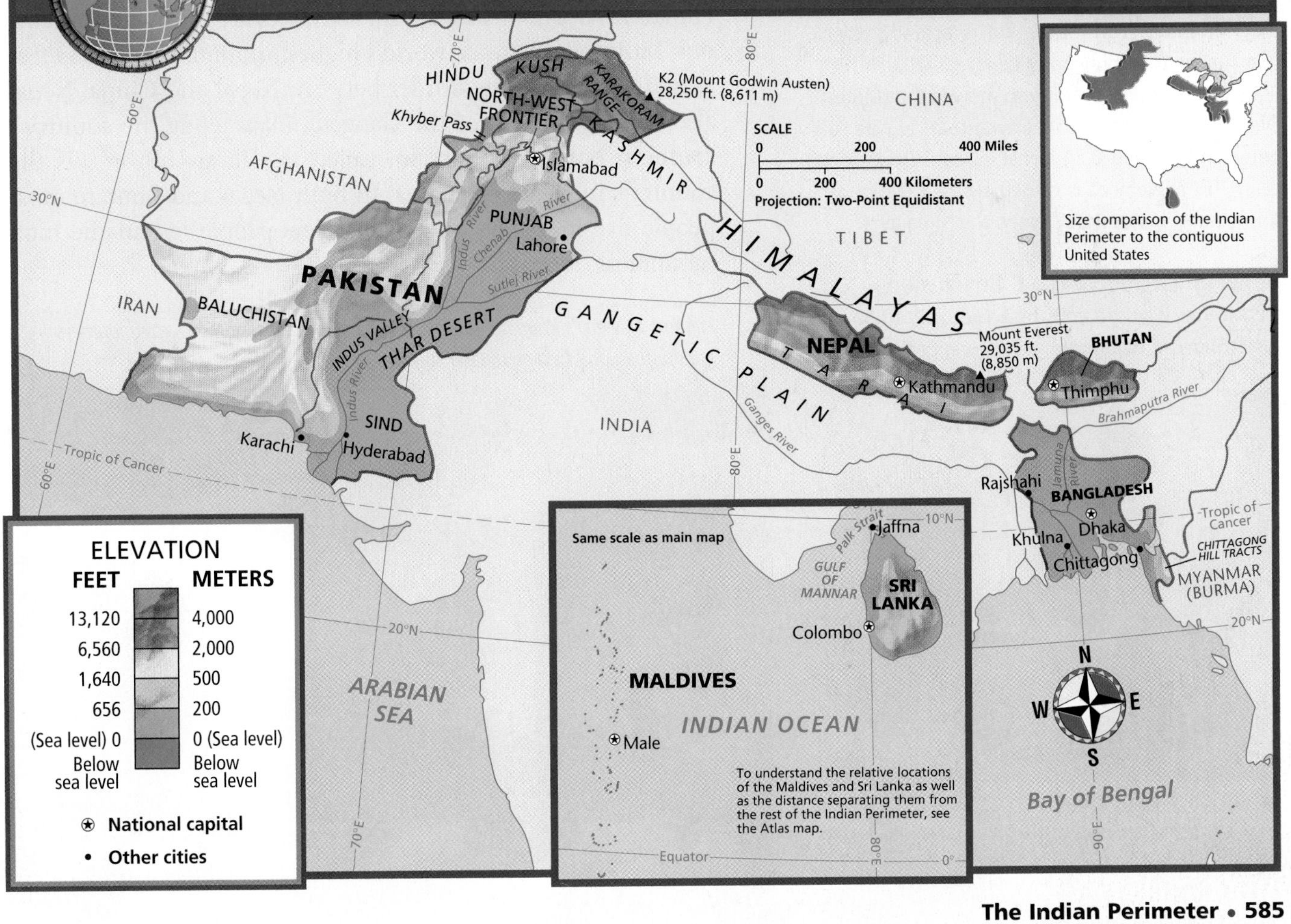

HINDU KUSH
KARAKORAM RANGE
K2 (Mount Godwin Austen) 28,250 ft. (8,611 m)
NORTH-WEST FRONTIER
Khyber Pass
KASHMIR
Islamabad
CHINA
AFGHANISTAN
PUNJAB
Lahore
Chenab River
Indus River
Sutlej River
HIMALAYAS
TIBET
PAKISTAN
IRAN
BALUCHISTAN
INDUS VALLEY
THAR DESERT
GANGETIC PLAIN
NEPAL
TARAI
Kathmandu
Mount Everest 29,035 ft. (8,850 m)
BHUTAN
Thimphu
Brahmaputra River
Ganges River
SIND
Karachi
Hyderabad
INDIA
Tropic of Cancer
Rajshahi
Jamuna River
BANGLADESH
Dhaka
Khulna
Chittagong
CHITTAGONG HILL TRACTS
MYANMAR (BURMA)
ARABIAN SEA
Bay of Bengal

SCALE
0 200 400 Miles
0 200 400 Kilometers
Projection: Two-Point Equidistant

Size comparison of the Indian Perimeter to the contiguous United States

Same scale as main map
Palk Strait
Jaffna
GULF OF MANNAR
SRI LANKA
Colombo
MALDIVES
Male
INDIAN OCEAN
Equator

To understand the relative locations of the Maldives and Sri Lanka as well as the distance separating them from the rest of the Indian Perimeter, see the Atlas map.

ELEVATION

FEET	METERS
13,120	4,000
6,560	2,000
1,640	500
656	200
(Sea level) 0	0 (Sea level)
Below sea level	Below sea level

⊛ National capital
• Other cities

Connecting to TECHNOLOGY

Measuring Mount Everest

Technology often changes what we know about Earth. For example, in 1954 surveyors measured the elevation of Mount Everest at 29,028 feet (8,848 m) above sea level. Then, in 1999, researchers funded by the National Geographic Society used Global Positioning System (GPS) technology to measure Everest. Special GPS equipment on Earth's surface sends information to orbiting satellites. Researchers can then use the data that is generated to find the exact location and elevation of any place on Earth's surface.

Getting GPS equipment to the top of Everest would have been impossible only a few years before. An early GPS unit might have weighed 50 pounds (23 kg). These early units were too heavy to carry in the thin air and dangerous conditions of Everest. By 1999 a GPS unit was about the size of a sandwich. The researchers also used a special radar device that could "see" through snow because the summit of Everest is generally snow-covered.

After the equipment was in place, the researchers studied the new GPS data to determine the exact elevation of Mount Everest. They learned that Mount Everest's elevation is actually seven feet higher than previously believed—29,035 feet (8,850 m). The 1999 survey also showed that Mount Everest is moving northeastward at a rate of between 0.12 and 0.24 inches (3 and 6 mm) per year. This movement is caused by the collision of the tectonic plates that created the Himalayas.

Drawing Inferences and Conclusions In what other ways do you think Global Positioning Systems might be helpful in studying Earth?

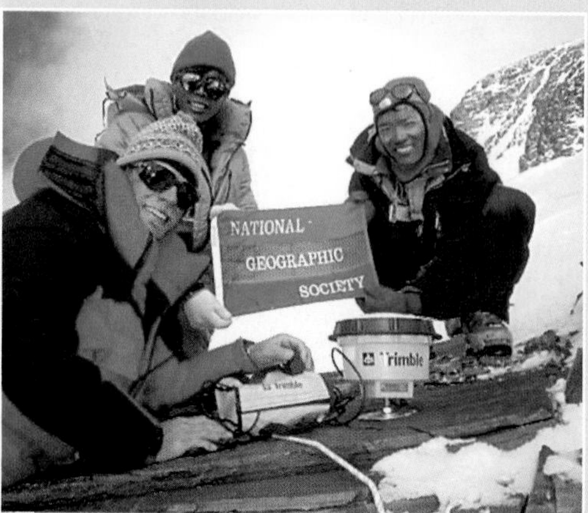

Physical Features

As the name of this region suggests, the countries of the Indian Perimeter border India. They are Pakistan, Nepal, Bhutan, Bangladesh, Sri Lanka, and the Maldives. These countries have greatly varying physical features. Landforms range from Earth's highest mountain range to islands that barely rise above the sea. Follow along on the chapter map as we look more closely at the natural environments of these countries.

Pakistan and the Himalayan Countries To the northwest of India is Pakistan. About twice the area of California, Pakistan is the largest country of the Indian Perimeter. The towering mountains of the Himalayas, Karakoram Range, and Hindu Kush cover the northern part of Pakistan. Four mountain peaks in Pakistan are higher than 26,000 feet (7,925 m).

South of these mountains lies a region of hills and plateaus, some of which are good for farming. The country's main farming region, however, is the Indus Valley. This large fertile plain was formed from massive amounts of sediment deposited by the Indus River. Although the area receives little rainfall, it is made productive with irrigation. West of the Indus Valley lies Baluchistan, an arid and lightly populated plateau. To the east of the Indus, the Thar Desert stretches into India.

Nepal and Bhutan are landlocked countries in the Himalayas north of India. Both countries are mostly mountainous. Mount Everest, the world's highest mountain at 29,035 feet (8,850 m), lies on the border between Nepal and China. Nepal also includes a strip of the Gangetic Plain along the country's southern border. This plain, called the Tarai (tuh-RY), is the country's main farming area. In both Nepal and Bhutan, most people live in valleys and plains. Few people live in the high mountains.

Pilgrims in Nepal's Khumbu Valley stack rocks as thanks for their sacred journeys and for the good fortune of later travelers.

Bangladesh and the Island Countries South of Bhutan and across a narrow extension of India lies Bangladesh. Most of this country is a broad flat alluvial plain. The Ganges and Brahmaputra Rivers join and flow southward through a huge river delta before draining into the Bay of Bengal. This delta is crisscrossed by many smaller rivers and waterways. The rivers overflow their banks every year. These floods deposit new layers of silt on the land, ensuring its fertility. Yet depending on their timing, floods often damage crops. Floods caused by tropical storms can also destroy villages and kill people and livestock.

Floods are also constantly reshaping the topography of Bangladesh. Floods can change the courses of rivers, sweep away years' worth of soil deposits, and form new islands. For this reason, no map of Bangladesh stays accurate for long.

South of India is Sri Lanka, a beautiful tropical island country in the Indian Ocean. The Palk Strait separates Sri Lanka from India. A coastal plain surrounds the mountainous center of the country.

The Maldives lie to the southwest of India in the Indian Ocean. This tropical country is made up of a chain of about 1,200 small coral islands. They have a total land area of just 115 square miles (298 sq km). None of the islands rises more than 8 feet (2.4 m) above sea level.

Our Amazing Planet

Mount Everest is named after Sir George Everest, a British surveyor general of India. In Nepal, Mount Everest is called Sagarmatha, which means "Goddess of the Sky." In Tibet it is called Chomolungma, or "Mother Goddess of the Universe."

TEKS **READING CHECK:** ***Physical Systems*** What role have rivers played in creating good farmland in the region?

Climates, Plants, and Animals

The climates of the Indian Perimeter range from arid and semiarid to tropical humid and highland. The peaks of the Himalayas are among the coldest places on Earth. Yet Sri Lanka and the Maldives, which are much lower and lie close to the equator, have warm tropical climates.

As in India, monsoons greatly affect the weather and vegetation of the Indian Perimeter countries. (See the climate graph.) The wet summer monsoon brings rain to most of the subcontinent, including lowland areas of Bhutan and Nepal. The rains support lush tropical rain forests from Bangladesh through the Himalayan lowlands. In fact, much of Bangladesh is covered with lush vegetation. Mango, bamboo, coconut, and date palm are common. About 15 percent of the country is forested. These areas support a variety of wildlife, including the Bengal tiger. Herds of wild elephants live in the Chittagong Hill Tracts in southeastern Bangladesh.

Powerful tropical cyclones often occur in the Bay of Bengal. These storms are called hurricanes in North America and typhoons in the western Pacific. They bring heavy rains and winds that can reach 150 miles (241 km) per hour. The strong winds can cause water to "pile up" in the Bay of Bengal. In what is called a **storm surge**, waters then wash ashore like a very high tide. Waves as high as 20 feet (6 m) may crash onshore. Because coastal areas are flat and low, these storm surges can move far inland and cause severe flooding. Since the early 1700s, when weather records were first kept, more than 1 million people have died because of these storms.

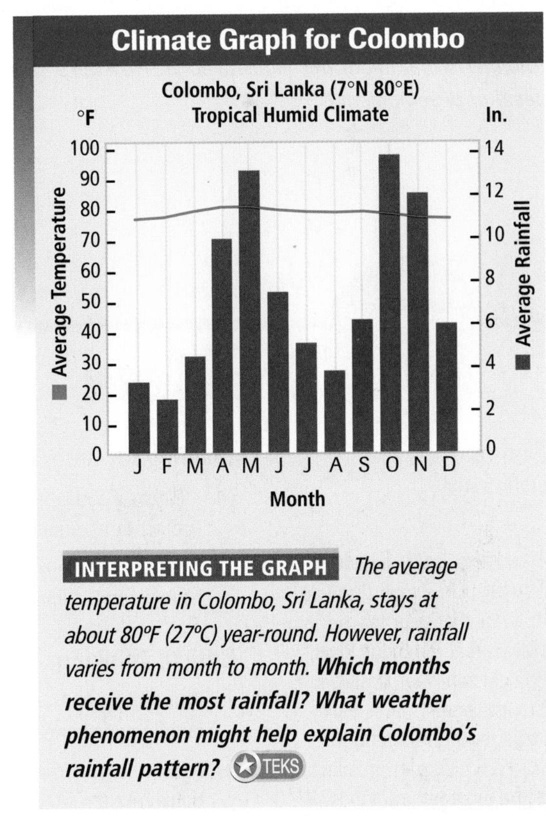

INTERPRETING THE GRAPH *The average temperature in Colombo, Sri Lanka, stays at about 80°F (27°C) year-round. However, rainfall varies from month to month.* ***Which months receive the most rainfall? What weather phenomenon might help explain Colombo's rainfall pattern?*** TEKS

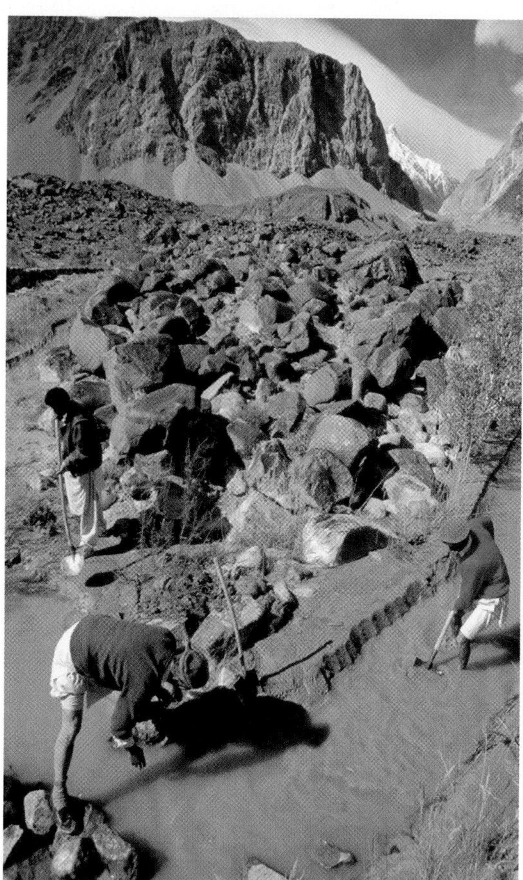

INTERPRETING THE VISUAL RECORD *Members of the Aga Khan Rural Support Program build an irrigation channel from the Batura Glacier to new orchards and fields outside a village in northern Pakistan. Irrigation has long been important in Pakistan.* ***How would you describe the level of technology being used to build this irrigation channel? What might this indicate about Pakistan's level of development?*** TEKS

In contrast to Bangladesh, most of Pakistan has dry climates. Much of southern and central Pakistan has an arid climate. Other areas have semiarid climates with cold winters. In the north, the foothills of the Himalayas have a humid subtropical climate. Because of the rain-shadow effect, south-facing slopes of hills are often wooded, while north-facing slopes are often bare and dry.

Mountainous parts of Nepal, Bhutan, and northern Pakistan have highland climates. Forests once covered large areas of these mountains. However, parts of these forests have been cleared. Bears, deer, snow leopards, wild goats, and many other animals live where forests remain. A valley in south-central Nepal is one of the last homes of the Indian rhinoceros, which is threatened by poaching.

READING CHECK: *Physical Systems* What factor greatly influences the region's climates and weather?

Natural Resources

Overall, the countries of the Indian Perimeter are not rich in natural resources. Pakistan and Sri Lanka have the most significant mineral deposits. Pakistan's minerals include iron and copper. Limestone for cement production is also mined there. The country has only small amounts of oil, but it has rich deposits of natural gas. Sri Lanka has deposits of several minerals, such as gemstones, iron, and salt.

Bangladesh has few minerals. However, it has small deposits of oil as well as some coal and natural gas. Rivers and good soil are the country's most important resources. Bhutan and Nepal also have few mineral deposits. However, both of these mountain countries have rich forests, and both have rivers that have potential for generating hydroelectric power.

READING CHECK: *Physical Systems* What are the most important natural resources of Bangladesh?

TEKS Working with Sketch Maps, Questions 1, 2, 3, 4, 5

Section 1 Review

Define

storm surge

Working with Sketch Maps

On a map that you draw or that your teacher provides, label the Indus River, Baluchistan, Thar Desert, Mount Everest, Tarai, Palk Strait, and Chittagong Hill Tracts. Which part of Pakistan is the most mountainous?

Reading for the Main Idea

1. *Physical Systems* Why is the topography of Bangladesh subject to frequent change?
2. *Places and Regions* Why do the Maldives and Sri Lanka have warmer climates than Nepal and Bhutan?

Critical Thinking

3. **Analyzing Information** In what ways is periodic flooding in the region beneficial? How can it be a problem?
4. **Analyzing Information** From which direction do you think Pakistan's moisture-bearing winds come? Why?

Organizing What You Know

5. Create a chart like the one below. Place an X in the box if the landform can be found in that country.

	Mountains	Alluvial plain	Desert
Bangladesh			
Bhutan			
Maldives			
Nepal			
Pakistan			
Sri Lanka			

Geography for Life

Environmental Change and the Maldives

Global warming and environmental change present serious threats to the Maldives. The islands are very low in elevation. As a result, just a slight melting of the polar ice caps—resulting in rising sea levels—could spell disaster. In fact, if the sea level of the Indian Ocean rose just 20 inches (51 cm), it could cover roughly 80 percent of the Maldives. Some scientists predict that in the next 100 years, water will completely cover the Maldives.

Recent changes in ocean levels and weather patterns point to other problems. For example, some residents believe the number and intensity of storms in the region has increased. In the past 15 years alone, two powerful storms have caused major damage to the islands. In addition, fishers have complained of a decline in the area's marine life. Residents say that the rising temperature of the ocean waters has killed sea life and damaged coral reefs.

Because of the seriousness of the problems facing the Maldives, its citizens are trying to prepare for the future. For example, the Maldives and other small island countries are pushing for worldwide reductions in greenhouse gases. Many scientists believe that these gases cause global warming. At an early age, students in the Maldives learn about global warming. Television and radio stations also teach the public about the greenhouse effect. With help from Japan, the Maldives built a 6-foot-high (1.8 m) concrete wall around its capital, Male. The island country's government hopes this seawall will protect the city from future storms. However, at $4,700 per foot, this type of wall is too expensive to build on each of the country's 200 inhabited islands. Some government officials have proposed another idea. They have suggested gathering the people of the smaller islands onto the three largest ones and defending them with seawalls.

Applying What You Know 

1. **Summarizing** How might global warming and rising sea levels affect the environment of the Maldives?
2. **Problem Solving** What do you think should be done to protect the Maldives from the threat of rising sea levels?

History and Culture

READ TO DISCOVER

1. What is the history of the countries of the Indian Perimeter?
2. What are some characteristics of the region's cultures?

DEFINE

protectorate

LOCATE

Indus Valley

WHY IT MATTERS

India, Pakistan, and Bangladesh were once united as one British colony. Use CNNfyi.com or other **current events** sources to find out about other countries that have split to form new countries in recent years.

History

The Harappan civilization grew along the Indus Valley in ancient times. This society built well-planned cities such as Mohenjo Daro. By about 1500 B.C. Aryans from central Asia had moved into what is now Pakistan. In time, they spread across northern India. Their culture mixed with the Dravidian culture already present in the area.

Empires—Ancient and Modern Other outsiders eventually took over the northwestern part of the subcontinent. These included the Persians and the armies of Alexander the Great. Two Indian empires, the Mauryan and the Gupta, pushed into parts of what is today the Indian Perimeter. Starting about A.D. 1000, Turkic Muslims from Afghanistan came to the subcontinent, bringing their religion. In time, Islam became

Mohenjo Daro was the largest city of the Harappan civilization, which flourished in the Indus Valley from about 2500 to 1500 B.C. This civilization was the first to use fire-hardened bricks extensively in construction, possibly because of a lack of building stone in the region. Although archaeologists have unearthed many examples of Harappan writing, scholars have been unable to decipher their meaning.

Crowds in Bangladesh celebrate Victory Day on December 16, the date on which Bangladesh achieved independence from Pakistan in 1971.

the main religion in the areas that are now Pakistan, Bangladesh, and the Maldives. The Muslim Mughal Empire flourished from the 1500s to the 1700s. About 1500, Europeans began sailing into the Indian Ocean, first to trade and later to set up colonies. Over time, the British came to control most of the subcontinent.

The Modern Period The British granted their Indian colony independence in 1947. India was divided into two countries, India and Pakistan. India was mostly Hindu. Pakistan was mostly Muslim and included present-day Pakistan and Bangladesh. These two parts of the country, separated by hundreds of miles, were known as West Pakistan and East Pakistan. While both were mostly Muslim, West and East Pakistan had important cultural differences. For example, their main languages were different. The new government was centered in West Pakistan, and many people in East Pakistan felt they had no real power. Then, in a war in 1971, East Pakistan broke away from West Pakistan and became the independent country of Bangladesh.

Sri Lanka became a colony of the British in 1802. They called their island colony Ceylon. In 1948 Sri Lanka became independent. The Maldives were a British **protectorate**. A protectorate gives up certain decision-making powers in exchange for protection by a stronger country. The Maldives gained full independence in 1965. Bhutan also was once a British protectorate. It became fully independent in 1949. However, India guides Bhutan's foreign policy.

Nepal was ruled by a series of dynasties until reforms were begun in 1951. Today the country is a constitutional monarchy. However, ethnic troubles, illiteracy, and poverty still make it hard for Nepal's people to build a strong democracy.

READING CHECK: *Places and Regions* Why did East and West Pakistan split into modern Pakistan and Bangladesh?

Durbar Square in Patan, Nepal, is home to many Buddhist monuments and Hindu temples. Patan is famous for its artisans and metalworkers, whose work can be seen in the town's many bronze gateways, guardian statues, and other carvings.

INTERPRETING THE VISUAL RECORD

A wide variety of foods are produced and sold in Bangladesh. In addition to staples like rice and fish, Bangladeshi markets may include beans, eggplant, lemons, plantains, red onions, and a wide variety of spices. ***Based on these photos, what type of economic system do you think Bangladesh has?***

Culture

All the countries of the Indian Perimeter are multiethnic. Three major religions dominate the region, and many languages are spoken.

People and Languages Pakistan's traditional regions—Baluchistan, the North-West Frontier Province, Punjab, and Sind—are each culturally distinct. However, there are also some common features, such as the importance of Islam. Urdu is Pakistan's official language and is taught in schools along with regional languages. An Indo-European language, Urdu is similar to Hindi, which is widely spoken in India. However, Urdu is written in a form of Persian script and, unlike Hindi and English, is read from right to left. Some Pakistanis also speak English.

In southern Nepal, most ethnic groups are of Indian Aryan ancestry. They speak Indo-European languages, such as Nepali, the country's official language. Others in the north are related to the people of Tibet (now part of China). They speak Sino-Tibetan languages.

Bhutan's population includes three main ethnic groups. The largest group, the Bhote, came from Tibet starting in the A.D. 800s. The second-largest group is made up of more recent immigrants from Nepal. Discrimination against the Nepalese is an important problem. The third group is made up of tribal peoples in eastern Bhutan. They are related to the peoples just across the border in India. Bhutan's official language, Dzongkha, is a Sino-Tibetan language. However, English is widely used in schools.

Most of the people in Bangladesh are Bengalis. The Bengalis are a mix of the region's early settlers with Turks and other Southwest Asians who came as merchants as early as the 1200s. Most speak Bengali.

Nearly 75 percent of Sri Lanka's population are Sinhalese. The Tamils, originally from southern India, make up most of the rest of the population. Tamils generally live in the northern and eastern parts of the country. In recent decades there have been bloody conflicts between the two ethnic groups.

READING CHECK: *Human Systems* How does Urdu differ from Hindi? Where are these languages widely spoken?

Education Education levels and literacy rates in the Indian Perimeter are generally low. In recent decades these countries have tried to increase literacy. However, there are too few schools and teachers. In general, women are less likely than men to be able to read. This is largely because of cultural attitudes that emphasize women's role in the home.

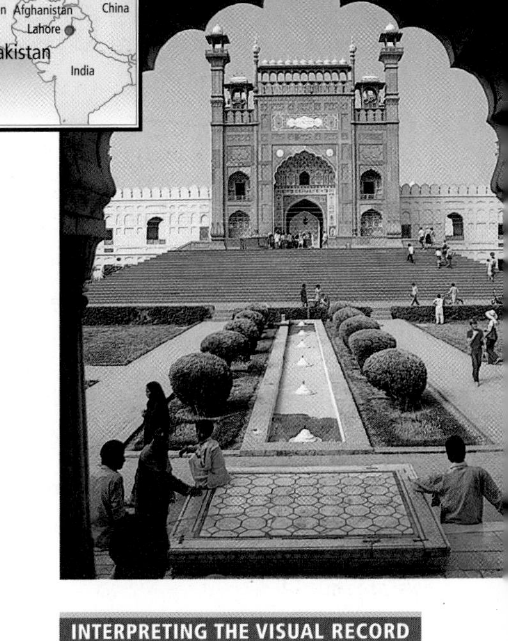

INTERPRETING THE VISUAL RECORD

A gateway leads to Badshahi Mosque in Lahore, Pakistan. This large mosque symbolizes the importance of Islam in Pakistan, a country where some 97 percent of the population is Muslim. Islam is the main link among the many different cultural groups that make up Pakistan. ***How might Islam help connect different cultural groups in Pakistan?***

Religion The region's main religions are Buddhism, Hinduism, and Islam. The religious makeup of each country reflects past events and historical migration patterns.

Pakistan, Bangladesh, and the Maldives are mostly Muslim. Small numbers of Hindus and Christians also live in these countries. At the time of India's partition, the new borders placed districts with Muslim majorities into East and West Pakistan. The number of Hindus in these areas fell sharply when millions fled to India. At the same time, millions of Muslims left India and settled in East and West Pakistan.

In Nepal and Bhutan, most Indian Aryan peoples are Hindu, while peoples of Tibetan origin are generally Buddhist. Bhutan's population is about 75 percent Buddhist and about 25 percent Hindu. The kingdom of Nepal is the world's only officially Hindu state. In Sri Lanka most Sinhalese are Buddhist and most Tamils are Hindu.

The region's religions are reflected in its traditions and customs. For example, Islamic influences shape life throughout Pakistan. People stop to pray several times a day, and Muslim holidays such as Ramadan are important. Midday prayers on Friday draw large numbers of people to the country's many mosques. Also in Pakistan, women often wear veils in public. In Sri Lanka, Buddhist festivals are held throughout the year. For example, the traditional Poya Days mark the phases of the Moon. The most important festival in Sri Lanka is held in the mountain city of Kandy each August. According to tradition, a temple there holds the country's most sacred relic, a tooth of the Buddha. During the annual August celebration, a beautifully decorated elephant is paraded through the streets accompanied by dancers and acrobats.

TEKS **READING CHECK:** ***Human Systems*** What are the three main religions of the Indian Perimeter?

TEKS Working with Sketch Maps, Questions 1, 2, 3, 4, 5

Section 2 Review

go.hrw.com **Homework Practice Online**
Keyword: SW3 HP26

Define protectorate

Working with Sketch Maps

On the map you created in Section 1, label the Indus Valley. Why do you think the region's first great civilizations grew there?

Reading for the Main Idea

1. ***Human Systems*** What people brought Islam to the region? What Muslim empire ruled the region?
2. ***Human Systems*** How did the partition of India affect the countries of this region?
3. ***Human Systems*** What are some examples of the ways religion is reflected in the region's customs?

Critical Thinking

4. **Analyzing Information** Which country of the region is the most ethnically and linguistically unified?

Organizing What You Know

5. Copy the time lines below. Then identify important events that occurred on or about the dates listed on the time lines.

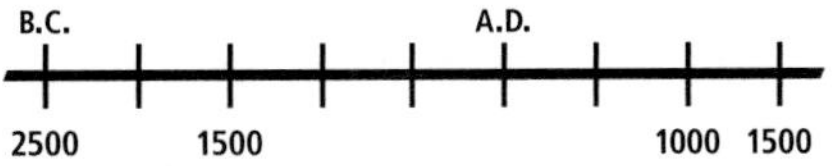

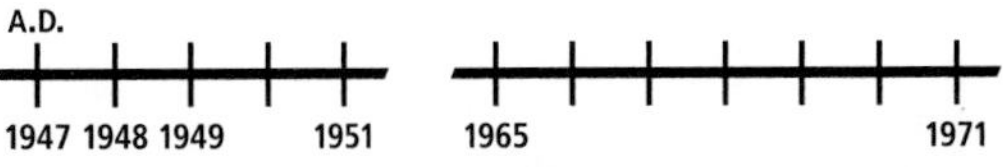

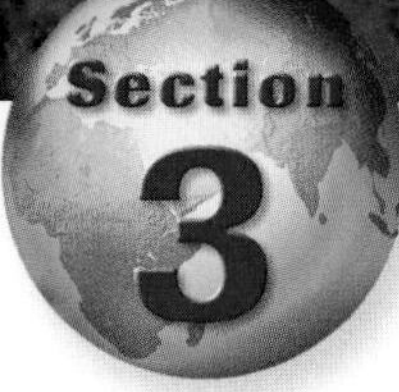

Section 3 The Region Today

READ TO DISCOVER

1. What is the basis of the economies of the Indian Perimeter?
2. What are settlement patterns like in the region?
3. What challenges will these countries face in the future?

WHY IT MATTERS

India and Pakistan both have nuclear weapons. Use CNNfyi.com or other **current events** sources to learn about the relationship between these two countries today.

DEFINE

graphite

LOCATE

Karachi
Lahore
Islamabad
Kathmandu
Thimphu
Dhaka
Colombo
Male

Economy

The level of economic development in the Indian Perimeter is generally low. The countries depend heavily on agriculture. Most have tried to build new industries. However, these efforts have been slowed by a lack of natural resources. Despite these similarities, there is quite a range between the richest and poorest countries of the region. For example, Nepal and Bhutan are among the poorest countries in the world. Yet Pakistan, while still a poor country, has experienced growth in GDP since independence.

Rice is an important crop in Sri Lanka. Since independence in 1948, the country has steadily increased the amount of land devoted to rice and has adopted new technologies to increase yields.

Pakistan, Bangladesh, and Sri Lanka Pakistan has used its significant mineral resources to help develop its manufacturing industries. However, nearly half the labor force still works in agriculture. Although Pakistan's economy has grown, so has its population. This population growth strains the country's ability to provide basic services to its people.

Bangladesh is overwhelmingly agricultural. More than half of the people work in farming. Jute, rice, and tea are the most important crops. Farming depends on the monsoon. Variations in the timing and intensity of the monsoon rains make the difference between a good harvest or a poor one. In recent times people have built a number of irrigation projects to control floods and conserve water for the dry months. These projects, the increased use of fertilizers, and new crops have increased farm output.

The government of Bangladesh has tried to industrialize the country. Textile and clothing makers are among the country's largest employers. However, a lack of mineral resources has made other development difficult. A dam in the Chittagong Hill Tracts produces hydroelectric power. Fishing and logging are also important. One common type of bamboo forms the basis of the country's paper industry. Construction is also a growing industry—a result of the country's fast-growing population.

Farming remains important to Sri Lanka's economy. Tea, rubber, and coconut are the main export crops. The country mines a variety of minerals, including gemstones and **graphite**. Graphite is a form of carbon that is used in pencil leads. Manufacturing is growing and now rivals agriculture in importance. Processed foods and textiles are leading manufactured products.

Bhutan's rulers banned foreigners from the country until well into the 1900s. This historical isolation, along with Bhutan's inaccessible location in the Himalayas, makes economic development difficult today.

Nepal and Bhutan Nepal and Bhutan are very poor and are still mainly agricultural. Timber is an important resource, with most being exported to India. Tourism is a growing part of the economies of the mountain countries. Many visitors come to hike in the Himalayas. However, some people fear the negative effects of tourism on the environment and cultures of Nepal and Bhutan. Their isolated locations have left many of their ecosystems and traditional ways of life largely intact. Bhutan, in particular, has taken a careful approach to tourism. The country was almost closed to outsiders until the 1970s. Today tourists must pay a fee and follow certain restrictions to limit their effect on the country.

Much of the development in Nepal and Bhutan has been helped by aid from other countries and the United Nations. Both these mountain countries have rivers that can generate hydroelectric power. However, they lack the resources needed to build dams. Aid from India did help Bhutan build one such project. Bhutan now sells extra electricity to India.

TEKS **READING CHECK:** ***Environment and Society*** How have Bangladeshis dealt with the effects of variations in monsoon rainfall on farming?

Cities, Settlement, and Land Use

Pakistan and Bangladesh each have more than 130 million people. In fact, they are 2 of the 10 most populous countries in the world. The other countries of the region have far smaller numbers of people. The population throughout the region is mostly rural. This pattern reflects the importance of agriculture there.

About a third of Pakistan's population lives in cities. Karachi is the largest city and main seaport. Lahore is the second-largest city. Islamabad, far to the north, is the capital. Pakistan's population is concentrated in the Indus Valley. Most other areas have

Kathmandu, Nepal, is a city of contrasts. Urban growth, tourism, and increasing connections with the outside world are causing rapid changes throughout the city. Kathmandu's old city is home to many old temples and shrines, but automobile exhaust, construction, and growing shantytowns are putting stress on the city's environment.

few people. The rural population lives mostly in small villages. In northwestern Pakistan, villages are sometimes laid out in a ring. The outer walls have no doors or windows, giving the village the appearance of a fortress.

In both Nepal and Bhutan, population density is low overall. However, the average is much higher in the lowlands and valleys where farming is possible. Nepal's most crowded area is the Kathmandu Valley, near the center of the country. The capital and largest city, Kathmandu, is located there. Bhutan had no cities at all until the 1960s. Even today Thimphu, the capital, is a town of only about 30,000 people.

Bangladesh and Sri Lanka are mostly rural and are densely populated. Bangladesh has about 2,361 people per square mile. This is a very high population density by world standards. For comparison, the United States has about 76 people per square mile. In and around Dhaka, in the country's most fertile region, there are more than 2,800 people per square mile. Sri Lanka's population density is much lower, but there are still some 766 people per square mile. The country's population is heavily concentrated along the fertile coastal plain.

In both countries most people live in villages. In good farming areas these villages are located close together. One village merges with the next in many areas. Bangladesh has two major cities. Dhaka is the country's capital, and Chittagong is the major port. Sri Lanka's largest city is Colombo, the capital and leading industrial area.

READING CHECK: ***Human Systems*** How are settlement patterns similar in Bangladesh and Sri Lanka?

Nepal's mountainous landscapes are a major destination for trekkers and mountain climbers from other countries. However, these international tourists have added to the country's pollution problems. Hikers in Nepal leave behind an estimated 110,000 pounds (50,000 kg) of garbage each year. Here, hikers at Everest Base Camp crush aluminum cans, which are taken back to Kathmandu and recycled.

Economic and Environmental Challenges

As in many other developing countries, the greatest challenge for the countries of the Indian Perimeter is poverty. Poor sanitation, disease, and poor nutrition cause major health problems. High population growth rates have made these problems worse. The largest cities, such as Karachi and Dhaka, are growing rapidly as people migrate from rural areas. Many people in these cities live in homemade shacks with no fresh water or electricity.

The region's countries also face environmental challenges. For example, deforestation is a serious problem in Nepal. Nepal and Bhutan are trying to limit tourism's destructive effects on their environment. In addition, flooding in Bangladesh and the Maldives, with their low flat terrain, might worsen if global warming causes ocean levels to rise. The threat of rising ocean levels has already affected the Maldives' capital of Male. (See Geography for Life: Environmental Change and the Maldives.)

In addition to problems between Pakistan and India over Kashmir, many countries in the region are struggling with political problems. For example, in Pakistan military leaders have overthrown the elected government three times. Rivalries between richer and poorer parts of society have also led to violence. Achieving democratic government and sharing economic benefits throughout society will be key challenges in the future.

Sri Lanka faces an ethnic conflict between the Hindu Tamil minority and the Buddhist Sinhalese majority. Tensions between the two groups have increased since the late 1970s. Some Tamils demand that they be allowed to form their own country. Fighting and violence continue today.

In Sri Lanka high quality "Ceylon" tea is grown on large plantations in the central highlands. Young tender leaves are picked by hand once a week and then quickly processed, producing teas of the highest quality. Women harvesting tea leaves hang large baskets from their heads so that both hands are free to pick leaves.

FOCUS ON GEOGRAPHY

Migration, Tea, and Ethnic Conflict in Sri Lanka The origins of multiethnic Sri Lanka date back more than 2,000 years. About that time, Hindu Tamils from southern India began trading in Sri Lanka. Those who settled in northern Sri Lanka became known as the Ceylon Tamil. Their descendants are usually considered natives of Sri Lanka, along with the Buddhist Sinhalese.

Another group, the Indian Tamil, have a different history. During the mid-1800s Tamil workers from southern India came to Sri Lanka's central plantation region to harvest the coffee crop. Then they would return to India. When a leaf disease destroyed the coffee business, farmers switched to tea. Tea is an evergreen plant that is harvested year-round. Therefore, the Tamil laborers settled in Sri Lanka permanently. Over time, they formed a poorly paid, overworked underclass.

Conflicts between Hindu Tamils and Buddhist Sinhalese have broken out from time to time over hundreds of years. Fighting in recent years has killed at least 55,000 people. Troubles between Indian Tamils and Ceylon Tamils have added to Sri Lanka's political crisis.

READING CHECK: *Human Systems* What role did the switch from coffee to tea play in Sri Lanka's modern ethnic conflicts?

TEKS Working with Sketch Maps, Questions 1, 2, 3, 4, 5

Section 3 Review

Homework Practice Online

Keyword: SW3 HP26

Define graphite

Working with Sketch Maps On the map you created in Section 2, label Karachi, Lahore, Islamabad, Kathmandu, Thimphu, Dhaka, and Colombo. Which of these are port cities?

Reading for the Main Idea

1. *Environment and Society* What is the general level of economic development in the region? What are some indicators of this?
2. *Human Systems* In which Indian Perimeter country is a civil war currently being fought?

Critical Thinking

3. **Analyzing Information** How are settlement patterns in Bangladesh related to the way most Bangladeshis make a living?
4. **Identifying Cause and Effect** What demographic trend has affected Pakistan's economic growth?

Organizing What You Know

5. Copy the following graphic organizer. Use it to identify the leading economic activities of each country.

	Economic activities
Bangladesh	
Bhutan	
Nepal	
Pakistan	
Sri Lanka	

CASE STUDY

Banking on Bangladesh's Female Entrepreneurs

Human Systems Bangladesh is one of the world's poorest countries. It is a young country that has experienced terrible cyclones and a costly struggle for independence, which have disrupted its economy. Economic development has not reached many small villages. As a result, Bangladesh's rural areas have a very low standard of living. For women in Bangladesh, raising the family's standard of living is a difficult task.

For several reasons, poverty often poses greater challenges for women than for men. In rural Bangladesh, begging and working as maids are the only jobs outside the home available to women. Culture traits can also keep women from working. Women in Bangladesh have traditionally had few contacts with anyone outside their families. Some are forbidden to travel far away from home. In addition, rural Bangladeshi women have little schooling. As a result of all these factors, women are limited in their ability to increase their families' income.

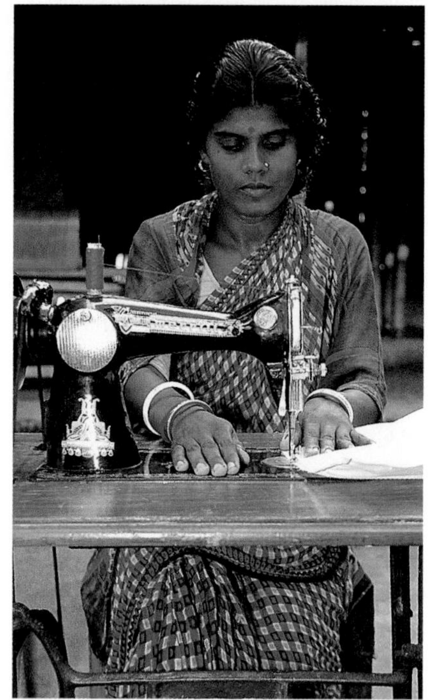

Rural development projects in Bangladesh are helping many women find jobs outside the home, which is raising their standard of living.

However, a creative lending program is showing that women may hold the key to economic development in rural Bangladesh.

Small Loans, Big Results

In the early 1980s the Grameen Bank was founded to offer small loans to rural poor people in Bangladesh. Known as microcredit, these loans typically range from about $100 to $300. Borrowers use the money to start small businesses in their villages. These businesses then provide a steady source of income. Soon after the bank started lending to poor families, officials noticed that the women were often better financial managers than the men. While men were more likely to spend the money on themselves, women used it to make long-range plans. Today about 95 percent of Grameen's borrowers are women.

One such woman is Mosammat Anowara Begum from the village of Chamurkhan. Anowara is a widow with several children. Even though she cannot read or write, the bank offered her a loan. She borrowed $390 from Grameen and used it to buy a cellular phone—the only telephone in her village. Anowara now sells telephone calls for 10 cents a minute. She hopes to pay back the loan in three years. Then Anowara will bring home a steady income of about $2 a day. While that amount may not seem like much, it is almost three times what the average Bangladeshi earns. Anowara's growing income has enabled her daughter to go to college.

Like Mosammat Anowara Begum, Noorjehan Begum borrowed about $320 from the Grameen Bank and bought a cellular phone, which she charges other villagers to use. With the money she earned, Noorjehan paid off her loan and bought a small piece of land and a house.

Women wait to apply for loans at a Grameen Bank weekly meeting. More than 2.3 million people from some 40,000 villages have borrowed money from the bank.

Anowara's story is just one example of how small loans are changing rural Bangladesh. Many other people have borrowed money from Grameen. They have started more than 400 different kinds of businesses. These include making ice cream, processing mustard seeds for oil, repairing radios, and weaving floor mats. The loans seem to be working. Some 125,000 families with Grameen loans pull themselves out of poverty each year. Researchers have noted that children's nutrition and education have significantly improved in households with female borrowers.

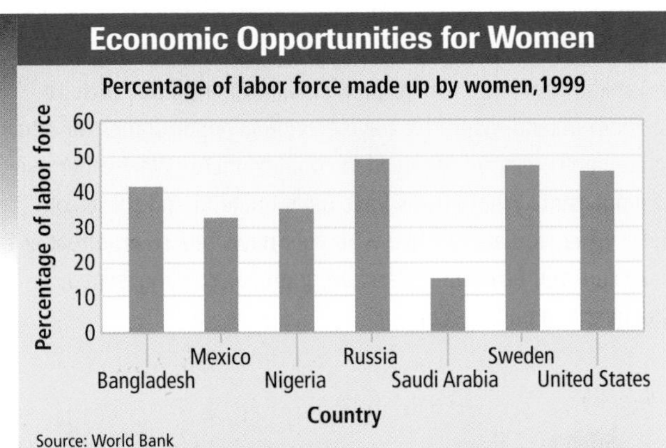

INTERPRETING THE GRAPH *The percentages of women in the labor forces of Bangladesh and the United States are nearly the same. However, most Bangladeshi women work in agriculture. Still, the Grameen Bank is helping many find other ways to make a living. Officials in other countries, such as Mexico, want to set up similar programs.* ***What might be some factors that explain differences in economic opportunities for women in these countries?*** TEKS

A New Approach

Microcredit projects have become so popular that many other countries, including the United States, have experimented with the idea. For example, the Women's Self-Employment Project in Chicago has loaned almost $1 million to a total of about 5,000 women. In all, more than 40 countries have applied the Grameen model, and it has reached nearly 22 million people.

Microcredit projects provide an alternative to conventional methods of helping the world's poor. Pouring large sums of money into industry has been the most common strategy for developing economies. However, these jobs then attract large numbers of people from rural areas to urban areas. By investing money in the rural areas one household at a time, development can be distributed throughout the country. With increasing opportunities for women to earn money, more and more households can lift themselves out of poverty.

Applying What You Know

1. **Summarizing** How does microcredit create economic opportunities that are not otherwise available to women in Bangladesh?
2. **Comparing and Contrasting** How is microcredit different from conventional forms of development aid?

Chapter 26 Review

Building Vocabulary

On a separate sheet of paper, explain the following terms by using them correctly in sentences.

storm surge   protectorate   graphite

Locating Key Places

On a separate sheet of paper, match the letters on the map with their correct labels.

Indus River	Mount Everest	Colombo
Baluchistan	Tarai	Male
Thar Desert	Dhaka	

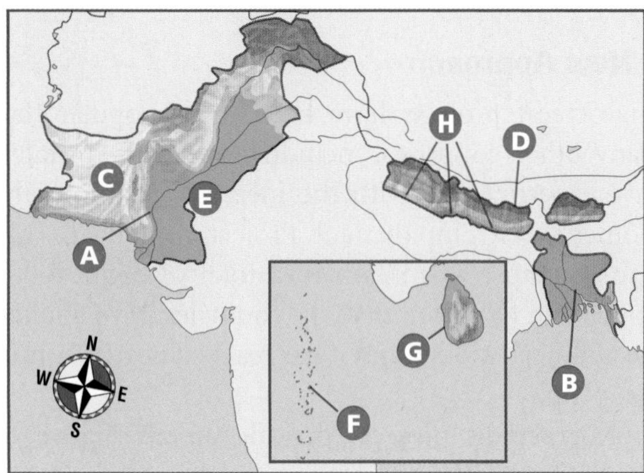

Understanding the Main Ideas

Section 1

1. *Physical Systems* Which countries of the Indian Perimeter include parts of the Himalayas? Which include parts of the plain of the Ganges River?

Section 2

2. *Human Systems* Where did Bhutan's largest ethnic group, the Bhote, come from?
3. *Human Systems* On what basis did the British divide their colony of India upon independence in 1947?

Section 3

4. *Human Systems* What are the major industries of Bangladesh?
5. *Environment and Society* What is the greatest potential energy resource of Nepal and Bhutan?

Thinking Critically for TAKS

1. **Analyzing Information** What combination of factors make Bangladesh so vulnerable to storm surges?
2. **Identifying Cause and Effect** What political factors might make further development difficult in some countries of the region?
3. **Comparing** What part of Pakistan bears a resemblance to the main environment of Bhutan? What part might resemble Bangladesh?

Using the Geographer's Tools

1. **Analyzing Maps** Study the physical and political map of the Indian Perimeter. Why do you think it would be difficult to govern modern-day Pakistan and Bangladesh as one country?
2. **Analyzing Tables** Study the information from the unit Fast Facts and Comparing Standard of Living tables. Then rank the countries according to their levels of economic development and standard of living. Explain why you have ranked the countries as you have.
3. **Preparing Charts** Create a two-column chart listing the benefits and drawbacks of tourism for Nepal and Bhutan. What economic benefits does tourism offer for these countries?

Writing for TAKS

Use classroom and library materials to research the ethnic and religious conflict in Sri Lanka. Write a news article that compares the Sri Lankan conflict to similar problems in other places, such as Northern Ireland. What role have economic opportunities for ethnic and religious minorities played in conflicts there? What other issues are important? What efforts have been made to end those conflicts? When you are finished with your news report, proofread it to make sure you have used standard grammar, spelling, sentence structure, and punctuation.

SKILL BUILDING

Geography for Life

Applying Geographic Models

Physical Systems Study the courses of the Indus, Ganges, and Brahmaputra Rivers. Consider the connection between erosion in the Himalayas and deposition of sediment downstream. Create a diagram showing how sediment is eroded and deposited by this process. Consider the interrelationships of climate, slope, and soil.

Building Social Studies Skills

TAKS PREP ONLINE
go.hrw.com
keyword: SW3 T26

Country	Births (per 1,000 people)	Deaths (per 1,000 people)
Bangladesh	25.44	8.73
Bhutan	36.22	14.32
Maldives	38.96	8.32
Nepal	33.83	10.41
Pakistan	32.11	9.51
Sri Lanka	16.78	6.43

Birthrates and Death Rates in Indian Perimeter Countries

Source: *The World Almanac and Book of Facts 2001*

Interpreting Charts

Study the chart above. Then use the information from the chart to help you answer the following questions.

1. Which country has the lowest birthrate?
   a. Pakistan
   b. Nepal
   c. Maldives
   d. Sri Lanka
2. Rank the six countries' birthrates and death rates from least to greatest. Which country has the highest birthrate but the second-lowest death rate?

Using Language

The following passage contains mistakes in grammar, punctuation, and usage. Read the passage and then answer the following questions.

(1) Family life in Bangladesh is different from Pakistan. (2) For example, many Bangladeshi women keep close ties to their own families after marriage. (3) However, as in other South Asian countries, marriages arranged by parents. (4) Most couples typically do not know each other prior to there wedding.

3. Which sentence contains a spelling error?
   a. 1
   b. 2
   c. 3
   d. 4
4. Which line is a fragment?
   a. 1
   b. 2
   c. 3
   d. All are complete sentences.
5. Rewrite sentence (1) to correct the mistake in comparison.

Alternative Assessment

PORTFOLIO ACTIVITY

Learning about Your Local Geography TEKS

Group Project: Presenting Information

Conduct research on natural hazards that can strike your area. Examples might be earthquakes, floods, forest fires, hurricanes, mudslides, or tornadoes. Analyze the way people prepare for and react to these hazardous events. Study some examples of how humans have used technology to adapt to the dangers of natural hazards. Compare and contrast the effects of local natural hazards with the effects of tropical cyclones and flooding in Bangladesh. How important is preparing for natural hazards in the daily lives of people in both regions? Use posters or another kind of display to present your findings to the class.

internet connect

Internet Activity: go.hrw.com
KEYWORD: SW3 GT26 TEKS

Choose a topic on the Indian Perimeter to:

- publish a poster on the landscapes and cultures of the Himalayas.
- learn the history of Sri Lanka.
- examine Pakistani history and culture.

Skill Building for TAKS

WORKSHOP 1

Using Media Services: The Internet

The Internet, or the Net, is made up of millions of computers linked worldwide through a telecommunications network. The Net provides access to a great deal of information in many formats and from many sources.

Developing the Skill An Internet service provider, or ISP, provides access to the Internet. A computer links to the ISP through a modem. Once you are connected, you can "surf" the Internet. Surfing the Internet lets you link to informational sites or send messages called e-mail. Many informational sites are locations on the World Wide Web, sometimes referred to as the WWW or simply the Web. You access the Web by using software called a Web browser. You can call up specific Web sites by entering a special address called a uniform resource locator, or URL. Such addresses typically begin with http:// and end with suffixes such as .com, .edu, .gov, or .org.

Most Web sites have a home, or main, page that will "link" you to additional information on other pages or sites through hyperlinks. These hyperlinks usually appear as underlined terms or phrases. For example, databases like the World Factbook from the U.S. Central Intelligence Agency can be found on the Web. The World Factbook includes hyperlinks to specific categories of information and countries.

You can also search the Web for sites that include information about a particular topic by using a search engine. Search engines will seek Web sites that include key words you have entered. You will need to experiment to find search engines that are most useful to you. Follow instructions on each search engine.

Users must be able to determine whether a Web page contains valuable information. Ask yourself these questions:

- Is a well-respected organization sponsoring the site? Whose views are represented?
- Is the information clear and easy-to-read? What has the designer done to attract your attention?

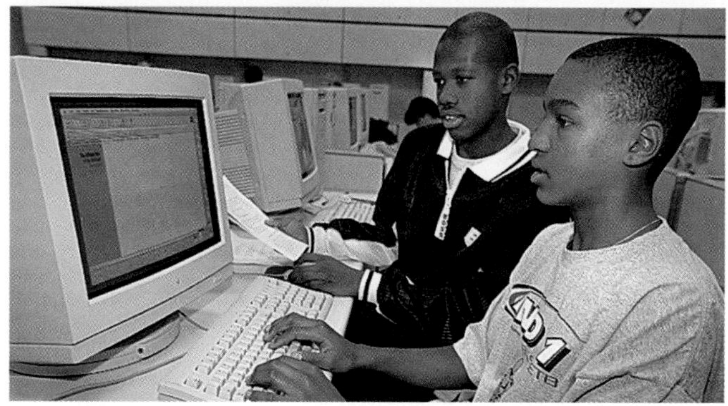

- Is the Web site easy to navigate, and are most of the page links active?
- Is the information current?
- Is the information biased? What information has been left out of the site? How might information there be interpreted in different ways by different people?

The Internet contains a huge amount of resources, making it an excellent tool for research. However, there is a lot of inappropriate material on the Web. Many well-respected sites have developed safe sites for children and teens. These sites are known as child-safe zones. Even so, you must be careful.

- Do not chat with or meet strangers.
- Do not give out any personal information.
- If you get into an unsecured area, get out immediately and tell your parent or teacher.
- Stay within well-respected search engines.

Practicing the Skill

1. Find Web sites operated by each of the governments of South Asia's countries. List the URL for what appears to be the main Web site for each government.
2. Locate a Web site that offers news from India or Pakistan. Identify the operator of the site. Is it a newspaper, television network, or other news provider?
3. Locate Web sites that provide statistics about countries in South Asia. What kind of data do you find on each site?

WORKSHOP 2

Solving Problems

Having effective problem-solving skills is important in geography. For example, some farmers must learn how to transform harsh environments into productive farmlands. Demographers work to find ways to collect more accurate information about populations. Governments study possible solutions to problems like rapid population growth, inadequate health care systems, and pollution.

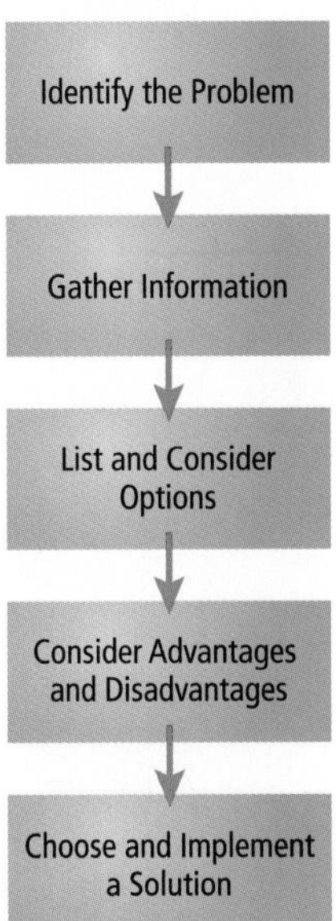

Developing the Skill The problems listed above are very different from each other. However, solving them can involve similar processes. Once you have identified a problem, these steps can help you find a solution:

- **Gather information** about the issue or problem. Libraries and reliable sites on the World Wide Web are common sources of information about geographic problems. The information you gather will help you understand a problem more clearly and identify possible solutions.
- **List and consider the options** you have. Brainstorm possible options and list them in a chart or idea web. List all the reasonable options that come to mind—you can narrow down your choices later.
- **Consider advantages and disadvantages** of each of your options. For complex problems, you might create a cost-benefits balance sheet for each option. List the disadvantages or additional problems associated with a particular option on the "cost" side of the balance sheet. On the "benefits" side list the advantages associated with that option. Follow the same process for all of your options. Then decide whether the advantages of each option outweigh the disadvantages.
- **Choose and implement the solution** you think is best from your list of options. The best solution may be clear once you have analyzed your cost-benefits balance sheets. If the best solution is not clear, you may need more information to help you decide. Then create an action plan for implementing your solution.
- **Evaluate the effectiveness** of your solution. What you learn from this evaluation can help you solve other problems in the future. It can also help you refine your own problem-solving process. Review your cost-benefits balance sheet for the option you chose. Note whether you anticipated all of the possible disadvantages and advantages associated with your solution.

Practicing the Skill

1. India and Pakistan have gone to war against each other twice since the late 1940s. Today both countries have nuclear weapons, making the possibility of war between them even more frightening. Imagine you are a member of a United Nations commission trying to forge a lasting peace between the two countries. Identify problems that divide the countries and then identify possible solutions.
2. Mount Everest and other mountains of the Himalayas attract many climbers and other tourists each year. Unfortunately, those visitors leave their garbage behind. How might the government of Nepal work to protect the natural environment from the problems visitors cause? How might the government implement your proposed solution?

East and Southeast Asia

Borobudur Temple, Indonesia

CONNECTING TO

Literature

"Thoughts of Hanoi" by Nguyen Thi Vinh

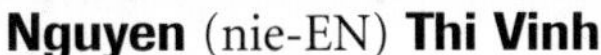

Nguyen (nie-EN) **Thi Vinh** (1924–) was born in northern Vietnam's Red River delta in Southeast Asia. A novelist and poet, she fled to South Vietnam after Communists took over the north in the 1950s. She remained in South Vietnam for a while after it fell to the Communists in 1975. In 1983 she moved to Norway. Her poem "Thoughts of Hanoi" reveals some of her memories of the land of her birth.

The night is deep and chill
as in early autumn. Pitchblack,
it thickens after each lightning
flash.
I dream of Hanoi:
Co-ngu Road
Ten years of separation
the way back sliced by a
frontier of hatred.
I want to bury the past
to burn the future
still I yearn
still I fear
those endless nights
waiting for dawn.

Brother,
how is Hang Dao now?
How is Ngoc Son temple?
Do the trains still run
each day from Hanoi
to the neighboring towns?
To Bac-ninh, Cam-giang,
Yen-bai,
the small villages, islands
of brown thatch in a lush
green sea?

The girls
bright eyes
ruddy cheeks
four-piece dresses
raven-bill scarves
sowing harvesting
spinning weaving
all year round,
the boys
plowing
transplanting
in the fields
in their shops
running across
the meadow at evening
to fly kites
and sing alternating
songs.

Stainless blue sky,
jubilant voices of children
stumbling through the alphabet,
village graybeards strolling to
the temple,
grandmothers basking in
twilight sun,
chewing betel* leaves
while the children run—

* **betel** (BEE-tuhl) **leaves:** *the leaves of a climbing pepper vine; chewed with betel nuts*

Analyzing the Primary Source

1. **Summarizing** What images does the writer remember from her homeland?
2. **Drawing Inferences** What do you suppose the writer means when she writes of her way back to Vietnam being "sliced by a frontier of hatred"?

UNIT 9 ATLAS

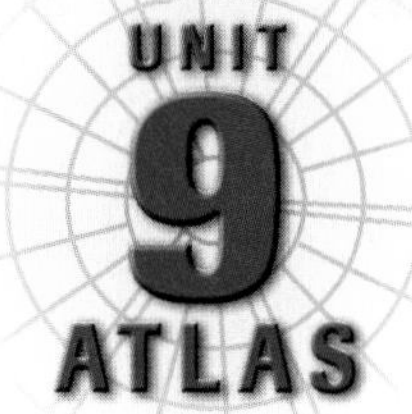

The World in Spatial Terms

East and Southeast Asia: Political

1. **Places and Regions** Which countries in the region are landlocked?
2. **Places and Regions** Compare this map to the physical map. Which four major islands make up Japan?

Critical Thinking

3. **Making Generalizations** How might Indonesia's island geography cause problems for its government? TEKS

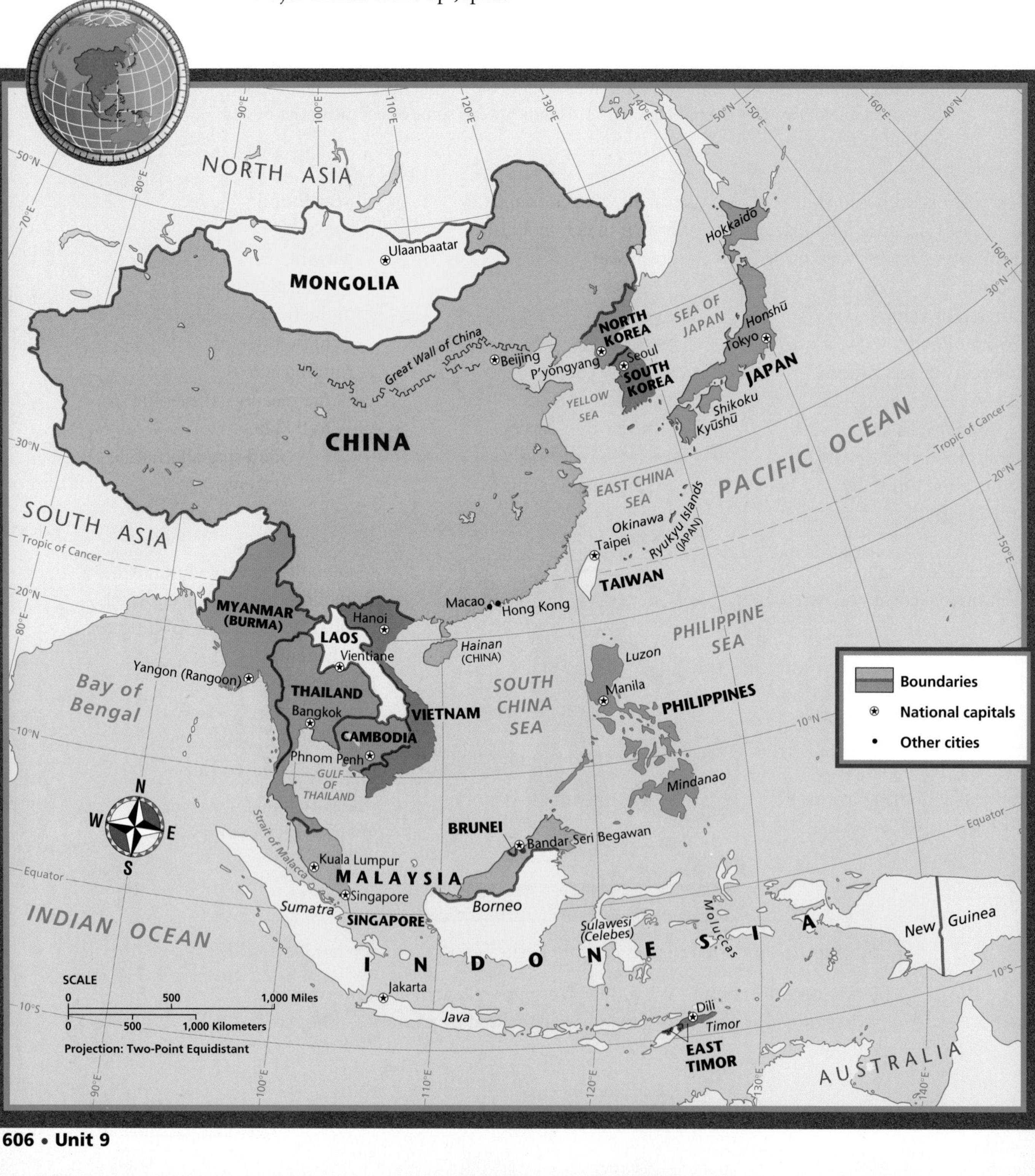

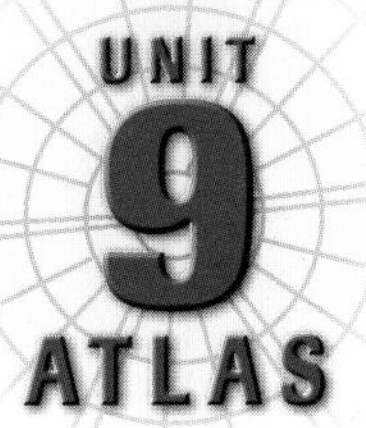

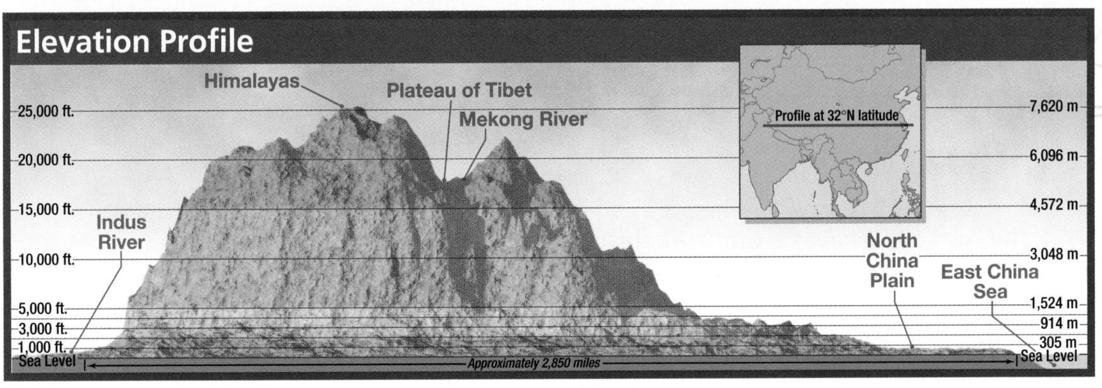

East and Southeast Asia: Physical

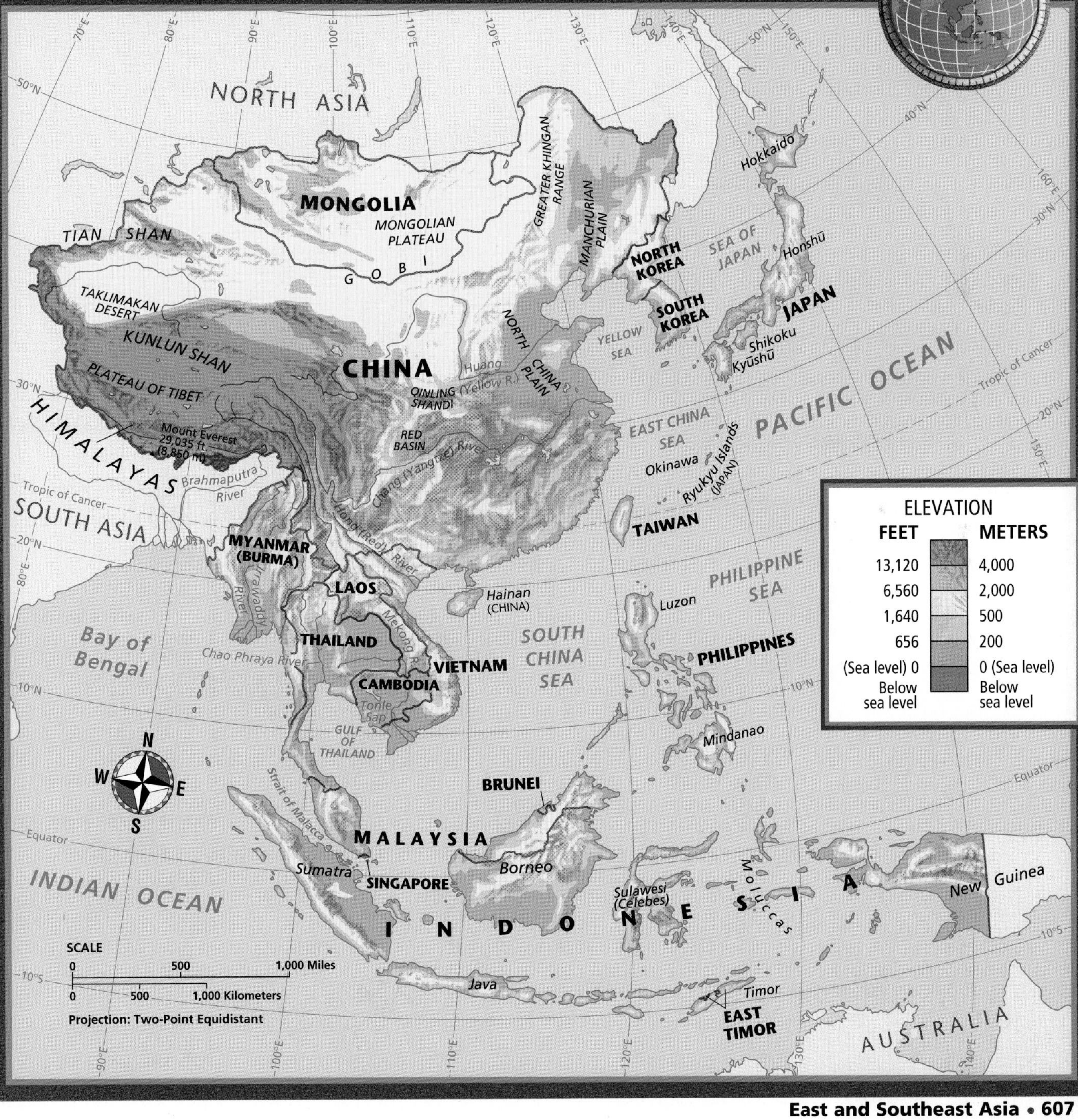

NORTH ASIA
MONGOLIA
MONGOLIAN PLATEAU
GOBI
TIAN SHAN
TAKLIMAKAN DESERT
KUNLUN SHAN
PLATEAU OF TIBET
HIMALAYAS
Mount Everest 29,035 ft. (8,850 m)
Brahmaputra River
SOUTH ASIA
CHINA
QINLING SHANDI
Huang (Yellow R.)
RED BASIN
Chang (Yangtze) River
Hong (Red) River
GREATER KHINGAN RANGE
MANCHURIAN PLAIN
NORTH CHINA PLAIN
NORTH KOREA
SOUTH KOREA
YELLOW SEA
SEA OF JAPAN
Hokkaido
Honshū
JAPAN
Shikoku
Kyūshū
PACIFIC OCEAN
EAST CHINA SEA
Okinawa
Ryukyu Islands (JAPAN)
TAIWAN
PHILIPPINE SEA
Luzon
PHILIPPINES
Mindanao
SOUTH CHINA SEA
Hainan (CHINA)
MYANMAR (BURMA)
Irrawaddy River
LAOS
THAILAND
Mekong R.
VIETNAM
CAMBODIA
Tonle Sap
Chao Phraya River
GULF OF THAILAND
Bay of Bengal
Strait of Malacca
MALAYSIA
BRUNEI
Borneo
Sumatra
SINGAPORE
INDONESIA
Sulawesi (Celebes)
Moluccas
New Guinea
Java
Timor
EAST TIMOR
AUSTRALIA
INDIAN OCEAN
Tropic of Cancer
Equator
50°N 40°N 30°N 20°N 10°N 10°S
70°E 80°E 90°E 100°E 110°E 120°E 130°E 140°E 150°E 160°E

ELEVATION

FEET	METERS
13,120	4,000
6,560	2,000
1,640	500
656	200
(Sea level) 0	0 (Sea level)
Below sea level	Below sea level

N E S W

SCALE
0 500 1,000 Miles
0 500 1,000 Kilometers
Projection: Two-Point Equidistant

East and Southeast Asia: Climate

1. **Places and Regions** Compare this map to the political map. Which climate types are found in China? TEKS
2. **Places and Regions** Compare this map to the political map. Which climate type dominates Indonesia, Malaysia, and the Philippines?

Critical Thinking

3. **Analyzing Information** Compare this map to the physical map. Which physical feature likely blocks the wet monsoon airflow from reaching the arid and semiarid regions of China? TEKS

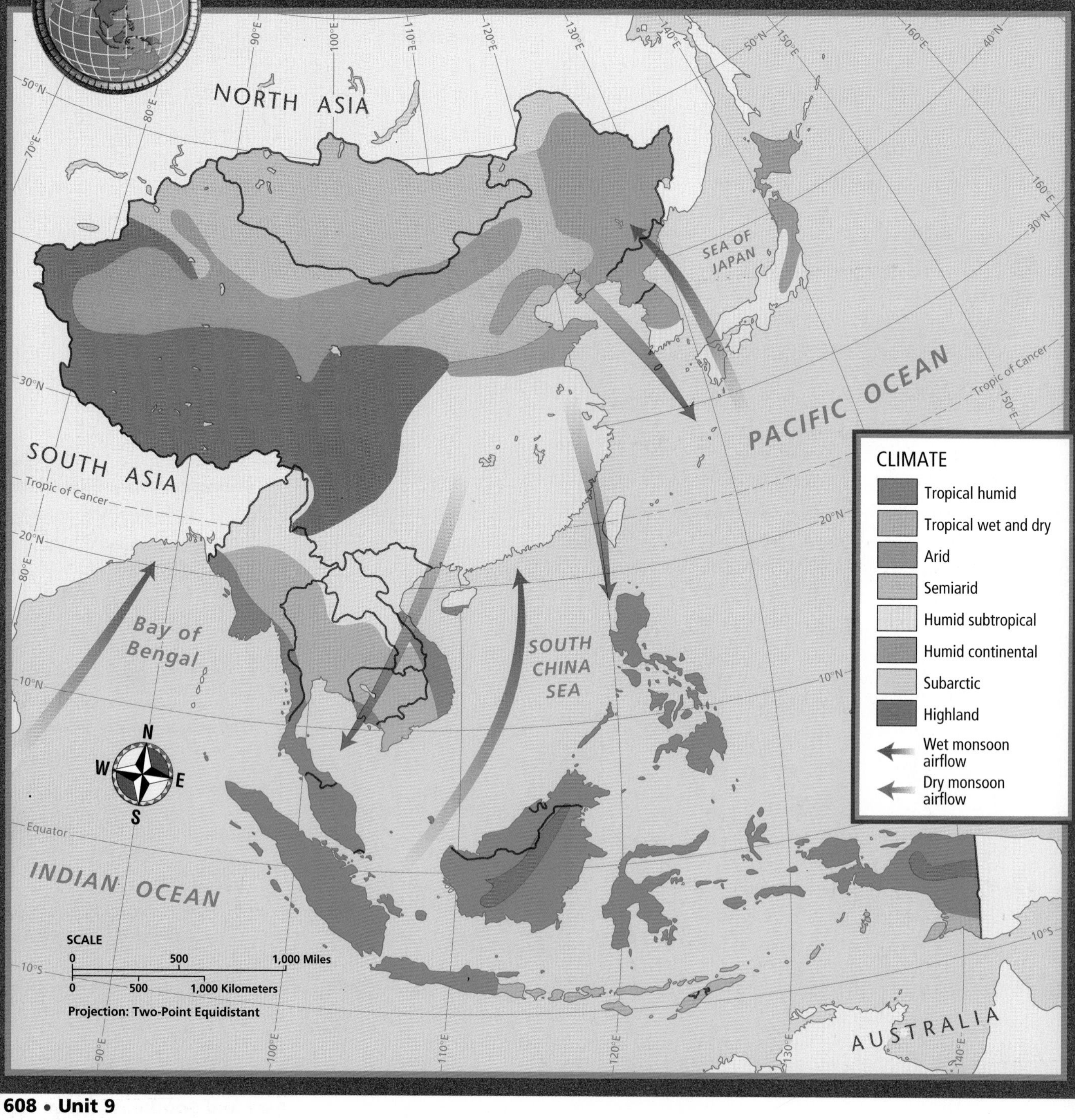

NORTH ASIA
SOUTH ASIA
SEA OF JAPAN
PACIFIC OCEAN
Bay of Bengal
SOUTH CHINA SEA
INDIAN OCEAN
AUSTRALIA
Tropic of Cancer
Equator

CLIMATE
- Tropical humid
- Tropical wet and dry
- Arid
- Semiarid
- Humid subtropical
- Humid continental
- Subarctic
- Highland
- Wet monsoon airflow
- Dry monsoon airflow

SCALE
0 500 1,000 Miles
0 500 1,000 Kilometers
Projection: Two-Point Equidistant

East and Southeast Asia: Precipitation

1. *Places and Regions* Compare this map to the population map. Which large metropolitan areas are located in areas that receive more than 80 inches (200 cm) of precipitation? TEKS
2. *Places and Regions* How much precipitation does most of Mongolia receive? TEKS

Critical Thinking

3. **Making Generalizations** How do you think the distribution of precipitation in East and Southeast Asia is related to latitude? TEKS

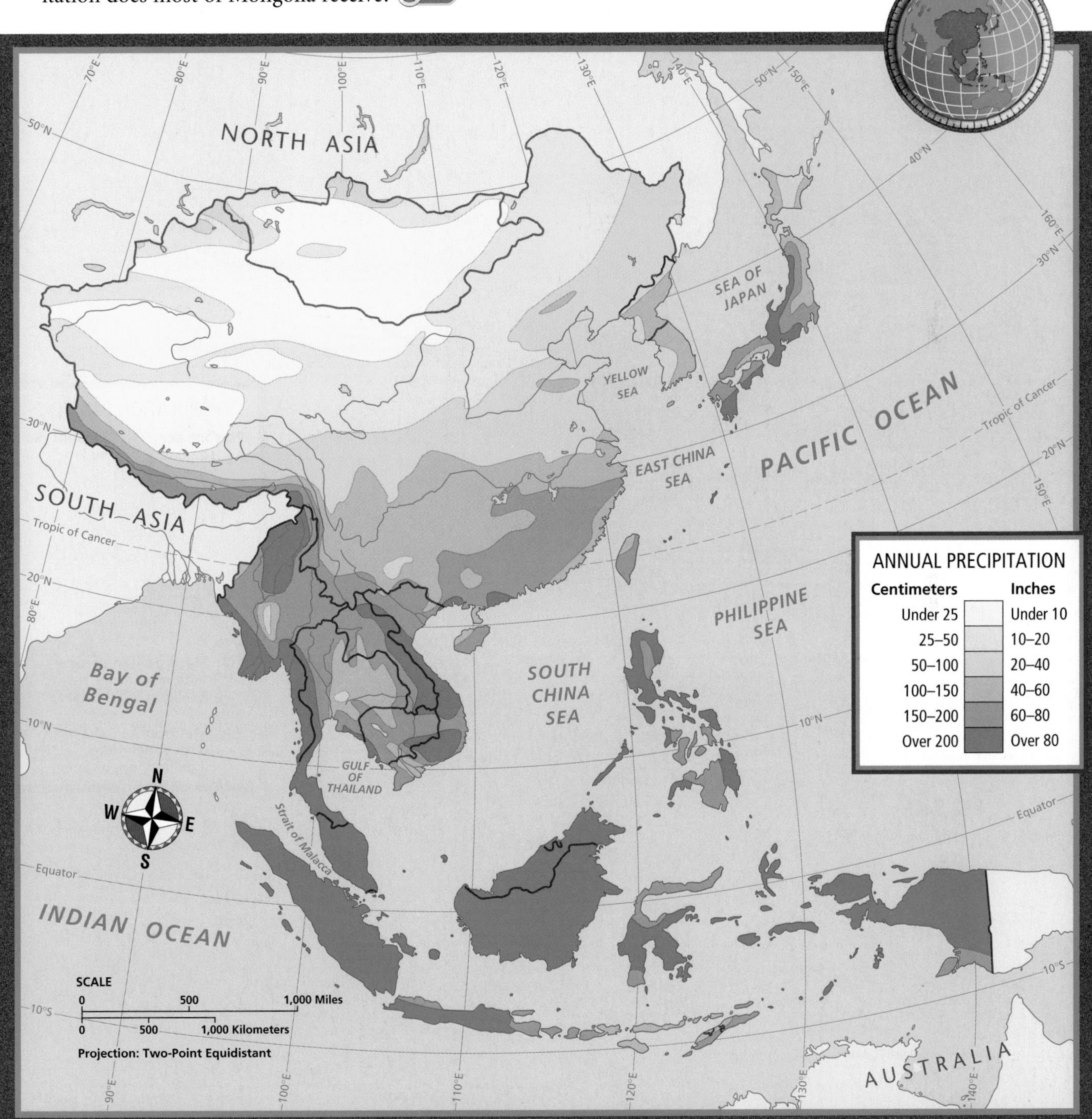

East and Southeast Asia: Population

1. *Places and Regions* Compare this map to the political map. Which country in the region has by far the largest number of metropolitan areas with more than 1 million inhabitants? TEKS
2. *Places and Regions* Compare this map to the physical map. Which large lowland region in northeastern China is densely populated? TEKS

Critical Thinking

3. **Analyzing Information** Compare this map to the physical map. If Indonesia's government wanted to lower the population density on Java, to which other islands in the country might they encourage people to move? TEKS

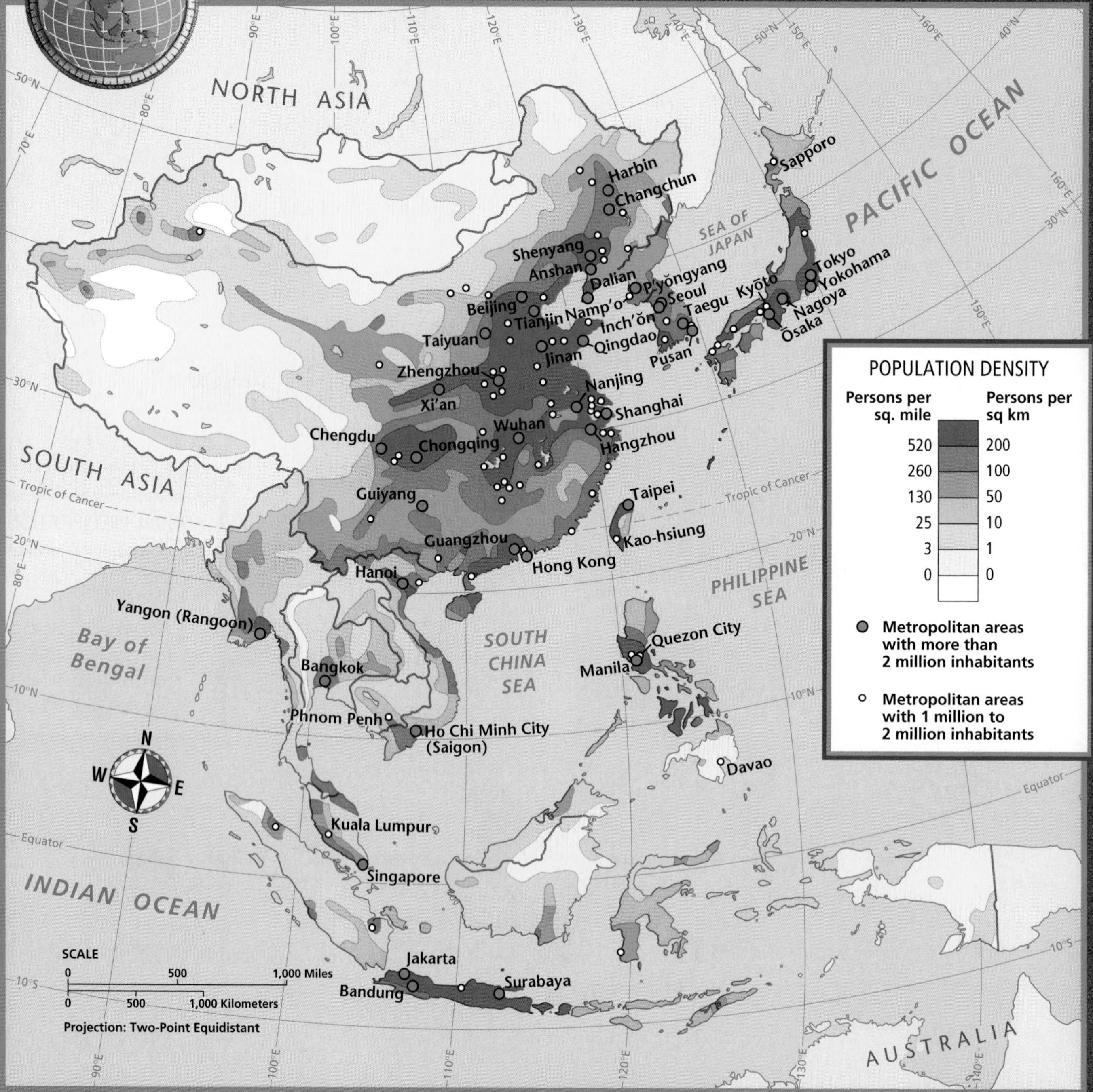

POPULATION DENSITY

Persons per sq. mile	Persons per sq km
520	200
260	100
130	50
25	10
3	1
0	0

● Metropolitan areas with more than 2 million inhabitants

○ Metropolitan areas with 1 million to 2 million inhabitants

SCALE: 0–500–1,000 Miles; 0–500–1,000 Kilometers

Projection: Two-Point Equidistant

UNIT 9 ATLAS

East and Southeast Asia: Land Use and Resources

1. **Places and Regions** What type of power is commonly generated in Japan? TEKS

2. **Environment and Society** Based on the map, which countries in the region do you think have the most oil? TEKS

Critical Thinking

3. **Analyzing Information** Compare this map to the population and political maps. How do you think Japan's location is advantageous for trade with other countries in the region? TEKS

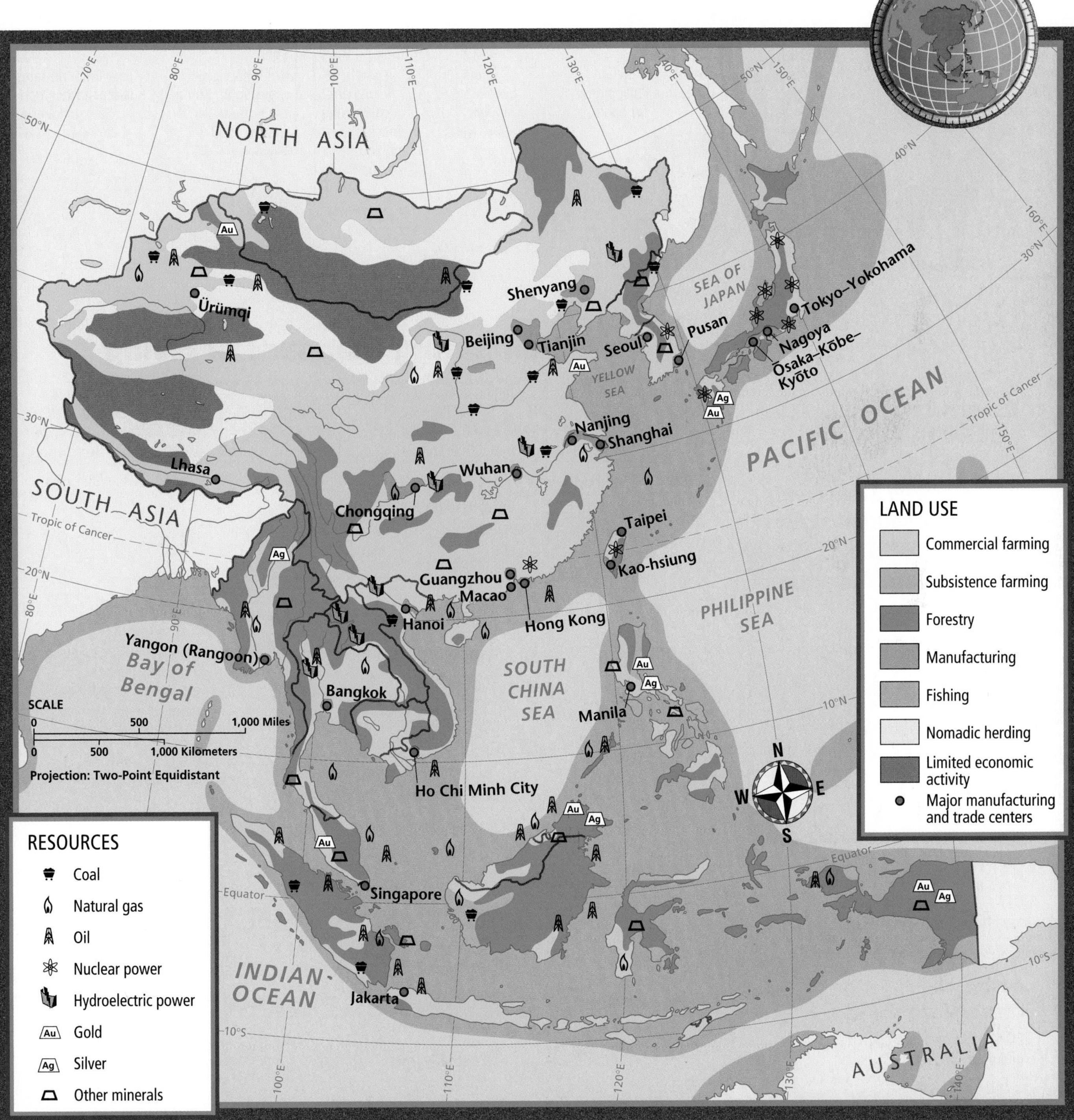

Time Line: East and Southeast Asia

108 B.C. Chinese invade Korea.

C. A.D. 500 Hindu and Buddhist states are established in Java and Sumatra.

1853 U.S. commodore Matthew C. Perry visits Japan.

1949 The Communists win China's civil war. Indonesia wins independence from the Netherlands.

1975 North Vietnamese and other Communist forces take over South Vietnam, ending a long war.

200 B.C. | A.D. 500 | 1200 | 1500 | 1800 | 1900

202 B.C. Han dynasty rises to power in China and spreads Chinese culture into southern China.

1279 Mongols rule all of China.

1521 Ferdinand Magellan claims the Philippines for Spain.

1912 China becomes a republic.

1945 Japan is defeated after conquering much of East and Southeast Asia in World War II.

1950 North Korea invades South Korea. The Korean War ends in 1953.

1999 East Timor declares independence from Indonesia.

Comparing Standard of Living

COUNTRY	LIFE EXPECTANCY (in years)	INFANT MORTALITY (per 1,000 live births)	LITERACY RATE	DAILY CALORIC INTAKE (per person)
Cambodia	55, male 59, female	65	35%	2,012
China	70, male 74, female	28	82%	2,741
Indonesia	66, male 71, female	41	84%	2,732
Japan	78, male 84, female	4	99%	2,887
North Korea	68, male 74, female	24	99%	2,360
Philippines	65, male 71, female	29	95%	2,395
Singapore	77, male 83, female	4	94%	Not available
South Korea	71, male 79, female	8	98%	3,268
Thailand	66, male 72, female	30	94%	2,296
Vietnam	67, male 72, female	30	94%	2,463
United States	74, male 80, female	7	97%	3,603

Sources: Central Intelligence Agency, *2001 World Factbook; Britannica Book of the Year, 2000*

The United States and East and Southeast Asia

Comparing Sizes

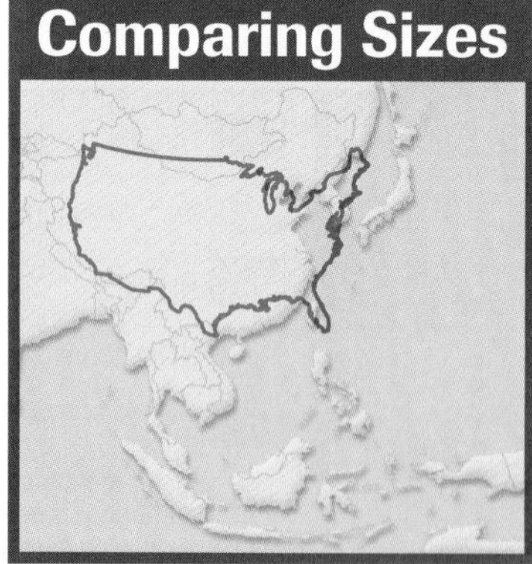

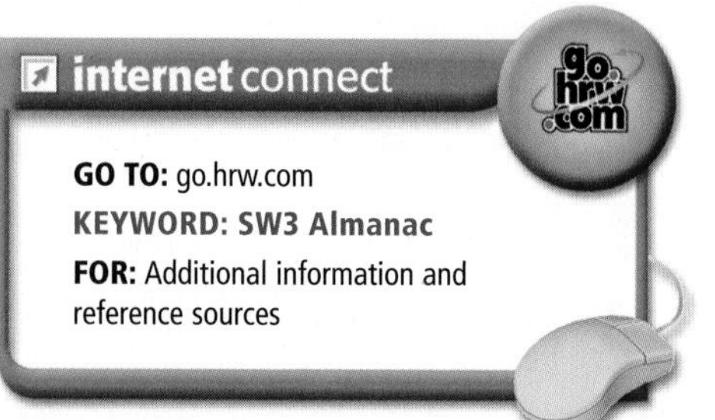

GO TO: go.hrw.com
KEYWORD: SW3 Almanac
FOR: Additional information and reference sources

Fast Facts: East and Southeast Asia

FLAG	COUNTRY Capital	POPULATION (in millions) / POP. DENSITY	AREA	PER CAPITA GDP (in US $)	WORKFORCE STRUCTURE (largest categories)	ELECTRICITY CONSUMPTION (kilowatt hours per person)	TELEPHONE LINES (per person)
	Brunei Bandar Seri Begawan	0.3 / 154/sq. mi.	2,228 sq. mi. 5,770 sq km	$ 17,600	48% government 42% oil, gas prod., services, const.	6,617 kWh	0.23
	Cambodia Phnom Penh	12.5 / 179/sq. mi.	69,900 sq. mi. 181,040 sq km	$ 1,300	80% agriculture	11 kWh	0.002
	China Beijing	1,273 / 344/sq. mi.	3,705,386 sq. mi. 9,596,960 sq km	$ 3,600	50% agriculture 26% services	851 kWh	0.11
	Indonesia Jakarta	228.4 / 308/sq. mi.	741,096 sq. mi. 1,919,440 sq km	$ 2,900	45% agriculture 39% services	320 kWh	0.02
	Japan Tokyo	126.8 / 869/sq. mi.	145,882 sq. mi. 377,835 sq km	$ 24,900	65% services 30% industry	7,470 kWh	0.48
	Laos Vientiane	5.6 / 62/sq. mi.	91,428 sq. mi. 236,800 sq km	$ 1,700	80% agriculture	31 kWh	0.004
	Malaysia Kuala Lumpur	22.2 / 175/sq. mi.	127,316 sq. mi. 329,750 sq km	$ 10,300	28% local trade, tourism 27% manufacturing	2,468 kWh	0.20
	Mongolia Ulaanbaatar	2.7 / 4/sq. mi.	604,247 sq. mi. 1,565,000 sq km	$ 1,780	Primarily herding, agricultural	1,042 kWh	0.04
	Myanmar (Burma) Yangon	42.0 / 160/sq. mi.	261,969 sq. mi. 678,500 sq km	$ 1,500	65% agriculture 25% services	107 kWh	0.006
	North Korea P'yŏngyang	22.0 / 472/sq. mi.	46,540 sq. mi. 120,540 sq km	$ 1,000	64% nonagricultural 36% agriculture	1,211 kWh	0.05
	Philippines Manila	82.8 / 715/sq. mi.	115,830 sq. mi. 300,000 sq km	$ 3,800	40% agriculture 19% gov., services	457 kWh	0.02
	Singapore Singapore	4.3 / 17,202/sq. mi.	250 sq. mi. 648 sq km	$ 26,500	38% fin., bus., serv. 21% manufacturing	5,921 kWh	0.45
	South Korea Seoul	47.9 / 1,260/sq. mi.	38,023 sq. mi. 98,480 sq km	$ 16,100	68% services 20% industry	4,859 kWh	0.50
	Taiwan Taipei	22.4 / 1,610/sq. mi.	13,892 sq. mi. 35,980 sq km	$ 17,400	55% services 37% industry	5,807 kWh	0.56
	Thailand Bangkok	61.8 / 311/sq. mi.	198,455 sq. mi. 514,000 sq km	$ 6,700	54% agriculture 31% services	1,359 kWh	0.09
	Vietnam Hanoi	79.9 / 628/sq. mi.	127,243 sq. mi. 329,560 sq km	$ 1,950	67% agriculture 33% ind., services	267 kWh	0.03
	United States Washington, D.C.	281.4 / 76/sq. mi.	3,717,792 sq. mi. 9,629,091 sq km	$ 36,200	30% manage., prof. 29% tech., sales, admin.	12,211 kWh	0.69

Sources: Central Intelligence Agency, *2001 World Factbook; The World Almanac and Book of Facts, 2001; Britannica Book of the Year, 2000;* U.S. Census Bureau The CIA calculates per capita GDP in terms of purchasing power parity. This formula equalizes the purchasing power of each country's currency. Data for East Timor was not available.

CHAPTER 27

China, Mongolia, and Taiwan

The People's Republic of China is the world's most populous country. The Chinese are experiencing great change as they develop a market economy. In this chapter you will learn about China, landlocked Mongolia, and the island of Taiwan.

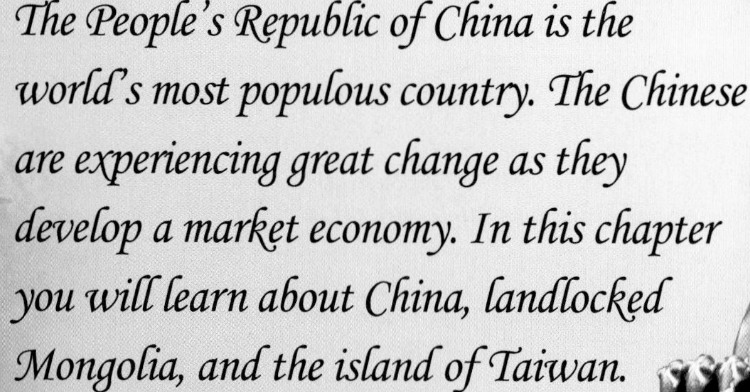

Man using cormorants to catch fish, China

Lion statue, Forbidden City, Beijing, China

Nihao! (How are you?) My name is Lu Hua, and I am 16. I live outside of Shanghai with my parents and my younger brother. My ancestors lived in our village already hundreds of years ago. My father is a clerk in a car factory. My mother works for a chemical factory. Behind our house is a kitchen and sheds for storing food, straw, and coal for cooking and heating.

I am in my last year at Jin Shan County High School. I am studying calculus, physics, chemistry, biology, Chinese, history, geography, and English. We also have a required class in communist philosophy and government. I am hoping to go to a university. In China only one or two out of a hundred kids can go to university.

After school, I help my family take care of the crops and animals. We have some small plots of land scattered around within a 15-minute walk of our house. We grow rice, wheat, and most of our own vegetables, like bok choy, which you would call Chinese cabbage. We also raise chickens and ducks. If I am not needed in the fields, I play basketball until it gets dark. I'm not on the school basketball team, but sometimes I play point guard for them. The best time of the week is Saturday night when my friends and I do whatever we want. Sometimes we go to a movie, or just talk. We also practice martial arts. Sunday morning I stay home with my family.

Natural Environments

READ TO DISCOVER

1. What are the major landforms and rivers in China, Mongolia, and Taiwan?
2. Which climate types are found in the region?
3. What are China's main natural resources?

DEFINE

paddy
intensive agriculture
double cropping
aquaculture

LOCATE

Himalayas
Kunlun Shan
Tian Shan
Altay Shan
Greater Khingan Range

Locate, continued

Plateau of Tibet
Tarim Basin
Turpan Depression
Huang (Yellow) River
Chang (Yangtze) River
Xi River
Red (Sichuan) Basin
North China Plain
Manchurian Plain
Mongolian Plateau
Gobi
Taiwan

WHY IT MATTERS

The construction of China's Three Gorges Dam, set for completion in 2009, is controversial both in China and in other countries around the world. Use CNNfyi.com or other **current events** sources to find the latest news about this dam and its effects on the area's people and environment.

China, Mongolia, and Taiwan: Physical-Political

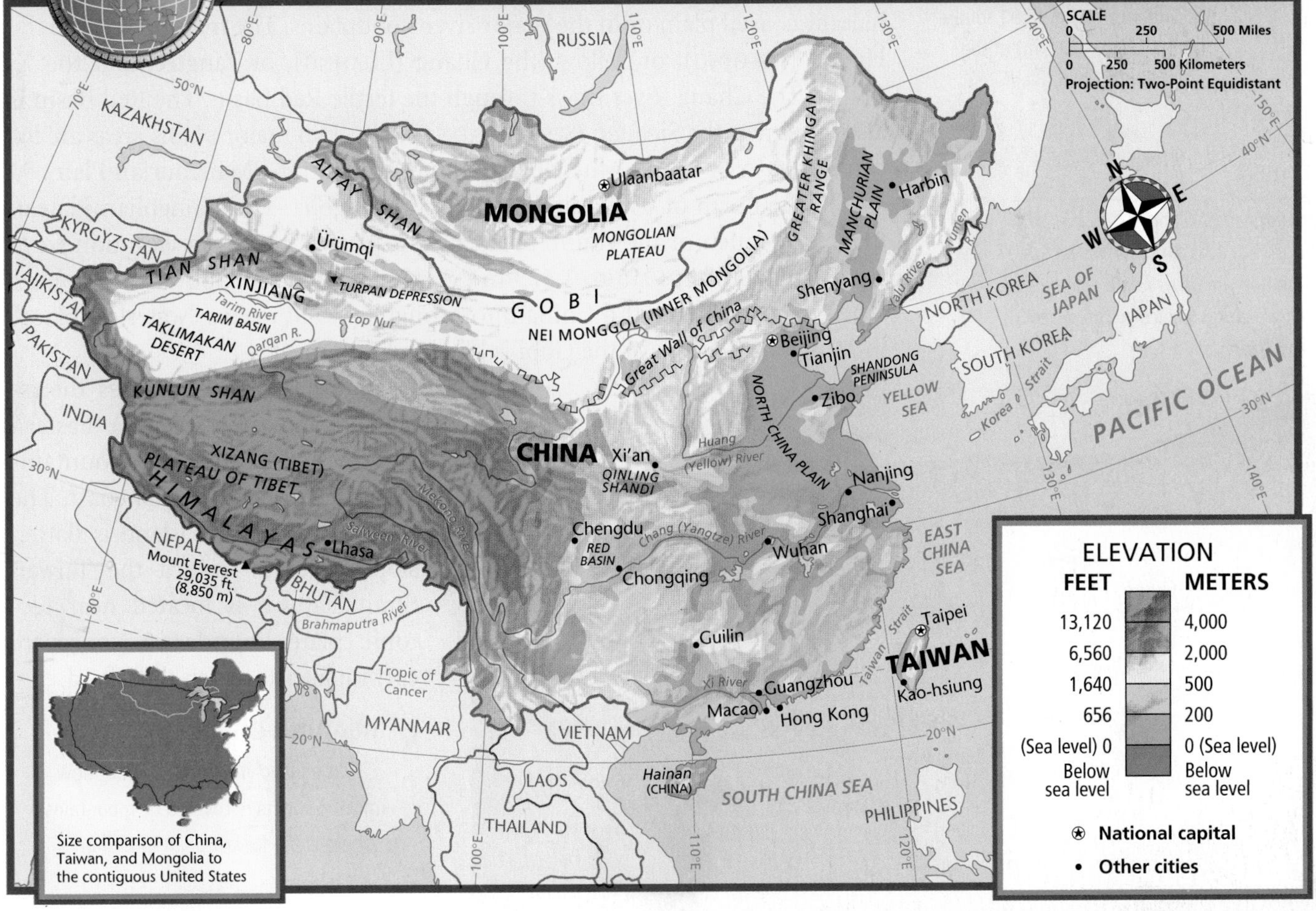

SCALE
0 250 500 Miles
0 250 500 Kilometers
Projection: Two-Point Equidistant

ELEVATION

FEET	METERS
13,120	4,000
6,560	2,000
1,640	500
656	200
(Sea level) 0	0 (Sea level)
Below sea level	Below sea level

⊛ National capital
• Other cities

Size comparison of China, Taiwan, and Mongolia to the contiguous United States

internet connect

GO TO: go.hrw.com
KEYWORD: SW3 CH27
FOR: Web sites about China, Mongolia, and Taiwan

Landforms and Rivers

In area, China is the world's third-largest country after Russia and Canada. China is slightly larger than the United States. From east to west, China and the 48 contiguous U.S. states cover about the same distance. However, from north to south, China extends farther in each direction. (See the size comparison on the chapter map.) China's natural environments include some of the world's driest deserts, highest mountains, and longest rivers.

Mountains cover more than 40 percent of China's land area. The Himalayas, the world's highest mountain range, are located along the southwestern border with Nepal and Bhutan. Mount Everest, the world's highest summit at 29,035 feet (8,850 m), is on the China-Nepal border. To the north are several other major mountain ranges. These include the Kunlun Shan (KOON-LOON SHAHN), Tian Shan, and Altay (al-TY) Shan. In the far northeast is the Greater Khingan (KING-AHN) Range. Much of southeastern China has a rugged terrain of low mountains and steep-sided river valleys.

Between the high mountain ranges of western China are large plateaus and basins. The Plateau of Tibet lies between the Himalayas and the Kunlun Shan. With an average elevation of about 16,000 feet (4,880 m), it is the world's highest plateau. To the north, between the Kunlun and Tian Shan, lies a lower area called the Tarim (DAH-REEM) Basin. The Turpan (toohr-PAHN) Depression, located in the northeastern part of the basin, is the lowest point in China. The Turpan Depression lies 426 feet (130 m) below sea level.

Our Amazing Planet

Built-up silt deposits on the bottom of the Huang River are so thick that in some areas the riverbed is more than 30 feet (10 m) above the surrounding plain. Levees keep the river from flooding more than 9,000 square miles (23,310 sq km) of the North China Plain.

The plains and river valleys of eastern China are the most densely populated areas of the country. Most of the flattest land is located along either the narrow coastal plain or in the major river floodplains. The major rivers are the Huang (HWAHNG), or Yellow; the Chang (CHAHNG), or Yangtze; and the Xi (SHEE). The Chang River flows through the fertile Red Basin. The Red Basin is also known as the Sichuan (SEE-CHWAHN) Basin. Two major plains areas are located in eastern China—the North China Plain and the Manchurian Plain.

To the north of China lies landlocked Mongolia. The Mongolian Plateau makes up much of the country's area. This plateau ranges in elevation from 3,000 to 5,000 feet (915 to 1,525 m). A desert called the Gobi (GOH-bee) extends from north-central China into Mongolia. *Gobi* means "waterless place" in Mongolian. Much of the Gobi is bare rock or gravel.

Southeast of China lies the island of Taiwan (TY-WAHN). Eastern Taiwan has high, steep mountains that rise from the Pacific coast. The western part of the island is flatter, sloping gently toward the Taiwan Strait. Taiwan is located near tectonic plate boundaries, so severe earthquakes threaten the island.

Nearly all of Mongolia is covered by desert or grassland. As this picture shows, the two are often found side by side. Overgrazing has led to the desert's expansion.

✓ READING CHECK:

Places and Regions About how much of China is covered by mountains? Where is the lowest point in China?

Climates, Plants, and Animals

The Asian monsoon system influences climates throughout most of China. (See the unit climate and precipitation maps.) Dry winter winds blow from the Asian interior toward the coast. In northern China these winds can be bitterly cold and dusty. Warm humid air from the ocean brings rain and sometimes typhoons during the summer. In general, the eastern third of the country receives the most rain. Rainfall in China tends to decrease to the north and west.

The color of Buddhist monastery ruins blends in with the dry soil and bare rock of Tibet. The river in the background gets its water from Himalayan snow and ice.

Southeastern China and Taiwan have a mild humid subtropical climate. Much of northeastern China has a humid continental climate. This area has warm humid summers and cold drier winters.

Dry highland climates dominate western China. The Plateau of Tibet has very cold and dry conditions. The high mountains surrounding the plateau have some of the severest weather conditions in the world. The sandy Taklimakan (tah-kluh-muh-KAHN) Desert occupies most of the Tarim Basin in northwestern China. The surrounding mountains create a rain shadow. These mountains block moist ocean air from bringing rain to the desolate basin. North-central China and Mongolia have semiarid and arid climates with the driest areas in the Gobi. Mongolia experiences extreme temperatures. Among the reasons for these extremes are the country's high elevation and its location in the Asian interior at high latitudes. Winter temperatures may drop to –50°F (–46°C), and severe blizzards occasionally hit the country. High temperatures during the dry summers set the stage for raging brush fires.

Because China is so large and has such diverse landforms and climates, the country has a huge number of plant and animal species. In fact, nearly all the major types of Northern Hemisphere plants grow in China. The main exceptions to this are polar tundra plants. China's plant life can be divided into two major regions—that of the dry northwest and the humid southeast. The dry conditions of the northwest produce large areas of steppe grasslands and scattered plants that are resistant to drought. China's humid regions include mangrove swamps along the South China Sea and tropical rain forests on the island of Hainan. China's forests include about 2,500 species of trees.

China's animal life is equally diverse, particularly in the isolated mountains and valleys of Tibet and Sichuan. Wildlife ranges from wild horses and bears to camels and wolves. Some animals that have become extinct in the rest of the world still survive in China. These include the giant panda, Chinese alligator, and Chinese paddlefish, which can grow up to 21 feet (6 m) in length. Many animals are abundant, such as antelope, pheasants, and carp.

READING CHECK: ***Physical Systems*** What factors influence Mongolia's extreme climate conditions?

INTERPRETING THE VISUAL RECORD

Fewer than 1,000 giant pandas remain in China's bamboo forests, which provide the animals' main source of food. Zoos have worked to increase this number, but pandas do not breed easily in captivity. ***How might the panda's limited diet have led to its decreased population?*** TEKS

INTERPRETING THE VISUAL RECORD

A coal miner drags his load from a tunnel. In some areas miners can use hand tools to remove coal from deposits near the surface. Small private mines like this one generally do not follow safety or environmental guidelines. ***What are some ways that government policies may have changed coal mining in China?*** 

Natural Resources

China has huge amounts of energy and mineral resources. It is the world's leading producer of coal, lead, tin, and tungsten. Other mineral resources include iron ore, bauxite, gold, and a wide range of metals. Coal deposits are found throughout the country, but the most important reserves are located in the north and northeast. The widespread use of coal has caused serious air pollution. Oil and natural gas are also found in many areas. China's government is working with Western oil companies to develop both offshore and interior areas for oil and gas production.

Hydropower is a major energy resource in China. Developing this resource further would help lessen dependence on coal and, in turn, reduce pollution. China is currently building the world's largest dam, which could provide vast amounts of hydroelectricity. The dam is located on the Chang River and is known as the Three Gorges Dam.

FOCUS ON GEOGRAPHY

Three Gorges Dam Detailed planning for the Three Gorges Dam began in the 1950s. After nearly 40 years of delays, dam construction finally began in 1993. It is scheduled to be completed in 2009. When it is finished, the Three Gorges Dam will generate as much power as 15 coal-burning power plants and create a reservoir about 370 miles (595 km) long. However, that reservoir will flood hundreds of towns and cities, forcing between 1 and 2 million people to move to higher ground.

One of the main reasons for building the dam is to control dangerous flooding along the Chang River. The Chinese also want to increase river traffic and trade and generate hydropower. However, the size and scale of the project have many people worried about its environmental impact. The dam will disrupt ecosystems along the river, affecting plants and animals, such as a rare freshwater dolphin species. As much as 240,000 acres (97,200 ha) of farmland will be lost, and more than 1,200 historical sites will be flooded. Researchers are racing to save as much as they can from these sites. Critics also worry about the cost. Originally estimated at $11 billion, the dam could cost up to $50 billion. Many critics argue that several smaller and cheaper dams could be built instead. They

In 1997 the waters of the Chang were diverted through a canal to allow construction of the Three Gorges Dam. More than 138.5 million cubic yards (102.6 million cubic m) of earth and stone will be moved before the dam is completed. Some 18,000 people are involved in the dam's construction.

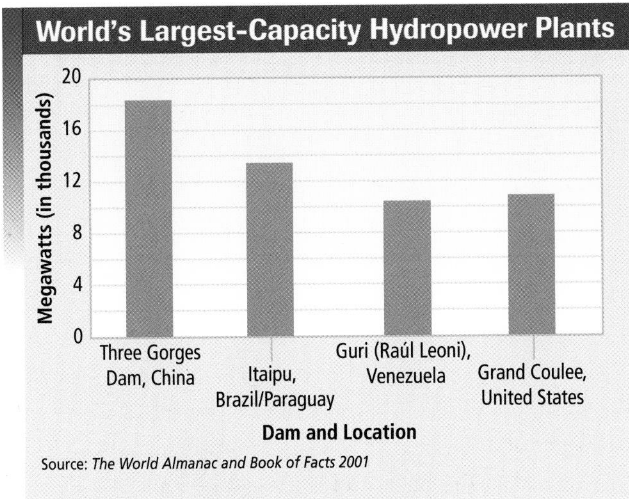

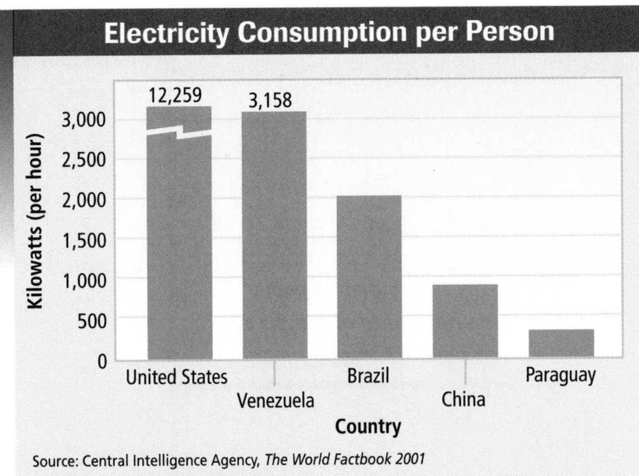

INTERPRETING THE GRAPHS *When fully operational, the Three Gorges Dam will generate more electricity than any other dam now in existence. China's population is more than four times the size of the U.S. population. Electricity consumption per person differs significantly, however.* ***How does China's electricity consumption compare to that of the United States? What factors may account for the difference?*** TEKS

argue that these dams would accomplish the same goals and affect the natural environment less.

READING CHECK: ***Human Systems*** How will the Three Gorges Dam affect settlement patterns and population distribution? TEKS

Agricultural Resources Because of its diverse climates and landform regions, China has a wide variety of soils. However, only about 10 percent of the country is arable. One important farming region is the fertile area along the Chang River. **Paddy** fields—wet lands where rice is grown—are common there. (See the map on the next page.) Because paddy agriculture uses so much human labor, it is a form of **intensive agriculture.** In many paddy areas, two crops are harvested each year from the same plot of land. This practice is known as **double cropping.**

Loess deposits in northern China are very fertile. However, they are easily eroded. The Huang River and its tributaries have eroded a huge loess plateau in the area. This erosion has created a distinctive landscape of steep-sided gullies. The Huang River often overflows its banks during the summer. Heavy rain swells the river, and the thick silt settles to the bottom, raising the level of the riverbed. The yellowish loess soil also gives the river its name, which means "yellow" in Chinese.

To feed its large population, China's farmers grow many different crops. Intensive farming methods boost production to high levels. In fact, China is a world leader in a wide range of farm products, such as ducks, peanuts, and rice. In general, southern China produces crops that need warmer temperatures, such as rice, citrus fruits, tea, and sugarcane. Northern China produces wheat, sorghum, millet, and soybeans. In the west, nomadic herding is common. (See the unit land use and resources map.)

China also has rich fishing resources. The Chinese have a long tradition of both ocean and freshwater fishing. Raising and harvesting fish and marine life in ponds or other bodies of water, known as **aquaculture**, is also important. This practice helps protect against taking too many fish from rivers and seas. It also provides valuable exports such as shrimp.

Like aquaculture, China's silk industry depends on raising small animals under controlled conditions. Silkworms feed on mulberry leaves. Their

INTERPRETING THE VISUAL RECORD

By the 200s B.C., Chinese farmers had developed an ox-drawn iron plow similar to the one used by the farmer shown. This innovation allowed them to increase production. ***How might this new technology have contributed to the growth of Chinese society?*** TEKS

INTERPRETING THE MAP *Chinese farmers have used paddy soils to grow rice for centuries. Over time, this rice farming has changed the chemical and physical properties of the paddies.* ***How could the distribution of paddy soils in China be used to divide the country into regions?*** TEKS

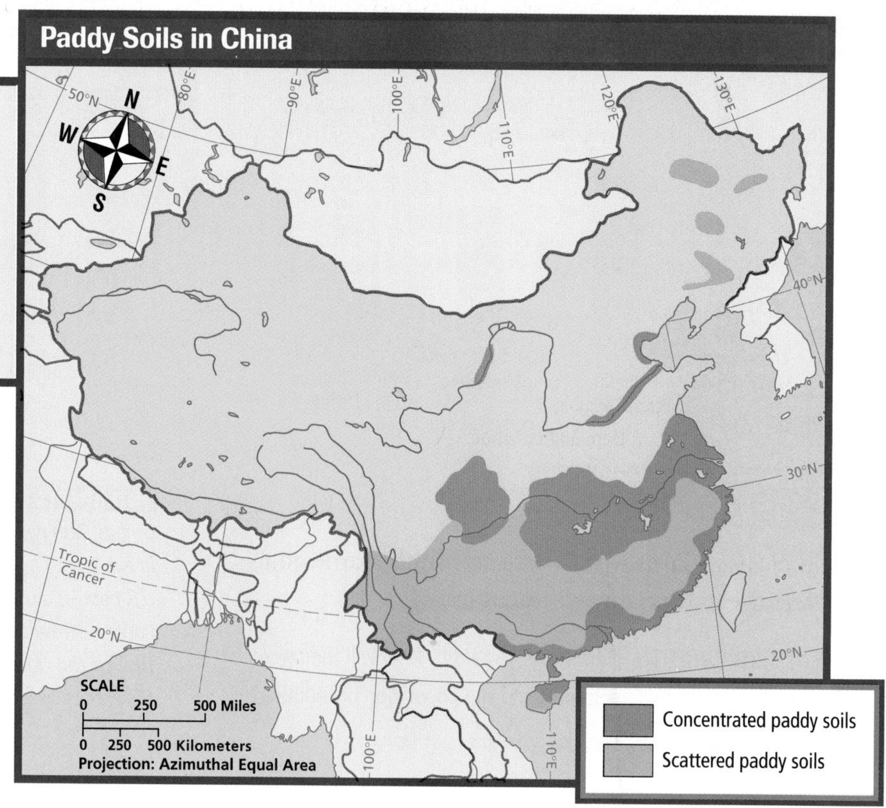

cocoons are unwound for the fiber. China produces more raw silk than any other country.

Population growth and the expansion of agriculture and industries are placing a strain on China's water resources. Droughts and mismanagement have resulted in water shortages across much of the country. Many geographers are concerned that China's water resources will not meet growing demands.

READING CHECK: *Human Systems* What factor has helped make China the world's leading producer of many farm products?

TEKS Working with Sketch Maps, Questions 1, 2, 3, 4, 5

Section 1 Review

Define paddy, intensive agriculture, double cropping, aquaculture

Working with Sketch Maps On a map of China, Mongolia, and Taiwan that you draw or that your teacher provides, label the Himalayas, Kunlun Shan, Tian Shan, Altay Shan, Greater Khingan Range, Plateau of Tibet, Tarim Basin, Turpan Depression, Huang (Yellow) River, Chang (Yangtze) River, Xi River, Red (Sichuan) Basin, North China Plain, Manchurian Plain, Mongolian Plateau, Gobi, and Taiwan. Which plateau is the world's highest?

Reading for the Main Idea

1. *Physical Systems* How do mountain barriers influence the climate of northwestern China?
2. *Physical Systems* How has erosion in northern China created a distinctive landscape along the Huang River?
3. *Environment and Society* How has the use of coal affected China's environment?

Critical Thinking

4. **Analyzing Information** Recall what you learned about tectonic forces in Chapter 4. How do you think tectonic forces have helped shape the landforms of southwestern China?

Organizing What You Know

5. Copy the chart below. Use it to describe the natural environments of the region's mountains, plateaus and basins, and plains and river valleys.

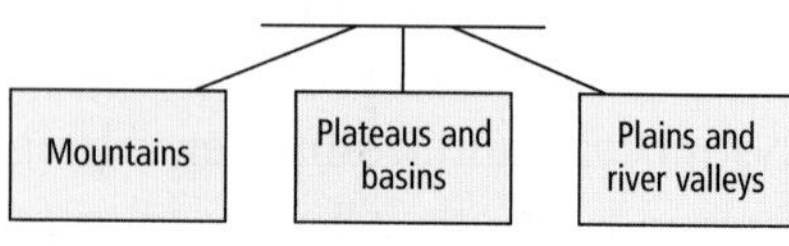

efficiently. They seized all private land and organized farmers and farmland into large, jointly run collective farms. The Communists forcibly relocated families that had owned the land. These families lost their personal property and individual freedoms.

INTERPRETING THE VISUAL RECORD

In the 1930s Mao Zedong and other Communists walked some 6,000 miles (9,650 km) across China, fighting Nationalist forces along the way. This event, known as the Long March, is shown in the painting above. The Long March inspired many young people to join the Communist Party. ***How are Mao and his followers portrayed in this painting?***

Communism also changed China's traditional family structure. In the past the oldest male had been the dominant family member. Under communism, women had equal status and worked in the fields along with the men. Later the government tried to slow the growth of its huge population by limiting families to one child. Despite opposition, the government still enforces this one-child policy in towns and cities.

The government also controlled the economy, setting fixed goals for agricultural and industrial production. For example, to improve China's economy, Mao launched the Great Leap Forward in 1958. The purpose of this program was to speed up industrialization. However, the program actually delayed the country's economic progress. The government organized workers to build backyard blast furnaces to make steel. Work groups built dams and dikes. The government also organized collective farms into larger **communes**. These agricultural communities organized farming and planned local public services.

Mao's program was a disaster. Poor planning and other problems led to the creation of large inefficient industries. In addition, an overemphasis on industry hurt agriculture and led to a famine that caused millions of deaths. The program was an environmental disaster as well. Forests were cut down to make charcoal to burn in the blast furnaces. No attempts at reforestation were made. The result was massive soil erosion.

The Cultural Revolution (1966–76) also delayed China's economic development and brought chaos to the country. During this period, Mao's followers tried to rid China of people they considered his enemies and critics. Anyone with an education was suspect, particularly intellectuals and scientists. Schools and universities were closed. Many old people and scholars were attacked, sent off to work in the countryside, or even killed.

A New China After Mao's death in 1976, Deng Xiaoping (DUHNG SHOW-PING) came to power. Deng realized that the Great Leap Forward and the Cultural Revolution had been mistakes. He pushed new policies to modernize China's agriculture, industry, and technology. He also worked to move the country toward a market economy. Agricultural production increased quickly. China's leaders continued Deng's market reforms after his death.

Teenagers take in the sights on Wang Fu Jing Street, one of Beijing's main shopping districts.

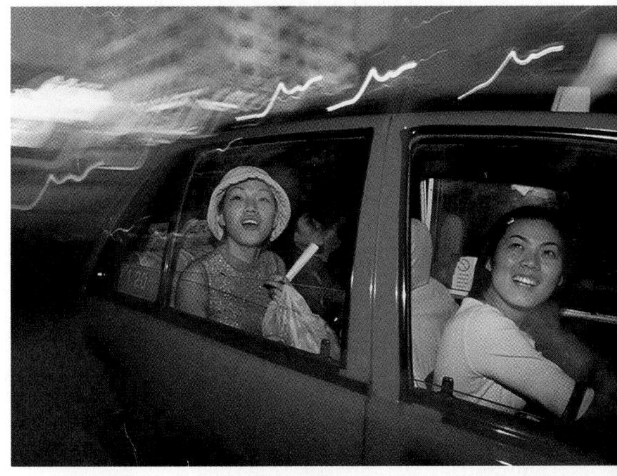

Today many farmers grow and sell their own crops and build their own homes. They do not own the land but lease it from the local government for long terms. Agricultural productivity has increased, and China is self-sufficient in food. A greater variety of fruits, vegetables, and meats has improved the diet of many Chinese. In addition, millions of town village enterprises (TVEs) produce textiles and other export goods. Industry has attracted investment from Chinese in

Scholars have studied how the Chinese written language reflects the country's geography. Here, you can see that mountain peaks are suggested in the character for mountain. *Similarly, the character for* tree *resembles the trunk and bare limbs of a tree. Many characters first recorded 3,000 years ago are still in use.*

Singapore and Taiwan. However, about half of China's industrial workers still work in inefficient state-owned industries.

Although China remains a poor country by Western standards, daily life has improved for many people. For example, more consumer goods such as televisions and refrigerators are now available. Nevertheless, challenges remain. One of the most serious is the worsening pollution of China's air and water. In fact, of the 10 cities with the worst air pollution in the world, 9 are in China. Also, the government still has considerable control over the lives of the Chinese people. While the government seems to be in favor of increasing economic freedoms, it does not want political reforms. The government continues to defend actions it took in 1989 to end pro-democracy demonstrations in Beijing (BAY-JING), China's capital. Army troops and tanks crushed peaceful protests, killing or injuring hundreds of people in Tiananmen Square.

READING CHECK: ***Places and Regions*** How has China's command economy changed since the death of Mao?

Culture

Today China has nearly 1.3 billion people, more than four times the population of the United States. The vast majority of the country's people are Han Chinese. The Han speak many Chinese dialects, but Mandarin is the official language. About 70 percent of the country's people speak it. Written Chinese uses symbols called characters. Some of these characters are **pictograms**, or simple pictures of the ideas they represent. (See the illustration.) In all, Chinese writing uses more than 50,000 characters. As you can imagine, learning to read and write Mandarin is a lengthy process. **Calligraphy**—artistic handwriting or lettering—developed along with the writing system. Japanese and Korean calligraphy evolved from Chinese styles.

Buddhism and **Taoism** are the major religions in China. Taoism originated in China in the 500s B.C. Its followers believe there is a natural order to the universe, called the Tao (DOW). A basic idea of Taoism is to live a simple life in harmony with nature. Many Chinese also follow the philosophy of **Confucianism**. Confucianism is based on the teachings of Confucius (551–479 B.C.), a Chinese philosopher. It is more a code of ethics than an organized religion. The Confucian code centers around family loyalty, duty, and education. Confucianism also spread to other Asian countries, particularly Japan, Korea, and Vietnam.

China has some 55 minority groups. Most members of these groups live along China's borders and in the western part of the country. Some of the largest minority groups in the north and northwest are the Mongols, Tibetans, Uighurs (WEE-goohrz), Kazakhs, and Kyrgyz. Ethnic groups in the southeast include the Yao, Zhuang, and Miao.

China's minority groups have distinct cultures and follow different religions. For example, Tibetans and Mongols are Buddhist, while the Uighurs, Kazakhs, and Kyrgyz are Muslim. Several million Han Chinese in the north are also Muslim. China's Communist government discourages religious practice. Nonetheless, it has recognized the religious freedom for some groups, such as the country's Muslims. However, this has not been the case with some other religious groups,

Confucius, or K'ung Ch'iu as he is known in Chinese, is revered for his teachings and the sayings recorded by his disciples.

These monks are students of the Tibetan form of Buddhism called Gelukpa, or the Yellow Hat sect. Before the Chinese conquest of the region, some 25 percent of Tibetan Buddhists belonged to religious orders.

including some Christian groups and, particularly, China's Tibetan Buddhists. The government has repressed Tibetan culture and religion to prevent any movement toward Tibetan independence.

READING CHECK: *Places and Regions* What is China's official and most common language?

Settlement and Traditions About two thirds of China's huge population lives in the eastern half of the country. The coast and major river valleys are densely populated. Most of the population in the east is concentrated in two areas. These are the North China Plain and the lower and upper basins of the Chang River. Other densely settled areas include the Xi and Zhu (JOO) River basins in southeastern China.

Today China is experiencing rapid migration from rural to urban areas. As a result, many cities are expanding rapidly. An estimated 85 million rural migrants live in China's cities. Many of these people are women from China's interior provinces who travel to work in factories along the coast. The money they send back home not only helps their families survive, but also earns respect for the women.

China's many artistic traditions include architecture, literature, music, painting, and pottery. Artists developed a style of landscape painting featuring towering mountains, clouds, and trees. Many Chinese landscapes include descriptive text written in calligraphy. Poetry flowered during the T'ang dynasty. The Chinese have long been innovators in pottery. They developed porcelain more than a thousand years ago. Traders around the world have long desired Chinese porcelain for its fine quality and beauty. Traditional Chinese architecture features wooden buildings on stone foundations. Large tiled roofs curve upward at the corners. These artistic traditions have influenced cultures throughout Asia.

Rice, noodles, and bread are basic Chinese foods. Tofu (soybean curd) and a wide range of vegetables, such as cabbage, are also common. Pork, poultry, and duck are popular, as are fish and other seafoods. Chinese cooking varies from region to region. Food from Sichuan is spicy with chilies, while northern China is famous for roasted duck. Tea is the most popular drink. Chinese food has diffused widely. For example, Chinese restaurants can be found in cities around the world.

READING CHECK: *Human Systems* What distinctive Chinese cultural patterns have spread to other parts of the world?

Chinese Jade

Jade is a hard, tough gemstone. It occurs naturally in many colors, but the most valuable jade is usually a brilliant emerald green. Since ancient times, the Chinese have valued jade much like Westerners have valued gold. In Chinese culture jade symbolizes ideas such as purity and indestructibility. Chinese artisans have long carved jade into jewelry, vases, weapons, and other objects. These objects are prized around the world for their beauty, strength, polished surface, and intricate details.

Drawing Inferences and Conclusions Why do you think jade symbolizes purity and indestructibility in Chinese culture?

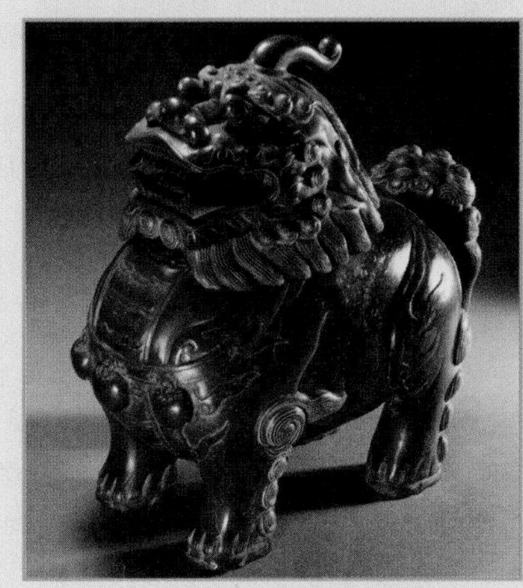

Places and Regions

Geography for Life

China's Karst Towers

Known for its massive karst towers, the area around Guilin (GWEE-LIN), China, has one of the most dramatic landscapes on Earth. The word *karst* refers to a limestone landscape with many caves, underground streams, and steep hills or mountains. There are karst landscapes all over the world, but few are as dramatic as the region near the Chinese city of Guilin.

The erosion of limestone by moving water creates karst landscapes. This process is particularly common in tropical areas with heavy precipitation and dense vegetation. Plants release mild acids and carbon dioxide that are absorbed and carried by water. The chemicals and gases in the water then slowly dissolve and carry away rock, sculpting hills and other features. When karst hills reach mountainous size—as they do in Guilin—they are referred to as tower karst.

China's karst towers have inspired the Chinese imagination for more than 1,000 years. According to a Chinese saying, Guilin's scenery is the "first under heaven." A Chinese scholar during the Sung dynasty (A.D. 960–1279) wrote, "I often sent pictures of the hills of Guilin which I painted to friends back home, but few believed what they saw." Many Chinese landscape paintings feature Guilin's karst towers covered with pine trees and shrouded in mist. In recent years the towers have attracted a growing number of tourists.

Karst towers covered with vegetation loom over the Li River, as farmers and their water buffalo cross rice paddies near Guilin.

Applying What You Know

1. **Summarizing** What physical processes have created the distinctive karst towers near Guilin?
2. **Comparing** How do you think the karst towers compare to physical features that have inspired photographers, writers, and artists in the United States?

TEKS Working with Sketch Maps, Questions 2, 3, 4, 5

Section 2 Review

go.hrw.com Homework Practice Online

Keyword: SW3 HP27

Identify Taoism, Confucianism

Define puppet government, communes, pictograms, calligraphy

Working with Sketch Maps

On the map you created in Section 1, label China, Mongolia, the Great Wall, Macao, Hong Kong, and Beijing. What European powers once controlled Macao and Hong Kong?

Reading for the Main Idea

1. *Human Systems* After which dynasty is China's main ethnic group named?
2. *The Uses of Geography* To where did China's Nationalists flee after they were defeated by the Communists in 1949?
3. *Places and Regions* How do China's religions make it a distinctive region? How does the government treat religion?

Critical Thinking

4. **Comparing** Why do you suppose China's Communist government has allowed more economic reform and freedom but continues to limit political rights and freedom?

Organizing What You Know

5. Copy the graphic organizer below. Use it to describe China under the leadership of Mao Zedong and China after Mao. In the center space, note features that remained the same.

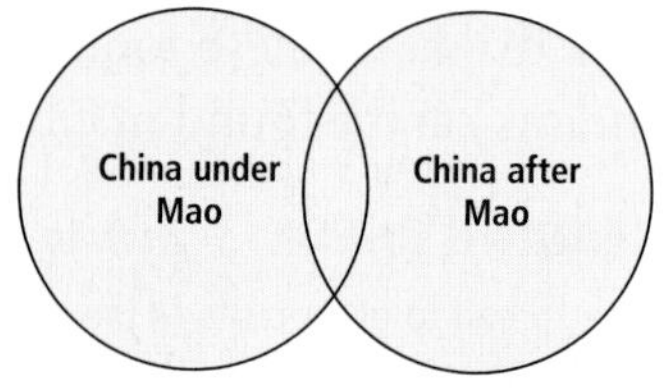

The Region Today

READ TO DISCOVER

1. What are China's major regions?
2. What is Mongolia like today?
3. What is Taiwan's relationship with China?

WHY IT MATTERS

International relations between China and Taiwan affect many countries in East Asia and U.S. foreign policy. Use CNNfyi.com or other **current events** sources to learn about recent developments between these two countries.

IDENTIFY

Special Economic Zones (SEZs)

DEFINE

martial law

LOCATE

Shanghai
Nanjing
Wuhan

Locate, continued

Chongqing
Guangzhou
Shenyang
Tibet (Xizang)
Xinjiang
Lhasa
Ürümqi
Ulaanbaatar
Taipei
Kao-hsiung

China

Today one out of five of the world's people lives in China. China is a huge country in both population and area. To study it more easily, we will look at China's four major geographic regions.

Southern China Southern China is bordered by the Qinling (CHIN-LING) Shandi range in the north and the Plateau of Tibet in the west. The East China and South China Seas lie off the eastern coast. Vietnam, Laos, and Myanmar (Burma) are located to the south.

Southern China is the country's most productive region economically. A large percentage of China's population lives there. Despite continuous cultivation for more than 4,000 years, the soil of southern China is still fertile. Alluvial deposits left by flooding rivers and careful farming renew the soils. The region is often called China's rice bowl.

The Chang Delta is a particularly important rice-growing area. Farmers there are able to grow two rice crops a year, plus a vegetable crop. Upstream

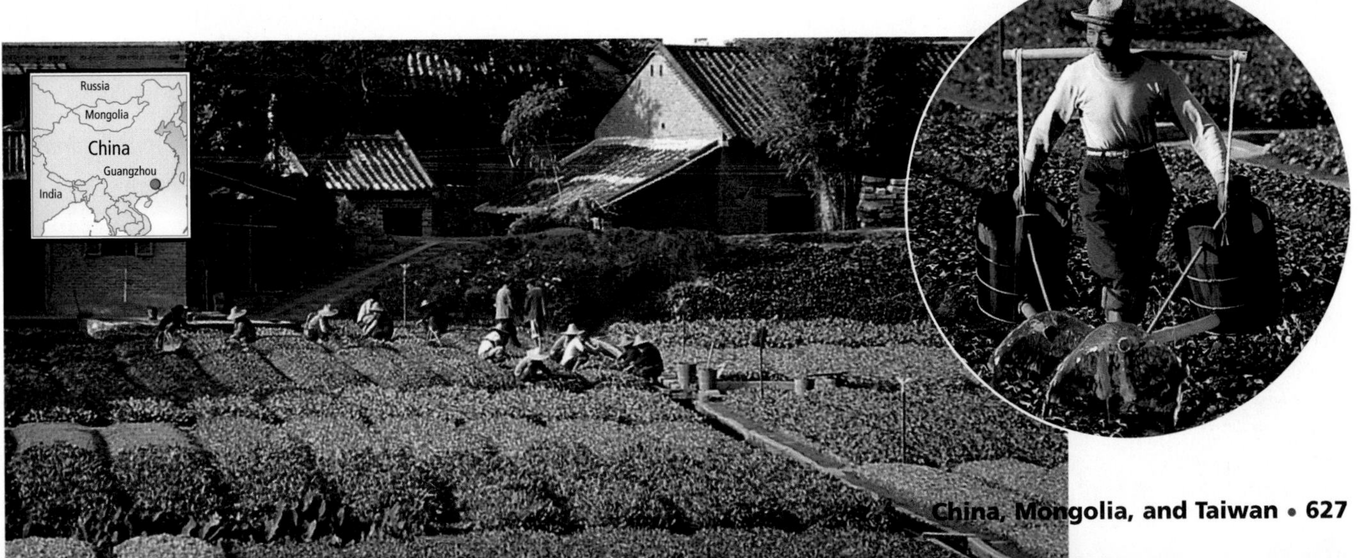

Farmers use traditional methods to work terraced vegetable gardens near Guangzhou. China is the world's leading producer of rice, ducks, hogs, chickens, and eggs, and a major producer of fish, tea, and wheat. Products you may not associate with China factor into its agricultural success. The country grows more than 20 percent of the world's corn, cotton, and peanuts, and more than 15 percent of its potatoes. It is the world's leading grower of apples.

INTERPRETING THE VISUAL RECORD

Hong Kong (shown at right) and Macao are classified as special administrative regions of China. They enjoy a high degree of autonomy in all areas except foreign policy and defense. Both have maintained capitalist economies. ***How may different points of view have influenced the relationship between China and these territories?*** TEKS

from the delta, the Chang River valley is the country's major cotton-growing area. Some farmers there use double cropping to grow cotton and a food crop. Cotton is the main raw material for China's textile industry.

The Chang Delta is also China's most populated and industrialized area. The country's largest city, Shanghai, is located there. Shanghai has some 13 million people. It is a huge industrial center and a major sea and river port. New skyscrapers are rising as the city rapidly spreads outward into farming areas. (See Cities & Settlements: Shanghai.) Two big industrial cities, Nanjing and Wuhan (WOO-HAHN), lie upriver. Industrialization there has been based on local iron ore and coal deposits. The city of Chongqing, farther inland along the Chang River, lies in the Red (Sichuan) Basin. That area has productive farms, coal, and minerals.

Farther south, at the mouth of the Xi River, is the famous trading center of Guangzhou (GWAHNG-JOH), once known as Canton. Guangzhou is the largest city south of Shanghai and has also attracted industrial and commercial development. Located on the coast south of Guangzhou is the former British colony of Hong Kong. Hong Kong's territory is about a third the size of Rhode Island. However, Hong Kong has about 7 million people, making it one of the world's most crowded places. Today it is China's major seaport and its banking and international trade center. It has a variety of industries, including textiles, shipbuilding, and manufacturing. Britain returned Hong Kong to China in 1997, and the Chinese government has granted the former colony a special status. This status includes local autonomy and the continuation of a capitalist economy.

Along the coast of southern China are **Special Economic Zones (SEZs)**. These zones are designed to attract foreign companies and investment to the country. The SEZs close to Hong Kong have seen rapid economic growth. Areas that were rice fields a few years ago now bustle with factories, high-rise buildings, and new freeways. Foreign goods and money circulate freely in these seaports.

TEKS **READING CHECK:** ***Places and Regions*** What has aided the growth of southern China's cities?

Tiananmen Square, in the center of Beijing, is one of the largest public squares in the world. At the north end of the square stand the gates to the Forbidden City (inset). The square has been the scene of both communist and anticommunist demonstrations.

Northern China Northern China lies north of the city of Nanjing and the Qinling Shandi range. This region includes the Huang River valley and the fertile North

China Plain. Chinese culture first developed in this area. It is a very densely populated region.

The Huang River begins in the Plateau of Tibet and flows across northern China to the Yellow Sea. It has changed course and flooded surrounding areas many times. Because so many people have died in floods along the Huang River, it has been called "China's sorrow." Dams along the river store water for irrigation and generating hydroelectricity.

Beijing is China's second-largest city, its cultural center, and its capital. In fact, *Beijing* means "northern capital." The city's history goes back about 2,700 years. The ancient walled part of Beijing is divided into two sections, the Outer City and Inner City. Within the Inner City lies the Imperial City, from which China's emperors once ruled. Within the Imperial City is the Forbidden City, an immense palace complex where the emperors lived. Beijing grew beyond its original walls and spread out across the plains. Today it is a modern city with industry, subways, department stores, hotels, and wide streets.

The Forbidden City contains some 800 buildings, which in turn have about 9,000 rooms. Yet the Great Hall of the People, built by the Communist government, has more floor space—6,060,000 square feet (563,000 sq m)—than all the Forbidden City's palaces put together.

Northeastern China Northeastern China includes three provinces once known as Manchuria. Oil, coal, iron ore, and other mineral resources are plentiful there. The oil fields have helped make China nearly self-sufficient in energy. The region also contains some of China's remaining forest resources. Important industries include iron and steel, chemicals, paper, textiles, and food processing. Shenyang is the region's largest city.

West of the Greater Khingan Range lies Nei Monggol, or Inner Mongolia. Mongols originally populated this area. Today the Han Chinese far outnumber the Mongols. Nomadic herding is common in this dry region at the edge of the Gobi. Irrigation allows some farming.

Western China Two large autonomous regions make up most of western China, Xizang (SHEE-DZAHNG) and Xinjiang (SHIN-JYAHNG). The land is generally too dry, high, and cold to support a large population. Most people there are either herders or irrigation farmers. These regions have some local government and retain their own culture and languages. Both areas were originally populated by people who were not Han Chinese. However, the number of Han Chinese colonists is rising.

The Potala Palace in Lhasa, Xizang (Tibet), is the traditional residence of the Dalai Lama and the seat of Tibet's government. It has more than 1,000 rooms. The current Dalai Lama, the fourteenth of the line, has been in exile since 1959, when the Tibetans unsuccessfully rebelled against the Chinese. Since that time he has led a peaceful campaign for Tibetan independence.

Xizang is the official Chinese name for the region of Tibet. The Chinese have occupied Tibet, once an independent kingdom, since 1950. Tibet is one of the highest and most barren regions on Earth. In fact, some people call it "the roof of the world." Most Tibetans practice a form of Buddhism called Lamaism. The huge Buddhist palace of the Dalai Lama (DAH-ly LAH-muh) towers above Lhasa (LAH-suh), Tibet's capital. The Dalai Lama is the spiritual leader of most Tibetans. In 1959 the Dalai Lama fled to India after an unsuccessful revolt against Chinese control. Tibetans have watched economic development draw many Han Chinese immigrants to the region. In fact, Tibetans fear that they will become an oppressed minority in their own land. Some hope that Tibet will one day again be independent.

North of Tibet, Xinjiang is populated mostly by Muslim Turkic people, particularly the Uighurs and Kazakhs. This dry region has coal, iron ore, copper, and oil. Ürümqi (ooh-ROOHM-CHEE) is the capital city and a manufacturing center. Some Muslims there have sought independence for the region.

READING CHECK: *Places and Regions* What religion do most Tibetans practice?

Mongolia

Mongolia is more than twice as big as Texas but has a population of less than 2.7 million. In fact, Mongolia is the least densely populated country in the world, and livestock outnumber people many times over. A large part of the population herd livestock to earn a living.

Mongolia's natural resources include coal, copper, and oil. Farming is limited, and the country faces food shortages. Water is also in short supply. The capital and only large city is Ulaanbaatar (oo-lahn-BAH-tawr). It has become more modern in recent years as the number of cars, cellular phones, and restaurants has increased. Industrial production includes processed foods, clothing and footwear, paper, and other products.

Mongolia held its first free elections in 1990. Before then, the country had been under the influence of the Soviet Union. The Communist government had suppressed the practice of Tibetan Buddhism, Mongolia's main religion. Mongolia has recently opened its economy to foreign aid and investment. However, its isolated and landlocked location limits Mongolia's economic opportunities. The growth of a market economy may mean a better future for the country.

READING CHECK: *Places and Regions* How have Mongolia's political and economic characteristics changed since 1990?

Left: In centuries past, Mongol warriors like the ones shown were known as superb horsemen and fighters. Right: Some Mongolians maintain their ancestors' way of life, following their herds and living in moveable homes called gers.

Taiwan

Taiwan is one of Asia's richest and most industrialized countries. It exports computers, scientific instruments, and sports equipment. Taiwan's major trading partners are the United States, Japan, and China. The per capita GDP of Taiwan is nearly five times that of China. (See the unit Fast Facts table.) Taipei (TY-PAY) is Taiwan's largest city, financial center, and capital. On the south end of the island is Kaohsiung (KOW-SHYOOHNG), the country's second-largest city. The southern city is a center of heavy industry and the main seaport.

Some 2 million Chinese Nationalists who escaped from mainland China after the Communists took over in 1949 settled in Taiwan. The Chinese Nationalist Party controlled Taiwan under **martial law**, or military rule, for 38 years. Only recently have democratic rights been expanded.

China's Communist government claims that Taiwan is really a province of China. On the other hand, Taiwan's government claims to be the legitimate government of China. This disagreement has caused tension throughout the region. It is possible that Taiwan and China may one day be reunited under one government. The two countries already have economic links. For example, the Taiwanese do invest in China's industrializing coastal regions. This increasing economic interdependence is, in some ways, drawing the two countries closer together. However, wide political and economic differences remain. In addition, some Taiwanese political parties oppose reuniting with China.

READING CHECK: *Places and Regions* How does Taiwan's per capita GDP compare to China's?

Taipei, the capital of Taiwan, is a bustling city of more than 2 million people. Between 1947 and 1967, the city more than quadrupled in size. High-rise buildings have replaced many more traditional structures.

TEKS Working with Sketch Maps, Questions 1, 3, 4, 5

Homework Practice Online

Keyword: SW3 HP27

Section 3 Review

Identify Special Economic Zones (SEZs)

Define martial law

Working with Sketch Maps

On the map you created in Section 2, label Shanghai, Nanjing, Wuhan, Chongqing, Guangzhou, Shenyang, Tibet (Xizang), Xinjiang, Lhasa, Ürümqi, Ulaanbaatar, Taipei, and Kao-hsiung. Which city is the largest in China?

Reading for the Main Idea

1. *Places and Regions* What is the most common economic activity in Mongolia? What are two problems facing the country?
2. *Places and Regions* What economic ties does Taiwan have with mainland China?

Critical Thinking

3. **Making Generalizations** In what ways could you describe Tibet as both a formal region and a perceptual region?
4. **Drawing Inferences and Conclusions** How do you think Taiwan's physical geography has allowed it to remain free of Communist control from mainland China?

Organizing What You Know

5. Copy the chart below. Use it to describe the major political, economic, social, and cultural features that characterize China's four main regions, Mongolia, and Taiwan.

Region/Country	Characteristics
Southern China	
Northern China	
Northeastern China	
Western China	
Mongolia	
Taiwan	

CITIES & SETTLEMENTS

Shanghai

Places and Regions Located where the Huangpu River flows into the Chang River delta, Shanghai is China's busiest port and largest city. Shanghai began as a quiet fishing village and grew to be a busy agricultural center. For a while it was the playground for Europeans in China. Today Shanghai is changing once again. The city is leading China's efforts to reform its communist economy. In the process the people of Shanghai are living through some of the most exciting times their city has ever seen.

Shanghai has been called the Pearl of the Orient. The Pearl of the Orient TV Tower contains 11 spheres of varying sizes, representing pearls. The tower houses exhibition space, a hotel, an observation deck, restaurants, and shops.

History Repeats Itself

Shanghai remained isolated for some 500 years until commercial agriculture developed in the region about A.D. 1000. Cotton production was the basis of Shanghai's economy until the 1840s, when China lost a war with Great Britain. Outside powers then forced China to open Shanghai to foreign development. Zones of the city came under the control of American, British, French, and Japanese companies. Each zone took on the culture of the country that controlled it. For example, the French zone became famous for its cafés and lively nightlife. As a result, in the 1920s and 1930s some people called Shanghai the Paris of the East. The city's international era ended in 1949. Communist forces took control of China that year and drove out all foreigners. For the next 40 years, Shanghai remained closed to outsiders.

Today many of the places that remind residents of their city's foreign past are disappearing in a frenzy of new construction. At the same time, new overseas influences have appeared. China's Communist government has reopened Shanghai to foreign businesses and investment. In the early 1990s the government created the Pudong Development Zone. This zone lies across the Huangpu River on the city's eastern edge. On what was recently farmland, a "mini-city" called the Pudong New Area has arisen. Within its 200 square miles are high-rise apartment houses, a financial district, and an industrial park. The government has also squeezed in an international airport.

Chinese leaders are using Pudong's high-tech facilities, tax breaks, and skilled workforce to lure foreign companies to Shanghai. More than 1,000 such companies began operations there in the 1990s. They include major corporations from Britain, Germany, Japan, South Korea, and the United States.

The Chinese completed construction of the country's tallest building, the 88-story Jin Mao Tower, in Pudong in 1999. The other notable feature on Pudong's skyline is the Pearl of the Orient TV Tower. The 1,535-foot (468 m) structure has been described as looking like a rocket ship poised for takeoff. The image seems to match the city's rise. By 1997 Pudong already had a GDP larger than that of many small countries. Its economy was growing at the rate of 15 percent per year.

Shanghai

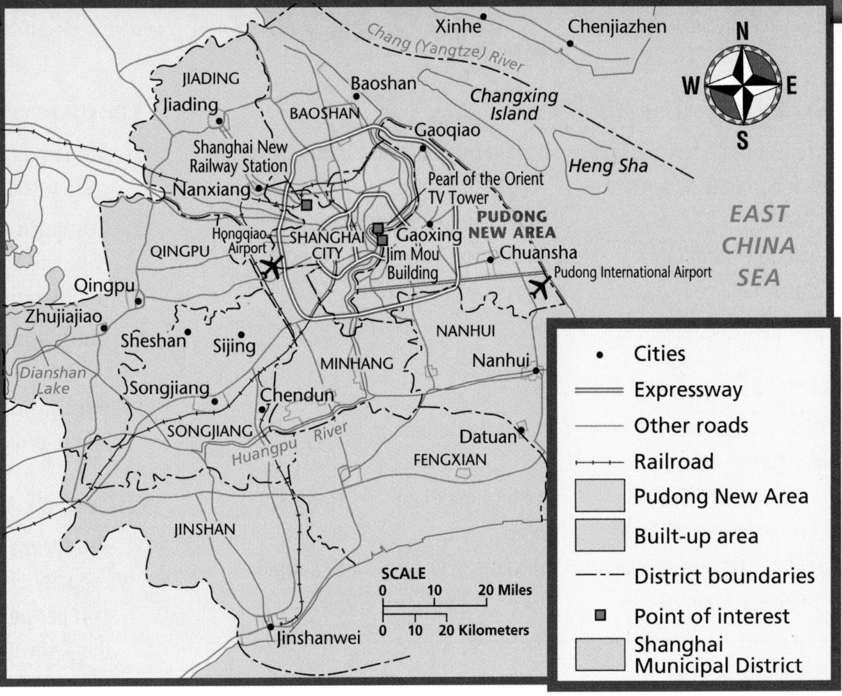

INTERPRETING THE MAP ***How does the Pudong New Area compare in size to Shanghai City, west of the Huangpu River? How might the Pudong New Area's location aid its development?*** TEKS

Open for Business

Back across the Huangpu River, the old city is also booming. In the process, entire neighborhoods are disappearing. Builders are leveling old British and French buildings from Shanghai's earlier period to make way for huge skyscrapers. In fact, more than 1,500 high-rise apartments and office buildings have been built in the city since 1990. New elevated freeways also criss-cross the growing urban area. Much of this work has been done by the rural Chinese who flood Shanghai in search of jobs.

These migrants make up almost 25 percent of the city's nearly 14 million people. They are only temporary residents, however. China's government is loosening its control of the country's economy, but it continues to keep tight control over the movement of China's citizens. One must get a permit to stay in Shanghai for longer than three days. For a two-year permit, a newcomer has to prove that he or she has a job and a place to live. Jobs are easy to find, but housing is in short supply. Nearly three fourths of the city's people live in downtown areas, which are particularly crowded. As a result, staying in Shanghai beyond two years is hard. A person must buy a house or apartment, be sponsored by an employer, or marry a resident. None of these alternatives is likely for most rural migrants. They tend to be poorly paid, poorly educated, and looked down upon by the city's permanent residents. This situation deepens the economic and social divisions that the boom has opened in the city.

Despite these problems, Shanghai's boom has radically changed daily life. Shanghai's residents can now enjoy French perfume and Swiss chocolates. They shop in Japanese department stores and luxury shops from New York's Fifth Avenue. Japanese cafés compete with American fast food and Tex-Mex restaurants. Huge German supermarkets provide groceries for meals at home. After dark, Shanghai now offers a scene that rivals the old days. Among the hot spots is Park 97, a nightclub in an old French mansion. Here, young urban professionals in suits or short skirts eat gourmet food and listen to French rap music. Today, however, the crowd is not European. The customers are nearly all Chinese.

Applying What You Know

1. **Summarizing** In what ways is history repeating itself in Shanghai?
2. **Comparing** How do you think life in Shanghai is like life in New York or Paris? What political, economic, and social similarities and differences would you expect to see between the cities? How might environmental change in those cities be similar?

CHAPTER 27 Review

Building Vocabulary TEKS

On a separate sheet of paper, explain the following terms by using them correctly in sentences.

paddy	communes	Confucianism
double cropping	pictograms	Special Economic Zones (SEZs)
aquaculture	calligraphy	martial law
puppet government	Taoism	

Locating Key Places TEKS

On a separate sheet of paper, match the letters on the map with their correct labels.

Plateau of Tibet	North China Plain	Hong Kong
Huang (Yellow) River	Gobi	Beijing
Chang (Yangtze) River	Taiwan	Shanghai
Red (Sichuan) Basin		

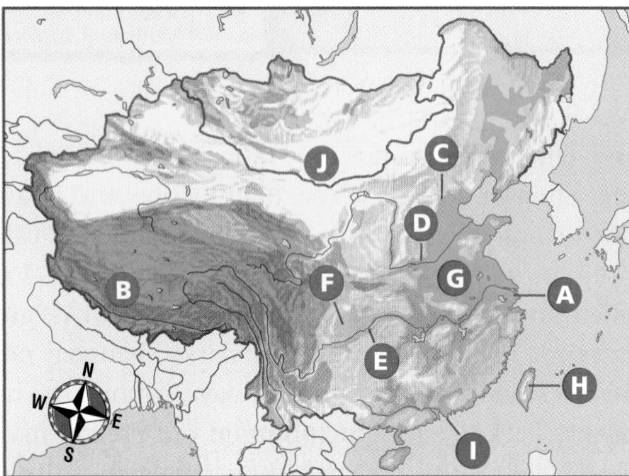

Understanding the Main Ideas TEKS

Section 1

1. *Places and Regions* How diverse are China's natural environments? Explain your answer.
2. *Environment and Society* Why is there a growing concern in China over a lack of water resources?

Section 2

3. *Human Systems* During which dynasty did distinctive Chinese culture traits such as the use of chopsticks and the early use of money develop?
4. *Environment and Society* How did Mao's Great Leap Forward cause an environmental disaster?

Section 3

5. *Places and Regions* Which region of China is known as China's rice bowl?

Thinking Critically for TAKS TEKS

1. **Identifying Cause and Effect** How does the existence of a certain soil type disrupt lives hundreds of miles away?
2. **Comparing** Compare population maps for Units 2 and 9. How does the distribution of China's population compare to the distribution of population in the United States?
3. **Comparing** How do you think the differing points of view held by China's and Taiwan's governments make unification difficult? Why do you suppose the two countries have increased economic ties despite these differing points of view?

Using the Geographer's Tools TEKS

1. **Analyzing Graphs** Review the graphs showing the generating capacity of the world's largest dams and electricity consumption per person. How might the data help explain why building the Three Gorges Dam is important to China's leaders?
2. **Interpreting Time Lines** Create a time line of major events and eras in China's history, using this chapter and the Unit 9 time line for dates. When did major cultural developments occur? When did China come into conflict with other countries?
3. **Preparing Maps** Work with a partner to research the Great Wall. Prepare a map showing dates of construction of various sections and their condition of preservation. Add information on sections of the wall recently discovered by remote imaging and why estimates of the wall's length vary so widely.

Writing for TAKS TEKS

Imagine that you are a researcher for an international magazine reporting on the effects of the Three Gorges Dam. How much power will the dam produce? How will it affect the physical environment? How will the distribution of resources and economic conditions in the region be affected? How will population distribution and patterns of settlement be changed? Summarize the answers to these questions in a benefits-cost balance sheet.

SKILL BUILDING

Geography for Life TEKS

Using Research Skills

Human Systems Conduct research on an aspect of Chinese culture that has diffused throughout Asia or the world. Examples include acupuncture, paper money, porcelain, or tea. Describe how the item spread beyond China and how its use has changed.

Building Social Studies Skills

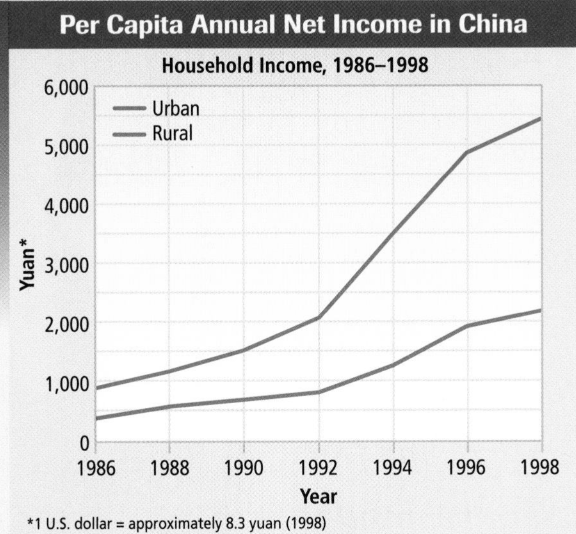

Interpreting Line Graphs

Study the line graph above. Then use the information from the graph to help you answer the questions that follow. Mark your answers on a separate sheet of paper.

1. The gap between urban and rural income was
   a. smallest in 1994.
   b. greatest in 1992.
   c. greatest in 1998.
   d. smallest in 1998.

2. How would you describe urban and rural income growth during the period shown on the graph?

Using Language

The following passage contains mistakes in grammar, punctuation, and usage. Read the passage and then answer the following questions. Mark your answers on a separate sheet of paper.

"(1) The Huang River rises on the eastern edge of the Plateau of tibet. (2) It flows eastward through the North China Plain and empties into the Yellow Sea. (3) Meaning "yellow river," the yellowish mud carried by the Huang gives it its name."

3. Which sentence contains an error in capitalization?
   a. 1
   b. 2
   c. 3
   d. all are correct

4. Rewrite sentence 3 to correct the dangling modifier.

Alternative Assessment

PORTFOLIO ACTIVITY

Learning about Your Local Geography TEKS

Group Project: Research

China is home to many endangered plant and animal species. Are there some endangered plants and animals in your state or region? Plan, organize, and complete a group research project about endangered species in your area. Contact your state department of wildlife to find out which plants and animals are considered endangered and where they are found. How many endangered species live in your state or area? What laws have been passed to protect them? What threatens their survival? What is being done to protect them? Is loss of habitat part of the reason they are threatened with extinction?

internet connect

Internet Activity: go.hrw.com
KEYWORD: SW3 GT27 TEKS

Choose a topic on China, Mongolia, and Taiwan to:

- learn about Chinese art.
- create a brochure on the Great Wall.
- write a report on diplomatic and trade issues between the United States and China.

CHAPTER 28

Japan and the Koreas

Japan lies off the Pacific coast of Asia. The Japanese call their country the land of the rising Sun. To the west is the Asian mainland. Stretching from the mainland is a rugged peninsula occupied by North and South Korea.

Ohaiyo (Hi!). I'm Akiko and I live in Japan. My dad works for Toyota, and my mom stays home. Every morning except Sunday I put on my school uniform and eat rice soup and pickles. I leave for school on the subway at 6:30 A.M. The subway is so crowded that I can't move. Special workers are hired to push more people into each car. At school, I study reading, math, English, science, and writing. I am still learning to write Japanese. I know 1,800 characters, but I need to know about 3,000 to pass the high school exams. For lunch, I eat rice and cold fish my mom packed for me. Before we can go home, we clean the school floors, desks, and windows. My dad usually isn't home until very late at night, so my mom helps me with my homework in our "big" room, which is 8 feet by 8 feet. In the evenings, I go to a special *juku* school to help me study harder for the important school exams. If I do not do well on these exams, I will not go to a good high school, and my whole family will be ashamed. On Sundays I sometimes go with my parents to visit my grandparents, who are rice farmers. We went to a baseball game once. I like rock music a lot, especially U2.

White-naped cranes

Geisha in Kyōto, Japan

Section 1 Natural Environments

READ TO DISCOVER

1. What are the major landforms of Japan and the Koreas?
2. Which climates are found in the region?
3. What are some important resources in Japan and the Koreas?

WHY IT MATTERS

Fishing fleets in the Pacific and around the world are facing major challenges from competition, overfishing, and pollution. Use CNNfyi.com or other **current events** sources to read about current issues in the fishing industry.

IDENTIFY

Chishima Current (Oyashio)
Japan Current (Kuroshio)

DEFINE

tsunamis
flyway

LOCATE

Hokkaidō
Honshū

Locate, continued

Shikoku
Kyūshū
Inland Sea
Ryukyu Islands
Okinawa
Kuril Islands
Japanese Alps
Fuji
Sea of Japan
Korea Peninsula

Japan and the Koreas: Physical-Political

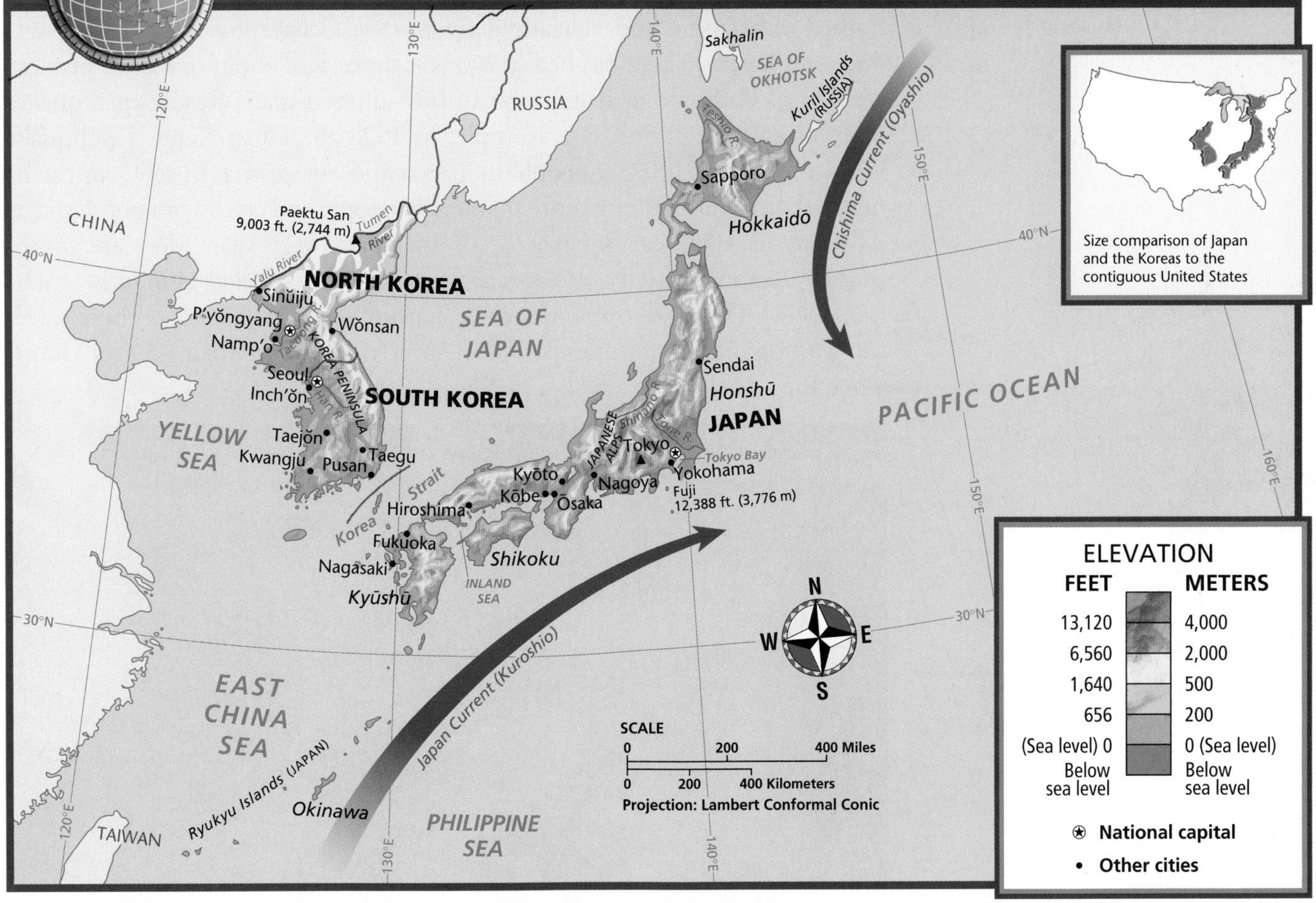

Size comparison of Japan and the Koreas to the contiguous United States

ELEVATION FEET	METERS
13,120	4,000
6,560	2,000
1,640	500
656	200
(Sea level) 0	0 (Sea level)
Below sea level	Below sea level

⊛ National capital
• Other cities

SCALE 0 200 400 Miles; 0 200 400 Kilometers
Projection: Lambert Conformal Conic

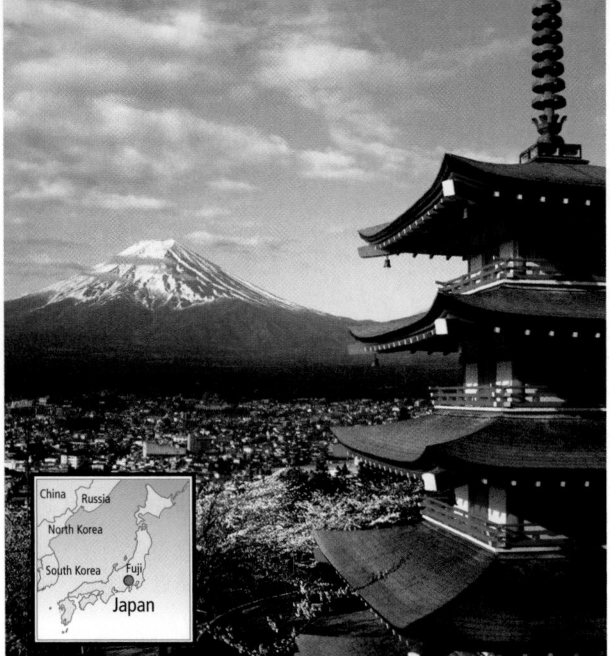

Fuji's volcanic cone is one of Japan's most recognized landmarks. Although Fuji has not erupted since 1707, geologists consider it an active volcano.

Landforms

Four main islands and thousands of smaller ones make up Japan. From north to south, the four main islands are Hokkaidō (hoh-KY-doh), Honshū (HAWN-shoo), Shikoku (shee-KOH-koo), and Kyūshū (KYOO-shoo). Honshū is the largest and most populous island. The Inland Sea separates the three main southern islands. The smaller Ryukyu (ree-YOO-kyoo) Islands to the south are also part of Japan. Okinawa is the largest of these. Japan also claims the Kuril (KYOOHR-eel) Islands to the north. The Soviet Union occupied the Kurils at the end of World War II. Russia now controls them.

More than 70 percent of Japan is mountainous. Japan's longest mountain range is the Japanese Alps. This range forms a rugged volcanic spine on the island of Honshū. Japan's highest peak, Fuji, rises to 12,388 feet (3,776 m) in central Honshū. Fuji's snow-capped volcanic cone has long been a symbol of Japan. The rest of the land—less than 30 percent—is made up of plains. Most are located on the Pacific coast of Honshū. The three most important plains include the Kanto Plain near Tokyo and the Nobi (NOH-bee) Plain near Nagoya. The third, in the Kansai region, lies near the cities of Kōbe (KOH-bay) and Ōsaka (oh-SAH-kah). All three coastal plains are densely populated.

Japan lies along a subduction zone. The Pacific plate dives under the Eurasian and Philippine plates in this zone. This location makes the islands a hotbed of tectonic and volcanic activity. (See Geography for Life: Tectonic Forces in Japan.) Japan has nearly 200 volcanoes, and about one third of these are active. Throughout history, Japan has suffered many deadly earthquakes and volcanic eruptions. For example, in 1923 the Great Kanto Earthquake killed more than 140,000 people in Tokyo and Yokohama. In 1995 an earthquake struck Kōbe, killing more than 6,400 people and causing major damage. As many as 1,500 earthquakes occur in Japan every year. Most are minor quakes. Tectonic activity also creates large sea waves called **tsunamis** (sooh-NAH-mees). The word *tsunami* means "harbor wave" in Japanese. These waves can travel hundreds of miles per hour. Some rise to more than 100 feet (30 m) when they reach shore.

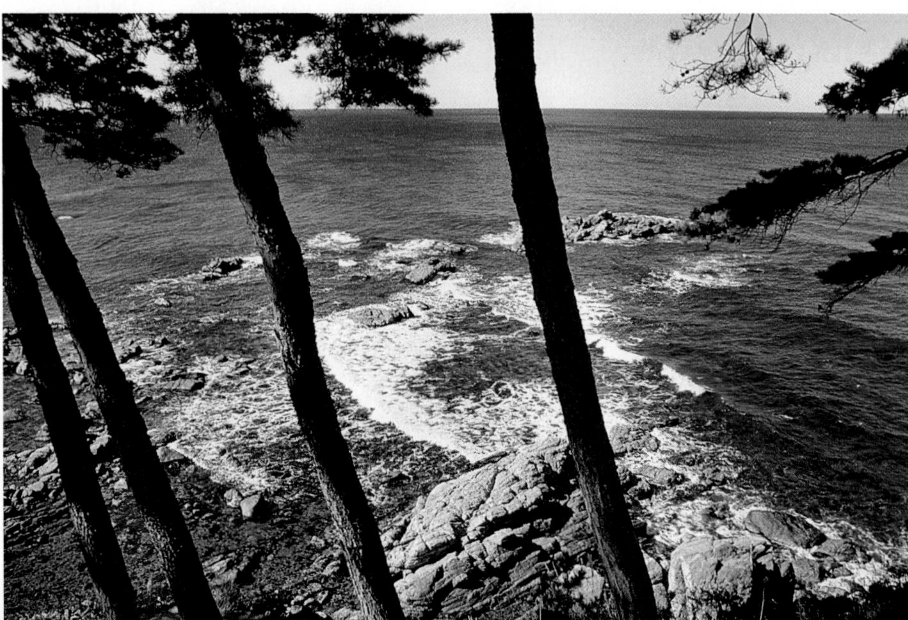

The rugged Korea Peninsula is made up of rocks such as granite and gneiss and has a rocky coastline.

West of Japan, across the Sea of Japan and Korea Strait, lies the Korea Peninsula. This 600-mile-long (965-km) peninsula is about the same size as the state of Utah. To the west is the Yellow Sea. Two countries, North Korea and South Korea, occupy the Korea Peninsula. North Korea borders China along the Yalu and Tumen Rivers. It also shares a short border with Russia in the far northeast.

The peninsula's landforms are mostly hills and low mountains. Unlike Japan, the Koreas do not have active volcanoes. The most mountainous region is in the northeast. Several peaks there rise higher than 8,000 feet (2,438 m). Steep mountains plunge into the sea along the east coast. A coastal plain is located on the west coast. This crowded plain contains the peninsula's best farmland and most of its population.

READING CHECK: *Physical Systems* How do tectonic forces affect Japan?

Our Amazing Planet

Japanese macaques (muh-KAKS), or snow monkeys, live on the northern tip of Honshū at 41° N latitude. This latitude marks the northern limit of monkey habitation in the world. Macaques survive 4 to 5 months of winter snow cover each year by eating tree bark.

Climates, Plants, and Animals

The Asian monsoon system influences the climates of Japan and the Korea Peninsula. During the summer, moist Pacific winds sweep across the region from the south. During the winter, dry northwest winds blow from Siberia in eastern Russia. Precipitation is greatest in the summer months. Most areas get between 40–60 inches (100–150 cm) of annual precipitation. (See the precipitation map in the unit atlas.) Typhoons from the tropical Pacific Ocean occasionally strike the Japanese and Korean coasts in the late summer and fall months.

Japan spans almost the same latitudes as the east coast of the United States. Therefore, the climates found in each place are similar. Hokkaidō and northern Honshū have a humid continental climate similar to that of the New England states. The cold **Chishima Current (Oyashio)** from the north cools summers. Cold winds blowing from the Asian mainland bring severe winters to northern Japan. Heavy snows fall in the mountains, particularly on Japan's western slopes. Southern Japan, including Kyūshū, Shikoku, and southern Honshū, has a humid subtropical climate similar to that of the southeastern United States. The warm **Japan Current (Kuroshio)** brings moist marine air to these areas. Summers are quite warm and humid, and winters are mild.

The Korea Peninsula has a similar climate pattern. The north has a humid continental climate, and the south has a humid subtropical climate. North Korea experiences very cold and snowy winters. South Korea's winters are milder. Summers are warm and humid across most of the peninsula.

Temperate and middle-latitude forests cover much of Japan and the Koreas. These forests contain camphor, oak, and pine trees. Maples are also common in Japan and are greatly admired for their beautiful fall colors. Some forests in the Koreas have been cleared. Deforestation and population growth have greatly limited the habitats of many large mammals there. Even in remote areas, bears, leopards, and tigers have almost disappeared. In Japan, however, many mammals are still common in forested mountain areas. These animals include antelope, bears, deer, and foxes. Both Japan and the Koreas are

INTERPRETING THE VISUAL RECORD

Tokyo's Tsukiji Market is one of the largest fish markets in the world. Every day, fresh fish are shipped in from around the world. However, even though Japan's fishing fleet is one of the most advanced and successful in the world, the country still imports seafood to meet its huge demand. ***How might improved technology have both positive and negative effects on Japan's fishing industry?***  TEKS

also on a major **flyway**—a migration route for birds. Hundreds of species of migratory birds pass through Japan and the Koreas on their journeys north and south.

 TEKS **READING CHECK:** *Physical Systems* How do the locations of warm and cold ocean currents affect the region's climates?

Natural Resources

Mineral and energy resources are quite limited in Japan. As a result, the Japanese rely heavily on oil and coal imports. Japan's industrial economy also depends on imports for many industrial metals and minerals. These imports include iron and aluminum. Because of its expensive oil imports, Japan has made efforts to conserve energy. Nuclear and hydropower plants have helped lower the country's dependence on imported oil. The Koreas are also oil-dependent and use nuclear power and hydropower. North Korea has deposits of iron ore, copper, lead, and coal.

More than 65 percent of Japan is forested—a very high percentage. In lowland areas, deciduous forests are common. Mountain regions have more evergreen trees. Commercial forestry in Japan is carefully controlled to limit soil erosion and protect plant and animal species. To further protect its forest resources, Japan imports much of its timber from Canada, Southeast Asia, and the United States.

Japan's island geography has helped create a culture that depends on the sea for much of its protein. Japanese waters are rich in marine life, and Japan has one of the world's largest fishing fleets. South Korea also has a large fishing fleet. Despite worldwide protests against whaling, the Japanese continue to hunt whales in international waters. In addition, aquaculture supplies fish, shellfish, seaweed, and pearls (from oysters) in Japan. Sometimes this practice is called sea farming.

TEKS **READING CHECK:** *Places and Regions* From where do the Japanese get much of the protein in their diet?

Section 1 Review

TEKS Questions 1, 2, 3, 4, 5

Identify Chishima Current (Oyashio), Japan Current (Kuroshio)

Define tsunamis, flyway

Working with Sketch Maps

On a map of Japan and the Koreas that you draw or that your teacher provides, label Japan, North Korea, South Korea, Hokkaidō, Honshū, Shikoku, Kyūshū, Inland Sea, Ryukyu Islands, Okinawa, Kuril Islands, Japanese Alps, Fuji, Sea of Japan, and the Korea Peninsula. Which islands are controlled by Russia but claimed by Japan?

Reading for the Main Idea

1. *Places and Regions* About how much of Japan is mountainous?
2. *Physical Systems* How are the climates of northern Japan and the Koreas different from climates in southern Japan?
3. *Environment and Society* What are some of the geographic and economic effects of Japan's forestry policies?

Critical Thinking

4. **Analyzing** How do you think Japan's dependence on ocean resources influences its culture? How might Japan's whaling policies affect its relations with other countries?

Organizing What You Know

5. Create a chart like the one shown below. Use it to compare the natural environments of Japan and the Koreas.

Japan	The Koreas

Physical Systems

Geography for Life

Tectonic Forces in Japan

Japan is located in one of the most tectonically active areas in the world. Stretching along the Pacific Ring of Fire, the country sits at the intersection of four major crustal plates. Tectonic forces along these plate boundaries have long shaped Japan's physical geography.

The Japanese islands are formed by the upper part of a large mountain chain that rises from the ocean floor. Volcanic processes created this mountain chain. These mountains are the result of the subduction of the Pacific and Philippine plates under the Eurasian and North American plates. This subduction created the Japan Trench as the heavier oceanic plates were pulled down below the continental plates. Material from the oceanic plates was then slowly heated and transformed into magma. Eventually, the magma rose back to the surface. There it formed the mountains and volcanoes that are now Japan.

Tectonic forces also created patterns in Japan's physical geography. Steep geologically young mountains are common. There has not been enough time for most of the land to be worn down by erosion. Also, Japan's many hot springs and most of its lakes are volcanic in origin.

Tectonic forces in Japan pose numerous hazards. Active volcanoes, earthquakes, and tsunamis are serious threats. For example, the Great Kanto Earthquake of 1923 wrecked Tokyo and Yokohoma. The quake struck just before noon. At that time, Japanese in thousands of homes and restaurants were using gas and wood-burning stoves to cook lunch. As wooden buildings collapsed and gas mains broke, fires raged out of control. Broken water mains made it impossible for firefighters to battle the blazes. The Great Hanshin Earthquake that struck Kōbe in 1995 was also a major disaster. It caused about $100 billion in damage and left thousands homeless.

Kōbe's Nagata Ward was devastated by the 1995 earthquake, which sparked raging fires.

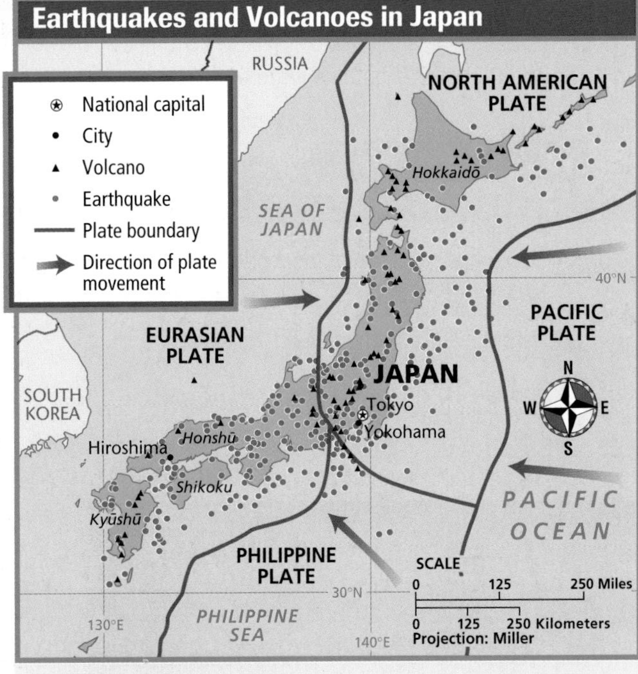

INTERPRETING THE MAP *Over the years, Japan has spent hundreds of millions of dollars on earthquake prediction research.* ***How might advanced warning of earthquakes help prevent disasters?*** TEKS

Dealing with and preparing for tectonic hazards has been a major concern in Japan. In fact, the Great Kanto Earthquake influenced engineering practices designed to make buildings earthquake-safe. For example, engineers began to focus on building structures flexible enough to withstand the violent shaking caused by earthquakes. Also, local governments made many changes to building codes. These changes included new rules limiting the height of buildings. Still, the 1995 damage in Kōbe shows that tectonic hazards remain a serious threat in Japan.

Applying What You Know 

1. **Summarizing** Why is Japan so tectonically active?
2. **Cause and Effect** How have tectonic hazards affected engineering practices in Japan? How might new engineering techniques help the Japanese adapt to their environment?

Section 2 History and Culture

READ TO DISCOVER

1. What cultures influenced the early history of Japan and the Koreas?
2. What were some major events in the modern history of the region?
3. What are some notable features of Japanese and Korean culture?

WHY IT MATTERS

For roughly 50 years the boundary between North and South Korea has been one of the world's most tense borders. Use CNNfyi.com or other **current events** sources to read about recent events on the Korea Peninsula.

IDENTIFY

Meiji Restoration
Diet

DEFINE

shogun
samurai
annex
armistice
demilitarized zone (DMZ)

LOCATE

Kyōto
Nagasaki
Tokyo
Hiroshima

In this woodcut, a samurai warrior orders farmers to kneel before their lord. During the rule of the shogun, Japanese society was divided into four main classes: artisans, farmers, merchants, and warriors.

Early History

Japan's early inhabitants were the Ainu (I-noo). They may be the descendants of people who migrated into Japan from northern Asia several thousand years ago. In about 300 B.C. invaders from Asia drove the Ainu into northern Japan. The new immigrants introduced rice farming to the islands. Within a few hundred years they had settled all the major Japanese islands. Today the Ainu number only about 20,000. Most live in northern Hokkaidō, where they fish and farm.

Korea's early people came from northern and central Asia. In 108 B.C. the Chinese invaded Korea. This event marked the beginning of a long period of Chinese influence on Korean culture. Eventually, the Koreans recaptured most of the peninsula. Korean culture flourished and became known in Asia for its architecture, ceramics, and painting.

China and Korea greatly influenced Japan's early culture. Over time, however, a distinct Japanese culture emerged. For example, Shintoism became Japan's main religion. Shintoism centers around *kami*. The *kami* are the spirits of natural places, sacred animals, and ancestors.

In the A.D. 700s Japan began to develop a unique political system. The Japanese established a capital at Kyōto (KYOH-toh) in central Honshū. A Japanese emperor was officially in control of Japan's political system. However, by the late 1100s real power rested with a powerful warlord called a **shogun**. Over time the power of the shogun grew. Eventually the shogun ruled over other wealthy landlords called *daimyo*. The *daimyo* controlled their own local territories. They also had professional warriors called **samurai** who protected them.

The Japanese political system was similar to the feudal system of medieval Europe. There was constant strife as local lords tried to invade each other's territories. However, when the Mongols tried to invade in 1274 and 1281, the Japanese put aside their rivalries. They united to defeat the invaders.

Portuguese traders arrived in Japan in the 1500s. Spanish and Dutch merchants soon followed. These Europeans introduced Christianity to Japan. However, the Japanese drove out European traders and missionaries in the 1600s. They allowed only a few foreigners, mostly Dutch traders, to remain in Japan. They restricted the Dutch traders to an island near the port of Nagasaki. Japanese leaders feared that foreign ideas might cause instability in Japanese society. Japan remained largely cut off from the world from the 1600s to the mid-1800s.

READING CHECK: ***Human Systems*** How did migration and the diffusion of ideas influence cultural change in the region?

Modern History

In 1853 U.S. Navy ships under Commodore Matthew C. Perry arrived in what is now called Tokyo Bay. This contact helped open Japan to foreign influences and trade. Over time these influences sparked change in the country. In 1868 a group of samurai demanding reforms overthrew the last shogun. They restored the emperor's power. This political revolution became known as the **Meiji Restoration**. *Meiji* means "enlightened rule." The Japanese then moved the capital from Kyōto to Tokyo. The emperor began to modernize Japan. Many ideas for modernization came from Europe and the United States. Over time the emperor pushed to reform the country's education system, government, industry, and laws. By 1890 Japan had a constitution and parliamentary system of government.

Japan soon became a world industrial and military power. To meet its growing need for natural resources and increase Japanese influence in Asia, Japan expanded its borders. In 1895 the Japanese took the island of Taiwan from China. In 1905 Japan defeated Russia in the Russo-Japanese War. The Japanese then gained control of the southern half of Sakhalin from Russia. In 1910 Japan annexed Korea. To **annex** an area means to formally join it to a country. Later the Japanese gained control of Manchuria, China's mineral-rich northeastern region. By 1920 Japan's empire also included many islands in the North Pacific. Then in 1937 Japan invaded the rest of China. Japan's growing empire provided the Japanese with valuable natural resources and military bases.

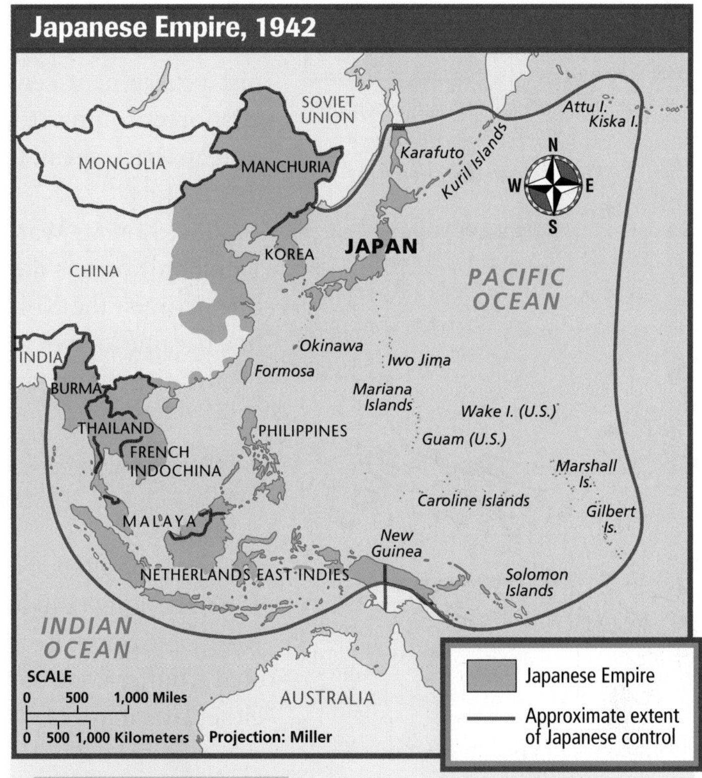

INTERPRETING THE MAP *By 1942 Japan had conquered Korea and large parts of China and Southeast Asia. The Japanese also established fortifications on many small islands in the Pacific.* ***What geographic factors led Japan to expand its territory?*** TEKS

Peace Memorial Park in Hiroshima marks the spot where the first atomic bomb was dropped on August 6, 1945. The blast killed more than 70,000 people and destroyed most of Hiroshima.

Japan signed an alliance with Germany and Italy in 1940. The next year Japan entered World War II by attacking the U.S. naval base at Pearl Harbor in Hawaii. Soon Japan controlled most of Southeast Asia and many Pacific islands. Later in the war, U.S. and Allied forces pushed the Japanese back. The United States dropped atomic bombs on the Japanese cities of Hiroshima and Nagasaki in August 1945. Japan then surrendered.

After World War II Japan set up a democratic government. (See Connecting to Government: Japan's Constitution.) That government includes an elected law-making body called the **Diet** (DEE-uht) and a prime minister. Japan's emperor is still the symbolic leader of the country but has no political power. With U.S. financial aid, Japan began to rebuild its economy and infrastructure.

Japan lost Korea at the end of World War II. The United States and Soviet Union then divided Korea along the 38th parallel (38° N latitude). The Soviets occupied the northern part of the peninsula, and the Americans occupied the south. In North Korea the Soviets set up a communist government called the Democratic People's Republic of Korea. In the south the United Nations supervised an election. South Korea then became the Republic of Korea. In 1949 the Soviet and U.S. occupation forces withdrew.

In 1950 North Korea invaded South Korea, sparking the Korean War. The UN sent troops to defend South Korea. Most were U.S. troops. UN forces drove the North Koreans back and nearly ended the war. However, China's communist government sent troops to support North Korea and again drove the UN forces southward. Eventually, the UN troops were able to push the North Koreans back again. In 1953 the two sides signed an **armistice**, or truce, to end the fighting.

The Korean War caused great damage across the peninsula. More than 1 million Koreans died. At the end of the fighting, the two sides set up a truce line between the North and South Korean forces. The strip of land along this line became known as the **demilitarized zone (DMZ)**. It stretches east-to-west near the old boundary at the 38th parallel.

FOCUS ON HISTORY

The DMZ Neither North nor South Korea governs the DMZ, which is about 150 miles long (240 km) and 2.5 miles (4 km) wide. At the end of the fighting, North and South Korean troops withdrew from the DMZ. This movement created a buffer zone between the two countries' armies. No military forces from either side may enter the area.

The DMZ between North and South Korea is one of the most heavily guarded borders in the world. The North Korean army keeps about 700,000 troops stationed near the DMZ. South Korea also has many troops stationed in the area, and some 37,000 American soldiers help to defend its border.

The Demilitarized Zone

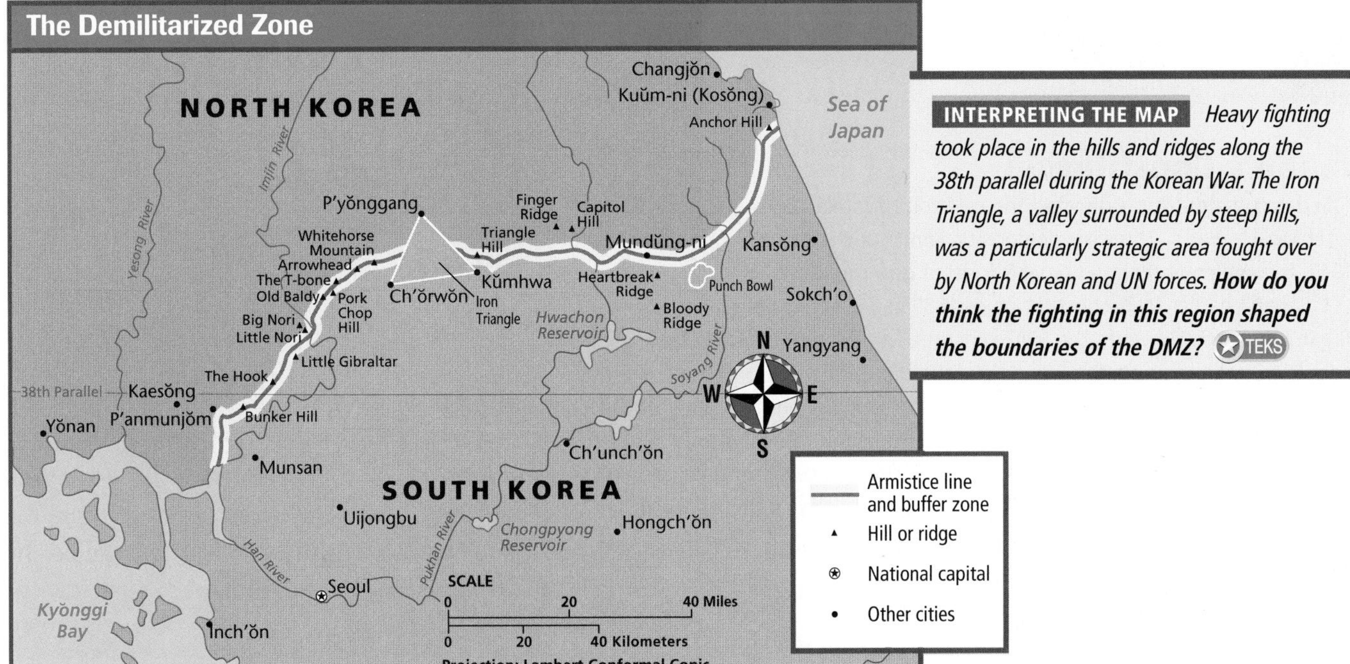

INTERPRETING THE MAP *Heavy fighting took place in the hills and ridges along the 38th parallel during the Korean War. The Iron Triangle, a valley surrounded by steep hills, was a particularly strategic area fought over by North Korean and UN forces.* ***How do you think the fighting in this region shaped the boundaries of the DMZ?*** TEKS

The DMZ has become known as one of the most fortified and tense borders in the world. North Korean troops face South Korean and U.S. troops across the divide. The landscape along the edges of the DMZ features barbed wire, concrete walls, guard towers, mines, and tank traps. However, the area between the edges is largely free of humans.

The almost total lack of human activity in the DMZ over the last 50 years has allowed a unique ecosystem to develop. Many rare and endangered animals have found refuge in the DMZ. These include Siberian tigers and Amur leopards. The red-crowned crane and the white-naped crane, two of the world's most endangered birds, also live there. Deforestation and urban growth have reduced the habitats of these animals in other parts of the peninsula.

A group is now working to turn the region into a system of nature reserves and protected areas. This group, called the DMZ Forum, hopes to create a future "peace park" out of the area. Conservation issues will likely remain unresolved for a while, however. First, the Koreas must resolve the political and military differences that separate them.

READING CHECK: ***Human Systems*** What processes created the unique ecosystem between North and South Korea?

Culture

Many elements of Japanese and Korean culture originated in China. For example, Chinese ideas and practices were the basis for Japan's early political systems. Chinese customs, food, and architecture have also greatly influenced the region.

People and Languages Japan and the Koreas are each dominated by a single major ethnic group with a common language. In Japan some 99 percent of the population are ethnic Japanese. The small minority groups are Koreans, Chinese, and Ainu. The Japanese language is spoken throughout the country,

Connecting to
GOVERNMENT

Japan's Constitution

Allied occupation forces directed the creation of Japan's current constitution after World War II. This new democratic constitution took effect on May 3, 1947. It completely changed Japan's form of government.

Japan's earlier Meiji constitution of 1889 had granted supreme power to Japan's emperor. However, the 1947 constitution gave power to the Japanese people through their elected legislature, the Diet. The emperor remained important in government only as "the symbol of the State and the unity of the people." Japan's new constitution also guaranteed individual freedoms, such as freedom of speech, religion, and the press. These individual rights had not been guaranteed under Japan's earlier constitution. Japan's new constitution also granted women the right to vote.

Analyzing Information How is the adoption of the 1947 constitution an example of the spread of culture traits and cultural convergence? TEKS

and several regional dialects exist. Japanese may be related to Korean. Written Japanese uses a combination of Chinese characters and Japanese symbols called kana.

Nearly the entire population of North and South Korea are ethnic Koreans. Chinese are the main minority group. The Korean language is spoken throughout the peninsula, and there are several major dialects. About half of all Korean words come from Chinese. However, the grammar of Korean is similar to Japanese. Written Korean uses a 24-letter alphabet called hangul.

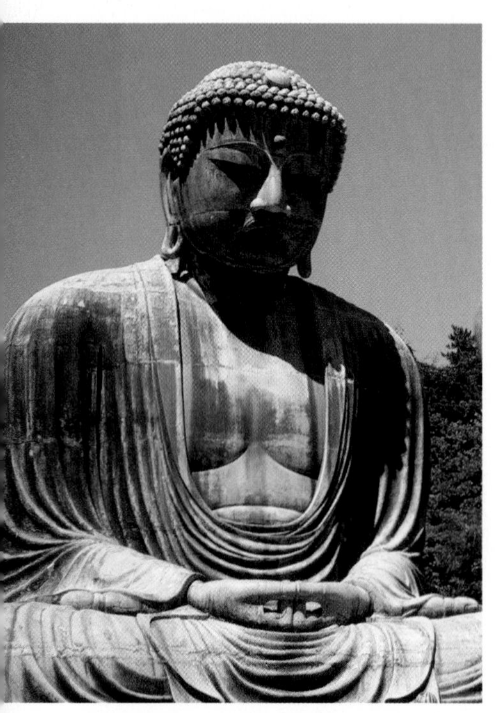

The Daibutsu, *or Great Buddha, in Kamakura, Japan, attracts pilgrims and tourists from all over the country. The hollow bronze statue is 37 feet (11 m) tall. Visitors can enter it through a door in the back.*

Religion Buddhism was introduced into Japan from China in the A.D. 500s. Today a combination of Shintoism and Buddhism dominates religion in Japan. Shintoism centers around the worship of natural spirits and ancestors. Buddhism stresses the unimportance of material goods. Both religions have greatly influenced Japan's traditional culture. For example, the same family might hold Shinto marriage ceremonies and Buddhist funerals.

Korea's major religion has historically been a blend of Buddhism and Confucianism. However, Christianity is now a major religion in South Korea. This is true even though Korea was never a European colony. About half of South Korea's population is Christian. North Korea's communist government officially allows freedom of religion. The state controls most religious activity, however.

Settlement and Land Use Japan is about the same size as California. Yet the country is home to nearly 127 million people—nearly four times as many people as there are in California. In fact, Japan is one of the world's most densely populated countries. On average there are 869 people per square mile in Japan. Most people live along Japan's narrow coastal plains, which are very crowded. The largest and most densely settled plains are in eastern Honshū. Japan's largest cities are also found there. (See the population map in the unit atlas.)

INTERPRETING THE VISUAL RECORD

Houses are packed together on this densely settled coastal plain near Tokyo. In areas close to cities, population densities can reach several thousand people per square mile. ***What human processes contribute to Japan's tremendous population density?*** TEKS

The Koreas are also densely populated. The most crowded areas lie along the west coast. South Korea's urban areas have grown tremendously in the last decades. About 81 percent of South Koreans now live in urban areas. In North Korea, only about 60 percent of the population is urban.

READING CHECK: ***Environment and Society*** Where do most Japanese live, and where are the country's largest cities?

Food Japan's main food is rice, which the Japanese eat at most meals. Fish are the major source of protein. Sushi (SOOH-shee)—vinegared rice and vegetables or raw fish—and sashimi (SAH-shee-mee)—thin slices of raw fish—are popular. Other common foods include cooked vegetables, tofu, and various types of noodles. Tea is the most popular drink. The diet of many Japanese has been changing since World War II. Fast food is more common, and people are eating more meat and dairy products.

Korea's staple food is also rice. Grilled meats and a wide range of vegetables are common. Barley, potatoes, and wheat are grown and used in many foods. One of the most popular dishes in Korea is kimchi (KIM-chee). This spicy dish is a mixture of Chinese cabbage, garlic, ginger, and other vegetables.

Education Japan has an excellent education system and one of the highest literacy rates in the world. This system has helped Japan become an industrial and economic power. Almost all children attend elementary and junior high schools, which are free. Students study many different subjects, such as art, mathematics, science, and social studies. Much time is spent learning how to read and write the Japanese language. There is intense competition among students to get into Japan's best universities. Many high school students attend special "cram" schools to prepare for difficult university entrance exams.

South Korea's education system is similar to Japan's. Most children attend free elementary, middle, and high schools. About one third of high school graduates go on to university. Admission is based on grueling entrance exams. In North Korea, teaching communist ideology is a major focus of the education system. The government also requires students to work as part of their

Korean women perform a traditional fan dance. The dancers' dresses are like those worn centuries ago by the Korean nobility. As they dance, the women make patterns with their fans.

schooling. Higher education in North Korea is much more limited than in South Korea and Japan. However, adult education is widespread. Many adults attend technical schools at night.

Traditions and Customs Western influences have greatly altered Japanese culture. Still, many traditions are important. Family ties are strong, and respect for elders is important. Traditional dress for both men and women is the kimono—a long robe with wide sleeves. Kimonos are mostly worn for special occasions. Removing one's shoes before entering a house is a common custom. Japanese wear slippers or socks instead of shoes inside the home.

The Japanese have a rich heritage in the arts. Traditional music is played on native instruments such as drums, flutes, and gongs. Japanese theater called Kabuki (kuh-BOO-kee) uses colorful costumes and makeup to portray historical events. In literature several types of Japanese poetry are widely popular. Painting, sculpture, and many other decorative arts also have a long history in Japan.

In the Koreas the central importance of family life has declined somewhat as the peninsula has developed and urbanized. However, many traditions and customs survive. For example, traditional clothing styles are still common, particularly in North Korea. Korean dance is popular in both North and South Korea. Dancers wear traditional clothes called *hanbok*.

TEKS **READING CHECK:** *Places and Regions* How is education in Japan and South Korea different from education in North Korea?

Working with Sketch Maps, Questions 1, 2, 3, 4, 5

Review

Identify
Meiji Restoration
Diet

Define
shogun, samurai,
annex, armistice, demilitarized zone (DMZ)

Working with Sketch Maps On the map you created in Section 1, label Kyōto, Nagasaki, Tokyo, and Hiroshima. Which city became Japan's capital after the Meiji Restoration?

Reading for the Main Idea

1. *The Uses of Geography* What events have created the boundary between North Korea and South Korea?
2. *The Uses of Geography* How did trade and the diffusion of ideas lead to cultural change in Japan in the last half of the 1800s?

Critical Thinking

3. **Drawing Inferences and Conclusions** What geographic factors may have influenced Japan's ability to control its territory and keep European traders out before the mid-1800s? How did this affect Japan's foreign policy at the time?
4. **Identifying Points of View** How do you think European traders and Japanese leaders viewed Japan differently?

Organizing What You Know

5. Copy the time line below. Use it to describe major events in the region's history.

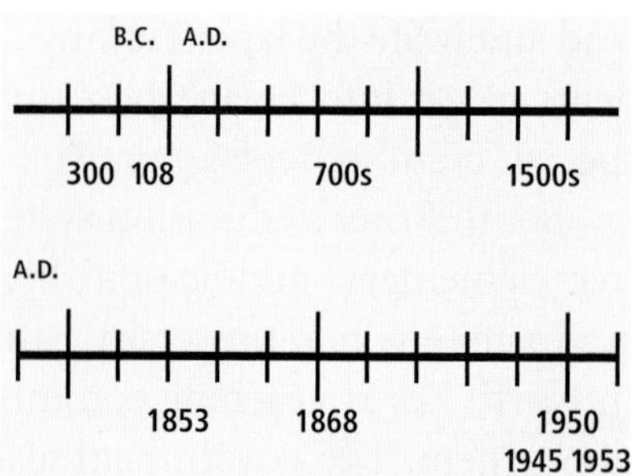

Section 3 The Region Today

READ TO DISCOVER

1. What are some characteristics of Japan today?
2. How is life in North Korea different from life in South Korea?

DEFINE

subsidies
work ethic
export economy
trade surplus
urban agglomeration

LOCATE

Yokohama
Nagoya
Ōsaka
Kōbe
P'yŏngyang
Seoul

WHY IT MATTERS

Japan is one of the world's leading economic powers. Use CNNfyi.com or other **current events** sources to read about current issues facing the Japanese economy.

Modern Japan

Despite modern influences, traditional Japanese culture remains an important part of Japanese society. Many Japanese still follow traditional ways of life, particularly in rural areas. A land of contrasts, Japan blends traditional and modern and East and West.

Agriculture About 11 percent of Japan is arable, and about 5 percent of Japanese workers are farmers. However, these farmers supply about 70 percent of the country's food needs. They have succeeded by using terraced cultivation and modern farming methods. Government **subsidies**, or financial support, protect Japanese farmers from foreign competition.

Farmers use more than half of Japan's farmland to grow rice. Tea, soybeans, fruits and vegetables, and mulberry trees (for silkworms) come from the south. Northern Japan is colder and has a shorter growing season. Farmers grow wheat, other grains, and vegetables there. Hokkaidō has a successful dairy industry.

Most Japanese farmers live in small villages. The average farm is only 2.5 acres (1 ha). However, expanding cities are taking over more and more good farmland.

INTERPRETING THE VISUAL RECORD

A farmer in central Japan harvests his rice crop. Farm machinery used in Japanese agriculture is often small and efficient. ***Why do you think many Japanese farmers prefer smaller farm machinery?*** TEKS

Industry Japan began its rapid economic growth in the 1950s. Soon the country became a model for economic success. One of the reasons for this success was Japan's culture. The Japanese have a strong **work ethic**—a belief that work itself has moral value. Long workdays and six-day workweeks are common. Also, most workers feel loyal to their companies. In turn, many companies have offered employees lifetime employment, although this has recently been changing. Japanese industry also benefited from the *keiretsu* system. This system brings together banking, big business, and government to set common goals. The combined efforts of these three groups helped Japanese industry grow.

In the 1990s, however, Japan faced stiff competition from the newly industrialized countries of East and Southeast Asia. The *keiretsu* system could not adapt quickly enough in the rapidly changing world economy. Many observers believe Japan will need to make some changes to regain its economic leadership in Asia. For example, many argue that Japan must allow more foreign competition and foreign investment into the country.

TEKS **READING CHECK:** ***Human Systems*** What is the *keiretsu* system?

International Economy Japan has been very successful at selling and manufacturing its products overseas. For example, Japanese companies have auto-manufacturing plants in the United States, which is a major market for Japanese cars. Japanese companies also invest in many other countries.

Japan's main economic competition comes from other Asian countries like China and South Korea. This competition has hurt some Japanese industries, such as shipbuilding and steel production. For example, South Korea is now the world's leading shipbuilder. Japan once dominated this industry. In the future, China may become the leader in shipbuilding because of its lower labor costs. However, Japan remains a world leader in quality manufactured products like automobiles, cameras, and electronics. It is also a leader in robotics and biotechnology.

Japan has an **export economy**. In this type of economy, goods are produced mainly for export rather than for domestic use. Japan imports raw materials and energy and exports high-quality manufactured goods. Japan's export economy is so strong that Japan has built up a huge **trade surplus**. A trade surplus exists when a country exports more than it imports. For example, Japan has a large trade surplus with the United States. This has become an important issue in U.S.-Japanese trade relations. The U.S. government wants Japan to open its markets to more American goods.

INTERPRETING THE VISUAL RECORD

Japan exports between 4 and 5 million automobiles each year and is the world's second-largest automobile producer after the United States. Consumers around the world buy Japanese cars for their high quality, excellent design, and competitive prices. ***How do you think the export of Japanese cars to the United States affects automobile production in this country?*** TEKS

Urban Geography Japan has many industrial regions. However, the country's industrial-urban core is located on Honshū's eastern coastal plains. Three large functional regions can be found there. The Keihin region includes Tokyo and Yokohama. To the southwest lies the Chukyo region, which includes the city of Nagoya. The cities of Ōsaka, Kōbe, and Kyōto are part of the Keihanshin region even farther west. A dense network of high-speed trains, highways, and air services connects these regions. (See Case Study: High-Speed Rail.)

Tokyo-Yokohama is the largest **urban agglomeration** in the world. An urban agglomeration is a densely populated region surrounding a central city. About 26 million people are jammed into the urban region of Tokyo-Yokohama. This region is Japan's center of commerce, education, entertainment, government, and trade. Because little land is available, real estate prices are among the highest in the world. Yokohama, just to the south on Tokyo Bay, is Japan's busiest seaport.

The Chukyo region around Nagoya is important in the production of textiles, ceramics, and motor vehicles. Kōbe and Ōsaka—Japan's second-largest city—are major seaports. Kyōto is Japan's ancient capital.

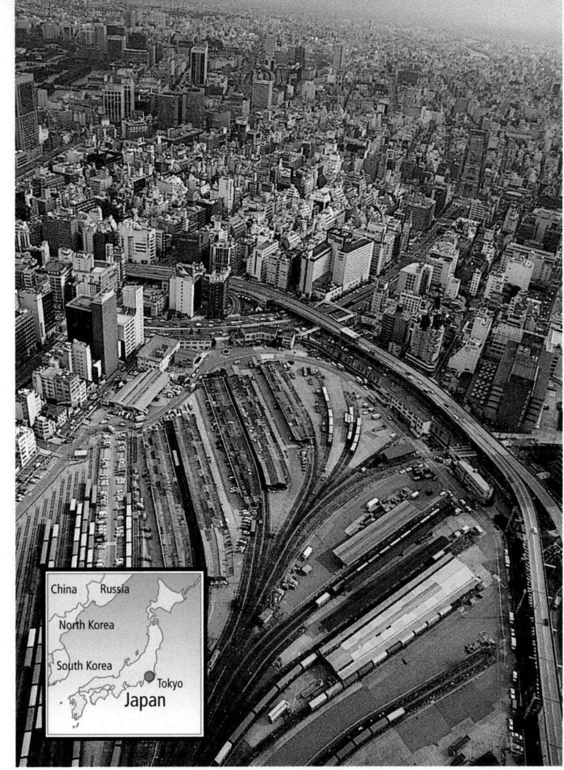

INTERPRETING THE VISUAL RECORD *Tokyo is one of the largest cities in the world and is the focus of Japan's transportation network.* ***What can you see in this photo that shows Tokyo is an important functional region?*** TEKS

Modern Ways of Life Most Japanese are middle-class. Families typically live in suburban areas. Parents often spend several hours each day commuting to and from work in the cities. Most Japanese families are small, with one or two children. Homes are also small and expensive. Sliding wooden screens separate sparsely furnished rooms. Families are used to tight spaces, and children often share bedrooms.

Japanese society has been greatly influenced by Western culture. For example, Western clothing, foods, and music are very popular. Baseball, golf, skiing, and soccer are common forms of recreation. At the same time, the Japanese have also influenced Western culture. Sushi bars and Japanese video games, landscaping, and cartoons can be found around the world.

Japan has one of the lowest birthrates in the world. In addition, life expectancy rates—84 years for women and 78 for men—are among the world's highest. As fewer people are born and people live longer, the average age of Japan's population is increasing. This "graying" population puts increased demands on health care and social services.

The role of women in Japanese society has also changed in recent decades. Women now receive more education than before and are an important part of the workforce.

INTERPRETING THE VISUAL RECORD

A family in Japan gets by with very little space when compared to a typical American family. ***What might be some reasons that Japanese houses are smaller on average than American houses?*** TEKS

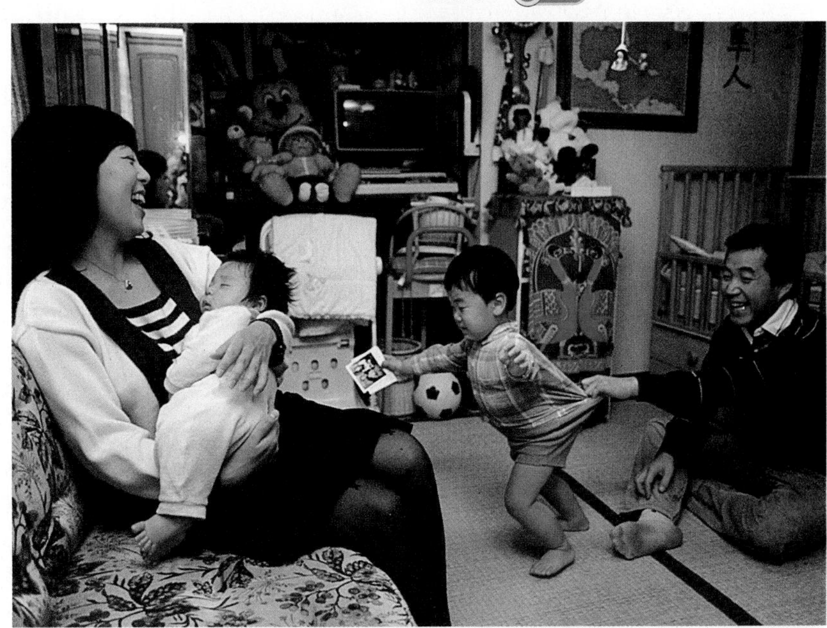

More than 50 percent of Japanese women are employed. As economic opportunities for women have grown, so have other opportunities. For example, women have become more involved in politics and government.

READING CHECK: ***Human Systems*** How have economic opportunities for women in Japan changed in recent decades?

The Two Koreas

The two Koreas share a common culture, but their level of political and economic development is very different. (See the unit Fast Facts table.) For example, the per capita GDP of South Korea is more than 16 times that of North Korea.

North Koreans honor a giant statue of Kim Il Sung, leader of North Korea from 1948 until his death in 1994. Kim set up a communist government in North Korea with close ties to the Soviet Union and China.

North Korea North Korea is an isolated country under a strict communist dictatorship. The country's leader, Kim Jong Il, came to power when his father died in 1994. North Korea has a large military and a command economy. The state controls most aspects of its citizens' lives. For example, the government limits and controls travel within the country. In general, government officials and high-ranking military officers live more comfortably than the average North Korean.

Outside the capital of P'yŏngyang (pyuhng-YANG), most North Koreans live in poverty. The country's farming system is based on inefficient communist state farms. The government lets some farmers have small garden plots for themselves. However, farmers cannot grow enough to feed themselves. Droughts and floods have caused severe food shortages. When the Soviet Union collapsed in the early 1990s, Soviet aid ended. The North Korean economy then collapsed. Food shipments from the United Nations have helped the country feed itself in recent years, but major problems still exist. Thousands of North Koreans have fled to China to seek a better life.

Seoul's Namdaemun Market features a wide range of everyday items and consumer goods.

South Korea For many years South Korea's military dominated the government. However, today the country has a democratically elected government. South Koreans enjoy many rights and freedoms not found in the north. In addition, South Korea used U.S. aid to rebuild its industries after the Korean War. It developed from a poor farming country into an industrial "Asian Tiger" in only decades. "Asian Tiger" is a term used to describe countries in East and Southeast Asia that have experienced rapid economic growth in recent decades. Like Japan, South Korea has an export economy. Its major industries are shipbuilding, steel, automobiles, textiles, and electronics.

South Korea's business model is similar to the *keiretsu* model in Japan. Huge family-owned conglomerates called *chaebol* dominate South Korea's economy and political system. Four *chaebol* control most of South Korea's manufacturing and exports. This situation has prevented competition and led to corruption and debt. The government has passed recent reforms

INTERPRETING THE VISUAL RECORD

In June 2000 Presidents Kim Jong Il of North Korea (left) and Kim Dae Jung of South Korea (right) met. The two leaders pledged to improve their countries' relations and to work toward the reunification of the peninsula. ***How do you think these meetings might affect relations between North and South Korea?*** TEKS

to try to improve South Korea's economy. They include opening the market to foreign investment and competition.

South Korea has a large urban middle class with access to consumer goods. The country's capital and largest city is Seoul (SOHL). Seoul is a huge city and a growing industrial and cultural center. South Korea's small farming villages are disappearing as the country continues to urbanize.

North Korea has almost no privately owned cars. The government owns nearly all cars and sets them aside for official use only.

Korean Reunification For decades Koreans have hoped to reunify their two countries. In recent years the two Koreas have had better relations. For example, South Korean tourists have been able to visit a few selected areas of North Korea. Talks and official visits continue, but the countries remain worlds apart. Fears of a North Korean invasion still remain in South Korea. Another concern is whether South Korea could afford the cost of reuniting with one of the world's poorest countries.

READING CHECK: *Places and Regions* How are the economies of North Korea and South Korea different?

TEKS Working with Sketch Maps, Questions 1, 2, 3, 4, 5

Section 3 Review

Homework Practice Online

Keyword: SW3 HP28

Define

subsidies
work ethic
export economy
trade surplus
urban agglomeration

Working with Sketch Maps

On the map you created in Section 2, label Yokohama, Nagoya, Ōsaka, Kōbe, P'yŏngyang, and Seoul. Which city is South Korea's capital and economic center?

Reading for the Main Idea

1. *Human Systems* How did Japan's culture help the country develop economically?
2. *Places and Regions* What are some important characteristics of the Japanese and South Korean market economies?
3. *Places and Regions* What are some obstacles to Korean reunification?

Critical Thinking

4. **Comparing** How are politics, economics, and society in North Korea related? How does this situation differ in South Korea?

Organizing What You Know

5. Copy the graphic organizer below. Use it to describe the three functional regions that are the industrial-urban core of Japan.

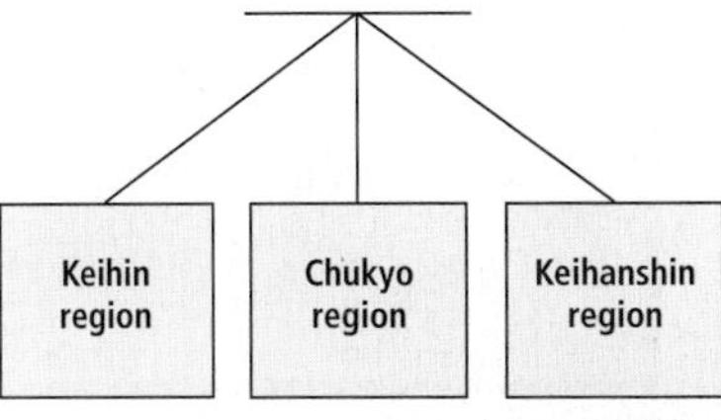

CASE STUDY

Japan's High-Speed Rail

Human Systems Japanese transportation engineers have developed one of the most sophisticated, efficient, and profitable rail networks in the world. For many Japanese travelers, railways are the preferred mode of transportation. A number of geographic factors have made train travel more attractive than road or air transportation.

Linking Major Urban Centers

The most advanced part of the Japanese rail network is the *Shinkansen*. This high-speed train links major urban centers such as Tokyo and Ōsaka. Only 320 miles (515 km) apart, Tokyo and Ōsaka lie on the southern coast of Japan's largest island, Honshū. These cities have easy access to deepwater ports. This access has helped them develop into the urban and industrial core of Japan. In 1964 the first *Shinkansen* line connected Tokyo and Ōsaka. It was named the "New Tokaido Line" after an ancient highway that once linked Kyōto and Tokyo. By 1975 the Japanese had extended the main rail line to Fukuoka. That city lies at the western edge of Japan's manufacturing belt on the island of Kyūshū. The Japanese have since added other lines radiating northward from Tokyo.

The most impressive aspect of the Japanese *Shinkansen* system is the incredible speeds that trains move. The fastest trains from Tokyo to Ōsaka travel at 160 miles per hour (257 kph). The trains can cover the 664-mile (1,070-km) journey from Tokyo to Hakata in less than seven hours and carry 1,000 passengers. In addition to being fast, the trains run often and on time. These qualities are very important for business passengers. For example, during the morning and evening rush hours, trains run every 7.5 minutes. On average, each train misses its schedule by only 36 seconds!

Factors Behind Success

Why has rail travel been so successful in Japan? Japan is a country where space is limited and land values are very high. As a result, trains have a distinct advantage over road and air travel. Why is this so? Road travel gobbles up larger chunks of real estate than rail networks. For example, cars and trucks need highways, access ramps, gas stations, and parking lots. In turn, increasing traffic constantly creates pressure to build even more roads. These roads then eat up even more space. High fuel costs can also make car and truck travel expensive. In fact, Japanese fuel prices are often about three times higher

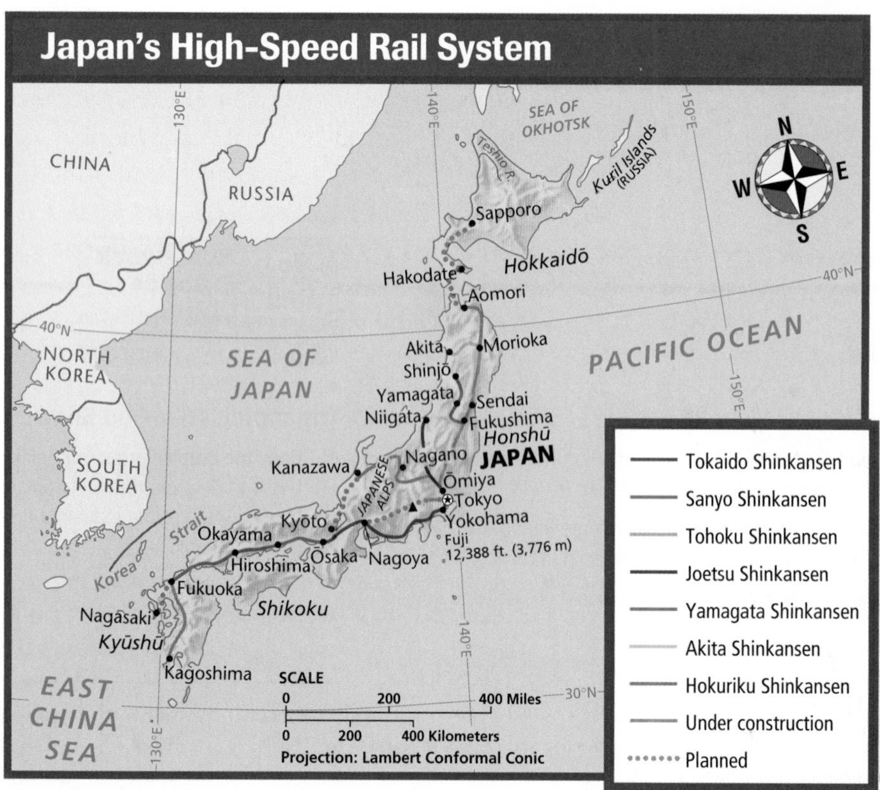

INTERPRETING THE MAP *Begun in the 1960s, the* Shinkansen *system now connects most of Japan's major cities, with even more routes planned or under construction.* ***Which areas on the map have lines planned or under construction? Why do you think these routes have not yet been built?*** TEKS

than those in the United States. It should be no surprise, then, that many Japanese do not want to take cars on long trips.

Airports present similar disadvantages for mass travel in a country like Japan. For one thing, they also require large areas of land. The most desirable airport locations are close to large urban areas. These locations are often in places where land prices are highest. In addition, the availability of land there is very limited. In Tokyo planners decided to locate the city's main airport an hour away—by train. Workers in Ōsaka built a new airport in the harbor on a human-made island. When you factor in the time it takes to get to and from these airports, trains can be faster than airplanes. Japanese travelers seem to agree. Even with expansion of their airports, Ōsaka and Tokyo are linked by six times as many trains as airline flights. In fact, train travel accounts for 84 percent of all trips between the two cities.

All Shinkansen *trains are monitored by computers in Tokyo. The computers control the trains' speeds and the distances between them.*

Spreading Success Around

Japan is not the only place where high-speed train travel makes sense. In 1998, 12 countries had passenger trains that average more than 125 miles per hour (200 kph). Sixteen other high-speed railways were in development in places such as South Korea and Sweden.

Originally designed to carry people over long distances, Shinkansen *trains have become popular among Japanese commuters.*

The same factors that led to the success of the *Shinkansen* are found where other such networks are in development. Many large cities are close together. In addition, road and air travel are stretched to the limit. As a result, high-speed rail is becoming more appealing. Even large countries have found high-speed rail attractive. In 1996, for example, Chinese leaders set a goal to launch a high-speed link between Beijing and Shanghai. Officials hope to cut the travel time between the cities from 17 hours to six or seven hours. The United States recently began operating high-speed train service between New York and Washington, D.C. Projects in Texas and California have also been proposed. Supporters of projects like these believe high-speed rail travel is likely to become an increasingly popular way to travel in densely populated regions. The Japanese *Shinkansen* set the standard for these future projects.

Applying What You Know

1. **Summarizing** What factors have made the *Shinkansen* a successful mode of transportation in Japan?
2. **Making Generalizations** What might be some other solutions to travel problems in countries like Japan?

Review

Building Vocabulary (TEKS)

On a separate sheet of paper, explain the following terms by using them correctly in sentences.

tsunamis	armistice
flyway	demilitarized zone (DMZ)
shogun	subsidies
samurai	work ethic
Meiji Restoration	export economy
annex	trade surplus
Diet	urban agglomeration

Locating Key Places (TEKS)

On a separate sheet of paper, match the letters on the map with their correct labels.

Hokkaidō	Kyūshū	P'yŏngyang
Honshū	Sea of Japan	Seoul
Shikoku	Tokyo	

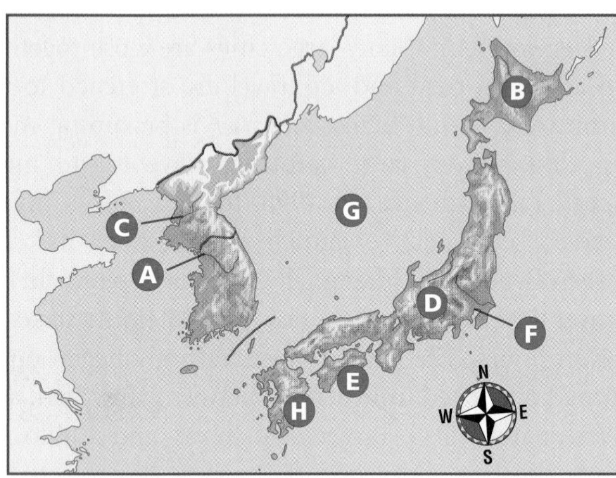

Understanding the Main Ideas (TEKS)

Section 1

1. *Places and Regions* What are Japan's four main islands? Which is the largest and most populated?
2. *Physical Systems* Why might you find so many different species of birds in Japan and the Koreas?

Section 2

3. *Places and Regions* During what period did Japan control the Korea Peninsula?
4. *Human Systems* How are the Japanese and Korean written languages different?

Section 3

5. *Environment and Society* What is agriculture like in North Korea?

Thinking Critically for TAKS (TEKS)

1. **Drawing Inferences and Conclusions** How did cultural patterns and attitudes in Japan influence innovation and the diffusion of ideas before the mid-1800s?
2. **Analyzing Information** Why do you suppose Japan's *keiretsu* system could not adapt quickly enough to the rapidly changing world economy of the 1990s?
3. **Evaluating** Why do you think the United States maintains a large military presence in South Korea today?

Using the Geographer's Tools (TEKS)

1. **Analyzing Maps** Study the map of the Japanese Empire in Section 2. How far did Japan expand its empire in the Pacific?
2. **Creating Charts** Create a chart describing and comparing the languages, religions, land use patterns, systems of education, and customs of Japan, North Korea, and South Korea.
3. **Preparing Maps** Use the information in Section 3 to prepare a map of Japan's three major urban-industrial regions. Do you think these are formal, functional, or perceptual regions? Why?

Writing for TAKS (TEKS)

Imagine you are a high school student in Japan. Write a letter to a friend in the United States in which you describe Japan's educational system. How is it similar to the education system in the United States? How is it different? When you are finished with your letter, proofread it to make sure you have used standard grammar, spelling, sentence structure, and punctuation.

SKILL BUILDING

Geography for Life (TEKS)

Field Work

Human Systems Conduct field work to find some Japanese products that affect your daily life. Start by creating a list of the products in your home that were made in Japan. What types of products are they? How much did they cost? Then interview friends and neighbors in your community to find out what Japanese-made products they find useful. What can these products tell us about Japan's level of development and economy? Why do you think Japanese products are so successful in the United States?

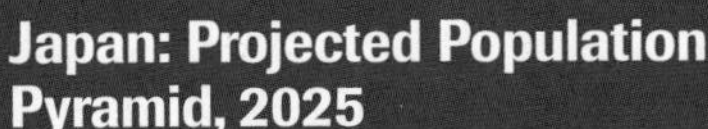

Japan: Projected Population Pyramid, 2025

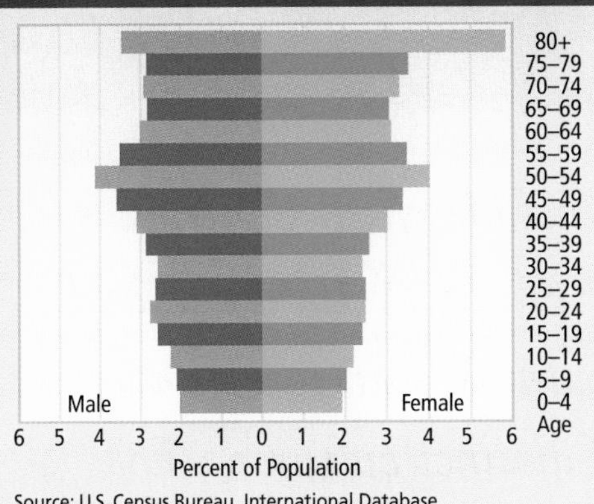

Source: U.S. Census Bureau, International Database

Interpreting Graphs

Study the projected population pyramid for Japan in 2025. Then answer the following questions.

1. According to the population pyramid, in 2025 Japan is projected to have
   - a. many more younger people than older people.
   - b. more people under the age of 4 than between 50–54.
   - c. more people over the age of 40 than below that age.
   - d. more men than women over the age of 80.
2. What factors might account for the way Japan's population pyramid is projected to look in 2025?

Analyzing Secondary Sources

Read the following passage and answer the questions that follow.

"Japan's earlier Meiji constitution of 1889 had granted supreme power to Japan's emperor. However, the 1947 constitution gave power to the Japanese people through their elected legislature, the Diet. The emperor remained important in government only as 'the symbol of the State and the unity of the people.' Japan's new constitution also guaranteed individual freedoms, such as freedom of speech, religion, and the press. These individual rights had not been guaranteed under Japan's earlier constitution."

3. Which of the following things did Japan's 1947 constitution *not* do?
   - a. guarantee individual freedoms
   - b. make the emperor more powerful
   - c. allow freedom of the press
   - d. take power from the emperor and give it to the people
4. What role did the emperor play in the government set up by the new constitution?

Alternative Assessment

PORTFOLIO ACTIVITY

Learning about Your Local Geography TEKS

Individual Project: Research

Like Japan and the Koreas, the United States is on a major flyway for migrating birds. How does this affect the wildlife in your area? Use your local library to learn about migratory birds that pass through your area each year. What are some of these types of birds? Where are they traveling from, and where are they going? At what time of the year do they travel? Write a short report answering these questions. You might want to create a map of the United States that notes important migratory bird routes.

internet connect

Internet Activity: go.hrw.com
KEYWORD: SW3 GT28 TEKS

Choose a topic on Japan and the Koreas to:

- create a newspaper on Japan and the Koreas.
- research active volcanoes along the Ring of Fire.
- learn about a typical school day in Japan.

CHAPTER 29

Mainland Southeast Asia

Mainland Southeast Asia includes five countries: Cambodia, Laos, Myanmar (Burma), Thailand, and Vietnam. The region's mountains, rain forests, and river valleys have nurtured distinct cultures. The entire region has tropical environments, with most of the population concentrated along major river valleys.

Sawaddee! (May you have good fortune!) I am Chosita, and I am 14 years old. I live in Bangkok with my parents and older sister. My mother is a colonel in the Royal Thai Air Force and my father is a government economist.

My sister and I get up very early for school because traffic in Bangkok is really bad. By 6:00 A.M. we are on the road. If traffic isn't too bad, we get to school in about 30 minutes. However, if traffic is heavy, it takes more than an hour. We go to school from June to September and from November to February. Our big "summer" vacation is from March to May.

Our school is near Jatujak Park, where the city government has relocated all the street vendors in Bangkok to help ease downtown traffic. However, people come to Jatujak to buy everything from food to shoes, clothing, and souvenirs, so traffic is still a big problem.

In my free time, I like to listen to music, watch TV, or go to the beach. The sea in the south is beautiful, and the beaches are not very crowded. I'm not sure what I want to be when I grow up—maybe an economist like my father or an accountant for a big firm. I still have quite a few more years to think about it.

Banana seller, Vietnam

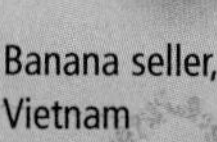

Detail of temple interior, Vietnam

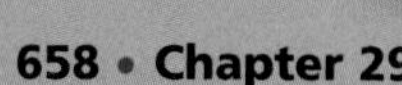

Natural Environments

READ TO DISCOVER

1. What are mainland Southeast Asia's major landforms and rivers?
2. Which climate and vegetation types are found in the region, and what animals live there?
3. What are some of mainland Southeast Asia's main resources?

DEFINE

arboreal

LOCATE

Bay of Bengal
Andaman Sea
Malay Peninsula
Gulf of Thailand
South China Sea
Gulf of Tonkin
Indochina Peninsula
Khorat Plateau
Irrawaddy River
Chao Phraya River
Mekong River
Hong (Red) River
Tonle Sap

WHY IT MATTERS

Biologists searching Earth's tropical regions, including those in mainland Southeast Asia, are still finding new plant and animal species. Use CNNfyi.com or other **current events** sources to read about some recent discoveries.

Mainland Southeast Asia: Physical-Political

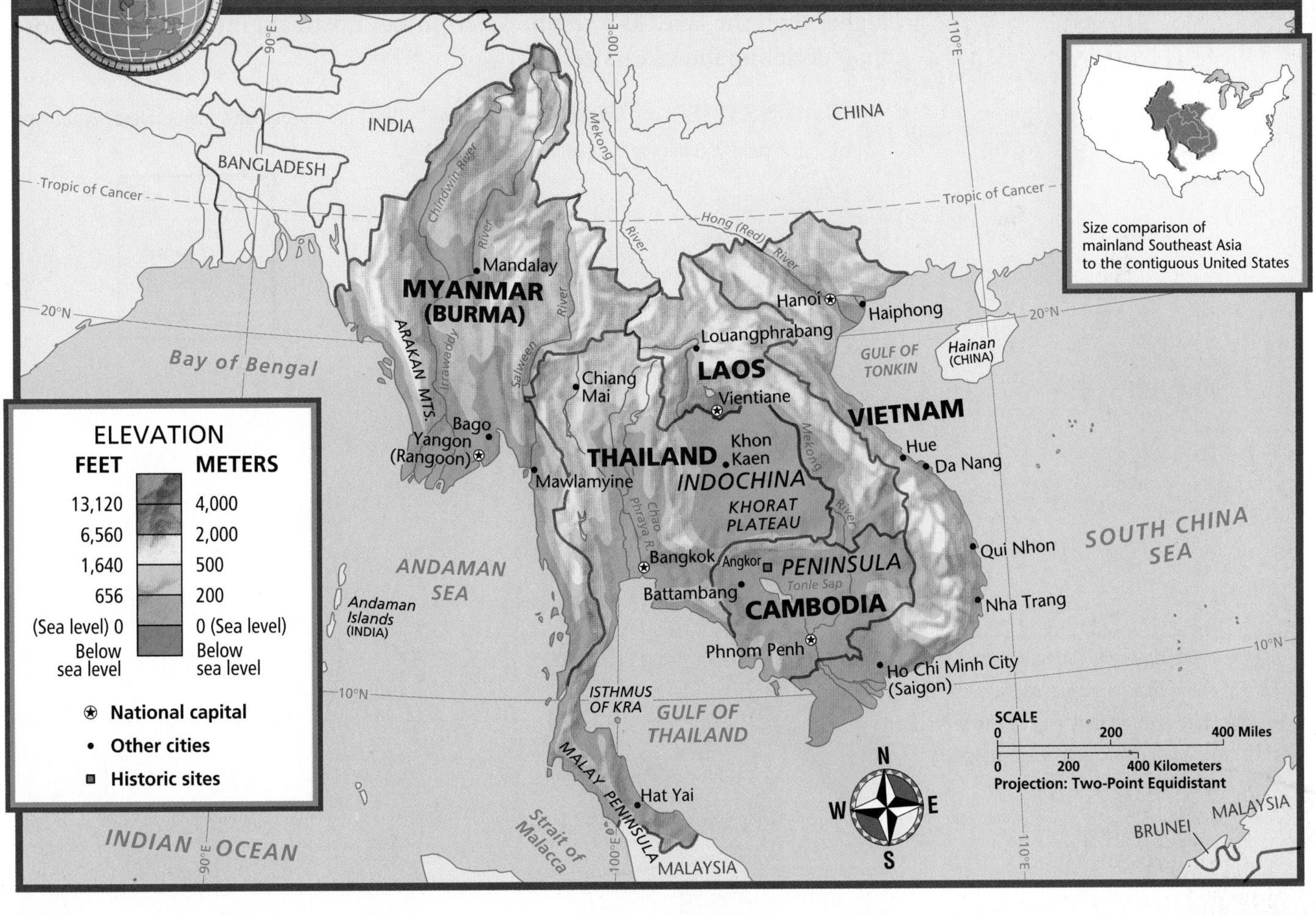

Size comparison of mainland Southeast Asia to the contiguous United States

ELEVATION

FEET	METERS
13,120	4,000
6,560	2,000
1,640	500
656	200
(Sea level) 0	0 (Sea level)
Below sea level	Below sea level

⊛ National capital
• Other cities
▫ Historic sites

SCALE: 0–200–400 Miles; 0–200–400 Kilometers
Projection: Two-Point Equidistant

internet connect

GO TO: go.hrw.com
KEYWORD: SW3 CH29
FOR: Web sites about mainland Southeast Asia

Landforms and Rivers

Mainland Southeast Asia stretches southward from the Asian landmass. To the west lie the Bay of Bengal and Andaman Sea. The narrow Malay Peninsula, which includes part of Thailand, extends to the south. In the east, the Gulf of Thailand, South China Sea, and Gulf of Tonkin surround the Indochina Peninsula. Laos is the region's only landlocked country.

The area has three main landform regions. In the north, rugged mountain ranges fan out from the Himalayas and the Plateau of Tibet. They stretch into Myanmar (MYAHN-mar), Thailand, Laos, and Vietnam. A central region of plains and low plateaus lies to the south in Thailand and Cambodia. Thailand's Khorat (koh-RAHT) Plateau is there. River valleys and deltas make up the third main landform region.

Four major rivers flow southward from Asia's mountainous interior. The Irrawaddy (ir-ah-WAH-dee) empties into the Bay of Bengal. The Chao Phraya (chow PRY-uh) flows into the Gulf of Thailand. The Mekong (MAY-KAWNG), the region's longest river, borders Thailand. It flows through Laos, Cambodia, and Vietnam to the South China Sea. The Hong (Red) River flows across northern Vietnam into the Gulf of Tonkin. The valleys and deltas of these large rivers have fertile alluvial soils, which support intensive farming and dense populations. The rivers have also been useful for local transportation. In addition, they have influenced the locations and growth of the region's cities.

Southeast Asia's largest freshwater lake is Tonle Sap (tohn-LAY SAP) in Cambodia. During the dry season, a river flows south from the lake into the Mekong River. However, during the wet season, the Mekong's water level is higher than the lake. As a result, water pushes upstream into Tonle Sap, more than doubling the lake's size.

TEKS **READING CHECK:** *Places and Regions* What are the region's major rivers? What type of soils would you find in their valleys and deltas?

Southern Thailand's Phi Phi Islands, or Ko Phi Phi in the Thai language, are famous for spectacular karst landscapes. Limestone cliffs have eroded to create these towering karst formations.

INTERPRETING THE VISUAL RECORD

A boat approaches a house isolated by the flooded Mekong River. Floodwaters can rise to 22 feet (7 m) along the Mekong during the rainy season.
How does this photo suggest that Cambodians have adapted to living in an area prone to floods? TEKS

Climates, Plants, and Animals

All of mainland Southeast Asia has tropical or subtropical climates. Precipitation is high throughout the region. Most coastal areas have a tropical humid climate with heavy rainfall. Interior plains and plateaus have a tropical wet and dry climate. The driest region is the Khorat Plateau in eastern Thailand. A tropical wet and dry climate extends across central Myanmar, Thailand, Cambodia, and southern Vietnam. A humid subtropical climate region is found in northern Myanmar, Laos, and northern Vietnam.

Mainland Southeast Asia is greatly affected by the monsoon wind system. This system causes a rainfall pattern with distinct wet and dry seasons. Summer winds from the southwest bring heavy rainfall. The mainland gets most of its rainfall during this season. Winter brings dry winds that blow from the northeast.

The monsoons create serious natural hazards. During the wet monsoon, severe flooding is common. On the other hand, fires can be a major problem in the dry season. Between monsoons in the fall, Vietnam is sometimes hit by powerful typhoons. Myanmar lies in the path of typhoons that come in from the Bay of Bengal.

In the 1990s biologists working in the remote mountains of Laos and Vietnam discovered at least five new mammal species. These species included a striped rabbit, a deer that barks like a dog, and a 200-pound wild ox.

The region's tropical and subtropical climates support a great number of plants and animals. While much of the environment has been greatly modified by people, large areas remain almost untouched. Rare and endangered plants and animals live in these areas. Some thickly forested mountain areas are so rugged and inaccessible that scientists are still discovering new species there. To protect the environment, some of the region's countries have created large national parks and reserves.

The thick tropical rain forests of many lowland and coastal areas support **arboreal**, or tree-dwelling, animals such as monkeys. A wide range of birds also live in the rain forests. Many fish species and other marine life live in coastal mangroves. These mangroves are particularly common in places where mud is deposited along the shore. Drier inland areas have monsoon forests. These forests of broadleaf trees are not as dense as tropical rain forests.

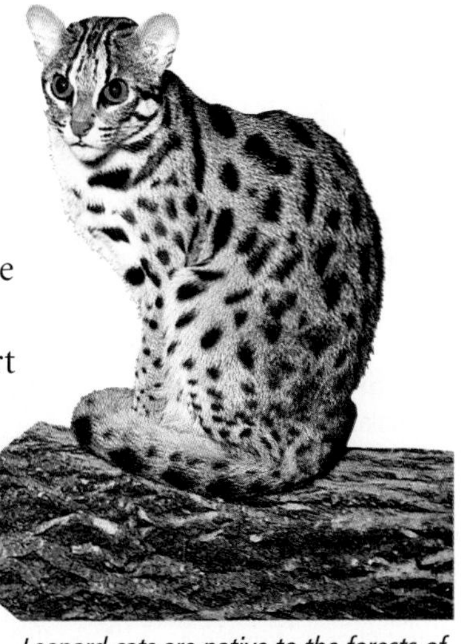

Leopard cats are native to the forests of Southeast Asia. Sometimes hunted for their pelts, these arboreal cats are now protected by law.

TEKS **READING CHECK:** ***Places and Regions*** What climates does the region have? Where is the driest region?

Natural Resources

Southeast Asia's tropical rain forests hold valuable hardwoods such as mahogany, teak, and ebony. Large areas have been logged for these woods or cleared by growing populations. Until recently there has been little concern for conservation or for replanting trees.

The continued clearing of tropical rain forests has damaged the habitat of endangered wildlife such as tigers and elephants. Removing trees and other plants has also led to flooding and soil erosion. In addition, a soil type common in the area, called laterite, hardens and becomes useless when the forest cover is removed. Some countries have tried to slow the deforestation. For example, Thailand has set up national parks to protect key areas and banned logging for export. However, in most countries logging and habitat loss continue at a rapid pace. (See Case Study: Studying Deforestation.)

Mainland Southeast Asia also has valuable minerals and fossil fuels. Iron, manganese, tin, and tungsten are found there. The region exports gems, including sapphires and rubies. In addition, Thailand has natural gas, Myanmar has oil, and Vietnam has coal. The large rivers have potential for the development of hydropower. Laos plans to build dams for hydropower in the near future. However, building dams may interfere with the natural soil-forming processes of the river valleys.

Although logging for export has been outlawed in Thailand since 1989, illegal operators continue to smuggle teak out of the country. Often elephants move the heavy logs. However, harsh treatment threatens the working elephants, and habitat destruction endangers the survival of those still living in the wild.

READING CHECK: *Environment and Society* How has clearing the region's rain forests affected the environment?

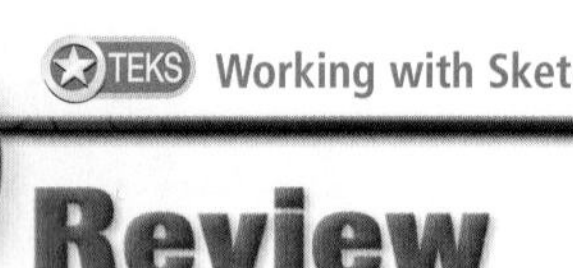

TEKS Working with Sketch Maps, Questions 1, 2, 3, 4, 5

Section 1 Review

Define arboreal

Working with Sketch Maps

On a map of mainland Southeast Asia that you draw or that your teacher provides, label the Bay of Bengal, Andaman Sea, Malay Peninsula, Gulf of Thailand, South China Sea, Gulf of Tonkin, Indochina Peninsula, Khorat Plateau, Irrawaddy River, Chao Phraya River, Mekong River, Hong (Red) River, and Tonle Sap. Which is mainland Southeast Asia's longest river?

Reading for the Main Idea

1. *Places and Regions* What are the major landform regions of mainland Southeast Asia?
2. *Physical Systems* How does the monsoon system influence the region's climates?
3. *Environment and Society* How may building dams affect the environment?

Critical Thinking

4. **Analyzing Information** How is the distribution of mangroves in the region related to soils?

Organizing What You Know

5. Copy the chart below. Use it to describe mainland Southeast Asia's four major rivers.

River	Flows through	Empties into
Irrawaddy		
Chao Phraya		
Mekong		
Hong (Red)		

History and Culture

READ TO DISCOVER

1. What are some major events in mainland Southeast Asia's history?
2. What are the main cultural features of the region?

DEFINE

domino theory

LOCATE

Angkor Wat

WHY IT MATTERS

Geographers, archaeologists, and other scientists study sites where plants and animals have been domesticated in mainland Southeast Asia and around the world. Use CNNfyi.com or other **current events** sources to find recent news on domestication.

History

Southeast Asia has been inhabited for a long time. The earliest ancestors of the region's peoples arrived at least 1.5 million years ago. Modern humans *(Homo sapiens)* have lived in Southeast Asia for at least 40,000 years.

Southeast Asia was an important center of plant domestication. Early peoples grew rice, citrus fruits, and bananas. By about 3000 B.C. rice farming had been established, and people had domesticated buffalo, pigs, and cattle. Evidence indicates that people from southern China began to migrate through the region at least 2,500 years ago. Over time, settlements headed by chiefs grew, and trade developed with China and India. Merchants from India probably introduced Sanskrit writing and Hinduism during this period.

Angkor Wat in Cambodia, built in the 1100s by King Suryavarman II as a Hindu temple, is the largest religious complex ever constructed. Inset: Statues line the road into the Bayon Temple near Angkor Wat.

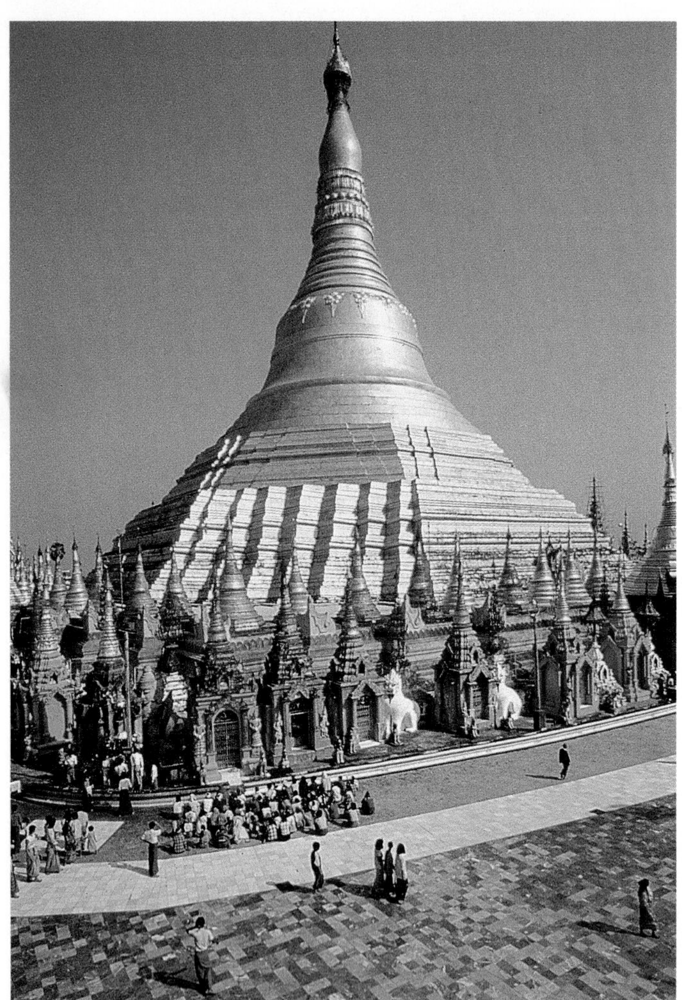
The Shwedagon Pagoda in Yangon, Myanmar, is a shrine sacred to the country's Buddhists. Tradition dates the gold-covered pagoda to the 500s B.C. The structure seen here, though, was rebuilt in the 1770s.

Early Cultures and Settlement Over the centuries various peoples moved into the region, particularly from China. The largest highly developed culture group in the region was the Khmer (kuh-MER). The Khmer dominated what is now Cambodia beginning in the A.D. 800s. By the end of the 1100s, their empire included most of mainland Southeast Asia. Angkor Wat, a huge temple complex built by the Khmer, reflects their advanced civilization and Hindu religion. In the 1200s the Thais (TYZ) migrated from southern China into the Khmer regions. Buddhism, introduced earlier from India and Sri Lanka, spread across mainland Southeast Asia, replacing Hinduism.

READING CHECK: *Human Systems* From where did Hinduism and Buddhism diffuse into the region?

Colonialism and Independence Europeans came to Southeast Asia in the early 1500s. Portuguese and Dutch sailors set up posts for trading in spices and other goods. Traders from Great Britain and France followed. Eventually, most of the region was colonized. Burma (now Myanmar) was a British colony. The French controlled Cambodia, Laos, and Vietnam, all of which they called Indochine, or French Indochina. Only Siam (now Thailand) was never colonized by Europeans.

During the 1800s the British and French set up plantations for growing export crops. They also built roads and railroads. They set up English- and French-language schools and introduced Christianity. Many Chinese and Indians migrated into the region during this period. They came to work on French and British plantations, mines, and railroads.

The Japanese invaded Southeast Asia during World War II. Only parts of Burma remained free of Japanese control. After the war, nationalist groups in the region tried to end colonialism. Over time four newly independent countries emerged. French Indochina was split into three countries—Vietnam, Cambodia, and Laos. Burma became independent from Britain. In 1989 Burma changed its name to Myanmar.

The transition to independence was not easy. In Vietnam, internal conflict led to a civil war involving the United States. (See Connecting to History: The Vietnam War.) The United States was involved because it wanted to stop the spread of communism in Asia. U.S. policy was based on the **domino theory**. This was the idea that if one country fell to communism, neighboring countries would follow like falling dominoes. The war in Vietnam also disrupted the neighboring countries of Laos and Cambodia.

In 1994 the U.S. space shuttle *Endeavour* scanned the Angkor area with radar. This and other remotely sensed images revealed the presence of previously unknown buildings.

READING CHECK: *Human Systems* Why did migrants from India and China come to the region during the European colonial era?

Culture

Each country in the region has one dominant cultural group. These are the Burmans, Thais, Khmer, Lao, and Vietnamese. However, there are differences within each group. For example, China has influenced culture in northern Vietnam, while Khmer influences are more pronounced in southern Vietnam. Each of the region's countries is also home to many minority groups. These groups have their own cultures and languages.

People, Languages, and Religion Mainland Southeast Asia has three main language families. They are spoken by the largest ethnic groups. In the west the Burmans speak a Sino-Tibetan language related to Chinese. Languages of the Tai family are spoken in Thailand and Laos. These languages—Thai and Laotian—may have originated in southwestern China. In the east the Vietnamese and Khmer peoples speak languages from the Austro-Asiatic family.

Many of the region's smaller ethnic groups live in mountain areas. In fact, more than 50 ethnic groups live in the Vietnamese highlands. They include the Cham, Yao, Hmong (MONG), and Muong. Many of these people have maintained their traditional ways of life. (See Geography for Life: Keeping Traditional Ways of Life in Myanmar.) For example, they practice animism, wear unique clothing and jewelry, and remain cut off from mainstream cultures.

Most of the region's major cities have large Chinese populations. About 14 percent of Thailand's population is ethnic Chinese. In addition, the colonial languages

Connecting to HISTORY

The Vietnam War

Vietnam's struggle for independence from France was long and hard. After years of fighting following World War II, the Vietnamese defeated the French in 1954. A peace settlement then divided Vietnam into two parts. North Vietnam had a Communist government, with its capital at Hanoi. In South Vietnam a pro-Western government was set up with its capital at Saigon. The two Vietnams could not come to an agreement about reunification, or rejoining as one country.

Communist groups supported by North Vietnam fought to control South Vietnam. The United States supported the South Vietnamese forces, entering directly into the conflict in late 1964. A long and bitter war followed. Nearly 2 million Vietnamese and more than 58,000 Americans lost their lives during the Vietnam War. The war even spread across the border into Laos and Cambodia. Most U.S. forces left South Vietnam in 1973. In 1975, North Vietnamese forces occupied Saigon, which they renamed Ho Chi Minh City in honor of their late leader.

More than 1 million South Vietnamese tried to escape the Communist takeover. Many became refugees in nearby countries. Some eventually settled in Western countries, particularly the United States, Australia, and France.

Drawing Inferences and Conclusions How did the Vietnam War affect the cultural geography of the United States? TEKS

In this 1975 photo, refugees flee to Saigon from cities taken by Communist forces. Local people offer food and drink.

INTERPRETING THE VISUAL RECORD

Thai Buddhists bring offerings of food and incense to a Bangkok temple. Theravada Buddhism has been Thailand's main religion since the 1300s. ***How might Buddhist traditions help unify the Thai people?***

of French and English are often spoken in the region. For example, English has become the language of international business in Southeast Asia. Some people in Vietnam, Cambodia, and Laos still speak French.

The region's dominant religion is Buddhism. Thailand, Cambodia, and Myanmar all have large Buddhist majorities. Most of the region's Buddhists practice a form of Buddhism called Theravada. They claim that Theravada, one of 18 major branches of Buddhism, is closest to the Buddha's original teachings. Buddhism and Hinduism coexist in a unique way in Thailand. There Hindu Brahmins lead most of the royal or official ceremonies.

Most Vietnamese practice a mix of Confucianism, which came from China, and Buddhism. Christianity and Islam are also present in the region. Animist religions are common in the highlands of Laos, Vietnam, and Myanmar.

TEKS **READING CHECK:** ***Human Systems*** What European languages are still spoken in mainland Southeast Asia? What religions are practiced there?

Food, Traditions, and Customs In the region's big cities, American fast food restaurants draw many new customers. Still, the most important food throughout mainland Southeast Asia is rice. In most places, rice is part of every meal. Other typical foods are fish and vegetables. Native tropical fruits like bananas, citrus, and durian are also available. Durian is a fruit known for its sweet flavor and unpleasant smell. The durian's odor is so strong, in fact, that some hotels and buses post signs forbidding the fruit. Spices such as ginger and chili peppers add flavor to regional dishes. Lime juice, lemon grass, and coriander add tang. Ground peanuts and coconut milk are other flavorings. Fish sauces are popular throughout the region. These sauces, which have different names in different countries, are based on the liquid poured off of salted, fermented fish. Sometimes chilies or sugar flavor the fish sauce. Spicy sauces with Indian origins called curries are popular in Thailand. Curries usually flavor rice or vegetables. Thais often wash down their spicy food with sugarcane juice. Vietnamese food is particularly varied, with nearly 500 traditional dishes. Some use exotic meats, such as bat or cobra. Plain white rice with various sauces is now typical fare, however.

Hmong women in Laos sew elaborate new clothes for the New Year to prevent bad fortune throughout the year.

Buddhism shapes people's lives in mainland Southeast Asia. For example, Thai men often spend time working and serving in monasteries. In Laos, all

Buddhist men have traditionally been expected to become monks for a while. One of Myanmar's main holidays is the Festival of Light, which marks an event in the Buddha's life. Paper lanterns light up the streets, and families visit the local shrine. Also in Myanmar, a water festival called Thingyan marks the country's Buddhist New Year. The festival celebrates the cleansing of the soul and washing away of the old year. During this mid-April festival, people in the streets soak each other with water. Since the event falls during the hottest season, the buckets of cold water may be welcome! Other countries celebrate the new lunar year in similar ways.

Throughout the region, cultures of urban and rural areas may differ widely. Many rural people follow the same practices generation after generation. For example, village religious festivals may celebrate local animist beliefs. Country people are more likely to wear garments such as the *panung* in Thailand and the *longyi* in Myanmar. A *panung* is a colorful cotton or silk cloth wrapped tightly around the body. A *longyi* is a long skirt. Western clothing is common in urban areas.

INTERPRETING THE VISUAL RECORD

Two different education systems are pictured in these photographs. At left, Thai boys study at the Wat Po monastery in Bangkok. At right, Vietnamese girls attend a state school in Ho Chi Minh City. ***What are some ways in which these schools might differ?*** TEKS

READING CHECK: *Places and Regions* How is Buddhism a unifying element of the region's culture?

TEKS Questions 1, 2, 3, 4, 5

Review

Homework Practice Online

Keyword: SW3 HP29

Define domino theory

Working with Sketch Maps On the map of mainland Southeast Asia that you created in Section 1, label the countries of mainland Southeast Asia and Angkor Wat. Shade in the countries that made up French Indochina. Which culture group built Angkor Wat?

Reading for the Main Idea

1. *Places and Regions* What were some of the first crops grown in mainland Southeast Asia?
2. *Places and Regions* Which two European countries had large colonies in mainland Southeast Asia? How did the U.S. military become involved in the region?
3. *Human Systems* What is the dominant religion throughout mainland Southeast Asia? What is unique about religious practices in Thailand?

Critical Thinking

4. **Analyzing** How has mainland Southeast Asia's location between India and China affected its culture?

Organizing What You Know

5. Create two time lines like the ones shown below. Use the time lines to trace the region's history through important periods and events.

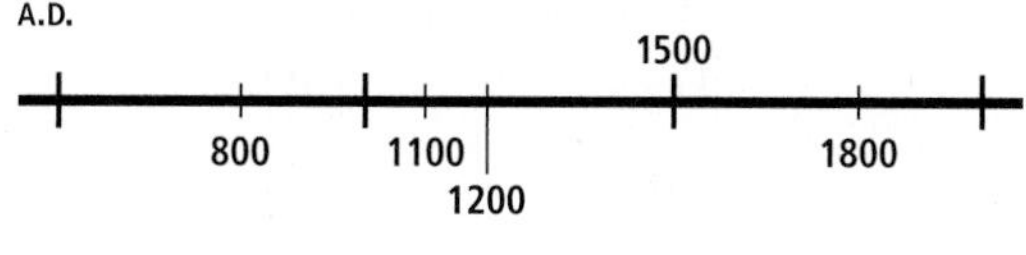

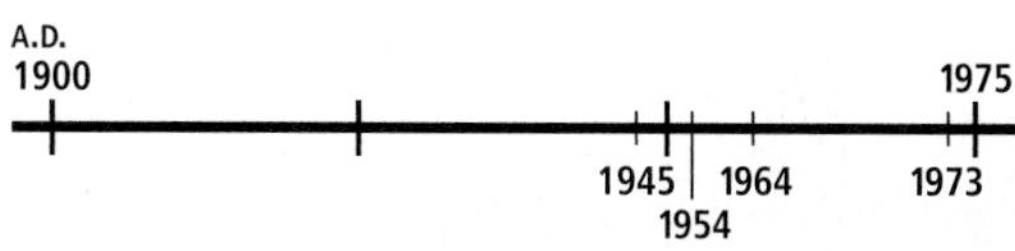

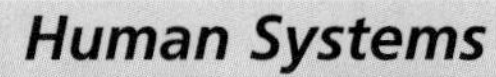

Geography for Life

Keeping Traditional Ways of Life in Myanmar

Traditions—established ways of thinking, acting, and behaving—are valued for many reasons. Traditional activities help people connect with their ancestors, land, and environment. In a world of rapid change, traditions offer a sense of stability. On the other hand, some traditions are controversial. For example, the shifting agriculture practiced by some peoples can damage the environment. Other customs may sharply limit the choices that women can make.

Indigenous cultures are particularly rooted in tradition. However, these cultures are under increasing pressure from the modern world. Modernization and economic development make it harder and harder for some to maintain their ways of life. These peoples are often viewed as backward by the majority culture in control of a country's politics, economic growth, and land use policies. The majority may dominate indigenous minority groups and force them to assimilate into modern society.

One example of the challenges facing traditional cultures is in Myanmar. Myanmar's largest ethnic group—the Burmans—make up about 68 percent of the country's population. The Burmans occupy the lowlands, river valleys, and large cities of Myanmar. As the country's largest ethnic group, they dominate Myanmar's government. However, Myanmar is also home to many smaller ethnic groups with different languages and cultures. Some of these groups include the Shan, Karen, Rakhine, Mon, Chin, and Kachin. In general, Myanmar's minority groups live in the hills and uplands near the country's borders. In fact, they are often referred to collectively as the hill peoples. Most of these peoples live in small villages and practice animism.

Since Myanmar became independent from Great Britain in 1948, many of the country's minority groups have struggled to maintain their identity and culture. Rapid economic and social change has altered old ways of life, particularly among younger generations. Minority groups have resisted government attempts at cultural assimilation. They have also demanded political freedom and basic human rights. Some have even taken up arms against the government. For example, the Kachin Independence Army (KIA) and Karen National Union (KNU) are fighting for independent states. Such conflicts are not uncommon in other parts of the world.

Applying What You Know

1. **Summarizing** How have the peoples who live in the hills and uplands of Myanmar struggled to maintain their traditional ways of life?
2. **Supporting a Point of View** Do you think younger generations are more or less likely to maintain traditional ways of life? Why?

Left: Heavy neck rings are common among the Kayan (Padaung). Right: Fishers in east-central Myanmar use traditional fishing nets. Many people in the area belong to the Mon ethnic group.

The Region Today

READ TO DISCOVER

1. What are the economies and politics of mainland Southeast Asia like?
2. What types of agriculture are practiced in the region?
3. What issues and challenges do the region's countries face?

IDENTIFY

ASEAN

DEFINE

wats
klongs

LOCATE

Bangkok
Ho Chi Minh City
Hanoi
Yangon (Rangoon)
Phnom Penh
Vientiane

WHY IT MATTERS

Some countries in mainland Southeast Asia still bear environmental scars left by years of warfare. Use CNNfyi.com or other **current events** sources to learn how these countries are working to repair the damage.

Economic and Political Development

The countries of mainland Southeast Asia have different levels of development. Some of them, such as Laos and Cambodia, are very poor and undeveloped. In contrast, Thailand has experienced rapid industrialization. In fact, it is one of the "Asian Tigers." This group of East and Southeast Asian countries includes Malaysia, Singapore, South Korea, and Taiwan. All have experienced rapid economic expansion since the 1980s.

Myanmar is rich in natural resources. However, the country's military government has kept its economy and people isolated. The Burmese have little freedom and remain poor. Both Laos and Vietnam have Communist governments. They have been trying to improve their economies by slowly accepting some features of capitalism. Vietnam has dynamic private businesses, but government policies still limit their growth. Still, Thailand is the region's leading economic power. It is a constitutional monarchy. Industry, agriculture, fishing, mining, and tourism drive Thailand's economy.

A Lao farming family loads hay into a cart.

INTERPRETING THE VISUAL RECORD

These shrimp were harvested from one of Thailand's seafood farms. Many farms release wastewater and fertilizer into coastal mangrove forests. This destroys the trees. ***How might damage to mangrove forests affect Thailand's coastal environment?*** TEKS

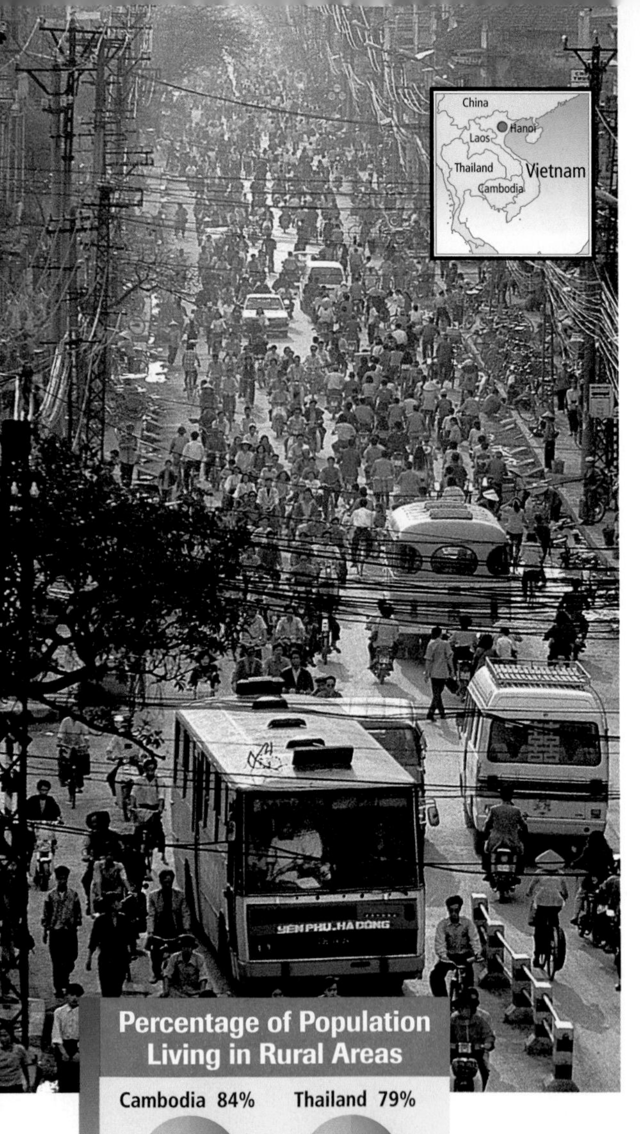

All the countries of Southeast Asia are members of the Association of Southeast Asian Nations, or **ASEAN**. ASEAN was founded in 1967. It promotes economic development as well as social and cultural cooperation among its members. The ASEAN countries have treaties of cooperation with each other. In the 1990s they moved to form a free-trade region.

READING CHECK: *Human Systems* Why was ASEAN formed?

Settlement and Cities Most people in mainland Southeast Asia live in rural areas and are subsistence farmers. In general, coastal and lowland areas are more densely populated, while highlands are less crowded. The most crowded regions are along the rivers. These regions have both the main rice-growing areas and the largest cities. Because of their location along riverbanks, however, the big cities are subject to occasional flooding.

The region's cities are growing rapidly and display many contrasts. For example, motorcycles speed past pedicabs. Pedicabs are three-wheeled carriages with a separate seat for a driver who pedals. In some cities skyscrapers tower over huts, and street vendors compete with shopping malls.

With more than 7 million people, Bangkok is by far the largest city of mainland Southeast Asia. Thailand's capital is one of the most dynamic cities in Asia, with its modern technology blending with older ways of life. For example, Bangkok has more than 400 ***wats***, which are Buddhist temples that also serve as monasteries. The *wats* are among the many attractions that draw thousands of tourists to the city each year.

Ho Chi Minh City, formerly called Saigon, is Vietnam's biggest city. Located in southern Vietnam, it is a thriving industrial center. It is also home to a busy social scene. Almost any type of entertainment, from disco to ancient forms of Vietnamese music, can be found in its nightspots. In the north, Hanoi is Vietnam's capital. Much of the country's history is evident in Hanoi's temples, monuments, and faded French colonial mansions. Hue has often been called the most beautiful city in Vietnam. Its historical attractions include a citadel surrounded by a wall 7 miles (11 km) long. Not far from Hue are tombs of the last Vietnamese royal family to rule the country.

The largest city in Myanmar is Yangon, the capital, also called Rangoon. It is a very green city, with so many trees that some travelers have compared it to a jungle. The city's skyline is dominated by the Shwedagon Pagoda, an enormous golden shrine which is said to hold eight of the Buddha's hairs. Gardens, museums, and a zoo also attract visitors.

The other national capitals are not as large as these cities. Phnom Penh (puh-NAWM PEN), Cambodia, saw most of its people driven out by the country's Communist government in the 1970s. The city's population is slowly increasing. Vientiane (vyen-TYAHN), also known as Viangchan, on the Mekong River, is the most important port for Laos.

READING CHECK: *Human Systems* How are ancient and modern times evident in the cities of mainland Southeast Asia?

Percentage of Population Living in Rural Areas

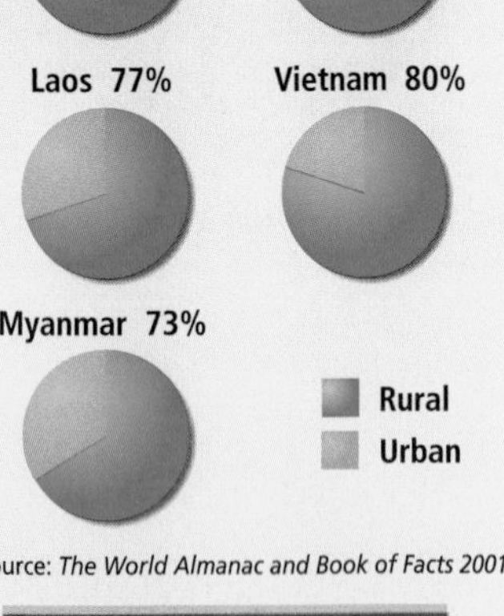

Source: *The World Almanac and Book of Facts 2001*

INTERPRETING THE GRAPH

Each country in Southeast Asia has only one or two large cities, like Hanoi, Vietnam, shown above. However, most people live in small villages. ***How do most rural residents make a living?***

Agriculture

Agriculture plays a central role in the economies of all the countries in mainland Southeast Asia. Farmers in the river lowlands grow rice on fertile slopes along the riverbanks. The wet tropical climate lets farmers grow two or three crops each year. Workers move the seedlings by hand to flooded paddy fields. Harvesting is also done by hand. Because paddy agriculture uses so much human labor, it is a form of intensive agriculture.

Paddy farming is particularly successful in areas with good water resources. In these areas farming supports dense populations. For example, the delta regions of the Mekong and Hong (Red) Rivers are very fertile, productive, and crowded. This is true also in the Irrawaddy and Chao Phraya valleys of Myanmar and Thailand. Farmers in mainland Southeast Asia produce enough rice to feed the local population and to export. In fact, Thailand and Vietnam are two of the world's leading exporters of rice. (See the graph.) Many farmers grow vegetables or fruit to sell or trade in local markets. Farmers also raise ducks, water buffalo, pigs, and fish.

The region's many subsistence farmers grow a variety of crops besides rice. These crops include cassava, yams, bananas, pineapples, beans, and sugarcane. In the rugged and forested mountains, poor soils make farming difficult. Farmers must leave their fields after only a few years when soil fertility decreases. Then they move to a new area. This shifting agriculture is common in Laos, northern Thailand, and northern Myanmar. Governments today discourage this practice because it can damage rain forests.

While many people are subsistence farmers, plantation agriculture is also important. For example, Thailand is the world's leading producer of natural rubber. Other plantation crops include sugarcane and pineapples.

Some people grow opium in the rugged mountains of Myanmar, Thailand, and Laos. Opium is a major crop for some poor mountain peoples. This illegal crop is processed to make the deadly and addictive drug heroin. Opium dealing produces riches and risks for corrupt officials, smugglers, and drug dealers. Opium is smuggled out of the region and into the United States, Europe, and other parts of Asia.

TEKS **READING CHECK:** ***Environment and Society*** Where is paddy farming most successful in the region?

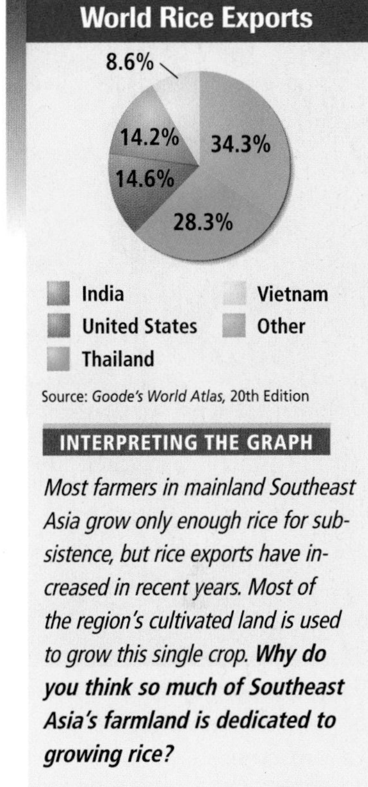

INTERPRETING THE GRAPH

Most farmers in mainland Southeast Asia grow only enough rice for subsistence, but rice exports have increased in recent years. Most of the region's cultivated land is used to grow this single crop. ***Why do you think so much of Southeast Asia's farmland is dedicated to growing rice?***

INTERPRETING THE VISUAL RECORD

Irrigation in this Vietnam rice field is a labor-intensive project. ***How does this irrigation system work?***

Vendors at Bangkok's floating markets sell fish, fruit, meat, rice, and vegetables to locals and tourists.

FOCUS ON GEOGRAPHY

Bangkok's *Klongs* The farm and the city meet on the canals of Bangkok, called ***klongs***. The people of Bangkok use this network of canals to travel around the city and to transport and sell goods. Some travelers have compared Bangkok to Venice, Italy, another city with a canal system.

In the 1900s Bangkok grew tremendously. New forms of transportation such as the automobile spread. Many *klongs* were filled in and paved to create space for cars and trucks. Today most of the canals have been converted to roads, and cars are more common than boats.

Despite these changes, *klongs* are still a part of life in Bangkok. Boats and river taxis ply the waters of Bangkok's remaining canals, transporting people around the city. As the city's traffic has grown, more people have been using the canals for local travel. Special commuter boats take people to and from work. The canals are also vital for moving fresh fish and produce around the city. In some places *klongs* are still the center of neighborhood life. Traditional houses built on stilts line the water. Families bathe on their front doorsteps. Boats loaded with mangoes, durians, and other produce or goods join together to form floating markets.

READING CHECK: ***Environment and Society*** What role have *klongs* played in the development of Bangkok?

Issues and Challenges

The major challenges in mainland Southeast Asia involve both political and economic issues. Corruption and dictatorial governments have caused serious problems for most of the region's countries. These political problems have in turn led to economic difficulties.

Cambodia has experienced terrible problems since independence. From 1975 to 1979, the country fell under the brutal rule of a Communist group called the Khmer Rouge ("Red Khmer"). Its leaders wanted to change Cambodia into a rural peasant society. To further their plan, the Khmer Rouge forced all citizens to work as field laborers. They emptied the cities, separated families, and targeted educated people for execution. The country's intellectual and skilled worker classes were practically wiped out. Perhaps 1 million Cambodians were killed. Starvation, disease, forced marches, and other hardships killed many more people. An invasion by Vietnamese forces finally ended the terror. However, millions of land mines are left over from decades of fighting. They maim or kill farmers and other civilians who accidentally disturb the hidden explosives. Today Cambodia has a stable elected government, but it remains a poor farming country with a troubled heritage.

A military government took control of Myanmar in 1962. Pro-democracy groups have tried to regain political freedoms but so far have failed. Their leaders and supporters have been jailed and harassed. Minority and rebel groups in the northern mountains have also battled Myanmar's government. The government controls many parts of the economy, such as the rice trade and heavy industry. Because of the political situation, many foreign investors have stayed away. As a result, the country remains poor, and living standards for most people have not improved.

Tourism might be a source of income for Myanmar. However, leaders of the democratic movement there favor a boycott on tourism. They say that most of the money spent by tourists goes directly to the military government.

Aung San Suu Kyi is the best known opponent of military rule in Myanmar. Her party, the National League for Democracy, won control of the country's parliament in 1990, but the military refused to give up power. The government has placed Aung San Suu Kyi under house arrest. Even when not detained, she was forbidden to leave Yangon. For her efforts to bring democracy to Myanmar, she received the Nobel Peace Prize in 1991.

Aung San Suu Kyi stands in front of the National League for Democracy flag.

Both Vietnam and Laos have Communist governments. They have been slowly moving toward market economies, but many challenges remain. Both countries are poor, with most people working in subsistence agriculture. For years Laos was almost cut off from the Western world. This lack of outside influence now draws tourists who want to see a glimpse of traditional Southeast Asian life. In the late 1990s Vietnam and the United States began forming closer relations. American businesses, products, and tourists are now increasing in Vietnam.

Thailand's economy grew by almost 9 percent a year between 1985 and 1995. It was the highest growth rate in the world during that time period. Investment from Japan helped Thailand develop textile, electronics, and automobile assembly plants. However, economic growth slowed in the late 1990s. Economic reforms are currently underway.

TEKS **READING CHECK:** *Human Systems* How has Myanmar's political situation hurt economic development there?

TEKS Questions 1, 3, 4

Section 3 Review

go.hrw.com **Homework Practice Online**
Keyword: SW3 HP29

Identify ASEAN

Define *wats, klongs*

Working with Sketch Maps

On the map of mainland Southeast Asia that you created in Section 2, label Bangkok, Ho Chi Minh City, Hanoi, Yangon (Rangoon), Phnom Penh, and Vientiane. Which city has been compared to Venice, Italy? Why?

Reading for the Main Idea

1. *Places and Regions* How is Thailand's economy unique in mainland Southeast Asia?
2. *Human Systems* What makes paddy farming a form of intensive agriculture?

Critical Thinking

3. **Analyzing Information** How do rivers influence the location and patterns of movement of products and people in the region? How do you think the rivers might be used to attract business investment?
4. **Making Generalizations** Why do you think the region's countries have worked to form closer economic and trading ties with each other? What might such ties mean for the economic and political futures of the region's countries?

Organizing What You Know

5. Copy the chart below. Use it to describe the major issues and challenges facing the countries of mainland Southeast Asia.

Country	Issues and challenges
Myanmar	
Thailand	
Cambodia	
Laos	
Vietnam	

CASE STUDY

Studying Deforestation

Environment and Society Environmental geographers around the world are concerned about deforestation in Southeast Asia. The region has about one fourth of the world's remaining tropical rain forests. These forests are being destroyed at an alarming rate. Some observers believe that most of the region's tropical rain forests could be gone in just 20 to 30 years.

Geographers bring a spatial and environmental perspective to studying this problem. They begin by asking geographic questions from a spatial perspective. Where is deforestation occurring? What are the patterns and causes of deforestation? How is deforestation in Southeast Asia related to economic development and world trade? Geographers also ask questions from an environmental perspective. How does deforestation affect the environment? How does it affect people?

Patterns and Causes

Geographers use satellite images and other tools to study the problem. They have found that Southeast Asia has one of the highest deforestation rates in the world. Between 1990 and 1995, about 56,255 square miles (145,700 sq km) of rain forest were lost. This figure represents an area about the size of Florida. The countries with the highest deforestation rates included the Philippines, Thailand, and Malaysia. (See the map.)

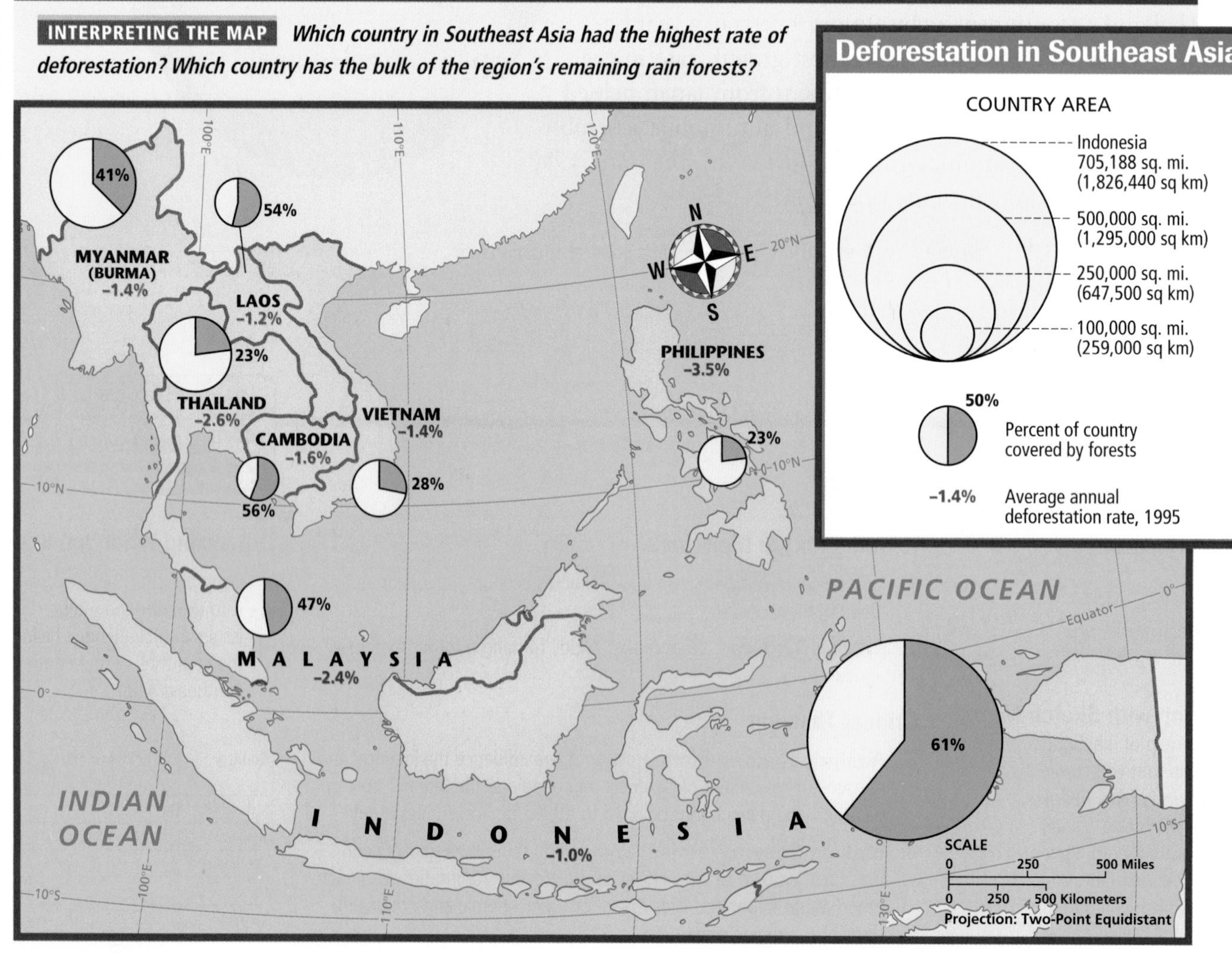

INTERPRETING THE MAP *Which country in Southeast Asia had the highest rate of deforestation? Which country has the bulk of the region's remaining rain forests?*

Left: In 1997, airplanes dropped water on the Indonesian fires. Right: A man looks out at the burning rain forest. For many years his people had used this part of Indonesia for hunting and for gathering medicinal plants.

In Southeast Asia there are many causes of deforestation. For example, people in many areas cut trees and use the wood for fuel. In addition, trees are cut and burned to clear fields for farming. As populations in Southeast Asia continue to grow rapidly, more forested areas are being used for farming. After nutrients in the soil are used up, farmers move on to other areas to grow crops. This pattern of shifting agriculture clears large forest areas.

Logging is another cause of deforestation. Valuable hardwood and wood-based products are important exports for countries like Malaysia, Indonesia, Myanmar, Cambodia, and Laos. Most logging is done by large corporations. Developed countries such as Japan import the timber. To limit deforestation, some Southeast Asian countries have enacted total or partial bans on logging. However, in many other countries, income from timber exports is used to develop national economies.

Environmental Effects of Deforestation

Geographers also study the effect of deforestation on people and environments. Some believe the most serious result of deforestation is the extinction of plant and animal species. A great diversity of plant and animal life thrives in the rain forests of Southeast Asia. For example, Indonesia may have about 11 percent of the world's plant species. About 12 percent of all mammal species and 17 percent of all bird species may also live there. Deforestation destroys the habitats of many of these rare plants and animals.

Deforestation also causes flooding and soil erosion, which can more directly affect humans. When areas are stripped of their forest cover, the soil is exposed directly to water's erosive power. As a result, the clearing of forests has led to erosion and flooding in many Southeast Asian countries, such as Thailand.

Smoke created by burning forests has been another problem. In 1997 and 1998, for example, a severe drought allowed fires to burn out of control, particularly in Indonesia. Some of the fires had been set to clear land. Smoke from these huge forest fires endangered people's health, shut down airports and schools, and disrupted the tourist industry.

Future Outlook

As you can see, deforestation in Southeast Asia is a very complicated problem. It is tied to the region's economies, world trade patterns, and people's need to use the resources around them to survive. To solve this problem, geographers and other scientists must learn more about the causes, rates, patterns, and environmental effects of deforestation.

Applying What You Know

1. **Summarizing** How is deforestation linked to economic development, population growth, and environmental change in Southeast Asia? What have some governments done to slow deforestation?
2. **Problem Solving** What do you think should be done to halt the destruction of Southeast Asia's tropical rain forests? Work with a partner to develop a proposed solution and present it to your class in a report.

Building Vocabulary

On a separate sheet of paper, explain the following terms by using them correctly in sentences.

arboreal | *wats*
domino theory | *klongs*
ASEAN

Locating Key Places

On a separate sheet of paper, match the letters on the map with their correct labels.

Malay Peninsula | Irrawaddy River | Mekong River
Indochina Peninsula | Chao Phraya River | Hong (Red) River
Khorat Plateau

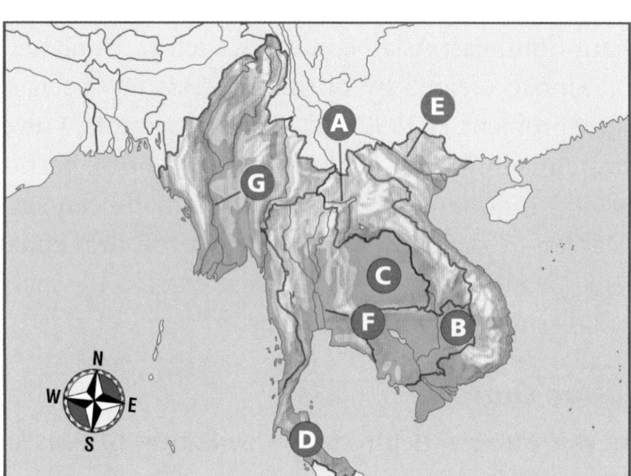

Understanding the Main Ideas

Section 1

1. *Places and Regions* What are the major climates in mainland Southeast Asia, and where are they found?
2. *Physical Systems* What type of soils have made the region's river valleys rich farming areas?

Section 2

3. *Places and Regions* Which of the region's countries was never a European colony? What two European countries had colonies in the region?
4. *Places and Regions* What three major language families are found in the region? Who are some of the primary speakers of the languages in each group?

Section 3

5. *Human Systems* In what areas do most of the region's people live? What are the large cities of the region like?

Thinking Critically for TAKS

1. **Drawing Inferences** Why do you think previously unknown animal species are still being discovered in the region?
2. **Analyzing Information** Where are the region's major cities located? What advantages might these locations offer?
3. **Comparing** Some crops, such as sugarcane and pineapples, are grown by both plantation owners and subsistence farmers. How are these two types of farming different in the region? How might the lives of subsistence farmers and plantation workers differ?

Using the Geographer's Tools

1. **Analyzing Statistics** Use the unit Fast Facts and Comparing Standard of Living tables to rank the region's countries in order, from most to least developed. Then note the political situation in the region's countries. How does the political geography affect the countries' economic geography?
2. **Creating Maps** Create a map of the region that shows rivers, deltas, mountains, plateaus, and plains. Then shade areas that have fertile soils and dense populations. Write a short paragraph noting the connections between the two.
3. **Creating Maps** Prepare a map showing past European colonies in the region. How might Thailand's (Siam's) relative location have helped it to remain independent?

Writing for TAKS

Write a newspaper article about life in Myanmar today. Be sure to discuss the country's recent political, economic, social, and environmental changes. How are these areas related? How do they affect life in Myanmar? Compare the situation in Myanmar to the United States. How are they similar? How are they different?

SKILL BUILDING

Geography for Life

Using Research Skills

Environment and Society Plan a research project on deforestation in mainland Southeast Asia. First, review the Case Study: Studying Deforestation. Then create three questions that you would like to research. For example, you might want to know more about how deforestation has affected a particular country. Use newspaper articles, documentaries, or statistics to research your questions. Then write a short report that answers your three questions.

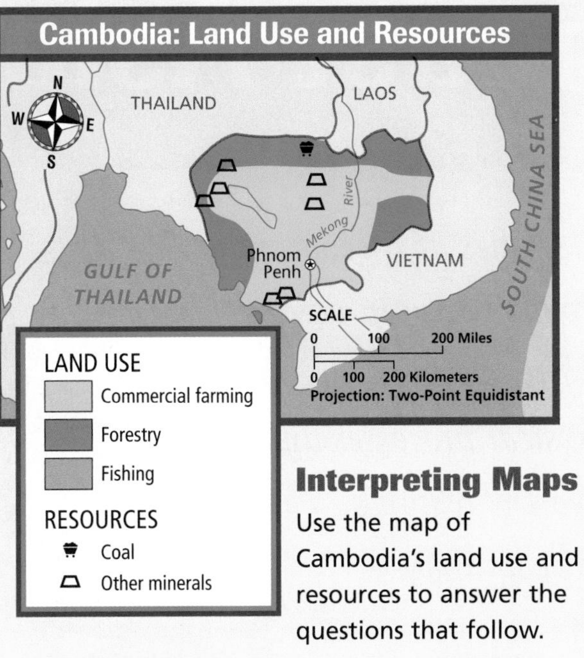

Interpreting Maps

Use the map of Cambodia's land use and resources to answer the questions that follow.

1. Which of the following can be concluded from the information shown on the map?
   a. The country has no major manufacturing and trade centers.
   b. Coal is found only along Cambodia's southern coast.
   c. Commercial farming is common in central areas of the country.
   d. Few forested areas are found in Cambodia.
2. Why do you think commercial farming is located where it is?

Analyzing Primary Sources

Karin Muller spent seven months hitchhiking through Southeast Asia. Read her observations about Vietnam and answer the questions.

> **"Rice in Vietnam . . . is not just that fluffy white stuff in a box that you dig out once a month as an accompaniment to your favorite lemon chicken. Rice is life. Almost no meal goes by without it . . . The poor mountain Vietnamese say "there is no money" and "there is no rice" interchangeably because if they had money, they would use it to buy rice. In many areas rice is money. It is the traditional currency."**

3. Based on what you have read, which of the following is true?
   a. Rice is Vietnam's least important food.
   b. Most of Vietnam's rice comes from mountain areas.
   c. Vietnamese eat rice mostly at special occasions.
   d. Rice is a very important part of the Vietnamese diet.
4. Why would a poor mountain Vietnamese person equate rice to money?

Alternative Assessment

PORTFOLIO ACTIVITY

Learning about Your Local Geography TEKS

Group Project: Research

Plan, organize, and complete a group research project on native arboreal animals in your area or state. First, use your local library to learn about some arboreal animals common near your community. Then have each individual in your group select one animal and learn more about it. In which tree species do these animals live? How are they adapted to life in the trees? What is their range of habitat in the United States and the rest of the world? To which other animals are they related? You might want to create maps and diagrams to illustrate the answers you find to these research questions. When you are done, share your results with the other group members.

internet connect

go.hrw.com

Internet Activity: go.hrw.com
KEYWORD: SW3 GT29 TEKS

Choose a topic on mainland Southeast Asia to:
- learn about the architecture of ancient and modern mainland Southeast Asia.
- create a brochure on Laos and Vietnam since the Vietnam War.
- understand the causes and effects of deforestation in mainland Southeast Asia.

CHAPTER 30

Island Southeast Asia

Island Southeast Asia lies at a geographic and cultural crossroads between oceans and continents. In this chapter you will read about how physical processes and location have shaped the region's history and human geography.

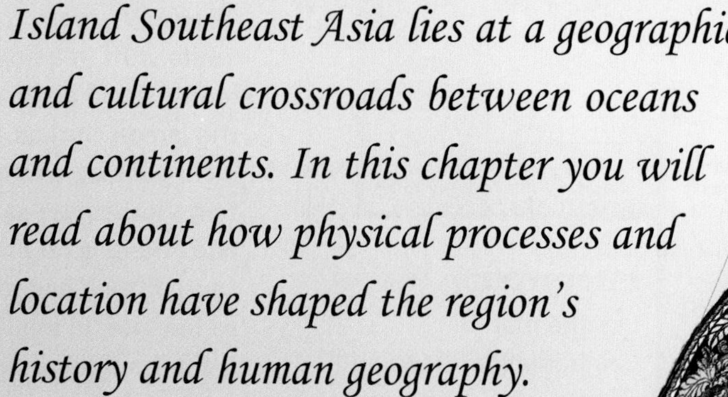
Malaysian kite

A Malaysian woman carries durian, a large fruit with a delicious taste but an unpleasant odor.

Hi! My name is Michelle-Anne Chan Yi Ping. I am 17 years old, and I live with my mum and grandmother in a house in Penang, Malaysia. I am in Form 5 now—what you call eleventh grade. I study six core subjects and three additional subjects. The core subjects are mathematics, English, Malay, science, morals (civics), and history. The three additional subjects are accounting, economics, and literature. In addition, there are four Houses, or school sports teams, that are identified by the colors red, blue, green, and yellow. I represent my class and Blue House in school netball. I am also a member of the Red Crescent (same as the Red Cross/First Aid in non-Islamic countries). I have joined the Anti-Crime Club too.

My grandmother prepares my lunch. She cooks mostly Chinese food. I eat at about 1:30 P.M., then take my bath after that. I eat dinner very early, around 5:00 P.M. The rest of my family eats between 7:30 and 8:30 P.M.

I really enjoy bowling, and I also collect stickers and stamps. Like many teenagers, I *lepak* (LUH-pak) a lot. That's a Malay word that means "loafing around or hanging out at malls." I can't afford to waste too much time, though! I hang out in the bowling alley even though sometimes I don't bowl. I surf the Internet every day and chat on-line. Chatting is something I love to do, and I'm known for my chatterbox behavior. Maybe that's why I want to become a famous lawyer!

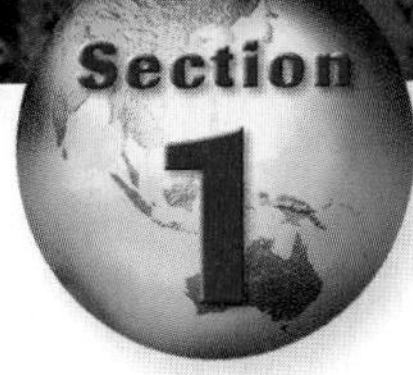

Natural Environments

READ TO DISCOVER

1. What landforms are found in island Southeast Asia, and what are some physical processes that have shaped them?
2. What climates, biomes, and natural resources does the region have?

DEFINE

archipelago
lahars
endemic species

WHY IT MATTERS

Earthquakes and volcanoes often shake the countries of island Southeast Asia. Heavy rains after a volcanic eruption can increase the amount of damage. Use CNNfyi.com or other **current events** sources to learn about the damage that these events can cause.

LOCATE

Malay Peninsula
Malay Archipelago
Sumatra
Irian Jaya
New Guinea
Borneo
Java Sea
South China Sea
Timor Sea
Strait of Malacca
Java
Spratly Islands

Island Southeast Asia: Physical-Political

SCALE 0 200 400 Miles; 0 200 400 Kilometers
Projection: Miller

ELEVATION

FEET	METERS
13,120	4,000
6,560	2,000
1,640	500
656	200
(Sea level) 0	0 (Sea level)
Below sea level	Below sea level

⊛ National capitals
• Other cities
▪ Historic sites

Size comparison of island Southeast Asia to the contiguous United States

Landforms

The Malay Peninsula and the Malay Archipelago (ahr-kuh-PEH-luh-goh) stretch from the Southeast Asian mainland almost to Australia. An **archipelago** is a large group of islands. We call this region island Southeast Asia. Look at the chapter map. The region sits between the Pacific and Indian Oceans. To the north and west lies the Asian mainland. Australia and the Pacific Ocean lie to the south and east.

The region's countries include Brunei, East Timor, Indonesia, Malaysia, the Philippines, and Singapore. East Timor, a former Portuguese and Indonesian territory, declared independence in 1999.

The Islands Island Southeast Asia contains more than 20,000 islands. These islands make up the Malay Archipelago. They extend along the equator from Sumatra to Irian Jaya (also called West Papua) on the island of New Guinea. Other major islands include Borneo and the islands of the Philippines. Some of the region's islands are among the largest in the world. Only the North Atlantic island of Greenland is larger than New Guinea. Borneo is the world's third-largest island. Malaysia's Malay Peninsula is almost an island. The Isthmus of Kra connects it to the Asian mainland.

The region's larger islands have high mountains. The highest peaks are in Irian Jaya on New Guinea. Peaks there rise to more than 16,000 feet (5,000 m). Some mountains are tall enough to have glaciers and snowfields.

Many seas and narrow straits separate the islands and the Malay Peninsula from one another. Some of these bodies of water are the Java Sea, South China Sea, and the Timor Sea. The Strait of Malacca lies between the island of Sumatra and the Malay Peninsula. This strait is located along a major shipping route. Trade along this route has benefited many nearby cities and countries, particularly Singapore.

INTERPRETING THE VISUAL RECORD

Mount Pinatubo, a volcano on the island of Luzon in the Philippines, erupted violently in June, 1991. It was the largest eruption anywhere in more than 50 years. Because scientists had predicted the event, rescuers were able to save thousands of lives. This photo was taken between eruptions. ***What evidence of erosion do you see in the photo?*** TEKS

Mount Pinatubo
Philippines
Malaysia
Indonesia
Australia

Tectonic Processes Tectonic activity is one of the physical processes that have shaped island Southeast Asia. The region is located in one of the world's most geologically active areas. It lies along and sits between several tectonic plate boundaries. Earthquakes and volcanic eruptions are common in areas near those boundaries. Volcanic islands have formed along deep ocean trenches at the edges of the Indian and Pacific Oceans. The Indonesian island of Java, for example, has about 13 active volcanoes.

Dangerous volcanic mudflows, called **lahars** (LAH-hahrz), sometimes rush down steep volcanic slopes. Lahars can bury river valleys and towns. However, volcanoes also provide Indonesia and the Philippines with rich soils. Undersea earthquakes and volcanoes make tsunamis an occasional threat to coastal areas in Indonesia and the Philippines.

A more stable area of Earth's crust lies under the shallow sea between the Malay Peninsula and Borneo. No plate boundaries exist in this area. As a result, Borneo, the Malay Peninsula, and eastern Sumatra do not have active volcanoes.

The volcano Krakatau (kra-kuh-TOW) is located between Sumatra and Java. Tsunamis caused by Krakatau's eruption in 1883 rose as high as 100 feet (30 m) and killed about 36,000 people.

READING CHECK: *Physical Systems* What physical processes created some of the region's islands?

Climates, Biomes, and Natural Resources

All of island Southeast Asia is located in the tropics. This tropical location strongly influences the islands' climates and biomes.

Climates Much of the region has a tropical humid climate. The weather is hot and damp all year. Rainfall is heavy most months—80 to 100 inches (200 to 250 cm) per year are normal. Some mountain areas receive even more rain. Afternoon thunderstorms are common.

The Philippines lie in the path of typhoons that sweep in from the Pacific Ocean. These huge storms can cause terrible destruction. Typhoons bring heavy rain and powerful winds. They also cause sea levels to reach dangerous heights. Most typhoons strike between August and October when ocean temperatures are warmest. Sometimes the storms are so severe that they cause long-lasting damage to the country's farmland and economy.

A few of the region's eastern islands have a tropical wet and dry climate. This climate region stretches from eastern Java to southern Irian Jaya. These areas have both very rainy and dry seasons. This wet-and-dry pattern results from the influence of the monsoon flow. Because this area is south of the equator, from November through March humid monsoon winds blow from the Indian and Pacific Oceans. The wet season occurs at this time. The dry season falls between June and September. At that time, monsoon winds blow from the dry interior of Australia.

Where would you expect to find the coolest climates in this tropical region? The only areas without tropical climates are found in the higher elevations of the region's mountains. Inland Borneo and Irian Jaya have highland climates. Resorts in the highlands offer an escape from the constant heat of the lower elevations.

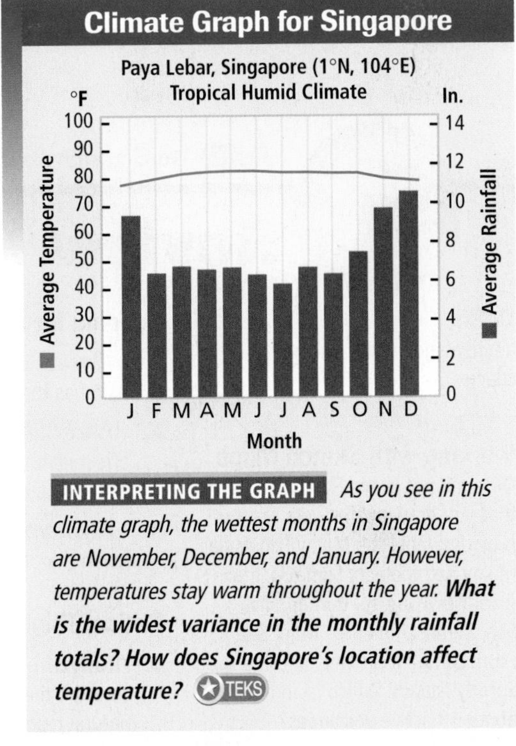

INTERPRETING THE GRAPH *As you see in this climate graph, the wettest months in Singapore are November, December, and January. However, temperatures stay warm throughout the year.* ***What is the widest variance in the monthly rainfall totals? How does Singapore's location affect temperature?*** TEKS

Plants and Animals The region's tropical climates support ancient tropical rain forests. Indonesia alone has about 10 percent of the world's remaining tropical rain forests. Thick mangrove forests grow in coastal areas. Mangroves are trees or shrubs that have exposed supporting roots. Birds, fish, and other small marine animals live in these tidal areas.

Today most of the forests and animals are at risk. Rain forests are being cut down at a fast rate. Loggers and developers are harvesting valuable hardwoods and clearing land for buildings. Farmers clear land for crops. Many indigenous peoples in Borneo, Irian Jaya, and the Philippines may also lose their forest homelands and traditional ways of life. Some countries have banned logging or created national parks to protect their tropical rain forests.

Tropical rain forests have the greatest variety of plant and animal life on Earth. In fact, island Southeast Asia has one of Earth's highest levels of biodiversity. There are many **endemic species** in the region. Endemic species are those native to a certain area. Southeast Asia's endemic species include the Komodo dragon, Javan rhinoceros, and orangutan. As people change the animals' habitats—by farming, logging, or setting fires—many of these animals are becoming endangered.

INTERPRETING THE VISUAL RECORD

With a length of about 10 feet (3 m) and weighing some 300 pounds (135 kg), the rare Komodo dragon is the biggest lizard on Earth. The Komodo tastes the air with its foot-long forked yellow tongue to locate prey. Komodos live only on a few small Indonesian islands. ***What factors might threaten the Komodo dragon's survival?*** TEKS

Natural Resources Tropical rain forests are among island Southeast Asia's many valuable natural resources. The region also has rich fisheries and volcanic soils that are good for farming. Rubber tree plantations are important in some countries, particularly Malaysia and Indonesia. The region produces many minerals, including copper, gold, and tin. Some countries, particularly Brunei and Indonesia, produce oil and natural gas. The tiny country of Brunei sits on top of a major oil-producing field.

Scientists believe that large oil and natural gas deposits lie near the Spratly Islands, southwest of the Philippines. China, Malaysia, the Philippines, Taiwan, and Vietnam all claim these uninhabited islands because they have valuable natural resources.

TEKS **READING CHECK:** *Physical Systems* What climates and biomes are distributed throughout the region?

Section 1 Review

TEKS Working with Sketch Maps, Questions 1, 2, 3, 4, 5

Define
archipelago, lahars, endemic species

Working with Sketch Maps
On a map of island Southeast Asia that you draw or that your teacher provides, label the Malay Peninsula, Malay Archipelago, Sumatra, Irian Jaya, New Guinea, Borneo, Java Sea, South China Sea, Timor Sea, Strait of Malacca, Java, and the Spratly Islands. Which island has about 13 active volcanoes?

Reading for the Main Idea
1. *Physical Systems* What area of island Southeast Asia has the least tectonic activity? Why?
2. *Places and Regions* How does the region's location affect its climate? What are two other factors that influence climates there?
3. *Physical Systems* What do the Komodo dragon, Javan rhinoceros, and orangutan have in common?

Critical Thinking
4. **Analyzing Information** Why might people consider the Strait of Malacca another important natural resource?

Organizing What You Know
5. Create a chart like the one shown below. Use it to identify the region's major natural hazards and their effects on the environment.

Natural hazards	Effects

Physical Systems

Geography for Life

Wallace's Line

Indonesia's great biodiversity results in part from its location. The country lies in a transition zone between ecosystems. Asian animal and plant species live in its western islands. Species related to those of Australia and New Guinea live in the eastern islands.

English naturalist and biogeographer Alfred Russel Wallace (1823–1913) studied the reasons for this distribution of species. Wallace spent eight years in the region. He noted the difference between the birds on the island of Bali and those just 20 miles east on Lombok. The birds on Bali appeared to be related to those on Java, mainland Malaysia, and Sumatra. The Lombok birds resembled those of New Guinea and Australia. Based on what he learned, Wallace drew a line across a map of the region. This line separated areas with Asian plants and animals from areas with Australian species. It eventually became known as Wallace's line. (See the map.)

What created this curious distribution of ecosystems? Recall what you learned about the theory of continental drift. Long ago, all of Earth's land surface was part of one huge landmass called Pangaea. Then Pangaea broke into what are today's continents and islands. As the pieces of Pangaea drifted apart, Australia, New Guinea, and nearby islands became isolated from other continents. Even during Earth's ice ages, lower sea levels failed to relink these regions. Thus, plant and animal life that evolved in Australia is very different from that in much of island Southeast Asia.

Many biogeographers today do not see the boundary between the two ecosystems as a simple line. They think of the island region between Java and New Guinea as an ecotone—a transition zone between adjoining ecosystems. Geographers refer to this transition zone as Wallacea, in honor of Alfred Wallace.

Applying What You Know

TEKS

1. **Summarizing** What past physical processes account for Indonesia's biodiversity today?
2. **Analyzing Information** Where do you think you might find other regions with high levels of biodiversity? Why?

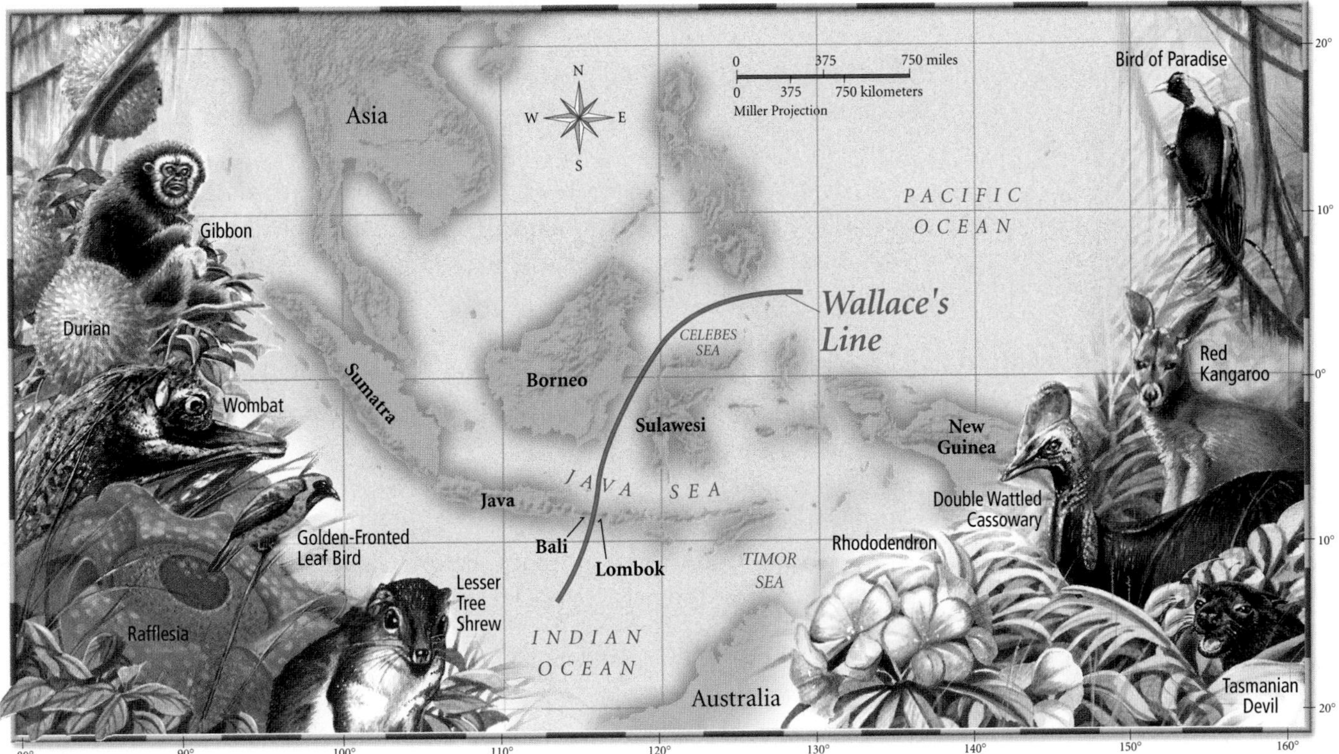

Section 2 History and Culture

READ TO DISCOVER

1. What early peoples migrated to island Southeast Asia?
2. How did colonialism affect the region's history?
3. What are the people and culture of island Southeast Asia like today?

DEFINE

homogeneous
slash-and-burn agriculture

LOCATE

Borobudur
Bali
Manila
Jakarta

WHY IT MATTERS

The Philippines was once the largest overseas possession of the United States. Use CNNfyi.com or other **current events** sources to learn about the relationship between the two countries today.

Early History

Island Southeast Asia is home to descendants of migrants from Asia and other places. Throughout the region's history, seafaring traders and various countries have tried to conquer the area. They wanted to control the region because it had rich resources and a useful location.

Human remains in the Philippines date back more than 30,000 years. The first Malay people from Asia probably migrated into the region about 2000 B.C. They mixed with the peoples who had long lived in the region. Over time, many other peoples came to island Southeast Asia.

Hindus from what is now India influenced the area early in its history. By about A.D. 700, Hindu and Buddhist kingdoms were well established in Java and Sumatra. These kingdoms controlled trade and built huge monuments, including Borobudur in central Java. This temple was completed about A.D. 850. Some of these monuments have been restored.

Majapahit (mah-jah-PAH-hit), the largest early kingdom, existed from the late 1200s to about 1500. Majapahit was centered on the islands of Java and Bali, but controlled many coastal areas. Trade and cities grew throughout the kingdom and the area.

Chinese merchants also sailed their ships long distances to trade among the islands. Some of the merchants then began to settle in coastal cities. Today ethnic Chinese are an important minority in the populations of many island Southeast Asian countries.

By the 1300s Arabs from Southwest Asia were also trading in the region. The Arabs introduced Islam to coastal peoples there, and the religion gradually spread. Areas in northern Sumatra and on the Malay Peninsula became early centers of Islam. Today Islam is island Southeast Asia's main religion.

READING CHECK: *Human Systems* What people began migrating to the region about 4,000 years ago?

INTERPRETING THE VISUAL RECORD

Indian influences on Indonesia appear in the movements of this Balinese dancer, who probably began training when she was four or five years old. ***Which religion came to Bali from India?*** 

Colonial Era and Independence

The Philippines were named for Philip II of Spain, pictured above. The first permanent Spanish settlement in the Philippines was established in 1565. The islands remained part of the Spanish Empire for more than 300 years.

As in many other parts of the world, island Southeast Asia came under the control of European colonial powers. (See Connecting to History: A Colonial History.) The Portuguese, who came in the 1500s, were the first Europeans to arrive. They were searching for spices such as cloves, nutmeg, and pepper and therefore called the area the Spice Islands. In the 1600s and 1700s the Dutch drove out the Portuguese. Portugal lost control of all its lands in the region except the island of Timor.

European Influence The explorer Ferdinand Magellan reached the Philippines in 1521 and claimed the islands for Spain. The Spaniards who followed wanted to Christianize and colonize the islands. Roman Catholicism, the religion brought by the Spaniards, is the main faith of the Philippines today. In addition, Manila became a major port for trade with China and the Spanish colonies in the Americas. In 1898, after the Spanish-American War, the United States took over the Philippines. These islands were the first large overseas U.S. territory.

The Dutch were much less interested than the Spaniards in converting the region's peoples to Christianity. In contrast, the Dutch went to the Spice Islands for commerce. They controlled the spice and tea trade of what became known as the Dutch East Indies. Today these islands make up Indonesia. The Dutch ruled from Batavia—now called Jakarta—their main port on Java.

The British set up colonies in Malaya on the Malay Peninsula and along the northern coast of Borneo. In 1819 they founded Singapore, which became a

INTERPRETING THE VISUAL RECORD

This design recalls the Portuguese arrival in the islands of Southeast Asia. The detail is from a piece of batik, a fabric from Indonesia. Batik is made by applying wax to fabric and then dyeing it with various colors. Areas covered by wax resist the dye. ***What elements of European culture are visible in the design?***

Connecting to HISTORY

A Colonial History

By the early 1900s foreign powers ruled all of island Southeast Asia. The United Kingdom, the Netherlands, Portugal, and the United States each controlled parts of the region. France ruled parts of nearby mainland Southeast Asia.

In some of the colonies, rebels fought for independence. In the Philippines, for example, rebels first fought against Spanish troops. When Filipino rebels declared independence in 1898, they fought against U.S. troops. The fighting was bitter and bloody. In 1946 the Philippines gained independence from the United States. Still, some Filipinos wanted fewer links to the United States. As a result, the United States closed its last military base in the Philippines in 1992. The Philippines remains a U.S. ally.

Making Generalizations What are some European and American cultural influences that you might expect to find in the region today? TEKS

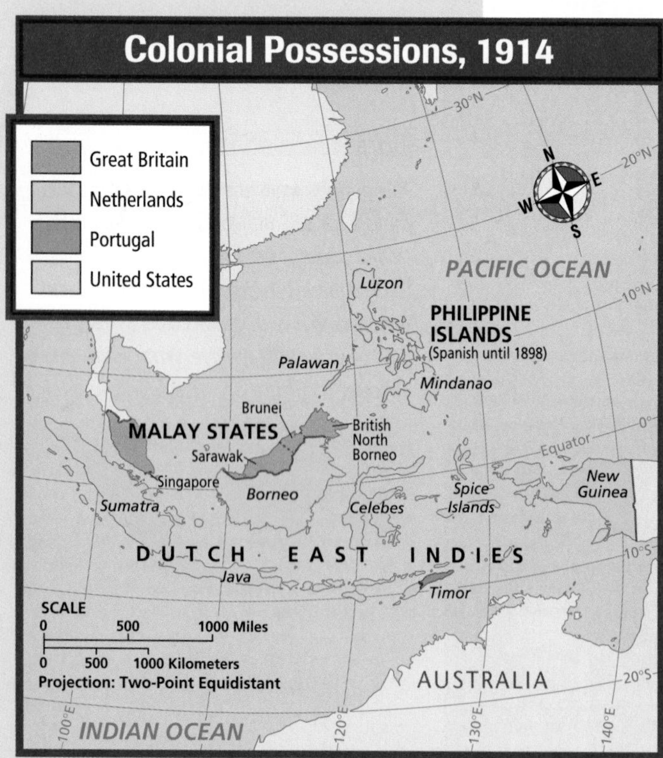

major port for British and Chinese trade. The British used many local workers to build roads, plantations, and schools. Workers from China and India also came to the plantations. Today Chinese and South Asians make up large ethnic groups in the region.

War and Independence In 1941 and 1942 Japan invaded European and U.S. territories in island Southeast Asia. During World War II the Japanese wanted the region's natural resources, particularly oil. They occupied much of the region until they surrendered at the end of World War II in 1945.

Soon after the war, the Philippines gained independence. The colonial system began to crumble throughout the area. The Dutch tried to reestablish their rule after the war, but the Indonesians resisted. As a result, the Dutch gave up the colony in 1949. Malaya won independence in 1957. Then in 1963 Malaya joined former British territories in Singapore and northern Borneo to form the Federation of Malaysia. Singapore later broke away from Malaysia in 1965. Brunei, a British colony, gained independence in 1984. This tiny country is ruled by a sultan. A sultan is the ruler of a Muslim country.

READING CHECK: *Human Systems* What European countries colonized the region?

Culture

How do you think island Southeast Asia's history has influenced its peoples and cultures? In what ways might you see these influences today?

People and Languages A history of migration and colonization has created a diverse population in island Southeast Asia. People from many different ethnic groups live in the region. In Indonesia, for example, no one ethnic group makes up a clear majority of the population. Nearly half of Indonesians are Javanese. Malays and others make up large minority groups.

Malaysia's population is somewhat less diverse. Nearly 60 percent of Malaysians are ethnic Malay. Still, this country also has large minority groups, like the descendants of Chinese migrants. In fact, ethnic Chinese and South Asians dominate much of Malaysia's economy.

Chinese live throughout the region, particularly in large cities. In Singapore, Chinese make up a majority—more than 75 percent of the population. Tensions between the city's mostly Chinese population and the Malays in Malaysia led Singapore to seek independence in 1965.

Each country in the region has one or more official languages. For example, Singapore has four official languages, including English and Malay. Chinese dialects are spoken in many large cities. In addition, indigenous peoples throughout the region speak local languages.

The Philippines is the region's most **homogeneous** (hoh-muh-JEE-nee-uhs) country. The word *homogeneous* means "of the same kind." About 95 percent of the country's people are ethnic Malays. Pilipino, which is based on a native language called Tagalog, is one of the Philippines' official languages. English is also an official language in this former U.S. territory.

Settlement and Land Use Island Southeast Asia's population is not evenly distributed. For example, Java has more than half of Indonesia's population of about 228 million. The Javanese live on an island smaller than New York State, which has fewer than 20 million people. The Indonesian government encourages citizens to move to less-populated islands. Between 1969 and 1994, some 8.5 million Indonesians were relocated. This policy has not been popular with the residents of those islands, however.

The country of Singapore, which occupies a small island, is almost completely urban. About 70 percent of tiny Brunei's people also live in cities. The larger countries are more rural. Many people are farmers. About a third of Indonesians and half of Malaysians and Filipinos live in cities. Still, many people are moving from rural areas to cities in search of work. Two of the most populous cities are Jakarta and Manila.

Religion As you have read, Buddhism, Hinduism, and Islam have long been practiced in the region. Indonesia is the world's most populous Islamic country. Nearly 90 percent of its people are Muslims. Hinduism is practiced in some areas, such as on the Indonesian island of Bali. Buddhism is most common in Malaysia and Singapore, where many Chinese live. Europeans brought Christianity, and today Christians live throughout the region. In the Philippines Christians make up more than 90 percent of the population.

INTERPRETING THE VISUAL RECORD

Traditional theater performances help preserve the Hindu-Buddhist heritage of Java. For example, wayang *puppets relate stories from Hindu epics.* Wayang *means "shadow." Elaborately painted leather puppets are placed behind a rice paper screen, with a candle or oil lamp as a light source behind them. The audience watches the shadows projected by the puppets. A* wayang *performance may last an entire night.* ***How might such a performance help unify a community?*** TEKS

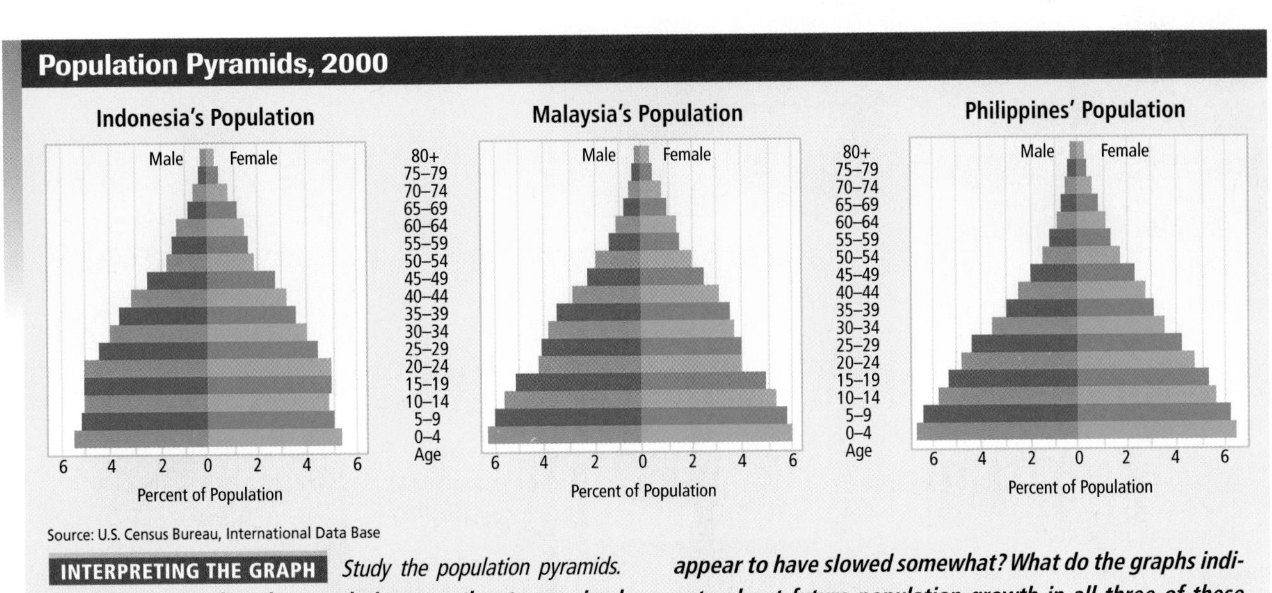

Source: U.S. Census Bureau, International Data Base

INTERPRETING THE GRAPH *Study the population pyramids.* ***In which country has the population growth rate remained the most constant? In which country does the growth rate appear to have slowed somewhat? What do the graphs indicate about future population growth in all three of these countries?*** TEKS

Food Farmers grow many kinds of foods in the region. However, rice is the main food crop, or staple, for most of the people. Rice is served with many other foods and spices, such as curries and chili peppers.

FOCUS ON GEOGRAPHY

Growing Rice Farmers in the region grow rice in three ways. Wet-rice, or paddy, cultivation is the most productive and common method. Rice paddies are constructed with dikes in lowland areas or with mud terraces in hilly areas. Water flow down steep slopes is controlled, and erosion is limited. This, along with the area's warm and wet climate, allows farmers to grow more than one rice crop each year.

Paddy cultivation is a form of intensive agriculture. Many workers are needed to harvest a particular crop. This type of rice farming supports large concentrated populations in island and mainland Southeast Asia. In addition, farmers can raise ducks, fish, and shrimp in the paddies. These food sources add protein to the diet. Paddy workers may use the water buffalo, a domesticated animal, for plowing and heavy farm duties.

In the tropical wet-and-dry climate areas, farmers use dry-rice cultivation. Farmers plow fields and plant rice seeds. Rivers may flood the area during the wet season. People practice a third type of rice cultivation in forested areas. This method is known as **slash-and-burn agriculture**, a form of shifting cultivation. Farmers clear or slash small areas of forest and burn the fallen trees. After a few years the soil's nutrients have been used up, and farmers move on to a new area. Dry farming and slash-and-burn agriculture are also used throughout the region to grow many other crops.

READING CHECK: *Environment and Society* How do farmers in the region grow rice?

INTERPRETING THE VISUAL RECORD

Rice paddies like these in Indonesia are common throughout island Southeast Asia. ***How have farmers limited erosion in the rice paddies pictured below?*** TEKS

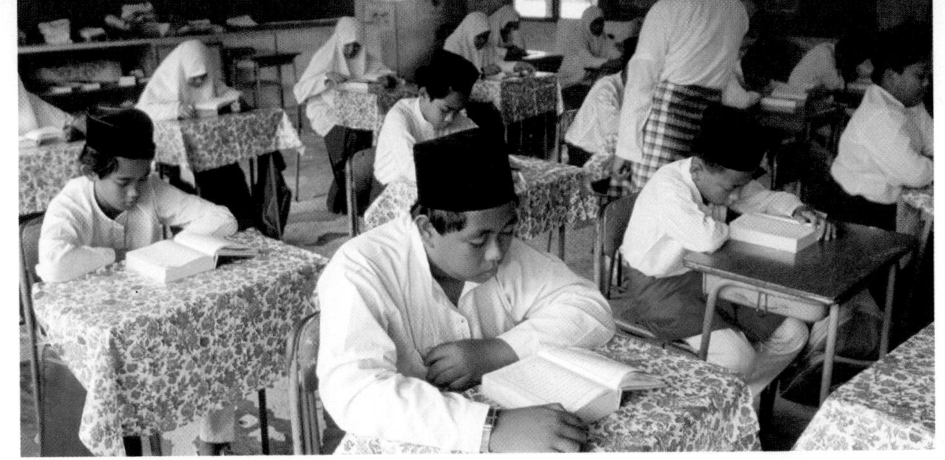

Bruneian students read in a religion class. The girls wear a traditional head covering called a tudong. *The boys wear the* songkok. *Men who have gone on a pilgrimage to Mecca (Makkah) wear a white* songkok.

Other Traditions and Education Many traditional clothing styles are still worn partly because they are ideal for the region's hot humid climates. For example, for business and other important occasions, Filipino men wear the *barong tagalog.* This light shirt is made from cotton or from fibers of the banana or pineapple plant. Malaysian men and women often wear sarongs. They wrap these long strips of cloth around their bodies.

People in the region use special methods to make some traditional clothing. For example, Indonesians create colorful fabrics called batiks (buh-TEEKS). Coating areas of the cloth with wax creates patterns on a batik. Uncoated areas can then be dyed with bright colors.

Education is a key to the region's future prosperity. To aid economic development, governments have tried to improve educational opportunities for all people. Schools have also been used to create a sense of national identity in the region's multiethnic countries. For example, in the Philippines the Pilipino language is used more and more in schools. Indonesian schools emphasize what are called the Pancasila, or "Five Principles." Indonesia's early leaders believed their new country should be based on those principles. One of the principles is the importance of national unity among the country's many ethnic groups. The others are belief in one God, a just and civilized humanity, democracy, and social justice.

READING CHECK: *Environment and Society* How does clothing help people adapt to the region's hot humid climate?

TEKS Questions 2, 3, 4, 5

Review

Define

homogeneous, slash-and-burn agriculture

Working with Sketch Maps

On the map you created in Section 1, label the countries of island Southeast Asia and Borobudur, Bali, Manila, and Jakarta. What was Jakarta called during the colonial era?

Reading for the Main Idea

1. *Human Systems* From where and when did Malays migrate to the region? What drew Europeans to the region during the colonial era?
2. *Environment and Society* How do farmers modify the region's environment to grow rice?

Critical Thinking

3. **Comparing and Contrasting** How are the people and culture of the Philippines unique in the region?
4. **Drawing Inferences and Conclusions** Why do you think the Indonesian government wants people to move to less-populated islands?

Organizing What You Know

5. Create a word web like the one shown below. Use the word web to describe the peoples and cultures of the region.

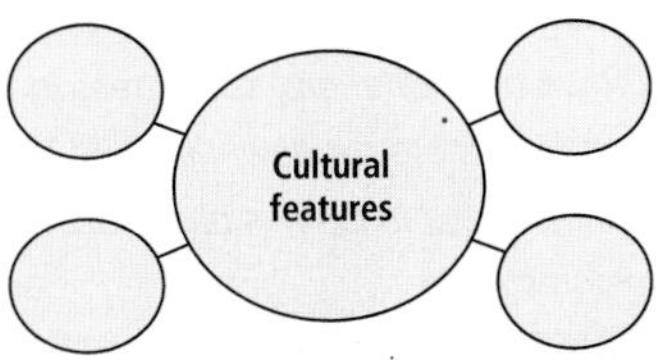

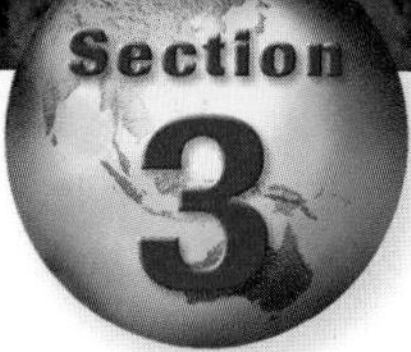

The Region Today

READ TO DISCOVER

1. How has island Southeast Asia's economy changed in recent decades?
2. What are the cities of the region like?
3. What challenges do the people of the region face?

DEFINE

kampongs

LOCATE

Kuala Lumpur
East Timor
Aceh
Moluccas
Mindanao
Sulu Archipelago

WHY IT MATTERS

Sometimes ethnic and religious conflicts have caused problems in island Southeast Asia. Use CNNfyi.com or other **current events** sources to identify international efforts to ease such conflicts around the world.

Economic Development

The countries of island Southeast Asia have mostly free-market economies. Governments have encouraged private industry and business. In addition, businesses in other countries have invested in the region.

Cooperation among the region's countries has helped economic development. All of the countries are members of the Association of Southeast Asian Nations (ASEAN). The region's leaders are also working to improve economic and trade ties to countries outside the region.

A swordfish adorns this Singaporean coin. Some coins display flowers to enhance Singapore's garden city image.

These boats bring teak and other types of lumber to Jakarta for distribution and export. Teak has long been valued for its beauty and durability. Although teak is grown on plantations, some trees in the rain forests are cut down illegally. This unauthorized logging is one cause of deforestation.

Economic Growth Rapid industrialization during the 1980s and 1990s brought great economic growth. Some countries became known as the Tigers of the Pacific Rim because of their spectacular growth. These newly industrialized countries included Singapore and Malaysia. Singapore was one of the earliest success stories. The city is located on the Strait of Malacca, a major shipping route. This location helped the city and its economy grow. Today Singapore is a major trade and industrial center. Many financial and high-technology companies have also opened offices in the country. This development has helped make Singapore's per capita GDP among the highest in the world.

Brunei also has a high GDP because of its oil and natural gas reserves. In fact, these reserves and the country's refineries account for more than half of the country's GDP. Brunei's government has used some of the oil income to benefit the country's citizens. For example, medical care is free. The government also helps pay food and housing costs.

A growing economy has also improved life for many Malaysians. Malaysia has tried to maintain its economic strength by selling many government-owned industries to private citizens. However, the Philippines and Indonesia have not done as well, because of their huge populations and recent political problems. Still, overall the region's future could be bright. Rich natural resources and a large and skilled labor force could fuel continued economic growth.

Island Southeast Asia	Major Exports
Brunei	crude oil
Indonesia	textiles/garments, wood products, electronics, footwear
Malaysia	electronic equipment, palm oil
Philippines	electronics and telecommunications, machinery and transport
Singapore	computer equipment, rubber

Source: *National Geographic Atlas of the World, Seventh Edition*

INTERPRETING THE CHART *Most countries of island Southeast Asia export electronics and products relating to technology.* ***What does this chart indicate about the level of economic development in these countries?*** TEKS

Agriculture Even with recent industrialization, agriculture has remained vital to the region. Wet-rice cultivation is the most common form of agriculture. The region's countries also produce and export coffee, fruit, spices, sugarcane, and tea. Rubber trees, which came to the region from South America, are also valuable. Malaysia and Indonesia, along with Thailand, are now the world's largest producers of natural rubber.

Fisheries provide this island region with seafood, the major source of protein. Traditional fishers have sailed nearby waters for thousands of years. Today, however, their small boats must compete with large commercial ships. Overfishing now poses a threat to local fisheries.

Tourism Tourism is also a big industry. For example, the Indonesian island of Bali is popular with tourists from around the world. The island's mixed Hindu-Buddhist culture and beautiful rice-paddy scenery attract thousands of visitors each year. Bali's skilled artists, dancers, and weavers help increase the island's popularity.

TEKS **READING CHECK:** ***Human Systems*** Which countries have experienced rapid industrialization?

INTERPRETING THE VISUAL RECORD

Lake Bratan and the Ulun Danu Temple on the Indonesian island of Bali typify the region's beautiful landscapes and architectural features. The temple, which includes Buddhist and Hindu elements, was built in 1633 and is dedicated to a Hindu water goddess. ***How does the temple's architecture compare to styles represented in other regions?*** TEKS

Kuala Lumpur gets its name from a Malay phrase meaning "muddy estuary." The city lies near the mouth of a river on the Malay Peninsula.

Urban Environments

The region's economic development has fueled the rapid growth of its cities. As you have read, many people are moving to cities to search for work. The largest cities are the capitals of the major countries. These include Jakarta in Indonesia, Kuala Lumpur in Malaysia, Manila in the Philippines, and Singapore.

All four cities have modern urban centers and government buildings. However, major differences exist among them. Singapore is orderly, wealthy, and very clean. Crime rates are low there. On the other hand, Kuala Lumpur, Manila, and Jakarta have smog and traffic problems. Manila and Jakarta have large slums.

Kampongs, which are villages built on stilts, make up the traditional Malay housing style. Today the term also refers to the crowded slums around Jakarta and other large cities.

Why is Singapore so different from its neighbors? The country's government has worked hard to clean up slums and provide better housing. (See Cities & Settlements: Singapore.) In addition, Singapore has strict laws against even minor offenses, such as littering. The government even outlaws chewing gum and certain kinds of music and movies. Some people believe that such strict limits on personal freedoms are a good trade-off for less crime, a clean city, and a strong economy. However, others believe Singapore could be just as successful with fewer rules.

READING CHECK: ***Human Systems*** How has economic growth affected the region's cities?

Kampong architecture reflects the region's physical environment. Stilts keep the houses safe from high tides or floods. The roofs are made of tiles woven from native plants. The Toraja people built these traditional houses in Sulawesi, one of Indonesia's larger islands.

Issues and Challenges

The region's countries face a number of issues. As you have read, the rapid growth of cities presents a major challenge. Other tasks include overcoming political problems and ethnic and religious tensions.

Political Challenges Protecting political and personal freedoms represents one of the region's challenges. The Philippines and Indonesia, for example, have held truly free elections only in recent years.

Indonesia faces another issue—the push by ethnic and religious groups for independence. In 1999 East Timor voted for independence. Indonesia had invaded this small Portuguese colony in 1975. Following the 1999 election, violence broke out between independence supporters and their opponents. The United Nations sent in troops to stop the fighting. East Timor's new government took power in 2002.

Other hot spots simmer. People in Aceh, on the northern tip of Sumatra, and Irian Jaya (also called West Papua) also want independence. At times, violence breaks out between independence supporters and government forces. Christian and Muslim residents of the Moluccas (moh-LUH-kuhs), another area of Indonesia, have also fought.

Religious differences have also led to conflict in the Philippines. The southern part of this mostly Roman Catholic country is home to many Muslims, some of whom want independence. Also in this area is a group of Muslim terrorists said to be linked to international terrorism. The United States has aided the Philippines by training Filipino troops to fight terrorism.

Other Challenges Environmental problems like deforestation, loss of wildlife diversity, overfishing, and air and water pollution present difficult challenges. In addition, with the exception of Singapore, many people in the region are still poor. A few business families, military officers, and politicians control most of the power and money. Some corrupt officials have managed to stay in power for decades. Poor workers from Indonesia have moved to richer Malaysia and Singapore to look for jobs. Raising the standard of living for all the region's people will be difficult.

TEKS **READING CHECK:** *Human Systems* What challenges face the region today?

Soaring 1,483 feet (452 m) into the air, the Petronas Towers in Kuala Lumpur ranked as the tallest buildings in the world when they were completed in 1998. This modern skyscraper of glass, steel, and concrete is a visual symbol of Malaysia's commitment to being modern.

Section 3 Review

TEKS Working with Sketch Maps, Questions 1, 2, 3, 4, 5

Define
kampongs

Working with Sketch Maps On the map you created in Section 2, label Kuala Lumpur, East Timor, Aceh, Moluccas, Mindanao, and Sulu Archipelago. Which island region declared independence in 1999?

Reading for the Main Idea

1. *Human Systems* Why are some of the countries in island Southeast Asia known as the Tigers of the Pacific Rim?
2. *Environment and Society* What factor helped make Singapore a major trade and industrial center?

Critical Thinking

3. **Comparing and Contrasting** What are some differences between the city of Singapore and other big cities in the region? What are some reasons for these differences?
4. **Making Generalizations** Why do you suppose Indonesia's various ethnic groups each view their country differently?

Organizing What You Know

5. Copy the table below. Use it to identify economic, environmental, and political challenges the region's countries face.

Challenges		
Economic	Environmental	Political

CITIES & SETTLEMENTS

Singapore

Human Systems Singapore has grown from a small port in the 1800s into a densely populated high-tech city. Efficient use of land, particularly for housing, has been crucial in planning the city's growth. Through housing policies, Singapore's government has tried to improve living conditions, encourage economic growth, and use land efficiently. It has also used housing policies to promote good relations among different ethnic groups.

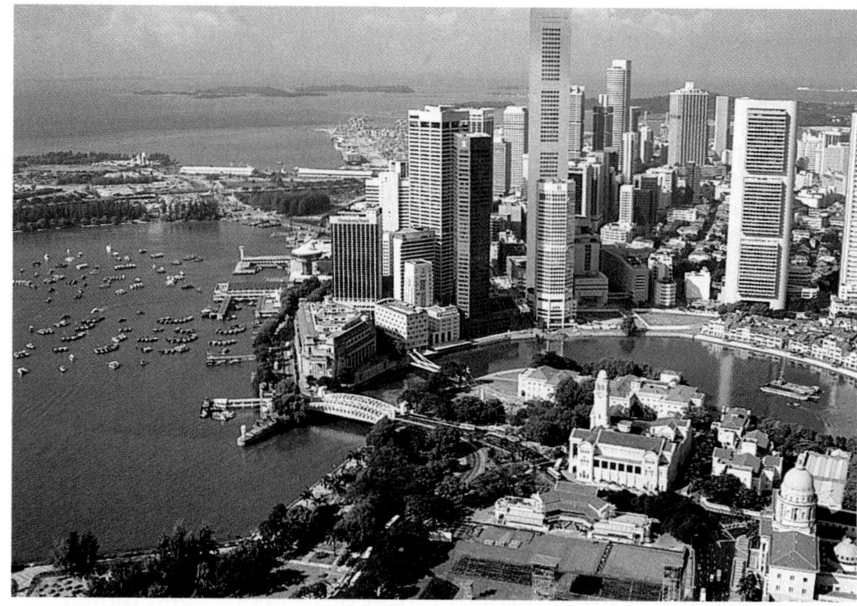

Singapore, also known as the Garden City because of its parks and tree-lined streets, began its modern history in the early 1800s. A representative of the British East India Company established the area as a trading site and gained possession of the harbor for the United Kingdom. Singapore's location on major shipping lanes has helped it become one of the world's busiest ports.

Development and Growth

Changes have been particularly dramatic in recent decades. In the early 1950s about 75 percent of Singaporeans lived in crowded slums on the island's southern shore. The other 25 percent lived in the more rural northern part of the island in traditional kampongs. In the 1960s providing low-cost housing became a major goal of Singapore's government. The newly created Housing and Development Board began to build high-rise high-density apartments for public housing. The government gave the board the power to clear slums and forcibly resettle residents.

The program was very successful. Thousands of people moved into the new apartments. Government loan programs helped even low-income families buy their apartments. By 1988, 2.3 million people—88 percent of Singapore's population—lived in apartments provided by the board. Most tenants owned their apartments. The government saw apartment ownership as giving people a "stake in Singapore" and encouraging political stability. At the same time, housing construction became a major industry, helping to fuel economic growth.

The government also built new roads and highways, along with a mass-transit rail system. Housing complexes rose where highways and the mass-transit railroad met. These new towns of up to 200,000 residents were meant to redirect urban growth away from the old city center. The government hoped these new towns would provide employment for most of their residents. In reality, many residents commuted to jobs either in the central business district or in Jurong, a huge industrial area west of the city center.

Housing and Ethnicity

Singapore's government has also tried to shape the role of ethnicity in society with its housing policies. Singapore has three major ethnic groups—Chinese, Indian, and Malay. In most cases, people of a single ethnic group had occupied Singapore's old slums and kampongs. The government used the resettlement program to break up the pattern of single-ethnic-group

Singapore

INTERPRETING THE MAP *Singapore's Mass Rapid Transit system is used daily by 62 percent of its population. The system eases transportation to and from work and home for the island's residents. In addition, the transit authority may add shopping centers to mass transportation centers. By building malls in the large stations, the transit authority could meet more of the commuters' needs.* ***Examine the map of Singapore's roads and mass transit systems. Where do you think new housing will be located?*** TEKS

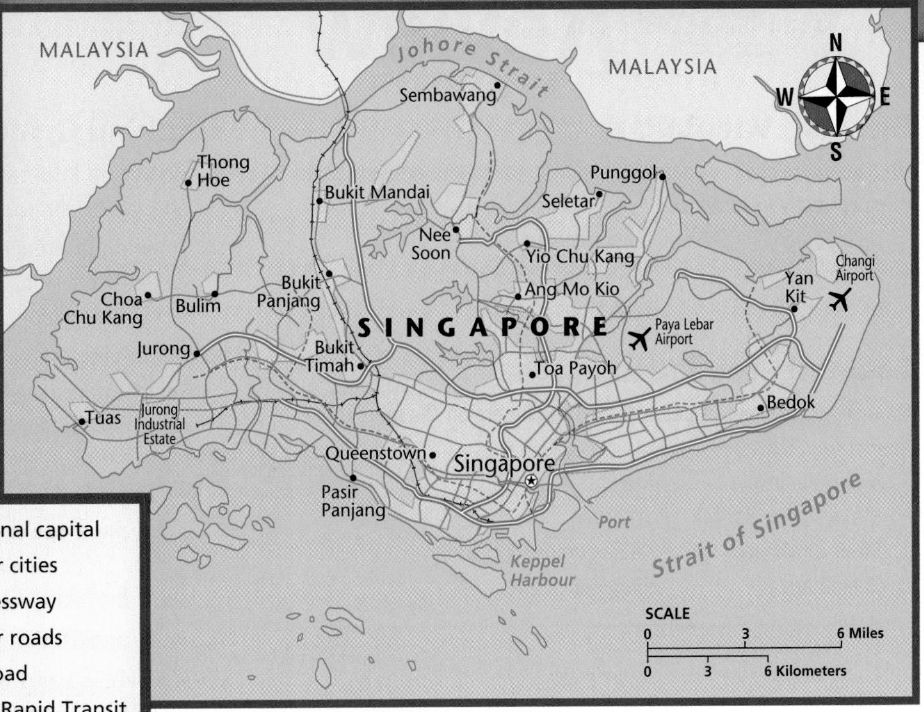

National capital
Other cities
Expressway
Other roads
Railroad
Mass Rapid Transit
Built-up area

neighborhoods. Singapore's leaders hoped that integrating the new apartment complexes would encourage tolerance of ethnic differences.

To accomplish this, a mix of Chinese, Indians, and Malays were assigned to every apartment block. The government also sponsored residents' committees and education and recreation programs for each apartment complex. These were designed to create a sense of community among the apartment residents.

Singapore's Ethnic Groups

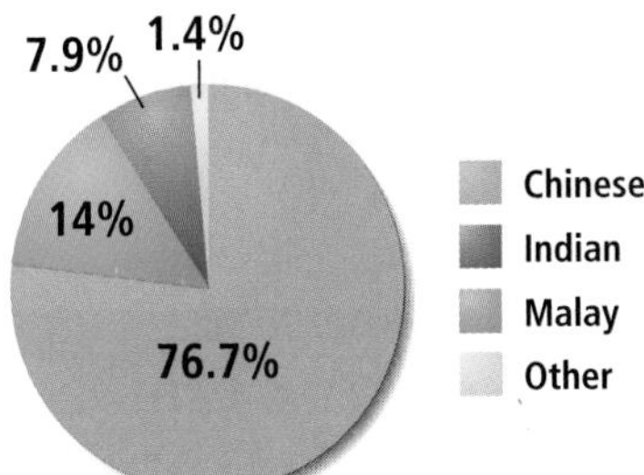

However, despite the hopes of the Housing and Development Board, few tenants thought of their apartment block as a real community. Most people maintained their ties to relatives and friends of the same ethnic group. Residents began selling their apartments to members of other ethnic groups so they could be near friends and family. The result was a trend toward apartment blocks of one ethnic group. In 1989 the government tried to halt this trend. It passed new rules to prevent people from selling their apartments or turning them over to members of other ethnic groups.

A Thriving City-State

Singapore's government has used housing policy to do more than simply alter settlement patterns. It has also tried to change attitudes. Not all of these goals have been met. However, the country has prospered in spite of its crowding and ethnic issues.

Applying What You Know

1. **Summarizing** What did Singapore's government hope its housing policies would accomplish?
2. **Analyzing Information** In what ways were Singapore's efforts to create new local communities successful? In what ways were they unsuccessful? Why?

Building Vocabulary

On a separate sheet of paper, explain the following terms by using them correctly in sentences.

archipelago	homogeneous
lahars	slash-and-burn agriculture
endemic species	kampongs

Locating Key Places

On a separate sheet of paper, match the letters on the map with their correct labels.

Malay Peninsula	Manila
Malay Archipelago	Jakarta
Strait of Malacca	Kuala Lumpur
Philippines	Singapore

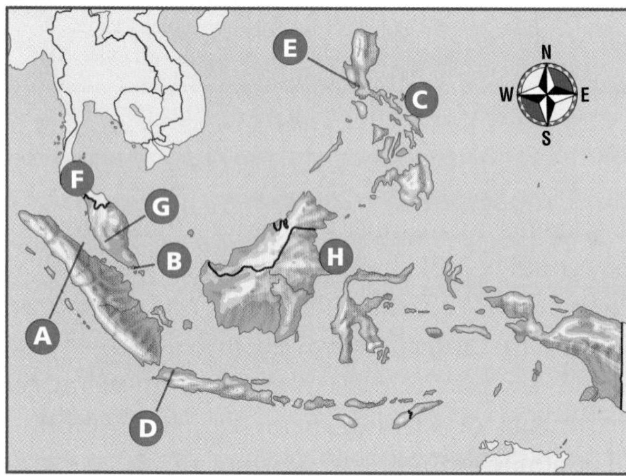

Understanding the Main Ideas

Section 1

1. *Physical Systems* Why is the land of island Southeast Asia so unstable in some areas?
2. *Places and Regions* How do the region's location and monsoons influence its climates?

Section 2

3. *Human Systems* Which ethnic group makes up the majority of Singapore's population? Do members of that ethnic group tend to live in urban or rural areas throughout island Southeast Asia?
4. *Places and Regions* What is the dominant religion of island Southeast Asia? What other religions are practiced there?

Section 3

5. *Places and Regions* What economic, environmental, and political challenges face the region today?

Thinking Critically for TAKS

1. **Analyzing Information** How has the region's cultural diversity affected the structure and goals of educational systems there?
2. **Drawing Inferences and Conclusions** Why do you think wet-rice cultivation is the most common method of growing rice in the region?
3. **Identifying Cause and Effect** How do you think island Southeast Asia might be different today if Europeans had never colonized the region?

Using the Geographer's Tools

1. **Analyzing Population Pyramids** Study the population pyramids in Section 2. What clues might these pyramids give us about the relative standard of living in the countries?
2. **Analyzing Statistics** Review the Fast Facts and Comparing Standard of Living tables at the front of this unit. Then go to **go.hrw.com** to access the World Factbook. Use information you find from these sources to rank the countries of island Southeast Asia by standard of living. Write a paragraph explaining your ranking.
3. **Preparing Maps** Draw an outline map of island Southeast Asia and identify areas that are experiencing ethnic and religious conflict.

Writing for TAKS

Review the various ways farmers in island Southeast Asia modify their physical environment to grow rice. Compare those methods with the ways farmers in dry climate regions of the United States modify their environment to grow wheat. You may want to use the Internet or consult library resources for more information. Write a short report about your findings. When you are finished with your report, proofread it to make sure you have used standard grammar, spelling, sentence structure, and punctuation.

SKILL BUILDING

Creating a Benefits-Cost Balance Sheet

Environment and Society As you have read, the tropical rain forests of island Southeast Asia are being cleared rapidly. What might be some advantages and disadvantages of clearing the forests for farmland, timber, or other purposes? Identify and evaluate those benefits and costs in the form of a two-column balance sheet.

Building Social Studies Skills

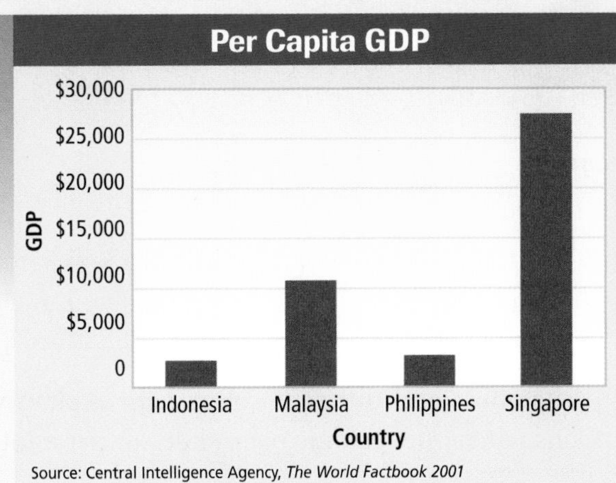

Interpreting Graphs

Study the bar graph above. Then use the information in the graph to help you answer the questions that follow. Mark your answers on a separate sheet of paper.

1. Singapore's per capita GDP is
   a. less than $10,000.
   b. more than five times the per capita GDP of Malaysia.
   c. more than $25,000.
   d. the lowest of the four countries.
2. Rank the countries from first to last by the value of each country's per capita GDP.

Analyzing Secondary Sources

Read the following passage and answer the questions. Mark your answers on a separate sheet of paper.

> **"Farmers in the region grow rice in three ways. Wet-rice, or paddy, cultivation is the most productive and common method. Rice paddies are constructed with dikes in lowland areas or with mud terraces in hilly areas. Water flow down steep slopes is controlled, and erosion is limited. This, along with the area's warm and wet climate, allows farmers to grow more than one rice crop each year."**

3. In the selection the word *terraces* refers to
   a. beautiful structures on the region's old colonial farmhouses.
   b. scenic spots along rivers in Malaysia.
   c. city gardens that have been transformed into rice farms.
   d. flat areas carved into hillsides so that rice can be grown there.
4. Why do you think being able to grow more than one rice crop a year is important in island Southeast Asia? Give specific reasons for your answer.

Alternative Assessment

PORTFOLIO ACTIVITY

Learning about Your Local Geography TEKS

Group Project: Field Work

Plan, organize, and complete a research project with a partner about the plant and animal life of your state. Identify important geographic questions about the species you identify in the area. What environmental conditions, such as climate, are common there? What species have adapted to that environment? What effect has interaction with humans had on the local environment and the species that live there? Organize your questions and answers in a table and identify connections between your local environment and that of island Southeast Asia. Finally, prepare a short report comparing the natural environments of your state and island Southeast Asia.

internet connect

Internet Activity: go.hrw.com
KEYWORD: SW3 GT30 TEKS

Choose a topic about island Southeast Asia to:
- test your knowledge of island Southeast Asia with an interactive map.
- research the rain forests of island Southeast Asia and make a poster.
- analyze tectonic forces that cause volcanoes and earthquakes.

Skill Building for TAKS

Creating a Diagram

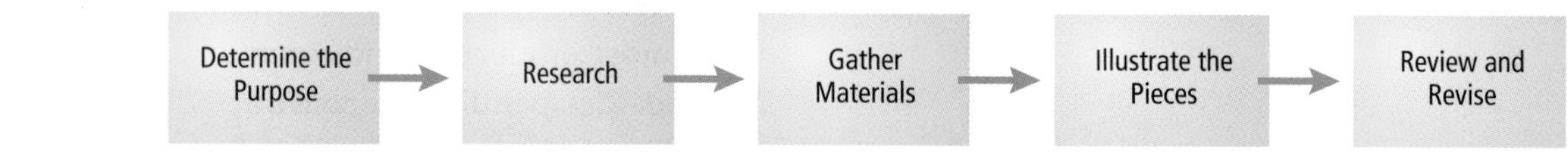

WORKSHOP 1

Using Diagrams

Some people are visual learners. They learn best when information is displayed in a visual format, such as a photograph, map, or diagram. Diagrams are illustrations that show us how to do something or how parts of a whole are related. You are probably familiar with many kinds of diagrams. For example, you may have used a diagram to put together a model airplane or ship. Car owner's manuals also have diagrams. They may show an owner how to accomplish important tasks, such as changing a tire. Public buildings often have diagrams that direct people to exits. Many kinds of diagrams are useful in geography. For example, the diagram on this page shows the kind of tectonic activity that makes parts of East Asia so prone to earthquakes.

Movement at Plate Boundaries

Labels, arrows, and different colors help explain what happens when one tectonic plate slides under another.

Developing the Skill Another useful diagram is a flowchart. A flowchart is a graphic representation of a series of related events or the steps in a process. Note the steps, or activities, in the flowchart at the top of this page. Arrows, or connectors, show the progress of activities. We can use this flowchart to help us plan steps for preparing other diagrams:

- Determine the purpose of your diagram. Are you showing a process, special features, a distribution, relationships, or something else?
- Conduct research to find the information you need. Research may also help you decide what kind of diagram best presents your information.
- Gather the materials you will need, such as paper, pencils, other art tools, or special software.
- Illustrate the pieces of the diagram, including labels that will help the reader follow along.
- Review and revise your diagram and labels for accuracy.

Practicing the Skill

1. Create a diagram that illustrates features in your school. For example, illustrate an evacuation plan for emergencies. You might also illustrate the suggested arrangement of a classroom or auditorium for a special event.
2. Research the process involved in producing rice. Then create a flowchart illustrating this process. Major steps will include planting, harvesting, and processing rice. You might want to create illustrations for each major step in the harvesting process.

Sample Population Data: Country A

E6 = 10.06

Workbook1

	A	B	C	D	E	F	G	H	I
1		1994	1995	1996	1997	1998	1999		
2									
3	Total Population (× 1,000)	1,198,500	1,211,210	1,223,890	1,236,260	1,248,100	1,259,090		
4	Birthrate (per 1,000)	17.70	17.12	16.98	16.57	16.03	15.23		
5	Death Rate (per 1,000)	6.49	6.57	6.56	6.51	6.50	6.46		
6	Natural Growth Rate (%)	11.21	10.55	10.42	10.06	9.53	8.77		
7	Total Number of Births (× 1,000)	21,040	20,630	20,670	20,380	19,910	19,090		
8									

Sheet1 / Sheet2 / Sheet3

WORKSHOP 2

Using Spreadsheets and Software

Sometimes you need to work with a large amount of numerical information in a simple and efficient manner. People use electronic spreadsheets, or worksheets, to do just that. You can use a spreadsheet for just text. However, a spreadsheet's real advantage is its ability to calculate values from preset formulas. It can even recalculate data automatically when entries change.

Developing the Skill All spreadsheets, regardless of the software, will follow the same design. Vertical columns are assigned letters in alphabetical order. Horizontal rows are arranged numerically. A cell is the intersection of one column and one row. Identifying a cell is similar to finding a location on a map using lines of latitude and longitude. The cell is named for its corresponding column and row. For example, the cell at the intersection of Column B and Row 2 is labeled B2. (See the spreadsheet above.)

To reach a specific cell you have several choices. You may use the mouse, arrow keys, or tab key. The computer will tell you which cell you are in by outlining the four sides of the cell. The cell will also be shown in a separate window at the top of the screen. In the example above you can see that cell E6 has a value of 10.06%. This is determined by studying the cell itself or by viewing the *status* or *formula line* found at the top of the worksheet.

Instead of calculating a value on a calculator, you can put a formula, such as C2 + C5, in a spreadsheet cell. Spreadsheets use standard formulas to create a simple mathematical equation. To create a formula, select the cell where you want to display the formula results. Type an equal sign (=) and then the rest of the formula. You may add, subtract, multiply, and divide. For example, the formula = *A1* + *A2* would add the value in cells A1 and A2. The formula = *C5* − *C4* would subtract the value in cell C4 from the value in cell C5. A slash (/) is used for division and an asterisk (*) for multiplication. Adding is the most common function of many spreadsheets. As a result, most spreadsheet software uses an *AutoSum* key (Σ) that you can place in a cell to add values automatically. For your convenience, many spreadsheet programs include a number of predefined formulas.

Another key feature of spreadsheets is the ability to calculate data from multiple worksheets. Suppose you are doing a report on precipitation around the world over a certain period of time. You could have separate worksheets for the continents, with precipitation amounts for individual countries on each continent. In addition, you could have a worksheet to show world precipitation changes over the same number of years. As you enter new country information, the numbers on the world worksheet change as well.

Practicing the Skill

Practice using a spreadsheet.

1. Open a new spreadsheet file.
2. Type the information as shown in the example on this page.
3. Estimate data you might expect to find for Country A in 2000 and place that data in Column H. Label this column "2000."
4. Label column I "Total." In cell I7 calculate the total births in Country A from 1994 to 2000. Use the formula = *B7* + *C7* + *D7* + *E7* + *F7* + *G7* + *H7* or use the *AutoSum* key (Σ).
5. Use a new spreadsheet to analyze data from a real country. For example, track annual growth in GDP, exports, or imports to analyze economic development in China in the 1990s.

The Pacific World

Green turtle, Great Barrier Reef

CONNECTING TO

Literature

A SECRET COUNTRY *by John Pilger*

John Pilger

(1939–) was born and educated in Sydney, Australia. Pilger is a journalist, filmmaker, and playwright. In this selection from his book *A Secret Country*, Pilger describes the beach in the Bondi neighborhood of Sydney, where he grew up. His reflections provide valuable insights into the human geography of his homeland.

By December, when the king tides have arrived from across the south Pacific, the salt spray blows up from the beach. It stiffens the air, covers windows with a stocky mist, corrodes paint on cars and mortar between bricks, and tastes like Bondi and summer. . . .

In Bondi, even the crankiest streets have a glimpse of the Pacific, if not of the beach itself. Whatever the state of life in the streets, the great sheet of dazzling blue-green is always there, framed between chimneys and dunnies [outhouses]. On weekdays my friend Pete and I would 'scale' a Bondi Beach tram if we spotted an old 'jumping jack' type, which had a peculiarly high suspension and could be bounced off the rails by a swarm of eleven-year-olds synchronising their efforts. And of course Len's [the tram conductor] apoplexy [great anger] was part of the fun. . . .

All principal beaches in Australia are public places. This is not so in the United States and Europe, where the private possession of land and sea is rightly regarded by visiting Australians as a seriously uncivilised practice. Although private property is revered by many Australians, there are no proprietorial [ownership] rights on an Australian beach. Instead, there is a shared assumption of tolerance for each other, and a spirit of equality which begins at the promenade steps. . . .

Australia, a society with a deeply racist past, has absorbed dozens of diverse cultures peacefully. The beach and the way of life it represents are central to this. A spectacle at Bondi in the 1950s . . . was the arrival on the beach of the first post-war immigrants. . . . Bolting lemming-like into a deceptively light surf, they would be duly rescued by lifesavers with a large trawling net. The ritual was repeated as each national group arrived. . . .

For most Australians, who live in congested coastal cities, the foreshore, the beach, is the one link with our ancient continent. . . . We see and understand little beyond the last of the urban red-tiled roofs, but many of us understand well the rhythm of water on sand, of wind on current. A Bondi child will know the feel of a westerly, a nor'easterly and a 'southerly buster.' There is a grace about this life.

Analyzing the Primary Source

1. **Comparing** How do beaches in Australia differ from beaches in the United States?
2. **Analyzing Information** According to Pilger, how have Australia's beaches shaped the country's society?

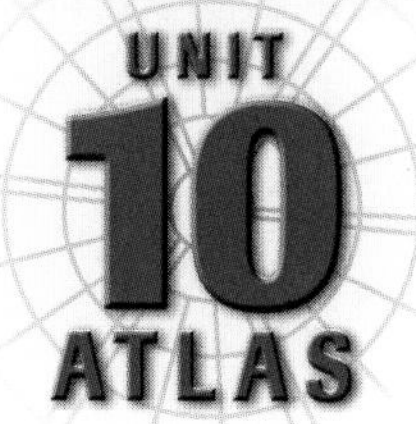

The World in Spatial Terms

The Pacific World: Political

1. *Places and Regions* Which of the region's capital cities is located farthest south? TEKS
2. *Human Systems* Which outside countries control islands in the Pacific region?

Critical Thinking

3. **Analyzing** Based on information on the map, which country would you expect to have the largest economy? Why?

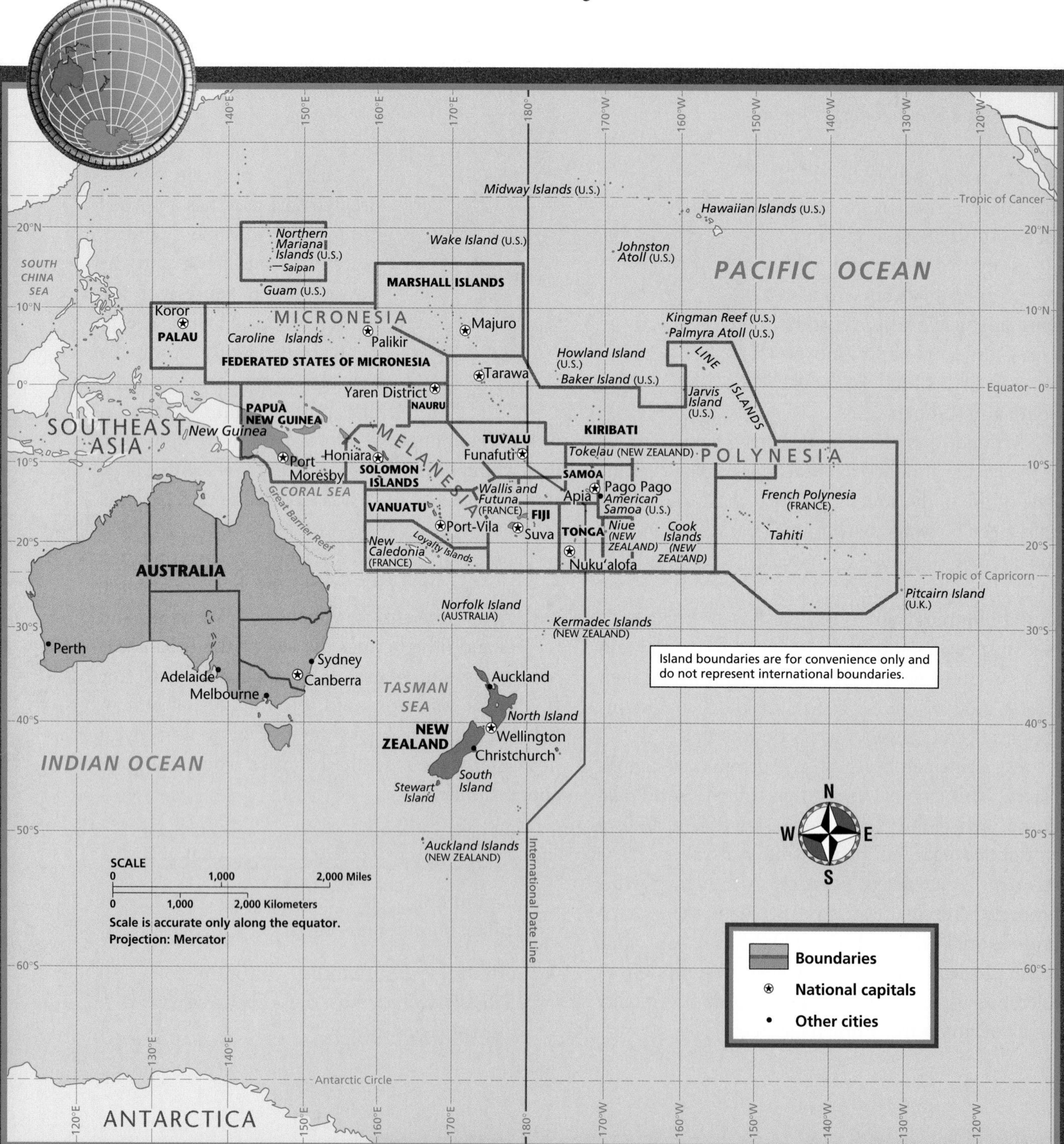

Midway Islands (U.S.)
Hawaiian Islands (U.S.)
Wake Island (U.S.)
Johnston Atoll (U.S.)
PACIFIC OCEAN
Northern Mariana Islands (U.S.)
Saipan
Guam (U.S.)
MARSHALL ISLANDS
Majuro
Koror
PALAU
MICRONESIA
Caroline Islands
Palikir
FEDERATED STATES OF MICRONESIA
Tarawa
Howland Island (U.S.)
Baker Island (U.S.)
Kingman Reef (U.S.)
Palmyra Atoll (U.S.)
LINE ISLANDS
Jarvis Island (U.S.)
Yaren District
NAURU
PAPUA NEW GUINEA
New Guinea
SOUTHEAST ASIA
SOUTH CHINA SEA
MELANESIA
KIRIBATI
TUVALU
Funafuti
Tokelau (NEW ZEALAND)
POLYNESIA
Port Moresby
Honiara
SOLOMON ISLANDS
CORAL SEA
Great Barrier Reef
SAMOA
Apia
Pago Pago
American Samoa (U.S.)
Wallis and Futuna (FRANCE)
FIJI
Suva
VANUATU
Port-Vila
TONGA
Nuku'alofa
Niue (NEW ZEALAND)
Cook Islands (NEW ZEALAND)
French Polynesia (FRANCE)
Tahiti
New Caledonia (FRANCE)
Loyalty Islands
Pitcairn Island (U.K.)
Tropic of Cancer
Equator
Tropic of Capricorn
AUSTRALIA
Perth
Adelaide
Melbourne
Sydney
Canberra
Norfolk Island (AUSTRALIA)
Kermadec Islands (NEW ZEALAND)
Island boundaries are for convenience only and do not represent international boundaries.
TASMAN SEA
Auckland
North Island
NEW ZEALAND
Wellington
Christchurch
South Island
Stewart Island
INDIAN OCEAN
Auckland Islands (NEW ZEALAND)
International Date Line
Antarctic Circle
ANTARCTICA

SCALE
0 — 1,000 — 2,000 Miles
0 — 1,000 — 2,000 Kilometers
Scale is accurate only along the equator.
Projection: Mercator

N S E W

Boundaries
National capitals
Other cities

UNIT 10 ATLAS

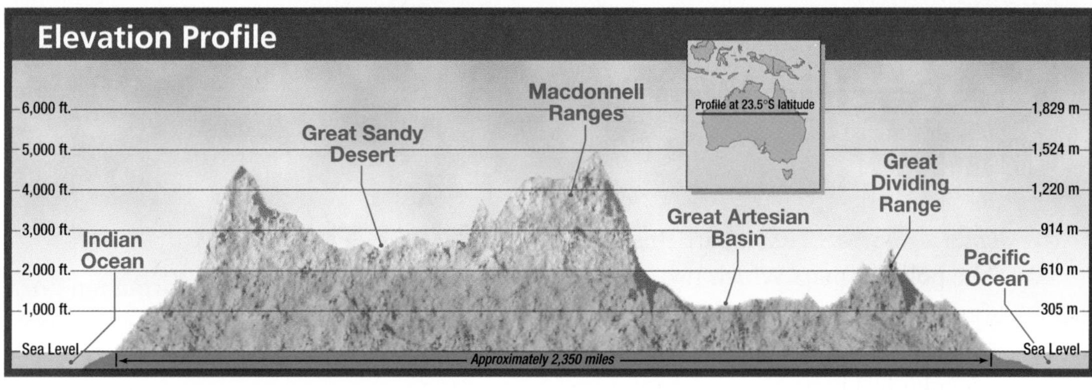

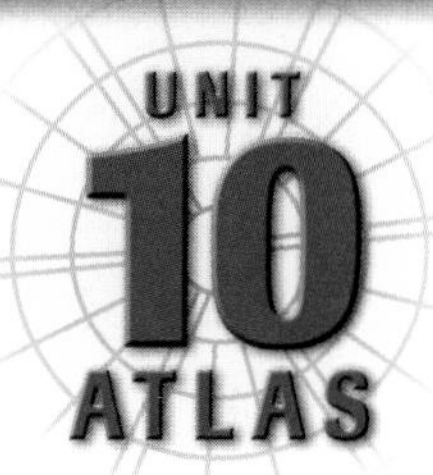

The Pacific World: Physical

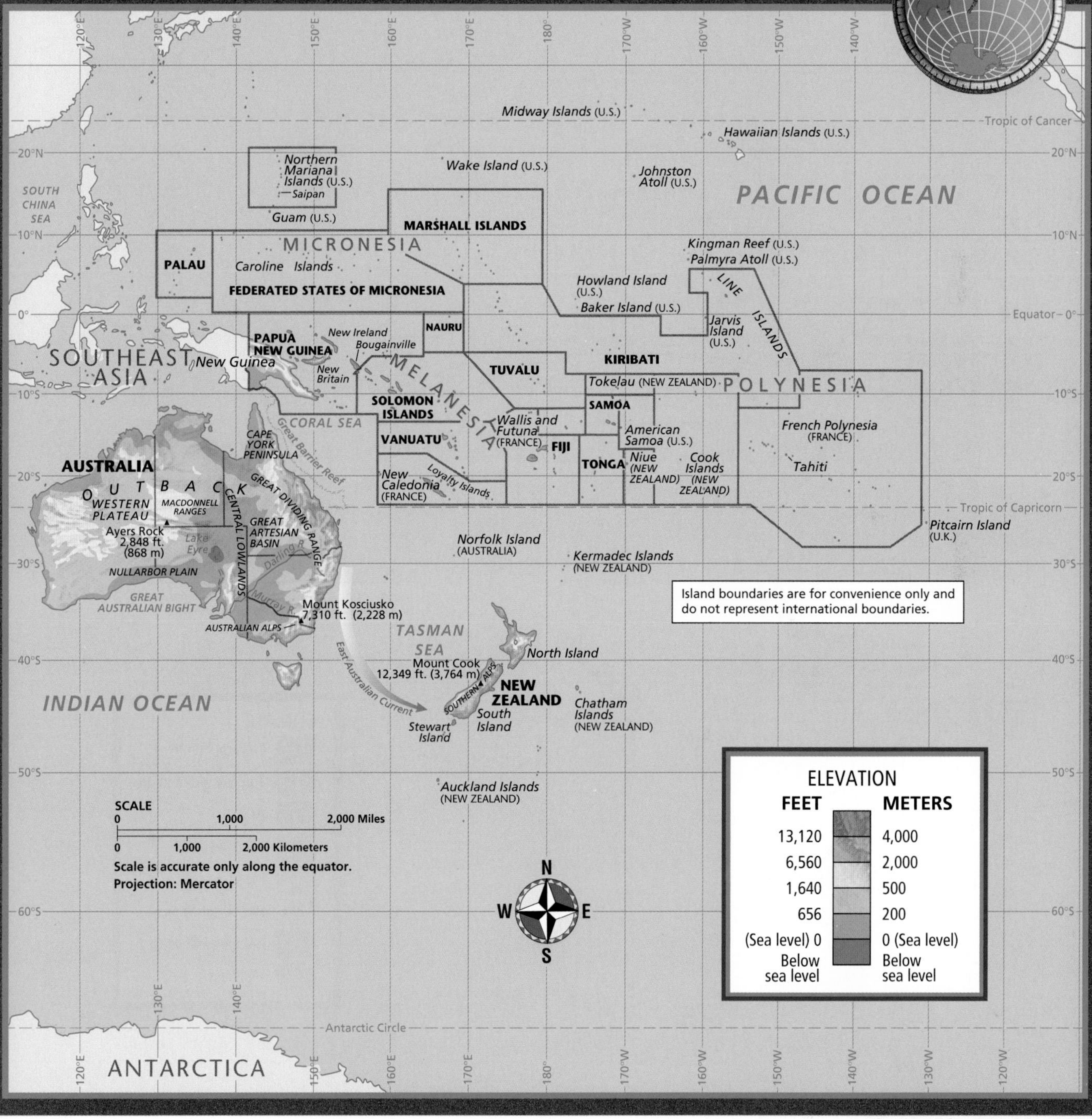

The Pacific World: Climate

1. **Physical Systems** Compare this map to the political map. Which two climates cover most of Australia? TEKS
2. **Physical Systems** Compare this map to the physical map. How might the East Australian Current affect New Zealand's weather? TEKS

Critical Thinking

3. **Analyzing** Based on information on the map, where would you expect most people in Australia to live? TEKS

140°E 150°E 160°E 170°E 180° 170°W 160°W 150°W 140°W 130°W 120°W

Tropic of Cancer

20°N

SOUTH CHINA SEA

PACIFIC OCEAN

10°N

Equator 0°

SOUTHEAST ASIA

10°S

CORAL SEA

Great Barrier Reef

20°S

Tropic of Capricorn

30°S

Island boundaries are for convenience only and do not represent international boundaries.

TASMAN SEA

40°S

INDIAN OCEAN

CLIMATE
Tropical humid
Tropical wet and dry
Arid
Semiarid
Mediterranean
Humid subtropical
Marine west coast
Highland

50°S

SCALE
0 1,000 2,000 Miles
0 1,000 2,000 Kilometers
Scale is accurate only along the equator.
Projection: Mercator

N W E S

60°S

Antarctic Circle

ANTARCTICA

120°E 130°E 140°E 150°E 160°E 170°E 180° 170°W 160°W 150°W 140°W 130°W 120°W

UNIT 10 ATLAS

The Pacific World: Precipitation

1. *Physical Systems* Compare this map to the political map. Which part of Australia receives the most precipitation?
2. *Places and Regions* Compare this map to the political and physical maps. How do the Southern Alps affect the distribution of precipitation in New Zealand? TEKS

Critical Thinking

3. **Making Generalizations** Compare this map to the political map. How do you think latitude affects the amount of precipitation in central Australia? TEKS

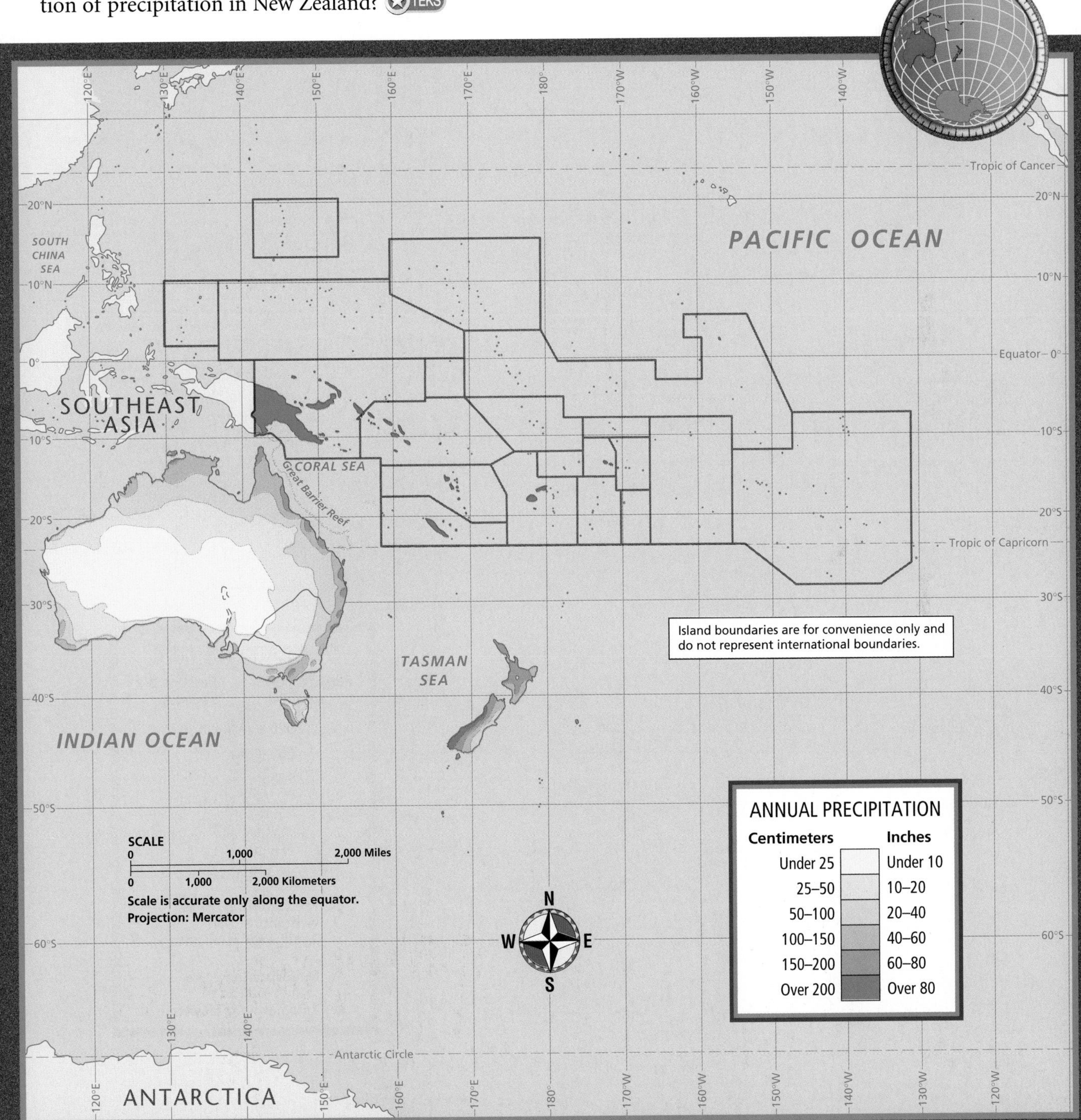

UNIT 10 ATLAS

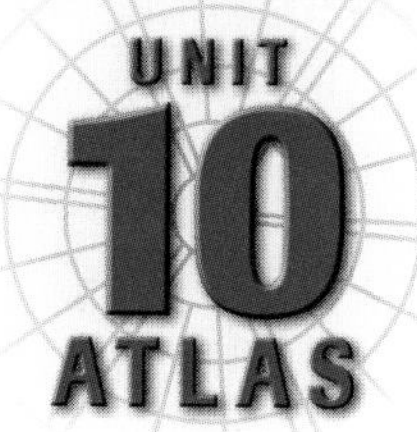

The Pacific World: Population

1. **Places and Regions** Compare this map to the political map. Which part of Australia has the highest population density? TEKS
2. **Environment and Society** Compare this map to the physical map. How do New Zealand's landforms appear to influence the distribution of its population? TEKS

Critical Thinking

3. **Comparing** How do you think the total population of the Pacific region compares to other major world regions? Why?

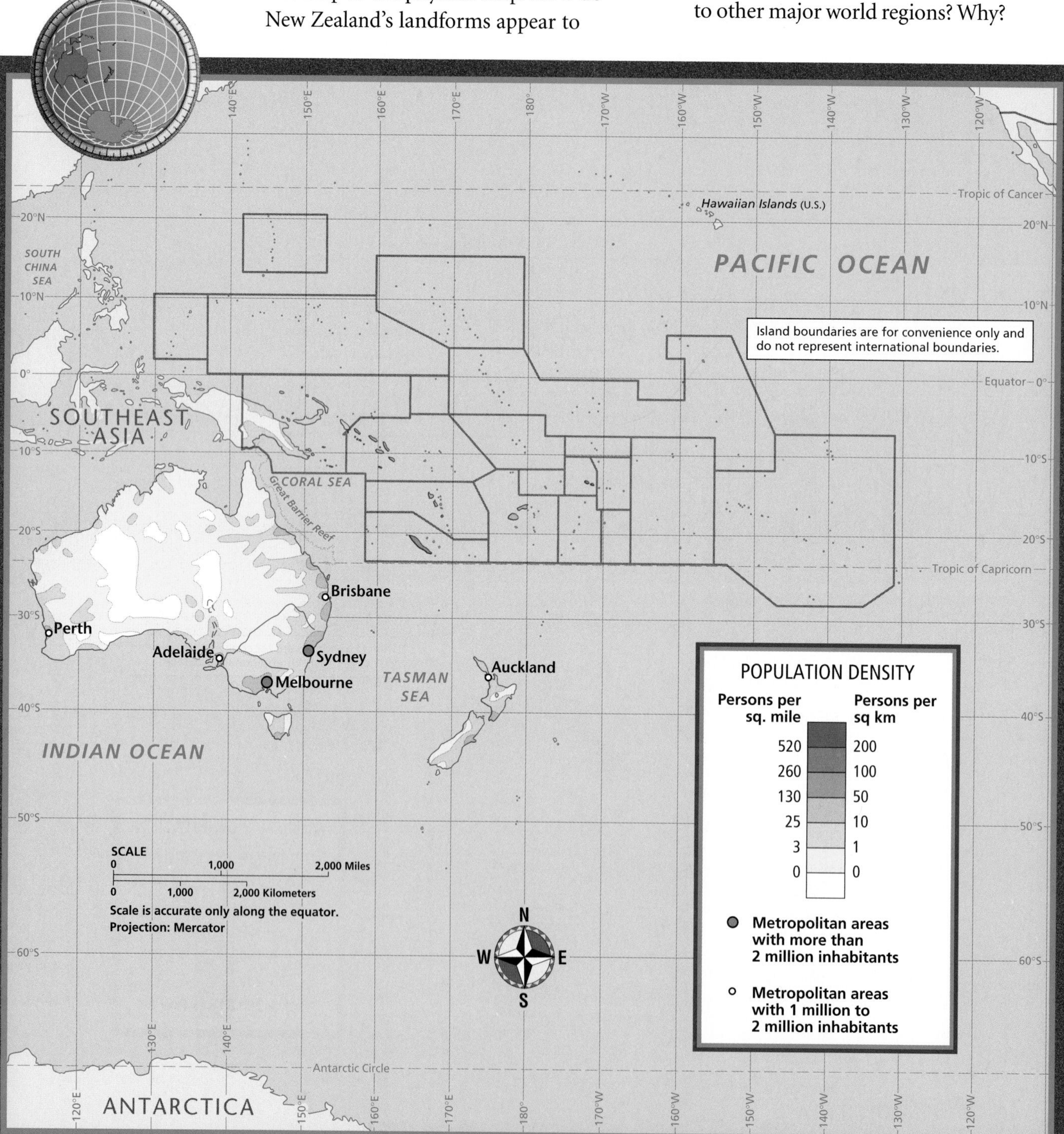

The Pacific World:
Land Use and Resources Map

1. *Places and Regions* Compare this map to the political map. Which natural resources are found in New Zealand?
2. *Places and Regions* Which resources are found in western Australia?

Critical Thinking

3. **Analyzing** Compare this map to the political map. How do you think the location of resources in interior Australia has affected the movement of products, capital, and people? TEKS

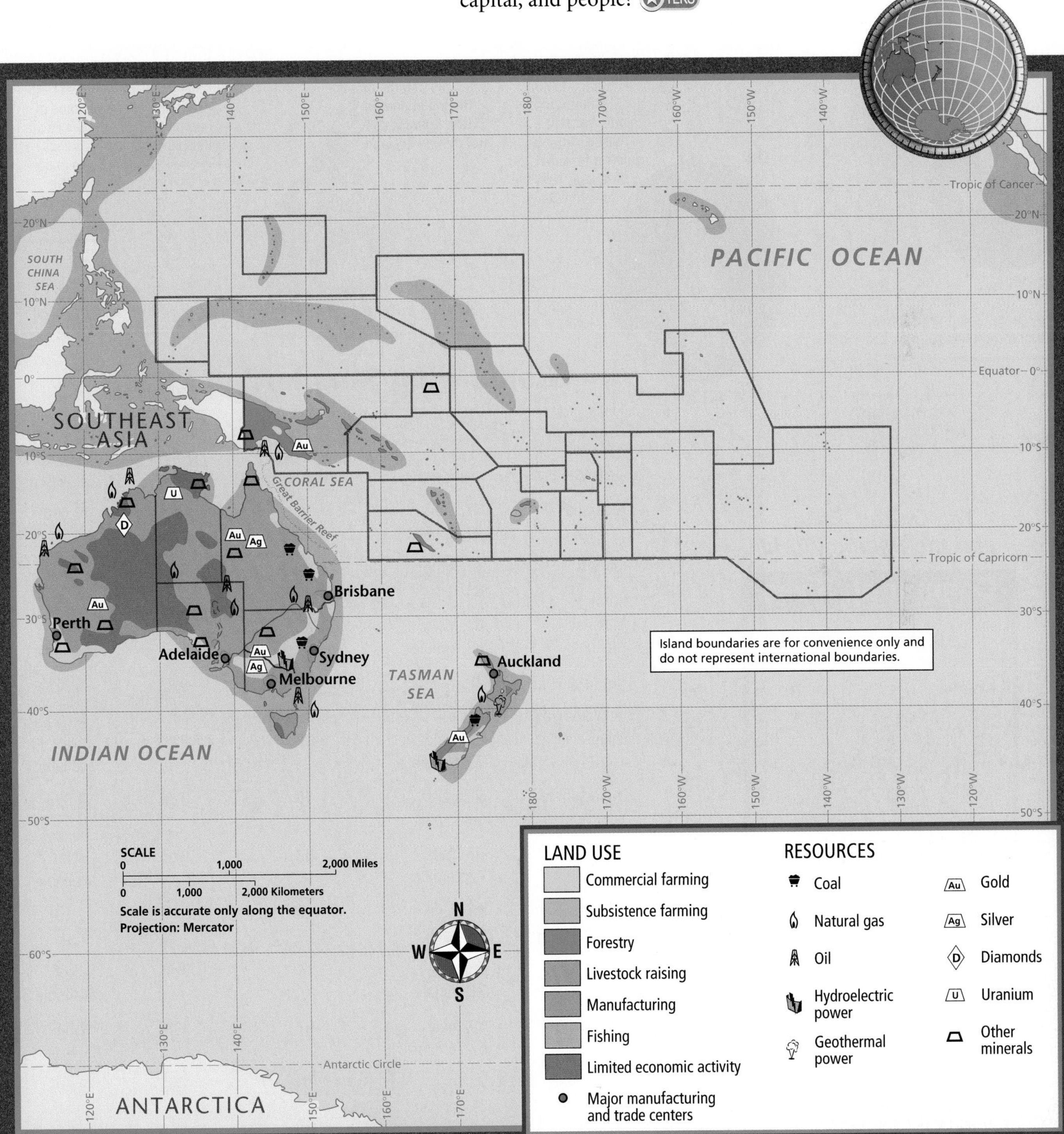

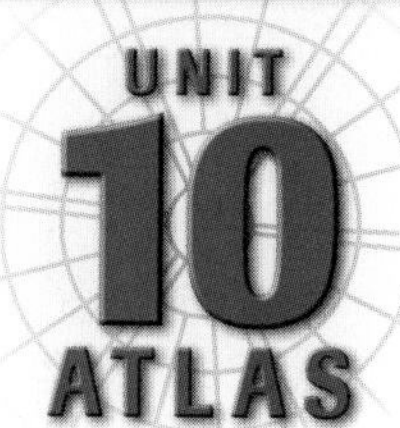

Time Line: The Pacific World

c. 1000
The Maori come to New Zealand by canoe from the Polynesian islands northeast of New Zealand.

1642
Dutch explorer Abel Tasman becomes the first European to reach New Zealand.

1770
James Cook explores Australia's east coast and claims it for Great Britain as New South Wales.

1788
Britain establishes its first prison colony in Australia.

1845–72
The Maori and the British fight for control of New Zealand.

1851
Gold is discovered in Australia.

1901
Australia becomes independent.

1941–45
Some of the Pacific Islands serve as major battlegrounds in World War II.

1967
Aborigines become legal citizens of Australia.

1000 | 1600 | 1700 | 1800 | 1900

Comparing Standard of Living

COUNTRY	LIFE EXPECTANCY (in years)	INFANT MORTALITY (per 1,000 live births)	LITERACY RATE	DAILY CALORIC INTAKE (per person)
Australia	77, male 83, female	5	100%	3,068
Kiribati	57, male 63, female	54	90%	2,772
Marshall Islands	64, male 68, female	40	93%	Not available
New Zealand	75, male 81, female	6	99%	3,379
Palau	66, male 72, female	17	92%	Not available
Papua New Guinea	61, male 66, female	58	72%	2,323
Samoa	67, male 72, female	32	97%	Not available
Solomon Islands	69, male 74, female	24	54%	2,131
Tonga	66, male 71, female	14	99%	Not available
Vanuatu	60, male 62, female	61	53%	2,542
United States	74, male 80, female	7	97%	3,603

Sources: Central Intelligence Agency, *2001 World Factbook; Britannica Book of the Year, 2000*

The United States and the Pacific World

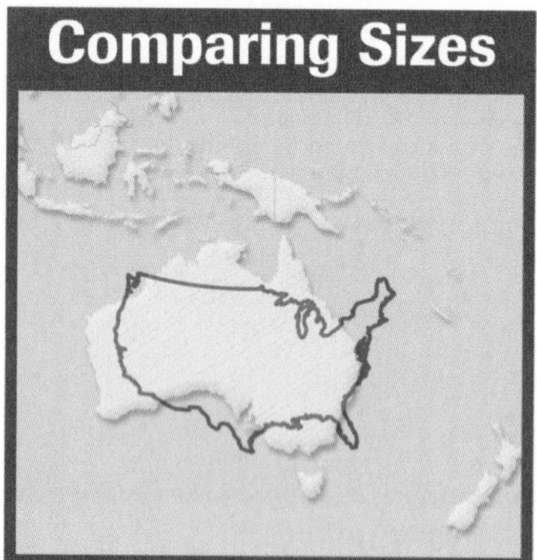

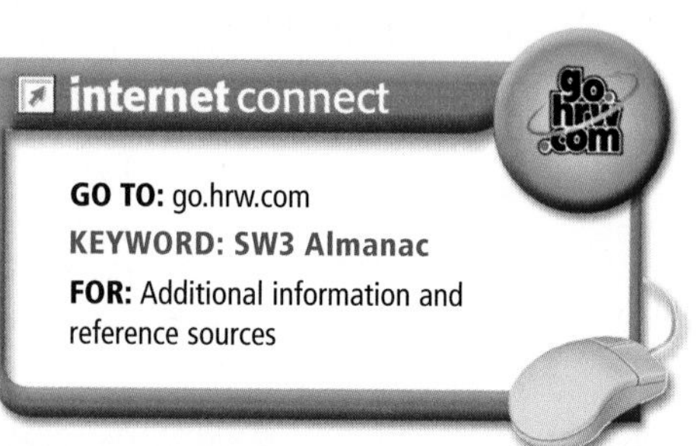

Fast Facts: The Pacific World

FLAG	COUNTRY Capital	POPULATION (in millions) POP. DENSITY	AREA	PER CAPITA GDP (in US $)	WORKFORCE STRUCTURE (largest categories)	ELECTRICITY CONSUMPTION (kilowatt hours per person)	TELEPHONE LINES (per person)
	Australia Canberra	19.4 7/sq. mi.	2,967,893 sq. mi. 7,686,850 sq km	$ 23,200	73% services 22% industry	9,211 kWh	.49
	Fiji Suva	0.8 120/sq. mi.	7,054 sq. mi. 18,270 sq km	$ 7,300	67% subsistence agriculture	562 kWh	0.09
	Kiribati Tarawa	0.09 340/sq. mi.	277 sq. mi. 717 sq km	$ 850	71% agriculture	69 kWh	0.02
	Marshall Islands Majuro	0.07 1,012/sq. mi.	70 sq. mi. 181.3 sq km	$ 1,670	26% services 19% agriculture	837 kWh	0.04
	Micronesia, Fed. States of Palikir	0.1 497/sq. mi.	271 sq. mi. 702 sq km	$ 2,000	67% government	Not available	0.08
	Nauru Yaren District	0.01 1,511/sq. mi.	8 sq. mi. 21 sq km	$ 5,000	Not available	2,308 kWh	0.17
	New Zealand Wellington	3.9 37/sq. mi.	103,737 sq. mi. 268,680 sq km	$ 17,700	65% services 25% industry	9,134 kWh	0.48
	Palau Koror	0.02 108/sq. mi.	177 sq. mi. 458 sq km	$ 7,100	26% services 19% trade	10,658 kWh	0.08
	Papua New Guinea Port Moresby	5.0 28/sq. mi.	178,703 sq. mi. 462,840 sq km	$ 2,500	85% agriculture	335 kWh	0.009
	Samoa Apia	0.2 162/sq. mi.	1,104 sq. mi. 2,860 sq km	$ 3,200	65% agriculture 30% services	519 kWh	0.04
	Solomon Islands Honiara	0.5 44/sq. mi.	10,985 sq. mi. 28,450 sq km	$ 2,000	27% agriculture 23% services	58 kWh	0.02
	Tonga Nuku'alofa	0.1 361/sq. mi.	289 sq. mi. 748 sq km	$ 2,200	65% agriculture	313 kWh	0.08
	Tuvalu Funafuti	0.01 1,099/sq. mi.	10 sq. mi. 26 sq km	$ 1,100	68% agric., fish. 22% services	277 kWh	0.09
	Vanuatu Port-Vila	0.2 41/sq. mi.	4,710 sq. mi. 12,200 sq km	$ 1,300	65% agriculture 32% services	169 kWh	0.02
	United States Washington, D.C.	281.4 76/sq. mi.	3,717,792 sq. mi. 9,629,091 sq km	$ 36,200	30% manage., prof. 29% tech., sales, admin.	12,211 kWh	0.69

Sources: Central Intelligence Agency, *2001 World Factbook; The World Almanac and Book of Facts, 2001; Britannica Book of the Year, 2000;* U.S. Census Bureau
The CIA calculates per capita GDP in terms of purchasing power parity. This formula equalizes the purchasing power of each country's currency.

CHAPTER 31

Australia and New Zealand

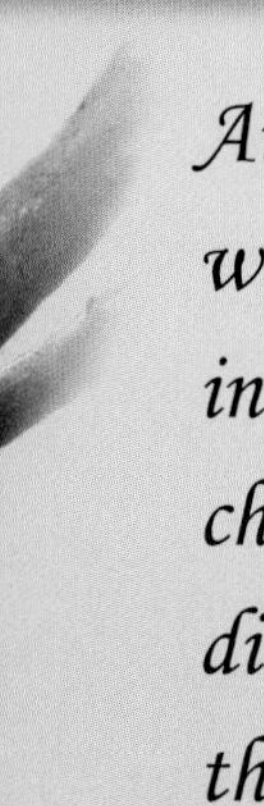

Australia and New Zealand are wealthy democratic countries located in the Southern Hemisphere. In this chapter you will learn about the distinctive and fascinating landscapes of these two countries.

Aboriginal art, *The Big Feast*

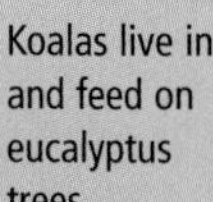

Koalas live in and feed on eucalyptus trees.

Hi! My name is Jared. My twin sister's name is Ashleigh. We live in Cooroy, Australia, about two hours north of Brisbane. We live with our mother. Our father doesn't live with us. He lives up north on the traditional lands of our people, the Djabugayndgi, at the edge of the Cape York Peninsula. Our house in Cooroy is on seven acres and has a big creek with a dam on it. The dam forms a big pond, or what we call a billabong. We can jump into the billabong from the trees along the edge. One tree used to have a swing, but it broke.

In the morning we shower and have orange juice, cereal, toast, and eggs before we catch the bus to school. We are in the ninth grade at Noosa High School. Roll call at school is at 8:30 A.M. Ashleigh and I are both studying math, science, English, history, and physical education. We also get to choose three electives. I'm taking art, speech, and woodworking. Ashleigh is taking art, music, and dance.

At 3:00 P.M. we catch the bus home. On Fridays we often stop at a friend's house and watch American and Australian TV shows. In my free time, I skateboard and play video games. On Saturdays I play sports. I have been invited to the state trials in rugby, softball, and three-on-three basketball. When I grow up, I'd like to be a rugby player. Ashleigh likes to play guitar and keyboard. She listens to rap music, pop, and reggae. She also designs houses. She wants to be an architect. We both like to go to the beach.

Australia

READ TO DISCOVER

1. What are the main features of Australia's natural environments?
2. What are Australia's history and culture like?
3. What are some important features of Australia's human systems?

WHY IT MATTERS

Protecting the environment from nonnative species is a challenge in Australia and many other countries. Use CNNfyi.com or other **current events** sources to find an issue involving nonnative species and environmental protection.

IDENTIFY

Aborigines

DEFINE

artesian wells
outback
marsupials
extensive agriculture
exotic species

LOCATE

Great Dividing Range
Central Lowlands
Western Plateau
Great Barrier Reef

Locate, continued

Cape York Peninsula
Tasmania
Murray River
Darling River
Great Artesian Basin
Lake Eyre
Sydney
Melbourne
Brisbane
Adelaide
Perth
Canberra

Australia and New Zealand: Physical-Political

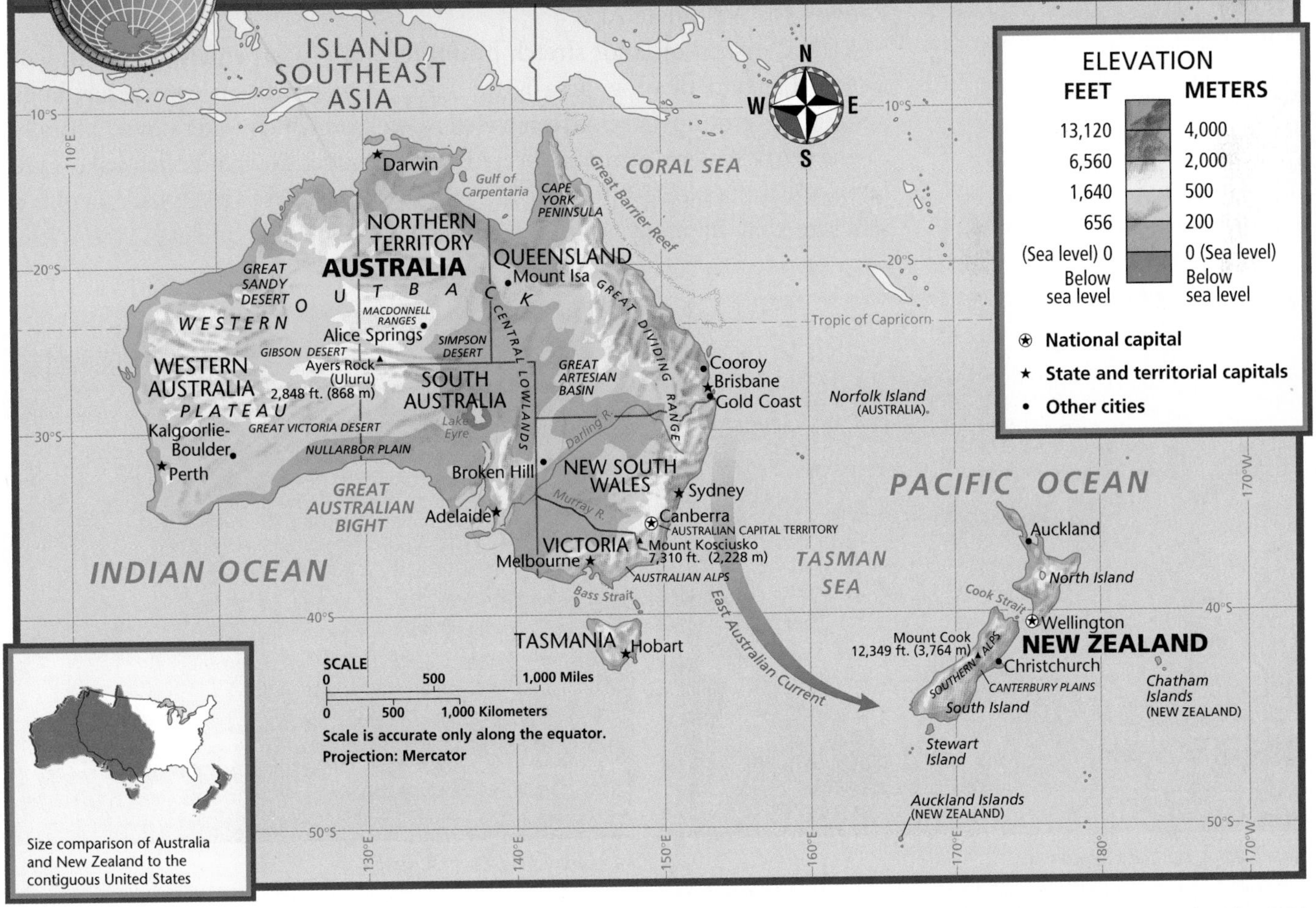

ISLAND SOUTHEAST ASIA
CORAL SEA
PACIFIC OCEAN
INDIAN OCEAN
TASMAN SEA
Darwin
Gulf of Carpentaria
CAPE YORK PENINSULA
Great Barrier Reef
NORTHERN TERRITORY
AUSTRALIA
QUEENSLAND
Mount Isa
GREAT SANDY DESERT
O U T B A C K
WESTERN PLATEAU
MACDONNELL RANGES
Alice Springs
SIMPSON DESERT
GIBSON DESERT
Ayers Rock (Uluru) 2,848 ft. (868 m)
WESTERN AUSTRALIA
SOUTH AUSTRALIA
CENTRAL LOWLANDS
GREAT ARTESIAN BASIN
GREAT DIVIDING RANGE
Cooroy
Brisbane
Gold Coast
Norfolk Island (AUSTRALIA)
Tropic of Capricorn
GREAT VICTORIA DESERT
Kalgoorlie-Boulder
NULLARBOR PLAIN
Lake Eyre
Darling R.
Perth
Broken Hill
NEW SOUTH WALES
Sydney
GREAT AUSTRALIAN BIGHT
Adelaide
Murray R.
Canberra
AUSTRALIAN CAPITAL TERRITORY
VICTORIA
Mount Kosciusko 7,310 ft. (2,228 m)
Melbourne
AUSTRALIAN ALPS
Bass Strait
East Australian Current
TASMANIA
Hobart
Auckland
North Island
Cook Strait
Wellington
NEW ZEALAND
Mount Cook 12,349 ft. (3,764 m)
SOUTHERN ALPS
Christchurch
CANTERBURY PLAINS
South Island
Chatham Islands (NEW ZEALAND)
Stewart Island
Auckland Islands (NEW ZEALAND)
10°S, 20°S, 30°S, 40°S, 50°S
110°E, 130°E, 140°E, 150°E, 160°E, 170°E, 180°, 170°W

ELEVATION

FEET	METERS
13,120	4,000
6,560	2,000
1,640	500
656	200
(Sea level) 0	0 (Sea level)
Below sea level	Below sea level

⊛ National capital
★ State and territorial capitals
• Other cities

SCALE
0 500 1,000 Miles
0 500 1,000 Kilometers
Scale is accurate only along the equator.
Projection: Mercator

Size comparison of Australia and New Zealand to the contiguous United States

internet connect

GO TO: go.hrw.com
KEYWORD: SW3 CH31
FOR: Web sites about Australia and New Zealand

Natural Environments

Australia is known as the Land Down Under because of its location south of, or "under" the equator. The name *Australia* comes from a Latin word that means "southern." The country is located between the Indian and Pacific Oceans. Almost the same size as the contiguous United States, Australia is the only country that is also a continent. It is also the smallest, flattest, and second-driest continent. As you will learn, flat topography and dry climates are major features of Australia.

Landforms and Rivers Geographers divide Australia into three main landform regions. In the east lies a highland region called the Great Dividing Range. The other two landform regions are the Central Lowlands and Western Plateau. In addition, the Great Barrier Reef lies off the northeast coast. This group of coral reefs is 1,250 miles (2,010 km) long. It is famous for its size and varied tropical sea life.

The Great Dividing Range stretches from the Cape York Peninsula to Tasmania. These highlands are the eroded remains of an old mountain range. They divide the flow of Australia's rivers. Those that flow down the eastern slopes empty into the Pacific Ocean. Those that flow west drain into the Central Lowlands. The Murray River and Darling River—Australia's major river system—flow west from this range.

The low Great Dividing Range is also Australia's main mountain system. Mount Kosciusko (kah-zee-UHS-koh), at only 7,310 feet (2,228 m), is the highest elevation on the continent. It is in the highest part of the Great Dividing Range—the Australian Alps.

The Central Lowlands stretch from the Gulf of Carpentaria to the Indian Ocean. Beneath this low area is the Great Artesian Basin, which has huge amounts of groundwater. **Artesian wells**—wells in which water flows naturally to the surface—are common here. In the Central Lowlands lies Lake Eyre (AYR), Australia's lowest point at 52 feet (16 m) below sea level. Lake Eyre is a salt lake that in many years is completely dry. North of Lake Eyre is the

Australia has many unusual landforms. This gorge at Alice Springs is near the country's center. Far to the south, near Adelaide, lies Kangaroo Island. Almost one third of the island is within national and conservation parks.

Alice Springs, Northern Territory

Kangaroo Island, South Australia

Simpson Desert. This desert has sand dunes that are from 70 to 120 feet (21 to 37 m) high. Winds move and shape these dunes.

The Western Plateau covers about two thirds of Australia. It has the oldest rocks on the continent. For millions of years, these rocks have been eroded. Deserts cover the central part of the area. In the south is the Nullarbor Plain—a dry flat limestone plateau.

READING CHECK: ***Places and Regions*** What are the three formal landform regions of Australia?

Climates Australia is a desert continent with green edges. About two thirds of Australia has an arid or semiarid climate. (See the unit climate map.) Most of the heart of Australia is extremely dry. Around these desert areas are ribbons of semiarid climate. Rainfall in these areas is often unreliable. Long droughts may be followed by short powerful storms and floods. During droughts, wildfires often sweep across the land. The dry interior of Australia is called the **outback**.

There are several reasons why Australia has such dry climates. Much of the country is between about 20° and 30° south latitude. These areas are often warm subtropical high-pressure zones with dry air. Another factor that contributes to Australia's dry climates is its generally low elevation. Only the Great Dividing Range is high enough to cause air to rise and cool, which creates rain. However, the mountains keep moisture from reaching much of the interior. West of the mountains is a rain shadow.

Temperatures in Australia are generally warm. Average January (summer) temperatures are higher than 85°F (29°C) in most of the interior and much of the north and northwest. Average July (winter) temperatures are warmest in the north and cool in the south. High elevations in the southeastern Great Dividing Range are the only areas that have cold weather. Winter skiing is possible there.

Much of northern Australia has a tropical wet and dry climate. This area has summer monsoons that bring heavy rainfall. Some parts of the Cape York Peninsula get more than 150 inches (381 cm) of rainfall per year. Winters are still warm, but may have long droughts. Along the east coast are humid subtropical and marine west coast climates. In the southeast, westerly winds bring winter storms and rain. Unlike most parts of Australia, the southeast is generally well watered. The mild climate and rainfall generate small streams and rivers. Two parts of the southern edge of Australia have a Mediterranean climate. Mild rainy winters and warm dry summers are normal there.

READING CHECK: ***Physical Systems*** How does the Great Dividing Range affect Australia's climates?

Plants and Animals Australia is known for its strange plants and animals. Most are endemic species. Because of past movements of Earth's tectonic plates, Australia has been separated from the other continents for about 35 million years. During this time, its plants and animals developed in isolation. This condition has caused a unique biogeography. For example, cats are native to every continent except Australia and Antarctica. Also, hoofed animals like deer and cattle are not native to Australia. Most of Australia's mammals are **marsupials**—mammals that have pouches to carry their young. These include the kangaroo, koala, and wallaby (WAH-luh-bee).

INTERPRETING THE VISUAL RECORD

These 300-foot cliffs stretch for nearly a hundred miles along the edge of the Nullarbor Plain. Nullarbor *comes from a Latin phrase that means "no tree."* ***What force is eroding these cliffs?*** TEKS

Rain forest, New South Wales

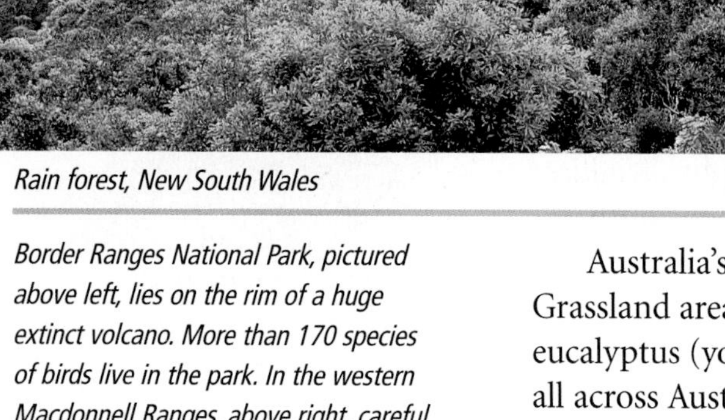

Semiarid landscape, Northern Territory

Border Ranges National Park, pictured above left, lies on the rim of a huge extinct volcano. More than 170 species of birds live in the park. In the western Macdonnell Ranges, above right, careful observers may see the rock wallabies that live there.

Australia's biomes mirror its climates. The interior is grassland and desert. Grassland areas have acacia (e-KAY-shuh) shrubs, bunch grasses, and scattered eucalyptus (yoo-kuh-LIP-tuhs) trees. About 500 kinds of eucalyptus are found all across Australia. Monsoonal northern Australia has large areas of savanna. Plants and animals there depend on seasonal rains. The Cape York Peninsula has tropical rain forests. Large trees and dense vegetation are common there. Animals include the tree kangaroo and many native birds, such as parrots and cockatoos. The south and southwest have Mediterranean scrub forests. The southeast and east coast have temperate forests.

READING CHECK: ***Places and Regions*** What is unique about Australia's plant and animal life?

Natural Resources Australia is rich in mineral and energy resources. As you have learned, water resources are scarce in many areas, particularly the arid interior.

Mineral resources include bauxite, copper, iron ore, and many other valuable minerals. (See the map of land use and resources in the unit atlas.) Many of these minerals are found in the dry interior. Mining centers such as Broken Hill and Mount Isa have been operating for many years. The Broken Hill mines in southeast Australia produce lead, silver, and zinc. Other mines yield valuable gems, such as diamonds, opals, and sapphires. Energy resources include coal, oil, and natural gas. Along Australia's east coast are large coal deposits. Most of the oil and natural gas comes from offshore fields. The main fields are in the Bass Strait near Tasmania and off the coast of western Australia.

The country's farming resources are more limited. Most areas have poor soils, and there is not much water. Only about 6 percent of the land is good for farming. The best farming areas are in the southeast. Because of the limited amount of good farmland, many areas are used for grazing. The many artesian wells and groundwater sources make this possible. Much of the water is too salty for people, but can be used for sheep.

READING CHECK: ***Environment and Society*** What factors affect the location of different types of economic activities in Australia?

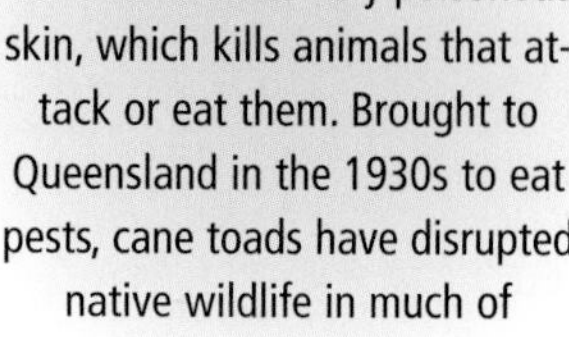

Cane toads have very poisonous skin, which kills animals that attack or eat them. Brought to Queensland in the 1930s to eat pests, cane toads have disrupted native wildlife in much of northern Australia.

History and Culture

Australia's native peoples have one of the world's oldest continuous cultures. However, Australian culture has been shaped by its history as a British colony. Although it is far from Europe, Australia's dominant culture is European.

Early History and Settlement Australia's first peoples were the **Aborigines** (a-buh-RIJ-uh-nees). They came to Australia from Southeast Asia at least 40,000 years ago. Early Aborigines lived a nomadic way of life. They hunted with spears, nets, and boomerangs—curved throwing sticks. There were many groups, each with a different name, speaking hundreds of different languages. Although estimates vary, at least 300,000 Aborigines probably lived in Australia when European settlers arrived in the late 1700s.

The British settled Australia as a prison colony. The first settlement, set up in 1788, later became the city of Sydney. By 1830, nearly 60,000 prisoners had been sent to Australia. Other people came to farm or raise sheep. In 1851 gold was discovered, attracting even more people. Many settlers forced Aborigines off their land. Aborigines had no resistance to the diseases brought by Europeans, and many of them died. In Tasmania, the Aborigines were completely wiped out.

In the mid-1800s more towns and colonies were founded. Eventually, six large colonies developed. (See Connecting to History: Australia's States and Territories.) In 1901 these six colonies joined to form the Commonwealth of Australia. The new country was a close ally of Great Britain. During World Wars I and II many Australians fought alongside British troops.

✓ **READING CHECK:** *Human Systems* About how many Aborigines lived in Australia when European settlers arrived?

Aborigine elders from Melville Island, off the north coast of Australia, display their spears.

Australia's States and Territories

Australia's first colony, New South Wales, was founded in 1788. In the 1820s Van Diemen's Land (later renamed Tasmania) and Western Australia were added. Later, colonies were created with land taken from New South Wales. These colonies included South Australia, Victoria, and Queensland. In 1901 Australia became independent from Great Britain. Soon after, the country's political geography looked much like it does today. The six colonies became states, and two territories—Northern Territory and the Australian Capital Territory—were created. (See the chapter map.)

Drawing Inferences and Conclusions Look at the chapter map. Which major cities do you think developed in each colony?

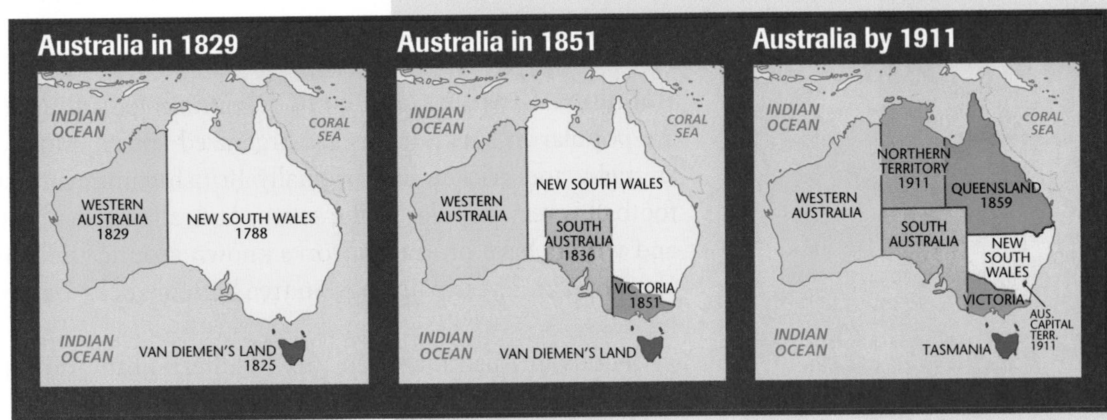

Some Common Australian Words	
Word	**Definition**
biscuit	cookie
clicks	kilometers (or miles) per hour
dunny	restroom with just a toilet
esky	ice chest
g'day	hello
mozzies	mosquitoes
Oz	Australia
roo	kangaroo
saltie	saltwater crocodile
shark biscuit	inexperienced surfer

People and Languages Australia's colonial history shaped its society. About 92 percent of Australia's 19 million people are of British or other European ancestry. English is the official language. However, it is spoken with a distinct "Aussie" accent and many special Australian words. (See the table.) Asians make up about 7 percent of the population. Many Asians began moving to Australia in the 1970s. They have added to the country's growing cultural diversity. Aborigines are a small but important group. About 200,000 Aborigines now live in Australia. Many of them have mixed European and Aboriginal ancestry.

Settlement and Land Use Most people live in cities along the southeastern coast. These include Sydney, Melbourne, Brisbane, and Adelaide. Perth is a large city in western Australia. All together, about 85 percent of Australians live in cities.

Australia's settlement pattern is tied to the country's colonial history. Major cities grew as ports during colonial times. Even today, each state and territory has just one major city. (See the chapter map.) The only large city not on the coast is the capital, Canberra. It was not founded until 1913. Inland settlements, such as Kalgoorlie-Boulder, are generally mining or agricultural towns. Alice Springs, the outback's major town, is important for transportation and tourism. Where people live is also tied to Australia's natural environment. Most people settled in the southeast. This area has pleasant climates and reliable rainfall. In the dry interior, a lack of water makes settlement and farming risky. Even today, very few people live there.

INTERPRETING THE VISUAL RECORD

Visitors can find fashionable galleries, hotels, and other establishments on Collins Street in Melbourne, Australia. ***How can you tell from the photograph that Melbourne has had a relatively long history of prosperity?***

Religion and Education The large majority of Australians are Christian. Asian immigration has brought many Buddhists and Muslims. Aborigines who follow traditional ways emphasize spiritual ties to the land. They believe that their ancestors created the world during what the Aborigines call the Dreamtime. These ancestors became part of nature. Such beliefs help Aborigines feel close to their ancestral lands.

Australia has a good education system. State and territorial governments run schools with help from the national government. In the outback, some students get lessons by radio, e-mail, video, and even satellite connections. One important national goal is to improve the situation of Aborigines. On average, they lag far behind other Australians in education. Historically, Aborigines have not had access to a good education in Australia. In fact, Aborigines did not even become legal citizens until 1967.

Traditions, Customs, and Food Swimming, surfing, and going to the beach are popular in Australia, as are organized sports. Most of these sports, such as rugby and cricket, are originally British games. Bruising Australian Rules football is very popular in the south. Australian artists, filmmakers, musicians, and writers have produced works known around the world. Aboriginal art is also popular. These paintings on tree bark or rocks feature human and animal figures.

Foods in Australia often mix Mediterranean and Asian styles. Common foods like beef and lamb are often grilled or roasted and served with potatoes and other vegetables. Italian and Greek foods are popular. Many immigrants to

Geography for Life

Australia's Changing Trade Patterns

Most of the world's developed countries have become rich through the production and export of manufactured goods. However, Australia's economy has long been based on the export of raw materials. The country's leading exports today include coal, gold, meat, wool, iron ore, and wheat. Australia ships these materials mainly to Japan, European Union (EU) countries, ASEAN (Association of Southeast Asian Nations) countries, and the United States. Factories in those countries make products using those raw materials.

While Australia exports mostly raw materials, it imports mostly manufactured goods. These imports include computers, machinery, office machines, and transportation equipment. Most of these imports come from the United States, Japan, ASEAN countries, and EU countries like the United Kingdom and Germany.

Since World War II Australia's government has tried to diversify the country's economy. Its goal has been to make and export more manufactured goods. Doing so would help the country become less dependent on the export of raw materials. Many factories now specialize in light manufacturing. Many also process farm products or minerals for sale overseas. In addition, Australia produces household appliances, paper products, processed foods, and textiles. The country also has a growing wine industry.

Just as Australia's economy has changed since World War II, so have its trading partners. Australia has long had strong trade relations with Britain. In fact, Britain was Australia's main trading partner for many years. However, these ties are not nearly as strong today. This change began after World War II as U.S. influence in the Pacific region began to grow. Trade and other ties between the United States and Australia strengthened. Trade with Japan and China also began to grow. Then in 1973 Britain joined what is now the EU and ended policies that favored trade with Australia and its other former colonies.

Wool from different breeds of sheep becomes products made by various segments of the wool industry, such as carpets, fine fabric, and upholstery. The fleece from the largest sheep may weigh 35 pounds (16 kg).

Today EU countries are significant importers of Australian goods. However, Japan and the United States are the largest single buyers of Australia's main exports. Japan, for example, imports Australian raw materials and natural resources that it lacks at home. In trade, as in other ways, Australia is increasingly looking toward Asia for its future.

Australia's Major Trading Partners

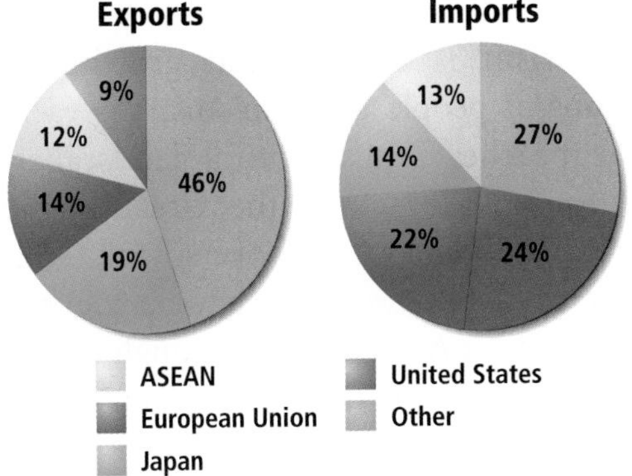

Source: Central Intelligence Agency, *The World Factbook 2001*

Applying What You Know

TEKS

1. **Summarizing** How and why have Australia's economy and trade patterns changed since World War II?
2. **Drawing Inferences and Conclusions** How might changing trade patterns increase cultural and other ties between Australia and Asia?

INTERPRETING THE VISUAL RECORD

Workers pick grapes near Adelaide, South Australia. ***Why do you think the southern coastal region is a good one for growing grapes?*** TEKS

Australia came from these countries. Asian foods have also become popular. Large cities now have many Indonesian, Vietnamese, and Chinese restaurants.

TEKS **READING CHECK:** ***Places and Regions*** What features of Australia's history and culture make it a distinctive region?

Australia Today

Australia is a developed country with a market economy. It has good transportation systems and health care and a tradition of stable democratic government. The per capita GDP is about $23,200—very high by world standards. Also, life expectancy and literacy rates are higher than in the United States.

Australia exports mostly raw materials and imports mostly manufactured goods. (See the chart.) Most people work in service industries like education, government, or tourism. Since World War II, Australia's major trading partners have changed. Australia used to trade mainly with Britain and other European countries. Today, however, the country trades mainly with Asia and the United States. (See Geography for Life: Australia's Changing Trade Patterns.)

Mining, Agriculture, and Tourism Mining is an important part of Australia's economy. In fact, Australia is the world's leading producer of bauxite, diamonds, opals, and lead. It is also a major producer of coal, copper, iron ore, silver, and many other minerals. Most minerals are exported to Japan or other countries in East Asia. Many of Australia's mineral resources are found in the dry interior.

Only 6 percent of Australia's land is good for farming. However, the country's large size and modern technology make it a major exporter of farm goods. The main products are wool, meat, and wheat. Wool supports the economy. The country has about 150 million sheep, or about 15 percent of all the sheep in the world. As a result, Australia is the world's leading producer of wool, supplying about 30 percent of the world's total. Most sheep are raised on the western slopes of the Great Dividing Range and around Perth. Cattle are raised in the north. Both sheep and cattle are raised on huge ranches called stations. (See Cities & Settlements: Life in the Outback.) These ranches are examples of **extensive agriculture**. This kind of agriculture uses much land but small inputs of capital and labor per unit area. Wheat is Australia's most important crop. It is concentrated in the southeast. Tropical crops like bananas, pineapples, and sugarcane grow along Queensland's wet coast. Australia produces many other fruits and vegetables also.

Tourism is a large and growing industry. Millions of people come to Australia each year to enjoy beaches and beautiful landscapes. Major vacation areas include the Great Barrier Reef, Queensland's Gold Coast, Ayers Rock (Uluru), and major cities like Sydney. Many people from other countries visit Australia even though they face long travel times to get there. The country is about a 15-hour flight from Los Angeles and a 24-hour flight from London.

Australia's Major Exports and Imports

Exports	Imports
dairy products, meat, fish, wool, forestry products, manufactures	machinery and equipment, vehicles and aircraft, petroleum, consumer goods, plastics

Source: Central Intelligence Agency, *The World Factbook 2001*

INTERPRETING THE CHART *Alumina is a component of bauxite, the main aluminum ore. It has other uses too, such as in ceramics and pigments. Because so many mineral deposits are far from population centers, mining these minerals can be expensive. Foreign investment is crucial.* ***What types of products does Australia export?***

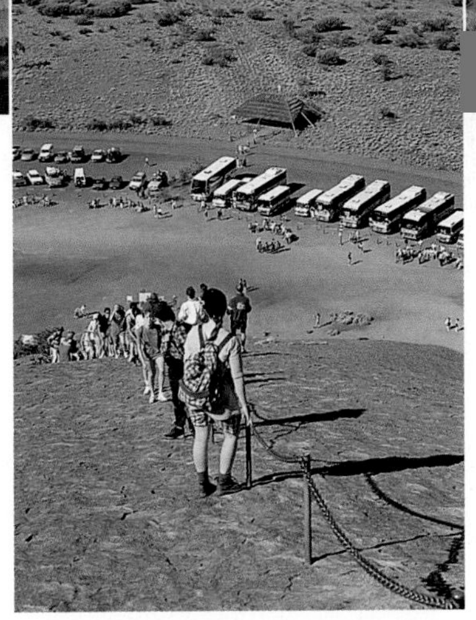

INTERPRETING THE VISUAL RECORD

Because of the rock's mineral content, Uluru changes color as the sun rises and sets. The Aborigines who own the site, the Anangu, believe that only two special men from the local group may climb Uluru. ***How can you tell from the photo that these tourists and the Aborigines view Uluru differently?*** TEKS

Issues and Challenges Important challenges include addressing Aboriginal claims to land and protecting the environment. In the late 1980s Aborigines began protesting mining on sacred lands. They also claimed that these lands belonged to them. This dispute has become a national issue. Judges have said that Aborigines have a right to claim traditional lands. As a result, many farmers, mining companies, and ranchers worry that they might lose control of the lands they use to earn a living.

Another challenge is protecting the environment. Australia's environment has changed greatly during the last 200 years. For example, more than a third of the country's woodlands have been cleared or altered. People usually cleared land to create areas for farming and raising animals. Hardest hit were the country's rain forests. About 75 percent of these forests were destroyed. This has reduced habitat for native wildlife.

Europeans brought many new plants and animals to Australia. New types of plants and animals that people introduce to an area are called **exotic species**. In Australia they include camels, cane toads, prickly pear cacti, and rabbits. Many have spread across the country and become problems. These new animals usually have no natural predators. For example, the British brought rabbits for sport hunting. However, the number of rabbits rose quickly. They damaged grazing lands, causing erosion. They became such a problem that people purposely introduced a disease among them. This succeeded in lowering their numbers. However, rabbits and other exotic species still cause problems.

READING CHECK: ***Human Systems*** What activities are important to Australia's market economy?

TEKS Questions 1, 2, 3, 4, 5

Section 1 Review

Homework Practice Online

Keyword: SW3 HP31

Identify

Aborigines

Define artesian wells, outback, marsupials, extensive agriculture, exotic species

Working with Sketch Maps On a map of Australia that you draw or that your teacher provides, label the Great Dividing Range, Central Lowlands, Western Plateau, Great Barrier Reef, Cape York Peninsula, Tasmania, Murray River, Darling River, Great Artesian Basin, Lake Eyre, Sydney, Melbourne, Brisbane, Adelaide, Perth, and Canberra. Which interior area has large amounts of groundwater?

Reading for the Main Idea

1. ***Physical Systems*** How do latitude and elevation influence Australia's climate regions?
2. ***Human Systems*** How have the human characteristics of Australia changed since the late 1700s?
3. ***Human Systems*** What political, economic, social, and demographic data give clues to Australia's level of development?

Critical Thinking

4. **Drawing Inferences and Conclusions** How is the distribution of plants and animals in Australia related to climate?

Organizing What You Know

5. Create a chart like the one shown below. Use it to describe Australia's three major landform regions.

Great Dividing Range	Central Lowlands	Western Plateau

CITIES & SETTLEMENTS

Life in the Outback

Places and Regions West of Australia's major mountains and rivers, a vast arid region extends across the country to the Indian Ocean. Some Australians call this desolate expanse the "back of the beyond," but it is better known as the outback. Temperatures here can reach 120°F (48°C) in the shade. Rainfall averages under 10 inches (25 cm) per year. The outback's riverbeds hold only rock and sand, except after heavy downpours. Dust storms are more common than rain! When it fills with water, Lake Eyre, at the edge of the Great Victoria Desert, is the country's largest lake. However, the lake fills completely only about twice in a century. Instead, most of the time the "lake" is a huge salt flat.

Adapting to the Environment

The outback covers about 75 percent of Australia. It is the ancestral home of Australia's first people, the Aborigines. Today only about 10 percent of Australians live on this desolate unforgiving land. Population density averages fewer than two people per square mile, and large areas are completely uninhabited. Many outback dwellers live 300 miles or more from the nearest store. Doctors and schools are even farther away. Such conditions have created a rugged and independent people. Yet the region defeats those who believe they can truly conquer it. The people of the outback survive only by adapting to the arid environment.

Coober Pedy, a town west of Lake Eyre in South Australia, provides a clear example of this adaptation. This isolated community is one of the few stops on the only paved road that crosses the entire outback. Called the Stuart Highway, this route connects the northern and southern coasts to Alice Springs, in the Northern Territory. With just 27,000 people, "the Alice" is one of the outback's largest towns.

Fewer than one percent of South Australia's 1.5 million people live in the state's interior. A visit to Coober Pedy quickly reveals why. The people of Coober Pedy have had to take extraordinary measures to live in such a harsh environment. In the Aborigine language, *Coober Pedy* means "White Man's Burrow." The name fits. Coober Pedy exists only because of the opal mines nearby. The town is the world's largest producer of these precious stones. In addition, much of Coober Pedy is itself underground. To escape the blistering heat, businesses, churches, and many homes are below the surface. These establishments, and about half the residents, occupy old mine shafts or specially dug homes called dugouts. In these underground places the temperature is 72°F (22°C) all year.

People shop for opals in one of the many belowground stores in Coober Pedy.

Adapting to Isolation

Along a rough dirt road between Coober Pedy and Lake Eyre lies the Anna Creek Station homestead. This small cluster of buildings is the station's headquarters. Covering some

Anna Creek family members and jackeroos pose for an informal portrait. The children's best friends may live hundreds of kilometers away. Still, some residents compare their far-flung community's closeness to a small town's.

18,600 square miles (30,000 sq km), Anna Creek is the largest cattle station in Australia. However, only 15 people live here. Besides the station manager and his family, the population includes a cook, pilot, and teacher. Several ranch hands—or "jackeroos" as cowboys are known in Australia—complete the group.

For those who are not used to it, station life can be uncomfortable. The ground is hot and dry. Snakes sun themselves on low sand dunes behind the buildings. Flies seem to be everywhere. The wind carries a fine red dust that covers everyone and everything. For some of the station's residents, dealing with the loneliness is difficult. After all, Coober Pedy, 100 miles to the west, is the closest town. Yet other station residents could not imagine living in a city again.

Few of the station's teachers have lasted more than a year before returning to the coast. Even when the station has an on-site teacher, basic education comes from the School of the Air. Using a high-frequency radio, the children take part in classes broadcast from the coastal city of Port Augusta. Every grade level has a half-hour class each day. Each student also receives an individual 10-minute radio session with his or her radio teacher once a week. Other lessons arrive on videotape. Students mail in their assignments. The teacher grades them and returns them by mail. Students work five or six hours a day on these assignments. A parent or a hired teacher, as at Anna Creek Station, supervises. Use of the Internet, e-mail, and video conferencing is increasingly important for instruction.

A worker watches a small part of the Anna Creek herd, which had been reduced by drought. Although horses are seldom used for roundups, many outback families keep them for other purposes.

March through November is the busiest time at Anna Creek. The cattle must be gathered before the heat and the flies make life miserable for people and animals. Calves must be tagged, horns cut, and some livestock shipped to market. The roundups are challenging. Some 16,000 head of cattle may be spread across an area larger than the state of Maryland. The jackeroos go out for weeks at a time to various parts of the station. Like education in the outback, this job also depends on technology. Jackeroos use motorcycles instead of horses on these roundups, and workers in airplanes spot cattle from the sky.

Applying What You Know 

1. **Analyzing Information** How have technological innovations allowed people to live and work in the outback?
2. **Comparing** How is education in the outback similar to education in the United States? How are the two systems different?

Section 2

New Zealand

READ TO DISCOVER

1. What are some important features of New Zealand's natural environment?
2. What are New Zealand's history and culture like?
3. On what is New Zealand's economy based, and what economic challenge does the country face?

WHY IT MATTERS

Depending heavily on world trade may cause problems for New Zealand. Use CNNfyi.com or other **current events** sources to investigate dependence on world trade in New Zealand or other countries.

IDENTIFY

Maori

DEFINE

economy of scale

LOCATE

North Island
South Island
Cook Strait
Southern Alps
Canterbury Plains
Auckland
Wellington

Natural Environments

The South Pacific island country of New Zealand is about 1,000 miles (1,609 km) southeast of Australia. Like Australia, New Zealand has a British colonial history and a high standard of living. However, New Zealand is also very different. It is much smaller, more mountainous, and has a wetter and milder climate. The country has two major islands—North Island and South Island. (See the chapter map.) They are separated by Cook Strait. Some smaller islands in the Pacific are also part of New Zealand.

North Island In the north are peninsulas with forests and fertile lowlands. New Zealand is located on the Pacific Ring of Fire, so it is a tectonically active country. The central and western parts of North Island have active volcanoes,

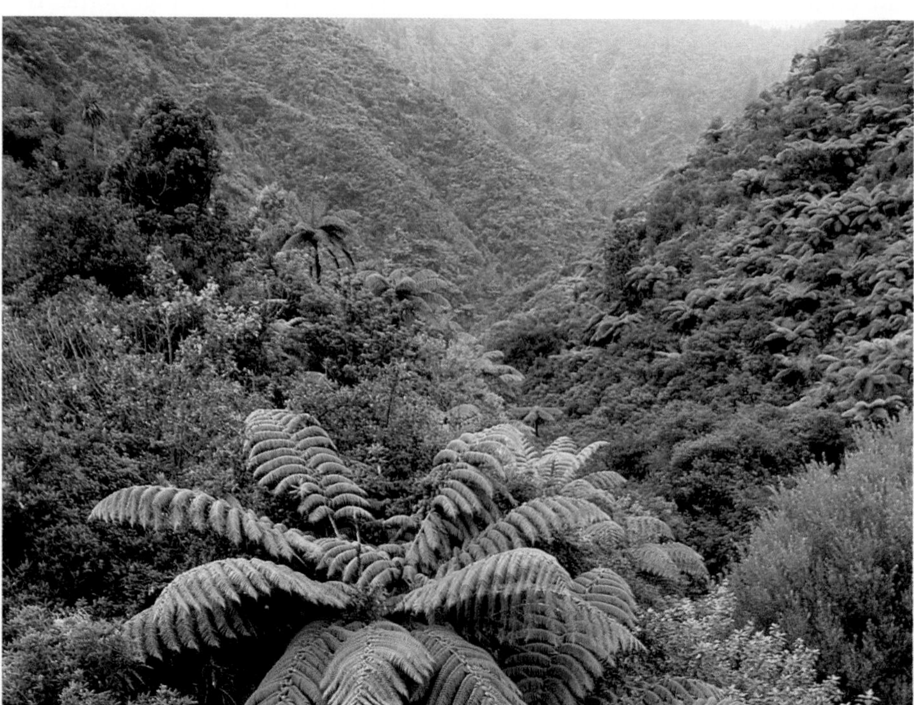

Tree ferns grow in Waitaanga Forest, North Island. These plants, which can be more than 30 feet (9 m) tall, have been called living fossils because the species has survived for some 150 million years. A fern leaf is one of New Zealand's national symbols.

geysers, and hot springs. These are created by the collision of the Pacific and Indo-Australian Plates. Subduction along this plate boundary causes earthquakes as well. The eastern part of North Island has rugged hills and small coastal lowlands.

The tuatara (TOO-uh-TAHR-uh), a small reptile, is found only on a few islands of New Zealand. Tuataras are the last living members of an ancient group of reptiles related to the dinosaurs. They appeared on Earth more than 200 million years ago. Some tuataras may live to be 100 years old.

South Island South Island is larger and has higher elevations than North Island. A steep mountain range called the Southern Alps runs along the west coast. The Southern Alps include Mount Cook, New Zealand's highest peak, which rises to 12,349 feet (3,764 m). The Southern Alps are famous for their beautiful scenery, with many glaciers and mountain lakes. Below are thick green forests. Along the east coast of South Island lie the Canterbury Plains. Different grain crops and livestock feeds are grown in this fertile lowland area.

READING CHECK: *Physical Systems* What physical processes have created volcanoes, geysers, and hot springs on North Island?

Climates and Biomes All of New Zealand has a mild marine west coast climate. However, temperatures and precipitation vary across the country. In general, North Island is warmer than South Island and rarely receives snow. Westerly winds bring moisture. When these winds hit the Southern Alps, they drop rain and snow. As a result, western South Island receives much more precipitation than the east, which is in a rain shadow. Most of the west coast averages more than 100 inches (254 cm) of rainfall each year. Some eastern areas receive less than 20 inches (51 cm).

Most of New Zealand has a temperate forest biome. Forests cover nearly 30 percent of the country. These forests have mostly evergreen trees. Plant and animal life includes many endemic species. The most well known of these are flightless birds like kiwis (KEE-weez), which live in forests. Kiwis sleep during the day and look for food at night. Moas, a much larger kind of flightless bird, became extinct several hundred years ago. Unlike Australia, New Zealand has no endemic mammals except bats. New Zealand also has many exotic species. European settlers brought cats, cattle, deer, and sheep.

TEKS **READING CHECK:** *Places and Regions* Which climate and biome are found throughout New Zealand?

Lake Matheson lies at the foot of Mount Cook and Mount Tasman on New Zealand's South Island. This tiny lake is protected from the wind, so its still surface reflects the mountains clearly.

This example of Maori folk art was created in 1888 for a visit by Te Kooti Rikirangi, a Maori guerrilla and religious leader. Maori artisans were highly skilled in wood carving and stone carving. This painted work, therefore, may show the influence of European art forms.

Natural Resources New Zealand's main resource is good land for farming. More than 50 percent of the land supports crops or livestock. Forests are also important. Pulp and paper are produced from the fast-growing *radiata* pine tree, an exotic species brought from California. Much hydroelectric power is produced from New Zealand's rivers. This kind of energy supplies 65 percent of the country's electricity. Geothermal power is also produced. New Zealand does not have many large mineral deposits. The most important are coal, gold, iron ore, and natural gas.

READING CHECK: ***Places and Regions*** What is New Zealand's main resource?

History and Culture

Like Australia, New Zealand was colonized by British settlers. What effects do you think this had on New Zealand?

Early History and Settlement The first people in New Zealand were the **Maori** (MOWR-ee). They came from Pacific islands to the north about 1,000 years ago. Anthropologists call the early Maori moa hunters because moas were their main prey. Other groups of Maori came later. Most settled on North Island. Their culture was based on farming, fishing, and hunting.

In 1642 Dutch explorer Abel Tasman became the first European to reach New Zealand. However, the Dutch did not return to settle the islands. In 1769 British explorer James Cook landed on North Island and made contact with the Maori. He explored both North Island and South Island.

The first European settlers in New Zealand came from the British colony in Australia. They were missionaries, traders, and whalers. In 1840, British settlers and Maori leaders signed a treaty that gave the British control of the islands. In exchange, the British agreed to protect Maori rights. The British set up several settlements on North Island in the 1840s. However, troubles between the Maori and British soon arose. Some British settlers began taking

Maori lands. Also, diseases introduced by Europeans killed many Maori. These problems led to the Maori Wars of 1845–72, which the Maori lost.

In 1907 New Zealand became an independent country within the British Empire. It continued to develop its farming economy. New refrigeration methods allowed farmers to ship meat and dairy products to Britain. Like Australia, New Zealand sent troops to fight alongside the British in World Wars I and II.

READING CHECK: ***Human Systems*** How did British settlement affect New Zealand's human geography?

People and Languages Most New Zealanders have British ancestors and speak English. Asians and Pacific Islanders are both small but growing minorities. New Zealand's largest minority group, the Maori, makes up nearly 10 percent of the population.

FOCUS ON CULTURE

The Maori The Maori are related culturally to other peoples of the Pacific Islands. According to Maori legend, they came to New Zealand in seven canoes. Then a Maori hero named Maui created North Island by fishing it from the sea. Maori society was made up of different tribes *(iwi)* ruled by chiefs *(ariki)*. Tribes lived in villages. Land belonged to smaller groups *(hapuu)* within each tribe. Maori artists decorated canoes and houses with beautiful designs. Tattooing was common. Chiefs and warriors had facial tattoos *(moko)* that symbolized their high place in society.

Maori culture has changed greatly since Europeans came. Today most Maori live in cities. Many have mixed Maori and European heritage. While Maori have adapted to Western society, many are behind other New Zealanders in education and employment.

INTERPRETING THE VISUAL RECORD

Left: Maori train for a celebration. Their war canoes are in the background. Right: New Zealand's national rugby team, the All Blacks, performs a Maori dance and chant called a haka *before every international match. A New Zealand team performed a* haka *overseas for the first time in 1888, in Great Britain.* ***By what process would the*** **haka** ***enter popular culture?*** TEKS

Many Maori have maintained traditional ways of life. For example, Maori often greet each other by pressing their noses together *(hongi)*. Dances called action songs are also common, as are traditional carved meeting houses. Although nearly all Maori speak English, some prefer to use the Maori language in their homes. In 1987 Maori became an official language of New Zealand.

READING CHECK: ***Human Systems*** How have the Maori maintained traditional ways since the arrival of Europeans?

Settlement Most of New Zealand's 3.9 million people live in lowland areas along the coast. About 75 percent live on North Island. More than 80 percent live in cities. New Zealand's primate city is Auckland. About 30 percent of all New Zealanders live in or around Auckland. Other major cities include Wellington, which is the capital, and Christchurch.

Traditions, Customs, and Food Like in Australia, outdoor activities are popular in New Zealand. A mild climate makes camping, hiking, and sailing possible year-round. Skiing is popular at resorts in the Southern Alps. Organized sports include rugby and cricket, both of which are played throughout the country. New Zealand also competes in international yachting races. In fact, practically every sport one can imagine is available. New Zealanders even enjoy creating new outdoor activities.

In New Zealand sheep outnumber people 13 to 1, so workers who can shear sheep quickly and well are valued. Recordholders can clip the wool from a sheep in less than a minute. The best shearers in the world come to New Zealand to compete against each other. Champions earn prize money and local fame.

Favorite foods in New Zealand include clam soup, lamb, and sweet potatoes. A meringue, fresh fruit, and cream dish called a pavlova is the national dessert. Tea is popular, reflecting the country's British heritage.

READING CHECK: ***Places and Regions*** Where do most New Zealanders live?

INTERPRETING THE VISUAL RECORD *Cable cars provide transportation for residents and visitors in Wellington, New Zealand.* ***Why do you think that cable cars are an efficient form of transportation in this city?***

Economy and Issues

New Zealand's economy is a combination of farming, manufacturing, and tourism. Major exports include wool, meat, fish, and dairy products. Historically, the country's economy has been based on the export of farm goods. However, manufacturing and services have become much more important in the past 20 years.

New Zealand's pastures support millions of sheep and cattle. Major crops include wheat, barley, fruits, potatoes, and vegetables. The country is the world's largest producer of kiwi fruit. Many industries are closely tied to agriculture. For example, factories make processed foods like butter and cheese. Most of these are exported. Other industries include wood and paper production, textiles, and machinery. The main industrial center is Auckland. New Zealand has a growing film industry. Many movies and television programs are filmed there. One reason is the wide range of settings the varied landscape provides. Tourism is also important to the economy.

One challenge with developing industries in New Zealand is the country's small population. It makes New Zealand a small market. Therefore, it is harder for industries to develop an **economy of scale**—a large production of goods that reduces the production cost of each item. For example, if a company produced machines but was only able to sell 10 per year, it might not make a profit. However, if the company sold 10,000 machines per year, it would lower the cost of producing each machine and make a large profit.

Another challenge facing New Zealand is diversifying its economy. The country depends heavily on exports. Therefore, changes in world markets can have important effects on its economy. For example, like Australia, Great Britain used to be New Zealand's main trading partner. In 1973 Britain joined what is now the European Union (EU). Britain then had to raise its tariffs on goods from non-EU trade partners. As a result, prices on products from New Zealand rose in Britain. This increase caused the number of goods New Zealand sold to Britain to fall. Since then, New Zealand has diversified its exports. It has also found new trading partners. New Zealand now trades with Australia, the United States, and Japan. However, global trade is still vital to the country.

INTERPRETING THE VISUAL RECORD *Sheep graze in a pasture on South Island.* ***How does the environment of this sheep station appear to differ from the one you read about in Cities & Settlements: Life in the Outback?***

TEKS **READING CHECK:** *Human Systems* How have changes in world trade patterns affected New Zealand's economy?

 TEKS Working with Sketch Maps, Questions 1, 2, 3, 4, 5

Section 2 Review

Identify
Maori

Define
economy of scale

Working with Sketch Maps
On a map of New Zealand that you draw or that your teacher provides, label North Island, South Island, Cook Strait, Southern Alps, Canterbury Plains, Auckland, and Wellington. What is New Zealand's primate city?

Reading for the Main Idea
1. *Physical Systems* How do the Southern Alps affect New Zealand's precipitation patterns?
2. *Environment and Society* How did technological changes in the early 1900s affect New Zealand's export economy?
3. *The Uses of Geography* In New Zealand's market economy, how are major commercial industries tied to its agricultural products?

Critical Thinking
4. **Analyzing Information** What do you think might have attracted early settlers to New Zealand?

Organizing What You Know
5. Create a chart like the one shown below. Use it to describe the natural environments of North Island and South Island.

North Island	South Island

CHAPTER 31 Review

Building Vocabulary TEKS

On a separate sheet of paper, explain the following terms by using them correctly in sentences.

artesian wells	extensive agriculture
outback	exotic species
marsupials	Maori
Aborigines	economy of scale

Locating Key Places TEKS

On a separate sheet of paper, match the letters on the map with their correct labels.

Western Plateau	Southern Alps
Tasmania	Sydney
North Island	Central Lowlands
Great Dividing Range	Great Barrier Reef

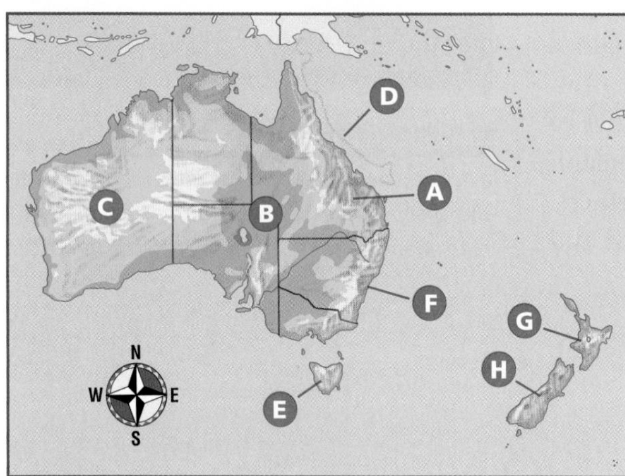

Understanding the Main Ideas TEKS

Section 1

1. *Physical Systems* What are some of the reasons Australia is dominated by arid and semiarid climates?
2. *Human Systems* What is the pattern of the distribution of major cities in Australia? Why do few people live in the continent's interior?

Section 2

3. *The Uses of Geography* How have the physical characteristics of New Zealand changed during the last 1,000 years?
4. *Places and Regions* What are some important characteristics of New Zealand's market economy?
5. *Places and Regions* How are the physical and human geography of Australia and New Zealand similar? How are they different?

Thinking Critically for TAKS TEKS

1. **Drawing Inferences and Conclusions** How do you think environmental conditions in Australia influenced where British colonists settled?
2. **Identifying Cause and Effect** How do you think the development of Australia's economy has affected the environment?
3. **Comparing** How are life and culture in New Zealand similar to life and culture in the United States?

Using the Geographer's Tools TEKS

1. **Analyzing Maps** Study the map of Australia's states and territories in Section 1. Why do you think the colony of New South Wales had land taken away to create other colonies?
2. **Analyzing Charts** Study the chart of Australia's major exports and imports in Section 1. What types of goods does Australia export? How do these exports relate to land use in Australia?
3. **Preparing Graphs** Use the information in Section 1 to create a pie graph of Australia's different ethnic groups. How do you think Australia's population will change during the next 20 years?

Writing for TAKS TEKS

Imagine you are a government consultant in Australia. You have been asked to write a report comparing different points of view on mining in lands the Aborigines hold sacred. You may want to use Internet or library resources to find more information. Write a short report. First, describe how Aborigines feel about mining operations in their sacred lands. Then describe a miner's point of view. Finally, suggest some ways that this difficult issue could be resolved. When you are finished with your report, proofread it to make sure you have used standard grammar, spelling, sentence structure, and punctuation.

SKILL BUILDING

Geography for Life TEKS

Drawing Sketch Maps

Places and Regions Review the unit political map. Then draw a sketch map of Australia showing its six states, two territories, five major cities, and lines of latitude and longitude. On what are the country's state and territorial boundaries based? Study the names of Australia's states and territories. How are they geographical names? What clues do they provide about Australia's culture and history?

Building Social Studies Skills

Interpreting Graphs

Study the pie graph below. Then use the information from the graph to help you answer the questions that follow.

1. What answer most accurately reflects the information from the graph?
   a. New Zealand produces more than half of the world's wool.
   b. China is the world's largest wool producer.
   c. Australia produces more of the world's wool than New Zealand and China combined.
   d. Australia produces more than four times the wool produced in China.
2. How do the countries shown in the graph rank, from the biggest to the smallest producers of the world's wool?

Analyzing Primary Sources

Read the following excerpt from an article by Bill Bryson. Then answer the questions.

> **"I first realized that I was going to like the outback when I read that the Simpson Desert, an area bigger than some European countries, was named in 1929 for a manufacturer of washing machines. . . . It wasn't so much the pleasingly unheroic nature of the name as the realization that an expanse of land of more than 50,000 square miles didn't even have a name until 70 years ago.**
>
> **But then that's the thing about the outback—it's so vast and forbidding that much of it has yet to be charted. . . ."**

3. Which of the following is *not* a reason that the writer likes the outback?
   a. Part of it was named for someone who was not a heroic explorer.
   b. The Simpson Desert is in Europe.
   c. Much of it has not been charted.
   d. Part of it was unnamed for a long time.
4. Which might the writer enjoy more—the Mojave Desert or Los Angeles? Why?

Alternative Assessment

PORTFOLIO ACTIVITY

Learning about Your Local Geography TEKS

Individual Project: Research

Plan, organize, and complete a research project on exotic plants and animals in your area. In your report answer the following questions: What are some of the plants and animals that have been introduced into your area? Where did they come from? Why were they introduced? Have any of these exotic species damaged the environment? If so, how? How have they affected endemic species? Prepare a report on your project and present it to your class.

internet connect

Internet Activity: go.hrw.com
KEYWORD: SW3 GT31 TEKS

Choose a topic on Australia and New Zealand to:

- explore Australia's Great Barrier Reef.
- meet the Aborigines of Australia.
- learn about tectonic processes in New Zealand.

CHAPTER 32

The Pacific Islands

A look at a map will tell you that the Pacific Islands region is characterized by vast expanses of water. The islands are scattered over thousands of miles of ocean, allowing different cultures to develop in isolation from each other.

A Huli man of Papua New Guinea applies face paint.

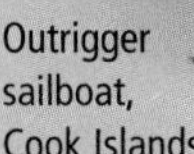
Outrigger sailboat, Cook Islands

Daba Namona! ("Good morning," in Motu.) My name is Jean Vanessa. I live in Vabukori, a village in Port Moresby, the capital of Papua New Guinea. I live with my mother, grandmother, four younger brothers, and my younger sister. My mother is a journalist, and my grandmother bakes bread for the family in a drum oven. Sometimes we sell the bread. My grandfather, whom I loved very much, passed away last year. He was Motu, but my grandmother's language is Susu.

Our house is big, built on stilts, and painted blue. It is a few yards away from the village square where we have meetings and church gatherings and play sports. Our house has one big bedroom. My mum, my sister, my three little brothers, and I sleep in this room. The others sleep in the main living room area.

For breakfast we usually have tea and bread with fillings such as Vegemite,® jam, or peanut butter, and sometimes spaghetti or baked beans. Cereal is expensive, so it's a real treat. On weekends, we often have fried flour (pancakes) for breakfast. After breakfast, I take a bus to school. All my classes are taught in English.

I really like Christmas and New Year's. In Vabukori, we celebrate with much feasting and singing. The villagers are divided into two groups. On Christmas Day one group cooks for the other one. The other group sings songs about Bible stories from dawn to dusk. On New Year's Day the groups exchange roles.

Natural Environments

READ TO DISCOVER

1. What are the main physical features of the Pacific Islands, and what physical processes affect them?
2. How is the Pacific Islands region divided into subregions?
3. What climates, biomes, and resources does the region have?

WHY IT MATTERS

Before air travel became common, travel across the Pacific Islands region was slow and difficult. Use CNNfyi.com or other **current events** sources to learn about how air travel has changed people's lives.

IDENTIFY

Intertropical Convergence Zone (ITCZ)

DEFINE

atoll

LOCATE

Tahiti
Melanesia
Micronesia
Polynesia

Locate, continued

Papua New Guinea
Solomon Islands
Vanuatu
New Caledonia
Fiji
Kiribati
Northern Mariana Islands
Marshall Islands
Guam
Marquesas Islands
Samoa

The Pacific Islands: Physical-Political

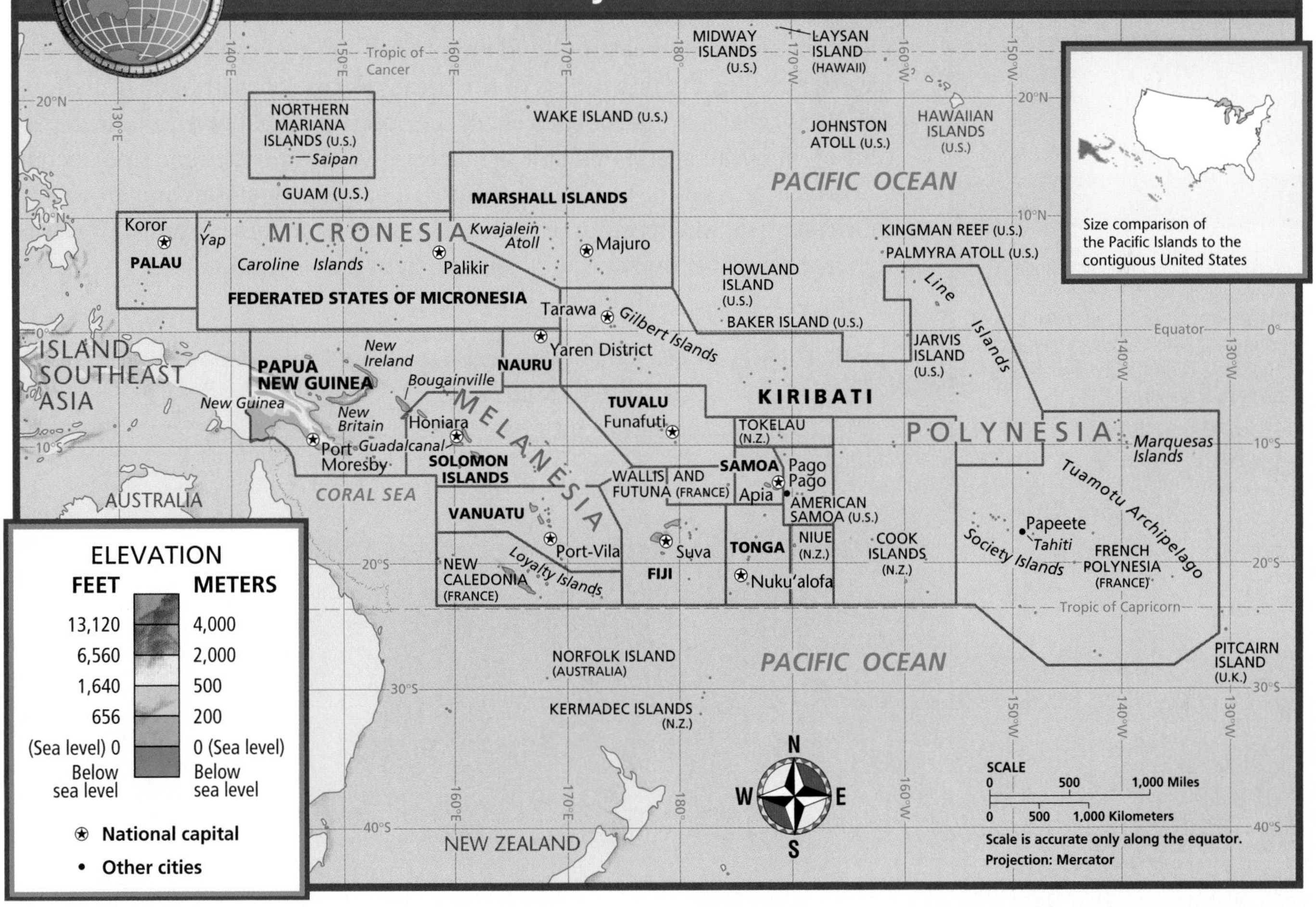

NORTHERN MARIANA ISLANDS (U.S.)
Saipan
GUAM (U.S.)
WAKE ISLAND (U.S.)
MIDWAY ISLANDS (U.S.)
LAYSAN ISLAND (HAWAII)
JOHNSTON ATOLL (U.S.)
HAWAIIAN ISLANDS (U.S.)
PACIFIC OCEAN
MARSHALL ISLANDS
Koror
PALAU
Yap
MICRONESIA
Kwajalein Atoll
Majuro
Caroline Islands
Palikir
FEDERATED STATES OF MICRONESIA
KINGMAN REEF (U.S.)
PALMYRA ATOLL (U.S.)
HOWLAND ISLAND (U.S.)
Line Islands
Tarawa
Gilbert Islands
BAKER ISLAND (U.S.)
Yaren District
NAURU
JARVIS ISLAND (U.S.)
ISLAND SOUTHEAST ASIA
PAPUA NEW GUINEA
New Ireland
Bougainville
New Guinea
New Britain
KIRIBATI
TUVALU
Funafuti
MELANESIA
Honiara
Guadalcanal
Port Moresby
SOLOMON ISLANDS
TOKELAU (N.Z.)
POLYNESIA
Marquesas Islands
SAMOA
Pago Pago
Apia
AMERICAN SAMOA (U.S.)
WALLIS AND FUTUNA (FRANCE)
Tuamotu Archipelago
AUSTRALIA
CORAL SEA
VANUATU
Port-Vila
Suva
FIJI
TONGA
Nuku'alofa
NIUE (N.Z.)
COOK ISLANDS (N.Z.)
Papeete
Tahiti
Society Islands
FRENCH POLYNESIA (FRANCE)
NEW CALEDONIA (FRANCE)
Loyalty Islands
Tropic of Cancer
Tropic of Capricorn
Equator
PITCAIRN ISLAND (U.K.)
NORFOLK ISLAND (AUSTRALIA)
KERMADEC ISLANDS (N.Z.)
NEW ZEALAND

Size comparison of the Pacific Islands to the contiguous United States

ELEVATION

FEET	METERS
13,120	4,000
6,560	2,000
1,640	500
656	200
(Sea level) 0	0 (Sea level)
Below sea level	Below sea level

⊛ National capital
• Other cities

SCALE
0 500 1,000 Miles
0 500 1,000 Kilometers
Scale is accurate only along the equator.
Projection: Mercator

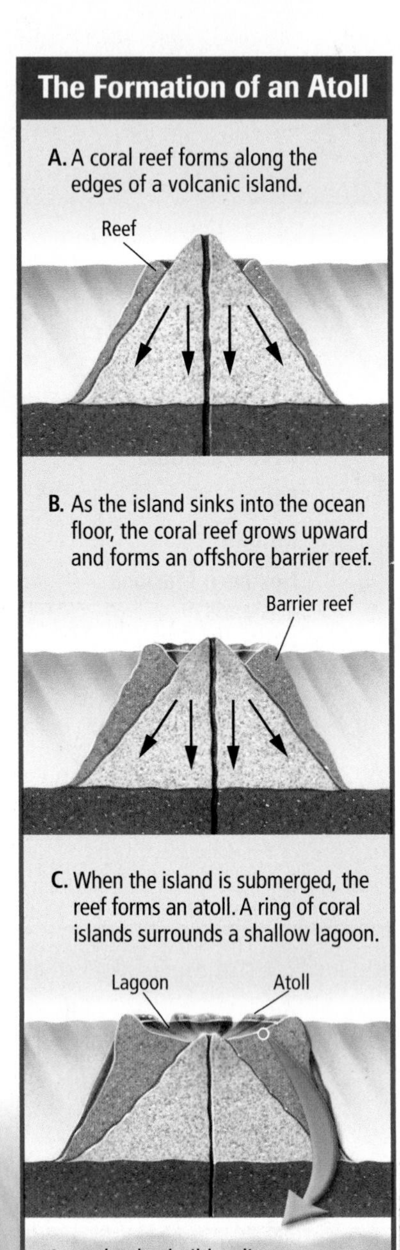

An Ocean Realm

The Pacific Ocean is the largest natural feature on Earth. It covers about a third of the world's surface. More than 10,000 islands lie within this region. Yet the total land area of the islands is very small. Most of them are tiny coral islands where no people live.

High and Low Islands Geographers classify the islands of the Pacific as either high islands or low islands. High islands can be further divided into continental and oceanic islands. Both types of high islands tend to be mountainous and rocky, and both may have volcanoes. Continental islands are formed from continental rock and lie on a shallow continental shelf such as that of Australia. New Guinea, the world's second-largest island, is a continental island. Oceanic islands such as Tahiti are simply volcanic mountains that have grown from the ocean floor to its surface.

Low islands form from coral. They are usually small and flat. Low islands also tend to have a characteristic ring shape. A ring-shaped coral island, or a ring of several islands linked by underwater coral reefs, is called an **atoll.** Within this ring lies a body of water called a lagoon. Kwajalein (KWAH-juh-luhn) Atoll in the Marshall Islands is about 80 miles (130 km) long and 20 miles (30 km) wide. However, most atolls are much smaller.

FOCUS ON GEOGRAPHY

Coral Reefs Coral is formed by colonies of tiny marine animals. Millions of these tiny creatures' skeletons build up into reefs on a volcanic base. Coral reefs have been called the rain forests of the ocean because of their biological diversity. Reefs cover less than 0.2 percent of the ocean floor. However, about 25 percent of all ocean species inhabit coral reefs, which provide food and shelter. Coral reefs around the world face natural threats like typhoons and crown-of-thorns starfish, which eat coral. Reefs can recover from these assaults. However, they are now also threatened by human activities. Pollution and destructive fishing practices are among the dangers.

READING CHECK: ***Physical Systems*** How might the destruction of the world's coral reefs affect the biological diversity of the oceans?

TEKS

Physical Processes High islands and low islands tend to have very different environments. The high islands' volcanic soils are usually rich. Many high islands support rain forests. Variations in elevation, rainfall, and soil lead to differences in plant life within one island as well as between islands. Low islands, on the other hand, are usually much less fertile. They often have no sources of freshwater.

The Pacific is the site of active tectonic processes. Many islands have active volcanoes, and many more volcanoes lie deep beneath the ocean's surface. Earthquakes are common in some areas. Tsunamis can present a danger as well to coastal areas.

Tectonic processes have also shaped the Pacific Ocean's floor. The Pacific's average depth is 14,000 feet (4,300 m). However, the deepest point, in the Mariana Trench, is more than 36,000 feet (11,000 m) below sea level. Oceanic trenches are the result of subduction. Volcanic ridges, visible on the map as chains of islands, often run parallel to the trenches. The Northern Mariana Islands and Tonga Islands are examples of volcanic island chains alongside deep trenches.

READING CHECK: *Physical Systems* Which type of Pacific island can often support rain forests, and why?

Three Island Groups

Geographers divide the Pacific Islands into three large subregions—Melanesia, Micronesia, and Polynesia. These groupings are based on the cultural variations and spatial arrangement of the islands.

Melanesia lies closest to Australia. It includes the eastern half of New Guinea, which makes up most of Papua (PA-pyooh-uh) New Guinea. From there Melanesia stretches east to include the Solomon Islands, Vanuatu (van-wah-TOO), New Caledonia, and Fiji. Most of the Melanesian islands have mountains and volcanoes.

Micronesia lies east of the Philippines, mostly north of the equator. Micronesia includes the Caroline Islands, the Gilbert Islands of Kiribati, the Northern Mariana Islands, and the Marshall Islands. These are a mix of high islands and low islands. The Northern Mariana Islands are a chain of volcanoes. Guam, the largest, is made up of a limestone plateau in the north and volcanic hills in the south. In contrast, the Gilbert Islands are all coral atolls.

Polynesia is the largest of the three subregions. It covers a huge triangle with its corners at Easter Island, the Hawaiian Islands, and New Zealand. See the unit atlas for Easter Island's isolated location about 2,000 miles (3,200 km) west of South America. Polynesia also includes the Cook, Marquesas, Samoa, Society, and Tonga Islands, and the Tuamotu Archipelago.

READING CHECK: *Places and Regions* What are the three subregions of the Pacific Islands?

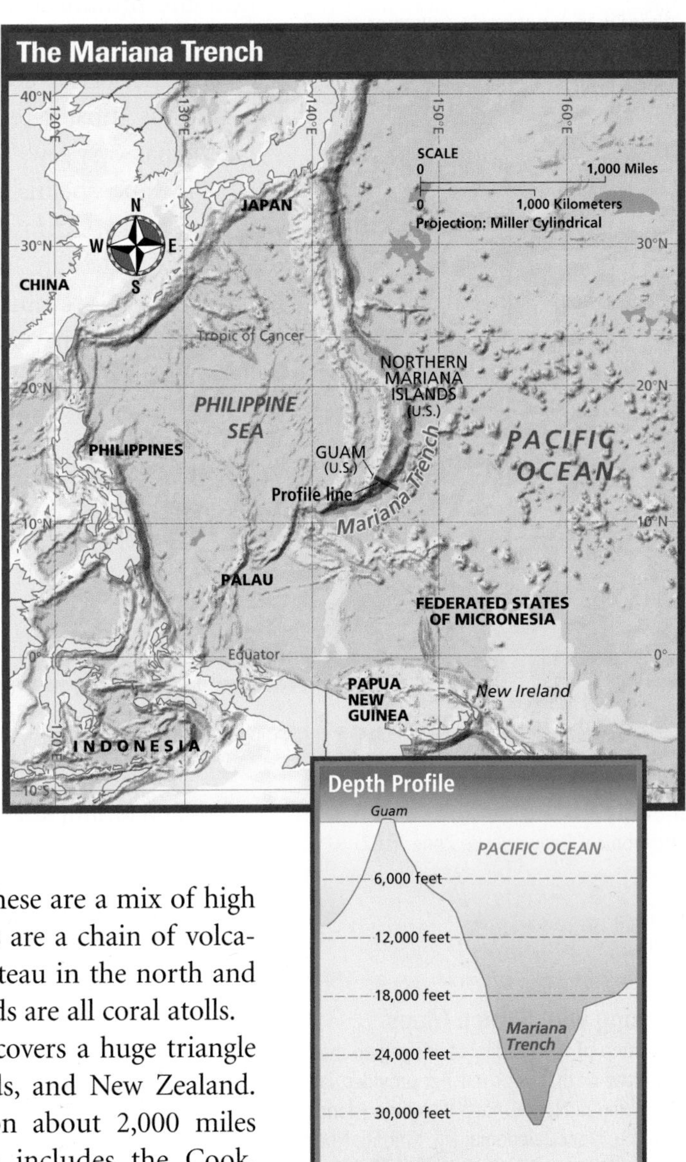

The Mariana Trench is the deepest known place on Earth's surface. The trench is more than 1,580 miles (2,550 km) long.

INTERPRETING THE VISUAL RECORD

The band of clouds in the photo marks the Intertropical Convergence Zone (ITCZ), which moves north or south of the equator throughout the year to the area with the warmest surface temperature. ***Does this photo show the ITCZ north or south of the equator? Which hemisphere was experiencing summer when this photo was taken? How can you tell?***

Climates, Biomes, and Resources

Most of the region's islands lie between the Tropic of Cancer and the Tropic of Capricorn. As you might expect, climates there are generally hot with high rainfall. Only Papua New Guinea, with its mountainous interior, has areas with highland climates. Here mountain peaks have tundra and alpine grasslands despite their location in low latitudes. Some parts of the region have distinct wet and dry seasons. For example, in the far western Pacific, wet and dry seasons are influenced by monsoons. In Papua New Guinea and some other areas, heavy rainfall supports thick tropical rain forests.

Trade winds play an important role in the climates of the islands. Northeast trade winds flow from the Tropic of Cancer. Southeast trade winds flow from the Tropic of Capricorn. The area where these prevailing winds meet near the equator is called the **Intertropical Convergence Zone (ITCZ)**. This area has generally calm humid air at low elevations. However, the warm, moist air of the ITCZ condenses and cools as it rises, causing heavy rainfall. Typhoons most often occur in the western Pacific, while the southeastern Pacific has very few. These swirling oceanic storms can be very destructive.

Fish and shellfish are important resources for the region. Lobsters, octopuses, sharks, shrimp, and tuna are just a few of the sea creatures people catch. Cultured pearls are harvested from oysters in French Polynesia and the Cook Islands. Other resources are less plentiful. Papua New Guinea and some other islands of Melanesia export timber. Only the large continental high islands have useful metal deposits. Papua New Guinea has major gold and copper deposits. New Caledonia is rich in nickel.

TEKS **READING CHECK:** *Places and Regions* What are the main natural resources of the Pacific?

TEKS Questions 1, 2, 3, 4

Section 1 Review

Identify
Intertropical Convergence Zone (ITCZ)

Define
atoll

Working with Sketch Maps
On a map of the Pacific Islands region that you draw or that your teacher provides, label Tahiti, Papua New Guinea, Solomon Islands, Vanuatu, New Caledonia, Fiji, Kiribati, Northern Mariana Islands, Marshall Islands, Guam, Marquesas Islands, and the Samoa Islands. Outline Melanesia, Micronesia, and Polynesia. Which of the three subdivisions lie mostly south of the equator? Which lies mostly north of the equator?

Reading for the Main Idea
1. *Places and Regions* What physical processes have shaped the region's islands? What environmental hazards affect the region?
2. *Physical Systems* What factors form the basis for dividing the Pacific Islands into three subregions?

Critical Thinking
3. **Analyzing Information** Why does New Guinea have highland climates? What factors affect the region's climates?
4. **Drawing Inferences and Conclusions** How do you think the distribution of mineral resources in the region relates to how different islands are formed?

Organizing What You Know
5. Create a chart like the one shown below. Use it to make generalizations about the two types of islands found in the region.

	High islands	Low islands
Surface landforms and elevation		
Soil-building process		
Soil fertility		
Vegetation		

Section 2 History and Culture

READ TO DISCOVER

1. What are some important events in the history of the region?
2. What are the traditions and culture of the region like?

DEFINE

trust territories
pidgin languages
matrilineal

WHY IT MATTERS

Early contacts with Europeans introduced new diseases into the Pacific Islands region. Use CNNfyi.com or other **current events** sources to learn about modern-day outbreaks of diseases.

LOCATE

Wake Island
American Samoa

History

Researchers have used archaeology and other evidence to learn that waves of people from Southeast Asia settled the region. Much later, Europeans and others arrived and brought many changes.

Migration Patterns Humans may have lived on New Guinea at least 33,000 years ago. Human migration into the Pacific may have begun even earlier than that. (See Geography for Life: Migration into the Pacific.) Over thousands of years, different peoples spread to different island groups. They sometimes mixed with earlier settlers. Over time, the peoples of Micronesia and Polynesia developed some distinct cultural features and became different in appearance. While the peoples of the two regions are different from each other, they share many cultural features. On the other hand the third region, Melanesia, is more distinct.

Polynesians created these stone figures, the tallest of which stands 37 feet (11 m) high, on Easter Island, or Rapa Nui (RAH-pah NOO-ee), as it is also known. After a period of peace and prosperity that lasted from about A.D. 1000 to 1500, rapid population growth and deforestation created an environmental crisis. Hunger and fighting over scarce resources devastated the island's people.

The French—like the British in Australia—used their colony in New Caledonia as a prison. They sent more than 22,000 French prisoners there between 1864 and 1897.

Cultural patterns, languages, and physical traits there all differ from those in Micronesia and Polynesia. Many Melanesians seem to be genetically linked to the Aborigines of Australia. However, peoples and cultures within Melanesia vary a great deal. Movement and mixing of peoples may have gone on for thousands of years. The picture is not complete, and more research may provide new details.

TEKS **READING CHECK:** ***Human Systems*** Where did the first settlers of the Pacific Islands come from?

European Arrival Europeans began to explore the Pacific Islands in the 1500s. Ferdinand Magellan, a Portuguese navigator working for Spain, sailed across the Pacific in 1520–21. Other explorers followed. Spanish, Dutch, English, and French sailors came to explore, trade, spread Christianity, and claim territory. Later, Germany, Japan, and the United States entered the race for colonies in the area. The United States captured Guam and the Philippines from Spain during the Spanish-American War in 1898. By the end of the 1800s, foreign powers controlled nearly the whole region.

At first, colonial control was limited. American and European whale hunters sailed into the Pacific, sometimes setting up small outposts. Before the discovery of petroleum, whale oil was very valuable for lighting and industrial uses. With little or no regulation by the colonial countries, the whalers and traders badly exploited the people of the islands. This disrupted traditional cultures in many areas. Some Pacific islanders were enslaved. Unknowingly, the Europeans also spread deadly diseases, including measles and influenza.

Over time, colonial rule became more organized. Colonial powers set up plantations and military bases. The British brought thousands of workers from India to work on sugar plantations in Fiji. Despite the colonial presence, the Pacific Islands remained uninvolved in global politics until the 1940s. World War II then brought sudden changes to the Pacific region. Many

INTERPRETING THE VISUAL RECORD

Captain James Cook, one of Europe's greatest explorers, led three voyages to the Pacific Ocean. William Hodges was the official artist on Cook's second trip. Hodges painted this view of Matavai Bay, in Tahiti, in the 1770s. ***How might this painting have influenced Europeans' perceptions of the Pacific Islands?*** TEKS

islands became battlefields between 1941 and 1945. Others were used as bases. Armies, planes, and ships moved through the region. Japan conquered many islands early in the war. Over time, the United States and its allies pushed back the Japanese and defeated them. At the end of the war, the United Nations made some islands **trust territories**. These were areas placed under the control of another country while they prepared for independence.

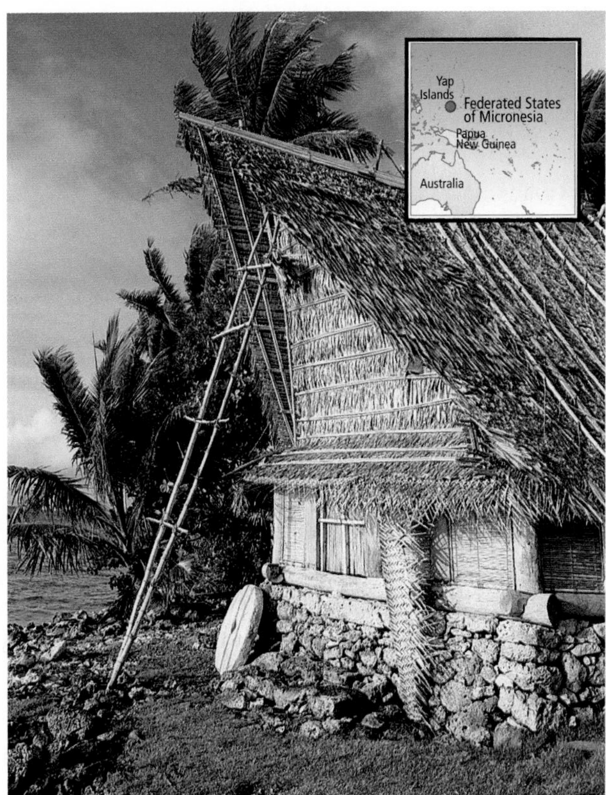

A piece of traditional stone money leans against the wall of this men's meeting house on the island of Yap. Although the U.S. dollar is the common currency in Yap, the stone disks are used for some major transactions, such as land purchases.

Independence Since the end of World War II, the islands of the Pacific have moved away from colonialism. Some have become fully independent. A few other islands are still colonies of or are otherwise associated with an outside country. Guam, Wake Island, and American Samoa are U.S. territories. The Northern Mariana Islands form a commonwealth with the United States, similar to Puerto Rico's situation. The Federated States of Micronesia are in free association with the United States. This status allows their citizens to work in this country. In return, the United States can keep military bases on the islands. Australia, Chile, France, New Zealand, and the United Kingdom also have territories in the Pacific.

READING CHECK: *Human Systems* Why did Europeans and other outsiders come to the Pacific Islands?

Traditions and Culture

Each of the societies of the Pacific developed its own culture. Despite the many variations, it is possible to make some generalizations about cultural features. Groups on the same island or within an island chain often shared cultural characteristics. Some similarities extended through a subregion or even through the entire region.

Today the Pacific Islands are home to a great number of different ethnic groups and languages. How did this happen? Huge stretches of ocean between islands allowed different cultures and languages to develop independently from one another. On New Guinea, thick rain forests and rugged mountains separated different groups of people in a similar way. Today the peoples of Papua New Guinea speak more than 700 different languages. Some of them are spoken by only a few hundred people.

Reflecting the lasting influence of colonialism, English and French are used in government and education in many parts of the region. To communicate within island groups where languages differ, some Pacific peoples have developed simplified languages based on English. These are called **pidgin languages**.

Education Almost all children in Polynesia and Micronesia now receive education through the high school level. Schooling is not yet available to everyone in Melanesia. There are several universities in the Pacific Islands. Most teach mainly in English, but a French-language university has locations in both New Caledonia and Tahiti.

Connecting to

TECHNOLOGY

Navigation Skills of the Pacific Islanders

To the untrained eye, the open ocean looks empty and featureless. How would you find your way from one tiny island to another? What if you were in a wooden sailboat without a compass, radio, or other modern instruments?

The sailors of the Pacific Islands used many skills to navigate the ocean. At the start of a trip, they would take their bearings from landmarks on shore to plot their direction. Then they would note the rising points of certain stars and steer by "star paths." Winds and currents were also familiar to the sailors. The flights of birds and the movements of clouds provided clues. The sailors noted the reflections of lagoons on the bottoms of clouds. They could even tell the way waves were deflected off of distant islands. Understanding all these signs helped these sailors reach their destinations, sometimes across thousands of miles of ocean. People of the Pacific Islands still use many of these skills.

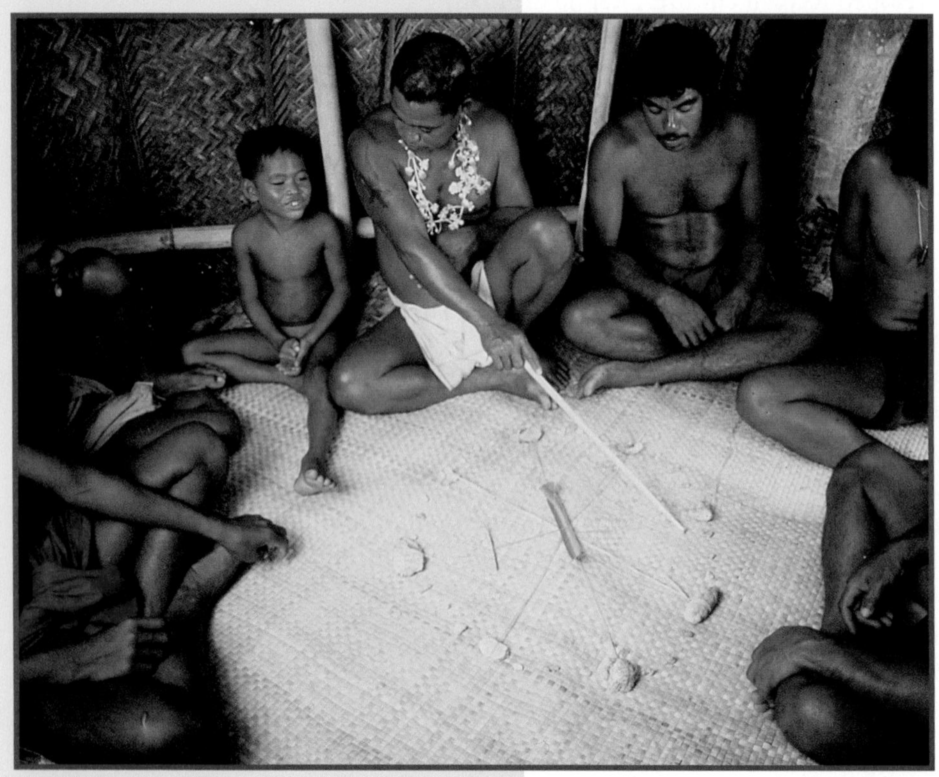

Expert navigators of the Pacific Islands teach others how to use a compass based on the locations of certain stars.

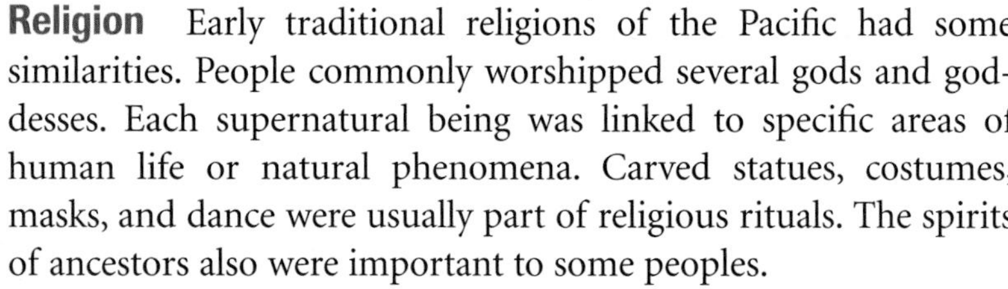

Religion Early traditional religions of the Pacific had some similarities. People commonly worshipped several gods and goddesses. Each supernatural being was linked to specific areas of human life or natural phenomena. Carved statues, costumes, masks, and dance were usually part of religious rituals. The spirits of ancestors also were important to some peoples.

Christianity is the main religion in the Pacific Islands today. Roman Catholic and Protestant missionaries spread their faiths through the area during the colonial period. Churches are important to community life in many parts of the region. However, people still practice local religions in parts of Melanesia, including Papua New Guinea.

READING CHECK: *Human Systems* How did French, English, and Christianity spread to the Pacific Islands?

Food Before Europeans came, three root crops—sweet potatoes, taro, and yams—were among the key foods in the Pacific Islands. Three tree crops—bananas, breadfruit, and coconut—were also important to the diet. The coconut palm was particularly useful. Pacific Islanders ate the soft flesh of the coconut and drank the milk. They used the shells as containers and got fibers from the coconut's outer husk. The people also made roofs and baskets from coconut palm tree fronds. Rice was the only grain grown in the region. It was grown only in parts of the Mariana Islands.

Around the Pacific, fish are the main source of protein. Before European contact, the domesticated animals of the Pacific Islands were limited to chickens, dogs, and pigs. People on some islands ate fruit bats. Not all these animals were found on all islands. For example, no dogs lived in the Marquesas Islands. Easter Island had only chickens. Many islands had no domesticated animals at all. In general, people served meat—especially pigs—only at special celebrations. In Samoa, for example, an occasion's importance could be measured by the number of pigs served.

Today imported foods such as canned meat, boxed cake mixes, and soft drinks are a growing part of people's diet. The need for money to buy imported goods has pushed many farmers to switch to growing cash crops such as sugarcane for export.

Traditional Societies Historically, people in the region tended to be organized into clans or tribes. They lived in places that ranged from small villages to communities with thousands of people. Patterns of social interactions within these groups had some features in common.

Polynesian groups often had complex rules of behavior and social ranks. On some islands, like Tahiti and Tonga, political status was kept within certain families. Individuals of high rank had great power over the common people. Chiefs distributed land and organized work such as the digging of irrigation systems. Competition for land, resources, and status often led to war.

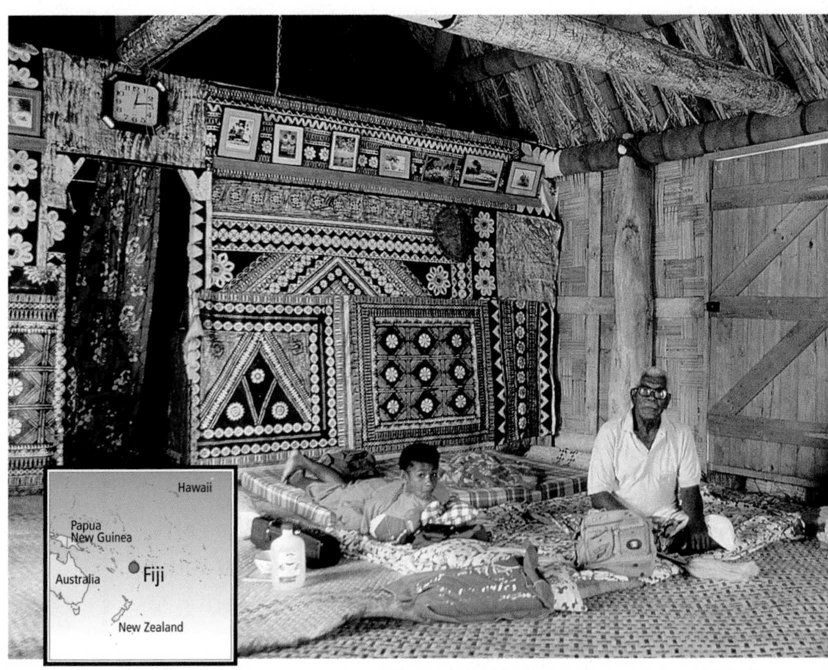

This Fijian chief's home contains traditional handicrafts. Fiji has a Great Council of Chiefs that elects the country's president.

Elsewhere in the Pacific, people placed less emphasis on inherited rank. People could gain status by giving feasts or presenting valuable gifts. They could also gain status by organizing trade with other groups or other islands. Many Melanesian peoples selected a leader through a kind of competition. A leader won supporters through his abilities or his wealth. Marriage or descent from an important family could also help determine leadership. The amount of power a leader held varied from group to group.

One interesting feature of Micronesia was that local groups were often **matrilineal**. That is, people traced kinship through the mother. When a marriage took place, the husband became part of the wife's clan, rather than the other way around. In some societies, women held high status. For example, Tongan women outranked men in several situations.

Art in the Pacific Islands was often connected to religion. Wood carvings, for example, usually showed gods and ancestors. Today island artists create similar carvings and sell them to tourists.

READING CHECK: *Human Systems* How did a Melanesian leader win his position?

TEKS Questions 1, 3, 4, 5

Review

Define

trust territories, pidgin languages, matrilineal

Working with Sketch Maps

On the map you created in Section 1, label Wake Island and American Samoa. In which of the Pacific Islands' three subregions is American Samoa located?

Reading for the Main Idea

1. *Environment and Society* What food crops did Pacific Islanders grow before contact with Europeans?
2. *Human Systems* What six foreign countries have territories in the Pacific?

Critical Thinking

3. **Analyzing Information** How did physical features affect migration and the distribution of culture groups in the region?
4. **Analyzing Information** How is the colonial history of the Pacific Islands reflected in the region's culture today? How has importing food changed the region?

Organizing What You Know

5. Create word webs like the one shown below to describe the people, languages, religions, educational systems, food, and traditional customs of the region.

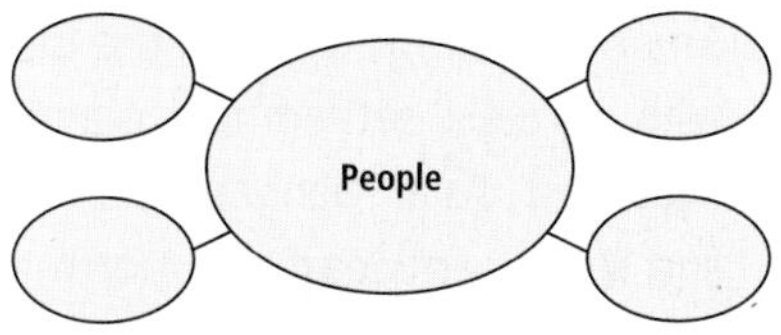

Geography for Life

Migration into the Pacific

To learn about the past, geographers study many clues. One great historical mystery is how the remote and numerous Pacific Islands were settled. Geographers have drawn from different areas of study to map the migration and settlement of the Pacific Islands.

One idea developed after a geographer drifted on a raft in the ocean for some time. He proposed that South Americans could have drifted west across the Pacific to reach Polynesia. Evidence from the field of botany may support this theory. Polynesians raise sweet potatoes. However, the plant is native to South America.

Geographers also study what cultures leave behind. For example, a type of pottery called Lapita ware is found throughout the western Pacific region. Lapita ware is named for a certain site in New Caledonia. These ceramic vessels are of many types and are highly decorated. The people who made this pottery had excellent navigation and canoe-building skills. They were also farmers. Archaeologists have compared samples of pottery from the region's different islands. They concluded that the Lapita culture spread from Fiji to Tonga and Samoa. Fiji and Tonga may have been settled by 1300 B.C.

Physical geography can also offer hints about the islands' settlement. For example, when ocean levels were lower during the ice ages some Pacific Islanders may have traveled across land bridges. At that time, New Guinea and Australia were joined by dry land.

Today most Polynesians believe their people came from places to the west—like Indonesia and the Philippines. The islands of Southeast Asia are large and close together. It is likely that mainland people who needed new fishing areas and land set off for the nearby islands. Overpopulation of their homeland was probably a major factor. After these travelers perfected their navigation skills, their voyages took them farther east. Eventually, they migrated through the Melanesian islands into Polynesia.

People from other parts of the world, such as South America, may have visited the islands. However, the people who actually settled the islands were most likely from Southeast Asia. Geographers think Polynesian migration and settlements east and north eventually formed a huge triangle in the Pacific Ocean. (See the map.) The cultures that evolved in central Polynesia had their beginnings in Tonga and Samoa. From there, some people moved eastward to the Marquesas Islands. During the next 200 years, one group went south to Easter Island and another to the Hawaiian Islands. By about A.D. 500, Polynesians had spread throughout the region, with the exception of New Zealand, which they reached in about A.D. 1000.

Migration and Settlement of the Pacific Islands

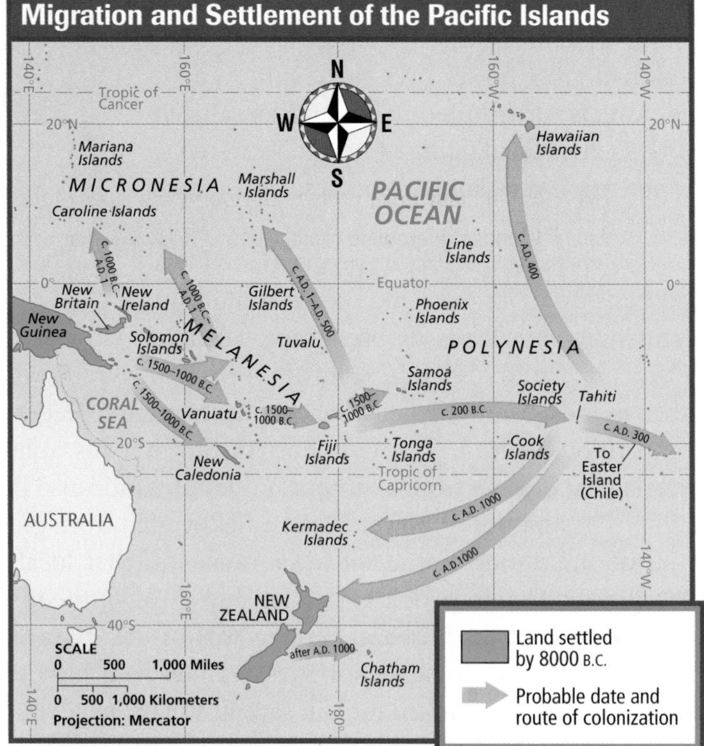

INTERPRETING THE MAP ***Which of the three subdivisions of the Pacific Islands was settled first? How might the winds in this area have affected early migration patterns?*** TEKS

Applying What You Know

1. **Summarizing** What are two theories about how the Pacific Islands were settled?
2. **Evaluating** Based on what you know, what geographic tools would you use to learn about how your state was settled?

Section 3 The Region Today

READ TO DISCOVER

1. What are the economies of the Pacific Islands region like?
2. What are some demographic characteristics of the region?
3. What challenges do the people of the region face?

WHY IT MATTERS

Like those in many other parts of the world, the cities of the Pacific Islands are growing rapidly. Use CNNfyi.com or other **current events** sources to learn more about why so many people are moving to cities and how this shift affects different countries.

IDENTIFY

Exclusive Economic Zone (EEZ)

DEFINE

copra
phosphates

LOCATE

Yap
Nauru
Port Moresby
Suva
Papeete

Economy

The region's economies have changed as they have become linked with the global economy. However, regional trade networks linked the islands long before European contact. The uneven distribution of resources, even on the larger islands, made trade essential. Trade networks not only stretched throughout island chains but also operated across wide areas of the Pacific. Traders carried goods such as feathers, food, mats, shells, spices, and wood in their canoes. Trading sometimes took on symbolic or political meanings. For example, many people in the Carolines offered yearly gifts to the chiefs of villages on the island of Yap. In exchange, the Yapese chiefs allowed people from the Carolines to continue to use land that the Yapese had claimed.

More recently, development in the region has been slow. Industries here face a number of hurdles. Local markets are small, and raw materials are limited. The need to import raw materials and export finished products adds to the cost of goods made in the region. For these reasons, the Pacific Islands region overall is poor. On many islands, people still rely on fishing and subsistence farming for food. Many of the main crops are those grown long ago. Coconut oil and **copra**—dried coconut meat—have been major exports. Some plantations grow introduced crops like cacao, pineapples, and vanilla.

Each country of the Pacific controls an **Exclusive Economic Zone** (**EEZ**). These zones stretch 200 nautical miles (370 km) from each country's shores. The countries can charge fees for economic activities within their EEZs. Fees paid by foreign businesses, particularly big tuna-fishing companies, provide much-needed income.

Mining has become important in Papua New Guinea, New Caledonia, and Nauru. The people of the Pacific may someday profit from the mining of metal deposits from the ocean floor. The technology needed to mine these minerals may not be practical for many years, however.

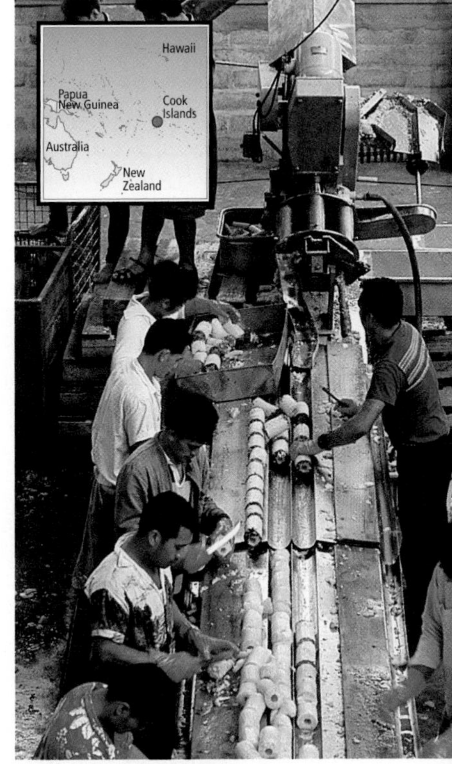

Workers process pineapple in a Cook Islands factory. Growing pineapples has caused extensive soil erosion on some of these islands.

Some islands have tried to move toward a manufacturing economy. The Northern Mariana Islands, Cook Islands, Fiji, and Tonga export textiles and clothing. Tourism provides more income than manufacturing, however. Clear blue water, white sand, and island culture draw visitors from around the world, particularly Japan and the United States. Tourism is a vital industry for some places, such as Tahiti and Tonga. Other islands have attracted few tourists.

READING CHECK: ***Places and Regions*** What role might technology play in the development of the region's resources?

Population and Migration Today

Overall, the population of the Pacific Islands is low. Even Papua New Guinea, which is larger than California, has only about 5 million people. However, some of the smaller island countries are very crowded.

There are few big cities in the region. Port Moresby in Papua New Guinea stands out as the largest city. It has a population of about 200,000. Suva, the capital of Fiji, is home to about 170,000 people. Papeete, a city on the island of Tahiti, has a population of about 24,000. It is the capital of French Polynesia and a regional center for tourism, transportation, and trade.

Although the cities of the Pacific are not large, the region saw rapid urban growth in the late 1900s. Population movement from island to island has also increased. People tend to move to cities and the more populated islands. At the same time, there is a great deal of emigration or movement out of the region. Many Pacific Islanders move to Australia, New Zealand, and the United States. The reasons for all three of these patterns are similar. The populations of most islands in the region have very high rates of natural increase. In fact, some of those rates are among the highest in the world. Birthrates in the Pacific have remained high while death rates have fallen. Populations have grown rapidly, straining resources. This is particularly true in the smaller islands. The search for jobs, education, and a better standard of living pushes people to move to other islands or to other regions.

While emigration may keep a population from growing too quickly, it can also cause problems. Young productive workers are often the ones who move away. As a result, some islands have a shortage of skilled labor.

READING CHECK: ***Human Systems*** What are the causes of migration within and out of the region?

Facing Challenges

In addition to rapid population growth, the region faces other challenges. These include concerns about how economic development will affect the environment. Political problems and issues of nuclear testing and climate also cloud the region's future.

With their small land area, many islands are particularly vulnerable to environmental destruction. For

Tourists relax at a hotel on a small island west of Tahiti. Visitors can enjoy a range of activities—from windsurfing to watching a guide feed sharks.

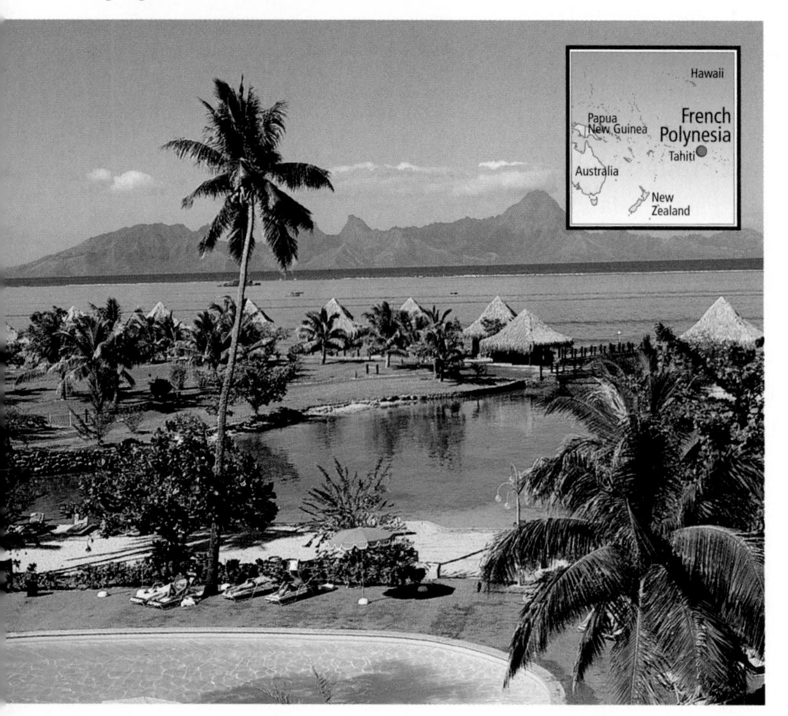

example, cutting forests in Melanesia could lead to rapid soil erosion. Also, mining has polluted some streams in Papua New Guinea. In addition, overfishing may reduce future catches.

The tiny island country of Nauru—less than 8 square miles (21 sq km) in area—displays an extreme case of environmental exploitation. Mining **phosphates**—chemicals used to make fertilizer—brings essential income to Nauru. The government shares some of the profits with Nauru's people. It has invested the rest of the money. Those investments may provide income after the supply of phosphates runs out. However, strip mining has steadily ruined most of the island's surface.

Phosphate mining on Nauru has left about 90 percent of the island a wasteland. In the photo above, ancient coral remains after the phosphate material has been removed.

Another concern involves past nuclear weapons testing. France, Great Britain, and the United States used their Pacific territories as nuclear testing grounds. They exploded bombs from the 1940s to the 1960s. France continued underground testing until the late 1990s, even though French Polynesians and other countries protested. Some people fear the radiation from the tests may cause health problems in the future.

Finally, the possibility of global warming is a special worry for the people of the Pacific. As you read earlier, many researchers believe worldwide temperatures are rising. As a result, melting polar ice might raise ocean levels. If the ocean rises even slightly, many low-lying islands will be submerged or become more vulnerable to storms.

Some islands have also suffered from political violence in recent years. Beginning in the 1970s, Papua New Guinea's military fought against a group demanding independence for the island of Bougainville. This fighting continued through the 1990s. Ethnic divisions have led to another conflict in Fiji. Indo-Fijians make up more than half of that country's population. They are descended from workers the British brought from India. Beginning in the 1970s, some ethnic Fijians became concerned about losing political control to Indo-Fijians. Tensions between the two groups have led to violence.

READING CHECK: *Environment and Society* How have humans modified the environment in parts of the region?

TEKS Questions 1, 2, 3, 4, 5

Section 3 Review

Identify Exclusive Economic Zone (EEZ)

Define copra, phosphates

Working with Sketch Maps On the map you created in Section 2, label Yap, Nauru, Port Moresby, Suva, and Papeete. Which is the region's largest city?

Reading for the Main Idea

1. *Human Systems* What two roles did traditional trade networks play in the Pacific?
2. *Places and Regions* What are five sources of income for the Pacific Islands region?
3. *Environment and Society* Why is the possibility of global warming particularly worrisome to the peoples of the Pacific?

Critical Thinking

4. **Evaluating** How have policies about resource development affected environments in Melanesia and Nauru? What economic impact have these policies had?

Organizing What You Know

5. Create a cause-effect diagram like the one below. Use it to identify the patterns of human movement in the Pacific Islands today.

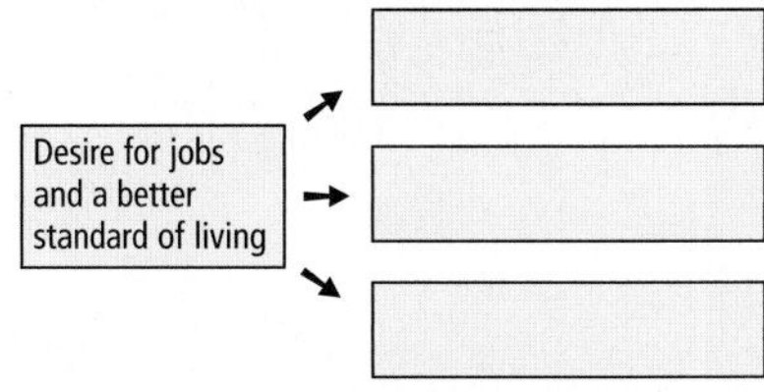

CASE STUDY

Exotic Invaders

Environment and Society The Pacific Islands are home to some of the most unusual creatures on Earth. Unfortunately, the introduction of exotic plants and animals threatens or endangers many of these species. When new plants or animals are brought to an island, disaster can result. In the most extreme cases, the original plants or wildlife die off and are replaced by the invaders.

The Brown Tree Snake

The brown tree snake provides a dramatic example of the effect an exotic invading animal can have on island wildlife. This snake was once found only in Australia, eastern Indonesia, New Guinea, and the Solomon Islands. During the 1940s and 1950s, the snakes probably stowed away in ships' cargoes and came to Guam accidentally. Soon they began to multiply and spread throughout the island. Today, in some of the island's forest areas, as many as 12,000 snakes live within just one square mile. In the 1960s local residents began to notice that there were fewer and fewer birds. Since then, native birds like the Guam flycatcher have practically disappeared from the island. Today native forest birds can only be found on the island's northernmost tip. Guam could become the first place on Earth to lose all its native birds.

A Detective Story

At first, researchers were not sure why the island's birds were disappearing. Possible reasons for the decline included disease, hunting, loss of habitat, pesticides, and predators. An introduced animal could also be responsible. Brown tree snakes, cats, dogs, and rats were all potential suspects.

Researchers used a geographic approach to solve the puzzle. First, they noticed that the snake was the only suspect not found on three nearby islands. These islands did not experience the same decline in bird populations as Guam. The scientists also noted that birds and bird eggs are important sources of food for the snake. The brown tree snake then became the number one suspect. Next, researchers gathered data about the date, location, and number of bird and snake sightings on Guam. Poultry owners helped by reporting when they first saw the snakes. Government biologists provided information about the number of birds on different parts of the island over previous years. Scientists then plotted this data on a map. They saw that as the snake's range moved north the range of native birds grew smaller and smaller. This evidence proved that the snake was the main culprit in the decline of native birds.

The Rabbits of Laysan

Laysan Island is another example of the disastrous effects introducing exotic animals can have on island habitats. About 1903, workers brought rabbits to this small sandy strip of land north of Hawaii. A year later, the workers left. In 1923, visitors returned and found Laysan Island a barren wasteland with only a few stunted trees. At some time the rabbits had multiplied to more than 5,000 and eaten most of the plants. Several types of native birds had also disappeared. Even the rabbits were dying out. With most of the plants gone, the rabbit population had shrunk to about 200. The remaining rabbits were removed. Within 10 years, plants had taken root again and several kinds of birds had returned to the island. Scientists now consider Laysan a success story.

The small photo opposite shows a brown tree snake. Goats, such as the feral animal in the inset photo below, are an exotic species in Hawaii. They have eaten practically all the native plants in some areas. In the background you can see what part of Kauai's coast looked like before (left) and after (right) the goats stripped the area's vegetation.

Islands of Trouble

Islands' small size and their isolation from other areas of land make them particularly vulnerable to invading species. Unlike plants and animals living on continents, those on islands quickly run out of new places to go when an exotic invader takes over. Islands are like fragile little rafts. They have limited space and a limited food supply. When one creature takes too much space and food, the other creatures often find themselves with nowhere to go.

Applying What You Know

1. **Summarizing** What steps did researchers take to solve the mystery of the disappearing birds?
2. **Identifying Cause and Effect** What role did people play in bringing the brown tree snake to Guam and rabbits to Laysan? What were the results of each event?

CHAPTER 32 Review

Building Vocabulary TEKS

On a separate sheet of paper, explain the following terms by using them correctly in sentences.

- atoll
- Intertropical Convergence Zone (ITCZ)
- trust territories
- pidgin languages
- matrilineal
- copra
- Exclusive Economic Zone (EEZ)
- phosphates

Locating Key Places TEKS

On a separate sheet of paper, match the letters on the map with their correct labels.

- Tahiti
- Papua New Guinea
- Solomon Islands
- Fiji
- Northern Mariana Islands
- Wake Island
- Nauru

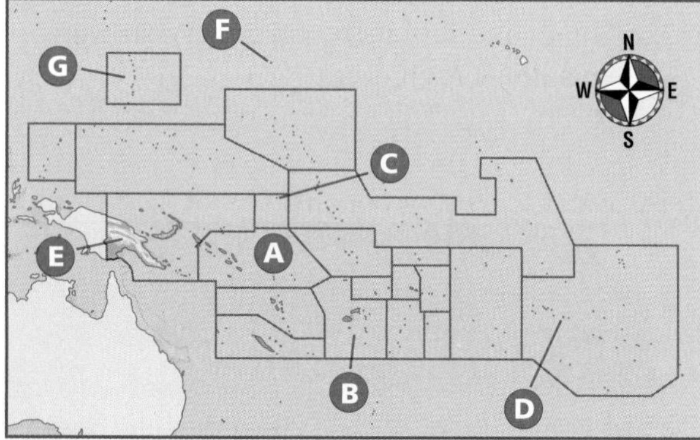

Understanding the Main Ideas TEKS

Section 1

1. *Places and Regions* What are the two types of islands in the Pacific? What are their origins?

Section 2

2. *Human Systems* What geographic factors helped a large number of languages to develop in the region?
3. *Environment and Society* What food crops have traditionally been important in the region? What uses did Pacific Islanders find for the coconut palm?

Section 3

4. *Human Systems* Why is manufacturing not a major factor in the region's economy?
5. *Human Systems* What are the three patterns of human movement in the Pacific Islands today?

Thinking Critically for TAKS TEKS

1. **Summarizing** What environmental hazards affect the Pacific Islands region? What impact might these hazards have on the land and people living there?
2. **Comparing** How do fishing, subsistence farming, and plantation agriculture play different roles in the region?
3. **Analyzing Information** How have migration, war, and trade affected the people and culture of the region?

Using the Geographer's Tools TEKS

1. **Interpreting Charts** Review the unit Fast Facts chart, the unit Comparing Standards of Living chart, and information from this chapter. Then use that information and the World Factbook (**go.hrw.com**) to rank the countries of this region by level of development. Write a short paragraph explaining why you ranked the countries as you have.
2. **Evaluating Maps** Look at the map of the ocean floor in Section 1. What does this map tell you about the region's geography? What are some clues that this is a tectonically active region?
3. **Creating and Interpreting Maps** Construct a map of the Pacific Islands region. Label the subregions of the Pacific Islands. Then use arrows to show the origin and direction of migration of people into the region over time. Label each arrow with the origin of migration. How would you expect these migrations to have influenced the region's cultures?

Writing for TAKS TEKS

Write a speech from the point of view of a resident of French Polynesia calling for greater political independence from France. Include specific reasons why you feel this way. When you are finished with your speech, proofread it to make sure you have used standard grammar, spelling, sentence structure, and punctuation.

SKILL BUILDING

Geography for Life TEKS

Gather and Organize Information from Maps

The World in Spatial Terms Using the physical-political map of the region, pick five Pacific islands. For each island, measure its approximate distance from Honolulu, Hawaii; Port Moresby, Papua New Guinea; Sydney, Australia; Suva, Fiji; and Papeete, Tahiti. Finally, create a chart to organize this information.

Building Social Studies Skills

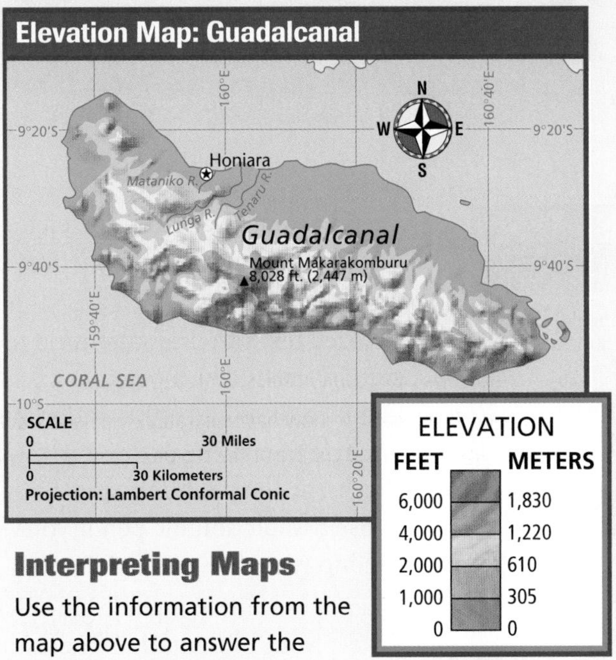

Interpreting Maps

Use the information from the map above to answer the questions that follow.

1. What is the elevation of Guadalcanal's highest area?
   a. less than 1,000 feet (305 m)
   b. between 1,000 and 2,000 feet (305 m and 610 m)
   c. above 9,000 feet (2,740 m)
   d. above 6,000 feet (1,830 m)
2. How does the elevation of the island change from north to south?

Analyzing Secondary Sources

Read the following passage and answer the questions.

"Today the Pacific Islands are home to a great number of different ethnic groups and languages. How did this happen? Huge stretches of ocean between islands allowed different cultures and languages to develop independently from one another. On New Guinea, thick rain forests and rugged mountains separated different groups of people in a similar way. Today the peoples of Papua New Guinea speak more than 700 different languages. Some of them are spoken by only a few hundred people."

3. Which feature was *not* a major factor in the development of the region's diverse cultural geography?
   a. ice cap
   b. rain forest
   c. mountain range
   d. ocean
4. What do the region's oceans, forests, and mountains have in common?

Alternative Assessment

PORTFOLIO ACTIVITY

Learning about Your Local Geography

Individual Project: Navigating with Landmarks

Just as Pacific Islanders navigated across vast ocean distances, you navigate from place to place every day. Like those sailors, you probably do not use a map or compass to find your way. Write detailed descriptions of some of your daily journeys, such as from home to school or from school to a job. Include how you get from place to place and what landmarks you use to find your destination.

internet connect

Internet Activity: go.hrw.com
KEYWORD: SW3 GT32 TEKS

Choose a topic about the Pacific Islands to:

- tour the South Pacific.
- learn the traditions of the Pacific Islands.
- explore the diversity of the region.

Skill Building for TAKS

WORKSHOP 1

Using Media Services: Overhead Transparencies

As you read in an earlier workshop, media services include a variety of tools that can be used to communicate information. These tools include sites on the World Wide Web, CD–ROMs, laser discs, DVDs, and a variety of software programs. Other tools also include VCRs, music CDs, and slides.

You have likely seen a teacher use a common media tool called an overhead projector. You can use an overhead projector to display information from transparencies. Transparencies allow you to quickly reorder your presentation or to repeat and emphasize something. All you need for an overhead presentation are the projector, transparency film, some transparency pens, a screen, and an electrical outlet.

Using Transparencies

1. Choose your main points. Keep it simple!
2. Create your transparencies by hand or computer.
3. Make it neat and easy to read.
4. Use color.
5. Include graphics and pictures.

Developing the Skill Putting together overhead presentations is not difficult. The following hints will help:

- Choose the main points and keep the number of words to a minimum. List just a few points on each transparency. You might list just one point on a transparency if it is particularly important or needs extended discussion.
- Either create your transparencies as you go, as if you were using a chalkboard, or use a computer to make professional-looking transparencies. If you use a computer, you will need to photocopy the original image onto the transparency.
- Make sure the lettering is big enough for people to see. If you choose to make handwritten transparencies, take your time. You want the writing to be clear and neat.
- Keep writing away from the edges of a transparency so that the text is visible on the screen.
- Colored text can draw attention to certain words. Use a pen of a different color to highlight, underline, or circle the text or to draw arrows, checks, or stars for emphasis. Be sure to use a water-soluble pen if you need to reuse the transparency.
- You might want to include graphics to keep your presentation interesting.

Before you begin a presentation, practice laying down the transparencies so they appear right-side up on the screen. You might want to number the transparencies in case you drop them. Also, darken the room slightly so your audience can see and read properly. Stand next to the screen after you place a transparency. Your audience will not have to look back and forth between you and the screen. Point to certain messages for emphasis by using a pen or pointer. You can also use a sheet of paper to reveal a portion of the transparency at a time. Doing so may allow you to use fewer transparencies. Finally, remember to talk to the audience, not to the screen.

Practicing the Skill

1. Create a transparency presentation about exotic species that have been introduced into the Pacific Islands region.
2. Use transparencies to summarize key points about the history of Australia and New Zealand.
3. Conduct research to learn how former Olympic cities have reused their Olympic sites. Then develop a proposal for reusing Sydney's 2000 Olympic site. Create transparencies to illustrate your proposal.

Parliament House in Canberra, Australia

WORKSHOP 2

Making Decisions

You make decisions every day. Some decisions are easy to make and take little time. More difficult decisions can take much longer to make. Perhaps you are deciding whether you should join an organization, buy a new CD player, or take a certain class. Such decisions often require that you gather information, identify options, predict consequences, and take an appropriate action to implement a decision.

The Decision-Making Process

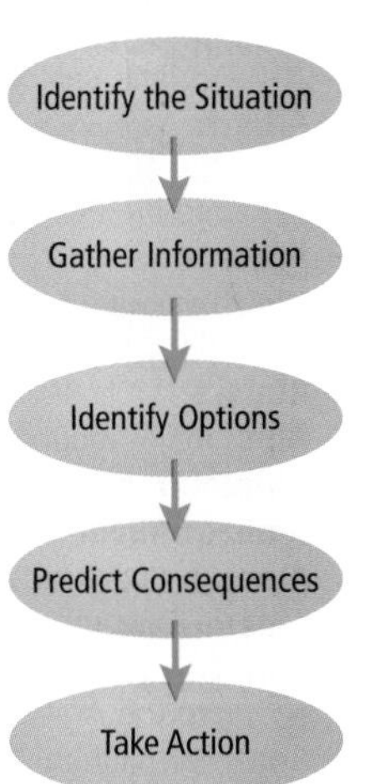

A similar process can be used to make decisions in geography. Suppose, for example, that a government in the Pacific Islands is debating whether to adopt new laws that would protect important natural resources and local ecosystems. Government officials would need to learn about the area, including local ecosystems and how human activities affect those ecosystems. Officials would also have to consider the kinds of protections that might be useful. In addition, they would need to predict how those laws might affect the local environment, economic development, and the ways of life for people there. Finally, the government would have to decide what laws to pass and then do so.

Developing the Skill As you can see, four key steps make up the decision-making process. You might want to copy the flowchart on this page and make notes about each step.

Gather Information To learn more about the issue, gather the necessary facts. Look for possible causes and solutions. Also, prioritize your needs and wants before you move to the next step of identifying your options. You might use a variety of databases, such as sites on the World Wide Web, to find needed information. You can also use newspapers, other periodicals, one or more questionnaires, and field work to gather information.

Identify Options Draw conclusions from the gathered information and identify your options. Then weigh the advantages and disadvantages of each. You might organize your options in a list or chart.

Predict Consequences It is important to understand the consequences of each option you have. Who would be affected by each option? What might those effects be? Which consequences are positive? Which are negative? You might rank the consequences from best to worst.

Take Action Follow through with your decision. Choose and implement the best option. You might want to monitor the effects of your actions. The information you learn could be useful in making decisions on future issues you might face.

Practicing the Skill

Consider a long-running debate in Australia about whether the country should become a republic or maintain ties to the British monarchy. In 1999 Australian voters rejected a proposal to make their country a republic. However, some Australians continue to work to replace the British monarch with a president or other head of state from Australia itself. Imagine that you have been asked to work on a new commission. That commission has been charged with recommending whether or not Australia should become a republic. Working with a small group of four or five students, use the decision-making process to develop a proposal for Australia's future political status.

Gazetteer

Phonetic Respelling and Pronunciation Guide

Many of the place-names in this textbook have been respelled to help you pronounce them. The letter combinations used in the respelling throughout the narrative are explained in the following phonetic respelling and pronunciation guide. The guide is adapted from *Merriam Webster's Collegiate Dictionary, Merriam Webster's Geographical Dictionary,* and *Merriam Webster's Biographical Dictionary.*

MARK	AS IN	RESPELLING	EXAMPLE
a	alphabet	a	*AL-fuh-bet
ā	Asia	ay	AY-zhuh
ä	cart, top	ah	KAHRT, TAHP
e	let, ten	e	LET, TEN
ē	even, leaf	ee	EE-vuhn, LEEF
i	it, tip, British	i	IT, TIP, BRIT-ish
ī	site, buy, Ohio	y	SYT, BY, oh-HY-oh
	iris	eye	EYE-ris
k	card	k	KAHRD
ō	over, rainbow	oh	OH-vuhr, RAYN-boh
u̇	book, wood	ooh	BOOHK, WOOHD
ȯ	all, orchid	aw	AWL, AWR-kid
ȯi	foil, coin	oy	FOYL, KOYN
au̇	out	ow	OWT
ə	cup, butter	uh	KUHP, BUHT-uhr
ü	rule, food	oo	ROOL, FOOD
yü	few	yoo	FYOO
zh	vision	zh	VIZH-uhn

*A syllable printed in small capital letters receives heavier emphasis than the other syllable(s) in a word.

Abu Dhabi (24°N 54°E) capital of the United Arab Emirates, **433**
Abuja (ah-BOO-jah) (9°N 7°E) capital of Nigeria, **499**
Acapulco (17°N 100°W) city on the southwestern coast of Mexico, **221**
Accra (6°N 0°) capital of Ghana, **499**
Addis Ababa (9°N 39°E) capital of Ethiopia, **517**
Adriatic Sea sea between Italy and the Balkan Peninsula, **349**
Aegean (ee-JEE-uhn) **Sea** sea between Greece and Turkey, **349**
Afghanistan landlocked country in Southwest Asia, **433**
Africa second-largest continent; surrounded by the Atlantic Ocean, Indian Ocean, and Mediterranean Sea, **S37–S38**
Ahaggar Mountains mountain range in southern Algeria, **483**
Albania country in the western Balkan region of Europe on the Adriatic Sea, **349**
Alberta province in western Canada, **189**
Aleutian Islands volcanic island chain extending from Alaska into the Pacific Ocean, **165**
Alexandria (31°N 30°E) city in northern Egypt, **483**
Algeria country in North Africa located between Morocco and Libya, **483**
Algiers (37°N 3°E) capital of Algeria, **483**
Alps major mountain system in south-central Europe, **349**
Altiplano broad, high plateau in Peru and Bolivia, **257**
Amazon River major river in South America, **257**
Amman (32°N 36°E) capital of Jordan, **449**
Amsterdam (52°N 5°E) capital of the Netherlands, **305**
Amu Dar'ya (uh-MOO duhr-YAH) river in Central Asia that drains into the Aral Sea, **403**
Amur (ah-MOOHR) **River** river in northeast Asia forming part of the border between Russia and China, **381**
Andes (AN-deez) great mountain range in South America, **257**
Andorra European microstate in the Pyrenees mountains, **S34**
Andorra la Vella (43°N 2°E) capital of Andorra, **S34**
Angkor ancient capital of the Khmer Empire in Cambodia, **659**
Angola country in southern Africa, **533**
Ankara (40°N 33°E) capital of Turkey, **449**
Antananarivo (19°S 48°E) capital of Madagascar, **533**
Antarctica continent around the South Pole, **S41**
Antarctic Circle line of latitude located at 66½° south of the equator; parallel beyond which no sunlight shines on the June solstice (first day of winter in the Southern Hemisphere), **S21–S22**
Antarctic Peninsula peninsula stretching toward South America from Antarctica, **S41**
Antigua and Barbuda island country in the Caribbean, **239**
Antwerp (51°N 4°E) major port city in Belgium, **305**
Apennines (A-puh-nynz) mountain range in Italy, **349**
Apia (14°S 172°W) capital of Western Samoa, **731**
Appalachian Mountains mountain system in eastern North America, **165**
Arabian Peninsula peninsula in Southwest Asia between the Red Sea and Persian Gulf, **433**
Arabian Sea sea between India and the Arabian Peninsula, **433**
Aral (AR-uhl) **Sea** inland sea between Kazakhstan and Uzbekistan, **403**
Arctic Circle line of latitude located at 66½° north of the equator; the parallel beyond which no sunlight shines on the December solstice (first day of winter in the Northern Hemisphere), **S21–S22**
Arctic Ocean ocean north of the Arctic Circle; world's fourth-largest ocean, **S21–S22**
Argentina second-largest country in South America, **257**

Gazetteer

Armenia country in the Caucasus region of Asia; former Soviet republic, **381**
Ashgabat (formerly Ashkhabad) (38°N 58°E) capital of Turkmenistan, **403**
Asia world's largest continent; located between Europe and the Pacific Ocean, **S35–S36**
Asmara (15°N 39°E) capital of Eritrea, **517**
Astana (51°N 71°E) capital of Kazakhstan, **403**
Astrakhan (46°N 48°E) old port city on the Volga River in Russia, **381**
Asunción (25°S 58°W) capital of Paraguay, **257**
Atacama Desert desert in northern Chile, **257**
Athens (38°N 24°E) capital and largest city in Greece, **349**
Atlanta (34°N 84°W) capital and largest city in the U.S. state of Georgia, **165**
Atlantic Ocean ocean between the continents of North and South America and the continents of Europe and Africa; world's second-largest ocean, **S21–S22**
Atlas Mountains African mountain range north of the Sahara, **483**
Auckland (37°S 175°E) New Zealand's largest city and main seaport, **711**
Augrabies (oh-KRAH-bees) **Falls** waterfalls on the Orange River in South Africa, **533**
Australia only country occupying an entire continent (also called Australia); located between the Indian Ocean and the Pacific Ocean, **S22**
Austria country in central Europe south of Germany, **327**
Azerbaijan country in the Caucasus region of Asia; former Soviet republic, **381**

Bab al-Mandab narrow strait that connects the Red Sea with the Indian Ocean, **433**
Baghdad (33°N 44°E) capital of Iraq, **433**
Bahamas island country in the Atlantic Ocean southeast of Florida, **239**
Bahrain country on the Persian Gulf in Southwest Asia, **433**
Baja California peninsula in northwestern Mexico, **221**
Baku (40°N 50°E) capital of Azerbaijan, **381**
Bali island in Indonesia east of Java, **679**
Balkan Mountains mountain range that rises in Bulgaria, **349**
Baltic Sea body of water east of the North Sea and Scandinavia, **305**
Baltimore (39°N 77°W) city in Maryland on the western shore of Chesapeake Bay, **165**
Bamako (13°N 8°W) capital of Mali, **499**
Bandar Seri Begawan (5°N 115°E) capital of Brunei, **679**
Bangkok (14°N 100°E) capital and largest city of Thailand, **659**
Bangladesh country in South Asia, **585**
Bangui (4°N 19°E) capital of the Central African Republic, **499**
Banjul (13°N 17°W) capital of Gambia, **499**
Barbados island country in the Caribbean, **239**
Barcelona (41°N 2°E) Mediterranean port city and Spain's second-largest city, **349**
Basel (48°N 8°E) city in northern Switzerland on the Rhine River, **327**
Basseterre (17°N 63°W) capital of St. Kitts–Nevis, **239**
Bay of Bengal body of water between India and the western coasts of Myanmar (Burma) and the Malay Peninsula, **563**
Bay of Biscay body of water off the western coast of France and the northern coast of Spain, **305**
Beijing (40°N 116°E) capital of China, **615**
Beirut (34°N 36°E) capital of Lebanon, **449**
Belarus country located north of Ukraine; former Soviet republic, **381**
Belém (1°S 48°W) port city in northern Brazil, **S32**
Belfast (55°N 6°W) capital and largest city of Northern Ireland, **305**
Belgium country between France and Germany in west-central Europe, **305**
Belgrade (45°N 21°E) capital of Serbia and Yugoslavia on the Danube River, **349**
Belize country in Central America bordering Mexico and Guatemala, **239**
Belmopan (17°N 89°W) capital of Belize, **239**
Benghazi (32°N 20°E) major coastal city in Libya, **483**
Benin (buh-NEEN) country in West Africa between Togo and Nigeria, **499**
Bergen (60°N 5°E) seaport city in southwestern Norway, **305**
Berlin (53°N 13°E) capital of Germany, **327**
Bern (47°N 7°E) capital of Switzerland, **327**
Bhutan South Asian country in the Himalayas located north of India and Bangladesh, **585**
Birmingham (52°N 2°W) major manufacturing center of west-central Great Britain, **305**
Bishkek (43°N 75°E) capital of Kyrgyzstan, **403**
Bissau (12°N 16°W) capital of Guinea-Bissau, **499**
Black Sea sea between Europe and Asia, **349**
Blue Nile East African river that flows into the Nile River in Sudan, **517**
Bogotá (5°N 74°W) capital and largest city of Colombia, **257**
Bolivia landlocked South American country, **257**
Bombay *See* Mumbai.
Bonn (51°N 7°E) city in western Germany; replaced by Berlin as the capital of reunified Germany, **327**
Borneo island in the Malay Archipelago in Southeast Asia, **679**
Bosnia and Herzegovina country in the western Balkans region of Europe between Serbia and Croatia, **349**
Bosporus a narrow strait separating European and Asian Turkey, **449**
Boston (42°N 71°W) capital and largest city of Massachusetts, **165**
Botswana country in southern Africa, **533**
Brahmaputra River major river of South Asia that begins in the Himalayas of Tibet and merges with the Ganges River in Bangladesh, **563**
Brasília (16°S 48°W) capital of Brazil, **257**
Bratislava (48°N 17°E) capital of Slovakia, **327**
Brazil largest country in South America, **257**
Brazilian Highlands region of old, eroded mountains in southeastern Brazil, **257**
Brazzaville (4°S 15°E) capital of the Republic of the Congo, **499**
Bridgetown (13°N 60°W) capital of Barbados, **239**
Brisbane (28°S 153°E) seaport and capital of Queensland, Australia, **711**
British Columbia province on the Pacific coast of Canada, **189**
British Isles island group consisting of Great Britain and Ireland, **305**
Brittany region in northwestern France, **305**
Brunei (brooh-NY) country on the northern coast of Borneo in Southeast Asia, **679**

Gazetteer

Brussels (51°N 4°E) capital of Belgium, **305**
Bucharest (44°N 26°E) capital of Romania, **349**
Budapest (48°N 19°E) capital of Hungary, **327**
Buenos Aires (34°S 59°W) capital of Argentina, **257**
Bujumbura (3°S 29°E) capital of Burundi, **517**
Bulgaria country on the Balkan Peninsula in Europe, **349**
Burkina Faso (boor-KEE-nuh FAH-soh) landlocked country in West Africa, **499**
Burma *See* Myanmar.
Burundi landlocked country in East Africa, **517**

Cairo (30°N 31°E) capital of Egypt, **483**
Calcutta *See* Kolkata.
Calgary (51°N 114°W) city in the western Canadian province of Alberta, **189**
Callao (kah-YAH-oh) (12°S 77°W) port city in Peru, **S32**
Cambodia country in Southeast Asia west of Vietnam, **659**
Cameroon country in Central Africa, **499**
Campeche (20°N 91°W) city in Mexico on the west coast of the Yucatán Peninsula, **221**
Canada country occupying most of northern North America, **189**
Canadian Shield major landform region in central Canada along Hudson Bay, **189**
Canberra (35°S 149°E) capital of Australia, **711**
Cancún (21°N 87°W) resort city in Mexico on the Yucatán Peninsula, **221**
Cantabrian (kan-TAY-bree-uhn) **Mountains** mountains in northwestern Spain, **349**
Cape Horn (56°S 67°W) cape in southern Chile; southernmost point of South America, **257**
Cape of Good Hope cape on the southwest coast of South Africa, **533**
Cape Town (34°S 18°E) major seaport city and legislative capital of South Africa, **533**
Cape Verde island country in the Atlantic Ocean off the coast of West Africa, **499**
Caracas (kuh-RAHK-uhs) (11°N 67°W) capital of Venezuela, **257**
Cardiff (52°N 3°W) capital and largest city of Wales, **305**
Caribbean Sea arm of the Atlantic Ocean between North America and South America, **239**
Carpathian Mountains mountain system in Eastern Europe, **327**
Casablanca (34°N 8°W) seaport city on the western coast of Morocco, **483**
Cascade Range mountain range in the northwestern United States, **165**
Caspian Sea large inland salt lake between Europe and Asia, **403**
Castries (14°N 61°W) capital of St. Lucia, **239**
Cauca River river in western Colombia, **257**
Caucasus Mountains mountain range between the Black Sea and the Caspian Sea, **381**
Cayenne (5°N 52°W) capital of French Guiana, **257**
Central African Republic landlocked country in Central Africa located south of Chad, **499**
Central America narrow southern portion of the North American continent, **239**
Central Lowlands area of Australia between the Western Plateau and the Great Dividing Range, **711**
Central Siberian Plateau upland plains and valleys between the Yenisey and Lena Rivers in Russia, **381**
Chad landlocked country in northern Africa, **499**
Chang (Yangtze) **River** major river in central China, **615**
Chao Phraya (chow PRY-uh) **River** major river in Thailand, **659**
Chelyabinsk (chel-YAH-buhnsk) (55°N 61°E) manufacturing city in the Urals region of Russia, **381**
Chernobyl (51°N 30°E) city in north-central Ukraine; site of a major nuclear accident in 1986, **381**
Chicago (42°N 88°W) major city on Lake Michigan in northern Illinois, **165**
Chile country on the west coast of South America, **257**
China country in East Asia; most populous country in the world, **615**
Chişinău (formerly Kishinev) (47°N 29°E) capital of Moldova, **349**
Chongqing (30°N 108°E) city in southern China along the Chang River, **615**
Christchurch (44°S 173°E) city on the eastern coast of South Island, New Zealand, **711**
Ciudad Juárez (32°N 106°W) city in northern Mexico opposite El Paso, **221**
Cologne (51°N 7°E) manufacturing and commercial city along the Rhine River in Germany, **327**
Colombia country in northern South America, **257**
Colombo (7°N 80°E) capital city and important seaport of Sri Lanka, **585**
Colorado Plateau uplifted area of horizontal rock layers in the western United States, **165**
Columbia River river that drains the Columbia Basin in the northwestern United States, **165**
Comoros island country in the Indian Ocean off the coast of Africa, **533**
Conakry (10°N 14°W) capital of Guinea, **499**
Congo Basin region in Central Africa, **499**
Congo, Democratic Republic of the largest and most populous country in Central Africa, **499**
Congo, Republic of the Central African country located along the Congo River, **499**
Congo River major navigable river in Central Africa that flows into the Atlantic Ocean, **499**
Copenhagen (56°N 12°E) seaport and capital of Denmark, **305**
Córdoba (30°S 64°W) large city in Argentina northwest of Buenos Aires, **S32**
Cork (52°N 8°W) seaport city in southern Ireland, **305**
Costa Rica country in Central America, **239**
Côte d'Ivoire (KOHT dee-VWAHR) (Ivory Coast) country in West Africa, **499**
Crimean Peninsula peninsula in Ukraine that juts southward into the Black Sea, **381**
Croatia country and former Yugoslav republic in the western Balkans region of Europe, **349**
Cuba country and largest island in the Caribbean, **239**
Cuzco (14°S 72°W) city southeast of Lima, Peru; former capital of the Inca Empire, **257**
Cyprus island republic in the eastern Mediterranean Sea, **449**
Czech Republic Central European country and the western part of the former country of Czechoslovakia, **327**

Dakar (15°N 17°W) capital of Senegal, **499**
Dallas (33°N 97°W) city in northern Texas, **165**

Gazetteer

Damascus (34°N 36°E) capital of Syria and one of the world's oldest cities, **449**
Danube River major river in Europe that flows into the Black Sea in Romania, **349**
Dardanelles narrow strait separating European and Asian Turkey, **449**
Dar es Salaam (7°S 39°E) capital and major seaport of Tanzania, **517**
Dead Sea salt lake on the boundary between Israel and Jordan in southwestern Asia, **449**
Deccan Plateau the southern part of the Indian subcontinent, **563**
Delhi (29°N 77°E) city in India, **563**
Denmark country in northern Europe, **305**
Detroit (42°N 83°W) major industrial city in Michigan, **165**
Devil's Island (5°N 53°W) French island off the coast of French Guiana in South America, **S31**
Dhaka (24°N 90°E) capital and largest city of Bangladesh, **585**
Dinaric Alps mountains extending inland from the Adriatic coast to the Balkan Peninsula, **349**
Djibouti country located in the Horn of Africa, **517**
Djibouti (12°N 43°E) capital of Djibouti, **517**
Dnieper River major river in Ukraine, **381**
Doha (25°N 51°E) capital of Qatar, **433**
Dominica Caribbean island country, **239**
Dominican Republic country occupying the eastern part of Hispaniola in the Caribbean, **239**
Donets Basin industrial region in eastern Ukraine, **373**
Douro River river on the Iberian Peninsula that flows into the Atlantic Ocean in Portugal, **349**
Drakensberg mountain range in southern Africa, **533**
Dublin (53°N 6°W) capital of the republic of Ireland, **305**
Durban (30°S 31°E) port city in South Africa, **533**
Dushanbe (39°N 69°E) capital of Tajikistan, **403**

Eastern Ghats mountains on the eastern side of the Deccan Plateau in southern India, **563**
Ebro River river in Spain that flows into the Mediterranean Sea, **349**
Ecuador country in western South America, **257**
Edmonton (54°N 113°W) provincial capital of Alberta, Canada, **189**
Egypt country in North Africa located east of Libya, **483**
Elburz Mountains mountain range in northern Iran, **433**
El Salvador country on the Pacific side of Central America, **239**
England southern part of Great Britain and part of the United Kingdom in northern Europe, **305**
English Channel channel separating Great Britain from the European continent, **305**
equator the imaginary line of latitude that lies halfway between the North and South Poles and circles the globe, **S21–S22**
Equatorial Guinea Central African country, **499**
Eritrea (er-uh-TREE-uh) East African country located north of Ethiopia, **517**
Essen (51°N 7°E) industrial city in western Germany, **327**
Estonia country located on the Baltic Sea; former Soviet republic, **327**
Ethiopia East African country in the Horn of Africa, **517**
Euphrates River major river in Iraq in southwestern Asia, **433**
Europe continent between the Ural Mountains and the Atlantic Ocean, **S33–S34**

Fergana Valley fertile valley in Uzbekistan, Kyrgyzstan, and Tajikistan, **403**
Fès (34°N 5°W) city in north-central Morocco, **483**
Fiji South Pacific island country; part of Melanesia, **731**
Finland country in northern Europe located between Sweden, Norway, and Russia, **305**
Florence (44°N 11°E) city on the Arno River in central Italy, **349**
France country in west-central Europe, **305**
Frankfurt (50°N 9°E) main city of Germany's Rhineland region, **327**
Freetown (9°N 13°W) capital of Sierra Leone, **499**
French Guiana French territory in northern South America, **257**
Funafuti (9°S 179°E) capital of Tuvalu, **731**

Gabon country in Central Africa located between Cameroon and the Republic of the Congo, **499**
Gaborone (24°S 26°E) capital of Botswana, **533**
Galway (53°N 9°W) city in western Ireland, **305**
Gambia country along the Gambia River in West Africa, **499**
Ganges River major river in India flowing from the Himalayas southeastward to the Bay of Bengal, **563**
Gangetic (gan-JE-tik) **Plain** vast plain in northern India, **563**
Gao (GOW) (16°N 0°) city in Mali on the Niger River, **499**
Gaza Strip area occupied by Israel from 1967 to 1994; partly under Palestinian self-rule since 1994, **449**
Geneva (46°N 6°E) city in southwestern Switzerland, **327**
Genoa (44°N 10°E) seaport city in northwestern Italy, **349**
Georgetown (8°N 58°W) capital of Guyana, **257**
Georgia (Eurasia) country in the Caucasus region; former Soviet republic, **381**
Germany country in central Europe located between Poland and France, **327**
Ghana country in West Africa, **499**
Glasgow (56°N 4°W) city in Scotland, United Kingdom, **305**
Gobi desert that makes up part of the Mongolian plateau in East Asia, **615**
Golan Heights hilly region in southwestern Syria occupied by Israel, **449**
Göteborg (58°N 12°E) seaport city in southwestern Sweden, **305**
Gran Chaco (grahn CHAH-koh) dry plains region in Paraguay, Bolivia, and northern Argentina, **257**
Great Artesian Basin Australia's largest source of groundwater; located in interior Queensland, **711**
Great Barrier Reef world's largest coral reef; located off the northeastern coast of Australia, **711**
Great Basin dry region in the western United States, **165**
Great Bear Lake lake in the Northwest Territories of Canada, **189**
Great Britain name for island and country of northern and western Europe, **291**
Great Dividing Range mountain range of eastern Australia, **711**

Gazetteer

Greater Antilles larger islands of the West Indies in the Caribbean Sea, including Cuba, Hispaniola, Jamaica, and Puerto Rico, **239**

Great Lakes largest freshwater lake system in the world; located in North America, **175**

Great Plains plains region in the central United States, **165**

Great Rift Valley valley system extending from eastern Africa to Southwest Asia, **517**

Great Slave Lake lake in the Northwest Territories of Canada, **189**

Greece country in southern Europe located at the southern end of the Balkan Peninsula, **349**

Greenland self-governing province of Denmark between the North Atlantic and Arctic Oceans, **305**

Grenada Caribbean island country, **239**

Guadalajara (21°N 103°W) city in west-central Mexico, **221**

Guadalquivir (gwah-dahl-kee-VEER) **River** river in southern Spain, **349**

Guam (14°N 143°E) South Pacific island and U.S. territory in Micronesia, **731**

Guatemala most populous country in Central America, **239**

Guatemala City (15°N 91°W) capital of Guatemala, **239**

Guayaquil (gwy-ah-KEEL) (2°S 80°W) port city in Ecuador, **214**

Guiana Highlands elevated region in northeastern South America, **257**

Guinea country in West Africa, **499**

Guinea-Bissau (GI-nee bi-SOW) country in West Africa, **499**

Gulf-Atlantic Coastal Plain North American landform region stretching along the Atlantic Ocean and Gulf of Mexico, **149**

Gulf of Bothnia part of the Baltic Sea west of Finland, **305**

Gulf of California part of the Pacific Ocean east of Baja California, Mexico, **221**

Gulf of Finland part of the Baltic Sea south of Finland, **S33**

Gulf of Guinea (GI-nee) gulf of the Atlantic Ocean south of western Africa, **499**

Gulf of Mexico gulf of the Atlantic Ocean between Florida, Texas, and Mexico, **165**

Gulf of St. Lawrence gulf between New Brunswick and Newfoundland Island in North America, **189**

Guyana (gy-AH-nuh) country in South America, **257**

Haiti country occupying the western third of the Caribbean island of Hispaniola, **239**

Halifax (45°N 64°W) provincial capital of Nova Scotia, Canada, **189**

Hamburg (54°N 10°E) seaport on the Elbe River in northwestern Germany, **327**

Hanoi (ha-NOY) (21°N 106°E) capital of Vietnam, **659**

Harare (18°S 31°E) capital of Zimbabwe, **533**

Havana (23°N 82°W) capital of Cuba, **239**

Helsinki (60°N 25°E) capital of Finland, **305**

Himalayas mountain system in Asia; world's highest mountains, **585**

Hindu Kush high mountain range in northern Afghanistan, **433**

Hispaniola large Caribbean island divided into the countries of Haiti and the Dominican Republic, **239**

Ho Chi Minh City (formerly Saigon) (11°N 107°E) major city in southern Vietnam; former capital of South Vietnam, **659**

Hokkaidō (hoh-KY-doh) major island in northern Japan, **637**

Honduras country in Central America, **239**

Hong Kong (22°N 115°E) former British colony in East Asia; now part of China, **615**

Hong (Red) River major river that flows into the Gulf of Tonkin in Vietnam, **659**

Honiara (9°S 160°E) capital of the Solomon Islands, **731**

Honshū (HAWN-shoo) largest of the four major islands of Japan, **637**

Houston (30°N 95°W) major port and largest city in Texas, **165**

Huang (Yellow) River one of the world's longest rivers; located in northern China, **615**

Hudson Bay large bay in Canada, **189**

Hungary country in central Europe between Romania and Austria, **327**

Iberian Peninsula peninsula in southwestern Europe occupied by Spain and Portugal, **349**

Iceland island country between the North Atlantic and Arctic Oceans, **305**

India country in South Asia, **563**

Indian Ocean world's third-largest ocean; located east of Africa, south of Asia, west of Australia, and north of Antarctica, **S22**

Indochina Peninsula peninsula in Southeast Asia that includes the region from Myanmar (Burma) to Vietnam, **659**

Indonesia largest country in Southeast Asia; made up of more than 17,000 islands, **679**

Indus River major river in Pakistan, **585**

Inland Sea body of water in southern Japan between Honshū, Shikoku, and Kyūshū, **637**

Interior Plains vast area between the Appalachians and Rocky Mountains in North America, **165**

Iran country in southwestern Asia; formerly called Persia, **433**

Iraq (i-RAHK) country located between Iran and Saudi Arabia, **433**

Ireland country west of Great Britain in the British Isles, **305**

Irian Jaya western part of the island of New Guinea that is part of Indonesia, **679**

Irish Sea sea between Great Britain and Ireland, **305**

Irrawaddy River important river in Myanmar (Burma), **659**

Islamabad (34°N 73°E) capital of Pakistan, **585**

Israel country in southwestern Asia, **449**

İstanbul (formerly Constantinople) (41°N 29°E) largest city and leading seaport in Turkey, **449**

Italy country in southern Europe, **349**

Jakarta (6°S 107°E) capital of Indonesia, **679**

Jamaica island country in the Caribbean Sea, **239**

Japan country in East Asia consisting of four major islands and more than 3,000 smaller islands, **637**

Java major island in Indonesia, **679**

Jerusalem (32°N 35°E) capital of Israel, **449**

Johannesburg (26°S 28°E) city in South Africa, **533**

Jordan Southwest Asian country stretching east from the Dead Sea and Jordan River into the Arabian Desert, **449**

Jordan River river in southwestern Asia that separates Israel from Syria and Jordan, **449**

Jutland Peninsula peninsula in northern Europe made up of Denmark and part of northern Germany, **305**

Gazetteer

Kabul (35°N 69°E) capital and largest city of Afghanistan, **433**

Kalahari Desert dry plateau region in southern Africa, **533**

Kamchatka Peninsula peninsula along Russia's northeastern coast, **381**

Kampala (0° 32°E) capital of Uganda, **517**

Kao-hsiung (23°N 120°E) Taiwan's second-largest city and major seaport, **615**

Karachi (25°N 69°E) Pakistan's largest city and major seaport, **585**

Karakoram Range high mountain range in northern India and Pakistan, **585**

Kara-kum (kahr-uh-KOOM) desert region in Turkmenistan, **403**

Kashmir mountainous region in northern India and Pakistan, **563**

Kathmandu (kat-man-DOO) (28°N 85°E) capital of Nepal, **585**

Kazakhstan country in Central Asia; former Soviet republic, **403**

Kenya country in East Africa south of Ethiopia, **517**

Khabarovsk (kuh-BAHR-uhfsk) (49°N 135°E) city in southeastern Russia on the Amur River, **381**

Khartoum (16°N 33°E) capital of Sudan, **517**

Khyber Pass major mountain pass between Afghanistan and Pakistan, **585**

Kiev (50°N 31°E) capital of Ukraine, **381**

Kigali (2°S 30°E) capital of Rwanda, **517**

Kilimanjaro (3°S 37°E) (ki-luh-muhn-JAHR-oh) highest point in Africa (19,341 ft.; 5,895 m); located in northeast Tanzania near the Kenya border, **517**

Kingston (18°N 77°W) capital of Jamaica, **239**

Kingstown (13°N 61°W) capital of St. Vincent and the Grenadines, **239**

Kinshasa (4°S 15°E) capital of the Democratic Republic of the Congo, **499**

Kiribati South Pacific country in Micronesia and Polynesia, **731**

Kjølen (CHUHL-uhn) **Mountains** mountain range on the Scandinavian Peninsula, **305**

Kōbe (KOH-bay) (35°N 135°E) major port city in Japan, **637**

Kolkata (23°N 88°E) giant industrial and seaport city in eastern India, **563**

Korea peninsula on the east coast of Asia, **637**

Koror (7°N 134°E) capital of Palau, **731**

Kosovo province in southern Serbia, **349**

Kuala Lumpur (3°N 102°E) capital of Malaysia, **679**

Kuril (KYOOHR-eel) **Islands** Russian islands northeast of the island of Hokkaidō, Japan, **381**

Kuwait country on the Persian Gulf in southwestern Asia, **433**

Kuwait City (29°N 48°E) capital of Kuwait, **433**

Kuznetsk Basin (Kuzbas) industrial region in central Russia, **381**

Kyōto (KYOH-toh) (35°N 136°E) city on the island of Honshū and the ancient capital of Japan, **637**

Kyrgyzstan (kir-gi-STAN) country in Central Asia; former Soviet republic, **403**

Kyūshū (KYOO-shoo) southernmost of Japan's main islands, **637**

Kyzyl Kum (ki-zil KOOM) desert region in Uzbekistan and Kazakhstan, **403**

Labrador region in the territory of Newfoundland, Canada, **189**

Lagos (LAY-gahs) (6°N 3°E) former capital of Nigeria and the country's largest city, **499**

Lahore (32°N 74°E) industrial city in northeastern Pakistan, **585**

Lake Baikal (by-KAHL) world's deepest freshwater lake; located north of the Gobi in Russia, **381**

Lake Chad shallow lake between Nigeria and Chad in western Africa, **499**

Lake Malawi (also called Lake Nyasa) lake in southeastern Africa, **533**

Lake Maracaibo (mah-rah-KY-buh) extension of the Gulf of Venezuela in South America, **257**

Lake Nasser artificial lake in southern Egypt created in the 1960s by the construction of the Aswān High Dam, **483**

Lake Nicaragua lake in southern Nicaragua, **239**

Lake Poopó (poh-oh-POH) lake in western Bolivia, **257**

Lake Tanganyika deep lake in the Great Rift Valley in Africa, **517**

Lake Titicaca lake between Bolivia and Peru at an elevation of 12,500 feet (3,810 m), **257**

Lake Victoria large lake in East Africa surrounded by Uganda, Kenya, and Tanzania, **517**

Lake Volta large artificial lake in Ghana, **499**

Laos landlocked country in Southeast Asia, **659**

La Paz (17°S 68°W) administrative capital and principal industrial city of Bolivia at an elevation of 12,001 feet (3,658 m); highest capital in the world, **257**

Lapland region extending across northern Finland, Sweden, and Norway, **305**

Las Vegas (36°N 115°W) city in southern Nevada, **165**

Latvia country on the Baltic Sea; former Soviet republic, **327**

Lebanon country in Southwest Asia, **449**

Lesotho country completely surrounded by South Africa, **533**

Lesser Antilles chain of volcanic islands in the eastern Caribbean Sea, **239**

Liberia country in West Africa, **499**

Libreville (0° 9°E) capital of Gabon, **499**

Libya country in North Africa located between Egypt and Algeria, **483**

Liechtenstein microstate in west-central Europe located between Switzerland and Austria, **327**

Lilongwe (14°S 34°E) capital of Malawi, **533**

Lima (12°S 77°W) capital of Peru, **257**

Limpopo River river in southern Africa forming the border between South Africa and Zimbabwe, **533**

Lisbon (39°N 9°W) capital and largest city of Portugal, **349**

Lithuania European country on the Baltic Sea; former Soviet republic, **327**

Ljubljana (lee-oo-blee-AH-nuh) (46°N 14°E) capital of Slovenia, **349**

Lomé (6°N 1°E) capital of Togo, **499**

London (52°N 0°) capital of the United Kingdom, **305**

Luanda (9°S 13°E) capital of Angola, **533**

Lubumbashi (loo-boom-BAH-shee) (12°S 27°E) industrial city in the Democratic Republic of the Congo, **477**

Gazetteer

Luxembourg small European country bordered by France, Germany, and Belgium, **305**
Luxembourg (50°N 7°E) capital of Luxembourg, **305**
Luzon chief island of the Philippines, **679**

Macao (22°N 113°E) former Portuguese territory in East Asia; now part of China, **615**
Macedonia Balkan country; former Yugoslav republic, **349**
Madagascar largest of the island countries off the eastern coast of Africa, **533**
Madrid (40°N 4°W) capital of Spain, **349**
Magdalena River river in Colombia that flows into the Caribbean Sea, **257**
Magnitogorsk (53°N 59°E) manufacturing city in the Urals region of Russia, **381**
Majuro (7°N 171°E) capital of the Marshall Islands, **731**
Malabo (4°N 9°E) capital of Equatorial Guinea, **499**
Malawi (muh-LAH-wee) landlocked country in Central Africa, **533**
Malay Archipelago (ahr-kuh-PE-luh-goh) large island group off the southeastern coast of Asia including New Guinea and the islands of Malaysia, Indonesia, and the Philippines, **679**
Malay Peninsula peninsula in Southeast Asia, **679**
Malaysia country in Southeast Asia, **679**
Maldives island country in the Indian Ocean south of India, **585**
Male (5°N 72°E) capital of the Maldives, **585**
Mali country in West Africa along the Niger River, **499**
Malta island country in southern Europe located in the Mediterranean Sea between Sicily and North Africa, **349**
Managua (12°N 86°W) capital of Nicaragua, **239**
Manama (26°N 51°E) capital of Bahrain, **433**
Manaus (3°S 60°W) city in Brazil on the Amazon River, **257**
Manchester (53°N 2°W) major commercial city in west-central Great Britain, **305**
Manila (15°N 121°E) capital of the Philippines, **679**
Manitoba prairie province in central Canada, **189**
Maputo (27°S 33°E) capital of Mozambique, **533**
Marseille (43°N 5°E) seaport in France on the Mediterranean Sea, **305**
Marshall Islands Pacific island country in Micronesia, **731**
Maseru (29°S 27°E) capital of Lesotho, **533**
Masqat (Muscat) (23°N 59°E) capital of Oman, **433**
Mato Grosso Plateau highland region in southwestern Brazil, **S31**
Mauritania African country stretching east from the Atlantic coast into the Sahara, **499**
Mauritius island country located off the coast of Africa in the Indian Ocean, **533**
Mazatlán (23°N 106°W) seaport city in western Mexico, **221**
Mbabane (26°S 31°E) capital of Swaziland, **533**
Mecca (21°N 40°E) important Islamic city in western Saudi Arabia, **433**
Mediterranean Sea sea surrounded by Europe, Asia, and Africa, **S22**
Mekong River important river in Southeast Asia, **659**
Melanesia island region in the South Pacific that stretches from New Guinea to Fiji, **731**
Melbourne (38°S 145°E) capital of Victoria, Australia, **711**
Mexican Plateau large, high plateau in central Mexico, **221**
Mexico country in North America, **221**
Mexico City (19°N 99°W) capital of Mexico, **221**
Miami (26°N 80°W) city in southern Florida, **165**
Micronesia island region in the South Pacific that includes the Mariana, Caroline, Marshall, and Gilbert island groups, **731**
Micronesia, Federated States of island country in the western Pacific, **731**
Milan (45°N 9°E) city in northern Italy, **349**
Minsk (54°N 28°E) capital of Belarus, **381**
Mississippi River major river in the central United States, **165**
Mogadishu (2°N 45°E) capital and port city of Somalia, **517**
Moldova European country located between Romania and Ukraine; former Soviet republic, **349**
Monaco (44°N 8°E) European microstate bordered by France, **305**
Mongolia landlocked country in East Asia, **615**
Monrovia (6°N 11°W) capital of Liberia, **499**
Montenegro *See* Yugoslavia.
Monterrey (26°N 100°W) major industrial center in northeastern Mexico, **221**
Montevideo (mawn-tay-bee-THAY-oh) (35°S 56°W) capital of Uruguay, **257**
Montreal (46°N 74°W) financial and industrial city in Quebec, Canada, **189**
Morocco country in North Africa south of Spain, **483**
Moroni (12°S 43°E) capital of Comoros, **533**
Moscow (56°N 38°E) capital of Russia, **381**
Mount Elbrus (43°N 42°E) highest European peak (18,510 ft.; 5,642 m); located in the Caucasus Mountains, **381**
Mount Everest (28°N 87°E) world's highest peak (29,035 ft.; 8,850 m); located in the Himalayas, **585**
Mozambique (moh-zahm-BEEK) country in southern Africa, **533**
Mumbai (Bombay) (19°N 73°E) India's largest city, **563**
Munich (MYOO-nik) (48°N 12°E) major city and manufacturing center in southern Germany, **327**
Murray-Darling Rivers major river system in southeastern Australia, **711**
Myanmar (MYAHN-mahr) (Burma) country in Southeast Asia between India, China, and Thailand, **659**

Nairobi (1°S 37°E) capital of Kenya, **517**
Namib Desert Atlantic coast desert in southern Africa, **533**
Namibia (nuh-MI-bee-uh) country on the Atlantic coast in southern Africa, **533**
Nanjing (32°N 119°E) city along the upper Chang River in China, **615**
Naples (41°N 14°E) major seaport in southern Italy, **349**
Nassau (25°N 77°W) capital of the Bahamas, **239**
Nauru South Pacific island country in Micronesia, **731**
N'Djamena (12°N 15°E) capital of Chad, **499**
Negev desert region in southern Israel, **449**
Nepal South Asian country located in the Himalayas, **585**
Netherlands country in west-central Europe, **305**
New Brunswick province in eastern Canada, **189**
New Caledonia French territory in the South Pacific Ocean east of Queensland, Australia, **731**
New Delhi (29°N 77°E) capital of India, **563**

Newfoundland province in eastern Canada including Labrador and the island of Newfoundland, **189**

New Guinea large island in the South Pacific Ocean north of Australia, **731**

New Orleans (30°N 90°W) major port city in Louisiana located on the Mississippi River, **165**

New York Middle Atlantic state in the northeastern United States, **165**

New York most populous city in the United States, **165**

New Zealand island country located southeast of Australia, **711**

Niamey (14°N 2°E) capital of Niger, **499**

Nicaragua country in Central America, **239**

Nice (44°N 7°E) city on the southeastern coast in France, **305**

Nicosia (35°N 33°E) capital of Cyprus, **449**

Niger (NY-juhr) country in West Africa, **499**

Nigeria country in West Africa, **499**

Niger River river in West Africa, **499**

Nile Delta region in northern Egypt where the Nile River flows into the Mediterranean Sea, **483**

Nile River world's longest river (4,160 miles; 6,693 km); flows into the Mediterranean Sea in Egypt, **483**

Nizhniy Novgorod (Gorki), Russia (56°N 44°E) city on the Volga River east of Moscow, **381**

North America continent including Canada, the United States, Mexico, Central America, and the Caribbean islands, **S21**

North China Plain region of northeastern China, **615**

Northern European Plain broad coastal plain from the Atlantic coast of France into Russia, **S33**

Northern Ireland the six northern counties of Ireland that remain part of the United Kingdom; also called Ulster, **305**

Northern Mariana Islands U.S. commonwealth in the South Pacific, **731**

North Island one of the two main islands of New Zealand, **711**

North Korea country on the northern part of the Korea Peninsula in East Asia, **637**

North Pole the northern point of Earth's axis, **S41**

North Sea major sea between Great Britain, Denmark, and the Scandinavian Peninsula, **305**

Northwest Highlands region of rugged hills and low mountains in Europe, including parts of the British Isles, northwestern France, the Iberian Peninsula, and the Scandinavian Peninsula, **291**

Northwest Territories division of a northern region of Canada, **189**

Norway European country located on the Scandinavian Peninsula, **305**

Nouakchott (nooh-AHK-shaht) (18°N 16°W) capital of Mauritania, **499**

Nova Scotia province in eastern Canada, **189**

Novosibirsk (55°N 83°E) industrial center in Siberia, Russia, **381**

Nuku'alofa capital of Tonga, **731**

Nunavut territory of northern Canada, **189**

Nuuk (Godthåb) (64°N 52°W) capital of Greenland, **305**

Ob River large river that drains Russia and Siberia, **381**

Oman country in the Arabian Peninsula; formerly known as Masqat (Muscat), **433**

Ontario province in central Canada, **189**

Orange River river in southern Africa, **533**

Orinoco River river in South America, **257**

Orizaba (19°N 97°W) volcanic mountain (18,700 ft.; 5,700 m) southeast of Mexico City; highest point in Mexico, **221**

Ōsaka (oh-SAH-kuh) (35°N 135°E) major industrial center on Japan's southwestern Honshū island, **637**

Oslo (60°N 11°E) capital of Norway, **305**

Ottawa (45°N 76°W) capital of Canada; located in Ontario, **189**

Ouagadougou (wah-gah-DOO-GOO) (12°N 2°W) capital of Burkina Faso, **499**

Pacific Ocean Earth's largest ocean; located between North and South America and Asia and Australia, **S21–S22**

Pakistan South Asian country located northwest of India, **585**

Palau South Pacific island country in Micronesia, **731**

Palikir capital of the Federated States of Micronesia, **731**

Pamirs mountain area mainly in Tajikistan in Central Asia, **403**

Panama country in Central America, **239**

Panama Canal canal connecting the Pacific Ocean and Caribbean Sea; located in central Panama, **239**

Panama City (9°N 80°W) capital of Panama, **239**

Papua New Guinea country on the eastern half of the island of New Guinea, **731**

Paraguay landlocked country in South America, **257**

Paraguay River river that divides Paraguay into two separate regions, **257**

Paramaribo (6°N 55°W) capital of Suriname in South America, **257**

Paraná River large river in southeastern South America, **257**

Paris (49°N 2°E) capital of France, **305**

Patagonia arid region of plains and windswept plateaus in southern Argentina, **257**

Peloponnese (PE-luh-puh-neez) peninsula forming the southern part of the mainland of Greece, **349**

Persian Gulf body of water between Iran and the Arabian Peninsula, **433**

Perth (32°S 116°E) capital of Western Australia, **711**

Peru country in South America, **257**

Philadelphia (40°N 75°W) important port and industrial center in Pennsylvania in the northeastern United States, **165**

Philippines Southeast Asian island country located north of Indonesia, **679**

Phnom Penh (12°N 105°E) capital of Cambodia, **659**

Phoenix (34°N 112°W) capital of Arizona, **165**

Plateau of Brazil area of upland plains in southern Brazil, **257**

Plateau of Tibet high plateau in western China, **615**

Poland country in central Europe located east of Germany, **327**

Polynesia island region of the South Pacific Ocean that includes the Hawaiian and Line Island groups, Samoa, French Polynesia, and Easter Island, **731**

Po River river in northern Italy, **349**

Port-au-Prince (pohr-toh-PRINS) (19°N 72°W) capital of Haiti, **239**

Port Elizabeth (34°S 26°E) seaport in South Africa, **533**

Portland (46°N 123°W) seaport and largest city in Oregon, **165**

Port Louis (20°S 58°E) capital of Mauritius, **533**

Port Moresby (10°S 147°E) seaport and capital of Papua New Guinea, **731**

Port-of-Spain (11°N 61°W) capital of Trinidad and Tobago, **239**

Porto-Novo (6°N 3°E) capital of Benin, **499**

Portugal country in southern Europe located on the Iberian Peninsula, **349**

Port-Vila (18°S 169°E) capital of Vanuatu, **731**

Gazetteer

Prague (50°N 14°E) capital of the Czech Republic, **327**
Praia (PRIE-uh) (15°N 24°W) capital of Cape Verde, **499**
Pretoria (26°S 28°E) administrative capital of South Africa, **533**
Prince Edward Island province in eastern Canada, **189**
Pripet Marshes (PRI-pet) marshlands in southern Belarus and northwest Ukraine, **381**
Puerto Rico U.S. commonwealth in the Greater Antilles in the Caribbean Sea, **239**
Pusan (35°N 129°E) major seaport city in southeastern South Korea, **637**
P'yŏngyang (pyuhng-YANG) (39°N 126°E) capital of North Korea, **637**
Pyrenees (PIR-uh-neez) mountain range along the border of France and Spain, **349**

Qatar Persian Gulf country located on the Arabian Peninsula, **433**
Qattara Depression lowland region in northern Egypt, **483**
Quebec province in eastern Canada, **189**
Quebec City (47°N 71°W) provincial capital of Quebec, Canada, **189**
Quito (0° 79°W) capital of Ecuador, **257**

Rabat (34°N 7°W) capital of Morocco, **483**
Rangoon *See* Yangon.
Red Sea sea between the Arabian Peninsula and northeastern Africa, **433**
Reykjavik (RAYK-yuh-veek) (64°N 22°W) capital of Iceland, **305**
Rhine River major river in west-central Europe, **327**
Riga (57°N 24°E) capital of Latvia, **327**
Río Bravo Mexican name for the river between Texas and Mexico, **221**
Rio de Janeiro (23°S 43°W) city in southeastern Brazil, **257**
Río de la Plata estuary between Argentina and Uruguay in South America, **257**
Riyadh (25°N 47°E) capital of Saudi Arabia, **433**
Rocky Mountains major mountain range in North America, **165**
Romania country in the eastern Balkans region of Europe, **349**
Rome (42°N 13°E) capital of Italy, **349**
Rosario (roh-SAHR-ee-oh) (33°S 61°W) city in eastern Argentina, **S32**
Roseau (15°N 61°W) capital of Dominica in the Caribbean, **239**
Ross Ice Shelf ice shelf in Antarctica, **S41**
Rub' al-Khali (Empty Quarter) uninhabited desert area in south-eastern Saudi Arabia, **433**
Russia world's largest country, stretching from Europe and the Baltic Sea to eastern Asia and the coast of the Bering Sea, **381**
Rwanda country in East Africa, **517**

Sahara desert region in northern Africa; world's largest desert, **483**
St. George's (12°N 62°W) capital of Grenada in the Caribbean Sea, **239**
St. John's (17°N 62°W) capital of Antigua and Barbuda in the Caribbean Sea, **239**
St. Kitts-Nevis Caribbean country in the Lesser Antilles, **239**
St. Lawrence River major river linking the Great Lakes with the Gulf of St. Lawrence and the Atlantic Ocean in southeastern Canada, **189**
St. Lucia Caribbean country in the Lesser Antilles, **239**
St. Petersburg (formerly Leningrad) (60°N 30°E) Russia's second-largest city and former capital, **381**
St. Vincent and the Grenadines Caribbean country in the Lesser Antilles, **239**
Sakhalin Island Russian island north of Japan, **381**
Salvador (13°S 38°W) seaport city of eastern Brazil, **257**
Salzburg city in Austria, **327**
Samarqand (40°N 67°E) city in southeastern Uzbekistan, **403**
Samoa South Pacific island country in Polynesia, **731**
Sanaa (15°N 44°E) capital of Yemen, **433**
San Diego (33°N 117°W) seaport and city in California, **165**
San Francisco (38°N 122°W) seaport and city in California, **165**
San José (10°N 84°W) capital of Costa Rica, **239**
San Juan (19°N 66°W) capital of Puerto Rico, **239**
San Marino microstate in southern Europe surrounded by Italy, **349**
San Marino (45°N 12°E) capital of San Marino, **S34**
San Salvador (14°N 89°W) capital of El Salvador, **239**
Santiago (33°S 71°W) capital of Chile, **257**
Santo Domingo (19°N 70°W) capital of the Dominican Republic, **239**
São Francisco River river in eastern Brazil, **257**
São Paulo (24°S 47°W) Brazil's largest city, **257**
São Tomé (1°N 6°E) capital of São Tomé and Príncipe, **499**
São Tomé and Príncipe island country located off the Atlantic coast of Central Africa, **499**
Sarajevo (sar-uh-YAY-voh) (44°N 18°E) capital of Bosnia and Herzegovina, **349**
Saskatchewan province in central Canada, **189**
Saudi Arabia country occupying much of the Arabian Peninsula in southwestern Asia, **433**
Scandinavia peninsula of northern Europe occupied by Norway and Sweden, **305**
Scotland northern part of the island of Great Britain, **305**
Sea of Azov sea in Ukraine connected to and north of the Black Sea, **S33**
Sea of Japan body of water separating Japan from mainland Asia, **637**
Sea of Marmara sea separating European and Asian Turkey, **449**
Sea of Okhotsk inlet of the Pacific Ocean on the eastern coast of Russia, **381**
Seattle (48°N 122°W) largest city in the U.S. Pacific Northwest located in Washington, **165**
Seine River river that flows through Paris in northern France, **305**
Senegal country in West Africa, **499**
Senegal River river in West Africa, **499**
Seoul (38°N 127°E) capital of South Korea, **637**
Serbia *See* Yugoslavia.
Seychelles island country located east of Africa in the Indian Ocean, **533**
Shanghai (31°N 121°E) major seaport city in eastern China, **615**

Shannon River river in Ireland; longest river in the British Isles, **305**
Shikoku (shee-KOH-koo) smallest of the four main islands of Japan, **637**
Siberia vast region of Russia extending from the Ural Mountains to the Pacific Ocean, **381**
Sierra Madre Occidental mountain range in western Mexico, **221**
Sierra Madre Oriental mountain range in eastern Mexico, **221**
Sierra Nevada one of the longest and highest mountain ranges in the United States; located in eastern California, **165**
Sinai (SY-ny) **Peninsula** peninsula in northeastern Egypt, **483**
Singapore small island country located at the tip of the Malay Peninsula in Southeast Asia, **679**
Skopje (SKAW-pye) (42°N 21°E) capital of Macedonia, **349**
Slovakia country in central Europe; formerly the eastern part of Czechoslovakia, **327**
Slovenia country in the western Balkans region of Europe; former Yugoslav republic, **349**
Sofia (43°N 23°E) capital of Bulgaria, **349**
Solomon Islands South Pacific island country in Melanesia, **731**
Somalia East African country located in the Horn of Africa, **517**
South Africa country in southern Africa, **533**
Southern Alps mountain range in South Island, New Zealand, **711**
South Island one of the two main islands of New Zealand, **711**
South Korea country occupying the southern half of the Korea Peninsula, **637**
South Pole the southern point of Earth's axis, **S41**
Spain country in southern Europe occupying most of the Iberian Peninsula, **349**
Sri Lanka island country located south of India; formerly known as Ceylon, **585**
Stockholm (59°N 18°E) capital of Sweden, **305**
Strait of Gibraltar (juh-BRAWL-tuhr) strait between the Iberian Peninsula and North Africa that links the Mediterranean Sea to the Atlantic Ocean, **349**
Strait of Magellan strait in South America connecting the South Atlantic and South Pacific Oceans, **S31**
Sucre (19°S 65°W) constitutional capital of Bolivia, **257**
Sudan East African country; largest country in Africa, **517**
Suez Canal canal linking the Red Sea to the Mediterranean Sea in northeastern Egypt, **483**
Sumatra large island in Indonesia, **679**
Suriname (soohr-uh-NAH-muh) country in northern South America, **257**
Suva (19°S 178°E) capital of Fiji, **731**
Swaziland country in southern Africa, **533**
Sweden country in northern Europe, **305**
Switzerland country in west-central Europe located between Germany, France, Austria, and Italy, **327**
Sydney (34°S 151°E) largest urban area and leading seaport in Australia, **711**
Syr Dar'ya (sir duhr-YAH) river draining the Pamirs in Central Asia, **403**
Syria Southwest Asian country located between the Mediterranean Sea and Iraq, **449**
Syrian Desert desert region covering parts of Syria, Jordan, Iraq, and northern Saudi Arabia, **449**

Tagus River longest river on the Iberian Peninsula in southern Europe, **349**
Tahiti French South Pacific island in Polynesia, **731**
Taipei (25°N 122°E) capital of Taiwan, **615**

Gazetteer

Taiwan (TY-WAHN) island off the southeastern coast of China, **615**
Tajikistan (tah-ji-ki-STAN) country in Central Asia; former Soviet republic, **403**
Taklimakan Desert desert region in western China, **615**
Tallinn (59°N 25°E) capital of Estonia, **327**
Tampico (22°N 98°W) Gulf of Mexico seaport in central eastern Mexico, **221**
Tanzania East African country located south of Kenya, **517**
Tarai (tuh-RY) region in Nepal along the border with India, **585**
Tarawa capital of Kiribati, **731**
Tarim Basin arid region in western China, **615**
Tashkent (41°N 69°E) capital of Uzbekistan, **403**
Tasmania island state of Australia, **711**
Tasman Sea part of South Pacific Ocean between Australia and New Zealand, **711**
Tbilisi (42°N 45°E) capital of Georgia in the Caucasus region, **381**
Tegucigalpa (14°N 87°W) capital of Honduras, **239**
Tehran (36°N 52°E) capital of Iran, **433**
Tel Aviv (tehl uh-VEEV) (32°N 35°E) largest city in Israel, **449**
Thailand (TY-land) country in Southeast Asia, **659**
Thar (TAHR) **Desert** sandy desert of northwestern India and eastern Pakistan; also called the Great Indian Desert, **563**
Thessaloníki (41°N 23°E) city in Greece, **349**
Thimphu (28°N 90°E) capital of Bhutan, **585**
Tian Shan (TIEN SHAHN) high mountain range separating northwestern China from Russia and some Central Asian countries, **403**
Tiber River river that flows through Rome in central Italy, **349**
Tibesti Mountains mountain group in northwest Chad, **499**
Tierra del Fuego group of islands at the southern tip of South America, **257**
Tigris River major river in southwestern Asia, **433**
Tijuana (33°N 117°W) city in northwestern Mexico, **221**
Timor island in Indonesia, **679**
Tiranë (ti-RAH-nuh) (42°N 20°E) capital of Albania, **349**
Togo West African country located between Ghana and Benin, **499**
Tokyo (36°N 140°E) capital of Japan, **637**
Tombouctou (Timbuktu) (17°N 3°W) city in Mali and an ancient trading center in West Africa, **499**
Tonga South Pacific island country in Polynesia, **731**
Toronto (44°N 79°W) capital of the province of Ontario, Canada, **189**
Transantarctic Mountains major mountain range that divides Antarctica into East and West, **58**
Trinidad and Tobago Caribbean country in the Lesser Antilles, **239**
Tripoli (33°N 13°E) capital of Libya, **483**
Tropic of Cancer parallel 23½° north of the equator; parallel on the globe at which the Sun's most direct rays strike Earth during the June solstice (first day of summer in the Northern Hemisphere), **S21–S22**
Tropic of Capricorn parallel 23½° south of the equator; parallel on the globe at which the Sun's most direct rays strike Earth during the December solstice (first day of summer in the Southern Hemisphere), **S21–S22**
Tunis (37°N 10°E) capital of Tunisia, **483**
Tunisia country in North Africa located on the Mediterranean coast between Algeria and Libya, **483**
Turin (45°N 8°E) city in northern Italy, **349**

Gazetteer

Turkey country of the eastern Mediterranean occupying Anatolia and a corner of southeastern Europe, **449**
Turkmenistan country in Central Asia; former Soviet republic, **403**
Tuvalu South Pacific island country in Polynesia, **731**

Uganda country in East Africa, **517**
Ukraine country located between Russia and Eastern European; former Soviet republic, **381**
Ulaanbaatar (oo-lahn-BAH-tawr) (48°N 107°E) capital of Mongolia, **615**
United Arab Emirates country located on the Arabian Peninsula, **433**
United Kingdom country in Europe occupying most of the British Isles; Great Britain and Northern Ireland, **305**
United States North American country located between Canada and Mexico, **165**
Ural Mountains mountain range in west-central Russia that divides Asia from Europe, **381**
Uruguay South American country on the northern side of the Río de la Plata between Brazil and Argentina, **257**
Uzbekistan country in Central Asia; former Soviet republic, **403**

Vaduz (47°N 10°E) capital of Liechtenstein, **S34**
Valletta (36°N 14°E) capital of Malta, **349**
Valparaíso (33°S 72°W) Pacific port for the national capital of Santiago, Chile, **S32**
Vancouver (49°N 123°W) Pacific port in Canada, **189**
Vanuatu South Pacific island country in Melanesia, **731**
Vatican City (42°N 12°E) European microstate surrounded by Rome, Italy, **349**
Venezuela country in northern South America, **257**
Victoria (4°S 55°E) capital of the Seychelles, **533**
Vienna (48°N 16°E) capital of Austria, **327**
Vientiane (18°N 103°E) capital of Laos, **659**
Vietnam country in Southeast Asia, **659**
Vilnius (54°N 25°E) capital of Lithuania, **327**
Vinson Massif (78°S 87°W) highest mountain (16,066 ft.; 4,897 m) in Antarctica, **S41**
Virgin Islands island chain lying just east of Puerto Rico in the Caribbean Sea, **239**
Vistula River river flowing through Warsaw, Poland, to the Baltic Sea, **327**
Vladivostok (43°N 132°E) chief seaport of the Russian Far East, **381**
Volga River Europe's longest river; located in west-central Russia, **381**

Wake Island (19°N 167°E) U.S. South Pacific island territory north of the Marshall Islands, **731**
Wales part of the United Kingdom occupying the western portion of Great Britain, **305**
Warsaw (52°N 21°E) capital of Poland, **327**
Washington, D.C. (39°N 77°W) U.S. capital; located between Virginia and Maryland on the Potomac River, **165**
Wellington (41°S 175°E) capital of New Zealand, **711**
West Bank area of Palestine west of the Jordan River; occupied by Israel in 1967; political status is in transition, **449**
Western Ghats (GAWTS) steep, rugged hills facing the Arabian Sea on the western side of the Deccan Plateau in India, **563**
Western Plateau large plain covering more than half of Australia, **711**
Western Sahara disputed territory in northwestern Africa; claimed by Morocco, **483**
West Siberian Plain region with many marshes east of the Urals in Russia, **381**
White Nile part of the Nile River system in eastern Africa, **517**
Windhoek (22°S 17°E) capital of Namibia, **533**
Windsor (42°N 83°W) industrial city across from Detroit, Michigan, in the Canadian province of Ontario, **189**
Winnipeg (50°N 97°W) provincial capital of Manitoba in central Canada, **189**
Witwatersrand (WIT-wawt-uhrz-rahnd) a range of low hills in north central South Africa, **533**
Wuhan (31°N 114°E) city in south central China, **615**

Xi River river in southeastern China, **615**

Yamoussoukro (7°N 5°W) capital of Côte d'Ivoire, **499**
Yangon (Rangoon) (17°N 96°E) capital of Myanmar (Burma), **659**
Yaoundé (4°N 12°E) capital of Cameroon, **499**
Yekaterinburg (formerly Sverdlovsk) (57°N 61°E) city in the Urals region in Russia, **381**
Yemen country located in the southwestern corner of the Arabian Peninsula, **433**
Yenisey (yi-ni-SAY) major river in central Russia, **381**
Yerevan (40°N 45°E) capital of Armenia, **381**
Yucatán Peninsula peninsula in southeastern Mexico, **221**
Yugoslavia former country of six republics on Europe's Balkan Peninsula; now including only the republics of Serbia and Montenegro, **349**
Yukon Territory Canadian territory bordering Alaska, **189**

Zagreb (46°N 16°E) capital of Croatia, **349**
Zagros Mountains mountain range in southwestern Iran, **433**
Zambezi (zam-BEE-zee) **River** major river in central and southern Africa, **533**
Zambia country in southern Africa, **533**
Zimbabwe (zim-BAH-bway) country in southern Africa, **533**
Zürich (47°N 9°E) Switzerland's largest city, **327**

Glossary

This glossary contains terms you need to understand as you study world geography. A brief definition or explanation of the meaning of the term as it is used in *World Geography Today* follows each term. The page number refers to the page on which the term is introduced in the textbook.

Phonetic Respelling and Pronunciation Guide

Many of the key terms in this textbook have been respelled to help you pronounce them. The letter combinations used in the respelling throughout the narrative are explained in the following phonetic respelling and pronunciation guide. The guide is adapted from *Merriam-Webster's Collegiate Dictionary, Merriam-Webster's Geographical Dictionary,* and *Merriam-Webster's Biographical Dictionary.*

MARK	AS IN	RESPELLING	EXAMPLE
a	alphabet	a	*AL-fuh-bet
ā	Asia	ay	AY-zhuh
ä	cart, top	ah	KAHRT, TAHP
e	let, ten	e	LET, TEN
ē	even, leaf	ee	EE-vuhn, LEEF
i	it, tip, British	i	IT, TIP, BRIT-ish
ī	site, buy, Ohio	y	SYT, BY, oh-HY-oh
	iris	eye	EYE-ris
k	card	k	KAHRD
ō	over, rainbow	oh	OH-vuhr, RAYN-boh
u̇	book, wood	ooh	BOOHK, WOOHD
ȯ	all, orchid	aw	AWL, AWR-kid
ȯi	foil, coin	oy	FOYL, KOYN
au̇	out	ow	OWT
ə	cup, butter	uh	KUHP, BUHT-uhr
ü	rule, food	oo	ROOL, FOOD
yü	few	yoo	FYOO
zh	vision	zh	VIZH-uhn

*A syllable printed in small capital letters receives heavier emphasis than the other syllable(s) in a word.

abdicate Resign, **387**

Aborigines (a-buh-RIJ-uh-nees) Australia's first peoples, **715**

abyssal plains Areas of the ocean floor where rocks gradually sink because they have no supporting heat below them; the world's flattest and smoothest regions, **65**

acculturation Process in which an individual or group adopts some of the traits of another culture, **96**

acid rain Polluted rain that can damage trees and kill fish in lakes, **77**

African National Congress (ANC) A political organization that pushed for an end to apartheid in South Africa, **539**

Afrikaners (a-fri-KAH-nuhrz) Dutch, French, and German settlers and their descendants in South Africa, **538**

agribusiness The operation of specialized commercial farms for more efficiency and profits, **126**

alliances Agreements between countries to support one another against enemies, **328**

alluvial fan Fan-shaped deposit of mud and gravel often found along the bases of mountains, **69**

alluvial soils Soils deposited by streams or rivers, **158**

animist religions Religions in which people believe in the presence of the spirits and forces of nature, **102**

annex To formally join an area to a country, **643**

Antarctic Circle The parallel 66 1/2° south of the equator, **30**

apartheid (uh-PAHR-tayt) Official policies that forced black South Africans to live in separate areas and use separate facilities from white South Africans, **539**

aquaculture Raising and harvesting fish and marine life in ponds or other bodies of water, **619**

aqueducts Artificial channels for transporting water, **78**

aquifers Rock layers where groundwater is plentiful, **78**

arable Fit for growing crops, **175**

arboreal Tree-dwelling, **661**

archipelago (ahr-kuh-PEH-luh-goh) Large group of islands, **680**

Arctic Circle A parallel 66 1/2° degrees north of the equator, **31**

arid Dry, **54**

armistice Truce, **644**

artesian wells Wells in which water flows naturally to the surface, **712**

ASEAN Association of Southeast Asian Nations, an organization founded to promote economic development as well as political cooperation among its members, **670**

atlas A collection of maps in one book, **11**

atmosphere The envelope of gases that surrounds a planet like Earth, **35**

atoll A ring-shaped coral island or ring of several islands linked by underwater coral reefs, **732**

autarky (AW-tahr-kee) A system in which a country tries to produce all the goods that it needs, **388**

autonomy Self-government, **351**

ayatollahs Religious leaders of the highest authority among Shia Muslims, **443**

Glossary

balance of power Condition existing when countries or alliances have such equal levels of strength that war is prevented, **328**

barrier islands Coastal islands created from sand deposited by ocean waves and currents in shallow water, **150**

basins Low areas of land, often surrounded by mountains, **151**

bauxite Ore from which aluminum is made, **241**

Bedouins Nomadic herders in Southwest Asia, **441**

Berbers Cultural group that has lived in North Africa since long before waves of Arab armies crossed the continent, **489**

bilingual Able to speak two languages, **169**

biodiversity The level of variety of plants and animals, **534**

biosphere The part of Earth that includes all life forms, **35**

birthrate The number of live births each year for every 1,000 people living in a place, **89**

Bolsheviks A communist group that overthrew the government during the Russian Revolution in 1917, **387**

boycott Refusal to buy certain goods from a country or business, **570**

buffer state A small country between two larger, more powerful countries, **265**

cacao Type of tree from which we get cocoa beans, **246**

calligraphy Artistic handwriting or lettering, **624**

cantons Largely self-governing states within a country such as Switzerland, **336**

capitalism An economic system in which businesses, industries, and resources are privately owned, **115**

caravans Groups of people traveling together for protection, **408**

Caricom Caribbean Community and Common Market, **252**

cartography The study of maps and mapmaking, **5**

cash crops Crops grown for sale in a market, **230**

caste Group of people who are born into a certain position in society, as in Hinduism, that restricts the occupation and associations of their members, **571**

central business district (CBD) City center dominated by large stores, offices, and buildings, **123**

Chibcha Early people of the Colombian Andes who developed gold-working skills, **263**

Chishima Current (Oyashio) Cold ocean current that cools northern Japan in the summer, **639**

city-states Self-governing cities and their surrounding areas, as in ancient Greece, **360**

climate Weather conditions in a region over a long time, **41**

climate graphs Graphs showing average temperature and precipitation in a place, **18**

colonies Territories controlled by people from a foreign land, **167**

command economy An economic system in which the government decides what to produce, where to make it, and what price to charge, **115**

commonwealth Self-governing territory associated with another country, **250**

communes Collective farms grouped together to organize farming and plan public services, **623**

communism An economic and political system in which the government owns or controls almost all the means of production, **115**

compass rose A map element with arrows pointing in all four principal directions, **14**

complementary region A region formed by the combination of two areas with different activities or strengths, each of which benefits the other, **342**

condensation The process by which water vapor changes from a gas into liquid droplets, **46**

confederation A group of states joined together for a common purpose, **336**

Confucianism A philosophy based on the teachings of Confucius, an ancient Chinese philosopher, **624**

coniferous forests Forests of cone-bearing evergreen trees, **55**

conquistadores (kahn-kees-tuh-DAWR-ez) Spanish conquerors of foreign lands during the colonial era, **225**

consensus General agreement, **197**

constitutional monarchy A type of government with a king or queen as head of state and a parliament as the lawmaking branch, **307**

contiguous Connecting or bordering, **14**

Continental Divide The crest dividing North America's major river systems into those flowing eastward and those flowing westward, **153**

continental drift The process by which Earth's plates slowly move across the upper mantle, **63**

continental shelves Areas where continental surfaces extend under the shallow ocean water around the continents, **66**

continents Large landmasses on Earth's surface, **10**

contour plowing Plowing fields across a hill, rather than up and down the hill, **75**

copra Dried coconut meat, **741**

core Earth's center, where pressures and temperatures are very high, **63**

cork Bark that is stripped from the trunks of cork oaks, **353**

Corn Belt A region of the U.S. Midwest that specializes in growing corn, **175**

cosmopolitan Having many foreign influences, **316**

Cossacks People from the southern steppe frontiers of the Russian Empire who played an important role in that empire's expansion, **387**

cottage industries Small-scale industries based in the home, **576**

coup (KOO) A change of government caused by a group taking control by force, **266**

creole A language blending African, European, or indigenous Caribbean languages, **251**

crop rotation The practice of planting different crops in a field in alternating years, **76**

cultural boundaries Boundaries that are based on culture traits, **129**

culture All the features of a people's way of life, **94**

culture region An area in which people have many shared culture traits, **95**

culture traits Learned activities and behaviors that people often take part in, **94**

cyclones Winds around centers of low atmospheric pressure, **42**

czar (ZAHR) Title of the emperor of Russia before the Russian Revolution, **387**

Dairy Belt A region of the U.S. Midwest that specializes in dairy products, **175**

Dalits Hindus in India who are not part of any caste, **571**

death rate Total number of deaths each year for every 1,000 people in a place, **89**

deciduous forests Forests made up of trees that lose their leaves during part of the year, **55**

deforestation Destruction or loss of forests, **76**

degrees Units used to measure distances between parallels and between meridians, **10**

delta Accumulation of sediment at the mouth of a river, **69**

demilitarized zone (DMZ) A buffer zone into which opposing armies may not enter, **644**

democracy A system of government in which the people decide who will govern, **130**

demography Statistical study of human populations, **87**

depressions Large low areas, **484**

desalinization Process of removing salt from ocean water, **70**

desertification Spreading of desert conditions, **501**

developed countries Countries with high levels of industrialization and high standards of living, **117**

developing countries Countries with less productive economies than developed countries and low standards of living, **117**

dharma For Hindus, the importance of doing one's duty according to one's station in life, **571**

dialect A regional variety of a language, **100**

dictator A leader who rules with almost absolute authority, **226**

Diet (DEE-uht) Japan's elected lawmaking body, **644**

diffusion A process occurring when an idea or innovation spreads from one person or group to another and is adopted, **97**

dikes Walls built to hold back water, **293**

doldrums Areas with no prevailing winds, **44**

domestication The process in which people grow plants and tame animals for their own use, **120**

domino theory The idea that if one country fell to communism, neighboring countries would follow like falling dominoes, **664**

double cropping Harvesting two crops in the same plot each year, **619**

drainage basin A region drained by a river and its tributaries, **71**

dryland farming Growing crops that can rely on limited rainfall rather than irrigation, **412**

dual economies Economies in which some goods are produced for export while other goods are produced for local consumption, **509**

dynasty Line of hereditary rulers, **438**

economy of scale Large production of goods that reduces the production cost of each item, **727**

Glossary

ecosystems Communities of plants and animals, **50**

ecotourism Type of tourism focusing on guided travel through natural areas, **248**

edge cities Clusters of large buildings away from the central business district of a city, **124**

El Niño (ehl NEEN-yoh) Ocean and weather pattern in which the southeastern Pacific Ocean is warmer than usual, affecting regional climates, **260**

emigrants People who leave a country to live somewhere else, **89**

enclave a distinct territorial or cultural unit, such as an ethnic group, surrounded by a different territory or culture, **362**

endemic species Plants and animals native to a certain place, **682**

environment Combination of Earth's four spheres, including all of the biological, chemical, and physical conditions that affect life, **35**

equator An imaginary line that circles the globe halfway between the North Pole and South Pole, **10**

equinox The time when both of Earth's poles are at a 90 degree angle from the Sun and the Sun's direct rays strike the equator, **31**

erg A sea of sand created by high, shifting sand dunes, **484**

erosion Movement of surface material from one location to another by water, wind, and ice, **67**

escarpment Steep face at the edge of a plateau or other raised area, **534**

estuaries Semi-enclosed bodies of water, seawater, and freshwater formed where a river meets an inlet of the sea, **71**

ethnic groups Human populations that share a common culture or ancestry, **95**

ethnic religions Religions found among people of one ethnic group and that generally have not spread into other cultures, **102**

Eurasia The landmass of Europe and Asia combined, **382**

European Union Organization of European countries featuring close cooperation on trade, economic, political, and social issues, **315**

evaporation Process by which liquid changes to gas, **46**

exclave An area separated from the rest of a country by the territory of other countries, **338**

Exclusive Economic Zone (EEZ) Zone extending 200 nautical miles (370 km) from a country's shore that includes resources controlled by that country, **741**

exotic rivers Rivers that begin in humid regions and then flow across dry areas, **434**

exotic species New types of plants and animals that people introduce to an area, **719**

export economy A type of economy in which goods are produced mainly for export rather than for domestic use, **650**

extensive agriculture A kind of agriculture using much land but small inputs of capital and labor per unit area, **718**

fall line A natural boundary between two landform regions with different elevations marked by rapids and waterfalls, **150**

Glossary

famine A widespread shortage of food, **307**

faults Places where rock masses have been broken apart and are moving away from each other, **66**

favelas (fah-VE-lahs) Large slums around Brazilian cities, **270**

fellahin Peasant farmers of North Africa, **492**

fjords (fee-AWRDZ) Narrow deep inlets of the sea set between high rocky cliffs, **292**

flyway Migration route for birds, **640**

folds Places where rocks have been compressed into bends by colliding plates, **66**

formal region Region with one or more common features that make it different from surrounding areas, **6**

fossil fuels Energy resources—formed from the remains of ancient plants and animals—including coal, natural gas, and petroleum, **79**

fossil water Groundwater that is not replenished by rain, **78**

free enterprise System that lets competition among businesses determine the price of products, **115**

free port Port where almost no taxes are placed on the goods unloaded from other places, **492**

front Zone of contact between two air masses of widely different temperatures or moisture levels, **44**

functional region Region made up of different places that are linked and function as a unit, **6**

fundamentalism Movement in which people believe in strictly following certain established principles or teachings, **98**

genocide The intentional destruction of a people, **527**

gentrification Process of buying run-down homes in older areas of a community and restoring them, **181**

geography The study of the physical, biological, and cultural features of Earth's surface, **3**

geometric boundaries Boundaries that follow geometric patterns, **129**

geothermal energy Heat of Earth's interior, **81**

geysers Hot springs that shoot water into the air, **320**

ghetto Section of a city where a minority group is forced to live, **339**

glaciers Thick masses of ice, including great ice sheets and bodies of ice that flow down mountains like slow rivers, **68**

globalization The process in which connections around the world increase and cultures become more alike, **98**

global warming The process in which Earth grows warmer over a period of time, **41**

Gran Chaco (grahn CHAH-koh) Very flat plains between the Andes and Brazilian Highlands of South America, **258**

graphite A form of carbon that is used in pencil leads, **595**

great-circle route Shortest route between any two places on the planet, **12**

greenhouse effect The process in which Earth's atmosphere traps heat energy, **41**

grid Lines crisscrossing each other in uniform intervals to create a pattern of squares, **9**

gross domestic product (GDP) The total value of goods and services created within a country, **116**

gross national product (GNP) The total value of goods and services that a country produces in a year, **116**

groundwater Water found below ground, **72**

gulag Network of prison labor camps in the Soviet Union, **388**

Gulf Stream Warm ocean current that moves warm tropical water northward along eastern North America, **45**

gum arabic Sap of acacia trees; a sticky substance that binds the ingredients in many candies and medicines, **525**

haciendas (hah-see-EN-duhs) Large Spanish colonial estates usually owned by wealthy families but worked by many peasants, **225**

hajj Religious journey to Mecca required of Muslims, **102**

headwaters First and smallest streams formed from the runoff of a mountain, eventually forming rivers, **71**

heavy industry Industry focusing on manufacturing, usually based on metals, **393**

hemispheres Halves of the globe divided by the equator or a meridian, **10**

hieroglyphs Pictures and symbols used in the ancient Egyptian writing system, **487**

hinterland A region that lies far away from major population centers, **192**

Holocaust The murder of millions of Jews during World War II, **456**

homogeneous (hoh-muh-JEE-nee-uhs) Of the same kind, such as a population made up mostly of the same ethnic group, **687**

hot spot Place where magma wells up to the surface from Earth's mantle, **152**

humidity The amount of water vapor in the air, **46**

humus Broken-down plant and animal matter in soil, **75**

hurricanes The most powerful and destructive tropical cyclones, **48**

hydroelectric power Electricity produced by moving water, **81**

hydrologic cycle Movement of water through the hydrosphere, **71**

hydrosphere Earth's water in all of its forms, **35**

icebreakers Ships that can break paths through ice, **384**

imams Muslim religious leaders, **440**

immigrants People who come to a new country to live, **90**

Inca Founders of South America's greatest early civilization, **263**

indigenous Native to an area, **244**

industrialization The process by which manufacturing based on machine power becomes widespread in an area, **116**

informal sector Sector of an economy made up of people who do not work for formal businesses, including many self-employed people and small family-owned businesses, **543**

infrastructure System of roads, ports, and other facilities needed by a modern economy, **117**

innovation New idea that a culture accepts, **96**

intensive agriculture A form of agriculture using a great amount of human labor and/or capital to farm small amounts of land, **619**

Intertropical Convergence Zone (ITCZ) An area where northeast and southeast trade winds meet near the equator, creating humid, often rainy conditions, **734**

irrigation A process in which water is artificially supplied to the land, **76**

isthmus Narrow strip of land with water on either side connecting two larger land areas, **222**

ivory Cream-colored material that makes up the tusk of an elephant, **522**

Jainism Religion with origins in India that emphasizes a strict moral code based on preserving life, **572**

Japan Current (Kuroshio) Warm ocean current bringing moist marine air to Japan, **639**

jute Plant fiber used to make products like burlap and rope, **576**

kampongs Traditional villages usually built on stilts in parts of Southeast Asia, **692**

karma For Hindus, Buddhists, and Jains, the positive or negative force caused by a person's actions, **571**

klongs Canals of Thailand, **672**

Kurds A minority group in Turkey and neighboring countries, **457**

lahars (LAH-hahrz) Dangerous volcanic mudflows, **681**

landlocked Having no coast on the ocean, **271**

landscapes Scenery of places, including physical, human, and cultural features, **4**

La Niña An ocean and weather pattern that causes the eastern Pacific Ocean to be colder than normal, **260**

latifundia (lah-ti-FOOHN-dee-uh) Large estates in South America, **265**

latitude Lines on a globe drawn in an east-west direction and measured in degrees, **9**

leaching Downward movement of minerals and humus in soils, **75**

legend Map key, **14**

lichens Small plants that consist of algae and fungi, **157**

light industry Industry using low-weight materials and focusing on the production of consumer goods, **393**

lingua franca A language of trade and communication used by people who speak different languages, **101**

literacy rate Percentage of people who can read and write, **117**

lithosphere Solid crust of Earth, **35**

Llanos (YAH-nohs) Large plains area in northeastern Colombia and western Venezuela, **258**

loess Fine-grained, windblown soil that is very fertile, **299**

longitude Lines on a globe drawn in a north-south direction and measured in degrees, **9**

Glossary

magma Liquid rock within Earth, **63**

magnesium Light metal that is valuable in certain industries, such as aerospace, **451**

mandates Territories temporarily placed under another country's control, **455**

mangrove Type of tree with roots that grow in saltwater; found in tropical coastal areas around the world, **241**

manioc Tropical plant with starchy roots, **267**

mantle The section of Earth's interior that lies above the outer core and has the most mass, **63**

Maori (MOWR-ee) First peoples of New Zealand, **724**

map projections Different ways of representing a round Earth on a flat map, **11**

maquiladoras (mah-kee-lah-DOHR-ahs) Special factories in Mexico owned mainly by American companies, **232**

market economy An economic system in which people choose freely what to buy and sell, **115**

market-oriented agriculture Agricultural system in which farmers grow products to sell to consumers, **126**

marsupials Mammals that have pouches to carry their young, **713**

martial law Military rule, **631**

matrilineal A social system characterized by tracing kinship only through the mother, **739**

Megalopolis A group of cities in the northeastern United States that has grown into one large, built-up area; also used to describe similar areas around the world, **173**

Meiji Restoration Japanese political revolution that overthrew the shogun and restored the power of the emperor, **643**

Mercosur South American organization whose purpose is to expand trade, improve transportation, and reduce tariffs among member countries, **270**

meridians Lines of longitude, **10**

mestizos People of mixed European and American Indian ancestry, **226**

meteorology The study of weather, **5**

metropolitan area A city and its surrounding built-up areas, **174**

microstates Very small countries, such as Andorra, Monaco, and San Marino, **357**

migration The process of moving from one place to live in another, **89**

millet Type of drought-resistant grain, **507**

minifundia (mi-ni-FOOHN-dee-uh) Small South American farms created when large colonial estates were broken up, **268**

missionaries People whose goal it is to spread their religion, **102**

monoculture Cultivation of a single crop, **409**

monotheism The belief in one god, **102**

monsoon A wind system in which winds reverse direction and cause seasons of wet and dry weather, **51**

moons Smaller objects that orbit a planet, **26**

mosques Islamic houses of worship, **102**

Glossary

mulattoes People with both African and European ancestry, **245**
multilingual Able to speak two or more languages, **336**

nationalism A feeling of pride and loyalty for one's country or culture group, **130**
nationalized The process by which organizations or businesses become owned and operated by the government, **308**
natural boundaries Boundaries that follow a feature of the landscape, **129**
natural hazards Events in the physical environment that can destroy human life and property, **155**
navigable Able to be used for shipping, **294**
neutral Characterized by not taking sides in conflicts, **336**
newsprint Inexpensive paper used mainly for newspapers, **158**
nomads People who move often from place to place, **409**
North American Free Trade Agreement (NAFTA) An agreement between Canada, Mexico, and the United States to eliminate tariffs on many products flowing between these three countries, **183**
North Atlantic Drift Warm ocean current that moderates climates in northwestern Europe, **297**

oasis A place where a spring bubbles to the surface in desert areas and plants grow, **436**
OPEC Organization of Petroleum Exporting Countries, which influences oil prices by controlling supply, **443**
ore Mineral-bearing rock, **79**
orographic effect Cooling effect that occurs when air is forced to rise over a mountain, resulting in a wetter windward side and a drier leeward side, **47**
outback Dry interior of Australia, **713**

paddy Wet land where rice is grown, **619**
Pampas Wide fertile grasslands in southern South America, **258**
pantheon All the gods of a religion, **568**
parallels Lines of latitude, **10**
parliament A legislature lead by a prime minister or premier, **196**
partition Division of a place, as in the 1947 partition of India, **570**
pastoralism A type of subsistence agriculture involving herding animals, **125**
perception Our awareness and understanding of the environment around us, **6**
perceptual regions Regions that reflect human feelings and attitudes, **6**
permafrost Permanently frozen soil below the ground's surface, **56**
perspective The way a person looks at something, **4**
petrochemicals Certain products made from oil, **80**
pharaohs Ancient Egyptian monarchs, **487**
phosphates Chemicals used to make fertilizer, **743**
pictograms Simple pictures used to represent ideas, **624**
pidgin languages Simplified languages used to help people with different languages communicate, **737**
piedmont An area at or near the foot of a mountain region, **150**
planets Major bodies that orbit a star, **25**
plantations Large farms that produce one major crop often for export, **168**
plateau An elevated flatland that rises sharply above nearby land on at least one side, **69**
plate tectonics The theory that Earth's crust is divided into rigid plates that slowly move across the upper mantle, **63**
plaza Open space in towns in Mexico and other Spanish-speaking countries, often in front of a church, **225**
polar regions High-latitude, cold regions around the North and South Poles, **29**
polders Lands reclaimed from the sea in the Netherlands, **293**
polytheism The belief in many gods, **102**
population density The average number of people living in an area; usually expressed as persons per square mile or square kilometer, **88**
population pyramids Graphs showing the percentage of males and females by age group in a country's population, **18**
potash A mineral used to process wool and to make fertilizers, glass, and soft soaps, **451**
precipitation Condensed droplets of water that fall as rain, snow, sleet, or hail, **16**
prevailing winds Winds that blow in the same direction most of the time, **44**
primate city A city that ranks first and dominates a country in terms of population and economy, **312**
prime meridian Imaginary line drawn from the North Pole through Greenwich, England, to the South Pole, **10**
protectorate Country that forfeits certain decision-making powers in exchange for protection by a stronger country, **591**
provinces Government districts similar to states, **190**
pull factors Factors attracting people to a new location, **90**
puppet government Government controlled by outside forces, such as another country's government, **622**
push factors Factors causing people to leave a location, **90**

quotas In trade matters, limits on the amount of a product that can be imported, **130**

rain shadow A drier area on the leeward side of a mountain range, **47**
reforestation The replanting of a forest, **76**
refugees People who have been forced to leave an area, **90**
reg A gravel-covered plain in a desert area, **484**

region An area with one or more common features that make it different from surrounding areas, **6**
regionalism A feeling of strong political and emotional loyalty to one's own region, **200**
reincarnation For Hindus, the repeated rebirth of one's soul into different forms, **571**
Renaissance (re-nuh-SAHNS) A period from about the 1300s to the 1500s marked by a renewed interest in learning in Europe, **356**
revolution One elliptical orbit of Earth around the Sun, **27**
rift valleys Places on Earth's surface where the crust stretches until it breaks, **65**
rotation One complete spin of Earth on its axis, **27**
Rus (ROOS) Scandinavian traders who were some of Kiev's early leaders; also the word from which Russia gets its name, **386**

Sahel (sah-HEL) A region of semiarid climate extending from Senegal and Mauritania to Sudan in Africa, **501**
samurai Professional Japanese warrior employed by a shogun for protection, **643**
sanctions Penalties intended to force a country to change its policies, **539**
Sanskrit Ancient language from which Hindi developed, **568**
Santeria Originating in Cuba, a religion that blends African traditions and Christian beliefs, **251**
satellite A body that orbits a larger body, **26**
savannas Areas of tropical grasslands, scattered trees, and shrubs, **51**
secular Nonreligious, **463**
sediment Small particles of weathered rock, **67**
separatism The belief that certain parts of a country should be independent, **200**
sepoys Indian troops under the command of British officers during the colonial era in India, **569**
sequent occupance A process of settlement by successive groups of people in which each group creates a distinctive cultural landscape, **306**
serfs Poor peasant farmers in Russia who worked for lords and were bound to the land, **387**
shatter belt A zone of frequent boundary changes and conflicts, often located between major powers, **389**
shifting cultivation A process in which farmers clear trees or brush for short-term planting before moving on to clear another area, **125**
Shi'ism One of the two main branches of Islam, **440**
shogun A powerful warlord in early Japan, **642**
Sikhism A religion with origins in India that combines the Muslim belief in one God with the Hindu belief in reincarnation, **572**
Silicon Valley An area south of San Francisco, California, that became the leading center of computer technology in the United States, **178**
silt Fertile, finely ground soil deposited by flowing water, **487**
sinkholes Steep-sided depressions that form when the roof of a cave collapses, **222**
sisal (SY-suhl) Strong, durable plant fiber used to make rope and twine, **522**
slash-and-burn agriculture A form of shifting cultivation in which farmers slash small areas of forest and burn the fallen trees, **688**

Glossary

Slavs The main people to settle in what are now Russia, Ukraine, Belarus, and eastern parts of central and southern Europe, **386**
smelters Factories that process metal ores, **394**
smog Air pollution resulting from chemical reactions involving sunlight and automobile and industrial exhaust, **179**
socialism An economic system in which the government owns and controls the means of producing goods, **321**
soil exhaustion A condition in which soil has lost nutrients and becomes nearly useless for farming, **76**
soil salinization Salt buildup in the soil, **76**
solar energy Energy from the Sun, which reaches Earth as light and heat, **27**
solar system The Sun and the group of bodies that revolve around it, **25**
solstice The time that Earth's poles point at their greatest angle toward or away from the Sun, **30**
sorghum A type of drought-resistant grain, **507**
souks Open-air markets in the Arab world, **461**
soviets Local governing bodies of the Soviet Union, **388**
Special Economic Zones (SEZs) Zones designed to attract foreign companies and investment to areas of China, **628**
staple The main food of a region, **507**
storm surge Waters washed ashore by large storms, **587**
subcontinent Very large landmass that is smaller than a continent, such as the Indian subcontinent, **564**
subsidies Financial support from the government, **649**
subsistence agriculture Farming in which food is produced by a family just for its own needs, **125**
sultans Ottoman rulers, **455**
Sunni One of the two main branches of Islam, **440**
superpower A huge powerful country, **183**
Swahili (swah-HEE-lee) Language in East Africa with roots in the languages of the African coast with added Arabic words, **522**

taiga A forest of mainly evergreen trees that covers half of Russia, **384**
Taoism A religion that teaches there is a natural order to the universe, called the Tao, **624**
tariffs Taxes on imports and exports, **130**
tar sands Rock or sand layers that contain oil, **261**
temperature Measurement of heat, **41**
tepuís (tay-PWEEZ) A chain of high plateaus in the Guiana Highlands, **258**
terrorism The use of fear and violence as a political force, **271**
textiles Cloth products, **173**
theocracy A country governed by religious law, **443**
topography Elevation, layout, and shape of the land, **17**
tornadoes Violent twisting spirals of air, **48**
totalitarian government A government in which one person or a small group of people rule a country and in which the people have no say, **130**

Glossary

trade deficit A situation in which the value of a country's exports is less than the value of the country's imports, **183**

trade surplus A condition existing when a country exports more than it imports, **650**

traditionalism Following longtime practices and opposing many modern technologies and ideas, **98**

transhumance The practice of moving herds from mountain pastures in the summer to lowland pastures in the winter, **409**

tree line The line of elevation above which trees do not grow, **259**

trench A deep valley marking a collision of plates, where one plate slides under another, **66**

tributary Any smaller stream or river that flows into a larger stream or river, **71**

Tropic of Cancer The parallel 23 1/2° north of the equator, **31**

Tropic of Capricorn The parallel 23 1/2° south of the equator, **30**

tropics Warm, low-latitude areas near the equator, **29**

trust territories Areas placed, after World War II, under the control of another country while the territories prepared for independence, **737**

tsetse fly African fly carrying a human disease called sleeping sickness, **519**

tsunamis (sooh-NAH-mees) Large sea waves created by undersea tectonic activity like earthquakes, **638**

typhoons The term for hurricanes in the western Pacific Ocean, **48**

uninhabitable Unable to support human life and settlements, **319**

universalizing religions Religions that seek followers all over the world, **102**

urban agglomeration Densely populated region surrounding a central city, **651**

urbanization Growth in the proportion of people living in towns and cities, **120**

veld (VELT) Grassland regions of South Africa, **535**

voodoo Haitian version of traditional African religious beliefs that are blended with elements of Christianity, **251**

wadis Dry streambeds that fill with water only after rainfall, **484**

watershed The entire region drained by a river and its tributaries, **71**

water table The groundwater level at which all the cracks and spaces in rock are filled with water, **72**

wats Buddhist temples that also serve as monasteries, **670**

weather The condition of the atmosphere at a given time and place, **41**

weathering The process by which rocks break and decay over time, **67**

wetlands Landscapes that are covered with water for at least part of the year, **72**

Wheat Belt The region of the U.S. interior west specializing in growing wheat, **178**

work ethic The belief that work itself has moral value, **649**

world cities Huge urban areas that are the most important centers of economic power and wealth, **122**

yurts Movable, round houses of wool felt mats in Central Asia, **410**

zinc An element important in metal processing and other industries, **406**

Zionism The movement that called for Jews to set up their own country or homeland in Palestine, **456**

Spanish Glossary

abdicate/abdicar Abandonar un cargo de jefe de gobierno, **387**

Aborigines/aborígenes Primeros habitantes de Australia, **715**

abyssal plains/suelos abisales Áreas del fondo del océano en que las rocas se hunden gradualmente debido a que no existen fuentes de calor que las mantengan en su sitio; suelen ser las regiones más llanas del mundo, **65**

acculturation/aculturación Proceso en el cual un individuo o un grupo adopta algunas características de otra cultura, **96**

acid rain/lluvia ácida Lluvia contaminada que puede dañar a los árboles y matar a los peces de los lagos, **77**

African National Congress/Congreso Nacional Africano (**ANC**, por sus siglas en inglés; **CNA**, en español) Organización política que luchó por poner fin a la separación racial en Sudáfrica, **539**

Afrikaners/afrikaners Colonizadores holandeses, franceses y alemanes y sus descendientes establecidos en Sudáfrica, **538**

agribusiness/agroeconomía Granjas especializadas que operan con mayor eficacia y ganancias económicas, **126**

alliances/alianzas Acuerdos entre diferentes países para respaldarse y defenderse de sus enemigos, **328**

alluvial fan/abanico pluvial Depósito de lodo y grava en forma de abanico que con frecuencia se acumula en las faldas de las montañas, **69**

alluvial soils/suelos pluviales Depósitos de tierra arrastrada por la corriente de los ríos, **158**

animist religions/religiones animistas Religiones cuyos integrantes creen en la presencia de espíritus y en las fuerzas de la naturaleza, **102**

annex/anexar Unir formalmente un territorio a un país, **643**

Antarctic Circle/Círculo Polar Antártico Paralelo 66 1/2° al sur del ecuador, **30**

apartheid/apartheid Política oficial que obligaba a las personas de raza negra en Sudáfrica a vivir en áreas separadas y a usar instalaciones diferentes de las que usaban los de raza blanca, **539**

aquaculture/acuacultura Cultivo y recolección de peces y vida marina en estanques y otros depósitos de agua, **619**

aqueducts/acueductos Canales artificiales usados para transportar agua, **78**

aquifers/acuíferos Capas de roca en las que abunda el agua, **78**

arable/cultivable Apropiado para el cultivo, **175**

arboreal/arborícola Que vive en los árboles, **661**

archipelago/archipiélago Grupo grande de islas, **680**

Arctic Circle/Círculo Polar Ártico Paralelo 66 1/2° al norte del ecuador, **31**

arid/árido Seco, **54**

armistice/armisticio Tregua, **644**

artesian wells/pozos artesianos Pozos en los que el agua fluye de manera natural a la superficie, **712**

ASEAN/ASEAN (por sus siglas en inglés) Asociación de Naciones del Sureste Asiático, organización creada para promover el desarrollo económico y los acuerdos políticos de las naciones que la integran, **670**

atlas/atlas Colección de mapas reunidos en un solo libro, **11**

atmosphere/atmósfera Capa de gases que envuelve a un planeta como la Tierra, **35**

atoll/atolón Isla coralina en forma de anillo o anillo de varias islas menores unidas por un arrecife de coral submarino, **732**

autarky/autarquía Sistema en el que un país trata de producir todos los bienes que necesita, **388**

autonomy/autonomía Autogobierno, **351**

ayatollahs/ayatolas Líderes religiosos de la máxima autoridad entre los shiíes musulmanes, **443**

balance of power/equilibrio de poder Condición que surge cuando varios países o alianzas mantienen niveles tan similares de poder evitar guerras, **328**

barrier islands/islas de barrera Islas costeras formadas por depósitos de arena arrastrada por las mareas y las corrientes de aguas poco profundas, **150**

basins/cuencas Zonas de tierras bajas, con frecuencia rodeadas por montañas, **151**

bauxite/bauxita Mineral con el que se elabora el aluminio, **241**

Bedouins/beduinos Pastores nómadas del suroeste de Asia, **441**

Berbers/berberes Grupo cultural que habita en el norte de África desde mucho antes que las oleadas de ejércitos árabes llegaran al continente, **489**

bilingual/bilingüe Persona que habla dos idiomas, **169**

biodiversity/biodiversidad Grado de variedad de plantas y animales, **534**

biosphere/biosfera Parte de la Tierra donde habitan todas las formas de vida que hay en ella, **35**

birthrate/tasa de natalidad Número de nacimientos anuales por cada 1,000 habitantes de un lugar, **89**

Bolsheviks/bolcheviques Grupo comunista que derrocó al gobierno durante la Revolución Rusa de 1917, **387**

boycott/boicot Negativa a comprar bienes provenientes de un país o de una empresa, **570**

buffer state/estado tapón País pequeño que se ubica entre dos países más grandes y poderosos, **265**

cacao/cacao Árbol del que se obtienen las semillas de cacao, **246**

calligraphy/caligrafía Escritura o rotulación artística, **624**

Spanish Glossary

cantons/cantones Grandes estados con autogobierno dentro de un país, como Suiza, **336**

capitalism/capitalismo Sistema económico en el que los negocios, las industrias y los recursos son de propiedad privada, **115**

caravans/caravanas Grupos de personas que viajan juntas para protegerse, **408**

Caricom/Caricom (por su abreviatura en inglés) Comunidad del Caribe y Mercado Común, **252**

cartography/cartografía Estudio y elaboración de mapas, **5**

cash crops/cultivo para la venta Productos cosechados para su venta en el mercado, **230**

caste/casta Grupo de personas nacidas con cierta posición social, como en el hinduismo, que restringe la ocupación y asociación de sus miembros, **571**

central business district/distrito comercial (**CBD**, por sus siglas en inglés) Zona de una ciudad dominada por grandes comercios, oficinas y edificios, **123**

Chibcha/Chibcha Antigua civilización de los Andes colombianos que desarrolló grandes habilidades orfebres, **263**

Chishima Current/corriente de Chishima (Oyashio) Corriente oceánica fría que disminuye la temperatura del norte de Japón en el verano, **639**

city-states/ciudades estado Ciudades autogobernadas y sus territorios aledaños, como en la antigua Grecia, **360**

climate/clima Condiciones meteorológicas de una región durante un periodo largo, **41**

climate graphs/gráficas del clima Gráficas que muestran la temperatura promedio y la precipitación pluvial de una región, **18**

colonies/colonias Territorios controlados por personas de otro país, **167**

command economy/economía dirigida Sistema económico en el que el gobierno decide lo que produce, dónde lo produce y el precio de venta de esos productos, **115**

commonwealth/mancomunidad Territorio autogobernado asociado con otro país, **250**

communes/comunas Granjas colectivas agrupadas para organizar las cosechas y la prestación de servicios, **623**

communism/comunismo Sistema económico y político en que el gobierno posee y controla casi todos los medios de producción, **115**

compass rose/rosa de los vientos Elemento de un mapa que señala con una flecha la posición de los cuatro puntos cardinales, **14**

complementary region/región complementaria Región formada por la combinación de dos zonas con diferentes actividades y niveles de poder para el beneficio de ambas, **342**

condensation/condensación Proceso por el cual el vapor pasa del estado gaseoso al estado líquido y forma pequeñas gotas, **46**

confederation/confederación Grupo de estados unidos con un propósito común, **336**

Confucianism/confucianismo Filosofía basada en las enseñanzas de Confucio, un antiguo filósofo chino, **624**

coniferous forests/bosques de coníferas Bosques poblados por árboles de semillas de cono y hojas siempre verdes, **55**

conquistadores/conquistadores Españoles que conquistaron territorios extranjeros durante la época colonial, **225**

consensus/consenso Acuerdo general, **197**

constitutional monarchy/monarquía constitucional Tipo de gobierno con un rey o una reina como jefe de estado y un parlamento legislador, **307**

contiguous/contiguo Que conecta o limita, **14**

Continental Divide/división continental Cresta que divide los principales ríos de Estados Unidos en dos partes, los que fluyen al este y los que fluyen al oeste, **153**

continental drift/desplazamiento continental Movimiento lento de las placas de la Tierra sobre el manto superior, **63**

continental shelves/arrecifes continentales Áreas en que la superficie continental se extiende a las aguas poco profundas que rodean a los continentes, **66**

continents/continentes Grandes masas de tierra sobre la superficie terrestre, **10**

contour plowing/arado de contorno Arado de una colina realizado en forma horizontal y no en forma vertical, **75**

copra/copra Carne de coco deshidratada, **741**

core/núcleo Centro de la Tierra donde la temperatura y la presión son muy altas, **63**

cork/corcho Material extraído de la corteza de los árboles de corcho, **353**

Corn Belt/región maicera Región del medio oeste de Estados Unidos especializada en el cultivo de maíz, **175**

cosmopolitan/cosmopolita Que recibe influencia de muchas culturas extranjeras, **316**

Cossacks/Cosacos Pueblo de la frontera sur del imperio ruso que jugó un importante papel en la expansión del imperio, **387**

cottage industries/industrias caseras Industrias de pequeña escala establecidas en el hogar, **576**

coup/golpe de estado Cambio de gobierno que ocurre cuando otro grupo toma el poder por la fuerza, **266**

creole/criollo Idioma que combina varios idiomas de África, Europa y el Caribe, **251**

crop rotation/rotación de cultivos Práctica que consiste en sembrar diferentes cultivos en años alternados, **76**

cultural boundaries/fronteras culturales Fronteras basadas en las tendencias culturales de diferentes grupos, **129**

culture/cultura Todas las características de vida de un grupo de personas, **94**

culture region/región cultural Área cuyos habitantes comparten varias tendencias culturales, **95**

culture traits/tendencias culturales Actividades y conductas aprendidas en las que participan las personas, **94**

cyclones/ciclones Vientos que se desplazan alrededor de centros de baja presión atmosférica, **42**

czar/zar Título del emperador de Rusia antes de la Revolución Rusa, **387**

Dairy Belt/región lechera Región del medio oeste de Estados Unidos especializada en la elaboración de productos lácteos, **175**

Dalits/dalits Habitantes de clase social baja en la India que no pertenecen a ninguna casta, **571**

death rate/tasa de mortalidad Número de muertes anuales por cada 1,000 habitantes de un lugar, **89**

deciduous forests/bosques de hoja caediza Bosques poblados por árboles que pierden sus hojas en cierto periodo del año, **55**

deforestation/deforestación Destrucción o pérdida de los bosques, **76**

degrees/grados Unidades usadas para medir la distancia entre meridianos y entre paralelos, **10**

delta/delta Acumulación de sedimentos en la desembocadura de un río, **69**

demilitarized zone/zona desmilitarizada (**DMZ**, por sus siglas en inglés) Zona de protección en la que los ejércitos enemigos no pueden entrar, **644**

democracy/democracia Sistema de gobierno en el que los ciudadanos deciden quién los gobernará, **130**

demography/demografía Estudio estadístico de los asentamientos humanos, **87**

depressions/depresiones geográficas Grandes áreas de terreno localizadas a muy baja altitud, **484**

desalinization/desalinización Proceso de remoción de la sal del agua del océano, **70**

desertification/desertificación Propagación de condiciones desérticas, **501**

developed countries/países desarrollados Países con altos niveles de industrialización y elevados estándares de vida, **117**

developing countries/países en desarrollo Países con economías productivas menores que las de los países desarrollados y bajos estándares de vida, **117**

dharma/dharma Según las prácticas hinduistas, importancia de cumplir con los deberes personales de acuerdo con la etapa en que vive cada quien, **571**

dialect/dialecto Variación regional de un idioma, **100**

dictator/dictador Líder que gobierna con autoridad absoluta, **226**

Diet/Dieta Órgano legislador de Japón que opera por elección, **644**

diffusion/difusión Proceso que ocurre cuando una idea innovadora se extiende y es adoptada por otras personas o grupos, **97**

dikes/dique Enorme muro usado para almacenar agua, **293**

doldrums/depresiones barométricas Áreas en las que no existen vientos predominantes, **44**

domestication/domesticación Proceso mediante el cual las personas cultivan la tierra y doman animales para su beneficio, **120**

domino theory/teoría del dominó Idea que explica que si un país cae víctima del comunismo, los países que lo rodean también caerán como si fueran fichas de dominó, **664**

double cropping/doble cosecha Cosecha de dos cultivos diferentes en la misma tierra en un solo año, **619**

drainage basin/cuenca de drenado Región drenada por un río y sus tributarios, **71**

dryland farming/cultivo de sequía Cultivos que pueden desarrollarse con lluvia limitada en vez de riego, **412**

dual economies/economías duales Economías en las que ciertos bienes son producidos exclusivamente para su venta y otros para el consumo local, **509**

dynasty/dinastía Línea de gobernantes por herencia, **438**

Spanish Glossary

economy of scale/economía de escala Producción masiva que reduce el costo de producción de cada artículo, **727**

ecosystems/ecosistemas Comunidades de plantas y animales, **50**

ecotourism/ecoturismo Tipo de turismo enfocado en visitas guiadas a regiones de interés natural, **248**

edge cities/ciudades satélite Grupos de grandes edificaciones lejanas de los distritos comerciales de una ciudad, **124**

El Niño/El Niño Patrón climatológico y de corrientes marinas que origina un aumento en la temperatura de las corrientes del sureste del océano Pacífico y afecta las condiciones climatológicas de varios países, **260**

emigrants/emigrantes Personas que abandonan su país para irse a vivir a otro lugar, **89**

enclave/enclave unidad distinto en territorio o cultura, como un grupo étnico, rodeado por otro territorio o cultura, **362**

endemic species/especies endémicas Plantas y animales nativos de cierto lugar, **682**

environment/medio ambiente Combinación de las cuatro esferas de la Tierra y que incluyen todas las condiciones biológicas, químicas y físicas necesarias para la vida, **35**

equator/ecuador Línea imaginaria que rodea al globo terrestre justo a la mitad de la distancia entre los polos Norte y Sur, **10**

equinox/equinoccio Momento en que los polos de la Tierra forman un ángulo de 90 grados en relación con el Sol, lo cual ocasiona que los rayos solares caigan directamente sobre el ecuador, **31**

erg/erg Mar de arena creado por el desplazamiento de altas dunas de arena, **484**

erosion/erosión Desplazamiento de fragmentos de una superficie ocasionado por la acción del agua, el viento o el hielo, **67**

escarpment/acantilado Cara sumamente empinada de una placa o área elevada de la superficie terrestre, **534**

estuaries/estuarios Extensiones semicerradas de agua, agua salada de mar y agua dulce, que se forman donde un río desemboca en una caleta del mar, **71**

ethnic groups/grupos étnicos Poblaciones humanas que comparten una cultura o herencia ancestral, **95**

ethnic religions/religiones étnicas Religiones creadas por personas de un grupo étnico que por lo general no se extienden a otras culturas, **102**

Eurasia/Eurasia Masa continental formada por Europa y Asia juntas, **382**

European Union/Unión Europea Organización de países europeos que muestra gran, cooperación en asuntos comerciales, económicos, políticos y sociales, **315**

evaporation/evaporación Proceso mediante el cual un líquido pasa al estado gaseoso, **46**

Spanish Glossary

exclave/exclave Área separada del resto de un país por la presencia geográfica de otro país, **338**

Exclusive Economic Zone/zona de exclusividad económica (**EEZ**, por sus siglas en inglés) Zona de 200 millas náuticas (370 km) a partir de la costa que contiene los recursos naturales propiedad de ese país, **741**

exotic rivers/ríos exóticos Ríos que nacen en regiones húmedas y fluyen a regiones más secas, **434**

exotic species/especies exóticas Plantas y animales que los humanos llevan a una región donde antes no existían, **719**

export economy/economía de exportación Tipo de economía cuyos bienes se producen casi exclusivamente para su exportación y no para su consumo local, **650**

extensive agriculture/agricultura extensiva Procesos agrícolas que cuentan con grandes extensiones de tierra pero pocos recursos económicos y mano de obra, **718**

fall line/línea de caída Frontera natural entre dos regiones con diferente elevación marcada por rápidos y caídas de agua, **150**

famine/hambruna Condición provocada por una escasez extrema de alimentos, **307**

faults/fallas Lugares donde las masas rocosas se fragmentan y se separan unas de otras, **66**

favelas/favelas Grandes asentamientos humanos localizados en las afueras de ciudades brasileñas, **270**

fellahin/fellahin Nombre dado a los agricultores del norte de África, **492**

fjords/fiordos Estrechos y profundos brazos de mar que penetran en los acantilados, **292**

flyway/ruta de vuelo Camino que siguen las aves en su paso migratorio, **640**

folds/pliegues Lugares donde el desplazamiento de las placas comprime las rocas y forma dobleces en la superficie, **66**

formal region/región formal Región con una o más características comunes que la hacen diferente de las zonas que la rodean, **6**

fossil fuels/combustibles fósiles Recursos energéticos formados por los restos de plantas y animales a lo largo de la historia; entre ellos se incluyen carbón mineral, gas natural y petróleo, **79**

fossil water/agua fosilizada Agua subterránea no renovada por las lluvias, **78**

free enterprise/libre empresa Sistema en el que la competencia entre empresas determina el precio de los productos, **115**

free port/zona libre Zona en la que los productos importados están casi exentos de impuestos, **492**

front/frente Zona de contacto entre dos masas de aire con diferente temperatura o nivel de humedad, **44**

functional region/región funcional Región conformada por diferentes lugares que operan en conjunto, **6**

fundamentalism/fundamentalismo Corriente cuyos seguidores obedecen estrictamente ciertas enseñanzas o principios establecidos, **98**

genocide/genocidio Aniquilamiento intencional de un pueblo, **527**

gentrification/aburguesamiento Adquisición de hogares en ruinas, en las zonas viejas de una comunidad, con la finalidad de restaurarlos, **181**

geography/geografía Estudio de las características físicas, biológicas y culturales de la superficie de la Tierra, **3**

geometric boundaries/fronteras geométricas Fronteras que siguen patrones geométricos, **129**

geothermal energy/energía geotérmica Calor proveniente del interior de la Tierra, **81**

geysers/géiseres Manantiales que lanzan chorros de agua caliente al aire, **320**

ghetto/ghetto Sector de una ciudad donde se ven obligados a vivir los grupos minoritarios, **339**

glaciers/glaciares Gruesas masas de hielo conformadas por una placa superficial y un gran bloque inferior que fluyen lentamente en los ríos congelados, **68**

globalization/globalización Proceso mediante el que las comunicaciones alrededor del mundo se han incrementado haciendo a las culturas más parecidas, **98**

global warming/calentamiento global Aumento de la temperatura de la Tierra en cierto espacio de tiempo, **41**

Gran Chaco/Gran Chaco Gran planicie localizada entre los Andes y las mesetas de Brasil en América del Sur, **258**

graphite/grafito Tipo de carbón usado para fabricar la punta de los lápices, **595**

great-circle route/ruta del gran círculo Ruta más corta entre dos lugares cualesquiera de la tierra, **12**

greenhouse effect/efecto invernadero Proceso mediante el cual la atmósfera de la Tierra atrapa energía calorífica, **41**

grid/enrejado Líneas verticales y horizontales que se cruzan a intervalos regulares para formar un patrón cuadriculado, **9**

gross domestic product/producto interno bruto (**GDP**, por sus siglas en inglés; **PNB**, en español) Valor total de los bienes y servicios creados en un país, **116**

gross national product/producto nacional bruto (**GNP**, por sus siglas en inglés; **PIB**, en español) Valor total de los bienes y servicios producidos por un país en un año, **116**

groundwater/aguas subterráneas Mantos de agua que se localizan debajo de la tierra, **72**

gulag/gulag Red de campos de trabajos forzados de las prisiones de la antigua Unión Soviética, **388**

Gulf Stream/corriente del Golfo Corriente oceánica tibia que transporta las aguas tropicales hacia el norte a lo largo de la costa este de América del Norte, **155**

gum arabic/goma arábiga Sustancia espesa extraída de los árboles de acacia, usada para combinar los ingredientes de diversos dulces y medicamentos, **525**

haciendas/haciendas Grandes fincas en la colonia española pertenecientes a familias adineradas en cuyas tierras trabajaban los campesinos de la región, **225**

hajj/hajj Peregrinación religiosa a la Meca que realizan de manera obligatoria todos los musulmanes, **102**

headwaters/cabeceras Corrientes de poco caudal que nacen en las montañas y que conforme descienden se convierten en ríos, **71**

heavy industry/industria pesada Industria enfocada en la fabricación de productos, especialmente aquellos hechos con metales, **393**

hemispheres/hemisferios Cada una de las mitades en que se divide la Tierra a partir del ecuador o un meridiano, **10**

hieroglyphs/jeroglíficos Dibujos y símbolos usados en el sistema de escritura del antiguo Egipto, **487**

hinterland/interior Región muy alejada de los grandes centros de población, **192**

Holocaust/holocausto Asesinato de millones de judíos durante la Segunda Guerra Mundial, **456**

homogeneous/homogéneo Que posee elementos del mismo tipo, como una población conformada por habitantes de un solo grupo étnico, **687**

hot spot/zona caliente Lugar en el que el magma sale del interior de la Tierra a la superficie, **152**

humidity/humedad Cantidad de vapor de agua que hay en el aire, **46**

humus/humus Restos orgánicos de plantas y animales acumulados en el suelo, **75**

hurricanes/huracanes Los ciclones tropicales de mayor poder destructivo, **48**

hydroelectric power/energía hidroeléctrica Electricidad producida con la fuerza del agua, **81**

hydrologic cycle/ciclo hidrológico Movimiento del agua en la hidrosfera, **71**

hydrosphere/hidrosfera El agua que existe en la Tierra, en todas sus formas, **35**

icebreakers/rompehielos Barcos equipados con un casco metálico que les permite abrirse camino en el hielo, **384**

imams/imanes Líderes religiosos musulmanes, **440**

immigrants/inmigrantes Personas que llegan a vivir a un nuevo país, **90**

Inca/inca Fundadores de la civilización antigua más importante de América del Sur, **263**

indigenous/indígena Persona nacida en una región específica, **244**

industrialization/industrialización Proceso por el cual la fabricación de productos basada en el uso de máquinas se extiende en toda una región, **116**

informal sector/sector informal Sector de la economía conformado por personas que no tienen empleos formales, incluidos quienes generan empleos por su cuenta y los pequeños comercios familiares, **543**

infrastructure/infraestructura Sistema de carreteras, puertos y otras instalaciones necesarias en una economía moderna, **117**

innovation/innovación Idea nueva aceptada por toda una cultura, **96**

intensive agriculture/agricultura intensiva Forma de agricultura que usa una gran fuerza laboral y capital en pequeñas extensiones de terreno, **619**

Intertropical Convergence Zone/zona de convergencia intertropical (**ITCZ**, por sus siglas en inglés) Área cercana al ecuador en la que se encuentran los vientos provenientes del noreste y del sureste, generando condiciones de humedad y a menudo de lluvia, **734**

irrigation/riego Proceso mediante el cual el agua se hace llegar a los cultivos de manera artificial, **76**

Spanish Glossary

isthmus/istmo Franja estrecha de tierra rodeada de agua por ambos lados que une dos regiones mayores, **222**

ivory/marfil Material de color blanco del que están hechos los colmillos de los elefantes, **522**

Jainism/jainismo Religión originada en India que enfatiza un código moral estricto basado en la preservación de la vida, **572**

Japan Current/Corriente de Japón (Kuroshio) Corriente de aguas tibias que transporta la mayor parte de la humedad de origen marino que llega a Japón, **639**

jute/yute Planta fibrosa que sirve para elaborar diversos productos como redes y cuerdas, **576**

kampongs/kampongs Aldeas tradicionales de algunas partes del sudeste de Asia construidas generalmente sobre pilotes, **692**

karma/karma Para los hinduistas, budistas y jainistas, es la fuerza positiva o negativa que surge como consecuencia de las acciones de las personas, **571**

klongs/klongs Canales de Tailandia, **672**

Kurds/kurdos Grupo minoritario que habita en Turquía y en países vecinos, **457**

lahars/lahars Deslizamientos peligrosos de lodo en los volcanes, **681**

landlocked/lugar bloqueado Sitio que no tiene litorales, **271**

landscapes/paisajes Panorama de un lugar que abarca características físicas, humanas y culturales, **4**

La Niña/La Niña Patrón climatológico y oceánico que hizo disminuir la temperatura de las aguas del este del océano Pacífico, **260**

latifundia/latifundio En América del Sur, gran extensión de terreno, **265**

latitude/latitud Líneas de un mapamundi dibujadas en dirección este a oeste y que se miden en grados, **9**

leaching/lixiviado Desplazamiento de minerales y humus hacia la parte inferior del suelo, **75**

legend/leyenda Clave de un mapa, **14**

lichens/líquenes Plantas pequeñas formadas de algas y hongos, **157**

light industry/industria ligera Industria que usa materiales ligeros en la fabricación de diversos bienes de consumo, **393**

Spanish Glossary

lingua franca/lengua franca Lenguaje de intercambio y comunicación comercial que usan personas que hablan distintos idiomas, **101**

literacy rate/tasa de analfabetismo Porcentaje de la población que no sabe leer ni escribir, **117**

lithosphere/litosfera Parte sólida de la corteza terrestre, **35**

Llanos/llanos Grandes planicies del noreste de Colombia y el oeste de Venezuela, **258**

loess/limo Suelo de grano fino que arrastra el viento y es muy fértil, **299**

longitude/longitud Líneas de un mapamundi dibujadas en dirección de norte a sur y que se miden en grados, **9**

magma/magma Rocas en estado líquido del interior de la Tierra, **63**

magnesium/magnesio Metal ligero muy valioso en ciertas industrias, como la aeroespacial, **451**

mandates/mandatos Territorios de un país que permanecen temporalmente en control de otros países, **455**

mangrove/mangle Tipo de árbol con raíces que crecen en agua salada; suelen encontrarse en las costas de las regiones tropicales de todo el mundo, **241**

manioc/mandioca Planta tropical de raíces almidonadas, **267**

mantle/manto Sección del interior de la Tierra localizada entre el centro y la mayor parte de la masa terrestre, **63**

Maori/maoríes Primeros pobladores de Nueva Zelanda, **724**

map projections/proyecciones de mapa Diferentes maneras de representar la curvatura de la Tierra con dibujos planos, **11**

maquiladoras/maquiladoras Fábricas mexicanas cuyos propietarios son principalmente empresas estadounidenses, **232**

market economy/economía de mercado Sistema en el que las personas eligen libremente dónde comprar y dónde vender sus bienes, **115**

market-oriented agriculture/agricultura orientada al mercado Sistema agrícola en el que los agricultores siembran un producto para venderlo a los consumidores, **126**

marsupials/marsupiales Mamíferos que tienen un pequeño saco en su vientre para transportar a sus crías, **713**

martial law/ley marcial Ley militar, **631**

matrilineal/matriarcado Sistema social en el que la transferencia del reino se realiza sólo por medio de la madre, **739**

Megalopolis/megalópolis Grupo de ciudades del noreste de Estados Unidos que se ha desarrollado abarcando una enorme área común; este término también describe grupos similares en otras partes del mundo, **173**

Meiji Restoration/restauración Meiji Revolución política japonesa que derrocó al shogún y restauró en el poder al emperador, **643**

Mercosur/Mercosur Organización de América del Sur creada con el propósito de extender el comercio, mejorar las vías de transporte y reducir los aranceles entre los países que lo integran, **270**

meridians/meridianos Líneas de longitud del globo terrestre, **10**

mestizos/mestizos Raza creada con la mezcla de europeos y nativos de América, **226**

meteorology/meteorología Estudio del clima, **5**

metropolitan area/área metropolitana Ciudad muy grande y zonas urbanas que la rodean, **174**

microstates/miniestados Países muy pequeños como Andorra, Mónaco y San Marino, **357**

migration/migración Proceso en el que un grupo de personas abandonan su lugar de origen para irse a vivir a otro, **89**

millet/mijo Tipo de grano resistente a la sequía, **507**

minifundia/minifundios Pequeñas granjas de América del Sur que surgieron con la separación de las colonias, 268

missionaries/misioneros Personas que se dedican a difundir su religión, **102**

monoculture/monocultivo Cultivo de un solo producto, **409**

monotheism/monoteísmo Creencia que considera la existencia de un solo dios, **102**

monsoon/monzón Sistema en el que los vientos cambian de dirección y crean temporadas de clima muy húmedo y muy seco, **51**

moons/lunas Pequeños cuerpos que giran alrededor de un planeta, **26**

mosques/mezquitas Centros islámicos de adoración, **102**

mulattoes/mulatos Personas con ancestros europeos y africanos, **245**

multilingual/políglota Persona que habla varios idiomas, **336**

nationalism/nacionalismo Sentimiento de orgullo y lealtad por el país o grupo cultural propio, **130**

nationalized/nacionalizar Proceso por el cual organizaciones o comercios dejan de ser operados por particulares y pasan a manos del gobierno, **308**

natural boundaries/fronteras naturales Límites creados por las características de un paisaje, **129**

natural hazards/peligros naturales Sucesos de un medio ambiente natural que pueden destruir la vida y las propiedades de los humanos, **155**

navigable/navegable Que puede usarse como vía de transporte acuático, **294**

neutral/neutral Que no toma partido en ningún conflicto, **336**

newsprint/papel periódico Papel de bajo precio que se usa para imprimir los diarios, **158**

nomads/nómadas Personas que frecuentemente cambian de lugar de residencia, **409**

North American Free Trade Agreement/Tratado de Libre Comercio de América del Norte (**NAFTA**, por sus siglas en inglés; **TLC**, en español) Acuerdo entre Estados Unidos, México y Canadá que elimina los aranceles de muchos productos que circulan entre estos tres países, **183**

North Atlantic Drift/corriente del Atlántico Norte Corriente de aguas tibias que regula el clima de la parte noroeste de Europa, **297**

oasis/oasis Lugar donde un manantial brota a la superficie en un área desértica y plantas crecen, **436**

OPEC/OPEC (por sus siglas en inglés; **OPEP** en español) Organización de Países Exportadores de Petróleo, creada con la finalidad de regular el precio del petróleo mediante el control de suministros, **443**

ore/mena Roca con minerales, **79**

orographic effect/efecto orográfico Efecto de enfriamiento que ocurre cuando una corriente de aire sube por la cuesta de una montaña produciendo un clima húmedo en un lado de la montaña y seco en el otro, **47**

outback/llanura desértica Región árida del centro de Australia, **713**

paddy/arrozal Tierra húmeda en la que se cultiva el arroz, **619**

Pampas/las pampas Zona de pastizales muy fértiles localizada al sudeste de América del Sur, **258**

pantheon/panteón Todos los dioses de una religión, **568**

parallels/paralelas Líneas de latitud, **10**

parliament/Parlamento Legislatura encabezada por un Primer Ministro, **196**

partition/división División de una región, como la de India, en 1947, **570**

pastoralism/pastoral Tipo de agricultura de subsistencia complementada por el pastoreo, **125**

perception/percepción Visualización y comprensión del ambiente que nos rodea, **6**

perceptual regions/regiones perceptuales Regiones que reflejan el comportamiento y sentimiento humanos, **6**

permafrost/permafrost Suelo que permanece congelado bajo la superficie, **56**

perspective/perspectiva Punto de vista de una persona sobre cualquier tema, **4**

petrochemicals/petroquímicos Productos hechos con petróleo como materia prima, **80**

pharaohs/faraones Monarcas del antiguo Egipto, **487**

phosphates/fosfatos Sustancias químicas que sirven para elaborar fertilizantes, **743**

pictograms/pictogramas Imágenes simples que representan ideas, **624**

pidgin languages/lenguaje franco Versión simplificada del lenguaje creada para facilitar la comunicación entre personas que hablan diferentes idiomas, **737**

piedmont/tierras bajas Área localizada al pie de una región montañosa, **150**

planets/planetas Cuerpos que giran alrededor de una estrella, **25**

plantations/plantaciones Grandes extensiones en las que se produce un cultivo principal para su exportación, **168**

plateau/meseta Región plana que tiene una elevación pronunciada al menos en uno de sus lados, **69**

plate tectonics/tectónica de placas Teoría que explica que la corteza de la Tierra está dividida en placas rígidas que se desplazan lentamente sobre el manto superior, **63**

plaza/plaza En ciudades de México y otros países de habla hispana, espacio abierto, con frecuencia localizado frente a una iglesia, **225**

polar regions/regiones polares Zonas frías de mayor latitud localizadas en las cercanías de los polos Norte y Sur, **29**

polders/pólderes Tierras ganadas al mar en los Países Bajos, **293**

polytheism/politeísmo Creencia que considera la existencia de varios dioses, **102**

Spanish Glossary

population density/densidad de población Promedio de habitantes de una región; por lo general se expresa como un número de habitantes por milla o kilómetro cuadrado, **88**

population pyramids/pirámide de población Gráfica que muestra el porcentaje de hombres y mujeres por edad en la población de un país, **18**

potash/potasio Mineral que se usa para procesar la lana y en la fabricación de fertilizantes, vidrio y jabones suaves, **451**

precipitation/precipitación Gotas condensadas de agua que caen en forma de lluvia, nieve, aguanieve o granizo, **16**

prevailing winds/vientos dominantes Vientos que soplan en la misma dirección la mayor parte del tiempo, **44**

primate city/ciudad primaria La ciudad más importante de un país en términos de población y economía, **312**

prime meridian/primer meridiano Línea imaginaria entre el Polo Norte y el Polo Sur que pasa por el meridiano de Greenwich, Inglaterra, **10**

protectorate/protectorado País que cede la responsabilidad de la creación de leyes a un país más poderoso a cambio de protección, **591**

provinces/provincias Distritos de gobierno similares a los estados, **190**

pull factors/factores de atracción Características propicias que hacen que un grupo de personas emigre a otra región, **90**

puppet government/gobierno marioneta Gobierno controlado por fuerzas externas, **622**

push factors/factores de abandono Características que hacen que un grupo de personas abandone una región, **90**

quotas/cuotas En términos comerciales es el límite en la cantidad que puede importarse de un producto determinado, **130**

rain shadow/sombra de lluvia Área seca opuesta al extremo húmedo en cadena montañosa, **47**

reforestation/reforestación Plantación de nuevos árboles donde se han cortado otros, **76**

refugees/refugiados Personas obligadas a abandonar la región donde viven, **90**

reg/reg Planicie cubierta de grava en un área desértica, **484**

region/región Área con una o más características comunes que la hacen diferente de otras zonas cercanas, **6**

regionalism/regionalismo Sentimiento político de lealtad a la región donde se vive, **200**

reincarnation/reencarnación Según los hinduistas es la serie de nacimientos repetidos en los que el alma adopta una forma distinta cada vez, **571**

Spanish Glossary

Renaissance/Renacimiento Periodo entre los años 1300 y 1500, aproximadamente, marcado por un renovado interés de la humanidad en el aprendizaje, **356**

revolution/revolución Una vuelta a la órbita elíptica de la Tierra alrededor del Sol, **27**

rift valleys/valles de fisura Puntos de la superficie de la Tierra en los que la corteza se estira hasta romperse, **65**

rotation/rotación Giro completo de la Tierra sobre su eje, **27**

Rus/rus Comerciantes escandinavos que actuaron como los primeros líderes de Kiev; esta palabra es el origen del nombre de Rusia, **386**

Sahel/sahel Región de clima semiárido que se extiende de Senegal y Mauritania hasta Sudán en África, **501**

samurai/samurai Guerrero profesional japonés empleado por el shogún para su protección, **643**

sanctions/sanciones Castigos impuestos a un país para obligarlo a cambiar sus políticas, **539**

Sanskrit/sánscrito Antigua lengua que dio origen al Hindi, **568**

Santeria/santería Religión originada en Cuba que combina tradiciones africanas con creencias cristianas, **251**

satellite/satélite Cuerpo que gira alrededor de otro cuerpo de mayor tamaño, **26**

savannas/sabanas Áreas de pastizales tropicales, árboles dispersos y matorrales, **51**

secular/seglar Que no está ligado a la religión, **463**

sediment/sedimento Pequeñas partículas de rocas fragmentadas, **67**

separatism/separatismo Creencia de que ciertas partes de un país deben ser independientes, **200**

sepoys/cipayos Tropas indias comandadas por el ejército británico en India durante la época de la colonia, **569**

sequent occupance/ocupación sucesiva Establecimiento de grupos sucesivos en el que cada grupo crea un paisaje cultural que lo identifica, **306**

serfs/siervos Campesinos pobres en Rusia que trabajaban la tierra para los señores feudales sin tener otra alternativa, **387**

shatter belt/zona de conflictos Zona de cambios y conflictos frecuentes, por lo general localizada entre grandes potencias, **389**

shifting cultivation/cultivo de reemplazo Técnica agrícola en la que se cortan árboles o se podan para sembrar cultivos de temporada en una zona a la vez, **125**

Shi'ism/chiitas Una de las dos ramas principales del Islam, **440**

shogun/shogún Señor feudal poderoso en el antiguo Japón, **642**

Sikhism/sikismo Religión creada en India que combina la creencia musulmán en la existencia de un solo dios y el punto de vista hinduista sobre la reencarnación, **572**

Silicon Valley/Silicon Valley (Valle del silicón) Área localizada al sur de San Francisco, California, que se convirtió en el principal centro de tecnología de computadoras de Estados Unidos, **178**

silt/limo Suelo fértil cubierto de arena de grano fino, por lo general depositada por el flujo de agua corriente, **487**

sinkholes/hoyos de hundimiento Depresiones de paredes empinadas que se forman cuando la bóveda de una cueva se derrumba, **222**

sisal/sisal Fibra fuerte y duradera obtenida de una planta que se usa para fabricar cuerdas trenzadas y sogas, **522**

slash-and-burn agriculture/agricultura de corte y quema Forma del cultivo de reemplazo en el que se talan pequeñas áreas de bosque y se queman los troncos caídos antes de sembrar, **688**

Slavs/eslavos Los grupos más importantes de habitantes establecidos en las regiones actualmente conocidas como Rusia, Ucrania, Bielorrusia y otras regiones del centro y el sur de Europa, **386**

smelters/fundidoras Fábricas de procesamiento de minerales, **394**

smog/smog Contaminación del aire producida por la reacción química de la luz solar y los contaminantes emitidos por los automóviles y las fábricas, **179**

socialism/socialismo Sistema económico en el que el gobierno posee y controla los medios de producción, **321**

soil exhaustion/agotamiento del suelo Condición en la que el suelo pierde nutrientes y se convierte en tierra no apta para el cultivo, **76**

soil salinization/salinización del suelo Acumulación de sales en el suelo, **76**

solar energy/energía solar Energía que proviene del Sol y llega a la Tierra en forma de luz y calor, **27**

solar system/sistema solar El Sol y el grupo de cuerpos que giran a su alrededor, **25**

solstice/solsticio Momento en que los polos de la Tierra forman el mayor ángulo posible en relación con el Sol, **30**

sorghum/sorgo Un tipo de grano resistente a la sequía, **507**

souks/souks Mercados árabes al aire libre, **461**

soviets/soviéticos Cuerpos locales de gobierno de la antigua Unión Soviética, **388**

Special Economic Zones/zonas de economía especial (SEZ, por sus siglas en inglés) Zonas diseñadas para atraer la inversión de compañías extranjeras en ciertas áreas de China, **628**

staple/cultivo básico Cultivo principal de una región, **507**

storm surge/marejada de tormenta Aguas agitadas en la costa debido a la acción de una gran tormenta, **587**

subcontinent/subcontinente Grandes masas de tierra menores que un continente, como el subcontinente de la India, **564**

subsidies/subsidios Apoyo financiero del gobierno, **649**

subsistence agriculture/agricultura de subsistencia Tipo de agricultura en el que una familia cultiva un producto sólo para su propio consumo, **125**

sultans/sultanes Gobernantes otomanos, **455**

Sunni/sunitas Una de las dos ramas más importantes del Islam, **440**

superpower/superpotencia País muy poderoso, **183**

Swahili/suahili Lengua utilizada en África Occidental que tiene raíces en las lenguas de la costa africana y combina algunas palabras de origen árabe, **522**

taiga/taiga Bosque poblado por árboles siempre verdes que abarca la mitad de Rusia, **384**

Taoism/taoísmo Religión que explica que existe un orden natural llamado Tao en el universo, **624**

tariffs/aranceles Impuestos sobre importaciones y exportaciones, **130**

tar sands/arena alquitranada Capas de arena y roca que contienen aceite, **261**

temperature/temperatura Sistema de medición de calor, **41**

tepuís/tepuís Cadena de mesetas en las altiplanicies de las Guayanas, 258

terrorism/terrorismo Uso del temor y la violencia como fuerza política, **271**

textiles/textiles Productos elaborados con telas, **173**

theocracy/teocracia País gobernado por una ley religiosa, **443**

topography/topografía Elevación, características y forma del terreno, **17**

tornadoes/tornados Espirales de aire que giran violentamente, **48**

totalitarian government/gobierno totalitario Gobierno en el que una persona o pequeño grupo de personas gobiernan un país sin dar a los habitantes voz ni voto, **130**

trade deficit/déficit comercial Situación en la que el valor de las exportaciones de un país es menor que el valor de sus importaciones, **183**

trade surplus/excedente comercial Condición que resulta cuando un país exporta más productos de los que importa, **650**

traditionalism/tradicionalismo Seguimiento de prácticas antiguas y opuestas a ideas y tecnologías modernas, **98**

transhumance/trashumar Práctica que consiste en llevar los rebaños de los pastizales de las montañas donde se alimentaron en el verano a tierras más bajas durante el invierno, **409**

tree line/límite de los árboles Altura a partir de la cual ya no crecen los árboles, **259**

trench/trinchera Valle profundo formado por un choque de placas en el que una de ellas se desliza debajo de la otra, **66**

tributary/tributario Corriente menor que alimenta un río más caudaloso, **71**

Tropic of Cancer/Trópico de Cáncer Paralelo 23 1/2° al norte del ecuador, **31**

Tropic of Capricorn/Trópico de Capricornio Paralelo 23 1/2° al sur del ecuador, **30**

tropics/trópico Área de clima cálido y baja latitud, cercanas al ecuador, **29**

trust territories/territorios bajo administración fiduciaria Áreas cuyo control fue cedido a otro país después de la Segunda Guerra Mundial como preparativo para su independencia, **737**

tsetse fly/mosca tse tse Mosca africana portadora de una enfermedad llamada mal del sueño, **519**

tsunamis/tsunamis Enormes olas creadas por movimientos tectónicos como los terremotos submarinos, **638**

typhoons/tifones Término usado al oeste del océano Pacífico para nombrar a los huracanes, **48**

uninhabitable/inhabitable Que no presenta condiciones adecuadas para la vida humana, **319**

universalizing religions/religiones universales Religiones que buscan seguidores en todo el mundo, **102**

Spanish Glossary

urban agglomeration/aglomeración urbana Región densamente poblada en los alrededores de una ciudad central, **651**

urbanization/urbanización Crecimiento en la proporción de los habitantes de poblados y ciudades, **120**

veld/veld Regiones de pastizales localizadas en África del Sur, **535**

voodoo/vudú Versión haitiana de ciertas creencias tradicionales religiosas africanas combinadas con elementos cristianos, **251**

wadis/vados Cuencas vacías de ríos que sólo se llenan con la lluvia, **484**

watershed/cuenca Lecho que antes ocupaba un río y sus tributarios, **71**

water table/mesa de agua Nivel al que llegan las aguas subterráneas cuando llenan las grietas de las rocas, **72**

wats/wats Templos budistas que también sirven como monasterios, **670**

weather/tiempo Condiciones atmosféricas de un lugar y momento determinados, **41**

weathering/desgaste Proceso de descomposición de las rocas con el paso del tiempo, **67**

wetlands/terreno pantanoso Paisaje cubierto de agua al menos una parte del año, **72**

Wheat Belt/región triguera Región central de Estados Unidos especializada en el cultivo del trigo, **178**

work ethic/ética laboral Creencia de que el trabajo en sí tiene valor moral, **649**

world cities/urbes del mundo Grandes áreas urbanas en las que se encuentran los centros de poder económico más importantes, **122**

yurts/yurtas Casas móviles de lana con forma circular comunes en la parte central de Asia, **410**

zinc/zinc Elemento muy importante en el procesamiento de metales y otras industrias, **406**

Zionism/sionismo Movimiento que exhortaba a los judíos a formar su propia nación en Palestina, **456**

Index

Index

Index

Index

Index

Index

Index

Index

Index

Index

Index

Index

Index

Index

Index

Index

Index

Index

Index

Acknowledgments

For permission to reprint copyrighted material, grateful acknowledgment is made to the following sources:

***The Asia Society*:** "Thoughts of Hanoi" by Nguyen Thi Vinh from *A Thousand Years of Vietnamese Poetry,* translated by Nguyen Ngoc Bich. Copyright © 1974 by The Asia Society.

Luba Brezhneva c/o Tanner Propp, LLP: From *The World I Left Behind: Pieces of a Past* by Luba Brezhneva, translated by Geoffrey Polk. Copyright © 1995 by Luba Brezhneva.

Bill Bryson: From "Australian Outback" by Bill Bryson from *National Geographic Online Magazine,* accessed May 10, 2001, at http://www.nationalgeographic.com/traveler/outback2/html. Copyright © 2000 by Bill Bryson.

John R. Mullin: Quote by John R. Mullin from "Architects, Planners and Residents Wonder How to Fill the Hole in the City" by Kirk Johnson and Charles V. Bagli from *The New York Times,* September 26, 2001, p. B9. Copyright © 2001 by John R. Mullin.

National Council for Geographic Education: From "Preferred Locations in North America: Canadians, Clues, and Conjectures" by Herbert A. Whitney from *Journal of Geography,* vol # 83, no. 5, September-October 1984, p. 222. Copyright © 1984 by National Council for Geographic Education.

W.W. Norton & Company, Inc.: From *Nature's Metropolis: Chicago and the Great West* by William Cronon. Copyright © 1991 by William Cronon.

Harold Ober Associates Incorporated: From "On the Beach" from *A Secret Country: The Hidden Australia* by John Pilger. Copyright © 1989 by John Pilger.

Penguin Books Ltd.: From *The Vinland Sagas,* translated by Magnus Magnusson and Hermann Pálsson (Harmondsworth: Penguin Classics 1965). Copyright © 1965 by Magnus Magnusson and Hermann Pálsson. From *The Epic of Gilgamesh,* translated by N. K. Sandars (Penguin Classics 1960, Third Edition, 1972). Copyright © 1960, 1964, 1972 by N. K. Sandars.

The Estate of Richard Rive: From *African Songs* by Richard Rive. Copyright © 1963 by Richard Rive. Published by Seven Seas Publishers, Berlin.

Wendy Rose: From "Loo-Wit" from *Bone Dance: New and Selected Poems 1967–1992* by Wendy Rose (University of Arizona Press, 1994). Copyright © 1985 by Wendy Rose. First appeared in *The Halfbreed Chronicles & Other Poems* by Wendy Rose (West End Press 1985/1993).

Viking Penguin, a division of Penguin Putnam Inc.: From *The Incas: The Royal Commentaries of the Inca* by Garcilaso de la Vega, translated by Maria Jolas. Copyright © 1961 by The Orion Press, Inc.

Villard Books, a division of Random House, Inc.: From *Fifty Years of Europe* by Jan Morris. Copyright © 1997 by Jan Morris.

Wallace Literary Agency, Inc.: "Hundred Questions" from *The Mahabharata* by R. K. Narayan. Copyright © 1978 by R. K. Narayan. Published by Viking Penguin.

Sources Cited:

From "Vancouver: Open for Business" by Lindsay Elliott from *BC Business,* vol. 26, no. 3, March 1998, page 73. Published by Canada Wide Magazine Ltd.

From "Mali's Nomads Coaxed to Adopt Modern Life; After a five-year conflict, government aims to integrate wandering Tuaregs into society" by Jennifer Ludden from *The Christian Science Monitor,* March 6, 1996. Published by The Christian Science Publishing Society, Boston, MA.

From "Prince Edward Island" by Ian Darragh from *National Geographic,* May 1998. Published by The National Geographic Society, Washington, D.C.

From "Blue Nile" by Virginia Morell from *National Geographic,* vol.198, no. 6, December 2000. Published by The National Geographic Society, Washington, D.C.

From "India" by Geoffrey C. Ward from *National Geographic,* May 1997. Published by The National Geographic Society, Washington, D.C.

From "London" by Simon Worrall from *National Geographic,* June 2000. Published by The National Geographic Society, Washington, D.C.

From "A Spare and Separate Walk of Life; Nomads Survive the Sahara Amid Dwindling Resources" by Stephen Buckley from *The Washington Post,* December 8, 1996.

Abbreviations used: (t) top, (c) center, (b) bottom, (l) left, (r) right, (bkgd) background

ILLUSTRATIONS

All work, unless otherwise noted, contributed by Holt, Rinehart & Winston.

Skills Handbook: Page S13 (c), MapQuest.com, Inc.; (bc), Leslie Kell.

Atlas Maps: MapQuest.com, Inc.

Texas Geography Today: Page TX1 (b), MapQuest.com, Inc.; TX2 (br), Mark Heine; TX3 (tl), Ortelius Design; TX3 (br), TX6 (tl), TX9 (tr), MapQuest.com, Inc.; TX13 (tr), Leslie Kell; TX18 (l), Argosy (Illustrations), Leslie Kell (Chart); TX21 (tl), TX22 (tl), TX26 (tl), TX28 (tl), TX29 (br), TX30 (tl), MapQuest.com, Inc.

Chapter One: Page 3 (cl), 4 (tl), 6 (cl), 7 (tl), (tr), 9 (bl), (bc), 11 (br), 12 (tc), (c), (bc), 13 (bl), 16 (tl), (tr), 17 (tl), (tr), MapQuest.com, Inc.; 18 (tc), Ortelius Design; 18 (cl), (cr), MapQuest.com, Inc.; 19 (t), (bl), Leslie Kell; 20 (cr), MapQuest.com, Inc.; 21 (b), Ortelius Design; 22 (cl), 23 (tl), MapQuest.com, Inc.

Chapter Two: Page 26 (t), Don Dixon; 26 (br), Uhl Studios, Inc.; 29 (cr), 30 (b), Argosy; 32-33 (bc), 36 (tl), 37 (tr), 38 (cl), MapQuest.com, Inc.; 39 (tl), Leslie Kell.

Chapter Three: Page 41 (cr), 42 (bl), MapQuest.com, Inc.; 43 (bl), Uhl Studios, Inc.; 44 (tl), 47 (b), (tl), 50 (bl), MapQuest.com, Inc.; 51 (tl), John White/The Neis Group; 52–53, 54 (tl), 55 (tc), (bl), (bc), 58 (b), 60 (cl), 61 (tl), MapQuest.com, Inc.

Chapter Four: Page 63 (br), Uhl Studios, Inc.; 64 (t), (b), MapQuest.com, Inc.; 65 (t), 66 (tl), Mark Heine; 66 (br), 67 (cr), MapQuest.com, Inc.; 69 (tr), 70 (br), Mark Heine; 72 (bl), 73 (tl), MapQuest.com, Inc.; 75 (tr), Mark Heine; 76 (tc), 78 (cl), 79 (cr), 80 (b), 82 (b), MapQuest.com, Inc.; 83 (b), Leslie Kell; 84 (cl), Mark Heine; 85 (tl), Leslie Kell.

Chapter Five: Page 87 (bc), MapQuest.com, Inc.; 88 (tl), Leslie Kell; 88 (b), 89 (c), MapQuest.com, Inc.; 91 (br), 92 (tl), 93 (tr), Leslie Kell; 96 (tl), 98 (cl), MapQuest.com, Inc.; 100 (bl), Leslie Kell; 101 (t), 103 (t), 110 (cl), MapQuest.com, Inc.; 111 (cl), Leslie Kell.

Chapter Six: Page 113 (b), Leslie Kell; 114 (cl), 115 (cr), MapQuest.com, Inc.; 116 (bl), 117 (b), Leslie Kell; 118 (tl), (cl), 119 (b), MapQuest.com, Inc.; 121 (t), Mark Heine; 122 (bl), David Chapman; 123 (bc), MapQuest.com, Inc.; 124 (t), 125 (br), 127 (tr), David Chapman; 127 (c), 130 (br), 131 (tc), MapQuest.com, Inc.; 134 (cl), David Chapman; 135 (tl), Leslie Kell.

Unit Two: Page 140 (c), 141 (c), MapQuest.com, Inc.; 141 (t), Ortelius Design; 142 (c), 143 (c), 144 (c), 145 (c), MapQuest.com, Inc.; 146 (cl), Ortelius Design; 147 (tl), MapQuest.com, Inc.

Chapter Seven: Page 149 (b), 150 (cl), 153 (cr), 154 (cr), 156 (tc), (c), 159 (bl), 160 (tl), 161 (bl), 162 (cl), MapQuest.com, Inc.; 163 (tl), Leslie Kell.

Chapter Eight: Page 165 (b), MapQuest.com, Inc.; 165 (cr), Craig Attebery/Jeff Lavaty Artist Agent; 166 (bl), (tl), 168 (tr), 169 (br), 170 (bl), (cl), MapQuest.com, Inc.; 172 (tr), Mark Heine; 173 (bl), 175 (tl), MapQuest.com, Inc.; 175 (br), Leslie Kell; 175 (tr), Nenad Jakesevic; 176 (tl), 178 (tl), 179 (tc), 180 (cl), 181 (br), 182 (tr), 185 (tr), (bl), 186 (cl), MapQuest.com, Inc.; 187 (tl), Leslie Kell.

Chapter Nine: Page 189 (b), MapQuest.com, Inc.; 191 (cr), (br), Leslie Kell; 193 (cl), 194 (bc), 196 (tl), 198 (b), 200 (cl), 201 (c), 202 (cl), MapQuest.com, Inc.; 203 (tl), Leslie Kell.

Unit Three: Page 210 (c), 211 (c), MapQuest.com, Inc.; 211 (t), Ortelius Design; 212 (c), 213 (c), 214 (c), 215 (c), MapQuest.com, Inc.; 216 (cl), Ortelius Design; 217 (tl), (cl), (bl), 218 (tl), (cl), (tl), (cl), MapQuest.com, Inc.

Chapter Ten: Page 221 (b), 222 (tc), 223 (tr), 224 (bc), 227 (tr), 229 (bl), MapQuest.com, Inc.; 230 (tl), Leslie Kell; 230 (cr), 231 (bc), 232 (b), (tr), 233 (tr), 236 (cl), MapQuest.com, Inc.; 237 (tl), Leslie Kell.

Chapter Eleven: Page 239 (b), 240 (cl), MapQuest.com, Inc.; 241 (tr), Leslie Kell; 243 (bl), (tr), 244 (cl), MapQuest.com, Inc.; 245 (br), Leslie Kell; 249 (cr), 252 (cl), 254 (cl), 255 (tl), MapQuest.com, Inc.

Chapter Twelve: Page 257 (b), 258 (tr), MapQuest.com, Inc.; 259 (b), Mark Heine; 260 (tc), 261 (cl), MapQuest.com, Inc.; 262 (b), Mark Heine; 263 (bc), 264 (tr), 267 (tr), 270 (tl), 272 (bl), 273 (t), 274 (cl), MapQuest.com, Inc.; 275 (tl), Leslie Kell.

Unit Four: Page 280 (c), 281 (c), MapQuest.com, Inc.; 281 (t), Ortelius Design; 282 (c), 283 (c), 284 (c), 285 (c), MapQuest.com, Inc.; 286 (cl), Ortelius Design; 287 (tl), (c), (cl), (bl), 288 (tl), (cl), (bl), 289 (tl), (cl), MapQuest.com, Inc.

Chapter Thirteen: Page 291 (b), 292 (c), MapQuest.com, Inc.; 293 (b), Ortelius Design; 296 (cl), 297 (tr), 298 (c), 299 (cr), 300 (tl), 301 (tr), 302 (cl), MapQuest.com, Inc.; 303 (tl), Leslie Kell.

Chapter Fourteen: Page 305 (b), 306 (b), (tl), 307 (tr), 310 (bl), 312 (tl), MapQuest.com, Inc.; 313 (cr), (br), Leslie Kell; 316 (tl), 318 (bl), 319 (bl), 322 (cl), 323 (t), 324 (cl), 325 (tl), MapQuest.com, Inc.

Chapter Fifteen: Page 327 (b), 328 (cl), MapQuest.com, Inc.; 330 (bl), Leslie Kell; 330 (cl), 333 (tr), MapQuest.com, Inc.; 334 (tr), Leslie Kell; 335 (cr), 336 (bl), 339 (tr), MapQuest.com, Inc.; 340 (bl), Leslie Kell; 341 (tc), 342 (bc), 343 (t), (cr), 344 (bl), (tr), 346 (cl), MapQuest.com, Inc.; 347 (tl), Leslie Kell.

Chapter Sixteen: Page 349 (b), 350 (bc), 351 (tc), (c), MapQuest.com, Inc.; 352 (tl), Leslie Kell; 353 (tr), Mark Heine; 354 (cr), 355 (bc), 356 (bl), 358 (cl), MapQuest.com, Inc.; 359 (tr), Leslie Kell; 360 (bc), 361 (tr), 362 (b), 363 (tc), 365 (cr), 366 (cl), MapQuest.com, Inc.; 367 (c), Leslie Kell.

Unit Five: Page 372 (c), 373 (c), MapQuest.com, Inc.; 373 (t), Ortelius Design; 374 (c), 375 (c), 376 (c), 377 (c), MapQuest.com, Inc.; 378 (cl), Ortelius Design; 379 (tl), (cl), MapQuest.com, Inc.

Chapter Seventeen: Page 381 (b), 385 (cl), 388 (cl), MapQuest.com, Inc.; 389 (cr), Leslie Kell; 389 (tr), MapQuest.com, Inc.; 390 (bl), Leslie Kell; 394 (tc), 395 (bl), 396 (cl), 397 (tc), 400 (cl), MapQuest.com, Inc.; 401 (tl), Leslie Kell.

Chapter Eighteen: Page 403 (b), 407 (br), MapQuest.com, Inc.; 407 (bl), Leslie Kell; 408 (cl), MapQuest.com, Inc.; 410 (bl), Leslie Kell; 411 (cr), 413 (bc), 416 (bl), 418 (cl), 419 (tl), MapQuest.com, Inc.

Unit Six: Page 424 (c), 425 (c), MapQuest.com, Inc.; 425 (t), Ortelius Design; 426 (c), 427 (c), 428 (c), 429 (c), MapQuest.com, Inc.; 430 (cl), Ortelius Design; 431 (tl), (cl), (bl), MapQuest.com, Inc.

Chapter Nineteen: Page 433 (b), 435 (cr), 436 (c), 437 (c), 440 (tl), 442 (cl), MapQuest.com, Inc.; 443 (b), Leslie Kell; 446 (cl), MapQuest.com, Inc.; 447 (tl), Leslie Kell.

Chapter Twenty: Page 449 (b), 450 (bc), 451 (tc), 453 (tl), 454 (bc), 455 (c), 456 (bl), 459 (bl), 461 (cr), 463 (c), 465 (tl), 465 (tr), 466 (cl), MapQuest.com, Inc.; 467 (tl), Leslie Kell.

Acknowledgments

Unit Seven: Page 472 (b), 473 (c), MapQuest.com, Inc.; 473 (t), Ortelius Design; 474 (c), 475 (c), 476 (c), 477 (c), MapQuest.com, Inc.; 478 (cl), Ortelius Design; 479 (tl), (cl), (bl), 480 (tl), (cl), (bl), 481 (tl), (cl), (bl), MapQuest.com, Inc.

Chapter Twenty-One: Page 483 (b), 485 (tr), MapQuest.com, Inc.; 486 (tl), Leslie Kell; 490 (bl), 492 (bl), MapQuest.com, Inc.; 493 (tr), Leslie Kell; 495 (bl), (tr), 496 (cl), MapQuest.com, Inc.

Chapter Twenty-Two: Page 499 (b), 500 (tc), 503 (bl), 504 (tr), 505 (br), 508 (tr), MapQuest.com, Inc.; 508 (cl), 510 (tl), Leslie Kell; 510 (b), Mark Heine; 511 (tr), 514 (cl), 515 (tl), MapQuest.com, Inc.

Chapter Twenty-Three: Page 517 (b), 518 (cr), 521 (bc), 522 (tr), 523 (tl), MapQuest.com, Inc.; 524 (tl), Leslie Kell; 526 (c), 528 (tc), MapQuest.com, Inc.; 529 (bl), Leslie Kell; 530 (cl), 531 (tl), MapQuest.com, Inc.

Chapter Twenty-Four: Page 533 (b), 534 (br), 536 (cl), 537 (bc), MapQuest.com, Inc.; 540 (bc), Leslie Kell; 542 (c), MapQuest.com, Inc.; 543 (tr), Leslie Kell; 544 (tl), (br), 545 (cr), 546 (bl), MapQuest.com, Inc.; 547 (tl), Leslie Kell; 548 (cl), 549 (tl), MapQuest.com, Inc.

Unit Eight: Page 554 (c), 555 (c), MapQuest.com, Inc.; 555 (t), Ortelius Design; 556 (c), 557 (c), 558 (c), 559 (c), MapQuest.com, Inc.; 560 (cl), Ortelius Design; 561 (tl), 561 (cl), MapQuest.com, Inc.

Chapter Twenty-Five: Page 563 (b), 569 (r), 570 (bl), 572 (bl), 575 (cr), 577 (c), MapQuest.com, Inc.; 578 (tl), Leslie Kell; 578 (b), 579 (cr), 581 (bl), 582 (cl), MapQuest.com, Inc.; 583 (tl), Leslie Kell.

Chapter Twenty-Six: Page 585 (b), MapQuest.com, Inc.; 587 (br), Leslie Kell; 589 (bl), 590 (cl), 592 (cr), 593 (tr), 594 (cl), 595 (cr), MapQuest.com, Inc.; 599 (bl), Leslie Kell; 600 (cl), MapQuest.com, Inc.; 601 (tl), Leslie Kell.

Chapter Twenty-Seven: Page 615 (b), 618 (cr), MapQuest.com, Inc.; 619 (tr), (tl), Leslie Kell; 620 (tc), 621 (cr), MapQuest.com, Inc.; 624 (cl), The John Edwards Group; 627 (bl), 628 (c), 633 (tr), 634 (cl), MapQuest.com, Inc.; 635 (tl), Leslie Kell.

Chapter Twenty-Eight: Page 637 (b), 638 (tl), 640 (tl), 641 (tr), 643 (cr), 644 (cl), (bc), 645 (tl), 651 (c), 652 (tl), (c), 654 (bl), 656 (cl), MapQuest.com, Inc.; 657 (tl), Leslie Kell.

Chapter Twenty-Nine: Page 659 (b), 660 (cr), 661 (tc), 663 (bl), 666 (tl), MapQuest.com, Inc.; 670 (bl), Leslie Kell; 670 (tl), MapQuest.com, Inc.; 671 (cr), Leslie Kell; 674 (b), 676 (cl), 677 (tl), MapQuest.com, Inc.

Chapter Thirty: Page 679 (b), 680 (cl), MapQuest.com, Inc.; 681 (br), Leslie Kell; 682 (cl), MapQuest.com, Inc.; 683 (b), Ortelius Design; 683 (b), Mark Heine; 686 (bl), MapQuest.com, Inc.; 687 (b), Leslie Kell; 688 (cl), Nenad Jakesevic; 691 (tr), Leslie Kell; 691 (c), 693 (cr), MapQuest.com, Inc.; 695 (bl), Leslie Kell; 695 (t), 696 (cl), MapQuest.com, Inc.; 697 (tl), Leslie Kell.

Unit Ten: Page 702 (c), 703 (c), MapQuest.com, Inc.; 703 (t), Ortelius Design; 704 (c), 705 (c), 706 (c), 707 (c), MapQuest.com, Inc.; 708 (cl), Ortelius Design; 709 (tl), (cl), (bl), MapQuest.com, Inc.

Chapter Thirty-One: Page 711 (b), 713 (cr), 715 (br), MapQuest.com, Inc.; 716 (tl), 717 (bl), 718 (bl), Leslie Kell; 718 (tl), 719 (tl), 723 (bl), 726 (bl), 728 (cl), MapQuest.com, Inc.; 729 (tl), Leslie Kell.

Chapter Thirty-Two: Page 731 (b), MapQuest.com, Inc.; 732 (l), Mark Heine; 733(cr), 737 (tr), 739 (c), 740 (tr), 741 (cr), 742 (c), 743 (tr), 746 (cl), 747 (tl), MapQuest.com, Inc.

PHOTO CREDITS

Features, Icons and Backgrounds: *Unit Opener Backgrounds:* (texture) CORBIS Images; (globes) Mountain High Maps® Copyright ©2003; Digital Wisdom, Inc. *Chapter Opener and Section Backgrounds:* (passport) Sam Dudgeon/HRW Photo; (textures) CORBIS Images; (globes) Mountain High Maps® Copyright ©2003 Digital Wisdom, Inc.; *Geography for Life Backgrounds:* (map) Map drawn by J. Asherton, engraved by J. Shury, Oregon Historical Society; (compass rose) CORBIS Images; (compass) Digital Imagery® copyright 2003 PhotoDisc. Inc. *Case Study Icons:* (notebook, compass, magnifying glass) Digital Imagery® copyright 2003 PhotoDisc, Inc.; (map, pencil) HRW Library Photo; (computer monitor) Image Copyright ©2003 PhotoDisc, Inc.; *Cities & Settlements Icons:* (Kurdish woman) Phillis Picardi/International Stock Photography; (Young cowboy) Paul Chesley/National Geographic Society Image Collection; (Woman in turtle neck) Peter Griffith/Masterfile; (Buildings collage) R. Ian Lloyd/Masterfile, J. A. Kraulis/Masterfile, Ron Stroud/Masterfile, Miles Ertman/Masterfile, CORBIS Images. *Internet**connect** Mouse Icon:* Digital Imagery® copyright 2003 PhotoDisc, Inc. *Our Amazing Planet Icon:* Mountain High Maps® Copyright ©2003 Digital Wisdom, Inc.

Front Matter: iv (tl), VCG/FPG International; iv (tr), Carr Clifton/Minden Pictures; iv (br), Yvette Cardozo/STONE; v (b), Stone/Shaun Egan; vi (t), George Hunter/H. Armstrong Roberts; vi (bl), Greg Ryan/Sally Beyer/Positive Relections; vii, Richard Turpin/Corbis Stock Market; viii (t), Wayne Eastep; viii (br), Robert Frerck/Woodfin Camp & Associates; ix (tr), Albert Normandin/Masterfile; ix (cr), Frank Fournier/Woodfin Camp & Associates; ix (b), ©Kevin Schafer/kevinschafer.com; x (t), Miwako Ikeda/International Stock Photography; x (br), Blaine Harrington; xi (tl), Telegraph Colour Library/FPG International; xi (b), Leo de Wys/Leo de Wys; xii (t), Stone/Chad Ehlers; xiii (t), W.P. Fleming/Viesti Collection, Inc.; xiv (t), Stone/Bob Krist; xiv (tc), CORBIS/Roger Ressmeyer; xv (t), Photo ©Stefan Schott/Panoramic Images, Chicago 1998; xvi (t), Corbis Images; xvii (t), Carr Clifton/Minden Pictures; xxii (tl), CORBIS/Tiziana and Gianni Baldizzone; xix (br), Digital imagery® copyright 2003 PhotoDisc, Inc.; xxii (tr), Art Wolfe, Inc.; xxiii (tl), Tom Stewart/Corbis Stock Market; xxiii (br), Daniel J. Schaefer.; S1 (cl), John Langford/HRW Photo; S1 (bl), John Langford/HRW Photo; S2 (b), ©Nik Wheeler/CORBIS; S2 (br), Digital imagery® copyright 2003 PhotoDisc, Inc.; S3 (tl), Jeffrey Aaronson/Network Aspen; S3 (tc), Index Stock Imagery, Inc.; S3 (bl), Amit Bhargava/Newsmakers/Getty Images; S3 (c), Norman Owen Tomalin/Bruce Coleman, Inc.; S3 (cr), ©Fritz Polking/Peter Arnold, Inc.; S4 (br), Sam Dudgeon/HRW Photo; S4 (bl), Larry Kolvoord Photography; S4 (c), Runk/Schoenberger/Grant Heilman Photography; S4 (tr), Wolfgang Kaehler Photography; S5 (bl), Sam Dudgeon/HRW; S7 (tr), Sam Dudgeon/HRW Photo; S8 (br), Sam Dudgeon/HRW; S9 (l), Sam Dudgeon/HRW Photo; S10 (c), Sam Dudgeon/HRW; S10 (tr), John Langford/HRW; S12 (tr), Digital imagery® copyright 2003 PhotoDisc, Inc.; S12 (cl), David Phillips/HRW; S12 (bl), Kyodo News Service; S13 (t), Digital imagery® copyright 2003 PhotoDisc, Inc.; S14 (cl), Sam Dudgeon/HRW Photo; S15 (l), Sam Dudgeon/HRW Photo; S16 (br), Sam Dudgeon/HRW Photo; S17 (bl), Sam Dudgeon/HRW; S20 (br), Digital imagery® copyright 2003 PhotoDisc, Inc.

Texas Geography Today: TX1 (tl), Corbis Images; TX4 (bl), Eleanor S. Morris; TX5 (br), Richard Cummins/Viesti Collection, Inc.; TX7 (tr), Wyman Meinzer Photography; TX7 (b), ©Annie Griffiths Belt/CORBIS; TX8 (bl), Earl Nottingham/AP/Wide World Photos; TX8 (tl), CORBIS/D. Robert & Lorri Franz; TX9 (br), Joe Viesti/Viesti Collection, Inc.; TX10 (tl), Archive Photos; TX11 (tl), Pat Sullivan/AP/Wide World Photos; TX12 (bl), Richard Erdoes/SuperStock; TX13 (br), Bob Daemmrich/Stock Boston; TX14 (tl), Kevin Stillman/TxDOT; TX14 (br), Laurence Parent; TX15 (cr), ©Morton Beebe, S.F./CORBIS; TX15 (tr), Laurence Parent; TX16 (b), David J. Phillip/AP/Wide World Photos; TX17 (t), Rebecca McEntee/AP/Wide World Photos; TX17 (br), Eric Gay/AP/Wide World Photos; TX18 (br), Don Couch; TX19 (t), Don Couch; TX20 (bl), David Muench Photography; TX23 (tr), Laurence Parent; TX24 (tl), Bob Daemmrich/Stock Boston; TX24 (br), Eleanor S. Morris; TX24 (bc), Courtesy of Texas Highways Magazine; TX25 (t), Laurence Parent; TX26 (bl), Texas Historical Commission; TX27 (tl), Ronald Martinez/AP/Wide World Photos; TX29 (tr), Victoria Smith/HRW Photo; TX30 (br), Laurence Parent; TX31 (tl), Laurence Parent.

Unit One: 1, CORBIS/Keren Su. **Chapter One:** 2 (bl), Victor Boswell/National Geographic Society Image Collection; 2 (r), Corbis Images; 2 (cl), ©Digital Vision/HRW; 3 (bl), Yvette Cardozo/STONE; 4 (tl), Boyd Norton/Evergreen Photo Alliance; 4 (br), Carr Clifton/Minden Pictures; 5 (br), CORBIS/Roger Ressmeyer; 6 (cl), David Muench Photography; 7 (tl), Stone/Earth Imaging; 7 (tr), Don Couch/HRW Photo–Postcard reprinted by permission from Jeffery Stanton; 8 (tr), William Campbell/Peter Arnold, Inc.; 11 (tl), Erich Lessing/Art Resource, NY; 15 (tl), The Art Archive/Dagli Orti (A); 15 (br), The Art Archive/Dagli Orti (A). **Chapter Two:** 24 (r), Stone/Earth Imaging; 24 (bl), CORBIS/Roger Ressmeyer; 25 (cr), Digital imagery® copyright 2003 PhotoDisc, Inc.; 27 (tr), NASA Image eXchange/www.nix.nasa.gov; 28 (bl), Space Telescope Science Institute/NASA/Science Photo Library/Photo Researchers, Inc.; 28 (bc), Jack Zehrt 1993/FPG International; 31 (tr), NASA's Global Hydrology and Climate Center, Huntsville, AL.; 34 (br), ©Douglas Peebles; 35 (atmosphere), VCG/FPG International; 35 (lithosphere), ©Douglas Peebles; 35 (hydrosphere), H. Richard Johnston/FPG International; 35 (biosphere), Nawrocki Stock Photo; 36 (tl), Calvin Larsen/Photo Researchers, Inc. **Chapter Three:** 40 (cl), Alan R. Moller/STONE; 40 (bl), Sepp Seitz /Woodfin Camp & Associates; 40 (r), Greg Gawlowski/Photo 20-20, Inc.; 41 (br), Larry Kolvoord Photography; 42 (br), ©Ned Gillette/Adventure Photo & Film; 44 (tr), ©2000 Kent Wood, All Rights Reserved; 46 (bl), W. Scarberry/The Image Works; 47 (tl), David Madison/Bruce Coleman, Inc.; 47 (tr), ©John Elk III; 48 (br), Reuters/Juan Carlos Ulate/Archive Photos; 49 (tl), NOAA (National Oceanic & Atmospheric Administration); 49 (tc), NOAA (National Oceanic & Atmospheric Administration); 49 (tr), NOAA (National Oceanic & Atmospheric Administration); 50 (b), Stone/James Martin; 54 (tr), Kal Muller/Woodfin Camp & Associates; 55 (tr), ©Dimaggio/Kalish/Peter Arnold, Inc.; 55 (bl), Bruce Coleman, Inc.; 55 (bc), Jane Van Eps/SuperStock; 56 (b), Calhoun/Bruce Coleman, Inc.; 57 (tl), ©David Ryan/Photo 20-20; 59 (tr), ©2003 Galen Rowell/Mountain Light Photography; 59 (tl), ©2003 Galen Rowell/Mountain Light Photography. **Chapter Four:** 62 (r), Carr Clifton/Minden Pictures; 62 (cl), Corbis Images/HRW; 62 (bl), Dorling Kindersley; 65 (br), "World Ocean Map", Bruce C. Heezen and Marie Tharp, 1977, ©Marie Tharp 1977. Reproduced by permission of Marie Tharp, 1 Washington Avenue, South Nyack, NY 10960; 66 (bl), CORBIS/Tom Bean; 66 (br), CORBIS/Dave Bartruff; 67 (br), Carr Clifton/Minden Pictures; 67 (cr), Wendell Metzen/Bruce Coleman, Inc.; 68 (b), Bill Hatcher/National Geographic Society Image Collection; 68 (tl), Andrew Miles; 68 (inset), Carr Clifton/Minden Pictures; 71 (tr), Joseph Hutchins Colton, Johnson's New Illustrated Family Atlas with Physical Geography (New York, 1864), pp. 10–11.; 72 (b), Frans Lanting/Minden Pictures; 73 (tl), Digital image ©2003 CORBIS; Original image courtesy of NASA/CORBIS; 74 (br), Walter H. Hodge/Peter Arnold, Inc.; 76 (tr), D. Wiggett/Natural Selection; 77 (br), David R. Frazier Photolibrary; 77 (t), www.visiblearth.nasa; 78 (tr), CORBIS/Ecoscene/Andrew Brown; 79 (tr), CORBIS/AFP Photo/Vanderlei Almeida; 81 (tr), Photo ©Stefan Schott/Panoramic Images, Chicago 1998. **Chapter Five:** 86 (r), Stone/Chad Ehlers; 86 (bl), Michael & Barbara Reed/SuperStock; 87 (bl), Erica Lansner/Black Star; 87 (br), Jeffrey Aaronson/Network Aspen; 89 (tl), Stone/Sally Mayman; 90 (b), Peter Turnley/Black Star; 94 (b), Bill Bachmann/Network Aspen; 95 (tr), Eric Sanford/International Stock Photography; 96 (t), Wolfgang Kaehler Photography; 97 (bl), Laura J. Hannam/Uniphoto Picture Agency; 98 (tl), Stone/Yann Layma; 99 (bl), Galen Rowell/Mountain Light Photography; 102 (bl), ©Ulrike Welsch; 104 (tc), Jewish Museum, London/Bridgeman Art Library; 104 (c), Richard Nowitz/Words & Pictures/Picture Quest; 104 (b), CORBIS/Annie Griffiths Belt; 104 (tr), Scala/Art Resource, NY; 104 (tl), Lisa Quinones/Black Star; 105 (tr), Mark Downey/Viesti Collection, Inc.; 105 (c), CORBIS/Lindsay Hebberd; 105 (b), CORBIS/Gian Berto Vanni; 105 (tl), ©Dinodia/TRIP Photo Library; 105 (tc), Milind A. Ketkar/Dinodia Picture Agency; 106 (tl), CORBIS/AFP; 106 (tc), CORBIS/Luca I. Tettoni; 106 (tr), Josef Beck/FPG International; 106 (cl), Dinodia/SOA; 106 (b), ©ML Sinibaldi/Corbis Stock Market; 107 (tc), CORBIS/Bohemian Nomad Picturemakers; 107 (tr), Courtesy of Information Division, Taipei Economic and Cultural Office in Chicago; 107 (c), Stone/Alan Thornton; 107 (tl), Bridgeman Art Library, New

York/London; 107 (b), Courtesy of Information Division, Taipei Economic and Cultural Office in Chicago; 108 (tr), Erich Lessing/Art Resource, NY; 108 (bc), Image Copyright ©2003 PhotoDisc, Inc.; 108 (c), CORBIS/Elio Ciol; 108 (tl), Scala/Art Resource, NY; 108 (b), CORBIS/Mark Thiessen; 109 (tl), Ronald Sheriidan/Ancient Art & Architecture Collection, Ltd.; 109 (tr), Bonhams, London/Bridgeman Art Library, New York/London; 109 (tc), Werner Forman/Art Resource, NY; 109 (cr), CORBIS/AFP; 109 (b), CORBIS/Reuters NewMedia, Inc. **Chapter Six:** 112 (cl), Stone/Shaun Egan; 112 (r), Stone; 112 (bl), Runk/Schoenberger/Grant Heilman Photography; 114 (br), Carolyn Schaefer/SCHAE/Bruce Coleman, Inc.; 115 (br), Stone/Cosmo Condina; 115 (cr), Archive Photos; 118 (tl), Stone/Glen Allison; 118 (cl), John Elk III/Bruce Coleman, Inc.; 120 (br), ©Werner Forman/CORBIS; 123 (bl), David R. Frazier Photolibrary; 124 (bl), Stone/Pete Seaward; 125 (tr), Stone/Doris DeWitt; 126 (tr), Arthur C. Smith III/ Grant Heilman Photography; 127 (bl), Stone/Paula Bronstein; 128 (br), Popperfoto/ Archive Photos; 129 (br), Stone/Tom Bean; 130 (br), Amit Bhargava/Newsmakers/ Getty Images; 131 (tl), Christine Osborne Pictures/MEP; 132 (br), Agencia Estado/ AP/Wide World Photos; 133 (bl), E. R. Degginger/Bruce Coleman, Inc.; 133 (t), www.visibleearth.nasa.gov; 137 (t), CORBIS/Lowell Georgia.

Unit Two: 138-139 (all), CORBIS/Michael S. Yamashita; 139 (tr), Science Photo Library/Photo Researchers, Inc.; 139 (c), Ray Manley/SuperStock; 146 (tl), Collection of The New-York Historical Society; 146 (soldier), Collection of The New-York Historical Society; 146 (Washington), White House Collection, copyright White House Historical Association; 146 (settlers), ©Shelburne Museum, Shelburne, Vermont, detail of the painting "Conestoga Wagon" by Thomas Birch; 146 (prospector), CORBIS/ Bettmann; 146 (uniform), The Granger Collection, New York; 146 (Uncle Sam), The Granger Collection, New York; 146 (Bush), AP/Wide World Photos; 146 (br), CORBIS/David Muench; 147 (b), CORBIS/Michael T. Sedam. **Chapter Seven:** 148 (cr), Raymond Gehman/CORBIS; 148 (bl), Michio Hoshino/Minden Pictures; 148 (cl), Chuck Place Photography; 151 (br), Chase Swift/CORBIS; 152 (bl), Ansel Adams Publishing Rights Trust/CORBIS; 152 (br), Roger Ressmeyer/CORBIS; 153 (tr), CORBIS/NASA; 154 (b), CORBIS/Robert Estall; 155 (tr), CORBIS/Jim Zuckerman; 156 (tr), Carr Clifton/Minden Pictures; 156 (bl), Carr Clifton/Minden Pictures; 157 (tr), Philip Rosenberg Photography; 158 (bl), Stone/Bruce Forster; 159 (tr), CORBIS/Jeffrey L. Rotman; 160 (tr), CORBIS/Bill Ross. **Chapter Eight:** 164 (bl), Stone/Walter Hodges; 164 (tr), CORBIS/Buddy Mays; 164 (cl), The Image Bank; 166 (t), Chuck Place Photography; 168 (bl), CORBIS; 169 (br), Mark Gibson Photography; 170 (tl), ©Christian Heeb/Gnass Photo Images; 171 (tr), CORBIS/Bettmann; 173 (b), CORBIS/Michael S. Yamashita; 174 (bl), CORBIS/Kelly-Mooney Photography; 176 (tl), David R. Frazier Photolibrary; 176 (bl), CORBIS/Robert Holmes; 177 (tr), Stone/Bruce Hands; 177 (b), ©Ric Ergenbright; 178 (tl), Tim Fitzharris/Minden Pictures; 179 (tr), Carr Clifton/Minden Pictures; 180 (cl), ©Ric Ergenbright; 181 (br), David R. Frazier Photolibrary; 181 (tr), CORBIS/Roger Ressmeyer; 183 (tl), CORBIS/James A. Sugar; 184 (tr), CORBIS/Bettmann. **Chapter Nine:** 188 (bl), Eastcott/Momatiuk/Woodfin Camp & Associates; 188 (tl), Jim Brandenburg/Minden Pictures; 188 (tr), The Lowe Art Museum, The University of Miami/SuperStock; 190 (t), The Huntington Library, Art Collections and Botanical Gardens, San Marino, CA/SuperStock; 192 (cl), George Hunter/SuperStock; 192 (tl), Stephan Poulin/SuperStock; 193 (bl), George Hunter/ H. Armstrong Roberts; 194 (br), M. Antman/The Image Works; 195 (br), Mauritius/ SuperStock; 196 (tl), George Hunter/H. Armstrong Roberts; 197 (cr), John Eastcott/ YVA Momatiuk/DRK Photo; 199 (br), Michael S. Yamashita/CORBIS; 200 (b), Nik Wheeler; 201 (tr), Walter Bibikow/Viesti Collection, Inc.; 204 (tr), Enrique Shore/Reuters/ TimePix; 204 (cr), Mike Segar /Reuters/TimePix; 205 (bl), Jeff Christensen/Reuters/ TimePix; 205 (tr), ©Monika Graff/The Image Works.

Unit Three: 208-209 (all), CORBIS/Kevin Schafer; 209 (tr), CORBIS/Wolfgang Kaehler; 209 (bc), SuperStock; 216 (cl), Museo Civico, Como, Italy/Art Resource, NY; 216 (c), Museo Nacional de Historia, Castillo de Chapultepec, Mexico City, Mexico/Art Resource, NY; 216 (tc), CORBIS/Lake County Museum; 216 (cr), CORBIS/Bettmann; 216 (tr), CORBIS/Dave G. Houser; 218 (b), Stone/Renee Lynn; 219 (b), Steve Vidler/ SuperStock. **Chapter Ten:** 220 (bl), Steve Ewert Photography/HRW Photo; 220 (tl), Stewart Aitchison/D. Donne Bryant Photography; 220 (tr), Scala/Art Resource, NY; 222 (tl), ©Robert Frerck/Odyssey/Chicago; 223 (tr), George Grall/National Geographic Society Image Collection; 223 (tc), James Hanken/Bruce Coleman, Inc.; 224 (br), D. Donne Bryant/D. Donne Bryant Photography; 225 (bl), Giraudon/Art Resource, NY; 226 (tl), J. Sarapochiello/Bruce Coleman, Inc.; 226 (bl), Nik Wheeler; 228 (tr), Eleanor S. Morris; 228 (bl), Alon Reininger/Woodfin Camp & Associates; 229 (cr), Sam Dudgeon/HRW Photo; 230 (br), Steve Vidler/SuperStock; 231 (tr), CORBIS/Jorge Uzon; 231 (b), ©Robert Frerck/Odyssey/Chicago; 232 (tr), Bob Daemmrich/The Image Works; 233 (tr), Joel Sartore/National Geographic Society Image Collection; 234 (t), ©Robert Frerck/Odyssey/Chicago; 234 (bl), ©Robert Frerck/Odyssey/Chicago; 235 (t), taken from Travel Guide Mexico ©Dorling Kindersley. **Chapter Eleven:** 238 (bl), Steve Ewert Photography/HRW Photo; 238 (tl), Karl Kummels/SuperStock; 238 (tr), CORBIS/Kevin Schafer; 240 (tl), Steve Vidler/SuperStock; 241 (br), Stone/Bob Krist; 242 (tr), ©Kevin Schafer/kevinschafer.com; 243 (tr), Rob Huibers/Panos Pictures; 244 (b), Stone/David Hiser; 245 (tr), SuperStock; 246 (b), ©2003 Greg Johnston; 247 (tr), August Upitis/SuperStock; 247 (br), ©Kevin Schafer/kevinschafer.com; 248 (tr), ©Kevin Schafer/kevinschafer.com; 249 (b), Wolfgang Kaehler Photography; 250 (tr), Stone/Nick Dolding; 250 (bl), Catherine Karnow/Woodfin Camp & Associates; 250 (tl), Wolfgang Kaehler Photography; 251 (cr), Michael K. Nichols/National Geographic Society Image Collection; 252 (tl), ©2003 Greg Johnston; 252 (b), Steve Vidler/ SuperStock; 253 (t), Steve Vidler/SuperStock. **Chapter Twelve:** 256 (bl), Steve Ewert Photography/HRW Photo; 256 (tl), ©Kevin Schafer/kevinschafer.com; 256 (tr), ©Gianni Dagli Orti/CORBIS; 258 (tr), ©Kevin Schafer/kevinschafer.com; 258 (bl), Angelo Cavalli/SuperStock; 260 (inset), ©Kevin Schafer/kevinschafer.com; 260 (tl), Norman Owen Tomalin/Bruce Coleman, Inc.; 261 (cl), D. Donne Bryant Photography; 261 (cr), D. Donne Bryant Photography; 262 (tc), NASA/Science Photo Library/Photo Researchers, Inc.; 263 (bl), ©Robert Frerck/Odyssey/Chicago; 263 (cr), The Art Archive/Museum of Mankind London/Eileen Tweedy; 264 (cl), The Art Archive/Simon Bolivar Amphitheatre/Dagli Orti (A) ; 265 (br), RAGA/The Stock Market; 266 (bl), Robert Perron; 267 (tr), Robert Frerck/Odyssey/Chicago; 268 (bl), ©Peter Lang/The PhotoWorks/D. Donne Bryant Photography; 268 (b), Pedro Raota/SuperStock; 269 (br), Still Pictures/Peter Arnold, Inc.; 269 (cr), Juca Martins/f4/D. Donne Bryant

Acknowledgments

Photography; 270 (tl), ©Mauricio Simonetti/f4/D. Donne Bryant Photography; 270 (br), Auscape (Parer-Cook)/Peter Arnold, Inc.; 271 (tr), ©Peter Wilson/CORBIS; 272 (b), Jeremy Homer/CORBIS; 273 (br), CORBIS/Yann Arthus-Bertrand.

Unit Four: 278-279 (all), Stone/Bob Krist ; 279 (tr), CORBIS/Werner Forman; 279 (bc), AKG, Berlin/SuperStock; 286 (cl), The Art Archive; 286 (tl), Alinari/Art Resource, NY; 286 (tc), The Art Archive/Bibliotheque des Arts Decoratifs Paris/Dagli Orti (A); 286 (cr), CORBIS/Hulton-Deutsch Collection Limited; 286 (tr), AP/Wide World Photos; 289 (b), SuperStock. **Chapter Thirteen:** 290 (tl), Miles Ertman/Masterfile; 290 (bl), CORBIS/Macduff Everton; 290 (tr), CORBIS/David Lees; 292 (bl), Corbis Images; 293 (c), Benjamin Rondel/CORBIS/Stock Market; 294 (tl), Pierre Boulat/Woodfin Camp & Associates; 295 (t), Bill Brooks/Masterfile; 295 (b), Stone/George Grigoriou; 296 (b), Ron Sanford/Tony Stone Images; 297 (br), CORBIS/Frank Lane Picture Agency; 298 (tl), Len Kaufman; 299 (br), Peter Weimann/Animals Animals/Earth Scenes; 300 (t), Richard Turpin/Corbis Stock Market; 301 (tl), Gaetano Barone/SuperStock. **Chapter Fourteen:** 304 (bl), Steve Ewert Photography/HRW Photo; 304 (tl), Steve Vidler/ Leo de Wys; 304 (tr), London, British Museum/AKG Photo, London; 306 (tl), Telegraph Colour Library/FPG International; 308 (bl), AKG Photo, London; 309 (tr), David Lomax/ Robert Harding; 310 (b), Gerry Johansson/Leo de Wys; 311 (bl), Musée d'Orsay/AKG Photo, London; 312 (tl), Picture Finders Ltd./Leo de Wys; 312 (br), Jean Paul Nacivet/ Leo de Wys; 313 (tl), Michael Jenner/Robert Harding; 314 (tl), Natalie B. Fobes/Tony Stone Images; 315 (cr), Steve Vidler/Leo de Wys; 317 (tr), Gavin Hellier/Robert Harding; 318 (br), Everett C. Johnson/Leo de Wys; 319 (tr), Vladpans/Leo de Wys; 319 (b), Tom Stewart/CORBIS/Stock Market; 320 (br), Photographers Library LTD/Leo de Wys; 320 (tl), CORBIS/Hubert Stadler; 321 (tr), Jan Kofod Winther; 322 (bl), Telegraph Coulour Library/FPG International. **Chapter Fifteen:** 326 (bl), Steve Ewert Photography/HRW Photo; 326 (cl), Dankwart Von Knobloch/SuperStock; 326 (tr), Kunsthistorisches Museum, Vienna, Austria/Erich Lessing/Art Resource, NY; 327 (tr), Wolfgang Kaehler Photography; 328 (t), Steve Vidler/SuperStock; 329 (bl), Frank Rumpenhorst/AP/Wide World Photos; 330 (tl), Wolfgang Kaehler Photography; 331 (tr), Argus Fotoarchiv/ Peter Arnold, Inc.; 332 (bl), SuperStock; 333 (tl), CORBIS/Reuters NewMedia Inc.; 335 (br), Wolfgang Kaehler Photography; 336 (tl), Robin Smith/SuperStock; 337 (tl), SuperStock; 338 (cl), Fitzwilliam Museum, University of Cambridge, UK/Bridgeman Art Library; 339 (br), Horst Schafer/Peter Arnold, Inc.; 340 (tr), Kurt Scholz/SuperStock; 341 (tr), Fred Bruemmer/Peter Arnold, Inc.; 342 (bl), Hubertus Kanus/SuperStock; 342 (tr), ©J. Icachsen/Art Directors & TRIP Photo Library; 343 (br), Wolfgang Kaehler Photography; 345 (cr), Sovfoto/Eastfoto. **Chapter Sixteen:** 348 (bl), MercuryPress.com; 348 (tl), Grant V. Faint/The Image Bank; 348 (tr), Erich Lessing/Art Resource, NY; 350 (br), Nathan Benn/Woodfin Camp & Associates; 351 (br), ©Ulrike Welsch; 351 (tl), ©Daniel Aubry/Odyssey/Chicago; 355 (bl), ©Louis Goldman/FPG International LLC; 357 (bl), Dorling Kindersley; 360 (br), ©Scala/Art Resource; 362 (tl), Giraudon/Art Resource, NY; 363 (tl), Mario Corvetto/Evergreen Photo Alliance; 364 (br), Peter Schmid/SuperStock; 365 (t), ©Figaro Magazine/Robert Tixador/Liaison Agency.

Unit Five: 370–371 (spread), Stone/David Sutherland; 371 (tr), CORBIS/Marc Garanger; 371 (c), SuperStock; 378 (cl), Giraudon/Art Resource, NY; 378 (c), Victoria and Albert Museum, London/Art Resource, NY; 378 (tr), Giraudon/Art Resource, NY; 378 (cr), CORBIS/Bettmann; 379 (b), Vladimir Jirnov/Bruce Coleman, Inc. **Chapter Seventeen:** 380 (bl), Steve Ewert Photography/HRW Photo; 380 (tr), SCALA/Art Resource, NY; 380 (tl), Mark Wadlow/Russia and Eastern Images; 382 (b), Randall Hyman; 383 (br), Sovfoto/Eastfoto; 384 (bl), Sovfoto/Eastfoto; 385 (tr), Galen Rowell/ Peter Arnold, Inc.; 385 (t), Sovfoto/Eastfoto; 386 (bl), The Art Archive/Hermitage Museum Saint Petersburg/Dagli Orti(A); 387 (b), Hulton Getty; 387 (t), CORBIS/ Bettmann; 388 (b), Deborah Harse/Peter Arnold, Inc.; 388 (t), CORBIS/UPI/Bettmann; 389 (tr), Sovfoto/Eastfoto; 391 (t), Marka/Leo de Wys; 393 (b), Stone/Paul Chesley; 394 (t), CORBIS/Steve Raymer; 394 (b), CORBIS/Fabian/Sygma; 395 (tr), AP/Wide World Photos; 395 (bl), Stone/Jerry Alexander; 396 (bl), CORBIS/David and Peter Turnley; 397 (tl), Sovfoto/Eastfoto; 398 (tc), Helga Lade/Peter Arnold, Inc.; 399 (tr), The Art Archive/Tretyakov Gallery Moscow/Dagli Orti (A); 399 (tl), IFA/Peter Arnold, Inc.; 399 (tc), Jeff Greenberg/Peter Arnold, Inc. **Chapter Eighteen:** 402 (bl), Brian Vikander/Vikander Photography; 402 (tr), Wayne Eastep; 402 (tl), CORBIS/David & Peter Turnley; 404 (t), Wayne Eastep; 405 (tr), Digital Image ©2003 CORBIS; Original image courtesy of NASA/CORBIS. ; 405 (br), ©Wolfgang Kaehler/CORBIS; 406 (tl), Hope Ryden/National Geographic Society Image Collection; 408 (bl), Wolfgang Kaehler Photography; 409 (tr), ©2003 BRIAN A. VIKANDER; 409 (tr), ©David Samuel Robbins/ CORBIS; 410 (t), Caroline Penn/Panos Pictures; 410 (cl), Nevada Wier; 411 (tr), Panos Pictures; 411 (cr), David Samuel Robbins; 412 (bl), Nevada Wier; 413 (tr), CORBIS/ Bettmann; 413 (br), ©2003 BRIAN A. VIKANDER; 414 (tr), Marcus Rose/Panos Pictures; 414 (tl), Gregory Wrona/Panos Pictures; 415 (tr), Brian Goddard/Panos Pictures; 416 (tr), @ I. Burgandinov/Art Directors & TRIP Photo Library; 417 (tl), ©P. Bucknall/Art Directors & TRIP Photo Library.

Unit Six: 422–423 (all), CORBIS/Yann Arthus-Bertrand; 423 (c), Louvre, Paris, France/Giraudon/Art Resource, NY; 423 (tr), Nik Wheeler/Black Star Publishing/ Picture Quest; 430 (cl), CORBIS/Bettmann; 430 (tc), CORBIS/Araldo de Luca; 430 (c), CORBIS/Burstein Collection; 430 (cr), CORBIS/Bettmann; 430 (tr), CORBIS/Francoise de Mulder. **Chapter Nineteen:** 432 (bl), Steve Ewert Photography/HRW Photo; 432 (tl), Adam Woolfitt/CORBIS; 432 (tr), Archivo Iconografico, S.A./CORBIS; 436 (tl), Wolfgang Kaehler Photography; 437 (bl), CORBIS; 437 (br), CORBIS; 438 (bl), University Library Istanbul/The Art Archive; 439 (br), AFP/CORBIS; 440 (tl), ©Abbas/ Magnum; 441 (br), Nik Wheeler; 442 (tl), ©Alexandra Avakian /Contact Press Images;

Acknowledgments

444 (tl), Wolfgang Kaehler Photography; 445 (bl), Burnett Moody/Bruce Coleman, Inc.; 445 (bc), Bill Foley/Bruce Coleman, Inc.; 445 (br), Bill Foley/Bruce Coleman, Inc. **Chapter Twenty:** 448 (cl), Richard T. Nowitz; 448 (tr), The Art Archive/British Museum; 448 (bl), ©2003 Zafer KIZILKAYA; 450 (bl), Steve Vidler/SuperStock; 450 (cl), ©Look GMBH/eStock Photography/Picture Quest; 451 (tl), ©1998 Peter Armenia; 452 (tr), D. Wells/The Image Works; 452 (bl), @J. Wakelin/Art Directors & TRIP Photo Library; 454 (br), Robert Frerck/Woodfin Camp & Associates; 455 (tr), ©Robert Frerck/Odyssey/Chicago; 456 (br), ©1998 Peter Armenia; 457 (cr), Rich Pomerantz; 457 (br), Stone/Robert Frerck; 458 (tl), Christine Osborne Pictures/MEP; 459 (br), CORBIS/Richard List; 460 (cl), Nik Wheeler; 461 (tr), Nik Wheeler; 461 (b), ©Alexandra Avakian/Contact Press Images; 462 (b), Wolfgang Kaehler Photography; 462 (cl), Rich Pomerantz; 463 (tl), Nik Wheeler; 464 (tc), Gut Marche/FPG International; 469 (tl), http://edcwww.cr.usgs.gov/earthshots.

Unit Seven: 470-471 (spread), Steve Vidler/SuperStock; 471 (c), Picture Finders Ltd./eStock Photography/Picture Quest; 471 (tr), Richard S. Durrance/National Geographic Society Image Collection; 478 (tc), CORBIS/Gallo Images; 478 (c), The Art Archive/British Museum/Harper Collins Publishers; 478 (diamond), CORBIS/D. Boone; 478 (tr), The Art Archive/London Museum/Eileen Tweedy; 478 (cr), CORBIS/Walter Dhladhla; 478 (cl), The Metropolitan Museum of Art, The Michael C. Rockefeller Memorial Collection, Gift of Nelson A. Rockefeller, 1972. (1978.412.323) Photograph by Malcolm Varon. Photography ©1986 The Metropolitan Museum of Art. **Chapter Twenty-One:** 482 (bl), Steve Ewert Photography/HRW Photo; 482 (tl), Stone/Sylvain Grandadam; 482 (tr), The Art Archive/Musee des Arts Africains et Oceaniens/Dagli Orti; 484 (br), ©Victor Englebert; 484 (bl), SuperStock; 485 (tl), ©Victor Englebert; 487 (br), ©Planet Art; 488 (t), Mimmo Jodice/CORBIS; 488 (br), Giorgio Ricatto/SuperStock; 488 (bl), Thomas J. Abercrombie/National Geographic Society Image Collection; 489 (cl), TRIP/Viesti Collection, Inc.; 489 (tr), AP/Wide World Photos; 490 (br), SuperStock; 491 (tr), Stone/Ben Edwards; 492 (bl), Stone/Michael Shopenn; 494 (tr), Stone/John Lamb; 495 (tr), AP/Wide World Photos; 497 (cl), GEOPIC/Earth Satellite Corporation. **Chapter Twenty-Two:** 498 (bl), Steve Ewert Photography/HRW Photo; 498 (cl), Lineair/R. Giling/Peter Arnold, Inc.; 498 (cr), M & E Bernheim/Woodfin Camp & Associates; 500 (tr), Wolfgang Kaehler Photography; 501 (tr), ©BIOS/Peter Arnold, Inc.; 502 (inset), ©Still Pictures/M. Edwards/Peter Arnold, Inc.; 502 (tr), ©Still Pictures/M. Edwards/Peter Arnold, Inc.; 503 (cr), Frank Fournier/Woodfin Camp & Associates; 504 (tl), Frank Fournier/Woodfin Camp & Associates; 504 (tr), Martha Cooper/Peter Arnold, Inc.; 506 (b), ©Victor Englebert; 507 (tl), Mark Edwards/Still Pictures /Peter Arnold, Inc.; 509 (cr), ©Lineair (R. Giling)/Peter Arnold, Inc.; 509 (br), ©Wendy Stone/Odyssey/Chicago; 511 (tr), AP/Wide World Photos/David Guttenfelder; 512 (tl), Bob Burch/Bruce Coleman, Inc.; 512 (br), M. & E. Bernheim/Woodfin Camp & Associates; 513 (t), Wolfgang Kaehler Photography. **Chapter Twenty-Three:** 516 (bl), Betty Press/Picture Group/Panos Pictures; 516 (tl), Daniel Westergren/National Geographic Society Image Collection; 516 (tr), The Art Archive; 518 (b), CORBIS/Sharna Balfour; 519 (tr), Boyd Norton; 520 (inset), Jason Lauré/Lauré Communications; 520 (c), Jason Lauré/Lauré Communications; 521 (bl), Dominic Harcourt-Webster/Panos Pictures; 521 (tr), CORBIS/Morton Beebe; 522 (cl), Jason Lauré/Lauré Communications; 522 (tr), Boyd Norton; 523 (t), Betty Press/Woodfin Camp/Picture Quest; 524 (cl), CORBIS/James A. Sugar; 525 (br), Gerald Cubitt; 526 (t), B. Brander/Photo Researchers, Inc.; 527 (br), Boyd Norton; 528 (tr), Crispin Hughes/Panos Pictures; 528 (cl), ©Charles Henneghien/Bruce Coleman, Inc.; 529 (br), ©2000 Alison M. Jones. **Chapter Twenty-Four:** 532 (cl), Friedrich Von Horsten /Animals Animals/Earth Scenes; 532 (bl), ©Jan Butchofsky/Houserstock; 532 (tr), Clem Haagner/Bruce Coleman, Inc.; 534 (t), John Elk III/Bruce Coleman, Inc.; 534 (b), Gerald Cubitt; 535 (br), Burnett Moody/Bruce Coleman, Inc.; 535 (tr), Albert Normandin/Masterfile; 536 (tr), Patti Murray/Animals Animals/Earth Scenes; 537 (br), S. Trevor/D.B./Bruce Coleman, Inc.; 538 (tr), Martin Gostelos/Bruce Coleman, Inc.; 538 (cl), Hulton Getty Collection/Archive Photos; 539 (br), A. Ramey/Woodfin Camp & Associates; 540 (tl), Piero Guerrini/Woodfin Camp & Associates; 541 (tl), Wolfgang Kaehler Photography; 541 (cr), Volkmar Wentzel/National Geographic Society Image Collection; 542 (bl), Gerald Cubitt; 543 (br), William R. Curtsinger/National Geographic Society Image Collection; 544 (br), Gerald Cubitt; 545 (tr), ©Victor Englebert; 551 (tr) Jason Lauré/Lauré Communications.

Unit Eight: 552-553 (all), ©Earl Bronssteen/Panoramic Images, Chicago 2003; 553 (tr), SuperStock; 553 (c), CORBIS; 560 (cl), Borromeo/Art Resource, NY; 560 (c), The Art Archive/Victoria and Albert Museum Lodon/Eileen Tweedy; 560 (tr), Hulton Getty; 560 (cr), @ H. Rogers/Art Directors & TRIP Photo Library; 561 (b), CORBIS/Tom Brakefield. **Chapter Twenty-Five:** 562 (bl), Steve Ewert Photography/HRW Photo; 562 (tr), Floyd Holdman/International Stock Photography; 562 (cl), SuperStock; 564 (b), Steve McCurry/Magnum Photos; 565 (tr), Earth Satellite Corporation/Science Photo Library/Photo Researchers, Inc.; 566 (inset), CORBIS/Jayanta Shaw/Reuters; 566 (tl), SuperStock; 567 (bl), Joel Simon/Stone; 567 (tr), CORBIS/©Jeremy Horner; 570 (bl); SuperStock; 570 (tl), ©Stone/Hulton Getty; 571 (tr), SuperStock; 572 (br), Gerald Brimacombe/International Stock Photography; 573 (tr), Miwako Ikeda/International Stock Photography; 574 (br), Ric Ergenbright; 575 (b), SuperStock; 576 (tl), SuperStock; 578 (br), James Strachan/Robert Harding; 580 (tr), Jeremy Bright/Robert Harding; 580 (c), CORBIS/Savita Kirloskar; 581 (t), Chris Stowers/Panos Pictures. **Chapter Twenty-Six:** 584 (bl), Steve Ewert Photography/HRW Photo; 584 (tr), Holton Collection/SuperStock; 584 (tl), CORBIS/Tom Owen Edmunds; 586 (b), Hugh Burden/SuperStock; 586 (bl), Charles Corfield; 588 (tl), CORBIS/Jonathan Blair; 589 (b), ©geyer@malediven.at; 590 (b), ©R. Graham/Art Directors & TRIP Photo Library; 591 (tr), CORBIS/Agence France Presse/Jewel Samad; 591 (bl), William McCloskey; 592 (tr), Baldev Kapoor/SuperStock; 592 (tl), SuperStock; 593 (tr), Peter Symasko/International Stock Photography; 594 (b), Miwako Ikeda/International Stock Photography; 595 (br), Karl Kummels/SuperStock; 596 (cl), Binod Joshi/AP/Wide World Photos; 597 (tr), CORBIS/Jeremy Horner; 597 (tl), Steve Vidler/SuperStock; 598 (bl), Shehzad Noorani/Woodfin Camp & Associates; 598 (tc), Fred Hoogervorst/Panos Pictures; 599 (t), Zed Nelson/Panos Pictures; 602 (tr), Jeff Greenberg/Photo Researchers, Inc.

Unit Nine: 604-605 (spread), Stone/David Ball; 605 (c), Wolfgang Kaehler Photography; 605 (tr), Ed Gifford/Masterfile; 612 (tl), CORBIS/©Jack Fields; 612 (tc), CORBIS/Bettmann; 612 (tr), Stone/Margaret Gowan; 612 (cl), Reunion des Musees Nationaux/Art Resource, NY; 612 (c), CORBIS/Gianni Dagli Orti (A). **Chapter Twenty-Seven:** 614 (bl), Steve Ewert Photography/HRW Photo; 614 (tl), Keren Su/China Span; 614 (tr), ©2003 Brian A. Vikander; 616 (bl), CORBIS/Dean Conger; 617 (tr), Thomas Laird/Peter Arnold, Inc.; 617 (br), Lynn M. Stone/Bruce Coleman, Inc.; 618 (b), Liu Liqun/China Stock; 618 (tl), Keren Su/China Span; 619 (br), ©Stone/Yann Layma; 621 (cr), Stone/Nigel Hicks; 621 (inset), Keren Su/China Span; 622 (t), James Montgomery/Bruce Coleman, Inc.; 622 (bl), China Stock; 622 (inset), Joe McNally/The Image Bank; 623 (tr), ©2003 Brian A. Vikander; 623 (br), Paul Chesley/Network Aspen; 624 (bl), ©Keren Su/FPG International LLC; 625 (t), Xue Jun Yuan/The Image Bank; 625 (br), CORBIS/Asian Art & Archaeology, Inc.; 626 (tr), Dennis Cox/China Stock; 627 (bl), Cameramann International, Ltd.; 627 (inset), Cameramann International, Ltd.; 628 (tr), Jeffrey Aaronson/Network Aspen; 628 (bl), Dennis Cox/China Stock; 628 (inset), ©2003 Brian A. Vikander; 629 (b), Suolang Luobu/Sovfoto/Eastfoto; 630 (br), Mahaux Photo/The Image Bank; 630 (bl), Mongolian Eight flags soldiers from Ching's military forces, engraved by R.Rancati (colour engraving)/Private Collection/Bridgeman Art Library, London/New York; 631 (tr), Jeffrey Aaronson/Network Aspen. **Chapter Twenty-Eight:** 636 (bl), Steve Ewert Photography/HRW Photo; 636 (tr), Wolfgang Kaehler Photography; 636 (tl), Wolfgang Kaehler Photography; 638 (tl), CORBIS/John & Dallas Heaton; 638 (br), H. Edward Kim/National Geographic Society Image Collection; 639 (cr), Stephen Walker/Peter Arnold, Inc.; 640 (tl), Paul Chesley/Network Aspen; 641 (bl), Michael Yamashita; 642 (bl), CORBIS/Asian Art & Archaeology, Inc.; 644 (tl), Bruno P. Zehnder/Peter Arnold, Inc.; 644 (br), Kyodo News Service; 646 (tr), Kyodo News Service; 646 (bl), CORBIS/John Dakers; Eye Ubiquitous; 647 (cr), Dr. Darlyne A. Murawski/National Geographic Society Image Collection; 647 (tl), Kyodo News Service; 648 (tr), Blaine Harrington; 649 (br), Jeffrey Aaronson/Network Aspen; 650 (b), Kyodo News Service; 651 (tr), Michael Yamashita; 651 (br), David A. Harvey/National Geographic Society Image Collection; 652 (bl), Catherine Karnow/Woodfin Camp & Associates; 652 (tl), Jeffrey Aaronson/Network Aspen; 653 (tl), CORBIS/AFP; 655 (tr), Kyodo News Service; 655 (bl), CORBIS/Reuters NewMedia Inc. **Chapter Twenty-Nine:** 658 (bl), Steve Ewert Photography/HRW Photo; 658 (r), Dennie Cody/FPG International; 658 (cl), Radhika Chalasani/Network Aspen; 660 (b), Stone/Connie Coleman ; 661 (br), Michael Dick/Animals Animals/Earth Scenes; 661 (tl), Mike Yamashita/Woodfin Camp & Associates; 662 (cl), Robert Frerck/Woodfin Camp & Associates; 663 (cr), Harvey Lloyd/FPG International; 663 (b), Jeffrey Aaronson/Network Aspen; 664 (tl), Jeffrey Aaronson/Network Aspen; 665 (b), CORBIS/Bettmann Archive; 666 (bl), Jeffrey Aaronson/Network Aspen; 666 (tl), Paul Chesley/Network Aspen; 667 (tl), Wolfgang Kaehler Photography; 667 (tr), Stone/Owen Franken ; 668 (br), Stone/James Strachan ; 668 (bl), Liverani-UNEP/Peter Arnold, Inc.; 669 (bl), Brecelj & Hodalic/Peter Arnold, Inc.; 669 (cr), Fritz Prenzel/Peter Arnold, Inc.; 670 (tl), Stone/Paul Chesley; 671 (bl), Dennie Cody/FPG International; 672 (tl), Telegraph Colour Library/FPG International; 673 (cr), Jeffrey Aaronson/Network Aspen; 675 (tl), Mike Yamashita/Woodfin Camp & Associates; 675 (tr), Mark Edwards/Still Pictures/Peter Arnold, Inc. **Chapter Thirty:** 678 (tl), Mike Yamashita/Woodfin Camp & Associates; 678 (bl), @ D. Saunders/Art Directors & TRIP Photo Library; 678 (tr), Don Couch/HRW Photo - Mustapha Mazurki's Wau Bulan kite from www.SKY-DANCER.com; 680 (b), CORBIS/Les Stone/Sygma; 682 (cl), Picture Finders Ltd./Leo de Wys; 684 (cl), Wolfgang Kaehler Photography; 685 (tr), SuperStock; 685 (bl), Werner Forman Archive Private Collection, Prague/Art Resource, NY; 687 (tr), Werner Forman Archive/Private Collection/Art Resource, NY; 688 (b), Stone/Denis Waugh; 689 (tl), CORBIS/Michael S. Yamashita; 690 (cr), SuperStock; 690 (bl), John Zoiner; 691 (bl), Stone/T Resource; 692 (b), Wolfgang Kaehler Photography; 693 (tr), Cesar Pelli & Associates/Photo by Jeff Goldberg/Esto; 694 (tr), Stone/Dave Saunders.

Unit Ten: 700-701 (all), Jeff Hunter/The Image Bank; 701 (tr), Peter Arnold, Inc.; 701 (c), David Burnett/Contact Press Images/Picture Quest; 708 (c), The Art Archive/Harper Collins Publishers; 708 (cr), CORBIS/Historical Picture Archive; 708 (cl), CORBIS/Werner Forman; 708 (tr), CORBIS/Charles & Josette Lenars; 708 (tl), CORBIS/Michael Maslan Historic Photographs. **Chapter Thirty-One:** 710 (bl), Penny Tweedie/HRW Photo; 710 (tl), Joseph Van Os/The Image Bank; 710 (tr), Leo de Wys/Leo de Wys; 712 (bl), Gisela Damm/Leo de Wys; 712 (br), R. Ian Lloyd/Masterfile; 713 (tr), ©David Doubilet; 714 (tl), Brian Sytnyk/Masterfile; 714 (tr), R. Ian Lloyd/Masterfile; 715 (tr), ©Bill Bachman; 716 (bl), Wes Thompson/CORBIS/Stock Market; 717 (tr), Robin Smith/SuperStock; 718 (tl), Jean-Paul Ferrero/Auscape International; 719 (cr), Steve Vidler/Leo de Wys; 719 (t), Corbis Images/HRW; 722 (br), Brian Sytnyk/Masterfile; 723 (bl), ZEFA GmbH/Leo de Wys; 724 (tr), Macduff Everton; 725 (bl), Andris Apse/APSEA/Bruce Coleman, Inc.; 725 (br), Michael Steele/Allsport; 726 (bl), CORBIS/Paul A. Sounders; 727 (tr), Picture Finders Ltd./Leo de Wys. **Chapter Thirty-Two:** 730 (bl), ©2000 Zafer A. Kizilkaya; 730 (tl), CORBIS/Dennis Marsico; 730 (tr), CORBIS/Wolfgang Kaehler; 732 (b), CORBIS/Australian Picture Library; 734 (tl), www.visibleearth.nasa.gov; 735 (b), Peter Hendrie/The Image Bank; 735 (inset), S. ACHERNAR/The Image Bank; 736 (b), The Art Archive/Eileen Tweedy; 737 (tr), Steve Vidler/Leo de Wys; 738 (bl), Steve Thomas The Last Navigator; 739 (tr), CORBIS/Wolfgang Kaehler; 741 (r), William Albert Allard/National Geographic Society Image Collection; 742 (bl), Steve Vidler/Leo de Wys; 743 (tr), Carl N. McDaniel; 744 (inset), Mike Tinsley/Auscape International Pty Ltd; 744 (b), David Alan Harvey/National Geographic Society Image Collection; 745 (inset), Chris Johns/National Geographic Society Image Collection; 745 (b), Chris Johns/National Geographic Society Image Collection; 749 (tr), SuperStock.